CLIMATE CHANGE IMPACTS ON THE UNITED STATES

The Potential Consequences of Climate Variability and Change

Foundation

Humanity's influence on the global climate will grow in the 21st century. Increasingly, there will be significant climate-related changes that will affect each one of us.

We must begin now to consider our responses, as the actions taken today will affect the quality of life for us and future generations.

A Report of the National Assessment Synthesis Team

US Global Change Research Program

CAMBRIDGE
UNIVERSITY PRESS

PUBLISHED BY THE PRESS SYNDICATE OF THE UNIVERSITY OF CAMBRIDGE
The Pitt Building, Trumpington Street, Cambridge, United Kingdom

CAMBRIDGE UNIVERSITY PRESS
The Edinburgh Building, Cambridge, CB2 2RU, UK
40 West 20th Street, New York, NY 10011-4211, USA
10 Stamford Road, Oakleigh, VIC 3166, Australia
Ruiz de Alarcón 13, 28014 Madrid, Spain
Dock House, The Waterfront, Cape Town 8001, South Africa

http://www.cambridge.org

First published 2001

Printed in the United States of America

ISBN 0-521-00075-0 paperback

This report was produced by the National Assessment Synthesis Team, an advisory committee chartered under the Federal Advisory Committee Act to help the US Global Change Research Program fulfill its mandate under the Global Change Research Act of 1990. The report was turned in to the Subcommittee on Global Change Research on October 31, 2000. The National Science and Technology Council has forwarded this report to the President and Congress for their consideration as required by the Global Change Research Act.

Administrative support for the US Global Change Research Program is provided by the University Corporation for Atmospheric Research, which is sponsored by the National Science Foundation. Any opinions, findings and conclusions or recommendations expressed in this publication are those of the authors and do not necessarily reflect the views of the National Science Foundation or the University Corporation for Atmospheric Research.

Comments on this report should be addressed to:
Office of the US Global Change Research Program
400 Virginia Avenue SW, Suite 750, Washington DC 20024
http://www.usgcrp.gov

The recommended citation of this report is as follows:
National Assessment Synthesis Team
Climate Change Impacts on the United States:
The Potential Consequences of Climate Variability and Change,
Report for the US Global Change Research Program,
Cambridge University Press, Cambridge UK, 620pp., 2001.

FOREWORD

The National Assessment of the Potential Consequences of Climate Variability and Change is a landmark in the major ongoing effort to understand what climate change means for the United States. Climate science is developing rapidly and scientists are increasingly able to project some changes at the regional scale, identifying regional vulnerabilities, and assessing potential regional impacts. Science increasingly indicates that the Earth s climate has changed in the past and continues to change, and that even greater climate change is very likely in the 21st century. This Assessment has begun a national process of research, analysis, and dialogue about the coming changes in climate, their impacts, and what Americans can do to adapt to an uncertain and continuously changing climate. This Assessment is built on a solid foundation of science conducted as part of the United States Global Change Research Program (USGCRP).

This document is the Foundation report, which provides the scientific underpinnings for the Assessment. It has been prepared in cooperation with independent regional and sector assessment teams under the leadership of the National Assessment Synthesis Team (NAST). The NAST is a committee of experts drawn from governments, universities, industry, and non-governmental organizations. It has been responsible for preparing an Overview report aimed at general audiences and for broad oversight of the Assessment along with the Federal agencies of the USGCRP. These two national-level, peer-reviewed documents synthesize results from studies conducted by regional and sector teams, and from the broader scientific literature.

This Assessment was called for by a 1990 law, and has been conducted under the authority of the USGCRP in response to a request from the President s Science Advisor. The NAST developed the Assessment s plan, which was then approved by the National Science and Technology Council, the cabinet level body of agencies responsible for scientific research, including global change research, in the US government. We would like to acknowledge their contributions to this effort. The agencies and their representatives are listed in the appendix to this volume. Of particular note have been Rosina Bierbaum and Peter Backlund of the Office of Science and Technology Policy, who provided consistent and helpful guidance throughout, and who organized our Oversight Board. In addition, Robert Corell (now a NAST Member), Aristides Patrinos, Paul Dresler, Richard Ball, Joel Scheraga, and Tom Spence, along with many additional individuals, have played major roles on behalf of the Subcommittee on Global Change Research, its National Assessment Working Group, and the ten cooperating agencies.

These assessment reports could not have been prepared without the extraordinary efforts of a large number of people. In addition to the members of the NAST, a number of individuals were entrained into development of the content and findings of the report, both as lead authors for the Overview and as lead and contributing authors for the chapters in this Foundation report. We want to express our sincere gratitude to these authors, the many names of whom are listed in the Overview report and in the chapter headings of this book. Those playing particularly important roles in the preparation of major sections of the Foundation report included Susan Bernard, Lynne Carter, David Easterling, Benjamin Felzer, John Field, Paul Grabhorn, Susan Jay Hassol, Schuyler Houser, Michael MacCracken, Michael McGeehin, Jonathan Patz, John Reilly, Joel Smith, Melissa Taylor, and Tom Wilbanks.

The report itself is based in large part on workshops and assessment efforts of five sector teams and teams in 20 regions across the US. Each of these groups has in turn involved many more experts from universities, governments at various levels, public and private organizations, and others interested in or affected by the changing global environment. All of these individuals have played an important role in developing and expanding the dialogue on the potential impacts of climate change. We want to especially thank the various regional and sector team leaders who are listed along with their team members in the chapters of this report.

In addition, we benefited from the comments of hundreds of reviewers, who helped encourage new insights and new ways of thinking about and presenting the results of these studies. Of particular help were the members of the Independent Review Board that was established by the President s Committee of Advisers on Science and Technology. Co-chaired by Peter Raven and Mario Molina, the board included Burton Richter, Linda Fisher, Kathryn Fuller, John Gibbons, Marcia McNutt, Sally Ride, William Schlesinger, James Gustave Speth, and Robert White. They provided cogent and helpful comments throughout the many drafts of the assessment documents.

The complexity of coordinating the activities of far-flung authors, providing background data, managing the inputs and responses from hundreds of reviewers, designing the reports to be accurate, accessible, and appealing, and ensuring that the final products were printed under tight timetables was very challenging. Many people devoted their personal and professional attention to those tasks without asking for credit. Here we acknowledge their contributions and dedication to seeing this job through, and thank them, most assuredly less than they deserve. Paul Grabhorn kept us focused on effectively communicating our message, helped us appreciate the importance of design, and he, Melody Warford, and their staff carried this through with an inspired design implemented through layout, graphics, and production of the documents. Susan Joy Hassol, with the gracious cooperation of the Aspen Global Change Institute, played a major role in making complex scientific issues more easily understood and helping our convoluted prose speak more clearly.

The staff of the National Assessment Coordination Office (NACO) played an important role in facilitating the entire assessment process by supporting the activities not only of the NAST, but also by coordinating the efforts of the regions, sectors, and agencies. Under the leadership of Michael MacCracken, the coordination and logistics associated with this very distributed effort came together. Melissa Taylor served as executive secretary to the NAST through March 2000. Lynne Carter served as NACO liaison to the regions, Justin Wettstein and LaShaunda Malone served as liaison to the sectors, and LaShaunda Malone also served as liaison with agencies and as coordinator for the various peer reviews. Thomas Wilbanks of Oak Ridge National Laboratory (ORNL) served as chair of the Inter-regional Forum that helped to encourage and coordinate regional activities. In addition, Forrest Hoffman of ORNL handled the Web site through which much of our information was distributed. The NACO staff were also assisted in their efforts by staff of the Global Change Research Information Office, including Robert Worrest, Annie Gerard, and Robert Bourdeau, who have helped in the posting of the full report for public comment and access.

The assessment studies are based on extensive data sets of various types. Benjamin Felzer, with assistance of staff at the National Center for Atmospheric Research (NCAR), assembled and analyzed the data from climate models and prepared most of the climate graphics. David Easterling, Byron Gleason, and other staff at the National Climatic Data Center provided databases describing past changes in the climate. Tim Kittel at the National Center for Atmospheric Research was instrumental in carrying through the processing of the climatic data to provide consistent sets for use across the US. We also very much appreciate the willingness of colleagues at the various modeling centers to provide results of their simulations, including particularly David Viner at the University of East Anglia, Francis Zwiers and George Boer at the Canadian Centre for Climate Modelling and Analysis, and John Mitchell, Ruth Carnell, and Jonathan Gregory at the Hadley Centre of the United Kingdom Meteorological Office. The availability of data for the assessment teams was made possible by Ben Felzer of NCAR and Annette Schloss and Denise Blaha of the University of New Hampshire.

Baseline distributions and simulations of changes in ecosystems were made available through the Vegetation/Ecosystem Modeling and Analysis Project (VEMAP) and their many team members. Tim Kittel of NCAR graciously served as coordinator of our links to this effort. The social science data sets were provided by Nestor Terlickij of NPA Data Associates through an agreement with the Oak Ridge National Laboratory based on the efforts of David Vogt and Thomas Wilbanks. In addition, Robert Chen at the Consortium for International Earth Science Information Networks (CIESIN) provided very helpful data sets on population and other social measures.

Many individuals have played important roles in carrying through the administrative aspects of this effort. We want to graciously acknowledge the contributions of Mary Ann Seifert of the Marine Biological Laboratory, Gracie Bermudez of the World Resources Institute, Rosalind Ledford of the National Climatic Data Center, Nakia Dawkins and Robert Cherry of NACO, and Susan Henson, Karen York, and Matt Powell of the National Science Foundation, all of whom assisted in making possible our many meetings and exchanges of reports, among many other tasks. In addition, the staff of the University Corporation for Atmospheric Research (UCAR) provided invaluable assistance with travel and contractual issues associated with the assessment process. Those playing particularly helpful efforts have been Gene Martin, Kyle Terran, Tara Jay, Amy Smith, Chrystal Pene, James Menghi, and Brian Jackson.

Finally, as co-chairs of the National Assessment Synthesis Team, we would like to thank the other members of this team. We have had quite an adventure, working to develop and analyze information, working with fellow NAST members and leaders of assessment teams around the country, considering and coming to agreement on findings, and writing and rewriting text in response to internal and external comments. Throughout there has been great comity, and we are very proud to have come to full consensus on all of the findings. We want to thank all of you especially for devoting your time and effort to this important effort; we know it has involved much more than any of you first thought, but we believe the product is also a very significant contribution to the Nation s future.

Jerry Melillo
Anthony Janetos
Thomas Karl

TABLE OF CONTENTS

ABOUT THE ASSESSMENT PROCESS

What is the purpose of this Assessment?

The Assessment's purpose is to synthesize, evaluate, and report on what we presently know about the potential consequences of climate variability and change for the US in the 21st century. It has sought to identify key climatic vulnerabilities of particular regions and sectors, in the context of other changes in the nation s environment, resources, and economy. It has also sought to identify potential measures to adapt to climate variability and change. Finally, because present knowledge is limited, the Assessment has sought to identify the highest priority uncertainties about which we must know more to understand climate impacts, vulnerabilities, and our ability to adapt.

How did the process involve both stakeholders and scientists in this Assessment?

This first National Assessment involved both stakeholders and scientific experts. Stakeholders included, for example, public and private decision-makers, resource and environmental managers, and the general public. The stakeholders from different regions and sectors began the Assessment by articulating their concerns in a series of workshops about climate change impacts in the context of the other major issues they face. In the workshops and subsequent consultations, stakeholders identified priority regional and sector concerns, mobilized specialized expertise, identified potential adaptation options, and provided useful information for decision-makers. The Assessment also involved many scientific experts using advanced methods, models, and results. Further, it has stimulated new scientific research in many areas and identified priority needs for further research.

What is the breadth of this Assessment?

Although global change embraces many interrelated issues, this first National Assessment has examined only climate change and variability, with a primary focus on specific regions and sectors. In some cases, regional and sector analyses intersect and complement each other. For example, the Forest sector and the Pacific Northwest have both provided insights into climate impacts on Northwest forests.

The regions cover the nation. Impacts outside the US are considered only briefly, with particular emphasis on potential linkages to the US. Sector teams examined Water, Agriculture, Human Health, Forests, and Coastal Areas and Marine Resources. This first Assessment could not attempt to be comprehensive: the choice of these five sectors reflected an expectation that they were likely to be both important and particularly informative, and that relevant data and analytic tools were available — not a conclusion that they are the only important domains of climate impact. Among the sectors considered, there was a continuum in the amount of information available to support the Assessment, with some sectors being at far earlier stages of development. Future assessments should consider other potentially important issues, such as Energy, Transportation, Urban Areas, and Wildlife.

Each regional and sector team is publishing a separate report of its own analyses, some of which are still continuing. The Overview and Foundation reports consequently represent a snapshot of our understanding at the present time.

After identifying potential impacts of climate change, what kinds of societal responses does this report explore?

Responses to climate change can be of two broad types. One type involves adaptation measures to reduce the harms and risks and maximize the benefits and opportunities of climate change, whatever its cause. The other type involves mitigation measures to reduce human contributions to climate change. After identifying potential impacts, this Assessment sought to identify potential adaptation measures for each region and sector studied. While this was an important first step, it was not possible at this stage to evaluate the practicality, effectiveness, or costs of the potential adaptation measures. Both mitigation and adaptation measures are necessary elements of a coherent and integrated response to climate change. Mitigation measures were not included in this Assessment but are being assessed in other bodies such as the United Nations Intergovernmental Panel on Climate Change (IPCC).

Does the fact that this report excludes mitigation mean that nothing can be done to reduce climate change?

No. An integrated climate policy will combine mitigation and adaptation measures as appropriate. If future world emissions of greenhouse gases are lower than currently projected, for whatever reason, including intentional mitigation, then the rate of climate change, the associated impacts, and the cost and difficulty of adapting will all be reduced. If emissions are higher than expected, then the rate of change, the impacts, and the difficulty of adapting will be increased. But no matter how aggressively emissions are reduced, the world will still experience at least a century of climate change. This will happen because the elevated concentrations of greenhouse gases already in the atmosphere will remain for many decades, and because the climate system responds to changes in human inputs only very slowly. Consequently, even if the world takes mitigation measures, we must still adapt to a changing climate. Similarly, even if we take adaptation measures, future emissions will have to be curbed to stabilize climate. Neither type of response can completely supplant the other.

How are computer models used in this Assessment?

State-of-the-science climate models have been used to generate climate change scenarios. Computer models of ecological systems, hydrological systems, and various socioeconomic systems have also been used in the Assessment to study responses of these systems to the scenarios generated by climate models.

What additional tools, besides models, were used to evaluate potential climate change impacts?

In addition to models, the Assessment has used two other ways to think about potential future climate. First, the Assessment has used historical climate records to evaluate sensitivities of regions and sectors to climate variability and extremes that have occurred in the 20th century. Looking at real historical climate events, their impacts, and how people have adapted, gives valuable insights into potential future impacts that complement those provided by model projections. In addition, the Assessment has used sensitivity analyses, which ask how, and how much, the climate would have to change to bring about major impacts on particular regions or sectors. For example, how much would temperature have to increase in the South before agricultural crops such as soybeans would be negatively affected? What would be the result for forest productivity of continued increases in temperature and leveling off of the CO_2 fertilization effect?

Has this report been peer reviewed?

This Overview and the underlying Foundation document have been extensively reviewed. More than 300 scientific and technical experts have provided detailed comments on part or all of the report in two separate technical reviews. The report was reviewed at each stage for technical accuracy by the agencies of the US Global Change Research Program. The public also provided hundreds of helpful suggestions for clarification and modification during a 60-day public comment period. A panel of distinguished experts convened by the President's Committee of Advisors on Science and Technology has provided broad oversight and monitored the authors' responses to all reviews.

ABOUT SCENARIOS AND UNCERTAINTY

Many of the maps in this document are derived from the two primary climate model scenarios. In most cases, there are three maps: one shows average conditions based on actual observations from 1961-1990; the other two are generated by the Hadley and Canadian model scenarios and reflect the models projections of change from present day conditions.

What are scenarios and why are they used?

Scenarios are plausible alternative futures — each an example of what might happen under particular assumptions. Scenarios are not specific predictions or forecasts. Rather, scenarios provide a starting point for examining questions about an uncertain future and can help us visualize alternative futures in concrete and human terms. The military and industry frequently use these powerful tools for future planning in high-stakes situations. Using scenarios helps to identify vulnerabilities and plan for contingencies.

Why are climate scenarios used in this Assessment and how were they developed?

Because we cannot predict many aspects of our nation's future climate, we have used scenarios to help explore US vulnerability to climate change. Results from state-of-the-science climate models and data from historical observations have been used to generate a variety of such scenarios. Projections of changes in climate from the Hadley Centre in the United Kingdom and the Canadian Centre for Climate Modeling and Analysis served as the primary resources for this Assessment. Results were also drawn from models developed at the National Center for Atmospheric Research, NOAA's Geophysical Fluid Dynamics Laboratory, and NASA's Goddard Institute for Space Studies.

For some aspects of climate, virtually all models, as well as other lines of evidence, agree on the types of changes to be expected. For example, all climate models suggest that the climate is going to get warmer, the heat index is going to rise, and precipitation is more likely to come in heavy and extreme events. This consistency lends confidence to these results.

For some other aspects of climate, however, the model results differ. For example, some models, including the Canadian model, project more extensive and frequent drought in the US, while others, including the Hadley model, do not. The Canadian model suggests a drier Southeast in the 21st century while the Hadley model suggests a wetter one. In such cases, the scenarios provide two plausible but different alternatives. Such differences can help identify areas in which the models need improvement.

Many of the maps in this document are derived from the two primary climate model scenarios. In most cases, there are three maps: one shows average conditions based on actual observations from 1961-1990; the other two are generated by the Hadley and Canadian model scenarios and reflect the models projections of change from those average conditions.

What assumptions about emissions are in these two climate scenarios?

Because future trends in fossil fuel use and other human activities are uncertain, the Intergovernmental Panel on Climate Change (IPCC) has developed a set of scenarios for how the 21st century may evolve. These scenarios consider a wide range of possibilities for changes in population, economic growth, technological development, improvements in energy efficiency, and the like. The two primary climate scenarios used in this Assessment are based on one mid-range emissions scenario for the future that assumes no major changes in policies to limit greenhouse gas emissions. Some other important assumptions in this scenario are that by the year 2100:

- world population will nearly double to about 11 billion people;
- the global economy will continue to grow at about the average rate it has been growing, reaching more than ten times its present size;
- increased use of fossil fuels will triple CO_2 emissions and raise sulfur dioxide emissions, resulting in an atmospheric CO_2 concentration of just over 700 parts per million; and
- total energy produced each year from non-fossil sources such as wind, solar, biomass, hydroelectric, and nuclear will increase to more than ten times its current amount, providing more than 40% of the world s energy, rather than the current 10%.

The Assessment's Emissions Scenario Falls in the Middle of the other IPCC Emissions Scenarios

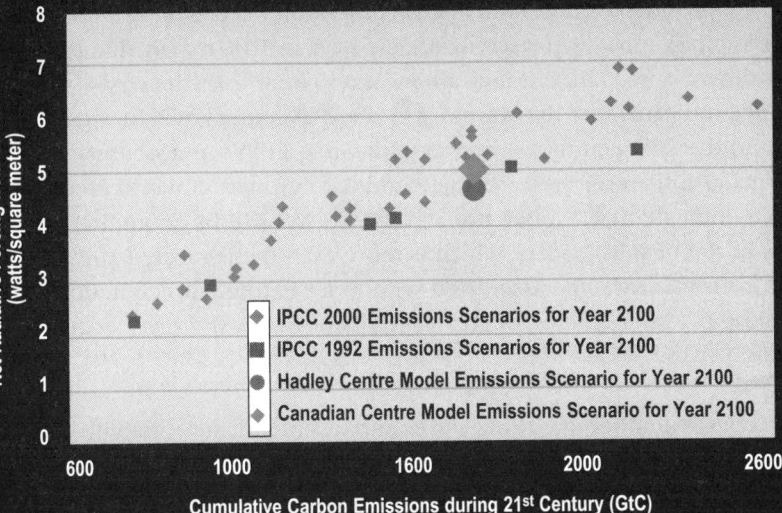

IPCC 2000 Emissions Scenarios for Year 2100
IPCC 1992 Emissions Scenarios for Year 2100
Hadley Centre Model Emissions Scenario for Year 2100
Canadian Centre Model Emissions Scenario for Year 2100

Net Radiative Forcing from 2000 to 2100 (watts/square meter)

Cumulative Carbon Emissions during 21st Century (GtC)

The graph shows a comparison of the projections of total carbon dioxide emissions (in billions of metric tons of carbon, GtC) and the human-induced warming influence due to all the greenhouse gases and sulfate aerosols for the emissions scenarios prepared by the IPCC in 1992 and 2000. As is apparent from the graph, both the emissions scenario and the human-induced warming influence assumed in this Assessment lie near the mid-range of the set of IPCC scenarios. Further detail can be found in the Climate chapter . See color figure section, page 546.

Both the emissions scenario and the human-induced warming influence assumed in this Assessment lie near the mid-range of the set of IPCC scenarios.

How is the likelihood of various impacts expressed?

To integrate a wide variety of information and differentiate more likely from less likely outcomes, the NAST developed a common language to express the team's considered judgement about the likelihood of results. The NAST developed their collective judgements through discussion and consideration of the supporting information. Historical data, model projections, published scientific literature, and other available information all provided input to these deliberations, except where specifically stated that the result comes from a particular model scenario. In developing these judgements, there were often several lines of supporting evidence (e.g., drawn from observed trends, analytic studies, model simulations). Many of these judgements were based on broad scientific consensus as stated by well-recognized authorities including the IPCC and the National Research Council. In many cases, groups outside the NAST reviewed the use of terms to provide input from a broader set of experts in a particular field.

Language Used to Express Considered Judgement

Common Language

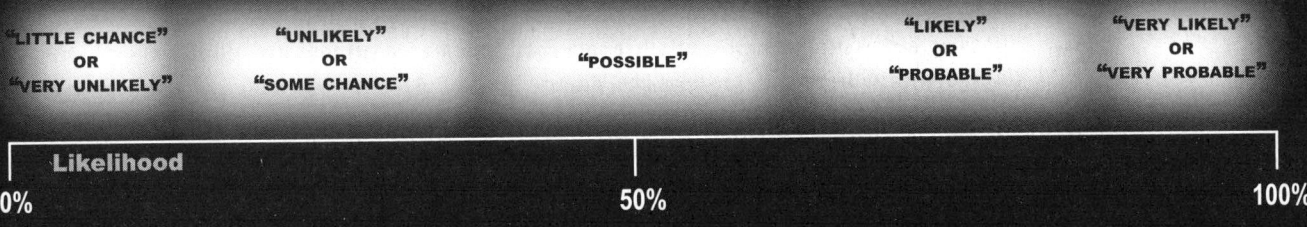

"LITTLE CHANCE" OR "VERY UNLIKELY"

"UNLIKELY" OR "SOME CHANCE"

"POSSIBLE"

"LIKELY" OR "PROBABLE"

"VERY LIKELY" OR "VERY PROBABLE"

Likelihood

0%   50%   100%

SUMMARY

CLIMATE CHANGE AND OUR NATION

The findings in this report are based on a synthesis of historical data, model projections, published scientific research, and other available information, except where specifically noted.

Long-term observations confirm that our climate is now changing at a rapid rate. Over the 20th century, the average annual US temperature has risen by almost 1°F (0.6°C) and precipitation has increased nationally by 5 to 10%, mostly due to increases in heavy downpours. These trends are most apparent over the past few decades. The science indicates that the warming in the 21st century will be significantly greater than in the 20th century. Scenarios examined in this Assessment, which assume no major interventions to reduce continued growth of world greenhouse gas emissions, indicate that temperatures in the US will rise by about 5-9°F (3-5°C) on average in the next 100 years, which is more than the projected *global* increase. This rise is very likely to be associated with more extreme precipitation and faster evaporation of water, leading to greater frequency of both very wet and very dry conditions.

This Assessment reveals a number of national-level impacts of climate variability and change including impacts to natural ecosystems and water resources. Natural ecosystems appear to be the most vulnerable to the harmful effects of climate change, as there is often little that can be done to help them adapt to the projected speed and amount of change. Some ecosystems that are already constrained by climate, such as alpine meadows in the Rocky Mountains, are likely to face extreme stress, and disappear entirely in some places. It is likely that other more widespread ecosystems will also be vulnerable to climate change. One of the climate scenarios used in this Assessment suggests the potential for the forests of the Southeast to break up into a mosaic of forests, savannas, and grasslands. Climate scenarios suggest likely changes in the species composition of the Northeast forests, including the loss of sugar maples. Major alterations to natural ecosystems due to climate change could possibly have negative consequences for our economy, which depends in part on the sustained bounty of our nation's lands, waters, and native plant and animal communities.

A unique contribution of this first US Assessment is that it combines national-scale analysis with an examination of the potential impacts of climate change on different regions of the US. For example, sea-level rise will very likely cause further loss of coastal wetlands (ecosystems that provide vital nurseries and habitats for many fish species) and put coastal communities at greater risk of storm surges, especially in the Southeast. Reduction in snowpack will very likely alter the timing and amount of water supplies, potentially exacerbating water shortages and conflicts, particularly throughout the western US. The melting of glaciers in the high-elevation West and in Alaska represents the loss or diminishment of unique national treasures of the American landscape. Large increases in the heat index (which combines temperature and humidity) and increases in the frequency of heat waves are very likely. These changes will, at minimum, increase discomfort, particularly in cities. It is very probable that continued thawing of permafrost and melting of sea ice in Alaska will further damage forests, buildings, roads, and coastlines, and harm subsistence livelihoods. In various parts of the nation, cold-weather recreation such as skiing will very likely be reduced, and air conditioning usage will very likely increase.

Highly managed ecosystems appear more robust, and some potential benefits have been identified. Crop and forest productivity is likely to increase in some areas for the next few decades due to increased carbon dioxide in the atmosphere and an extended growing season. It is possible that some US food exports could increase, depending on impacts in other food-growing regions around the world. It is also possible that a rise in crop production in fertile areas could cause prices to fall, benefiting consumers. Other benefits that are possible include extended seasons for construction and warm weather recreation, reduced heating requirements, and reduced cold-weather mortality.

Climate variability and change will interact with other environmental stresses and socioeconomic changes. Air and water pollution, habitat fragmentation, wetland loss, coastal erosion, and reductions in fisheries are likely to be compounded by climate-related stresses. An aging populace nationally, and rapidly growing populations in cities, coastal areas, and across the South and West, are social factors that interact with and alter sensitivity to climate variability and change.

There are also very likely to be unanticipated impacts of climate change during the 21st century. Such "surprises" may stem from unforeseen changes in the physical climate system, such as major alterations in ocean circulation, cloud distribution, or storms; and unpredicted biological consequences of these physical climate changes, such as massive dislocations of species or pest outbreaks. In addition, unexpected social or economic changes, including major shifts in wealth, technology, or political priorities, could affect our ability to respond to climate change.

Greenhouse gas emissions lower than those assumed in this Assessment would result in reduced impacts. The signatory nations of the Framework Convention on Climate Change are negotiating the path they will ultimately take. Even with such reductions, however, the planet and the nation are certain to experience more than a century of climate change, due to the long lifetimes of greenhouse gases already in the atmosphere and the momentum of the climate system. Adapting to a changed climate is consequently a necessary component of our response strategy.

The warming in the 21st century will be significantly greater than in the 20th century.

Natural ecosystems, which are our life support system in many important ways, appear to be the most vulnerable to the harmful effects of climate change...

Major alterations to natural ecosystems due to climate change could possibly have negative consequences for our economy, which depends in part on the sustained bounty of our nation s lands, waters, and native plant and animal communities.

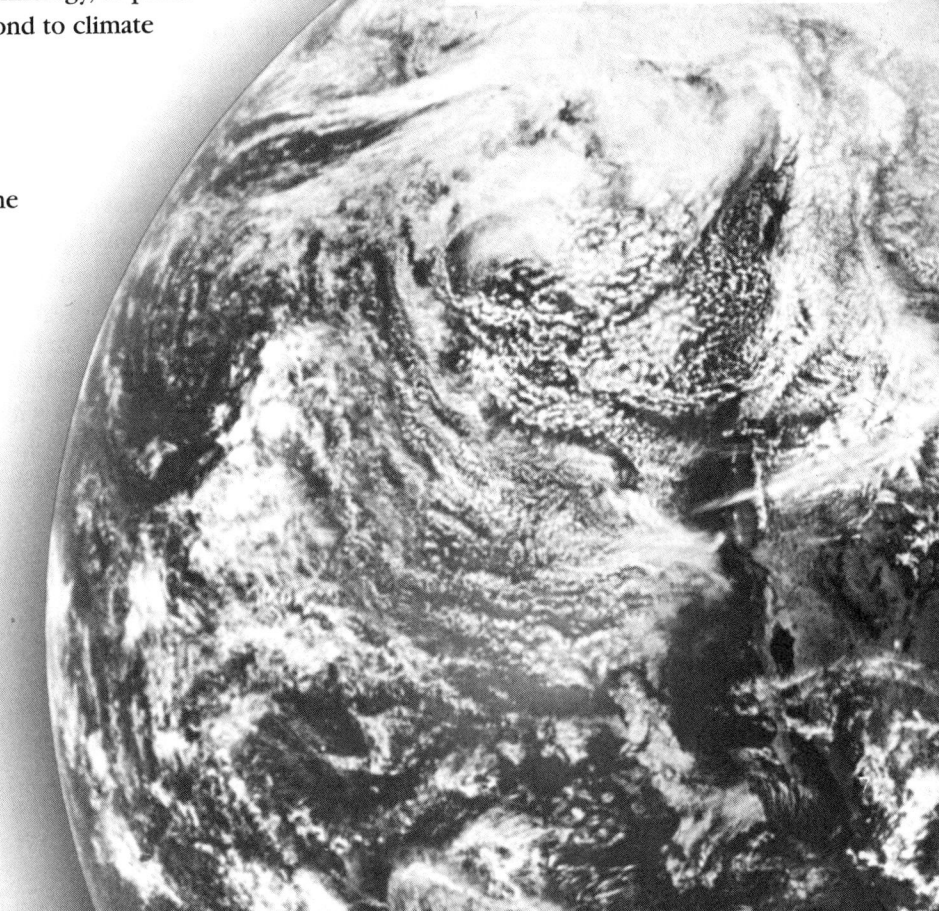

SUMMARY

CLIMATE CHANGE AND OUR NATION

The magnitude of climate change impacts depends on time period and geographic scale. Short-term impacts differ from long-term impacts, and regional and local level impacts are much more pronounced than those at the national level.

For the nation as a whole, direct economic impacts are likely to be modest, while in some places, economic losses or gains are likely to be large. For example, while crop yields are likely to increase at the national scale over the next few decades, large increases or decreases in yields of specific crops in particular places are likely.

Through time, climate change will possibly affect the same resource in opposite ways. For example, forest productivity is likely to increase in the short term, while over the longer term, changes in processes such as fire, insects, drought, and disease will possibly decrease forest productivity.

Adaptation measures can, in many cases, reduce the magnitude of harmful impacts or take advantage of beneficial impacts. For example, in agriculture, many farmers will probably be able to alter cropping and management practices. Roads, bridges, buildings, and other long-lived infrastructure can be designed taking projected climate change into account. Adaptations, however, can involve trade-offs, and do involve costs. For example, the benefits of building sea walls to prevent sea-level rise from disrupting human coastal communities will need to be weighed against the economic and ecological costs of seawall construction. The ecological costs could be high as seawalls prevent the inland shifting of coastal wetlands in response to sea-level rise, resulting in the loss of vital fish and bird habitat and other wetland functions, such as protecting shorelines from damage due to storm surges. Protecting against any increased risk of water-borne and insect-borne diseases will require diligent maintenance of our public health system. Many adaptations, notably those that seek to reduce other environmental stresses such as pollution and habitat fragmentation, will have beneficial effects beyond those related to climate change.

Vulnerability in the US is linked to the fates of other nations, and we cannot evaluate national consequences due to climate variability and change without also considering the consequences of changes elswhere in the world. The US is linked to other nations in many ways, and both our vulnerabilities and our potential responses will likely depend in part on impacts and responses in other nations. For example, conflicts or mass migrations resulting from resource limits, health, and environmental stresses in more vulnerable nations could possibly pose challenges for global security and US policy. Effects of climate variability and change on US agriculture will depend critically on changes in agricultural productivity elsewhere, which can shift international patterns of food supply and demand. Climate-induced changes in water resources available for power generation, transportation, cities, and agriculture are likely to raise potentially delicate diplomatic issues with both Canada and Mexico.

This Assessment has identified many remaining uncertainties that limit our ability to understand fully the spectrum of potential consequences of climate change for our nation. To address these uncertainties, additional research is needed to improve our understanding of ecological and social processes that are sensitive to climate, ways of applying climate scenarios and reconstructions of past climates to the study of impacts, and assessment strategies and methods. Results from these research efforts will inform future assessments that will continue the process of building our understanding of humanity's impacts on climate, and climate's impacts on us.

KEY FINDINGS

1. Increased warming

Assuming continued growth in world greenhouse gas emissions, the primary climate models used in this Assessment project that temperatures in the US will rise 5-9°F (3-5°C) on average in the next 100 years. A wider range of outcomes is possible.

2. Differing regional impacts

Climate change will vary widely across the US. Temperature increases will vary somewhat from one region to the next. Heavy and extreme precipitation events are likely to become more frequent, yet some regions will get drier. The potential impacts of climate change will also vary widely across the nation.

3. Vulnerable ecosystems

Many ecosystems are highly vulnerable to the projected rate and magnitude of climate change. A few, such as alpine meadows in the Rocky Mountains and some barrier islands, are likely to disappear entirely in some areas. Others, such as forests of the Southeast, are likely to experience major species shifts or break up into a mosaic of grasslands, woodlands, and forests. The goods and services lost through the disappearance or fragmentation of certain ecosystems are likely to be costly or impossible to replace.

4. Widespread water concerns

Water is an issue in every region, but the nature of the vulnerabilities varies. Drought is an important concern in every region. Floods and water quality are concerns in many regions. Snowpack changes are especially important in the West, Pacific Northwest, and Alaska.

5. Secure food supply

At the national level, the agriculture sector is likely to be able to adapt to climate change. Overall, US crop productivity is very likely to increase over the next few decades, but the gains will not be uniform across the nation. Falling prices and competitive pressures are very likely to stress some farmers, while benefiting consumers.

6. Near-term increase in forest growth

Forest productivity is likely to increase over the next several decades in some areas as trees respond to higher carbon dioxide levels. Over the longer term, changes in larger-scale processes such as fire, insects, droughts, and disease will possibly decrease forest productivity. In addition, climate change is likely to cause long-term shifts in forest species, such as sugar maples moving north out of the US.

7. Increased damage in coastal and permafrost areas

Climate change and the resulting rise in sea level are likely to exacerbate threats to buildings, roads, powerlines, and other infrastructure in climatically sensitive places. For example, infrastructure damage is related to permafrost melting in Alaska, and to sea-level rise and storm surge in low-lying coastal areas.

8. Adaptation determines health outcomes

A range of negative health impacts is possible from climate change, but adaptation is likely to help protect much of the US population. Maintaining our nation's public health and community infrastructure, from water treatment systems to emergency shelters, will be important for minimizing the impacts of water-borne diseases, heat stress, air pollution, extreme weather events, and diseases transmitted by insects, ticks, and rodents.

9. Other stresses magnified by climate change

Climate change will very likely magnify the cumulative impacts of other stresses, such as air and water pollution and habitat destruction due to human development patterns. For some systems, such as coral reefs, the combined effects of climate change and other stresses are very likely to exceed a critical threshold, bringing large, possibly irreversible impacts.

10. Uncertainties remain and surprises are expected

Significant uncertainties remain in the science underlying regional climate changes and their impacts. Further research would improve understanding and our ability to project societal and ecosystem impacts and to provide the public with additional useful information about options for adaptation. However, it is likely that some aspects and impacts of climate change will be totally unanticipated as complex systems respond to ongoing climate change in unforeseeable ways.

IMPACTS OF CLIMATE CHANGE

It is very likely that the US will get substantially warmer. Temperatures are projected to rise more rapidly in the next one hundred years than in the last 10,000 years. It is also very likely that there will be more precipitation overall, with more of it coming in heavy downpours. In spite of this, some areas are likely to get drier as increased evaporation due to higher temperatures outpaces increased precipitation. Droughts and flash floods are likely to become more frequent and intense.

SPECIES DIVERSITY

While it is possible that some species will adapt to changes in climate by shifting their ranges, human and geographic barriers, and the presence of invasive non-native species will limit the degree of adaptation that can occur. Losses in local biodiversity are likely to accelerate towards the end of the 21st century.

PERMAFROST AREAS

It is very probable that rising temperatures will cause further permafrost thawing, damaging roads, buildings, and forests in Alaska.

FORESTRY

Timber inventories are likely to increase over the 21st century. Hardwood productivity is likely to increase more than softwood productivity in some regions, including the Southeast.

WATER SUPPLY

Reduced summer runoff, increased winter runoff, and increased demands are likely to compound current stresses on water supplies and flood management, especially in the western US.

ISLANDS

Sea-level rise and storm surges will very likely threaten public health and safety and possibly reduce the availability of fresh water.

CORAL REEFS

Increased CO_2 and ocean temperatures, especially combined with other stresses, will possibly exacerbate coral reef bleaching and die-off.

FRESHWATER ECOSYSTEMS

Increases in water temperature and changes in seasonal patterns of runoff will very likely disturb fish habitat and affect recreational uses of lakes, streams, and wetlands.

FOREST ECOSYSTEMS

Forest growth is likely to increase in many regions, at least over the next several decades. Over the next century, tree and animal species' ranges will probably shift in response to the changing climate. Some forests are likely to become more susceptible to fire and pests.

AGRICULTURE

The Nation's food supply is likely to remain secure. The prices paid by consumers and the profit margins for food producers are likely to continue to drop.

HUMAN POPULATIONS

Heat waves are very likely to increase in frequency, resulting in more heat-related stresses. Milder winters are likely to reduce cold-related stresses in some areas.

COASTAL ECOSYSTEMS

Sea-level rise is very likely to cause the loss of some barrier beaches, islands, marshes, and coastal forests, throughout the 21st century.

EXTREME EVENTS

It is very likely that more rain will come in heavy downpours, increasing the risk of flash floods.

COASTAL COMMUNITIES AND INFRASTRUCTURE

Coastal inundation from storm surges combined with rising sea level will very likely increase threats to water and sewer systems, transportation and communication systems, homes, and other buildings.

RARE ECOSYSTEMS

Alpine meadows, mangroves, and tropical mountain forests in some locations are likely to disappear because the new local climate will not support them or there are barriers to their movement.

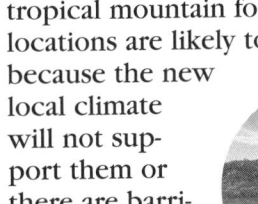

Adaptation

There are substantial opportunities to minimize the negative impacts and maximize the benefits of climate change through adaptation. Examples include cultivating varieties of crops, trees, and livestock that are better suited to hotter conditions. This report includes an initial identification of potential adaptation strategies, but an analysis of their effectiveness, practicality, and costs was not considered in this Assessment.

11

CHAPTER 1

SCENARIOS FOR CLIMATE VARIABILITY AND CHANGE

Michael MacCracken[1,2], Eric Barron[3], David Easterling[1,4], Benjamin Felzer[1,5], and Thomas Karl[4]

Contents of this Chapter

[1]Coordinating author for the National Assessment Synthesis Team; [2]Lawrence Livermore National Laboratory, on assignment to the National Assessment Coordination Office of the US Global Change Research Program; [3]Pennsylvania State University; [4]National Climatic Data Center; [5]National Center for Atmospheric Research

CHAPTER SUMMARY

Climate Context

Climate[1] provides the context for the environment and for many human activities — changes in the climate will thus have consequences for the environment and for human activities. While solar radiation is the primary energy source for maintaining the Earth's temperature, the atmospheric concentrations of water vapor, carbon dioxide (CO_2), methane (CH_4), and other gases determine the intensity of the natural greenhouse effect that currently keeps the Earth's surface temperature at about 58°F (14°C). Without this natural greenhouse effect, the Earth's surface temperature would be about 0°F (about -18°C), a temperature that would make the Earth uninhabitable for life as we know it. Over the last 150 years, combustion of coal, oil, and natural gas (collectively called fossil fuels), deforestation, plowing of soils, and various industrial activities have led, among other changes, to increases in the atmospheric concentrations of critical greenhouse gases. In particular, the CO_2 concentration has increased by about 30% and the CH_4 concentration by about 150%. The warming influence of these changes, amplified by associated increases in the atmospheric water vapor concentration, have intensified the natural greenhouse effect and initiated changes in the climate.

[1] Throughout the National Assessment reports, the term "climate" is intended to include both climate variability and climate change. "Climate change" refers to long-term or persistent trends (over decades or more) or shifts in climate, while "climate variability" refers to short-term (generally decadal or less) climate fluctuations.

Climate of the Past Century

- Since the mid-1800s, the global average temperature has warmed by about 1°F (about 0.6°C). The Northern Hemisphere average temperature during the 1990s is almost 1.5°F (about 0.9°C) warmer than during the few centuries prior to the Industrial Revolution. While some of this warming may be due to an intensification of solar radiation and a small portion due to urban warming, a variety of analyses indicate that the current warming is too large to be explained by natural fluctuations alone. The observed magnitude, pattern, and timing of the global warming indicate that the rising concentrations of CO_2 and other greenhouse gases caused by human activities are contributing significantly to the recent warming.

- During the 20th century, the average temperature over the US increased by about 1°F (0.6°C), with some regions warming as much as 4°F (about 2.4°C) and some other regions showing slight cooling. In general, nighttime minimum temperatures rose more than daytime maximums, and wintertime temperatures rose more than those of summertime. Total annual precipitation also increased, with most of the increase occurring in heavy precipitation events.

- Reconstructions of the climate of the past thousand years using ice cores, tree rings, vegetation types, and other proxy measures suggest that the warming of the 20th century is unprecedented compared to natural variations prior to this century that were presumably caused by solar, volcanic, and other natural influences. In addition, the current warming is much more extensive and intense than the regional scale warming that peaked about 1000 years ago in Europe during what is referred to as the Medieval Warm Period. The recent warming is also far more than can be characterized as a recovery from the cool conditions centered in Europe and the North Atlantic region a few hundred years ago that are often referred to as the Little Ice Age. Looking back over the few thousand years for which we are able to provide some reconstruction of the temperature record, the current global warmth appears unprecedented.

- An ice-core record from Antarctica covering the past 420,000 years indicates that temperatures in that region have been up to about 10°F (6°C) colder than present values for about 90% of the

420,000-year period. During these cold periods, massive glaciers covered much of the land area of the Northern Hemisphere (e.g., covering eastern North America with roughly a mile of ice to south of the Great Lakes), even though global temperatures were only several degrees colder. Evidence suggests that these variations have been driven primarily by changes in the seasonal and latitudinal distribution of solar radiation caused by cyclic variations in the Earth's orbit around the Sun, but amplified by a number of factors. These additional factors include changes in glacial height and extent, in ocean circulation, and in the atmospheric CO_2 and CH_4 concentrations that were apparently driven by the initial temperature change.

- The geological record indicates that the global climate has varied markedly over the past billion or more years. It appears that these natural variations resulted from changes in identifiable factors that still determine climatic conditions today. These factors include the amount of solar radiation and shape of the Earth's orbit around the Sun, the gas and particle composition of the atmosphere (which determines the efficiency of the absorption and reflection of incoming solar energy), the geographical pattern of land and ocean, the heights of mountains, the direction and intensity of ocean currents, the chaotic nature of the interactions among the atmosphere, land, and oceans, and more. The geological record clearly indicates that changes in these factors can cause significant changes in climate.

Climate of the Coming Century

- Projections of the expanding uses of coal, oil, and natural gas as sources of energy indicate that human activities will cause the atmospheric CO_2 concentration to rise to between 2 and 3 times its preindustrial level by the end of the 21st century unless very significant control measures are initiated. The concentrations of CH_4 and some other greenhouse gases are also projected to rise, whereas controls on chlorofluorocarbon emissions are expected to allow their concentrations to fall.

- The ongoing effects of past increases in the concentrations of greenhouse gases and the changes projected for the 21st century are very likely to cause the world to warm substantially in comparison to natural fluctuations that have been experienced over the past 1000 years. Model-based projections for a mid-range emissions scenario are that the global average temperature is likely to rise by about 2 to 6°F (about 1.2 to 3.5°C), with a central estimate of almost 4°F (2.4°C), by the end of the 21st century. The range of these estimates depends about equally on ranges in the estimates of climate sensitivity and of growth in fossil fuel emissions.

- For the mid-range emissions scenario, the projected warming is likely to be greater in mid and high latitudes than for the globe as a whole, and warming is likely to be greater over continents than over oceans. For this mid-range emissions scenario, the models used for this Assessment project that the average warming over the US would be in the range of about 5 to 9°F (about 2.8 to 5°C). However, given the wide range of possible emissions scenarios and uncertainties in the sensitivity of the climate to emissions scenarios, it is possible that the actual increase in US temperatures could be higher or lower than indicated by this range.

- A warming of 5 to 9°F (2.8 to 5°C) would be approximately equivalent to the annual average temperature difference between the northern and central tier of states, or the central and southern tier of states. Wintertime warming is projected to be greater than summertime warming and nighttime warming greater than daytime warming.

- Even though less warming is projected in summertime than in wintertime, the summertime heat index, which combines the effects of heat and humidity into an effective temperature, is projected to rise anywhere from 5 to 15°F (or even more for some scenarios) over much of the eastern half of the country, especially across the southeastern part of the country. If the projected rise in the heat index were to occur, summertime conditions for New York City could become like those now experienced in Atlanta, those in Atlanta like those now experienced in Houston, and those in Houston like those in Panama.

- The amount of rainfall over the globe is also very likely to rise because global warming will increase evaporation; however, the pattern of changes is likely to vary depending on latitude and geography as storm tracks are altered. Model projections of possible changes in annual precipitation across the US are generally mixed. Results from the two models used in the National Assessment tend to agree that there is likely to be an increase in precipitation in the southwestern US as Pacific Ocean temperatures increase, but do not provide a clear indication of the trend in the southeastern US.

- It is likely that the observed trends toward an intensification of precipitation events will continue. Thunderstorms and other intensive rain events are likely to produce larger rainfall totals. While it is not yet clear how the numbers and tracks of hurricanes will change, projections are that peak windspeed and rainfall intensity are likely to rise significantly.

- Although overall precipitation is likely to increase across the US, the higher temperatures will increase evaporation. Even with a modest increase in precipitation, the increase in the rate of evaporation is expected to cause reductions in summertime soil moisture, particularly in the central and southern US.

- Sea level, which has risen about 4 to 8 inches (10-20 cm) over the past century, is projected to increase by 5 to 37 inches (13-95 cm) over the coming century, with a central estimate of about 20 inches (50 cm). The range is so broad because of uncertainties concerning what might happen to the Antarctic and Greenland ice caps. To determine the amount of sea-level rise in particular regions, the global rise in sea level must be adjusted by the local rise or sinking of coastal lands.

- Limitations in scientific understanding mean that the potential exists for surprises or unexpected events to occur, for thresholds to be crossed, and for nonlinearities to develop. Such surprises have the potential of either amplifying projected changes or, in rarer cases, moderating the potential changes in climate. Examples might include amplified rates of sea-level rise if deterioration of the Greenland or Antarctic ice caps is accelerated; limited warming or perhaps even cooling in some regions if ocean currents and deep ocean overturning is suppressed; disappearance of Arctic sea ice over a few decades; sufficient warming of methane trapped in frozen soils to allow its release and subsequent amplification of the warming rate, etc. While such possibilities could cause large impacts, estimating the likelihood of their occurrence is presently highly problematic, making risk assessments quite difficult.

SCENARIOS FOR CLIMATE VARIABILITY AND CHANGE

INTRODUCTION

This National Assessment is charged with evaluating and summarizing the potential consequences of climate variability and change for the United States over the next 100 years (Dresler et al., 1998). Studies of the interactions of climate with both the environment and with societal activities show clearly that there are important interconnections. The very hot and dry conditions of the 1930s, coupled with poor land management practices, not only created Dust Bowl conditions on the Great Plains, but also led large numbers of people to migrate from the central US to settle in the Southwest and California. Drought conditions in 1988 and flood conditions in 1993 had devastating effects on many regions in the upper Mississippi River basin. Climate variations along the West Coast have led to years of drought (with subsequent fires) and of flood (with subsequent mudslides). It is these many interactions that have led to the focus on what will happen in the future as climate variations continue and as human activities believed to be capable of altering the climate continue.

The hypothesis that human activities could be influencing the global climate was first postulated more than a century ago (Arrhenius, 1896) and has become much better developed during the 20th century (e.g., beginning with papers by Callendar, 1938; Manabe and Wetherald, 1975; Hansen et al., 1981 and continuing to include thousands of additional scientific papers). Assessments of the scientific literature to evaluate the basis for postulating that human activities are affecting the global climate have been undertaken by many groups, including the Intergovernmental Panel on Climate Change (IPCC, 1990, 1992, 1996a), eminent advisory groups (PSAC, 1965; NRC, 1979, 1983; NAS, 1992), government agencies (e.g., USDOE, 1985a, 1985b), professional societies (most recently, the American Geophysical Union, see Ledley et al., 1999), and prominent scientific researchers (e.g., Mahlman, 1997). All of these analyses have come to similar conclusions, indicating that human activities are changing atmospheric composition in ways that are very likely to cause significant global warming during the 21st century. Results presented in this chapter draw upon the basis of scientific understanding described in these and related reports and the recent scientific literature, providing a limited set of citations that can be expanded upon by reference to these assessments.

Although these scientific studies indicate that the future will be different from the past, determining how different it will be and the significance of these differences presents a tremendous scientific challenge. The future will be affected by how the climate varies due to natural and human influences, how the environment may respond to climate change and to other factors, and how society may evolve due to a myriad of influences, including climate variability and change. Quite clearly, definitive predictions cannot be made, being too dependent on factors ranging from uncertainties introduced by our growing, but limited, understanding of the climate system to the complexities introduced by the pace of technological development and social evolution.

Given the seriousness and strength of the projections of climate change arising from the scientific community and from careful assessments, prudent risk management led Congress in 1990 to call for assessments of the potential impacts of climate change. During the 1990s, scientific assessments have focused on the global-scale consequences of human activities, leading to the conclusion that "the balance of evidence suggests a discernible human influence on the global climate" (IPCC, 1996a). IPCC assessments of the consequences of climate change have also indicated that potentially important consequences could arise (IPCC, 1996b, 1996c). It was these global-scale findings that indicated both the need for and the possibility of being able to conduct an assessment of the potential consequences of climate variability and change for the United States.

As a basis for this Assessment, and in the context of the uncertainties inherent in looking forward 100 years, Assessment teams are pursing a three-pronged approach to considering how much the climate may change. The three approaches involve use of: (1) historical data to examine the continuation of trends or recurrence of past climatic extremes; (2)

comprehensive, state-of-the-science, model simulations to provide plausible scenarios for how the future climate may change; and (3) sensitivity analyses that can be used to explore the resilience of societal and ecological systems to climatic fluctuations and change. This chapter provides background and information concerning past and projected changes in climate needed to carry through the National Assessment goal of analyzing potential consequences for society and the environment.

It should be emphasized that this chapter does not attempt a full scientific review of the adequacy or accuracy of climate observations or climate simulations of the past or future. For such a review, this Assessment relies on the very comprehensive, international assessments being undertaken by the IPCC (e.g., IPCC 1996a and the report now in preparation for release in 2001). Rather, this chapter provides information needed to understand and explain the analyses of the regional to national scale impact studies that are described in this National Assessment report and the supporting regional and sector reports. In presenting the needed background information, this chapter summarizes the strengths and weaknesses of the various approaches that need to be considered in interpreting the results of the impact analyses. This consideration includes balancing the many limitations that preclude making accurate specific predictions with the need for providing the best available information for conducting a risk-based analysis of the potential consequences of climate change.

CLIMATE AND THE GREENHOUSE EFFECT

The ensemble of weather events at any location defines the climate in that place. The climate is described by such measures as the averages of temperature, precipitation, and soil moisture as well as the magnitude and frequency of their variations, the likelihood of floods and droughts, the temperature of the oceans, and the paths and intensities of the winds and ocean currents. In contrast to climate's focus on average conditions over seasons to centuries and longer, weather describes what is happening at a particular place and time (e.g., when and where a thunderstorm occurs). Although the weather is constantly changing, the time- and space-averaged conditions making up the climate can also vary from season to season or decade to decade and can change significantly over the course of decades or centuries and beyond. While a slowly warming climate may seem hardly noticeable, the record of the Earth's environmental history indicates that seemingly small changes in climate (e.g., changes in the long-term average temperature of a few degrees) can have quite noticeable consequences for society and the environment.

Many factors determine the Earth's weather and climate, including the intensity of solar radiation, concentrations of atmospheric gases and particles, interactions with the oceans, and the changing character of the land surface. The predominant source of warming is energy received from the Sun in the form of solar radiation. Energy from the Sun enters the top of the atmosphere with an average intensity of about 342 watts per square meter. About 25% of this energy is immediately reflected back to space by clouds, aerosols (micron-sized particles and droplets, including sulfate aerosols), and other gases in the atmosphere; an additional 5% is reflected back to space by the surface, making the overall reflectivity (or albedo) of the Earth about 30%. Of the other 70% of incoming solar radiation, about 20% is absorbed in the atmosphere and the rest is absorbed at the surface. Thus, 70% of incoming solar energy is the driving force for weather and climate (Kiehl and Trenberth, 1997).

Studies of the Earth's climatic history extending back hundreds of millions of years indicate that there have been global-scale climate changes associated with changes in the factors that affect the Earth's energy balance. Factors that have exerted important influences include changes in: solar irradiance, the Earth's orbit about the Sun, the composition of the atmosphere, the distribution of land and ocean, the extent and type of vegetation, and the thickness and extent of snow and glaciers. Records of global glacial extent derived from ocean sediment cores (e.g., see Imbrie et al., 1992, 1993) and of temperature and atmospheric composition derived from deep ice cores drilled in Greenland and Antarctica (e.g., Petit et al., 1999) provide strong indications of the interactions and associations of these various influences. The Antarctic record (Figure 1), for example, indicates that the atmospheric CO_2 concentration can be changed by up to 100 parts per million by volume (ppmv)[2] as a result of the climate changes that occur due to the glacial-interglacial cycling over the past 420,000 years (Petit et al., 1999). While explanations of the relationships among orbital forcing, atmospheric concentrations of GHGs, and glacial extent are not yet fully quantified, it is clear that the Earth's climate has been dif-

---

[2] Parts per million by volume (ppmv) is equivalent to the number of molecules of CO_2 to the number of molecules of air, which is made up mostly of nitrogen and oxygen.

ferent when atmospheric composition has been different. Analyses indicate that these natural changes in atmospheric composition are being driven mainly by the initial changes in climate due to the orbital changes, and are then acting as feedbacks that amplify or moderate the initial changes in the climate. Given the evidence that changes in atmospheric composition have been a factor in determining climatic conditions over the Earth's history, human-induced changes in atmospheric composition (particularly greenhouse gas concentration) would also be expected to have an important influence on the climate. Scientific understanding of the changes in climates of the geological past would be significantly compromised if the Earth's climate were not now responding to changes in atmospheric composition.

Changes in the Earth's orbit around the Sun occur quite slowly, with periods ranging from about 20,000 to 400,000 years (Berger, 1978; Berger and Loutre, 1991). While these long periods mean that changes will be slow, their influences are steady and the changes, along with other factors, seem to cause trends in temperature evident in records of a few centuries or more in length (Berger, 1999). On the time scale of many centuries to millennia, observations from Antarctic ice cores (Petit et al., 1999; Imbrie et al., 1989) suggests that these orbital changes cause changes in climate that lead to changes in the amount of carbon dioxide in the atmosphere. These changes in the CO_2 concentration, working in parallel with the dynamics of ice sheets and their underlying geological substrate, then seem likely to have reinforced the glacier-inducing and melting influences of the changes in solar radiation caused by the orbital variations (Pisias and Shackleton, 1984; Shackleton et al., 1992; Petit et al., 1999; Clark et al., 1999). Following the end of the last glacial period about 10,000 years ago, orbital changes appear to have contributed to a

Northern Hemisphere warming that peaked about 6,000 years ago when the Earth was closer to the Sun during the Northern Hemisphere summer. Subsequent to this peak, a slow and sometimes intermittent cooling of the Northern Hemisphere started that seems to have continued until overwhelmed by the warming effects of the recent increases in the CO_2 concentration due to human activities (Thomson, 1995).

The amount of solar radiation reaching a given location on the Earth can also be changed by changes in solar output (irradiance). Satellite observations of solar irradiance over the past 20 years indicate that the amount of energy put out by the Sun varies by about 0.1% over the 11-year sunspot cycle, with more energy coming out at sunspot maximum and less at solar minimum (Willson, 1997). Analyses of records of atmospheric conditions indicate that stratospheric temperatures do vary somewhat with

Figure 1: Changes in the global average concentration of carbon dioxide (light) and the local surface air temperature (dark) have been reconstructed for the past 420,000 years using information derived from an ice core drilled at the Vostok station in Antarctica (Petit et al., 1999). The local temperature record is derived from measurements of oxygen-18 isotope concentrations in the water frozen as snow. The record shows a series of long-term variations in the lower tropospheric (above the inversion layer) temperature that are similar to changes in solar radiation caused by changes in the Earth's orbit around the Sun. For most of past 420,000 years, temperatures in Antarctica (and by implication the globe) have been lower than recent values. Independent geological evidence indicates that glacial ice amounts peaked on Northern Hemisphere continents during these cold periods, most recently about 20,000 years ago. The very brief warm periods coincide with interglacial periods over the world's continents, with the Eemian interglacial of about 120,000 years ago being the last warm period until the present interglacial started about 10,000 years ago. In the absence of human influences on the climate, models of the advance and retreat of glaciers that include representations of changes in the Earth's orbit, natural variations in atmospheric composition, effects of climate change on land cover, sinking and rising of land areas due to the presence or absence of glaciers, and other factors suggest that the Earth would not return to glacial conditions for many thousands of years (Berger et al., 1999). These studies also suggest that global-scale glaciation would be unlikely if the CO_2 concentration is above about 400 ppmv.

400,000 Years of Antarctic CO_2 and Temperature Change

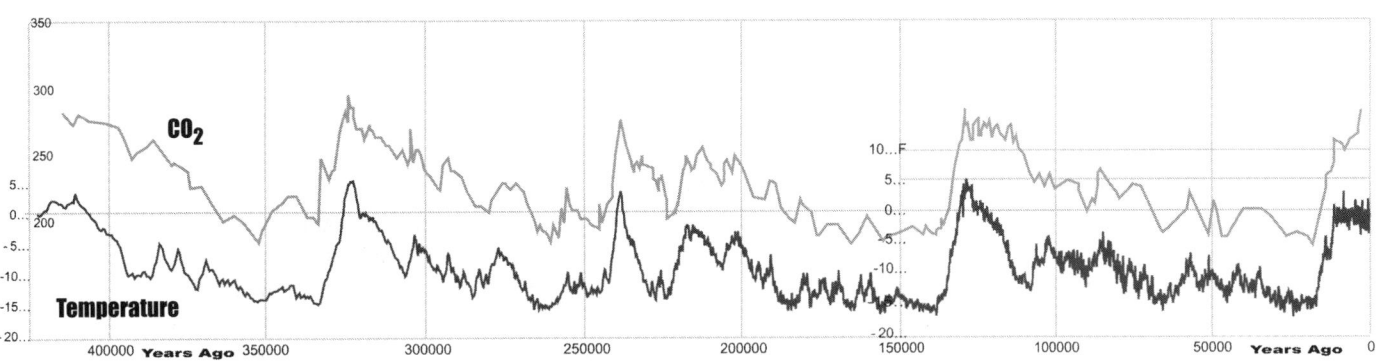

the sunspot cycle, but most scientists believe that these variations are too small to have caused a detectable impact on global average surface temperatures, especially with the thermal buffering provided by the global ocean. However, over the longer-term, reconstructions of changes in solar irradiance suggest that there may have been an increase of 0.24% to 0.3% in solar output over the past several centuries (Lean et al., 1995; Hoyt and Schatten, 1993). Calculations indicate that this increase in solar energy may have created a global warming of as much as 0.4°F (about 0.2°C) from the 17th to early 20th century, and perhaps contributed to a small cooling influence since solar irradiance peaked near the middle of this century (Lean and Rind, 1998). Over hundreds of millions of years, astronomical studies indicate that the amount of solar energy emitted by the Sun has been slowly increasing, but that these changes are too slow to be inducing noticeable climate change during human existence.

For global average temperatures to be relatively stable over time, there must be a balance of incoming solar energy and outgoing energy radiated away as heat (or infrared) energy. Observations from satellites confirm that the amount of outgoing energy is in close balance with the amount of absorbed solar energy. However, the observations of the amount of energy being emitted are consistent with a celestial body (like the Moon) that has an average temperature close to 0°F (about −18°C). Were 0°F really the surface temperature, the Earth's surface would be covered with snow and ice and it would be too cold for life as we know it. Observations indicate, however, that the Earth's atmosphere acts to warm the surface in a manner similar in effect (but different in detail) to the glass panels of a greenhouse. The Earth's natural "greenhouse" effect occurs because only a small fraction of the infrared radiation emitted by the surface and lower atmosphere is able to move directly out to space. Most of this heat radiation is absorbed by gases in the atmosphere and then, along with other contributions of energy to the atmosphere (e.g., from absorption of solar energy or heat released by the condensation of precipitation) is re-emitted, either out to space or back toward the surface. Because the downward emitted energy is available to further warm the surface, this blanketing effect raises the average surface temperature of the Earth to about 58°F (about 14°C) (Jones et al., 1999).

The gases that absorb and reemit infrared radiation are called greenhouse gases (GHGs). The set of GHGs includes water vapor (the most important

greenhouse gas), carbon dioxide (the most important greenhouse gas whose concentration is being directly influenced by human activities), methane, nitrous oxide, chlorofluorocarbons, stratospheric and tropospheric ozone, and others. Most of the GHGs occur naturally in the atmosphere, contributing to the natural greenhouse effect that acts to keep the Earth at a higher temperature than it otherwise would be were these gases not present. Observations and laboratory experiments indicate that as the amount of these GHGs is increased, more of the infrared radiation emitted upward from the surface and lower atmosphere is absorbed before being lost out to space. This process intensifies the natural greenhouse effect, trapping more energy near the surface and causing the temperatures of the surface and atmosphere to rise (e.g., see Goody and Yung, 1989).

Small particles or droplets (known collectively as aerosols) and changes in cloudiness and land reflectivity can affect how much energy is absorbed by the Earth, creating a warming influence if the overall reflectivity decreases, or a cooling influence if overall reflectivity increases. For example, aerosols can result from major volcanic eruptions or burning of sulfur-laden coals or vegetation (e.g., both natural and human-induced fires). Cooling can result when light colored aerosols (such as sulfate aerosols or volcanically injected aerosols) increase the amount of solar energy reflected back to space and thereby decrease the amount of energy absorbed by the atmosphere and surface. In addition to their direct effect, it is possible that sulfate aerosols exert an indirect cooling influence by increasing the reflectivity, extent, and character of clouds. By contrast, carbonaceous aerosols, such as organic compounds and soot that are injected by fires and inefficient combustion can increase solar absorption by the atmosphere, thereby creating a warming influence by adding to the amount of energy that can be recycled by the greenhouse effect. Changes in the vegetation cover can themselves affect the energy balance, changing surface reflectivity, evapotranspiration rates, wind drag, and the amount by which snow cover can increase surface reflectivity in winter (Pitman et al., 1999). Unfortunately, the understanding of these direct and indirect influences is quite limited, although they are not thought to be dominant (IPCC, 1996a).

While the large-scale, long-term climate of the Earth as a whole is determined by the balance of incoming solar radiation and outgoing infrared radiation (moderated by the movement of energy within the Earth system), the climate at a particular place depends on interactions of the atmosphere, land surface (including its latitude, altitude, type, and vegetative cover), and oceans. The atmosphere and oceans transport energy from place to place, store it in the upper ocean, transform the form of energy from heat to water vapor through evaporation and back through condensation, and create the climate experienced at particular places. Some of the interactions are very rapid, as in the creation and movement of storms that have important local influences. Others, however, are quite slow, as in the several year cycle of El Niño (warm) and La Niña (cold) events in the tropical eastern and central Pacific Ocean that influence the weather around much of the world. Changes in land cover also cause changes in the amount of energy absorbed or emitted. Such changes can occur as a result of deforestation, changes in snow cover, growth or decay of glaciers, or other factors. Thus, changes in the processes that determine how energy is absorbed, moved around, and stored cause the climate to fluctuate or even change over long periods.

HUMAN ACTIVITIES AND CHANGES IN ATMOSPHERIC COMPOSITION

Observations from the Vostok ice core record and other ice core records (e.g., Petit et al., 1999; Neftel et al., 1994) indicate that, until the last couple of centuries, the atmospheric CO_2 concentration had varied between about 265 and 280 ppmv over the past 10,000 years (Indermuehle et al., 1999). Even though the average atmospheric concentration varied over this time by only a few ppmv, exchanges of carbon were occurring among the atmosphere, oceans, and vegetation (each referred to as being a reservoir for carbon, in that carbon comes in and goes out over time). For example, carbon was being taken up by vegetation into living plants and being returned to the atmosphere as soil carbon decayed. Carbon dioxide was also being released into the atmosphere as cold, upwelling ocean waters warmed in low latitudes, and CO_2 was being taken up in the cold waters sinking in high latitudes. Estimates of the annual fluxes (transfers) of carbon between the atmosphere and ocean (and back), and the atmosphere and vegetation (and back), suggest that transfers of 60 to 90 billion metric tons of carbon (abbreviated as GtC, for gigatonnes of carbon) per year have been taking place for each pathway for thousands of years (Schimel et al., 1995). The relatively stable atmospheric concentration of CO_2

in the 10,000 years prior to the start of human contributions suggests that the fluxes tended to be in balance, with the amounts of carbon (or CO_2) in any particular reservoir not changing significantly over time.

Over the past few hundred years, evidence clearly indicates that human activities have started to change the balance. The lower curve in Figure 2 provides the best available reconstruction of carbon emissions to the atmosphere (as CO_2) since about 1750 (Marland et al., 1999). Deforestation and the spread of intensive agriculture initiated a growth in emissions of CO_2 in the mid-18th century that has moved about 130 GtC from the biosphere into the atmosphere (updated from Houghton, 1995) since that time. Starting in the 19th century and accelerating in the 20th century, combustion of coal, oil, and natural gas has led to emissions totaling more than 270 GtC (extended from data presented in Andres et al., 2000). These fuels are collectively referred to as fossil fuels because they were formed many millions of years ago from the fossil remains of plants and animals. The effect of combustion of fossil fuels is to add carbon to the atmosphere that has been isolated in geological formations for many millions of years. Combustion of fossil fuels is currently adding more than 6 GtC per year to the atmosphere.

As indicated in the middle curve of Figure 2, the atmospheric concentration of CO_2 has been responding to these additions. The concentrations shown here are derived from air bubbles trapped in ice cores (Neftel et al., 1994) and since 1957 from direct measurements taken at the Mauna Loa Observatory in Hawaii (Keeling and Whorf, 1999; Conway et al., 1994). These observations, and others from around the world, provide convincing evidence that there has been an increase in the atmospheric CO_2 concentration from historical levels of about 270-280 ppmv in the early 19th century to over 365 ppmv at present. Many types of studies confirm that it has been the rise in CO_2 emissions from land clearing and fossil fuel use that have caused the rise in the atmospheric CO_2 concentration over the last 200 years (e.g., Wigley and Schimel, 2000).

Although the natural fluxes of carbon being exchanged each year between the atmosphere and the oceans and between the atmosphere and vegetation are at least 10 times larger than the 6 GtC/yr from fossil fuel emissions, only about half of the fossil fuel carbon can be taken up by the vegetation and oceans. The other half of the atmospheric increase, for reasons that relate to the slow over-

Figure 2: Records of CO_2 emissions, CO_2 concentrations, and Northern Hemisphere average surface temperature for the past 1000 years: (a) Reconstruction of past emissions of CO_2 as a result of land clearing and fossil fuel combustion since about 1750 (in billions of metric tons of carbon per year) [data from CDIAC, 2000; Andres et al., 2000; Marland et al., 1999; Houghton, 1995; Houghton and Hackler, 1995]; (b) Record of the CO_2 concentration for the last 1000 years, derived from measurements of CO_2 concentration in air bubbles in the layered ice cores drilled in Antarctica, a location that has been found to be representative of the global average concentration [data from Etheridge et al., 1998; Keeling and Whorf, 1999]; (c) Reconstruction of annual-average Northern Hemisphere surface air temperatures based on paleoclimatic records (Mann et al., 1999). For the Mann et al. data, the zero change baseline is based on the average conditions over the period 1902-80. The error bars for the estimate of the annual-average anomaly increase somewhat going back in time, with one standard deviation being about 0.25˚F (0.15˚C). Although this record comes mostly from the Northern Hemisphere, it is likely to be a good approximation to the global anomaly based on comparisons of recent patterns of temperature fluctuations. See Color Plate Appendix.

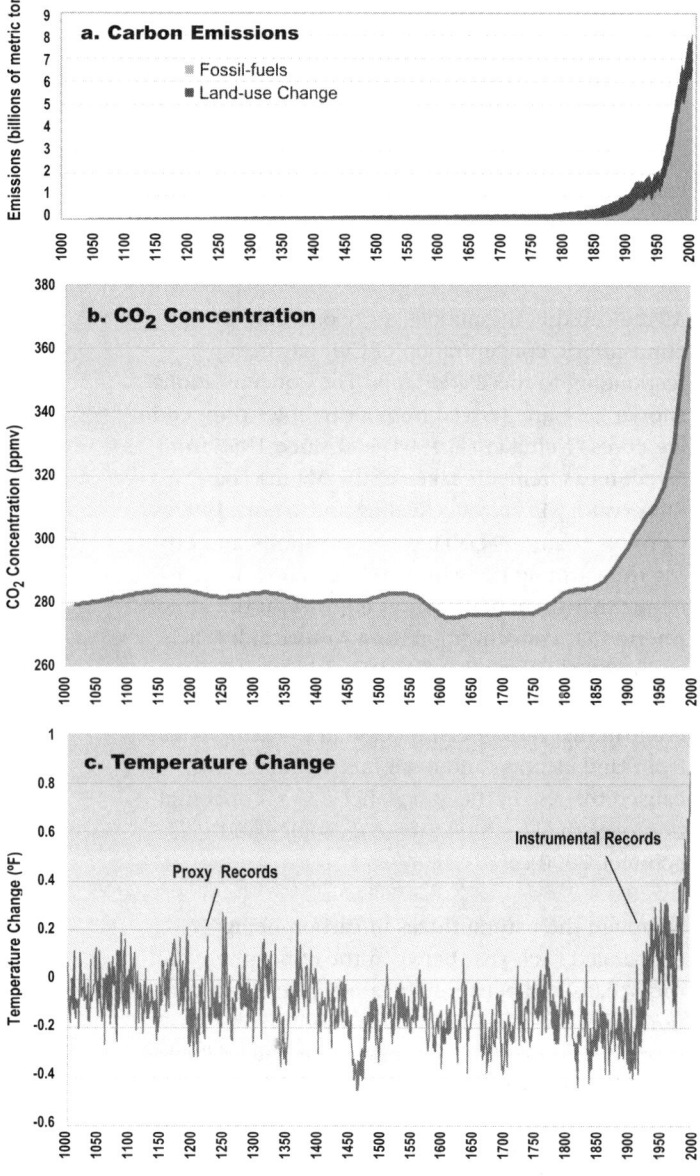

1000 Years of Global CO_2 and Temperature Change

turning rate of the oceans and limits on how much vegetation can accumulate, is destined to remain in the atmosphere for at least 100 years, even if global emissions are substantially reduced. Just as adding water to a multi-pool fountain raises its level even though the amount of water being pumped through is many times larger than the amount of water being added, adding carbon (as CO_2) from geological storage to the amount being exchanged among the atmosphere, ocean, and vegetation reservoirs causes a rise in the atmospheric concentration (as well as in ocean and vegetation levels).

HUMAN ACTIVITIES AND CLIMATE CHANGE

Based on scientific understanding of the greenhouse effect, increasing the atmospheric composition of greenhouse gases should cause the global temperature to rise. The top curve of Figure 2 presents a reconstruction of the annual-average near surface air temperature for the last 1000 years for the Northern Hemisphere (Mann et al., 1999); Crowley (2000) finds similar results. Because instrumental data are sparse or non-existent before the mid-19th century, these estimates of temperature are based on such proxy indicators as widths of tree-rings, types of vegetation, amounts of snowfall as recorded in ice cores, etc. While these measures are not as precise as thermometers, such indicators have proven to be reasonably accurate for reconstructing the fluctuations in Northern Hemisphere average temperature, providing a good indication of the variations that have occurred prior to the start of instrumental data in the mid 19th century. Although not as precise in their time resolution, records of subsurface ground temperatures also confirm that long-term warming is occurring (Huang et al., 2000).

These proxy data suggest that for most of the past 1000 years, the Northern Hemisphere average temperature had been slowly cooling at about –0.03˚C/century (Thomson, 1995; Mann et al., 1999). Then, starting in the late 19th century, the temperature started to rise, and has risen especially sharply during the latter part of the 20th century. This 20th century warming appears to be unprecedented compared to natural variations prior to this century that were presumably caused by solar, volcanic, and other natural influences. In addition, the current warming is much more extensive and intense than the regional scale warming that peaked about 1000 years ago in Europe during what is referred to as the Medieval Warm Period (Mann et al., 1999; Crowley, 2000). The recent warming is also far more than can be characterized as a recovery from the cool conditions centered in Europe and the North Atlantic region a few hundred years ago that are often referred to as the Little Ice Age (Crowley and North, 1991; Mann et al., 1999; Crowley, 2000). Overall, looking back over the few thousand years for which we can reconstruct estimates of large-scale temperatures, the current warmth of global conditions appears unprecedented.

Figure 3 presents the instrumental records of temperature change for the globe and for the US. The global results indicate that the annual average temperature has risen about 1.0°F (about 0.6°C) since the mid-19th century, with sharp rises early and late in the 20th century and a pause in the warming near the middle of the century. Sixteen of the 17 warmest years this century have occurred since 1980, and, counting the projected temperature for 1999, the seven warmest years in the instrumental record have all occurred in the 1990s. The global average temperature in 1998 set a new record by a wide margin, exceeding that of the previous record year, 1997, by about 0.3°F (Karl et al., 2000). Higher latitudes have warmed more than regions nearer the equator and nighttime temperatures have warmed more than daytime. To the extent that available data are globally representative, the 1990s are the warmest decade in the last 1000 years (the period for which we have adequate data, see Mann et al., 1999). A recent report by the National Research Council (NRC, 2000) confirms that, although satellite-measured temperatures of the lower atmosphere since that record began in 1979 are rising more slowly than surface temperatures, the two measures of the global climate have been rising at similar rates over the four-decade long record of balloon measurements (Angell, 2000). The NRC report also confirms that there is good reason to accept the evidence that the increase in the surface temperatures is real and has become relatively rapid compared to the rates of warming earlier in the 20th century.

Of course, distributions of temperature change around the world are more varied, with some regions warming at a rate substantially greater than the global average and others even experiencing a modest cooling. Observations derived from the United States Historical Climatology Network (USHCN) for 1200 of the highest quality observing stations in the US indicate that surface temperatures have increased over the past century at near to the global average rate. As is the case around the world, the largest observed warming across the US has occurred in winter. Note that it is generally not appropriate to

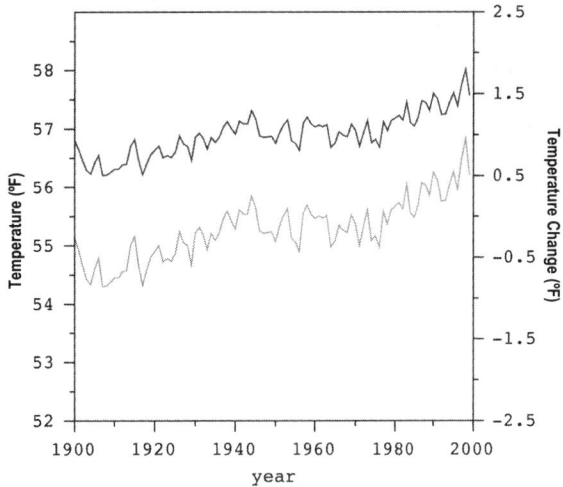

Global 20th Century Temperature

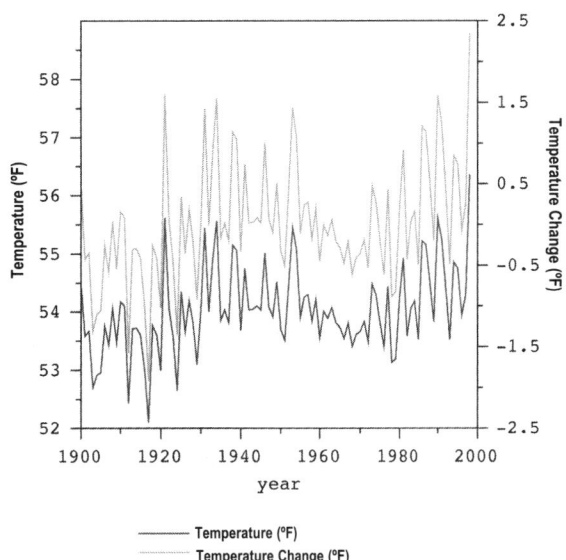

U.S. 20th Century Temperature

— Temperature (°F)
······ Temperature Change (°F)

Figure 3: (a) Global annual-average surface temperature and temperature change for combined land and ocean regions for the period 1900-1999 based on the method of Quayle et al. (1999); (b) US annual-average surface temperature and temperature change for the period 1900-1999 using the USHCN data set (Easterling et al., 1996). See Color Plate Appendix.

compare the spatial patterns of the satellite observations with the spatial patterns of the surface temperature record because, for example, the atmosphere redistributes temperature anomalies, near surface inversions disconnect surface and atmospheric temperature changes, and forcings such as by volcanic eruptions and ozone changes have different effects on the surface and atmospheric temperature trends. However, other measures of climate change across the US indicate that changes are indeed occurring.

An increasing number of studies indicates that the time histories of greenhouse gas emissions, concentrations, and surface temperature are closely related rather than just random correlations (IPCC 1996a; Tett et al., 1999). Each type of factor that could contribute to the observed warming of the climate would have a distinctive character or "fingerprint" that can be searched for in the observations. For example, an increase in solar radiation would be expected to warm both the lower and upper atmosphere, yet the lower atmosphere has warmed while the upper atmosphere has cooled. Although there is some evidence that some of the warming in the first half of the 20th century may have been due to an increase in the intensity of solar radiation, major warmings like that of the 20th century have not been evident in the records of the past thousand years (and likely much longer), suggesting that an increase in solar radiation is unlikely to be the primary cause of the recent warming (IPCC, 1996a).

It is also becoming more clear that the change is not due to a diminution of the influence of major volcanic eruptions, especially because the relatively recent El Chichón (in 1983) and Pinatubo (in 1991) eruptions injected very large amounts of aerosol into the stratosphere and yet, although there was some cooling, global average temperatures remained well above temperatures following the Krakatoa eruption in 1883 and major eruptions during the first decade of the 20th century (see Figure 3). Third, were the warming due mainly to a change in the coupling of the atmosphere and oceans, we would expect to see variations of this size and rate in the past. However, such variations do not appear to have occurred, except perhaps as the world was emerging from the last glacial period when large ice sheets were melting. Lastly, the possibility of urban heating contaminating the temperature record has been examined in numerous studies and in each case only about 0.1˚C or less of the observed 0.6˚C warming over the 20th century can be linked to urban contamination of temperature records (Karl et al., 1988, Jones et al. 1990, Easterling et al., 1997). Based on the inadequacy of natural factors to explain the recent change, the IPCC (1996a) concluded that

> "the probability is very low that these correspondences [i.e., the observed time history of the geographical, seasonal and vertical patterns of atmospheric temperature change] could occur by chance as a result of natural internal variability only. The vertical patterns of change [i.e., with stratospheric cooling and tropospheric and surface warming] are also inconsistent with those expected for solar and volcanic forcing."

More recent studies are confirming these findings

(e.g., see Hegerl et al., 1997; Barnett et al., 1999; Knutson et al., 1999).

Climatic changes due to factors being influenced by human activities also have characteristic fingerprints. Because greenhouse gases are essentially transparent to solar radiation, yet absorb infrared radiation, increasing the concentrations of greenhouse gases creates a warming influence at the surface and a cooling influence in the stratosphere (which is consistent with what has been occurring). Increases in sulfate aerosols that have occurred over the 20th century as a result of sulfur dioxide emissions resulting from coal combustion would be expected to have led to a surface cooling that would be greater in the Northern Hemisphere than in the Southern Hemisphere and most dominant in the mid-20th century. Depletion of stratospheric ozone as a result of the emissions of chlorofluorocarbons would be expected to have led to surface warming and cooling of the lower stratosphere.

Accounting for the effects of the increases in greenhouse gas and aerosol concentrations and the changes in stratospheric ozone, the time and space patterns of temperature changes are consistent with a strong warming during the 20th century caused by the changes in greenhouse gas concentrations and a cooling influence due to aerosols that grew in strength in the middle of the 20th century. Based on these diverse results, the IPCC (1996a) concluded that "the balance of evidence suggests a discernible human influence on global climate." Since that assessment, an increasing number of studies are providing more quantitative information indicating that the 20th century warming is unlikely to be due to solely to changes in solar radiation and is likely to be a result of the increasing concentrations of greenhouse gases and aerosols, especially during the latter half of the 20th century (Tett et al., 1999; Stott et al. 2000).

APPROACHES FOR ASSESSING THE IMPACTS OF CLIMATE CHANGE

Because it would be very disruptive to rapidly terminate use of fossil fuels around the world[3], it is clear that the atmospheric CO_2 concentration will continue to increase for many decades into the future. In addition, the concentrations of other greenhouse gases are increasing, and limitation of emissions of these gases would require implementing significant emission control measures. Theoretical analyses, measurements in laboratory and field experiments, and knowledge of the processes determining the temperatures of the Earth, Mars, and Venus all indicate that increasing the concentrations of GHGs in the atmosphere will increase the natural greenhouse effect, causing the world to warm. Given the weight of evidence provided by assessments of the potential for climate change, prudent risk management demands an assessment of the potential impacts of climate changes that will occur over the 21st century.

Early attempts to investigate the potential consequences of climate change often simply assumed that climate would change by an arbitrary amount (e.g., temperatures increase by 5 °F, or precipitation goes up or down by 20%). For other studies (e.g., USEPA, 1989), results from model simulations with a doubled CO_2 concentration were all that were available. Such studies, however, could only be used to investigate the potential sensitivity of existing systems to a different climate, rather than to explore how such changes might evolve over time and in so doing how these changes might spur natural and societal adaptations that could moderate the potential consequences.

In this Assessment, our goal is to examine the consequences of time-dependent climatic change. Doing this requires a two-step process. First, estimates of how the climate may change in the future must be developed. The development of scenarios of climate change that can be used in this effort is the primary subject of this chapter. Second, estimates must be developed of how the climate will affect the environment and society, and of how society might respond. These topics are the subject of subsequent chapters, but are coupled to this chapter in that the potential impacts often depend on having certain types of information available about how the weather or climate will change. To assist in these analyses, this chapter summarizes our understanding about a number of the particular climatic influences that may occur.

Three approaches have been used to develop the information base needed to evaluate the potential consequences of climate change on the US:

- Carefully checked historical data are being used to examine the potential consequences of the continuation of past climatic trends and weather and climate extremes in order to evaluate the consequences of recurrences of the types of climate fluctuations and variations that occurred in the past (e.g., the Dust Bowl period);
- Results from general circulation model simulations extending out to the year 2100 are being used to generate plausible quantitative estimates of the combined influences on climate of projected changes in greenhouse gas and aerosol concentrations; and
- Sensitivity analyses are being encouraged to facilitate exploration of the limits of vulnerability (both strengths and weaknesses) for particular regions, sectors, societal activities, and ecosystems.

This strategy has several advantages in that it serves multiple purposes and addresses several needs. These include: (a) providing a historical basis for assessing the significance of potential changes in the climate; (b) providing a range of plausible future climatic conditions as a means of recognizing the limitations and degree of uncertainty in the model formulations or assumptions; (c) incorporating the range and character of natural variability for consideration, given its importance for human and natural systems; (d) providing opportunities to compare model simulations with observations in order to evaluate model capability; and (e) ensuring opportunities to include sensitivity analyses to explore the implications of thresholds or limits in human and ecosystem adaptability. While not all groups have been able to pursue all approaches fully, having a variety of approaches has helped broaden the approach and served many of these purposes.

Thus, in this multi-pronged approach, climate models provide the Assessment process with physically consistent projections that are sufficiently plausible and quantitative to investigate the potential impacts of climate change on water, health, ecosystems, food production, and coastal areas, among other types of consequences. Use of the model projections is guided by knowledge of the climate of the last century and sensitivity analyses, and experience with the weather and climate during the historical record provides a benchmark, a personal and national context for assessing the future.

---

3 Estimates are that it would take a reduction in cumulative emissions of somewhat over 50% during the 21st century to stop any further rise in CO_2 concentration, with virtually no emissions allowed thereafter. Even stabilizing the atmospheric concentration at twice its preindustrial level would require limiting average emissions over the 21st century to about 130% of current levels, as opposed to the projected tripling of CO_2 emissions by the year 2100 as projected by the mid-range emissions scenario (IPCC, 1996a).

1. Use of Historical Records[4]

Records of how the climate has actually changed over the past century and over earlier times provide an important context for evaluating the potential consequences of future changes in climate. Climatologists have used two types of data to identify changes and variations in climate. The first consists of actual observations made over the 20th century of temperature, precipitation, and other weather-related variables that have been routinely measured at thousands of locations across much of the globe, including the US. These data have been supplemented over the past few decades by space-based measurements. Because the observing methods, instruments, and station locations have changed over time, climatologists have used various methods to assess and correct for the non-climate related factors that can affect these data. The second type is "paleoclimate" data: physical, biological, and chemical indicators recorded in rocks, ice, trees, and sediments that can be used to infer past climate conditions. Examples include the width or density of tree rings, ice cores containing air that has been trapped inside the ice for thousands and even hundreds of thousands of years, sediment at the bottom of lakes and the ocean, and others. These data are calibrated against modern-day climate measurements, and indicate that, on a global scale, climate fluctuations have been at most several tenths of a degree F (few tenths of a degree C) over the past several thousand years, indicating a quite stable climate compared to conditions occurring over the past million years.

For the US, carefully documented data records exist for the 20th century that provide climate information for most of the inhabited areas of the country. Data from the United States Historical Climatology Network (USHCN), which has been developed from a carefully selected and processed set of observations from the US Cooperative Observing Network, have been thoroughly quality controlled (Easterling et al., 1996). The data set was developed and is maintained by NOAA's National Climatic Data Center (NCDC). It contains monthly averaged maximum, minimum, and mean temperature and total precipitation data for 1200 of the highest quality observing stations in the continental United States for the period 1895 to 1997. These data have been carefully screened for recording errors and, based on well-defined procedures, adjusted for long-term variability or trends that might be introduced by changes in instrumentation, station location, urban warming, or other factors that can cause small, but

important, contamination of temperature and precipitation observations. A similar high-quality data set has been developed for Alaska, but the spatial representativeness of these data sets is not as high due to the sparsity of stations.

In addition to the monthly average station data, data sets of daily maximum and minimum temperature and precipitation have been used to examine variability and trends in climatic parameters. The Daily Historical Climatology Network data set contains observations for 187 high-quality stations in the contiguous US for the period 1910-1997 and observations for 1000 stations in the contiguous US for the period 1948-1997. An additional data set ("Probabilities of Temperature Extremes in the U.S.A." CD-ROM, available from NCDC) has been developed that includes observations of daily maximum and minimum temperatures for 300 stations in the contiguous US, Alaska, and Hawaii for the period 1948-1996. The software on this CD-ROM uses a statistical model described in Karl and Knight (1997) to provide probability estimates of how daily extreme temperatures and heat waves may change under various warming scenarios. This CD-ROM also contains software to allow the user to examine probabilities of extreme daily temperatures under the observed climate and how they might change with climate change.

An important additional data set for sensitivity studies in examining ecosystem impacts has been provided by the Vegetation-Ecosystem Modeling and Analysis Project (VEMAP Members, 1995; Kittel et al., 1995, 1997). The VEMAP data set extends from 1895-1993. This record was created by using statistical models that could link data from long-term stations to help fill in records at stations spanning only part of the period 1895-1993. The statistical methods allowed information for missing periods to be inferred and provided a spatially and temporally uniform data set for driving ecosystem and agricultural models, for example. The VEMAP record is based on USHCN stations plus USDA-Natural Resources Conservation Service Sno-Tel stations for high elevation precipitation. Altogether, the data set draws on information from about 8000 stations. The processing algorithm for deriving a high spatial resolution data set accounts for elevation and slope changes. The primary data set provides gridded monthly average data for minimum and maximum temperature, precipitation, humidity (both relative and absolute) and solar radiation at a 0.5° x 0.5° latitude-longitude spacing (about 27 miles or 43 kilometers in longitude and 35 miles or 55 km in latitude). Because some ecosystem and agricultural models require

[4]The data sets described in this section are available at http://www.nacc.usgcrp.gov/scenarios/

estimates of daily projections of these variables, a statistically based "weather-generator" technique has been used to provide estimates of daily temperature and precipitation for each grid location.

2. Use of Climate Model Simulations[5]

As a second approach, physically consistent projections of future climatic conditions derived from climate models provide an important tool for investigating the potential consequences of climate change. Climate models have been developed and are used because the Earth's atmosphere/ocean/land/ice system is far too complex to reproduce in a laboratory and simple extrapolations of past changes in climate cannot account for the rapid changes in human influences on the climate. These mathematical representations of the Earth atmosphere/ocean/land/ice system rely on the well-established laws for conservation of mass, momentum, and energy, and on empirical relationships derived from observations of how particular processes work, to specify transfers of these conserved quantities among latitude/longitude/altitude grid boxes that cover the Earth like tiles. The typical size of the grid boxes that cover the Earth in current atmospheric models is about the size of a modest sized US state and these boxes average several thousand feet (about a kilometer) thick; ocean models tend to have finer grid sizes to represent the smaller ocean eddies.

Developing models that can be used to project possible future climatic conditions requires incorporating the most important physical principles and processes that determine climatic conditions. The most comprehensive models of Earth's climate system to date are called General Circulation Models or GCMs[6] (e.g., see Nihoul, 1985; Washington and Parkinson, 1986; Mote and O'Neil, 2000). The domains for these models include the global atmosphere (up to mid-stratospheric altitudes), the oceans (from surface to the bottom), the land surface (although with limited detail in mountainous regions), and sea ice and snow cover (with Greenland and Antarctic ice caps assumed to be present). The processes represented include solar

and infrared radiation, transfer and transformation of energy, evaporation and precipitation, winds and ocean currents, snow cover and sea ice, and much more. While full detail cannot always be included, present models are constructed so as to represent key processes with sufficient detail that the large-scale climate and its sensitivity to potential changes by human activities can be self-consistently calculated. Tests are performed to determine the ability of the models to simulate the evolution of temperature, rainfall, snow cover, winds, soil moisture, sea ice, ocean circulation, and other key variables over the entire globe through the seasons and over periods of decades to centuries (e.g., Gates et al., 1999; Meehl et al., 2000a).

The advantages of using model simulations are that they are quantitative and are based on the fundamental laws of physics and chemistry, often affected and moderated by biological interactions. However, while attempts are made to ensure climate models are adequately comprehensive, such models are obviously simplified versions of the real Earth that, in their current versions, cannot capture its full complexity, especially at regional and smaller scales. The level of confidence that can be placed in such models can be evaluated by testing their ability to simulate past and present climate conditions. Among the tests that have been used to evaluate the skill of climate models have been comparisons of model simulations of the weather (to the limit that it is predictable), the cycle of the seasons, climatic variations over the past 20 years when globally complete data sets are available, climatic changes over the past 150 years during which the world has warmed, and climatic conditions for periods in the geological past when the climate was quite different than at present. Studies on comparisons of model simulations of paleoclimatic variations also suggest that models can simulate some of the types of changes that have been reconstructed from the geological records (e.g., COHMAP, 1988; Kutzbach et al., 1993; Joussaume et al., 1999). Beyond studies of particular periods, only quite simplified models have been able to be tested on their simulations of the onset, duration, and termination of the glacial periods of the past million years, and these results suggest that the GCMs likely do not adequately include all of the feedback processes that may be important in determining the long-term climate (Berger, 1999; Berger et al., 1999).

The capabilities of the most developed of these models have been carefully reviewed by the IPCC and as part of other national and international scientific efforts to evaluate their ability to represent

---

[5]Results from the models described in this section are available at http://www.nacc.usgcrp.gov/scenarios/

[6] Some studies refer to these models of the global climate system as Global Climate Models (also condensed to GCMs). However, technically, it is only the atmospheric and oceanic components of such models that are actually considered to be General Circulation Models in that they calculate how air and oceans move. Because the atmospheric and oceanic parts of global climate models are so dominant and so widely discussed, we have chosen to refer to the overall climate system models as General Circulation Models.

most aspects of the present and historical climates (e.g., see discussions in IPCC, 1996a; Gates et al., 1999; Meehl et al., 2000a). These evaluations indicate that climate models represent many, but not all, of the important large-scale aspects of the global climate quite well. The evaluations also show, however, that there are important limitations of their simulations of regional conditions, particularly in and downwind of mountainous regions, because important local influences are not well represented in the models. Model capabilities for representing natural climate variations over periods of years (e.g., the El Niño/La Niña fluctuations) to several decades (e.g., over the Pacific and Atlantic oceans) are only beginning to show success. Basically, the model evaluations indicate that the models can be used to provide important and useful information about potential long-term climate changes over periods of up to a few centuries on hemispheric scales and across the US, but care must be taken in interpreting regionally specific and short-term aspects of the model simulations. Rather than repeat the full analysis of model results being undertaken as part of the ongoing IPCC assessments, this chapter focuses on the performance of the two selected models over the US, while the Assessment as a whole focuses on determining how the selected climate change scenarios may impact human and natural systems.

Because these models are based on quantitative, physically based relationships governing, to the extent of current understanding, the global distributions of air pressure, heat, moisture, and momentum, climate models can be used to investigate how a change in greenhouse gas concentrations, or a volcanic eruption, may modify the Earth's climate. Using models in this way enables the generation of information that can potentially be used in assessment of impacts across the regions and sectors of the country. Because of continual efforts at improvement over the last several decades, these models provide a state-of-the-science glimpse into the climate of the 21st century and represent a growing capability to learn how climate change may impact the nation. However, real uncertainty remains in the ability of models to simulate many aspects of the future climate such that the model results must be viewed as providing a view of future climate that is physically consistent and plausible, but incomplete.

To convey the importance of the limitations, assumptions, and uncertainties in the model results, the IPCC has adopted the terms "projection" and "scenario" rather than "prediction" or "forecast" to refer to the results of climate model simulations of the future. This choice is meant to emphasize that we must recognize that climate model simulations do not provide precise forecasts, but rather are best used to develop insights about plausible climate changes resulting from specific assumptions such as about how energy technologies and emissions will evolve. Relying on this approach, even with the recognized uncertainties, can be useful, just as it is in other cases where individuals and organizations make use of information, even if it is associated with some level of uncertainty. For example, many people plan their days around weather forecasts with uncertainty conveyed both in words and numbers, e. g., a 30% chance of rain, or snow likely with a probability of 70%, etc. Others invest financial resources based on economic trends or decide to purchase a new home based on interest rate analyses. Understood in this light, the model-based scenarios can help to provide useful insights about the consequences of climate variability on the US, but the model results should be considered as plausible projections rather than specific predictions.

3. Use of Vulnerability Analyses

The third approach to exploring potential impacts of future climate change is to ask what degree of change would cause significant impacts in areas of critical human concern, and then to seek to determine the likelihood that such changes might occur (based on the historical record, model simulations, etc.). This approach is a form of "sensitivity analysis" conducted to determine under what conditions and to what degree a system might be sensitive to change. Such analyses are not predictions that such changes will occur; rather, they examine what the implications would be if the specified changes did occur.

For example, questions that might be explored could include: What would happen to weather conditions over the US if El Niño conditions occurred more frequently or more intensely? What would be the implications if there were simultaneous droughts in the US and in other grain-growing regions? What if the 1980s California drought lasted ten years instead of six? What if the deepwater circulation of the North Atlantic Ocean were disrupted and colder conditions prevailed from New England across to Europe? Alternatively, such questions could be phrased: How large would climate change have to be in order to cause a particular impact? How dry would conditions need to be for fire frequency and extent to increase significantly in the southeastern US? How high do ocean temperatures have to become for coral reefs to be seriously threatened? How low does river flow in the Mississippi-Missouri basin have to become for extensive areas of hypoxia (lack of oxygen) to occur in the Gulf of Mexico?

While there are always values for which one could get disastrous consequences, this approach is most useful when it focuses on basing the questions on types of climatic fluctuations, changes, or conditions that might have occurred before the instrumental record began. For example, a recent study by Woodhouse and Overpeck (1998) suggests that the 1930s drought in the Great Plains, while severe, was much shorter than earlier droughts that have occurred in the past several hundred years. They also found that droughts of similar magnitude to the 1930s drought are expected to occur about once or twice a century. Thus, a return of the 1930s drought, perhaps even lengthened, seems a plausible scenario for the future (Stahle et al., 2000 report similar findings). Similarly, various proxy records indicate that droughts in California have lasted much longer than the 1980s drought. The fact that such conditions have occurred suggests that they could occur again, and that it would be prudent to think about the impacts such climate fluctuations might have, given the way society has developed.

Because generating scenarios for sensitivity analyses necessarily focuses on considering particular conditions in particular places inducing particular types of impacts, the details of this approach are not developed in this chapter. Instead, the region and sector chapters pose the questions and contain the information underpinning these analyses and their application. This chapter is instead devoted to building the base of national-scale information that these studies have used.

TRENDS IN CLIMATE OVER THE US DURING THE 20th CENTURY

The climate of the United States contains an incredible variety of climatic types. It ranges from the high latitude Arctic climate found in northern Alaska, to tropical climates in Hawaii, the Pacific Islands and Caribbean, with just about every climate regime in between. Because of this wide array of climate, and the large area involved, the interannual variations (year-to-year variability) of climate in different parts of the country are affected differently by a variety of external forcing factors. Perhaps the most well-known of these factors is the El Niño-Southern

Oscillation (ENSO) which has an irregular period of about 2-7 years. ENSO has reasonably well-known effects in different parts of the country. In the El Niño phase, which involves unusually high sea surface temperatures (SSTs) in the eastern and central equatorial Pacific from the coast of Peru westward to near the international date line, effects include more winter-time precipitation in the southwestern and southeastern US, and above average temperatures in the Midwest that, with a strong El Niño, can extend into the northern Great Plains. The La Niña phase, which involves unusually low SSTs off the west coast of South America, often leads to higher winter-time temperatures in the southern half of the US, with more hurricanes in the Atlantic and more tornadoes in the Ohio and Tennessee valleys (Bove et al., 1998; Bove, personal communication). Furthermore, in the summertime, La Niña conditions may contribute to the occurrence of drought in the eastern half of the country (Trenberth and Branstator, 1992).

Other factors that affect the interannual variability of the US climate include the Pacific Decadal Oscillation (PDO), and the North Atlantic Oscillation (NAO). The PDO is a phenomenon similar to ENSO, but is manifest in the SSTs of the North Pacific Ocean (Mantua et al., 1997). The PDO has an irregular period that is on the order of decades, and like ENSO, has two distinct phases, a warm phase and a cool phase. In the warm phase, SSTs are higher than normal in the equatorial Pacific, and lower than normal in the northern Pacific, leading to a deepening of the Aleutian Low, higher winter temperatures in the Pacific Northwest, and relatively high SSTs along the Pacific coast. This condition also leads to dry winters in the Pacific Northwest, and wetter conditions both north and south of there. Essentially, the opposite conditions occur in the cool phase. The NAO is a phenomenon that displays a seesaw in temperatures and atmospheric pressure between Greenland and northern Europe. However, the NAO also includes effects in the US such that when Greenland is warmer than normal, the eastern US is usually colder, particularly in winter, and vice versa (Van Loon and Rogers, 1978).

As context for evaluating the importance of climate change during the 21st century, it is useful to review how the climate over the US has changed over the 20th century. Whereas Figure 3b showed the results for the US as a whole, Figure 4a displays the spatial pattern of the trend in annual average temperature across the US for the past 100 years calculated using the USHCN data set. Over most areas of the US, except for the Southeast, there has been warming of

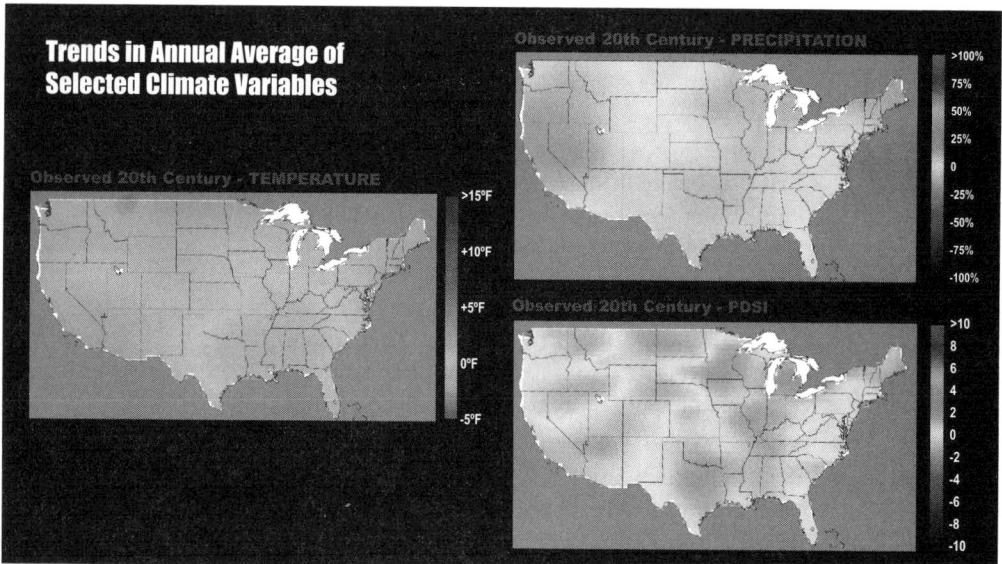

Figure 4: Trends in the annual average of selected climatic variables over the US during the 20th century as derived from observations compiled in the USHCN data set (Easterling et al., 1996). (a) Temperature (°F/century); (b) Precipitation (percent change/century); (c) Palmer Drought Severity Index (percent change/century). See Color Plate Appendix.

Observed US Trends in Daily Precipitation Intensity

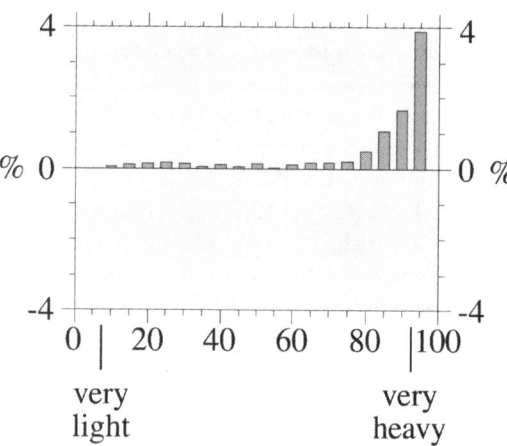

Figure 5: US trends (1910-1996) in mean precipitation (in percent change per century) for various categories of daily precipitation intensity. Values are plotted for each 5%, such that 5 represents from the lowest to 5th percentile and 95 represents the 95th to highest values of precipitation intensity. The lowest to 5th percentile are the lightest daily precipitation amounts and the 95th to highest are the heaviest daily amounts (Karl and Knight, 1998).

strong. The Southeast is one of the handful of places in the world indicating some cooling, due perhaps to the increased presence of sulfate aerosols, changes in atmospheric circulation regimes, and/or changes in cloud cover (Karl et al., 1996). Locally in some areas, interannual variability is high enough and trends small enough that some trends are not statistically significant. However, wherever the absence of statistically significant results occurs, significant trends are found nearby, reinforcing the overall observed pattern of warming for the US.

Not only are average temperatures changing, but the variability of the global climate also seems to be changing. For example, Parker et al. (1994) compared spatially averaged variances of annual temperature anomalies between the two periods 1954-1973 and 1974-1993. An increase in temperature variability of between 4 and 11% was found for the latter period. In some areas, such as North America, the increase was even larger. However, Karl et al. (1995a) analyzed changes in variability over the 20th century on a variety of time scales, from 1-day to 1-year for most of the Northern Hemisphere. They found evidence of a decrease in variability on shorter time scales (e.g., 1-day), but no broad scale patterns for longer scale variability. Thus, it appears that, for example, temperature variability on longer time scales (e.g., year-to-year variability) is increasing, but variability on shorter time scales (e.g., day-to-day or month-to-month variability) is decreasing.

Recent analysis of changes in the number of days where the minimum temperature drops below freezing indicates that the frequency of such conditions is changing across the US. Over the 20th century, averaged over the country, there has been a decline of about two days per year (i.e., -2 days/100 years). The spatial pattern of the change mirrors the changes in average annual temperature, showing cooling in the southeastern US and warming everywhere else. Thus, the Southeast has experienced an increase in the number of days below freezing while the western portion of the country has experienced strong decreases, with moderate declines or no change elsewhere. Seasonally, this change is most apparent in winter and spring, with little change in the autumn. Examination of changes in the dates of the first autumn frost and the last spring frost shows

more than 1°F, which is consistent with the observed warming of the world as a whole (Karl et al., 1996). In some regions, particularly in the Northeast, the Southwest, and the upper Midwest, the warming has been greater, in some places such as the northern Great Plains, reaching as much as 3°F. Warming in interior Alaska has also been quite

a similar pattern, with little change in autumn, but a change to earlier dates of the last spring frost. This shift has resulted in a lengthening of the frost-free season over the country with a trend of 1.1 days per decade (Easterling, 2000).

Observations indicate that total annual precipitation is increasing for both the globe and over the US. Although global precipitation has only increased by about 1%, the increase north of 30°N has been significantly larger, estimated to be 7-12% (IPCC, 1996a). For the conterminous US, the increase in precipitation during the 20th century is estimated to be 5-10% (Karl and Knight, 1998), which is broadly consistent with the global-scale changes in mid-latitudes. Although there is more spatial variation across the US for precipitation trends than for temperature trends, and although the high year-to-year variability means that small changes are not as likely to be statistically significant, there is an overall increasing trend that is highly significant, both statistically and practically (see Figure 4b). Across the US, most regions have experienced increased precipitation with the exception of localized decreases in the upper Great Plains, the Rocky Mountains, and parts of Alaska. Recent analyses suggest that much of this increase in precipitation is due to increases in heavier precipitation events (see Figure 5) and an increase in the number of rain-days (Karl and Knight, 1998). Not only is this trend evident in daily (24-hour) precipitation events, but the frequency of heavy multi-day (7-day) precipitation events is also increasing (Kunkel et al., 1999). Trends in additional types of variability and extreme events are only starting to become available (Smith, 1999; Easterling et al., 2000a).

Soil moisture is a function of how much precipitation falls and when, as well as how much evaporation and runoff occur. Figure 4c shows the trends in soil moisture across the US during the 20th century, calculated using a Palmer Drought Severity Index model (Palmer, 1965). Overall, there has been relatively little change, except for some areas of the Rocky Mountains and northern Great Plains that have become somewhat drier, and for the Mississippi River Valley, which, the way this index is calculated, tends to show a recovery from the drought years of the 1930s.

CLIMATE MODEL SIMULATIONS USED IN THE NATIONAL ASSESSMENT

Over the past decade, models have been developed that can quite reasonably simulate the climatic conditions of the 20th century and that can be used to simulate the climatic effects of changes in atmospheric composition in the 21st century. These models offer simulations of time-dependent scenarios, based on quantitative relationships, grounded in observational evidence and theoretical understanding. Around the world, there are more than two dozen groups that are developing models to simulate the climate (Gates et al., 1999; Meehl et al., 2000a). However, the various models are in various stages of development and validation, and their treatments of greenhouse gases, aerosols, and other natural and human-induced forcings continue to evolve. The various models that have been used to simulate the climate of the 21st century have been used in various types of simulations, including equilibrium and time-dependent simulations (IPCC, 1996a). The most important characteristics of the models that were considered as possible choices for use in the National Assessment are summarized in Table 1.

For the purposes of the National Assessment, to ensure use of up-to-date results, and to promote a helpful degree of consistency across the broad number of research teams participating in this activity, the National Assessment Synthesis Team (NAST) developed a set of guidelines to aid in narrowing the set of simulations to be considered for use by the regional and sector teams. To build the basis for its set of guidelines, the NAST developed a set of objectives for the characteristics of model simulations that would be most desirable. The criteria for making the selections, which included aspects concerning the structure of the model, the character of the simulations, and the availability of the needed results, included that the models must, to the greatest extent possible:

- be coupled atmosphere-ocean general circulation models that include comprehensive representations of the atmosphere, oceans, and land surface, and the key feedbacks affecting the simulation of climate and climate change;
- simulate the evolution of the climate through time from at least as early as the start of the detailed historical record in 1900 to at least as far as into the future as the year 2100 based on a

well-documented scenario for changes in atmospheric composition that takes into account time-dependent changes in greenhouse gas and aerosol concentrations (equilibrium simulations assuming a CO_2 doubling were excluded) [7];

- provide the highest practicable spatial and temporal resolution (roughly 200 miles [about 300 km] in longitude and 175 to 300 miles [about 275 to 425 km] in latitude over the central US);
- include the diurnal cycle of solar radiation in order to provide estimates of changes in minimum and maximum temperature and to be able to represent the development of summertime convective rainfall;
- be capable, to the extent possible, of representing significant aspects of climate variations such as the El Niño-Southern Oscillation cycle;
- have completed their simulations in time to be processed for use in impact models and to be used in analyses by groups participating in the National Assessment;

Table 1: Characteristics of Global Models

Model Component or Feature	Characteristics of Climate Models Recommended for Use in the National Assessment		Characteristics of Climate Models for which Some Results Were Available for the National Assessment				
	Canadian Climate Centre (CGCM1)	Hadley Centre, United Kingdom (HadCM2)	Max Planck Institute, Germany (ECHAM4/ OPYC3)	Geophysical Fluid Dynamics Laboratory (GFDL)	National Center for Atmospheric Research (NCAR CSM)	Parallel Climate Model (PCM)	Hadley Centre, United Kingdom (HadCM3)
Atmospheric resolution in horizontal (latitude-longitude) and vertical	3.75° by 3.75° (spectral T32) 10 layers	2.5° by 3.75° (grid) 19 layers	2.8° by 2.8° (spectral T42) 19 layers	3.75° by 2.25° (spectral R30) 14 layers	2.8° by 2.8° (spectral T42) 18 layers	2.8° by 2.8° (spectral T42) 18 layers	2.5° by 3.75° (grid) 19 layers
Treatment of land surface, evaporation and evapotranspiration	Modified bucket for soil moisture	Soil layers, plant canopy, and leaf stomatal resistance included	Soil layers, plant canopy, and leaf stomatal resistance included	Simplified bucket for soil moisture	Soil layers, plant canopy, and leaf stomatal resistance included	Soil layers, plant canopy, and leaf stomatal resistance included	Soil layers, plant canopy, stomatal resistance, and CO_2 processes included
Includes diurnal cycle	Yes	Yes	Yes	No	Yes	Yes	Yes
Oceanic resolution in horizontal (latitude-longitude) and vertical	1.8° by 1.8° 29 layers (based on GFDL MOM 1.1)	2.5° by 3.75° 20 layers	2.8° by 2.8° 9 layers	1.875° by 2.25° 18 layers (GFDL MOM 1.1)	2.4° by 1.2° (variable) 45 layers	0.66° by 0.66° (variable) 32 layers	1.25° by 1.25° 20 layers
Treatment of sea ice	Thermodynamic only	Dynamic and thermodynamic	Dynamic and thermodynamic	Dynamic and thermodynamic	Dynamic and thermo-dynamic	Dynamic and thermodynamic	Dynamic and thermodynami
Treatment of atmosphere-ocean coupling	Flux-adjusted	Flux-adjusted	Flux-adjusted	Flux-adjusted	Not flux-adjusted	Not flux-adjusted	Not flux-adjusted
Treatment of multiple greenhouse gases	No, CO_2 used as surrogate	No, CO_2 used as surrogate	No, CO_2 used as surrogate	No, CO_2 used as surrogate	Yes	Yes	Yes
Treatment of sulfate chemistry	Albedo change only	Albedo change only	Albedo change only	Albedo change only	Yes, with reduced sulfur emissions	Sulfate loading specified from NCAR CSM	Yes
Equilibrium temperature response of system model to CO_2 doubling	3.5°C 6.3°F	2.6°C (4.1°C for AGCM with simple ocean) 4.7°F (7.4°F)	2.6°C 4.7°F	3.4°C 6.1°F	2.0°C 3.6°F	2.0°C 3.6°F	3.3°C 5.9°F
Year when results from 1900 to 2100 simulation were made available	1998	1998	1998 (but only through 2049)	1999	1999	1999	2000

[7] Note that although vegetation is an important feature of the land surface that can affect the climate, human-induced changes in future vegetation cover and changes in vegetation due to changes in climate are not yet being treated in these climate models.

- be models that are well-documented and whose groups are participating in the development of the Third Assessment Report of the Intergovernmental Panel on Climate Change (IPCC) in order to ensure comparability between the US efforts and those of the international community;

- provide a capability for interfacing their results with higher-resolution regional modeling studies (e.g., mesoscale modeling studies using resolutions finer by a factor of 5 to 10); and

- allow for a comprehensive array of their results to be provided openly over the World Wide Web.

Including at least the 20th century in the simulation adds the value of comparisons between the model results and the historical record and can be used to help initialize the deep ocean to the correct values for the present-day period. Having results from models with specific features, such as simulation of

the daily cycle of temperature, which is essential for use in cutting edge ecosystem models, was important for a number of applications that Assessment teams were planning. Despite uncertainties surrounding available emissions scenarios, using results with consistent assumptions about increases in greenhouse gases and sulfate aerosols helps to ensure that the assessment efforts of the various regional and sector teams can be combined into a consistent national synthesis and could then be interfaced with international assessments.

These restrictions led to a decision to consider mainly model simulations that used emissions scenarios that were close to the IPCC's "IS92a" scenario (see IPCC, 1992) (see box, "What Does the IS92a Scenario Assume?") so that there could be ready comparison with international studies and analyses. As shown in Figure 6, the net radiative forcing for

What Does the "IS92a" Scenario Assume?

To prepare a projection of future changes in climate, a scenario of future concentrations of greenhouse gases must be developed. This is often done by starting with a scenario for changes in emissions of greenhouse gases. The future emissions scenario most used for analysis throughout the 1990s, including to drive model simulations of climate change, has been the IS92a scenario. This scenario is near the middle of the range of six peer-reviewed scenarios of possible alternative futures published by the IPCC in 1992 (IPCC, 1992). Based on calculations done with models of greenhouse gas and aerosol concentrations, the IS92a scenario results in a climate forcing that is similar to that used in the two models chosen for primary use by the National Assessment. The recently published set of IPCC 2000 emissions scenarios finds that the net radiative forcing of this emissions scenario (i.e., greenhouse gas induced warming minus aerosol induced cooling influence) is still well within the range of what the IPCC has recently concluded are plausible scenarios for how energy technologies, energy use, economic development, and population growth of the 21st century may evolve (IPCC, 2000).

The IS92a scenario makes a number of assumptions based on current and projected trends and expectations. Like each of the IPCC's 1992 scenarios, it assumes that the nations of the world will implement no major changes in their policies that would limit the growth of activities that are contributing to climate change. The scenario also assumes that global population will approximately double over the 21st century and that continued economic growth at rates typical of the recent past will raise total economic output by a factor of about 10; growth by this amount would mean that global average per capita economic activity would go up by a factor of 5. Because of increasing efficiencies and new technologies, the scenario assumes world energy growth will, however, only need to increase by about a factor of 4. To meet this increase in energy demand, energy derived from fossil fuels (coal, oil, and natural gas) is projected to more than double (increased use of coal, however, would increase CO_2 emissions by a factor of about 3). To provide the rest of the energy, the IS92a scenario assumes that energy derived from non-fossil fuel energy sources (e.g., solar, wind, biomass, hydroelectric, and nuclear) will increase by a factor of about 15. The scenario assumes that this growth in non-fossil energy sources will occur without any implementation of climate-specific policies because the costs of these energy sources will decline relative to fossil fuels. If this scenario comes to pass, it would mean that the fraction of energy coming from non-fossil sources would rise from just over 10% of all energy now to over 40% by 2100. Scenarios forecasting less rapid availability of non-fossil technologies would lead to greater CO_2 emissions to meet the same growth in population and economic activity; scenarios leading to reduced CO_2 emissions would require some combination of more rapid increases in efficiency improvements, faster development of non-fossil technologies, a slower rate of economic development, and reduced population growth.

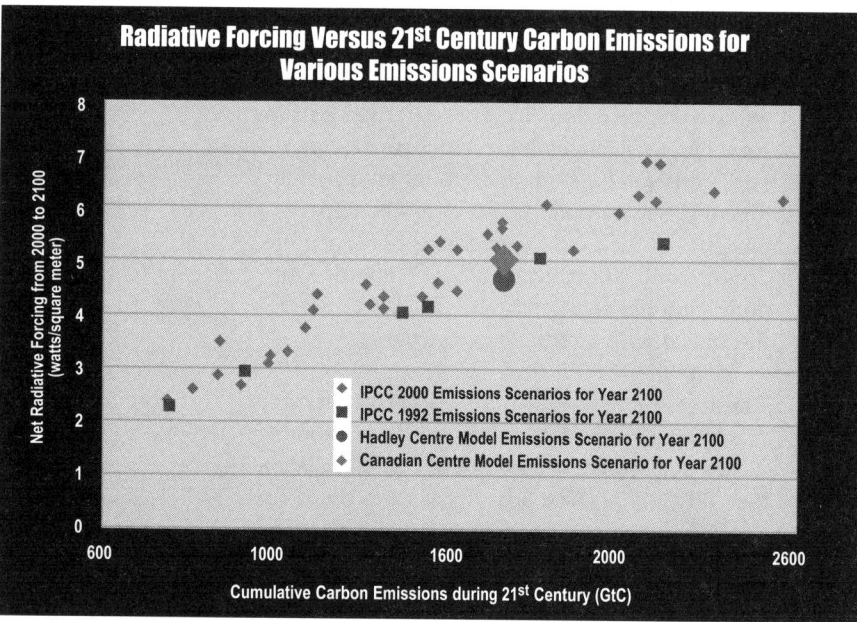

Figure 6: Comparison of the projections of total carbon emissions and overall human-induced radiative forcing for the six emissions scenarios prepared by the IPCC in 1992 (IS92 scenarios; IPCC, 1992) and the 35 emissions scenarios prepared by the IPCC in 2000 for which radiative forcing could be estimated (SRES scenarios; IPCC, 2000). These scenarios are based, although in different ways, on projected changes in emissions resulting from changes in population, economic development, energy use, efficiency of energy use, the mix of energy technologies, etc. The horizontal axis gives the total emissions of fossil fuel-derived carbon dioxide projected for the 21st century (in billions of tonnes of carbon, GtC). For reference, if the current level of global carbon emissions is maintained from 2000 to 2100, cumulative emissions over the 21st century would be roughly 650 GtC. Assuming no climate-related controls on emissions are introduced, this value is near the lowest value projected by any of the scenarios for the 21st century. The vertical axis gives the projected change in net radiative forcing at a pressure level approximating the tropopause (in watts per square meter) for all human-induced changes in greenhouse gases and aerosols (both direct and indirect contributions) over the 21st century using relationships employed in the IPCC Second Assessment Report (IPCC, 1996a; Smith et al., 2000), including the uptake of CO_2 by the oceans and land. Radiative forcing is important because it is the driving force for global warming; for reference, the projected change in radiative forcing up to the year 1992 is about 1.6 watts per square meter (IPCC, 1996a). The figure also shows the net radiative forcing and the approximate emissions of carbon used in the Hadley and Canadian scenarios. For these scenarios, which increase the equivalent CO_2 concentration by 1% per year, the carbon emissions are estimated by calculating the emissions needed to match the net radiative forcing after subtracting the radiative effects of other greenhouse gases and aerosols based on the average of IS92a and IS92f scenarios, and is an amount between the IS92a and IS92f scenarios. Based on these calculations, the Canadian and Hadley scenarios lie near the mid-range of the proposed scenarios in terms of both carbon emissions and net radiative forcing. See Color Plate Appendix.

the 21st century for the IS92a emissions scenario is near the mid-range of radiative forcing scenarios constructed based on the new set of emissions sce-

narios prepared by the IPCC (2000). The new scenarios suggest that the upper limit of possible increases in radiative forcing by 2100 is greater than for the IS92 emissions scenarios due to the recent recognition that significantly intensified use of fossil fuels could lead to substantial increases in emissions of methane, carbon monoxide, nitrogen oxides, and volatile organic compounds that would significantly increase the concentration of tropospheric ozone, a strong greenhouse gas. What is clear from this diagram, and is discussed more fully later in the text for the particular models used, is that the IS92a emissions scenario is a quite plausible choice for consideration if the results from only one emissions scenario are available. However, it must be emphasized that the climate model results that are available are simply one representation of what could happen, and are not predictions or forecasts of what might actually happen. This restriction could start to be relaxed in future assessments by considering results from a wider range of climate models and a wider range of emissions scenarios.

In the selection of the particular set of model results to be used for the Assessment, a number of additional constraints were also considered. For example, time and computer resource constraints generally prevented the completion of a new set of model simulations with these models specifically designed for this Assessment. Given the limited duration of the Assessment, and the desire to process the GCM results through the VEMAP processing package in order to better account for changes in mountainous regions, it was essential that scenarios be completed early in the assessment process (i.e., mid to late 1998) in order to enable timely availability of processed model results. In addition, the limitations in capabilities and resources have meant that the set of cases and situations that all teams would be asked to use needed to be kept to a minimum. For these reasons, it was necessary to limit the selection to a minimum, but representative, set of model simulations.

Given these guidelines and considerations, the results from particular simulations of two models were selected to be the primary sources of simulation-based projections for this first National Assessment. The specific simulations selected were those runs that are closest to the IS92a emissions scenario from the GCMs developed by the Canadian Centre for Climate Modelling and Analysis (henceforth referred to as the "Canadian model scenario") and the Hadley Centre for Climate Prediction and Research of the Meteorological Office of the United Kingdom ("Hadley model scenario," specifically the simulation using the HadCM2 GCM). Although careful consideration was given, the timing and types of simulations available from US modeling centers did not meet as many of the important criteria as the models selected (see NRC, 1998), although results from US modeling groups were able to be used by some regional teams and for some types of investigations.

Using the results from more than one major modeling center helps to capture a sense of the range of conditions that may be plausible in the future, even though the range of possible futures is likely to be broader due to the wide range of possible emissions scenarios as well as uncertainties arising from model limitations. Both of the models selected are coupled ocean-atmosphere models that are well documented and have been peer-reviewed by the scientific community (Boer et al., 1984, 2000b; Johns et al., 1997). Both models include the day-night cycle, which enables them to provide estimates of changes in minimum and maximum temperature. Both models reasonably represent the broad scale features of the global climate, including the major high and low pressure centers and the major precipitation belts that generate the weather. Even though each simulation can take several hundred hours on the fastest supercomputers that are available (Karl and Trenberth, 1999), both models have available ensembles of simulations (Mitchell et al., 1995; Mitchell and Johns, 1997; Boer et al., 2000a).

Although the fundamental physical principles driving these models are similar, there are differences in how, and even whether, the models incorporate some important processes. Therefore, there are some differences in the results of these models. One important factor in causing these differences is the uncertainty remaining in how best to represent such processes as changes in cloud cover in response to global climate change (e.g., see Mitchell et al., 1987). Because of such uncertainties, it is considered important to use models representing a range of possible values in impact studies. In addition, it needs to be noted that none of the model projections consider the potential influences of changes in natural forcings, even though it is likely that fluctuations will continue to occur as a result of variations in solar forcing and occasional volcanic eruptions (Hyde and Crowley, 2000).

Global Mean Temperature Anomalies (a)

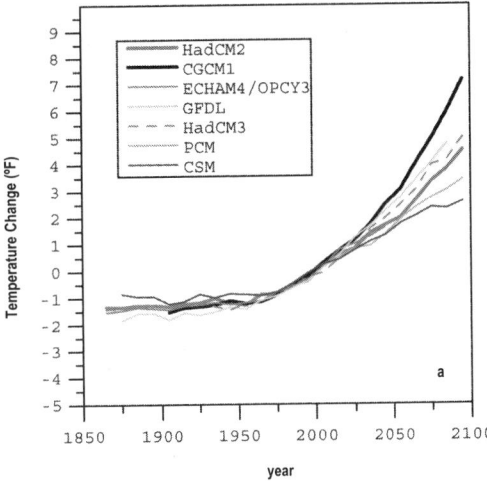

U.S. Mean Temperature Anomalies (b)

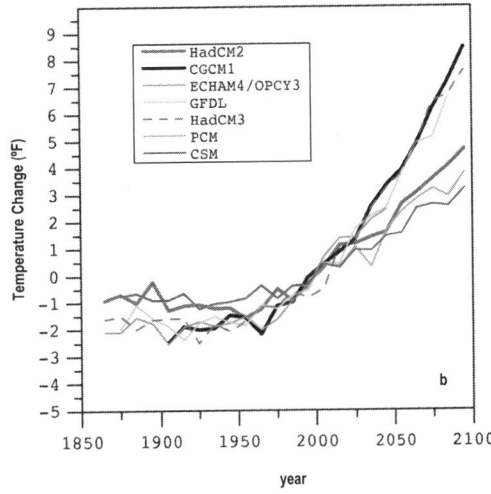

Figure 7: Comparison of the annual average changes in (a) global average surface air temperature (°F), and (b) US average surface air temperature (°F) from the Canadian model scenario and Hadley model scenario simulations used in the National Assessment and from the simulations of other modeling groups, including a very recent result from the Hadley Centre model version 3, Germany's Max Planck Institute/German Climate Computing Center (DKRZ), NOAA's Geophysical Fluid Dynamics Laboratory, and from the Parallel Climate and the Climate System models from the National Center for Atmospheric Research (which used a slightly lower greenhouse gas emission scenario and a significantly lower sulfate emissions scenario than the other models). Decadal means have been plotted to suppress the natural year-to-year variability. The baseline period is 1961-1990. The anomalies are with respect to the year 2000, calculating the values from a 2nd order polynomial fit over adjacent decades. See Color Plate Appendix.

Figure 7 and Table 2 provide a comparison of the projected changes in annual average surface temperature for the globe and for the US based on results from the Canadian and Hadley models. Results are

35

Table 2: Model-Simulated Changes in 20th and 21st Century Surface Temperatures for the US

Model simulated changes in annual-average surface temperature for the 20th and 21st centuries based on linear fits to the decadal average values derived from the model simulations with comparison to estimates of observed changes for the 20th century and the range of warming projected by the IPCC (1996a) for the various emission scenarios and climate model sensitivities.

Source of Estimate	Simulated Change in Global Average Surface Temperature		Simulated Change in Average Surface Temperature for Conterminous US	
	20th Century	21st Century	20th Century	21st Century
Hadley - Version 2	1.0°F 0.55°C	4.7°C 2.6°C	0.8°F 0.4°C	4.7°F 2.6°C
Canadian Centre	1.2°F 0.7°C	7.5°F 4.2°C	1.9°F 1.05°C	9.0°F 5.0°C
Max Planck Institute (MPI)	1.0°F 0.55°C	3.4°F* 1.9°C	1.6°F 0.9°C	4.1°F* 2.3°C
Geophysical Fluid Dynamics Laboratory (GFDL)	1.4°F 0.8°C	5.7°F* 3.2°C	1.65°F 0.9°C	7.8°F* 4.3°C
Hadley - Version 3	1.1°F 0.6°C	5.6°F 3.1°C	1.4°F 0.8°C	8.85°F 4.9°C
Parallel Climate Model	0.9°F 0.5°C	3.7°F 2.0°C	0.7°F 0.4°C	4.1°F 2.3°C
Climate System Model	0.9°F 0.5°C	2.8°F 1.5°C	0.7°F 0.4°C	3.3°F 1.8°C
Observed (Quayle, 1999 and Karl et al., 1995b)	0.7-1.4°F 0.4-0.8°C		0.5-1.4°F 0.3-0.8°C	
IPCC (1996a) for 1990 to 2100 (uncontrolled sulfur emissions)		1.6-6.3°F^ 0.9-3.5°C^		
IPCC (1996a) for 1990-2100 (level sulfur emissions)		1.4-8.1°F^ 0.8-4.5°C^		

*Estimates for less than the full 21st century have been linearly extrapolated to develop an estimate for change over the full century (MPI from 2049 to 2100; GFDL from 2090 to 2100).
^For estimates for just the 21st century, about 0.2-0.3°F (0.1-0.2°C) must be subtracted, depending on scenario considered.

also provided for a set of simulations done with other models, some of which became available after processing of results for use in impacts studies had been completed. The Canadian and Hadley simulations each use an emissions scenario for changes in greenhouse gas and aerosol concentrations over the 21st century that is designed to represent the IS92a (or no policy intervention) case of the IPCC (1992). New simulations are being carried out by the world's modeling groups for the new range of climate scenarios developed by the IPCC (2000). As an example of the results of this type of simulation, the figure also includes the newer simulations with the NCAR CSM and PCM models that use a lower

emissions scenario than IS92a for greenhouse gases and aerosols (ACACIA-BAU, see Dai et al., 2001) and that are carried out with a model with a climate sensitivity in the lower part of the range of 2.7 to 8.1°F (1.5 to 4.5°C). It is important to note that these model results also indicate that substantial warming occurs even assuming that emissions are reduced significantly below the IS92a scenario.

Although the emissions scenarios are the same for the Canadian and Hadley simulations, the Canadian model scenario projects that the world will warm more rapidly than does the Hadley model scenario. This greater warming in the Canadian model scenario occurs in part because the Hadley model scenario projects a wetter climate at both the national and global scales, and in part because the Canadian model scenario projects a more rapid melting of Arctic sea ice than the Hadley model scenario. Results from other models, with the exception of the latest results from the National Center for Atmospheric Research (Dai et al., 2001), are generally within or slightly below the lower bound of this range. The larger reduction in the NCAR model results from the slower rise in greenhouse gas concentrations that is assumed and due to a projected increase of low cloud cover that is not evident in simulations by other models, although these effects are somewhat offset by reduced loadings of sulfate aerosols. Compared to the range suggested for the year 2100 in the IPCC results (IPCC, 1996a), the Hadley model scenario projects warming for the 21st century that is slightly above the central IPCC estimate of about 4˚F (2.4˚C) after adjusting for the change in baseline years. The Canadian model scenario projects global average warming that is slightly above the high-end of the IPCC suggested range if sulfur emissions are not controlled, but within the range if they are assumed to be controlled. The greater warming for the Canadian model (Hengeveld, 2000), as for the Hadley-3 model, is likely a result of their higher climate sensitivity. While neither the Hadley nor Canadian model scenarios projects a rate of warming coincident with the low end of the IPCC range, this lower bound is also generally not consistent with estimates of climate sensitivity derived from comparison of model simulations with the paleoclimatic record or with the extent of warming that has occurred over the last two hundred years.

All of the models, with the exception of the Hadley version 2 GCM (HadCM2), project greater warming over the US than for the globe as a whole. The variation of results among model results is also greater for the US than for the globe. It is especially interesting that the projected warming due to these changes in greenhouse gas concentrations is very rapid after the mid-1970s, when much of the recent warming began. As an indication of how the sequential improvement of models by the various groups may change the results, it is instructive to compare the results from the HadCM2 that were used in this Assessment, with the results from the Hadley version 3 GCM (HadCM3) that were not available in time for full use in this Assessment. The more recent Hadley model results suggest significantly more warming over the US than the Hadley model selected for this Assessment. Recognizing that all model results are plausible projections

rather than specific quantitative predictions, the primary models used for this Assessment project that the average warming over the US will be in the range of about 5 to 9˚F (about 2.8 to 5˚C). However, given the wide range of possible emissions scenarios and uncertainties in model simulations, it is possible that the actual increase in US temperatures could be higher or lower than indicated by this range. Such a warming is approximately equivalent to the annual average temperature difference between the northern and the central tier of states, or the central and the southern tier of states.

Figure 8 provides similar information for projected changes in precipitation. For the globe, the two primary Assessment models represent a range of plausible conditions that are typical of results from other climate models that have used the same emissions scenario, although the simulation of NOAA's Geophysical Fluid Dynamics Laboratory (GFDL) does suggest an even greater increase in global precipitation than either of these primary models. Over the US, the spread among model results is greater than over the globe due to the patchier nature of precipitation and changes in precipitation. The Hadley model scenario projects a very large increase in precipitation (which is one reason its temperature increase is lower than for other models) whereas the Canadian model scenario results show an increase mainly in the second half of the 21st century. The greater variability of the precipitation results, compared to the temperature results, reflects the larger natural variability of precipitation. By using the selected results from the Canadian and Hadley models, we are not only capturing results for differing model sensitivities, but also, to a large extent, for much of the wet/dry and hot/warm range of future climate conditions generated by the wider set of climate models. As such, these cases seem quite representative of the types of conditions that could occur.

While the available information provides quite plausible estimates for the future, there are important limitations that need to be recognized:

- Each model simulation provides a snapshot of the temporal and spatial variations of the climate as the global climate is evolving through time in response to changes in greenhouse gases and aerosols. Because of inherent variability in the model that results from small differences in the initial model conditions, only by employing an ensemble of simulations would we be able to assess the statistical significance of the model results for any decade over this interval. When an ensemble of simulations is analyzed, the long-term trends in variables have been found to be generally consistent across multiple simulations, but quite variable for particular years, decades, and locations.

• The particular simulations we have selected reflect only one particular emissions scenario rather than a range of emission scenarios (the

Global Precipitation Anomalies

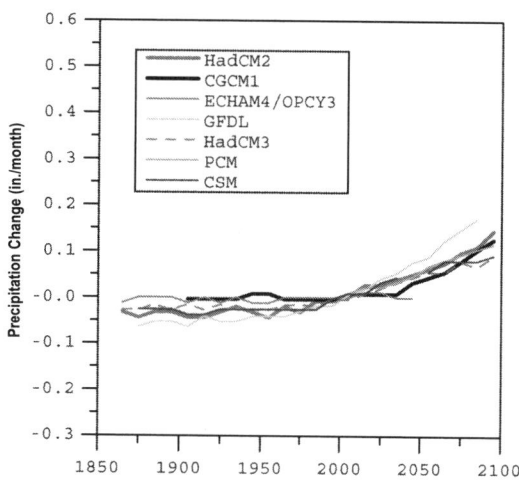

US Precipitation Anomalies

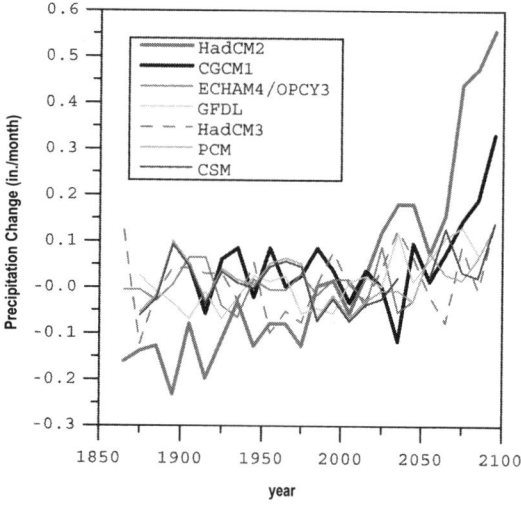

Figure 8: Comparison of the annual average changes in (a) global average precipitation (inches per month), and (b) average precipitation over the US from the Canadian model scenario and Hadley model scenario simulations used in the National Assessment and from the simulations of other groups (same as for Figure 7). The baseline period is assumed to be 1961-1990. Although decadal means have been applied to suppress year-to-year fluctuations, the greater variability of precipitation than temperature still reveals significant variations due to natural factors; the magnitude, although not the timing, of the remaining fluctuations may be considered plausible. The anomalies are with respect to the year 2000, calculating the values from a 2nd order polynomial fit. See Color Plate Appendix.

emissions scenario being used for the model runs we are using is described in a subsequent section). For the future, the actual emissions of greenhouse gases and aerosols are likely to be different than the baseline used. For example, it is quite possible that emissions of both greenhouse gases and aerosols may be lower (as a result of societal development, control measures, etc.) or higher if oil shale and coal become the fuels of choice throughout the world. Changing the emissions scenario would give different results, although the rate of climate change over the next few decades is not likely to differ significantly from the model results because of the momentum created by climate and global energy systems.

• Use of only two model simulations provides a limited opportunity to investigate the consequences of climate variability and change. To help overcome this limitation, regions and sectors have been asked, as explained earlier, to look both at the historical record and to consider cases that reflect educated guesses based on the nature and importance of specific regional and sector sensitivities. One tool developed for use in the sensitivity analyses is the "Probabilities of Temperature Extremes" CD-ROM that has been developed by NCDC. Other approaches focus on drawing information from the regional paleoclimatic record.

Recognizing the limitations in the minimum strategy approach that could be proposed for the entire set of Assessment teams, some groups have had the resources available to carry through additional impact studies using results from the models developed at the National Center for Atmospheric Research (NCAR), NOAA's Geophysical Fluid Dynamics Laboratory (GFDL), NASA's Goddard Institute for Space Studies (GISS), and the Max Planck Institüt/Deutsches Klimarechencentrum (MPI/DKRZ, referred to simply as MPI) in Germany. To support this extended effort, access to the wider set of climate information is provided through the National Assessment web site.

While GCMs have shown significant improvement over recent decades, and the models used in the Assessment are considered among the world's best, there are a number of shortcomings that arise in applying the models to study potential regional-scale consequences of climate change. For this Assessment, several types of effort have been used to start to address these problems. Of most importance for the analyses done as part of the National Assessment, the results of the GCMs have been passed through the VEMAP processing algorithms so that information could be provided at a scale that

Normalization of Results from Climate Models

While the Canadian and Hadley models both provide reasonable simulations of the large-scale features of the 20th century climate over the US, there are differences in the absolute values of temperature and precipitation that could affect many types of impact studies if adjustments were not made. For the 20th century, this process is accomplished by simply driving the impact models with the observed climatic conditions rather than the model-generated conditions. For studying the 21st century, this procedure is not possible as observations of the future are not available. Instead, the assumption is made that the differences between models and observations for the 20th century are systematic – that is, that the differences between models and observations are a result of limitations in the model formulation and will be present in simulations for both the 20th and 21st centuries. If this assumption is valid, then the changes in climate due to human activities can be determined by taking the difference between a model simulation with increasing concentrations of greenhouse gases and one simulation without such changes and adding the difference to the observations for the 20th century to yield plausible estimates for the changing climatic conditions of the 21st century. Although this assumption is certainly not completely valid, it is likely to be sufficiently valid that the uncertainties introduced in making this assumption will, for many types of situations, be of less importance than uncertainties resulting from other factors (e.g., the differences between models, uncertainties in climate sensitivity to changes in greenhouse gases, uncertainties in impact models, etc.).

To carry out this normalization of the model results using the differencing approach, and to provide improved spatial resolution of key climate variables, the VEMAP methodology applied initially to the observed station data was used to process the Canadian and Hadley model scenarios of climatic changes during the 21st century. This procedure was done by interpolating the monthly average changes calculated by the models to the VEMAP grid and then basing the scenario for the 21st century on the model calculated increment to the 20th century climate baseline. In the case of temperature, the adjustment was carried out by adding the model estimate for the monthly average change in temperature from the model's 1961-90 baseline to the local value of the observed monthly baseline temperature for the same period. For precipitation, the adjustment was made based on the multiplying by the ratio (percentage) change calculated by the model. In this way, projections for the 21st century were made for changes in mean maximum surface air temperature, mean minimum surface air temperature, and total precipitation on a monthly basis. A weather generator was used to derive daily values for these variables, and incoming solar radiation and humidity were then derived from these variables. These data sets are available at http://www.nacc.usgcrp.gov/scenarios and values for particular regions or time periods can be extracted by going to http://eos-webster.sr.unh.edu.

While this application of the VEMAP technique of using the changes calculated by the models to simulate the changes to the historical record provides a practical way of accounting for the systematic offsets between modeled and locally observed conditions, this technique is not without its limitations. For example, care must still be taken when analyzing the effects of special situations where thresholds effects might occur (e.g., the presence or absence of snow cover in mountainous regions) resulting in projected changes that may be too strong or too weak. Also, assuming that the temperature changes will be the same in valleys and on mountaintops fails to deal with the effects of inversions and the special weather conditions of mountain regions. Using the ratioing approach to estimate precipitation change also assumes, at least to some extent, that weather systems will be of the same type, being different only in overall intensity or number, while not recognizing that changes in storm direction into mountainous regions could have a large effect. Darwin (1997) argues that at least some of these limitations can be reduced, especially in desert regions, by using absolute amounts of precipitation to make the adjustment; however, in mountainous regions, this approach seems to fail to deal with the strong gradients in precipitation with altitude. Alcamo et al. (1998) have compared the risk for worldwide natural vegetation using the two approaches, and find a lower risk using the ratioing approach that is used in the Assessment than the difference-adjustment technique, suggesting that the conclusions drawn in this Assessment may be somewhat conservative, although this uncertainty is likely less than the uncertainty resulting from the differences in the model projections.

What is most clear is that, for future assessments, meso-scale models need to be used to more rigorously and accurately simulate regional patterns of changes in precipitation (and such efforts are already underway in a couple of the regions).

was comparable to the information in the data sets for the historical period and in a way that accounted for at least some of the shortcomings and biases in the models. In particular, the model scenario results used in the impact assessments were adjusted to remove the systematic differences with observations that are present in the GCM calculations in particular regions due to mountainous terrain and other problems. The VEMAP normalization process is described in the box "Normalization of Results from Climate Models."

In addition, some regional teams have applied other types of "down-scaling" techniques to the GCM results in order to derive estimates of changes occurring at a finer spatial resolution. One such technique has been to use the GCM results as boundary conditions for mesoscale models that cover some particular region (e.g., the West Coast with its Sierra Nevada and Cascade Mountains). These models are able to represent important processes and mountain ranges on finer scales than do GCMs. However, these simulations are very computer intensive and it has not yet been possible to apply the techniques nationally or for the entire 21st century. With the rapid advances in computing power expected in the future, this approach should become more feasible for future assessments. To overcome the computational limitations of mesoscale models, other participants in the Assessment have developed and tested empirically based statistical techniques to estimate changes at finer scales than do the GCMs, and these efforts are discussed in the various regional assessment reports. These techniques have the important advantage of being based on observed weather and climate relationships, but have the shortcoming of assuming that the relationships prevailing today will not change in the future.

CLIMATE MODEL SIMULATIONS OF THE 20th CENTURY FOR THE US

An important measure of the adequacy of the applicability of these models for simulation of future climatic conditions is to compare their results for simulation of the climate of the 20th century over the US with observations[8]. In conducting these simulations, the models are driven by observations and, particularly for aerosols, reconstructions of the changing composition of the atmosphere. While one might want the simulations to match observa-

tions very accurately, several complications must be accounted for in making the comparisons. First, the model simulations have not been designed to, and cannot be designed to, exactly reproduce the climate of the 20th century. One reason that reproduction of the 20th century climate is not possible is that observations are poor or entirely lacking of changes in some of the factors that could lead to part of the naturally induced fluctuations in the climate. These factors include changes in solar radiation,[9] injection of volcanic aerosols into the stratosphere, and the state of the global ocean and ice sheets at the start of the century. Over the long term, omitting such natural forcing factors should tend to average out to a near zero net effect on global average temperatures. For this reason, the effect of these omissions is often assumed to be small over periods of many decades compared to the steady and long-term growth of the greenhouse effect. Second, because of the chaotic nature of the climate, we cannot expect to match the year-by-year or decade-by-decade fluctuations in temperature that have been observed during the 20th century. Third, these particular model simulations do not yet include consideration of all of the effects of human-induced changes that are likely to have influenced the climate, including changes in stratospheric and tropospheric ozone and changes in land cover (and associated changes relating to biomass burning, dust generation, etc.). Finally, while it is desirable for model simulations not to have significant biases in representing the present climate, having a model that more accurately reproduces the present climate does not necessarily mean that projections of changes in climate developed using such a model would provide more accurate projections of climate change than models that do not give as accurate simulations. This can be the case for at least two reasons. First, what matters most for simulation of changes in future climate is proper treatment of the feedbacks that contribute to amplifying or limiting the changes, and accurate representation of the 20th century does not guarantee this will be the case. Second, because projected changes are calculated by taking differences between perturbed and unperturbed cases, the effects of at least some of the systematic biases present in a model simulation of the

---

[8]It should be noted that while we are interested in changes over the US, these changes are in many cases determined by how well the model represents changes in the global scale features of the climate that in turn then affect what is happening over the US and in particular regions. Although the models selected do include flux adjustments to reduce drift in global average temperatures, these flux adjustments have only a limited influence on determining the patterns of continental-scale climate simulated by these models. It should also be noted that models not including flux adjustments give a generally similar pattern and range of model projected changes in climate.
[9]The newest GCM simulations are beginning to investigate the effects of past variations in solar radiation on climate, even though reconstructions of past levels of solar output are uncertain.

present climate can be eliminated. While potential nonlinearities and thresholds make it unlikely that all biases can be removed in this manner, it is also possible that the projected changes calculated by such a model could turn out to be more accurate than simulations with a model that provided a better match to the 20th century climate.

Recognizing these many limitations, evaluation of the simulations of the Canadian and Hadley models are presented here to give an indication of the general adequacy of the models for use in these studies. Analyses at the global scale by the two modeling groups indicate that there is general agreement with the observed long-term trend in temperature over the 20th century, although there is significant variation over decadal time scales (e.g., Johns et al., 1997; McFarlane et al., 1992; Flato et al., 2000). As shown by Stott et al. (2000), simulations with the Hadley model also show that, by accounting for changes in greenhouse gas concentrations, sulfate aerosols, and solar forcing, there is a close similarity between the observed and the modeled climates, with both model simulations warming about 1°F during the 20th century and showing a roughly similar temporal pattern even though not all influences were considered.

Few of these comparisons have focused on the character of the simulations at the continental and national scale that are of interest in this Assessment, and so this section presents a selection of these model results. At these scales, so many types of comparisons can be made, and there are so many ways to display and interpret the results, that the set of comparisons included here is augmented by additional comparisons available on the Web site[10] to provide the interested reader the opportunity to gain a more complete perspective. The set of figures here have been chosen to illustrate that results from these models, while not predictions, are plausible and suitable for use in investigating the potential consequences of climate variability and change for the US.

Figure 9 compares the Canadian and Hadley model scenarios to observations, presenting results for annual average temperature and for seasonal temperature range[11] (summer average temperature minus winter average temperature) for the period 1961-1990; this period, by common convention, is considered the baseline climate period. For annual average temperature, the model results and observa-

tions have quite similar values and distributions across the US, with average temperatures exceeding 80°F (about 28°C) along the southeastern edges of the US and near 40°F (about 5°C) across the north-central US. The maps of the seasonal range in temperature across the US (summer minus winter) show that the seasonal ranges of temperature for the models extend from about 5°F (about 3°C) near southern and southwestern coastal regions to over 50°F (about 28°C) in the northern Great Plains, in reasonable concurrence with observations.

The comparisons also show that the models are a bit warmer than observations along mountain ridges (e.g., the Sierra Nevada and Cascade Mountains) and a bit colder than observed over mountain basins. Doherty and Mearns (1999) report that both models exhibit large year-round cold biases over mountainous regions of the West when compared to the Legates and Wilmott (1990a) climatology. However, that climatology likely has a warm bias (making the models look cold) because most observing stations are located in valleys in mountainous regions. The VEMAP surface climatology used in the National Assessment comparisons improved on the Legates and Wilmott climatologies by adding in information from a large number of high altitude stations and otherwise accounting for the effects of mountains. Compared to this presumably more accurate representation of the observed conditions, the model differences with observations are smaller, but not eliminated.

Differences with observations remain particularly large over the southern Rocky Mountains and Great Basin (see Web site for a map of actual differences). These differences are most likely due to the effects of smoothing the mountain ridges and uplifting the mountain valleys to match the relatively coarse resolution available in current climate models (figures of the differences in topographic height of models and observations are also shown on the Web site). Both primary models also exhibit a warm bias over Hudson Bay during winter that extends southward into the northern US. This bias may be partly due to insufficient observational measurements over Hudson Bay itself, so that the observed surface temperature is likely more representative of cold land areas than of water bodies covered by sea ice (Doherty and Mearns, 1999). Other biases may well reflect the limited spatial resolution and representation of climatic processes in the models. For example, both models also have a warm bias during summer in the central Great Plains and Midwest that probably reflects inadequate treatment of summer convection and soil moisture processes. This bias

---

[10]See additional figures at www.cgd.ucar.edu/naco/found/figs.html.
[11]Figures showing the model projections of temperature for the summer and winter seasons and for differences between simulations and observations are available on the Web site.

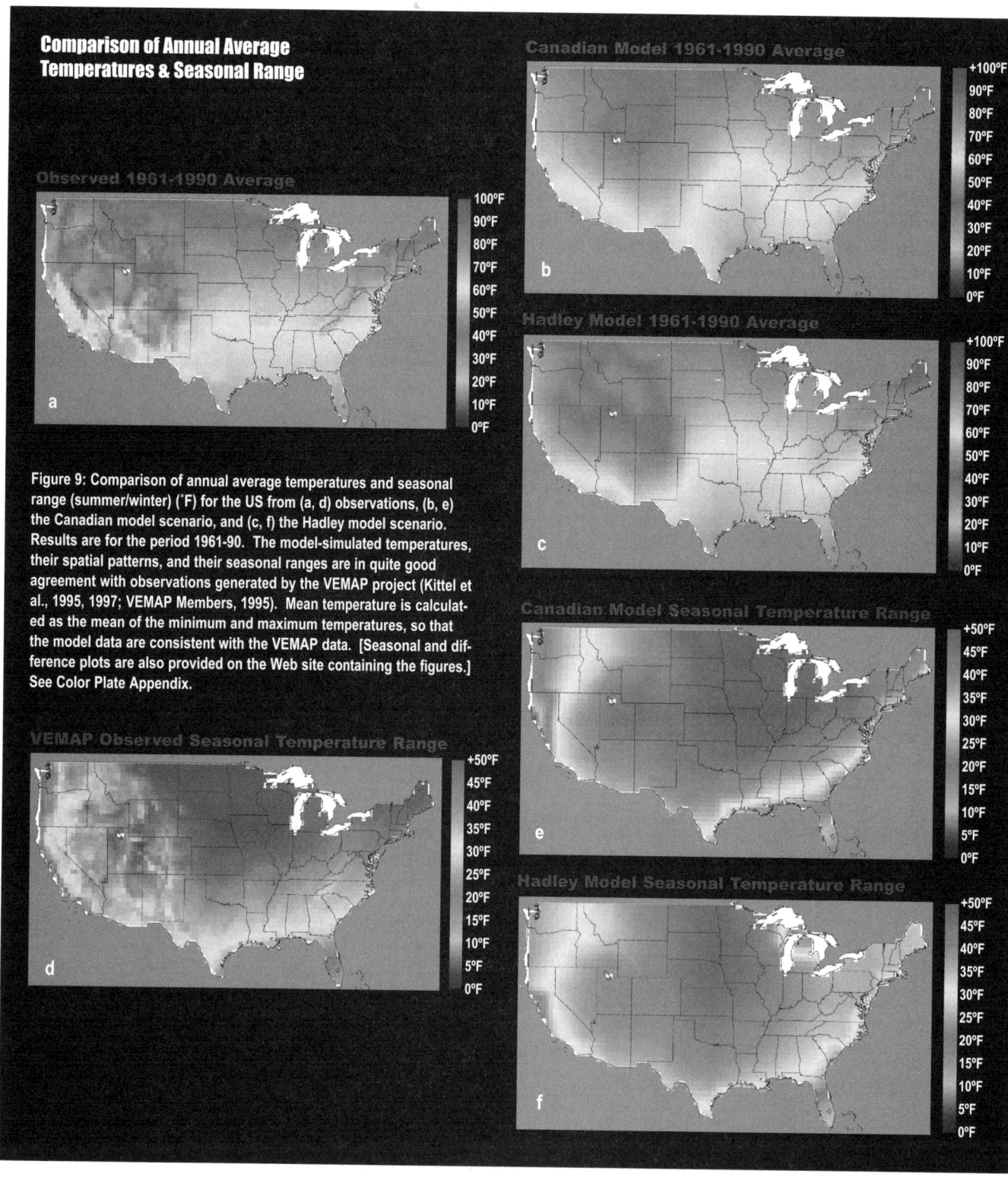

Comparison of Annual Average Temperatures & Seasonal Range

Figure 9: Comparison of annual average temperatures and seasonal range (summer/winter) (°F) for the US from (a, d) observations, (b, e) the Canadian model scenario, and (c, f) the Hadley model scenario. Results are for the period 1961-90. The model-simulated temperatures, their spatial patterns, and their seasonal ranges are in quite good agreement with observations generated by the VEMAP project (Kittel et al., 1995, 1997; VEMAP Members, 1995). Mean temperature is calculated as the mean of the minimum and maximum temperatures, so that the model data are consistent with the VEMAP data. [Seasonal and difference plots are also provided on the Web site containing the figures.] See Color Plate Appendix.

extends further into the eastern US in the Canadian model scenario than in the Hadley model scenario. Over adjacent ocean areas, the Canadian model also indicates temperatures slightly above observations whereas the Hadley model indicates temperatures are slightly below observations, likely reflecting remaining problems with representation of coastal ocean areas (Doherty and Mearns, 1999). These differences of several degrees can create problems in the direct application of model results, but the agreement of the overall patterns and seasonal ranges provides considerable confidence that the projected changes in temperature due to human influences are plausible for use in impacts studies.

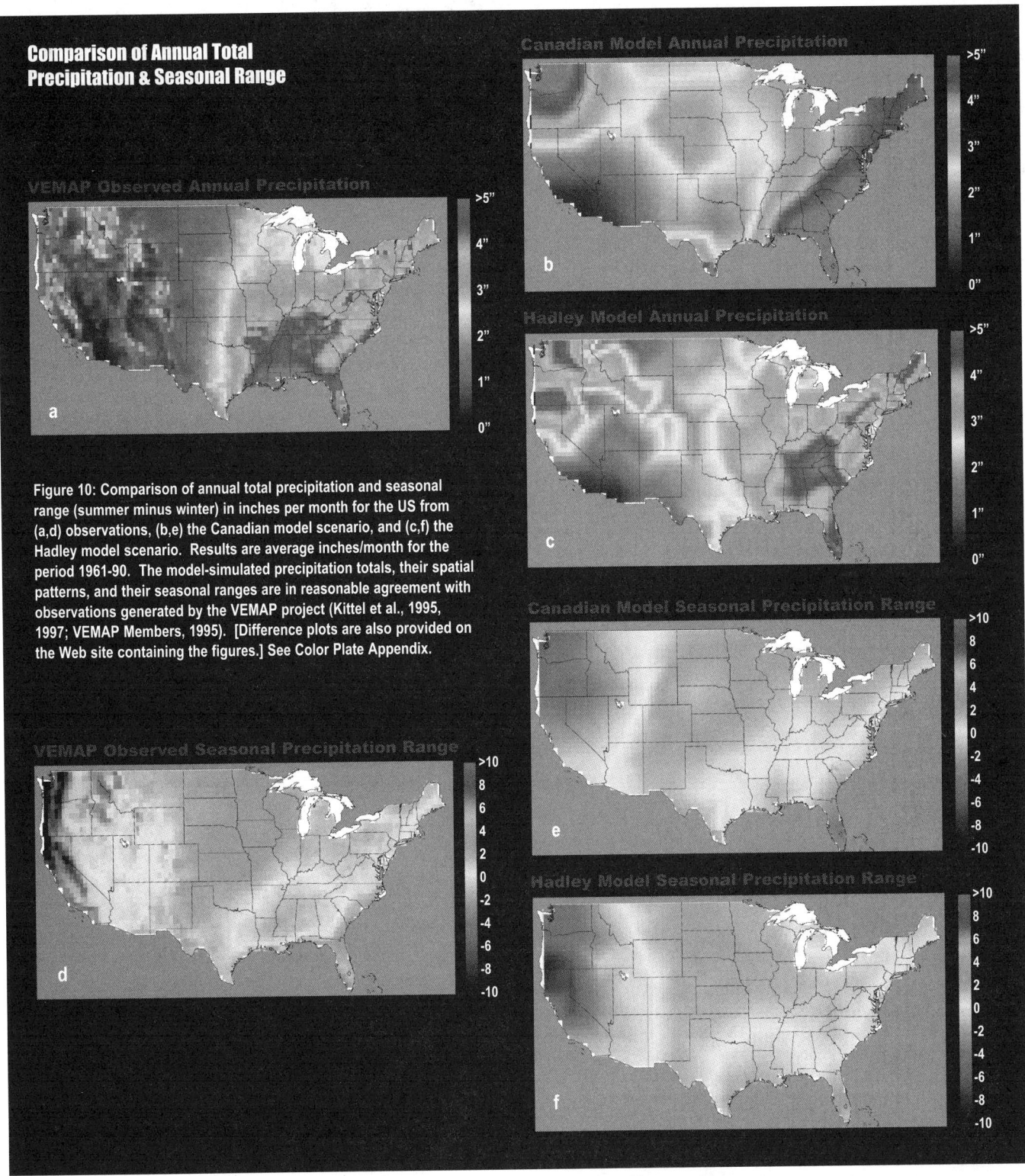

Comparison of Annual Total Precipitation & Seasonal Range

Figure 10: Comparison of annual total precipitation and seasonal range (summer minus winter) in inches per month for the US from (a,d) observations, (b,e) the Canadian model scenario, and (c,f) the Hadley model scenario. Results are average inches/month for the period 1961-90. The model-simulated precipitation totals, their spatial patterns, and their seasonal ranges are in reasonable agreement with observations generated by the VEMAP project (Kittel et al., 1995, 1997; VEMAP Members, 1995). [Difference plots are also provided on the Web site containing the figures.] See Color Plate Appendix.

Figure 10 presents similar results for annual total precipitation and seasonal range (summer minus winter, in inches/month). Precipitation amounts in complex terrain are highly variable as a result of the local interaction of storms with mountains and local variations in the surface warming that drives convective rain systems (Legates and DeLiberty, 1993;

Legates 1997). The relative coarseness of the model resolution means, therefore, that agreement is not likely to be as good, especially over the western US. Both models and observations (from VEMAP and Legates and Wilmott, 1990b) show a similar range from a minimum in the dry areas of the Southwest to much larger amounts over other parts of the

43

Mean Temperature Change

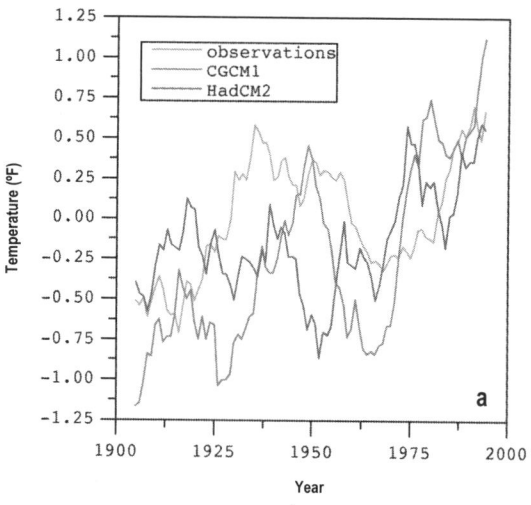

Precipitation Change

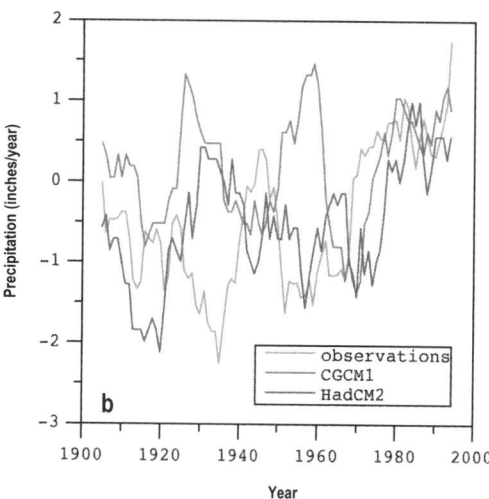

Figure 11: Time histories of the changes in (a) annual average temperature (˚F), and (b) annual total precipitation (inches per year) for the 20th century based on observations and on simulations from the Canadian and Hadley models, calculated as 10-year running means from 1900 to 2000. Mean temperature is the actual mean temperature from the models, rather than the mean of the minimum and maximum temperatures. Anomalies are shown with respect to 1961-1990. In these simulations, unlike in intercomparisons of the atmospheric models as in the AMIP project (Gates et al., 1999), the ocean temperatures are freely calculated and the concentrations of greenhouse gases and aerosols are imposed; natural forcings, such as changes in solar radiation and volcanic eruptions that are likely affecting the observed climate are not, however, being treated in the models because observations of their precise radiative influences are not available. See Color Plate Appendix.

country. Although the very broad-scale patterns are similar, the role of mountain chains in concentrating precipitation into particular locations is much more evident in the observations than in the models with their very smoothed representation of mountain ranges. The pattern of the seasonal range is also

plausibly represented, with the indication that precipitation in the West occurs much more in winter than summer whereas over the rest of the country there tends to be a modestly larger amount in summer. Overall, the model results show broad agreement with observations, except in the Canadian model over Florida. Similarities, however, are evident in the simulation of high amounts of precipitation in the West in winter and low amounts in summer giving a negative seasonal range, and a smoother seasonal cycle in the eastern US. However, there are important differences, especially in the regions of mountainous terrain where observations of precipitation are also problematic due to the great spatial variability.

As for the temperature differences, differences with observations arise because the models do not fully represent the high reach of mountain ridgelines. Because of this discrepancy, the models do not create as much precipitation along the Pacific coast ridgelines as is observed, allowing more precipitation further inland. For the rest of the country, comparisons by Doherty and Mearns (1999) indicate that the models have a wet bias over northeastern North America in spring and summer, and a dry bias in southern North America in both summer and winter. In our comparisons, the Canadian model (but not the Hadley) shows a wet bias in the northeastern US, but a dry bias when integrated over the whole country. The biases in coastal regions may result from the relatively coarse resolution of the models, which does not allow adequate representation of the relatively small-scale spatial patterns of the sea breeze and other coastal meteorology. Also, the tropical rainbelt created by the Intertropical Convergence Zone (ITCZ) does not extend far enough northward in either model, creating dry biases in some of the equatorial regions of the Northern Hemisphere. That there are differences that must be accounted for in the analyses becomes especially clear when focusing on very particular regions (e.g., Florida), and the Web site provides difference maps for simulations of the total and seasonal precipitation.

Because some impact studies require scenarios of changes in day-to-day variability in the weather, comparison should also be made over this time scale, considering, for example, the adequacy of model simulations of the frequency, intensity, and amounts of precipitation. Unfortunately, such detailed comparisons are only beginning to be carried out and so caution must be exercised in interpretations than depend on these results. Nonetheless, as for temperature, if account is taken of systematic differences, the model results would seem to give a plausible set of baseline conditions to use in estimating changes to temperature and precipitation that could occur as the climate changes.

In evaluating model performance, it is also important to look at how well the models simulate the temporal variations of climatic conditions in the immediate past. Figure 11 shows a comparison over the US of the observed and modeled time histories of changes in annual average temperature and annual total precipitation during the 20th century. Remembering that complete agreement of each climate fluctuation should not be expected due to the natural variability of the climate, these plots indicate that the models generally have the right magnitude and duration of natural climate anomalies and that, with the exception of the start of rapid warming late in the century in the Canadian model scenario, the trends are plausibly similar. The Web site provides diagrams that go beyond these comparisons to provide estimates of the actual values of temperature and precipitation, thereby illustrating the systematic differences that are present between the models and observations. These differences arise both because of limitations in the models (e.g., inadequate resolution, inadequate representation of various processes, etc.) and shortcomings in the monitoring network (e.g., few stations at high latitudes, etc.). To the extent that these differences are systematic, the model projections of changes can be used if care is taken in working near thresholds such as the freeze line. To the extent that the differences are inherent in the treatment of climate processes and how they might respond with a different climate, uncertainties are introduced into the climate scenarios, again emphasizing that these result must be viewed as scenarios rather than predictions.

While these analyses indicate that the model results are generally similar to observations, it is clear that systematic errors are present, especially in mountainous areas. To account for these differences, historical analyses have generally been based on compilations of observational data, such as the USHCN or VEMAP data sets, rather than numerical model results, and appropriate adjustments need to be made when applying model results for the future (as explained in the box on page 28 on Normalization of Results).

SCENARIOS FOR CHANGES IN ATMOSPHERIC COMPOSITION AND RADIATIVE FORCING FOR THE 21st CENTURY

Projecting changes in climate for the 21st century requires not only a tested climate model, but also a scenario for the development and evolution of the human activities that are expected to affect the climate. In particular, projections of climate change require a projection of how atmospheric composition will be changing in the 21st century as a result of the ongoing use of fossil fuels and the release of other greenhouse gases[12]. To provide the basis for such estimates, scenarios of societal and technological evolution during the 21st century must be developed; these in turn can be used to develop emissions scenarios. The accuracy of these scenarios is necessarily limited by uncertainties in insights and assumptions about what will happen many decades into the future. Because of the resulting uncertainties, the concentration scenarios that are used, like the climate scenarios, cannot be viewed as predictions of the future. Instead, they must be treated as plausible estimates of future conditions that are appropriate for use in exploring vulnerabilities through analysis and assessment.

A range of scenarios has been developed by a number of groups to describe how atmospheric concentrations of CO_2, other GHGs, and aerosols may change in the future. These scenarios are generally based on projections of future changes in population, energy technology, economic development, environmental controls, and other factors. The 1992 scenarios proposed by the Intergovernmental Panel on Climate Change (IPCC, 1992) have become widely used because of the international effort that went into their consideration[13]. The set of 1992 IPCC greenhouse-gas emission scenarios was based upon six plausible demographic and socioeconomic scenarios that spanned a wide range of possibilities for population growth, types of energy use, and rates of economic growth. The range of projected emissions for the 21st century is quite broad (see IPCC, 1992 and Figure 6).

The central baseline (sometimes called "business-as-usual") estimate from the set of IPCC 1992 scenarios is closely comparable to the radiative forcing scenario represented by a 1% per year compounded

[12]Note that for the purposes of these studies, the level of solar insolation and the occurrence of volcanic eruptions are assumed to remain as they were for the 20th century. Even though changes are likely (e.g., see Hyde and Crowley, 2000), the net effect of these changes are likely to be small in comparison to the human-induced influences on radiative forcing.

[13]These scenarios are presently being updated as part of the effort leading up to the IPCC's Third Assessment Report (IPCC, 2000). The newer scenarios tend to span a similar range to the 1992 scenarios.

increase in the equivalent CO_2 concentration that has been used by most climate modeling groups to generate their central estimates of potential climate change for the 21st century. This scenario has been taken as the baseline scenario for this study because of its wide use, because it represents neither maximum nor minimum emissions projections (see Figure 6), and because this Assessment did not have the resources to either construct better alternative scenarios or ensure that such scenarios would be used by the climate modeling groups for calculations that would be available in time for impact evaluation as part of this Assessment. Although the IS92a emissions scenario tended to overestimate greenhouse gas emissions during the 1990s (Hansen et al., 1998), it is not clear that the recent tendency toward lower emissions compared to IS92a will persist as the global economy recovers from its recent recession. In particular, the new IPCC (2000) scenarios suggest a wide range of possible future emissions scenarios, some higher and some lower than the IS92a scenario (see Figure 6). To the extent that actual greenhouse gas emissions might be greater or less than this central scenario over the long term, the climatic changes at a given time would be greater or less. Alternatively, the climatic changes that are projected with this scenario would be projected to occur either earlier or further in the future, although the difference would likely be less than one or two decades. Although the potential consequences of a somewhat faster or slower rise in greenhouse gas emissions has not yet been evaluated, it seems likely that such changes in emissions scenarios would have a relatively small influence over the climate changes projected for the first half of the 21st century.

Figure 12 shows the projected changes in CO_2 and equivalent CO_2 concentration for the IS92a scenario (projected changes in the concentrations of other greenhouse gases are described in IPCC, 1992) and the 1% per year change in equivalent CO_2 concen-

Forcing Scenarios

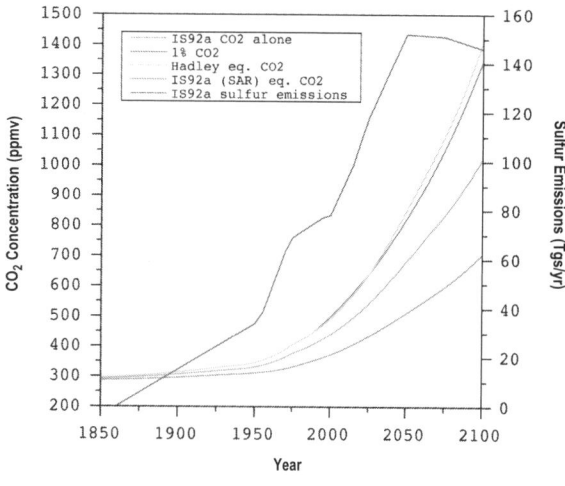

Radiative Forcing

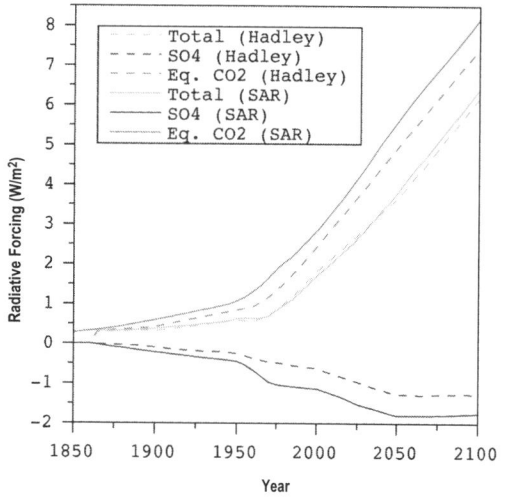

Figure 12: Comparison of different projections for aerosol effects and for (a) the CO_2 and equivalent CO_2 concentrations, and (b) the associated radiative forcings for the period 1850-2100. In the top figure, the lavender line shows the IPCC's IS92a scenario estimate of the CO_2 concentration; values prior to 1990 are based on observations. Based on this projection, the CO_2 concentration would rise to about 705 ppmv in 2100 from a level of about 353 ppmv in 1990. Because many of the climate models treat the effects of the set of human-affected greenhouse gases by use of an equivalent CO_2 concentration, the green line shows the scenario for the equivalent CO_2 concentration, which rises to about 1022 ppmv in 2100 from a value of about 410 ppmv in 1990. For this curve, the equivalent CO_2 concentration is calculated so as to incorporate the radiative effects of changes in the concentrations of all greenhouse gases using the IPCC radiative forcing equivalents (the conversion factor is 6.3 based on Appendix 2 in IPCC, 1997). The light blue line shows the equivalent CO_2 concentration that results from using the Hadley radiative forcing equivalents to approximate the IS92a scenario; the conversion factor used is 5.05 (John Mitchell, personal communication). Using the Hadley conversion factor, the equivalent CO_2 concentration for the IS92a scenario would rise to about 1409 ppmv in 2100. The red line shows that the Hadley IS92a equivalent CO_2 scenario is quite well fitted by use of a 1% per year compounded increase in the Hadley equivalent CO_2 concentration. In this case, the CO_2 equivalent concentration in 2100 reaches about 1346 ppmv. The deep blue line shows the IPCC IS92a scenario for sulfur emissions, which shows a rise until about 2050, when emissions roughly level off. While there are some differences in the projected concentrations of equivalent CO_2 between the IPCC (1996a) and the Hadley model scenario, the bottom figure shows that these differences are mostly overcome when comparing the radiative forcings that are projected by the IPCC and are actually used in the Hadley model scenario. The red and blue lines, respectively show the radiative forcings as projected by the IPCC (solid lines) and as included in the Hadley model (dotted lines). For both forcings, the Hadley model projects slightly less influence than the projections using the IPCC conversion factors. When these forcings are combined, as shown by the green lines, the net radiative forcings projected by the IPCC and used in the Hadley model 1% per year scenario are very close. See Color Plate Appendix.

tration. For the National Assessment, the IS92a time history of the CO_2 concentration (which rises to about 708 ppmv by 2100) has been used in the studies of the consequences of a rising CO_2 concentration for plants, coral, etc. However, because concentrations of CH_4, N_2O, and CFCs are also changing, the model simulations need to be forced by the net radiative effect of these greenhouse gas changes. This consideration is implemented in the Hadley and Canadian models by increasing the equivalent CO_2 concentration by 1% per year (compounded) starting in 1990 to account for the combined radiative effects of all the greenhouse gases (Boer et al., 2000a). Thus, while the 1990 concentration of CO_2 is about 354 ppmv, the model uses an equivalent CO_2 concentration of about 420 ppmv to account for the effects of the other greenhouse gases. For the year 2100, rather than reaching a CO_2 concentration of about 708 ppmv, an equivalent CO_2 concentration of about 1055 ppmv is reached (note that the 1% per year case is slightly higher than this concentration, actually being closer to case IS92f). In terms of radiative forcing, the 1% simplification overestimates the net forcing in 2100 of all greenhouse gases by about 10% compared with IS92a (of course, there are other scenarios that have higher forcing than IS92a). As shown in Figure 12, the IS92a scenario used by the models also significantly increases sulfate emissions (and therefore sulfate aerosol loadings) until about 2050, after which levels are projected to remain roughly constant. The net changes in forcing for both the Hadley and Canadian model scenarios are, as indicated in Figure 6, near the middle of the range for all emissions scenarios.

In that sulfate aerosols contribute significantly to air pollution and acid rain, the newer scenarios in IPCC (2000) suggest that sulfate aerosol levels (and so their cooling influence) will be lower than in IS92a, thereby raising the overall warming influence. In addition, the new scenarios suggest that significant increases in the use of fossil fuels will lead to increased emissions of methane, carbon monoxide, nitrogen oxides, and volatile organic compounds, which will lead to significant increases in both regional and hemispheric levels of tropospheric ozone, a strong greenhouse gas.

What is quite clear from Figure 6 is that the radiative forcing could be either higher or lower than the case that has been the most frequent reference case for the modeling groups and is being used in this Assessment. The normal way to treat such a range of possible futures would be to treat a range of possible future conditions rather than rely on only one case. In this first National Assessment, constraints on

time and resources, however, have forced a limitation to considering the consequences of only one concentration scenario. While this approach is a limitation that should be relaxed in future assessment efforts, the constraints of this limitation are reduced by the recognition that much of the climate change over the next few decades will be due to already recorded changes in atmospheric composition. In addition, with the momentum created by the world's present use of fossil fuel energy, deviations in the concentration scenario for the various GHGs are likely to have only a limited influence on the climate over the next few decades. To explore this issue further, a later section of this chapter does summarize the climatic consequences of using a scenario that moves toward stabilization of the atmospheric CO_2 concentration at double its preindustrial value. However, even if such a stringent emissions limitation were imposed now, the effect on CO_2 concentrations and climate would be relatively modest during the early 21st century before increasing and becoming quite significant during the 22nd century.

CLIMATE MODEL SCENARIOS FOR CHANGES IN TEMPERATURE, PRECIPITATION, SOIL MOISTURE, AND SEA LEVEL OVER THE US FOR THE 21st CENTURY

Temperature and Heat Index

All climate models project significant warming for the 21st century. Results shown in Figure 7 clearly indicate that the global warming projected for the 21st century will be significantly greater than during the 20th century. This increase in the rate of warming is due to both the continuing rise in the CO_2 concentration projected for the 21st century and the continuing response of the climate system to the increasing rate of rise in the CO_2 concentration in the second half of the 20th century. Figure 7 also demonstrates that the projections for warming over the US are very likely to be greater than for the global average, both because warming is greater over land areas than over ocean areas and because the US is located in mid-latitudes. This figure also shows that, although the rate of warming is not like-

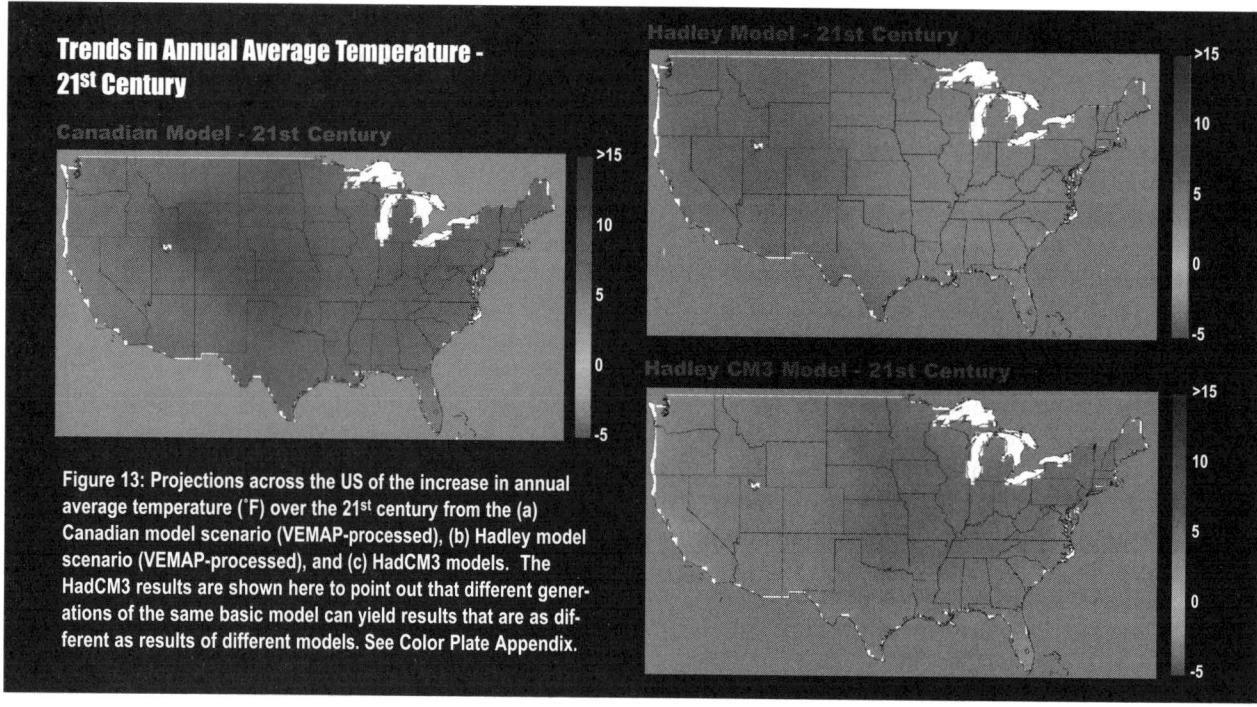

Trends in Annual Average Temperature – 21st Century

Canadian Model - 21st Century

Hadley Model - 21st Century

Hadley CM3 Model - 21st Century

Figure 13: Projections across the US of the increase in annual average temperature (°F) over the 21st century from the (a) Canadian model scenario (VEMAP-processed), (b) Hadley model scenario (VEMAP-processed), and (c) HadCM3 models. The HadCM3 results are shown here to point out that different generations of the same basic model can yield results that are as different as results of different models. See Color Plate Appendix.

ly to be uniform over this period, the average rate warming rate is very likely to increase during the 21st century. This change in rate may occur in an uneven way, with some very warm years, and then some not-so-warm or even cooler years. Although we do not yet have the ability to forecast these short-term fluctuations precisely, the model scenarios clearly show that the long-term rate of warming

is very likely to increase substantially over coming decades.

Figure 13 shows the annual average geographic patterns of the projected warming across the US as calculated by the Canadian and Hadley models[14]. The trends are in degrees (°F) of warming per century and represent the expected warming for the several decades around 2100[15]. In the Canadian model scenario for the next 100 years, increases in annual average temperature of 10°F (5.6°C) are projected across the central US, with changes about half this large projected along the East and West Coasts. The projections indicate that the changes will be particu-

Figure 14: Time histories of (a) maximum and (b) minimum temperature over the US (°F). The values prior to the present are based on observations from 1900-1998 (the HCN data set) and values for the future are based on the VEMAP version of the Canadian and Hadley model scenarios (i.e., in the VEMAP data sets, model projections of climate change are added to the observed 1961-90 baseline climate). See Color Plate Appendix.

Maximum Temperature in the US (annual average)

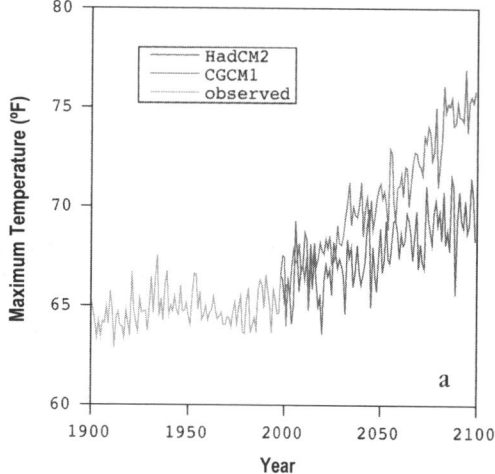

Minimum Temperature in the US (annual average)

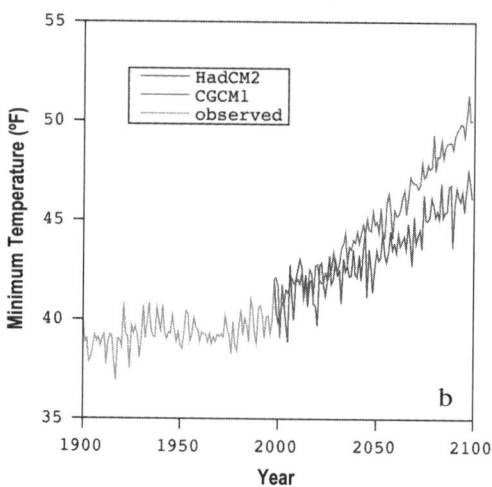

[14] Maps of projected changes for the winter and summer seasons are available on the Web site.

larly large in winter, with minimum temperatures rising more than maximum temperatures. Large increases in temperature are also projected over much of the South in summer. In the Hadley model scenario, temperatures in the eastern US are projected to increase by 3 to 5°F (2-3°C) by 2100. Regions across the rest of the nation are projected to warm by up to about 7°F (4°C). Such changes would be equivalent to shifting the climate of the southern US to the central US and the climate of the central US to the northern US.

Model results from the HadCM3 model are also shown in Figure 13. This model is a more recent version of the Hadley Centre model than was available at the start of this Assessment. This model shows greater warming in the eastern half of the US. This reinforces the point that although all the models agree that there will be a strong warming trend, projections of the spatial and temporal pattern of the warming differ among the models.

While the maps included in this chapter provide results for the conterminous 48 states, model results are also available for Alaska and for the Pacific and Caribbean Islands on the Web site. Both models project that Alaska will experience even more intense warming than the conterminous US[16]. In contrast, Hawaii and the Caribbean islands are likely to experience somewhat less warming than the continental US, because they are at lower latitudes and are surrounded by ocean, which warms more slowly than land. These results are shown in maps appearing in the respective chapters of this report. Although the details of the projected climate fluctu-

ations over time are less reliable than the projections of the overall trends, it is useful to examine the projected time histories of the changes. Figure 14 shows the time histories for the projected changes over the US in the annual averages of minimum and maximum temperature for the two models. As is suggested by the maps, the time series show that the warming is projected to be greater in the Canadian model scenario than in the Hadley model scenario. The larger increase in minimum than maximum temperature indicates that nighttime temperatures are projected to increase more than daytime temperatures. Factors causing this difference could include the increase in downward infrared radiation, the increase in the dew point temperature, changes in cloud cover, changes in soil moisture, and changes in snow and ice cover, each of which would act to raise nighttime temperatures more than daytime temperatures. In addition, an increase in sulfate concentrations or increases in cloud cover might act to limit daytime warming by reflecting more solar radiation back to space. That both models suggest that minimum temperatures will rise more rapidly than maximum temperature is consistent with what has been observed over the past century (Easterling et al., 1997).

Although the two primary models used here project that the temperature increase will be greater in the western than in the eastern US, the intensification of the hydrologic cycle caused by the warming will also cause an increase in the amount of moisture in the air. This increase is particularly important for the southeastern and eastern US, where humidity is relatively high and upward trends in temperature

July Heat Index Change - 21st Century

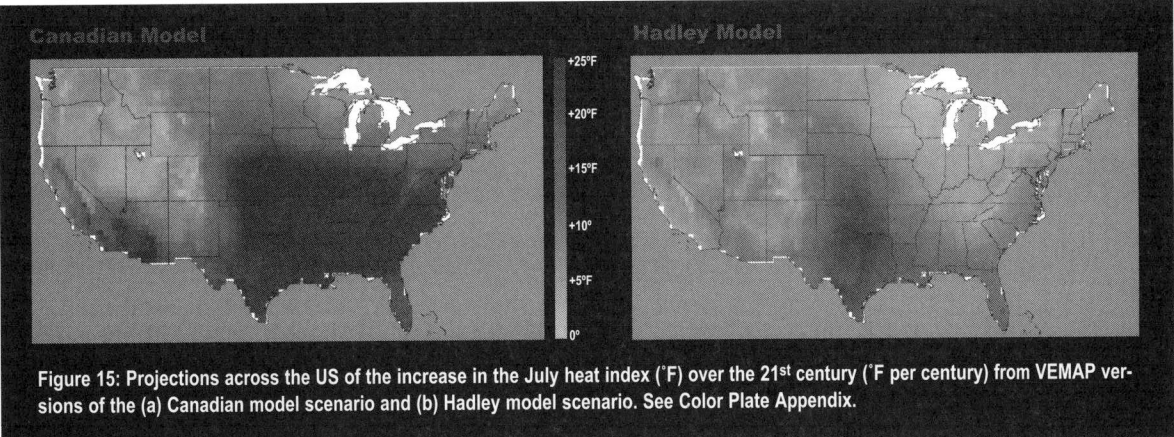

Figure 15: Projections across the US of the increase in the July heat index (°F) over the 21st century (°F per century) from VEMAP versions of the (a) Canadian model scenario and (b) Hadley model scenario. See Color Plate Appendix.

[15]Because the model simulations are most valid over long-periods of time, these results are based on a linear fit to the model projected changes for the 21st century rather than being based on the differences between particular years or decades at the beginning and end of the century. This choice is intended to make clear that it is the overall century-long rate of change that is the result in which we can have the most confidence. Because of the long-term warming, great care should be taken in comparing this projected rate of change to observed changes over shorter periods because it is widely recognized that there will be considerable natural variability through the century as a result of the effects of natural influences such as solar variations, volcanic eruptions, and the ocean-atmosphere interactions that create such fluctuations as ENSO events.

[16]For results for Alaska, see http://www.cgd.ucar.edu/naco/alaska/tx.html.

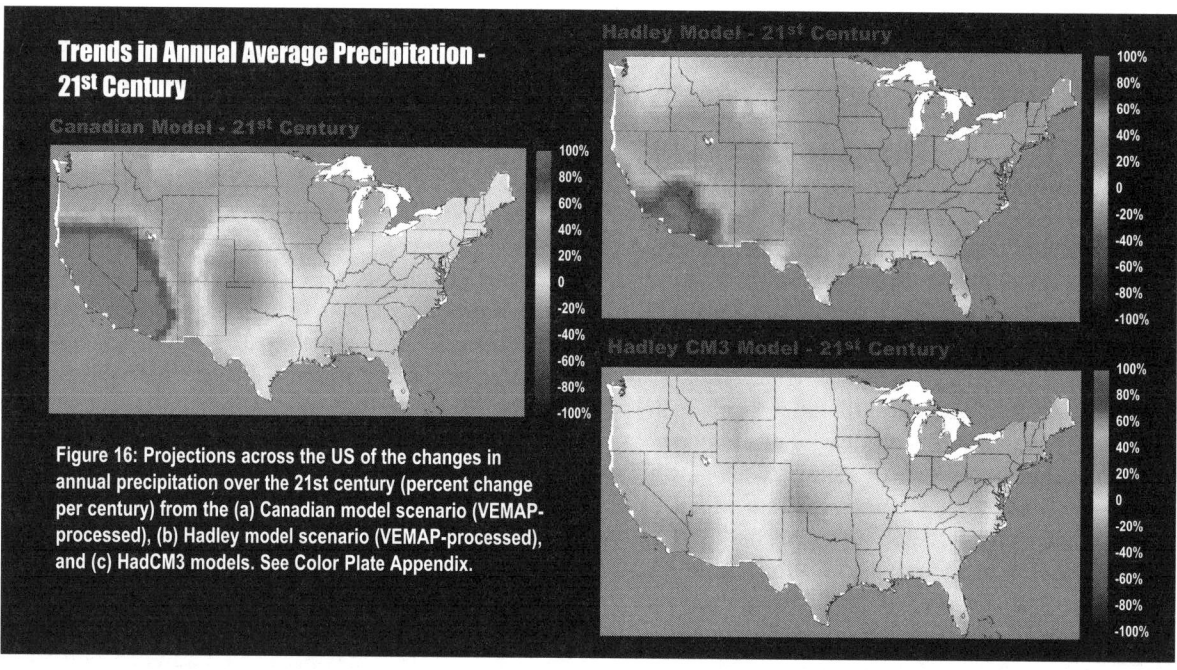

Figure 16: Projections across the US of the changes in annual precipitation over the 21st century (percent change per century) from the (a) Canadian model scenario (VEMAP-processed), (b) Hadley model scenario (VEMAP-processed), and (c) HadCM3 models. See Color Plate Appendix.

are quite large (Karl and Knight, 1997). Figure 15 shows the projected increases in the heat index across the US based on the model projections for the changes in maximum temperature; similar results have been reported from the GFDL model (Delworth et al., 1999). The heat index is a measure of the rise of apparent temperature and is a good measure of discomfort because it combines both heat and humidity effects (Steadman, 1979). These results indicate that, even though the relative humidity may drop slightly (not shown), the rise in the heat index will be more than double the actual rise in temperature across much of the South and East, making the projected warming in these parts of the country feel particularly significant. By the end of the 21st century, the heat index of the Northeast is likely to feel more like that of the Southeast today; the Southeast is likely to feel more like today's south Texas coast; and the south Texas coast is likely to feel more like the hottest parts of Central America today.

Precipitation

Figure 16 shows the projected pattern of changes in precipitation across the contiguous US, expressed as a percentage change from the present amount[17]. The most noticeable feature in both models is a projected increase in precipitation in California and the southwestern US. The projected increase is larger in the Canadian than in the Hadley model scenario. This feature is a result of a warmer Pacific Ocean causing an increase mainly in wintertime precipitation. Although the projected changes over the

Pacific Ocean are not well-established, particularly with regard to how El Niño conditions may change, both primary models project Pacific Ocean warming and a southward movement of the storm-generating Aleutian Low which would together lead to increased precipitation along the West Coast. For these conditions, a greater fraction of the increased wintertime precipitation would be expected to fall as rain rather than snow, causing, on average, a reduction in mountain snow pack. These changes are likely to increase wintertime and decrease summertime river flows in the West. Even with an accurate projection of Pacific Ocean changes, the regional pattern of this precipitation increase could only be roughly estimated due to the limited representation of the region's mountains (e.g., see Mearns et al., 1999). As global scale models improve, mesoscale models will be able to be used to explore this issue further.

Across the Northwest and over the central and eastern parts of the US, the precipitation projections from the models are in less agreement. The differences between model projections are likely a result of a number of factors. For example, the two models show different positions and intensities of the storm tracks in the Southeast during winter in their simulations of recent decades. The Canadian model scenario projects that there will be a decrease in annual precipitation across the southern half of the nation east of the Rocky Mountains. Decreases are projected to be particularly large in eastern Colorado and western Nebraska in the west central Plains, and in the southern states in an arc from

[17]Changes in the absolute amount of precipitation are shown in figures available on the web site.

Louisiana to Virginia. These projected decreases in precipitation are largest in the Great Plains during summer and in the East during both winter and summer. In the Hadley model scenario, virtually the entire US is projected to experience increases in precipitation, with the exception of small areas along the Gulf Coast and in the Pacific Northwest. Precipitation is projected to increase in the eastern half of the nation and in southern California and parts of Nevada and Arizona in summer, and in every region except for the Gulf States and northern Washington and Idaho during the winter. However, while the Hadley (HadCM2) scenario used in this Assessment suggests greater precipitation in the Southwest, the more recent HadCM3 model suggests that there will be less rainfall in the Southwest; the projected pattern of change is similar to the Canadian model scenario in parts of the Southeast. Because of the differences among these results, the projected direction of the trend for changes in precipitation in any given region needs to be viewed as uncertain, although continuation of the increasing precipitation trend for the US as a whole seems plausible. Resolving these differences in precipitation projections will occur only by increasing resolution and implementing other improvements in the climate models.

Figure 17 provides the time histories of the projected changes in precipitation for the US. Both models project a long-term increase in total annual precipitation across the US. However, the time histories clearly indicate that the very large variability that currently exists is likely to continue, with the possibility of periods of both increased and even reduced precipitation within the overall upward trend.

Annual Precipitation in the US

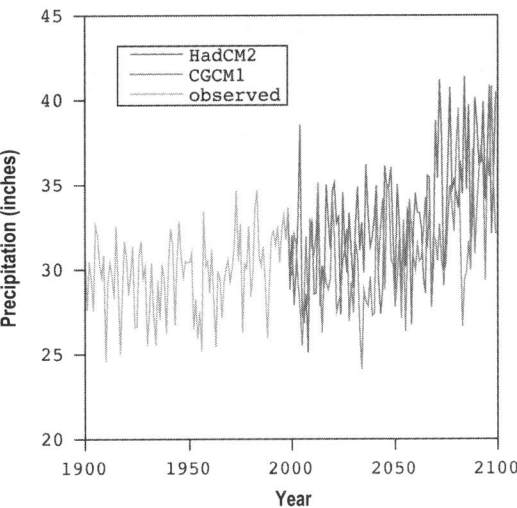

Figure 17: Time history of model projected changes in precipitation over the US (inches per year). The values prior to the present are based on observations from 1900-1998 (the HCN data set) and values for the future are based on the VEMAP version of the Canadian and Hadley model scenarios. See Color Plate Appendix.

Soil Moisture

Projections of changes in soil moisture depend on the balance between precipitation, evaporation, runoff, and soil drainage. By itself, an increase in precipitation would tend to increase soil moisture. However, higher air temperatures increase the rate of evaporation and may remove moisture from the soil faster than it can be supplied by precipitation. Under these conditions, some regions are likely to become drier even though rainfall increases. In fact, soil moisture has already decreased in portions of the Great Plains and Eastern Seaboard, including in some locations where precipitation has increased

Summer Soil Moisture - 21st Century

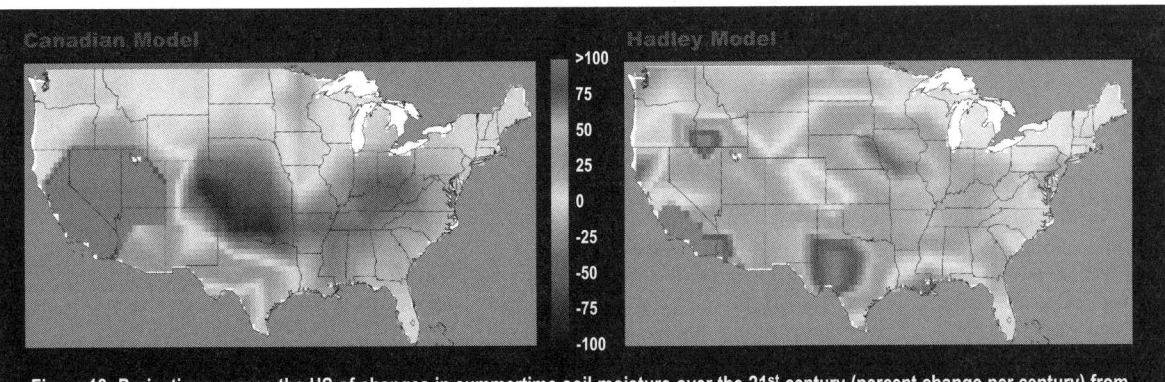

Figure 18: Projections across the US of changes in summertime soil moisture over the 21st century (percent change per century) from the (a) Canadian model scenario, and (b) Hadley model scenario. Figure prepared by the National Climatic Data Center. See Color Plate Appendix.

but air temperature has risen. Figure 18 shows the projected changes in the summer soil moisture across the US. In the Canadian model scenario, the Southeast and the region extending through the central US to just east of the Rocky Mountains are projected to experience the largest decreases in soil moisture. Increases in soil moisture are projected for the areas surrounding Iowa and from Utah to California. In the Hadley model scenario, summer soil moisture is projected to increase in the eastern half of the US and is generally unchanged or slightly decreased from the Rocky Mountains westward, except for Southern California.

Increased drought becomes a national problem in the Canadian model scenario and is also found in the GFDL model (Wetherald and Manabe, 1999). Intense drought tendencies occur in the region east of the Rocky Mountains and throughout the Mid-Atlantic-Southeastern states corridor. Increased tendencies toward drought are also projected in the Hadley model scenario for the regions immediately east of the Rocky Mountains. California and Arizona, as well as the region from eastern Nebraska to the Virginia coastal plain are projected to have reduced drought tendency. The differences in soil moisture

Sea Level Rise

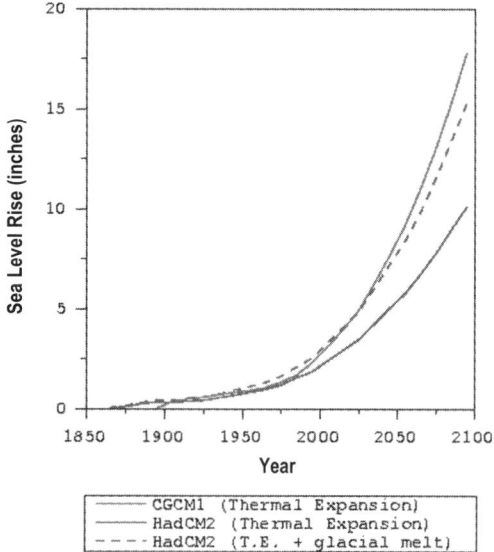

Figure 19: Historic and projected changes in sea level (inches above baseline) based on the Canadian and Hadley model scenarios. The Canadian model projection includes only the effects of thermal expansion of warming ocean waters (F. Zwiers, personal communication). The Hadley model simulation adds on the sea level increment of melting of mountain glaciers (Gregory and Oerlemans, 1998). Neither model includes consideration of possible changes of sea level (upward or downward) due to melting or accumulation of snow on Greenland and Antarctica. See Color Plate Appendix.

and drought tendencies are likely to be the most critical for agriculture, forests, water supply, and lake levels.

Sea Ice and Sea Level

The two primary model scenarios include projections of a decline in sea ice cover and a rise in sea level, both of which are of particular importance for assessing the potential consequences of climate change along coastlines. The Canadian model scenario projects that sea ice in the Arctic Ocean is very likely to melt completely each summer and be significantly reduced in winter thickness and extent by the end of the 21st century, whereas the Hadley model scenario envisions a slower process of melting. Observations indicate that the average depth of sea ice in the Arctic has dropped by 40%, from about 10 feet (3.1 meters) to about 6 feet (1.8 meters) over the past three decades (Rothrock et al., 1999); this suggests that the model projections of melting of sea ice in the future are likely to be quite plausible. Sea ice is particularly important for coastal regions because its presence suppresses waves from wintertime storms that erode coastlines. In addition, some marine species depend on the protection or convenience of sea ice to feed and reproduce, making the meltbacks ecologically important.

Because sea ice floats, its melting does not affect sea level. However, the melting of glaciers on land and the warming of ocean waters do cause sea level to rise. Over the 20th century, observations indicate that sea level has risen about 4 to 8 inches (10-20 cm). In estimating the potential rise in sea level during the 21st century, the Canadian model scenario includes consideration of only the sea-level rise caused by the warming of ocean waters (the thermal expansion effect), whereas the Hadley model scenario also includes consideration of the rise caused by the melting of mountain glaciers. Although melting of the polar ice sheets may contribute to sea level rise in the long-term, neither of the model estimates includes consideration of the changes in sea level caused by the accumulation or melting of snow on Greenland and Antarctica[18]. The global climate models also do not include the local, but significant, component of sea-level change

18 In the IPCC 1996 report, the accumulation and melting of Antarctica and Greenland were assumed to balance to give no net contribution to sea level change over the 21st century; more recent studies are finding that some parts of Greenland are melting while others seem to be accreting, and that the global warming at the end of the last glacial period apparently has initiated deterioration of parts of the West Antarctic Ice Sheet. Due to limitations in the observations of these ice sheets, however, projections included here do not include contributions from changes in the size of these polar ice sheets, even though strong warming seems likely to contribute to their melting and thence to sea-level rise.

caused by changes in the heights of coastlines as they rise or fall due to regional or even local effects (e.g., the pumping out of groundwater, earthquakes, isostatic adjustment from the last glacial period, etc.). Given these caveats, Figure 19 presents the model projections for the estimated rise in global sea level from the two models used in the National Assessment. Over the next hundred years, a rise of sea level of about one and a half feet (about 0.5 meters) is considered likely based on these projections. Maps of the regional pattern of sea-level change around the US are presented in the Coastal chapter, where the effects of local changes in the level of the coastline are also considered. A rise of this amount would be several times as much as occurred during the 20th century. As indicated in the Coastal chapter, even a relatively modest rise can cause extensive coastal erosion (e.g., see Leatherman et al., 2000).

CLIMATE MODEL SCENARIOS OF CHANGES IN CLIMATIC PATTERNS, VARIABILITY, STORMS, AND EXTREMES FOR THE 21st CENTURY

Changes in Climate Patterns, Variability, and Storms

Examination of the patterns of global-scale climate change provides the broader context needed to understand the changes in storm tracks, precipitation belts, and other variations over the US. Such analyses can be undertaken because the simulation of large-scale natural variability by the climate models is generally reasonable (e.g., Stouffer et al., 2000). As for virtually all global models, both models used in the National Assessment project that, in comparison to global average changes, warming at high latitudes will be greater during winter and warming of the land will be greater than of the ocean (Figure 20). The dramatic wintertime warming in high latitudes is very likely due to feedbacks involving the reduction in the reflectivity of the surface as sea ice melts and because weakening of the near-surface inversion allows a relatively large temperature change to occur. Land warms more than ocean because of the oceans' greater ability to limit and redistribute the trapped energy by evaporating moisture, mixing heat downward, transporting heat

around by ocean currents, and the ocean's larger heat capacity. In addition, the warming of land areas increases as soil moisture is reduced, which reduces the potential for evaporative cooling. Although not shown in these figures, another robust feature of global warming is greater warming at upper levels of the tropical atmosphere. This warming occurs because of the way that the vertical atmospheric structure is determined through convection and the

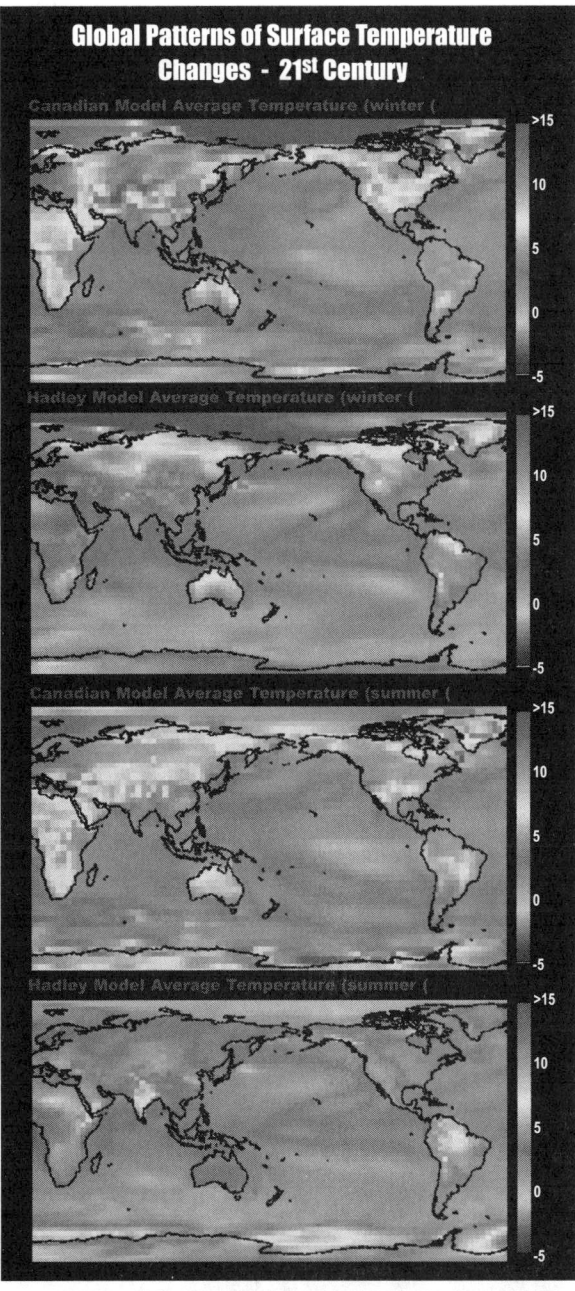

Figure 20: Global patterns of projected changes in surface temperature (°F) over the 21st century [future (2090-2099) and modern (1961-1990)] for (a) December, January, February (DJF) from the Canadian model scenario, (b) DJF from the Hadley model scenario, (c) June, July, August (JJA) from the Canadian model scenario, and (d) JJA from the Hadley model scenario. See Color Plate Appendix.

removal of moisture with altitude. The upper atmosphere warming affects how the atmospheric circulation changes, the generation and intensity of convective rainfall (rainfall resulting from vertical motion in the atmosphere), and the development of tropical storms.

Model projected changes in regional temperatures over the Pacific Ocean indicate greater warming over the equatorial and northern East Pacific Ocean, and that these changes extend to the West Coast of

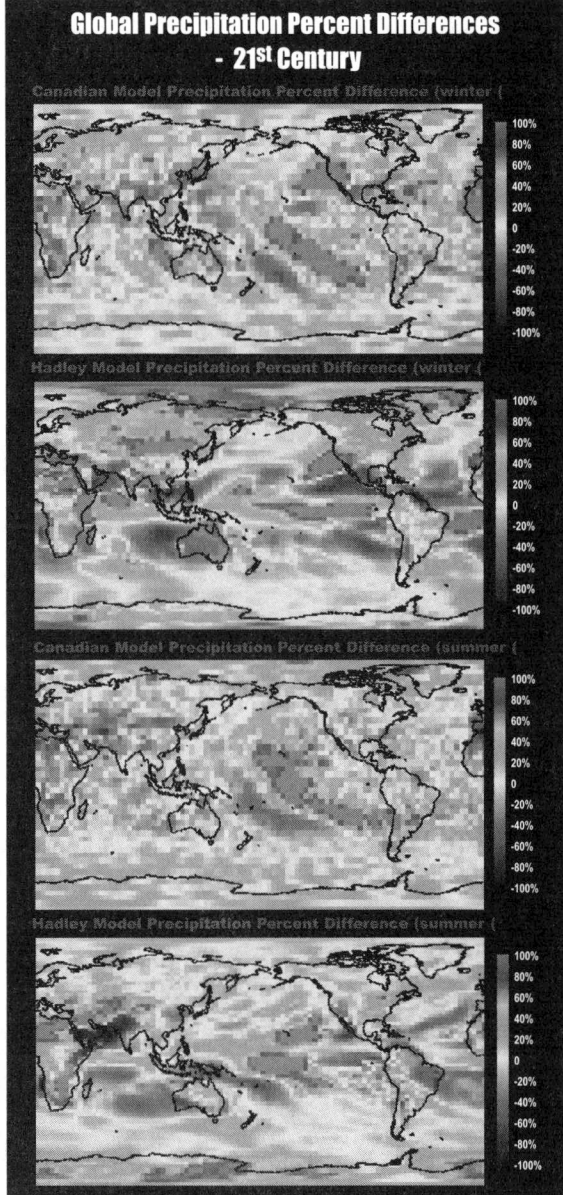

Figure 21: Global precipitation percent differences [(future - modern)/modern) x 100] for (a) December, January, February (DJF) from the Canadian model scenario, (b) DJF from the Hadley model scenario, (c) June, July, August (JJA) from the Canadian model scenario, and (d) JJA from the Hadley model scenario. See Color Plate Appendix.

the US in both models (Figure 20). This pattern of warming resembles an El Niño pattern of sea surface temperature (SST) anomalies, and so it would seem very likely to lead to an El Niño-like wind and precipitation response. Other models (Meehl et al., 2000b) also show this type of response (Meehl and Washington, 1996; Knutson and Manabe, 1995, 1998; Timmermann et al., 1999), although some models show a La Niña-like (Noda et al., 1999), or an initial La Niña-like, pattern that transitions into an El Niño-like pattern (Cai and Whetton, 2000). This response appears to be highly dependent upon how cloud feedbacks are represented by the models, so remains quite uncertain (Meehl et al., 2000b).

The global precipitation anomalies (Figure 21) projected by the models show increased precipitation coinciding with the region of these warm anomalies. The increased precipitation in the Southwest appears to be largely a result of the warmer SSTs in the Pacific Ocean off the coast of North America. During winter, decreased precipitation along the northern branch of the Hadley Circulation (the atmospheric circulation with rising air near the equator and sinking air near 30° latitude, resulting in the trade winds and subtropical dry regions) extends over the eastern US in the Canadian model scenario, but not in the Hadley model scenario. During summer, the Hadley model scenario shows a large area of decreased precipitation in the eastern Pacific and Atlantic Oceans, whereas the Canadian model scenario projects decreased precipitation over land areas. Recent analyses of the 6-hourly data from the Hadley model indicate that the model is accurately reproducing the Southwest monsoon during summer. In a simulation with increased greenhouse gases (although without sulfates), there are indications of a strengthening of the monsoon (Arritt et al., 2000), which correlates with the region of increased summer precipitation in the Southwest.

Because the Northern Hemisphere's atmospheric circulation is more vigorous during winter, examining the winter circulation pattern provides an indication of the causes of these precipitation changes. The polar jet stream is known to be dependent upon both global and local temperature gradients. While the reduced pole-to-equator temperature gradient at the surface suggests a weaker or northward-shifted jet stream, the increased pole-to-equator temperature gradient in the upper troposphere suggests the reverse. The models calculate the relative influence of each factor and provide a result that is a physically and quantitatively consistent representation of how temperatures, winds, and other atmospheric features might change in the future.

As shown in Figure 22, both models project the strengthening and southward shift in the region of maximum upper atmospheric winds in the eastern Pacific and across the West Coast (Sousounis, 1999; Felzer, 1999). The changes in these winds, which seem in the models to be a combination of the polar and subtropical jets, are indicative of a deepened and southward-shifted Aleutian Low in both models, especially for the Hadley model scenario. The Aleutian Low is a center for storms coming into North America off the Pacific, so a deepening and southward-shift in the Aleutian Low would allow more storms to penetrate further southward towards the California coast, helping to explain the precipitation increases projected for that region. The projected weakening of the Pacific Subtropical High (centered near Hawaii) would reduce upwelling of colder ocean waters, allowing SSTs to rise and enabling more storms to penetrate into the Southwest. Storm counts (Figure 23) confirm that the models are projecting more storms associated with the stronger Aleutian Low. Although the Hadley model scenario shows a slight decrease in storms over the Southwest, there is more moisture in the atmosphere, resulting from the higher SSTs (Felzer and Heard, 1999). As a result, the amount of precipitation is actually projected to increase. Other models, however, show reduced storm activity along the Pacific coast (Christoph et al., 1997), so that these results are apparently model dependent.

The region of storm formation off the East Coast of the US is locally dependent upon the land-sea temperature gradient. Warm Gulf Stream waters and a cold land surface in winter provide ideal conditions for generating storms (e.g., nor'easters). With warming of the land surface, the land-sea contrast is reduced and the intensity of these storms could be reduced. The storms in the Hadley model scenario start in the Mid-Atlantic region and track north and east over the Atlantic Ocean; in contrast, the storms in the Canadian model scenario track closely along the East Coast (Figure 23). Observations indicate that present storm tracks extend along the southeastern coast of the US (Klein, 1957), so, in this particular region, the storm tracks are better located in the Canadian model scenario than in the Hadley model scenario, although they are over-represented to the south. Both models project a decrease in the number of storms along this predominant East Coast storm track (Figure 23), although some individual storms appear to be more intense (Felzer and Heard, 1999; Carnell and Senior, 1998; Lambert, 1995). Because of the different baseline positions of the storm tracks, however, the effect of the reduced number of storms is felt over the US only in the

Wintertime Changes in Jet Stream and Atmospheric Circulation

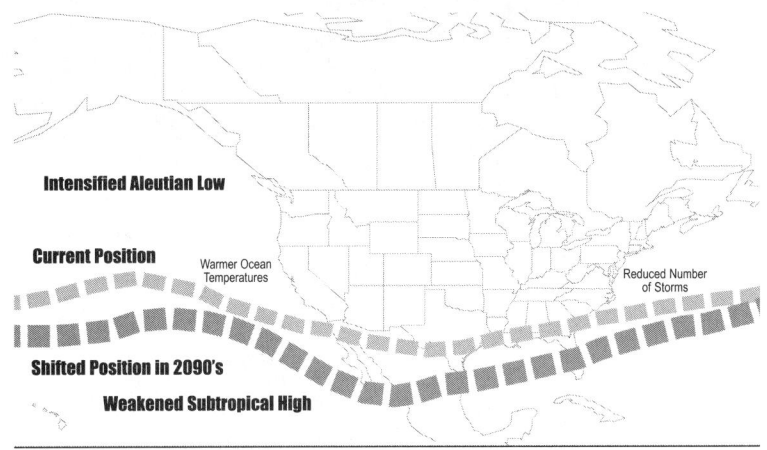

Figure 22: Schematic illustrating wintertime changes in the jet stream, pressure systems, sea surface temperatures, and storm tracks over and adjacent to North America. The Canadian and Hadley model scenarios both show: a southward-shifted jet stream over the eastern Pacific and Southwest; a southward-shifted and intensified Aleutian Low and weakened subtropical High in the West; and warmer ocean surface temperatures off the coast of California. The Canadian model scenario also shows a reduction in the number of storms along the East Coast storm track; however, the Hadley model scenario does not show this reduction nor did it develop this observed storm center in its control simulation. For more details, see Sousounis (1999). See Color Plate Appendix.

Canadian model scenario (Felzer and Heard, 1999). Note that the increase in the number of storms over East Coast land areas in the Hadley model scenario is probably not statistically significant because there are very few storms there to begin with. A decreasing number of storms would be a change from the historical pattern, which does not show any decrease in East Coast storms over the past 100 years, but instead shows an increase during the 1960s (Hayden, 1999). A separate study of the results from the Canadian model also indicates that a higher CO_2 concentration will alter wintertime variability and the behavior of the Arctic Oscillation, affecting primarily the North Atlantic and European regions (Monahan et al., 2000). Other studies show an entire range of possibilities for storm changes in the North Atlantic (Meehl et al., 2000b), including more intense storms (Lunkeit et al., 1996), less intense storms (Beersma et al., 1997), and a shift in storm tracks towards the northeast with no change in intensity (Schubert et al., 1998).

Changes in the tracks of storms and jet streams may also be the result of changes in tropical circulation due to changes in the model projections for the El Niño-Southern Oscillation (ENSO). ENSO is presently a major cause of inter-annual variations in tropical

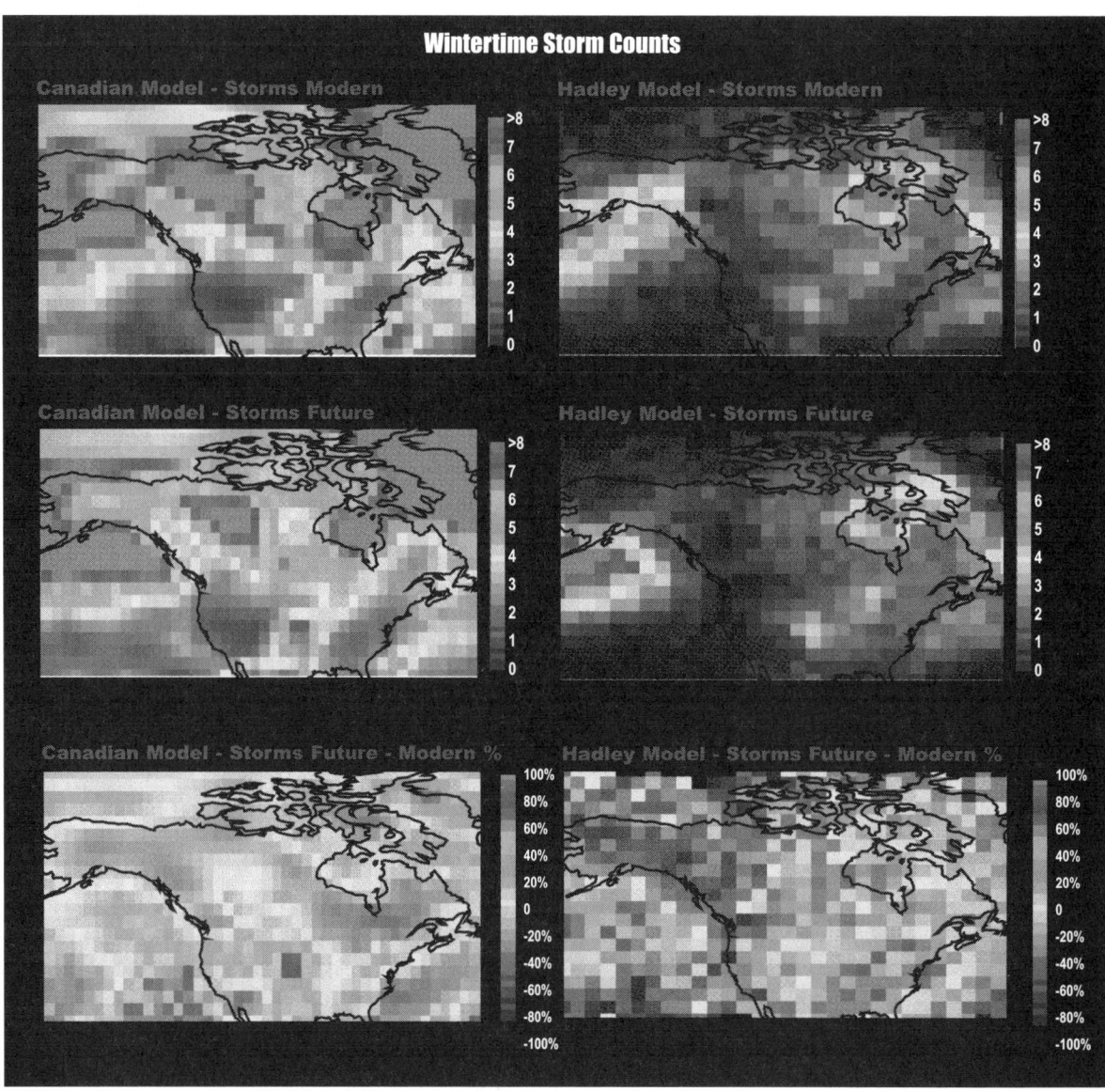

Figure 23: Wintertime (DJF) storm counts (Carnell and Senior, 1998; Lambert, 1995) from the (a) Canadian model scenario (1901-1910 total); (b) Hadley model scenario (1990-2110 mean from unforced control run); (c) Canadian model scenario (2091-2100 total); (d) Hadley model scenario (2070-2100 mean from transient run); (e) Canadian model scenario delta (c-a); and (f) Hadley model scenario delta (d-b). Units are number of winter storms per 145,000 km^2. See Color Plate Appendix.

and global circulation. During warm ENSO events (El Niño), the waters in the eastern and central equatorial Pacific Ocean warm, changing the atmospheric and oceanic circulations in the Pacific region, which affects global weather patterns and the position of the jet stream over North America. The potential effects of global warming on ENSO are not yet established with confidence, in part because of the limited ability of GCMs to simulate ENSO variations over the 20th century. However, both oceanic and atmospheric indices can be used to provide some indications of the types of changes the models are projecting. In particular, although the findings must be considered uncertain, the Niño SST-based indices and the Southern Oscillation Index (SOI) do provide an indication of how atmospheric pressure patterns may shift.

Indices for the Niño-3 and 4 regions in the Pacific Ocean, which record changes in the SST, show ENSO cycles continuing to occur in both models as the world warms, although around a higher average oceanic temperature (D. Legler and J. O'Brien, personal communication; see http://www.coaps.fsu.edu/~legler/NAST/Assess_ENSO.html). Examining the SOI results, the Canadian model scenario projects a shift towards a more persistent set of conditions that is similar to an El Niño state (He and Barnston, personal communication), while the Hadley model scenario shows no change. In neither case do these models project a significant change in the frequency or amplitude of ENSO variability (Collins, 2000). A recent study using a model with sufficient tropical resolution to more accurately reproduce ENSO variability has shown increased ENSO amplitude (El Niños and La Niñas are both more intense) as a result of greenhouse warming (Timmermann et al., 1999). Other studies (Meehl et al., 2000b), however, show little change (or even a slight reduction) in ENSO amplitude (Knutson et al., 1997), while the Hadley model scenario shows an increase in amplitude only after CO_2 levels have been quadrupled (Collins, 2000). Because there are several frequencies of variability within the ENSO signal (Meehl et al., 2000b; Zhang et al., 1997; Lau and Weng, 1999; Allan et al., 1996; Knutson et al., 1997), it is often difficult to determine how ENSO is changing, even with a century-long time series. Given these model results, the stronger Aleutian Low and weaker subtropical high (Trenberth and Hurrell, 1994) that both Assessment models project over the Pacific seem likely to result from either the El Niño-like response in the two models or from the warm ENSO phase (El Niño) response evident in the Canadian model scenario.

Many of the precipitation changes in the GCMs, particularly during summer when the atmospheric circulation is weaker, appear to be the result of feedbacks involving the land surface. During winter, snow cover is the mechanism for this interaction, while during summer, soil moisture is most important. As warming occurs over land areas, evaporation of available moisture increases and the soil moisture decreases; as the land dries out, this soil moisture leads to a decrease in overall evaporation and therefore of the amount of precipitable water in the atmosphere. This decrease, in turn, results in fewer clouds, less precipitation, and increased warming, completing the positive feedback loop. Thus, while increased warming over the ocean is projected to result in increased precipitation, the increased warming over land is projected to lead to less precipitation because of the limited moisture-holding capacity of the land. Differences in model projections of changes in precipitation over land during summer may therefore result from differences in the respective land surface models used in each GCM. Soil moisture trends generally correlate with precipitation anomalies during winter. Although soil moisture trends (Figure 18) during summer also correlate with the precipitation anomalies, there are even broader areas of decreased soil moisture due to the large increases in evapotranspiration. For example, both models show drying in the western Great Plains during summer. Another example is Alaska, where increases in evaporation due to increased summer temperatures are projected to lead to decreased soil moisture even though precipitation increases (Felzer and Heard, 1999).

Snow cover also plays an important role in wintertime changes in climate. Given the degree of warming across the US, the models project that the extent of snow cover is very likely to be significantly reduced (Figure 24). As the snow line retreats poleward, a larger surface area is exposed to a lower albedo surface, which increases the amount of warming, creating a large positive feedback. While both the Canadian and Hadley models show the snowline retreating towards the end of the 21st century, the reduction in snow cover over the US is projected to be particularly dramatic in the Canadian

Winter Average Snow Cover Difference - 2090s

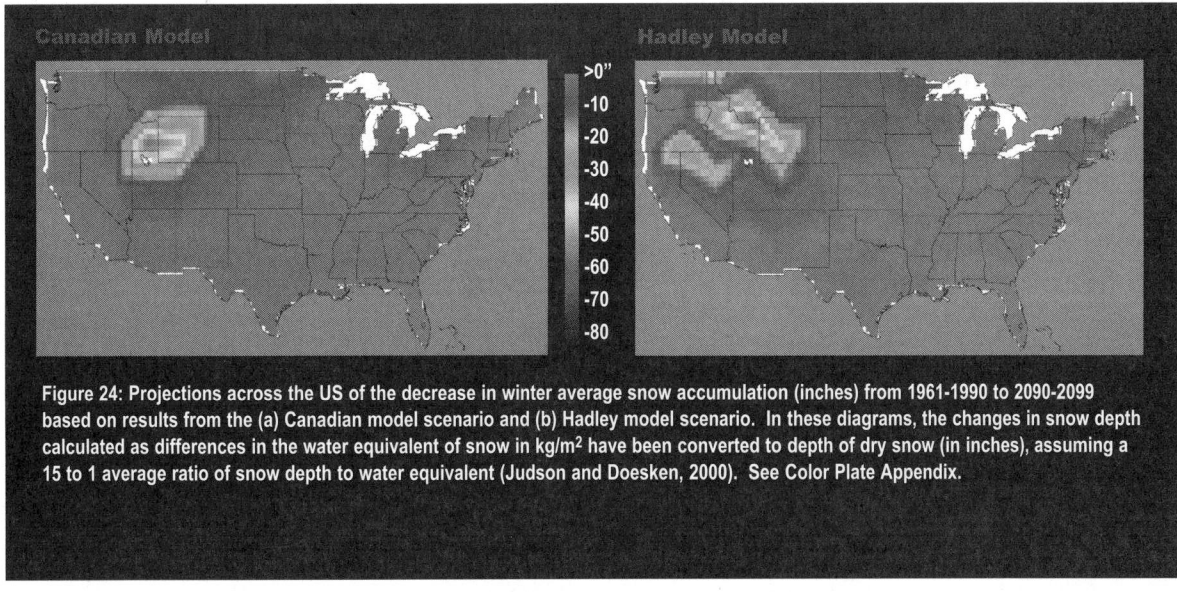

Figure 24: Projections across the US of the decrease in winter average snow accumulation (inches) from 1961-1990 to 2090-2099 based on results from the (a) Canadian model scenario and (b) Hadley model scenario. In these diagrams, the changes in snow depth calculated as differences in the water equivalent of snow in kg/m² have been converted to depth of dry snow (in inches), assuming a 15 to 1 average ratio of snow depth to water equivalent (Judson and Doesken, 2000). See Color Plate Appendix.

Projected Changes in Intensity of National Daily Precipitation

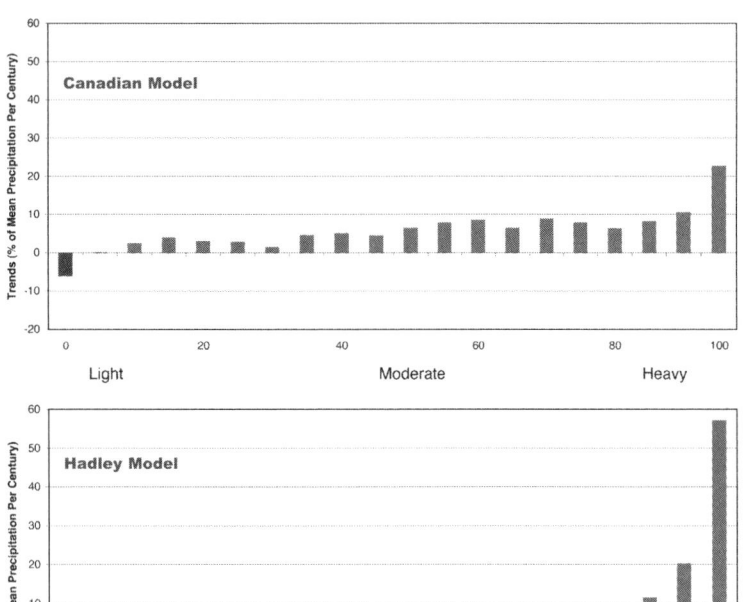

Figure 25: Bar chart showing projected changes in frequency of various types of precipitation. Both the (a) Canadian and (b) Hadley model scenarios project increases in the frequency of heavy precipitation events, intensifying the trend observed for the 20th century. Figure prepared by Byron Gleason of the National Climatic Data Center based on the methods described in Karl and Knight (1998).

model, where mean wintertime snow cover exists only in the northern Rocky Mountains and northern Great Plains. Although snow cover still remains in the northern Rocky Mountains, both models project that the amount of snow in this region will be dramatically reduced. The sharply reduced extent of snow cover in the Canadian model scenario, which may have been initiated by the more zonal flow conditions leading to higher wintertime temperatures and fewer outbreaks of Arctic storms, allows more absorption of solar radiation, especially early in the year. This effect further diminishes snow cover and increases the warming in the Canadian model simulation.

Climate Extremes

While changes in average conditions and inter-annual variability are expected to have significant effects for some ecosystems and some parts of the economy, other parts of the economy are projected to be affected more by potential changes in the frequency or intensity of extreme events. It is likely that the

frequency of occurrence of exceeding certain thresholds and the intensity of extreme events might change because temperatures and absolute humidity are projected to increase, as is the hydrologic cycle of evaporation and precipitation. Many models project a greater frequency of extreme high temperatures and decrease in frequency of extreme low temperatures as a result of increased greenhouse gas concentrations (Giorgi et al., 1998; Meehl et al., 2000b). Increased daily temperature variability in summer and decreased daily temperature variability in winter is also likely (Mearns et al., 1995; Gregory and Mitchell, 1995; Zwiers and Kharin, 1998; Meehl et al., 2000b). As observed (Karl and Knight, 1998) and modeled (Meehl et al., 2000b), reduced diurnal temperature range may result from a greater increase in minimum temperatures than maximum temperatures. Both observations (Gaffen and Ross, 1998) and model results (Delworth et al., 1999), including the current scenarios (Figure 15), show an increase in the heat index, which is a measure of the discomfort level due to warming.

Trends in one-day and multi-day precipitation events over the US and other countries show an increase in the number of days with the heaviest amounts of precipitation (Karl and Knight, 1997, 1998). The number of days annually with precipitation exceeding 2 inches (about 5 cm) has been increasing in the US (Karl et al., 1995a) and the frequency of the highest 1- to 7-day precipitation totals has also been increasing (Kunkel et al., 1999). Increases have been largest for the Southwest, Midwest, and Great Lakes regions of the US. Projections from the Hadley and Canadian model scenarios show an increase of heavy precipitation events as the climate warms (Figure 25). Hulme et al. (1998) found some agreement between the projected precipitation changes and recently observed trends. In reviewing the Canadian model results, Zwiers and Kharin (1998) found that extreme temperature and precipitation events are very likely to occur more frequently. Many other modeling studies also show an increase in the heaviest precipitation events (Meehl et al., 2000b; Kothavala, 1997; Hennessy et al., 1997; Durman et al., 2000; Giorgi et al., 1998). Studies on changes in climate extremes are summarized in recent workshop proceedings (Karl and Easterling, 1999; AGCI, 1999; Easterling et al., 2000b) and in Meehl et al. (2000b). Summer drying in mid-continental regions due to increased evaporation, sometimes coupled with decreased precipitation, has also been projected in many models (Haywood et al., 1997; Gregory et al., 1997; Wetherald and Manabe, 1999; Meehl et al., 2000b).

Studies with the GFDL hurricane model (Knutson et al., 1998; Knutson and Tuleya, 1999) also suggest that the rate of precipitation during tropical storms could increase due to the warmer conditions and the increased amount of water vapor in the atmosphere. Other studies confirm these results (Krishnamurti et al., 1998; Walsh and Ryan, 1999; Meehl et al., 2000b). Two additional studies show a decrease in the frequency of hurricanes as a result of global warming (Bengtsson et al., 1996; Yoshimura et al., 1999). Both the Canadian and Hadley model scenarios project an increase of heavy precipitation events as the climate warms. Ultimately there is a strong dependence of hurricanes on ENSO (Meehl et al., 2000b; Knutson et al., 1998; Knutson and Tuleya, 1999), indicating that how ENSO changes is likely to be an important indicator of how hurricanes will vary, especially for the southeastern US.

Precipitation is the driving factor affecting streamflow (Langbein, 1949; Karl and Reibsame, 1989) so the observed and projected increase in the intensity and frequency of heavy (the upper 5% percentiles of all precipitation events) and extreme precipitation (the highest annual 1-day precipitation events) have the potential to increase inland flooding. Higher temperatures, conversely, have the potential for exacerbating drying of the soil and, over time, of increasing drought frequency and intensity. Separating these two influences is challenging. Nonetheless, analyses of changes in drought frequency and intensity (Karl et al., 1995a) reveal no trend in drought frequency, but they do reveal an increase in the area affected by severe and extreme moisture surplus. Streamflow data analyzed by Lins and Slack (1999) also reveal an increase in low-stream flows, adding more confidence to the notion that drought frequency and intensity has not become more severe, despite the increase in US average temperature. On the other hand, Lins and Slack (1999) do not find an unusual number of statistically significant increases of streamflow, despite the fact that Karl and Knight (1998) show statistically significant increases of precipitation, including heavy and extreme events. New analyses indicate a strong relation between multi-decadal increases in heavy and extreme precipitation events and high and low streamflows, but with considerable variability (Groisman et al., 1999, 2000). These results indicate that part of this variability is related to reductions in snow cover extent in the West, which have modified the peak stream flows and ameliorated the effect of increased heavy precipitation. In these results, Groisman et al. (2000) find that, when averaged across watersheds and across the country, a clear relationship between heavy precipitation and high streamflow events emerges.

THE CLIMATIC EFFECTS OF STABILIZING THE CARBON DIOXIDE CONCENTRATION

The objective of the Framework Convention on Climate Change (FCCC), of which over 160 countries including the US are signatories, is to stabilize the atmospheric concentration of greenhouse gases "at a level that would prevent dangerous anthropogenic interference with the climate system." Precise goals for stabilization of the CO_2 concentration have not been established. To provide information for the negotiating process, the IPCC considered stabilization at concentrations of 350, 450, 550, 650, 750, and 1000 ppmv, plus a variety of temporal pathways to reach these goals (Wigley et al., 1997). Many alternative carbon emission and CO_2 concentration pathways have been evaluated for achieving stabilization at 550 ppmv, which represents an approximate doubling of the pre-industrial CO_2 concentration. Different end points and emission pathways arise from different assumptions about the speed at which emissions can or will be reduced based on views about feasibility or optimality of policies, measures, and technological changes.

To provide an estimate of the reduction in climate change that might occur with CO_2 stabilization, the NAST asked the National Center for Atmospheric Research (NCAR) to use new models to carry out special simulations to provide an indication of the size of the climatic change that would result from stabilizing the CO_2 concentration. NAST and NCAR scientists chose to examine the climatic consequences of a reduced emission growth scenario involving eventual stabilization at 550 ppmv. This emissions path would allow continued growth in emissions for a few decades into the 21st century, followed by rapid decreases in emissions. While this emission path is a plausible alternative for investigation of potential climatic impacts, it should not be interpreted as the only way to achieve stabilization, as a prediction of what is most likely to happen, or as a preferred policy alternative. Reductions in the projected warming would be greater from scenarios that begin reducing emissions earlier in the 21st century than is assumed in the stabilization scenario used here.

To carry out these simulations, two different climate models were used (Boville and Gent, 1998; Washington et al., 2000). Having the results of only one modeling group (albeit with two similar models) is somewhat limiting, especially because the baseline simulation reported here was not the same baseline used in the Canadian and Hadley model scenarios. However, these calculations do provide interesting insights[19]. Similar stabilization runs have now also been completed by the Hadley Centre (Mitchell et al., 2000).

Global Mean Temperature Anomolies

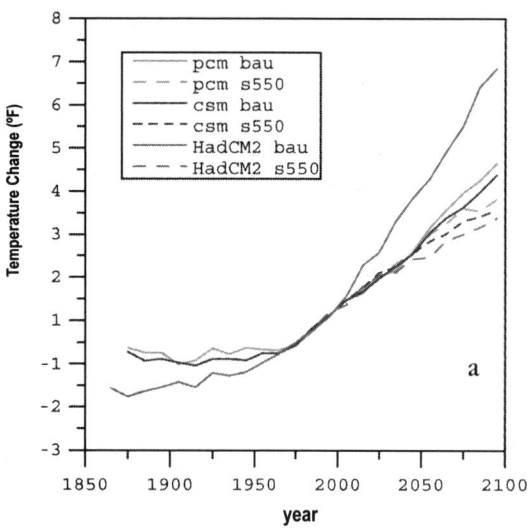

US Mean Temperature Anomolies

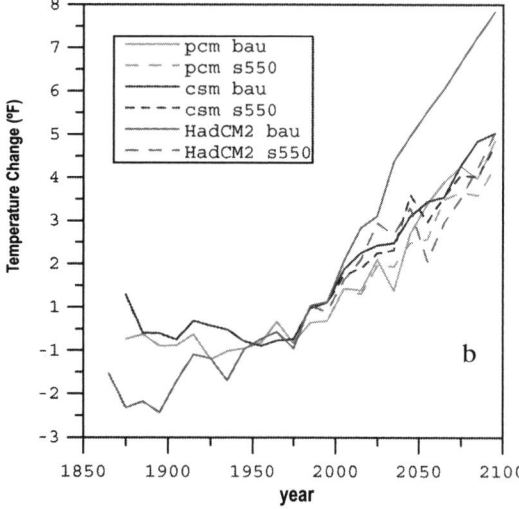

Figure 26: Comparison of the time history of the increase in annual-average surface temperature for (a) the globe and (b) the US as projected by two related models developed at the National Center for Atmospheric Research for an emission scenario where the greenhouse gas concentrations are allowed to rise without restriction (baseline) and for a case (stabilization) where steps are taken to limit the rise in the CO$_2$ concentration to 550 ppmv (Dai et al., 1999; Washington et al., 2000). Results are also shown for a recent Hadley model simulation (Mitchell et al., 2000). See Color Plate Appendix.

While most of the differences in climatic conditions that would result from moving toward stabilization would occur in the 22nd century, modest effects do become apparent in the latter half of the 21st century (or could occur earlier if earlier actions are taken to reduce the rate of rise of emissions). As indicated in Figure 26, if the emissions pathways were to occur as projected, global and US average temperatures would likely continue to rise significantly during the 21st century, even if actions were taken starting in the near future to limit the growth of emissions in order to move toward stabilizing atmospheric concentrations at 550 ppmv[20]. Basically, the NCAR results (Dai et al., 2001) suggest that, even if such actions are taken, the warming in 2100 would still likely be several degrees Fahrenheit (and other more responsive models would suggest even more). With movement toward stabilization, the warming is projected to be about half a degree Fahrenheit less (or about 10-15% lower) in 2100 than for the "no climate policy" scenario used in the Hadley and Canadian models used in the Assessment. Figure 27 shows the effects of the move toward stabilization on temperature and precipitation patterns over the US. In these model simulations (and other simulations may give different effects), the emissions cutback begins to reduce the warming across the southern US and to change the resulting precipitation pattern slightly.

It should also be noted that the reduction in the rate of rise of the CO$_2$ concentration itself would also have important effects. For forests and agriculture, increased CO$_2$ stimulates growth and improves water use efficiency under a range of conditions, so that a lesser rise in the CO$_2$ concentration would likely reduce the increase in crop production and growth of natural biomass (see chapters on Agriculture and Forests) as well as reduce climatic stress on the various ecosystems. For coral reefs, the acidifying effects of CO$_2$ cause reduced alkalinity of ocean waters, reducing calcification and weakening corals; therefore, limiting the rate of CO$_2$ increase would help to ameliorate this situation (see Coastal chapter and Kleypas et al., 1999).

---

[19]Detailed results from these model simulations are available at http://www.nacc.usgcrp.gov/scenarios/.

[20]Stabilizing the atmospheric CO$_2$ concentration at 550 ppmv over the 21st century would require keeping global average per capita emissions of CO$_2$ at roughly their present level of 1 tonne of carbon per year as global population increases by about 50% and developing nations raise their energy levels to enhance their standard-of-living. Accomplishing this would require that all energy needs for the growing population would have to be met by an appropriate combination of reducing CO$_2$ emissions (e.g., through higher efficiencies, use of less carbon intensive fuels, etc.), switching to energy sources not based on fossil fuels (e.g., wind, solar, hydro, biomass, nuclear, etc.), providing more efficient energy services, or reducing the emissions of other greenhouse gases.

CRUCIAL UNKNOWNS AND RESEARCH NEEDS

While much has been learned about the types of climate changes that could occur over the 21st century as atmospheric concentrations of CO_2 increase, much remains to be learned, especially about how the variability and extremes of the climate will change. Although the similarities in how the Canadian and Hadley models represent changes in global scale features are encouraging, the differences between their results on regional scales suggest that significant uncertainties remain. For example, even though the Canadian model scenario produces a reasonable response to El Niño occurrences across North America, problems with the way these GCMs simulate ENSO variability suggest that the projected pattern of changes may not be definitive. Also, as illustrated by the different projections of changes in summer precipitation in the Southeast, there are often several processes that contribute to the pattern of change that is seen, and these may progress differently. As illustrated by the discussion about changes in storm tracks, often the same process can lead to different projections of changes when imposed on a slightly different base state of the climate. In addition, the different representations of land surface processes (as well as other parameterizations) included in different GCMs can have an important impact on projections of changes in regional precipitation. This dependence occurs because precipitation, unlike atmospheric dynamics, is a highly localized feature of the climate, depending on the interaction of many processes, some of which are still represented in quite schematic ways. Given these many limitations, it is important to mention again that the model projections are not predictions, but that they instead should be viewed as internally consistent scenarios of climatic changes that might occur over the 21st century. As a result, they can, as indicated earlier, only provide indications of the types of consequences that might result.

To build confidence in the projections, much remains to be done. Further improvements in climate models are needed, especially in the representations of clouds, aerosols (and their interactions with clouds), sea ice, hydrology, ocean currents, regional orography, and land surface characteristics. Improving projections of the potential changes in atmospheric concentrations of greenhouse gases and aerosols is underway under the auspices of the IPCC (IPCC, 2000) and model simulations based on

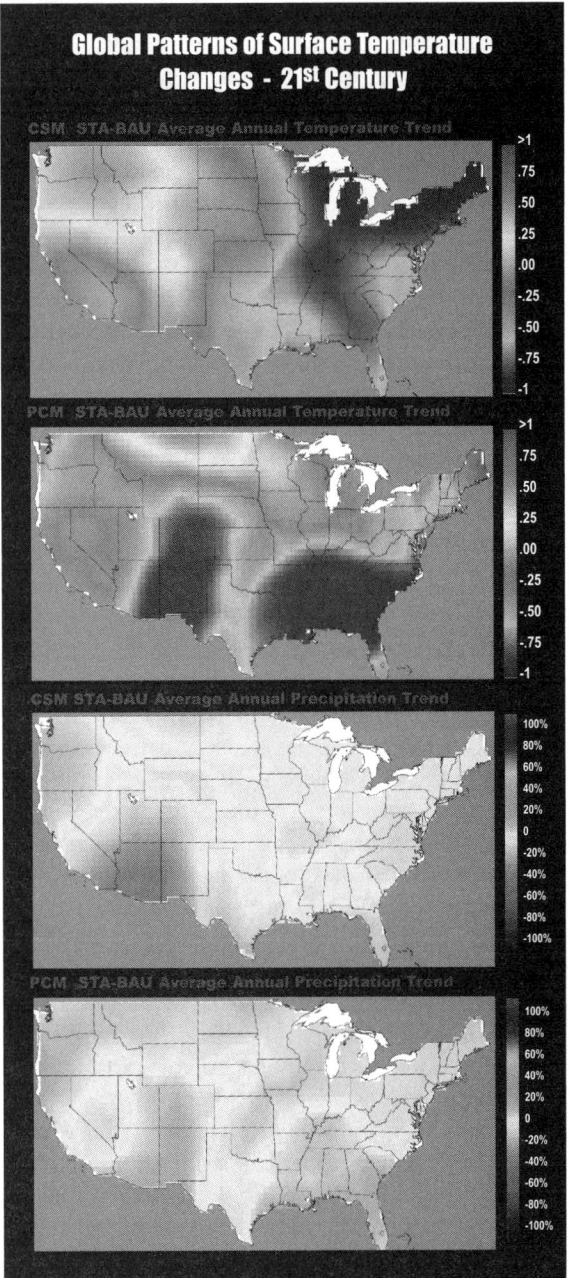

Figure 27: Patterns across the US of projected changes in the trends of annual mean surface temperature and precipitation for the 21st century assuming an emissions profile that moves toward stabilization of the CO_2 concentration at 550 ppmv in the 22nd century (STA) as compared to the baseline case (roughly case IS92a, or BAU, except projections in sulfur emissions are reduced in the CSM scenario). The projected differences in the changes that would generally be projected (case STA minus BAU) are based on results from: (a) NCAR CSM for annual mean temperature; (b) PCM for annual mean temperature: (c) NCAR CSM annual average monthly precipitation; and (d) PCM annual average monthly precipitation. Temperature trend differences are given as ßF per 100 years. Precipitation trend differences are given in percent, with both trends calculated using a 1980-1999 baseline. Trends are derived based on a linear regression through each grid point. Results are described in Dai et al. (1999) and Washington et al. (2000). See Color Plate Appendix.

these revised emissions forecasts are expected to provide improved estimates of future change. In addition to having results from more models available, ensembles of simulations from several model runs are needed so that the statistical significance of the projections can be more fully examined. As part of these efforts, it is important to develop greater understanding of how the climate system works (e.g., of the role of atmosphere-ocean interactions and cloud feedbacks), to refine model resolution, to more completely incorporate existing knowledge into climate models, to more thoroughly test model improvements, and to augment computational and personnel resources in order to conduct and fully analyze a wider variety of model simulations, including mesoscale modeling studies.

While much remains to be done that will take significant time, much can also be done at present to improve the use and understanding of potential climate change scenarios. For example, an intensified analysis program is needed to provide greater understanding of the changes and the reasons they occur. New efforts to examine the synoptic patterns of the changes in the global models were started at the national level in the analyses presented here. Such effort are also starting through region-specific studies that combine analysis of the model results with the insights available from analysis of historical climatology and past weather patterns (i.e., synoptic conditions). For example, Risbey et al. (1999) have constructed regional climate scenarios for two study regions in North America (Chesapeake Bay and Oklahoma/Great Plains) using a combination of GCM output and dynamical reasoning. Other approaches being pursued involve use of mesoscale models that provide higher resolution of spatial conditions even though they can only provide simulations of shorter periods of time and smaller spatial scales.

SUMMARY

There are clear indications that the atmospheric concentrations of CO_2 and other greenhouse gases and aerosols are being increased by human activities. These changes in atmospheric composition, combined with the influences of other natural and human-induced changes, are changing the climate. Available model simulations, combined with our understanding of the factors that have changed the Earth's climate in the past, provide clear evidence that global warming is occurring and that additional warming will result. There is clear evidence on the global scale that this warming is occurring, with the

warming during the 20th century in reasonable accord with model simulations. Because climate affects the environment and many of our natural resources, considering how such changes might affect us will provide important information on how we will need to adapt to such changes.

Use of historical climate data provides one basis for exploring the environment and society's vulnerability to a changing climate. Records of the US climate indicate that the climate is apparently starting to change in a manner consistent with the observed global-scale changes. These records also indicate that there have been significant variations in the climate, which have in turn had important effects on agriculture, water resources, and public health. There is no reason to believe from the historical or paleoclimatic record or from model results that such changes will not recur in the future. To help in analyzing societal vulnerability to ongoing climate variations, a range of historical information about the climate of the 20th century has been provided for the Assessment.

To explore how future climate may be affected by the rising concentrations of greenhouse gases, model simulations have been used to provide quantitative estimates. Although emission scenarios are uncertain and model simulations are still imperfect (e.g., due to limitations in the representations of important processes and feedbacks), two sets of model results have been assembled to provide plausible projections of how conditions may change over the US during the 21st century. These models project quite significant warming across the US and substantial stress on water resources in several regions. While the particular sets of model results do not fully bound all of the possible futures, they do provide a range of possible future conditions that can be used to start to explore the potential consequences of climate change for the US.

The results available for this Assessment thus provide the basis for the most complete analysis yet undertaken and point to pathways for future analysis and research. Current understanding clearly indicates that the climate is changing and is very likely to change significantly more in the future. At the same time, much more work is needed over the coming years to improve global and mesoscale projections of future changes in CO_2 concentration and of climate, to improve simulation of climate variability, to develop the means to project changes in extreme events, and to expand the statistical analysis and interpretation of existing and planned model simulations. Practical efforts should continue at the community level to interpret the scientific data and climate scenarios with the aim of providing usable information that integrates current needs with planning for future community development.

LITERATURE CITED

AGCI (Aspen Global Change Institute), *Elements of Change 1998*, edited by S. J. Hassol and J. Katzenberger, Aspen Colorado, 1999.

Alcamo, J., E. Kreileman, M. Krol, R. Leemans, J. Bollen, J. Van Minnen, M. Schaeffer, S. Toet, and B. De Vries, Global modelling of environmental change: An overview of Image 2.1., pp. 3-94 in *Global Change Scenarios of the 21st Century: Results from the Image 2.1 Model*, edited by J. Alcamo, R. Leemans, and E. Kreileman, Pergamon Press, Oxford, United Kingdom, 1998.

Allan, R. J., J. Lindesay, and D. Parker, *El Niño Southern Oscillation and Climate Variability*, CSIRO Publications, Melbourne, Australia, 405 pp., 1996.

Andres, R. J., G. Marland, T. Boden, and S. Bischoff, Carbon dioxide emissions from fossil fuel combustion and cement manufacture, 1751 to 1991, and an estimate for their isotopic composition and latitudinal distribution, in *The Carbon Cycle*, edited by T. M. L. Wigley and D. Schimel, Cambridge University Press, Cambridge, United Kingdom, 312 pp., 2000.

Angell, J. K., Difference in radiosonde temperature trend for the period 1979-1998 of MSU data and the period 1959-1998 twice as long, *Geophysical Research Letters*, 27, 2177-2180, 2000.

Arrhenius, S., On the influence of carbonic acid in the air upon the temperature of the ground, *Philosophical Magazine*, 41, 237, 1896.

Arritt, R. W., D. C. Goering, and C. J. Anderson, The North American monsoon system in the Hadley Centre coupled ocean-atmosphere GCM, *Geophysical Research Letters*, 27, 565-568, 2000.

Barnett, T. P. K., et al., Detection and attribution of recent climate change: A status report, *Bulletin of the American Meteorological Society*, 80, 2631-2659, 1999.

Beersma, J. J., K. M. Rider, G. J. Komen, E. Kaas, and V. V. Kharin, An analysis of extratropical storms in the North Atlantic region as simulated in a control and 2 x CO_2 time-slice experiment with a high-resolution atmospheric model, *Tellus*, 49A, 347-361, 1997.

Bengtsson, L., M. Botzet, and M. Esch, Will greenhouse gas-induced warming over the next 50 years lead to higher frequency and great intensity hurricanes?, *Tellus*, 48A, 57-73, 1996.

Berger, A., Long-term variation of daily insolation and Quaternary climatic changes, *Journal Atmospheric Sciences*, 35, 2362-2367, 1978.

Berger, A., The role of CO_2, sea-level and vegetation during the Milankovitch forced glacial-interglacial cycles, in *Geosphere-Biosphere Interactions and Climate*, edited by L. Bengtsson, 1999.

Berger, A., and M. F. Loutre, Insolation values for the climate of the last 10 million years, *Quaternary Science Reviews*, 10, 297-317, 1991.

Berger, A., M. F. Loutre, and J. L. Melice, *The 100 kyr period in the astronomical forcing, Scientific Report 1999/6*, Institut D'Astronomie et de Geophysique G. Lemaitre, Université Catholique de Louvain, 1999.

Boer, G. J., G. M. Flato, and D. Ramsden, A transient climate change simulation with historical and projected greenhouse gas and aerosol forcing: Projected climate for the 21st century, *Climate Dynamics*, 16, 427-450, 2000a.

Boer, G. J., G. M. Flato, M. C. Reader, and D. Ramsden, A transient climate change simulation with historical and projected greenhouse gas and aerosol forcing: Experimental design and comparison with the instrumental record for the 20th century, *Climate Dynamics*, 16, 405-425, 2000b.

Boer, G. J., N. A. McFarlane, R. Laprise, J. D. Henderson, and J. P. Blanchet, The Canadian Climate Centre spectral atmospheric general circulation model, *Atmosphere-Ocean*, 22(4), 397-429, 1984.

Bove, M. C., J. B. Elsner, C. W. Landsea, X. Niu, and J. J. O'Brien, Effect of El Niño on US landfalling hurricanes, revisited, *Bulletin of the American Meteorological Society*, 79, 2477-2482, 1998.

Boville, B. A., and P. R. Gent, The NCAR Climate System Model, Version One, *Journal of Climate*, 11(6), 1115-1130, 1998.

Cai, W., and P. H. Whetton, Evidence for a time-varying pattern of greenhouse warming in the Pacific Ocean, *Geophysical Research Letters*, 27, 2577-2580, 2000.

Callendar, G. S., The artificial production of carbon dioxide and its influence on temperature, *Quarterly Journal of the Royal Meteorological Society*, 64, 223, 1938.

Carnell, R. E., and C. A. Senior, Changes in mid-latitude variability due to increasing greenhouse gases and sulphate aerosols, *Climate Dynamics*, 14, 369-383, 1998.

CDIAC (Carbon Dioxide Information Analysis Center), Oak Ridge National Laboratory, 2000. (Available at http://cdiac.esd.ornl.gov:80/cdiac/)

Christoph, M., U. Ulbrich, and P. Speth, Midwinter suppression of Northern Hemisphere storm track activity in the real atmosphere and in GCM experiments, *Journal of the Atmospheric Sciences, 54*, 1589-1599, 1997.

Clark, P. U., R. B. Alley, and D. Pollard, Northern Hemisphere ice-sheet influences on global climate change, *Science, 286*, 1104-1111, 1999.

COHMAP (Climates of the Holocene Mapping Project), Climatic changes of the last 18,000 years: Observations and model simulations, *Science, 241*, 1043-1052, 1988.

Collins, M., The El-Niño Southern Oscillation in the second Hadley Centre coupled model and its response to greenhouse warming, *Journal of Climate, 13*, 1299-1312, 2000.

Conway, T. J., P. P. Tans, and L. S. Waterman, Atmospheric CO_2 records from sites in the NOAA/CMDL air sampling network, in *Trends '93: A Compendium of Data on Global Change*, edited by T. A. Boden, et al., ORNL/CDIAC-65, Carbon Dioxide Information Analysis Center, Oak Ridge National Laboratory, US Department of Energy, Oak Ridge Tennessee, 1994.

Crowley, T. J., Causes of climate change over the past 1000 years, *Science, 289*, 270-277, 2000.

Crowley, T. J., and G. R. North, *Paleoclimatology*, Oxford University Press, New York, 339 pp., 1991.

Dai, A., K. E. Trenberth, and T. R. Karl, Effects of clouds, soil moisture, precipitation, and water vapor on diurnal temperature range, *Journal of Climate, 12*, 2451-2473, 1999.

Dai, A., T. M. L. Wigley, B. A. Boville, J. T. Kiehl, and L. E. Buja, Climates of the 20th and 21st centuries simulated by NCAR Climate System Model, *Journal of Climate, 14*, 485-519, 2001.

Darwin, R. F., World agriculture and climate change: Current questions, *World Resource Review, 9*, 17-31, 1997.

Delworth, T. L., J. D. Mahlman, and T. R. Knutson, Changes in heat index associated with CO_2-induced global warming, *Climatic Change, 43*, 369-386, 1999.

Doherty, R., and L. O. Mearns, A comparison of simulation of current climate from two coupled atmosphere-ocean global climate models against observations and evaluation of their future climates, National Institute for Global and Environmental Change (NIGEC), Boulder, Colorado, 1999.

Dresler, P. V., M. C. MacCracken, and A. Janetos, National assessment of the potential consequences of climate variability and change for the United States, *Water Resources Update, 112*, 16-24, 1998.

Durman, C. F., J. M. Gregory, D. C. Hassell, and R. G. Jones, The comparison of extreme European daily precipitation simulated by a global and a regional climate model for present and future climates, Quarterly Journal of the Royal Meteorological Society, in press, 2000.

Easterling, D. R., Variability and trends in temperature threshold exceedances and frost dates in the United States, *Bulletin of the American Meteorological Society*, in review, 2000.

Easterling, D. R., T. R. Karl, E. H. Mason, P. Y. Hughes, D. P. Bowman, R. C. Daniels, and T. A. Boden, United States Historical Climatology Network (USHCN) Monthly Temperature and Precipitation Data, Publication No. 4500, Carbon Dioxide Information and Analysis Center, Oak Ridge National Laboratory, Oak Ridge, Tennessee, 83 pp. with appendices, 1996.

Easterling, D. R., et al., Maximum and minimum temperature trends for the globe, *Science, 277*, 364-367, 1997.

Easterling, D. R., J. L. Evans, P. Ya. Groisman, T. R. Karl, K. E. Kunkel, and P. Ambenje, Observed variability and trends in extreme climate events: A review, *Bulletin of the American Meteorological Society, 81*, 417-425, 2000a.

Easterling, D. R., G. A. Meehl, C. Parmesan, S. A. Changnon, T. R. Karl, and L. O. Mearns, Climate extremes: Observations, modeling, and impacts, *Science, 289*, 2068-2074, 2000b.

Etheridge, M., L. P. Steele, R. L. Langenfelds, R. J. Francey, J.-M. Barnola, and V. I. Morgan, Historical CO_2 records from the Law Dome DE08, DE08-2, and DSS ice cores, in *Trends: A Compendium of Data on Global Change*, Carbon Dioxide Information Analysis Center, Oak Ridge National Laboratory, US Department of Energy, Oak Ridge Tennessee, 1998.

Felzer, B., Hydrological implications of GCM results for the U.S. National Assessment, pp. 69-72 in *Proceedings of the Specialty Conference on Potential Consequences of Climate Variability and Change to Water Resources of the United States, May 10-12, 1999, Atlanta, Georgia*, American Water Resources Association, Middleburg, Virginia, 1999.

Felzer, B., and P. Heard, Precipitation differences amongst GCMs used for the U.S. National Assessment, *Journal of the American Water Resources Association, 35*(6), 1327-1339, 1999.

Flato, G. M., G. J. Boer, W. G. Lee, N. A. McFarlane, D. Ramsden, M. C. Reader, and A. J. Weaver, The Canadian Centre for Climate Modeling and Analysis global coupled model and its climate, *Climate Dynamics, 16*, 451-467, 2000.

Gaffen, D. J., and R. J. Ross, Increased summertime heat stress in the U.S., *Nature, 396,* 529-530, 1998.

Gates, W. L., et al., An overview of the results of the Atmospheric Model Intercomparison Project (AMIP I), *Bulletin of the American Meteorological Society, 80,* 29-55, 1999.

Giorgi, F., L. Mearns, C. Shields, and L. McDaniel, Regional nested model simulations of present day and 2 x CO_2 climate over the central Great Plains of the United States, *Climatic Change, 40,* 457-493, 1998.

Goody, R. M., and Y. L. Yung, *Atmospheric Radiation,* Oxford University Press, Oxford, United Kingdom, 1989.

Gregory, J. M. and J. F. B. Mitchell, Simulation of daily variability of surface temperature and precipitation over Europe in the current and 2 x CO_2 climates using the UKMO climate model, *Quarterly Journal of the Royal Meteorological Society, 121,* 1451-1476, 1995.

Gregory, J. M., J. F. B. Mitchell, and A. J. Brady, Summer drought in northern midlatitudes in a time-dependent CO_2 climate experiment, *Journal of Climate, 10,* 662-686, 1997.

Gregory, J. M., and J. Oerlemans, Simulated future sea-level rise due to glacier melt based on regionally and seasonally resolved temperature changes, *Nature, 391,* 474-476, 1998.

Groisman, P. Ya., et al., Changes in the probability of heavy precipitation: Important indicators of climatic change, *Climatic Change, 42,* 243-283, 1999.

Groisman, P. Ya., R. W. Knight, and T. R. Karl, Heavy precipitation and streamflow in the United States: Trends in the 20th century, *Bulletin of the American Meteorological Society,* in press, 2000.

Hansen, J., D. Johnson, A. Lacis, S. Lebedeff, P. Lee, D. Rind, and G. Russell, Climatic impact from increasing carbon dioxide, *Science, 213,* 957-966, 1981.

Hansen, J., M. Sato, A. Lacis, R. Ruedy, I. Gegen, and E. Matthews, Climate forcings in the industrial era, in *Proceedings of the National Academy of Sciences, 95,* 12753-12758, 1998.

Hayden, B. P., Extratropical storms: Past, present, and future, pp. 93-96 in *Proceedings of Specialty Conference on Potential Consequences of Climate Variability and Change to Water Resources of the United States,* May 1999, American Water Resources Association, 1999.

Haywood, J. M., R. J. Stouffer, R. T. Wetherald, S. Manabe, and V. Ramaswamy, Transient response of a coupled model to estimated changes in greenhouse gas and sulfate concentrations, *Geophysical Research Letters, 24*(11), 1335-1338, 1997.

Hegerl, G. C., K. Hasselmann, U. Cubasch, J. F. B. Mitchell, E. Roeckner, R. Voss, and J. Waszkewitz, Multi-fingerprint detection and attribution of greenhouse gas-and aerosol-forced climate change, *Climate Dynamics, 13,* 613-634, 1997.

Hengeveld, H. G., Projections for Canada's Climate Future, Special report CCD-00-01, Environment Canada, Downsview, Ontario, Canada, 2000.

Hennessy, K. J., J. M. Gregory, and J. F. B. Mitchell, Changes in daily precipitation under enhanced greenhouse conditions, *Climate Dynamics, 13,* 667-680, 1997.

Houghton, R. A., Land-use change and the carbon-cycle, *Global Change Biology, 1,* 275-287, 1995.

Houghton, R. A., and J. L. Hackler, Continental scale estimates of the biotic carbon flux from land cover change, 1850-1980, ORNL/CDIAC-79, NDP-050, Oak Ridge National Laboratory, 144 pp., 1995.

Hoyt, D. V., and K. H. Schatten, A discussion of plausible solar irradiance variations: 1700-1992, *Journal of Geophysical Research, 98,* 18895-18906, 1993.

Huang, S., H. N. Pollack, and P. Y. Shen, Temperature trends over the past five centuries reconstructed from borehole temperatures, *Nature, 403,* 756-758.

Hulme, M., T. J. Osborn, and T. C. Johns, Precipitation sensitivity to global warming: Comparison of observations with HadCM2 simulations, *Geophysical Research Letters, 25*(17), 3379-3382, 1998.

Hyde, W. T., and T. J. Crowley, Probability of future climatically significant volcanic eruptions, *Journal of Climate, 13,* 1445-1450, 2000.

Imbrie, J., et al., On the structure and origin of major glaciation cycles 1. Linear responses to Milankovitch forcing, *Paleoceanography, 7,* 701-738, 1992.

Imbrie, J., et al., On the structure and origin of major glaciation cycles 2. The 100,000 year cycle, *Paleoceanography, 8,* 699-735, 1993.

Imbrie, J., A. McIntyre, and A. Mix, Oceanic response to orbital forcing in the late Quaternary: Observational and experimental strategies, pp. 121-164 in *Climate and the Geosciences,* edited by A. Berger et al., Kluwer Academic Publishers, Boston, Massachusetts, 1989.

Indermuehle, A., et al., Holocene carbon-cycle dynamics based on CO_2 trapped in ice at Taylor Dome, Antarctica, *Nature, 398,* 121-126, 1999.

IPCC (Intergovernmental Panel on Climate Change), *Climate Change: The IPCC Scientific Assessment,* edited by J. T. Houghton, G. J. Jenkins, and J. J. Ephraums, Cambridge University Press, Cambridge United Kingdom, 365 pp., 1990.

IPCC (Intergovernmental Panel on Climate Change), *Climate Change 1992: The Supplementary Report to the IPCC Scientific Assessment,* edited by J. T. Houghton, B. A. Callander, and S. K. Varney, Cambridge University Press, Cambridge, United Kingdom, 200 pp., 1992.

IPCC (Intergovernmental Panel on Climate Change), *Climate Change 1995: The Science of Climate Change,* edited by J. T. Houghton, L. G. Meira Filho, B. A. Callander, N. Harris, A. Kattenberg, and K. Maskell, Cambridge University Press, Cambridge, United Kingdom, 572 pp., 1996a.

IPCC (Intergovernmental Panel on Climate Change), *Climate Change 1995: Impacts, Adaptations, and Mitigation of Climate Change: Scientific-Technical Analyses,* edited by R. T. Watson, M. C. Zinyowera, and R. H. Moss, Cambridge University Press, Cambridge, United Kingdom, 879 pp., 1996b.

IPCC (Intergovernmental Panel on Climate Change), *Climate Change 1995: Economic and Social Dimensions of Climate Change,* edited by E. J. Bruce, Hoesung Lee, and E. Haites, Cambridge University Press, Cambridge, United Kingdom, 464 pp., 1996c.

IPCC (Intergovernmental Panel on Climate Change), *An Introduction to Simple Climate Models Used in the IPCC Second Assessment Report,* by D. Harvey, J. Gregory, M. Hoffert, A. Jain, M. Lal, R. Leemans, S. Raper, T. Wigley, and J. deWolde, and edited by J. T. Houghton, L. G. Meira Filho, D. Griggs, and K. Maskell, World Meteorological Organization/United Nations Environment Programme, 50 pp., 1997.

IPCC (Intergovernmental Panel on Climate Change), *Special Report on Emissions Scenarios,* N. Nakicenovic (lead author), Cambridge University Press, Cambridge, United Kingdom, 599 pp., 2000.

Johns, T. C., R. E. Carnell, J. F. Crossley, J. M. Gregory, J. F. B. Mitchell, C. A. Senior, S. F. B. Tett, and R. A. Wood, The second Hadley Centre coupled ocean-atmosphere GCM: Model description, spinup and validation, *Climate Dynamics, 13,* 103-134, 1997.

Jones, P. D., M. New, D. E. Parker, S. Martin, and I. G. Rigor, Surface air temperature and its changes over the past 150 years, *Reviews of Geophysics, 37,* 173-199, 1999.

Jones, P. D., P. Ya. Groisman, M. Coughlan, N. Plummer, W.-C. Wang, and T. R. Karl, Assessment of urbanization effects in time series of surface air temperature over land, *Nature, 347,* 169-172, 1990.

Joussaume, S., et al., Monsoon changes for 6000 years ago: Results of 18 simulations from the Paleoclimate Modelling Intercomparison Project (PMIP), *Geophysical Research Letters, 26,* 859-862, 1999.

Judson, A., and N. Doesken, Density of freshly fallen snow in the central Rocky Mountains, *Bulletin of the American Meteorological Society, 81,* 1577-1587, 2000.

Karl, T. R., and D. R. Easterling, Climate extremes: Selected review and future research directions, *Climatic Change, 42*(10), 309-325, 1999.

Karl, T. R., and R. W. Knight, The 1995 Chicago heat wave: How likely is a recurrence?, *Bulletin of the American Meteorological Society, 78,* 1107-1119, 1997.

Karl, T. R., and R. W. Knight, Secular trends of precipitation amount, frequency, and intensity in the United States, *Bulletin of the American Meteorological Society, 79,* 231-241, 1998.

Karl, T. R., and W. E. Reibsame, The impact of decadal fluctuations in mean precipitation and temperature on runoff: A sensitivity study over the United States, *Climatic Change, 15,* 423-447, 1989.

Karl, T. R., and K. Trenberth, The human impact on climate, *Scientific American, 281,* 100-105, 1999.

Karl, T. R., H. F. Diaz, and G. Kukla, Urbanization: Its detection and effect in the United States climate record, *Journal of Climate, 1,* 1099-1123, 1988.

Karl, T. R., R. W. Knight, and B. Baker, The record breaking global temperatures of 1997 and 1998: Evidence for an increase in the rate of global warming?, *Geophysical Research Letters, 27,* 719-722, 2000.

Karl, T. R., R. W. Knight, D. R. Easterling, and R. Quayle, Indices of climate change for the United States, *Bulletin of the American Meteorological Society, 77,* 279-292, 1996.

Karl, T. R., R. W. Knight, and N. Plummer, Trends in high-frequency climate variability in the twentieth century, *Nature, 377,* 217-220, 1995a.

Karl, T. R., R. W. Knight, D. R. Easterling, and R. G. Quayle. Indices of climate change for the United States, *Bulletin of the American Meteorological Society, 77,* 279-292, 1995b.

Keeling, C. D., and T. P. Whorf, Atmospheric CO_2 records from sites in the SIO air sampling network, in *Trends: A Compendium of Data on Global Change*, Carbon Dioxide Information Analysis Center, Oak Ridge National Laboratory, US Department of Energy, Oak Ridge, Tennessee, 1999.

Kiehl, J. T., and K. E. Trenberth, Earth's annual global mean energy budget, *Bulletin of the American Meteorological Society*, 78, 197-208, 1997.

Kittel, T. G. F., N. A. Rosenbloom, T. H. Painter, D. S. Schimel, and VEMAP participants, The VEMAP integrated database for modeling United States ecosystem/vegetation sensitivity to climate change, *Journal of Biogeography*, 22, 857-862, 1995.

Kittel, T. G. F., J. A. Royle, C. Daly, N. A. Rosenbloom, W. P. Gibson, H. H. Fisher, D. S. Schimel, L. M. Berliner, and VEMAP participants, A gridded historical (1895-1993) bioclimate dataset for the conterminous United States, pp. 219-222 in *Proceedings of the 10th Conference on Applied Climatology*, American Meteorological Society, Boston, Massachusetts, 1997.

Klein, W. H., Principal tracks and mean frequencies of cyclones and anticyclones in the Northern Hemisphere, *Research Paper 40*, US Weather Bureau, Washington DC, 60 pp., 1957.

Kleypas, J. A., R. W. Buddemeier, D. Archer, J. P. Gattuso, C. Langdon, and B. N. Opdyke, Geochemical consequences of increased atmospheric carbon dioxide on coral reefs, *Science, 284*, 118-120, 1999.

Knutson, T. R., and S. Manabe, Time-mean response over the tropical Pacific to increased CO_2 in a coupled ocean-atmosphere model, *Journal of Climate, 8*, 2181-2199, 1995.

Knutson, T. R., and S. Manabe, Model assessment of decadal variability and trends in the tropical Pacific Ocean, *Journal of Climate, 11*, 2273-2296, 1998.

Knutson, T. R., and R. E. Tuleya, Increased hurricane intensities with CO_2-induced warming as simulated using the GFDL hurricane prediction system, *Climate Dynamics, 15*, 503-519, 1999.

Knutson, T. R., S. Manabe, and D. Gu, Simulated ENSO in a global coupled ocean-atmosphere model: Multidecadal amplitude modulation and CO_2 sensitivity, *Journal of Climate, 10*, 138-161, 1997.

Knutson, T. R., T. L. Delworth, K. W. Dixon, and R. J. Stouffer, Model assessment of regional surface temperature trends (1949-1997), *Journal of Geophysical Research, 104*, 30981-30996, 1999.

Knutson, T. R., R. E. Tuleya, and Y. Kurihara, Simulated increase of hurricane intensities in a CO_2-warmed climate, *Science, 279*, 1018-1020, 1998.

Kothavala, Z., Extreme precipitation events and the applicability of global climate models to study floods and drought, *Mathematics and Computers in Simulations, 43*, 261-268, 1997.

Krishnamurti, T. N., R. Correa-Torres, M. Latif, and G. Daughenbaugh, The impact of current and possibly future sea surface temperature anomalies on the frequency of Atlantic hurricanes, *Tellus, 50A*, 186-210, 1998.

Kunkel, K. E., K. Andsager, and D. R. Easterling, Long-term trends in extreme precipitation events over the conterminous United States and Canada, *Journal of Climate, 12*, 2515-2527, 1999.

Kutzbach, J. E., P. J. Guetter, P. J. Behling, and R. Selin, Simulated climatic changes: Results of the COHMAP climate-model experiments, in *Global Climates Since the Last Glacial Maximum*, edited by H. E. Wright, Jr., et al., pp. 24-93, University of Minnesota Press, Minnesota, 1993.

Lambert, S. J., The effect of enhanced greenhouse warming on winter cyclone frequencies and strengths, *Journal of Climate, 8*(5), 1447-1452, 1995.

Langbein, W. B., Annual Runoff in the United States, *US Geological Survey Circular No. 5*, Department of the Interior, Washington, DC, 1949 (reprinted 1959).

Lau, K. M., and H. Weng, Interannual, decadal-to-inter-decadal and global warming signals in sea surface temperature during 1955-1997, *Journal of Climate, 12*, 1257-1267, 1999.

Lean, J., and D. Rind, Climate forcing by changing solar radiation, *Journal of Climate, 11*, 3069-3094, 1998.

Lean, J., J. Beer, and R. S. Bradley, Reconstruction of solar irradiance since 1620: Implications for climate change, *Geophysical Research Letters, 22*(23), 3195-3198, 1995.

Leatherman, S. P., K. Zhang, and B. C. Douglas, Sea level rise shown to drive coastal erosion, *Transactions of the American Geophysical Union, EOS, 81*, 55-57, 2000 (also see responses and reply in *EOS, 81*, 436-437 and 439-441, 2000).

Ledley, T. S., E. T. Sundquist, S. E. Schwartz, D. K. Hall, J. D. Fellows, and T. L. Killeen, Climate Change and Greenhouse Gases, *EOS, Transactions of the American Geophysical Union*, 80(39), September 28, 1999, p. 453, 1999. (Available at http://www.agu.org/eos_elec/99148e.html)

Legates, D. R., Global and terrestrial precipitation: A comparative assessment of existing climatologies: A reply, *International Journal of Climatology*, *17*, 779-783, 1997.

Legates, D. R., and T. L. DeLiberty, Precipitation measurements biases in the United States, *Water Resources Bulletin*, *29*, 855-861, 1993.

Legates, D. R., and C. J. Wilmott, Mean seasonal and spatial variability in global surface air temperature, *Theoretical and Applied Climatology*, *41*, 11-21, 1990a.

Legates, D. R., and C. J. Wilmott, Mean seasonal and spatial variability in gauge-corrected global precipitation, *International Journal of Climatology*, *10*, 111-127, 1990b.

Lins, H. F., and J. R. Slack, Streamflow trends in the United States, *Geophysical Research Letters*, *26*(2), 227-230, 1999.

Lunkeit, F., M. Ponater, R. Sausen, M. Sogalla, U. Ulbrich, and M. Windelband, Cyclonic activity in a warmer climate, *Contributions to Atmospheric Physics*, *69*, 393-407, 1996.

Mahlman, J. D., Uncertainties in projections of human-caused climate warming, *Science*, *278*, 1416-1417, 1997.

Manabe, S., and R. T. Wetherald, The effects of doubling the CO_2 concentration on the climate of a general circulation model, *Journal of the Atmospheric Sciences*, *32*, 3, 1975.

Mann, M. E., R. S. Bradley, and M. K. Hughes, Northern hemisphere temperatures during the past millennium: Inferences, uncertainties, and limitations, *Geophysical Research Letters*, *26*(6), 759-762, 1999.

Mantua, N., S. Hare, Y. Zhang, J. Wallace, and R. Francis, A Pacific interdecadal climate oscillation with impacts on salmon production, *Bulletin of the American Meteorological Society*, *78*, 1069-1079, 1997.

Marland, G., T. A. Boden, R. J. Andres, A. L. Brenkert, and C. Johnston, Global, regional, and national CO_2 emissions, in *Trends: A Compendium of Data on Global Change*, Carbon Dioxide Information Analysis Center, Oak Ridge National Laboratory, US Department of Energy, Oak Ridge, Tennessee, 1999.

McFarlane, N. A., G. J. Boer, J. P. Blanchet, and M. Lazare, The Canadian Climate Centre second-generation general circulation model and its equilibrium climate, *Journal of Climate*, *5*, 1013-1044, 1992.

Mearns, L. O., I. Bogardi, F. Giorgi, I. Matyasovszky, and M. Palecki, Comparison of climate change scenarios generated from regional climate model experiments and statistical downscaling, *Journal of Geophysical Research*, *104*(D6), 6603-6621, 1999.

Mearns, L. O., F. Giorgi, L. McDaniel, and C. Shields, Analysis of the diurnal range and variability of daily temperature in a nested modeling experiment: Comparison with observations and 2 x CO_2 results, *Climate Dynamics*, *11*, 193-209, 1995.

Meehl, G. A., and W. M. Washington, El Niño-like climate change in a model with increased atmospheric CO_2 concentrations, *Nature*, *382*, 56-60, 1996.

Meehl, G. A., G. J. Boer, C. Covey, M. Latif, and R. J. Stouffer, Meeting summary: The Coupled Model Intercomparison Project (CMIP), *Bulletin of the American Meteorological Society*, *81*, 313-318, 2000a.

Meehl, G. A, F. Zwiers, J. Evans, T. Knutson, L. Mearns, and P. Whetton, Trends in extreme weather and climate events: Issues related to modeling extremes in projections of future climate change, *Bulletin of the American Meteorological Society*, *81*, 427-436, 2000b.
Mitchell, J. F. B., and T. C. Johns, On modification of global warming by sulfate aerosols, *Journal of Climate*, *10*(2), 245-267, 1997.

Mitchell, J. F. B., T. C. Johns, J. M. Gregory, and S. Tett, Climate response to increasing levels of greenhouse gases and sulphate aerosols, *Nature*, *376*, 501-504, 1995.

Mitchell, J. F. B., T. J. Johns, W. J. Ingram, and J. A. Lowe, The effect of stabilising atmospheric carbon dioxide concentrations on global and regional climate change, *Journal of Geophysical Research*, *27*, 2977-2980, 2000.

Mitchell, J. F. B., C. A. Wilson, and W. M. Cunnington, On CO_2 climate sensitivity and model dependence of results, *Quarterly Journal of the Royal Meteorological Society*, *113*, 293-322, 1987.

Monahan, A. H., J. C. Fyfe, and G. M. Flato, A regime view of Northern Hemisphere atmospheric variability and change under global warming, *Geophysical Research Letters*, *27*, 1139-1142, 2000.

Mote, P., and A. O'Neill, *Numerical Modeling of the Global Atmosphere in the Climate System*, Kluwer Academic Publishers, Boston, Massachusetts, 517 pp., 2000.

NAS (National Academy of Sciences), *Policy Implications of Greenhouse Warming: Mitigation, Adaptation, and the Science Base, Panel on Policy Implications of Greenhouse Warming*, National Academy Press, Washington, DC, 1992.

Neftel, A., H. Friedli, E. Moor, H. Lötscher, H. Oeschger, U. Siegenthaler, and B. Stauffer, Historical CO_2 record from the Siple Station ice core, in *Trends: A Compendium of Data on Global Change*, Carbon Dioxide Information Analysis Center, Oak Ridge National Laboratory, US Department of Energy, Oak Ridge Tennessee, 1994.

Nihoul, J. C. J. (Ed.), *Coupled Ocean-Atmosphere Models*, Elsevier Science, Amsterdam, 767 pp., 1985.

Noda, A., K. Yamaguchi, S. Yamaki, and S. Yukimoto, Relationship between natural variability and CO_2-induced warming pattern: MRI AOGCM experiment, pp. 359-362 in *Preprints volume, 10th Symposium on Global Change Studies*, American Meteorological Society, Boston, Massachusetts, 1999.

NRC (National Research Council), *Carbon Dioxide and Climate: A Scientific Assessment*, National Academy of Sciences, Washington, DC, 22 pp., 1979.

NRC (National Research Council), *Changing Climate: Report of the Carbon Dioxide Assessment Committee*, National Academy Press, Washington, DC, 496 pp., 1983.

NRC (National Research Council), *Capacity of US Climate Modeling*, National Academy Press, Washington, DC, 1998.

NRC (National Research Council), *Reconciling Observations of Global Temperature Change*, National Academy Press, Washington, DC, 85 pp., 2000.

Palmer W., Meteorological Drought, *Research Paper No. 45*, US Weather Bureau, NOAA Library and Information Services Division, Washington, DC, 58 pp., 1965.

Parker, D. E., T. P. Legg, and C. K. Folland, Interdecadal changes of surface temperatures since the late 19th century, *Journal of Geophysical Research*, *99*, 14373-14399, 1994.

Petit, J. R., et al., Climate and atmospheric history of the past 420,000 years from the Vostok ice core, Antarctica, *Nature*, *399*, 429-436, 1999.

Pisias, N. G., and N. J. Shackleton, Modelling the global climate response to orbital forcing and atmospheric carbon dioxide changes, *Nature*, *310*, 757-759, 1984.

Pitman, A., R. Pielke, Sr., R. Avissar, M. Claussen, J. Gash, and H. Dolman, The role of the land surface in weather and climate: Does the land surface matter?, *IGBP Newsletter*, 39, pp. 4-9, September 1999.

PSAC (President's Science Advisory Council), Atmospheric Carbon Dioxide, Appendix Y4 in *Restoring the Quality of Our Environment*, Report of the Environmental Pollution Panel, The White House, Washington, DC, 1965.

Quayle, R. G, T. C. Peterson, A. Basist, and C. Godfrey, An operational near-real-time global temperature index, *Geophysical Research Letters*, *26*, 333-335, 1999.

Risbey, J. S., P. J. Kushner, P. J. Lamb, R. Miller, M. C. Morgan, M. Richman, G. Roe, and J. Smith, Generating regional climate scenarios by combining synoptic-climatological guidance and GCM output, US EPA Research Report, Washington DC, 88 pp., 1999 (also submitted to *Journal of Climate* by a subset of the authors and with different title).

Rothrock, D. A., Y. Yu, and G. A. Maykut, Thinning of the Arctic sea-ice cover, *Geophysical Research Letters*, *26*(23), 3469-3472, 1999.

Schimel, D, I. G. Enting, M. Heimann, T. M. L. Wigley, D. Raynaud, D. Alves, and U. Siegenthaler, CO_2 and the carbon cycle, in *Climate Change 1994, Radiative Forcing of Climate Change and an Evaluation of the IPCC 1992 Emission Scenarios*, edited by J. T. Houghton, et al., Cambridge University Press, Cambridge, United Kingdom, 1995.

Schubert, M., J. Perlwitz, R. Blender, K. Fraedrich, and F. Lunkeit, North Atlantic cyclones in CO_2-induced warm climate simulations: Frequency, intensity, and tracks, *Climate Dynamics*, *14*, 827-837, 1998.

Shackleton, N. J., J. Le, A. C. Mix, and M. Hall, Carbon isotope records from Pacific surface waters and atmospheric carbon dioxide, *Quaternary Science Reviews*, *11*, 387-400, 1992.

Smith, E., Atlantic and East Coast hurricanes 1900-98: A frequency and intensity study for the twenty-first century, *Bulletin of the American Meteorological Society*, *80*, 2717-2720, 1999.

Smith, S. J., N. Nakicenovic, and T. M. L. Wigley, Radiative forcing in the IPCC SRES scenarios, *Nature*, in review, 2000.

Sousounis, P., A synoptic assessment of climate change model output: Explaining the differences and similarities between the Canadian and Hadley climate models, in *Eleventh Symposium on Global Change Studies, 80th American Meteorological Society Annual Meeting, January 2000*, American Meteorological Society, Boston, Massachusetts, 1999.

Stahle, D. W., E. R. Cook, M. K. Cleaveland, M. D. Therrell, D. M. Meko, H. D. Grisino-Mayer, E. Watson, and B. H. Luckman, Tree-ring data document 16th century megadrought over North America, *Transactions of the American Geophysical Union (EOS), 81,* 121 and 125, 2000.

Steadman, R. G., The assessment of sultriness, part I: A temperature-humidity index based on human physiology and clothing science, *Journal of Climate and Applied Meteorology, 18,* 861-873, 1979.

Stott, P. A., S. F. B. Tett, G. S. Jones, M. R. Allen, W. J. Ingram, and J. F. B. Mitchell, Attribution of twentieth century temperature change to natural and anthropogenic causes, *Climate Dynamics,* in press, 2000.

Stouffer, R. J., G. Hegerl, and S. Tett, A comparison of surface air temperature variability in three 1000-year coupled ocean-atmosphere model integrations, *Journal of Climate, 13,* 513-537, 2000.

Tett, S. F. B., P. A. Stott, M. R. Allen, W. J. Ingram, and J. F. B. Mitchell, Causes of twentieth century temperature change, *Nature, 399,* 569-572, 1999.

Thomson, D. J., The seasons, global temperature, and precession, *Science, 268,* 59-68, 1995.

Timmermann, A., Oberhuber, J., Bacher, A., Esch, M., Latif, M., and E. Roeckner, Increased El Niño frequency in a climate model forced by future greenhouse warming, *Nature, 398,* 694-697, 1999.

Trenberth, K. E., and G. Branstator, Issues in establishing causes of the 1988 drought over North America, *Journal of Climate, 5,* 159-172, 1992.

Trenberth, K. E. and J. W. Hurrell, Decadal atmosphere-ocean variations in the Pacific, *Climate Dynamics, 9,* 303-319, 1994.

USDOE (US Department of Energy), *Projecting the Climatic Effects of Increasing Carbon Dioxide,* edited by M. C. MacCracken and F. M. Luther, US Department of Energy, Washington DC, 1985a.

USDOE (US Department of Energy), *Detecting the Climatic Effects of Increasing Carbon Dioxide,* edited by M. C. MacCracken and F. M. Luther, US Department of Energy, Washington DC, 1985b.

USEPA (US Environmental Protection Agency), *The Potential Effects of Global Climate Change on the United States,* EPA-230-05-89-050, edited by J. Smith and D. Tirpak, Washington DC, 1989.

Van Loon, H., and J. Rogers, The seesaw in winter temperatures between Greenland and Northern Europe. Part 1: General description, *Monthly Weather Review, 106,* 296-310, 1978.

VEMAP (Vegetation/Ecosystem Modeling and Analysis Project) Members: Comparing biogeography and biogeochemistry models in a continental-scale study of terrestrial ecosystem responses to climate change and CO_2 doubling, *Global Biogeochemical Cycles, 9,* 407-437, 1995.

Walsh, K. J. E., and B. F. Ryan, Idealized vortex studies of the effect of climate change on tropical cyclone intensities, pp. 403-404 in *Preprints volume, 23rd Conference on Hurricanes and Tropical Meteorology,* American Meteorological Society, Boston, Massachusetts, 1999.

Washington, W. M., and C. L. Parkinson, *An Introduction to Three-Dimensional Climate Modeling,* University Science Books, Mill Valley, California, 422 pp., 1986.

Washington, W. M., et al., Parallel Climate Model (PCM) control and 1%/year CO_2 simulations with a 2/3° ocean model and a 27 km dynamical sea ice model, *Climate Dynamics,* in press, 2000.

Wetherald, R. T., and S. Manabe, Detectability of summer dryness caused by greenhouse warming, *Climatic Change, 43,* 495-511, 1999.

Wigley, T. M. L., and D. Schimel (Eds.), *The Carbon Cycle,* Cambridge University Press, Cambridge United Kingdom, 312 pp., 2000.

Wigley, T. M. L., A. K. Jain, F. Joos, B. S. Nyenzi, and P. R. Shukla, Implications of proposed CO_2 emissions limitations, Technical paper for the Intergovernmental Panel on Climate Change, Geneva, Switzerland, 41 pp., 1997.

Willson, R. C., Total solar irradiance trend during solar cycles 21 and 22, *Science, 277,* 1963-1965, 1997.

Woodhouse, C., and J. Overpeck, 2000 years of drought variability in the central United States, *Bulletin of the American Meteorological Society, 79,* 2693-2714, 1998.

Yoshimura, J., M. Sugi, and A. Noda, Influence of greenhouse warming on tropical cyclone frequency simulated by a high-resolution AGCM, pp. 555-558 in *Preprints volume, 10th Symposium on Global Change Studies,* American Meteorological Society, Boston Massachusetts, 1999.

Zhang, Y., J. M. Wallace, and D. S. Battisti, ENSO-like interdecadal variability: 1900-93, *Journal of Climate, 10,* 1004-1020, 1997.

Zwiers, F. W., and V. Kharin, Changes in the extremes of the climate simulated by CCC GCM2 under CO_2 doubling, *Journal of Climate, 11,* 2200-2222, 1998.

ACKNOWLEDGEMENTS

Many of the materials for this chapter are based on contributions from participants on and those working with the

Climate Scenario Team

Richard Ball, Department of Energy (retired)

Tony Barnston, NOAA National Centers for Environmental Prediction, Climate Prediction Center

Eric Barron, Pennsylvania State University

Denise Blaha, University of New Hampshire

George Boer, Canadian Centre for Climate Modelling and Analysis, Victoria, BC

Ruth Carnell, Hadley Centre, Meteorological Office, Bracknell, UK

Aiguo Dai, National Center for Atmospheric Research

Christopher Daly, Oregon State University

David Easterling, NOAA National Climatic Data Center

Benjamin Felzer, National Center for Atmospheric Research

Hank Fisher, National Center for Atmospheric Research

Greg Flato, Canadian Centre for Climate Modelling and Analysis

Byron Gleason, NOAA National Climatic Data Center

Jonathan Gregory, Hadley Centre, Meteorological Office, Bracknell, UK

Yuxiang He, NOAA National Centers for Environmental Prediction, Climate Prediction Center

Preston Heard, Indiana University - Bloomington

Roy Jenne, National Center for Atmospheric Research

Dennis Joseph, National Center for Atmospheric Research

Tom Karl, NOAA National Climatic Data Center

Tim Kittel, National Center for Atmospheric Research

Richard Knight, NOAA National Climatic Data Center

Steven Lambert, Canadian Centre for Climate Modelling and Analysis, Victoria, BC

Michael MacCracken, USGCRP/National Assessment Coordination Office

Linda Mearns, National Center for Atmospheric Research

John Mitchell, Hadley Centre, Meteorological Office, Bracknell, UK

James Risbey, Carnegie Mellon University

Nan Rosenbloom, National Center for Atmospheric Research

J. Andy Royle, U. S. Fish and Wildlife Service, Laurel MD

Annette Schloss, University of New Hampshire

Joel B. Smith, Stratus Consulting

Steven J. Smith, Battelle Pacific Northwest National Laboratory

Peter Sousounis, University of Michigan

David Viner, Climatic Research Unit, Norwich, UK

Warren Washington, National Center for Atmospheric Research

Tom Wigley, National Center for Atmospheric Research

Francis Zwiers, Canadian Centre for Climate Modelling and Analysis, Victoria, BC

71

CHAPTER 2

VEGETATION AND BIOGEOCHEMICAL SCENARIOS

Jerry Melillo[1,2], Anthony Janetos[3,2], David Schimel[4,5], and Tim Kittel[5]

Contents of this Chapter

[1]The Ecosystems Center, Marine Biological Laboratory; [2]Coordinating author for the National Assessment Synthesis Team, [3]World Resources Institute, [4]Max-Planck Institut für Biogeochemie, [5]National Center for Atmospheric Research

CHAPTER SUMMARY

Ecosystems are communities of plants and animals and the physical environment in which they exist. Ecologists often categorize ecosystems by their dominant vegetation – the deciduous broad-leafed forest ecosystems of New England, the short-grass prairie ecosystems of the Great Plains, the desert ecosystems of the Southwest. Concerns for continued ecosystem health and performance stem from two primary issues. Ecosystems of all types, from the most natural to the most extensively managed, produce a variety of goods and services that benefit humans. Examples of ecosystem services include modification of local climate, air and water purification, landscape stabilization against erosion, flood control, and carbon storage. Ecosystems are also valued for recreational and aesthetic reasons. Climate change has the potential to affect the structure, function, and regional distribution of ecosystems, and thereby affect the goods and services they provide.

For this Assessment, the Vegetation/Ecosystem Modeling and Analysis Project (VEMAP) was used to generate future ecosystem scenarios for the conterminous United States based on model-simulated responses to the Canadian and Hadley scenarios of climate change. The ecosystem scenarios were then shared with Assessment participants to assist them in their evaluations of the potential sensitivities of ecosystems and ecosystem goods and services to climate change.

Key Findings

Some of the key results from VEMAP for ecosystems in the absence of land-cover and land-use changes are as follows:

- Over the next few decades climate change is very likely to lead to increased plant productivity and increased terrestrial carbon storage for many parts of the country, especially those that get moderately warmer and wetter. Areas where soils dry out during the growing season, such as the Southeast for the climate simulated with the Canadian model, are very likely to see reduced productivity and decreases in carbon storage.

- By the end of the 21st century, many regions of the country are likely to have experienced changes in vegetation distribution. Areas in which soil moisture increases are likely to maintain or exhibit an increased woody component of vegetative cover. Areas in which soil moisture decreases are likely to lose woody vegetation. For example, in the Southeast, the climate simulated by the Canadian model causes soil drying that would lead to forest losses and savanna and grassland expansion.

- Modeling of vegetation responses to climate change is in the early stages of development. No single model simulates all of the important factors affecting vegetation responses to climate; model results must therefore be viewed with caution. The complex, non-linear nature of ecosystems almost certainly means that we will be surprised by some of the changes in ecosystem function and structure that climate changes set in motion. Keeping the magnitude of climate change as small as possible and slowing its rate are the two things we can do to minimize the negative impacts on natural ecosystems.

VEGETATION AND BIOGEOCHEMICAL SCENARIOS

INTRODUCTION

This chapter is designed to report the results of the Vegetation/Ecosystem Modeling and Analysis Project II (VEMAP II); a project that has provided data about terrestrial ecosystem responses to climate change to Assessment participants. The chapter is not meant to be a comprehensive, in-depth analysis of climate impacts on all aspects of terrestrial ecosystem structure and function. The chapter has two focus areas – biogeochemistry and plant biogeography in natural terrestrial ecosystems. While animal communities are mentioned briefly in the chapter, they were not considered in the VEMAP II analysis and so are not focused on in this chapter. These scenarios for vegetation and biogeochemical change can serve as background for analyses of changes to fauna and biological diversity by contributing broad-scale information on habitat changes.

VEMAP II is an international, collaborative effort supported by several US Global Change Research Program agencies and sponsored by the International Geosphere-Biosphere Program (IGBP) to conduct an analysis of the potential effects of climate change on ecosystem processes and vegetation distribution within the continental United States. Modeling results to date indicate that natural terrestrial ecosystems are sensitive to changes in global surface temperature, precipitation patterns, atmospheric carbon dioxide (CO_2) levels, and other climate parameters. Major ecological characteristics to be affected include the geographic distribution of dominant plant species, productivity of plants, biodiversity within natural ecosystems, and basic ecological processes and their feedbacks to the climate system.

Two types of models that have been used in VEMAP II to examine the ecological effects of climate change are biogeochemistry models and biogeography models. Biogeochemistry models project changes in basic ecosystem processes such as the cycling of carbon, nutrients, and water (ecosystem function), and biogeography models simulate shifts in the geographic distribution of major plant species and communities (ecosystem structure).

VEMAP II involves a comparison of three biogeochemistry and three biogeography models (Schimel et al., 2000; Neilson et al., 2000). The models use a common "baseline" data set and two potential climate scenarios. Common data are used to ensure that any variability in predicted responses is attributable to the different structures and formulations of individual ecological models rather than to input data.

For the National Assessment, the focus is on model outputs for two time periods: 2025-2034 (near term) and 2090-2099 (long term). Outputs of the biogeochemistry models are used to consider near-term ecological impacts, while outputs of the biogeography models are used to consider longer-term impacts. This is based on the team's expert judgement that biogeochemical changes will dominate ecological responses to climate change in the next few decades, while species shifts will dominate ecological responses to climate change towards the end of the 21st century, as organisms attempt to migrate to occupy "optimal" climate space.

RESEARCH APPROACH

Biogeochemistry Models

The biogeochemistry models simulate the cycles of carbon, nutrients (e.g., nitrogen), and water in terrestrial ecosystems which are parameterized according to life form (VEMAP Members, 1995, Schimel et al., 2000). The models consider how these cycles are influenced by environmental conditions including temperature, precipitation, solar radiation, soil texture, and atmospheric CO_2 concentration. These environmental variables are inputs to general algorithms that describe plant and soil processes such as carbon capture by plants with photosynthesis, decomposition, soil nitrogen transformations mediated by microorganisms, and water flux between land and the atmosphere in the processes of evaporation and transpiration. Common outputs from biogeochemistry models are estimates of net primary productivity, net nitrogen mineralization, evapotranspiration fluxes (e.g., PET, ET), and the storage of carbon and nitrogen in vegetation and soil. In the VEMAP II activity, three biogeochemistry models were used: BIOME-BGC (Hunt and Running, 1992; Running and Hunt, 1993), CENTURY (Parton et al., 1987, 1988, 1993), and the Terrestrial Ecosystem Model (TEM)

(Melillo et al., 1993; McGuire et al., 1997; Tian et al., 1999). The similarities and differences among the models are summarized in Table 1. A detailed intercomparison of these biogeochemistry models has recently been published (Pan et al., 1998). The capabilities and limitations of the models are identified in this intercomparison. A comparison of model results to field data for the Mid-Atlantic region of the northeastern US is presented in Jenkins, et al. (2000).

BIOME-BGC

The BIOME-BGC (BioGeochemical Cycles) model is a multi-biome generalization of FOREST-BGC, a model originally developed to simulate a forest stand development through a life cycle (Running and Coughlan, 1988; Running and Gower, 1991). The model requires daily climate data and the definition of several key climate, vegetation, and site conditions to estimate fluxes of carbon, nitrogen, and water through ecosystems (Table 4 in VEMAP Members, 1995). Allometric relationships are used to initialize plant and soil carbon (C) and nitrogen (N) pools based on the leaf pools of these elements (Vitousek et al., 1988). Components of BIOME-BGC have previously undergone testing and validation, including the carbon dynamics (McLeod and Running, 1988; Korol et al., 1991; Hunt et al., 1991; Pierce, 1993; Running, 1994) and the hydrology (Knight et al., 1985; Nemani and Running, 1989; White and Running, 1995).

CENTURY

The CENTURY model is a general model of plant-soil nutrient cycling which has been used to simulate carbon and nutrient (nitrogen, phosphorus, and sulfur) dynamics for different types of ecosystems including grasslands, agricultural lands, forests, and savannas (Parton et al., 1987, 1993; Metherell, 1992). For VEMAP, only carbon and nitrogen dynamics are included. The model uses monthly temperature and precipitation data as well as atmospheric CO_2 and N inputs to estimate monthly stocks and fluxes of carbon and nitrogen in ecosystems. The CENTURY model also includes a water budget submodel which calculates monthly evapotranspiration, transpiration, water content of the soil layers, snow water content, and saturated flow of water between soil layers. The CENTURY model incorporates algorithms that describe the impact of fire, grazing, and storm disturbances on ecosystem processes (Ojima et al., 1990; Sanford et al., 1991; Holland et al., 1992; Metherell, 1992).

TEM

The Terrestrial Ecosystem Model (TEM version 4.1) describes carbon and nitrogen dynamics of plants and soils for non-wetland ecosystems of the globe (Tian et al., 1999). This model requires monthly climatic data along with soil and vegetation-specific parameters to estimate monthly carbon and nitrogen fluxes and pool sizes. The model includes algorithms from the water balance model of Vörösmarty et al. (1989) to calculate potential and actual evapotranspiration, soil moisture, and drainage. Estimates of net primary production and carbon storage by this version of TEM have been evaluated in previous applications of the model at both regional and global scales (Xiao et al., 1998; Tian et al., 1998, 1999, 2000; Kicklighter et al., 1999; Prinn et al., 1999; Reilly et al., 1999; McGuire et al., 2000).

BIOGEOGRAPHY MODELS

The models used to estimate biogeographic responses to climate change in VEMAP II include LPJ, MAPSS (Mapped Atmosphere-Plant-Soil System) and MC1. These three models project the local dominance of various terrestrial vegetation forms based on (1) ecophysiological constraints, which determine the broad distribution of major categories of woody plants, and (2) response limitations, which determine specific aspects of community composition, such as the competitive balance of trees and grasses. Though similar in some respects, these models simulate potential evapotranspiration and direct CO_2 effects differently, and as a result they show varying sensitivities to temperature, CO_2 levels, and other factors. Two of the models, LPJ and MC1 have biogeochemistry modules while the third, MAPSS, does not. Both LPJ and MC1 are dynamic vegetation models, while MAPSS is an equilibrium model.

LPJ

The LPJ-Model was constructed in a modular framework. Individual modules describe key ecosystem processes, including vegetation establishment, resource competition, growth, and mortality (Sitch, 2000). Vegetation structure and composition is described by nine plant functional types (PFTs) which are distinguished according to their plant physiological (C_3, C_4 photosynthesis), phenological (deciduous, evergreen) and physiognomic (tree, grass) attributes. The model is run on a grid cell basis with input of soil texture, monthly fields of temperature, precipitation, and percentage sunshine hours. Each grid cell is divided into fractions covered by the PFTs and bare ground. The presence and fractional coverage of an individual PFT depends on its specific environmental limits, and on the outcome of resource competition with the other PFTs.

Table 1. Key Characteristics of the Three Biogeochemical Models used in VEMAP II.

References	Biome-BGC Running and Hunt (1993)	Century Parton et al. (1994)	TEM Tian et al. (1999)
Responses of Plant Physiology			
CO_2	Reduction in canopy conductance and leaf N concentration; and increases in intercellular CO_2 concentration, production and water-use efficiency	Reductions in transpiration and leaf N concentration; and prescribed increases in potential production	Increases in intercellular CO_2 concentration and production
Temperature	Optimum temperature for photosynthesis; maintenance respiration increases with temperature; growth respiration increases with photosynthesis	Optimum temperature of production	Optimum temperature of gross primary production (GPP); maintenance respiration increases with temperature; growth respiration increases with GPP
Moisture regime	Canopy conductance increases with enhanced soil moisture and reduced vapor pressure deficit	Potential production increases with enhanced soil moisture	GPP increases with enhanced evapotranspiration; phenology modified with enhanced evapotranspiration
Solar radiation	Photosynthesis increases with enhanced photosynthetically active radiation (PAR)	None	GPP increases with enhanced PAR
Responses of Soil Processes			
CO_2	Soil moisture increases with reduced canopy conductance; decomposition decreases with lower N concentration in litterfall	Soil moisture increases with reduced transpiration; decomposition decreases with lower N concentration in litterfall	Decomposition decreases with lower N concentration in litterfall
Temperature	with increases in temperature: 1) decomposition increases; 2) soil moisture decreases: and 3) net N mineralization increases	with increases in temperature: 1) decomposition increases; 2) soil moisture decreases: and 3) net N mineralization increases	with increases in temperature: 1) decomposition increases; 2) soil moisture decreases: and 3) net N mineralization increases
Precipitation	Soil moisture increase with enhanced precipitation; optimum soil moisture for decomposition	Soil moisture increase with enhanced precipitation; optimum soil moisture for decomposition	Soil moisture increase with enhanced precipitation; optimum soil moisture for decomposition
Solar radiation	Soil moisture decreases with enhanced solar radiation	None	Soil moisture decreases with enhanced solar radiation
Disturbance Regimes	Prescribed mortality	Scheduled fire regimes	Implicitly implemented through litterfall fluxes

The two-layer soil water balance model is based on Haxeltine and Prentice (1996). Moisture in each layer, expressed as a fraction of water holding capacity, is updated daily. Percolation from the upper to the lower layer, and absolute water holding capacity are soil texture dependent.

Establishment and mortality are modeled on an annual basis. Plant establishment, in terms of additional PFT individuals, depends on the fraction of bare ground available for seedlings to successfully establish. Natural mortality is taken as a function of PFT vigor, and corresponds to an annual reduction in the number of PFT individuals. Dead biomass enters the litter pool, and the soil pools. Mortality also occurs due to disturbance (Thonicke et al., 2000).

MAPSS

The MAPSS (Mapped Atmosphere-Plant-Soil System) model begins with the application of ecophysiological constraints to determine which plant types can potentially occur at a given location. A two-layer hydrology module (including gravitational drainage) with a monthly time step then allows simulation of leaf phenology, leaf area index (LAI) and the competitive balance between grass and woody vegetation. A productivity index is derived based on leaf area duration and evapotranspiration. This index is used to assist in the determination of leaf form, phenology, and vegetation type, on the principle that any successful plant strategy must be able to achieve a positive Net Primary Production (NPP) during its growing season.

The LAI of the woody layer provides a light-limitation to grass LAI. Stomatal conductance is explicitly included in the water balance calculation, and water competition occurs between the woody and grass life forms through different canopy conductance characteristics as well as rooting depths. The direct effect of CO_2 on the water balance is simulated by reducing maximum stomatal conductance. The MAPSS model is calibrated against observed monthly runoff, and has been validated against global runoff (Neilson and Marks, 1995). A simple fire model is incorporated to limit shrubs in areas such as the Great Plains (Neilson, 1995).

The forest-grassland ecotone is reproduced by assuming that closed forest depends on a predictable supply of winter precipitation for deep soil recharge (Neilson et al., 1992). An index is used that decrements the woody LAI as the summer dependency increases.

MC1

MC1 consists of three linked modules simulating biogeography, biogeochemistry, and fire disturbance (Lenihan et al., 1998; Daly et al., 2000). The main functions of the biogeography module are: (1) to simulate the composition of deciduous/evergreen, needleleaf/broadleaf tree and C_3/C_4 grass life-form mixtures from climatic thresholds; and (2) to classify those woody and herbaceous life forms into different vegetation classes based on their biomass (or leaf area index) simulated by the biogeochemistry module.

The biogeochemistry module, which is based on the CENTURY model (Parton et al., 1987), simulates monthly carbon and nutrient dynamics for a given life-form mixture. It was configured to always allow tree-grass competition. Above- and below-ground processes are modeled in detail, and include plant production, soil organic matter decomposition, and water and nutrient cycling. Nitrogen (N) demand is always assumed to be met in this study and never limited by local conditions since there were no soil N data available to initialize and calibrate the model.

Parameterization of this module is based on the life form composition of the ecosystems, which is updated annually by the biogeography module. The fire module simulates the occurrence, behavior and effects of severe fire. Allometric equations, keyed to the life-form composition supplied by the biogeography module, are used to convert above-ground biomass to fuel classes. Fire effects (i.e., plant mortality and live and dead biomass consumption) are estimated as a function of simulated fire behavior (i.e., fire spread and fire line intensity) and vegetation structure. Fire effects feed back to the biogeochemistry module to adjust the levels of the carbon and nutrient pools. A detailed description of the model can be found in Daly et al. (2000).

Simulated grazing is species-independent and only occurs in the model between April and September. Only grasses are consumed and there is no tree death assumed due to either consumption or trampling by herbivores. A fraction of the material consumed by the grazers (C and N) is returned to the site.

DATABASES

To meet the various input requirements of the biogeochemistry and biogeography models and ensure a common starting point for the VEMAP II simulations, the "baseline" database was created to incorporate "current" climate parameters (including atmospheric CO_2 of 354 ppmv in 1990), existing soil properties, a uniform vegetation classification, and two climate-change scenarios. Key database design criteria include temporal consistency, with daily and monthly climate sets having the same monthly average. The database is also spatially consistent with, for example, climate and vegetation reflecting topographic effects. And finally, the database is physically consistent, with relations maintained among climate variables and among soil properties in soil profiles.

The database covers the coterminous United States with a spatial resolution of 0.5°. The coterminous United States is made of about 3100 of the 0.5° x 0.5° grid cells. The baseline vegetation is assumed to be in equilibrium under current climate. The current vegetation distribution is determined by first defining a "potential" vegetation distribution based on ecophysiological and resource constraints.

VEGETATION IN THE FUTURE

Current cropland and urban areas are defined and a cropland and urban "mask" is applied to the potential vegetation distribution to define the extent of current natural vegetation. This same unchanged cropland and urban mask is used in throughout the

21st century in VEMAP II and so shifts in cropland areas and expansion of urban areas is not included.

Climate change scenarios are based on two atmospheric general circulation model (GCM) experiments – one conducted at the Hadley Centre for Climate Prediction and Research of the Meteorological Office of the United Kingdom (HadCM2 version) (henceforth, Hadley) and the other at the Canadian Centre for Climate Modelling and Analysis (henceforth, Canadian). These scenarios were selected because they are representative of the higher and lower halves of the range of temperature sensitivity among the "transient" GCMs available at the beginning of VEMAP II.

Because elevated CO_2 may directly affect plants independently of whether it causes any change in climate, VEMAP II included a partial factorial experimental design in which simulations were run with both climate and CO_2 changing through time and then only climate changing through time. Both the biogeochemistry and biogeography models were run with both transient climate and CO_2 and with transient climate alone.

For the biogeochemistry models, several aspects of carbon cycle changes were analyzed including changes in annual net primary production, and in annual net carbon storage. For the biogeography models, the focus was on changes in the area of major vegetation assemblages.

Table 2. Simulated Changes in Annual Net Primary Production due to Changes in Climate plus CO_2 and Climate only in the Conterminous United States.

Models	Modeled Current NPP	Factors affecting NPP	Changes Hadley Climate Simulation	Changes Canadian Climate Simulation
Biome-BGC	2800	Climate + CO_2	+439 (15.7%)	+222 (7.9%)
		Climate	+98 (3.6%)	-274 (-10.0%)
CENTURY	3300	Climate + CO_2	+177 (7.1%)	+72 (2.9%)
		Climate	+109 (4.4%)	+3 (0.1%)
TEM	3500	Climate + CO_2	+539 (15.4%)	+397 (11.3%)
		Climate	+221 (6.6%)	-102 (-3.1%)

Changes are given as deviations from "current" NPP as both absolute (Tg C/yr) and relative (%) values. Simulations are for the period 2025-2034.

BIOGEOCHEMICAL SIMULATION RESULTS

The three biogeochemistry models estimate continental scale Net Primary Production (NPP) in natural ecosystems for contemporary climate and CO_2. For NPP, estimates range from 2.8 Pg C/yr to 3.5 Pg C/yr. In the near term (2025-2034), all three models project small increases in continental NPP for both climate simulations when climate and CO_2 effects are considered (Table 2). For the scenario used, the CO_2 concentration in 2025-2034 averaged about 425 ppmv. The magnitude of the CO_2 fertilization effect in the decade 2025-2034 ranges from a low of about 3% in CENTURY to a high of 18% in Biome-BGC. These sensitivities to CO_2 differ from experimental results, in part, because most field experiments are done at doubled pre-industrial CO_2 (about 300 ppmv CO_2), higher than the projected levels in 2025-2034 in the mid-range IPCC emissions scenario used in this assessment. For the near-term climate simulated by the Canadian model, both Biome-BGC and TEM suggest that without a CO_2 fertilization effect, average annual NPP for the period 2025-2034 would decline relative to current average annual NPP. This is an important point since the exact magnitude of the CO_2 fertilization effect on NPP is uncertain for many natural ecosystems, especially forests.

Annual net carbon storage at the continental level is projected by all three biogeochemistry models to increase in the near term for both climate simulations, when climate and CO_2 effects are considered (Table 3). The biogeochemistry models estimate

Changes in Vegetation Carbon

Figure 1. The maps above show projections of relative changes in vegetation carbon between 1990 and the 2030s for two climate scenarios. Under the Canadian model scenario, vegetation carbon losses of up to 20% are projected in some forested areas of the Southeast in response to warming and drying of the region by the 2030s. A carbon loss by forests is treated as an indication that they are in decline. Under the same scenario, vegetation carbon increases of up to 20% are projected in the forested areas in the West that receive substantial increases in precipitation. Output from TEM (Terrestrial Ecosystem Model) as part of the VEMAP II (Vegetation Ecosystem Modeling and Analysis Project) study. See Color Plate Appendix

Table 3. Simulated Annual Net Carbon Storage due to Changes in Climate and CO_2 in the Major Regions of the Conterminous United States for "Today" and the Period 2025-2034 in Tg C/yr.

Region	Current	Future (2025-2034)	
		Canadian	Hadley
Northeast	3	9	13
Southeast	14	-4	34
Midwest	6	17	27
Great Plain	14	16	16
West	22	41	16
Northwest	7	17	11
Conterminous US Total	66	96	117

Results are the mean of three biogeochemistry models.

that today, the average carbon storage rate of 66 Tg/yr. For the climate simulated with the Hadley model over the period 2025-2034, the biogeochemistry models estimate an average carbon storage rate of 117 Tg/yr, almost a 100% increase relative to present. For the climate simulated with the Canadian model for the same period, the biogeochemistry models estimate an average carbon storage rate of 96 Tg/yr. One particularly interesting result becomes apparent when the annual carbon storage data are analyzed by regions (Table 3). For the climate simulated with the Canadian model, the mean projection of the biogeochemistry models is that the southeastern ecosystems will loose carbon in the near term (Figure 1). This ecological response is consistent with the hot, dry climate conditions the model projects for this region during the period of 2025-2034.

BIOGEOGRAPHY SIMULATION RESULTS

For both the Hadley and Canadian climate scenarios, the biogeography models project shifts in the distribution of major vegetation types as plant species move in response to climate change (Figure 2). An implicit assumption in the biogeography models is that vegetation will be able to move freely from location to location; an assumption that may be at least in part unwarranted because of the barriers to plant migration that have been put in place on landscapes through agricultural expansion and urbanization.

The projected changes in vegetation distribution with climate change vary from region to region (Figure 3a-f; Tables 4-9). Some of the major changes as simulated by the biogeography models for the six National Assessment regions of the coterminous US can be summarized as follows:

Northeast

- Under both simulated climates, forests remain the dominant natural vegetation, but the mix of forest types changes. For example, winter-deciduous forests expand at the expense of mixed conifer-broadleaf forests.
- Under the climate simulated by the Canadian model, there is a modest increase in savannas and woodlands.

Southeast

- Under the climate simulated by the Hadley model, forest remains the dominant natural vegetation, but once again the mix of forest types changes.
- Under the climate simulated by the Canadian model, all three biogeography models show an expansion of savannas and grasslands at the expense of forests. For two of biogeography models, LPJ and MAPSS, the expansion of these non-forest ecosystems is dramatic by the end of the 21st century. Both drought and fire play an important role in the forest breakup.

Midwest

- Under both simulated climates, forests remain the dominant natural vegetation, but the mix of forest types changes.
- One biogeography model, LBJ, simulates a modest expansion of savannas and grasslands.

Great Plains

- Under the climate simulated by the Hadley model, two biogeography models project an increase in woodiness in this region, while the third projects no change in woodiness.
- Under the climate simulated by the Canadian model, the biogeography models project either no change in woodiness or a slight decrease.

West

- Under the climate simulated by both the Hadley and Canadian models, the area of desert ecosystems shrinks and the area of forest ecosystems grows.

Northwest

- Under both simulated climates, the forest area grows slightly.

Ecosystem Models

Current Ecosystems

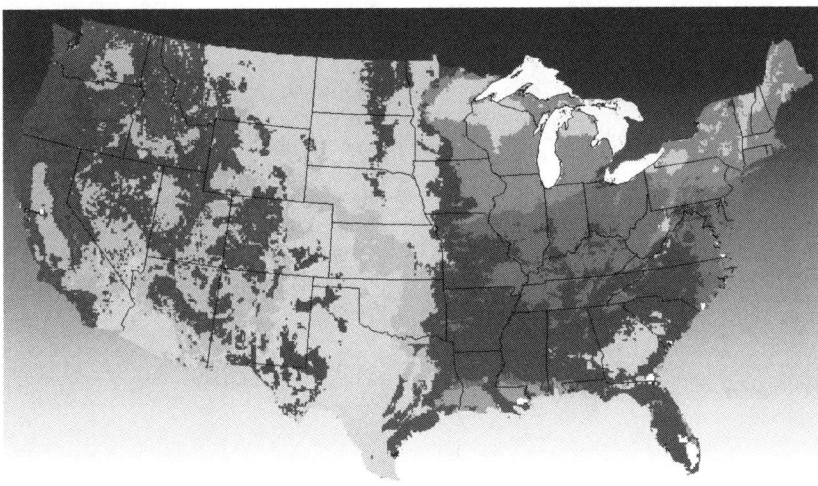

Canadian Model

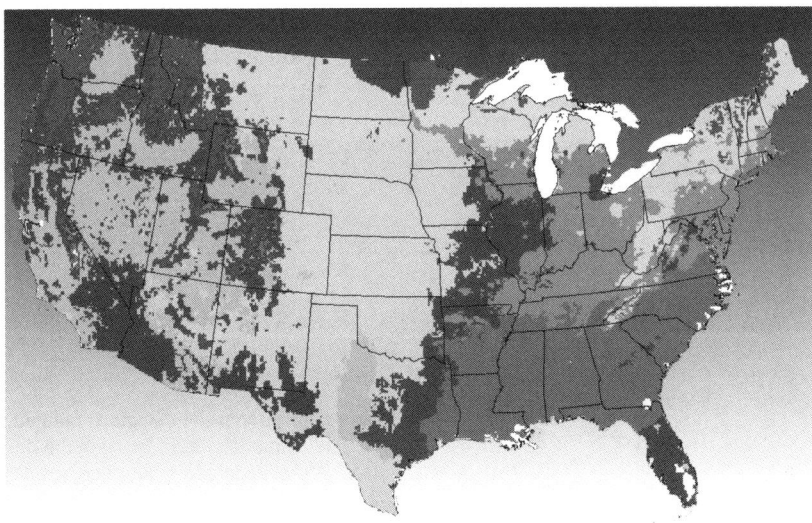

Tundra
Taiga / Tundra
Conifer Forest
Northeast Mixed Forest
Temperate Deciduous Forest
Southeast Mixed Forest
Tropical Broadleaf Forest
Savanna / Woodland
Shrub / Woodland
Grassland
Arid Lands

Hadley Model

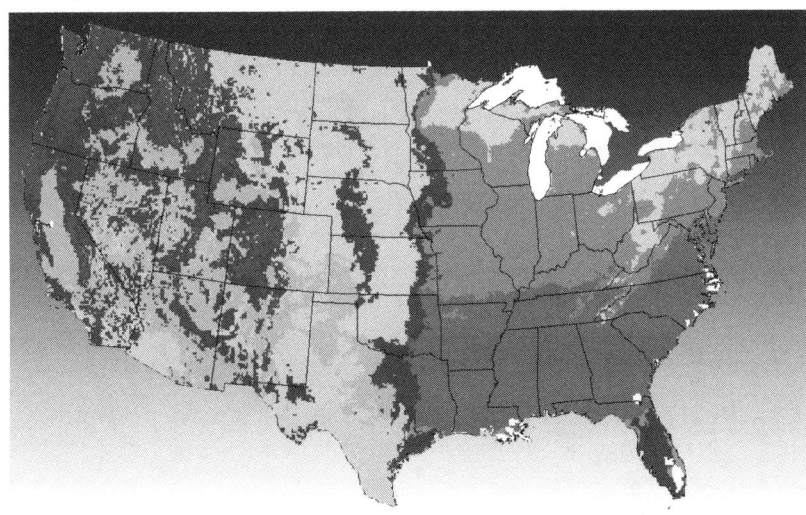

Figure 2. The models used to estimate biogeographic responses to climate change in VEMAP II include LPJ, MAPSS, and MC1. These three models predict the local dominance of various terrestrial vegetation forms based on: (1) ecophysiological constraints, which determine the broad distribution of major categories of woody plants; and (2) response limitations, which determine specific aspects of community composition, such as the competitive balance of trees and grasses. Though similar in some respects, these models simulate potential evapotranspiration and direct CO_2 effects differently, and as a result they show varying sensitivities to temperature, CO_2 levels, and other factors. Two of the model models, LPJ and MC1 have biogeochemistry modules, while the third, MAPPS, does not. For both the Hadley and Canadian climate scenarios, the biogeography models project shifts in the distribution of major vegetation types as plant species move in response to climate change. The projected changes in vegetation distribution with climate change vary from region to region. (Source: VEMAP, 1998). See Color Plate Appendix

LPJ, MC1 and MAPSS Estimates

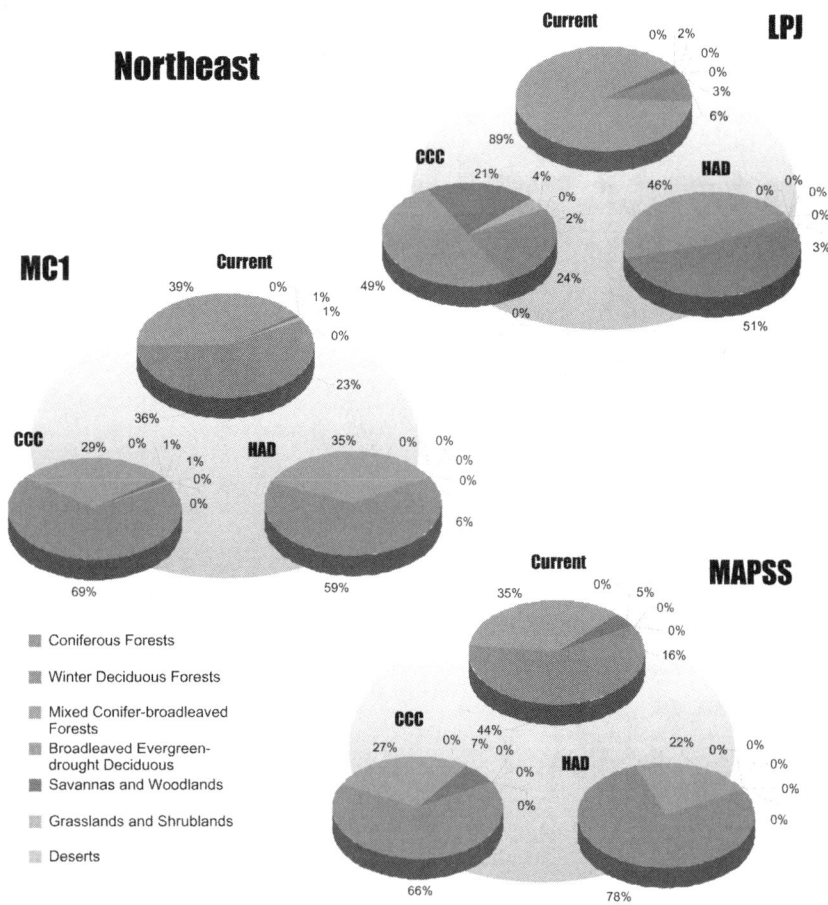

Northeast

Figure 3(a) Under both simulated climates, forests remain the dominant natural vegetation, but the mix of forest types changes. For example, winter-deciduous forests expand at the expense of mixed conifer-broad-leaved forests. Under the climate simulated by the Canadian model, there is a modest increase in savannas and woodlands. See Color Plate Appendix.

- Coniferous Forests
- Winter Deciduous Forests
- Mixed Conifer-broadleaved Forests
- Broadleaved Evergreen-drought Deciduous
- Savannas and Woodlands
- Grasslands and Shrublands
- Deserts

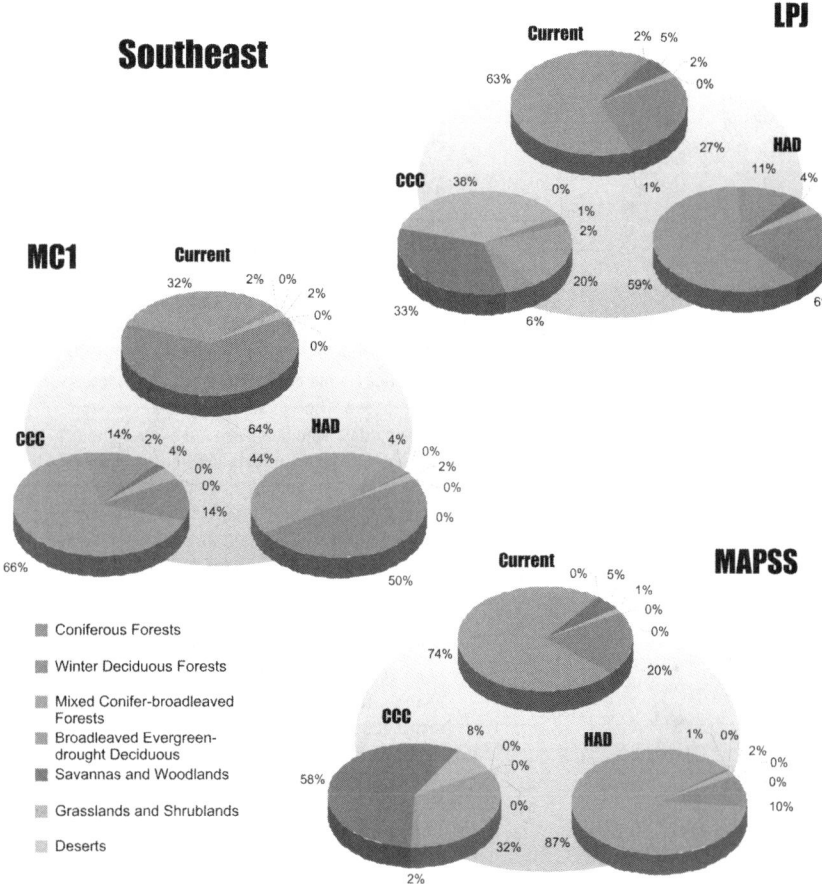

Southeast

Figure 3(b) Under the climate simulated by the Hadley model, forest remains the dominant natural vegetation, but once again the mix of forest types changes. Under the climate simulated by the Canadian model, all three biogeography models show an expansion of savannas and grasslands at the expense of forests. For two of biogeography models, LPJ and MAPSS, the expansion of these non-forest ecosystems is dramatic by the end of the 21st century. Both drought and fire play an important role in the forest breakup. See Color Plate Appendix.

- Coniferous Forests
- Winter Deciduous Forests
- Mixed Conifer-broadleaved Forests
- Broadleaved Evergreen-drought Deciduous
- Savannas and Woodlands
- Grasslands and Shrublands
- Deserts

Mid-West

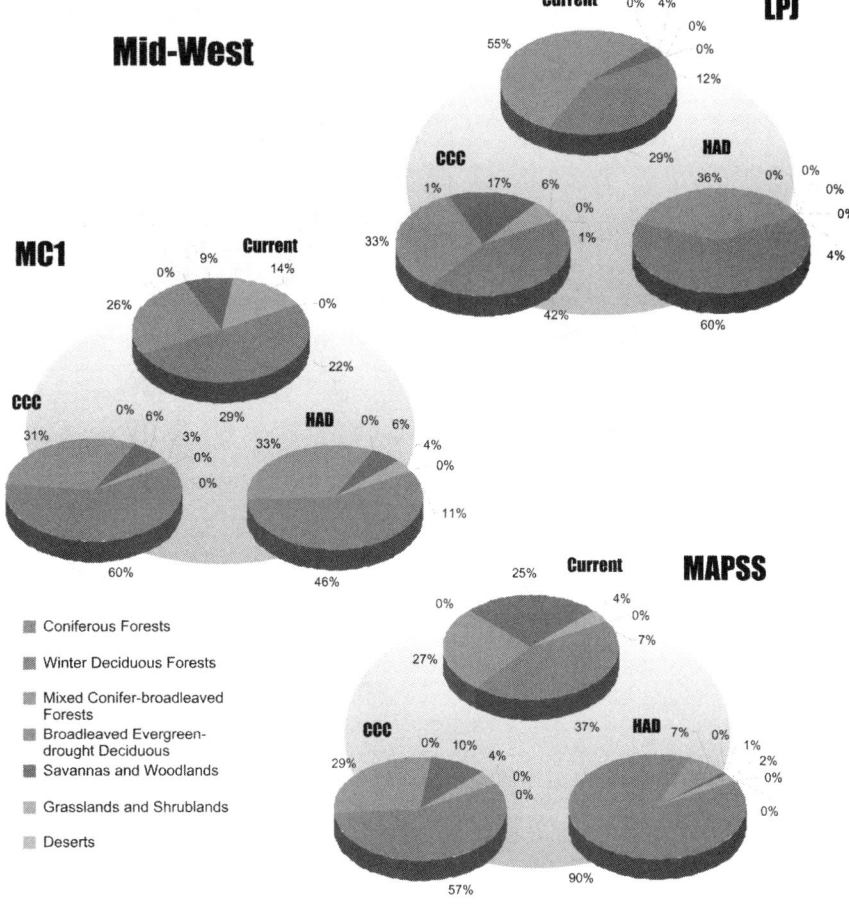

Figure 3(c) Under both simulated climates, forests remain the dominant natural vegetation, but the mix of forest types changes. One biogeography model, LBJ, simulates a modest expansion of savannas and grasslands. See Color Plate Appendix.

Coniferous Forests

Winter Deciduous Forests

Mixed Conifer-broadleaved Forests

Broadleaved Evergreen-drought Deciduous

Savannas and Woodlands

Grasslands and Shrublands

Deserts

Great Plains

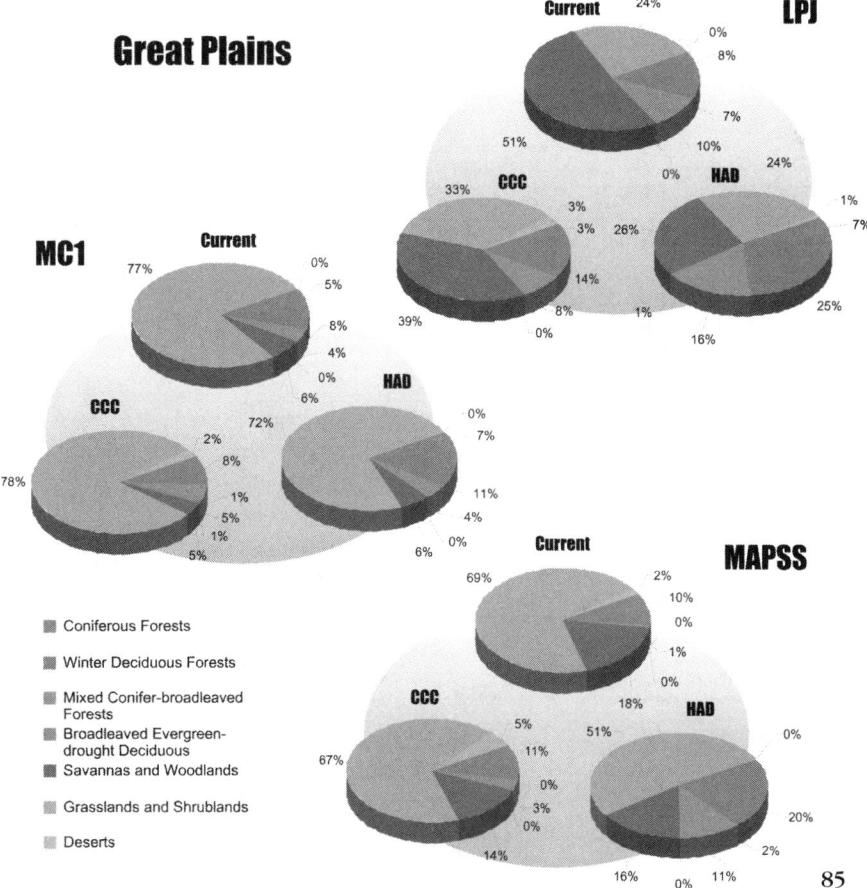

Figure 3(d) Under the climate simulated by the Hadley model, two biogeography models project an increase in woodiness in this region, while the third projects no change in woodiness. Under the climate simulated by the Canadian Model, the biogeography models project either no change in woodiness or a slight decrease. See Color Plate Appendix.

Coniferous Forests

Winter Deciduous Forests

Mixed Conifer-broadleaved Forests

Broadleaved Evergreen-drought Deciduous

Savannas and Woodlands

Grasslands and Shrublands

Deserts

85

Northwest

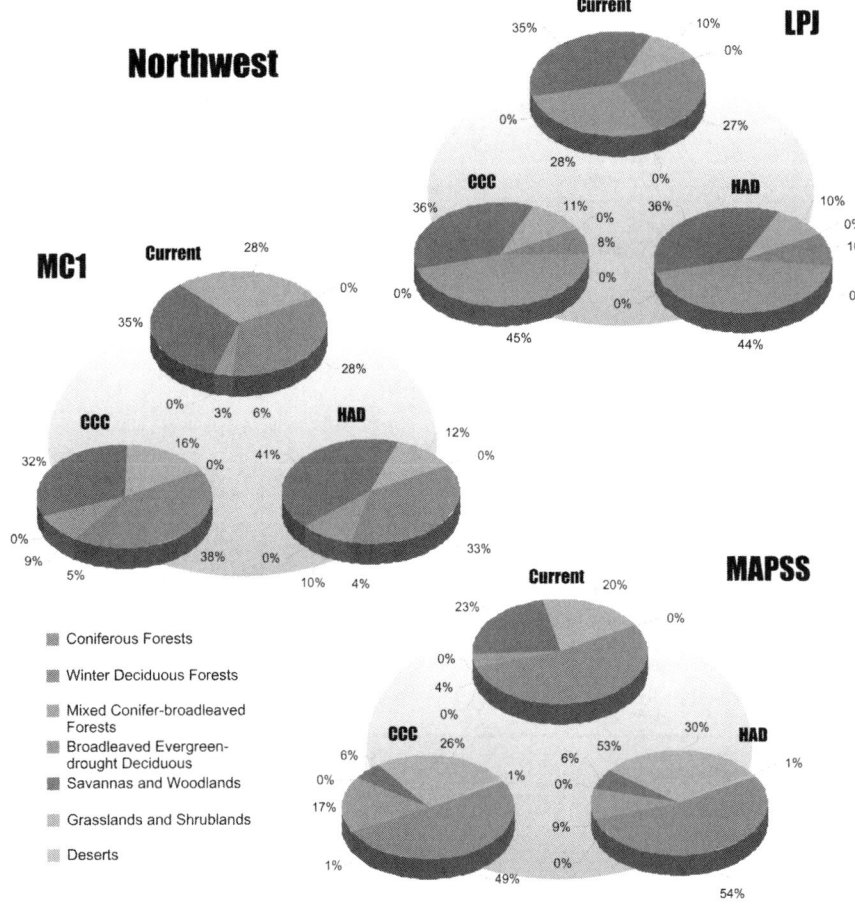

Figure 3(e): Under both simulated climates, the forest area grows slightly. See Color Plate Appendix.

Coniferous Forests

Winter Deciduous Forests

Mixed Conifer-broadleaved Forests

Broadleaved Evergreen-drought Deciduous

Savannas and Woodlands

Grasslands and Shrublands

Deserts

West

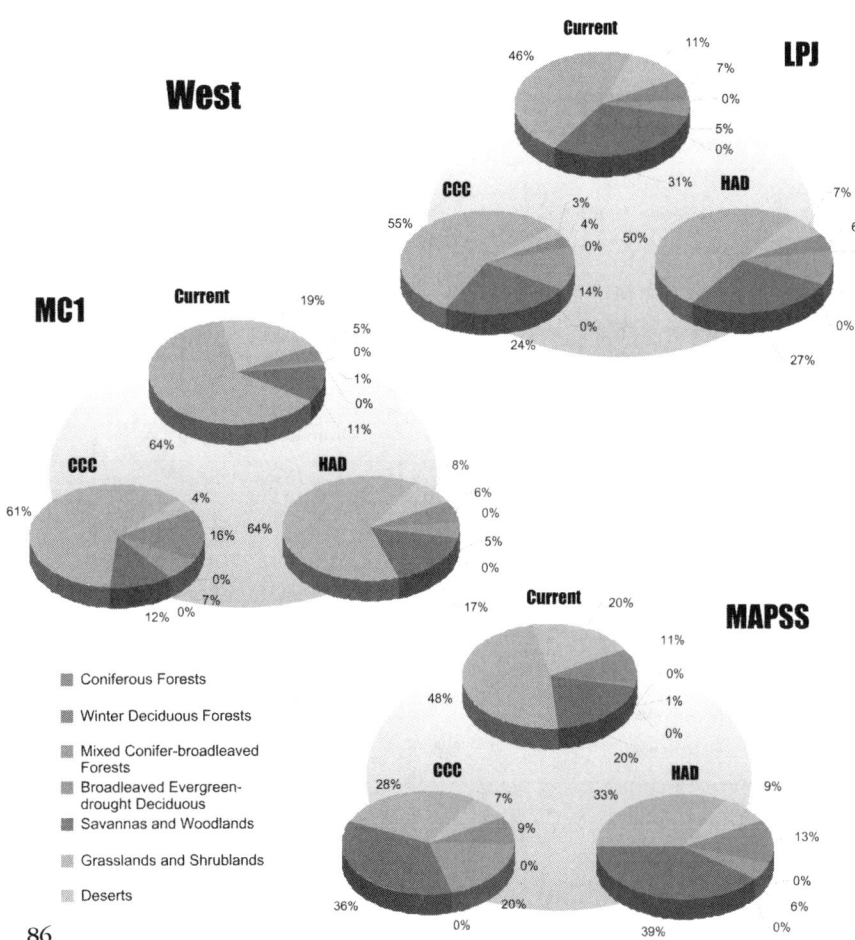

Figure 3(f). Under the climate simulated by both the Hadley and Canadian models, the area of desert ecosystems shrinks and the area of forest ecosystems grows. See Color Plate Appendix.

Coniferous Forests

Winter Deciduous Forests

Mixed Conifer-broadleaved Forests

Broadleaved Evergreen-drought Deciduous

Savannas and Woodlands

Grasslands and Shrublands

Deserts

Table 4. Vegetation redistribution associated with climate change as estimated by LPJ using Hadley Climate. Current and future (2090-2099) distributions are presented as percentages of the total non-agricultural area in the region. (Neilson et al., 2000; EOS Webster, 2000)

Vegetation Type	Northeast		Southeast		Mid-west		Great Plains		Northwest		West	
	Current	Future	Current	Future	Current	Future	Current	Future	Current	Future	Current	Future
Coniferous Forests	3.3	2.9	26.8	16.5	11.9	4.2	7.6	6.6	26.7	10.2	6.9	5.5
Winter Deciduous Forests	6.1	51.3	0.5	6.4	29.0	60.0	7.1	25.2	0.0	0.0	0.0	0.1
Mixed Conifer-broadleaved Forests	88.3	45.8	63.4	59.5	54.9	35.8	10.4	15.8	27.8	44.3	5.3	10.2
Broadleaved Evergreen-drought Deciduous	0. 0	0.0	2.0	10.6	0.0	0.0	0.0	0.7	0.0	0.0	0.0	0.0
Savannas and Woodlands	2.3	0.0	5.3	4.0	4.3	0.0	51.3	26.7	35.1	35.7	30.5	27.2
Grasslands and Shrublands	0.0	0.0	2.0	3.0	0.0	0.0	23.7	24.0	10.5	9.9	45.9	50.0
Deserts	0.0	0.0	0.0	0.0	0.0	0.0	0.0	0.9	0.0	0.0	11.3	7.0
Total	100	100	100	100	100	100	100	100	100	100	100	100

Table 5. Vegetation redistribution associated with climate change as estimated by LPJ using Canadian Climate. Current and future (2090-2099) distributions are presented as percentages of the total non-agricultural area in the region. (Neilson et al., 2000; Neilson and Drapek, 1998)

Vegetation Type	Northeast		Southeast		Mid-west		Great Plains		Northwest		West	
	Current	Future	Current	Future	Current	Future	Current	Future	Current	Future	Current	Future
Coniferous Forests	3.3	1.9	26.8	1.1	11.9	0.5	7.6	3.4	26.7	8.3	6.9	3.6
Winter Deciduous Forests	6.1	23.9	0.5	1.6	29.0	42.9	7.1	13.7	0.0	0.0	0.0	0.0
Mixed Conifer-broadleaved Forests	88.3	0.0	63.4	20.2	54.9	32.5	10.4	8.4	27.8	45.3	5.3	13.7
Broadleaved Evergreen-drought Deciduous	0.0	49.6	2.0	6.3	0.0	1.1	0.0	0.0	0.0	0.0	0.0	0.0
Savannas and Woodlands	2.3	20.8	5.3	33.2	4.3	17.0	51.3	38.0	35.1	35.3	30.5	24.1
Grasslands and Shrublands	0.0	3.9	2.0	37.5	0.0	6.0	23.7	33.4	10.5	10.5	45.9	55.3
Deserts	0.0	0.0	0.0	0.0	0.0	0.0	0.0	3.1	0.0	0.0	11.3	3.2
Total	100	100	100	100	100	100	100	100	100	100	100	100

Table 6. Vegetation redistribution associated with climate change as estimated by MC1 using Hadley Climate. Current and future (2090-2099) distributions are presented as percentages of the total non-agricultural area in the region. (Neilson et al., 2000; Bachelet, et al., 2000)

Vegetation Type	Northeast		Southeast		Mid-west		Great Plains		Northwest		West	
	Current	Future	Current	Future	Current	Future	Current	Future	Current	Future	Current	Future
Coniferous Forests	22.8	6.1	0.0	0.0	22.2	10.8	5.3	6.5	28.1	32.8	5.0	6.4
Winter Deciduous Forests	35.9	58.5	63.9	49.4	28.3	46.7	7.8	10.8	6.4	4.2	0.0	0.2
Mixed Conifer-broadleaved Forests	39.5	35.3	31.7	44.1	26.2	32.7	3.8	4.3	3.3	9.5	1.4	4.7
Broadleaved Evergreen-drought Deciduous	0.0	0.0	2.3	4.5	0.0	0.0	0.0	0.0	0.0	0.0	0.0	0.0
Savannas and Woodlands	0.7	0.0	0.0	0.4	8.9	5.9	6.3	6.0	33.9	41.8	11.6	17.1
Grasslands and Shrublands	1.1	0.0	2.1	1.7	14.4	3.9	76.8	72.3	28.3	11.6	63.7	63.4
Deserts	0.0	0.0	0.0	0.0	0.0	0.0	0.0	0.0	0.0	0.0	19.7	8.3
Total	100	100	100	100	100	100	100	100	100	100	100	100

Table 7. Vegetation redistribution associated with climate change as estimated by MC1 using Canadian Climate. Current and future (2090-2099) distributions are presented as percentages of the total non-agricultural area in the region. (Neilson et al., 2000)

Vegetation Type	Northeast		Southeast		Mid-west		Great Plains		Northwest		West	
	Current	Future	Current	Future	Current	Future	Current	Future	Current	Future	Current	Future
Coniferous Forests	22.8	0.0	0.0	0.0	22.2	0.0	5.3	7.9	28.1	37.6	5.0	15.8
Winter Deciduous Forests	35.9	68.8	63.9	14.0	28.2	60.3	7.8	1.2	6.4	5.4	0.0	0.3
Mixed Conifer-broadleaved Forests	39.5	29.0	31.7	66.4	26.2	30.8	3.8	4.8	3.3	9.1	1.4	6.8
Broadleaved Evergreen-drought Deciduous	0.0	0.0	2.3	13.8	0.0	0.0	0.0	1.1	0.0	0.0	0.0	0.1
Savannas and Woodlands	0.7	1.4	0.0	2.3	8.9	6.2	6.3	4.6	33.9	31.8	11.6	11.8
Grasslands and Shrublands	1.1	0.7	2.1	3.5	14.4	2.7	76.8	78.2	28.3	16.1	63.7	61.3
Deserts	0.0	0.0	0.0	0.0	0.0	0.0	0.0	2.3	0.0	0.0	19.7	3.9
Total	100	100	100	100	100	100	100	100	100	100	100	100

Table 8. Vegetation redistribution associated with climate change as estimated by MAPSS using Canadian Climate. Current and future (2090-2099) distributions are presented as percentages of the total non-agricultural area in the region. (Neilson et al., 2000; Neilson and Drapek, 1998)

Vegetation Type	Northeast		Southeast		Mid-west		Great Plains		Northwest		West	
	Current	Future	Current	Future	Current	Future	Current	Future	Current	Future	Current	Future
Coniferous Forests	16.0	0.0	0.0	0.0	7.0	0.2	9.8	11.1	53.7	49.7	10.9	8.8
Winter Deciduous Forests	44.2	66.1	20.0	0.0	37.1	56.8	0.0	0.0	0.0	0.6	0.0	0.0
Mixed Conifer-broadleaved Forests	35.3	26.6	73.6	32.3	26.8	28.5	0.8	3.3	3.7	17.4	0.8	20.5
Broadleaved Evergreen-drought Deciduous	0.0	0.0	0.0	2.0	0.0	0.0	0.0	0.0	0.0	0.0	0.0	0.0
Savannas and Woodlands	4.5	7.3	5.2	57.3	25.1	10.2	18.2	14.0	2.3	5.9	20.2	35.7
Grasslands and Shrublands	0.0	0.0	1.3	8.4	3.9	4.3	69.6	66.9	19.6	25.7	48.2	27.7
Deserts	0.0	0.0	0.0	0.0	0.0	0.0	1.6	4.7	0.0	0.7	19.8	7.4
Total	100	100	100	100	100	100	100	100	100	100	100	100

Table 9. Vegetation redistribution associated with climate change as estimated by MAPSS using Hadley Climate. Current and future (2090-2099) distributions are presented as percentages of the total non-agricultural area in the region. (Neilson et al., 2000; Bachelet, et al., 2000)

Vegetation Type	Northeast		Southeast		Mid-west		Great Plains		Northwest		West	
	Current	Future	Current	Future	Current	Future	Current	Future	Current	Future	Current	Future
Coniferous Forests	16.0	0.0	0.0	0.0	7.0	0.4	9.8	19.5	53.7	53.0	10.9	13.4
Winter Deciduous Forests	44.2	77.9	20.0	10.0	37.1	88.8	0.0	2.3	0.0	0.3	0.0	0.0
Mixed Conifer-broadleaved Forests	35.3	22.1	73.6	87.1	26.8	7.4	0.8	11.2	3.7	9.4	0.8	5.6
Broadleaved Evergreen-drought Deciduous	0.0	0.0	0.0	0.8	0.0	0.0	0.0	0.0	0.0	0.0	0.0	0.0
Savannas and Woodlands	4.5	0.0	5.2	0.1	25.1	1.1	18.2	15.5	23.0	6.4	20.2	39.2
Grasslands and Shrublands	0.0	0.0	1.3	2.0	3.9	2.2	69.6	51.1	19.6	29.8	48.2	32.8
Deserts	0.0	0.0	0.0	0.0	0.0	0.0	1.6	0.3	0.0	1.1	19.8	8.9
Total	100	100	100	100	100	100	100	100	100	100	100	100

LITERATURE CITED

Bachelet, D., R. P. Neilson, J. M. Lenihan, and R. J. Drapek, Climate change effects on vegetation distribution and carbon budget in the US, *Ecosystems*, in review, 2000.

Daly C., D. Bachelet, J. M. Lenihan, R. P. Neilson, W. Parton, and D. Ojima, Dynamic simulation of tree-grass interactions for global change studies, *Ecological Applications*, 10(2) 449-469, 2000.

EOS Webster, University of New Hampshire, http://eos-webster.sr.unh.edu/, 2000.

Haxeltine, A., I. C. Prentice, BIOME3: An equilibrium terrestrial biosphere model based on ecophysiological constraints, resource availability, and competition among plant functional types, *Global Biogeochemical Cycles, 10*(4), 693-709, 1996.

Holland, E. A., W. J. Parton, J. K. Delting, and D. L. Coppock, Physiological response of plant population to herbivory and their consequences for ecosystem nutrient flow, *American Naturalist, 140*, 685-706, 1992.

Hunt, E. R., Jr, F. C. Martin, and S. W. Running, Simulating the effect of climatic variation on stem carbon accumulation of a ponderosa pine stand: Comparison with annual growth increment data, *Tree Physiology, 9*, 161-172, 1991.

Hunt, E. R., Jr., and S. W. Running, Simulated dry matter yields for aspen and spruce stands in the North American boreal forest, *Canadian Journal of Remote Sensing, 18*, 126-133, 1992.

Jenkins, J. C., R. A. Birdsey, and Y. Pan, Biomass and NPP estimations for the mid-Atlantic region (USA) using plot-level forest inventory, *Ecological Applications,* in review, 2000.

Kicklighter, D. W., et al., A first-order analysis of the potential role of CO_2 fertilization to affect the global carbon budget: A comparison study of four terrestrial biosphere models, *Tellus ,51B*, 343-366, 1999.

Knight, D. H., T. J. Fahey, and S. W. Running, Factors affecting water and nutrient outflow from lodgepole pine forests in Wyoming, *Ecological Monographs, 55*, 29-48, 1985.

Korol, R. L., S. W. Running, K. S. Milner, and E. R. Hunt, Jr., Testing a mechanistic carbon balance model against observed tree growth, *Canadian Journal of Forest Research, 21*, 1098-1105, 1991.

Lenihan, J. M., C. Daly, D. Bachelet, and R. P. Neilson, Simulating broad-scale fire severity in a dynamic global vegetation model, *Northwest Science, 72*, 91-103, 1998.

McGuire, A. D., J. M. Melillo, D. W. Kicklighter, Y. Pan, X. Xiao, J. Helfrich, B. Moore, III, C. J. Vorosmarty, and A. L. Schloss, Equilibrium responses of global net primary production and carbon storage to doubled atmospheric carbon dioxide: Sensitivity to changes in vegetation nitrogen concentration, *Global Biogeochemical Cycles*, 11, 173-189, 1997.

McGuire A. D., J. S. Clein, J. M. Melillo, D. W. Kicklighter, R. A. Meier, C. J. Vorosmarty, and M. C. Serreze, Modeling carbon responses of tundra ecosystems to historical and projected climate: Sensitivity of pan-Arctic carbon storage to temporal and spatial variation in climate, *Global Change Biology*, in press, 2000.

McLeod, S., and S. W. Running, Comparing site quality indices and productivity of ponderosa pine stands in western Montana, *Canadian Journal of Forestry Research*, 18, 346-352, 1988.

Melillo, J. M. et al. Global climate change and terrestrial net primary production, *Nature* 63, 234-240, 1993.

Metherall, A. K., Simulation of soil organic matter dynamics and nutrient cycling in agroecosystems, Ph.D. dissertation, Colorado State University, Ft. Collins, Colorado, 1992.

Neilson, R. P., and R. J. Drapek, Potentially complex biosphere responses to transient global warming, *Global Change Biology*, 4, 505-521, 1998.

Neilson, R. P., G. A. King, and G. Koerper, Toward a rule-based biome model, *Landscape Ecology*, 7, 27-43, 1992.

Neilson, R. P., A model for predicting continental-scale vegetation distribution and water balance, *Ecological Applications*, 5, 362-385, 1995.

Neilson, R. P., and D. Marks, A global perspective of regional vegetation and hydrologic sensitivities from climate change, *Journal of Vegetation Science* 5, 715-730, 1995.

Neilson, R. P., D. Bachelet, J. M. Lenihan and R. J. Drapek, The VEMAP Models and Potentially Complex Biosphere—Atmosphere Feedbacks: Is There a Threshold Global, paper presented at AGU, Spring 2000.

Nemani, R. R., and S. W. Running, Testing a theoretical climate-soil-leaf area hydrologic equilibrium of forests using satellite data and ecosystem simulation, *Agriculture and Forest Meteorology*, 44, 245-260, 1989.

Ojima, D.S., W. J. Parton, D. S. Schimel, and C. E. Owensby, Simulated impacts of annual burning on prairie ecosystems, in *Fire in North American Tallgrass Prairies*, edited by S. L. Collins and L. L. Wallace, 175 pp., University of Oklahoma Press, Norman, Oklahoma, 1990.

Pan, Y., et al., Modeled responses of terrestrial ecosystems to elevated atmospheric CO_2: A comparison of simulations by the biogeochemistry models of the Vegetation/Ecosystem Modeling and Analysis Project (VEMAP), *Oecologia*, 114, 389-404, 1998.

Parton, W. J., D. S. Schimel, C. V. Cole, and D. S. Ojima, Analysis of factors controlling soil organic levels of grasslands in the Great Plains, *Soil Science Society of America Journal*, 51, 1173-1179, 1987.

Parton, W. J., J. W. B. Stewart, and C. V. Cole, Dynamics of C, N, P, and S in grassland soils: A model, *Biogeochemistry*, 5, 109-131, 1988.

Parton, W. J., et al., Observations and modeling of biomass and soil organic matter dynamics for the grassland biome worldwide, *Global Biogeochemical Cycles*, 7, 785-809, 1993.

Pierce, L. L., Scaling ecosystem models from watersheds to regions: Tradeoffs between model complexity and accuracy, Ph.D. dissertation, School of Forestry, University of Montana, 146 pp., 1993.

Prinn, R., et al., Integrated global system model for climate policy assessment: Feedbacks and sensitivity studies, *Climatic Change*, 41(3/4), 469-546, 1999.

Reilly, J., R. Prinn, J. Harnisch, J. Fitzmaurice, H. Jacoby, D. Kicklighter, P. Stone, A. Sokolov, and C. Wang, Multi-gas assessment of the Kyoto Protocol, *Nature*, 401, 549-555, 1999.

Running, S. W., and J. C. Coughlan, A general model of forest ecosystem processes for regional applications, I. Hydrologic balance, canopy gas exchange, and primary production processes, *Ecological Modelling*, 42, 125-154, 1988.

Running, S. W., and G. T. Gower, FOREST-BGC, a general model of forest ecosystem processes for regional applications, II. Dynamic carbon allocation and nitrogen budgets, *Tree Physiology*, 9, 147-160, 1991.

Running, S. W., and E. R. Hunt, Jr., Generalization of a forest ecosystem process model for other biomes, BIOME-BGC, and an application for global-scale models, in *Scaling Processes Between Leaf and Landscape Levels*, edited by J. R. Ehleringer and C. Field, pp. 141-158, Academic Press, San Diego, California, 1993.

Running, S. W. Testing FOREST-BGC ecosystem process simulations across a climatic gradient in Oregon, *Ecological Applications* 4, 238-247, 1994.

Sanford, R. L., Jr., W. J. Parton, D. S. Ojima, and D. J. Lodge, Hurricane effects on soil organic matter dynamics and forest production in the Luquillo Experimental Forest, Puerto Rico: Results of simulation modeling, *Biotropica*, 23, 364-372, 1991.

Schimel, D., et al., Contribution of increasing CO_2 and climate to carbon storage by ecosystems in the United States, *Science*, 287, 2004-2006, 2000.

Sitch, S., The role of vegetation dynamics in the control of atmospheric CO_2 content, doctoral dissertation, Department of Ecology, Plant Ecology, Lund University, Lund, Sweden, 2000.

Thonicke, K. S., S. Venevsky, S. Sitch, and W. Cramer, The role of fire disturbance for global vegetation dynamics: Coupling fire into a dynamic Global Vegetation Model, *Global Change Biology*, in review, 2000.

Tian, H., J. M. Melillo, D. W. Kicklighter, A. D. McGuire, J. V. K. Helfrich, III, B. Moore, III, and C. J. Vörösmarty, Effect of interannual climate variability on carbon storage in Amazonian ecosystems, *Nature*, 396, 664-667, 1998.

Tian, H., J. M. Melillo, D. W. Kicklighter, A. D. McGuire, and J. Helfrich, The sensitivity of terrestrial carbon storage to historical climate variability and atmospheric CO_2 in the United States, *Tellus* 51B, 414-452, 1999.

Tian, H., J. M. Melillo, D. W. Kicklighter, A. D. McGuire, B. Moore, III, and C. J. Vorosmarty, Climatic and biotic controls on interannual variations of carbon storage in undistubed ecosystems of the Amazon Basin, *Global Ecology and Biogeography*, in press, 2000.

VEMAP Members, Vegetation/ecosystem modeling and analysis project: Comparing biogeography and biogeochemistry models in a continental-scale study of terrestrial ecosystem responses to climate change and CO_2 doubling, *Global Biogeochemical Cycles*, 4, 407-437, 1995.

Vitousek, P. M., T. Fahey, D. W. Johnson, and M. J. Swift, Element interactions in forest ecosystems: Succession, allometry, and input-output budgets, *Biogeochemistry*, 5, 7-34, 1988.

Vörösmarty, C. J., et al., Continental scale model of water balance and fluvial transport: An application to South America, *Global Biogeochemical Cycles* 3, 241-265, 1989.

White, J. D., and S. W. Running, Testing scale dependent assumptions in regional ecosystem simulations, *Journal Vegetation Science*, 5, 687-702, 1995.

Xiao, X., J. M. Melillo, D. W. Kicklighter, A. D. McGuire, R. G. Prinn, C. Wang, P. H. Stone, and A. Sokolov, Transient climate change and net ecosystem production of the terrestrial biosphere, *Global Biogeochemical Cycles*, 12, 345-360, 1998.

ACKNOWLEDGMENTS

Many of the materials for this chapter are based on contributions from participants on and those working with the

Ecosystem Scenario Team
Timothy G. F. Kittel*, National Center for
    Atmospheric Research
Jerry Melillo*, Marine Biological Laboratory
David S. Schimel*, Max-Planck-Institute for
    Biogeochemistry, Jena, Germany
Steve Aulenbach, National Center for Atmospheric
    Research
Dominique Bachelet, Oregon State University
Sharon Cowling, Lund University, Sweden
Christopher Daly, Oregon State University
Ray Drapek, Oregon State University
Hank H. Fisher, National Center for Atmospheric
    Research
Melannie Hartman, Colorado State University
Kathy Hibbard, University of New Hampshire
Thomas Hickler, Lund University, Sweden
Cristina Kaufman, National Center for Atmospheric
    Research
Robin Kelly, Colorado State University
David Kicklighter, Marine Biological Laboratory
Jim Lenihan, Oregon State University
David McGuire, U.S. Geological Survey and
    University of Alaska, Fairbanks
Ron Neilson, USDA Forest Service
Dennis S. Ojima, Colorado State University
Shufen Pan, Marine Biological Laboratory
William J. Parton, Colorado State University
Louis F. Pitelka, University of Maryland Appalachian
    Laboratory
Colin Prentice, Max-Planck-Institute for
    Biogeochemistry, Jena, Germany
Brian Rizzo, University of Virginia
Nan A. Rosenbloom, National Center for
    Atmospheric Research
J. Andy Royle, U. S. Department of the Interior
Steven W. Running, University of Montana
Stephen Sitch, Potsdam Institute for Climate Impact
    Research, Germany
Ben Smith, Lund University, Sweden
Thomas M. Smith, University of Virginia
Martin T. Sykes, Lund University, Sweden
Hanqin Tian, Marine Biological Laboratory
Justin Travis, Lund University, Sweden
Peter E. Thornton, University of Montana
F. Ian Woodward, University of Sheffield, UK

* Assessment Team chair/co-chair

CHAPTER 3

THE SOCIOECONOMIC CONTEXT FOR CLIMATE IMPACT ASSESSMENT

Edward A. Parson[1] and M. Granger Morgan[2] served as Coordinating Authors for the National Assessment Synthesis Team with contributions from: Anthony Janetos[3], Linda Joyce[4], Barbara Miller[5], Richard Richels[6], and Tom Wilbanks[7]

Contents of this Chapter

[1] John F. Kennedy School of Government, Harvard University; [2] Dept. of Engineering and Public Policy, Carnegie-Mellon University; [3] World Resources Institute; [4] US Forest Service; [5] World Bank; [6] EPRI; [7] Oak Ridge National Laboratory

CLIMATE IMPACTS AND THEIR ASSESSMENT

It is obvious, from history and everyday observation, that weather and climate can have impacts on people. Human impacts can arise from weather and climate events at many scales: from individual extreme events such as hurricanes or ice storms; from anomalous seasons such as an unusually cold winter or dry summer; or from multi-year departures from normal climate conditions, such as the drought of the 1930s.

Although particular climate impacts may be clear, their mechanisms of causation can be complex and the degree of influence climate has on human affairs in aggregate remains controversial. The view that climate determines major historical events and the character of societies and economies, which has been periodically expressed since antiquity and enjoyed perhaps excessive respect in the early 20th century (e.g., Huntington, 1915), has fallen into perhaps excessive disrepute, although it has never been fully refuted. More persuasive arguments for significant climatic influence on particular historical events or characteristics of societies continue to be advanced (e.g., Myrdal, 1972; Bryson et al., 1974; Lambert, 1975; Schneider, 1984; Diamond, 1997; Sachs, 1999), but it remains the case that the aggregate degree and mechanisms of climatic influence on human affairs are not fully understood (Riebsame, 1985).

Given an assumed state of America's society and economy, the *impacts* of a specified weather or climate event are the changes it induces in matters of human concern. Defining climate impacts as changes implies an alternative, the baseline climate against which changes are measured. For studying the impacts of climate change, the baseline is normally assumed to be continuation of the climate of the past few decades. Describing impacts also requires specifying the perturbed climate, whose effects relative to the baseline are to be measured. Methods for specifying such hypothesized changes in climate, through model projections and historical analogs, are discussed in Chapter 1.

A specified climate change may have multiple impacts. For example, an unusually warm winter can have diverse impacts on home heating bills, driving safety, recreational opportunities, ski area profitability, and the over-wintering of household or crop pests. Impacts may be beneficial or harmful, with most climate scenarios bringing mixed effects: benefits to some people, places, and sectors, and harm to others. A system is more or less *sensitive* to climate depending on whether a specified change in climate brings large or small impacts.

The simplest framework for assessing climate impacts involves specifying the climate change and climate baseline, and attempting to infer impacts directly. The state of the society or economy that bears the climate change is not considered (Kates, 1985). Although this framework has been widely criticized as too simplistic, it is adequate for studies of some important impacts, which can be described without detailed or explicit consideration of socioeconomic conditions. In particular, assessments that only describe climate's first-order effects on environmental characteristics, or biological or physical resources whose importance to society is clearly evident, can be conducted without explicit consideration of socioeconomic context. Assessments of this type might, for example, attempt to calculate the effects of specified climate change on the range of sugar maple trees in New England, the productivity of loblolly pine forests in Georgia, the expected wheat yield in Kansas, the mean annual runoff in the Colorado basin, the average July heat index in Chicago, or the expected frequency and intensity of storms in North Carolina. Most assessments of climate impacts conducted to date have followed this framework. With a few exceptions, this Assessment has from necessity followed the same practice.

Conducting assessments using this approach is difficult. It requires projecting future behavior of the climate system, and of managed and unmanaged ecological systems. These projections are challenging because the systems are highly complex, interactive, and uncertain, and because we do not understand all the factors that control their operation.

But this approach, challenging as it is and useful as it can be, is not sufficient for a full assessment of climate impacts that seeks to identify, describe, and value their effects on people, economies, and societies. Climate variability and change occur in a social and economic context that contributes to determining impacts. In some cases, socioeconomic conditions may mediate or alter even first-order biophysical impacts such as the examples listed above, so socioeconomic information will be necessary to describe and assess even these impacts. The effect of a specified climate change on wheat yields or pine productivity will depend on how the farm or forest is managed, as well as on how the climate changes. The heat index in Chicago is strongly influenced by the urban heat island effect, which depends on the size, density, and surface characteristics (e.g., building, roofing, and paving materials) of the city. Runoff in the Colorado basin under a specified climate can be altered by large-scale land-use change in the basin, as well as by water engineering projects. A specified runoff event may cause a flood or may not, depending on the infrastructure present. Winnipeg survived the Red River flood that destroyed Grand Forks, because a large emergency flood channel had been constructed around Winnipeg decades earlier.

More fundamentally, the impacts of climate change that matter to people are not limited to direct biophysical impacts, but can also include many indirect effects on such factors as health, income, and employment; the price, availability, and quality of goods and services; property values and losses; recreational opportunities; the character of the landscape; and the political, social, and economic character of their community – as well as the direct effects of weather and climate on people's experience. Such impacts are not exclusively caused by weather or climate, but are mediated by many characteristics of the economy and society. They can only be meaningfully defined relative to specified individual and collective perceptions, interests, and values, which in turn may themselves be subject to change. For example, what is the value of fall foliage in New England, and what would be the impacts if it changed? The settlement patterns and demographic structure of the population, the prosperity and structure of the economy, the technologies available and in use, the patterns of land and natural resource use, and the institutions and policies in place will all contribute to how – and how much – climate will matter to people, and what they can and might wish to do about it. Climate conditions and societal conditions jointly cause climate impacts (Kates, 1985).

Because of this joint causation, making a coherent assessment of climate impacts requires careful, systematic assumptions about future socioeconomic conditions as well as future climatic conditions. However challenging it is to model and project future climate, projecting future socioeconomic conditions is even more so. As is the case for climate and ecosystems, the nation's economy, society, patterns of resource use, technology, and land use, are shaped by highly complex, interactive, and uncertain processes. But while most aspects of climate projection are based on well understood physical processes, our understanding of the basic structure and causal factors operating in socioeconomic systems and their evolution is vastly more limited.

Reasonable judgments can be drawn about what kinds of futures are more or less likely, but causal laws of society and history – if they should exist at all – are not known.

The central place of socioeconomic conditions in determining impacts requires that they be considered, and for many analyses, be explicitly projected. But the profound limits to our knowledge of the factors that determine socioeconomic change require that explicit acknowledgment of uncertainty be central to such projections. This requirement cannot be met by assuming any single socioeconomic future. Rather, multiple scenarios representing a plausible range of alternative socioeconomic futures are needed, ideally with explicit quantification of judgments about uncertainty. The sensitivity of results to alternative assumptions should also be examined. In particular, the charge to not assume just one socioeconomic future applies to the widespread practice of studying the impacts of future climate changes as if they were imposed on today's society. Although it has long been recognized that this practice introduces serious biases to impact assessment, and several major studies have demonstrated the alternative of explicit, coherent socioeconomic projections (e.g., Rosenberg, 1993), the practice remains widespread. This practice, often advocated in order to avoid criticism for engaging in speculation, is equivalent to assuming that the future society that will bear the impacts of climate change will resemble the present in all relevant ways – an assumption that may be acceptable for near-term assessments, but grows increasingly unacceptable as the time horizon lengthens. To see how wrong this assumption is likely to be, one need only compare America's society and economy of today to that of 100, 50, or even 25 years ago.

CLIMATE IMPACTS IN SOCIOECONOMIC CONTEXT

Lessons from History

Looking backward a century underscores the extent to which impacts of climate depend on socioeconomic conditions. It also shows the severity of the challenge posed by attempting to project socioeconomic conditions up to a century in the future. At the turn of the 20th century, most of the US workforce was employed on farms; aircraft, electronics, and antibiotics had not been invented; aluminum was a semi-precious metal; the automobile existed only as a primitive novelty; and the predominant form of transportation – and the predominant urban environmental problem – was the horse. Over the intervening century, the population of the United States nearly quadrupled, from 76.2 million to 275 million (US Bureau of the Census, 1998), while US real GDP increased more than thirty-fold, from just under $300 Billion to about $9.5 Trillion (1996 dollars, Bureau of Economic Analysis, 2000) – corresponding to a nearly ten-fold increase in real per capita income.

These increases in material welfare, and the process of industrial transformation that drove them, have had profound effects on the nation's relationship and sensitivity to climate. As first the industrial sector and later the services sector grew to dominate the American economy, fewer Americans' livelihoods have been directly tied to climate. Moreover, wealthier nations – like wealthier individuals – are in general better able to cope with the negative impacts of climate variability and change, and better able to take advantage of the opportunities they present. Wealthy societies can spare resources to support adaptation, can better afford to make required changes in technology and infrastructure, and can more easily endure climate-related losses. Within societies, climatic harms and opportunities will not be equally distributed among individuals and communities: some will face greater burdens than others. Moreover, high rates of economic and population growth can themselves impose stresses on natural systems, through rising pollution (including greenhouse gases), congestion, and demands for land and resources, potentially increasing these systems' vulnerability to climatic stresses.

Much of our recent prosperity has been fueled by new technology. Although technological change can also carry significant social and environmental costs, in aggregate it has greatly contributed to Americans' increased material well-being over the 20th century. For example, in the past decade, computers and new communication technologies have transformed many activities, bringing increases in productivity as well as new products and services.

Technology affects society's relationship to climate in many ways. Technological change will strongly influence the success of future efforts to control greenhouse gas emissions. Many technological changes, large and small, have reduced Americans' vulnerability to weather and climate in a host of ways. A striking example has been weather and climate forecasting, which with increasing understanding of large-scale patterns of variability is now developing substantial predictive skill on weekly and even seasonal intervals. Other examples include better roads and automobiles, navigation and instrument systems for aircraft and shipping, broadcasting and other forms of wireless communication, air conditioning and improved heating technology, new construction materials and techniques that have allowed construction of huge indoor spaces, and technologies that have made many forms of outdoor sport and recreation (e.g., skiing and climbing) safer and more accessible.

Technology can also increase society's vulnerability to climate, particularly to extreme climate or weather events. This can happen because modern societies are organized around the available technologies, and become dependent on them. Contemporary American society relies in critical ways on electric power, transportation, and communications systems, all of which can be disrupted by extreme events if systems have not been adequately designed to deal with them. Large-scale loss of power lines in an ice storm can have catastrophic effects on a modern industrial society, even though all societies, including early industrial ones, functioned without widespread electrical service only a century ago.

US population has not simply grown in the past century, it has also shifted markedly in its demographic structure and its distribution around the country. These trends have also shaped patterns of sensitivity to climate. For example, the US population is growing older. The fraction of Americans aged 65 or over has increased from 1 in 25 in 1900 to 1 in 8 in 2000. Older people are physiologically more vulnerable to heat stress. Without adaptive measures, a more aged society will be more vulnerable to increases in heat-related illness and death under a warmer climate. A warmer climate may also bring a reduction in cold-related mortality, a trend that will also interact with the aging of the population, although the effect of temperature changes on mortality appears to be weaker for cold conditions than for hot. Recent migration to the South and Southwest demonstrates that many older Americans prefer warmer climates, although the nearly universal spread of one technology – air conditioning – has played an essential role in allowing the rapid growth of these regions. At the same time, rapid population and economic growth in arid parts of the Southwest has sharply increased vulnerability to water shortages.

America is also becoming more urban. Over the 20th century the fraction of Americans living in cities increased from 40% to more than 75% (US Bureau of the Census, 1999). Urbanization affects climate vulnerabilities and capacity for adaptation in multiple and complex ways. City dwellers depend less on climate-sensitive activities for their livelihoods, and have more resources and social support systems close at hand. But the dense concentration of people and property in coastal or riverside metropolitan areas, dependent on extensive fixed infrastructure such as water, sewer, and energy utilities, and roads, tunnels, and bridges (which are aging and overburdened in many US cities), can increase vulnerability to extreme events such as floods, storm surges, and heat waves. Combined with other urban stresses such as congestion, pollution, and the urban heat island effect, climate change could significantly harm urban quality of life and health.

Americans are also moving to the coasts. Some 53% of the total US population now live in the 17% of land area that comprises the coastal zone, and the largest continuing population increases for several decades are projected to be in coastal areas. This trend is exacerbating wetland loss and coastal-zone pollution. In addition, locating more people and more valuable property in low-lying coastal areas increases vulnerability to storms, storm surges, coastal erosion, and sea-level rise – as severe recent losses in Florida, Georgia, and the Carolinas, as well as a century of damage trends, all confirm (Changnon et al., 2000).

Observing past patterns of climate impacts reveals how America's vulnerability to climate and its capacity for adaptation have depended on many highly detailed and specific characteristics of its economy and society. For particular communities or activities, the most important factors shaping climate vulnerability might be as diverse as local zoning ordinances, housing styles, or building codes; popular forms of recreation; the age and degree of specialization of capital in particular industries; world market conditions; and the distribution of income. For example, the vulnerability of American agriculture to past climate extremes has been shaped by a host of socioeconomic factors, including the size and structure of farm families, agricultural practices and available technologies, markets for alternative crops, available capacity for storage and transport, groundwater accessibility, local and nationwide markets for capital and labor, bank lending practices and the nationwide organization of banking and capital markets, global trade rules, and public policies.

Over the 21st century, population and demographic structure, settlement patterns, economic output and structure, technology, policy, and other social and economic factors will continue to affect the ease with which American society can adapt to, or take advantage of, climate variability and change. Continuing income growth and continuing development of new technologies remain likely, in aggregate, to reduce our vulnerability to climate. But as in the 20th century, specific climate impacts and vulnerabilities in the 21st century are likely to remain dependent on many detailed and specific characteristics of America's society, with the particular factors that turn out to be most important not evident in advance. Moreover, the changes in these factors over the 21st century are likely to be at least as great, and at least as unpredictable in their details, as the changes that took place over the 20th century.

ADAPTATION AND VULNERABILITY

People need not merely suffer the climate conditions they face, but can change their practices, institutions, or technology to take maximum advantage of the opportunities the climate presents and to limit the harms they suffer from it. Through such *adaptations*, people and societies (like ecosystems) adjust to the average climate conditions, and the variability of conditions they have experienced in the recent past. Present climates are not tuned to maximize human welfare, of course, so some potential changes might be purely benign (e.g., if there was a reduction in maximum hurricane wind speeds). But when habits, livelihoods, capital stock, and management practices are finely tuned to current climate conditions, the direct effect of many types of change in these conditions, particularly if the change occurs rapidly, is more likely to be harmful and disruptive than beneficial.

But just as societies adapt to the present climate, they can also adapt to changes in it. Adaptation can be intentional or not, and can be undertaken either in anticipation of projected changes or in reaction to observed changes. Society's capacity to adapt to future climate change is a crucial uncertainty in determining what the actual consequences of climate change will be. Societies and economies are *vulnerable* to climate change if they face substantial unfavorable impacts, and have limited ability to adapt. Like impacts themselves, the set of options and resources available to adapt to change, and the ability of particular individuals, communities, and societies to adopt them, depend on complex sets of linked social and economic conditions. Such factors as wealth, economic structure, settlement patterns, and technology play strong roles in determining vulnerability to specified climate conditions (Downing et al., 2000).

Human societies and economies have demonstrated great adaptability to wide-ranging environmental and climatic conditions found throughout the world, and to historical variability. Wealthy industrial societies like the US function quite similarly in such divergent climates as those of Fairbanks, Alaska and Orlando, Florida. While individual adaptability also contributes, it is principally social and economic adaptations in infrastructure, capital, technology, and institutions that make life in Orlando and Fairbanks so similar that individual Americans can move between them (in either direction) with at most moderate discomfort.

But adaptability has limits, for societies as for individuals, and individuals' ability to move through large climate differences tells us little about these limits. Moving between Fairbanks and Orlando may only be uncomfortable, but rapidly imposing the climate of either place on the other would be very disruptive. The countless ways that particular local societies have adapted to current conditions and their history of variability can be changed, but not without cost, not all with equal ease, and not overnight. The speed of climate change, and its relationship to the speed at which skills, habits, resource-management practices, policies, and capital stock can change, is consequently a crucial contributor to vulnerability. Moreover, however wisely we may try to adjust long-lived decisions to anticipate coming climate changes, we will inevitability remain limited by our imperfect projections of the coming changes. Effective adaptation may depend as much on our ability to devise responses that are robust to various possible changes, and adjustable as we learn more, as on the quality of our projections at any particular moment. While societies have shown substantial adaptability to climate variability, the challenge of adapting to a climate that is not stable, but evolving at an uncertain rate, has never been tested in an industrialized society.

While adaptation measures can help Americans reduce harmful climate impacts and take advantage of associated opportunities, one cannot simply assume that adaptation will make the aggregate impacts of climate change negligible or beneficial. Nor can one assume that all available adaptation measures will necessarily be taken. Even for such well-known hazards as fire, flood, and storms, people often fail to adopt inexpensive and easy risk-reduction measures in their choices of building sites, standards, and materials – sometimes with grave consequences. In this first National Assessment, potential climate adaptation options were identified, but their feasibility, costs, effectiveness, and the likely extent of their actual implementation were not assessed. Careful assessment of these will be needed.

SOCIOECONOMIC SCENARIOS IN IMPACT ASSESSMENT

Coping with Complexity

One way to assemble the socioeconomic assumptions needed for impact assessment is to construct scenarios. Scenarios are coherent, internally consistent, and plausible descriptions of possible future states of the world, used to inform investigations of future trends, potential decisions, or consequences (IPCC, 1994). Scenarios can be simple or complex, quantitative or qualitative, stochastic or deterministic, and can provide variable levels of detail according to their purpose. In most usage, scenarios are exogenous to the analysis: they describe aspects of the world that must be specified for the analysis to proceed, but which are simply assumed, not calculated within the analysis.

In assessments of climate change, the craft of developing and applying scenarios is most advanced for the scenarios of future greenhouse-gas emissions used to drive climate models. Scenarios for the largest sources of emissions can be developed by projecting a few aggregate characteristics of the nation or region being considered, such as population, economic growth, and changes in the energy intensity and carbon intensity of economic output (Nakicenovic and Stewart, 2000). While projections of these variables may have wide uncertainty ranges, they can be based on widely available consistent historical data and their complexity is not overwhelming. Moreover, because emission trends depend jointly on trends in population, economic growth, and technological change, it is possible to generate a wide range of emissions futures while considering only a narrow, largely benign range of population and economic futures, by making widely divergent assumptions of technical change.

Developing scenarios for assessment of impacts is a fundamentally different and more complex problem, on which less experience is available. No simple aggregate technical coefficients are known, analogous to energy intensity or carbon intensity in emissions scenarios, which would largely define impacts. Indeed, impacts and vulnerability are likely to depend on highly specific, detailed, often local characteristics of particular communities or activities, for which reliable consistent data are unlikely to be available – if we even knew what the relevant characteristics were. Also in contrast to emission scenarios, scenarios for impact assessment must consider the possibility of sustained low economic growth as well as high, since income and wealth are likely to be important determinants of vulnerability and adaptive capacity.

A working group of the NAST was charged with developing scenarios for the socioeconomic assumptions necessary for the Assessment. Because of the complexity and diversity of the socioeconomic characteristics that might be important determinants of impacts and vulnerability, and because of the highly decentralized nature of the National Assessment process, this working group judged it infeasible to attempt to develop fully detailed socioeconomic scenarios centrally. To do so would amount to trying to predict a century of American history. Moreover, such an attempt would be inappropriate because the determinants of impacts are likely to vary among regions, and identifying the most important ones is likely to require detailed regional expertise. Rather, the working group attempted to balance the Assessment's competing needs – to reflect regional concerns and expertise while maintaining enough consistency to allow national-level synthesis – by recommending a two-tracked approach to scenario development, partly centralized and partly decentralized.

The centralized track comprised a few key socioeconomic variables likely to influence many domains of impact, such as population, economic output, and employment. For these, where nationwide consistency was most important, the working group developed three internally consistent socioeconomic scenarios, which were used in all region and sector studies in the Assessment. The three scenarios spanned a wide range of high- and low-growth futures. Projections of population, income, and employment were provided in substantial detail through 2050 – by county and by thirteen economic sectors – and at the national level through 2100. These scenarios are described in Appendix 1 of this chapter.

The decentralized track was to be used when particular analyses required specifying future values of more specific or local socioeconomic characteristics. In such cases, the relevant assessment teams were asked to develop and document the required assumptions themselves. A common template was provided to guide teams in developing scenarios, which involved identifying two or three key characteristics they judged to have the most direct effects on the impact of interest, constructing uncertainty ranges for these characteristics, and varying them jointly through their ranges. In addition, two background papers were provided that reviewed alternative methods and attempts at projecting future trends in technology and institutions (Patt et al., 1998; Wilbanks, 1998). The template for the decentralized track is described in Appendix 2 of this chapter.

Teams were also requested to attempt an alternative, exploratory approach to projecting impacts in 2100, which would avoid the need for 100-year socioeconomic projections. This exploratory approach involved reversing the relationship between assumed socioeconomic futures and climate impacts. The standard approach used throughout the Assessment involved assessing the impact of a specific climate scenario under a specific future socioeconomic scenario. Instead, this alternative, exploratory approach involved specifying *only* a future climate scenario, and trying to identify plausible socioeconomic conditions that would make for large variation in the impacts of this specified climate. For the region or sector in question, what potential future socioeconomic conditions might make this climate change seriously harmful? What conditions might make it insignificant? What conditions might make it greatly beneficial? The purpose of this alternative approach was to engage teams in a more open-ended process of thinking through potential socioeconomic futures, to scout for potential vulnerabilities and opportunities that might escape notice in a more conventionally structured inquiry.

In this first Assessment, the region and sector teams made very limited use of the socioeconomic scenarios and template provided. In some cases, such as the Human Health sector, teams judged the state of knowledge in their domain insufficient to support any prospective, scenario-based analysis. In several other cases, analyses projected only first-order biophysical impacts such as changes in forest productivity or streamflow, for which no socioeconomic assumptions were needed. The few analyses that attempted to project impacts further down the causal chain to effects on humans only required, or could only effectively use, the scenarios of economic and population growth specified by the centralized track. No analysis in this Assessment used the full template for socioeconomic scenario development discussed in Appendix 2. The limited use of socioeconomic scenarios in this first Assessment has limited the extent to which impacts can be described or assessed in terms of human relevance – e.g., in terms of monetary loss or gain, valuation of non-market changes, or the incidence of such extreme events as bankruptcy, property loss or abandonment, or regional economic booms or busts. Further developing, testing, and applying such methods for constructing scenarios sufficiently rich and detailed to do impact assessment, but that still sufficiently limit complexity and maintain enough consistency to permit aggregation, will be a key methodological and research challenge for subsequent assessment of climate impacts.

The approach taken in this Assessment to projecting socioeconomic futures has obvious limitations. On the one hand, it represents a vast simplification of the linked climatic, ecological, economic, and social processes that will actually determine climate impacts and adaptive capacity. On the other hand, it is so complex and difficult to implement, that no analysis undertaken as part of this first Assessment was able to follow the template fully. Still, this general approach of combining central guidance on over-arching assumptions with structured use of decentralized expertise for other assumptions has allowed us to make a start. It has allowed this Assessment to take a first look at climate impacts, more detailed and consistent than has hitherto been conducted, which can be refined and extended as substantive knowledge and assessment methods are progressively improved.

MULTIPLE STRESSES

Human society has imposed various stresses on the environment, at diverse spatial scales, for centuries. Over the 21st century, climate change will occur in parallel with, and be jointly determined with, many other environmental stresses and many other forms of change. The same social, economic, and ecological systems that will bear the stress of climate change will also often be bearing other simultaneous stresses. These will include environmental stresses such as air pollution, acid deposition, coastal and estuarine pollution, loss of habitat and natural ecosystems, and unsustainable exploitation of marine resources. They will also include broader

socioeconomic stresses such as rapid shifts in technology and world market conditions, potential increases in migration and in economic inequality, and overloading of infrastructure in rapidly growing metropolitan and coastal regions. Other technological, economic, institutional, or social trends may help to increase systems' adaptability and mitigate the effects of climate change and other stresses. As climate varies and changes, so will these other factors. The aggregate impacts on ecological, economic, and social systems will reflect the joint application of multiple environmental and other stresses, as well as potential interactions between them.

For most US ecosystems, other stresses currently greatly exceed those arising from climate. Pacific Salmon populations are predominantly stressed by fishing, dams, and watershed alteration. Maple trees in New England are predominantly stressed by pests and air pollution. Endangered species are predominantly stressed by loss of habitat. Over the coming decades, some non-climatic stresses are likely to decline while others increase. For example, increasingly strict emission controls are likely to reduce acidifying pollution, while larger and wealthier populations are likely to increase the stresses that development, land-use conversion, pollution, and recreation impose on forests, mountain regions, wetlands, and coastlines. Although non-climatic stresses exceed climatic ones for most systems at present, one cannot assume that this will remain so – particularly for natural ecosystems, which in general are much more dependent on climate than socioeconomic systems. People may move with little discomfort between Alaska and Florida, but species adapted to the climate of one of these States could not survive in the other: a Martin or an Arctic Tern could not live in the wild in Florida, nor a Great White Heron or a Manatee in Alaska. Moreover, climate variability is already a discernible stress for some systems. A changing climate, interacting with other environmental and socioeconomic trends, is likely to become an increasingly important stress for many more systems. A system already bearing multiple stresses at high levels is likely to be less able, other factors being equal, to adapt to climate change. This observation is likely to apply not just to natural ecosystems, but also to managed ecosystems and communities, such as marginal agricultural lands or resource-dependent communities suffering job loss and out-migration.

Although it is likely that interactions among multiple stresses will be key dimensions of socioeconomic and ecological vulnerability, our current knowledge of how stresses interact is very limited. This first National Assessment has only been able to undertake the most preliminary investigation of interactions and multiple stresses. Several specific examples of multiple-stress effects have been identified as high priority needs for research and analysis, in order to improve our ability to analyze and respond to multiple stresses in future assessments.

THRESHOLDS, BREAKPOINTS, AND SURPRISES

The response of many systems to external changes is continuous: if you touch the accelerator, the car speeds up a little; if you touch the brake, it slows down a little. Many of the analyses of climate impacts discussed in this Assessment assume such continuous responses, so the projected impacts are often extensions of processes and trends that are already underway today.

Sometimes, however, systems can respond in highly discontinuous or nonlinear ways: if you tighten the propeller of the rubber-band airplane by one more turn, it can break into a dozen pieces. While many natural and social systems are likely to respond gradually to climate change, responses can also be sudden if a small additional stress pushes the system over a threshold or breakpoint.

Such discontinuities or surprises can be seen clearly after they happen, and attempting to explain them often generates important advances in our understanding, but they are extremely difficult to predict. It is imperative to remember that complex climatic, ecological, and socioeconomic systems might surprise us – by sudden or discontinuous response, or by evolving in some other way quite different from what we expect. We have been surprised by environmental and socioeconomic changes many times. Environmental examples include the failure of rain to "follow the plow" in the 19th Century American West, the appearance (and cessation) of the 1930s drought, and the 1980s appearance of the Antarctic ozone hole. Several possible surprises and discontinuities have been suggested for the Earth's atmosphere, oceans, and ecosystems.

Equivalently high-consequence surprises could also arise in socioeconomic systems, causing emissions, impacts, or vulnerability to be markedly different than we expect. Potential candidates for such surprises might include rapid development and deploy-

ment of technologies for non-fossil energy or carbon sequestration – which could greatly reduce future carbon dioxide emissions – or for coal-based synthetic fuels, which would greatly increase emissions. Other candidates include exhaustion of major reinsurance pools from high weather-related casualty losses, leading to financial destabilization; or climate-related emergence of new epidemic diseases. Still more potential for surprise arises from the intrinsic unpredictability of human responses to the challenges posed by climate change.

Even if the probability of any particular surprise occurring is low – which is widely assumed, but may or may not be true – potential surprises are so numerous and diverse that the likelihood of at least one occurring is much greater. We do not know how far the climate system, or the systems it affects, can be perturbed before they respond in quite unexpected ways. As with multiple stresses, in this first Assessment we have only been able to identify this possibility and conduct some preliminary speculation. Potential large-consequence surprises present some of the more worrisome concerns raised by climate change, and pose some of the greatest challenges for policy and research.

By their very nature, surprises are unpredictable. But two broad approaches can help us prepare to live with a changing and uncertain climate, even considering the possibility of surprise. First, some of our assessment effort can be devoted to identifying and characterizing potential large-impact events, even if we presently judge their probability to be very small. Second, society can maintain a diverse and advancing portfolio of scientific and technical knowledge, and conditions that encourage the creation and use of new knowledge and technology. Continually advancing knowledge and technology, and the social, economic, and policy conditions that support them, provide a powerful foundation for adapting to whatever climate changes might come.

INTEGRATED ASSESSMENT
Thinking About the Future

Multiple climatic and socioeconomic characteristics jointly determine climate impacts. Further complexity arises from the fact that the multiple socioeconomic factors likely to determine impacts and adaptive capacity all influence each other, and are in turn influenced by patterns of environmental change. Patterns of population growth, technological change, economic growth, and structural change all affect each other, and collectively determine the character and degree of environmental stresses society imposes, including the emissions that contribute to climate change. Public policies will also contribute to this complex mix, both those directed toward climate change and others. Tax policy can influence investments in research. Immigration policy can influence the rates of both population and economic growth, and the cultural, educational, and economic mix of the population. Reliable prediction of such complex and uncertain processes is not possible.

Since the early 1990s, research groups have sought to represent linked processes of global environmental change and associated ecological, economic, and social processes in "Integrated Assessment" models of global climate change. These models are intended to allow consistent examination of the human contributions to climate change and ways to mitigate them, with the human consequences of climate change and ways to adapt to them. They consequently allow coherent assessment of possible responses including both emissions reduction and adaptation to resultant change, and the tradeoffs between them. They also thereby allow consistent comparison of uncertainties in all domains of the climate issue – future emissions and means to reduce them, the responses of the climate system at global and regional scales, and the resultant impacts and means to adapt to them (Weyant et al., 1996; Parson and Fisher-Vanden, 1997; Rotmans and Dowlatabadi, 1998).

Over the past ten years this work has yielded significant insights into the economic determinants of emission trends, potential feedbacks between climate change and managed and unmanaged terrestrial ecosystems, and the relative contributions of atmospheric, ecological, and socioeconomic uncertainty to uncertainties in future climate impacts and advantageous responses. While the promise of important further insights from such work remains substantial, the characterization of impacts and adaptation remains the weakest element of integrated assessment at present.

APPENDIX 1:

Three Scenarios of Future Socioeconomic Conditions

The three socioeconomic scenarios developed by the NAST working group all assumed that broad 20th century trends of population and economic growth are likely to persist, to varying degrees, in the 21st

century. Barring major wars or other catastrophes, US population growth is likely to continue, though at a declining rate, moving toward a stable or nearly stable population in the second half of the 21st century. The population is also likely to continue to grow older for several decades or more, depending principally on the balance between immigration and increased life expectancy, and is projected to continue present trends of moving to metropolitan areas and the coast. Income and employment growth are projected to move with the people to the cities and coasts, and to continue the long-standing shift among sectors away from agriculture, resources, and primary industry and toward technology, trade, and services.

The Assessment focused on two target dates, 2030 and 2100. While both are far in the future, 2030 lies within the range of some projection models and strategic-planning tools, while 2100 lies beyond the useful range of nearly all such tools. Consequently, the Assessment provided socioeconomic projections with substantial spatial and sector detail for 2030, but only aggregate national projections of a few key variables for 2100.

For 2030, the Assessment provided detailed high, medium, and low scenarios of population and economic growth. These three scenarios were based on alternative assumed trends in fertility, mortality, and migration, labor-force participation by age group, and labor productivity, and were implemented using a commercial regional economic growth model. (Terleckyj, 1999a, 1999b). This model provided annual projections of population, by sex and by five-year age cohort, for each state, county, and metropolitan area in the United States. The NAST working group specified the assumed trends in fertility, mortality, and migration that determined nationwide population trends, while the model's demographic module calculated the resultant age structure of the population and its economic module distributed people around the nation.

In specifying these aggregate demographic trends, the working group used the Census Bureau's assumptions for future trends in US age-specific fertility and mortality (US Bureau of the Census, 2000), but applied a wider range of assumptions for future immigration. The low scenario followed the Census Bureau's low immigration assumption, which reduces net immigration from roughly 750,000 per year in the mid-1990s (0.3% of population) to 300,000 per year in 2000, and holds it numerically constant thereafter. The middle and high scenarios each projected that recent trends of increasing immigration will continue, using growth trends derived from two different recent periods. The middle scenario took the average trend in the ratio of annual net immigration to current population that has prevailed over the past thirty years (1967-1997), and projected this trend forward until 2025. Projected in this way, net immigration reaches 0.46% of the population in 2025, and is held constant at this fraction of population thereafter. The high scenario differed from the middle only in that it calculated the trend in the ratio of immigration to population over the most recent ten years (1987-1997), a period of particularly rapid immigration growth. Projecting this steeper trend forward, net immigration reaches 0.86% of population in 2025, and is held at that fraction of population thereafter. In all three scenarios, the aging of the post-war baby boom generation brings sharp increases in the fraction of older Americans. The fraction of Americans aged 65 or over begins to surge after 2010 from its present value of 12.5%, reaching 20% by 2030. Still greater increases are projected in the fraction of Americans in the oldest age groups. The fraction of Americans aged 85 and over is projected to triple to 4.5% by 2050, while those 100 and over are projected to increase seven-fold, to 0.2%.[1]

The three scenarios also provided projections of employment and income for thirteen major economic sectors (including three government sectors), with the same level of spatial detail – by state, metropolitan area, and county. As in the case of population, the NAST-specified assumptions determined nationwide economic trends – in this case, trends in nationwide labor productivity and age-specific labor-force participation rates – while the distribution of employment and income among locations and economic sectors was calculated internally by the model.

Rates of labor-force participation were varied only for older workers. For workers under 55, participation rates were held at present levels. For workers aged 55 through 64, all three scenarios project increases in participation that extend recent trends, reaching a higher, constant level in 2025. Only the high scenario differs, in projecting an increase in participation for workers 65 and over, which also levels off in 2025. For productivity, the middle scenario assumes a continued constant increase of 1.2% annually in real economic output per worker, equal to the average annual increase over the 20th century. The high and low scenarios double and halve this

rate of productivity growth respectively, to 2.4% and 0.6% per year.

Considering three alternative trends for population and productivity growth and two for labor-force participation yields eighteen possible combinations. If practicality dictates using only a few scenarios, only a small subset of these possible combinations can be considered. The recent IPCC scenario exercise faced a similar but much more complex problem of selecting scenarios from a large set of combinatorial possibilities, because their scenarios included diverse growth trends for multiple world regions. To reduce this complexity, they constructed narrative storylines that provided broad political and social context for particular worldwide patterns of population and economic growth (Nakicenovic and Stewart, 2000).

In constructing the three scenarios for this Assessment, high population growth and high economic growth were combined in the high scenario, while low population and economic growth are combined in the low scenario. This combination of high population with high economic growth would not be appropriate in constructing scenarios for the world as a whole, because historical evidence and demographic theory both suggest that higher rates of economic growth are associated with lower rates of population growth. This situation is reversed, however, for projections of American growth in the 21st century, because most of the variation in population growth arises from variation in the assumed level of immigration. In contrast with natural population increase, immigration tends to follow economic opportunity, so the pairing of high population growth with high economic growth is more plausible than the reverse.

For 2100, the more distant target date of the Assessment, a much less detailed set of socioeconomic projections was specified. Over this time horizon, the likelihood of fundamental changes in economic structure, technology, and culture are likely to render the incremental methodology of the near-term regional economic model invalid. Indeed, any attempt to specify century-scale socioeconomic trends with the spatial and sector detail the NPA model provides for the near term is likely to be ludicrous. No detailed assumptions for the regional and sector distribution of population, employment, or income were specified for the second half of the 21st century.

Rather, a simple aggregate set of nationwide projections of population, employment, and GDP for the US were developed. As in the case of the more detailed

---

[1] These figures are for the middle scenario. In the Census Bureau's highest scenario – which is not identical to the Assessment high scenario because the Census assumes less immigration and consequently an older population – centenarians reach 0.75% of the total population by 2050.

projections through 2030, three scenarios were developed that combined high, medium, and low population and economic growth. Each of these scenarios was constructed to track the growth of national population and output in the corresponding more detailed scenario for the near term, then to converge in growth rates of both population and productivity over the second half of the 21st century. The scenarios were constructed using a simple reduced-form integrated-assessment model (Scott et al, 1999), and were broadly consistent with three of the "marker" scenarios developed for the IPCC Third Assessment Report (Nakicenovic and Stewart, 2000).[2]

The assumptions specified for these long-run scenarios are as follows. Population growth rates in the three scenarios converge beginning in 2050, until they become equal in 2075 and follow a common path thereafter, declining from 0.35% per year in 2080 to 0.15% in 2100. Aggregate rates of labor-force participation also converge after 2050, but not to full equality, reaching 75%, 77.5%, and 80% in the three scenarios by 2100. Finally, annual growth rates of economic output per worker begin converging and declining after 2050, reaching 1.12% per year in 2075 and remaining at that level through 2100.

The consequences of these assumptions for US population and GDP are shown in Tables 1 and 2, and

[2] The correspondence with IPCC scenarios is not exact, in part because of differences in time-steps and the definition of regions (the US is not a complete region in the IPCC scenarios).

Figures 1 and 2, for both target years 2030 and 2100. US population is projected to reach 356 million by 2030 in the middle scenario (corresponding to an average growth rate of 0.86% per year between 1995 and 2030), with a range of 305 to 398 million in the low and high scenarios (0.39% to 1.21% annual growth). By 2100, the US population has reached 494 million in the middle scenario (average annual growth of 0.47% from 2030 to 2100), with a range from 353 to 640 million (0.21% to 0.68% average annual growth). United States GDP grows from its present $7.2 trillion to $14.4 trillion by 2030 (range $10.3 to $24.1 trillion), and to $39.2 trillion by 2100 (range $19.2 to $114.7 trillion). In terms of GDP per person, these scenarios give a range from $33,800 to $60,600 in 2030 ($40,500 in the middle scenario), and from $54,400 to $179,200 in 2100 ($79,400 in the middle scenario).[3]

APPENDIX 2:

Template for developing socio-economic scenarios

If a team required more detailed or specific socioeconomic assumptions to conduct an analysis than were provided in the centrally defined scenarios, they were asked to develop and document them using a common template, as follows.

[3] All monetary figures are expressed in 1992 dollars.

Table 1: Scenarios of US Population (Millions)

	1997	Growth rate, 1995-2030	2030	Growth rate, 2030-2100	2100
Present Population	268				
Low Scenario		0.39%	305	0.21%	353
Middle Scenario		0.86%	356	0.47%	494
High Scenario		1.21%	398	0.68%	640

Table 2: Scenarios of US GDP (Trillions of 1992 Dollars)

	1997	Growth rate, 1995-2030	2030	Growth rate, 2030-2100	2100
Present GDP	$7.2				
Low Scenario		1.1%	$10.3	0.9%	$19.2
Middle Scenario		2.1%	$14.4	1.4%	$39.2
High Scenario		3.7%	$24.1	2.25%	$114.7

First, each team was to select *a few key issues* they judged would be most important for their region or sector, or would best illustrate important patterns of impact. These are the "key issues" discussed in each of the regional and sector chapters. Second, for each key issue the team was asked to identify *one or two key socioeconomic factors*, such as specific aspects of development patterns, land use, technologies, or market conditions, that they judged likely to have the *most direct influence* on climate impacts, capacity for adaptation and vulnerability for that issue. In choosing their key issues and key socioeconomic factors, each team was requested to use whatever combination of preliminary analysis, expert judgment, and stakeholder consultation they judged most appropriate. They were then to examine the impacts of specified climate-change scenarios on their key issues, under a range of values for their chosen socioeconomic factors. If they identified more than one key socioeconomic factor, they were asked to construct a few alternative socioeconomic scenarios by varying the factors jointly between high and low values. Other than the few key factors they chose to vary, any other required socioeconomic assumptions were to be fixed at baseline or best-guess values.

The ranges chosen for key socioeconomic factors were intended to reflect all sources of socioeconomic uncertainty *except* climate change itself and US policy responses to climate change. Since the purpose of the Assessment was to examine the effects of climate explicitly, these did not need to be embedded in variation of socioeconomic input assumptions. In contrast, the ranges were to include climate-related uncertainty outside the US, if the team judged such uncertainty to matter for US impacts. This situation might arise, for example, in estimating demand for US grain exports or immigration to the US, either of which could be influenced by climate-related impacts abroad.

The template also provided some guidance in deciding how wide a range of values to assume for the key socioeconomic factors. In general terms, teams were asked to make the range wide enough to generate instructive variation in impacts, but to remain within their judgment of plausibility. Specifically, the range chosen for any factor should correspond to roughly a 10% chance that the true value would lie above the upper end of the range, and a 10% chance that it would lie below the lower end. The NAST working group followed this same guideline, constructing ranges to capture the true value with 80% confidence, in developing the three scenarios of aggregate US population and economic growth dis-

Scenarios of 21st Century Growth in America

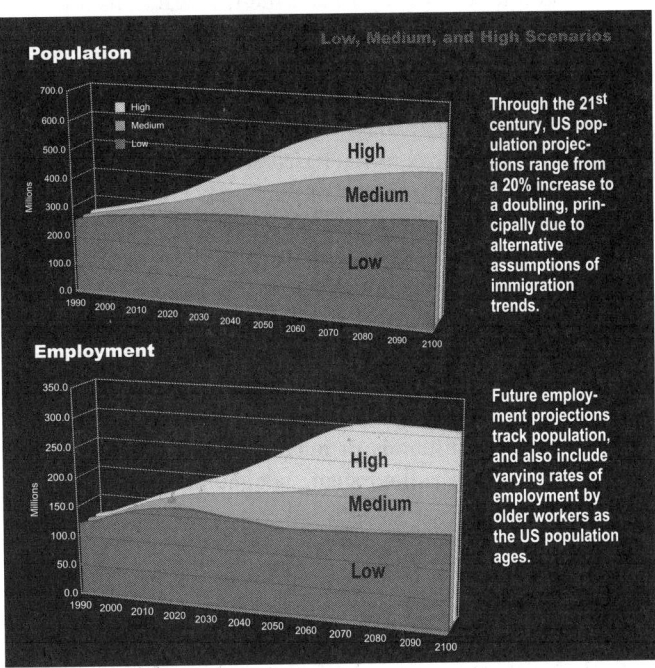

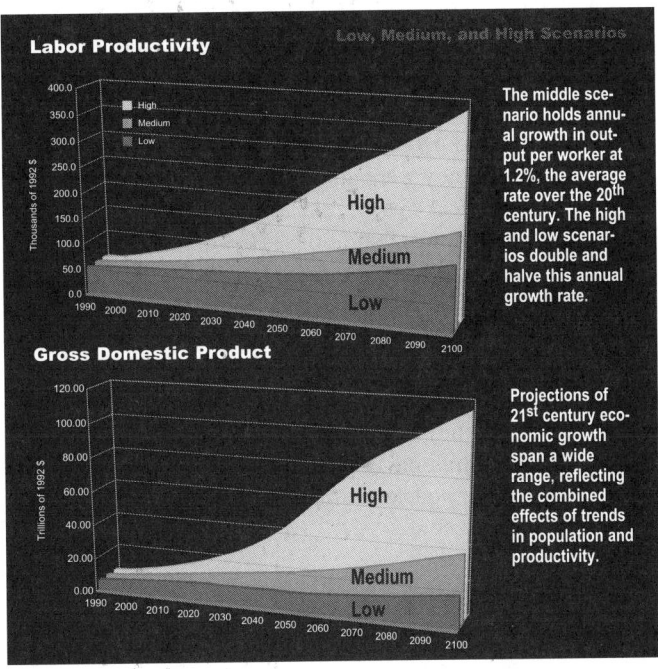

Figure 1. The Assessment considered high, medium, and low scenarios of future US population and economic growth. Future trends in population, economic growth, and technological change will all shape our contribution to climate change, our vulnerability to it, and our ability to adapt.

cussed above. In constructing these ranges, participants were cautioned to draw them wide, seeking to mitigate people's widely known tendency to be too confident in estimating unknown quantities (Kahneman et al., 1982; Morgan and Henrion, 1990).

The combination of climate scenarios and socioeconomic scenarios provides the raw materials for assessing climate impacts. Imposing a specified climate scenario – whether derived from historical experience or from a model projection – on a specified socioeconomic scenario, and examining its effects relative to the baseline climate, provides a first-order illustration of potential climatic impacts. How this difference varies among alternative climate scenarios illustrates how impacts depend on uncertainty in climate. How the difference varies among alternative socioeconomic scenarios illustrates how socioeconomic factors shape vulnerability to impacts. How it varies with specific hypothesized responses (such as changes in management, policy, institutions, or infrastructure), illustrates key decisions that may shape vulnerability and adaptive capacity.

For example, in the Pacific Northwest, the projected climate scenario had warmer wetter winters and drier summers. As one of their key issues, the Northwest team identified the impacts of this change on seasonal streamflow and freshwater supplies. In two separate analyses, they examined impacts of climate variability and change on the risk of winter flooding, and on the reliability of summer water and hydroelectric supplies. For the flooding analysis, they were able to define flooding purely by hydrological characteristics, without regard to socioeconomic conditions. However, more detailed assessment that considered not just occurrence of floods but the damages they cause would clearly have to consider settlement patterns, construction standards, emergency response measures, and other socioeconomic conditions as well. For the shortage analysis, they examined impacts of summer shortages on multiple uses under two alternative sets of operating policies to manage flows and allocate supply (Mote et al., 1999).

LITERATURE CITED

Bryson, R. A., H. H. Lamb, and D. L. Donley, Drought and the decline of Mycenae, *Antiquity*, 48, 46, 1974.

Bureau of Economic Analysis (BEA), Survey of current business, US Treasury Department, January 2000.

Carter, T. R., M. L. Parry, H. Harasawa, and S. Nishioka, *IPCC Technical Guidelines for Assessing Climate Change Impacts and Adaptations*, National Institute for Environmental Studies, Tsukuba, Japan, 1994.

Changnon, S. A., R. A. Pielke, Jr., D. Changnon, R. T. Sylves, and R. Pulwarty, Human factors explain the increased losses from weather and climate extremes, *Bulletin of the American Meteorological Society, 81*, 437-442, 2000.

Diamond, J., *Guns, Germs, and Steel: The Fates of Human Societies*, Norton, New York, 1997.

Downing, T. E, R. Butterfield, S. Cohen, S. Huq, R. Moss, A. Rahman, Y. Sokona, and L. Stephen, Climate change vulnerability: Toward a framework for understanding adaptability to climate change impacts, Report to UNEP from the Environmental Change Institute, University of Oxford, Oxford, England, May 2000.

Glantz, M. T. (Ed.), *Societal Responses to Regional Climatic Change: Forecasting by Analogy, Westview Press*, Boulder, Colorado, 1988.

Huntington, E., *Civilization and Climate*, Yale University Press, New Haven, Connecticut, 1915.

Kahneman, D. A., P. Slovic, and A. Tversky (Eds.), *Judgment under Uncertainty: Heuristics and Biases*, Cambridge University Press, New York, 1982.

Kates, R. W., The interaction of climate and society, in *Climate Impact Assessment*, edited by R. W. Kates, J. H. Ausubel, and M. Berberian, John Wiley and Sons, Chichester, UK, SCOPE/ICSU, chap. 1, 3-36, 1985.

Krunkel, K. E., R. A. Pielke, Jr., and S. A. Changnon, Temporal fluctuations in weather and climate extremes that cause economic and human health impacts: A review, *Bulletin of the American Meteorological Society, 80*, 1077-1098, 1999.

Lambert, L. D., The role of climate in the economic development of nations, *Land Economics, 47*, 339, 1975.

Morgan, M. G., Uncertainty analysis in risk assessment, *Human and Ecological Risk Assessment, 4*(1), 25-39, 1998.

Morgan, M. G., and M. Henrion, *Uncertainty: A Guide to Dealing with Uncertainty in Quantitative Risk and Policy Analysis*, Cambridge University Press, New York, 1990.

Mote, P. W., et al., Impacts of climate variability and change: Pacific Northwest, A report of the Pacific Northwest Regional Assessment Group for the US Global Change Research Program, Climate Impacts Group, University of Washington, Seattle, Washington, November 1999.

Myrdal, G., *Asian Drama: An Inquiry into the Poverty of Nations*, Random House, New York, 1972.

Nakicenovic, N., and R. Stewart (Eds.), *Emissions Scenarios: Special Report of the Intergovernmental Panel on Climate Change*, Cambridge University Press, New York, 2000.

Parson, E. A., and K. Fisher-Vanden, Integrated assessment models of global climate change, *Annual Review of Energy and the Environment, 22*, 589-628, 1997.

Patt, A., B. Tomm, A. Wolfe, and T. Wilbanks, *Forecasts of Institutional Change in the United States: A Summary*, prepared for the U.S. National Assessment of Possible Consequences of Climate Variability and Change, Oak Ridge National Laboratory, Oak Ridge, Tennessee, 1998.

Riebsame, W. E., "Research in climate-society interaction," in *Climate Impact Assessment*, edited by R. W. Kates, J. H. Ausubel, and M. Berberian, SCOPE/ICSU, Wiley, Chichester, UK, chap. 3, 69-84, 1985.

Rosenberg, N. J. (Ed.), Towards an integrated assessment of climate change: The MINK study, *Climatic Change (special issue), 24*, 1-173, 1993.

Rotmans, J., and H. Dowlatabadi, Integrated Assessment Modeling, in *Human Choice and Climate Change*, edited by S. Rayner and E. Malone, Vol. 3, chap. 5, 291-377, Battelle Press, Columbus, Ohio, 1998.

Sachs, J., T. Panayotou, and A. Peterson, Developing countries and the control of climate change: A theoretical perspective and policy implications, CAERII discussion paper no. 44, Cambridge, MA, November 1999.

Schneider, S. H., and R. Londer, *The Coevolution of Climate and Life*, Sierra Club Books, San Francisco, 1984.

Scott, M. J., R. D. Sands, J. Edmonds, A. M. Liebertrau, and D. W. Engel, . Uncertainty in integrated assessment models: Modeling with MiniCAM 1.0, *Energy Policy*, 27, 855-879, 1999.

Terleckyj, N. E., Analytic documentation of three alternate socioeconomic projections, 1997-2050, NPA Data Services, Inc., Washington, DC, May 1999a.

Terleckyj, N. E., Development of three alternate national projection scenarios, 1997-2050, NPA Data Services, Washington DC, July 26, 1999b.

Weyant, J., O. Davidson, H. Dowlatabadi, J. Edmonds, M. Grubb, E. A. Parson, R. Richels, J. Rotmans, P. R. Shukla, R. S. J. Tol, W. Cline, and S. Fankhauser, Integrated assessment of climate change: An overview and comparison on approaches and results, in *Climate Change 1995: Economic and Social Dimensions of Climate Change*, edited by J. P. Bruce, H. Lee, and E. F. Haites, Report of Working Group 3, Intergovernmental Panel on Climate Change (IPCC), chap. 10, 367-396, Cambridge University Press, New York, 1996.

Wilbanks, T. J., Forecasts of technological change in the United States: a summary, prepared for the U.S. National Assessment of Possible Consequences of Climate Variability and Change, Oak Ridge National Laboratory, Oak Ridge, Tennessee, 1998.

Wilhite, D. A. (Ed.), Drought: A global assessment in *Hazards and Disasters*, Vol. 2. Routledge Publishers, New York, 2000.

US Bureau of the Census, *Current Population Reports, Series P23-194, "Population Profile of the United States: 1997,"* US Government Printing Office, Washington, DC, 1998.

US Bureau of the Census, "Aging in the United States," http://www.census.gov/population/www-socdemo/age.html#elderly, 1999.

US Bureau of the Census, Population Projections through 2050, Document NP-T1, February 11, 2000.

CHAPTER 4

POTENTIAL CONSEQUENCES OF CLIMATE VARIABILITY AND CHANGE FOR THE NORTHEASTERN UNITED STATES

Eric Barron[1,2]

Contents of this Chapter

[1]Pennsylvania State University; [2]Coordinating author for the National Assessment Synthesis Team

CHAPTER SUMMARY

Regional Context

The Northeast is characterized by diverse waterways, extensive shorelines, and a varied landscape in which weather and the physical climate are dominant variables. The contrasts, from mountain vistas and extensive forests to one of the most densely populated corridors in the US, are noteworthy. The Northeast includes the largest financial market in the world (New York City), the nation's most productive non-irrigated agricultural county (Lancaster, PA), and the largest estuarine region (the Chesapeake Bay) in the US. The Northeast is dominated by managed vegetation, with much of the landscape covered by a mosaic of farmland and forest. The varied physical setting of the Northeast is matched by its highly diversified economy and by the character of its human populations. The majority of the population is concentrated in the coastal plain and piedmont regions, and within major urban areas. The economic activities within the region range from agriculture to resource extraction (forestry, fisheries, and mining), to major service industries highly dependent on communication and travel, to recreation and tourism, to manufacturing and transportation of industrial goods and materials. Assessment of the impacts of climate change is based on observed climate trends, climate simulations, and the importance of past extreme weather events.

Climate of the Past Century

- The Northeast has been prone to natural disasters related to weather and climate, including floods, droughts, heat waves, and severe storms.
- Temperature increases of as much as 4°F (2°C) over the last 100 years have occurred along the coastal margins from the Chesapeake Bay through Maine.
- Precipitation shows strong increases, with trends greater than 20% over the last 100 years occurring in much of the region. Precipitation extremes appear to be increasing while the amount of land area experiencing drought appears to be decreasing.
- The period between the first and last dates with snow on the ground has decreased by 7 days over the last 50 years.

Climate of the Coming Century

- The Northeast has among the lowest rates of projected future warming in comparison with the other regions of the US.
- Winter minimum temperatures are likely to show the greatest change, with models projecting increases ranging from 4-5°F (2-3°C) to as much as 9°F (5°C) by 2100, with the largest increases in coastal regions. Maximum temperatures will possibly increase much less, but again the largest changes are likely to occur in winter.
- For precipitation, model scenarios offer a range of potential future changes, from roughly 25% increases by 2100 on average for the entire region, to little change.
- The variability in precipitation in the coastal areas of the Northeast is likely to increase.
- Models provide contrasting scenarios for changes in the frequency and intensity of winter storms.

Key Findings

- On the time scale of a century, winter snowfalls and periods of extreme cold are very likely to decrease. In contrast, heavy precipitation events have been increasing and warming is very likely to continue this trend. Potential changes in the intensity and frequency of hurricanes are a major concern.

- Climate change is very likely to exacerbate current stresses on estuaries, bays, and wetlands in the Northeast with rising sea level and increasing water temperatures, with significant effects on fish populations, productivity, and human and ecosystem health.

- Decreased snowfalls and more moderate winter temperatures are very likely to lower winter stresses in major northeastern urban areas. However, climate change has greater potential to add to existing stresses in urban areas due to the impact of rising sea level and elevated storm surges on transportation systems, increased heat-related mortality and morbidity associated with temperature extremes, increased ground-level ozone pollution problems associated with warming, and the impact of precipitation and evaporation changes on water supply.

- Typical summer recreational activities involving beaches or freshwater reservoirs are likely to experience extended seasons and the region's diverse waterways are very likely to become havens for escape from increasing summer heat. In contrast, negative impacts are likely to include inability of ski areas to maintain snow pack, muting of fall foliage colors, displacement of maple-sugaring, increases in insect populations, accelerated beach erosion due to sea-level rise, and worsening ground-level ozone pollution prob-

lems, even in the mountains of New England.

- The complex institutional framework of communities, municipal, county, regional, and statewide formal and informal governing bodies that characterize the Northeast is likely to limit the region's ability to deal with extremes in water supply.

- Infectious disease vectors, such as ticks and mosquitoes, are likely to be altered by warmer and wetter conditions. Increased rainfall and flooding have a historical association with contamination of public and private water supplies (e.g., with *Cryptosporidium*).

- Agriculture in the Northeast is very likely to be relatively robust to climate change although the crop mix may change. The ability to change crop types and to take advantage of hybrids limits vulnerability. Northern cool weather crops are a possible exception.

- The species composition of northeastern forests is very likely to change under the climate scenarios examined in this Assessment, with significant northward migration of forest types.

- Projected changes in temperature and precipitation are likely to have a direct impact on species distribution mix or an indirect impact associated with changing predator-prey relationships or changes in pests or disease. In many cases, the species effected may be truly characteristic of a region or may be of economic significance (e.g., lobster, migratory birds, and trout).

- Climate change, resulting in higher temperatures and poorer air quality, could lead to increases in heat-related mortality and morbidity and respiratory illness. The elderly, children, those already ill, and lower-income residents are groups most at risk.

POTENTIAL CONSEQUENCES OF CLIMATE VARIABILITY AND CHANGE FOR THE NORTHEASTERN UNITED STATES

PHYSICAL SETTING AND UNIQUE ATTRIBUTES

The northeastern United States is dominated by diverse waterways, extensive shorelines and a varied landscape in which weather and climate are dominant variables. The regional contrasts and strengths are noteworthy. Most mental images of New England focus on an environment of quaint villages, mountain vistas, extensive forests, and brilliant fall colors, maple sugaring, and skiing through forested glades. The Northeast also includes the most densely populated corridor in the US and is intimately connected to the North Atlantic coastline and ocean-accessible waterways. Of these waterways, the Chesapeake Bay is the largest estuarine region in the US and is unmatched in terms of its importance for recreation, fish, and wildlife. The Northeast is also crossed by a remarkable network of streams and rivers superimposed on the mountainous terrain of the Appalachians. The vegetative landscape has changed dramatically over the last 100 years, in part because of significant areas of forest re-growth.

SOCIOECONOMIC CONTEXT

The varied physical setting of the Northeast is matched by its highly diversified socioeconomic characteristics (NPA Data Services, 1998; Bureau of Economic Analysis, 2000; Polsky et al., 2000; Rose et al., 2000). The Northeast includes the largest financial market in the world (New York City), as well as the most productive non-irrigated agricultural county (Lancaster, PA) in the US. The economic activities in the region range from resource extraction (forestry, fisheries, and mining) and agriculture, to major service industries highly dependent on communication networks and travel, to recreation and tourism, to manufacturing and transportation of industrial goods and materials. The human populations are largely concentrated in coastal and urban areas. The region contains some of the most densely populated counties in the United States.

The socioeconomic characteristics of the Northeast vary considerably across the region. In the Mid-Atlantic region, more than 50% of the 35 million people and associated jobs are concentrated in six urban areas, for the most part within the coastal plain and piedmont regions. The Mid-Atlantic population increased nearly 20% during the last three decades, somewhat less than the 33% population growth experienced by the nation as a whole. The working age population increased by about 34%, the over-65 population by 72%, and the 0-19 age group declined by 16%. Per capita income increased by 82%, with the largest growth in total income in the service sector (300%). Farm employment declined by almost one-half, reflecting the national trend. The economy is diverse and substantial. The Mid-Atlantic region alone accounts for 13% of the total US economy, with sizeable export and import flows. Agriculture, forestry, and mining comprise about 2% of the region's economy, while manufacturing and service comprise 26% and 20% respectively.

The Northeast is characterized by a megalopolis that extends from Boston to Washington, D.C. The greater New York City area alone includes parts of three states, 31 counties, and nearly 1,600 cities, towns, and villages with more than 19 million inhabitants. At the heart of this metropolis is a city with more than 7 million people. This population places tremendous demands on land and water resources. Approximately 30% of the 31 counties comprising the greater New York City land area is fully converted to urban uses, and the rate of conversion has accelerated even though the rate of population growth has slowed. As an example of water demand, the amount supplied by the New York City water system, serving the city, most of Westchester, and some additional communities, is approximately 1.3 billion gallons a day. The area's development is intimately connected with its 1,500 miles of coastline and a complex transportation infrastructure. The general economy of this metropolitan area is mostly based on service industries, of which the economic heartbeat is finance, corporate headquarters, and trade centers.

New England is a study in contrasts with a fascinating land use history (Cronon, 1983). Most of the region's almost 13 million inhabitants live in the densely populated coastal segments in the south and east. To the north and west, New England is heavily forested and mountainous with more isolated smaller urban concentrations. The New England economy is dominated by the service sector (individual state averages range from 27 to 35% of their economies), followed by manufacturing of durable goods, finance, insurance and real estate, and trade. The service sector and the finance, insurance, and real estate sector, tend to be the fastest growing segments of the economy (ranging from 5 to 10% growth over the decade). New England also includes the state (Connecticut) with the highest per capita personal income in the nation but, within the region, only Massachusetts and Connecticut currently have per capita personal incomes that are growing at rates above the national average (NPA Data Services, 1998). Data from the 1979-1989 Census placed Connecticut, Massachusetts, and New Hampshire with the highest percentage increases in median household income of US states for the decade of comparison.

The large urban areas of the Northeast include some of the oldest metropolitan centers in the nation, and are often characterized by aging infrastructure and a wide variety of stresses. The Boston to Washington corridor is largely an urban landscape, with the densest population in the nation.

CLIMATE VARIABILITY AND CHANGE

Historically, the Northeast has been prone to natural disasters related to weather and climate. Floods, droughts, heat waves, and severe storms are characteristic. For example, seven major tropical storms crossed the Mid-Atlantic region since 1986 and six years of the last 20 have been characterized by significant drought in some parts of the region. In addition, the major cities of the Northeast have experienced episodes of increased morbidity and mortality during heat waves (Kalkstein and Greene, 1997). The 1990s have been characterized by a number of significant winter precipitation events including a number of heavy snowfalls and major ice storms, while winter temperatures have tended to be mild. Both the weather events and mild winters raised the regional consciousness about climate and climate change.

The climate of the northeastern US experienced significant changes in temperature and precipitation during the 20th century (Hughes et al., 1992; Karl et al., 1996; Karl and Knight, 1998). Based on observations derived from the highest quality US observing stations as a part of the US Historical Climatology Network, temperature increases of as much as 4°F (2°C) over the last 100 years occurred along the coastal margins from the Chesapeake Bay to Maine. Within the Northeast, the only area with a decreasing historical trend in temperature is a small portion of southern Pennsylvania. Analysis of extreme temperatures shows little change in the number of days exceeding 90°F or below freezing.

Precipitation trends also show strong increases over the last 100 years, with trends greater than 15-20% occurring in Pennsylvania, western New York, and from northern New York across the middle of New England (Karl et al., 1996; Karl and Knight, 1998). Precipitation extremes show increases of 12% in highest annual 1-day precipitation total, and 3% in the number of days per year exceeding 2 inches of precipitation over the 20th century. The region of the Northeast adjacent to the Great Lakes had the largest annual increase in precipitation over the period 1931-1996 and the largest increases in very extreme precipitation events (Kunkel et al., 1999a).

For the region as a whole, the period between the first and last dates with snow on the ground has decreased by 7 days over the last 50 years, resulting in a shorter snow season (Karl et al., 1993; Groisman et al., 2000). The Palmer Drought Severity Index (PDSI), which indicates trends in drought and wet periods, confirms the precipitation patterns by demonstrating a tendency toward more wet periods in Pennsylvania and southern New England, and tendencies toward drier periods in northern New England and around the Chesapeake Bay. On average, the amount of land area experiencing drought appears to have decreased over the last half century.

Projections of future climate change in the Northeast based on the Hadley climate model (Mitchell et al., 1995; Mitchell and Johns, 1997; Johns et al., 1997) and Canadian climate model (Boer et al., 1984; McFarlane et al., 1992; Boer et al., 1999; Boer et al., 1999; Flato et al., 1999) suggest lower than average temperature increases in comparison with other regions of the US. The Hadley model yields a trend for the 21st century of a 5°F (2.6°C) average temperature increase while the Canadian climate model indicates a 9°F (5°C) increase. Winter minimum temperatures increase by 4 to 12°F (2 to 7°C) in the Canadian model with the greatest increases in the western part of the region and the smallest increases in southern New England. In the Hadley model, winter minimum temperatures increase by approximately 7°F (4°C) in the entire region north of Maryland, with somewhat smaller increases immediately to the south. Summertime increases in minimum temperatures are projected to be greater than 5°F (3°C) in both models. Maximum temperatures also increase, with the largest changes in winter (from more than 12°F in West Virginia to less than 5°F in New England in the Canadian model, and from 3 to 5°F in the Hadley model). Climate models differ substantially in their projections of the summertime increase in maximum temperatures, with projections ranging from 2-3°F (1.3°C) to as much as 7-11°F (4-6°C), between the Hadley and Canadian models respectively. Both models used in this Assessment generally indicate small decreases in the variability of temperature (the exception is an increase in summer temperature variance in the Canadian model). The more limited warming in the Northeast relative to much of the rest of the US may be partially attributed to the cooling effect of sulfate aerosols that are concentrated in the Northeast and the maritime influences in the coastal regions, offsetting some of the warming.

The Northeast currently has more total precipitation than all other regions except the Southeast. The model simulations offer rather different scenarios for future changes. In the Hadley model, regional

precipitation is projected to increase by as much as 25% by 2100. In summer, the greatest increases are in western portion of the region, while in winter the largest increases are in New England. In contrast, the Canadian model projects little change in precipitation or small regional decreases approaching 5-10%. The largest decreases are in the Mid-Atlantic region, during both winter and summer. Only small regions have a projected increase in precipitation. The variability in precipitation in the Northeast is projected to increase in both model projections (slightly in winter and substantially in summer in the Canadian model and substantially in both winter and summer in the Hadley model).

Severe storms are a major issue in the Northeast throughout the year. Given the spatial resolution of global climate models, neither thunderstorm activity nor hurricanes are simulated by the models. An analysis of sea-level pressure patterns in the Hadley and Canadian models, which provides some indication of the path of hurricanes if they form, suggests little reason to expect changes in the average track of hurricanes over the 21st century. Changes in frequency and intensity of hurricanes under future climate conditions remains a topic of considerable debate. Global climate models are capable of resolving and simulating mid-latitude cyclones responsible for winter storms, although the simulations are imperfect. For example, Hayden (1999) indicates that the Atlantic coast storm track in the Hadley model is displaced offshore in simulations of current conditions. In the climate projections the Northeast continues to be a location of winter storms, as an analysis of the winter storm variability and locations does not shift over the 21st century. However, an analysis of storm counts and intensities in the Hadley and Canadian climate scenarios yields some differences. The Canadian model produces decreased counts over much of the eastern seaboard, with the exception of small increases in parts of the Mid-Atlantic region. In contrast, the Hadley model indicates an increase over the coastal region with slightly stronger storms. The differences in these two scenarios reflect the position of the jet stream. The more zonal jet stream in the future simulated by the Canadian model would mean fewer cold-air outbreaks in the Northeast. The more north-south jet stream in the Hadley model results in an increase in east-coast storms. Carnell et al. (1996) and Carnell and Senior (1998) describe results from a storm analysis. Storm tracks in the Atlantic are weakened, but there is a statistically significant increase in storm counts across the Mid-Atlantic region. Further, they find a shift toward deeper low-pressure systems, and hence stronger storms. The

Changes in Storm Tracks

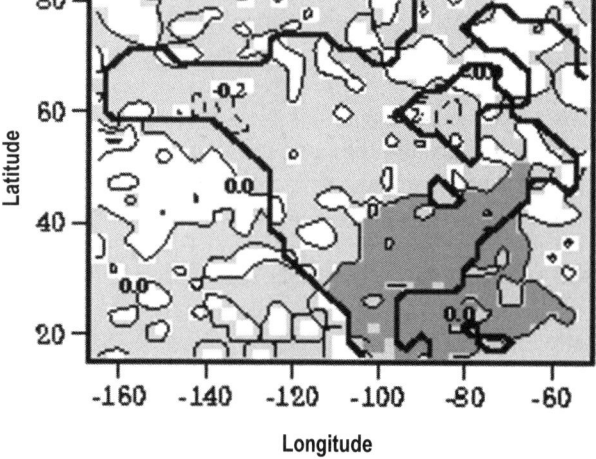

Figure 1. A storm track analysis from the Hadley climate model scenario projects a slightly strengthened wintertime storm track through the Northeast, in the 2020s, because the jet stream has a more north-to-south position along the East Coast. This scenario projects a slightly stronger winter storm area (dark shaded region). The Canadian climate scenario has a more east-west jet, and in general indicates slightly weaker storminess.

strength of the storms is dependent on two factors. First, as the continental regions become warmer in winter in the future, the decreased temperature contrast between land and sea in the region tends to produce weaker storms. However, increased heating associated with higher atmospheric water vapor tends to counteract this effect to produce deeper low-pressure storms.

The Palmer Drought Severity Index derived from the model simulations provides two very different pictures for the Northeast. The Hadley model projects less drought tendencies in the Northeast while the Canadian model projects tendencies for severe drought to increase over the 21st century. This result follows from the precipitation and temperature projections in the two models, with the Canadian model projecting larger temperature increases but with smaller precipitation changes. Increased evaporation with the warmer temperatures yields a greater drought tendency. In contrast, the Hadley model projects a smaller warming, and with regional precipitation increases of as much as 24% by 2100.

For perspective, the specific conditions projected by the climate models for the end of the 21st century can be matched with areas in the US that currently experience these specific conditions. Such a comparison provides perspective on the magnitude and nature of the projected climate changes. For summer temperature and precipitation, the Hadley model projects that New York State will have summer conditions similar to present day Maryland and southern Pennsylvania by the end of the 21st century. In contrast, the higher temperatures and smaller change in precipitation found in the Canadian model yield a summer climate regime for NY by the end of the 21st century that is closer to present day central Illinois or Missouri.

ECOLOGICAL PERSPECTIVE

A mosaic of farmland and forest covers much of the landscape in the Northeast. The southern part of the region (including the Appalachian Mountains of West Virginia, Pennsylvania, and Maryland) is characterized by higher percentages of corn, cotton, soy, and grasses with forests, than the rest of the region. These forests include oak, oak-hickory-pine, mixed hardwoods, mixed pines, maple-ash-beech, and limited spruce-fir forests. The most northern states of the region exhibit northern hardwoods, red spruce-balsam fir, white pine-hemlock, oak, hickory, oak-pine, elm-red maple, and cool mountain spruce-fir

Palmer Drought Severity Index Change

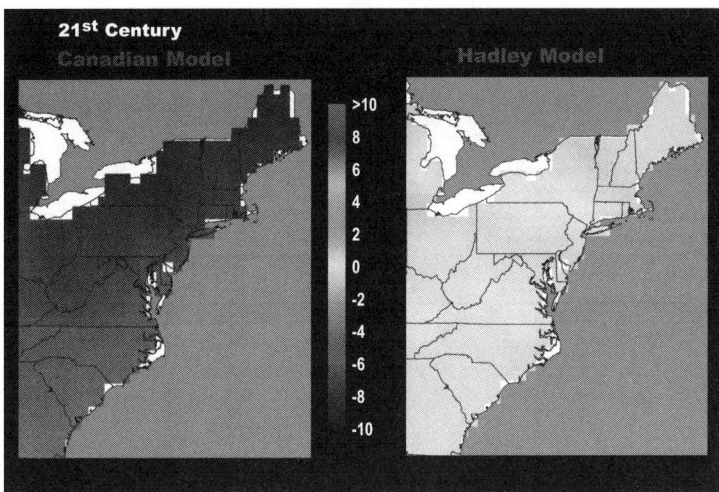

Figure 2. The projected trends in the Palmer Drought Severity Index (PDSI) are dependent on the projections of temperature and precipitation. Large increases in drought tendencies occur in the Northeast in the Canadian model associated with substantial warming and small changes in precipitation. In contrast, the Hadley model yields larger increases in precipitation and a more modest warming, conditions under which the drought tendency tends to decline. See Color Plate Appendix.

forests. Human activities converted significant fractions of forested land to agriculture. However, since 1900 there has been extensive abandonment of agricultural land, and land covered by forest has increased significantly. One exception is in urban areas where significant conversion of land to human use continues. The forests of the Northeast are still about 70% of the level of the 1600s if USDA (1998) data are compared with early analyses (see Forest sector foundation chapter). Forest area has been relatively stable over much of the region during the last few decades, although biomass increased with the maturity of second growth forests (Powell et al., 1996). Estimates of the percentage of forest cover in the region range from 34% in Delaware to 88% in Maine and New Hampshire (Klopatek et al., 1979).

Moderate species richness characterizes the northern half of the Northeast region, while the southern half exhibits moderately-high to high richness (Ricketts et al., 1999). For example, tree species richness in the Northeast region ranges from about 60 species in northern Maine to more than 140 species in parts of Pennsylvania, West Virginia, and Maryland (Currie, 1991). Reptile species richness varies from fewer than 10 species in northern Maine to about 40 in Maryland and Delaware, while amphibian species richness ranges from fewer than 20 to more than 30 species per state.

Ecosystem Models

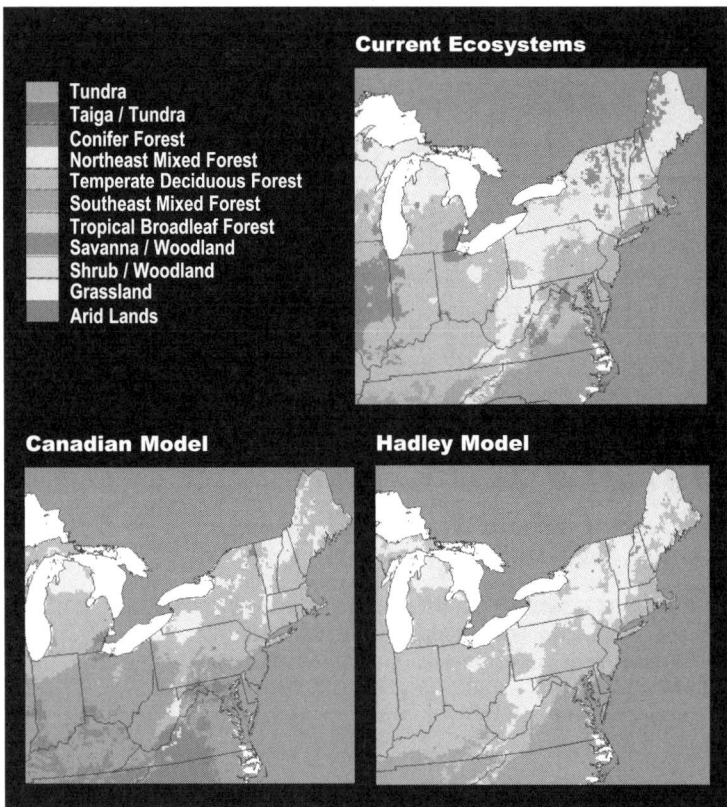

Figure 3. The projected changes in vegetation character using output of the Canadian (a) and the Hadley (b) models indicates a substantial northward shift in the vegetation types. These changes are significantly larger in the Canadian model scenario, which projects a greater warming trend with little change or a decrease in precipitation. Based on the model of Neilson and Drapek (1998). See Color Plate Appendix.

The Vegetation/Ecosystem Modeling and Analysis Project (VEMAP) provides the basis for an assessment of changes in vegetation cover (VEMAP members, 1995; Kittel et al., 1997) and primary productivity. The specific forest character is also projected using the MAPSS vegetation model (Neilson and Drapek, 1998). Climate projections from the Canadian model yield substantial changes in the nature of the forests in the Northeast. The conifer forest of northern New England and much of the northeast mixed forest of New England, New York and Western Pennsylvania are projected to change to a temperate deciduous forest similar to southeastern Pennsylvania, Maryland, and northern Virginia today. The area of southeast mixed forest, today characteristic of the region south of Virginia, would become compressed into a small area of West Virginia, southern Pennsylvania, and the coastal plain of Virginia, Delaware, and New Jersey, while much of Virginia would become savanna/woodland.

The changes based on the Hadley model are less dramatic but still noteworthy. The conifer forest of northern New England is replaced by northeast mixed forest. The area of temperate deciduous forest in New England and Pennsylvania/West Virginia grows slightly. The area of southeast mixed forest grows in Virginia. The differences in vegetation projections are a strong reflection of the differences in moisture and temperature projections in the two climate model scenarios.

KEY ISSUES

Regional perspectives on the potential impacts of climate change are naturally influenced by personal experience related to historical weather and climate and their associated impacts. The key issues also reflect perceptions of current stresses and problems. In some cases, the issues are not directly related to climate change, but are likely to be exacerbated by climate change.

Four key issues are identified that are of major importance for the Northeastern US:

- Vulnerability to changes in extreme weather events;
- The compounding of climate change with other stresses for important ecosystems such as the Chesapeake Bay and other bays and estuaries;
- The impact of climate change on major urban environments; and
- The potential changes in recreation due to climate change.

Two additional issues are also noteworthy:

- Significant change in the character of forests in the Northeast; and
- Limited vulnerability in the agricultural sector

1. Vulnerability to Changes in Extreme Weather Events

The fact that the Northeast is prone to natural weather disasters and weather extremes figures strongly in the specific examples identified by stakeholders in each area of the Northeast. The reasoning is clear. Severe weather presents threats to both safety and property. During the period 1950-1989, storms caused more than $12 billion in damages in the Northeast (Changnon and Changnon, 1992). Twenty-two events each caused more than $200 million in damages, with hurricanes causing the most damage, followed by thunderstorms, winter storms, and wind. A 1996 blizzard in the Mid-Atlantic and New England region, followed by flooding, caused an estimated $3 billion in damages and 187 deaths according to the National Oceanic and Atmospheric Administration. Changnon and Changnon (1992) and Agee (1991) note some correlation between higher historical temperatures and increased cyclonic and anticyclonic activity when five-year averages from 1950 to 1989 are analyzed. The strongest relationships were with thunderstorm activity and winter storms. The Northeast had relatively few "weather disasters" during the cooler 1960s, followed by increased numbers of events during the warming trend that followed. However, changes in severe weather are widely regarded as one of the most uncertain aspects of future climate projection (USGCRP, 1995; Barron, 1995). Further, the nature of these events is likely to change significantly as climate evolves over the 21st century. In other words, some of the tendencies for changes in severe weather are likely to be short term or transient features of a changing climate. Northeast stakeholders tend to focus attention on historical events with significant impacts, including ice storms, severe flooding, nor'easters, hurricanes and other tropical storms, and severe or persistent drought.

Ice storms. The ice storm of January 1998 caused substantial environmental, economic, and societal damage. This series of devastating ice storms hit northern New York and New England, along with portions of southeastern Canada, causing extensive damage to forests and energy and transportation infrastructure, as well as impacting human health. The magnitude of the storm in terms of measurements of the number of hours of persistent freezing precipitation was unprecedented (DeGaetano, 2000). While a number of significant ice storm events have occurred over the 20th century, the extensive area of impact (37 counties were declared Federal disaster areas) was also unusual.

A conservative estimate of the damage approaches 1 billion dollars for the US, with insured losses exceeding 200 million dollars (DeGaetano, 2000). Many people across the region were without power for up to three weeks in mid-winter. Seventeen deaths occurred, primarily associated with carbon monoxide poisoning and hypothermia associated with the power failure. In Maine, 70% of the state's population of 1.2 million people were without power for at least some period of time. Over the entire region (portions of New York, Vermont, New Hampshire, and Maine) approximately 1.5 million

Figure 4. The Northeast is prone to a wide variety of natural weather disasters and weather extremes including the 1998 ice storm illustrated.

people were without electricity for up to three weeks. In addition, nearly 17 million acres of rural forests and urban trees were affected, with five million acres classified as severely damaged. Hardwood species were most heavily damaged with trees bent and limbs and branches broken under the weight of the ice coating. The longer-term ecological impacts of severe tree damage from the storm are not yet clear, especially as the ice storm was followed by the 1998-1999 drought.

Although the 1998 storm was extreme in terms of persistence and extent, other severe events have occurred in the 20th century. Changes in the frequency, intensity, or path of ice storms are not evident in the historical record. A primary concern is the potential for such storms to become more frequent. Ice storms occur when warm moist air masses are uplifted over cold polar air masses or move over cold surfaces. Such conditions will possibly become more common if the milder winters projected by the climate models increase the frequency of northern displacement of warm moist air masses as occurred in 1998. However, these effects, if they do occur, are likely to be transitory. In the Canadian scenario, winter precipitation decreases over much of the region and minimum winter temperatures eventually increase significantly (by 7 to more than 10°F above present values), reducing the occurrence of subfreezing temperatures.

Severe flooding. The impacts of severe flooding (such as occurred during tropical storm Agnes in 1972 and Floyd in 1999) are amply demonstrated by the historical record from the Northeast. Records of severe floods reveal a diverse set of responsible meteorological conditions, including:

117

- rapid melting of snow with warming events following a major nor'easter;
- spring snow melt following heavy winter snowfall;
- heavy rainfall (as opposed to snow) as warm air masses move over a frozen ground that limits percolation and drainage;
- major summer thunderstorm systems; and
- major precipitation events associated with hurricanes or tropical depressions.

The frequency and occurrence of future flooding in the Northeast will depend on how this diverse set of meteorological conditions changes. Several elements of the historical and model-derived future climate projections raise flooding as an increased concern. These elements are:

- the historical trends that illustrate increases in extreme precipitation events through the latter half of the 20th century (Groisman et al., 2000),
- the Hadley model's tendency to simulate wetter conditions in summer and winter, and
- the uncertainties associated with projecting how the intensity and frequency of major hurricanes may change in the future.

Other elements of model-derived future climate projections suggest that winter and spring flooding increases are possibly transient in nature, or are likely to decline in the future. These elements are:

- the Canadian model's tendency to simulate drier conditions in summer and winter,
- the model simulations indicating milder winters and hence the potential for northward movement of warm air masses in winter producing rainfall over frozen ground as the climate warms, then with decreased flooding as continued warming substantially reduces the length of time the ground is frozen, and
- the potential for warming events during winter with associated higher snowfall creating rapid melting periods as a transient effect as the climate warms, followed by decreased snowmelt events as the climate continues to warm and precipitation tends to fall increasingly as rain.

Nationally, annual flood damages increased steadily over the period 1903 to 1997, and flood-related fatalities have been high since the 1970s (Kunkel et al., 1999b). Although societal growth is certainly a factor in the increase in flood damages, it is an insufficient explanation. More heavy precipitation events (Karl and Knight, 1998) have also been suggested as a factor in this increase. Since 1983, flood damages

in the river-rich Mid-Atlantic states total 4.7 billion dollars (US Army Corp of Engineers, 1998). Flooding also disrupts water supplies and is a significant health risk (Solley et al., 1998; Yarnal et al., 1997). Several water-borne diseases present risks even in wealthier countries when flood waters compromise water systems. These include viruses (e.g., rotovirus), and bacteria-borne (e.g., *Salmonella*) or protozoan-borne (e.g., *Giardia* and *Cryptosporidium*) diseases.

Nor'easters. Major nor'easters produce significant precipitation accumulations and cause significant coastal damage, in terms of beach erosion and structural damage, and thus are of major interest to the Northeast. This is particularly true because five states of the Northeast (New York, Massachusetts, Connecticut, New Jersey, and Maryland, in respective order) represent five of the top six states along the Atlantic and Gulf coasts in terms of value of insured coastal property (Insurance Research Council, 1995). The climate model projections are divergent with regard to nor'easters. The shift to deeper winter cyclones in the western North Atlantic with stronger winds for doubled carbon dioxide concentrations found by Carnell et al. (1996) indicates the potential for increased property damage. In contrast, Stephenson and Held (1993) found little change in the North Atlantic using the NOAA Geophysical Fluid Dynamics Laboratory model, and thus future increases in storm damages would more reflect development of coastal property rather than climate change.

The climate models used in this Assessment also provide different scenarios. The Mid-Atlantic region is the only area of the east coast in which both the Canadian and the Hadley climate models indicate slight increases in the frequency and intensity of winter storms with little change in storm track. The increases are more significant in the Hadley model with the north-south shift of the jet stream under future carbon dioxide conditions, while the Canadian model suggests decreases in storm counts with the exception of the Mid-Atlantic region.

A significant hazard to coastal areas stems from changes in flood levels superimposed on a more gradual rise in sea level. Return periods of coastal flood events will shorten considerably, even in the absence of any change in storm climatology. For example, by 2100, the 100-year flood event (the flood height that occurs on average once every 100 years) in New York City is likely to occur much more frequently (e.g., every 19 years in the Hadley model scenario) because of the sea-level increase (Rosenzweig and Solecki et al., 2000).

Interestingly, many of the severe winter weather conditions predicted for the Northeast may seem counter-intuitive. For example, the Great Lakes are very likely to experience decreased ice cover or a shorter season of ice cover with climate warming, yet a transient increase in the frequency and intensity of lake effect snows is possible. Specifically, the lack of ice cover allows increased lake effect snows as cold polar air masses move southward. Hence, the lake effect snows in areas such as Buffalo, NY could break records even though the climate warms. This result depends on the nature of cold air outbreaks that initially may not be substantially different from today. The effect is likely to be temporary or transient however, because as climate warms substantially, the precipitation increasingly falls as rain according to both climate model scenarios.

Hurricanes. Major tropical storms also remain a significant concern for several reasons:

- Hurricanes moving inland across the southern states have produced historically high precipitation extremes in the Northeast, and both wind and rainfall damage in New England and Long Island. Hurricanes rank first in terms of severe weather damage in the Northeast (Changnon and Changnon, 1992).
- Considerable debate is on-going as to whether hurricane intensity may increase with warmer tropical and extra-tropical temperatures, particularly resulting in greater precipitation, higher winds, or both (see for example Emanuel, 1988; Idso et al., 1990; Lighthill et al., 1994; Bengtsson et al., 1996; Knutsen et al., 1998).
- The debate about potential increases in the frequency of Atlantic hurricanes is also tied to whether warming will result in an increase or decrease in the tendency for El Niño-like conditions. Historically, during El Niño events, the probability of US land-falling hurricanes is reduced, while during La Niña events, the probability of US land-falling hurricanes increases (Bove et al., 1998). One modeling study, by Timmermann et al. (1999), suggests more frequent El Niño-like conditions and stronger La Niña events as climate warms.
- East Coast population growth substantially increases the potential health effects and property damage associated with hurricane events.

In 1995, the Insurance Research Council estimated that a category 4 hurricane making land-fall in Asbury Park, NJ, New York City, or Long Island had the potential to cause insurance losses of $40 to $52 billion. Conservatively, total damages can easily exceed twice the level of insured damages. In short, the potential for hurricane damage in the Northeast from a single storm far exceeds the region's total damages from hurricanes over the last 40 years. Much of the vulnerability stems from a remarkable increase in coastal property values. A comparison of storm intensity, frequency, and damages during this century (Kunkel et al., 1999b) indicate that large increases in damages are associated with increasing value of property exposed to weather risk. For example, the value of New York insured property doubled from 1988 to 1993. Unfortunately, hurricanes' spatial scales prevent their simulation as features of most global climate models. The Assessment climate models do not indicate any systematic change in the steering forces that might govern the path of future hurricanes compared to the present. Potential changes in intensity and frequency remain highly uncertain.

Drought. The differences in precipitation and temperature projected for the Northeast indicate substantial difficulty in determining how drought tendencies may change in the future. There are significant reasons for concern, but also the potential for little change from present conditions. Although the Northeast is on average "water-rich" in comparison with precipitation levels for the rest of the nation, drought is a significant concern for three reasons:

- Six of the last 20 years were characterized by drought in some part of the region and even a single year drought can result in water restrictions in many counties.
- The increased warming associated with smaller precipitation changes in the Canadian model provides a scenario for the Northeast characterized by a strong tendency toward frequent extreme droughts.
- The lack of water storage in the Northeast is a significant factor in creating vulnerability. In addition, the regions water withdrawals are highly dependent on surface flow. For example, in the Mid-Atlantic region 95% of the water withdrawals are from surface flows (Neff, 2000). Drought is a frequently cited potential impact of global warming because of increased evaporation rates associated with rising air temperatures.

In contrast, drought tendencies could change very little or decrease given that:

- The Hadley model, with a smaller temperature increase and increases in precipitation, yields a tendency toward neutral changes in drought

119

Potential Changes In Severe Weather

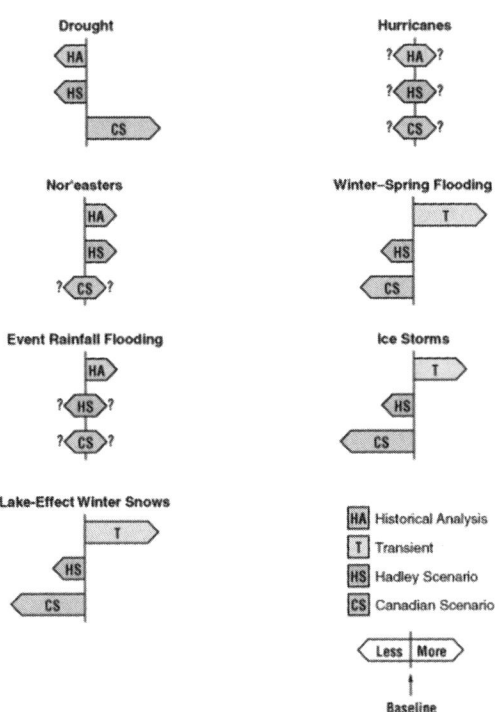

Figure 5. Schematic of the potential changes in severe weather for the Northeast based on historical data (H), the Hadley model scenario (HS), the Canadian model scenario (CS) or an assessment of possible transient effects (T). See Color Plate Appendix.

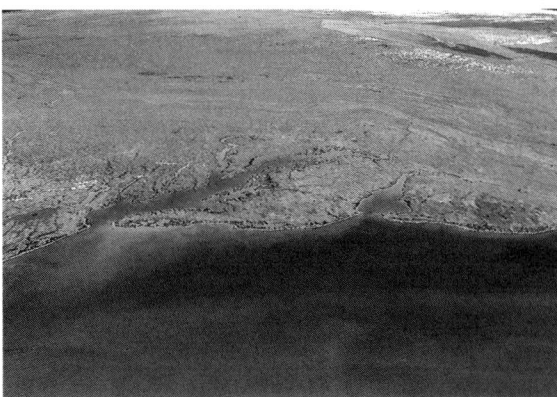

Figure 6. The coastal regions of the Northeast are dominated by extensive estuaries and bays.

severity or slightly decreased drought tendencies,

• Historical analyses indicate that the extent of area experiencing drought in the Northeast declined somewhat.

The Northeast is currently prone to weather-related natural disasters. However, historical analysis and climate model projections present a range of possibilities, including the potential that such weather

disasters could increase in both summer and winter. The historical cases of large-scale damages associated with these events even under current climate conditions add a perspective of significant vulnerability. Many human structures are designed based on historical climate records. If these structures are already vulnerable, then this argues for adaptation strategies that focus on "over-designing" critical structures to add margins of safety and more frequent design-criteria review based on updated climate projections. The potential for changes in frequency, path, and intensity of hurricanes, and in the nature of severe winter storms, becomes a key uncertainty in assessing climate impacts. Coping with substantial increases in severe storms, or even repeat of historical events coupled with higher sea level, might necessitate relocation of infrastructure away from high-risk zones. Historically, negative economic impacts of severe weather on forestry and agriculture resulted in different planting and harvesting methods. The impact of changes in severe weather on ecosystems is a significant unknown.

2. Compounding Stresses on Major Estuaries and Bays

Major estuaries and bays, such as the Long Island Sound, Delaware Bay, and the Chesapeake Bay, characterize the coastal regions of the Northeast. These coastal embayments represent unique ecosystems and unique resources for fisheries and recreation (Fisher et al., 2000). Importantly, these geographically defined features are unable to "migrate" in response to climate change. In addition, growth in coastal populations has added substantial stresses to these environments. The Chesapeake Bay, the nation's largest estuary, is a key example.

Chesapeake Bay is characterized by multiple stresses with significant combined impact on water characteristics and ecosystems (Funderbunk et al., 1991; Kearney and Stevenson, 1991; Drake et al., 1996; Perry and Deller, 1995; 1996; Jones et al., 1997; Abler and Shortle, 2000; Walker et al., 2000). Human land-use practices and upstream industry have a marked impact on water quality and pollutant levels. High nutrient and particulate loading from agricultural and urban runoff and air-borne pollutants reduces oxygen levels, which reduces productivity and organism habitat area. High nutrient loading is very likely to increase algal blooms, shading deeper water and limiting submerged aquatic vegetation. Increased land use and growth in the area covered by impervious surfaces has resulted in "flashier" streams, meaning that any rainfall causes a rapid peak flow response. In naturally vegetated areas such as forests, a significant fraction of the rain percolates into the ground. As the soil becomes saturated, more water flows into the streams. It thus takes some time for rainfall to cause stream levels to rise. When land uses cover much of the ground with streets, buildings, and parking lots, less rainfall can percolate into the ground; thus it runs off quickly, causing the streams to "flash." Increased population pressure has also altered the boundary between the land environment and the Bay, providing fixed or "hardened" margins and decreasing the area of wetlands.

Climate change adds an additional stress. The most important influences reflect the potential to change water temperatures and freshwater inputs (hence salinity). Warming of the atmosphere is very likely to have a direct impact on Bay temperatures. Changes in the frequency and intensity of precipitation events, changes in frequency and strength of hurricanes, and any change in the strength and frequency of droughts would substantially influence freshwater inputs. With "flashier" streams, any increase in high precipitation events is very likely to have a marked impact. These changes would then influence salinity in the Bay and stratification of the water mass. Sea-level rise is likely to contribute to these changes. Local rates of sea-level rise in the Chesapeake (close to 4mm/year, or 0.16 inches per year yielding 16 inches in 100 years) are anomalously high for the Atlantic Coast, due to regional subsidence and other factors (Gornitz, 1999). The global average increase in sea level is closer to 1.2mm/year (0.048 inches per year) according to Gornitz and Lebedoff (1987). Both temperature and salinity are significant environmental controls on organism character, influencing fish populations, productivity, and human and ecosystem health. For example, cholera bacteria are present in the Chesapeake Bay. Increased cholera risk is associated with rising water temperatures, however water and waste treatment practices should prevent US epidemics (Colwell et al., 1998).

Gibson (1999) and Najjar et al. (2000) used a water balance model to project a 24% increase in runoff in the Susquehanna River Basin under the Hadley model scenario and a 4% decrease for the Canadian model scenario. Gibson and Najjar (2000) then analyzed changes in Chesapeake salinity as a function of these changes in runoff. The increased runoff in the Hadley model resulted in a 20% decrease in the surface salinity within the northern segment of the Chesapeake, with as much as a 4% change penetrating to deeper waters within the southern segment of the Bay. The Canadian model scenario results in changes of 3% or less.

The potential for significant changes in temperature

and salinity raises several important concerns about species composition in the Chesapeake and other bays. Any significant changes in salinity and temperature are very likely to result in the migration or loss of key species. The introduction of opportunistic invasive species during changing conditions is also very likely to change predator-prey relationships influencing the character of ecosystems, or result in elimination of key species that may not be vulnerable to the direct effects of climate change. This type of indirect change is already evident in Chesapeake waterfowl populations in which declines of submerged aquatic vegetation, attributed to excessive nutrients and sedimentation, are associated with dramatic declines in some species of waterfowl unable to adapt to changing food sources (Perry and Uhler, 1988). The Chesapeake and Delaware Bays are stopover points for millions of bird species. Numerous waterfowl (the eight most dominant species have been numbered at 700,000) winter in the Chesapeake. If affected species have economic significance or if they are connected to the uniqueness of a region, the impact is very likely to be significant.

Sea-level change substantially influences wetlands through inundation, saltwater intrusion into fresh and brackish marshes, and erosion. Marshes are already estimated to have lost one-third of their original area.

Percent Salinity Change in the Chesapeake Bay

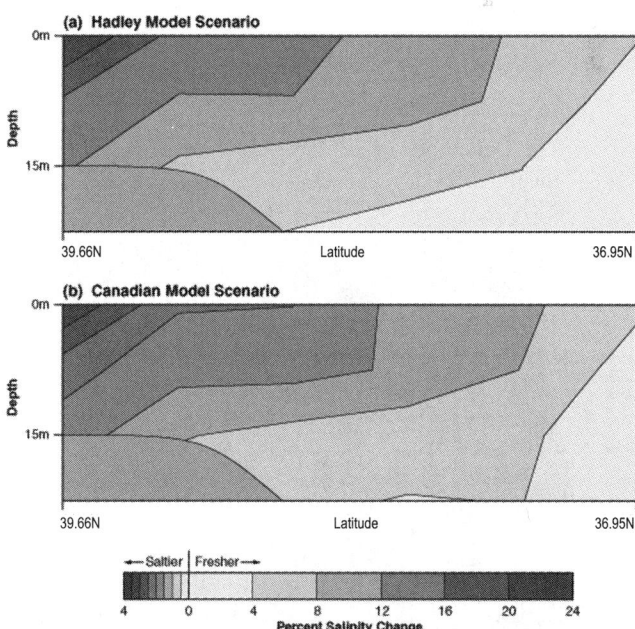

Figure 7. Calculated salinity within the Chesapeake given the runoff calculated from the Hadley (top) and Canadian (bottom) climate scenarios by Gibson and Najjar (2000). The distribution of salinities ranges from the upper reaches of the Bay (39.66N) to the Lower Chesapeake near its Atlantic opening (36.95N). See Color Plate Appendix.

Sea-level rise can also submerge protective barrier islands or cause them to retreat landward onto marsh and lagoonal areas. The hardening of the bay boundaries (through construction of impoundments, retaining walls, dockage, etc.) adds an additional limitation to the landward movement of wetlands as sea level rises. Marshes and lagoons are significant stopover habitats for migratory birds.

Adaptation strategies are governed by three factors. First, estuaries and bays are characterized by a number of compounding stresses. Second, there are limited avenues for protecting critical ecosystems that are geographically fixed and therefore cannot "migrate." Third, the uncertainties in projecting the water balance for the Northeast region under future climate conditions results in substantial uncertainty in determining the future water properties of the Bay and other estuaries in the region. One of the few adaptation strategies available may be to limit the non-climate stresses on the region in order to minimize any climatic impacts. This may argue for greater control of land-use at the boundaries of estuaries and increased concern over nutrient and pollutant fluxes into estuaries and bays. Humans are also gaining some experience in constructing artificial wetlands. However, we know little about the long-term viability of these constructed wetlands.

3. Multiple Stresses on Major Urban Areas

The major urban areas of the Northeast are stressed even without climate change (Rosenzweig and Solecki et al., 2000). In the Northeast, there have been major investments in the infrastructures of system elements such as roads, water supply, communication, energy delivery, and waste disposal. Still, for many large urban areas this infrastructure is characterized by aging, problems with under-capacity, overuse, and deferred maintenance. Cities are associated with a host of continuing problems, some related to climate like air quality, and others that are non-climate related such as crime and poverty. The complex web of institutional relationships among communities, municipalities, regions, states, the federal government, and the public, private, and non-profit sectors often make consideration of overarching issues, such as the environment, problematic. Scenarios from climate models suggest that climate change intersects with a significant number of other stresses, all with implications for overall quality of life.

Sea Level and Storm Surge. One of the more significant potential climate change impacts in urban coastal communities is rising sea level (Bloomfield, 1999) and elevated storm surge levels. For example, historical events suggest that the metropolitan areas are particularly vulnerable. By using daily tide gauge

Nor'easter of December 1992

The December 11-12, 1992 nor'easter produced some of the worst flooding and strongest winds on record for the area. It resulted in a near shutdown of the New York metropolitan transportation system and evacuation of many seaside communities in New Jersey and Long Island. This storm should have provided a "wake-up" call, heralding the vulnerability of the transportation system to major nor'easters and hurricanes. Had flood levels been only 1 to 2 feet above the actual high water level of 8.5-foot above mean sea level, massive inundation of rail and subway tunnels could have resulted in loss of life. With rising sea levels, even a weaker storm would produce comparable damage. While hurricanes are much less frequent than nor'easters in this area, they can be even more destructive because the geometry of the New Jersey and Long Island coasts amplifies surge levels toward the New York City harbor. For a worst-case scenario category 3 hurricane, surge levels could rise 25 feet above mean sea level at JFK airport and 21 feet at the Lincoln tunnel entrance.

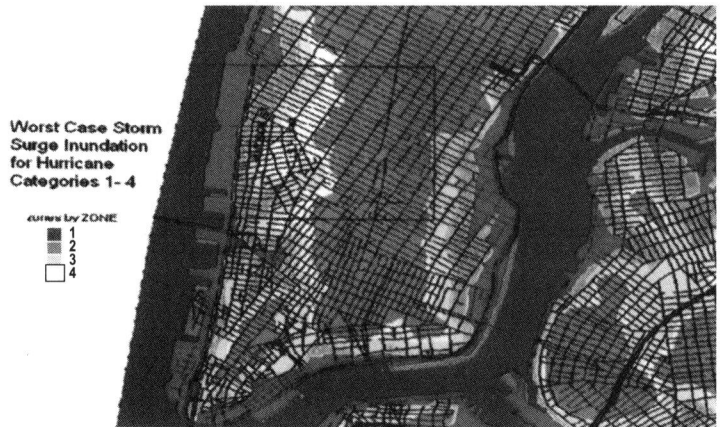

Worst Case Storm Surge Inundation for Hurricane Categories 1-4

curves by ZONE
1
2
3
4

Figure 8. Vulnerable coastal areas for Manhattan, based on a 20 foot high flooding zone for the year 2100, derived by Klaus Jacob of Lamont-Doherty Earth Observatory for the Metroeast workshop. See Color Plate Appendix.

data and statistics on extreme events (Ebersole, 1982), the Metro-East Assessment team (Rosenzweig and Solecki et al., 2000) calculated a range of scenarios delineating significant coastal vulnerability, particularly in urban areas. Critical points for flooding for many of the region's vital transportation systems (including airports, subways, highways, and major road and railroad tunnels) are located at elevations between 7 and 20 feet above current sea level and are very likely to be inundated by coastal storm surges with estimated recurrence periods of about 100 years. Taking into account models of sea-level rise (IPCC, 1996) of 9 inches to 3 feet (23 to 96 cm) in the next 100 years, these current recurrence periods are likely to be shortened by factors of 3 to 10 before the year 2100.

Water Supply. Water supply systems in the major northeastern cities also exhibit substantial vulnerability to climate change. For example, the New York City water system is large and relatively inflexible in terms of demands on the system, and therefore susceptible to large changes in the water balance. A few key examples illustrate the nature of the problem. New York City's water supply is derived from water collected from a 2,000 square mile area, stored in three upland reservoir systems. This is an ecosystem service that the City has secured by making a capital investment of about $1 billion in the surrounding communities. The City's water supply is sensitive to climate variability and change given the demands on the current system, and its dependence on annual precipitation levels. Even with increased winter precipitation, rapid run-off (rather than accumulation as snow) has resulted in significant regional water supply problems in recent decades (McCabe and Ayers, 1989). The climate stresses from current conditions are already evident. The climate vulnerability is compounded by two factors. First, the upstate communities have experienced substantial growth. Second, these upstate communities have a legal right to water in times of low supply. In addition, New York City has a legal obligation to provide water to the Delaware Basin because it has access to water in the river headlands. In the Delaware Basin, additional water releases from the reservoir systems might be required if sea-level rise advances the salt water front up the Delaware and Hudson rivers (Alpern, 1996). A prolonged drought is likely to force the city to seek alternative water sources.

Soil Moisture. The combination of future precipitation changes and an increase in evaporation associated with higher future temperatures is expected

to produce a decrease in summer soil moisture (Broccoli, 1996). The climate model scenarios used in this Assessment both project little trend in soil moisture in the region immediately adjacent to the city. However, significant climate change is likely to result in the need for new large-scale investments in order to replace the current ecosystem service from the growing surrounding communities.

Heat-related Illness and Death. Warmer summers are likely to be associated with higher maximum temperatures and then with increased heat-related morbidity and mortality in major cities of the Northeast (Kalkstein and Greene, 1997; Chestnut et al., 1998; Kilbourne et al., 1982). By 2090, increases in maximum temperatures from 1-2°F to 5°F are projected for the coastal Northeast. An associated increase from the current 13 days to a projected 16-32 days above 90°F might result in a five-fold increase in heat-related mortality in New York City according to the Metroeast regional assessment, if no adaptation occurs. Kalkstein and Greene (1997) and Kalkstein and Swift (1998) utilized three different climate models to examine winter and summer mortality for 2020 and 2050. In all three models, increases in summer mortality exceeded decreases in winter mortality for Baltimore, Philadelphia, Pittsburgh, and Washington D.C.

Air Pollution. Higher summer temperatures also increase photochemical reaction rates leading to an increase in ground-level ozone (smog) and other pollutants. The New York City area already has one of the nation's highest rates of respiratory disease associated with airborne pollutants. This is likely to be exacerbated by increases in the urban heat island

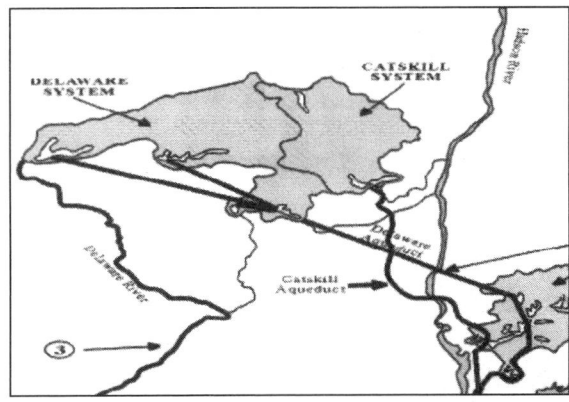

Figure 9. Water is one of the major ecosystem services provided to New York City by the surrounding land areas (Metroeast Workshop). Population growth in the region, and legal rights to the water by upstate communities, place the adequacy of the water supply at greater risk under some scenarios of future climate change.

effect and conditions of persistent elevated summer temperatures.

In summary, the major urban areas of the Northeast are characterized by multiple stresses. Climate change has the potential to exacerbate many of these problems. The most significant climate-related stresses in the urban regions of the Northeast are increased heat mortality, greater ozone pollution associated with higher temperatures, infrastructure risk due to higher sea level, water availability problems associated with significant changes in regional water balance, and increased vulnerability to severe storms.

The addition of significant multiple stresses to the urban environment argues for a variety of adaptation strategies. Some of the direct health effects of heat can be mitigated with active warning systems (heat alerts, opening of shelters, spraying water on dark building tops) such as those currently in place in Philadelphia, structural adaptations (construction that promotes air-flow, reduction in area of black roofing, etc.) and "cool community" measures such as increased planting of trees. Some of the projected mortality studies do not reflect potential changes in air conditioning use or other adaptations, which offer substantial protection against heat waves.

Figure 10. Outdoor recreation is of major economic importance in the Northeast, and it is tightly coupled to climatic conditions.

Indirect air pollution problems can be addressed through stricter controls on pollutants and ozone precursors. Construction of dike systems or relocation of critical infrastructure can reduce some of the vulnerability to sea-level rise.

Significant financial investment would be required to produce an urban system that is more robust under conditions of higher storm surges or increased tendency toward drought. Increased flooding and a significant increase in severe storms would argue for establishment of set back zones, re-zoning, buyouts of high risk areas, relocation of structures at risk, or altered management. Investing in new reservoirs, limiting growth in water source regions, and increasing water costs to promote conservation are the primary coping strategies for water-related problems, but major reservoir construction projects have associated environmental consequences. A more comprehensive regional water management strategy with increased emphasis on safe water supplies could substantially mitigate major water supply and quality problems.

4. Recreation Shifts

Changes in warmth and the seasonal characteristics of precipitation are likely to have substantial impacts on recreation in the Northeast. In particular, this has been identified as a key issue for New England (Rock and Moore, 2000). These impacts differ widely with the type of recreation and season. The winter ski industry is particularly vulnerable. Increases in minimum nighttime temperatures, periodic warm spells, and increased occurrence of winter precipitation as rain are likely to limit the ability of ski areas to maintain adequate snow pack. Current skiing locations with marginal climate characteristics are likely to become untenable.

The recent tendency for mild winters in the Northeast, coupled with climate model projections indicating significant increases in winter minimum temperatures, points to possible benefits. Mild winters and extended periods of warmth in fall and spring may encourage new recreational activities in the forested mountains of the region.

Higher sea level coupled with increased winter storms is likely to result in loss of beachfront property and destruction of barrier islands, decreasing the opportunities for beach recreation during warm months. In highly populated areas, prime recreational beaches can probably be maintained by more frequent episodes of beach nourishment, but costs could increase substantially (Valverde et al., 1999).

Wet springs and mild winters are likely to lead to increased populations of insects and high pollen counts. Increases in disease-bearing vectors, such as mosquitoes and ticks, add increased health risks. For example, mosquito populations are tightly connected to minimum temperature characteristics and water availability. Increases in pest populations are likely to adversely impact recreation. The climate model projections offer different scenarios; the increased drought frequency in the Canadian model yields a different pest population than the warmer and wetter projection from the Hadley model.

The New England region currently experiences summer air quality and ozone pollution problems, both of which are exacerbated by warming. Ground-level ozone across New England may reach unhealthful levels during summer months, especially during humid periods when maximum temperatures exceed 90°F. The combination of high temperatures and full sunlight, coupled with nitrogen oxides generated primarily by automobile traffic, and volatile organic compounds from both natural and human sources, result in elevated levels of ground-level ozone, a form of air pollution often called smog. Exposure to elevated levels of ozone has a negative impact on both forest health and human health. Due to the topographic variability typical of New England and upstate New York, and the fact that the region is typically downwind from major urban centers, high-elevation areas (above 3,000 feet) are likely to have unhealthful levels of ozone. The prospect of air quality alerts for hikers in the mountains of New England is likely to have a negative effect on tourism.

In contrast, typical summer recreational activities involving beaches or freshwater reservoirs are very likely to experience extended seasons. However, sea-level rise is projected to lead to increased beach erosion and loss due to the impact of both storm surges and permanent inundation. Due to the large amount of human development abutting beach areas, establishment of new beaches further inland is difficult in many cases or could result in significant costs to land owners or both. Adaptation measures including beach replenishment or hard structures such as sea walls and groins are costly and often ineffective except in the short-term. The diverse waterways of the Northeast, including lakes, rivers, beaches, and estuaries, are likely to continue to be havens for escape from the summer heat.

In autumn, a major recreational draw for the Northeast is the display of fall foliage. However, increased autumn warmth is associated with muting

Figure 11. On warm humid days when temperatures exceed 90°F, ozone problems are exacerbated across the region. The top figure shows the view on a clear day at the Great Gulf of Mount Washington, New Hampshire. The bottom figure shows the same view when temperatures exceed 90°F and air quality problems occur.

of fall foliage colors. Drought decreases leaf color and changes the timing of leaf drop. Both factors detract from a major tourist attraction in the Northeast. The two climate models used in this Assessment offer different recreational scenarios as the Canadian model projects increased drought while the Hadley model projects little change in drought risk.

The key outcome of an analysis of recreational activities is that the extent of potential impacts is highly dependent on the type of activity and on the differences between the model results. Many of the activities are likely to "migrate" out of portions of the Northeast or will move northward (such as the ski industry). Ski resorts are likely to be required to continue current trends toward development of year-round attractions. An extended warm weather season is very likely to make waterways, mountains, and forests greater attractions. Undoubtedly, humans will make trade-offs in terms of type and location of recreational activities. The uncertainties associated with the water balance projections for the Northeast contribute to uncertainties about the impact of climate change on recreation in the region.

New England Maple Syrup

A successful maple syrup season in New England depends on the proper combination of freezing nights and warm daytime temperatures, along with prolonged cold temperatures (resulting in a recharge of sugar to the sap) during the months of February and March. When the right combination of these climatic conditions occurs, the sugar maple tree produces a sap containing 2-5% sugar. In addition, the first flow of sap in a given season generally produces the highest quality maple syrup. A sustained, early flow heralds a good year for the maple industry in an area. If the initial flow occurs too early, before many of the producers have tapped their trees, they will miss this profitable opportunity. The maple industry in New England depends to a large extent on the timing of these critical climate events. Due to changes in both technology (the advent of tubing) and climate (very early initial flows and a reduction in freeze/thaw cycles and cold recharge periods), the maple syrup industry is moving from New England into Canada.

In the past, the success of the maple syrup industry in Canada was limited by deep snow cover (limiting access to individual trees) and fewer freeze/thaw cycles due to prolonged periods of low nighttime and daytime temperatures. The development of tubing-based sap collection methods, which provide easier access to trees and eliminate the need to make frequent collections, has allowed the Canadians to become more competitive in the past several decades. Changes in climate over the past several decades also allowed the Canadians to collect more sap over a longer "sugar season" than in the past. Conditions for sustained sap flow now mark the Canadian season while higher temperatures have led to fewer freeze/thaw cycles and reduced cold recharge periods in New England. This, coupled with earlier and earlier initial flows over the past two decades, has resulted in a shift in the volume of syrup production from the US to the Gaspe Peninsula of Quebec. It is interesting to note that in 1928, the major syrup production center in the US was located in Garrett County, Maryland.

If wintertime minimum temperatures continue to increase more rapidly than the maximum temperatures (as indicated in the climate models used in this Assessment) then the current northward shift in maple syrup production is very likely to continue. In the long term, change in the range of the maple tree is very likely to completely dominate the ability to produce maple syrup. The climate model scenarios, when coupled with assessments of the species composition of forests under the Canadian and Hadley model climate projections, project a substantial northward displacement in the distribution of maple trees (see Chapter 17 on Forests). It is likely under these scenarios that maple syrup production will not be possible in many regions of the Northeast because conditions for tree growth are unsuitable.

A key uncertainty is whether climate change will yield an increased tendency toward drought, which will impact fall colors, forest health, and the nature of water-related recreational activities.

ADDITIONAL ISSUES

Two additional issues are likely to be of considerable significance for the Northeast:

Forests

The species composition of the forests of the Northeast is very likely to change dramatically under both the Hadley and Canadian climate scenarios. Two approaches are available for assessing the forest changes in the Northeast. The MAPSS vegetation analysis (Neilson and Drapek, 1998) provides an analysis of the distribution changes in large-scale for-

est types (e.g., the Northeast mixed forest or the conifer forests of New England). The model of Iverson and Prasad (1998) examines changes in dominant forest species (e.g., maple-beech-birch, oak-hickory, elm-ash-cottonwood, and oak-gum-cypress). Although both vegetation models indicate substantial changes in forest character in response to the climate model scenarios, there are also some differences between the models as might be expected from two different approaches.

Based on the Canadian model scenario and the MAPSS vegetation analysis, the conifer forest of northern New England and much of the northeast mixed forest of New England, New York, and western Pennsylvania changes to a temperate deciduous forest similar to southeastern Pennsylvania and northern Virginia today. The area of southeast mixed forest, today characteristic of the region south of Virginia, becomes compressed into a small area of

West Virginia, southern Pennsylvania, and the coastal plain of Virginia, Delaware, and New Jersey, while much of Virginia becomes savanna/woodland. The Hadley model predicts less dramatic changes but still the conifer forest of northern New England is replaced by northeast mixed forest. The area of temperate deciduous forest in New England and Pennsylvania/West Virginia grows slightly. The area of southeast mixed forest grows in Virginia. The northeast mixed forest declines dramatically in total area (72% loss of area in the US according to the Forest Sector Foundation report) and specific species, for example the sugar maple (*Acer saccarum*), are lost entirely from the US (Watson et al., 1998).

An additional analysis by the Mid-Atlantic Assessment (Fisher et al., 2000) examined the distribution of dominant forest types based on the model of Iverson and Prasad (1998). Oak-hickory forests (46%) and maple-beech-birch (37%) dominate the region today. In both climate model scenarios, oak and hickory forests replace the maple-beech-birch forests of western and northern Pennsylvania, West Virginia, and southern New York. These large changes have significant potential to affect aesthetics, tourism, fall colors, and to cause people concern. Although in the short-term, forests are regarded as only moderately vulnerable to the specific climate changes projected, changes in timber, and non-timber values such as recreation, scenic views, and wildlife habitat may be long-term issues.

Forest issues extend well beyond the species composition of the forests. These issues include (a) the potential for alteration of tree resistance to insects associated with changes in temperature and water availability (Roth et al., 1997), (b) increased fire occurrence associated with increased tendency toward drought, (c) reduction in primary productivity and carbon storage despite carbon dioxide fertilization if drought becomes as significant as projected by the Canadian model, (d) changes in forest disturbance as a function of changes in hurricane frequency and intensity, or changes in wind, ice, or heavy precipitation events. Climate change can also compound the impacts of invasive species, as invasive species tend to have high reproductive rates, good dispersal ability, and rapid growth rates (Williamson, 1999). Many of these impacts are associated with aspects of climate model projections that are associated with uncertainty – e.g., changes in the character of extreme weather and the nature of the future water balance in the Northeast under climate change conditions.

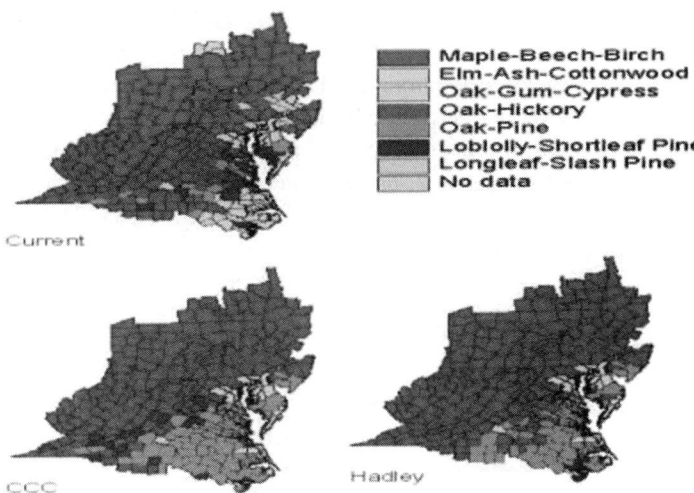

Figure 12. Dominant forest types for the mid-Atlantic region for current climate, and the potential distribution of these forest types for the Canadian and Hadley climate scenarios based on the Mid-Atlantic Assessment. Based on the model of Iverson and Prasad (1998). See Color Plate Appendix.

Agriculture

Although crop production is tied to climate, agriculture is a relatively small and declining fraction of economic output of the Northeast. Further, most studies indicate that the agriculture in the Northeast is relatively robust to climate change even though the crop mix may change (Abler et al., 1999). The ability to change crop types and to take advantage of hybrids limits vulnerability. The most frequently cited concern related to agriculture is associated with market forces, some of which are likely to be climate-related, generated by agricultural changes outside of the Northeast region. Some cool weather crops and many small family farms are exceptions to these conclusions. Farmers in the Northeast region make a significant contribution to the national supply of dairy products and food crops such as apples, grapes, potatoes, sweet corn, onions, cabbage, and maple syrup. In addition, small family farms throughout the Northeast are vital to the economy of rural areas, and they fill an important market niche for fresh, high quality, affordable local produce. These farms will be particularly sensitive to climate change due to the cost of adaptation, especially if compounded by other market forces. For example, the Northeast dairy industry is already quite fragile. Milk production by dairy cows is optimal at cool temperatures, so an increase in temperatures will require substantial increases in air conditioning costs. Most research on climate change impacts on agriculture has focused on major world trade crops such as wheat, soybeans, and corn.

Information from these studies has only very limited application to the Northeast where dairy and high value horticultural crops dominate the economy. Overall, agriculture in the Northeast is likely to survive a climate change, and may even benefit relative to some other regions of the US. However, the costs of adaptation could be high, and the vulnerability of small farms is likely to increase.

ADDED INSIGHTS

The results of assessment research in the Northeast yield several additional insights into the impacts of climate change on the region. Substantive issues include, but are not limited to the following:

Institutional Complexity. The complex institutional framework of community, municipal, county, regional, and statewide formal and informal governing bodies that characterize the Northeast have the potential to limit the region's ability to deal with extremes. For example, the complex array of watersheds, small management units, and urban dependencies on broad surrounding regions with different institutional characteristics all have the potential to limit drought planning and response management. Climate change is likely to cause a change in political focus and management of critical ecosystem services like water. Similar issues arise for weather disaster social services and energy distribution. There are signs of innovative management even with complex institutional structures in the Northeast transportation systems (e.g., introduction of electronic fare collection systems and EZ Pass). The ability of the Northeast to adapt to extreme situations will depend upon the ability of institutions to identify and prioritize vulnerable facilities and populations. Targeting and flexibility in the use of resources among the many institutions is needed to adapt more effectively to climate change.

Infectious Diseases. Infectious disease vectors are often strongly influenced by climate. For example, the primary Lyme disease vector is the deer tick. The tick population is governed by the size of both mouse and deer populations. Increased acorn mast production directly influences rodent populations. Milder winters contribute to a larger survival rate for deer. Larger deer populations increase the human contact with deer and deer ticks, increasing the possibility of Lyme disease infections. Milder winters are projected by virtually all climate models (the two primary models used in this Assessment both indicate large increases in winter minimum temperatures). Although climate has a strong impact on Lyme disease vectors, the complexity of the relationships makes changes in the distribution and frequency of the disease under altered climate difficult to predict (Martens, 1999). Changes in mosquito populations and survival are also possible with warmer and wetter conditions. Recent examples of outbreaks of West Nile Virus and equine encephalitis in Northeast urban areas have substantially raised concerns about vector-borne diseases and illustrate that improved monitoring and better understanding of these diseases are important to the region. Increased temperatures in coastal bays are likely to increase algal blooms, which have a chance of harboring cholera. Increased rainfall and flooding, if severe, have historically caused contamination of public and private water supplies (e.g., with *cryptosporidium*) and models project an increase in very heavy rainfall events and the potential for greater flood risks. However, in large measure, US public health infrastructure and response capabilities, if vigorously sustained, are likely to limit the potential impacts.

Species Changes. Although ecosystem character was not selected as one of the major topics for the first US National Assessment, regional assessment teams raised a significant number of issues concerning changes in species composition beyond forest character. Changes in temperature and precipitation will have direct impacts on species distribution as well as indirect impacts associated with changing predator-prey relationships and/or changes in pests and disease. In many cases, these species may be truly characteristic of a region or may be of economic significance. For example, lobster populations are associated with cooler waters and warming may promote northward migration of the population — a key issue for New England. Coastal population pressures combined with sea-level rise are very likely to limit habitat regions along the Atlantic Flyway for migratory birds. Warming is likely to substantially limit trout populations — a key issue for Pennsylvania. Changes in species mix and introduction of climate-driven invasive species are likely to also have unanticipated feedbacks on ecosystems.

Differential Human Impacts. The large differences in economic status and the aging of the population in the Northeast is likely to be associated with differential impacts based on the ability to respond to climate change. Where impacts are significant, climate change is likely to have greater impact on lower-income residents and the elderly, as well as children and the ill (e.g., those with chronic respiratory ailments). The key concerns are heat mortality, susceptibility to disease, and changes in air quality factors such as ozone. For example, lower income residents, often including the elderly, may have additional burdens because of lack of air conditioning, poor housing conditions, and unsafe neighborhoods leading to unwillingness to open windows or seek relief out of doors or away from the city.

ADAPTATION STRATEGIES

The most important elements of adaptation strategies proposed for the Northeast include:

- relocating structures at risk from severe weather (e.g., hurricanes) and flooding (both in coastal regions and in river systems prone to high water levels);
- strengthening design criteria for critical infrastructure (e. g., power supply) to ensure robust operation under possible changes in weather and climate extremes;
- increasing reservoir construction and improving management of water supplies to increase robustness of the water systems under conditions of flooding or drought, recognizing the potential for other negative consequences;
- greater emphasis on water quality and air quality controls to minimize the compounding of climate impacts, including stricter adherence to existing regulations and potentially stricter controls on pollutants and their precursors;
- incorporating active warning systems (e.g., Philadelphia's heat wave warning system) and structural adaptations (e.g., construction that promotes air-flow, and reduction in area of black roofing) to limit the potential for increased heat mortality and morbidity;
- limiting the non-climate stresses in order to minimize impacts on critical regions that are geographically fixed and therefore can't "migrate" such as the Chesapeake Bay;
- limiting agricultural and forestry economic impacts by introducing different planting and harvesting methods based on historical responses to weather and climate; and
- limiting coastal development through existing regulatory frameworks to protect coastal regions, increasing the focus on "smart" land use to reduce vulnerability to floods and storm damage, and limiting pollution delivered to coastal regions by water run-off from the adjacent land area.

The social and economic framework of the future is likely to be substantially more advanced than at present. As a result, the adaptation strategies may become more numerous, more effective, and more easily implemented.

CRUCIAL UNKNOWNS AND RESEARCH NEEDS

Seven critical factors limit our ability to assess the potential importance of climate change on the Northeast:

- For the Northeast, changes in extreme weather are viewed as a critical issue in assessing potential impacts. A key issue is an ability to assess the potential changes in severe storms, including hurricane intensity, path, and frequency and the character and frequency of nor'easters. Changes in the frequency and intensity of extreme weather events with climate change remain one of the most uncertain aspects of future climate.
- Improved projections of the spatial and temporal distribution of future temperature and precipitation would enable a more robust assessment. The difference in the projections of changes in the Palmer Drought Severity Index between the Canadian and Hadley models presents significantly different scenarios, with dramatically different implications for ecosystems, water, recreation, and agriculture.
- The ability to combine multiple stresses and to simulate the resultant effects on the environment is severely limited in many specific environments (e.g., the Chesapeake Bay). Fully integrated observational networks for multiple variables and comprehensive models will be required to address multiple stresses.
- Changes in population and land use patterns will have dramatic effects on ecosystems and on the nature and magnitude of climate impacts, yet the ability to project these changes is limited. Continued growth in coastal populations in conjunction with sea-level rise introduces substantial additional risk due to severe weather and greater concerns about habitat loss.
- Understanding of the potential change in species composition and character in response to climate change is in its infancy. This includes changes in economically important species, pests, invasive species, and predator-prey relations. Validation of biological response models is a major problem.
- Understanding of how people and societies will adapt, due to uncertainties in overall changes in

the economy, technology, and societal values, is a major area of uncertainty.

- Uncertainties in estimating sea-level rise substantially limit the ability to assess the height of storm surges and the recurrence interval of flooding events with substantial economic impacts.

LITERATURE CITED

Agee, E. M., Trends in cyclone and anticyclone frequency and comparison with periods of warming and cooling over North America, *Journal of Climate*, 4, 263-267, 1991.

Abler, D. G. and J. S. Shortle, Climate change and agriculture in the Mid-Atlantic region, *Climate Research*, 14, 185-194, 2000.

Alpern, R., Impact of global warming on water resources: Implications for New York City and the New York metropolitan region, in The Baked Apple? *Metropolitan New York in the Greenhouse*, Annals of the New York Academy of Sciences, vol. 790, edited by D. Hill, 86, 1996.

Barron, E. J., Climate models: How reliable are their predictions?, *Consequences*, 1, 16-27, 1995.

Bengtsson, L., M. Botzet, and M. Esch, Will greenhouse-gas-induced warming over the next 50 years lead to higher frequency and greater intensity of hurricanes?, *Tellus*, 48, 57-73, 1996.

Bloomfield, J., Hot nights in the city: global warming, sea-level rise and the New York metropolitan region, Environmental Defense Fund, New York, 1999.

Boer, G. J., N. A. McFarlane, R. Laprise, J. D. Henderson, and J. P. Blanchet, The Canadian Climate Centre spectral atmospheric general circulation model, *Atmosphere-Ocean*, 22(4), 397-429, 1984.

Boer, G. J., G. M. Flato, M. C. Reader, and D. Ramsden, A transient climate change simulation with historical and projected greenhouse gas and aerosol forcing: experimental design and comparison with the instrumental record for the 20th century, *Climate Dynamics*, 16, 405-426, 1999.

Boer, G. J., G. M. Flato, and D. Ramsden, A transient climate change simulation with historical and projected greenhouse gas and aerosol forcing: projected climate for the 21st century, *Climate Dynamics*, 16, 427-450, 1999.

Bove, M. C., J. B. Elsner, C. W. Landsea, X. Niu, and J. J. O'Brien, Effect of El Niño on U.S. land falling hurricanes, revisited, *Bulletin of the American Meteorological Society*, 79, 2477-2482, 1998.

Broccoli, A. J., The greenhouse effect: The science base, in *The Baked Apple? Metropolitan New York in the Greenhouse*, edited by D. Hill, Annals of the New York Academy of Sciences, vol. 790, 23, 1996.

Bureau of Economic Analysis, Survey of current business, June 2000, U.S. Government Printing Office, 2000. Available online at http://www.bea.doc.gov/bea/regional/data.htm.

Carnell, R. E., and C. A. Senior, Changes in mid-latitude variability due to increasing greenhouse gases and sulphate aerosols, *Climate Dynamics*, 14, 368-383, 1998.

Carnell, R. E., C. A. Senior, and J. F. B. Mitchell, An assessment of measures of storminess: simulated changes in northern hemisphere winter due to increasing CO_2, *Climate Dynamics*, 12, 467-476, 1996.

Changnon, S. A., and J. M. Changnon, Temporal fluctuations in weather disasters: 1950-1989, *Climatic Change*, 22, 191-208, 1992.

Chestnut, L. G., W. S. Freffle, J. B. Smith, and L. S. Kalkstein, Analysis of differences in hot-weather-related mortality across 44 US metropolitan areas, *Environmental Sciences Policy*, 1, 59-70, 1998.

Colwell, R. P. Epstein, D. Gubler, M. Hall, P. Reiter, J. Shukla, W. Sprigg, E. Takafuji, and J. Trtanj, Global climate change and infectious diseases, *Emerging Infectious Diseases*, 4, 451-452, 1998.

Cronon, W., *Changes in the Land*, Hill and Wang, New York, 241 pp., 1983.

Currie, D. J., Energy and large-scale patterns of animal- and plant-species richness, *The American Naturalist*, 137, 27-49, 1991.

DeGaetano, A. T., Climate perspective and impacts of the 1988 northern New York and New England ice storm, *Bulletin of the American Meteorological Society*, 81, 237-254, 2000.

Drake, B. G., G. Peresta, E. Beugeling, and R. Matamala, Long-term elevated CO_2 exposure in a Chesapeake Bay wetland: Ecosystem gas exchange, primary production, and tissue nitrogen, in *Carbon Dioxide and*

Terrestrial Ecosystems, edited by G.W. Koch and H.A. Mooney, Academic Press, San Diego, California, 197-214, 1996.

Ebersole, B.A., Atlantic coast water-level climate, WIS Report 7, US Army Corps of Engineers, Washington, DC, 1982.

Emanuel, K.A., The maximum intensity of hurricanes, *Journal of the Atmospheric Sciences*, 45, 1143-1154, 1988.

Fisher, A., et al., Preparing for a changing climate: The potential consequences of climate variability and change, Mid-Atlantic foundations, US National Assessment, http://www.essc.psu.edu/mara/results/foundations_report, in review, 2000.

Flato, G. M., G. J. Boer, W. G. Lee, N. A. McFarlane, D. Ramsden, M. C. Reader, A. J. Weaver, The Canadian Centre for Climate Modelling and Analysis Global Coupled Model and its climate, *Climate Dynamics*, 16, 451-468, 1999.

Funderbunk, S. L., S. J. Jordan, J. A. Mihursky, and D. Riley (Eds.), *Habitat Requirements for Chesapeake Bay Living Resources*, 2nd edition, Chesapeake Research Consortium, Inc., Solomons, Maryland, 1991.

Gibson, J. R., and R. G. Najjar, Modeling Chesapeake Bay salinity under climate change, *Limnology and Oceanography*, in review, 2000.

Gibson, J. R., Modeling Chesapeake Bay salinity and phytoplankton dynamics in response to varying climate, M.S. thesis, Penn State University, University Park, PA, 1999.

Gornitz, V., Regional sea-level variations in eastern North America: A geological perspective, *Transactions. American Geophysical Union*, 80, (17) S85, 1999.

Gornitz, V., and S. Lebedeff, Global sea-level changes during the past century, in *Sea-Level Rise and Coastal Subsidence*, edited by D. Nummedal, O. Pilkey, and J. Howard, Kluwer Academic Publishers, Dordrecht, Netherlands, 357-364, 1987.

Groisman, P.Ya , R.W. Knight, T.R. Karl, and H.F. Lins, Heavy precipitation and streamflow in the United States: Trends in the 20th century, *Bulletin of the American Meteorological Society*, in press, 2000.

Hayden, B. P., Climate change and extratropical storminess in the United States: An assessment, *Journal of the American Water Resources Association*, 35, 1387-1398, 1999.

Hughes, P. Y., E. H. Mason, T. R. Karl, and W. A. Brower, United States historical climatology network daily temperature and precipitation data, Environmental Sciences Division Publication 3778, Carbon Dioxide Information and Analysis Center, Oak Ridge National Laboratory, Oak Ridge, Tennessee, 55 pp., 1992.

Idso, S.B., R.C. Balling, Jr., and R.S. Cerveny, Carbon dioxide and hurricanes: implications of northern hemispheric warming for Atlantic/Caribbean storms, *Meteorology and Atmospheric Physics*, 42, 259-263, 1990.

Insurance Research Council, Coastal exposure and community protection, Insurance Research Council, Wheaton, Illinois, 45 pp., 1995.

Iverson, L. R., and A. M. Prasad, Predicting abundance of 80 tree species following climate change in the eastern United States, *Ecological Monographs*, 68(4), 465-485, 1998.

IPCC (Intergovernmental Panel on Climate Change), *Climate Change 1995: The IPCC Second Scientific Assessment*, Cambridge University Press, Cambridge, UK, 1996. Johns, T. C., R. E. Carnell, J. F. Crossley, J. M. Gregory, J. F. B. Mitchell, C. A. Senior, S. F. B. Tett, and R. A. Wood, The second Hadley Centre coupled ocean-atmosphere GCM: Model description, spinup, and validation, *Climate Dynamics*, 13, 103-134, 1997.

Jones, K. B., et al., *An Ecological Assessment of the United States Mid-Atlantic Region: A Landscape Atlas*, US EPA, Office of Research and Development: Washington, DC, 1997 (Document #EPA/600/R-97/130).

Kalkstein, L. S., and J. S. Greene, An evaluation of climate/mortality relationships in large U.S. cities and the possible impacts of climate change, *Environmental Health Perspectives*, 105, 84-93, 1997.

Kalkstein, L. S., and J. J. Swift, An evaluation of climate/mortality relationships in the Mid-Atlantic region and the possible impacts of climate change, Prepared for inclusion in the Mid-Atlantic Regional Assessment, 1998.

Karl, T. R., P. Ya. Groisman, R. W. Knight, and R. R. Heim, Jr., Recent variations of snow cover and snowfall in North America and their relation to precipitation and temperature variations, *Journal of Climate*, 6, 1327-1344, 1993.

Karl, T. R., and R. W. Knight, Secular trends in precipitation amount, frequency, and intensity in the United States, *Bulletin of the American Meteorological Society*, 79, 231-241, 1998.

Karl, T. R., R. W. Knight, D. R. Easterling, and R. G. Quayle, Trends in U.S. climate during the Twentieth century, *Consequences*, 1, 2-12, 1996.

Kearney, M. S., and J. C. Stevenson, Island land loss and marsh vertical accretion rate evidence for historical sea-level changes in Chesapeake Bay, *Journal of Coastal Research*, 7, 403-415, 1991.

Kilbourne, E. M., K. Cho, T. S. Jones, S. B. Thacker, and F. I. Team, Risk factors for heat stroke: A case central study, *JAMA*, 247, 3332-3336, 1982.

Kittel, T. G. F., J. A. Royle, C. Daly, N. A. Rosenbloom, W. P. Gibson, H. H. Fisher, D. S. Schimel, L. M. Berliner, and VEMAP2 participants, A gridded historical (1895-1993) bioclimate dataset for the conterminous United States, in *Proceedings of the 10th Conference on Applied Climatology*, October 20-24, 1997, Reno, Nevada, American Meteorological Society, Boston, Massachusetts, 219-222, 1997.

Klopatek, J. M., R. J. Olson, C. J. Emerson, and J. L. Jones, Land-use conflicts with natural vegetation in the United States, *Environmental Conservation*, 6, 191-200, 1979.

Knutson, T. R., R. E. Tuleya, and Y. Kurihara, Simulated increase of hurricane intensities in a CO2-warmed climate, *Science*, 279, 1018-1020, 1998.

Kunkel, K. E., K. Andsager, and D. R. Easterling, Long-Term trends in extreme precipitation events over the conterminous United States, *Journal of Climate*, 12, 2515-2527, 1999.

Kunkel, K. E., R. A. Pielke Jr., and S. A. Changnon, Temporal fluctuations in weather and extremes that cause economic and human health impacts, *Bulletin of the American Meteorological Society*, 80, 1077-1098, 1999.

Lighthill, J., G. Holland, W. Gray, C. Landsea, G. Craig, J. Evans, Y. Kurihara, and C. Guard, Global climate change and tropical cyclones, *Bulletin of the American Meteorological Society*, 75, 2147-2157, 1994.

Martens, P., How will climate change affect human health?, *American Scientist*, 87, 534-541, 1999.

McCabe, G. J., Jr., and M. A. Ayers, Hydrologic effects of climate change in the Delaware River Basin, *Water Resources Bulletin*, 25(6), 1989.

McFarlane, N. A., G. J. Boer, J. P. Blanchet, and M. Lazare, The Canadian Climate Centre second-generation general circulation model and its equilibrium climate, *Journal of Climate*, 5, 1013-1044, 1992.

Mitchell J. F. B., T. C. Johns, J. M. Gregory, and S. Tett, Climate response to increasing levels of greenhouse gases and sulphate aerosols, *Nature*, 376, 501-504, 1995.

Mitchell J. F. B., and T. C. Johns, On modification of global warming by sulfate aerosols, *Journal of Climate*, 10, 245-267, 1997.

Najjar, R. G., et al., The potential impacts of climate change on the mid-Atlantic coastal region, *Climate Research*, 14, 219-233, 2000.

Neff, R., H. Chang, C. G. Knight, R. G. Najjar, B. Yarnal, and H. A. Walker, Impact of climate variation and change on mid-Atlantic hydrology and water resources, *Climate Research*, 14, 207-218, 2000.

Neilson, R. P., and R. J. Drapek, Potentially complex biosphere responses to transient global warming, *Global Change Biology*, 4, 505-521, 1998.

NPA Data Services, Inc., Regional Economic Projection Series, Washington, DC, 1998.

Perry, M. C., and A. S. Deller, Review of factors affecting the distribution and abundance of waterfowl in shallow-water habitats of Chesapeake Bay, *Estuaries*, 19, 272-278, 1996.

Perry, M. C., and A. S. Deller, Waterfowl population trends in the Chesapeake Bay area, in *Toward a Sustainable Coastal Watershed: The Chesapeake Experiments. Proceedings of a Conference*, Chesapeake Research Consortium Publication No. 149, edited by P. Hill and S. Nelson, Chesapeake Research Consortium, Inc., Edgewater, Maryland, 1995.

Perry, M. C., (Ed.), F. M. Uhler, Food habits and distribution of wintering canvasbacks, Aythya valisinerina, on Chesapeake Bay, *Estuaries*, 11, 57-67, 1988.

Polsky, C., J. Allard, N. Currit, R. G. Crane, and B. Yarnal, The mid-Atlantic region and its climate: past, present, and future, *Climate Research*, 14, 161-173.

Powell, D. S., J. L. Faulkner, D. R. Darr, Z. Zhu, and D. W. MacCleery, *Forest Resources of the United States, 1992*, USDA Forest Service, 133 pp. (General Tech Report, RM-243), 1996.

Ricketts, T. H., E. Dinerstein, D. M. Olson, and C. Loucks, Who's where in North America? Patterns of species richness and the utility of indicator taxa for conservation. *BioScience, 49*, 369-381, 1999.

Rock, B., and B. Moore III et al., The New England regional assessment of the potential consequences of climate variability and change, US National Assessment, in prep. 2000.

Roth, S., E. P. McDonald, and R. L. Lindroth, Atmospheric CO_2 and soil water availability consequences for tree-insect interactions, *Canadian Journal of Forest Research*, 27, 1281-1290, 1997.

Rose, A., Y. Cao, and G. Oladosu, Simulating the economic impacts of climate change in the Mid-Atlantic region, *Climate Research*, 14, 175-183, 2000.

Rosenzweig, C., and W. Solecki, et al., Climate change and a global city: an assessment of the metropolitan East Coast region, US National Assessment, in review, 2000.

Solley, W. B., R. R. Pierce, and H. A. Perlman, Estimated water use in the United States in 1995, *US Geological Survey Circular 1200*, Washington, DC, 1998.

Stephenson, D. B., and I. M. Held, GCM response of northern winter stationary waves and storm tracks to increasing amounts of carbon dioxide, *Journal of Climate*, 6, 1859-1870, 1993.

Timmermann, A., J. Oberhuber, A. Bacher, M. Esch, M. Latif, and E. Roeckner. 1999. Increased El Niño frequency in a climate model forced by future greenhouse warming. *Nature*, 398, 694-696.

US Army Corps of Engineers, *Annual Flood Damage Report to Congress for Fiscal Year 1997*, 1998. Available online at http://www.usace.army.mil/inet/functions/cw/cecwe/flood.htm.

USDA Forest Service, *Report of the Forest Service. Fiscal Year 1998*, Washington, DC, 1998.

US Global Change Research Program Forum on Global Change Modeling. Washington, D.C. USGCRP Report 95-02, Washington, DC, 1995.

US Historical Climatology Network: Available on line at http://www.ncdc.noaa.gov/ol/climate/research/ushcn/daily.html

Valverde, H. R., A. C. Trembanis, and O. H. Pilkey, Summary of beach nourishment episodes on the U.S. East Coast barrier islands, *Journal of Coastal Research*, 15, 1100-1118, 1999.

VEMAP members, Vegetation/Ecosystem Modeling and Analysis Project (VEMAP): Comparing biogeography and biogeochemistry models in a continental-scale study of terrestrial ecosystem responses to climate change and CO_2 doubling, *Global Biogeochemical Cycles*, 9, 407-437, 1995.

Walker, H. A., J. S. Latimer, and E. H. Dettman, Assessing the effects of natural and anthropogenic stressors in the Potomac Estuary: Implications for long-term monitoring, *Environmental Monitoring and Assessment*, in press, 2000.

Watson, R. T., M. C. Zinyowera, and R. H. Moss (eds). *The Regional Impacts of Climate Change: An Assessment of Vulnerability*. Cambridge University Press, New York, 1998.

Williamson, M., Invasions, *Ecography*, 22, 5-12, 1999.

Yarnal, B., D. L. Johnson, B. Frakes, G. I. Bowles, and P. Pascale, The flood of '96 in the Susquehanna River basin, *Journal of the American Water Resources Association*, 33, 1299-1312, 1997.

ACKNOWLEDGMENTS

Many of the materials for this chapter are based on contributions from participants on and those working with the

Metropolitan East Coast Workshop and Assessment Teams
Cynthia Rosenzweig*, National Aeronautics and Space Administration, Goddard Institute for Space Studies, and Columbia University
William Solecki*, Montclair State University
Carli Paine, Columbia University
Peter Eisenberger, Columbia University Earth Institute
Lewis Gilbert, Columbia University Earth Institute
Vivien Gornitz, Columbia University Center for Climate Systems Research
Ellen K. Hartig, Columbia University Center for Climate Systems Research
Douglas Hill, State University of New York, Stony Brook
Klaus Jacob, Lamont-Doherty Earth Observatory of Columbia University
Patrick Kinney, Columbia University Joseph A. Mailman School of Public Health
David Major, Columbia University Center for Climate Systems Research
Roberta Balstad Miller, Center for International Earth Science Information Network (CIESIN)
Rae Zimmerman, New York University Institute for Civil Infrastructure Systems, Wagner School

Mid-Atlantic Workshop and Assessment Teams
Ann Fisher*, Pennsylvania State University
David Abler, Pennsylvania State University
Eric J. Barron, Pennsylvania State University
Richard Bord, Pennsylvania State University
Robert Crane, Pennsylvania State University
David DeWalle, Pennsylvania State University
C. Gregory Knight, Pennsylvania State University
Ray Najjar, Pennsylvania State University
Egide Nizeyimana, Pennsylvania State University
Robert O'Connor, Pennsylvania State University
Adam Rose, Pennsylvania State University
James Shortle, Pennsylvania State University
Brent Yarnal, Pennsylvania State University

New England and Upstate New York Workshop and Assessment Teams
Barry Rock*, University of New Hampshire
Berrien Moore III*, University of New Hampshire
David Bartlett, University of New Hampshire
Paul Epstein, Harvard School of Public Health
Steve Hale, University of New Hampshire
George Hurtt, University of New Hampshire
Lloyd Irland, Irland Group, Maine
Barry Keim, New Hampshire State climatologist
Clara Kustra, University of New Hampshire
Greg Norris, Sylvatica Inc., Maine
Ben Sherman, University of New Hampshire
Shannon Spencer, University of New Hampshire
Hal Walker, EPA, Atlantic Ecology Division, Rhode Island

* Assessment Team chair/co-chair

CHAPTER 5

POTENTIAL CONSEQUENCES OF CLIMATE VARIABILITY AND CHANGE FOR THE SOUTHEASTERN UNITED STATES

Virginia Burkett[1,2], Ronald Ritschard[3], Steven McNulty[4], J. J. O'Brien[5], Robert Abt[6], James Jones[7], Upton Hatch[8], Brian Murray[9], Shrikant Jagtap[7], and Jim Cruise[3]

Contents of this Chapter

[1]US Geological Survey; [2]Coordinating author for the National Assessment Synthesis Team; [3]University of Alabama in Huntsville; [4]USDA Forest Service; [5]Florida State University; [6]North Carolina State University; [7]University of Florida; [8]Auburn University; [9]Research Triangle Institute

CHAPTER SUMMARY

Regional Context

The Southeast "sunbelt" is a rapidly growing region with a population increase of 32% between 1970 and 1990. Much of this growth occurred in coastal counties, which are projected to grow another 41% between 2000 and 2025. The number of farms in the region decreased 80% between 1930 and 1997 as the urban population expanded, but the Southeast still produces roughly one quarter of US agricultural crops. The Southeast has become America's "woodbasket," producing about half of America's timber supplies. The region also produces a large portion of the nation's fish, poultry, tobacco, oil, coal, and natural gas. Prior to European settlement, the landscape was primarily forests, grasslands, and wetlands. Most of the native forests were converted to managed forests and agricultural lands by 1920. Although much of the landscape has been altered, a wide range of ecosystem types exists and overall species diversity is high.

Climate of the Past Century

- Southeastern temperature trends varied between decades over the past 100 years, with a warm period during the 1920s-1940s followed by a downward trend through the 1960s. Since the 1970s, temperatures have been increasing, with 1990's temperatures reaching peaks as high as those of the 1920s-1940s.

- Annual rainfall trends show very strong increases of 20-30% or more over the past 100 years across Mississippi, Arkansas, South Carolina, Tennessee, Alabama, and parts of Louisiana, with mixed changes across most of the remaining area. The percentage of the Southeast landscape experiencing severe wetness increased approximately 10% between 1910 and 1997.

Climate of the Coming Century

- Climate model simulations provide plausible scenarios for both temperature and precipitation over the 21st century in the Southeast. Both of the principal climate models used in the National Assessment suggest warming in the Southeast by the 2090s, but at different rates.

- The Canadian model scenario shows the Southeast experiencing a high degree of warming, which translates into lower soil moisture as higher temperatures increase evaporation. The Hadley model scenario simulates less warming and a significant increase in precipitation by 2100 (about 20% on average, but with some differences within the region).

- Both climate model simulations indicate that the heat index (a measure of comfort based on temperature and humidity) will increase more in the Southeast than in other regions. Heat index in the Southeast is projected to rise by as much as 8 to 15°F (4 to 8°C) in the Hadley model and by over 15°F (8°C) across the entire region in the Canadian model simulation by 2100.

Key Findings

- The increase in the southeastern summer heat index simulated by both the Hadley and Canadian climate models would likely affect human activity and possibly demographics in the Southeast during the 21st century.
- Agriculture could possibly benefit from increased CO_2 and modest warming (up to 3 to 4°F, or 2°C) as long as rainfall does not decline, but there are differences in individual crop responses. Management adaptations could possibly offset potential losses in individual crop productivity due to increased evapotranspiration.
- Biological productivity of pine and hardwood forests will likely move northward as temperatures increase across the eastern US. Hardwoods are more likely to benefit from increases in CO_2 and modest increases in temperature than pines. Physiological forest productivity and ecosystem models suggest that, without management adaptations, pine productivity is likely to increase by 11% by 2040 and 8% by 2100 across the Southeast compared to 1990 productivity. These models suggest that hardwood forest productivity will likely increase across the region, by 25% by 2090 compared to 1990 regional hardwood productivity.
- Under the Hadley model scenario, the region's land use change in the next century is likely to

be dominated by non-climate factors such as commodity prices and demographic forces. Urbanization will likely continue to convert forest and agricultural land, while continued movement of land from agriculture to forest is also expected. Under more extreme climate scenarios, land reallocation could possibly be more dramatic as the productivity of land-based activities such as forestry and agriculture is more profoundly influenced by climate.
- During the 21st century, the IPCC projects that sea-level rise will likely accelerate 2- to 5-fold compared to the global average rate during the 1900s (which was 4-8 inches). This would very likely have dramatic effects on population centers, infrastructure, and natural ecosystems in the low-lying Gulf and South Atlantic coastal zone.
- Water and air quality are concerns given the changes in temperature and precipitation that are simulated by climate models.
- Changes in minimum temperature, rainfall, and CO_2 will likely alter ecosystem structure, but interactions are difficult to model or predict, particularly relative to disturbance patterns.
- Changes in fresh water and tidal inflows into coastal estuaries will likely alter the ecological structure and function of these highly productive and valuable ecosystems.

POTENTIAL CONSEQUENCES OF CLIMATE VARIABILITY AND CHANGE FOR THE SOUTHEASTERN UNITED STATES

PHYSICAL SETTING

The Southeast region (Fig. 1) represents 15.4% of the land area of the US and 22.6% of its citizenry (US Bureau of the Census, 1994). Although the Southeast has considerable variation in landforms, it is possible to divide it into five fairly distinct regions based on physical geography. The lower third of the low, flat Florida peninsula is a sub-tropical province with unique features such as the Everglades and the Florida Keys. The Coastal Plain, which dominates the region, is a broad band of territory paralleling the Gulf and South Atlantic seacoast from Virginia to Texas, with a deep extension up the lower Mississippi River valley. The Coastal Plain is relatively flat, with broad, slow-moving streams and sandy or alluvial soils. The Piedmont is a slightly elevated plateau that begins at the "fall line," where rivers cascade off the eastern edge of the plateau onto the Coastal Plain, and ends at the Appalachian Mountains. This land is rolling to hilly, with many streams. The Highlands comprise the inland mountain regions and include the southernmost Appalachian Mountains in the east and the Ozark and Ouachita Mountains in the west. The Interior Plains stretch into the north-central portion of the region, including parts of Tennessee and Kentucky. The southern portions of the State of Virginia are included in aspects of this regional analysis. The northern portions are included in the Northeast chapter with a discussion of the Chesapeake Bay watershed.

SOCIOECONOMIC CONTEXT

The 544,000 square mile southeastern "sunbelt" is one of the fastest growing regions in the US. The southeastern population increased by 21%, more than double the national rate, between 1970 and 1980 and another 11% between 1980 and 1990. Much of the historical population growth from Texas to North Carolina occurred in the 151 counties within the southeastern coastal zone, which are projected to grow another 41% between 2000 and 2025, compared to the projected national average of about 25% (NPA, 1999). However, warming at higher latitudes combined with increased heat stress in the southeastern US may serve to decrease population migration towards the Southeast.

Based on the 1990 Census, about 61% of the Southeast's population is considered urban. The number of farms in the region decreased 80% between 1930 and 1997 (USDA, 1999). The 20th century was one of dramatic transition from an agrarian economy to one based on a combination of natural resources, manufacturing and trade, technology, and tourism. Roughly one half of 1990 employment fell in the categories of manufacturing and wholesale/retail trade, compared with an average of less than 5% in agriculture (US Bureau of the Census, 1994). Prior to 1950, corn and cotton were the most important crops. Today agriculture is more varied; soybean and hay outweigh cotton in acreage harvested and rice has become increasingly important. The Southeast still produces roughly one quarter of the nation's agricultural crops, but timber harvests are more valuable in terms of annual economic impact in most states. Forest products industries were among the top four manufacturing employers in Mississippi, Alabama, North Carolina and Georgia in 1997. The Southeast has become America's "woodbasket," producing about half of America's timber supplies. The region is also responsible for a large portion of the nation's fisheries, poultry, tobacco, oil, natural gas, bauxite, coal, and sulfur production.

According to the 1990 census, 18% of the population of the Southeast region lives below the poverty level. The most poverty-prone areas include the Lower Mississippi River Valley and parts of Appalachia. While certain measurements, such as per capita income, have moved in the direction of the national averages, poverty rates in some areas are as much as two and a half times the national average. Levels of education of the population in some areas also lag behind national standards. Some of the smaller, more remote and geographically isolated areas of the region suffer from a lack of economic opportunities, have significant dependent populations, and lack the public institutions needed to support progressive development (Glasmeier, 1998). These distressed counties present a profound challenge to policy makers concerned about climate change mitigation strategies and issues, particularly in the Appalachian coal-producing and Gulf Coast oil-producing regions.

ECOLOGICAL CONTEXT

Prior to European settlement, the Southeast was dominated by upland forests, grasslands, and wetlands. Nearly one-third of the region may have been wetland (Dahl, 1990), but by 1990, wetlands had been reduced to about 16% of the southeastern landscape (Hefner, et. al., 1994). A wide range of ecosystem types is presently found in the region, ranging from coastal marshes to high-elevation spruce fir forests. Diversity of both plant and animal species is high compared to other regions considering the extent of landscape alteration that has occurred. On an area basis, the Southeast has relatively high overall species richness indices (Ricketts, et. al., 1999). Vascular plant diversity is second only to Puerto Rico. Tree species richness is greatest in the Southeast, with approximately 180 tree species found in parts of South Carolina and more than 140 tree species identified in most of the remainder of the region.

Forests still dominate parts of the Southeast; the share of forestland in each state averages about 30%. About 20% of the present forests exist as pine plantations. Native longleaf pine was the predominant species in the Coastal Plain in the late 1800s but less than 3 million of the 60 million acres of southeastern longleaf pine remain today (Boyer, 1979). More than 60 species of mammals occur in a relatively small area of the southern Appalachian mountains, while 40 or fewer mammal species are found in the Coastal Plain (Currie, 1991). The region has very high diversity of amphibians and reptiles. Roughly half of the remaining wetlands in the US are located in the Southeast, and more than three-quarters of the nation's annual wetland losses over the past 50 years occurred in the region.

Two types of ecosystem models (biogeochemical and biogeographical) show a wide range of potential changes in vegetation in the Southeast during the 21st century, depending upon the climate scenario selected. One of the biogeographical models (MAPSS) projects significant shifts in major biomes under the Canadian climate scenario, but not under the Hadley climate scenario. Under the Hadley climate scenario, the Southeast mixed forest retains its

southern boundaries while expanding west into parts of eastern Oklahoma and Texas, and north into parts of Missouri, West Virginia, Kentucky, and Virginia. Water stress and increased fire disturbance restricts the Southeast forest under the Canadian climate scenario, and large areas of the Southeast are converted to savanna (grasslands with scattered trees and shrubs) and the Southeast forest moves into the northcentral part of the US.

One of the biogeochemistry models (TEM) used in the National Assessment projects large differences in carbon storage for the Southeast depending upon the climate scenario used. Under the Canadian climate model scenario, vegetation is projected to lose up to 20% of its carbon mass by 2030. However, under the Hadley climate scenario the same biogeochemistry model indicates that vegetation will add between 5 and 10% to its carbon mass over the next 30 years. These differences in carbon storage reflect the differences in climate scenario projections for temperature and precipitation that are greatest in the southeastern part of the US (Felzer and Heard, 1999).

Ecological models used in the National Assessment do not simulate species-level response, nor do they simulate land use changes, invasive species impacts, or other influences on ecosystems that cannot be effectively modeled based on historical or empirical evidence, unless the ecological models are linked

Land Cover Map of the Southeast Region

Figure 1. The Southeast region includes all of nine states (Alabama, Florida, Georgia, Kentucky, Louisiana, North Carolina, Mississippi, South Carolina, and Tennessee), the southern portion of Virginia, and 50 counties in east Texas. Four subregional workshops were conducted in the Southeast region. See Color Plate Appendix

with other process models, both biological and economic. For example, Harcombe and others (1998) observed that Chinese tallow, a freeze-intolerant non-native tree species, increased dramatically in southeastern Texas over the past few decades. Chinese tallow increased by a factor of 30 between 1981 and 1995, often out-competing native species when canopy gaps form in mesic (medium moisture) and wet sites (Harcombe, et. al., 1998). These kinds of interactions and changes in forest dynamics are difficult to simulate.

Mixed responses among species to fertilization effects of elevated atmospheric CO_2 further confound our ability to model ecosystem structure and productivity. Several studies showed that elevated CO_2 increased photosynthesis rates and improved water use efficiency in many forest species and agricultural crops (Acock, et. al., 1985; Allen, et. al., 1989; Nijs, et. al., 1988). Two reviews of CO_2 exposure studies with deciduous and coniferous species found that increases in growth rates varied widely, but that generally tree growth was stimulated by increases in CO_2 (Eamus and Jarvis, 1989; NCASI, 1995). However, limits on the availability of soil nutrients and water in many natural or semi-natural ecosystems can severely constrain the potential improvement in water use efficiency due to suppressed transpiration induced by enhanced CO_2 levels, thereby offsetting potential gains in productivity (Lockwood, 1999). Temperature, plant pests, air pollution, and light availability could also limit the potential enhancement of growth by elevated CO_2 (NCASI 1995). Hence, one should be very cautious

in assuming what the net effects of CO_2 enrichment might be across a region or biome.

CLIMATE VARIABILITY AND CHANGE

Past Century

The Southeast has some of the warmest conditions in the US. However, it is the only region to show widespread but discontinuous cooling periods of 2 to 3.5°F (1 to 2°C) over almost the entire area during the past 100 years (Figure 2). Peninsular Florida, North Louisiana, and a few small areas in the Appalachian Mountains have shown a modest warming of around 2°F (1°C) since 1900. The reason the Southeast temperature record shows a net cooling trend over the past 100 years is that there was a warm period between the 1920s and 1940s, then a significant downward trend through the late 1960s. The mid-1900s cooling trend may have been due to natural variation. Human-caused sulfate aerosol emissions during this period may have also played some role. Sulfate aerosols reflect some sunlight back into space, thereby cooling the atmosphere (Kiehl, 1999). Since 1970, the average annual temperature increased, with the most significant increases occurring during the 1990s. Trends in temperature extremes over the past one hundred years exhibit a decrease of about 5 days in the number of days per year exceeding 90°F (32.2°C), and an increase of 6 days in the number of days below freezing over the entire region. However, over the past fifty years the average annual length of the snow season decreased by 4 days.

The Southeast receives more rainfall than any other region. Annual precipitation trends show increases of 20-30% or more over the past 100 years across Mississippi, Arkansas, South Carolina, Tennessee, Alabama, and parts of Louisiana, with mixed changes across most of the remaining area. The southern mountains of North Carolina, the southern tip of Texas, and a couple of other small areas have slightly decreasing trends in annual precipitation. Much of the increase in precipitation was associated with more intense events (rainfall greater than 2 inches or 5 cm per day). A small percentage of the increased precipitation was associated with moderate rainfall events, which are generally beneficial to agriculture and water supply. Analysis of stream flow trends during 1944-1993 showed little change in annual maximum daily discharge, but significant increases in annual median and minimum flows in

Southeast US Annual Mean Temperature

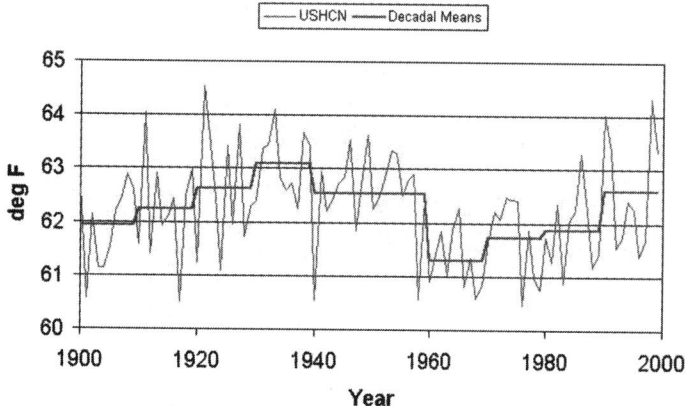

Figure 2. Decadal average temperatures in the Southeast.
Source D. Easterling, NOAA National Climatic Data Center. See Color Plate Appendix

the lower Mississippi Valley, and decreases in these categories in parts of Georgia and North Carolina (Linns and Slack, 1999). Increased precipitation intensity in extreme events during the next century is suggested by climate models under doubled CO_2 for the US (Mearns, et. al., 1995) and there is evidence that moisture in the atmosphere is increasing over the Caribbean region (Trenberth, 1999). Heavy rains are less efficient (more water runs off into the sea) and are more likely to cause flooding, which is a serious problem in the region.

Trends in wet and dry spells during the 20th century, as indicated by the Palmer Drought Severity Index (PDSI), are spatially consistent with the region's annual precipitation trends, showing a strong tendency to more wet spells in the Gulf Coast states, and a moderate tendency in most other areas. The percentage of the southeastern landscape experiencing "severe wetness" (periods in which the PDSI averages more than +3) increased approximately 10% between 1910 and 1997.

Effects of El Niño on Climate in the Southeast

The El Niño/Southern Oscillation (ENSO) phenomenon contributes to variations in temperature and precipitation that complicate longer-term climate change analysis in certain parts of the country, particularly the Southeast. ENSO is an oscillation between warm and cold phases of sea-surface-temperature (SST) in the tropical Pacific Ocean with a cycle period of 3 to 7 years. US climate anomalies (departures from the norm) associated with ENSO extremes vary both in magnitude and spatial distribution. El Niño events (the warm phase of the ENSO phenomenon) are characterized by 2 to 4°F (about 1 to 2°C) cooler average wintertime air temperatures in the Southeast. During the spring and early summer months, the region returns to near-normal temperatures. Precipitation anomaly patterns following warm events indicate that Gulf Coast states encounter wetter than normal winters (by about 1 to 2 inches per month). By the spring, the entire eastern seaboard shows increased precipitation. In summer, climate impacts of warm events are more localized; for example, drier conditions are found in eastern coastal regions, and from north Texas to northern Alabama. El Niño events also create upper atmospheric conditions that tend to inhibit Atlantic tropical storm development, resulting in fewer hurricanes, while La Niña events have the opposite effect, resulting in more hurricanes. Figure 3 depicts US Gulf of Mexico hurricane landfall trends and the probability of hurricane landfall during El Niño and La Niña years.

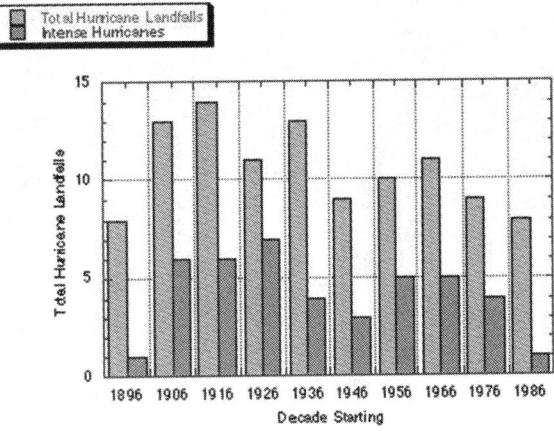

Gulf Landfalling Hurricanes by Decade

Figure 3(a). US Hurricane Landfall Trends in the Gulf of Mexico. This figure shows the number of US hurricanes making landfall in the Gulf of Mexico by decade for the past 100 years. There were peaks in activity during the 1910s and 20s, as well as a lower peak in the 1960s. The past 30 years have shown a decrease in the number and intensity of Gulf hurricanes making landfall. See Color Plate Appendix

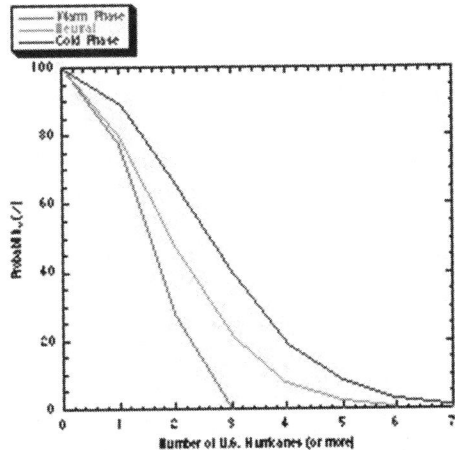

US Hurricanes

Figure 3(b). Effect of ENSO Phase on Hurricane Landfall This figure shows the probability of the number of hurricane landfalls on the US in a given hurricane season and ENSO phase (El Niño, Neutral, La Niña). Based on the past 100-year record, the probability of at least 1 hurricane landfall is similar for all three phases, with probabilities ranging from 78% for El Niño to 90% for La Niña. For multiple landfalls, however, the differences caused by ENSO phase become apparent. The probability of at least 2 landfalls during El Niño is 28%, but is 48% in neutral years, and 66% during La Niña. The probability of at least 3 landfalling hurricanes is near 0% for El Niño, 20% for neutral years, and 50% for La Niña. It is clear that El Niño years have few multiple hurricane strikes on the US, while neutral years and La Niña years often see multiple hurricane strikes on the US coast. (Source: Florida State University, Center for Ocean-Atmosphere Prediction). See Color Plate Appendix

July Heat Index Change - 21st Century

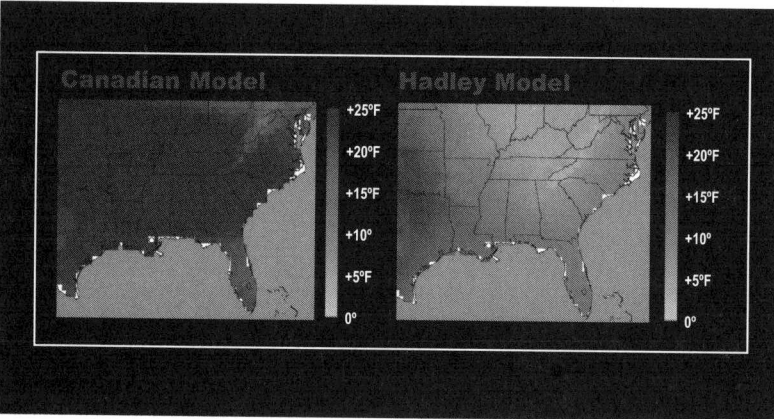

Figure 4. The changes in the simulated heat index for the Southeast are the most dramatic in the nation with the Hadley model suggesting increases of 8 to 15°F for the southern-most states, while the Canadian model projects increases above 20°F for much of the region. Heat indices simulated for the Southeast by 2100. (source, NOAA National Climatic Data Center). See Color Plate Appendix

During La Niña events (the cold phase of ENSO), the anomalies are sometimes reversed from those associated with warm events, but not everywhere. Above-average wintertime temperatures are present East of the Mississippi. By spring, the warmer anomalies in the east are focused in the Ohio Valley and northern Florida, Georgia, and South Carolina. Wintertime precipitation patterns associated with cold events show increases (1-2 inches or 2.5-5 cm per month) in the band stretching from northern Mississippi to southwestern Pennsylvania. In the spring, Gulf Coast areas have increased precipitation. In summer, the extreme southern US is colder than normal and greater precipitation is evident in the Southeast. Dry to very dry conditions are found in parts of Texas and Louisiana. Thus, as suggested by these results, the climate anomalies associated with opposite phases of ENSO are not in direct opposition. For example, climate anomalies in Florida are nearly opposite (for cold and warm events) while those in many of the midwestern states are of the same sign for both precipitation totals and temperatures during both warm and cold events. Further evidence demonstrates that climate anomalies associated with strong warm events are not amplifications of normal warm events (Rosenberg, et al. 1997).

Scenarios for the Future

The Hadley Centre climate model projects that by 2030, maximum summer temperatures in the

Southeast will increase by about 2.3°F (1.3°C) while maximum winter temperatures will increase by 1.1°F (0.6°C). The projected increase in mean annual temperature of 1.8°F (1.0°C) by 2030 and 4.1°F (2.3°C) by 2100 represents a smaller degree of projected warming than for any other region. The smaller simulated warming rate is possibly due to the buffering affects of the oceans, large amounts of surface water for evaporative cooling, and the sulfate aerosol emissions that are prevalent throughout the eastern US. Sulfate aerosols may help explain the mid-20th century cooling trend in parts of the US; however, over the past two decades, sulfate emissions decreased, and the future cooling affect of sulfate aerosols is not expected to be as important due to Clean Air Act restrictions. Although the increase in temperature under the Hadley model is small compared to other regions, the resulting increase in the summer heat index by 8 to15°F (4 to 8°C) (calculated from monthly maximum temperature and relative humidity) would likely affect human activity and, possibly, demographics in the 21st century.

The Canadian Centre climate model projects higher temperature scenarios and higher southeastern heat indices by the end of the 21st century than does the Hadley model. The Canadian model simulates an increase in mean annual southeastern temperature of about 3°F (1.7°C) by 2030 and 10°F (5.5°C) by 2100. In the Canadian model, increases in maximum summer temperature are the highest in the Nation for both 2030 (5°F or 2.8°C) and 2100 (12°F or 6.5°C). The Canadian climate model simulates an increase in average summer heat index above 15°F (8°C) across the entire region by 2100 (Figure 4).

Another important difference between the two models for the Southeast lies with the simulated changes in rainfall; the Canadian climate model simulates reduced average annual precipitation (10% less than present by 2090) while the Hadley model simulates more precipitation than present (20% more by 2090). These differences have important implications for hydrologic impacts on the Southeast, because the Canadian model simulates decreased soil moisture, while the Hadley model simulates increased soil moisture (Felzer and Heard, 1999). Differences between the two models are illustrated in Figure 5, which depicts the simulated summer climate in Georgia in 2030 and 2090. According to the Hadley climate model scenario, the Southeast will remain the wettest region of the US for the next 100 years. The precipitation changes projected by the Hadley model by 2100 are consistent with other parts of the eastern and midwestern US.

The Max Planck Institute climate model (ECHAM4/OPYC3), one of a few models with sufficient resolution in the tropics to adequately simulate narrow equatorial upwelling and low frequency waves, simulates more frequent El Niño-like conditions and stronger La Niñas under a doubling of CO_2, which is consistent with the Hadley model projections with a doubling of CO_2. The Max-Planck model also suggests that the mean climate in the tropical Pacific region will shift toward a state corresponding to present-day El Niño conditions (Timmermann, et. al., 1999). McGowan, Cayan, and Dorman (1998) showed that the frequency of warm sea surface events off the western coast of North America increased since 1977, but relationships between this trend and reduced hurricane landfall in the Gulf Coast region have not been established.

Summer Climate Changes from Hadley Centre Scenario

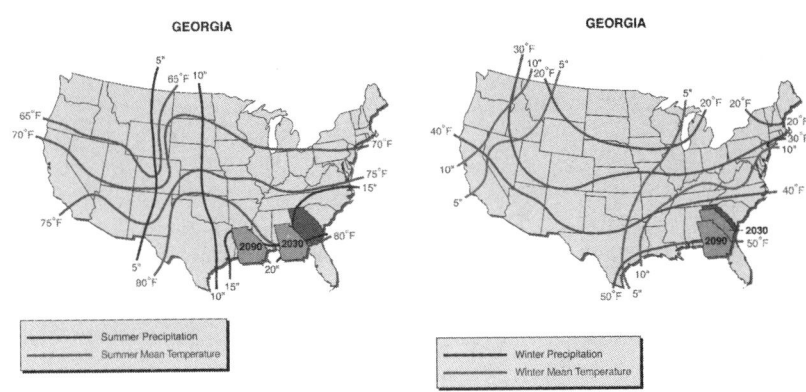

Figure 5. Illustration of how the summer and winter climates in Georgia would shift under the Hadley climate scenario (HADCM2). For example, the summer climate in Georgia in the 2030s would be more like the current climate of the Florida panhandle. Source: NOAA, National Climatic Data Center. See Color Plate Appendix

Table 1: Billion-Dollar Southeast Weather Disasters, 1980-1999

Disaster	Year	Estimated Damages/ Costs*	Estimated Deaths**
AR/TN Tornadoes	1999	$ 1.3 billion	17
TX Flooding	1998	$ 1.0 billion	31
Hurricane Georges	1998	$ 5.9 billion	16
Hurricane Bonnie	1998	$ 1.0 billion	3
Southern Drought/Heat Wave	1998	$ 6.0-9.0 billion	200
El Niño/Tornadoes and floods	1998	$ 1.0 billion	132
MS/OH Valley Floods/Tornadoes	1997	$ 1.0 billion	67
Hurricane Fran	1996	$ 5.0 billion	37
Hurricane Opal	1995	$ 3.0 billion	27
TX/OK/LA/MS Severe Weather	1995	$ 5.0-6.0 billion	32
TX Flooding	1994	$ 1.0 billion	19
Tropical Storm Alberto	1994	$ 1.0 billion	32
Southeast Ice Storm	1994	$ 3.0 billion	9
Summer Drought/Heat Wave	1993	$ 1.0 billion	***
Hurricane Andrew	1992	$ 27.0 billion	58
Hurricane Bob	1991	$ 1.5 billion	18
TX/OK/LA/AR Flooding	1990	$ 1.0 billion	13
Hurricane Hugo	1989	$ 9.0 billion	57
Hurricane Juan	1985	$ 1.5 billion	63
Hurricane Elena	1985	$ 1.3 billion	4
Florida Freeze	1985	$ 1.2 billion	0
Florida Freeze	1983	$ 2.0 billion	0
Hurricane Alicia 1999	1983	$ 3.0 billion	21
Total		$ 83.7-87.7 billion	856

* not adjusted for inflation, ** US only, *** undetermined (Source: National Climatic Data Center, 1999)

KEY ISSUES

1. Weather-related Stresses on Human Populations
2. Agricultural Crop Yields and Economic Impacts
3. Forest Productivity Shifts
4. Water Quality Stresses
5. Threats to Coastal Areas

Changes in average climate and weather extremes have important economic implications in the Southeast. There are several reasons why this region is of relatively high interest and concern. First, there is a strong ENSO signal, primarily in the Gulf Coast states, that results in seasonal and year-to-year variations in temperature and precipitation. Understanding potential future climate change in the context of current natural variability can provide an important contribution to ongoing discussions of mitigation options. A second consideration is that the Southeast experiences many extreme climate events such as hurricanes, heat waves, tornadoes, ice storms, floods, and lightning storms that cause significant economic losses to industry and local communities. The agriculture and forestry sectors make substantial contributions to the regional and national economy and these sectors are quite vulnerable to climate variability. Water resources, air quality, coastal resources, and land use are other important regional issues that may be strongly influenced by climatic trends and variability.

1. Weather-related Stresses on Human Populations

The US experienced 42 weather-related disasters over the past 20 years that resulted in damages/costs in excess of $1 billion each; 23 of these disasters occurred in Southeast states, resulting in total damages/costs of about $85 billion. Most of the property damages were associated with floods and hurricanes. Low-lying Gulf and South Atlantic coastal counties are particularly vulnerable to storm surge. Between 1978 and 1998, 56% of the National Flood Insurance Program (NFIP) policies in force and 74% of total NFIP claim payments occurred in southeastern coastal counties (Heinz Center, 1999).

In addition to the projected shift towards more frequent El Niño-like conditions in the Southeast, some climate models suggest that rainfall associated with El Niño events will increase as atmospheric CO_2 increases. Increased flooding in low-lying coastal counties from the Carolinas to Texas could possibly have adverse effects on human health. Floods are the leading cause of death from natural disasters in the Southeast and nationwide. Flooding, however, is not the only problem stemming from unusual meteorological events. The southern heat wave and drought of 1998 resulted in damages in excess of $6 billion.

El Niño and the 1998 Florida Wildfires
Florida consistently ranks among the top five states in terms of wildfire frequency and acreage affected, due largely to frequent thunderstorms and a warm moist climate that promotes lush growth of volatile understory plants. To limit the accumulation of fuels that promote disastrous wildfires, landowners in Florida routinely treat close to 2 million acres a year with prescription fire. The unseasonably warm weather and copious rainfall brought about by El Niño conditions during the winter of 1997-98 resulted in even higher plant growth than usual and high soil moisture that limited the acreage that could be treated effectively with dormant-season prescription fire. As El Niño conditions began to subside in March, 1998, record breaking rainfall changed to record-setting drought. Many prescription burns were postponed further because of the increasing probability of crown fires and root damage to trees. Lightning activity picked up as usual in May, but with lower than average rainfall. By June 1, drought indices reached record heights. During the ensuing six weeks, more than 2,500 fires burned roughly 500,000 acres in Florida, destroying valuable timber and damaging roughly 350 homes and businesses. Predicting El Niño conditions holds obvious benefits for fire preparedness and prevention. Changes in El Niño patterns would have both ecological and economic implications.

Also of concern in the Southeast are the effects that elevated surface temperatures have on human health as a result of prolonged or persistent periods of excessive summertime heat events coupled with droughty conditions. For example, it is known that urban surface temperatures in cities in the Southeast can be elevated as much as 5 to 10°F (approximately 3 to 5°C) over non-urbanized areas (Lo, et. al., 1997; Quattrochi and Luvall, 1999). These elevated urban surface temperatures are a heat stress to humans and can significantly contribute to increasing both the duration and magnitude of photochemical smog, particularly ozone concentrations (Southern Oxidants Study, 1995; Quattrochi, et. al., 1998). Increases in maximum summer temperatures are of particular concern among lower income households that lack sufficient resources to improve insulation and install and operate air conditioning systems.

Adaptation Options

Understanding the risks and vulnerability of communities to weather-related hazards (considering hidden and reported costs and the actual frequency with which these disasters occur in the Southeast) is important to the quality of adaptation strategies. Across the region, intense precipitation has increased over the past 100 years and some models suggest that this trend will continue as the atmosphere warms. Traditional approaches to mitigation such as flood proofing, elevated structures, and building codes, are no longer adequate in themselves, particularly in the coastal zone. Even if storms do not increase in frequency or intensity, sea-level rise alone will increase the propensity for storm surge flooding in virtually all southeastern coastal areas.

The National Oceanic and Atmospheric Administration (NOAA) and the Federal Emergency Management Agency (FEMA) commissioned a study on the true costs and mitigation of coastal hazards in 1996. The report of this study calls for a strategic shift in hazard mitigation and focuses on model state programs developed in Florida and others parts of the country to foster more disaster-resilient communities. Recommendations include improvements in disaster cost accounting and risk assessment, insurance/mitigation policy linkages, integrated approaches to coastal management/development, and community-based mitigation planning (Heinz Center, 1999). Changes in climate and sea-level rise should be an integral consideration as Southeast coastal communities develop strategies for hazard preparedness and mitigation. Several states have implemented permanent "set backs" or "rolling easements" to prevent further development in areas that will become more flood-prone as sea level rises (see Coastal and Marine Resources chapter).

Health advisory systems, community-wide heat emergency plans, improved weather prediction capabilities, and other adaptations that would likely reduce urban heat stress and air pollution-related health effects are presented in the Northeast and Health Chapters.

2. Agricultural Crop Yields and Economic Impacts

Current Conditions

Great agricultural changes have taken place in the Southeast over the past 150 years. In 1849, the South produced more corn than the Midwest; Southeast acreage in corn was higher than in cotton. Cotton production expanded greatly after the Civil War, and by the late 1920s, cotton was more dominant in the South than it was a century before. There was complete mechanization of crop production in the Southeast after World War II and millions of sharecroppers moved to the big cities in the North. This had an important impact, not only on labor requirements, but on the whole economic structure of agriculture. There has also been a shifting cropland base in this region. For example, over the last 50 years soybeans changed from a minor forage crop to an agricultural staple second only to corn in value of production. As soybeans and rice replaced corn and cotton, farmers chose soils most suitable for the new crops. Drained wetland soils in Arkansas were more productive in soybeans than the old Piedmont soils abandoned by cotton farmers.

In terms of agricultural potential, one of the Southeast's most important assets is its potential to expand the acreage devoted to crops beyond the current level. The land from which new cropland can be drawn is currently about evenly divided between pasture and forestland. Although the Southeast could substantially increase acreage devoted to agriculture, it fares poorly in terms of native

Table 2: Principal Crops in the Southeast

$(10^3$ acres)
(source, USDA, Census of Agriculture, 1996).

	1929	1949	1969	1987	1996
Corn	23,940	20,417	7,896	4,309	5,005
Cotton	23,228	13,031	4,711	3,345	5,931
Peanuts	2,207	2,348	1,046	971	927
Rice	598	1,011	1,194	1,654	2,156
Soybeans	1,321	2,599	13,894	25,645	12,303

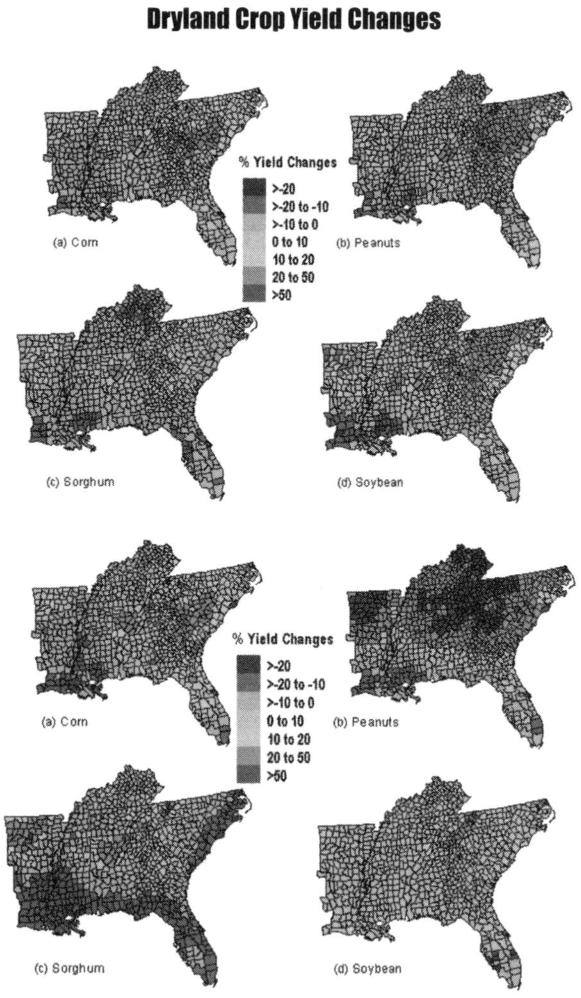

Dryland Crop Yield Changes

Figure 6. Dryland crop yield changes in 2030 (a) and 2090 (b) without adaptation for various climate sensitivity scenarios (source: Auburn University, Global Hydrology and Climate Center; University of Florida, Agricultural and Biological Engineering Department). See Color Plate Appendix

soil fertility. In addition to having low fertility, millions of acres of soils in the Southeast are moderately to severely eroded, the result of decades of continuous corn and cotton production under poor soil management. However, another of the region's agricultural assets is its latitude and proximity to the warm Gulf and Caribbean maritime influence. Overall, the Southeast has a consistent 30- to 90-day longer growing season than the Corn Belt and the Great Plains. The Southeast has enormous supplies of fresh water in the form of rainfall, surface water flowing through streams and creeks, and groundwater. Water availability gives the Southeast some substantial advantages, including irrigation possibilities that have barely been exploited.

The Southeast's mild climate and frequent rainfall predispose the region to an array of agricultural pest problems more serious than anywhere else in the nation. Agricultural pests can reduce crop yields and raise production costs. Another consequence of the region's pest problems has been relatively high use of pesticides. Although the Southeast accounts for only 14% of the nation's cultivated cropland, it consumes 43% of the insecticides and 22% of the herbicides used by farmers (USDA, Census of Agriculture 1994).

Potential Impact of Climate Change on Crop Yield and Water Use at the Field Scale

Two families of mechanistic crop simulation models CROPGRO (for soybean and peanut) and CERES (for corn, wheat, rice, and sorghum) in DSSAT V 3.5 (Tsuji et al., 1998; Jones et al., 1999) were used to simulate potential dryland and irrigated crop production using state- and crop-specific management practices throughout the Southeast. The Hadley Centre model was chosen for this analysis by the Southeast working group because this model was gridded properly for the region at the time the analysis was begun, and because the model performed best, among those tested, at hindcasting southeastern ENSO-related climate anomalies.

Crop yield changes simulated under dryland (non-irrigated) conditions suggest that yields were sensitive to the Hadley climate change and CO_2 fertilization and that the response varied by crops and locations (Figure 6). Dryland crop yields generally decreased along the Gulf Coast by 1 to 10% due to water stress. Furthermore, increased demand for water by other sectors under higher temperatures is likely to amplify these impacts. Increases in atmospheric CO_2 concentrations may reduce crop water use to some extent, due to increases in leaf stomata resistance, but cannot fully compensate for increases in crop water demand due to the higher temperatures in climate change scenarios.

In the crop simulation model, dryland corn yield increased by 1 to 30% in the Coastal Plain and decreased up to 10% in Louisiana and large parts of Mississippi, Arkansas, and Kentucky. The shorter growing cycle reduced yield while increased CO_2 and rainfall boosted yield. Due to lower water use efficiency, the model suggests that it could possibly become uneconomical to irrigate corn, prompting farmers to increase dryland corn production. Sorghum yields increased in the model results by 1 to 30% in parts of Alabama, Georgia, South Carolina, and North Carolina where seasonal rainfall increased by 5 to 15%. Simulated yields from irrigated sorghum were 4 to 7% lower almost everywhere, even where higher yields were predicted under dryland conditions, largely due to shorter growing seasons.

148

Table 3: Temperature Tolerance Limits for Various Crops

(based on the projected rainfall changes in 2030s and 2090s)

	445 ppmv CO_2 as in 2030s Change in Current Rainfall			680 ppmv CO_2 as in 2090s Change in Current Rainfall		
	-20%	0%	20%	-20%	0%	20%
	Temperature Tolerance					
Soybean	+0	+1	+3	+1	+3	+4
Peanuts	+0	+2	+3	+2	+3	+4
Corn	+0	+0	+1	+2	+3	+3
Sorghum	+0	+0	+1	+1	+1	+1
Wheat	+0	+1	+2	+3	+3	+3

Simulated soybean and peanut yields increased by 1 to 30% mostly within the Coastal Plain and mid-south and more than 30% in parts of North and South Carolina. Yields also increased, in parts of Arkansas and upper Mississippi, where dryland corn and sorghum yields decreased. Irrigated yields increased 1 to 10% over almost all of the region including where losses had been predicted under dryland conditions. The models simulated decreases of up to 30% in dryland peanut yields in the lower Delta and along the Gulf Coast, but when irrigation was added in the same areas, yield increased by over 30%. If the model-simulated changes were to occur slowly over next 25 to 45 years, farmers would be likely to slowly increase irrigation as the marginal value of irrigation increased. The spatial variation in simulated yield induced by climate change suggests that many farmers in the lower Delta and Gulf Coast may drop dryland production of peanuts while production of these crops may expand in other parts of the region.

Simulated winter-wheat yield increased in all regions except in the lower Delta and parts of Arkansas. Simulated irrigated yields increased following a similar trend as the dryland yield. Demand for irrigation increased by 20 to 50% in the Delta where rainfall decreased, and evaporation increased due to higher temperatures. Over the same period, irrigated yields declined. In parts of Arkansas and Louisiana, where irrigated rice dominates, the model simulated 1 to 10% yield losses by the 2030s and 3 to 39% increases by the 2090s. One of the major threats to rice production is increasing competition for and increasing costs of irrigation water.

Sensitivity Analysis of Crop Response to Climatic Change

To avoid the narrow range of the climate conditions simulated by the Hadley model, we conducted a sensitivity analysis in 10 agriculturally-dominant southeastern areas using 25 combinations of anticipated temperature and rainfall at 445 and 680 ppm CO_2 levels superimposed on the current climate. The sensitivity analysis identified the climate conditions that would be particularly damaging to the dryland production (see Table 3) and allowed us to consider to what extent yields can be maintained with current management practices. Results indicate that yield would likely decrease compared to the current values if temperature exceeds the current value more than the corresponding value for change in rainfall amount.

Figure 7 shows that under simulated 2030s CO_2 levels, +1°C (1.8°F) change in temperature increased dryland production of soybean, peanuts, and wheat under current rainfall, while yields of corn and sorghum declined. A 2°C (3.6°F) increase in temperature in 2030 resulted in further yield losses for all crops simulated. In contrast, the effects of 2°C (3.6°F) temperature increase in 2090 with no change in rainfall suggested a generally positive effect on crop yields. However, decreases in rainfall by 20% accompanied by temperature increases of 1 to 2°C (1.8-3.6°F) almost doubled yield losses for all dryland crops studied. Under conditions of lower or the same growing season rainfall amounts, yields of all crops increased more due to increased CO_2 levels than due to higher seasonal rainfall. For all crops and combinations of temperature and CO_2 changes, decreases in precipitation resulted in differential decreases in crop yields. Changes in yields with 20% lower rainfall were of similar magnitude at all other temperatures and CO_2 levels simulated. Furthermore, results showed that crop yields were much less sensitive to changes in temperature compared to changes in precipitation. An increase in growing season rainfall of about 20% almost completely offset the negative effect of temperature increase. Irrigated yields were simulated assuming current rainfall by varying only the temperatures

Simulated Changes in Dryland Yields for Southeastern Crops based on the Hadley (HADCM2) Scenario

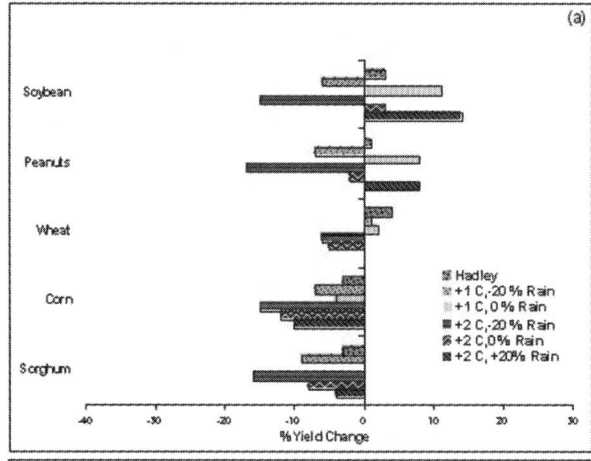

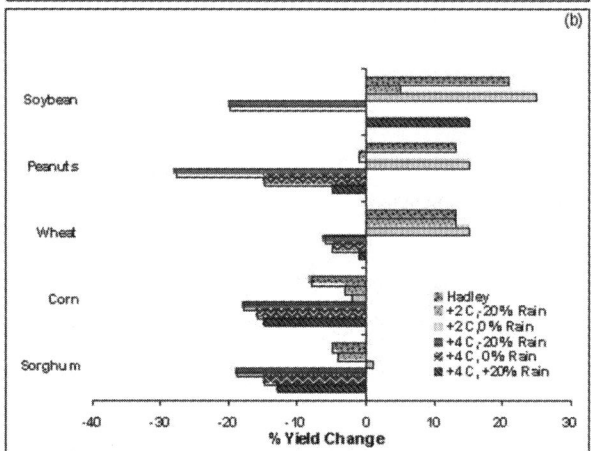

Figure 7. Dryland yields changes in 2030(a) and 2090(b) without adaption for various climate sensitivity scenarios. (source: Auburn University, Global Hydrology and Climate Center; University of Florida, Agricultural and Biological Engineering Department). See Color Plate Appendix

from +1 to +5°C (1.8 to 9°F). If future rainfall is higher, then irrigation requirements will likely decrease, and if it is lower, irrigation needs will likely increase. Most irrigated crops were sensitive to simulated climate changes and CO_2 fertilization.

Implications for Field Scale Adaptation

Future climate change may strongly affect agriculture in the Southeast. In cropping systems, a wide range of low cost adaptations to climate change may exist to maintain or even increase crop yields under future climates. Farmers may be able to respond to changes in environmental conditions by choosing the most favorable crops, cultivars, and cropping systems. Our findings indicate that changes in planting dates and maturity groups could possibly increase yield under the climate change conditions studied. In the simulations, all crops benefited from

variety and planting date adaptations either by shifting from a decreased yield without management changes to an increased yield when management was changed. For most dryland crops, adaptation did not completely eliminate yield loss under a 20% less rainfall condition. For corn, peanuts, sorghum, and wheat, there may be little need to change currently adapted varieties under all combinations of temperature and rainfall changes, in part due to strong CO_2 fertilization effects on crop yields.

Sub-regional Impacts on Productivity and Profitability

The largest threat to crop production in the Southeast appears to occur where decreases in precipitation coincide with higher temperatures. This combination of climate change would likely increase evapotranspiration demands in a climate with lower water availability, which would increase water stress and reduce crop growth and yields. In many areas of the Southeast, however, projected temperature rises of no more than 3 or 4°F (2°C or less) together with declining needs for irrigation, should enhance crop production. The Hadley model scenario for 2030 indicates improved conditions for water availability in the Tennessee Valley, coastal North and South Carolina, and the lower Mississippi Valley. Water supplies are projected to be much worse in the Mississippi Delta and slightly worse in Louisiana, Southern Alabama, the Florida panhandle, and the Coastal Plain of Georgia. By 2090 they also become much worse in Northern Mississippi, Southern Alabama, and Southwestern South Carolina.

A farm management model was used to simulate changes in crop mix, water use, and farm income associated with the climate-induced yield changes described above (Hatch et al., 1999). Of the major crop growing areas of the Southeast, the southern Mississippi Delta and Gulf Coast areas are more negatively impacted, while the northern Atlantic Coastal Plain is more positively impacted. Analyses indicate that farmers could possibly mitigate most of the negative effects and possibly benefit from changes in CO_2 and moisture that benefit crop growth. The discussion that follows is organized around the two principal row crop growing areas of the Southeast, the Mississippi Delta and Atlantic Coastal Plain.

The southern portion of the Mississippi Delta is expected to endure the severest negative impacts with the northern portion relatively less impacted. In both 2030 and 2090, simulated crop yield, water use, and income all are relatively worse off in the southern area of the Delta, particularly Louisiana,

Mississippi, and Arkansas. This picture contrasts rather sharply with the largely beneficial impacts in much of the Coastal Plain, especially the northern tier. Southern Alabama, the Panhandle of Florida, and southwest Georgia, the crop growing areas in proximity to the Gulf coast, are the areas of the Coastal Plain that are negatively impacted. The rest of this important crop growing area, that stretches from central Georgia to North Carolina, is expected to see beneficial impacts from the climate-induced yield changes and the resultant changes in farm management.

Simulated changes in water use for irrigation of row crops show a distinct north-south pattern. That is, the southern tier of the Southeast is expected to increase its needs for irrigation water whereas the northern tier is expected to decrease its relative need for irrigation water. This pattern is somewhat evident in the 2030 simulation and very pronounced in the 2090 crop and management simulations.

Economic sensitivity to increased temperatures was also investigated. Two sensitivity scenarios were analyzed to provide an indication of sensitivity to increased temperature without any changes in precipitation. The sensitivity scenarios were "hot" and "very hot;" the former was an increase of 1°C in 2030 and 2°C in 2090, and the very hot scenario increased temperature by 2°C in 2030, and 5°C in 2090. These temperature changes were selected because they roughly reflect the temperature changes associated with the Hadley and Canadian models, respectively, without the simulated changes in precipitation. The "hot" scenario had a slightly more negative impact than the Hadley scenario in many areas of the Southeast because the hot scenario did have the Hadley's accompanying increase in moisture. The "very hot" scenario produced rather dramatic negative effects, again because these were not mitigated by additional moisture.

The heterogeneous growing conditions of the Southeast and the great diversity of crops and management systems used in the region make broad generalizations about regional climate effects on agriculture very difficult. The Southeast is one region of the nation that is very likely to experience changes in the mix of crops that can be profitably grown. As a result, the Southeast is a region that will gain from improved information on climate effects and on improved dissemination of this information to farm managers. Improvements in understanding climate and forecasting weather will improve the ability of managers to deal effectively with these

and future changes, for example by providing them with forecasts based on ENSO phase (Legler, et. al., 1999).

Additional Adaptation Options
Expected changes in productivity and profitability will very likely stimulate adjustments in management strategies. As yields change, commodity prices will also change. Producers have several options by which they can adapt to changes in yield and price expectations. As previously pointed out, they can change to alternative crops. They can also grow the same crops, but adjust cultural practices, varying planting dates, seeding rates, row spacing, patterns of water usage, crop rotations, and the amounts, timing, and application methods for crop nutrients and pesticides.

Technology can also be expected to respond with new products and methods to optimize production under changing climatic conditions. Plant breeders will very likely respond by developing new varieties to accommodate climatic changes. Combinations of technological advances and adaptive management practices could very likely minimize the potential adverse effects and amplify the potential positive effects of climate change on agricultural productivity in the Southeast.

3. Forest Productivity Shifts

Current Conditions
Most of the Southeast's native forests were converted to farmland by 1920, with a large percentage of this conversion occurring prior to the Civil War. By 1860, about 43% of the total land area in the Southeast was reported as farmland, but a substantial part of the farm holdings remained in forest, which was often used as a place for grazing livestock (USFS, 1988). With continued expansion of settlements, timberland continued to decline until the early 1920s. Significant changes in agriculture took place after 1920 that caused abandonment of large areas of crop and pasturelands. These included the boll weevil infestation, which made cotton growing unprofitable in many parts of the Southeast. Some of the abandoned land was planted with trees, but the majority reverted naturally to forest leading to increases in timberland acreage (USFS 1988).

By the late 1950s and early 1960s, the decline in timberland began again in the Southeast, caused primarily by the clearing of forest for soybean and other crop production. Much of this timberland reduction occurred in the bottomland hardwood forest areas of the Mississippi Delta. Forest reductions were further fueled by growth in urban areas, highways, power lines, and related development. Throughout the 1970s, timberland was cleared for agricultural use and for an expanding export market. In the decade 1982-92, the National Resources Inventory reports roughly a half million-acre loss (less than 1%) in forestland in the Southeast.

Land use changes in the region are sensitive to any projected changes in the value of agricultural and forest lands. Expansion of urban and built-up areas in the Southeast also represents a significant demand for land, but one that will continue to be small relative to the total land base. For example, although developed land increased about 27% in the decade 1982-92, the total land use in this category represents only 8% of the total land use in the Southeast. Future land use changes are likely to have major impacts on things which do not have market prices: wildlife and habitat, topsoil, aesthetics, pollution of groundwater by agricultural chemicals, soil erosion, sedimentation, and loss of wetlands. The management of these natural resources

Potential Southern Pines and Hardwoods Net Primary Productivity (NPP)

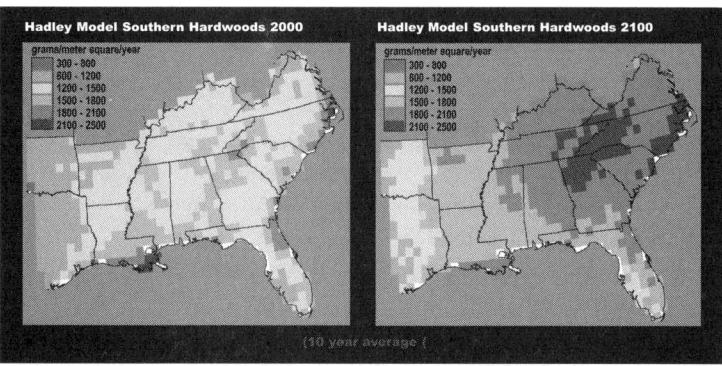

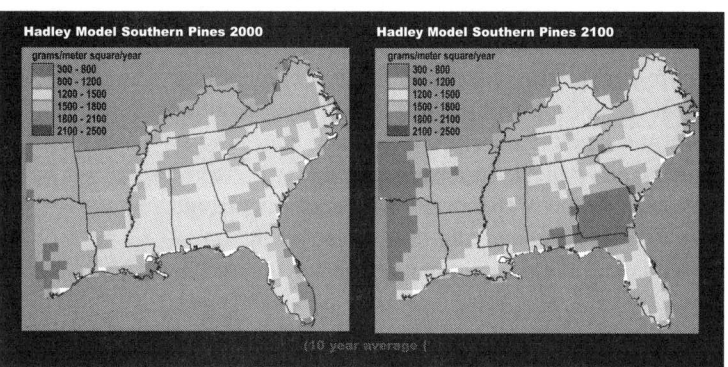

Figure 8. Potential net primary productivity (NPP) of loblolly pine and southern hardwoods simulated by the PnET model with the Hadley climate scenario (HADCM2). (Source: USDA Forest Service, Southern Global Change Program). See Color Plate Appendix

and their relationships with climate variability and change remains an integral part of the economic well being of the Southeast.

Potential Impacts of Climate Change
As part of this Assessment, PnET-II, a physiologically-based forest process model, combines climate, soil, and vegetation data to simulate annual soil water stress, drainage and biological productivity in southern US forests (Aber, et. al., 1993; McNulty, et. al., 1994). Using Forest Inventory and Analysis data from the USDA Forest Service, predictions of forest productivity per unit area were projected into regional growth using current volume for southern pine forests (USDA, 1988). Model projections of future forest productivity included the influence of doubled CO_2. The Hadley climate scenario (the wetter of the principal climate models used in the National Assessment) was used for reasons cited earlier. Total changes in standing volumes were calculated by multiplying growth per unit area by the total area of pine forest across the region.

The PnET-II model indicated a 12% increase in overall southeastern forest productivity by 2100, but there were important differences between hardwoods and pines. In model simulations using the Hadley scenario, southern pine plantations experienced an 11% increase in productivity by 2040 and an 8% increase by 2100, while hardwood and mixed pine hardwood forest (which represent 64% of the total forest area) experienced a 22% increase in productivity by 2040 and a 25% increase by 2100 compared to 1990 productivity estimates. This difference would likely be accentuated under the warmer temperatures simulated by the Canadian model. This is significant because pines (used for pulp and paper), presently account for almost two-thirds of the region's forest industry land and about half of the nation's softwood inventory. Climate models used as input in both the forest and VEMAP models suggest a northward shift in forest productivity over the next century, but they do not consider changes in management that could potentially ameliorate adverse effects. At least two ecosystem models run with the Canadian climate scenario suggest that there will be a 25 to 50% increase in fires and that part of the southeastern pine forest will be replaced by pine savannas and grasslands due to increased moisture stress (see Vegetation and Biogeochemical Scenarios Chapter).

Figure 8 shows PnET-II model outputs for southern forest net primary productivity in the Southeast at present (baseline), and at 2040 and 2090 decadal averages under the Hadley model climate scenario.

The principal factor influencing the lower increase in pine relative to hardwood productivity by 2090 is the fact that pines have greater water demands than do hardwoods on a year-round basis. Even with the increased water use efficiencies associated with increased atmospheric CO_2, the southern pines are limited by water as evapotranspiration rates increase with air temperature.

The impact of climate change on the distribution and impact of forest pests and diseases remains uncertain. Southern pine beetles caused over $900 million in damage to southern US pine forests between 1960 and 1990. Higher winter air temperatures will increase over-wintering beetle larva survival rate, and higher annual air temperatures will allow the beetles to produce more generations per year. Both of these factors could increase beetle populations. On the one hand, field research has demonstrated that moderate drought stress can increase pine resin production and therefore reduce the colonization success rate of the beetle. However, severe drought stress reduces resin production and greatly increases the susceptibility of trees to beetle infestation. Insufficient evidence currently exists to predict which of these factors will control future beetle populations and impacts (McNulty, et. al., 1998).

Potential Effects on Timber Markets
The Sub-Regional Timber Supply (SRTS) model (Abt, et. al., 2000) was developed to link forest inventory models with timber market models. The model uses estimated relationships between prices, harvest, and inventory to model market impacts of shifts in supply or demand. The SRTS model uses the spatially explicit and species specific growth changes from PnET-II to modify inventory accumulation. The cumulative nature of inventory tends to dampen the market impacts of the variability found in annual growth rates.

This analysis of the future of the forest sector comes at an important turning point in historical trends. Since the turn of the century, southern inventories have been increasing due to recovery from exploitation in the 1920s and the emergence of industrial forestry in the 1950s. During the last decade, removals of both hardwoods and softwoods have increased rapidly and are approximately equal to growth. This implies that even subtle climate change impacts may influence the direction of future inventory changes. Overall, the SRTS model (using the Hadley climate scenario) indicated that climate change would more likely favor the Mid-Atlantic over the East Gulf, and hardwoods over softwoods, and that growth over the 2000-2020 period would be sig-

nificantly lower than over the 2020-2050 time period. This, along with currently favorable growth/removal ratios in the Mid-Atlantic region, led to shifts northward in pine and hardwood harvest in all model runs.

Beyond the spatial and market adjustments to climate change within the forest sector, land-use feedbacks from the agricultural sector, discussed below, also tend to move inventory and harvest to areas with comparative advantages. Sensitivity analysis to higher temperatures (Hadley +2°C, or 3.6°F) indicated that the northern shift in inventories and markets became more pronounced and regional prices increased as the mid-Gulf region experienced significant growth declines.

Potential Effects on Forestland Area
Although forests and agriculture dominate the Southeastern landscape, the effect of a changing climate on the relative productivity of these activities is just one of many factors that will determine how the region's land will be used in the 21st century. Urban and other developed uses, while currently a relatively small part of the regional land base, have expanded substantially in the last two decades and are likely to continue to do so in the future. In recent years, much of the forest area lost to development in the region was about equally offset by gains from forest establishment on previously agricultural land due to the decline in agricultural returns. This has tended to stabilize net forest area trends while exacerbating losses in agricultural land.

Without accounting for climate change, forest area is projected to remain fairly stable in aggregate to 2040. But within the region, there are expected to be areas with substantial land use change (Figure 9). Urbanized areas in the North Carolina and Georgia piedmont and southern Florida are projected to continue the conversion of forestland and agricultural land to developed use, but on a regional basis, these losses are expected to be offset by movement of land from agriculture to forest in other areas, such as the Mississippi delta.

Relative changes in forest and agricultural returns brought on by climate change could possibly change the pattern of stable forest areas in the future if, as some scenarios suggest, agriculture can adapt to climate change in some parts of the region better than forests can. Under the Hadley base climate scenario, our model simulations suggest relatively little change in the way that land is allocated between forests and other uses between now and 2040, though some northern migration of forest area

153

Timberland Acreage Shift

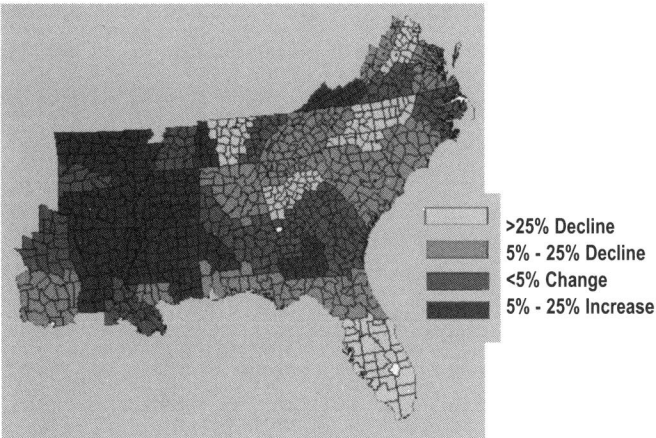

Figure 9 (a). Changes in land use based on timberland acreage shift for 1993-2040: baseline without climate change. Forestland losses are projected in the more urbanized areas of the Southeast, from northern Virginia through the Georgia piedmont and southern Florida. The movement of land from agriculture to forest is projected in many parts of the mid-South. See Color Plate Appendix

> □ >25% Decline
> ▨ 5% - 25% Decline
> ▨ <5% Change
> ■ 5% - 25% Increase

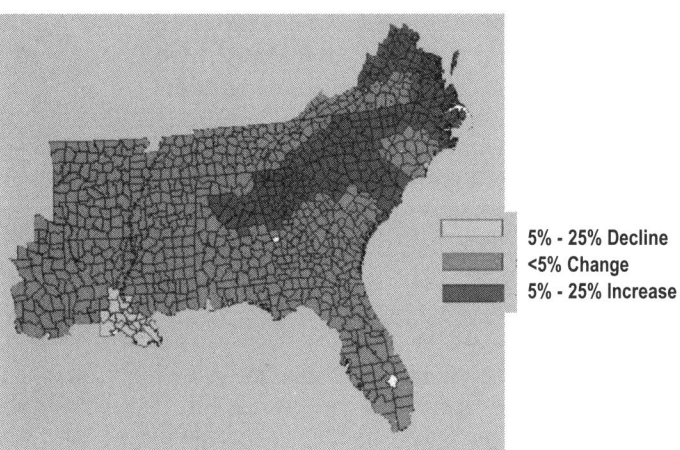

> □ 5% - 25% Decline
> ▨ <5% Change
> ■ 5% - 25% Increase

Figure 9(b). Timberland acreage shifts by 2040 due to Hadley climate change. In 2040, forestland is projected to be slightly higher with Hadley base climate change than without climate change in some of the northern reaches of the Southeast, but slightly lower under climate change in parts of the deep South. Year 2040 land allocation effects in most of the region are fairly neutral. (Source: North Carolina State University, Department of Forestry; Research Triangle Institute, Center for Economics Research). See Color Plate Appendix

is expected (Figure 9). However, sensitivity analysis reveals that substantial variation from the Hadley base scenario (e.g., +2 or 4°C, 3.6°F or 7.2°F) would likely have more dramatic effects on land allocation and lead to larger net losses in forest area. These projections should be evaluated with caution, however, as the land use model has employed more limited information on the sensitivity of agricultural economic returns to climate change than on on economic returns to forestry.

Adaptation Options

In general, the biological productivity of southeastern forests will likely be enhanced by atmospheric carbon enrichment, as long as precipitation does not decline or air temperature increase soil moisture stress to a level that would offset potential CO_2 benefits on productivity. The modeling system employed to analyze regional impacts of climate change captures some adaptation through its modeling of economic behavioral responses to changes in the biophysical conditions in forestry and agriculture. For instance, the northward shift in forest productivity is projected to lead to relative increases in the proportion of regional timber harvests that come from the northern reaches of the region. This will tend to compensate for the biophysical effects of climate by reducing harvest pressures in the more negatively affected southern parts and increasing pressure in the more positively affected northern parts. In addition, landowners are projected to shift land between forests and agriculture in places and at times where the change in relative productivity warrants it.

Other potential adaptation strategies not modeled in this Assessment include genetic and silvicultural system improvements that increase water use efficiency or water availability. The knowledge of the role of fire, hurricanes, droughts, and other natural disturbances will be important in developing forest management regimes and increasing stand productivity in ways that are sustainable over the long term. Under a hotter, drier climate, an aggressive fire management strategy could prove to be very important in this region.

Timber productivity associated with increased temperature, growing season length, and carbon enrichment may be further enhanced by improved genetic selection, bio-engineering, use of marginal agricultural land for tree production, and more intensive forest management. Reduction of air pollutants (e.g., ozone, nitrogen oxides) could also be an important strategy for increasing forest productivity.

4. Water Quality Stresses

Current Conditions

The Southeast has abundant surface water resources, most of which are intensively managed. Almost all major river systems in the region have been dammed and there are few minor streams that have not been affected by landscape alteration, channelization, surface or ground water withdrawals, or other human activities. Based on 50-year or longer streamflow records at 395 stations, Lins and Slack (1999) found that the conterminous US is getting wetter. Exceptions were noted in parts of the Southeast and the Pacific Northwest, where some stream gauges showed a decrease in minimum daily discharge. Parts of Florida and Georgia appear to be experiencing a trend towards decreased minimum flows, while the Lower Mississippi River Valley stations showed an increase in both annual median daily and annual minimum daily discharge (Lins and Slack, 1999). In Louisiana, when Keim and others (1995) modeled historical streamflow based on precipitaion, streamflow per unit drainage area increased significantly since 1900.

Potential Impacts of Climate Change

Changes in climate that result in decreased runoff during early summer generally reduce water quality in the Southeast (Mulholland, et. al., 1997). Summer low flows occur when water quality (particularly dissolved oxygen) of many southeastern streams and rivers is at its lowest (Meyer, 1992). Reduced dissolved oxygen during summer months can result in massive fish kills and harmful algal blooms in both coastal and inland waters.

An assessment of southeastern water quality associated with changes in climate was conducted using EPA's GIS-based BASINS model to evaluate current and future water quality conditions under both mean and extreme hydrologic conditions in the Southeast. Analyses were conducted for each of the US Geological Survey's eight-digit Hydrologic Unit Codes (HUCs). Water quality indices included dissolved oxygen, nitrogen and nitrates, and pH. The assessment included three steps: identification of basins with current and potential water quality problems, prediction of general change in stream flow conditions under scenarios of future climate, and re-evaluation of affected basins using the Hadley climate model scenarios for 2030 and 2100.

While water quality problems across the Southeast are not critical under current conditions, quality attainment status is not met in several cases during the majority of the year, and can become critical under extreme low flow conditions during some portions of the year. Stresses on the water quality of the Southeast appear to be associated with intensive agricultural practices, urban development, coastal processes, and possibly mining activities. As might be expected, the impacts of these stresses appear to be more frequent during extreme conditions, probably associated with dry weather. Analysis of the current status (based on 1990-97 observations) of the watersheds in the Southeast for dissolved oxygen revealed few problems under average conditions.

However, it must be recognized that because the BASINS database is indexed to major USGS hydrologic units, it necessarily consists of observations of conditions on the larger streams in the regions. Thus, smaller tributaries may exhibit water quality degradation that was not apparent at the larger scale of this analysis.

The analysis suggested that only scattered HUC watersheds in a few states currently exhibit dissolved oxygen (DO) conditions below, or nearly below the recommended 5 mg/l during average conditions. However, dissolved oxygen problems under extreme low flows do arise at a few locations in most, if not all, states. Nitrate levels in streams in the Southeast are used as an indication of nutrient content in these HUCs. While some nutrients are essential for ecosystem health, excessive levels can result in harmful conditions such as algal blooms, which can negatively impact DO levels. The current status (1990-97) based on observations for total average nitrate nitrogen content for streams of the Southeast reveals that many exhibit levels above 0.5 mg/l and in some cases above 4 to 5 mg/l, which is 3- to 4-times higher than levels common in most southeastern streams.

One interesting observation is that streams that currently exhibit low dissolved oxygen levels do not correspond to those basins where nutrient levels appear to be high. However, in both cases, the problems are most prevalent in watersheds with intensive forestry or agricultural operations.

Climate scenarios for the southeastern US provide contrasting results in terms of temperature and precipitation estimates over the region, so that in some cases conditions may improve while in others they may degrade. The Canadian model results show little overall change until 2030, followed by drier weather in most of the region over the next seventy years. On the other hand, the Hadley Centre model predicts a slight decrease in overall precipitation over the region during the next 30 years, after which precipitation increases significantly. The Hadley model results also show significantly decreased precipitation during the first six months of the year with rainfall returning to normal, or near normal, for the last six months, particularly by the end of the century. These results are particularly striking for the immediate Gulf Coast region and indicate that this area may be exceptionally vulnerable to degraded conditions. Intensive agricultural activity including disking and planting in the early spring, fertilizer application in the late spring/early summer, and harvesting in the fall may significantly

155

exacerbate water quality conditions during this period.

Preliminary hydrologic analyses based on the Hadley scenario suggest that streamflow in the Southeast (particularly along the Gulf Coast) may decline by as much as 10% during the early summer months over the next 30 years (Cruise, et. al., 1999; Ritschard, et. al., 1999). These results lead to the conclusions that water quality conditions may become critical during more frequent periods of extreme low flow. Correlation of the hydrologic analyses with the land use in basins where water quality problems already exist suggests that the problems may be most acute in areas of intensive agricultural activity, in coastal areas, or near coastal streams (Cruise et al., 1999).

Many of the basins with high nitrate levels form the boundary between two states. The Chattahoochee boundary between Georgia and Alabama and the Tombigbee boundary between portions of Alabama and Mississippi are two outstanding examples. The Hadley scenario suggests decreased water availability throughout much of this region over the next 50 years. As streamflow and soil moisture decrease, intensity of fertilizer application may increase and irrigation needs may become critical. These issues would likely lead to intense competition for scarce water resources and conflicts between these states over runoff treatment and water quality.

Water quality is also a concern in nearshore marine environments. Both the Canadian and Hadley climate models suggest an increase in rainfall in the Upper Mississippi Valley (see Great Plains and Midwest chapters). A large (8,000 to 18,000 km^2 during 1985-1997) zone of oxygen-depleted (hypoxic) coastal waters is found in the north-central Gulf of Mexico and is influenced in its timing, duration, and extent by Mississippi River discharge and nutrient flux (Rabalais, et. al., 1999; Justic et al., 1997). Nitrate delivered from the Mississippi Basin to the Gulf of Mexico, principally from non-point agricultural sources, is now about three times larger than it was 30 years ago as a result of increases in nutrient loading per unit discharge (Goolsby, et. al., 1999). Hypoxia, which is most prevalent in the lower water column, can adversely affect marine life and is a growing concern to those who harvest and manage Gulf fisheries. An increase in Upper Mississippi Basin streamflow, where the majority of the nitrogen and phosphorus loading occurs, portends an increase in the hypoxic zone offshore.

5. Threats to Coastal Areas

Current Conditions

Few regions have the combination of special characteristics and vulnerabilities found in southeastern coastal areas. The interaction of sea-level rise, storms, beach erosion, subsidence, salt water intrusion, urban development, and human population growth, and shifts in the transition zone where land meets ocean, creates conditions for potential adverse effects on the largest segments of the southeastern population. Large cities located in the coastal zone (such as Houston, Charleston, and New Orleans) already suffer frequent and severe flood damages.

Potential Impacts of Climate Change

Sea-level rise is regarded as one of the more certain consequences of increased global temperature, and sea level has been rising gradually over the past 15,000 years. Globally, average sea level rose 4 to 8 inches (10 to 20 cm) during the past 100 years and this average rate is projected to accelerate 2 to 5-fold over the next 100 years (IPCC, 1998). Parts of the City of New Orleans that are presently 7 feet below mean sea level may be 10 or more feet below sea level by 2100, due to a combination of rising sea level and subsidence of the land surface.

Low-lying marshes and barrier islands of the southeastern coastal margin are considered particularly vulnerable to sea-level rise, but all are not equally vulnerable. Cahoon and others (1998) found that some Gulf coastal marshes and one mangrove site in south Florida are being gradually submerged because they do not accumulate sediment quickly enough to keep up with present rates of sea-level rise. In coastal Louisiana, landforms created by Mississippi River sediment deposition over the past 8,000 years are naturally de-watering and compacting. As sea level rises, inundation or displacement of coastal wetlands and barrier islands is occurring. The impacts of subsidence (the lowering of the land relative to sea level) are aggravated by human activities such as levee construction along the Mississippi River, ground water withdrawals, and canal dredging though marshes, passes, and barrier islands. Changes in tidal amplitude have caused salt-water intrusion into many formerly fresh and brackish water habitats. Roughly one million acres of south Louisiana wetlands have been converted to open water since 1940, and Louisiana's barrier islands have eroded to two-thirds of the size they were in 1900.

If sea-level rise accelerates during the 21st century as predicted by the Intergovernmental Panel on Climate Change (IPCC), many other southeastern coastal areas will experience shoreline retreat and coastal land loss. Under the IPCC's "best estimate" of average global sea-level rise over the next 100 years, the Big Bend area of the Florida Gulf coast will likely undergo extensive losses of salt marsh and coastal forest (Doyle, 1998, Figure 11). Since 1980, losses of coastal forests in parts of Florida, South Carolina, and Louisiana have been attributed to salt water intrusion and/or subsidence. Since 1991, landowners and public land managers in Florida have observed extensive die-offs of Sabal palm along a 40-mile stretch of coast between Cedar Key and Homosassa Springs. Williams and others (1999) attribute the forest decline to salt water intrusion associated with sea-level rise. Since 1852, when the first topographic charts were prepared of this region, high tidal flood elevations have increased approximately 12 inches.

Rising sea levels due to climate warming may also affect estuaries and aquatic plant communities. Sea-level rise reduces the amount of light reaching sea-grass beds (light penetration decreases exponentially with water depth), thereby reducing growth rates. Some marine grass beds may be eliminated because their shoreward migration is impeded by shoreline construction and armoring (Short and Neckles, 1999). Increased tidal range associated with sea-level rise may have deleterious effects on estuarine and fresh water submerged aquatic plants by altering both salinity and water depth. (See Compounding Stresses on Major Estuaries and Bays in the Northeast chapter for a discussion of sea-level rise effects on salinity in the Chesapeake Bay.)

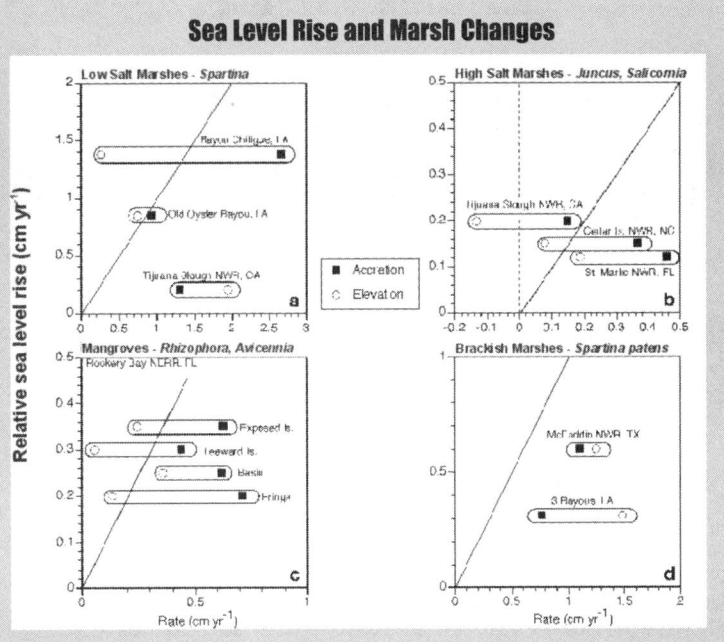

Figure 10. Relationship of vertical accretion and marsh surface elevation change with local relative sea level-rise for sites located in a) low salt marsh, b) high salt marsh, c) mangrove forest and d) brackish marsh. The diagonal line indicates parity between accretion or elevation change and sea-level rise. (source: Cahoon et al., 1998)

Coastal Wetland Vulnerability

Thresholds at which sea-level rise results in coastal wetland loss vary among sites due to differences in rates of vertical accretion (sediment build up) and local subsidence or uplift processes. Cahoon and others (1998) estimated the potential for submergence of 10 southeastern wetlands by simultaneously measuring surficial sediment accretion and soil surface elevation changes, and then comparing these rates to observed sea-level change from tide gauges. Three of the 10 sites are experiencing a net elevation deficit relative to sea-level rise. The other sites are presently accumulating enough sediment to keep pace with sea-level rise. If sea-level rise were to accelerate 4-fold, the Oyster Bayou site would be submerged by about the year 2045. The Oyster Bayou site would not be submerged if sea-level rise increases 3-fold or less, unless the site is impacted by a hurricane or other disturbance. Both long-term processes (e.g., accretion, compaction, and decomposition) and episodic events (e.g., hurricanes) affect the threshold at which coastal wetlands are submerged by sea-level rise. (See Figure 10)

Changes in Florida's Big Bend

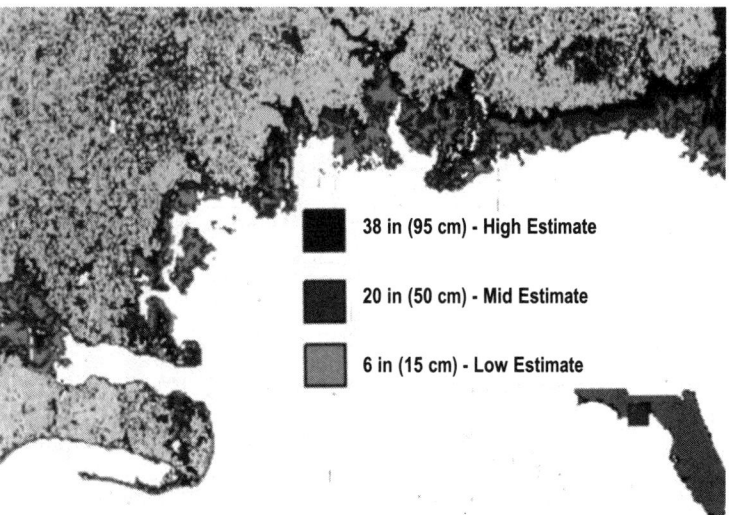

38 in (95 cm) - High Estimate

20 in (50 cm) - Mid Estimate

6 in (15 cm) - Low Estimate

Figure 11. Changes in Florida s Big Bend region forest, marshes, and open water under IPCC (1998) sea-level rise scenarios. (Source: Doyle, 1998). See Color Plate Appendix

Storm surge is intensified as sea level rises and natural coastal defenses deteriorate. It is important to note that even if there is no significant climate-change-driven increase in the frequency or intensity of Atlantic hurricanes, these storms will be more damaging when making landfall on coastal regions as sea level rises and coastal landforms erode. South Atlantic and Gulf coastal populations increased 15% between 1980 and 1993. During this period, the value of insured coastal property in the US increased by 179%(Insurance Institute for Property Loss Reduction and Insurance Research Council, 1995). Florida topped the list of coastal states with potential hurricane damages, with $872 billion in insured properties. Insurance Research Council demographic projections suggest that the number of persons living in the most hurricane prone counties will have increased to 73 million by 2010, a doubling since 1995 (Insurance Institute for Property Loss Reduction and Insurance Research Council, 1995).

During the past 30 years, the Gulf of Mexico has seen a decrease in the number of hurricanes making landfall. Hurricanes Andrew, Hugo, and Brett are the only Category 4 storms to make landfall since 1969, but since 1900, 35% of all hurricanes hit Florida and along the middle Gulf Coast, and 50% or more of all hurricanes making landfall are Category 3 or higher (Heinz Center, 1999). Property losses due to hurricanes increased from less than $5 billion per decade between 1900 and 1940 to about $15 billion per decade during the 1960s to 1980s; however, the

number of deaths attributed to hurricanes declined dramatically since the 1950s (Pielke and Pielke, 1997).

Adaptation Options
There are few practical options for protecting coastal communities and ecosystems as a whole from increased temperature, changes in precipitation, or rapidly rising sea level. Still, a variety of management measures could be applied on a site-by-site basis to increase the resiliency of specific communities and ecosystems or to reduce or partially compensate for impacts. Many of these measures could be justified based solely upon non-climate threats to coastal regions. For example, increased protection for existing coastal wetlands and removal of other stresses (such as dredge-and-fill activities and water pollution) may not only reduce the sensitivity of coastal communities, wetlands, and barrier islands to small changes in average sea level but also achieve broader conservation goals (Burkett and Kusler, 2000).

Other no-risk measures for achieving broader objectives and reducing climate change impacts include: limiting construction in areas where coastal wetlands may be displaced as sea level rises; installing sediment diversions for dams; linking presently fragmented wetlands and waters to provide the corridors needed for plant and animal migration; using water control structures for some wetlands to enhance particular functions and address decreased precipitation and/or increased evaporation; increasing management programs for invasive species control; and implementing various coastal restoration measures (Burkett and Kusler, 2000).

ADDITIONAL ISSUES

Six additional climate-related issues for the Southeast region are:

- Climate Model Limitations: Existing general circulation models cannot adequately resolve some components of climate or certain geographic or topographic features that are important because of their interaction with regional climate features. For coastal regions, much uncertainty exists about the effects of global climate change and variability on tropical storms, the most important natural hazard affecting regional vulnerability. Effects of climate change on areas of hurricane origination and threshold for hurricane formation, intensity, frequency, and tracks are poorly understood.

• Water Resources: Fresh water plays an important role in many sectors including coastal resources, health, agriculture, estuarine fisheries, and forestry. Competing demands from urban development, agriculture, and recreation for already stressed ground water systems would likely be exacerbated by changes in precipitation and salt water intrusion due to sea-level rise.

• Impacts on Coastal Ecosystems and Services: Sea-level rise, changes in fresh water delivery to coastal estuaries, and increased atmospheric temperature and CO_2 all portend changes in the structure and function of coastal and estuarine systems. Losses of coastal marshes and submerged aquatic vegetation will have impacts at higher trophic levels. Gulf coastal states currently produce most of the Nation's shrimp, oysters, and crabs, and each of these estuarine fisheries is dependent upon the primary productivity of coastal ecosystems.

• Health Issues related to Water Quality: The effects on surface waters of changes in precipitation have important health implications in the region. Increased precipitation promotes the transportation of bacteria as well as other pathogens and contaminants by surface waters throughout the region. Health consequences may range from shellfish infections transmitted to humans to ground water contamination associated with saltwater intrusion.

• Socioeconomic and Insurance Issues: In the ultimate analysis, the issue of climate change and the need for an assessment of potential consequences will be relevant to the degree that it is placed on a human scale. To this end, the potential societal impacts of climate change must be identified and understood. Insurance exposure and/or the insurability of coastal and island facilities are issues that should be examined in the context of climate change and variability.

• Urban Issues: A distinctive characteristic of southeastern coastal regions is their current high level of urbanization and rate of population growth, which could possibly be affected by changes in climate that are presented in this Assessment. The urban environment will have its own responses to the impacts of climate change and variability. While responses will be driven by stakeholder and policymaker decisions, they need to be evaluated and understood within the framework of different potential regional scenarios of climate change consequences. Design and construction factors, building code issues, infrastructure and lifelines, energy use, structural vulnerability to natural hazards, land use and zoning issues, traffic patterns, and evacuation/shelter infrastructure are all important areas for potential mitigation and future research. The potential impacts of climate change on oil, natural gas, and navigation infrastructure is also of concern in the region.

CRUCIAL UNKNOWNS AND RESEARCH NEEDS

Precipitation Uncertainties

The Southeast is the only region for which current climate models simulate large and opposing changes in precipitation patterns over the next 100 years. The range of differences is so great that it is difficult to state with any degree of confidence that precipitation will increase or decrease in the Southeast over the next 30-100 years as atmospheric CO_2 increases. Until climate models are improved (or until there is a way to validate and compare the accuracy of the existing models), people in this region must consider a wide range of potential future changes in soil moisture and runoff, as we have in this Assessment.

Human Health

There are serious human health concerns related to the plausible increases in maximum temperature for this region, particularly among lower income households. Similarly, the flood prone nature of coastal counties from the Carolinas to Texas could have significant human health implications in addition to the economic losses discussed earlier. Approaches for modeling human health effects should include these aspects of the population in addition to the climatic science. Water quality degradation could possibly become a more serious human health problem in the region; improvements in our capability to model streamflow and water quality in both inland and coastal waters are needed.

Agriculture and Forestry

Our current understanding of the potential consequences of climate change on agriculture in the Southeast is focused on a few key row crops. In the future, it will be important to include the affects of climate variations on other high value crops (e.g., citrus and vegetables) and on animal management practices. Furthermore, the role of climate on pests and pest management systems needs to be included in future assessments. It will also be necessary to

159

develop, validate, and evaluate new technology capabilities such as new genetic varieties that will help farmers cope with any changes in future climate. The effects of biotechnology (e.g., transgenic crops) on future agricultural productivity in this region, including the benefits of using less fertilizer, pesticides, or water, need to be evaluated in light of plausible climate scenarios. Because the incorporation of new climate-related technological capabilities into agriculture is relatively new and yet unproven, future pilot studies should explore the communication of such information to the agriculture community.

The potential effects of climate change on agricultural prices are an equally complex interaction of physical effects and managerial responses worldwide. Spatial equilibrium economic models that would address these market issues require that the information from all regions be reasonably similar. In the case of climate change, detailed information from one region set against very general information from many important competing crop growing areas would not provide a consistent framework for understanding worldwide response in the agriculture sector. Thus, it would be very useful to investigate in greater detail climate-induced production effects in major international crop areas to integrate such farm management results from important growing areas worldwide to address potential climate-induced price effects.

Although extensive laboratory and field research has been completed on the individual impacts of changing air temperature, precipitation, ozone, carbon dioxide, and nitrogen availability on forest productivity, water use, and carbon sequestration, there is still little understanding of the synergistic impacts of environmental change on southern forests. Field experiments with multiple treatment factors (e.g., variables) are quite costly, and there are scaling problems associated with laboratory experiments. Therefore, improved development, testing, and validation of integrated stress impacts through computer modeling are crucial future research needs. Models can provide a mechanism for examining changing atmospheric and socioeconomic impacts on forest structure and function. However, before any confidence can be given to such model projections, priority needs to be given to testing, model verification, and analysis.

Aquatic and Coastal Resources

If precipitation patterns continue to change on a scale similar to that observed over the past 100 years, many southeastern aquatic ecosystems, including estuaries, will be affected by changes in streamflow. There are several additional unknown variables corresponding to future conditions that might affect the quality of southeastern water resources. They fall into two categories: future pollutant loadings (natural and anthropogenic) and biophysical reactions. The pollutant loadings, both point and non-point, will be directly related to changes in land use activity including the presence of confined animal systems and growing population centers. Also, atmospheric deposition of nitrogen will be tied to continued emissions of nitrogen oxides (NOx). The biophysical reactions on the land surface that might serve to uptake nitrogen and other constituents will be associated with land cover conversion and vegetation. Future research programs that include these three critical unknowns seem crucial to gaining a clearer understanding of the relationship between variations in climate and water quality in the Southeast.

Quantitative data describing the response of native southeastern plant communities to atmospheric carbon enrichment, water quality (e.g., salinity), and changes in temperature and soil moisture are limited to a few key species. Moreover, very few studies have addressed the potential interactions among climate variables and between plant species, and even fewer studies deal with climate effects at secondary and higher levels in the food web. Several recent studies suggest that a number of invasive species will be favored by climatic change in the Southeast, such as the freeze-intolerant Chinese tallow, which is now a serious invader in the Gulf coastal plain. Models that integrate environmental change, species responses, and interactions among species are needed to describe pathways that are likely to alter plant and animal community structure. Research is also needed to determine secondary and higher-order effects on ecosystem goods and services. Carbon sequestration is one ecological function that has been poorly described in this region of abundant wetlands and forests that play potentially significant roles as carbon sinks.

Research and demonstration projects are needed to identify and prioritize methods that may be implemented to minimize the adverse ecological effects of climatic change on native southeastern flora and fauna. Monitoring is needed to evaluate long-term trends in the abundance and distribution of native and non-native species, focusing on species and groups that are considered highly sensitive to the range of predicted climatic changes in the region (e.g., amphibians in pine flatwood ponds, benthic invertebrates in coastal estuaries, and salt-tolerant invasive aquatic plants that could out-compete native plant species that are important wildlife food sources).

Extreme Events/Disturbance Patterns

Changes in disturbance patterns (e.g., hurricanes, floods, droughts) are possibly more significant in terms of potential economic losses than longer-term changes in precipitation or temperature. Ecosystems are also impacted by climate-related disturbance. Disturbance is a natural process that, in many cases, not only structures ecosystems but sustains them as well. Our limited understanding of the role of disturbance in natural ecosystems and our inability to predict climate extremes is problematic for those interested in mitigating the potential adverse impacts of climate change. Research should be undertaken to examine potential changes in disturbance regimes that may be expected as the climate warms and precipitation patterns change. Disturbance topics should not be limited to weather events, however. Fire, harmful algal blooms, and insect outbreaks are ecological disturbances that may be heavily influenced by climatic conditions. The ecological effects of these types of disturbances are difficult to model or predict because they are often poorly understood. Basic information is needed to identify ecological changes that are likely to occur as the type, frequency, and spatial patterns of disturbance are altered as the climate changes.

LITERATURE CITED

Aber, J. D., C. Driscoll, C. A. Federer, R. Lathrop, G. Lovett, J. M. Melillo, P. Steudler, , and J. Vogelmann, A strategy for the regional analysis of the effects of physical and chemical climate change on biogeochemical cycles in northeastern (U.S.) forests, *Ecological Modeling, 67*, 37-47, 1993.

Abt, R. C., F. W. Cubbage, and G. Pacheco, Forest resource assessment using the subregional timber supply model (SRTS), *Forest Products Journal, 50*(4), 25-33, 2000.

Acock, B., V. R. Reddy, H. F. Hodges, D. N. Baker, and J. M. McKinion, Photosynthetic response of soybean canopies to full-season carbon enrichment, *Agronomy Journal, 77*, 942-47, 1985.

Allen, L. H., K. J. Boote, J. W. Jones, A. J. Rowland-Bamford, G. Bowes, D. A. Graetz, K. R. Reddy, Temperature and CO_2 effects on rice: 1988. US Department of Agriculture, Carbon Dioxide Research Division, Washington, DC, 1989.

Boyer, W. D., Regeneration of natural longleaf pine forest, *Journal of Forestry, 77*, 572-575, 1979.

Burkett, V., and J. Kusler, Climate change: Potential impacts and interactions in wetlands of the United States, *Journal of American Water Resources Association, 36*(2), 313-320, 2000.

Cahoon, D. R., J. W. Day, D. J. Reed, and R. S. Young, Global climate change and sea-level rise: estimating the potential for submergence of coastal wetlands, in *Vulnerability of Coastal Wetlands in the Southeastern United States: Climate Change Research Results*, US Geological Survey, Biological science report USGS/BRD/BSR 1998-0002, pp. 21-34, 1998.

Cruise, J. F., A. S. Limaye, and N. Al-Abed, Assessment of the impacts of climate change on water quality in the southeastern United States, *Journal of American Water Resources Association, 35*(6), 1539-1550, 1999.

Currie, D. J., Energy and large-scale patterns of animal- and plant-species richness, *American Naturalist, 137*(1), 27-49, 1991.

Dahl, T. E., Wetland losses in the United States: 1780's to 1980's, US Department of Interior, Fish and Wildlife Service, Washington DC, 21 pp., 1990.

Doyle, T. W., Modeling global change effects on coastal forests, in *Vulnerability of Coastal Wetlands in the Southeastern United States: Climate Change Research Results*, US Geological Survey, Biological Science Report USGS/BRD/BSR 1998-0002, Lafayette, Louisiana, pp. 67-80, 1998.

Eamus, D., and P.G. Jarvis, The direct effects of increases in the global atmospheric CO_2 concentration on natural and commercial temperate trees and forests, *Advances in Ecological Research*, *19*, 1-55, 1989.

Felzer, B., and P. Heard, Hydrological implications of GCM results for the US national assessment, *Journal of American Water Resources Association*, *35*(6), 1327-1339, 1999.

Glasmeier, A., The history of Central and Southern Appalachia: A socioeconomic analysis, paper presented at the Central and Southern Workshop on Climate Variability and Change, Morgantown, West Virginia, May 27-28, 1998.

Goolsby, D. A., W. A. Battaglin, G. B. Lawrence, R. S. Artz, B. T. Aulenbach, R. P. Hooper, D. R. Keeney, and G. J. Stensland, Flux and source of nutrients in the Mississippi-Atchafalaya River Basin, In Review, White House Office of Science and Technology Policy, Committee on Environment and Natural Resources, Hypoxia Working Group, May 1999, pp. 13-37, 2000.

Harcombe, P. A., R. B. W. Hall, J. S. Glitzenstein, E. S. Cook, P. Krusic, and M. Fulton, Sensitivity of Gulf Coast forests to climate change, in *Vulnerability of Coastal Wetlands in the Southeastern United States: Climate Change Research Results*, US Geological Survey, Biological Science Report, USGS/BRD/BSR 1998-0002, Lafayette, Louisiana, pp. 47-67, 1998.

Hatch, U., S. Jagtap, J. Jones, and M. Lamb, Potential effects of climate change on agricultural water use, *Journal of American Water Resources Association*, *35*(6), 1551-1562, 1999.

Hefner, J. M., Southeast wetlands: Status and trends, mid-1970's to mid-1980's, US Department of Interior, Fish and Wildlife Service, Atlanta, Georgia, 32 pp., 1994.

Heinz Center, *The Hidden Costs Of Coastal Hazards: Implications For Risk Assessment And Mitigation*, Island Press, Washington, DC, 220 pp., 1999.

Insurance Institute for Property Loss Reduction and Insurance Research Council, Community exposure and community protection: Hurricane Andrew's legacy (IIPLR and IRC). Wheaton, IL (IILPR) and Boston, MA (IRC), 1995.

Intergovernmental Panel on Climate Change, *The Regional Impacts of Climate Change: An Assessment of Vulnerability*, Cambridge University Press, United Kingdom, 514 pp., 1998.

Jones, J. W., S. S. Jagtap, and K. J. Boote, Climate change: implications of soybean yield and management in the USA, paper presented at the World Soybean Research Conference VI, Chicago, Illinois, August 7, 1999.

Justic´, D., N. N. Rabalais, and R. E. Turner, Impacts of climate change on net productivity of coastal waters: Implications for carbon budget and hypoxia, *Climate Research, 8*, 225-237, 1997.

Keim, B. D., G. E. Faiers, R. A. Muller, J. M. Grymes, and R. V. Rohli, Long-term trends of precipitation and runoff in Louisiana, USA, *International Journal of Climatology*, *15*, 531-541, 1995.

Kiehl, J. T., Climate change enhanced: Solving the aerosol puzzle, Science, *283*(5406), 1273-1275, 1999.

Legler, D. M., K. J. Bryant, and J. J. O'Brien, Impact of ENSO-related climate anomalies on crop yields in the US, *Climatic Change, 42*(2), 351-375, 1999.

Lins, H. F., and J. R. Slack, Streamflow trends in the United States, *Geophysical Research Letters*, *26*(2), 227-230, 1999.

Lo, C. P., D. A. Quattrochi, and J. C. Luvall, Application of high-resolution thermal infrared remote sensing and GIS to assess the urban heat island effect, *International Journal of Remote Sensing*, *18*, 287-304, 1997.

Lockwood, J. G., Is potential evapotranspiration and its relationship with actual evapotranspiration sensitive to elevated CO_2 levels?, *Climatic Change, 41*(2), 193-212, 1999.

McGowan, J. A., D. R. Cayan, and L. M. Dorman, Climate-ocean variability and ecosystem response in the northeast Pacific, *Science, 281*, 210-217, 1998.

McNulty, S. G., P. L. Lorio, Jr., M. P. Ayres, and J. D. Reeve, Predictions of Southern Pine Beetle populations using a forest ecosystem model, in *The Productivity and Sustainability of Southern Forest Ecosystems in a Changing Environment*, edited by R. Mickler and S. Fox, Springer Publishing, pp. 617-634, 1998.

McNulty, S. G., J. M. Vose, W. T. Swank, J. D. Aber, and C. A. Federer, Landscape scale forest modeling: Data base development, model predictions, and validation using a GIS, *Climate Research, 4*, 223-231, 1994.

Mearns, L. O., F. Giorgio, L. McDaniel, and C. Shields, Analysis of daily variability of precipitation in a nested regional climate model: Comparison with observations and doubled CO_2 results, *Global Planetary Change, 10*, 55-78, 1995.

Meyer, J. L., Seasonal patterns of water quality in black-water rivers of the coastal plain, southeastern United States, in *Water Quality in North American River Systems*, edited by C. D. Becker and D. A. Neitzel, Battelle Press, Columbus, Ohio, pp. 249-276, 1992.

Mulholland, P. J., G. R. Best, C. C. Coutant, G. M. Hornberger, J. L. Meyer, P. J. Robinson, J. R. Stenberg, R. E. Turner, F. Vera-Herra, and R. G. Wetzel, Effects of climate change on freshwater ecosystems of the southeastern United States and the Gulf Coast of Mexico, *Hydrological Processes, 11*, 949-970, 1997.

National Climatic Data Center, Billion dollar US weather disasters, 1980-1999, PDF version, www.ncdc.noaa.gov/ol/reports/billions.html. July 20, 1999.

National Council of the Paper Industry for Air and Stream Improvement (NCASI), Global change and forest responses: theoretical basis, projections, and uncertainties, NCASI, New York, Technical Bulletin No. 690, pp. 11-18, 1995.

Nijs, I., I. Impens, and T. Behacghe, Leaf and canopy responses of *Lolium perenne* to long-term elevated atmospheric carbon-dioxide concentration, *Planta 177*, 312-320, 1988.

Pielke, Jr., R. A., and R. A. Pielke, Sr., Vulnerability to hurricanes along the US Atlantic and Gulf coasts: Considerations of the use of long-term forecasts, in Hurricanes: *Climate and Socioeconomic Impacts*, edited by H. F. Diaz and R. S. Pulwarty, Springer Publishing, New York, 147-184, 1997.

Quattrochi, D. A., J. C. Luvall, M. G. Estes, Jr., C. P. Lo, S. Q. Kidder, J. Hafner, H. Taha, R. D. Bornstein, R. R. Gillies, and K. P. Gallo, Project ATLANTA (ATlanta Land use ANalysis: Temperature and Air quality): A study of how the urban landscape affects meteorology and air quality through time, Preprints, *American Meteorological Society*, 104-107, 1998.

Quattrochi, D. A., and J. C. Luvall, Urban sprawl and urban pall: Assessing the impacts of Atlanta's growth on meteorology and air quality using remote sensing and GIS, *Geographical Information Systems, 9*, 26-33, 1999.

Rabalais, N. N., R. E. Turner, D. Justic´, Q. Dortch, and W. J. Wiseman, Characterization of hypoxia, *Topic #1, Gulf of Mexico Hypoxia Assessment*, NOAA Coastal Program Decision Analysis Series, National Oceanic and Atmospheric Administration, Coastal Ocean Office, Silver Spring, Maryland, 144 pp. plus 3 appendices, 1999.

Ricketts, T. H., E. Dinerstein, D. M. Olson, and C. Loucks, Who's where in North America? Patterns of species richness and the utility of indicator taxa for conservation, *BioScience, 49*(5), 369-381, 1999.

Ritschard, R. L., J. F. Cruise, and L. U. Hatch, Spatial and temporal analysis of agricultural water requirements in the Gulf Coast of the United States, *Journal of American Water Resources Association, 35*(6), 1585-1596, 1999.

Rosenberg, N. J., R. C. Izaurralde, M. Tiscareño-Lopéz, D. Legler, R. Srinivasan, R. A. Brown, and R. D. Sands, Sensitivity of North American agriculture to ENSO-based climate scenarios and their socio-economic consequences: Modeling in an integrated assessment framework, Pacific Northwest National Laboratory, Seattle, Washington, 146 pp., 1997.

Short, F. T., and H. A. Neckles, The effects of global climate change on seagrasses, *Aquatic Botany, 63*, 169-196, 1999.

Southern Oxidants Study Team. 1995. The state of the Southern Oxidants Study: Policy-relevant Findings in ozone pollution research, 1988-1994. Southern Oxidants Study, Raleigh, North Carolina, 94 pp.

Timmermann, A., J. Oberhuber, A. Bacher, M. Esch, M. Latif, and E. Roeckner, Increased El Niño frequency in a climate model forced by future greenhouse warming, *Nature, 398*, 694-697, 1999.

Trenberth, K. E., Conceptual framework for changes of extreme of the hydrological cycle with climate change, *Climatic Change, 42*, 327-339, 1999.

Tsuji, G., G. Hoogenboom, P. Thornton, (Eds.), *Understanding Options for Agricultural Production*, Kluwer Academic Publishers, 399 pp., 1998.

US Bureau of the Census, County and City Data Book, US Government Printing Office, Washington, DC, GPO #003-024-08753-7, 1994.

US Department of Agriculture, Census of Agriculture, National Agricultural Statistics Service, Washington, DC, 1994.

US Department of Agriculture, Census of Agriculture, National Agricultural Statistics Service, Washington, DC, 1996.

US Department of Agriculture, Census of Agriculture, National Agricultural Statistics Service, Montgomery, Alabama, 1999.

US Forest Service, The South's fourth forest: alternatives for the future, US Department of Agriculture, Forest Service, Washington DC, Forest Resource Report No. 24, 512 pp., 1998.

Williams, K., K. C. Ewel, R. P. Stumpf, F. E. Putz, and T. W. Workman, Sea-level rise and coastal forest retreat on the west coast of Florida, *Ecology, 80*(6), 2045-2063, 1999.

ACKNOWLEDGMENTS

Dedication
This chapter is dedicated to Dr. Ronald Ritschard, a good friend, colleague, and scientist, without whose efforts this Assessment would not have been possible. Ron will be deeply missed.

Many of the materials for this chapter are based on contributions from participants on and those working with the

Central and Southern Appalachians Workshop Team
William T. Peterjohn (PI), West Virginia University
Richard Birdsey, USDA Forest Service
Amy Glasmeier, Pennsylvania State University
Steven McNulty, USDA Forest Service
Trina Karolchik Wafle, West Virginia University

Gulf Coast Workshop and Assessment Team
Zhu Hua Ning*, Southern University and A & M College
Kamran Abdollahi*, Southern University and A & M College
Virginia Burkett, USGS National Wetlands Research Center
James Chambers, Louisiana State University
David Sailor, Tulane University
Jay Grymes, Southern Regional Climate Center
Paul Epstein, Harvard University
Michael Slimak, US Environmental Protection Agency

Southeast Workshop and Assessment Teams
Ron Ritschard*, University of Alabama – Huntsville
James O'Brien*, Florida State University
James Cruise*, University of Alabama – Huntsville
Robert Abt, North Carolina State University
Upton Hatch, Auburn University
Shrikant Jagtap, University of Florida
James Jones, University of Florida
Steven McNulty, USDA Forest Service
Brian Murray, Research Triangle Institute

* signifies Assessment team chairs/co-chairs

CHAPTER 6

POTENTIAL CONSEQUENCES OF CLIMATE VARIABILITY AND CHANGE FOR THE MIDWESTERN UNITED STATES

David R. Easterling[1,2] and Thomas R. Karl[1,2]

Contents of this Chapter

[1]NOAA National Climatic Data Center; [2]Coordinating author for the National Assessment Synthesis Team

CHAPTER SUMMARY

Regional Context

The Midwest is characterized by farming, manufacturing, and forestry. The Great Lakes form the world's largest freshwater lake system, providing a major recreation area as well as a regional water transportation system and access to the Atlantic Ocean via the St. Lawrence Seaway. The region encompasses the headwaters and upper basin of the Mississippi River and most of the length of the Ohio River, both critical water sources and means of industrial transportation providing an outlet to the Gulf of Mexico. The Midwest contains some of the richest farmland in the world and produces most of the nation's corn and soybeans. It also has important metropolitan centers, including Chicago and Detroit. The largest urban areas in the region are found along the Great Lakes and major rivers. The "North Woods" are a large source of forestry products and have the advantage of being situated near the Great Lakes, providing for easy transportation.

Climate of the Past Century

Over the 20th century, the northern portion of the Midwest, including the upper Great Lakes, has warmed by almost 4°F, while the southern portion along the Ohio River valley has cooled by about 1°F.

Annual precipitation has increased, up to 20% in some areas, with much of this coming from more heavy precipitation events.

Climate of the Coming Century

During the 21st century, it is highly likely that temperatures will increase throughout the region, likely at a rate faster than that observed in the 20th century, with models projecting a warming trend of 5 to 10°F over 100 years.

Precipitation is likely to continue its upward trend, with 10 to 30% increases across much of the region. Increases in the frequency and intensity of heavy precipitation events are likely to continue in the 21st century.

Despite the increase in precipitation, rising air temperatures and other meteorological factors are likely to lead to a substantial increase in evaporation, causing a soil moisture deficit, reduction in lake and river levels, and more drought-like conditions in many areas.

Key Findings

- A reduction in lake and river levels is likely to occur as higher temperatures drive increased evaporation, with implications for transportation, power generation, and water supply.
- Agriculture as a whole in this region is likely to be able to adapt and increase yields with the help of biotechnology and other developments.
- For both humans and other animals, a reduction in extremely low temperatures is likely to reduce cold-weather stress and mortality due to exposure to extreme cold, while an increase in extremely high temperatures is likely to decrease comfort and increase the likelihood of heat stress and mortality in summer.
- Preventative measures such as adequate storm-water discharge capacity and water treatment could help offset the likely increased incidence of water-borne diseases due to an increase in heavy precipitation events. As temperatures increase there is also a chance of more pest-borne diseases, such as St. Louis encephalitis.
- Changes in seasonal recreational opportunities are likely, with an expansion of warm weather activities during spring and fall, and a reduction during summer due to excessively hot days. Cold weather activities are likely to decline as warmer weather encroaches on the winter season.
- Boreal forest acreage is likely to be reduced under projected changes in climate. There is also a chance that the remaining forestlands would be more susceptible to pests, diseases, and forest fires.
- Major changes in freshwater ecosystems are likely, such as a shift in fish composition from cold water species, such as trout, to warm water species, such as bass and catfish.
- Higher water temperatures are likely to create an environment more susceptible to invasions by non-native species.
- The current extent of wetlands is likely to decrease due to declining lake levels.
- Changes in bird populations and other native wildlife have already been linked to increases in temperature and more changes are likely in the future
- Eutrophication of lakes is likely to increase as runoff of excess nutrients due to heavy precipitation events increases and warmer lake temperatures stimulate algae growth.

POTENTIAL CONSEQUENCES OF CLIMATE VARIABILITY AND CHANGE FOR THE MIDWESTERN UNITED STATES

PHYSICAL SETTING AND UNIQUE ATTRIBUTES

The Midwest region is dominated by the Great Lakes, two major river systems, and large tracts of both forests and agricultural lands. The landscape of most of this region was scoured by a series of continental glaciers resulting in thousands of lakes, wetlands, and huge expanses of relatively flat land. The last of these glaciers retreated about 10,000 years ago and glacial meltwater carved many important river valleys. As the glaciers retreated, the lands that were once covered by ice up to two miles deep developed some of the richest and most productive agricultural soils in the world. In addition to the Great Lakes, they also formed tens of thousands of small- to moderate-size lakes, which have become a characteristic of the region, the "land of 10,000 lakes," as Minnesota license plates proclaim (Lew, 1998).

The Great Lakes are one of the world's largest inland lake systems, containing 20% of the world's freshwater reserve (Botts and Krushelnicki, 1987). They also provide a regional water transportation system and access to the Atlantic Ocean via the St. Lawrence Seaway. This is a vital competitive advantage for transporting manufactured goods and agricultural products produced in this region. This region also contains the headwaters and upper basin of the Mississippi River and most of the length of the Ohio River, both critical water sources and means of industrial transportation providing an outlet to the Gulf of Mexico.

The Midwest region is comprised of the area north of the Ohio River, including the Great Lakes region, and west to the states adjacent to the Mississippi River (Figure 1). It contains 12% of the US land area and 22% of its population. This area covers a wide range of ecosystems with some extensive urban/suburban development. The region can be divided into three subregions: the rolling forested landscapes of southern Missouri, Illinois, Indiana, and Ohio; the relatively flat farmland in the northern portions of these states as well as in Iowa; and the heavy evergreen and deciduous forests of Wisconsin, Minnesota, and Michigan. Native vegetation ranges from mixed deciduous forests in the south to the so-called Prairie Peninsula, (a region of tall grass prairie in southwestern Minnesota, Iowa, and large portions of Missouri, Illinois, and southern Wisconsin), giving way to the deciduous and coniferous forests in the northern Great Lakes. The flora of the region are accustomed to periodic extreme droughts, flooding, late-spring or early-autumn frost, low minimum temperature, high maximum temperature, and severe storms.

Figure 1. Map of Midwest region.

SOCIOECONOMIC CONTEXT

The Midwest is a combination of the Manufacturing and Corn Belts, a region of manufacturing and agricultural production on which the entire country depends. Shaped by surface water systems and other emerging transportation networks, it developed rapidly in the latter half of the 19th century with the arrival of both large numbers of settlers from eastern US and immigrants from Europe. It has relatively high population density, and numerous pockets of national excellence such as the Mayo Clinic in Minnesota. The region provides more than 40% of the nation's industrial output and is responsible for 30% of the nation's foreign agricultural exports.

Historically, the Midwest's image as the heart of the Manufacturing Belt has been closely associated with the automobile age, and its prosperity has traditionally been tied to the fortunes of the automobile industry. Other heavy industry has been important as well, including the production of chemicals, steel, paper, and medical products. Due to foreign competition, an aging industrial infrastructure, and environmental issues, the Manufacturing Belt declined in competitiveness during the 1970s, coming to be referred to as the Rust Belt. However, the region responded with vigorous industrial restructuring related to modern technologies and components such as electronics, and also to services and financial industries. Meanwhile, continued improvements in farming methods and seed stock from new research and development has pushed the yields of corn, soybeans, fruits, vegetables, and other crops up to previously inconceivable levels. This revitalized economic base is expected to lead to continued economic and demographic growth (see Figures 2 and 3), but still lagging behind the rest of the country (with the notable exception of Minnesota).

ECOLOGICAL CONTEXT

Land use in the Midwest region is dominated by managed ecosystems such as farmland. However, the natural land cover of the region is characterized by three prominent environmental gradients. First, a southwest to northeast gradient from prairie to forest in Minnesota is largely a function of water availability. Second, a south to north gradient from Eastern deciduous (oak-hickory) to Northern mixed hardwood forests (beech, maple, hemlock) in Michigan and Wisconsin is a prominent landscape feature. These patterns correspond to climatic and soil gradients and a steep south-north land-use gradi-

Midwest Population Estimates and Projections

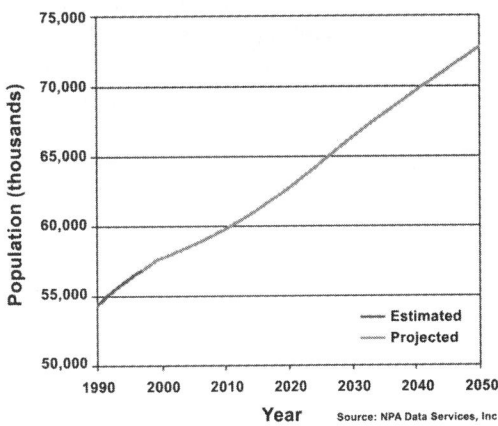

Figure 2. Population trend estimate for the Midwest region using the baseline assumptions from the NPA Data Services estimates. Under this scenario, the population of the Midwest is expected to increase by about 30% by 2050.

Midwest Industry Income

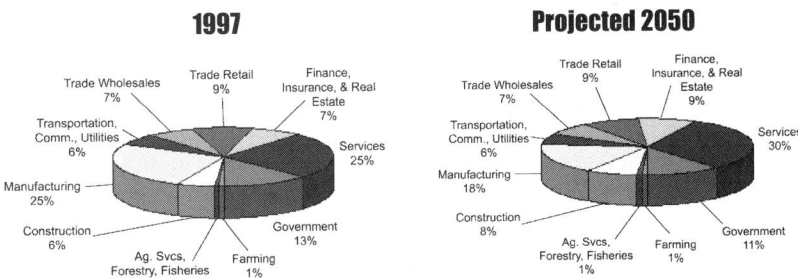

Figure 3. Percentage of economy by sector for (a) 1997, and (b) 2050, estimated from the NPA Data Services using baseline assumptions (NPA 1999b). Under these estimates, by 2050 the manufacturing percentage of the economy decreases by 7%, and the service sector increases by 5%. See Color Plate Appendix

ent from predominantly agriculture to predominantly forested. Third, the region is at the southern margin of the boreal forest (spruce-fir) and northern portions of the region include boreal species as locally dominant, especially on wetter sites.

In 1992, the Upper Great Lakes region (Michigan, Minnesota, and Wisconsin) was about 42% (over 50 million acres) forestland. Over 90% of the forestland is used for commercial forestry, and more than half of the commercial forest land is owned by the non-industrial private sector. The forestry sector employs about 200,000 people and produces over $24 billion dollars a year in forest products. Expectations in the industry are for sustained or increased output of forest products, particularly given increasing demand for forest products, decreasing supply from the Pacific Northwest, and

171

the already high production from the neighboring southern and southeastern regions of the US. The second and third-growth forests of the Upper Great Lakes are maturing, and recent forest inventories report substantial increase in the amount of forested land and in stocking on those lands. The majority of Americans, including those in the region, express a desire for increased emphasis on non-commodity values in forest management (e.g., recreation, aesthetics, and biodiversity). This desire often conflicts with the dependence of rural landowners on forests for employment and community development. While both standing volume and demand for forest products continue to increase in the Upper Great Lakes, the amount of land available for timber production continues to decrease due to conversion to urban and industrial uses, and development of seasonal and retirement homes.

Two trends in land use should be considered and are very likely to continue for the short term. Declines in the amount of farmland in Michigan, Minnesota, and Wisconsin (a 5% decline between 1998 and 1997) was observed in the Census of Agriculture. Forest cover increased by 3% between 1980 and 1993 according to the USDA Forest Service forest inventory, and urban sprawl has accelerated, both replacing important agricultural land use. Although the pressures causing these changes are still in place (declining agricultural productivity and increasing demand for recreational and aesthetic uses of land), it seems unlikely that the trends can continue long term. Increasing development and declining rates of agricultural abandonment are likely to lead to declines in forest area in the longer term (Warbach and Norberg, 1995). Furthermore, large-scale management of forests on private lands is becoming increasingly difficult as ownership is becoming increasingly fragmented among many more and smaller parcels (Norgaard, 1994; Brown and Vasievich, 1996). Between 1960 and 1990, average private parcel sizes declined by an average of 1.2% per year across the region. While this "parcelization" associated with recreational and seasonal home development doesn't necessarily result in forest clearing, it does affect the management of forests and, therefore, the ability of foresters and thus forests to respond to changing climatic conditions.

Given the substantial potential expansion of the temperate deciduous forests and savannas (oak and hickory dominant) it is important to consider two limiting factors. First, between two-thirds and three-quarters of these two communities are under active human management for agriculture and/or develop-

ment. This can affect the availability of seed sources and, therefore, slow the migration of species northward. This delay can contribute to dieback as communities make the transition from one type to another. Second, the northern forests are strongly influenced not only by climate, but also by the soils present, with conifers tending to dominate on the sandy soils, such as found in the north of the region, especially in lower Michigan. Sandy soils are more prone to drought conditions. Although vegetation models consider this influence, the scale of the variation in soil effects is much finer than can be represented in the models. Therefore, soil effects contribute to uncertainties in projections.

CLIMATE VARIABILITY AND CHANGE

The climate of this region is typical of an interior continental climate, although the Great Lakes exert a strong influence on nearby areas for both precipitation and temperature. Total annual precipitation varies from a low of about 25 inches in western Minnesota and Iowa to more than 40 inches per year along the Ohio River. Rainfall in most of the region is highly seasonal, with most falling in the summer. Only in the Ohio River Valley is precipitation more consistent throughout the year. Temperatures range widely from winter to summer, year-to-year, and even decade-to-decade, with reduced variability in areas adjacent to the Great Lakes.

Annual mean temperature trends in this region over the 20th century indicate that the northern portion, including the upper Great Lakes, is warming at a rate of almost 4°F/100 years (2°C), but the southern portion along the Ohio River valley has a slight cooling trend of around 1°F/100 years (0.6°C). By the end of the 21st century, the projected changes in mean annual temperature for much of the Midwest are between 5 and 10°F (3-6°C) with about twice the rate of temperature increase in the Canadian climate model scenario than in the Hadley scenario. In both scenarios, the mean daily minimum temperature rises more than the maximum temperature, often by a degree or two F by the end of the 21st century, a characteristic also observed during the 20th century.

The decrease in the annual mean temperatures over the 20th century along the Ohio River is also associated with a reduction in the number of days in the late 20th century exceeding 90°F (32°C). Decreases of up to 14 days per year exceeding 90°F in the Ohio Valley have been observed, owing in part to the extreme heat in the first half of the century associated with the intense droughts during the 1930s. This is in contrast to the projected increase in the probability of temperatures exceeding 90°F during the 21st century. For example, in Cincinnati, the probability of more than half the days in July having temperatures at or above 90°F increases from 1 in 20 now, to 1 in 2 by 2030 in the Canadian climate scenario and to 1 in 10 for the Hadley scenario.

Changes in extreme temperature can be even more dramatic. For example, the Canadian scenario suggests that for places like Chicago, the probability of three consecutive days with nighttime temperatures remaining above 80°F (27°C) and daytime temperatures exceeding 100°F (38°C) increases from about a once in 50 years occurrence today, to approximately once every 10 years by the decade of the 2030s. By the decade of the 2090s this probability increases to one year in two or 50% for any given year. Analysis of the historical data shows no change during the 20th century in the number of days below freezing for the southern portion of the region, and only slight decreases of around 2 days per 100 years around the Great Lakes. Again, this is expected to change considerably by the end of the 21st century. For example, in southern Wisconsin, the Canadian scenario projects an increase in the number of wintertime nights remaining above freezing from the current 5 nights per winter to 15 of the nights by the end of the 21st century.

The length of the snow season over the last 50 years increased by about 6 days per year in the northern Great Lakes area, but decreased by as much as 16 days in the Ohio River valley and adjacent areas. Both climate scenarios project a decrease of up to 50% in the length of the snow cover season by the end of the 21st century. Although it is possible that during the next few decades lake-effect snows might initially increase with a reduction in lake ice cover, it is highly likely that by the end of the 21st century, a reduction in the conditions favorable for lake-effect snows would cause the frequency of lake-effect snows to decrease (Kunkel et al., 2000). Therefore, it is likely that by the end of this century, sustained snow cover (more than 30 continuous days of snow cover) could disappear from the entire southern half of the region.

Observed trends for annual precipitation over the 20th century show moderate to strong increases almost everywhere in the region, often exceeding 20% per century. On an annual basis, precipitation is projected to increase by 20 to 40% by the end of the 21st century in the upper Midwest and decrease by up to 20% along the Ohio River in the Canadian scenario. However, these changes are not uniform throughout the year. For the Hadley scenario, increases in precipitation occur everywhere in the Midwest with increases of 20 to 40% common. However, the magnitude of precipitation increases in the Hadley scenario is unusual compared to other $2xCO_2$ equilibrium models run over the past several years in that it does produce more precipitation at the end of the 21st century (Quinn, et al., 1999).

Observed changes in soil moisture calculated from the Palmer Drought Severity Index indicate moderate to very strong increases in wetness in the eastern portions of the region. In contrast, even with the temperature increases there is a strong enough increase in precipitation to outweigh increased evaporation in the Hadley scenario which leads to small positive changes of soil moisture content by 2100. In the Canadian scenario however, despite the increase in annual precipitation, a reduction in summer precipitation coupled with the larger increase in temperature relative to the Hadley scenario leads to increased evaporation and reduced soil moisture content, especially during summer. The frequency and intensity of droughts increase in the Canadian scenario, but decrease slightly in the Hadley scenario by 2100.

Changes in variability related to the changes in extremes are of particular interest. The interannual variability of the annual mean temperature decreases slightly for both model scenarios by the end of the 21st century, but the magnitude of this decrease is very small compared to the increase in the mean. As a result, new record high extremes of temperature are common in both climate scenarios throughout the 21st century (Karl and Knight, 1998; National Climatic Data Center, 1999).

The interannual (year-to-year) variability of precipitation in both scenarios shows no significant changes, even by the end of the 21st century. But an increase in the mean annual precipitation would likely be accompanied by even greater changes in heavy daily precipitation events (Groisman et al., 1999a). Both climate scenarios show increases in precipitation for the Midwest region, and analysis indicates that this increase is occurring due to increases in the highest daily precipitation amounts (Figure 4). However, in the Canadian scenario, even with projected increases in precipitation, the increased evaporation due to

Midwest Daily Precipitation/HadCM2

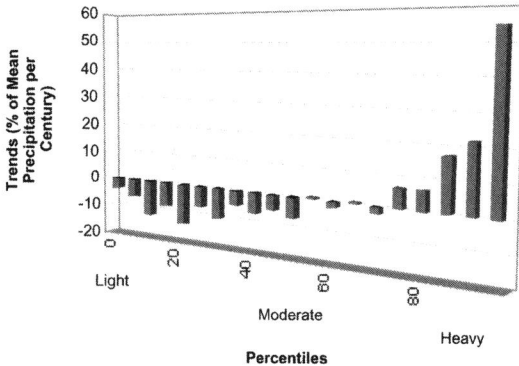

Midwest Daily Precipitation/CGCM1

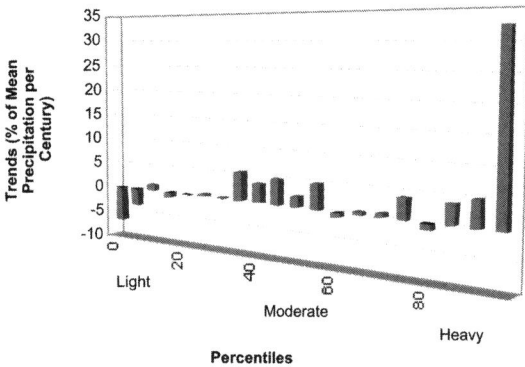

Figure 4. Annual trends in daily precipitation by percentile for the (a) CGCM1 (Canadian model) and (b) HadCM2 (Hadley model) scenarios. Notice the largest trend is in the heaviest daily precipitation amount for both model simulations indicating that most of the increase in annual precipitation is due to an increase in precipitation on days already receiving large amounts (analysis based on method in Karl and Knight, 1998). See Color Plate Appendix

rising temperatures would result in both decreased runoff and increased direct lake evaporation, resulting in decreased lake levels (see water resources section below).

KEY ISSUES

1. Water Resources
2. Agricultural Ecosystems
3. Natural and Semi-natural Ecosystems
4. Quality of Life

The key issues discussed here were chosen to be representative of those issues that are potentially affected directly by climate change, in either positive or negative ways. Moreover, they are issues of key economic and/or environmental importance. It is clear that secondary and even tertiary effects due

to climate change among the various sectors are possible, however these types of effects are very difficult to quantify and are not considered here.

1. Water Resources

Some climate model scenarios project a reduction in available water with increasing atmospheric CO_2 even with increased precipitation, due to enhanced evaporation with increased temperatures outweighing the precipitation increase (Meehl et al., 2000). Decreased lake levels and river flow would pose special problems for the region, affecting commerce and recreation and altering entire ecosystems. Understanding potential impacts of climate change on water resources is a key element in understanding impacts on a variety of other sectors. Issues with the region's many lakes including the Great Lakes, and the river systems including the Ohio and Mississippi basins, are critical to the region's economy and ecosystems. These issues include water levels, water temperature, ice cover, and water quality.

Lake levels. For the Great Lakes, the potential impacts related to the various scenarios of climate change span a wide range. The Canadian climate scenario projects a reduction in the levels of the Great Lakes by as much as 4 to 5 feet because of decreases in the net basin supply of water caused by reduced land surface runoff and increased evaporation of lake water. These reductions are projected to occur in spite of the projected increases in total precipitation. Should these changes occur, they would be 3 to 6 times larger than the seasonal variation of Lake levels, which normally reach a minimum during winter and a maximum in late summer, unlike many smaller lakes which reach their minimum in late summer. The reduction in water levels projected by the Canadian scenario would result in a 20-40% decrease in outflow in the St. Lawrence River. The Hadley scenario, with its smaller temperature increases and greater precipitation increases, leads to smaller changes (a slight increase of 1 foot) in Lake levels during the 21st century (see box, Quinn et al., 1999). However, Chao (1999) presents Great Lakes level changes calculated using a number of both transient and equilibrium climate change simulations. For each lake the results show lake level declines ranging from less than 1 foot (0.3 meter) to more than 5 feet (1.8 meters) (see Figure 5). The differences in the results from various model projections illustrate the difficulty in planning long-term adaptation strategies because they require consideration of a broad range of possibilities.

The Drought of 1988

Along with the droughts of the 1930s, the 1988 Midwest drought was one of the worst in the previous 100 years and brought home the socioeconomic impacts of these types of short-period climate fluctuations. Major impacts occurred in most sectors of the economy including agriculture, recreation, and transportation. The National Climatic Data Center (1999) lists this drought and associated heat waves as one of the most expensive natural disasters in US history, with costs of over $30 billion.

One of the unforeseen impacts was on river-borne transportation on the lower Mississippi River. The drought reduced water levels in the Mississippi River to the point that barge traffic was restricted, causing bulk commodity shipping to be shifted to more expensive rail transport. A controversial proposed response was to increase the diversion of Lake Michigan water into the Mississippi River system via the Chicago River. The diversion is limited to 3,200 cubic feet per second (cfs) by a US Supreme Court decision and the proposed increase was to 10,000 cfs. However, a number of factors led to a decision not to implement this increased diversion. First, it was determined to be too politically controversial, particularly because Great Lake water levels were rapidly falling from their record levels of the previous few years and the increased diversion could accelerate this fall. Furthermore, hydrologic studies indicated that the increased diversion would be insufficient to improve conditions anyway.

As with many such impacts, there were winners and losers. Losers included producers such as farmers, petroleum companies, the coal industry, and others who had to pay more to ship their products. In addition to the river-based shipping industry itself, other losers were consumers of the commodities normally shipped via river traffic that had to pay more due to higher shipping costs. There were also negative ecological impacts due to low water levels including fish kills, wetland damage, and salt-water intrusion up the lower Mississippi River past New Orleans. The major winners economically were the alternative shippers such as the rail and trucking industries.

If reductions in runoff as large as those projected by the Canadian climate scenario occur, this would have a substantial effect on all the major rivers of the region. During the 1988 drought, hundreds of millions of dollars were lost due to transportation inefficiencies (Changnon, 1989). On an annual basis, over $3 billion in business revenue and personal income, and 60,000 jobs relate to the movement of goods on the Great Lakes-St. Lawrence transportation system (Allardice and Thorp, 1995). Low water levels often lead to gouged ship hulls or damaged propellers. Dredging operations can be used to offset these problems. In addition to the added cost of dredging, another risk is the possibility of re-suspending human-made inert toxins and heavy metals lying within the lake bottom sediments, significantly impacting water quality. Such "surprises" are often difficult to anticipate. There is the potential that the competitive advantage this region has had due to reliable and efficient transportation of goods in and out of the area could be lost.

This event illustrated the high likelihood for future controversy if droughts of this magnitude become more frequent due to climate change. Clearly if this occurs the barge industry would be severely impact-ed by a change in shipping patterns favoring railroads and trucking.

Lake ice cover. Some offsetting changes related to the water transportation problem are projected in both the Hadley and Canadian climate scenarios. For example, with water temperature increases there would be a longer ice-free season (see box) resulting in a one to two month extension of the shipping season by the end of the 21st century. This could translate into hundreds of millions of dollars of additional business revenue, however the ship-

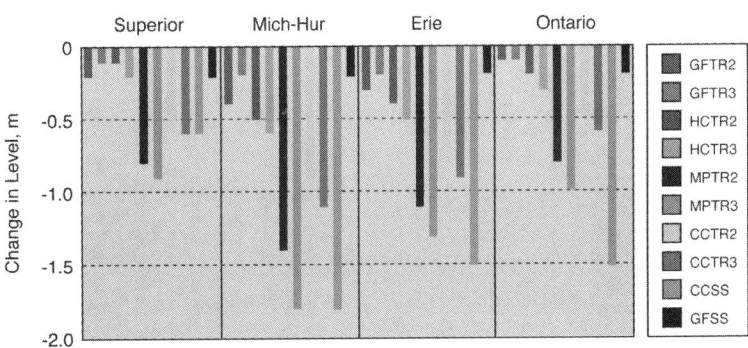

Figure 5. Change in water level for each of the Great Lakes under a number of climate change scenarios, from Chao (1999). See Color Plate Appendix

ping cost per ton would likely increase for reasons previously discussed due to low water levels (Chao, 1999, see Figure 6).

Improved understanding of Great Lakes ice cover, its climatic variability, and the ensuing economic and ecological impacts is needed for the development of strategies for adaptation to potential climate change. Reduced ice cover duration over the 21st century, as recent studies suggest (see box, Quinn et al., 1999), would have feedbacks on lake evaporation, lake levels, and even lake-effect snowfall (by possibly affecting the seasonality of lake effect snowstorms). Ecosystem services that could possibly be lost with a reduction of lake ice include: 1) storage of airborne atmospheric particulates until their rapid release in the spring, 2) enhancing overwinter survival of fish and fish eggs, and 3) protecting the shore against erosion.

For the thousands of smaller lakes in the region, ice cover is projected to form every year only in lakes in Minnesota, upper Michigan, and Wisconsin by the end of the 21st century (Fang and Stefan, 1998), but shortened by nearly half of its normal duration. These changes could eliminate fish winterkill in most shallow lakes, but possibly can endanger traditional recreational users of these lakes because of reduced ice thickness.

Storm-water routing and flood plains. Despite the lower water levels as temperatures increase, it is likely that a continuation of the increase in heavy precipitation events would occur as the climate warms (IPCC, 1996; Groisman et al., 1999a; Groisman et al., 1999b; Karl and Knight, 1998). Parts of the region, including Indiana and much of the area around Chicago are already at a high risk related to the number of residents living in the 500-year flood plain. The addition and expansion of impervious surfaces such as pavement can compound the

difficulties related to flash flooding events, taxing existing storm water routing infrastructure. Already states like Illinois have updated their 100- and 500-year rainfall return periods in recognition of the changing climate (Angel and Huff 1997). Furthermore, Olsen et al. (1999) found evidence for increased flood risk over the most recent decades in the lower part of the Missouri basin, and on the Mississippi River at St. Louis.

With a warmer climate that is drier due to enhanced evaporation, but also experiencing an increase in heavy precipitation events, it is likely that there would be changes in eutrophication incidence in lakes. Increased surface water temperature reduces vertical mixing of nutrient rich water between the hypolimnion (the colder bottom layer in a thermally stratified lake) and the epilimnion (the warmer, top layer) resulting in a reduction in the intensity of summer algae blooms. However, heavier rain events with associated greater runoff would likely increase eutrophication because heavy rain events are most important in transporting nutrients such as fertilizer from the watershed to the lake (Stephan et al., 1993). Also under this scenario it is likely that groundwater could also be affected through a reduction in water tables due to both increased irrigation and a reduction in recharge with enhanced evaporation; there would also be an increased risk of wellhead contamination during flooding events (Rose et al., 1999).

2. Agricultural Ecosystems

Agriculture in this region is vitally important to the nation and the world. With continued intensive application of technological advances, the impact of projected climate changes on agricultural yields would likely increase the high productivity of the region. Larger impacts could be related to changes in growing conditions abroad affecting crop prices.

Under climate change scenarios, a longer growing season would likely translate into increased farm production. Although only 4% of the farms in the Upper Great Lakes region have irrigation capabilities, even non-irrigated crops are likely to increase yields with an increased growing season length. Increasing irrigation capabilities would allow farmers to take advantage of the increased growing season length and push yields even higher. Although soil-types and farming practices in the Great Lakes region are more suitable for current crop types than those grown further south, these limitations are likely to be outweighed by increases in growing season length and warmer summer temperatures (Andresen

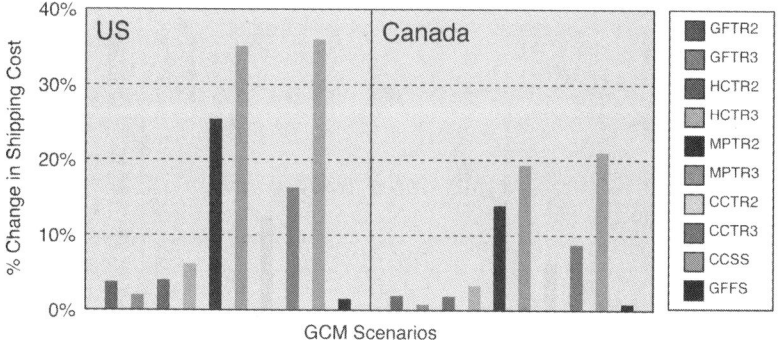

Figure 6. Change in shipping costs under a number of climate change scenarios, from Chao (1999). See Color Plate Appendix

Great Lakes Water Diversion

In 1900, the city of Chicago built the Chicago Sanitary and Ship Canal to keep sewage from contaminating the Chicago water supply intakes in Lake Michigan. The flow of water down the Chicago River was reversed. Sizeable amounts of water were diverted from Lake Michigan. This diversion launched a series of continuing legal controversies involving Illinois as a defendant against claims by the federal government, various Lake States, and Canada (which wanted the diversion stopped or drastically reduced). During the past 96 years, extended dry periods lowered the lake levels. Using these dry periods as surrogates for future conditions, their effects on the past controversies were examined as analogs for what might occur as a result of climate change. The results suggest that changing socioeconomic factors, including population growth, will likely cause increased water use, with Chicago seeking additional water from the Great Lakes. New priorities for water use will emerge as in the past. Future reductions in available water could lead to increased diversions from the Great Lakes to serve interests in and outside the basin. Lower lake levels in the future could lead to conflicts related to existing and proposed diversions, and these conflicts would be exacerbated by the consequences of global warming. Costs of coping with the new water levels could also be significant. Should Lake Michigan levels drop as much as 5 feet by the end of the 21st century as predicted by the Canadian climate scenario, it is possible that a lowering of the level of the Canal would be needed. To lower the canal by 4 feet, at least 30 miles of the canal would need to be dredged, and 15-17 miles would be rock excavation at huge financial costs (Injerd, 1998). A warmer climate, even with modest increases in precipitation, will likely lead to a drier climatic regime and will tax the economy and challenge existing laws and institutions for dealing with Great Lakes water issues.

Great Lakes Physical Impacts

Hydrologic impact analyses for the Great Lakes were developed for 20-year periods centered about 2030, 2050, and 2090 using the Canadian and Hadley climate scenarios. These scenarios were used with the Great Lakes Environmental Research Laboratory's Advanced Hydrologic Prediction suite of models (Quinn et al., 1999) and freezing degree-day ice cover models to assess impacts to Great Lakes water supplies, lake levels, ice cover, and tributary river flows for the 121 Great Lakes major tributary basins. Relative changes in hydrologic factors predicted by the two scenarios compared to the 1961-90 base period are summarized below.

1. Increases in precipitation are offset by increases in evaporation with higher temperatures in the Canadian climate scenario resulting in decreases in Lake levels of between 4 and 5 feet by the year 2090. The Hadley scenario shows modest increases in Lake levels, generally less than 1 foot, largely due to smaller increases of temperature accompanying the increase in precipitation.

2. The Canadian climate scenario leads to a decrease in mean annual outflow from each lake of -20 to -30% by 2090, whereas the Hadley scenario produces a modest increase in outflow of between 2 and 7%.

3. Both scenarios produce increases in water surface temperatures between 4.5°F (2.5°C) and 9°F (5°C).

4. Ice cover duration for Lakes Superior and Erie show decreases ranging from 29 to 65 days depending on the model scenario used.

et al., 1999). Figure 7 shows the projected changes in the climate of Illinois during the 21st century using the two climate scenarios. For the Hadley climate scenario, farming conditions during the summer will be like those in West Virginia by 2030 and like the eastern part of North Carolina by the end of the 21st century. In contrast, the much drier Canadian scenario projects Illinois' climate to be similar to that of Missouri and Arkansas by 2030 and eastern Texas and Oklahoma by the end of the 21st century. Simulation results by Izaurralde et al. (1999) suggest that the Corn Belt in general, is likely to experience an increase in corn yields due to both the CO_2 fertilization effect, and decreased plant stress due to low temperatures. In particular, results from the Hadley scenario show a simulated warm-

Summer Climate Shifts

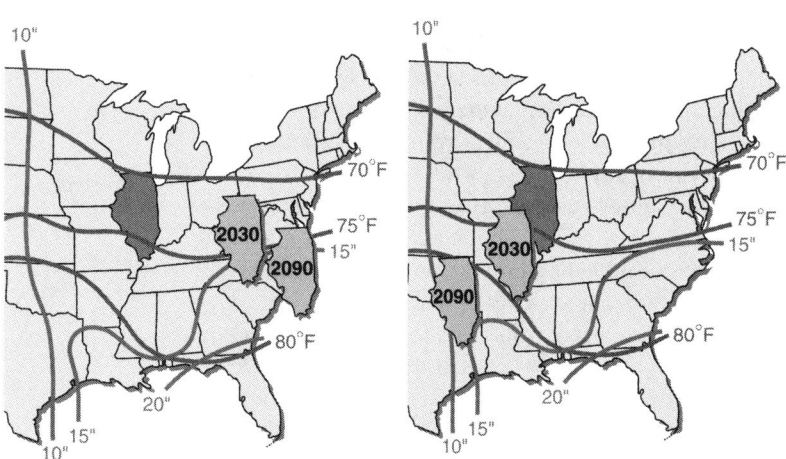

Figure 7. Illustration of how the summer climate of Illinois would shift under the (a) CGCM1 (Canadian model) scenario, and (b) HadCM2 (Hadley model) scenarios. For example, under the CGCM1 Canadian scenario, the summer climate of Illinois would become more like the current climate of southern Missouri in 2030 and more like Oklahoma s current climate in 2090. See Color Plate Appendix.

Midwest Soybean Yield and Precipitation

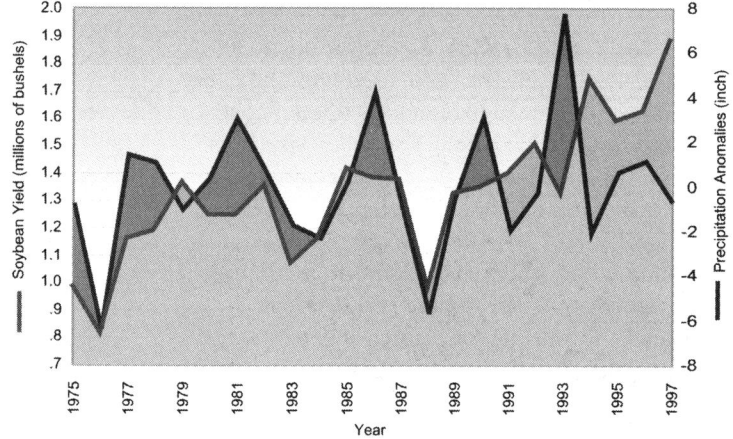

Figure 8. The relationship between Midwest soybean yield and precipitation is shown here. Soybean yields in thousands of bushels are shown as the differences from the average yield in recent decades. Precipitation is the difference from the 1961-90 average precipitation. Note that lower yields result from both extreme wet and extreme dry conditions. Soybean yields from National Agricultural Statistics Service, USDA. See Color Plate Appendix.

ing in minimum temperatures of over 3°C (5.5°F) by the end of this century that helps contribute to simulated corn yield increases (Izaurralde et al., 1999).

The year-to-year variability found in corn and soybean yields is primarily driven by growing season weather conditions. The climate variability of the Midwest since 1960 has been rather small and yields in most years have been within 10 to 15% of a run-

ning 5-year mean. Occasionally, severe drought, heat, or flooding can cause larger yield reductions, such as in 1983, 1988, and 1993. It is possible that severe drought will pose the biggest threat to future crop production. Severe droughts, such as in 1988, can cut production by 30% or more, and heavy flooding events can have a similar effect (see Figure 8). If more droughts or wet spells do become more common, adaptations in farming practices and crop selection would be necessary to offset yield reductions due to these types of climate events.

Dairy productivity is also directly affected by temperature. For example, optimal ranges for dairy cattle are between 40 and 75°F (4 and 24°C); dairy cattle are sensitive to heat stress and high humidity (Wolfe, 1997). As the climate warms, it is likely additional measures such as artificial cooling methods will be necessary to ensure that the productivity of livestock is not reduced by extremes of heat. On the other hand, with warmer winters it is likely that productivity will be enhanced by a reduction in cold stress. It is uncertain whether enhanced productivity with the reduction in cold stress will offset the costs of dealing with increased heat stress.

The types of grasses, including crops that grow in the Midwest, are a mixture of C3 and C4 type grasses. The C4 grasses include warm weather crops, such as corn, and the weedy grasses, such as crabgrass and many other creeping perennials. The C3 grasses, which dominate in the northern portions of the region, include cultivated grasses such as wheat and ryegrass that thrive in cooler growing season conditions. Total biomass transient climate change experiments, examining the response of grasslands to gradual changes in climate, show that total biomass in grassland regions of the western Midwest that are initially dominated by C3 perennial grasses would be replaced by C4 perennial grasses (Coffin and Lauenroth, 1996). The change in grass types is strongly determined by increased temperature throughout the year. Because many weedy creeping perennials are C4-like, this could pose problems for agricultural areas in the western and northern Midwest where the current dominant natural vegetation is C3 grasses. Agriculture has contributed to pollution in lakes and rivers of the region including bioaccumulation of fertilizers and pesticides in fish. The use of herbicides is likely to increase as temperatures increase and the dominant grass type becomes the C4 grasses (Allardice and Thorpe, 1995). Conflicting priorities between agricultural yields and water quality are likely to be exacerbated.

Soils in the Midwest region contain significant amounts of carbon and the management of the vegetation above the soil strongly influences the carbon that is removed from and is stored in those soils. Land management activities that influence soil carbon include deforestation and afforestation, biomass burning, cultivation crop residue management, application of inorganic fertilizer and organic manures, and the various farming systems (Lal et al., 1998). On agricultural lands, soil management practices such as conservation tillage, mulch farming, water management, soil fertility management, liming, and acidity management have been shown to increase the carbon stored in agricultural soils. However, the climate change effects on soil, particularly soil carbon storage, are difficult to predict. For example, in agricultural areas where crop residue is available and incorporated into the soil system, there tends to be a build up of soil carbon. This is especially evident in the productive soybean/corn rotations where nitrogen fixation by the beans and the additional residue input from corn rapidly builds soil carbon.

Adaptation Options
Agriculture has exhibited a capacity to adapt to moderate differences in growing season climate. It is likely that agriculture would be able to adapt under the moderate climate change scenarios produced by many GCMs. There are several adjustment possibilities that are already used and could be employed in the future to adapt to climate change. Some of these possibilities are farm practices (e.g., earlier planting and harvests, changed planting densities, integrated pest management, conservation tillage) and development of new varieties and farming technologies. There are already examples of individual farmers getting the benefit of double cropping (planting a second crop after the first is harvested) as the growing season has noticeably increased in just the past few years, but these are isolated cases at present. The

potential for double cropping soybeans with a warmer climate was examined using the two Assessment scenarios. If only temperature is considered, simulation results using both the Canadian and Hadley scenarios indicate that the potential exists for double cropping even in the northern parts of the region. However, the limiting factor appears to be a lack of adequate soil moisture. If a second crop of soybeans is grown, particularly in more northern sites, adequate yields would likely be dependent on the use of irrigation (Garrity and Andresen, 1999).

If extreme climate conditions like the droughts of the 1930s or 1988 become more common, then additional adaptation measures would be necessary. Some possibilities include shifting the mix of crops, increasing irrigation, and removing marginal lands from production. Hybrid strains of corn, for example, have been developed that allow the corn to be grown under a wide range of conditions. If climate conditions in the Midwest become unsuitable for the current hybrids to be grown, one adaptation mechanism would be a switch to a hybrid more suited to future climate conditions. Furthermore, there is evidence that carbon dioxide fertilization by itself can enhance crop production. Studies indicate that this effect, coupled with a warmer climate, outweighs poor soil quality in the northern parts of the region and could allow corn and soybean crops to be grown in areas where they are not currently found (Andresen et al., 1999).

3. Natural and Semi-natural Ecosystems

Both natural and semi-natural ecological systems on land and in water will be affected by the sustained warming projected by the climate models. Water-based ecosystems are currently under multiple

Agriculture and Drought

The agricultural impacts of the drought of 1988 illustrate potential impacts of one possible future climate scenario, one that is warmer and much drier. The temperature and precipitation conditions in 1988 were similar to conditions in the 1930s. Overall grain production was down by 31%, with corn production down by 45%. The supply of grain was adequate to meet demand because of large surpluses from previous years; however, the drought reduced surpluses by 60%. Interestingly, overall farm income was not reduced because grain prices increased substantially (35% for corn, 45% for soybeans); however, these overall figures mask large losses in the heart of the drought region where yield reductions were much larger than the national average. The reduced production caused slightly higher domestic food prices, estimated at a 1% increase in 1988 and a 2% rise in 1989 (in 1989, the drought persisted in some areas and surpluses were much reduced following 1988). In summary, the drought of 1988 demonstrated that the agribusiness sector remains vulnerable to severe climatic anomalies, despite decades of advances in agricultural technology.

stresses such as eutrophication, acid precipitation, toxic chemicals, and the spread of exotic organisms. Climate change will interact with these existing stresses, often in not very well understood ways.

Forest types in the Midwest range from the deciduous forests of Missouri, southern Illinois, Indiana, and Ohio, to coniferous forests in northern Minnesota, Wisconsin, and Michigan. The upper Midwest has a unique combination of soil and climate that allows for abundant coniferous tree growth. For example, Michigan is second only to Oregon in Christmas tree production. It is possible that regional climate change will displace boreal forest acreage, and make current forestlands more susceptible to pests and disease. A decrease in soil water content as projected by the Canadian scenario would increase the potential for more drought and excessive wet periods, increasing forest fire potential. Higher temperatures could also increase growth rates of marginal forestlands that are currently temperature limited, but can also reduce stands of aspen, the major hardwood harvested in the northern Midwest. The southern transition zone of the boreal forest is susceptible to expansion of temperate deciduous forests, which in turn would have to compete with other land use pressures. This could have major impacts on the boreal forest industry, such as Christmas tree production.

Oak/hickory forests in the southern Midwest are also a major resource for the forestry industry. Oak decline or oak dieback is caused by a complex interaction of environmental stresses. Trees weakened by environmental stresses such as drought, flooding, or insect defoliation could then be invaded and killed by insects or diseases that could not successfully attack a healthy tree (Wargo et al., 1983). Although commonly called oak decline, this problem is not confined to oak or other deciduous trees (see box). Because the initial stresses leading to this problem are either drought or excess wetness, the Canadian scenario's combination of temperature and precipitation change would likely reduce soil moisture leading to more drought-like conditions and increasing the likelihood of this kind of decline in both deciduous and coniferous species.

Forests

Another area of concern is the urban forest. These are tracts of forest-type land in urban settings such as parks or trees lining streets and are particularly important for large urban areas such as Chicago and Detroit. Besides beauty, these urban forests provide practical benefits such as shade, wind breaks, habitat for urban wildlife, and help in mitigating water, air,

and noise pollution, and in reducing storm water runoff. Furthermore, urban forests help reduce urban air temperatures through increased shade and evapotranspiration, which reduces air conditioning needs and summer-time energy demand (McPherson et al., 1997). The more than 70 million acres of urban forests in the US are dwindling (American Forests, 1998, Wong, 1999). With almost 80% of the US population living in urban areas, the health of these small pockets of forest is a growing concern (Sadof and Raupp, 1999). Because of their urban environment, these forests are already stressed beyond what a comparable rural forest would be in the same region, and climate changes that could affect the health of rural forests would likely be magnified in the urban setting. Encouraging the planting of appropriate tree species in urban areas wherever possible would be beneficial.

Fragmentation of large forest tracts in the region is another major stress on the forest industry, and this could become an even greater problem if scenarios of vegetation change are correct. Competing land use, particularly between forest and agriculture, could become a major problem, possibly significantly reducing the current extent of forestland. Forest fragmentation also severely affects wildlife, by reducing protective cover and natural migration stopover habitat. Furthermore, there is concern that natural responses to climate change are likely to be nonlinear, with an apparent resistance to change up to a certain threshold, beyond which time a rapid or catastrophic transition could occur (Moll, 1998).

Model simulations under both the Hadley and Canadian climate scenarios project the disappearance of the boreal forest from the region, and a consistent and substantial reduction in the amount of area covered by both the temperate continental coniferous forest and cool temperate mixed forest types as well. This suggests that the northern hardwood forests that sustain the regions forest products industry are likely to undergo substantial conversion to temperate deciduous forest and temperate deciduous savannas. The results for the grasslands are mixed, depending on the moisture projections in the climate scenarios and the assumptions about water use in vegetation models. The fact is, however, that very little natural grassland remains and the fate of the grasslands has more to do with agricultural policies and economic conditions than climate (Brown , 1999).

Fish and Wildlife

Wildlife impacts are a critical issue. For example, since the mid-20th century songbird populations have declined in the Midwest for a variety of reasons including habitat destruction and fragmentation (Robinson, 1997). However, bird populations, like other wildlife, are particularly susceptible to extreme climate events. Periods of excessive wetness and particularly drought affect both habitat and food supplies, resulting in diminished reproduction and a decline in populations the next year. Furthermore, the locations of the climate extremes affecting migratory Midwest bird populations could very well be outside the borders of the US, for example in Central America, where the birds spend the winter (Robinson, 1997).

Climate changes could have large and unpredictable impacts on aquatic ecosystems. Along with concerns regarding changes in water availability and ice cover, changes to other characteristics such as water temperatures are a concern. For example, Great Lakes surface water temperatures are likely to increase by as much as 5°C by the end of the 21st century (see box). Water temperature increases in deep inland lakes can lead to a decrease in dissolved oxygen content and primary productivity, and a decline in cold-water fish populations. A rise in water temperature would change the thermal habitats for warm-, cool-, and cold-water fish. The thermal habitat for warm- and cool-water fishes would likely increase in size, but could be eliminated for cold-water fish in lakes less than 13 meters (43 feet) deep (especially those in eutrophic states) and in many streams (Magnusson, 1997). Species such as brown and rainbow trout are at risk (EPA, 1995; 1997) as well as other cold water fish found in shallow lakes. Furthermore, Chao (1999) used a number of GCM-based scenarios of climate change to examine changes in cold water habitat in Lake Erie, (defined as a well-oxygenated hypolimion), and found reductions of 50 to 80% depending on the scenario.

Wetlands

Because Great Lakes' water levels have an unusual seasonal cycle (related to winter precipitation being locked up as snow/ice cover with frozen soils), reduced ice cover and higher temperatures are likely to affect the winter Great Lakes ecosystem by changing the annual cycle. The annual and interannual water level fluctuations act as a perturbation to wetland biophysical systems and maintain a more productive intermediate stage of wetland development. Changes in the annual cycle could be compounded due to substantially lower water levels, which would lead to more pressure on wetlands. Projected lake-level changes would likely affect wetland distributions, resulting in a displacement or loss of current wetlands (Mortsch, 1998). These wetlands serve a variety of important functions including waterfowl habitat, fish breeding areas, and natural improvement of water quality. Wetlands particularly at risk are those in areas where the topography inhibits successful wetland migration, such as areas with irregular topography.

Invasive Species

Invasive species are another current stress that could be exacerbated under climate change. For example, the zebra mussel has become a major problem in the Great Lakes system. Zebra mussels are small, fingernail-sized mussels native to the Caspian Sea region of Asia. It is believed that they were transported to the Great Lakes via ballast water from a transoceanic vessel. The ballast water, taken on in a freshwater European port, was subsequently discharged into Lake St. Clair near Detroit, where the mussel was discovered in 1988. Since that time, it has spread rapidly to all of the Great Lakes and waterways in many US states, as well as Ontario and Quebec. Hydroelectric power plants, municipal drinking water facilities, and other water-

Drought, Insects, and the Decline of Forest Species

In the late 1930s, many hemlock trees showed signs of deterioration in the Menominee Indian Reservation in east-central Wisconsin. Hemlock is an important forest tree at Menominee, which is located about 30 miles (50 km) north of the southwestern range limit of the tree. By the late 1930s, the hemlock borer, ordinarily not a problem, reached epidemic proportions. Careful examination by Secrest et al. (1941) revealed that extensive root damage had occurred during the drought years 1930-1937. Borer attacks in 1938 were successful only on trees that had 10% or less of their root system still alive. In 1939, attacks were successful only on trees with less than half of the main lateral roots alive. Obviously hemlock borer attacks were successful only on trees that were already heavily damaged by unfavorable climatic conditions. The 1930s droughts were ultimately responsible for the loss of trees near the range limit, but insect attack was the proximate cause of death of trees already weakened by drought. The implication of this example is that unfavorable climatic conditions may not kill trees outright, but by stressing the trees, climate can contribute to death by insect attack.

using industries are most heavily impacted by zebra mussel populations. Mussels colonize the surfaces of pipes, diminishing the flow rate through water intake pipes. Unless preventive measures are taken, larval zebra mussels colonize the interior parts of turbines and other equipment, leading to costly repairs. On the other hand, zebra mussels do act to improve water quality through their natural filtration of nutrients.

Temperature can limit the extent of zebra mussel colonization and has likely kept populations in Lake Superior small. Each mature female produces several hundred thousand eggs during the breeding season, which occurs when the water temperature is above 54°F (12°C). The longer this period, the more successful colonization is likely to be. Adults are unable to survive prolonged exposure to temperatures above 90°F (32°C) and they can tolerate temperatures as low as 32°F (0°C), provided they do not freeze. With warmer summer water temperatures in northern lakes, such as Lake Superior, the small current infestation is likely to become as widespread as it is in the lakes in warmer parts of the region.

4. Quality of Life

The Great Lakes and surrounding region is the industrial heartland of the United States. It is a leader in automobile, paper products, medicine, chemical, and pharmaceutical production. Recreation is also an important part of the regional economy. Although most of these industries are not directly vulnerable to climate change, severe indirect effects are possible. For example, changes in governmental policy designed to address climate change issues have the potential to dramatically impact certain industries (e.g., the automobile industry) affecting the overall economic health of the region.

Human Health
It is possible that human health will be affected by climate change in the Midwest in a number of ways. Although our understanding of the relationship between cold weather and mortality is weaker than the relationship between heat and mortality, milder winters are likely to have beneficial effects on cold-related mortality. For example, Changnon (1999) argues that 850 lives were saved across the US during the record warm winter of 1997-1998. But, even with generally milder winters, temperatures are still expected to go below 0°F every other year even out to the end of the 21st century in places like Chicago. On the other hand, increases in dangerously high day and nighttime

temperatures (at or above 100°F during the day and 80°F during the night) are likely to increase in frequency, with 3-day consecutive heat waves occurring every other year by the end of the 21st century. If heat waves do increase in frequency and severity this would likely lead to more temperature-related mortality and morbidity. This is particularly true in most large Midwest cities that experience these kinds of events relatively infrequently and where the population is not accustomed to these events (Kalkstein and Smoyer, 1993). Studies have shown that during the first heat wave of a season, most heat-related deaths begin to occur on the second or third day of the heat wave (Kalkstein and Greene, 1997). With a diverse population that ranges from crowded urban settings to dispersed farming communities, impacts are likely to vary considerably among different social and demographic groups. For example, under current climate change scenarios, more frequent extremely hot days are likely to occur more often in urban areas where the "heat island effect" reduces nighttime cooling. The recent susceptibility to heat-related mortality in places like Chicago has been attributed to factors such as a growing elderly lower income population that cannot afford or would not use air conditioning (Kunkel et al., 1999). However, recent adaptation and responses to potentially deadly heat waves, such as enhanced warning and educational programs, have been markedly improved and would likely help reduce severity of the impact on human mortality in the future. Other concerns for both urban and rural areas of the region include a possible increase in respiratory disease due to excess air pollution. Poor dispersion and high temperatures are closely related to elevated levels of ozone and other air pollutants, potentially compounding the health effects of extremely high temperatures (Karl, 1979, see Figure 9).

Indirect effects of climate variability on infectious diseases are another possible area of health impact, but whether any change in disease rates might occur is uncertain and highly specific to each disease. An increase in heavy precipitation events could also increase the potential for water-borne disease outbreaks similar to the one in Milwaukee during the 1993 flood (Rose et al., 1999, see sidebar story). Increases in water-borne infections such as *Cryptosporidiosis* are possible if heavy precipitation events become more frequent resulting in greater cattle- and human-derived *Cryptosporidium* contamination of surface water in combined sewer and storm water drainage systems. The water-associated disease known as "Swimmers itch" could become more prevalent if lakes are more often contaminated by larval *Schistosoma* from birds or mammals, and swimming becomes more frequent with hotter days. Various vector-borne diseases caused by infectious agents transmitted by mosquitoes or ticks might also be effected, but the complex and multiple impacts of climate on transmission make prediction difficult. For example, increases in temperature and precipitation extremes can affect mosquito ecology and potentially the transmission of viral agents such as Lacrosse or St. Louis encephalitis, however climate predictors of encephalitis outbreaks in the US are still unclear (see Human Health, Chapter 15).

For most of the health effects discussed above, there appears to be considerable opportunity for adaptation, either through increased prevention and/or greater education. The development of better educational and monitoring programs and increased support for public health systems would help reduce weather-related mortality from heat waves or increases in vector-borne diseases.

Recreation

Recreation is clearly sensitive to climate change. Ice fishing and snowmobiling are favorite pastimes in the upper Midwest, and higher winter temperatures coupled with a possible reduction in lake-effect snows are a direct threat to these activities. Reductions in lake-ice cover would significantly reduce the length of the season for recreational activities dependent on ice cover. Changes in the seasonal characteristics of precipitation resulting in more winter precipitation falling as rain than snow and increased numbers of days above freezing would affect both snowmobiling and skiing opportunities.

Temperature and Stagnation

(Analysis Period: 1948 to 1998, Summer)

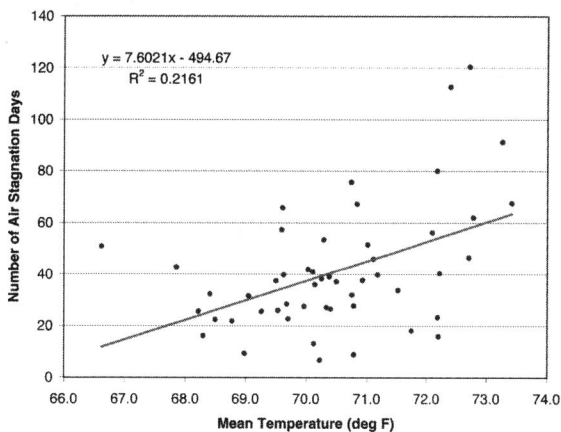

Figure 9. Relationship between average summer air temperature and the number of air stagnation days for the Midwest showing that as the mean summer temperature increases, the number of air stagnation days also increases (air stagnation data from Wang and Angell, 1999).

Reductions in cold-season recreational opportunities would likely be offset by increases in opportunities for more warm-season recreational activities such as golfing and boating in the fall and spring seasons as the length of the warm season increases. But as the climate continues to warm, the number of desirable days for outdoor summer activities is likely to be reduced. Clearly, habits and preferences for outdoor recreation options will need to change, and commercial interests will need to keep abreast of these changes.

1993 Midwest Flooding and Water-borne Disease

The heavy rainfall and major flooding on the upper Mississippi River basin in the spring and summer of 1993 was an exceptional and unprecedented event in modern times. By any measure this was a climate anomaly unseen in the modern historical record for this region. The flooding resulted from pre-existing conditions that created circumstances ripe for flooding, and day after day of heavy rainfall over the basin.

One unforeseen impact of the flooding was an outbreak of water-borne *Cryptospiridiosis* in the city of Milwaukee in the early spring of 1993. Unusually heavy spring rainfall and snowmelt washed fecal material from agricultural non-point sources into streams. This fecal material is often infected with *Cryptospiridium* oocysts. The streams emptied their infected load into Lake Michigan near the intakes for the Milwaukee water supply that made its way into the water treatment system. Engineering malfunctions combined with the massive flooding resulted in an outbreak of water-borne *Cryptosporidiosis* that infected over 400,000 people and caused over 100 deaths. As a result of this outbreak, the Milwaukee water system was forced to extend the water intake pipes much further from the shoreline, and improve water treatment controls.

ADDITIONAL ISSUES

A number of additional consequences are also likely to be of importance for the Midwestern US:

- Warmer winters are likely to result in savings in winter heating bills, potentially offsetting increased cooling bills during summer. Fewer lake-effect snows would reduce transportation problems, and could reduce building costs.
- Great Lakes issues include impacts on water supply intake pipes due to reduced water levels, and increased temperatures resulting in reduced water cooling efficiencies for power plants.
- Erosion of agricultural soil is currently a major problem, which could be exacerbated by the potential for increased heavy downpours.
- Quality of cool season vegetable crops could be reduced with brief high temperature events at critical stages in crop development.
- With climate changes and changes in natural ecosystems, there are concerns for proper management of game and wildlife and their respective habitat resources in order to maintain current levels of hunting and other outdoor recreational activities that contribute to the regional economy.
- Climate impacts on agriculture will have secondary impacts on local economies and land use, resulting in winners and losers, and potentially large community changes.
- With a drier climate, freshwater ecosystems may be more susceptible to the effects of such current problems as acid rain increasing surface water acidity, and the effects of increased development of dams for reservoirs.
- Exotic species will likely continue to invade this region and some native species will start or continue to decline. The climate change effects on these kinds of biological changes are unknown.

ADAPTATION STRATEGIES

Perhaps the most important approach to adapting to the potential effects of climate change is to develop and maintain flexibility in vulnerable activities and sectors. This would include, for example, developing water resource policies that are flexible enough to adapt to the potential for either an increase or decrease in available surface and ground water. Other strategies include: improving forecasts and warnings of extreme precipitation events and the related impacts; as well as designing large infrastructure projects to provide increased capability to cope with climate extremes. This might include rethink-

ing the construction of dams and levees on river systems, or the development of new hydroelectric generation facilities in the face of potential changes in lake levels and river flow (Smith, 1997).

On the other hand, a number of activities and sectors will likely be able to adapt easily with no outside intervention. Agriculture and many types of businesses and industries have benefited and will continue to benefit from consistent technological advances and will easily adapt as conditions change. Advances in plant genetics suggest that most cropping activities will be able to cope with most foreseeable climate shifts. Furthermore, the turnover rate in most large capital items, such as manufacturing facilities, is rapid enough that these types of facilities will have turned over or been remodeled at least once, if not two or more times by the year 2050 (Ausubel, 1991).

Another critical adaptation strategy is to develop a public education program regarding the potential risks and consequences associated with rapid climate change. For example, the potential for increasing fire danger associated with warmer and drier conditions should be communicated to homeowners in high fire-risk ecosystems. The increased potential for flooding with increases in the frequency of heavy rain events should be communicated to flood plain landowners. With better information, the residents of the region would be better prepared to respond to a less certain climate.

CRUCIAL UNKNOWNS AND RESEARCH NEEDS

1. Perhaps one of the most important unknowns with regard to future climate is reconciling projections of temperature and precipitation change from various climate models. Both model scenarios used here project warming and increased precipitation. However, in the Canadian scenario, warming was great enough to increase evaporation to the point that decreased soil moisture becomes a critical issue. Since some climate model simulations even project a decrease in precipitation for the region, this could severely impact water levels in both the Great Lakes and Mississippi/Ohio River systems, ground water, and soil moisture levels. However, other simulations project increases with only modest effects on water levels, which leaves this question an important one for future research.

2. Another crucial unknown is how extreme weather events might change. If precipitation variability increases, even with no change in average annual amount, this would have large implications for drought and wet-spell occurrence. It is clear that extreme events such as droughts and heat waves, flooding, and severe winter storms all can have dramatic effects on agriculture, transportation, human health, etc. This is perhaps one of the most difficult but critical issues to address as far as future climate is concerned, particularly since relatively little is known about past changes in extreme events, and the current generation of climate models still does not resolve many of these types of events well.

3. Also at issue are effects of CO_2 fertilization on crops, forests, and other flora and the ability of farmers to adapt to a changing, but uncertain, future climate in an economically competitive world agricultural market. For example, significant increases in growing season temperatures will require shifts to new varieties that are more heat tolerant, do not mature too quickly, and have a higher temperature optimum for photosynthesis. This may be achieved through plant genetics, but it is uncertain as to how the public in the US and abroad will react to these new genetically modified food sources. Moreover, new crop types may require abandoning traditional crops and this may be difficult to accomplish. For example, an apple grower may need to change varieties, taking years to grow new trees.

LITERATURE CITED

Allardice, D. R., and S. Thorp, A changing Great Lakes economy: Economic and environmental linkages, SOLEC working paper presented at State of the Lakes Ecosystem Conference, U.S. Environmental Protection Agency, , EPA 905-R-95-017, 1995.

American Forests (AF), Regional Ecosystem Analysis, Puget Sound Metropolitan Area, Final Report, Urban Forestry Center, American Forests, http://www.amfor.org/ufc/uea/report.html, 8 pp., 1998.

Andresen, J. A., W. B. Sea, G. Alagarswamy, and H. H. Cheng, National Assessment, Upper Great Lakes Final Report, Agricultural Sector, US Global Change Research Program, 10 pp., 1999.

Angel, J.R., and F.A. Huff, Changes in heavy rainfall in Midwestern United States, *J. Water Resources Planning. Management*, 123, 246-250, 1997.

Ausubel, J. H., Does climate still matter?, *Nature, 350,* 649-652, 1991.

Botts, L., and B. Krushelnicki, *The Great Lakes: An Environmental Atlas and Resource Book.* US Environmental Protection Agency and Environment Canada, Downsview, Ontario, 44 pp., 1987.

Brown, D. G., and J. M. Vasievich, A study of land ownership fragmentation in the Upper Midwest, *Proceedings, GIS/LIS 96 Conference,* Denver, Colorado., pp. 1199-1209, 1996.

Brown, D. G. US National Assessment, Great Lakes regional summary report, *Land Ecology,* US Global Change Research Program, pp. 51-57, 1999.

Changnon, D., 1997: Damaging storms in the United States: Selection of quality data and monitoring indices, in *Preprints of the Workshop on Indices and Indicators for Climate Extremes,* NOAA/National Climatic Data Center, Asheville, North Carolina, USA, June 3-6, 1997, 24 pp., 1997.

Changnon, S.A., The drought, barges, and diversion, *Bulletin of the American Meteorological Society,* 70(9) 1092-1104, 1989.

Changnon, S.A., Impacts of 1997-98 El Nino-generated weather in the United States, *Bulletinof the American Meteorological Society,* 80, (9), 1819-1928, 1999.

Chao, P., Great Lakes water resources: Climate change impact analysis with transient GCM scenarios, *Journal of American Water Resources Association, 35,* 1499-1507, 1999.

Coffin D. P., and W. K. Lauenroth, Transient responses of North American grasslands to changes in climate, *Climatic Change, 34,* 269-278, 1996.

EPA, *Climate Change and Wisconsin* (230-F-97-008 ww), *Climate Change and Michigan* (230-F-97-008 v), *Climate Change and Minnesota* (230-F-97-008 w), 1997. available at http://www.epa.gov/globalwarming/publications/impacts/state/index.html

EPA, *Ecological Impacts from Climate Change: An Economic Analysis of Freshwater Recreational Fishing* (220-R-95-004), 1995. available at http://www.epa.gov/globalwarming/publications/

Fang, X., and H. G. Stefan, Potential climate warming effects on ice covers of small lakes in the contiguous US, *Cold Regions Science and Technology, 27,* 119-140, 1998.

Garrity, C. M., and J. A. Andresen, Climatological constraints of a double-cropping system in the Upper Great Lakes region: An assessment of risk and potential, paper presented at 1999 Annual Workshop of the US National Assessment: The Potential Consequences of Climate Variability and Change, Atlanta, Georgia, April 12-15, 1999.

Groisman, P. Ya., et al., Changes in the probability of heavy precipitation: Important indicators of climatic change, *Climate Change, 42,* 243-283, 1999.

Groisman, P. Ya., R. W. Knight, T. R. Karl. Relations between changes of heavy precipitation and streamflow in the United States, in press, *Bulletin of the American Meteorological Society,* 2000.

Injerd, D., Panel Presentation, Impacts and risks of climate change and variability: stakeholder perspectives in *Adapting to Climate Change and Variability in the Great Lakes-St. Lawrence Basin: Proceedings of a Binational Symposium,* edited by L. D. Mortsch, S. Quon, L. Craig, B. Mills, and B. Wrenn, Environment Canada, Downsview, Ontario, 1998.

IPCC (Intergovernmental Panel of Climate Change), *Climate Change 1995: The Science of Climate Change,* edited by J. Y. Houghton, F. G. Meira Filho, B. A. Callander, N. Harris, A. Kattenberg, and K. Maskell, Cambridge University Press, Cambridge, United Kingdom, 570 pp., 1996.

Izaurralde, R. C., R. A. Brown, and N. J. Rosenberg,. US regional agricultural production in 2030 and 2095: Response to CO_2 fertilization and Hadley Climate Model (HADCM2) projections of greenhouse-forced climatic change, Pacific Northwest National Laboratories, Richland, Washington, Rep. No. PNNL-12252, 42 pp., 1999.

Kalkstein, L., and J. S. Greene, An evaluation of climate/mortality relationships in large US cities and the possible impacts of a climate change, *Environmental Health Perspectives,* 105(1), 84-93, 1997.

Kalkstein, L., and K. Smoyer, The impacts of climate change on human health: Some international implications, *Experientia, 49,* 969-979, 1993.

Karl, T. R., Potential application of model output statistics (MOS) to forecasts of surface ozone concentrations, *Journal of Applied Meteorology,* 18, 254-265, 1979

Karl, T. R., and R. W. Knight, Secular trends of precipitation amount, frequency, and intensity in the USA, *Bulletin American Meteorological Society, 79,* 231-241, 1998.

Kunkel, K. E., R. A. Pielke, Jr., and S. A. Changnon, Temporal fluctuations in weather and climate extremes that cause economic and human health impacts: A review, *Bulletin of the American Meteorological Society, 80,* 1077-1098, 1999.

Kunkel, K. E., N. Westcott, and D. Kristovich, Assessment of potential effects of climate change on heavy lake-effect snowstorms near Lake Erie, preprints, *11th Symposium on Global Change Studies,* Long Beach, California, January 10-13, 2000, American Meteorological Society, Boston, Massachusetts, pp. 50-53, 2000.

Lal, R., J. M. Kimble, R. F. Follett, and C. V. Cole, The Potential of US Cropland to Sequester Carbon and Mitigate the Greenhouse Effect, Sleeping Bear Press. Ann Arbor, Michigan, 1998.

Lew, A. A., The Midwest: The Nation's Heartland, in *Geography USA: A Virtual Textbook,* (http://www/for.nau.edu/~alew/midwest7.html)., 1998.

Magnuson, J. J., Potential effects of climate changes on aquatic systems: Laurentian great lakes and Precambrian shield region, chap. 2, in *Assessment,* edited by C. E. Cushing, John Wiley & Sons, Chichester, England, pp. 7-53, 1997. [Papers in this volume were originally published in *Hydrological Processes,* 11(8), 819-1067, 1997.

McPherson, E. G., D. Nowak, G. Heisler, S. Grimmond, C. Souch, R. Grant, and R. Rowntree, Quantifying urban forest structure, function, and value: The Chicago Urban Forest Climate Project, *Urban Ecosystems,* 1(1), 49-61, 1997.

Meehl, G.A., F. Zwiers, J. Evans, T. Knutson, L. Mearns, and P. Whetton, Trends in extreme weather and climate events: Issues related to modeling extremes in projections of future climate change, *Bulletin of the American Meteorological Society,* 81, V. 81, No. 3 pp. 427-436.

Moll, R., Reports of the working groups: Ecosystems health in *Adapting to Climate Change and Variability in the Great Lakes-St. Lawrence Basin: Proceedings of a Binational Symposium,* edited by L. D. Mortsch, S. Quon, L. Craig, B. Mills, and B. Wrenn, Environment Canada, Downsview, Ontario 1998.

Mortsch, L. D., Assessing the impact of climate change on the Great Lakes shoreline wetlands, *Climatic Change, 40,* 391-416, 1998.

National Climatic Data Center (NCDC), Billion Dollar US Weather Disasters 1980-1999, Asheville, North Carolina, US Department of Commerce, NOAA, Website, (http://www.ncdc.noaa.gov/ol/reports/billionz.html), 1999.

Norgaard, K. J., Impacts of the Subdivision Control Act of 1967 on land fragmentation in Michigan's townships, Ph.D. dissertation, Michigan State University, East Lansing, Michigan, 1994.

NPA Data Services, 1999a, *Demographic Databases, Household Databases Three Growth Projections 1967-2050,* 1424 16th Street, NW, Suite 700, Washington, DC 20036

NPA Data Services, 1999b, *Economic Databases Three Growth Projections 1967-2050,* 1424 16th Street, NW, Suite 700, Washington, DC 20036

Olsen, J., J. Stedinger, N. Matalas, and E. Stakhiv, Climate variability and flood frequency estimation for the upper Mississippi and lower Missouri rivers, *Journal of American Water Resources Association, 35,* 1509-1523, 1999.

Quinn, F.H., B. M. Lofgren, A. H. Clites, R. A. Assel, A. Eberhardt, and T. Hunter, Great Lakes water resources sector report, Great Lakes Environmental Research Laboratory, Ann Arbor, Michigan, 6 pp., 1999.

Robinson, S. K., The case of the missing songbirds, *Consequences, 3,* 2-15, 1997.

Rose, J. B., S. Daeschner, D. R. Easterling, F. C. Curriero, S. Lele, and J. A. Patz, . Climate and waterborne outbreaks in the US, *Journal of American Water Works Association,,* 77-87, Sept. 2000.

Sadof, C., and M. Raupp, Biological controls in the urban forest, *Midwest Biol Cont. News,* (5)12, 1-5, 1999.

Secrest, H. C., H. J. Mac Aloney, and R. C. Lorenz, Causes of the decadence of hemlock at the Menominee Indian Reservation, Wisconsin, *Journal of Forestry, 39,* 3-12, 1941.

Smith, J. B., Setting priorities for adapting to climate change, *Global Environmental Change,* 7(3), 251-264, 1997.

Stefan, H.G., M. Hondzo, and X. Fang, Lake water quality modeling for projected future climate scenarios, *Journal of Environmental Quality,* 22, 417-431, 1993.

Wang, J. X. L., and J. K. Angell, Air stagnation climatology for the United States (1948-1998), NOAA/Air Resources Laboratory ATLAS No. 1, Silver Spring, Maryland, 73 pp., 1999.

Warbach, J. D., and D. Norberg, *Michigan Society of Planning Officials Trend Futures Project: Public Lands and Forestry Trends Working Paper,* Michigan Society of Planning Officials, Rochester, Michigan, 1995.

Wargo, P.M., D. R. Houston, L. A. LaMadeleine, Oak decline, *Forest Insect & Disease Leaflet 165,* US Department of Agriculture, Forest Service, 8 pp., 1983.

Wolfe, D. W., Potential climate change impacts on New England agriculture, in *Workshop Summary Report, New England Regional Climate Change Impacts Workshop,* Institute for the Study of Earth, Oceans, and Space, University of New Hampshire, Durham, New Hampshire, 1997.

Wong, K., A pixel is worth 1000 words: Satellite images reveal startling tree loss in American cities, *US News and World Report,* July 19, 1999.

ACKNOWLEDGMENTS

Many of the materials for this chapter
are based on contributions from participants
on and those working with the

Eastern Midwest Workshop Team
J. C. Randolph, Indiana University
Otto Doering, Purdue University
Mike Mazzocco, University of Illinois, Urbana -
    Champaign
Becky Snedegar, Indiana University

Great Lakes Workshop and Assessment Team
Peter J. Sousounis*, University of Michigan
Jeanne Bisanz*, University of Michigan
Gopal Alagarswamy, Michigan State University
George M. Albercook, University of Michigan
J. David Allan, University of Michigan
Jeffrey A. Andresen, Michigan State University
Raymond A. Assel, Great Lakes Environmental
    Research Laboratory
Arthur S. Brooks, University of Wisconsin-Milwaukee,
    Wisconsin
Michael Barlage, University of Michigan
Daniel G. Brown, Michigan State University
H.H. Cheng, University of Minnesota
Anne H. Clites, Great Lakes Environmental Research
    Laboratory
Thomas E. Croley II, Great Lakes Environmental
    Research Laboratory
Margaret Davis, University of Minnesota
Anthony J. Eberhardt, Buffalo District, Army Corps of
    Engineers
Emily K. Grover, University of Michigan
Galina Guentchev, Michigan State University
Vilan Hung, University of Michigan
Kenneth E. Kunkel, Illinois State Water Survey
David A. R. Kirstovich, Illinois State Water Survey

John T. Lehman, University of Michigan
John D. Lindeberg, Center for Environmental Studies,
    Economics & Science
Brent M. Lofgren, Great Lakes Environmental
    Research Laboratory
James R. Nicholas, USGS, Lansing, Michigan
Jamie A. Picardy, Michigan State University
Jeff Price, American Bird Conservancy
Frank H. Quinn, Great Lakes Environmental Research
    Laboratory
Paul Richards, University of Michigan
Joe Ritchie, Michigan State University
Terry Root, University of Michigan
William B. Sea, University of Minnesota
David Stead, Center for Environmental Studies,
    Economics & Science
Shinya Sugita, University of Minnesota
Karen Walker, University of Minnesota
Eleanor A. Waller, Michigan State University
Nancy E. Westcott, Illinois State Water Survey
Mark Wilson, University of Michigan
Julie A. Winkler, Michigan State University
John Zastrow, University of Wisconsin

Additional Contributors
Stanley Changnon, Illinois State Water Survey
Byron Gleason, National Climatic Data Center

* signifies Assessment team chairs/co-chairs

CHAPTER 7

POTENTIAL CONSEQUENCES OF CLIMATE VARIABILITY AND CHANGE FOR THE GREAT PLAINS

Linda A. Joyce[1,2], Dennis Ojima[3], George A. Seielstad[4], Robert Harriss[5], and Jill Lackett[3]

Contents of this Chapter

[1]Coordinating author for the National Assessment Synthesis Team; [2]USDA Forest Service, [3]Colorado State University, [4]University of North Dakota, [5]National Center for Atmospheric Research

CHAPTER SUMMARY

Regional Context

The Great Plains produces much of the nation's grain, meat, and fiber, and in addition provides recreation, wildlife habitat, and water resources. Though more rural than the rest of the United States, the urban areas of the Great Plains provide housing and jobs for two-thirds of the people of the Great Plains. Soil organic matter is a major resource of the Great Plains as it provides improved soil water retention, soil fertility, and the long-term storage of carbon.

Climate of the Past Century

Over the 20th century, temperatures in the Northern and Central Great Plains have risen more than 2°F (1°C), with increases up to 5.5°F (3°C) in some areas. There is no evidence of a trend in the historical temperature record of the Southern Great Plains. Over the last 100 years, annual precipitation has decreased by 10% in the lee of the Rocky Mountains. Texas has seen significantly more high intensity rainfall.

Climate of the Coming Century

Air temperatures will likely continue to rise throughout the region, with the largest increases in the northern and western parts of the Plains. Seasonally, more warming is projected to occur in winter and spring than in summer and fall.

A pattern of decreasing precipitation appears likely in the lee of the Rocky Mountains while other sections of the Great Plains may experience slight increases. Although precipitation increases are projected for parts of the Great Plains, increased evaporation due to rising air temperatures are projected to surpass these increases, resulting in net soil moisture declines for large parts of the region.

Key Findings

- Productivity of crops and grasses of the region will likely respond positively to additional atmospheric carbon dioxide, especially those systems with adequate water and nitrogen such as alfalfa or soybeans.
- The warmer and predicted longer growing seasons will likely change the life cycles of all biological organisms and these changes will have profound impacts on the ecology of the Great Plains native ecosystems.
- The projected climate changes will likely alter the current biodiversity, resulting in a new composition of plant and animal species that may or may not be detrimental to society. A possible migration of invasive species across the Great Plains is a concern to stakeholders in the region because the rapid rate of change in climate may be disadvantageous to native species.
- Extreme temperatures and heat stress events, where the temperature remains over 90°F (32°C) for 3 consecutive days, will likely increase in the Southern Great Plains. These events will increase the heat stress on humans and livestock.
- Changes in the demand for irrigation water vary by crop type and the changes in the seasonality of precipitation. For example, in northeastern Colorado the consumptive demand for water for perennial crops such as grass and alfalfa was estimated to increase at least 50% over current use, however, consumptive water use for corn was not expected to increase substantially.
- Increased air temperatures may reduce soil organic matter affecting soil fertility, water-holding capacity, and the storage of carbon in the soil.
- The intensity of rainfall events may increase in the Southern Great Plains, resulting in more rainfall in shorter periods of time, with implications to urban and rural flood controls and soil erosion.
- Rural communities, already stressed by their declining populations and shrinking economic bases, are dependent on the competitive advantage of their agricultural products in domestic and foreign markets. A changing climate will likely bring additional stresses that will disproportionately impact family farmers and ranchers.
- Stakeholders in the region thought that community-based adaptive management was an important component for future planning.

POTENTIAL CONSEQUENCES OF CLIMATE VARIABILITY AND CHANGE ON THE GREAT PLAINS

PHYSICAL SETTING AND UNIQUE ATTRIBUTES

The Great Plains is often pictured as an agricultural landscape dotted with many small rural towns. In fact, the Great Plains is more rural than the rest of the US; nearly 40% of the region's counties have populations less than 2,500 people in contrast to the 20% of rural counties in the remainder of the US (Skold, 1997; Gutmann, 1999). These landscapes of grasses, forbs, and shrubs, interspersed with a variety of crops, give the Plains a sense of wide-open space. Increasing sizes and declining numbers of farms and ranches, fewer rural trade centers, and declining rural populations are recent socio-economic trends that further contribute to the remoteness of some rural counties. In contrast, other Great Plains' counties with large urban centers or with scenic amenities are experiencing population increases and economic growth (Drabenstott and Smith, 1996).

Distributions of the naturally occurring vegetation and the planted agricultural crops (Figure 1) are strongly linked to the gradients of temperature (north to south) and precipitation (west to east) within the Great Plains (Figure 2). Cool-season grasslands in the North give way to warm-season grasslands in the central and southern parts of the region, which give way to drought-adapted shrubs in the southwestern parts and trees in the southeastern parts. As precipitation increases from west to east across the Great Plains, the native vegetation includes more mixed-grass and tall-grass species, and finally tree species. This precipitation gradient influences the use of irrigation in agriculture with a higher percentage of crops grown under irrigation in the western Great Plains and more crops dependent on growing season precipitation in the eastern Great Plains. The extreme western Great Plains is dominated by dryland cropping because of the reduced availability of water for irrigation. Crop types tend to follow the temperature gradient, with cool-season grains such as wheat, barley and oats dominating in the north and warm-season crops such as corn, sorghum, sugar beets, and cotton dominating in the south. However, the ability of the agricultural sector to expand across climatic zones is seen in that both corn and wheat can be found from North Dakota to Texas. Both natural and human systems cope with the natural variability in climate that characterizes the Great Plains. Periods of drought, a result of variability in climate, heighten the importance of water in this region.

Though dominated by grasslands, the Great Plains is also home to a diversity of plants and animals in shrublands, wetlands, woodlands, and forest communities. Riparian vegetation including deciduous forests, woodlands, and shrublands trace the Mississippi, Missouri, Platte, Kansas, Arkansas, Colorado, and Rio Grande Rivers. Juniper woodlands and conifer forests are found on the escarpments of the South Dakota badlands. Wind-deposited material forms extensive sand dune systems in

Great Plains Vegetation Map

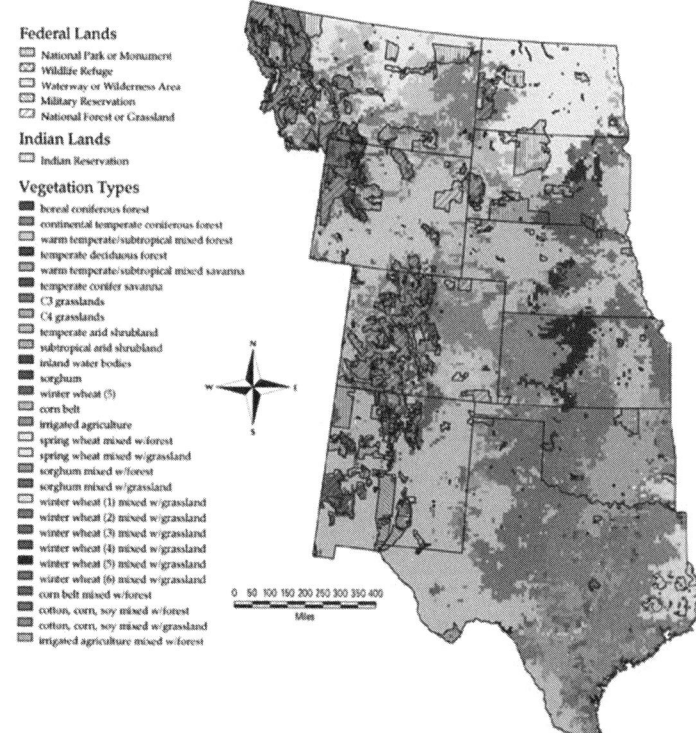

Federal Lands
- National Park or Monument
- Wildlife Refuge
- Waterway or Wilderness Area
- Military Reservation
- National Forest or Grassland

Indian Lands
- Indian Reservation

Vegetation Types
- boreal coniferous forest
- continental temperate coniferous forest
- warm temperate/subtropical mixed forest
- temperate deciduous forest
- warm temperate/subtropical mixed savanna
- temperate conifer savanna
- C3 grasslands
- C4 grasslands
- temperate arid shrubland
- subtropical arid shrubland
- inland water bodies
- sorghum
- winter wheat (5)
- corn belt
- irrigated agriculture
- spring wheat mixed w/forest
- spring wheat mixed w/grassland
- sorghum mixed w/forest
- sorghum mixed w/grassland
- winter wheat (1) mixed w/grassland
- winter wheat (2) mixed w/grassland
- winter wheat (3) mixed w/grassland
- winter wheat (4) mixed w/grassland
- winter wheat (5) mixed w/grassland
- winter wheat (6) mixed w/grassland
- corn belt mixed w/forest
- cotton, corn, soy mixed w/forest
- cotton, corn, soy mixed w/grassland
- irrigated agriculture mixed w/forest

0 50 100 150 200 250 300 350 400
Miles

Figure 1. Distributions of the naturally occurring vegetation and the current planted agricultural crops are strongly linked to the gradients of temperature (north to south) and precipitation (west to east) within the Great Plains. Outlines show the federal land holdings in the region (vegetation map from Natural Resource Ecology Lab, Colorado State University. Potential natural vegetation according to VEMAP members, 1995). See Color Plate Appendix

Great Plains Climate

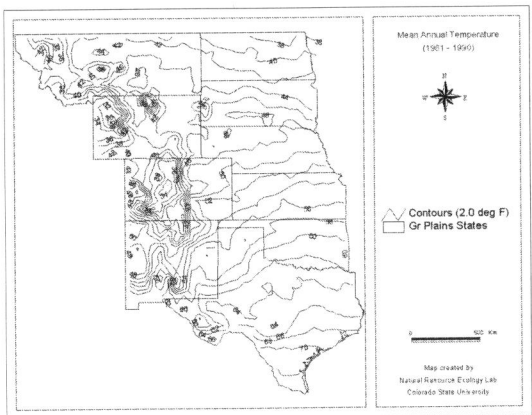

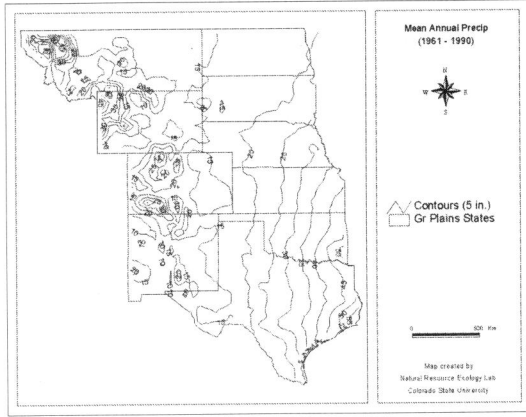

Figure 2. Great Plains climate is characterized by a strong north-south temperature gradient and a strong east-west precipitation gradient (averages based on the 1961-1990 period; data from Dennis Ojima, VEMAP climate). See Color Plate Appendix

the Great Plains. The Nebraska Sandhills contain the largest dune system covered by grassland in the United States (Ostlie et al., 1997). The rich grasslands of the region have been the basis of a large grazing system for thousands of years and currently support a diversity of cattle, bison, and other mammals, as well as a diversity of insect and bird species. Nearly 60% of the bird species that breed within the US are considered regular breeders within the Great Plains (Ostlie et al., 1997). Endemic plants and animals are found throughout unique habitats in the Great Plains. The wetland basins in the Prairie Pothole region of the Northern Great Plains and the playa lakes of the Central and Southern Great Plains are important habitat and breeding grounds for migratory waterfowl. Fish habitats include large streams with erratically variable flow, prairie ponds, marshes and small streams, and residual pools of highly intermittent streams. The largest and most diverse class of animals in the Great Plains is insects (Ostlie et al., 1997).

Throughout this region, soil organic matter is a major resource as it provides improved soil water retention, soil fertility, and the long-term storage of carbon. Soil organic matter integrates climate, geology, topography, and ecosystem dynamics (Jenny, 1980). The historical levels of grassland productivity and soil organic matter varied widely from north to south across the Great Plains, reflecting rainfall and temperature gradients (Jenny, 1980; Parton et al., 1987; Schimel et al., 1990; Peterson and Cole, 1995). Ecosystem processes associated with productivity and decomposition influence the net changes in soil carbon (C) and these net changes serve as an index

of soil fertility. Soil organic matter in grassland ecosystems is estimated to be an important reservoir of terrestrial carbon (Anderson, 1991). For example, the average aboveground plant biomass production for a cool-season grassland in Havre, MT is 33.9 g m^{-2}. For a short-grass steppe ecosystem at the Central Plains Experimental Range in Colorado, aboveground plant biomass production is 45.9 g m^{-2} (Haas et al., 1957; Cole et al., 1990). Soil C values are 3230 and 2310 g m^{-2}, for Havre and the Central Plains Experimental Range, respectively. The conversion of Great Plains grasslands to croplands has resulted in major changes in soil organic matter and nutrient supplying capacity on these lands (Haas et al., 1957; Tiessen et al., 1982; Burke et al., 1990).

SOCIOECONOMIC CONTEXT

More than 70% of the Great Plains landscape (Figure 1) is used to produce a large proportion of the nation's food. Over 60% of the nation's wheat is produced in Montana, North Dakota, South Dakota, Nebraska, Kansas, Oklahoma, and Texas (Skold, 1997). The states of Texas, Oklahoma, New Mexico, Nebraska, Kansas, and Colorado produce 87% of the nation's grain sorghum. Over 54% of the nation's barley and 36% of the nation's cotton are produced in the region (Skold, 1997). Livestock in the Great Plains constitutes over 60% of the nation's total, including both grazing and grain-fed cattle operations. Nearly 75% of the grain-fed cattle in the US are from the Great Plains, using the readily available supply of feed grains, over 50% of which is also produced in the region. Great Plains' farmers and ranchers have excelled by being adaptive and by incorporating new technologies to buffer their production against the highly variable climate. One tenth of Great Plains' cropland is irrigated. New

Great Plains Agicultural Exports

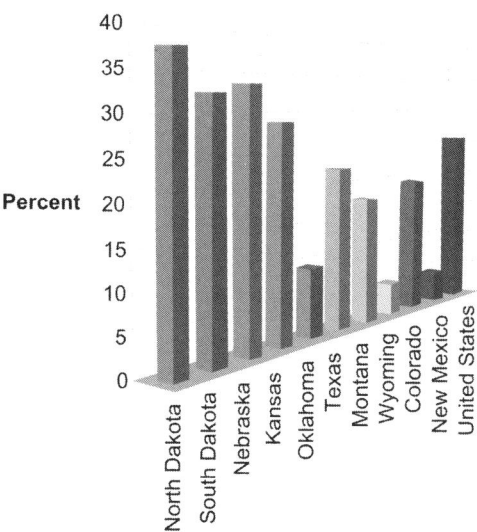

Figure 3. Agricultural exports are an important percentage of the total agricultural production within each state. (USDA, 1997 Census of Agriculture). See Color Plate Appendix

When the Great Plains was being settled over 100 years ago, county seats were established about 30 miles apart in the eastern portion and somewhat farther apart in the central and western parts of the region (Drabenstott and Smith, 1996). These county seats became the centers for government, commerce, and finance. Consolidation in agriculture, expansion of telecommunications, improved transportation, and discount retailing in rural areas are trends that have significantly reduced the number of viable rural economic centers (Drabenstott and Smith, 1996), placing an increasing burden on rural governments to provide health care and education with dwindling financial resources. In contrast, those rural counties with large trade centers or scenic attractions, as in western Montana and Colorado, have seen growth in their local economies.

technologies in agriculture, crop genetics, and livestock production have facilitated the expansion and diversification of Great Plains agriculture.

However, Great Plains' communities are undergoing dramatic changes as the industries of agriculture, commerce, and finance consolidate (Drabenstott and Smith, 1996; Barkema and Drabenstott, 1996).

Agriculture, nationally and regionally, is becoming a farm-to-grocery integrated industry in response to consumer's demands for conveniently prepared and highly nutritious food products (Barkema and Drabenstott, 1996). While the total area in farmland has remained fairly stable in the Great Plains, farms are getting bigger and the number of farmers fewer. Within the Central Great Plains, the total number of farms has gone from nearly 200,000 in 1930 to less than 100,000 in 1990 and big farms have increased from 10% to over 30% of the total number of farms (Gutmann, 1999). More crops are grown under contracts with rigid production guidelines. Contract

Multiple Stresses on Urban and Natural Environments

The Rio Grande River is life itself to cities, industries, wildlife, and rare vegetation on both sides of the US-Mexico border. However, environmental stresses and socioeconomic changes are overwhelming the urban infrastructure in this area of the Southern Great Plains. The Rio Grande supplies water to the rapidly expanding human population as well as the rapidly expanding manufacturing facilities. Total population in cities along the US-Mexico border is projected to double in less than 30 years. More than 60% of all US-Mexico trade passes through Laredo, Texas in trucks, making it the largest land port in the US. Mexico will soon displace Japan as the number two trading partner of the US, despite the fact that Mexico's economy is one-twelfth the size of Japan's.

The rapid increase in trade flowing through border cities, following passage of the North American Free Trade Agreement (NAFTA), has resulted in a complex array of costs and benefits to border cities. Industrial point source pollution as well as automobile emissions are a significant problem in the industrialized areas of the Lower Rio Grande Valley and Northern Mexico. In the unincorporated shanty-towns along both sides of the Rio Grande border, infrastructure for water supply and wastewater treatment is generally absent or totally inadequate. Reported cases of hepatitis and other viral diseases are typically two times greater in border counties compared to the statewide average.

The state of human health and environmental quality are very closely tied to land use, climate variability, air quality, and water supply. Climate change is likely to exacerbate these multiple stresses on human and natural systems; both by changes to local weather and to the flow regime of the river.

production varies across crop types, but even the small proportion of contracted wheat and feed grain is rising. Cattle feeders without contracts report that their markets have shrunk (Barkema and Drabenstott, 1996).

One consequence of changing agricultural economics is that larger farms' volume of business often justifies searching greater distances to seek lower prices on purchased items or higher prices on items sold (Barkema and Drabenstott, 1996). Local markets decline as the large rural centers or urban areas attract consumers.

Great Plains' agriculture has also been limited by two other global trends. First, the largest increase in the agricultural market is in food products that are ready or nearly ready for human consumption, such as cereals and snack foods, meats, fruits, and vegetables. The main agricultural crops of the Great Plains are bulk commodities such as wheat and corn. With the exception of meat processing, minimal food processing is done within the Great Plains because of prohibitive transportation costs to reach the major markets. Secondly, the international markets for US agricultural production have shifted from the grain purchases of the Soviet Union and the European Union to the more diversified markets of Asia. In 1996, over 40% of US agricultural exports went to Asian markets (Barkema and Drabenstott, 1996). From 4 to 38% of each Great Plains state's crop production — grain and livestock — is exported outside of the US (Figure 3). Thus, Great Plains agriculture is highly sensitive to changing consumer preferences and the global economy.

Although agriculture dominates land use in the Great Plains, the percent share of agriculture is small, 2% of the 1997 gross state product of all Great Plains states (US Department of Commerce 1998). The Northern Great Plains states are more dependent on agriculture than the Central Great Plains states, which are more dependent on agriculture than the Southern Great Plains states (Figure 4). Agriculture, forestry, and fisheries comprise 11% of North Dakota's gross state product, in contrast to 1% in Texas. In 1996, North Dakota, South Dakota, and Nebraska ranked one, two, and three nationally in terms of the farm share of the gross state product.

Changes in agricultural economies and lifestyles in the Great Plains are altering demographic patterns of the region. Populations are declining in most rural areas. In North Dakota, over 65% of the counties have fewer than 6 residents per square mile. The average age of a farm operator in the Central

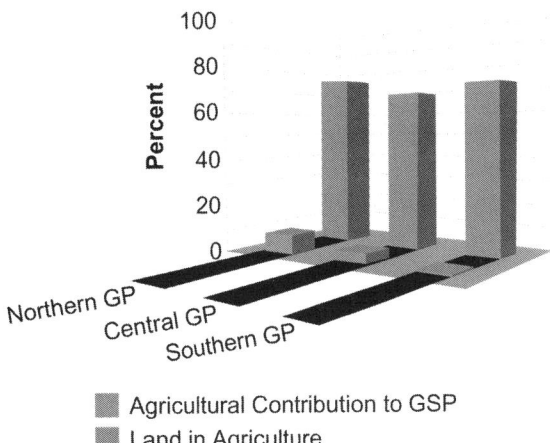

Agricultural Land Comparisons

Figure 4. The Northern Great Plains are more dependent on agriculture than the Central which is more dependent on agriculture than the Southern Great Plains, yet agriculture dominates land use in all regions of the Great Plains. (Economic data from US Dept of Commerce, Bureau of Economic Analysis, Regional Economic Analysis Division, June 1998, and land use data from USDA, 1997 Census of Agriculture.) See Color Plate Appendix

Great Plains is nearly 54 years, 52 years in the Northern Great Plains, and over 56 years in the Southern Great Plains (USDA Census of Agriculture, 1997). Urban population in the Central Great Plains has gone from 27% of the total population in 1930 to 67% in 1990 (Gutmann, 1999). Some 80% of the people in Texas live in cities. This shift in population increases the demand for services in the urban areas and introduces urban problems such as deteriorating air quality. In addition, agricultural to municipal water transfers are becoming increasingly common as more water is needed for urban inhabitants (NRC, 1992).

ECOLOGICAL CONTEXT

Water and temperature strongly influence the structure and function of the Great Plains grasslands. The transition from short-grass steppe in the lee of the Rocky Mountains to mixed prairie and finally tall-grass prairie at the eastern edge of the Great Plains corresponds to a precipitation gradient of low rainfall east of the Rockies to relatively high and more evenly-distributed rainfall in eastern parts of the region. The temperature gradient from north to south in the Great Plains represents another important gradient that determines plant type distribution and local abundance (e.g., warm-season versus cool-season grasses).

Organic matter in Great Plains' soils also varies with temperature and precipitation gradients (Jenny, 1980; Parton et al., 1987; Schimel et al., 1990; Peterson and Cole, 1995). The net changes in soil C integrate ecosystem processes associated with productivity and decomposition rates that change with weather and with land management. Changes in soil organic matter serve as an index of soil fertility. Soil carbon storage is the largest carbon pool in the grassland ecosystems. Soil carbon storage and fluxes are influenced by vegetation, characteristic soil properties, inherent climate regime, and land use practices. The current carbon level is not only determined by the current state of these factors but is also dependent on the land use history. Changes in rates of decomposition, plant production, and respiration from climate and land management affect the storage and flux of carbon from the soil. Rainfall amounts tend to control the amount of plant production that occurs whereas temperature determines decomposition rates. However, both rainfall and temperature interact in both processes to a certain degree. Changes in soil carbon are responsive to agricultural practices that alter the amount of plant material entering the soil. Changes in tillage practices, grazing patterns, and manure distribution all affect the storage of soil carbon. Conversion of grasslands or forests to cropland can result in a rapid decline in carbon stores. Up to 50% of the soil carbon and the woody biomass of the forest can be lost due to cropland conversion (Haas et al., 1957; Cole et al., 1989, 1990). Ecosystems are often sensitive to changes in extreme events. Changes in low or high temperature extremes or in drought occurrence can rapidly change the structure of an ecosystem, especially when these modifications in extremes coincide with a disturbance event, such as fire, that resets the ecosystem to a different vegetation composition. In these instances, large fluxes of carbon and long-term changes in carbon storage may result.

Three environmental parameters of global change — increased carbon dioxide, increased temperature, and altered precipitation — will likely affect Great Plains ecosystems primarily through their combined effects on plant and soil water interactions, photosynthesis, and other aspects of plant metabolism. Responses will result from direct effects of the changing environment on individual plants (e.g., productivity), as well as from changes in the mix of plant communities and cropping systems that occur due to different sensitivities of individual species. In order to assess these changes at the regional scale, the ecosystem model CENTURY (Parton et al., 1987, 1994) was used to simulate

plant productivity and carbon storage in natural and in agricultural ecosystems.

Both of the two climate scenarios used in this assessment have a marked increase in temperatures, with the Canadian scenario being the warmer of the two scenarios (see Climate chapter). Both scenarios are associated with increases in precipitation, however, the regional pattern and the magnitude of the increase in precipitation is highly varied. Of the two, the Hadley scenario simulates a slightly wetter growing season compared to the Canadian scenario. The timing of precipitation, seasonal changes in temperature, and the increased atmospheric carbon dioxide concentrations determine the impact of the two climate scenarios on the plant productivity and soil carbon levels of the Great Plains.

In order to capture the impact of land use practices on carbon storage and fluxes, the dominant natural or managed vegetation type was assigned to individual grid cells across the region. The CENTURY model simulated net carbon storage in these grid cells for each climate scenario. These simulations were based on current dominant agricultural practices and do not represent adaptive agricultural management practices being tested for maximum carbon storage. Recent no-till or reduced-till practices have been developed to lessen the impact on the soil carbon losses from croplands. The use of adaptive cropping management strategies in the US is estimated to account for 0.08-0.2 PgC/y (Lal et al., 1998), although these cropland areas are vulnerable to large carbon losses due to removal of vegetation and soil disturbances on an annual basis. The historical land management usage greatly determines the current carbon storage and potential fluxes from these managed systems.

For both climate scenarios, the increased atmospheric carbon dioxide associated with the climate changes partially ameliorates the negative warming trends on crop and grassland productivity (Figure 5). Productivity of most crops and grasses responds positively to additional carbon dioxide, especially those with adequate nitrogen such as the nitrogen-fixing alfalfa or soybean systems. These cropping systems, when simulated under dryland management, perform much better with increased carbon dioxide since these cropping systems are able to use the available soil moisture more effectively than under conditions in the absence of elevated carbon dioxide.

Soil organic matter responds to changes in vegetation as well as atmospheric carbon dioxide levels

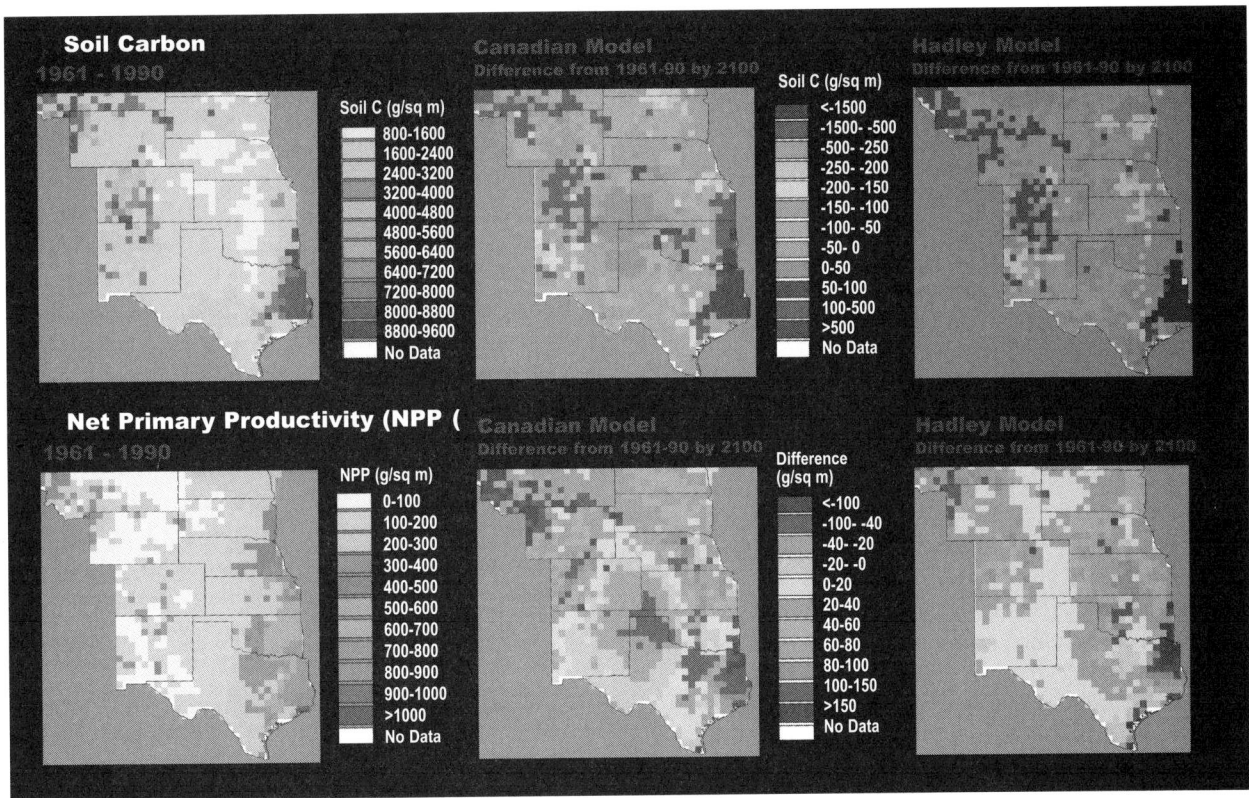

Figure 5. The productivity of the Great Plains increases from west to east and from north to south, following the precipitation and the temperature gradients. Land uses are strongly influenced by productivity. Both climate scenarios increase the moisture stress on the central parts of the Great Plains and productivity declines in this region. Soil organic matter in the Great Plains is an important reservoir of terrestrial carbon. The amount of carbon stored in the soil is strongly influenced by past and present land management practices and weather patterns. Where moisture levels and productivity decline, soil carbon may actually increase as decomposition processes become limiting. Where soil moisture levels increase from increased water use efficiency, soil carbon levels may decline. (CENTURY results from VEMAP analysis, Natural Resource Ecology Lab, Colorado State University.) – See Color Plate Appendix

(Figure 5). Where moisture levels and productivity decline, soil carbon may actually increase as decomposition processes become limiting as seen along the lee of the Rocky Mountains and especially in the Southern Great Plains. The southeastern parts of the Great Plains lose soil organic matter due to the increased soil moisture levels resulting from increased water use efficiency associated with the elevated carbon dioxide levels. In the dryland systems with little excess crop residue, soil organic matter continually declines over the duration of the climate change scenario. Systems such as wheat-fallow continue to mine soil carbon as temperature and moisture conditions facilitate release of carbon by microorganisms during the fallow period. In systems where excess crop residue is available and incorporated into the soil system, there tends to be a build up of soil carbon. This is especially evident in the productive irrigated soybean/corn rotations where nitrogen fixation by the beans and the additional residue input from corn rapidly builds soil carbon.

Some changes, such as enhanced forage production in response to elevated carbon dioxide, may be beneficial. However, plant growth under elevated carbon dioxide conditions directly influences impor-

tant metabolic responses that bear on forage quality, including reduced crude protein content and increased total non-structural carbohydrates. This direct effect on forage quality may be overshadowed by quality changes due to altered plant communities, such as changes in the balance of dominant warm- and cool-season grasses, more legumes, or more shrubs. The warmer and predicted longer growing seasons will likely alter life cycles of all biological organisms and these changes will likely have profound impacts on the ecology of the Great Plains native ecosystems.

CLIMATE VARIABILITY AND CHANGE

The Great Plains climate is characterized by a strong north-south temperature gradient and a strong east-west precipitation gradient (Figure 2). Annual precipitation ranges from less than 7.8 inches (200 mm) on the western edge to over 43 inches (1,100 mm) on the eastern edge of the Great Plains. Average annual temperature is less than 39°F (4°C) in the Northern Great Plains and exceeds 72°F (22°C) in the Southern Great Plains.

The spring and summer peaks in precipitation provide growing season moisture for the prairies. For example, 75% of the annual precipitation falls during the growing season at the Konza Prairie, Kansas (Hayden, 1998). Growing seasons range from 110 days in the Northern Great Plains to 300 days in the Southern Great Plains (Donofrio and Ojima, 1997). Monthly average maximum temperatures exceed 91°F (33°C) in most places on the Great Plains during one of the summer months. Three consecutive days of temperature over 90°F (32°C) can signal heat stress for both humans and livestock.

The variability of weather in the Great Plains is a characterizing feature, as "normal" years are rare and extremes are most often the common experience. Blizzards, floods, droughts, tornadoes, hail storms, thunderstorms, high winds, severe cold, and extreme heat often arrive suddenly, disrupt normal daily activities, and can be life-threatening.

Analysis of weather data for the last 100 years in the Great Plains indicates a warming pattern in the Northern and Central Great Plains, but no evidence of a trend in the historical temperature record for the Southern Great Plains (see Climate Chapter). Temperatures have risen in the Northern and Central Great Plains by about 2°F (1°C) in the 20th century (Karl et al., 1999), with increases of 5°F (3°C) in Montana, North Dakota, and South Dakota. This warming trend in the Northern Great Plains is reflected in 6 fewer days with temperatures less than 32°F (0°C). Meteorologists use indices, such as heating degree days and cooling degree days, to describe likely changes that would be required to maintain the human quality of life — additional heating needed in the winter or cooling in the summer. Heating Degree Days (HDD) are the degrees the average daily temperature is below 65°F (18°C). Cooling Degree Days (CDD) are the degrees the average daily temperature is above 65°F (18°C). The Northern Great Plains has had significantly fewer heating degree days (-628 HDD) and significantly more cooling degree days (40 CDD) over the 20th

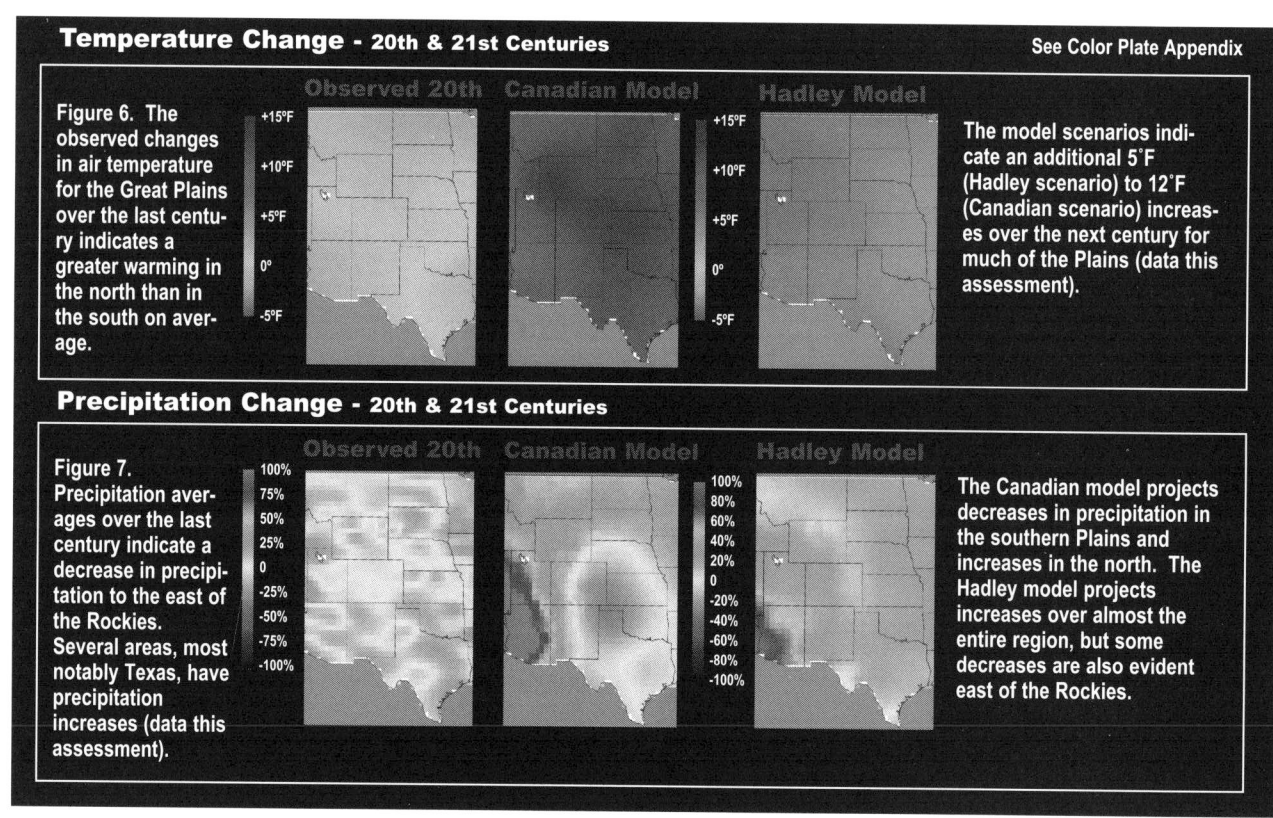

Temperature Change - 20th & 21st Centuries See Color Plate Appendix

Observed 20th Canadian Model Hadley Model

Figure 6. The observed changes in air temperature for the Great Plains over the last century indicates a greater warming in the north than in the south on average.

The model scenarios indicate an additional 5°F (Hadley scenario) to 12°F (Canadian scenario) increases over the next century for much of the Plains (data this assessment).

Precipitation Change - 20th & 21st Centuries

Observed 20th Canadian Model Hadley Model

Figure 7. Precipitation averages over the last century indicate a decrease in precipitation to the east of the Rockies. Several areas, most notably Texas, have precipitation increases (data this assessment).

The Canadian model projects decreases in precipitation in the southern Plains and increases in the north. The Hadley model projects increases over almost the entire region, but some decreases are also evident east of the Rockies.

century (Chapter 1). Daily minimum temperatures throughout the year have increased more than maximum temperatures, indicating greater nighttime warming. In the Northern Great Plains, over the last 50 years the mean date of the last measurable snow (greater than 1 inch on the ground) has occurred 4 days earlier. This significant change indicates a warming trend in the winter-spring months. Over the last 100 years, annual precipitation has decreased by 10% in eastern Montana, eastern Wyoming, western and central North Dakota, and Colorado and has increased by 5 to 20% in South Dakota, Oklahoma, Texas, and parts of Kansas (Karl et al., 1999). Texas has seen an increase in high intensity precipitation events, with significantly more area reported in severe wetness and significantly less area reported in drought conditions over the last 100 years (see Chapter 1).

The two climate scenarios used in this Assessment (see descriptions in Climate Chapter) project a continuation of the trends seen in Great Plains historical climate: higher temperatures, and for some areas, greater precipitation. The Canadian scenario projects greater increases in temperature than the Hadley scenario (Figure 6). In both scenarios, the annual average temperature rises more than 5°F (3°C) by the 2090s. Increases in temperature are greatest along the eastern edge of the Rocky Mountains. Minimum temperatures rise more in the winter (December-January-February) than the summer (June-July-August). By the 2090s winter temperatures increase 14°F (7°C) compared to 9°F (5°C) for summer in the Canadian scenario, and 9°F (5°C) versus 7°F (4°C) for the Hadley scenario.

Great Plains' annual precipitation increases by at least 13% in both scenarios by the 2090s (Figure 7). A pattern of decreasing precipitation trends appears in the lee of the Rocky Mountains and is greatly accentuated in the Canadian scenario. The annual increases are greatest in the eastern and northern parts of the Great Plains. Winter precipitation increases slightly more than summer precipitation in both scenarios. Precipitation is likely to occur in more intense rainfall events, especially in the Southern Great Plains. Although precipitation increases are projected for parts of the Great Plains, increased evaporation from rising air temperatures will outweigh the surplus of moisture from increased precipitation and soil moisture will likely decline for large parts of the region.

Drought conditions within 8 climate divisions in the Great Plains have been described using the Palmer Drought Severity Index (PDSI): values between 1.99

Annual Average Palmer Drought Severity Index (PDSI)

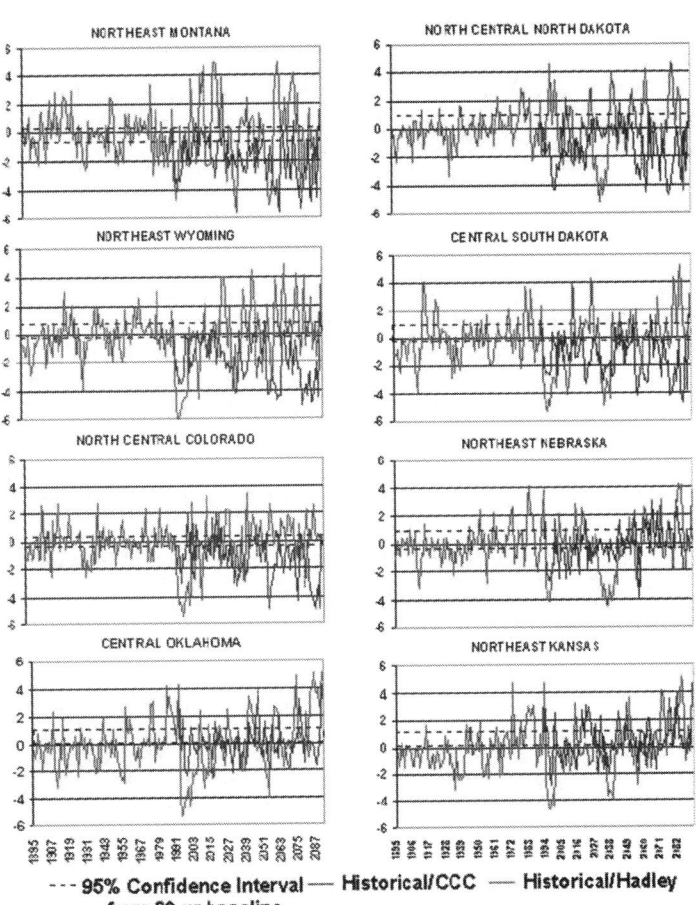

Figure 8. The droughts of the 1930s and 1950s are shown as years or periods where the Palmer Drought Severity Index (PDSI) was less then –2 and both climate scenarios (CCC = Canadian, Hadley) suggest future periods in each of these 8 climate divisions in the Great Plains where drought conditions appear likely. The 95% confidence interval for the historical period of 1960 to 1990 is shown as two dashed lines (VEMAP data, Natural Resource Ecology Lab, Colorado State University.) See Color Plate Appendix

and –1.99 are considered near normal. Values of –2 to –2.99 are considered moderate drought; values of –3 to –3.99, severe drought, and values less then –4 are considered extreme drought. The droughts of the 1930s and 1950s are shown as years or periods where the PDSI was less then –2 and both climate scenarios suggest future periods where drought conditions appear likely (Figure 8). If vegetation cover is not maintained in the sand dune areas of the Great Plains, there is some chance that increased droughtiness may result in mobilizing sand dunes (Muhs and Maat, 1993).

The number of times that a climate division in each of 8 Great Plains states experiences three consecutive days exceeding 90°F (32°C) also increases in

3+ Consecutive Days exceeding 90°F (32° C)

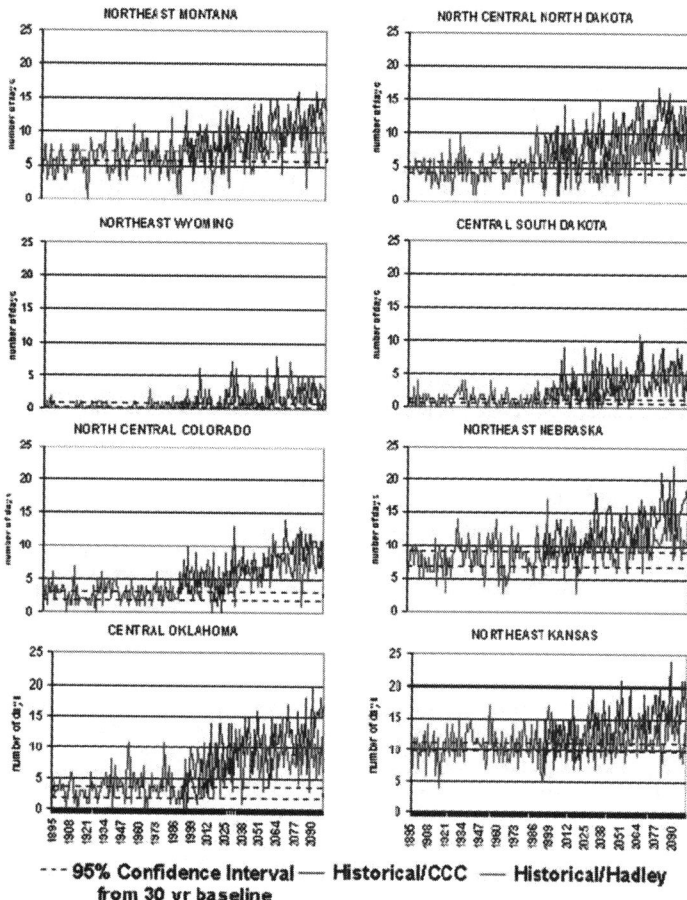

- - - 95% Confidence Interval — Historical/CCC — Historical/Hadley
from 30 yr baseline

Figure 9. Heat stress events can be triggered for livestock and for humans when the temperature exceeds 90°F (32°C) for three or more consecutive days. The number of times that a climate division in each of the 8 Great Plains states experiences three consecutive days where temperatures exceed 90°F (32°C) increases in both scenarios. The 95% confidence interval for the historic period of 1960 to 1990 is shown as two dashed lines (VEMAP data, Natural Resource Ecology Lab, Colorado State University). See Color Plate Appendix

both scenarios (Figure 9). For the climate divisions in Colorado and Oklahoma, this represents more than a doubling of the number of times such heat stress would occur.

KEY ISSUES

Workshops held in the Great Plains identified stakeholders' concerns about climate change and climate variability in the context of other current stresses on the environment and society (Ojima et al., 1997; Seielstad, 1998; National Aeronautics and Space Administration and University of Texas at El Paso, 1998). The key issues explored in detail below apply equally to the Northern, Central, and Southern

Great Plains. Additional issues were identified and these are discussed as a group following the sections below. The key issues impact the full spectrum of industries and ecosystems in the Great Plains and were chosen partly because they are common to a great deal of the economy and ecology of the region; for example, all of the key issues relate to agriculture.

- Changes in the timing and quantity of water could exacerbate the current conflicts surrounding water allocation and use in the Great Plains.
- Potential shifts in climate variability may increase the risks associated with farming, ranching, and wildland management.
- Invasive species may have unanticipated indirect impacts on the Great Plains ecology and economy.
- Rural communities, already stressed by their declining populations and shrinking economic base, are dependent on the competitive advantage of their agricultural products in domestic and foreign markets. A changing climate will bring additional stresses that will disproportionately impact family farmers and ranchers.
- Soil organic matter is a critical resource of the Great Plains as it provides improved soil water retention, soil fertility, and the long-term storage of carbon. (see Agriculture Solutions box)

1. Changes in Timing and Quantity of Water

Current Issues

Water supply and demand, allocation and storage, and quality are all climate-sensitive issues affecting the region's economy. Competing uses for water include agriculture, domestic and commercial uses, recreation, natural ecosystems, and industrial uses including energy and mining. Texas leads the Great Plains states in the total amount of water withdrawn from either surface or groundwater sources (Figure 10) although Wyoming has the highest per capita use of water, the result of a small human population and a large agricultural use. Agriculture (irrigation and livestock) withdraws the largest share of water in every Great Plains state, except North Dakota (Figure 10). "Consumptive use" refers to that part of water withdrawn that is evaporated, transpired, incorporated into products and crops, consumed by humans or livestock, or otherwise removed from the immediate water supply (Solley, 1997). Agriculture is the largest consumptive use category in every state (Figure 10), accounting for over 40% of the total water used in most states.

Sources of water include precipitation, surface water in rivers, streams and lakes, and groundwater in aquifers. Surface water supplies most of the water withdrawals in Montana, North Dakota, and Wyoming whereas groundwater sources are greatest for Nebraska (Solley, 1997). Seasonality of precipitation also influences user dependency on these sources (Miller, 1997). Irrigated agriculture along the eastern edge of the Rocky Mountains is more dependent upon snowmelt runoff from the Rocky Mountains than on spring and early summer rains. Non-irrigated agriculture in the eastern parts of the Great Plains is less dependent upon snowmelt runoff and more dependent upon the spring and early summer rains.

Water quality is a current constraining factor in the productive use of water. The management of water quality problems, such as salinity, nutrient loading, turbidity, and siltation of streams, is tied to the availability of water to accommodate agricultural and human demands. Dams, diversions, channelizations, and groundwater pumping have influenced nearly all freshwater ecosystems in the Great Plains by altering riparian habitats, aquatic ecosystems, hydrological cycles, and recreational opportunities.

In each state of the region, the allocation of water among competing uses depends on the ownership of water rights, and on the contracts and operating rules governing federal and other public water projects. Initial allocations can be modified by market transactions, but the cost of transferring water or water rights through markets varies considerably from state to state. For example, laws restrict transfers from agricultural to non-agricultural uses in some states such as Nebraska. Water apportionment decision-making among the various sectors is a major challenge and is currently marked by conflict, negotiation, or cooperation, depending on the setting. While interstate allocation issues have generated significant conflict, there are promising signs of cooperation in the tri-state effort between Colorado, Wyoming, and Nebraska to address endangered species preservation on the Platte River. In addition, negotiations between urban centers and irrigators who are willing to sell their water rights or rent water owned by the city are becoming commonplace in some states. Management of both surface water and groundwater to meet diverse and increasing water needs is a major political and management concern in the region.

Potential Impacts
The projected increases in temperature and droughts are expected to exacerbate the current

Great Plains Water Use

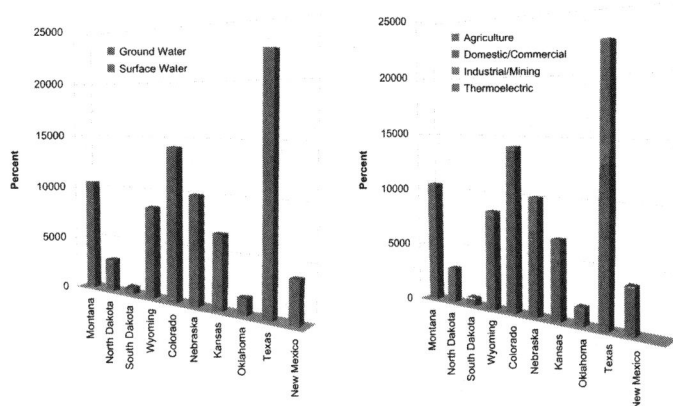

Figure 10. Surface waters are important sources for the western and northern Great Plains. Ground water, such as the Ogalalla aquifer, supplies large shares of the water for Nebraska, Kansas, and Oklahoma. Although the total amount of water withdrawal varies across the Great Plains, agriculture is the dominant consumptive use in all states (Solley, 1997). See Color Plate Appendix

competition for water among the agricultural sector, urban and industrial users, recreational users, and natural ecosystems, as well as within each user community. As water needs and available resources differ across use categories and within categories, changes such as a shift in the seasonality of precipitation will impact users differently. For example, alteration in the timing of snowmelt runoff from the Rocky Mountains would impact the current system of water management in the Platte Basin. In the South Platte Basin, water is taken from the streams/rivers in the spring and pumped strategically into nearby shallow groundwater aquifers such that the release of the recharge meets downstream user needs at the appropriate time. Winter warming coupled with an increase in winter precipitation could result in earlier and greater snowmelt from the Rocky Mountains. This storage system along the South Platte is adapted to the current climate and temporal needs of water users; under projected changes, the release of recharge may not meet the timing or amount of the downstream users' water needs.

Irrigated cropland has allowed for a diversification in agriculture in the Great Plains. Lack of soil moisture can greatly reduce yield of crops and forage. Ojima et al. (1999) found that, under the Canadian climate scenario, consumptive demand for water for perennial crops such as grass and alfalfa would increase at least 50% by the 2090s over current use (average of 1981-90). However, because growing season precipitation increases, consumptive water

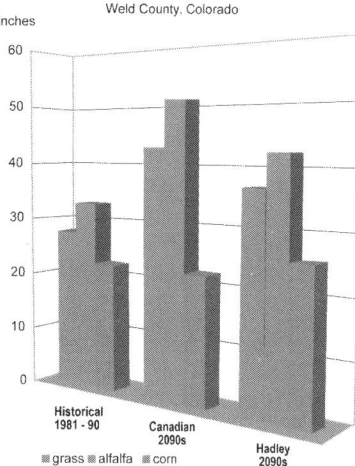

Consumptive Water Use

Weld County, Colorado

inches

Historical 1981 - 90 · Canadian 2090s · Hadley 2090s

■ grass ■ alfalfa ■ corn

Figure 11. Lack of soil moisture can greatly reduce yield of crops and forage. Under both climate scenarios, the consumptive demand for water on grass pasture increases more than 50% while the water needs for irrigated corn change little. Perennial crops experience an increase in consumptive demand for water; the size of the increase depends on the climate scenario (Ojima et al., 1999). See Color Plate Appendix

use for corn does not increase substantially (Figure 11). Thus, depending upon the future changes in temperature and precipitation, both irrigated lands and farmed acreage not currently irrigated could compete for scarce water resources in the future.

Potential increases in drought and/or storm intensity could severely impact water quality. Non-point source pollution can contain contaminants from fertilizers, herbicides, pesticides, livestock wastes, salts, and sediments that reduce the quality of both surface water and groundwater drinking water supplies. Many small towns in the Great Plains struggle to meet current drinking water standards. The projected increase in intense rainfall in the Southern Plains may increase problems with runoff in urban areas or runoff of livestock wastes from feedlots.

Adaptation Options

There are existing strategies that deal with droughts, chronic water shortages, extreme weather events and year-to-year climate variability. These strategies favor improvement and maintenance of soil, water, biotic, and land resources. Numerous cultural, economic, political, and social factors often inhibit a rapid and widespread adoption of more sustainable practices (Wilhite, 1997). These existing practices and new technologies are possibly only marginal in effecting the climate change impacts.

Water availability in both dryland and irrigated systems could be improved by existing and new technologies for residue management and tillage practices. Various techniques have been used to increase storage and availability of water: management of groundwater aquifers; enhanced snowpack storage in mountains through forest management; crop management practices that enhance soil moisture retention through crop stubble, wind breaks, and mulches; and snow management strategies on the Plains. Such techniques could increase the quantity of stored water to provide resilience to a changing climate. Irrigation scheduling, adjusting yield target to match available water, and/or changing cropping systems or land use in the event that irrigation costs exceed the worth of increased production are other options. More efficient irrigation application methods in agriculture (such as precision farming) could decrease water consumption. The need for better, non-consumptive water use in urban areas could be achieved through conservation, xeriscaping (low water-use landscaping), and the use of gray water systems for landscapes. However, the effectiveness of such measures and distribution of impacts across various water users would depend on a number of both natural and institutional factors. For example, more efficient non-consumptive water use in one location does not necessarily result in less overall water demand as diminished use at one point may simply allow increased use downstream. In addition, potential water quality issues may arise with repeated use.

Water trading is another drought management response that is more developed in some states than in others. In Colorado, for example, there are active water rental markets along the Front Range of the Rocky Mountains, and many cities that have acquired water rights in advance of need routinely rent them back to agricultural users in normal water years. This practice provides a drought buffer for the urban uses because the city can decide not to lease the water during drought years. "Water banking" is a term applied either to conjunctive use of groundwater and surface water supplies or to a formal mechanism to facilitate voluntary water transfers. This is a relatively new concept in the Great Plains region, but both Texas and Kansas have instituted programs to encourage water banking.

2. Weather Extremes

Current Issues

Extreme weather events include severe winter snow storms, ice storms, high winds, hail, tornadoes, lightning, drought, intense heavy rain, floods, heat waves, extreme cold snaps, and unexpected frosts. Natural systems have adapted to this variability, but climate extremes have significant economic impacts on farmers and ranchers as well as the human communities in the Great Plains. For example, in May of 1999, an outbreak of F4-F5 tornadoes hit the states of Oklahoma, Texas, Kansas, and Tennessee, resulting in at least $1 billion in damages and 54 deaths. In fall 1998, severe flooding in southeast Texas from 2 heavy rain events with 10-20 inch rainfall totals caused approximately $1 billion in damages and 31 deaths. The severe drought from fall 1995 through summer 1996 in the agricultural regions of the Southern Great Plains resulted in about $5 billion in damages.

Urban and industrial infrastructures have also been impacted by extreme weather events. For example, the summer 1998 heat wave and drought severely impacted roads and pipelines in Texas. In addition, this extreme event resulted in over $6 billion in damages from Texas/Oklahoma eastward to the Carolinas and at least 200 deaths. The April 1997 flood put nearly 90% of Grand Forks, North Dakota under water and caused over $1 billion in damages.

The extremes of hot and cold, as well as wet and dry, pose challenges for livestock enterprises. In the winter of 1996-97, eight blizzards in North Dakota resulted in the deaths of over 120,000 cattle, 9,500 sheep, and several thousand hogs and poultry (Junkert, personal communication via Seielstad) with direct losses of $250 million. Heat, particularly hot-humid conditions, impacts the performance of intensive livestock operations more than cold, and cattle are impacted to a greater degree than sheep (Hahn and Morgan, 1999). In August 1992, a 3-day heat wave after a relatively cool period in central and eastern Nebraska caused several hundred feedlot cattle deaths. In a July 1995 heat wave, over 4,000 feedlot cattle died in the central US (Hahn and Morgan, 1999). Weather is the primary factor in management decisions on the timing of calving seasons because newborns and neonatal animals are vulnerable to extremes of both heat and cold.

Potential Impacts

Analyses of the historical record (1895-1995) identify increases in high intensity rainfall events (greater than 2 inches/day) in the Southern Great Plains (Karl et al., 1999). Model results suggest that the frequency of high intensity rainfall will continue to increase in the Southern Great Plains, resulting in more rainfall in shorter periods of time. The histori-

THI and Wind Speed at Rockport, MO., in July, 1995

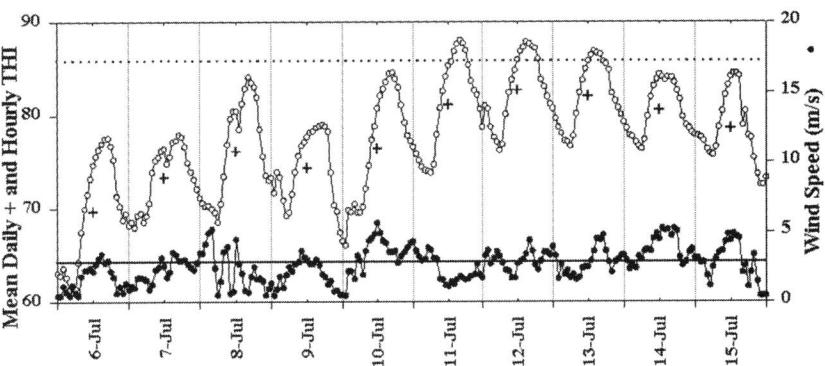

Figure 12. The hourly temperature-humidity index (THI) values and windspeeds at a mid-central US location in the highest-risk area of an early summer, 1995 heatwave which caused large-scale livestock deaths and performance losses (Hahn and Mader, 1997). The temperature-humidity index averaged above 80 for more than 3 days here, and nighttime relief (THI less than 74) from the extreme daytime heat did not occur naturally. *See Color Plate Appendix*

cal record also indicates that 7-day extreme precipitation totals are increasing in the Great Plains (Kunkel et al., 1999). In both the Canadian and the Hadley scenarios, the frequency of very high temperatures and heat stress events appear likely to increase.

Stakeholders in the region view slow changes in climate averages as less of an issue than the possibility of greater or more frequent extremes. Hail suppression programs, rainfall enhancement programs, and drought mitigation programs currently operate in different areas in the Plains. These current concerns about extreme events were also highlighted as concerns under a changing climate in the Great Plains.

Adaptation Options

Stakeholder groups recognized that climate in the Great Plains is inherently variable, and that decisions regarding land use, land management, and development are made with this variability in mind. This variability underscores the need for adaptation strategies that reduce risk and uncertainty through better access to more timely, accurate, easily accessible information about near-term weather (weeks to growing season), extreme weather events, and forecasts for weather 6 to 18 months in the future. Physically, emotionally, and psychologically, people cope better with a disaster or an abrupt change if they are prepared or if a response plan is in place.

Making decisions under uncertainty requires that strategic (long-term) planning decisions be made so that a tactical (short-term) response can be initiated when needed. For example, the stress of a heat event can be minimized if livestock operators have 3-4 days to reduce feed intake (and therefore, metabolic heat production) in cattle. Livestock can cope with high heat during the day if nighttime temperatures are sufficiently low for the animals to cool down. In the 1995 event, the heat wave was sufficiently extreme (Figure 12) and extensive to cause heat loads which exceeded the stress threshold and led to impaired performance for many animals and death for those most vulnerable (Hahn and Morgan, 1999). If livestock operators knew that the heat index was going to exceed a certain threshold more often, operators could make strategic decisions to prepare for extreme heat events through some type of cooling management, such as fans, shade, or the use of water sprays. Without such forethought, an immediate response to cool animals is difficult to arrange. Information about climate and real-time weather would allow farmers and ranchers to make strategic decisions and be able to respond quickly when necessary.

In the Northern Great Plains, the stakeholders discussed how they could build learning communities in which stakeholders share information with each other; in other words, information flows in all directions (Seielstad 1998). Information outreach could include short television segments containing information pieces on climate and other aspects of the environment, a web site with interactive information, newsletters, and partnerships with the National Park Service to develop information kiosks about environmental changes.

Nearly all Great Plains states have drought management plans in place or are developing them. These plans tend to be reactive, focusing on the emergency response to drought rather than a mitigation plan (Wilhite, 1997). Such plans may need to be updated to reflect the changing demographic and economic conditions within Great Plains states. Colorado's drought plan incorporates a monitoring system, an impact assessment, and a response system. The monitoring system allows for the early detection of a drought. In the assessment plan, the potential impact of a drought on 8 sectors is evaluated: municipal water, wildfire protection, agricultural industry, tourism, wildlife, economics, energy loss, and health. The response phase deals with the unmet needs identified by the assessment and assists local communities when their capabilities are exceeded. In the aftermath of a drought, an evaluation of the response provides suggestions to revise and improve the drought response system. The impact of a drought—economic, social or environmental—is the combination of the meteorological

event and the vulnerability of the human community (Wilhite, 1997). If increasing demand for water in the Great Plains increases the social vulnerability to water supply disruptions, then the impact of future droughts will be greater even if climate patterns were to remain the same. Thus, policies that develop regionally appropriate drought mitigation measures today will likely reduce the social impacts associated with future droughts, whether or not they are the result of climate change (Wilhite, 1997).

3. Invasive Species and Biodiversity

Current Issues
Natural systems in the Great Plains are currently stressed by a variety of agents including fragmentation of grasslands through land conversions to agriculture, cities, and roads; sedimentation and water pollution from fertilizer and pesticide runoff; introduction of invasive species through human activities and natural encroachment; altered hydrology due to the impoundment and diversion of water; and changes to natural runoff from watersheds by human activities that alter the natural efficiency of watersheds and the permeability of soil surfaces. Increasing human demands on natural systems for consumable wildlife opportunities (such as hunting and fishing) and other recreational opportunities are also likely to continue to stress natural systems.

The pattern of persistence of native species in the Great Plains is likely associated with the regional pattern of agriculture and urban development. In the agriculturally rich, eastern portions of the Great Plains, the native habitats are absent or highly fragmented. In the central portion of the Plains, these habitats may still be present but are largely disconnected. In the western edges, the native habitats are often more continuous although the recent rapid expansion of urban areas along the Rocky Mountains is increasingly fragmenting these native communities. Persistence of species depends on sufficient habitat, sufficient core areas, or sufficient connectivity between habitat patches.

Invasive species are currently a significant issue on the Great Plains. Invasive species are plant or animal species that have been introduced into an environment in which they did not evolve and they usually have no natural enemies present to limit their reproduction and spread (Westbrooks, 1998). Invasive species typically have high reproductive rates, fast growth rates, and good dispersal mechanisms. The costs and weed-associated losses in

Risk Management in the Northern Great Plains: Predicting Wheat Yield and Quality

Farmers have always monitored their crops closely for signs of nutrient deficiency, heat and water stress, insect infestation, and disease. With large acreages, however, it can be difficult to find the time to monitor all fields and even more difficult to follow change over a growing season and from year to year. Remote sensing has the advantage of being able to view large areas over one or several growing seasons. It can also detect radiation—for example, near-infrared (IR) radiation—which is invisible to the human eye. Plants absorb radiation in the visible red (R), but radiate it strongly in the near-infrared; how strongly depends on the plants' vigor and health. When vegetation indices, calculated from IR and R observations, are measured frequently over a season, the progression of canopy emergence, maturity, and senescence can be seen over large areas. The indices are also related to crop yields and, for grains, protein content. Real-time access to such data could be useful in agricultural and ranching operations.

Dryland wheat farmers and researchers in the Northern Great Plains used one vegetation index, the Normalized Difference Vegetation Index (NDVI), to see if protein content of wheat could be determined remotely during the 1998 growing season (Seielstad, 1999). Protein content, a measure of wheat quality, is related to existing soil nutrients, chemical inputs, plant biomass, and weather. The amount of nitrogen fertilizer required to produce a specific grain protein content is related to rainfall. In a wet year, early rapid plant growth will deplete soil nitrogen, resulting in a deficit late in the season and if additional nitrogen is not applied, the high yield will be of low quality. Conversely, late-season nitrogen application would not likely improve a crop's quality in a dry year.

With early information on weather and protein content of the crop, farmers could apply part of the total fertilizer typically needed for the entire growing season. The decisions about further fertilizer applications could be made as the growing season progresses. If less than average precipitation occurs or appears likely to occur, the initial fertilizer application may be sufficient. If growing season precipitation is predicted to be adequate for wheat production, then additional fertilizer to meet the growing needs could be applied. This management practice could save money, reduce runoff of unnecessary nitrogen, and lower the eventual emission of nitrogen into the atmosphere as a greenhouse gas.

Producer response was positive. In an article published in the Precision Agriculture Research Association (PARA) Newsletter, cooperating producers Carl and Janice Mattson write "...these NDVI images showed us the broad picture and gave us a great deal of optimism when it comes to satellite images. Perhaps with the more detailed images possible with the new satellites going up, there may be a vast amount of knowledge that we can apply to our management decisions." Several other producers also had positive responses. A similar study was applied to rangelands to develop efficient methods of assessing forage quantity and quality.

crop and forage production in the agricultural sector are nearly $15 billion annually. Introduced in 1827 as a contaminant in seed, leafy spurge occurs in all of the Great Plains states except Oklahoma and Texas. Grazing capacity of areas with more than 10 to 20% leafy spurge cover is significantly reduced (Westbrooks, 1998). The direct and secondary economic impacts of leafy spurge infestations on over 1.6 million acres in North and South Dakota, Montana, and Wyoming were approximately $129 million (Leitch et al., 1994). Crop losses in Kansas are annually $40 million from field bindweed (Westbrooks, 1998). With an ability to reduce wheat yield by 25%, jointed goatgrass has infested 5 million acres of winter wheat and is

spreading at a rate of 50,000 acres or more a year (Westbrooks, 1998).

In native ecosystems, invasive species may compromise the ecosystem's ability to maintain its structure or function (Stohlgren et al., 1999, Mack and D'Antonio, 1998, Vitousek et al., 1996). In grassland ecosystems, riparian areas next to streams and rivers are rich in native plant species and highly productive. These riparian areas are also critical habitat for species associated with the surrounding drier ecosystems, offering shelter and forage. These riparian ecosystems may be easily invaded by invasive species and may facilitate the establishment and migration of exotic plant invasions to upland sites in

Agriculture Solutions to Global Warming
By Martin Kleinschmit, Farmer and Rancher, Bow Valley, Nebraska

Farmers have a lot at risk as global climate heats up, but they also have a lot to gain by participating in the solution to climate change. By conserving soil organic matter, farmers can improve soil health and productivity as well as capture and store (sequester) carbon in the extensive crop and rangelands of the Great Plains. The higher temperatures and greater numbers of droughts and floods projected for the region could threaten crops, raise production expenses, and increase the risk of failure. To protect our food supply, healthy soils able to withstand erratic weather patterns are needed. Increasing the carbon content of the soil will help to mitigate global warming by keeping carbon dioxide out of the atmosphere, but it will do even more to buffer the soil against the threats of climate change. Presently most US farmland has only half or less of its historical level of organic matter. Soil scientists have established that a 6-inch (15 cm) block of soil with 1-2% organic matter can hold only about one inch (2.5 cm) of rain before it runs out of the bottom. With 4-5% organic matter, that same soil can hold 4-6 inches (10-15 cm) of rain before it leaves the root zone and takes with it the water-soluble nutrients. Increasing soil organic matter also reduces the risks of flooding and erosion, and retains moisture longer so plants have access to it during periods of dry weather. It lessens the need for (and expense of) irrigation, reduces ground water pollution, and reduces the amount of run-off, lessening the threat of stream pollution. It also lowers the cost of fertilization since nutrients not lost to erosion and leaching need not be replaced. Agricultural incentives that encourage carbon sequestration in soil provide an opportunity to promote food security in a changing climate and reduce the threat of climate change at the same time.

---

grassland ecosystems (Stohlgen et al., 1998). Natural disturbance regimes such as fire and grazing are seen as important in maintaining the native species. Alteration of the frequency, intensity, spatial pattern, or scale of these disturbances may expedite the replacement of native species with exotics (Stohlgen et al., 1999). The loss of native species and the increasing presence of exotic or invasive species are current and continual challenges for natural resource managers of the Great Plains.

Potential Impacts
The projected climate changes will likely alter the biodiversity of the Great Plains. The "new" composition of species that might arise in response to the "new" climate may or may not be detrimental to society, but the rapid rate of change in climate is likely to be disadvantageous to native species. Invasive species exploit habitats left vacant by native species susceptible to multiple stresses. As climate changes, the indirect impacts of weeds and pests are likely to bring surprising challenges.

Some native species will be unable to adapt fast enough to new climate regimes, resulting in a lowered competitive edge and weakened resistance to infestations by invasive plants and animals. Potential impacts include shifts in the relative abundance and distribution of native species, significant changes in species richness and assemblages, and local extinctions of native species.

Subtle changes in the diurnal or seasonal patterns of temperature have been shown to affect plant community composition. In the short-grass steppe, for instance, the slight warming of nighttime temperatures seen in the last 20 years has been linked to the decline of blue grama grass, the dominant grass of the short-grass prairie (Alward et al., 1999). Increased average temperature and annual precipitation in the Central Great Plains may make it possible for invasive plants, such as kudzu and Johnson grass, now found further south, to migrate north. Additional land-use pressure on these native systems is likely if agricultural practices extend into these areas as a result of more favorable climate or demands for agricultural products.

Changes, such as nutritional value of plants and changes in timing of insect emergence, may imply a decline in avian populations in the Great Plains, while longer growing seasons and the possibility of increased productivity may mitigate the declines (Larson, 1994). Grassland bird species are currently declining, a function of loss of habitat. Further alteration of their habitat from climate change would be a likely additional stress on these declining populations.

Changes in precipitation, temperature, and the hydrological cycle are likely to impact aquatic systems and the terrestrial animals dependent on these ecosystems. The abundance of wetlands is closely tied to the interannual variation in climate, both precipitation and temperature (Malcolm and Markham, 1998). Changes in annual precipitation in the northern regions are likely to have significant impacts on the breeding duck populations (Sorenson et al,. 1998; Bethke and Nudds, 1995). Changes in precipitation and riverine flow regimes are likely to exacerbate land-use conflicts and competition for water supply including conservation needs.

Maintaining natural biodiversity—the full array of native plants, animals, natural communities, and ecosystems that occur within the Great Plains—may be difficult as climate changes. In the Northern Great Plains, the relatively undisturbed landscapes of the Indian reservations contain a variety of micro-environments supporting a wide range of indigenous plant and animal species. The scattered federal land holdings within the Northern and Central Great Plains also offer refuges for species and opportunities to enhance native vegetation. In the Southern Plains, however, there is very little protected land (Figure 1). The challenging issue in this area relates to private sector land management for ecotourism, and whether this is an appropriate route to maintaining some degree of biodiversity.

Adaptation Options
Rather than identify specific strategies, stakeholders in the Central Great Plains proposed a set of general principles to guide strategic development of social responses to climate change. These five principles were articulated as follows:

First, "no regrets" strategies, those that respond to existing stresses while making the system more resilient to climate change without incurring significant costs (OTA, 1993), should be vigorously explored. There is a high level of uncertainty in regional climate projections and even greater uncertainty associated with how natural systems will respond to those changes. Developing detailed adaptation strategies based on predictions of future behavior of natural systems is currently not tenable. Instead, no-regrets strategies are particularly appropriate for natural systems because current environmental stresses could be addressed and mitigated through such strategies. The implementation of beneficial strategies today could have a positive influence on future stresses/impacts that may accrue from a changing climate.

Second, the key to developing effective coping strategies for present and future stresses is to provide organisms with alternatives for adaptation, such as landscape heterogeneity and high levels of connectivity in aquatic and terrestrial systems.

Landscape heterogeneity is the diversity among ecosystems, such as grasslands and forests, and within ecosystems, such as different successional stages. This latter type of diversity depends on maintaining appropriate disturbances to the ecosystems, such as periodic fire. In many cases, disturbance activities need to be created by management actions, such as livestock grazing and prescribed burning, in order to compensate for the loss of natural disturbance regimes, such as buffalo herds and wildfire. The idea of enhancing land stewardship by private landowners is central to the success of this management principle.

The third principle focuses on preserving current land uses that promote integrity in natural systems. This would entail, to the extent possible, encouraging conservation and restoration through proper land management. A fundamental need in implementing this principle is to identify actions that foster long-term economic vitality while at the same time enhancing ecosystem resiliency.

The fourth principle is adaptive management. The stakeholders felt that it is critically important to learn by doing and to evaluate what works and what fails to work in an attempt to lessen the impact(s) of climate change on natural systems. There will be surprises no matter how well prepared stakeholders may be; therefore management that is flexible and responds quickly will be most effective for dealing with uncertainties.

Finally, effective coping strategies depend on informing the public and decision makers about the implications of climate change for natural systems, and what these effects mean to the quality of human life. For example, why is the role of wetlands in flood control important to society? What could changes in this natural system mean to a community or to natural systems on local and regional bases?

The stakeholders saw these principles as fundamental to the discussions of climate change in all regions and an effective way to educate the general public and decision-makers about the related issues involved with climate change.

4. Quality of Rural Life on the Great Plains

Current Issues
Current demographic changes are imposing challenges to the rural areas of the Great Plains. Declining rural populations, the aging of remaining

residents, and the increased remoteness of neighbors place rural communities at increased socioeconomic risk. The growing urban areas are magnets for jobs and people. This shift in population increases the demand for services in urban areas, while increasing the burden on rural governments to provide health and education services with a declining economic base. The vulnerability of the rural human populations on the Great Plains will affect their ability to marshal resources, both natural and societal, to cope with increased risk and uncertainty. As the urban centers continue to grow, problems such as air quality will likely compromise the quality of life in urban areas, including the health-related aspects (see Additional Issues).

The consequences of weather and change on agricultural economics, beef cattle production, and grassland use can be subtle and complex due to indirect impacts from international trade, cost of feed, and markets. Most agricultural commodities are subject to production/price cycles, with the time between peaks and troughs of production controlled largely by the producer's ability to respond to price signals and consumer behavior. However, climate variability, especially drought, can significantly modify the dynamics of cattle inventories and production/price cycles, resulting in losses to producers. For example, the 1995-96 Texas drought resulted in larger numbers of cattle being sent to market due to poor range condition, increased corn feed prices, and the largely diminished winter wheat feed crop.

Potential Impacts

The projected changes in climate – increases in temperature, reductions in soil moisture, and more intense rainfall events – will likely require changes in crop and livestock management in the Great Plains. Because rural populations and their communities are highly dependent on the natural resources of the Great Plains, they are at risk from climate change, and from potential increases in climate variability. Rural economies in semi-arid regions are economically vulnerable due to lower marginal economies (lower profits and tax bases, fewer resources available) and their reliance on livestock and cropping systems that are often stressed. International exports will reflect the climate change impacts on the global agricultural markets.

Increases of warm-season forages may be a welcome addition to a forage mix in the Northern Great Plains, but the loss of the current diversity of warm- and cool-season forages in the Central Great Plains may pose limitations in grazing management. The elevated atmospheric concentration of carbon dioxide will possibly lower forage quality of native grasses. Legumes, a potential source of nitrogen, could be a new and important part of farm and ranch management. Changes in the seasonality of precipitation, particularly the growing season precipitation,

Table 1. Soil and Water Conservation Strategies and their Benefits

(Soils Working Group, Ojima et al., 1997).

Adaptation Action	Benefits to Farmer/Rancher	Benefits to Climate Change Issue
Soil organic matter management	Increase in water-holding capacity Increase in soil fertility	Increase in soil carbon storage
Precision Farming Targeted fertilizer application Targeted water application Targeted pesticide application	Cost savings Cost savings/reduced salinization Cost savings/reduced toxification	Reduction in N$_2$O emissions
Energy from biomass	Diversified	Reduction in carbon dioxide emissions from fossil fuel burning
Managing livestock wastes to capture methane	Usable energy	Reduction in methane emissions

will likely impact plant growth of native vegetation and crops with and without irrigation in the Great Plains. Warmer winters will mean some chance of more rain than snow with resultant deeper recharge, enhancing the competitive advantages of shrubs. More intensive storm activity and an increased frequency of heat waves will likely be an increasing problem for the Southern Great Plains. Whether or not the plant community will be able to accommodate changes in growing season climate or hydrological patterns is a matter of concern among stakeholders who depend on these weather patterns for their livelihood.

Adaptation Options
The stakeholders in the Great Plains proposed that the most effective adaptation strategies would be "no regrets" actions, developed from the bottom-up through community-based efforts, with an emphasis on risk reduction and increasing diversification. Because each community has different needs and values, a community-based approach was strongly supported to address issues related to adaptation and mitigation of climate change and variability. Stakeholders identified that any government policy directed towards responding to climate change at the national scale should focus on the long-term, not short-term economic incentives; should be flexible, allowing for local implementation and short response times; and should promote adaptation strategies that are sustainable and economically viable.

In the agricultural sector, various strategies have evolved to cope with drought and soil erosion (Ojima et al., 1997). Many of these coping strategies not only provide direct benefits to the farmer or rancher, but are also beneficial to the environment (Table 1). The loss of carbon and water from croplands can be minimized through practices such as reduced tillage. Cover crops and residue management can facilitate soil conservation by suppressing soil loss from wind and water erosion. Precision agricultural practices that integrate specialized crop varieties, fertilizer inputs, and irrigation schedules into crop management may provide technology to cope with climate changes.

Stakeholders also spoke about diversification, how this strategy had helped in coping with other climate or economic events in the Great Plains, and what institutional factors limited the ability of people in the Great Plains to diversify their operations and their local and regional economies. Livestock enterprises are often a mix of range management and planted forage or crop activities. Stakeholders

identified diversification of land use as an important strategy to increase profits and/or reduce risk. Examples included a strategy that some operators have already adopted such as diversifying ranch operations to include recreation, or a new strategy that policy makers could implement such as carbon credits. In addition, stakeholders identified the importance of diversification within a land use, such as ranching, to cope with the effects of climate change. Diversification strategies included 1) mix or change animal species to fit the new environment, 2) change genetics of the animal species, 3) change seasonality of production (e. g., calving, lambing, weaning), or 4) reduce production practices in stressful environments.

ADDITIONAL ISSUES

- Industrial and urban infrastructures designed for historical climate extremes may be inadequate under a different climate. As the Great Plains continues to change due to social and economic factors, demands on the existing water resource structures and other social infrastructures will also change and further challenge resource allocations in the future for land and water. In addition, climate change will affect long-term planning regarding current capacity or future design of additional infrastructure needs to sustainably utilize water and land resources. For example, in the Southern Great Plains, researchers have been working with urban engineers to assess unexpected impacts on urban infrastructures from recent extreme events, and to develop adaptation strategies for future events.

- Across the Great Plains, the poor living in communities of all sizes are disconnected from some of the most essential safety nets to cope with natural hazards. The human mortality that results from severe heat waves, floods, and natural hazards is a problem that eludes the best-intentioned policies. For example, in the Southern Plains, the poor along the Lower Rio Grande are extremely vulnerable to both droughts and flash floods. The relationship between poverty and carrying capacity of the environmental system is not a typical part of a community's approach to developing social service systems. The current struggle to achieve a sustainable economy that properly takes account of ecosystem services and social services will likely become an increasingly important issue.

- Rising air temperatures will exacerbate the current air quality issues in the urban areas of the

Great Plains. Denver (and much of the Colorado Rocky Mountain Front Range) suffers serious air quality problems on a seasonal basis. Every major city in Texas is out of compliance with current air quality standards. Houston is the second most polluted city in the nation. Typically, in Texas, a longer warm season will produce more days exceeding air quality standards. This is an immediate issue in Texas because these cities will either lose federal funds in the next decade due to their inability to control emissions, or will spend millions of dollars in courts trying to challenge regulatory penalties. The problems are exacerbated by rising urban populations from population growth as well as rural migration to cities for jobs. The issue of air quality has the potential to be an international issue if situations such as the 1998 wildfires in Mexico that inundated the Great Plains with smoke and dust become more frequent under an altered climate. Major air quality impacts occurred in the Rio Grande and Southern Great Plains and plumes from these wildfires were transported as far north as Canada.

- Advancements in technology have improved farming capability. Partnership in global change research and advancement of these technologies should provide a greater buffer against perturbations to the agricultural ecosystems and economy.
- Dissemination of information on adaptive varieties of crops and livestock suitable for changing conditions is needed from reliable sources for stakeholders.

CRUCIAL UNKNOWNS AND RESEARCH NEEDS

Research needs identified for the Great Plains focused on improvement of weather forecasts, enhancement of diversification in agriculture, management of the biodiversity of native ecosystems, improved water allocation decision-making tools, improved pest management strategies, carbon sequestration techniques, and human dimensions research.

Monitoring, early detection, and distributed warning of extreme weather events would allow for preparation and responses to minimize damages. Interannual and seasonal forecasts of weather would improve advanced planning of many activities: what crops to plant, how many cattle to graze, and other management activities. Increased research on the incidence and possible timing of hail in the Great

Plains would benefit farmers by helping them make more informed decisions about their cropping plans under the forecasted climatic conditions.

For native species, there is a need to synthesize currently available information about the potential impacts of climate change on native species and natural systems. This would include quantifying the ecological and physiological thresholds of native organisms and their tolerances to changes in environmental factors such as temperature, salinity, and sedimentation, and developing coordinated cropping/grazing systems that minimize impacts at critical periods for wildlife reproduction. Given that many natural systems in the region are substantially altered by human activities, research is needed on restoration techniques that will be effective to restore biological diversity and ecosystem services to degraded systems.

A better understanding of the relationship between livestock dynamics and rangeland condition, and the role that the diversity of both plant and animal components of rangeland ecosystems play in maintaining good rangeland condition is a needed area of research. Studies of climate change and elevated carbon dioxide levels on vegetation and animal dynamics are needed to understand ecosystem-level responses. Climate change will interact with the other current stresses on these native ecosystems and it is the cumulative effects of multiple stresses on natural systems that needs greater understanding and the development of management tools.

For agriculture, research is needed on the best feasible methods of diversification under certain climatic conditions and in specific localities. Soil types, rainfall, and growing seasons limit agricultural diversification. For example, including leguminous crops, like field peas, lentils and Austrian winter peas, in crop rotations has been successful in the Northern Plains but of limited value in the Central Plains.

Research on new crop and crop variety development for a future climate involves long lead times. Waiting until the seeds are actually marketable will mean that there will be crop failure in the first few years of a changed climate until new seeds which are better adapted to the new climate are produced, tested, and marketed. With new crops come new pests, and these need to be taken into consideration when new seed is developed.

The attainment of viable management practices for storing carbon and conserving Great Plains lands and aquatic areas is a complex issue, and will require new knowledge. Research is needed to understand the carbon cycle in the context of the ecology of Great Plains' ecosystems and agro-ecosystems, with the objective of using that information to develop conservation systems that lead to carbon storage and soil conservation. The role of wetlands in sequestering carbon needs to be better understood. Economic research is also needed to start to understand how best to achieve the goal of promoting sustainable management practices including carbon sequestration.

Research Needs Related to Humans in the Great Plains

All of the previously mentioned research will also enhance human-managed systems in the Great Plains, as the health of natural systems is vital for human quality-of-life and for recreational activities. Water quality issues are important for rural settlements, as are water supply issues. Diversification may be an important tool for survival of many family farms and ranches in the Plains, and better forecasts may encourage human preparedness for extreme events in the Plains.

There are however, several additional research needs that focus on the human populations of the region (cf Stern and Easterling, 1999). These include comprehending the perception of, understanding of, and awareness of climate change and its impacts in different parts of the Great Plains. A survey should be conducted to determine respondents' perceptions of whether and how climate has changed, how they monitor it, how it has affected their livelihoods, and if they have made any changes as a result.

Involving local people in designing and implementing the monitoring of climate change is important. Stakeholders can provide input to scientists regarding what types of information and forecasting are needed at the local level. This can make research more valuable to the farmer/rancher and provide an opportunity for scientist-practitioner interaction. Variables that may require study include soil organic matter, soil moisture, plant productivity, and extreme event frequency.

Rural residents other than farmers or ranchers, those living in urban areas, and those that depend on the Plains for products may also be affected by climate change. These affects should be a further area of research on human dimensions. How climate change affects demographic patterns in the region, both urban and rural (specifically age structure and employment) would also be an important research topic. Many people enjoy recreation in the Plains, and people all over the country depend on farm products from the Plains. Therefore, changes that affect the Plains will have indirect effects spread widely throughout the US and the world.

LITERATURE CITED

Alward, R. D., J. K. Detling, and D. G. Milchunas, Grassland vegetation changes and nocturnal global warming, *Science*, 283, 229-231, 1999.

Anderson, J. M., The effects of climate change on decomposition processes in grassland and coniferous forests, *Ecological Applications*, 1, 326-347, 1991.

Barkema, A., and M. Drabenstott, Consolidation and changes in heartland agriculture, in *Economic Forces Shaping the Rural Heartland*, Federal Reserve Bank of Kansas City, http://www.kc.frb.org/publicat/heartlnd/hrtmain.htm, 1996.

Bethke, R. W., and T. D. Nudds, Effects of climate change and land use on duck abundance in Canadian Prairie-park lands, *Ecological Applications*, 5, 588-600, 1995.

Burke, I. C., T. G. F. Kittel, W. K. Lauenroth, P. Snook, C. M. Yonker, and W. J. Parton, Regional analysis of the Central Great Plains: Sensitivity to climate variability, *BioScience*, 41, 685-692, 1990.

Cole, C. V., J. W. B. Stewart, D. S. Ojima, W. J. Parton, and D. S. Schimel, Modeling land use effects on soil organic matter dynamics in the North American Great Plains, in *Ecology of Arable Land - Perspectives and Challenges, Developments in Plant and Soil Sciences*, edited by M. Clarholm and L. Bergstrom, Kluwer Academic Publishers, Dordrecht, the Netherlands, 39, pp. 89-98, 1989.

Cole, C. V., I. C. Burke, W. J. Parton, D. S. Schimel, D. S. Ojima, and J. W. B. Stewart, Analysis of historical changes in soil fertility and organic mater levels of the North American Great Plains, in *Challenges in Dryland Agriculture, A Global Perspective, Proceedings of the International Conference on Dryland Farming*, August 1988, edited by P. W. Unger, T. V. Sneed, W. R. Jordon, and R. Jensen, Texas A&M University, College Station, Texas, pp. 436-438, 1990 .

Donofrio, C., and D. S. Ojima, The Great Plains today, in *Climate Change Impacts on the Great Plains*, Office of Science and Technology Policy and United States Global Change Research Program Workshop, compiled by D. S. Ojima, W. E. Easterling, and C. Donofrio, Colorado State University, Fort Collins, Colorado, pp. 1-24, 1997.

Drabenstott, M., and T. R. Smith, The changing economy of the rural heartland, in *Economic Forces Shaping the Rural Heartland*, Federal Reserve Bank of Kansas City, http://www.kc.frb.org/publicat/heartlnd/hrtmain.htm, 1996.

Gutmann, M., Social and demographic changes in the Central Great Plains, paper presented at the Central Great Plains Regional Assessment of Climate Change Impacts Workshop, Loveland, Colorado, March 22-24, 1999, Sponsored by the Natural Resource Ecology Laboratory, Colorado State University, and the US Department of Energy, 1999.

Haas, H. J., C. E. Evans, and E. F. Miles, Nitrogen and carbon changes in Great Plains soils influenced by cropping and soil treatments, USDA Technical Bulletin 1164, 1957.

Hahn, G. L., and T. L. Mader, Heat waves in relation to thermoregulation, feeding behavior, and mortality of feedlot cattle, Proceedings of 5th International Livestock Environment Symposium. Vol. I, pp. 563-571, American Society of Agricultural Engineers, Minneapolis, Minnesota, 1997.

Hahn, G. L., and J. A. Morgan, Potential consequences of climate change on ruminant livestock production, paper presented at the Climate Change Impacts on Rangeland Systems Symposium, Society for Range Management annual meeting, Omaha, Nebraska, February 25-26, 1999, Sponsored by the Central Great Plains Assessment and the Department of Energy, 1999.

Hayden, B. P., Regional climate and the distribution of tall grass prairie, in *Grassland Dynamics*, edited by A. K. Knapp, J. M. Briggs, D. C. Hartnett, and S. L. Collins, Oxford University Press, Inc, New York, New York, US, p. 19-34, 1998.

Jenny, H., *The Soil Resource, Origin and Behavior*, Ecological Studies, vol. 37, Springer-Verlag, New York, 1980.

Karl et al., Trends in average annual temperature and average precipitation over the 1895 to 1995 period. Data are the US Historical Climatology Network stations averaged in each of the 344 Climate Divisions. Analysis prepared for National Assessment Synthesis Team, 1999.

Kunkel, K., K. Andsager, and D. R. Easterling, Long-term trends in extreme precipitation events over the conterminous United States and Canada, *Journal of Climate*, 12, 2515-2527, 1999.

Lal, R., J. M. Kimble, R. F. Follett, and C. V. Cole, The Potential of US cropland to Sequester Carbon and Mitigate the Greenhouse Effect, Sleeping Bear Press, Chelsea, Michigan, 1998.

Larson, D. L., Potential effects of anthropogenic greenhouse gases on avian habitats and populations in the Northern Great Plains, *American Midland Naturalist, 131, 330-346,* 1994.

Leitch, J. A., F. L. Leistritz, and D. A. Bangsund, Economic effect of leafy spurge in the Upper Great Plains: Methods, models, and results, Agricultural Economics Report No. 316, North Dakota State University, Agricultural Experiment Station, 1994.

Mack, M., C., and C. M. D'Antonio, Impacts of biological invasions on disturbance regimes, *Trends in Ecology and Evolution, 13,* 195-198, 1998.

Malcolm, J. R., and A. Markham, Climate change threats to the National Parks and protected areas of the United States, World Wildlife Fund, Washington, DC, 1998.

Miller, K., Climate variability, climate change, and western water, in *Report to the Western Water Policy Review Advisory Commission,* National Technical Information Service, Springfield, Virginia, 1997, see also http://www.den.doi.gov/wwprac/reports/west.htm.

Muhs, D. R., and P. B. Maat, The potential response of Great Plains eolian sands to greenhouse warming and precipitation reduction on the Great Plains of the USA. *Journal of Arid Environments, 25,* 351-361, 1993.

National Aeronautics and Space Administration and University of Texas at El Paso, Tilting the balance: climate variability and water resource management in the Southwest, Southwest Border Regional Workshop on Climate Variability and Change, University of Texas at El Paso, March 2-4, 1998, University of Texas at El Paso, El Paso, Texas, 1998.

National Research Council, *Water Transfers in the West: Efficiency, Equity, and the Environment,* National Academy Press, Washington, DC, 1992.

Natural Resource Conservation Service, Northern Plains Regional Office, *State of the Land for the Northern Plains Region,* USDA NRCS, Northern Plains Regional Office, Lincoln, Nebraska, 1996.

Office of Technology Assessment, *Preparing for an Uncertain Climate,* US Government Printing Office, Washington, DC, 1993.

Ojima, D. S., L. Garcia, E. Eigaali, K. Miller, T. G. F. Kittel, and J. Lackett, Potential climate change impacts on water resources in the Great Plains, *Journal of the American Water Resources Association, 35,* 1443-1454, 1999.

Ojima, D. S., W. E. Easterling, C. Donofrio (Compilers), Climate change impacts on the Great Plains, Workshop Report, May 27-29, 1997, Sponsored by Department of Energy, Office of Science and Technology Policy, US Global Change Research Program, and Colorado State University, College of Natural Resources, Colorado State University, Fort Collins, Colorado, 1997.

Opie, J., *The Law Of The Land: Two Hundred Years Of American Farmland Policy,* University of Nebraska Press, Lincoln, Nebraska, 1987.

Ostlie, W. R., R. E. Schneider, J. M. Aldrich, T. M. Faust, R. L. B. McKim, S. J. Chaplin, *The Status of Biodiversity in the Great Plains,* The Nature Conservancy, Arlington, Virginia, 1997.

Parton, W. J., D. S. Schimel, C. V. Cole, and D. S. Ojima, Analysis of factors controlling soil organic matter levels in Great Plains grasslands, *Soil Science Society of America Journal, 51,* 1173-1179, 1987.

Parton, W. J., D. S. Schimel, D. S. Ojima, and C. V. Cole, A general model for soil organic matter dynamics: Sensitivity to litter chemistry, texture, and management, in *Quantitative Modeling of Soil Forming Processes,* edited by R. B. Bryant and R. W. Arnold, pp. 147-167, Soil Science Society of America Special Publication 39, Soil Science Society of America, Madison, Wisconsin, 1994.

Peterson, G. A., and C. V. Cole, *Productivity of Great Plains Soils: Past, Present, and future, in Conservation of Great Plains Ecosystems,* edited by S. R. Johnson, and A. Bouzaher, pp. 325-342, Kluwer Academic Publishers, Dordrecht, the Netherlands, 1995.

Schimel, D. S., W. J. Parton, T. G. F. Kittel, D. S. Ojima, and C. V. Cole, Grassland biogeochemistry: Links to atmospheric processes, *Climate Change, 17,* 13-25, 1990.

Schimel, D. S., T. G. F. Kittel, and W. J. Parton, Terrestrial biogeochemical cycles: global interactions with the atmosphere and hydrology, *Tellus, 43AB,* 188-203, 1991a.

Schimel, D. S., T. G. F. Kittel, A. K. Knapp, T. R. Seastedt, W. J. Parton, and V. B. Brown, Physiological interactions along resource gradients in a tallgrass prairie, *Ecology, 72,* 672-684, 1991b.

Seielstad, G., (Ed.), *Proceedings, Northern Great Plains Regional Workshop on Climate Variability and Climate Change,* University of North Dakota, Upper Midwest Aerospace Consortium, 1998.

Seielstad, G., Progress report for a public access resource center (PARC) empowering the general public to use EOSDIS – Implementation, University of North Dakota, Grand Forks, North Dakota, 1999.

Skold, M. D., Agricultural systems and economic characteristics of the Great Plains, in *Climate Change Impacts on the Great Plains,* compiled by D. S. Ojima, W. E. Easterling, and C. Donofrio, Office of Science and Technology Policy and US Global Change Research Program Workshop, pp. 85-88, Colorado State University, Fort Collins, Colorado, 1997.

Solley, W. B., Estimates of water use in the western United States in 1990 and water-use trends 1960-90, in *Report to the Western Water Policy Review Advisory Commission,* Western Water Policy Review Advisory Commission, National Technical Information Service, Springfield, Virginia, 1997.

Sorenson, L.G., R. Goldberg, T. L. Root, and M. G. Anderson, Potential effects of global warming on waterfowl populations breeding in the northern Great Plains, *Climatic Change, 40,* 343-369, 1998.

Stern, P.C., and W. E. Easterling, (Eds.), *Making Climate Forecasts Matter,* National Academy Press, Washington, DC, 1999.

Stohlgren, T. J., K. A. Bull, Y. Otsuki, C. A. Villa, and L. Lee, Riparian zones as havens for exotic plant species in the central grasslands, *Plant Ecology, 138,* 113-125, 1998.

Stohlgen, T. J., D. Binkley, G. W. Chong, M. A. Kalkhan, L. D. Schell, K. A. Bull, Y. Otsuki, G. Newman, M. Bashkin, and Y. Son, Exotic plants species invade hot spots of native plant diversity, *Ecological Monographs, 69,* 25-46, 1999.

Tiessen, H., L. W.B. Stewart, and J.R. Bettany, Cultivation effects on the concentration and amounts of carbon, nitrogen, and phosphorus in grassland soils, *Agronomy Journal, 74,* 831-835, 1982.

USDA Census of Agriculture 1997, http://www.nass.usda.gov/census/

US Department of Commerce, Bureau of Economic Analysis, Regional Economic Analysis Division, June 1998

VEMAP members, Vegetation/ecosystem modeling and analysis project: comparing biogeography and biogeochemistry models in a continental-scale study of terrestrial ecosystem responses to climate change and CO_2 doubling, Global Biogeochemical Cycles, 9, 477-437, 1995.

Vitousek, P. M., C. M. D'Antinio, L. L. Loope, and R. Westbrooks, Biological invasions as global environmental change, *American Scientist, 84,* 468-478, 1996.

Westbrooks, R. G., Invasive plants: changing the landscape of America: fact book, Federal Interagency Committee for the Management of Noxious and Exotic Weeds, Washington, DC, 1998.

Wilhite, D., Improving drought management in the West, in *Report to the Western Water Policy Review Advisory Commission,* National Technical Information Service, Springfield, Virginia, see also http://www.den.doi.gov/wwprac/reports/adrought.htm, 1997.

ACKNOWLEDGMENTS

Materials for this chapter are based in large part
On the products of workshops and assessments
carried out by the

Central Great Plains Assessment Team
Dennis Ojima, Colorado State University, co-chair
Jill Lackett, Colorado State University, co-chair
Lenora Bohren, Colorado State University
Alan Covich, Colorado State University
Dennis Child, Colorado State University
Celine Donofrio, Colorado State University
William Easterling, Pennsylvania State University
Kathy Galvin, Colorado State University
Luis Garcia, Colorado State University
Jim Geist, Colorado Corn Administrative Committee
Myron Gutmann, University of Texas at Austin
Tom Hobbs, Colorado State University/Colorado
    Division of Wildlife
Tim Kittel, National Center for Atmospheric
    Research
Martin Kleinschmit, Center for Rural Affairs
Kathleen Miller, National Center for Atmospheric
    Research
Jack Morgan, USDA Agricultural Research Service
Gary Peterson, Colorado State University
Bill Parton, Colorado State University
Keith Paustian, Colorado State University
Rob Ravenscroft, Rancher, Nebraska
Lee Sommers, Colorado State University
Bill Waltman, Natural Resources Conservation
    Service

Northern Great Plains Assessment Team
George Seielstad, University of North Dakota, co-
    chair
Leigh Welling, University of North Dakota, co-chair
Kevin Dalsted, South Dakota State University
Jim Foreman, Ten Sleep, Wyoming
Bob Gough, Intertribal Council on Utility Policy
James Rattling Leaf, Sinte Gleska University
Janice Mattson, Precision Agriculture Research
    Association
Patricia McClurg, University of Wyoming
Gerald Nielsen, Montana State University
Gary Wagner, Climax, Minnesota
Pat Zimmerman, South Dakota School of Mines and
    Technology

**Southern Great Plains Workshop Steering
    Committee**
Robert Harriss, Texas A&M University (currently
    NCAR), chair
Tina Davies, Houston Advanced Research Center
David Hitchcock, Houston Advanced Research
    Center
Gerald North, Texas A&M University

**Southwest-Rio Grande Workshop Steering
    Committee**
Charles Groat, University of Texas-El Paso (currently
    USGS), chair
Honorable Silvestre Reyes, US House of
    Representatives, Texas, honorary chair

CHAPTER 8

POTENTIAL CONSEQUENCES OF CLIMATE VARIABILITY AND CHANGE FOR THE WESTERN UNITED STATES

Joel B. Smith[1], Richard Richels[2,3], and Barbara Miller[3,4]

Contents of this Chapter

[1]Stratus Consulting, Inc.; [2]EPRI; [3]Coordinating author for the National Assessment Synthesis Team; [4]World Bank

CHAPTER SUMMARY

Regional Context

The West is characterized by variable climate, diverse topography and ecosystems, an increasing human population, and a rapidly growing and changing economy. Western landscapes range from the coastal areas of California, to the deserts of the Southwest, to the alpine tundra of the Rocky and Sierra Nevada Mountains. Since 1950, the region's population has quadrupled, with most people now living in urban areas. The economy of the West has been transformed from one dominated by agriculture and resource extraction to one dominated by government, manufacturing, and services. National parks attract tourists from around the world. The region has a slightly greater share of its economy in sectors that are sensitive to climate than the nation as a whole; these include agriculture, mining, construction, and tourism, which currently represent one-eighth of the region's economy.

As a result of population growth and development, the region faces multiple stresses. Among these are air quality, urban sprawl, and wildfires. Perhaps the greatest challenge, however, is water, which is typically consumed far from where it originates. Competition for water among agriculture, urban, recreation, environmental, and other uses is intense, with water supplies already oversubscribed in many areas.

The combination of continued development of the West and climate change is likely to introduce some new stresses, exacerbate some existing stresses, and ease other stresses.

Climate of the Past Century

- In the 20th century, temperatures in the West rose 2 to 5°F.
- The region generally became wetter, with some areas having increases in precipitation greater than 50%. A few areas, such as portions of Arizona, became drier and experienced more droughts. The length of the snow season in California and Nevada decreased by about 16 days from 1951 to 1996.

Climate of the Coming Century

- During the 21st century, temperatures are very likely to increase throughout the region, at a rate faster than that observed, with the Hadley and Canadian General Circulation Models (GCMs) projecting increased temperatures of about 3 to over 4°F by the 2030s and 8 to 11°F by the 2090s.
- The two climate model scenarios project increased precipitation, particularly during winter, and especially over California. However, parts of the Rocky Mountains are projected to get drier and the Canadian model projects most of the region getting drier by the 2030s. Other changes in climate are possible and there is some chance that that climate over much of the West could become generally drier during the 21st century.
- Under the Hadley and Canadian scenarios, runoff is estimated to double in California by the 2090s, though the climate models also suggest the potential for more extreme wet and dry years in the region.
- This chapter considers the effects of warmer and wetter conditions, based on the climate model scenarios used in this Assessment. It also considers a scenario of generally warmer and drier conditions.

Key Findings

Water Resources

- The potential for flooding is very likely to increase because of earlier and more rapid melting of the snowpack and more intense precipitation. Even if total precipitation increases substantially, snowpacks are likely to be reduced. However, it is possible that more precipitation would also create additional water supplies, reduce demand and ease some of the competition among competing uses.

- In contrast, a drier climate is very likely to decrease water supplies and increase demand for such uses as agriculture, recreation, aquatic habitat, and power, thus increasing competition for scarcer supplies.

- Improved technology, planting of less water-demanding crops, pricing water at replacement cost, and other conservation efforts can help reduce demand and vulnerability to drought. Advanced planning for potentially larger floods is needed to reduce flood risks.

Natural Ecosystems

- Vegetation models estimate that under wetter conditions there is likely to be an increase in biomass, a reduction in desert areas, and a shift toward more woodlands and forests in many parts of the West. However, should the climate become drier, forest productivity would likely be reduced and arid areas would expand. It is possible that fire frequency could increase whether the region gets wetter or drier.

- Human development of the West has resulted in habitat fragmentation, creation of migration barriers such as dams, and introduction of invasive species. The combination of development, presence of invasive species, complex topography, and climate change is likely to lead to a loss of biodiversity in the region. However, it is probable the mountains will enable some species to migrate to higher altitudes. It is also possible that some ecosystems, such as alpine ecosystems, would virtually disappear from the region.

- Human interventions to aid adaptation by species will be challenging, but reducing the pressures of development on ecosystems and removing barriers to migration could be the most effective strategies.

Agriculture and Ranching

- Higher CO_2 concentrations and increased precipitation are likely to increase crop yields and decrease water demands while milder winter temperatures are likely to lengthen the growing season and result in a northward shift in cropping areas. However, there is some chance that higher temperatures will inhibit growth of certain fruits and nuts that require winter chilling, and changes in the rainfall and humidity can harm some crops, such as grapes, by increasing potential for disease.

- It is possible that higher temperatures and increased precipitation will increase forage production and lengthen the growing and grazing season for ranching, but flooding and increased risk of animal disease can adversely affect the industry.

- Increasing crop diversity can improve the likelihood that some crops will fare well under variable conditions, while switching to less water-demanding crops and improving irrigation efficiency would conserve water. Improved weather forecasting could aid farmers in selecting crops, timing harvests, and increasing irrigation efficiency; and aid ranchers in timing cattle sales and breeding.

Tourism and Recreation

- Higher temperatures are very likely to result in a longer season for summer activities such as backpacking, but a shorter season for winter activities, such as skiing. Ski areas at low elevations and in more southern parts of the region are very likely to be at particular risk from a shortening of the snow season and rising snowlines.

- Adaptation strategies for tourism and recreation involve diversification of income sources.

POTENTIAL CONSEQUENCES OF CLIMATE VARIABILITY AND CHANGE FOR THE WESTERN UNITED STATES

PHYSICAL SETTING AND UNIQUE ATTRIBUTES

The West region spans from California to the Rocky Mountains in Colorado and south to the Mexican border. The region contains 19% of the land area and 17% of the population in the United States. On average, the West has low precipitation, although some parts are quite wet. It also has some of the greatest variance in topography and climate in the lower 48 states. The West includes the lowest point (Death Valley, which is 282 feet below sea level) and the highest point (Mt. Whitney, 14,494 feet above sea level) in the lower 48 states. Among its major mountain ranges are the Sierra Nevada, the Wasatch, and the Rockies. The region also contains the Great Basin in Nevada and Utah; in which most of the rivers do not run to the sea. Especially because of its varied topography, climate zones in the West range from deserts to alpine.

Historic and Estimated Population for the West

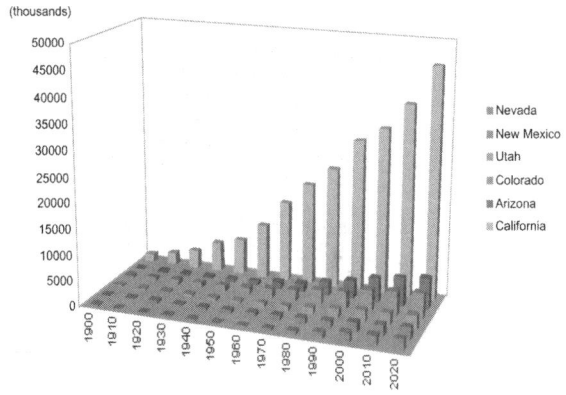

Figure 1: The West's population grew from less than 10 million in 1940 to 46.2 million in 1998 (US Census Bureau, 1998). California's population mushroomed from less than 7 million in 1940 to more than 33 million in 1998 (California Trade and Commerce Agency, 1997; California Department of Finance, 1998). Although more than two-thirds of the West's population lives in California, in recent decades, the intermountain states have become the fastest-growing in the nation. For example, Arizona's population grew from 1.3 million in 1960 to 4.5 million in 1998 (CLIMAS, 1998). Six of the 10 fastest-growing states in the US are projected to be in this region, with Arizona, Nevada, and Utah being the fastest. California's population is projected to rise from its 1998 level of 33 million to about 45 million (NPA Data Services, Inc., 1999). See Color Plate Appendix.

SOCIOECONOMIC CONTEXT

The West underwent a dramatic transformation in the 20th century in its human population, economy, and landscape. Since the middle of the century, the population has increased fourfold (see Figure 1). Although more than two-thirds of the West's 46 million people live in California, more recently the intermountain states have become one of the fastest-growing areas in the nation. Most people in the West live in urban areas. To the large cities of California — San Francisco, Los Angeles, San Diego, and Sacramento — the West has now added Denver, Salt Lake City, Albuquerque, Phoenix, and Las Vegas as major metropolitan areas (see Figure 2). Thus, once predominantly rural states are now among the most urban in the country. The regional population is projected to grow by about one half, reaching 60 to 74 million people, by 2025 (NPA Data Services, Inc., 1999).

The economy of the West has been transformed from one dominated by agriculture and resource extractive industries in the 19th century to one dominated by government, manufacturing, and services such as tourism. Figure 3 displays the relative value of all goods and services produced in the region in 1996. About 11% of the region's output is currently in sectors considered relatively sensitive to climate, including agriculture, mining, construction, and the tourism related sectors of hotels and amusement/recreation. This share of the region's output in these sectors is projected to increase to 12% by 2045, mainly because of increases in tourist related activities, but also because of increases in agricultural services. The share of total output in agriculture is projected to decrease, although the total value of agricultural production is projected to increase (US BEA, 1999a).

ECOLOGICAL CONTEXT

Although much of the West is semi-arid grassland or shrubland, the region's diverse ecosystems contain alpine tundra, coniferous and mixed forests, chaparral, wetland, and coastal and estuarine areas (USGS, 1993).

Water and land in the West have been substantially altered by people. In the West, water is typically consumed far from where it originates. For California users, water is extracted from natural systems primarily in the northern part of the state, and from the Colorado River. More than one-third of the water Arizona uses is from the Colorado River (CLIMAS, 1998). Western water tends to be subsidized (by the federal government and states) and sold to consumers at prices effectively below what it costs to make supplies available. Irrigation is the major consumer of Western water (see Figure 4).

The federal government owns more than half of the land in the West, including 83% of Nevada. Most of the federally owned land is managed by the Bureau of Land Management, Forest Service, Park Service, and Department of Defense (Riebsame, 1997). Indian reservations are scattered throughout the region, and are most concentrated in Arizona, where they comprise about one-third of the state's land area (estimated based on Riebsame, 1997). Between two-thirds and three-quarters of the land in the West is used for pasturelands, agriculture, and forests, with ranching using most of that land (USGS, 1999). However, the amount of land used for farming (including cultivated and non-cultivated land such as pastureland) in the West decreased by 8% between 1992 and 1997 (USDA, 1997).

Continued population and economic growth could result in more demand for water, wood products, and minerals; more roads, and conversion of land to urban uses (which could increase runoff and, in coastal areas, vulnerability to sea-level rise); potentially more automobile emissions (although this depends on future technology and transport practices); and increased demands for recreation. All of these could put more pressure on the remaining undeveloped areas. However, protecting open space could ease the current pressures of development on ecosystems and enhance the ability of species to cope with climate change.

CLIMATE VARIABILITY AND CHANGE

The West experiences great temporal and spatial variation in precipitation and temperature. Temperature regimes range from hot desert environments to cold alpine environments. Precipitation ranges from up to 40 inches per year in northern California to less than 10 inches in the deserts of Nevada, southeastern California, and western Arizona. Although many parts of the region, particu-

Urban Population Growth in the West

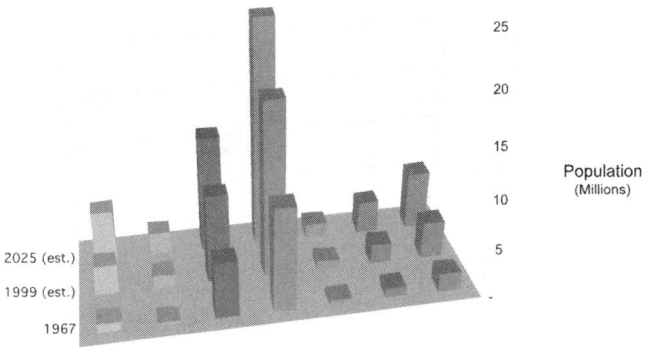

- Denver-Boulder-Greeley; CO-KS-NE
- Albuquerque; NM-AZ
- San Francisco-Oakland-San Jose; CA
- Phoenix-Mesa; AZ
- Salt Lake City-Ogden; UT-ID
- Los Angeles-Riverside-Orange County; CA-AZ
- Las Vegas; NV-AZ

Figure 2: Over 93% of California's residents live in cities, including San Francisco, Los Angeles, San Diego, and Sacramento, and their surrounding metropolitan areas. In intermountain areas, population growth is also largely concentrating in cities, such as Denver, Salt Lake City, Albuquerque, Phoenix, Las Vegas, Santa Fe and Provo. Much of the future population growth is expected to occur in urban areas. Source: NPA Data Services, 1999. See Color Plate Appendix.

The Relative Value of Economic Activitiy in the West

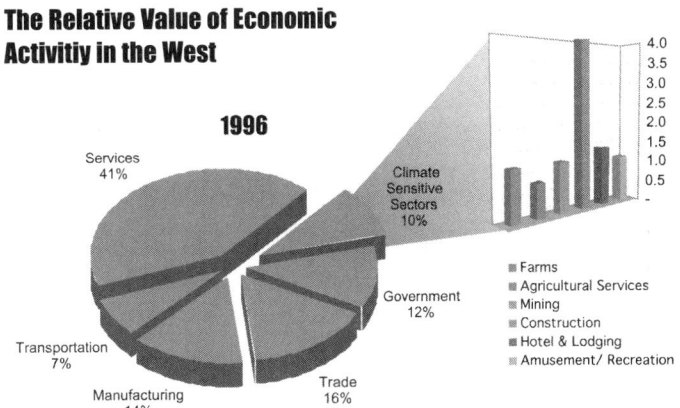

- Farms
- Agricultural Services
- Mining
- Construction
- Hotel & Lodging
- Amusement/ Recreation

Figure 3: The West produces 18% of US Gross National Product. The region has a slightly greater share of its economy in relatively climate-sensitive sectors such as agriculture, mining, construction, and tourism, than the nation as a whole. While 1.8% of the nation's economic output is from agriculture (which includes forests and fisheries), 2.0% of the West's economic output is from the agriculture sector. The West has 4.1% of its gross product from hotels, amusement/recreation, restaurants, and museums, which are strongly affected by tourism, while the nation as a whole has 1.6% (US BEA, 1999a). With its Gross State Product of $962 billion, California comprises 72% of the total Regional Product of $1.3 trillion in 1996 (US BEA, 1999a). Ranked as a nation, California would be the seventh largest economy in the world (California Trade and Commerce Agency, 1997). See Color Plate Appendix.

larly in the Southwest, receive most of their precipitation from summer monsoons, highly variable winter precipitation provides most of the annual runoff in the rest of the region (Bales and Liverman, 1998).

Relative Water Use in the West

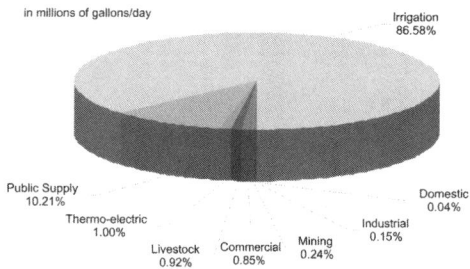

in millions of gallons/day

Irrigation 86.58%

Public Supply 10.21%

Thermo-electric 1.00%

Livestock 0.92%

Commercial 0.85%

Mining 0.24%

Industrial 0.15%

Domestic 0.04%

Figure 4: In 1995, 87% of the water consumed in the West was for irrigation (Solley et al., 1998; see Figure 4). However, water use for irrigation has declined slightly since 1980, while municipal uses have grown (Diaz and Anderson, 1995). For example, agriculture accounts for 81% of all water used in Arizona, down from 93% in 1963, while municipal demand currently accounts for 14% of water used, up from 5% in 1963 (CLIMAS, 1998). In addition, irrigated land in the region fell by 8% from 1982 to 1992, although acreage may have increased in recent years (USDA, 1997). Total water use in the region appears to have been declining since 1980 (Templin, 1999). See Color Plate Appendix.

El Niño and Events 1997-1998

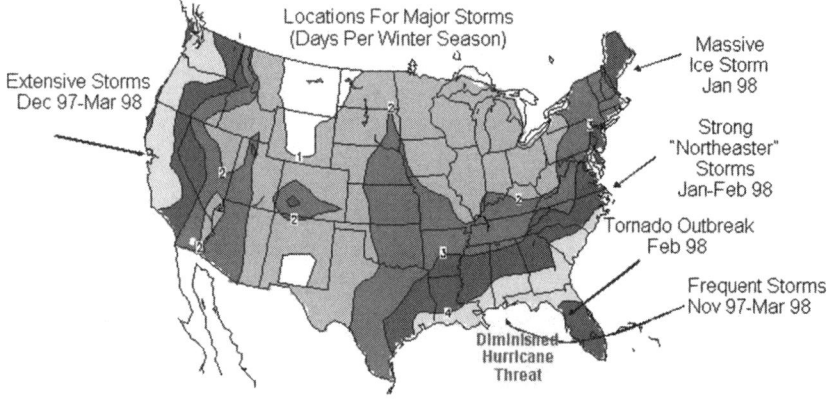

Locations For Major Storms (Days Per Winter Season)

Extensive Storms Dec 97-Mar 98

Massive Ice Storm Jan 98

Strong "Northeaster" Storms Jan-Feb 98

Tornado Outbreak Feb 98

Frequent Storms Nov 97-Mar 98

Diminished Hurricane Threat

Figure 5: The 1997-1998 El Niño had quite strong effects in the West, with particularly large winter precipitation events. The heavy precipitation lead to such localized consequences as flooding and landslides. See Color Plate Appendix.

In many areas of the West, paleoclimatic data suggest that on some occasions droughts and floods were more extreme over the past few thousand years than was observed during the 20th century (Bales and Liverman, 1998). Since 1900, temperatures in the West have been rising, with increases of 2 to 5°F per 100 years in all areas except southern Colorado, western New Mexico, and eastern Arizona (See Climate Chapter). Averaged over the region, the number of days with high temperatures over 90°F increased in the 20th century while days below freezing decreased (David Easterling, National Climatic Data Center, personal communication, 1999).

Over the 20th century, annual precipitation over most of the region generally increased 10 to 40%. However, precipitation in the Central Valley of California, southeastern California, south-central Utah, northeastern Arizona, and western Colorado decreased and some areas have experienced more drought (Karl et al., 1990; USHCN, 1999). The length of the snow season decreased by about 16 days from 1951 to 1996 in California and Nevada, and stayed about the same elsewhere (David Easterling, National Climatic Data Center, personal communication, 2000). Since the late 1940s, snowmelt has come earlier in the year in many northern and central California river basins (Dettinger and Cayan, 1995). The proportion of annual precipitation from heavy storm events has increased in the 20th century (Karl and Knight, 1998).

The region is quite vulnerable to climate variability, as the 1998 El Niño event demonstrated, particularly in California. El Niño storms during February 1998 brought as much as three times the average rainfall for the month, causing numerous deaths in addition to damages to homes, businesses, roads, utilities, and crops (Willman, 1998). On the other hand, an advanced forecast for El Niño resulted in many protective measures being undertaken (see Figure 5).

With its complex topography, developing reliable projections of climate change in the West is particularly difficult. General Circulation Models (GCMs) tend to be least reliable projecting changes in coastal areas and in mountains, two features prevalent in the West. However, it is possible to develop GCM-based scenarios that give an indication of how increased greenhouse gas concentrations could change the climate. The limitations of GCMs are discussed in more detail in Chapter 1.

Average annual outputs from the Hadley and Canadian GCMs are shown in Figure 6. The Hadley model projects a 3.8°F (2.1°C) winter warming and a 3.1°F (1.7°C) summer warming by the 2030s[1] over 1961-1990 temperatures and an 8.8°F (4.9°C) winter and an 8.3°F (4.6°C) summer increase by the 2090s. The Canadian model projects more winter warming, with a 4.8°F (2.7°C) winter and a 2.5°F (1.4°C) increase in summer temperature by the 2030s and a 12.8°F (7.1°C) winter and 7.7°F (4.3°C) summer increase by the 2090s (NCAR, 1999a).

Both models project a doubling of winter precipitation over California. However, the Hadley and

---

[1] The results for the 2030s are an average for 2025-2034.

Canadian models also show the potential for decreased precipitation in some parts of the Rocky Mountains. The Canadian model shows no change in summer precipitation, while the Hadley model projects that summer precipitation would decrease.

The models do not project a significant change in interannual variation of precipitation. Should inter-annual variation of precipitation increase, there would be more extreme wet years and more extreme dry years. It is likely that many areas in the West could have wetter winters and drier summers. It is very unlikely that changes in precipitation will be uniform across the West; some areas will likely be wetter while it is possible that others will be drier. Wet periods will very likely be followed by dry periods because, even with climate change, there will still be variability — seasonally, from year to year, and from place to place.

California has experienced relatively less sea-level rise than the eastern United States because many areas are being uplifted by moving of geological plates (Neumann et al., 2000). The coast south of La Jolla, California has been experiencing a relative sea-level rise of approximately 8 inches (20 cm) per century; the coast from Los Angeles to San Francisco has had a 0 to 6 inches (15 cm) per century of sea-level rise; and the coast in far northern California has experienced a relative reduction in sea level of 2 to 6 inches (5 to 16 cm) per century. The Intergovernmental Panel on Climate Change estimates that sea level will rise 6 to 37 inches (15 to 95 cm) by 2100 (Houghton et al., 1996), which would result in net sea-level rise for the entire California coast.

KEY ISSUES

The key issues in the West involve those systems that are sensitive to climate and, in a number of cases, are already stressed by current development patterns. All of these systems will be affected by climate change.

1. Changes in seasonality and amount of water resources
2. Plant and animal changes in natural ecosystems
3. Changes in agricultural crop productivity
4. Precipitation and forage changes for ranching
5. Sea-level rise effects on coastal resources
6. Changes in tourism and recreation

Changes in Annual Mean Temperature and Precipitation

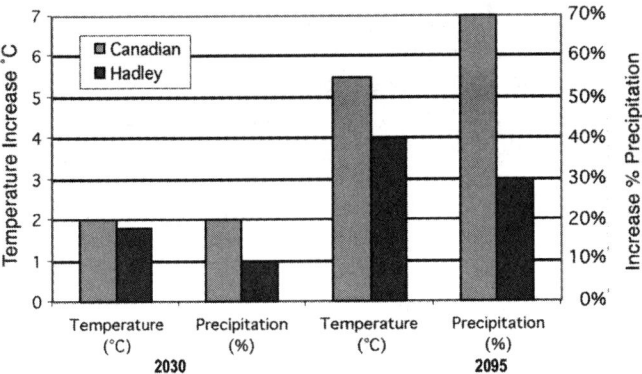

Figure 6: Changes in annual mean temperature and precipitation for the West as projected by the Hadley and Canadian models compared to 1961-90 Base Period.

1. Water Resources

The more than fourfold increase in the population of the West since the middle of the 20th century has dramatically changed the use of natural resources in the West and imposed stresses on these resources. One of the more stressed resources is water. Although agricultural water use is declining,[2] water supplies are tight because of growth in environmental, municipal, and industrial demands and could become tighter as the population and economy continue to grow and unresolved water rights claims are settled. For example, over the last ten years, California consumed more than its normal year apportionment of Colorado River water, but surplus water and water unused by Arizona and Nevada was available to meet California's needs (US Bureau of Reclamation, 1997; US Bureau of Reclamation, 1999).[3] Meanwhile, rapidly growing urban areas such as Las Vegas are demanding more water. In addition, many aquifers are being depleted at rates faster than their recharge, and high-volume groundwater mining has caused land subsidence (sinking) and fissuring (cracking).

[2] Although total water use for irrigation is declining, agricultural production is sensitive to changes in precipitation and subsequent changes in water allocation. For example, in 1991, during the fifth year of a drought, water supplies to California agriculture were severely curtailed. Overall economic losses were approximately $400 million – about 2% of total agricultural revenues. In spite of the drought, agricultural revenues in 1991 reached an all-time high (Gleick and Nash, 1991).

[3] Use of Colorado River water is allocated between the Upper Basin and the Lower Basin. The Lower Division states of California, Arizona, and Nevada are guaranteed a delivery of 75 maf (million acre feet) in each 10 year period. Also, the Upper Division states (Colorado, New Mexico, Utah, and Wyoming) are to supply one-half of the water required to be delivered by treaty to Mexico, that is, 0.75 mafy (million acre feet per year), if waters over and above the quantities of use apportioned to the Upper Basin (7.5 mafy) and the Lower Basin (8.5 mafy) are insufficient. (House Document No. 717, 1948) Nevada's apportionment of Colorado River water is 0.3 mafy 4 percent of the surplus water made available. The Upper Basin states receive the following shares: Arizona 0.05 mafy, Colorado 3.855 mafy, Utah 1.713 mafy, Wyoming 1.043 mafy, and New Mexico 0.84 mafy (NYT, 1999).

Brown (2000) forecast that by 2040 net water withdrawals in the region will increase, with withdrawals for domestic and public use increasing most, and irrigation withdrawals declining slightly except in the Upper Colorado Basin. As water use shifts from agriculture to municipal uses, the ability to reduce withdrawals during droughts declines.

It has become increasingly difficult to build any significant new water resources infrastructure because of economic, environmental, and social constraints. In addition, institutional factors such as water rights, local planning and zoning, and regulations influence and can limit the nature and level of response that water managers can make to changes in supply or demand. Reserved and Native American water rights claims are senior to those of many other water consumers, and many of these rights are not currently being exercised (see box on Native American water claims).

Because of its semiarid climate, water supplies in the West are considered to be more vulnerable to climate change than water supplies in other regions (Gleick, 1990; Hurd et al., 1999a). Detailed hydrologic modeling conducted for the western US projects

Hypothetical Change in Runoff for a Western Snowmelt Basin

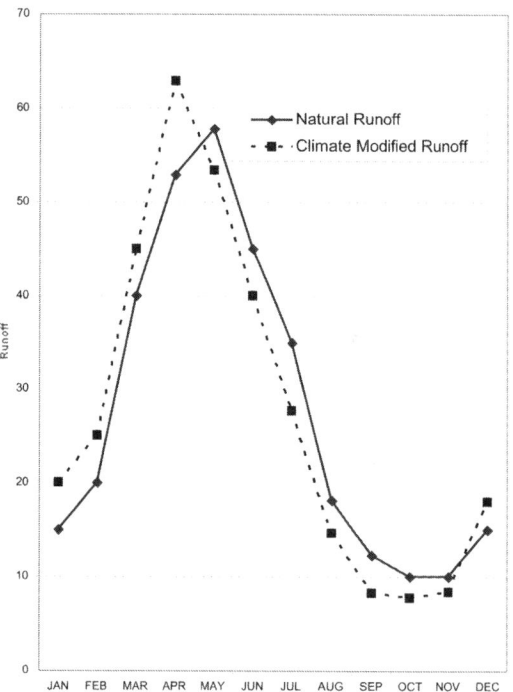

Figure 7: Natural runoff (solid line) peaks in May as winter snow melts. Under conditions of climate change (dashed line), runoff peaks earlier and higher, but is lower in the summer. Source: Gleick and Chalecki, 1999.

a significant change in snowfall and snowmelt dynamics because of higher temperatures. Rising temperatures are likely to shorten the snowpack season by delaying the autumnal change from rainfall to snow and advancing the spring snowmelt. A larger proportion of winter precipitation in mountainous areas is also very likely to fall as rain rather than snow, even if overall precipitation amounts do not change. McCabe and Wolock (1999) found that under the two GCM scenarios, April 1 snowpack in the major western mountain ranges would be reduced, except that under the Hadley scenario, snowpack in the Rocky Mountains would have little change.[4] Peak runoff is very likely to occur earlier in the year (see Figure 7) (Gleick and Chalecki, 1999). Jeton et al. (1996) found that snowmelt would occur more than two weeks earlier than currently in the East Fork of the Carson River and North Fork of the American River in the Sierra Nevada under a 2.2°C (4°F) warming, which the Hadley and Canadian scenarios suggest would occur by the 2030s.[5]

Wolock and McCabe (1999) projected changes in runoff for the region using the Hadley and Canadian climate models (see Table 1). They estimate that California runoff will increase by the 2030s by about three-fifths and double by the 2090s.[6] Their study projected small changes in runoff in the rest of the West by the 2030s, and no change to approximately 30% increases in runoff outside of California by the 2090s. The changes in runoff for the areas outside California are not considered to be statistically significant because there is so much variance in year to year runoff. Soil moisture under both scenarios is projected to increase, but in many locations outside of California conditions could be drier during some periods, particularly in the summer (NCAR, 1999b).

These changes in runoff have important consequences for water management. Any changes in runoff timing or variability could possibly cause problems (Gleick, 1987). Earlier spring runoff is likely to increase risk of spring flooding, complicate seasonal allocation schedules, and create problems for matching supply and demand and meeting envi-

---

[4]The article does not state at what altitude snowpack is measured.
[5]Jeton et al. (1996) also found that total annual flow was insensitive to changes in temperature and much more sensitive to changes in precipitation.
[6]In contrast, Miller et al. (1999a, 1999b) found that total streamflow in the Russian River in northern California, which is not snowmelt driven, would not change significantly under the Hadley 2090s scenario, but peak runoff may occur one month earlier because of a potential change in winter storms. In contrast, snowmelt driven streamflow in the Sierra Nevada would likely happen earlier and peak streamflow would rise. Miller et al. (2000) found that the American River in the Sierra Nevada, which is snowmelt driven, showed both an increase in magnitude and earlier peak flow (see also Hay et al., 2000).

ronmental in-stream flow requirements in the summer. It is likely to be problematic for the current reservoir system to store earlier spring runoff for use in the summer unless new operating rules and regimes are implemented (Lettenmaier and Sheer, 1991), and it is not clear that such a change would be sufficient to reduce spring flooding and increase summer supplies. This may be especially true in California, where both climate models used in this Assessment show a substantial increase in runoff, particularly in the winter. In addition, more intense precipitation events (such as the extreme event in Las Vegas on July 8, 1999 that caused extensive flooding in the city) could increase flooding. The risk of increased flooding is exacerbated by continued urban development, which increases surface runoff during storms. Development in floodplains and expansion of areas that could be flooded because of increased runoff could result in more people and property at risk to the effects of climate change. In addition, higher runoff can increase mudslides.

On the other hand, it is possible that increased runoff would create more water supplies for the West. Presumably, this could contribute to an easing of many current stresses on the water management system because there would be relatively more water available for users. A wetter climate would also likely reduce the demand for surface water and groundwater for such purposes as irrigation and watering lawns.

There is some chance that higher runoff could ease water quality problems although it could also result in more runoff of pollutants from farms and streets, which can degrade water quality. It is likely that hydropower production would increase with more runoff. However, earlier runoff is likely to result in more electricity production in winter time, when demand for heating is very likely to be falling, and less electricity production in summer when demand for cooling is very likely to be rising.

If there is reduced or even only small increases of precipitation, runoff is very likely to be reduced. In addition, both groundwater recharge and reservoir supplies are very likely to be reduced as higher temperatures increase evaporation (Wilkinson and Rounds, 1998a).

Reduced runoff, particularly if combined with higher demands due to hotter and drier conditions, would very likely make allocation of water supplies a more critical issue for the West. It is likely that instream uses such as hydropower and recreation would be among those most affected by a reduction in runoff. It is also likely that urban and industrial users would be less vulnerable to supply reductions. Hurd et al. (1999b) found that urban and industrial users of Colorado River water would have very small reductions in supplies if runoff is reduced. In general, it is very likely that those with more junior water rights claims (those who receive their allocations after the senior claims are met) would be at greatest risk should runoff decline (Miller et al., 1997). In addition it is possible that Native Americans will more fully exercise their rights to water (see box). Furthermore, during droughts there is likely to be increased dependence on groundwater, causing increased overdraft, subsidence, and reduced baseflow of rivers. On the other hand, it is possible that drier conditions would result in a decrease in flood potential and mudslides in California.

With less runoff, water quality is likely to decline if stronger pollution control measures are not undertaken. Higher temperatures alone would decrease dissolved oxygen levels in water while lower streamflow would concentrate pollutants. Lower flows in the Colorado River are likely to result in increased salinity levels, unless additional steps are taken to control the problem (Gleick and Nash, 1991). Lower lake levels could also increase water quality problems. For example, salinity concentrations in the Great Salt Lake are likely to increase with lower lake levels (Grimm et al., 1997).

Table 1: Estimated Changes in Runoff

Current and Estimates Changes in Runoff from the Canadian and Hadley Models (mm)

Region	Historical Runoff 1961-90 (mm/yr)	Change in Annual Runoff 2025-2034 (mm/yr)		Change in Annual Runoff 2090-2099 (mm/yr)	
		Canadian	Hadley	Canadian	Hadley
Upper Colorado	43	-15	3	2	28
Lower Colorado	2	-1	6	0	33
Great Basin	21	-1	4	16	29
California	232	60	63	320	273

Native American Water Claims

Indian water rights remain an unresolved and important issue for water allocation in the West in a number of cases. Under the legal doctrine established by the 1908 Winters case (*Winters v. United States* [207 US 564 (1908)]), Indian tribes have reserved water rights that could amount to 45-60 million acre-feet (Western Water Policy Review Advisory Commission, 1998). However, the vast majority of those claims have never been clearly quantified or developed for the benefit of the tribes. In many cases, non-Indian water users have already fully appropriated and used the sources of water potentially available to satisfy tribal rights. Tribal efforts to protect and develop their water rights have encountered resistance from other water users and state water authorities. There is substantial ongoing litigation (approximately 60 pending cases as of 1995) and about 20 ongoing negotiation efforts aimed at achieving settlements of Indian water rights claims. The low availability of financial resources in certain cases makes it difficult for tribes to develop their water rights or to contest competing uses that interfere with Indian water rights, including instream flow rights for fishery purposes.

Historically, tribes often made significant concessions of their reserved water rights to obtain water development on reservations. Yet, many Indian irrigation projects have fallen into disrepair for lack of project funding. Some projects such as the Navajo Irrigation Project remain uncompleted, and others such as the Animas-La Plata Project have yet to be built despite Congressionally approved water settlements. Recently, the Secretary of the Interior promoted a comprehensive dialogue on a government-to-government basis with tribes in an attempt to develop a water rights negotiation process that responds to the concerns of tribes.

Adaptation Options

Although building additional flood controls or storage infrastructure to address the need to store earlier runoff for the summer may be more attractive under climate change, environmental and cost constraints could serve as impediments. Where both local and imported supplies are available, there will be greater flexibility to deal with changes in water supply availability. If groundwater supplies are maintained as a buffer against drought, local areas are likely to have better coping ability.

Adaptation to potentially increased demand and reduced supply may focus on the demand side of water use. Here too, the development path for the West is critical. Should the increased population continue to use water at the same or an increasing rate, agriculture water allocations could be further reduced. As noted above, this can make it more difficult to reduce demand during droughts.

One source of adaptation lies in changing water pricing structures. Pricing water closer to its replacement cost would discourage wasteful uses. While market-based solutions would increase efficiency, it is possible there will be equity problems: users with limited resources, such as the poor and some farmers, may have to cut back on water use more than others.

Water transfers (between users and across river basins) will almost certainly play some role in addressing future water demand. These transfers include water savings derived from system enhancement measures such as canal lining and other waste reduction measures, and transfer of water currently used in agriculture for use in urban areas. In addition, institutions to manage groundwater quantity and quality may need to be strengthened (Knox, 1991).

The efficiency of municipal and industrial water uses can be significantly improved. Increased application of conservation technologies such as ultra low flush toilets and landscaping practices such as xeriscaping can reduce the growth of urban demand for water and lower the vulnerability of urban areas to drought. Use of treated effluent could be increased (Wong et al., 1999). Municipalities near the ocean can also reduce water demand by desalting seawater, which is an expensive option. For example, Santa Barbara recently built a desalinization plant.

Increasing flood storage or flood control measures is likely to be an adaptation to increased risk of flooding. However, flood control management is shifting away from reliance on physical structures to effective management of floodplains, including restricting development, using wetlands, and trying to re-create the ability of rivers to spread floods to avoid concentrated downstream impacts (Wong et al., 1999). These adaptations may be effective if implemented in response to climate change, but would be more effective if implemented in anticipation of climate change. If annual precipitation increases, but summers become hotter and drier, there is likely to still be a need for additional storage to provide more water in the summer or for demand reduction measures to lessen the need for water in the summer.

2. Natural Ecosystems

The wide diversity of natural ecosystems in the West ranges from low-elevation deserts to alpine tundra (see Figure 8). In addition, productivity varies considerably. Most of the West is grassland, shrubland/grassland, and desert shrubland. The mountains contain coniferous forests, woodlands, deciduous forests (mostly aspen), and mixed forests. California has a wide diversity of ecosystems, including mostly coniferous forests in the north and in the Sierras, oak savanna and chaparral along the central coast, and shrubland and grassland along the southern coast and interior. The central and southern Rocky Mountains are dominated by ecosystems associated with mountains: alpine, coniferous forests interspersed with grasslands, and, at lower elevations, woodlands. The very dry environments in the Great Basin support shrublands, some grasslands, and deserts. The wetter parts of the Great Basin support woodland vegetation (USGS, 1993). Aquatic habitats range from cool mountain to desert streams and rivers, including reservoirs which have substantially altered the aquatic ecology of the West. In addition, wetlands in the west, particularly in arid areas, are important habitat for endangered species, fish rearing, and migratory waterfowl.

With this wide diversity of ecosystems and topography comes a wide diversity of species, many of which are in isolated habitats. California's climate zones, from coastal to desert to alpine regions, support a wide variety of plants and animals, as does the area near the New Mexico-Arizona-Mexico borders and Utah, with its deserts, canyons, and alpine peaks (Wilkinson and Rounds, 1998b; US EPA, 1998a and b).

Development has taken its toll on the natural ecosystems of the region. Dams and reservoirs have altered free-flowing streams, numerous plant and animal species have been eliminated or reduced to low numbers, and agriculture and ranching have transformed lowland ecosystems. By some estimates, 90% of California's wetlands have disappeared (Wilkinson and Rounds, 1998b). All of this alteration has made natural ecosystems vulnerable to invasion by hundreds of non-native species.

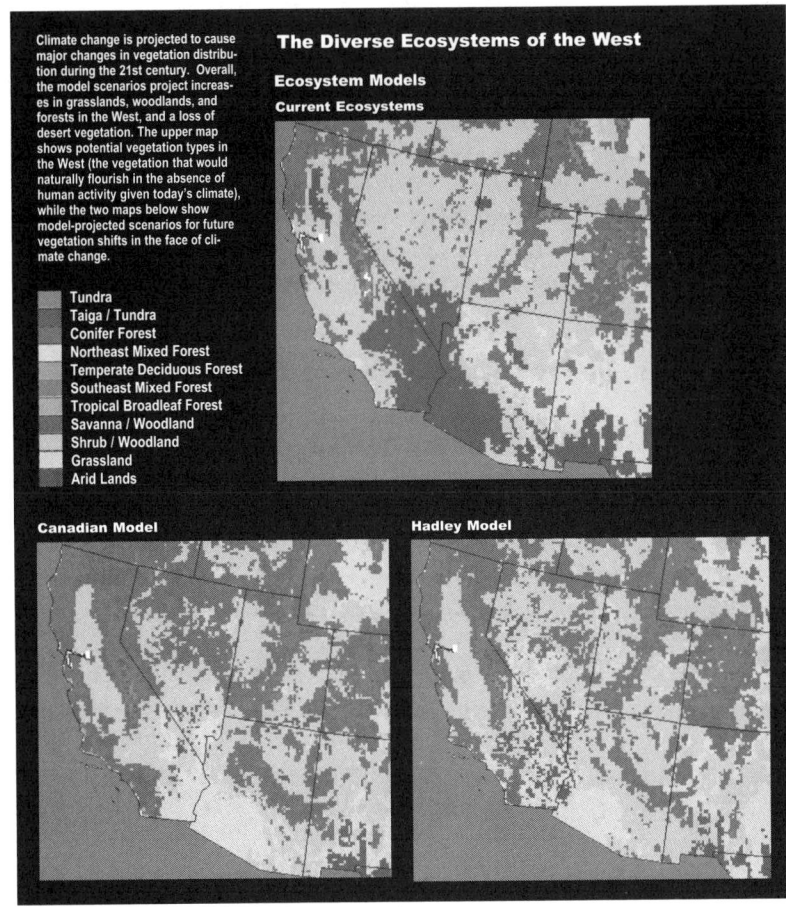

The Diverse Ecosystems of the West

Climate change is projected to cause major changes in vegetation distribution during the 21st century. Overall, the model scenarios project increases in grasslands, woodlands, and forests in the West, and a loss of desert vegetation. The upper map shows potential vegetation types in the West (the vegetation that would naturally flourish in the absence of human activity given today's climate), while the two maps below show model-projected scenarios for future vegetation shifts in the face of climate change.

Ecosystem Models
Current Ecosystems

- Tundra
- Taiga / Tundra
- Conifer Forest
- Northeast Mixed Forest
- Temperate Deciduous Forest
- Southeast Mixed Forest
- Tropical Broadleaf Forest
- Savanna / Woodland
- Shrub / Woodland
- Grassland
- Arid Lands

Canadian Model

Hadley Model

Figure 8: Currently the West has a large diversity of ecosystems. Under the two climate change scenarios, the area in arid and grassland ecosystems would decrease and the area in forest ecosystems would increase. See Color Plate Appendix.

California contains more threatened and endangered species (257) than any of the other lower-48 states (US Fish and Wildlife Service, 1999) and is second highest in rate of species extinction (The Nature Conservancy, 1999). Myers et al. (2000) consider the California Floristic Province as one of the 25 "hotspots" in the world that have exceptional diversity of species and are experiencing exceptional loss of habitat.

The rise in population has resulted in more urban development and development into wooded areas which among other things has exposed human settlements to wildfires (see box on fires). Fire is a natural part of the ecology of the West. However, fire suppression has resulted in an unnatural increase in the density of vegetation, thereby making the landscape more susceptible to severe fires. In addition, some invasive species, such as cheatgrass, have increased fire frequency, while species such as star thistle and *Tamarix* have reduced water supplies and increased flooding (Chapin et al., 2000).

Vegetation

Under both the Canadian and Hadley climate scenarios, using the VEMAP biogeography and biochemistry models, biomass is projected to increase and vegetation to shift from deserts and grasslands to woodlands and forests in many parts of the region. Forests are projected to expand in California, Utah, and Colorado, mostly in the mountains. Nevada, northern Arizona, and western New Mexico are projected to see a shift toward shrub woodland and savanna woodland, while southwestern Arizona and southeastern California are projected to shift from arid lands to grasslands (see Chapter 2: Vegetation and Biogeochemical Scenarios: Future Vegetation). Across the West, a wetter climate is likely to increase forest productivity, including shifting some conifer forests to broadleaf forests, although there could still

Fire in the West

The rise in population in the West has resulted in more development into wooded areas and increased exposure to fire risk, which was already high (see Figure 9). For example, there were major fires in recent years in urban areas, including Oakland, Santa Barbara, Malibu, and Los Alamos. The Oakland fire destroyed or damaged about six thousand structures. In addition, fire suppression, which has resulted in dense growth and invasion of non-native species such as cheat grass, have made many Western forests more vulnerable to major fires. Continued development into forested areas, along with continued suppression of fires and spread of non-native species, is likely to increase risks of severe fires.

Studies suggest there is a good chance that climate change will increase the risk of fire frequency, whether precipitation increases or decreases in the region. Lower precipitation renders montane forests more fire-prone. These forests are already at risk because of the massive fuel buildup and predisposition to uncontrollable crown fires. Torn et al. (1998) found that warmer and drier conditions could lead to a "dramatic" increase in land area burned and potentially catastrophic fires in California. Higher precipitation increases the fuel loads of sparse vegetation in arid areas. If interannual variability of precipitation does not decrease, wet periods will be followed by dry periods and there is a good chance fires would increase. Modeling with a dynamic global vegetation model (MC1) found that fires across the West could increase under such conditions. As temperatures continue to rise, so would evapotranspiration, which can lead to more drying and more fires (Neilson and Drapek, 1998). Under the Hadley and Canadian scenarios, the fire severity rating in the West increases 10%.

Increased fire could reduce the indigenous vegetation in some cases and promote conversion to nonnative weeds. More fire could degrade water quality because of increased runoff of sediments. Fires also add to air pollution. Should fire increase, there could be increased risks for human settlements within or close to forests and grasslands.

Fire Severity - July 1994

☐ Low
▨ Medium
■ High

Figure 9: Relative fire severity across the United States in July, 1994. All of the states with high fire severity were in the West. Source: Liverman, 1998. (see http://udallcenter.arizona.edu/publications/pdfs/swclimatereport-final.pdf, page 22.)

The risk of fire in urban areas and in heavily forested areas could be reduced through a number of measures. Restrictions can be placed on development in fire-prone areas. Building and landscape design criteria have been developed for fire-prone areas. Construction with nonflammable materials and installation of "firescape" landscape designs are also being used in high-risk areas. Controlled burns may also need to be used as part of a vegetation management strategy in urban areas. Many of these adaptations have been implemented in response to urban fires such as those in Oakland. Fires in natural areas should not be suppressed to the degree that a large amount of fuel buildup is allowed. These adaptations should be implemented in anticipation of climate change.

be a net increase in conifer forest cover (Neilson and Drapek, 1998). The higher temperatures, however, are likely to result in many alpine areas virtually disappearing from the West and being replaced by temperate forests (see Chapter 2: Vegetation and Biogeochemical Scenarios). Note that the projected changes do not show steady increases in biomass in all places at all times. Under the Hadley model, vegetation productivity declines in New Mexico and Arizona by the 2030s. One model result shows that in Colorado, forests first decrease in area by 2030, but expand by 2095 to cover an area larger than today.

There are a number of reasons for caution about these projections. First the CO_2 fertilization effect on plant growth and water use efficiency may not be as positive as assumed in the models (Walker and Steffen, 1997). Under the Canadian model, assuming no CO_2 fertilization effect, biomass is projected to decline in some parts of the West (Aber et. al., 2001). Modeling conducted for this Assessment and other studies discussed in the box on fire show an increased risk of fire in the West. Climate change could also make conditions more favorable for pest outbreaks and introduction and spread of invasive alien species (Dale et al., 2001). Should high levels of air pollution continue and wind storms increase, these would be additional stresses on forests. It is also uncertain whether transitions from one type of ecosystem to another would be smooth or involve disruptions.

Furthermore, as climate continues to change, the CO_2 fertilization effect (which increases growth and water use efficiency) becomes saturated and declines, and higher temperatures would impose more moisture stress on vegetation.

If conditions become drier, productivity of vegetation is likely to decrease (Neilson and Drapek, 1998). There could be a shift from forests, woodlands, and shrublands, to grasslands and deserts.

Biodiversity

As noted above, development has resulted in fragmentation of habitats, creation of barriers to migration, such as urban areas and dams, and introduction of invasive species. This, in combination with the complex topography and varied climate of the region, is likely to make it difficult for many species to adapt to climate change through migration. It is also likely that development would favor the spread of invasive and non-indigenous species because invasive species are generally better suited to chang-

ing conditions. Without development, the adverse impacts of climate change on biodiversity would likely be substantially reduced.

While the mountains of the West can serve as barrier to species migration, they also provide higher altitude and northern routes for migration as well as many microclimates that can create refugia for some species. But, migration upslope also means migrating to smaller and smaller areas of habitat, which would only support smaller and smaller populations. As climate change continues, species migrating upslope are very likely to be threatened as their habitats figuratively disappear off the tops of mountains.

The faster the rate of climate change, the greater the stress will be on many species and populations.

Terrestrial Species

Hansen et al. (2001) found there is a slight chance that Quaking Aspen and Engleman Spruce will not survive under projected climate change (however, this study did not account for the positive effects of CO_2 fertilization). Interestingly, paper birch is projected to expand southward in the Rocky Mountains. Hansen et al. (2001) also found that animal populations could change. It is possible that higher temperatures lead to a decrease in bird and mammal populations that are currently found in the region because they cannot tolerate higher temperatures. It is possible that higher temperatures could increase reptiles and amphibians in the southern Rocky Mountains because of their greater tolerance for heat.

Murphy and Weiss (1992) projected that a 5°F (3°C) warming would result in a substantial reduction in the area of the Great Basin suitable for boreal species. They estimated that plant species would be reduced from 305 to 254, four of nine mammals would be lost, and 23 to 30% of butterflies living in boreal areas in the Great Basin would become extinct. On the other hand, there is some chance that higher temperatures would enable some southwestern desert plants to invade the Great Basin (Neilson and Drapek, 1998), although such a large-scale change could take thousands of years to be realized.[7]

---

[7] In warm periods in the past, some species migrated to new locations, while others remained in the same general location (Tausch et al., 1995).

Observed Shift in Range of Edith's Checkerspot Butterfly: 1900 to 1990s

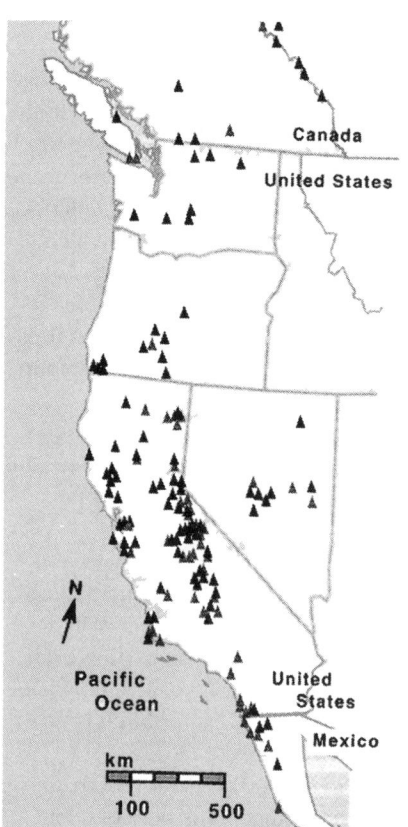

Figure 10: On this map of studied sites, the lighter triangles represent extinct populations of Edith s Checkerspot butterfly, while the darker triangles represent present populations. The mean location of populations of this butterfly has shifted northward by 57 miles (92 kilometers) and upward in altitude by 407 feet (124 meters) since 1900. This is an indication that climate change is already having an affect on the some species ranges. Source: Parmesan, 1996. See Color Plate Appendix.

Aquatic Species

Aquatic and riparian ecosystems in the West are also vulnerable to changing precipitation and runoff regimes. Wetlands may have some resiliency to climate change because they currently cope with highly variable climate conditions (Grimm et al., 1997). While wetter conditions are likely to alleviate some existing stresses, higher temperatures are likely to exceed the thermal tolerances of many fish species and lead to increased fragmentation of many cold water fish habitats particularly in mountains (Meyer et al., 1999). It is probable that some alpine and cold water fish species will not survive in the region (Grimm et al., 1997). In addition, higher temperatures are likely to allow for invasions by non-native fish species (Wagner and Barron, 1998). It is also possible that higher water temperatures would be a problem for salmonid populations, since these fish are near the southern end of their range now in

California and show signs of stress in the warmer years (Wilkinson and Rounds, 1998a). Drier conditions are likely to result in the loss of many small water bodies and aquatic ecosystems (Grimm et al., 1997).

In addition, the change in seasonality of runoff is likely to have adverse effects on many species. It is difficult to anticipate exactly how these changes in flow magnitude and timing would affect particular species or flow-dependent habitats. However, some general predictions can be made based on knowledge of species life history strategies in relation to hydrology. In general, climate-related hydrologic changes are very likely to favor some species more than others, resulting in decreased species diversity and altered composition of native biological communities. For example, it is possible that alterations to the timing and magnitude of spring flows will favor non-native riparian plants that would otherwise be suppressed by high runoff in spring (Kattelmann and Embury, 1996). Modified flow regimes are also very likely to affect populations of native fish species. For example, the distribution and abundance of the four seasonal runs of chinook salmon native to the Sacramento River drainage that are already in jeopardy are likely to be further altered by seasonal changes in the availability of spawning flows (Yoshiyama et al., 1996).

Observed Effects on Species

The effects of climate change on species are already being observed. Parmesan (1996) found that the mean location of populations of Edith's Checkerspot butterfly shifted northward and upward in elevation since the beginning of the century (Figure 10). She found that the southern boundary moved northward but was unable to determine if the northern boundary moved further northward (Camille Parmesan, University of Texas, personal communication.) These butterflies do not migrate; in fact, it is their relatively sedentary nature that makes them a good choice for tracking long term trends in wildlife range shifts in response to climatic warming. A range shift northward is a process which takes decades. In theory, as climate change makes the most southern regions less suitable and the far northern regions more suitable, populations at the southern end of the range go extinct while new populations are established northward of the previous boundary. However fragmentation of habitat and barriers to migration are likely to impede northward migration of many species, resulting in decreases in their total range.

Sagarin et al. (1999) found that in the past 50 years, the southern invertebrates have become more common and northern invertebrates have become less common in the rocky intertidal community in Pacific Grove, California. Both of these changes appear to be the result of higher temperatures.

Adaptation Options

A number of steps could be taken to at least help reduce some of the pressures of development on ecosystems and biodiversity and even anticipate the need for species to migrate in response to climate change. Urban development could be managed to better protect riparian areas and reduce habitat fragmentation. There could be concerted efforts to link habitats and even create migration corridors for species to migrate northward or upslope in response to climate change. The current trend toward reduced land for agriculture could present some opportunities if abandoned lands are used for habitat. Reducing offstream water use will also help improve aquatic habitats. These measures would need to be implemented in anticipation of climate change. It is not clear how effective many of these measures, particularly migration corridors, would be in averting negative effects of a warmer and wetter climate on natural systems. In addition, implementing these measures may be challenging, while continued urban and suburban development could result in increased stress on ecosystems and species diversity.

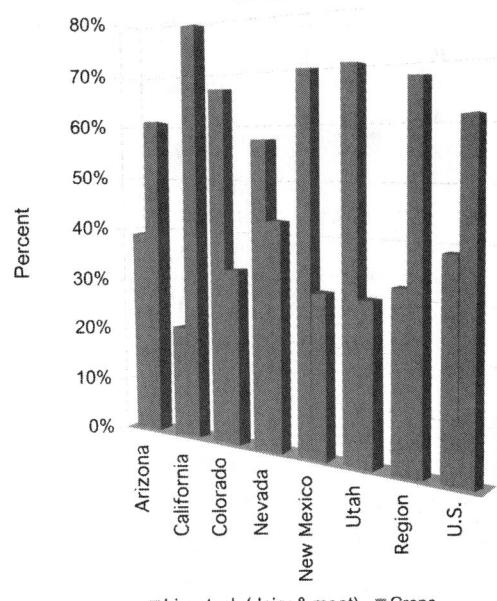

Relative Share of Crop and Livestock Output in the West.

Figure 11: For most of the states in the West, the majority of value-added agriculture production comes from livestock and dairy production. However, because California s agricultural production is dominated by crops (75% of total agricultural output for the state), and because California dominates regional agricultural output (84% of regional crop production, 51% of regional livestock and dairy production), the majority of the region s total agricultural production comes from crops. This difference between the dominant types of agricultural production on a state level and on a regional level highlights the heterogeneity of agriculture in the West. Source: USDA Economic Research Service State Farm Sector Value-Added Data; (http://www.econ.ag.gov/briefing/fbe/fi/fivadmu.htm). August 30, 1999. See Color Plate Appendix.

Table 2. Relative Share of Crop and Livestock Output in the West

State	Output	Percentage of Combined Crop and Animal Output
Arizona	Crop	60.94%
	Livestock and Dairy	39.06%
California	Crop	79.02%
	Livestock and Dairy	20.98%
Colorado	Crop	33.14%
	Livestock and Dairy	66.86%
Nevada	Crop	42.66%
	Livestock and Dairy	57.34%
New Mexico	Crop	30.22%
	Livestock and Dairy	39.78%
Utah	Crop	29.75%
	Livestock and Dairy	70.25%
Region	Crop	67.74%
	Livestock and Dairy	32.26%

3. Agriculture

The total value of crop and livestock production in the West in 1997 was $32 billion (US BEA, 1999a; see Figure 11. About two-thirds of the value of western agriculture is from crops, with the rest from livestock (Figure 11 and Table 2). Fruits, tree nuts, and vegetables comprise about two-thirds of the value of crop production, while seven-eighths of livestock production is from meat animals and dairy products. The West produces 17% of the nation's agricultural output, but three-fifths of the country's fruits and tree nuts, almost half of the vegetables, and almost one-quarter of the dairy products.

Higher CO_2 concentrations are likely to help increase crop yields and decrease water demand, although higher temperatures are also likely to hasten phenological development of crops (resulting in reduced yields) and increase demand for water. Higher precipitation can increase yields but can also cause flooding and waterlogging of crops. The net effect on yield will depend on relative changes in CO_2 concentrations, temperature, and precipitation.

Milder winter temperatures are likely to lengthen the growing season and result in a northward shift of where some crops are planted, assuming the land and infrastructure are available for such geographic shifts. In addition, there is some chance that frost-sensitive plants once grown primarily in areas such as the Imperial Valley of California will be grown in the state's Central Valley.

Conversely, it is possible that crops that prefer cold winters such as winter wheat and potatoes could be limited to more northern areas (although other wheat varieties could be grown). It is very likely to be more difficult to relocate perennial crops such as vineyards, fruits and nuts, than to relocate annual crops, because perennials can take many years to decades to get established. In addition, warmer temperatures can inhibit growth of certain fruit and nut crops that require chilling during the winter. It is also possible that warmer temperatures will increase heat stress, weeds, pests, and pathogens that affect plants, animals, and farm workers.

Changes in the seasonality of precipitation could cause some problems. There is some chance that vineyards, for example, could experience losses if rains increase near harvest time — unseasonable rain can cause molds, ruining the grapes. Higher air temperatures and humidity can increase risk of diseases that can harm vineyards. However, higher temperatures in the Sonoma and Napa Valleys since 1951, which is mainly the result of nighttime warming, improved the quality and yield of wines (Nemani et al., 2001). Cotton yields can also be reduced by rain at critical stages of growth.

Should the climate become hotter and drier, agriculture would be at particular risk. It is probable that the amount of water available for irrigation will be reduced substantially (Hurd et al., 1999b). Thus, agriculture could be squeezed between an increased need for water and less available water. If additional irrigation water is applied, there would be increased salinity in soils and rivers. Rural communities would be sensitive to declines in agriculture or ranching.

Estimated changes in irrigated crop yields using scenarios derived from the Hadley and Canadian climate models in the 2030s and 2090s for the "Pacific" and "Mountain" regions are displayed in Table 3. The Pacific region includes California, Oregon, and Washington, and the Mountain region contains all of the Rocky Mountain states. The 2030 results assume a CO_2 concentration of 445 parts per million (ppm), and the 2095 results assume CO_2 levels of 660 ppm (Francesco Tubiello, Goddard Institute for Space Studies, personal communication, 1999). The specific numerical results should be treated with caution since they include states outside the West as it is defined here and include optimistic assumptions about the CO_2 fertilization effect while not considering other effects such as pests and disease. The results show increases in yields for many crops, but decreases for some crops such as tomatoes in the Pacific and hard red spring wheat in the mountain states. Although not shown, the results tend to show small changes in demand for irrigation water for the major western crops, but in a few cases, significant decreases in demand. Crop production in the Pacific and Mountain states is projected to increase (see Chapter 13: Agriculture).

Adaptation Options
One strategy to adapt to the effects of climate change is to maintain and increase the diversity of crop types and varieties, because diversity increases the likelihood of having some crops that fare well under variable climate conditions. For example, in California, the artichoke crop was good in 1998, but the orange crop was devastated by freezing conditions. Farmers may also plant low-chill varieties of certain tree crops in anticipation of higher average temperatures. This adaptation is already under way

Table 3: Estimated Changes in Crop Production in the West

Estimated Percent Changes in Dryland Crop Production for the Mountain Region from the Canadian and Hadley Models (%)

Crop	2030s		2090s	
	Canadian	Hadley	Canadian	Hadley
Cotton	4.86	16.73	50.41	38.55
Hard Red Spring Wheat	12.92	16.90	-10.54	27.47
Hay	9.57	11.14	16.77	30.50
Tomatoes (processed)	21.92	23.59	-22.99	35.19
Oranges (processed)	66.90	69.90	114.60	111.60
Pasture	20.90	19.50	51.49	49.27

Estimated Percent Changes in Irrigated Crop Production for the Mountain Region from the Canadian and Hadley Models (%)

Crop	2030s		2090s	
	Canadian	Hadley	Canadian	Hadley
Cotton	74.22	92.11	188.24	170.36
Hard Red Spring Wheat	-16.98	-1.22	-29.62	-1.41
Hay	17.29	30.58	16.32	33.00
Tomatoes (processed)	21.92	23.59	-22.99	35.19
Oranges (processed)	66.90	69.90	114.60	111.60

Estimated Percent Changes in Dryland Crop Production for the Pacific Region from the Canadian and Hadley Models (%)

Crop	2030s		2090s	
	Canadian	Hadley	Canadian	Hadley
Cotton	6.58	22.63	68.22	52.17
Hard Red Spring Wheat	16.25	65.75	137.90	131.10
Rice	6.49	6.27	1.76	5.77
Hay	26.76	28.38	62.24	50.29
Tomatoes (processed)	-19.95	-9.14	-7.62	-19.54
Oranges (processed)	36.87	42.77	77.90	73.03
Pasture	47.53	58.83	102.12	92.55

Estimated Percent Changes in Irrigated Crop Production for the Pacific Region from the Canadian and Hadley Models (%)

Crop	2030s		2090s	
	Canadian	Hadley	Canadian	Hadley
Cotton	41.66	51.70	105.66	95.62
Hard Red Spring Wheat	0.25	4.60	4.80	11.75
Rice	6.49	6.27	1.76	5.77
Hay	38.26	61.06	52.94	70.33
Tomatoes (processed)	-19.95	-9.14	-7.62	-19.54
Oranges (processed)	36.87	42.77	77.90	73.03

and can be enhanced in response to climate change. Breeding crops better suited to take advantage of higher CO_2 levels and more heat may also make sense.

Development of drought- and heat-resistant crops will help reduce the vulnerability of the agriculture sector. Bioengineering could be helpful in this regard, but this is a complicated issue with advantages and disadvantages.

There is substantial potential to reduce current and future water use through less water demanding technologies and better water management practices. Agriculture could switch from high water use crops such as irrigated pasture, alfalfa, cotton, and rice, to less water demanding crops such as soybeans, wheat, barley, corn for grain, and sorghum (USDA, 1997; Gleick et al., 1995). Water-intensive crops grown in desert areas could possibly become uneconomic if water prices increase. More efficient irrigation technologies such as sprinklers or drip irrigation can reduce water demand. Crops may need to be planted earlier to take better advantage of earlier runoff (higher temperatures may also favor earlier planting of crops).

4. Ranching

Ranching is quite sensitive to climate variability. The cattle industry in Arizona reduced herd size by about 80,000 head during the 1994 to 1996 drought, but an increase in precipitation in New Mexico in the same period resulted in an increase of 100,000 head of cattle (McClaren and Patterson, 1998)

It is possible that an increase in temperature and precipitation could have the benefit of increasing forage production in many locations, and lengthening the growing and grazing season on native rangelands. Moreover, increased water supplies and longer growing seasons would make it possible to harvest more alfalfa crops per year (now typically two to three), increase hay supplies, and reduce prices.

A warmer and wetter climate can pose problems for dairy cattle. There is some chance that flooding could wash out holding ponds. If winters become wetter, it is possible dairy cattle will suffer. In the Chino, California area, which produces 25% of the state's milk, some 6,500 head of cattle died during El Niño conditions in February 1998. Cows and calves became mired in mud and weakened by the cold, succumbing to bacterial infec-

tions that breed in the muck. However, should conditions become generally wetter, it is likely vegetation will get more dense, which may reduce winter mud.

Ranching is extremely vulnerable to drought (Liverman, 1998) and should the climate become drier, vegetation productivity, water supplies, and the carrying capacity of land and, hence, livestock production, would be reduced. In addition, higher temperatures can increase livestock diseases and calving problems (Wagner and Baron 1998). The economic impact would be felt most strongly in the rural and intermountain areas.

Adaptation Options
Stakeholders identified improvement in weather forecasting to be the most important adaptation for ranching. The timing of cattle sales and breeding, and the range of management strategies that ranchers employ, depend on knowledge of anticipated and observed range conditions and long-term water availability. Consideration may be given to raising different species or breeds more suitable for hotter conditions (Wagner and Barron, 1998). Management practices should be adjusted to changes in conditions to reduce stress on ecosystems when appropriate.

5. Coastal Resources

Although a large portion of California's coast is made up of cliffs, many of the state's most populous coastal areas are vulnerable to sea-level rise, including the San Francisco Bay area and the coast south of Santa Barbara. If no protective measures are taken, sea-level rise will inundate hundreds of square miles of low-lying land in California (Gleick, 1988). Unless protected, coastal structures from harbors to houses could succumb to the ocean, as numerous California beachfront homes did in February 1998. Also, beaches will be flooded unless defensive actions are taken. Agricultural lands in the Sacramento-San Joaquin delta, some already as much as 25 feet below sea level, are threatened with inundation. As the ocean encroaches, some aquifers near the coast will become contaminated by saltwater intrusion. Rising sea level could inundate many coastal wetlands and unprotected development (see Figure 12). Should sea walls be used to protect coastal areas downslope, wetlands are likely to be blocked from migrating inland with the sea and could thus be lost.

A study of the costs of protecting the margins of San Francisco Bay from a 3.3-foot (1-meter) sea-level

[8]Gleick and Maurer (1990) also noted that many costs were not, or could not be, quantified.

rise concluded that more than $1 billion (1990$) would be needed for new or upgraded levees to protect existing industrial and commercial developments, with an additional annual maintenance cost exceeding $100 million (Gleick and Maurer, 1990).[8]

Adaptations Options

Strategies for protecting developed coastal areas include defending with engineered fortifications any assets of high economic value such as cities, airports, ports, and delta levees (for water supply security); relocating vital assets to higher ground (or engineering alternative solutions); and, for less economically valued areas of the developed coast (housing on coastal bluffs), retreating. Building coastal defenses can block inland migration of wetlands and result in loss of beaches. Advance planning can prevent new developments from being built in areas likely to be at risk in the future. Avoiding new construction is likely to prove far less costly than trying to protect such development in the future. For new development of any kind, local government agencies such as the Coastal Commission could be authorized to consider "risk of harm" from impacts of climate change. After consideration of risk of harm, developments would be approved only with no assured warranty of safety or loss, and private insurance would underwrite the risk or self-insurance would bear any costs or losses.

6. Tourism and Recreation

The spectacular scenery, favorable climate, and large amounts of public land, especially in national parks, have made the West a major destination for tourists from around the world. Billions of dollars have been invested in ski resorts in all of the region's states, with Colorado, Utah, and California having particularly extensive facilities which attract many visitors. Tourist expenditures in the West are growing. Hotels, lodging, amusement, and recreation provided $32 billion in revenues in 1996 and are projected to provide $52 billion in 2045 (US BEA, 1999a).

Since the tourism industry in the West is so strongly outdoors oriented, it is particularly sensitive to climate. The period for winter activities is likely to shrink, while the period for summer activities is likely to increase. Natural vegetation provides part of the aesthetic attraction, and significant climate-change effects on western ecosystems are very likely to change the distribution and abundance of vegetation and animals. Much of the attraction for tourists is associated with water: its inherent aesthetic appeal, and the growing water-related sports

of fishing, whitewater rafting, kayaking, and canoeing. Some of this recreation is on the many artificial lakes such as Lake Powell and Lake Mead. Increases in runoff could possibly enhance these sports while decreases could possibly reduce their attractiveness.

The skiing industry is at particular risk from higher temperatures. With rising temperatures, snowpack seasons are very likely to shorten. Moreover, snow-

Current and Projected Wetlands in South San Francisco Bay

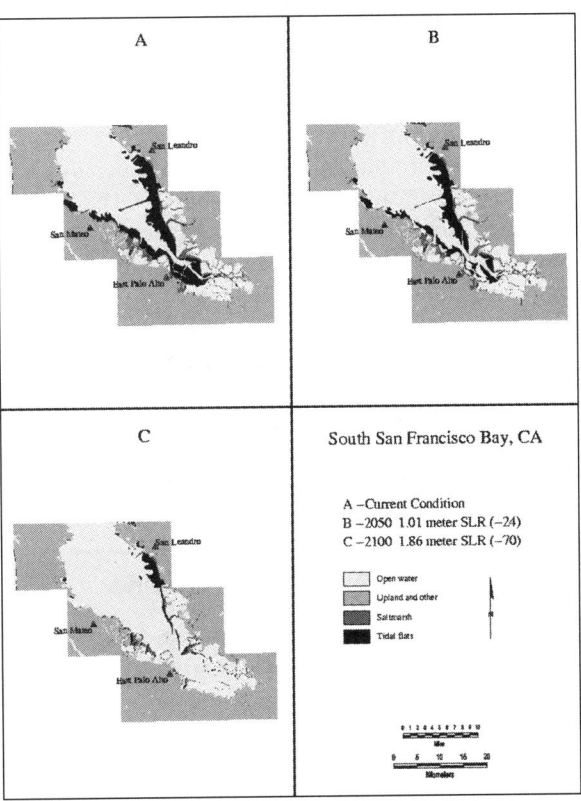

Figure 12: This figure shows the spatial extent and distribution of current and projected wetland habitat types in southern San Francisco Bay (derived from US Fish and Wildlife, National Wetlands Inventory data) following sea-level rise as calculated using the Sea Level Affecting Marshes Model (SLAMM4) (Galbraith et al., In prep.). The sea-level rise scenarios use historic rates that include local subsidence (obtained from tide gages at or close to each of the sites), superimposed on the median estimate of the likely rate of sea-level change due to climate change (Titus and Narayanan, 1996). The historic rate of sea-level rise in the southern part of San Francisco Bay is estimated to be 3.0 feet (0.9 meter) by 2050 and 5.3 feet (1.6 meter) by 2100. This could be due to tectonic movements resulting in land subsidence and/or crustal subsidence due to the depletion of subterranean aquifers. When combined with the projected median estimate of 13.4 inches (34 cm) eustatic (global) sea-level rise by 2100 from climate change, sea-level rise is estimated to be 3.3 feet (1.0 meter) by 2050 and 6.1 feet (1.9 meters) by 2100. The numbers shown in parenthesis on the figure indicate that approximately 57.7% of tidal flat habitat will be lost by 2050 and 62.1% by 2100, compared to the current condition. Using only the historic rate of local sea level rise, approximately 58.9% (2050) and 61.1% (2100) of tidal flat habit. See Color Plate Appendix.

line elevations will rise. Lower-elevation and more southern ski areas are likely to be at greatest risk.

On the other hand, rising temperatures are likely to result in a longer summer season for warm weather recreation activities. Backpacking, biking, mountain climbing, and rock climbing have been growing in popularity. For example, the number of backpackers in the Canyonlands of Utah rose sevenfold from the early 1970s to the mid-1990s (Riebsame, 1997). But there is some chance that increased precipitation could decrease the number of days desirable for summer recreation activities. Whether warmer and wetter conditions would result in a net increase or decrease in summer recreation is unclear.

Adaptations Options

Adaptations for tourism and recreation generally involve diversification of income sources. The larger, better-capitalized resorts such as Aspen and Vail have already adapted their facilities to serve as summer destination resorts with a range of warm-season recreational activities, conference facilities, and music and dance programs; those with private land have extensive, high-priced real estate development. The smaller areas may not be sufficiently capitalized or have the private land to achieve these forms of diversification. This strategy can be taken in response to climate change and can be done in anticipation of climate change only to the extent that current recreation patterns support it.

ADDITIONAL ISSUES

Mining

The mining industry is quite sensitive to climate variability and change because of the importance of water to its production processes, and the fact that environmental laws hold mines liable for the quality of effluent water. Water is needed for the concentration step of processing. In addition, a typical mining operation is required to collect and use or process all precipitation that falls within the limits of the facility or otherwise comes in contact with unnaturally exposed material. There is some chance that increased precipitation can result in more runoff of pollutants, while decreased precipitation could result in reductions in water supplies for processing. The mining industry is likely to adapt to climate variability by relying on short-term forecasts of precipitation in day-to-day operations, interannual forecasts of precipitation for temporary enhancement of water treatment facilities, and long-term climate outlooks to decide on capital improvements in water holding areas, mechanical pumps, and water treatment facilities.

Air Quality

Air quality is a significant problem in many parts of the West. For example, with 17 million inhabitants occupying a basin subject to many temperature inversions, the greater Los Angeles area has a particularly serious problem with ground-level ozone levels and particulate matter. In addition, San Francisco, Las Vegas, Phoenix, Reno, and Salt Lake City have problems meeting federal government standards for ozone levels, and many western cities have particulate matter concentrations close to or exceeding federal standards.

If precursors are not reduced and temperatures increase, it is possible that ozone levels, which are at their peak in the summer, will increase. Higher temperatures increase ozone formation when precursors are available. Should wetter conditions increase biomass, which emits ozone precursors, air quality could further decline. Fine particulate matter concentrations could also increase. This could lead to more health problems. On the other hand, increased El Niño conditions, which would result in more storms and precipitation in the winter, would be likely to reduce levels of winter air pollutants, such as carbon monoxide and particulates. Reducing emissions of air pollutants, which is needed anyway in many Western cities, may be even more necessary because of climate change.

Health Effects

Since the West is generally dry, it is likely to be at lower risk of increase in vector-borne infectious diseases than more humid regions. Should the West become warmer and significantly wetter, there is some chance that there could be an increase in the potential presence of disease vectors. In recent years, wetter conditions contributed to the outbreak of cases of Hantavirus in the region, particularly in the Four Corners area (Engelthaler et al., 1999). It is possible that wetter conditions would increase the potential for a Hantavirus outbreak and other climate sensitive diseases such as plague (Parmenter et al., 1999), assuming other control measures are not taken. But, because of the capability of the public health system, it is unlikely that there will be large outbreaks of infectious diseases in the West. It is more likely that if climate gets warmer and wetter, the potential for small outbreaks from people carrying the diseases from other countries into the region would increase. To keep health risks low, it is critical that the public health system be maintained.

The region currently has lower risk of heat stress mortality than Midwest and Northeastern cities. Kalkstein and Greene, 1997 found in San Francisco and Los Angeles, winter mortality would decrease, while in Los Angeles summer mortality would increase. The estimated net change in mortality across the nine large western cities studied is close to zero.

ADAPTATION STRATEGIES

For managed systems in the West, there appears to be significant potential to reduce negative consequences of climate change and take advantage of positive impacts. For example, wise water management can reduce the risks from droughts and floods. The potential for adaptation appears to be high in many of the other potentially affected sectors of the economy. And many of the measures mentioned above would have significant benefits regardless of climate change. Clearly though, these adaptations will involve costs, are not necessarily easy to implement, and can result in both winners and losers. The costs and feasibility of these adaptations were not assessed. Should there be sudden or extreme climate changes, it is not clear how effective adaptations would be in ameliorating adverse impacts.

Risks from climate change are likely to be greatest for those affected sectors or subsectors that lack the resources or capacity to adapt. For example, it is uncertain how effective the adaptations discussed above would be in reducing the vulnerability of natural ecosystems and biodiversity to climate change. Reducing current stresses on natural systems may help, but adverse impacts are still likely to occur. Poor or immobile people are likely to bear particular risks from climate change. In addition, activities that are fixed in place, such as national parks and Indian reservations, are at particular risk because they are unable to relocate in response to climate change. The development of adaptation strategies may need to pay particular attention to these types of situations.

Many development trends can increase vulnerability to climate change. But the development of the West also presents many opportunities to prepare for and thereby reduce the risks of climate variability and change in development plans and projects. For example, development can attempt to minimize water use and degradation of water and air quality. Coastal structures can be designed to minimize the risks of sea-level rise and harm to natural ecosystems. Development in flood plains can be reduced. The tourist industry can further diversify into both winter and summer recreation. The public health

system can be maintained and improved. Riparian areas can be protected, fragmentation of ecosystems reduced, and migration corridors developed or maintained. The capability of the poor and immobile to adapt can be enhanced. The effectiveness of these strategies in reducing the risks of climate change has not been assessed.

One strategy that should help virtually all affected sectors is improved forecasting of climate. In particular, improved seasonal and annual forecasting of climate would help water supply managers, farmers, ranchers, miners, health care professionals, and others plan for wet or dry seasons and extreme heat and cold episodes. Improved multidecadal forecasts of climate change would help infrastructure designers, land use planners, and others in identifying future directions of climate change.

CRUCIAL UNKNOWNS AND RESEARCH NEEDS

Clearly there are many uncertainties about how climate in the West will change and what the impacts of such changes will be, and there are many research needs that should be addressed to help resolve uncertainties. Improved research is a coping strategy itself, and many of the research areas will help improve the effectiveness of adaptations identified above. A number of general research needs cut across all sectors sensitive to climate change:

- Improve climate forecasts for the West: improve predictions of the sign, magnitude, and seasonality of change of important climate variables such as precipitation, and improve the estimation of probabilities.
- Seek a better understanding of the interrelationships between climate impacts and the institutional structures that facilitate or constrain effective action.
- Improve methods for involving the public in research and communicating research results to the public and decision makers.
- Conduct more research on adaptation, specifically to improve understanding of the potential effectiveness, costs, and impediments to adaptations.

Water
- Develop a better understanding of the human and ecological impacts of climate variability and change on water resources, particularly at the local and regional levels.
- Analyze all water resource options, including full

239

efficiency potential in all sectors, water transfer options, impacts of pricing changes on all sectors (including the impacts of different water price levels on the types of crops grown in different locations).

- Develop methodologies, analytical tools, and design criteria for incorporating increased climatic variability and change into hydraulic design and water resources planning and management.
- Develop effective long-term strategies for conservation.
- Study improvements in flood forecasting and response, improvements in reservoir management, and enhancement of other infrastructure that may be vulnerable to climate impacts.
- Improve understanding of groundwater resources in terms of amounts, locations, water quality, relationship to surface water, and potential for recharge, including effects of climate variability and altered precipitation. Develop an accurate and complete inventory of groundwater, ascertain the rates of use and potential for natural recharge, examine the extent to which it can be recharged by technology, and understand how all of these parameters would be affected by an increase or decrease in precipitation.
- Examine how to effectively transfer knowledge and technology from the research community to the public, particularly with regard to improving long-term planning and developing more realistic supply/demand water budgets.

Agriculture

Many of the research topics that apply to water resources are critical for agriculture. Additional research topics include the following:

- Improve understanding of the effect of climate change on plant yield and health.
- Enhance knowledge of how climate change and variability may affect pest and disease problems.
- Improve understanding of the effects of ENSO on agriculture.
- Examine the impact of climate change on the competitiveness of agriculture with other regions in the US and globally.
- Analyze the institutional obstacles to adaptation to climate change in agriculture (water laws, endangered species, etc.)

Ranching

- Examine how ranchers cope with climate variability and how their experience can be used to enhance their ability to adapt to climate change. Examine the interactions between urban development, climate change, and loss of land for ranching.

- Examine the impact of climate change on the competitiveness of ranching with other regions in the US and globally.

Coastal Issues

- Develop a statewide (California) map identifying the extent of sea-level rise. Certain areas have been mapped using a simple 1-meter demarcation, but the maps have not been based on the best available mapping technology, such as that used by NOAA and NASA.
- Analyze the impacts of sea-level rise and accelerated cliff erosion on buildings, energy, transportation, coastal infrastructure, and other features. The impacts of altered sediment flows along the coast may also have important implications for harbors and navigation.

Ecosystem Management

- Conduct extensive interdisciplinary ecosystem research, monitoring, and modeling in the region to provide an understanding of ecosystem structure and function on which sound land-management practices can be based.
- Improve understanding of CO_2 fertilization on natural ecosystems.
- Improve understanding of the effectiveness of possible adaptations for preserving biodiversity.

Fire

- Improve modeling and predictive capacity to allow fire personnel to deploy resources as needed.
- Link the remote sensing and GIS-based images being used with models to better understand fire risk and the dynamics of fire to increase ground-truthing. Additional work on the dynamics of fire and ecological communities would improve the modeling efforts.

Health

- Improve understanding of the vulnerability of the region to the spread of infectious diseases and heat waves.
- Improve understanding of the relationships between emissions of air pollutants, climate change, and resulting air pollution.

Landscape Processes

- Conduct more research on how climatic change will affect the land surface, in terms of erosion by wind and water, sediment discharge, and landslide potential.

LITERATURE CITED

Aber, J., R. Neilson, S. McNulty, J. Lenihan, D. Bachelet, and R. Drapek, Forest processes and global environmental change: Predicting the effects of individual and multiple stressors, BioScience, in press, 2001.

Bales, R. C., and D. M. Liverman, Climate patterns and trends in the Southwest, in *Climate Variability and Change in the Southwest: Final Report of the Southwest Regional Climate Change Symposium and Workshop,* edited by R. Merideth, D. Liverman, R. Bales, and M. Patterson, The University of Arizona, Tucson, Arizona, 1998.

Brown, T. C., Projecting US freshwater withdrawals, *Journal of Water Resources Research, 36,* 769-780.

California Department of Finance, California's population tops 33 million, press release, May 1998.

California Trade and Commerce Agency, California: An economic profile, September 1997. (http://commerce.ca.gov/california/economy).

Chapin, F. S., et al., Consequences of change biodiversity, *Nature 405,* 234-242, 2000.

CLIMAS, Climate variability in the Southwest region, (http://geo.ispe.arizona.edu/swclimate/water%context.htm.), 1998.

Dale, V. H., et al., Forest disturbances and climate change, *BioScience,* in press, 2000.

Dettinger, M. D., and D. R. Cayan, Large-scale atmospheric forcing of recent trends toward early snowmelt runoff in California, *Journal of Climate, 8,* 606-623, 1995.

Diaz, H. F., and C. A. Anderson, Precipitation trends and water consumption related to population in the southwestern United States: A reassessment, *Water Resources Research,* 31, 713-720, 1995.

Engelthaler, D. M., et al., Climatic and environmental patterns associated with Hantavirus pulmonary syndrome, Four Corners region, United States, *Emerging Infectious Diseases 5*(1), 87-94, 1999.

Galbraith, H., D. Park, R. Jones, J. Clough, B. Harrington, S. Herrod-Julius, and G. Page, Potential impacts of sea level rise due to global climate change on migratory shorebird populations at coastal sites in North America,

Report to US Environmental Protection Agency, Stratus Consulting Inc., Boulder, Colorado. Draft. April 19, 2000.

Gleick, P., Regional hydrologic consequences of increases in atmospheric CO_2 and other trace gases, *Climate Change, 10,* 137-161, 1987.

Gleick, P., Climate change and California: Past, present, and future vulnerabilities, *Societal Responses to Regional Climate Change: Forecasting By Analogy,* edited by M. H. Glantz, Westview Press, Boulder, Colorado, 1988.

Gleick, P. H., Vulnerability of water systems, in *Climate Change and US Water Resources,* edited by P. E. Waggoner, John Wiley & Sons, New York, pp. 223-240, 1990.

Gleick, P. H., and E. P. Maurer, Assessing the costs of adapting to sea level rise: A case study of San Francisco Bay, The Pacific Institute for Studies in Development, Environment, and Security and the Stockholm Environment Institute, Stockholm, Sweden, 1990.

Gleick, P. H., and L. Nash, The societal and environmental costs of the continuing California drought, Pacific Institute for Studies in Development, Environment, and Security, Berkeley, California, 66 pp., 1991.

Gleick, P. H., P. Loh, S. V. Gomez, and J. Morrison, "California Water 2020: A Sustainable Vision," Pacific Institute for Studies in Development, Environment, and Security, Oakland, California, 113 pp., 1995.

Gleick, P. H., and E. L. Chalecki, The impacts of climate changes for water resources of the Colorado and Sacramento-San Joaquin River basins, *Journal of the American Water Resources Association, 35*(6), 1429-1441, 1999.

Grimm, N. B., A. Chacon, C. N. Dahm, S. W. Hostetler, O. T. Lind, P. L. Starkweather, and W. W. Wurtsbaugh, Sensitivity of aquatic ecosystems to climatic and anthropogenic changes: The basin and range, American Southwest and Mexico, *Hydrologic Processes, 11,* 1023-1041, 1997.

Hansen, A. J., R. P. Neilson, V. Dale, C. Flather, L. Iverson, D. J. Currie, S. Shafer, R. Cook, and P. J. Bartlein, Global change in forests: Responses of species, communities, and biomes, *BioScience,* in press, 2000.

Hay, L. E., R. L. Wilby, and G. H. Leavesley, A comparison of delta change and downscaled GCM scenarios for three mountainous basins in the United States, *Journal of the American Water Resources Association, 36*(2), 2000.

Houghton, J. T., L. G. Meira Filho, B. A. Callander, N. Harris, A. Kattenberg, and K. Maskell, (Eds.), *Climate Change 1995: The Science of Climate Change, Contribution of Working Group I to the Second Assessment Report of the Intergovernmental Panel on Climate Change,* Cambridge University Press, Cambridge, England, 1996.

Hurd, B. H., N. Leary, R. Jones, and J. B. Smith, Relative regional vulnerability of water resources to climate change, *Journal of the American Water Resources Association, 35*(6), 1399-1410, 1999a.

Hurd, B. H., J. M. Callaway, J. B. Smith, and P. Kirshen, Economic effects of climate change on US water resources, in *The Economic Impacts of Climate Change on the US Economy,* edited by R. Mendelsohn and J. E. Neumann, Cambridge University Press, Cambridge, England, 133-177, 1999b.

Jeton, A. E., M. D. Dettinger, and J. L. Smith, Potential effects of climate change on streamflow, eastern and western slopes of the Sierra Nevada, California, and Nevada, *Water-Resources Investigations Report 95-4260,* US Geological Survey, Sacramento, California, 1996.

Kalkstein, L. S., and J. S. Greene, An evaluation of climate/mortality relationships in large US cities and the possible impacts of a climate change, *Environmental Health Perspectives, 105*(1), 2-11, 1997.

Karl, T. R., C. N. Williams, Jr., F. T. Quinlan, and T. A. Boden, 1990: United States Historical Climatology Network (HCN) Serial Temperature and Precipitation Data, Environmental Science Division, Publication No. 3404, Carbon Dioxide Information and Analysis Center, Oak Ridge National Laboratory, Oak Ridge, TN, 389 pp.

Karl, T. R., and R. W. Knight, Secular trends of precipitation amount, frequency, and intensity in the United States, *Bulletin of the American Meteorological Society, 79,* 231-241, 1998.

Kattlemann, R., and M. Embury, Riparian areas and wetlands, *Sierra Nevada Ecosystem Project: Final Report to Congress, Volume III, Assessments, Commissioned Reports, and Background Information,* Centers for Water and Wildland Resources, University of California, Davis, 1996.

Knox, J. B., Global climate change: Impacts on California, an introduction and overview, in *Global Climate Change and California,* edited by J. B. Knox, University of California Press, Berkeley, California, 1991.

Krieger, D. J., Saving open spaces: Public support for farmland protection, Center for Agriculture in the Environment, (http://farm.fic.niu.edu/cae/wp/99-1/wp99-1.html) December 29, 1999.

Lettenmaier, D. P., and D. P. Sheer, Climatic sensitivity of California water resources, *Journal of Water Resources Planning and Management, 117,* 108-125, 1991.

Liverman, D., Southwest overview, in *Climate Variability and Change in the Southwest: Final Report of the Southwest Regional Climate Change Symposium and Workshop,* edited by R. Merideth, D. Liverman, R. Bales, and M. Patterson, University of Arizona, Tucson, Arizona, 1998.

McCabe, G. J., and D. M. Wolock, General-Circulation-Model simulations of future snowpack in the western United States, *Journal of the American Water Resources Association, 35*(6), 1473-1484, 1999.

McClaren, M., and M. Patterson, Ranching, in *Climate Variability and Change in the Southwest: Final Report of the Southwest Regional Climate Change Symposium and Workshop,* edited by R. Merideth, D. Liverman, R. Bales, and M. Patterson, University of Arizona, Tucson, Arizona, 1998.

Meyer, J. L., M. J. Sale, P. J. Mulholland, and N. L. Poff, Impacts of climate change on aquatic ecosystem functioning and health, *Journal of the American Water Resources Association, 35*(6), 1373-1386, 1999.

Miller, K. A., S. L. Rhodes, and L. V. MacDonnell, Water allocation in a changing climate: Institutions and adaptation, *Climatic Change, 35,* 157-177, 1997.

Miller, N. L., J. Kim, R. K. Hartman, and J. Farrara, Downscaled climate and streamflow study of the southwestern United States, *Journal of the American Water Resources Association, 35*(6), 1525-1537, 1999a.

Miller, N. L., J. Kim, and M. D. Dettinger, California stream flow evaluation based on a dynamically downscaled eight year hindcast, observations, and physically based hydrologic models, *Eos, Transactions,* AGU, 80, F406, 1999b.

Miller, N. L., J. Kim, and M. D. Dettinger, Climate change sensitivity study for two California river basins, U.S. Department of Energy Workshop on the Climate Change Prediction Program, Sponsored by the U.S. Department of Energy, Bethesda, Maryland, March 27-29, 2000.

Murphy, D. D., and S. B. Weiss, Effects of climate change on biological diversity in western North America: Species losses and mechanisms, in *Global Warming and Biological Diversity,* edited by R. L. Peters and T. E. Lovejoy, Yale University Press, New Haven, Connecticut, pp. 355-368, 1992.

Myers, N., R. A. Mittermeier, C. G. Mittermeier, G. A. B. da Fonseca, and J. Kent, Biodiversity hotspots for conservation priorities, *Nature, 403,* 853-858, 2000.

NCAR (National Center for Atmospheric Research), VEMAP tables of means and variances, Also available on Web site (http://www.cgd.ucar.edu/naco/vemap/vemtab.html), October 11, 1999a.

NCAR (National Center for Atmospheric Research), Soil Moisture [Changes Projected by GCMs], Also available on Web site (http://www.cgd.ucar.edu/naco/gcm/sm.html), August 31, 1999b.

NCAR (National Center for Atmospheric Research), Foundation document figures, Also available on Web site (www.cgd.ucar.edu/naco/found/figs.html), November 5, 1999c.

Neilson, R. P., and R. J. Drapek, Potentially complex biosphere responses to transient global warming, *Global Change Biology, 4,* 505-521, 1998.

Nemani, R. R., M. A. White, D. R. Cayan, G. V. Jones, S. W. Running, and J. C. Coughlan, Asymmetric climate warming improves California vintages, *Climate Research,* in press, 2001.

Neumann, J. E., G. Yohe, R. Nicholls, and M. Manion, Sea-level rise and global climate change: A review of impacts to the US coasts, Pew Center on Global Climate Change, Arlington, Virginia, 2000.

NPA Data Services, Inc., *Demographic Databases; Three Growth Projections 1967-2025,* NPA Data Services, Inc., Washington, DC, 1999.

NPS, National Park Service, (http://www.nps.gov/grca/), March 17, 1999.

NYT, New rules sought on tapping the Colorado River, *The New York Times,* May 23, 1999.

Parmenter, R. R., E. P. Yadav, C. A. Parmenter, P. Ettestad, and K. L. Gage, Incidence of plague associate with increased winter-spring precipitation in New Mexico, *American Journal of Tropical Medical Hygiene, 61,* 814-821, 1999.

Parmesan, C., Climate and species' range, *Nature, 382,* 765-766, 1996.

Riebsame, W. E. (Ed.), *Atlas of the New West,* W. W. Norton & Company, New York, 1997.

Sagarin, R. D., J. P. Barry, S. E. Gilman, and C. H. Baxter, Climate related changes in an intertidal community over short and long time scales, *Ecological Monographs, 69,* 465-490.

Solley, W. B., R. R. Pierce, and H. A. Perlman, Estimated use of water in the United States in 1995, US Geological Survey Circular 1200, US Government Printing Office, Denver, Colorado, 1998.

Tausch, R. J., C. L. Nowak, and R. S. Nowak, Climate change and plant species responses over the Quaternary: Implications for ecosystems management, in *Interior West Global Change Workshop,* edited by R. W. Tinus, General Technical Report RM-GTR-262, USDA Forest Service, Fort Collins, Colorado, 1995.

Templin, W. E., California — Continually the Nation's leader in water use, (http://ca.water.usgs.gov/wuse/awra/), accessed May 13, 1999.

The Nature Conservancy, (http://www.consci.tnc.org/library/pubs/rptcard/map.html), accessed March 17, 1999.

Titus, J. G., and V. Narayanan, The risk of sea level rise, *Climatic Change, 33,* 151-212, 1996.

Torn, M., E. Mills, and J. Fried, Will climate change spark more wildfire damages?, Lawrence Berkeley Laboratory, Berkeley, California, (http://eande.lbl.gov/CBS/EMills/wild.html), 1998.

(US BEA) US Bureau of Economic Analysis, Gross State Product by component and industry 1977-1997." 1999a (http://www.bea.doc.gov/bea/regional/gsp/gsplist.html).

(US BEA), US Bureau of Economic Analysis, Regional Economic Information System (http://fisher.lib.virginia.edu/reis/), 1999b.

US Bureau of Reclamation, Updating the Hoover Dam documents, Denver, Colorado, 1978.

US Bureau of Reclamation, Compilation of records in accordance with Article V of the Decree of the Supreme Court of the United States in *Arizona versus California* dated March 9, 1964, for calendar years 1989-1997.
US Bureau of Reclamation, Estimate of 1999 Colorado River Use, Boulder City, Nevada, US Bureau of Reclamation Memorandum to All Interested Persons, May 26, 1999.

US Census Bureau, *Statistical Abstract of the United States: 1998* (118th edition), US Department of Commerce, Washington, DC, 1998.

US Department of Agriculture (USDA), Agricultural resources and environmental indicators, 1996-1997, *Agricultural Handbook No. 712,* USDA, Economic Research Service, Washington, DC, 1997.

US Environmental Protection Agency, Climate change and New Mexico, EPA 236-F-98-007p, US EPA, Washington, DC, 1998a.

US Environmental Protection Agency, Climate change and Utah, EPA 236-F-98-007z, US EPA, Washington, DC, 1998b.

US Fish and Wildlife Service, Listed species by state and territory, (http://www.fws.gov/r9endspp/listmap.html), Accessed March 17, 1999.

US Geological Survey (USGS), Seasonal land cover regions: scale: 1:7,500,000," map, US Geological Survey, Reston, Virginia, 1993.

US Geological Survey (USGS), (http://goechange.er.usgs.gov/sw/changes/anthropogenic/cropland/lu4592_lyt.gif), 1999.

US Historical Climatology Network (USHCN), 1999. (http://www.ncdc.noaa.gov/ol/climate/research/ushcn/ushcn.html)

Wagner, F. H., and J. Barron, Rocky Mountain/Great Basin regional climate-change workshop, Utah State University, Logan, Utah, 1998.

Walker, B., and W. Steffen, An overview of the implications of global change for natural and managed terrestrial ecosystems, *Conservation Ecology* Vol 1; Issue 2. (online) (http://www.consecol.org/Journal/vol1/iss2/art2/), 1997.

Western Water Policy Review Advisory Commission, *Water in the West: Challenge for the Next Century,* National Technical Information Service, Springfield, Virginia, 1998.

Willman, M. L., State may consider mudslide insurance for homeowners, *Los Angeles Times,* February 14, 1998.

Wilkinson, R., and T. Rounds, Potential impacts of climate change and variability for California: California regional workshop report, National Center for Ecological Analysis and Synthesis, University of California at Santa Barbara, 1998a.

Wilkinson, R., and T. Rounds, Climate change and variability in California: White paper for the California Regional Assessment, National Center for Ecological Analysis and Synthesis, University of California at Santa Barbara, 1998b.

Wolock, D. M., and G. J. McCabe, Simulated effects of climate change on mean annual runoff in the conterminous United States, *Journal of the American Water Resources Association, 35*(6), 1341-1350, 1999.

Wong, A. K., L. Owens-Viani, A. Steding, P. H. Gleick, D. Haasz, R. Wilkinson, M. Fidell, and S. Gomez, Sustainable use of water: California success stories, Pacific Institute for Studies in Development, Environment, and Security, Oakland, California, 1999.

Yoshiyama, R. M., E. R. Gerstung, F. W. Fisher, and P. B. Moyle, Historical and present distribution of Chinook salmon in the Central Valley drainage of California, *Sierra Nevada Ecosystem Project: Final Report to Congress, Volume III, Assessments, Commissioned Reports, and Background Information,* Centers for Water and Wildland Resources, University of California, Davis, 1996.

ACKNOWLEDGMENTS

Many of the materials for this chapter are based on contributions from participants on and those working with the:

California Workshop and Assessment Steering Committees
Robert Wilkinson, University of California-Santa Barbara, co-chair
Jeff Dozier, University of California-Santa Barbara, co-chair
Richard Berk, University of California, Los Angeles
Dan Cayan, Scripps Institution of Oceanography, University of California, San Diego
Keith Clarke, University of California, Santa Barbara
Frank Davis, University of California, Santa Barbara
James Dehlsen, Dehlsen Associates
Peter Gleick, Pacific Institute for Studies in Development, Environment, and Security
Michael Goodchild, University of California, Santa Barbara
Nicholas Graham, Scripps Institution of Oceanography / University of California, San Diego
William J. Keese, California Energy Commission
Charles Kolstad, University of California, Santa Barbara
Michael MacCracken, USGCRP and Lawrence Livermore National Laboratory
Jim McWilliams, University of California, Los Angeles
John Melack, University of California, Santa Barbara
Norman Miller, Lawrence Berkeley National Laboratory / University of California, Berkeley
Harold A. Mooney, Stanford University
Peter Moyle, University of California, Davis
Walter Oechel, San Diego State University
Larry Papay, Bechtel Group
Claude Poncelet, Pacific Gas and Electric Company
Thomas Suchanek, NIGEC / University of California, Davis
Henry Vaux, University of California Office of the President
James R. Young, Southern California Edison

Rocky Mountain/Great Basin Workshop and Assessment Teams
Frederic Wagner, Utah State University, co-chair
Thomas Stohlgren, US Geological Survey, co-chair
Connely Baldwin, Utah State University
Jill Baron, US Geological Survey, Fort Collins, CO
Hope Bragg, Utah State University
Barbara Curti, Nevada Farm Bureau, Reno, NV

Martha Hahn, U.S. Bureau of Land Management, Boise, ID
Sherm Janke, Sierra Club, Bozeman, MT
Upmanu Lall, Utah State University
Linda Mearns, National Center for Atmospheric Research, Boulder, CO
Hardy Redd, Private Rancher, Lasal, UT
Gray Reynolds, Sinclair Corporation, Salt Lake City, UT
David Roberts, Utah State University
Lisa Schell, Colorado State University
Susan Selby, Las Vegas Valley Water District
Carol Simmons, Colorado State University
Dale Toweill, Idaho Dept. of Fish and Game, Boise, ID
Booth Wallentine, Utah Farm Bureau Federation, Salt Lake City, UT
Todd Wilkinson, Journalist, Bozeman, MT

Southwest/Colorado River Basin Workshop and Assessment Teams
William A. Sprigg, University of Arizona, co-chair
Todd Hinkley, US Geological Survey, co-chair
Diane Austin, University of Arizona
Roger C. Bales, University of Arizona
David Brookshire, University of New Mexico
Stephen P. Brown, Federal Reserve Bank of Dallas
Janie Chermak, University of New Mexico
Andrew Comrie, University of Arizona
Prabhu Dayal, Tucson Electric Power Company
Hallie Eakin, University of Arizona
David C. Goodrich, US Department of Agriculture
Howard P. Hanson, Los Alamos National Laboratory
Laura Huenneke, New Mexico State University
William Karsell, WAPA
Korine Kolivras, University of Arizona
Diana Liverman, University of Arizona
Rachel A. Loehman, Sandia National Laboratories
Jan Matusak, Metropolitan Water District of Southern California
Linda Mearns, National Center for Atmospheric Research
Robert Merideth, University of Arizona
Kathleen Miller, National Center for Atmospheric Research
David R. Minke, ASARCO
Barbara Morehouse, University of Arizona
Dan Muhs, US Geological Survey
Wilson Orr, Prescott College
Thomas Pagano, University of Arizona
Mark Patterson, University of Arizona
Kelly T. Redmond, Desert Research Institute
Paul R. Sheppard, University of Arizona
Verna Teller, Isleta Pueblo
James R. Young, Southern California Edison

CHAPTER 9

POTENTIAL CONSEQUENCES OF CLIMATE VARIABILITY AND CHANGE FOR THE PACIFIC NORTHWEST

Edward A. Parson[1,2], with contributions from members of the Pacific Northwest Assessment Team: Philip W. Mote[3], Alan Hamlet[4], Nathan Mantua[5], Amy Snover[6], William Keeton[7], Ed Miles[8], Douglas Canning[9], Kristyn Gray Ideker[10]

Contents of this Chapter

[1]John F. Kennedy School of Government, Harvard University, [2]Coordinating author for the National Assessment Synthesis Team, [3]University of Washington (UW), Chair, Pacific Northwest Regional Assessment Team, [4]Dept of Civil and Environmental Engineering, UW, [5]Climate Impacts Group, UW, [6]Climate Impacts Group, UW, [7]College of Forest Resources, UW, [8]Climate Impacts Group, UW, [9]Washington State Dept of Ecology, Olympia, [10]Ross and Associates (work completed while at UW)

CHAPTER SUMMARY

Regional Context

The Northwest, which includes the states of Washington, Oregon, and Idaho, has a great diversity of resources and ecosystems, including spectacular forests containing some of the world's largest trees; abrupt topography that generates sharp changes in climate and ecosystems over short distances; mountain and marine environments in close proximity, making for strong reciprocal influences between terrestrial and aquatic environments; and nearly all the volcanoes and glaciers in the contiguous US. The region has seen several decades of rapid population and economic growth, with population nearly doubling since 1970, a growth rate almost twice the national average. The same environmental attractions that draw people and investment to the region are increasingly stressed by the region's rapid development. The consequences include loss of old-growth forests, wetlands, and native grass and steppe communities, increasing urban air pollution, extreme reduction of salmon runs, and increasing numbers of threatened and endangered species. Climate change and its impacts will interact with these existing stresses in the region.

Climate of the Past Century

- Over the 20th century, annual-average temperature in the Northwest rose 1 to 3°F (0.6 to 1.7°C) over most of the region, with nearly equal warming in summer and winter.
- Annual precipitation increased nearly everywhere in the region, by 11% on average, with the largest relative increases about 50% in northeastern Washington and southwestern Montana.
- Year-to-year variations in the region's climate show a clear correlation with two large-scale patterns of climate variation over the Pacific, the El Niño/Southern Oscillation (ENSO), and Pacific Decadal Oscillation (PDO). The region-wide pattern associated with these phenomena is that warm years tend to be relatively dry with low streamflow and light snowpack, while cool ones tend to be relatively wet with high streamflow and heavy snowpack. This has clear effects on important regional resources: warmer drier years tend to have summer water shortages, less abundant salmon, and increased risk of forest fires.

Climate of the Coming Century

- Regional warming is projected to continue at an increased rate in the 21st century, in both summer and winter. Average warming over the region is projected to reach about 3°F (1.7°C) by the 2020s and 5°F (2.8°C) by the 2050s.
- Annual precipitation changes projected through 2050 over the region range from a small decrease (-7% or 2") to a slightly larger increase (+13% or 4").
- Projected precipitation increases are concentrated in winter, with decreases or smaller increases in summer. Because of this seasonal pattern, even the projections that show increases in annual precipitation show decreases in water availability.

Key Findings

- Projected warmer wetter winters are highly likely to increase flooding risk in rainfed rivers, while projected year-round warming and drier summers are highly likely to increase risk of summer shortages in both rainfed and snowfed rivers, because of smaller snowpack and earlier melt.

- Salmon are likely to be harmed by increased winter flooding, reduced summer and fall flows, and rising stream and estuary temperatures. It is also possible that earlier snowmelt and peak streamflow will deliver juveniles to the ocean before there is enough food for them. Climate change is consequently likely to hamper efforts to restore depleted stocks, and to stress presently healthy stocks.

- The coniferous forests that dominate much of the Northwest landscape are sensitive to summer moisture stress. Their extent, species mix, and productivity are likely to change under projected 21st century climate change, but the specifics of these changes are not yet known.

- Sea-level rise will likely require substantial investment to avoid coastal inundation, especially in low-lying communities of southern Puget Sound where the coast is subsiding. Projected heavier winter rainfall is likely to increase soil saturation, landsliding, and winter flooding.

- El Niño events increase erosion both by raising sea level for several months and by changing the direction of winds and waves from westerly to southwesterly. Climate change is projected to bring similar changes and associated impacts, including severe storm surge and coastal erosion.

POTENTIAL CONSEQUENCES OF CLIMATE VARIABILITY AND CHANGE FOR THE PACIFIC NORTHWEST

PHYSICAL SETTING AND UNIQUE ATTRIBUTES

The Pacific Northwest region includes the states of Washington, Oregon, and Idaho, and for assessing impacts in the Columbia River basin, some areas in adjoining states and the Canadian province of British Columbia. The region has a great diversity of resources and ecosystems, including spectacular forests containing some of the world's largest trees; abrupt topography that generates sharp changes in climate and ecosystems over short distances; mountain and marine environments in close proximity, making for strong reciprocal influences between terrestrial and aquatic environments; and nearly all the volcanoes and glaciers in the contiguous US.

The region is divided climatically, ecologically, economically, and culturally by the Cascade Mountains. The low-lying areas west of the Cascades hold three quarters of the region's population, concentrated in the metropolitan areas of Tacoma-Seattle-Everett along the Puget Sound coast, and Portland in the Willamette Valley. Here, once-dominant forestry, fishing and agriculture have been overtaken by aero-

Figure 1: The Puget Sound area has experienced rapid growth over the past two decades. Source: P. Mote, University of Washington

space, computer software and hardware, trade and services, although the relatively declining resource sectors remain economically and culturally important. The Northwest still provides about a quarter of the nation's softwood lumber and plywood (Haynes et al., 1995, tables 18 and 27). Agriculture is now much more important in the east of the region. Thanks in part to massive water-management and irrigation projects, the fertile lowlands of eastern Washington are the "Fruit Bowl" of the nation, producing 60% of the nation's apples and large fractions of its other tree fruit, while Idaho produces about a quarter of the nation's potatoes (USDA, 2000).

SOCIOECONOMIC CONTEXT

The region has seen several decades of rapid population and economic growth. Population has nearly doubled since 1970, a growth rate almost twice the national average. Growth has been strongly concentrated in the major western metropolitan areas and in the smaller but fast-growing inland cities of Boise and Spokane (Jackson and Kimmerling, 1993). Federal lands comprise roughly half the region's land area.[1] The region's environment presents a great variety of outdoor recreational opportunities, and its moderate climate and quality of life contribute to its continuing attraction to so many newcomers. The region is projected to continue growing faster than the national average, its population increasing from the present 10.5 million to 19 million by 2050 (with a range of 14.5 million to 23 million) (Terleckyj 1999a, 1999b; US Census Bureau, 2000). Both recent and projected growth rates are similar east and west of the Cascades (very slightly higher on the west), so the west side is projected to continue to contain nearly three quarters of the region's population.

The same environmental attractions that draw people and investment to the region are increasingly stressed by the region's rapid development. The predominant current stresses arise from direct

---

[1] Federal lands are 30% of Washington, 48% of Oregon, and 64% of Idaho, or 48% of the region overall, with an additional 5% state-owned. (Jackson and Kimmerling 1993, p. 32).

human interventions in the landscape, through such activities as dam building, forestry (including replacement of natural forests by plantations), and land-use conversion from the original forests, wetlands, grasslands, and sagebrush to expansion of metropolitan areas, intensively managed forests, agriculture, and grazing. The consequences include loss of old-growth forests, wetlands, and native grass and steppe communities, increasing urban air pollution, extreme reduction of salmon runs, and increasing numbers of threatened and endangered species.

ECOLOGICAL CONTEXT

The Northwest has a great diversity of landscapes and ecosystems, reflecting the region's varied climate and topography. Dense, tall moist coniferous forests cover about 80% of western Washington and Oregon, with Douglas fir, western red cedar and western hemlock at most low-elevation locations, western hemlock and Pacific silver fir at middle elevations, and mountain hemlock at high elevations (Franklin and Dyrness, 1973). A century of commercial logging has cut nearly all this forest at some time, greatly altering its species and age distribution. Only 10 to 20% of the original extent remains as old-growth forest (Marcot et al., 1991; Kellogg, 1992). The west also includes oak forests and grasslands in low-lying river valleys, coastal salt marshes and freshwater wetlands, and in the Klamath Mountains of southern Oregon, a mixed forest of drought-resistant conifers and hardwoods (Mac et al., 1998, p. 646). East of the Cascade crest, the drier climate and more frequent fires generate an open, park-like forest of ponderosa pine and Douglas fir, with sub-alpine fir, Engelmann spruce and patches of alpine larch at higher elevations and whitebark pine especially prominent near the upper tree line. In the Rocky Mountains and the east slope of the Cascades, forest gives way at high elevations to alpine meadows, and at lower elevations to juniper woodlands, sagebrush steppe, and grasslands, as well as high desert and lava fields in Idaho.

As on the west side of the Cascades, human influence has greatly altered east-side ecosystems. Fire suppression, grazing, and selective cutting have transformed all but a few percent of the original ponderosa pine forest into overstocked mixed-species forests that are highly susceptible to fire, insects, and disease (Henjum et al., 1994). More than 99% of the prairie grasslands near the meeting point of Idaho, Oregon, and Washington have been converted to crops, mostly wheat, while about 90% of the sagebrush steppe on the Snake River plain in

Figure 2: Old-Growth Douglas Fir Forest in the Cascades. Source: T.B. Thomas, US Forest Service

Figure 3: The Columbia is one of the most intensively developed river systems in the world. Source: ©P. Grabhorn

Idaho has been converted to agriculture (Mac et al., 1998, p. 649; Noss et al., 1995). Grazing has transformed nearly all of the remaining grassland and sagebrush steppe, leading to large-scale replacement of native perennials with invasive annuals such as Cheatgrass, Medusahead and Yellow Starthistle, and to expansion of Juniper woodlands into former rangeland (Miller and Rose, 1995; Miller and Wigand, 1994; West and Hassan, 1985).

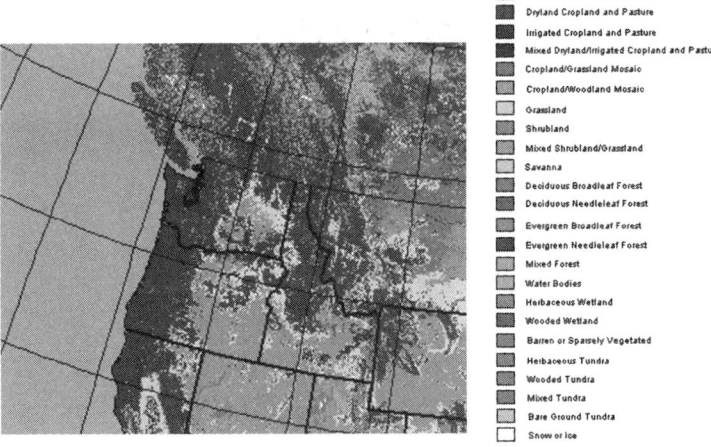

Urban and Built-Up Land
Dryland Cropland and Pasture
Irrigated Cropland and Pasture
Mixed Dryland/Irrigated Cropland and Pasture
Cropland/Grassland Mosaic
Cropland/Woodland Mosaic
Grassland
Shrubland
Mixed Shrubland/Grassland
Savanna
Deciduous Broadleaf Forest
Deciduous Needleleaf Forest
Evergreen Broadleaf Forest
Evergreen Needleleaf Forest
Mixed Forest
Water Bodies
Herbaceous Wetland
Wooded Wetland
Barren or Sparsely Vegetated
Herbaceous Tundra
Wooded Tundra
Mixed Tundra
Bare Ground Tundra
Snow or Ice

**Figure 4: Major ecological regions of the Pacific Northwest
Source: United States National Atlas – See Color Figure Appendix**

An additional stress on inland forests is the devastation of whitebark pine *(Pinus albicaulis)*, a dominant species near the upper tree-line in the Rockies and Cascades, by the introduced fungus, white pine blister rust. Throughout the species' range in northern Idaho more than half of the trees are dead, while infection rates of living trees are above 50% throughout Washington and Idaho, and above 20% in Oregon (Keane and Arno, 1993; Kendall, 1995).

The region is high in biodiversity across major taxa. In the wet west-side forests, more than 150 species of terrestrial snails and slugs have been identified, and 527 species of fungi, of which 234 are rare and occur nowhere else (FEMAT, 1993). It is estimated that these forests may support 50,000 to 70,000 species of arthropods, although only preliminary surveys of arthropods have been conducted. A survey of one experimental forest in the Oregon Cascades found more than 3,400 (Parsons et al., 1991). Oregon contains between 3,000 and 4,000 identified species of vascular plants, Idaho and Washington between 2,400 and 3,000, putting Oregon in the top six states for plant diversity and Washington and Idaho in the top 15. Of these plant species, about 8-12 are rare in Oregon, 5-8 in Washington and Idaho.[2] The 33 species of amphibians in the region include 17 that are endemic, but only one candidate for federal listing, the Oregon spotted frog (Bury, 1994). Although 67 populations of fish in the region are on either federal or state sensitive species lists, much more is known about salmon and trout, the most highly valued species in the region, than about other species. Of 58 distinct salmonid stocks in the region, 26 are now listed as

endangered or threatened under the Endangered Species Act (ESA), including the Puget Sound Chinook, the first ESA listing to affect a major metropolitan area (NMFS, 2000). Of roughly 450 species of birds identified in five sub-regions of the Northwest by the Breeding Bird Survey, from 10 to 35 species per sub-region show decreased numbers since the 1960s, while 3 to 25 show increases, with the largest net decreases in the coastal forests of southern Oregon (Carter and Barker, 1993). Among mammals, seven carnivores — the Grizzly Bear, Gray Wolf, Lynx, Wolverine, Fisher, Marten, and Kit Fox — have small and threatened regional populations, principally due to disturbance, loss of forest habitat, and the secondary effects of logging road construction (Weaver et al., 1996).

CLIMATE VARIABILITY AND CHANGE

West of the Cascades the climate of the Northwest is maritime, with abundant winter rains, dry summers, and mild temperatures year-round — usually above freezing in winter, so snow seldom stays on the ground more than a few days. Most places west of the Cascades receive more than 30 inches (75 cm) of precipitation annually, while some westward mountain slopes of the Olympics and Cascades receive more than 200 inches (500 cm). Although a mild maritime climate has prevailed in the region for several centuries, thousand-year records show substantial fluctuations. For example, from about 4,000 to 8,000 years ago in the Puget Sound area, dominance of dry vegetation types such as California chaparral suggests the region's climate was much warmer and drier, resembling the present climate of California's northern Central Valley (Detling, 1953, 1968).

East of the Cascade crest, the climate shifts sharply from abundant rainfall to abundant sunshine, with annual precipitation generally less than 20 inches (50 cm), as little as 7 inches (20 cm) in some places. These precipitation differences are most pronounced in winter: summer precipitation in the west is only slightly higher than in the east, while winter precipitation is four to five times higher. Figures 5 and 6 illustrate the large differences in annual and seasonal precipitation across the region. Even the inland mountain ranges receive much less precipitation than the western Cascades or Olympics. Though average temperatures are similar east and west, the east has larger daily and annual ranges, with hotter summers and colder winters.

---

[2]"Rare" means an international endangerment rank of G1 to G3 (Morse et al., 1995).

Observed Climate Trends

Over the 20th century, the Northwest has grown warmer and wetter. Annual-average temperature rose 1 to 3°F (0.6 to 1.7°C) over most of the region, with nearly equal warming in summer and winter. Annual precipitation also increased nearly everywhere in the region, by 11% on average, with the largest relative increases about 50% in northeastern Washington and southwestern Montana.[3]

In addition to this trend toward a warmer, wetter climate, the Northwest's climate also shows significant recurrent patterns of multi-year variability. These year-to-year variations tend to be consistent over the entire region, and are evident in both winter and summer. The predominant pattern is that warm years tend to be relatively dry with low streamflow and light snowpack, while cool ones tend to be relatively wet with high streamflow and heavy snowpack. Although the differences in temperature and precipitation are relatively small (differences in monthly-average temperature of up to 2 to 4°F or 1.1 to 2.2°C in winter), they have clearly discernible effects on important regional resources. Warmer drier years tend to have summer water shortages, less abundant salmon, and increased risk of forest fires (dell'Arciprete et al., 1996; Mantua et al., 1997; Hulme et al., 1999).

These year-to-year variations in the region's climate show a clear correlation with two large-scale patterns of climate variation over the Pacific, one more and one less well known. The El Niño/ Southern Oscillation (ENSO) is an irregular oscillation with a period of 2 to 7 years, which is widely known and intensively studied. ENSO's positive El Niño phase warms sea-surface temperature in the equatorial Pacific and cools it in the central North Pacific, deepening the winter low-pressure system off the Aleutians and bringing substantial changes in mid-latitude atmospheric circulation (Trenberth, 1997). A more recently identified pattern of longer-term variability is the Pacific Decadal Oscillation (PDO), defined in terms of changes in Pacific sea-surface temperature north of 20 degrees latitude. Like the warm El Niño phase of ENSO, the warm or positive phase of PDO warms the Pacific near the equator and cools it at northern mid-latitudes. But unlike ENSO, PDO's effects are stronger in the central and northern Pacific than near the equator, and its irregular period is several decades, tending to stay in one

Average Annual Precipitation, Pacific Northwest, 1961-1990

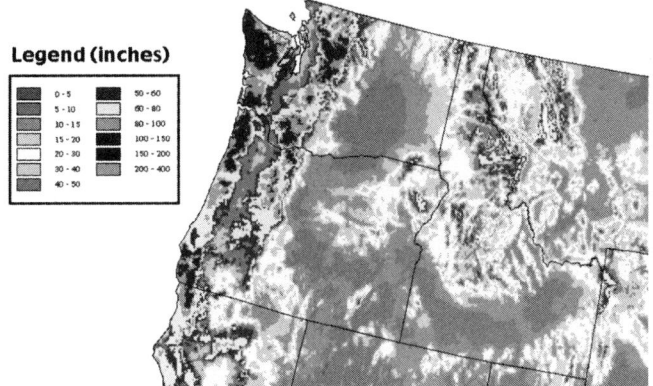

Legend (inches)

	0 - 5		50 - 60
	5 - 10		60 - 80
	10 - 15		80 - 100
	15 - 20		100 - 150
	20 - 30		150 - 200
	30 - 40		200 - 400
	40 - 50		

Figure 5: The Cascade mountains divide the wetter west from the drier east. Source: Mapping by C. Daly, graphic by G. Taylor and J. Aiken, copyright © 2000, Oregon State University. – See Color Figure Appendix

Average Monthly Precipitation in the Pacific Northwest

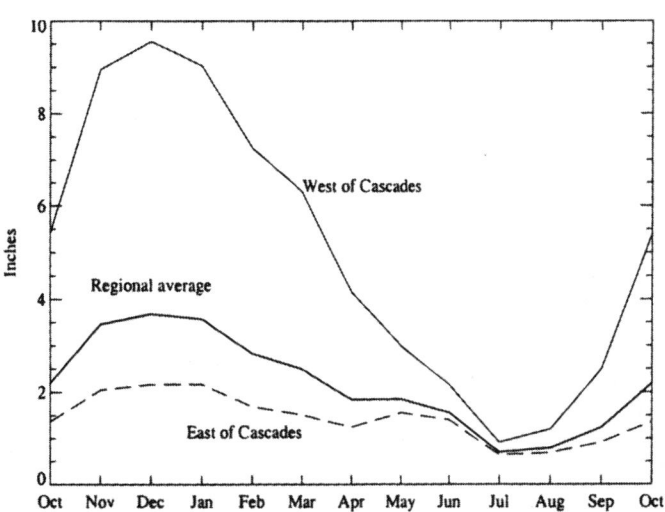

Figure 6: West of the Cascades is wetter than east, but nearly all the difference occurs in winter. Source: Mote et al. (1999), Figure 3, pg. 5.

phase or the other for 20 to 30 years at a time. PDO is also much less well understood than ENSO, in part because its period is so long relative to the history of reliable records that only two complete oscillations have been observed. The PDO was in its cool, or negative phase from the first sea-surface temperature records in 1900 (and possibly before) until 1925, then in warm or positive phase until 1945, cool phase again until 1977, and warm phase until the 1990s (Miller et al., 1994; 1998). Evidence is beginning to mount that another change to the

---

[3]Analyses of Historical Climate Network data by NCDC. A similar analysis of historical trends by UW JISAO (in Mote et al., 1999) using slightly different re-weighting algorithms and regional boundaries found a 14% average precipitation increase over the 20th century. The difference between these is not significant.

Northwest Average Temperature, Observed and Modeled

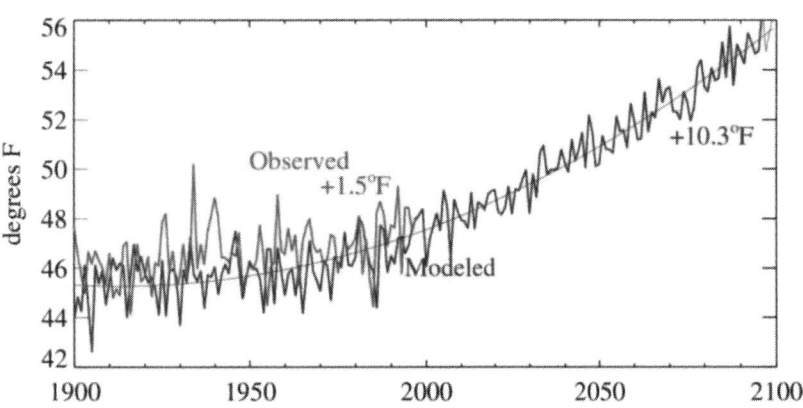

Figure 7: The red line shows annual-average temperature in the Northwest in the 20th century, observed from 113 weather stations with long records. The blue line shows the historical Northwest average temperature calculated by the Canadian model from 1900 to 2000, and projected forward to 2100. Source: Mote et al (1999), Summary (p. 6). – See Color Figure Appendix

cool phase of PDO likely occurred in the mid-1990s, but it is too early to tell with confidence. The warm phase of PDO, like El Niño, strengthens the Aleutian low, bringing warmer winter temperatures over western North America and warmer ocean temperatures along the coast. In these winters, the mid-lati-tude storm track tends to split, with one branch carrying storms south to California, the other north to Alaska. These winters consequently tend to be drier than normal in the Pacific Northwest and wetter than normal along the coasts to both the south and north. In contrast, years during the cool phase of PDO and during the cool, La Niña phase of ENSO are associated with a weaker Aleutian low, which tends to bring winters that are cooler and wetter than normal in the Pacific Northwest. The major exception to this cool-wet versus warm-dry pattern occurs during the strongest El Niño events, such as that of 1998. During these events, the Aleutian low is very strong and is also shifted to the Southeast, making winters on the Northwest coast warmer and wetter – i.e., while moderate El Niños tend to make Northwest winters warmer and drier, the strongest El Niños reverse the effect on precipitation and make the region warmer but with near normal precipitation.

Scenarios of Future Climate

Projections of climate change in the Northwest were conducted through 2100 using the Canadian and Hadley models (Boer et al., 1984, 1999a,b; McFarlane et al., 1992; Flato et al., 1999; Mitchell et al., 1995; Mitchell and Johns, 1997; Johns et al., 1997), and through 2050 with five additional general circulation models (GCMs), two of 1998 vintage and

Temperature Change 20th and 21st Centuries

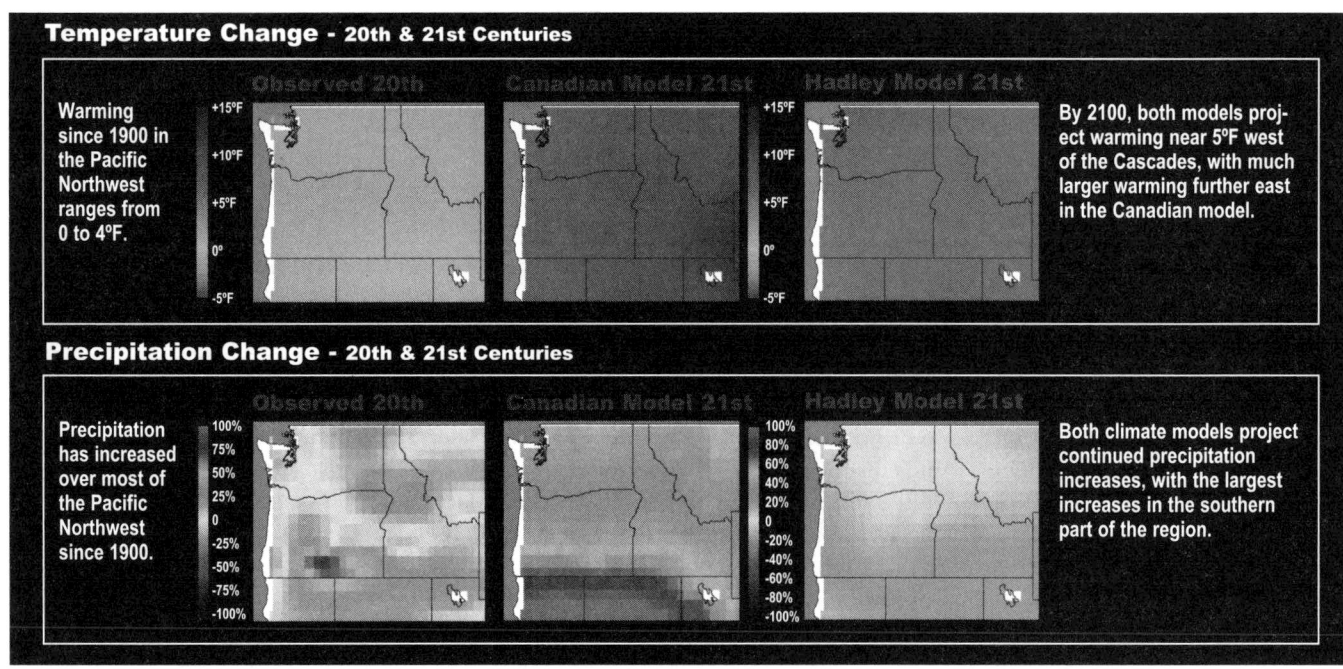

Figure 8: Temperature change observed in the 20th and projected for the 21st centuries. – See Color Figure Appendix

three of 1995 vintage. In addition, one analysis linked another GCM to a regional atmospheric model, to provide a climate projection with the fine spatial resolution (1.5 kilometers, about 1 mile) necessary for hydrological studies (Leung and Ghan 1999a, 1999b). In addition to the Canadian and Hadley models, the 1998 models included those developed by the Max-Planck Institut fur Meteorologie (MPI) and the National Oceanic and Atmospheric Administration's Geophysical Fluid Dynamics Laboratory (GFDL). These four models were run using the standard emission scenario, a 1% annual increase in equivalent atmospheric CO_2 concentration with trends in sulfate aerosol loading from the IPCC IS92a scenario. The 1995 models included earlier versions from the Hadley Center, MPI, and GFDL. Compared to the 1998 group, these models included simpler representations of sea, ice, and land surfaces. It is important to note that these earlier model runs also used a different emission scenario, which did not include aerosols. The finer-scale analysis used the Community Climate Model (CCM3) of the National Center for Atmospheric Research (NCAR) driving the Regional Climate Model of the Pacific Northwest National Laboratory (PNNL-RCM). This analysis used the same scenario as the 1995 models, 1% annual CO_2 –equivalent increase with no aerosols.

The coarse spatial resolution of GCMs is particularly troublesome for replicating the spatial character of climate in the Northwest, which is strongly shaped by the region's abrupt topography. Instead of the sharp Cascade crest, the models show a relatively smooth rise from the Pacific to the Rockies. Consequently, they simulate a climate for the region that is too "maritime" —milder in both winter and summer, with precipitation more evenly distributed across the region. One study comparing the Canadian model to a finer-scale regional climate model for western Canada (where GCMs have the same bias) suggested that these biases are more acute for precipitation than for temperature (Laprise et al., 1998). In projecting future climate, each model's bias relative to the present climate is removed. Despite their overall maritime bias for the Northwest, the Canadian and GFDL models (although not the Hadley model) do reproduce the Northwest region's observed 20th-century climate trends fairly well, particularly for temperature. Figure 7 shows the annual-average Pacific Northwest temperature calculated by the Canadian model through the 20th and 21st centuries, with a comparison to the observed record for the 20th century. The model matches the observed trend of the 20th century closely, and projects substantially more rapid warming through the 21st century. Figure 8 shows the changes in temperatue and precipitation over the region projected by the Canadian and Hadley models for the 21st century, and compares these to the observed pattern of changes over the 20th century.

These projections all show regional warming continuing at an increased rate in the next century, in both summer and winter. Average warming over the region is projected to reach about 3°F (1.7°C) by the 2020s and 5°F (2.8°C) by the 2050s. Annual precipitation changes projected through 2050 over the region range from a small decrease (-7% or 2") to a slightly larger increase (+13% or 4") (Hamlet and Lettenmaier, 1999), but these precipitation changes,

Table 1: Model Projections of Northwest Regional Climate

Model	2020s Temp change	2020s Precip change, inches Apr-Sep	2020s Precip change, inches Oct-Mar	2050s Temp change	2050s Precip change, inches Apr-Sep	2050s Precip change, inches Oct-Mar
Canadian	3.5°F	+0.2	+3.0	5.9°F	+0.3	+4.1
Hadley	3.2°F	+1.3	+3.5	4.8°F	+0.8	+2.5
MPI	3.7°F	-0.3	+0.6	5.3°F	-0.8	-0.4
GFDL	4.5°F	-0.2	+0.5	7.3°F	+0.4	+1.7
MPI 95	2.2°F	-2.5	+0.8	4.6°F	-1.8	+0.7
Hadley 95	2.8°F	-1.7	+2.5	5.4°F	-1.1	+2.5
GFDL 95	3.3°F	+0.4	+2.7	6.1°F	-0.8	+2.8
Average	3.1°F	-0.3	+1.6	5.3°F	-0.5	+1.4

Note: Results from 1995-vintage climate models were based on an emission scenario that did not include aerosols. Source: Table 3 (pg. 21), Mote et al., (1999b).

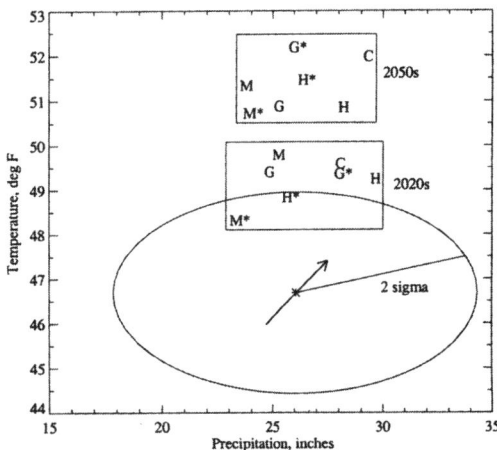

Projected Northwest Climate Change, Compared to 20th Century Variability

Figure 9: Climate change by the 2020s and 2050s over the Northwest Region from seven climate model scenarios. Any point on the graph shows a particular combination of regional annual-average temperature and total annual precipitation. The asterisk and arrow through it show the average climate over the 20th century and its trend, warming about 1.5°F (0.8°C) with a 2.5" (6 cm) precipitation increase. The oval illustrates how much the region's climate varied over the 20th century, enclosing all combinations of temperature and precipitation that were more than 5% likely to occur. Each letter shows one model's projection of the region's average climate, either in the 2020s or the 2050s. The models project that regional precipitation changes will lie within the range of 20th century variability, but projected temperature changes lie outside it. By the 2050s, all models project a climate so much warmer in the Northwest that it lies well outside the range of 20th century variability (*=1995-vintage model; H=Hadley; M=Max-Planck; G=GFDL; C=Canadian). Source: Regional report, Mote et al. (1999), fig. 12, pg. 19. – See Color Figure Appendix

unlike projected temperature changes, lie within the 20th century range of year-to-year variability. Projected precipitation increases are concentrated in winter, with decreases or smaller increases in summer. Because of this seasonal pattern, even the projections that show increases in annual precipitation show decreases in water availability. The 1995 and 1998 models show no systematic differences in their projections. These results are shown in Table 1, and are compared to 20th century climate variability in Figure 9. Most models by 2020, and all models by 2050, project a regional climate that is warmer than even the warmest years of the 20th century. The simulation that linked a GCM to a regional model showed similar results, with average annual warming of about 4°F (2.2°C) by 2050, wetter winters and drier summers, but showed larger seasonal differences – about 6°F (3.3°C) warming in winter and 2°F (1.1°C) in summer – due to its finer-scale representation of snow-albedo feedback (Leung and Ghan, 1999a, 1999b).

After 2050, the projected trend to a warmer, wetter regional climate continues in both the Hadley and Canadian models, with substantially more warming in winter than in summer in both models. By the 2090s, projected average summer temperatures rise by 7.3°F (4.1°C) in the Canadian model and 8.3°F (4.6°C) in the Hadley model, while winter temperatures rise 8.5°F (4.7°C) in the Hadley model and 10.6°F (5.9°C) in the Canadian model. Projected precipitation increases over the region range from a few percent to 20% (with a regional average of 10%) in the Hadley model, and from 0 to 50% (with a regional average of 30%) in the Canadian model.

These projected changes are associated with large-scale shifts in atmospheric circulation over the Pacific, especially in winter, which resemble the changes that occur during the strongest El Niño events. The Aleutian low is both strengthened and moved to the southeast, displacing the mid-latitude storm track southward and making the average winds on the Oregon and Washington coast stronger and more northward. Consequently, winters along the coast are warmer and wetter – both in total precipitation, and in the amount of rainfall in heavy storms, because warmer temperatures increase the quantity of water vapor the atmosphere can hold.

KEY ISSUES

Of the many potentially significant areas of climate impact in the region, the Northwest regional study selected four critical issues to examine:

Freshwater. Wetter winters are highly likely to increase flooding risk in rainfed and mixed rain/snow rivers, while year-round warming and drier summers are highly likely to increase risk of summer water shortages in rainfed, mixed, and snowfed rivers, including the Columbia. In the Columbia system, allocation conflicts are already acute, while a cumbersome network of overlapping authorities limit the system's adaptability, making it quite vulnerable to shortages.

Salmon. While non-climatic stresses on Northwest salmon presently overwhelm climatic ones, salmon abundances have shown a clear correlation with 20th century climate variations. Climate models cannot yet project the most important oceanic conditions for salmon, but the likely effects on their freshwater habitat, such as warmer water and reduced summer streamflow, are all highly likely to be unfavorable. Climate change is consequently likely to impede efforts to restore already depleted stocks, and to stress presently healthy stocks.

Forests. Northwest forests have been profoundly altered by timber harvesting and land-use conversion, both east and west of the Cascades. Whether Northwest forests will expand or contract under projected climate change, and what the effects on species mix and productivity will be, are highly likely to depend on assumptions related to plant response of water use efficiency to CO_2 enrichment, on which present evidence is incomplete. Models project several decades of forest expansion east of the Cascades and contraction to the west, with preliminary indications of larger forest contraction in the longer-term, as increased moisture stress overwhelms CO_2-induced increases in water-use efficiency.

Coasts. Sea level rise is likely to require substantial investments to avoid coastal inundation, and abandonment of some property, especially in low-lying communities of southern Puget Sound where the land is subsiding 0.3 to 0.8 inches per decade. Other likely effects include increased risk of winter landslides on bluffs around Puget Sound, and increased erosion on sandy stretches of the Pacific Coast.

1. Freshwater

Freshwater is a crucial resource in the Northwest, and climatic effects on water resources strongly influence and couple with many other domains of impact. Despite the region's reputation as a wet place, this only applies annually to the west slopes, and even they are dry in summer. Most of the region receives less than 20 inches (50 cm) of precipitation a year, and dry summers make freshwater a limiting resource for many ecosystems and human activities. Water supply, availability, and quality are already stressed by multiple growing demands.

The Cascades largely divide rivers partly or entirely controlled by rainfall, whose flow peaks in winter, from those controlled by snowmelt, whose flow peaks in late spring. The Columbia, a snowmelt-dominated river, is one of the nation's largest, draining roughly three-quarters of the region and carrying 55-65% of its total runoff. The Columbia is the region's primary source of energy and irrigation water, and is managed by multiple agencies for multiple, often conflicting values, including electricity, flood control, fish migration, habitat protection, water supply, irrigation, navigation, and recreation. Agriculture takes the largest share of present withdrawals, but other demands are growing, in particular, the recent demand for in-stream flow requirements to protect salmon. With more than 250 reservoirs and 100 hydroelectric projects, the Columbia system is among the most developed in the world and has little room for

further expansion, even as regional population growth and changing allocation priorities are intensifying competition for water. Because its watershed is so large, the Columbia's flow reflects an averaging of weather conditions over large areas and seasonal time-scales. Consequently, climatic effects on its flow can be detected and projected with more confidence than for smaller river systems, which respond most strongly to shorter-term and more local events.

Columbia basin hydrology shows a strong signal of both ENSO and PDO. Because the warm phases of these oscillations tend to make winters both warm and dry, their effects on snowpack and streamflow, and hence on regional water supply, are stronger than their effects on either temperature or precipitation. Warm-phase years accumulate less snowpack, and shift from snow accumulation to melting earlier in the season. In the Columbia, the warm phases of ENSO and PDO each reduce average annual flow by roughly 10% relative to the long-term mean, with larger reduction of peak spring flow. The effects of the two oscillations are nearly additive, so years with both in their warm phase have brought the lowest snowpack and streamflow, and the highest incidence of droughts. Five of the six extreme multi-year droughts since 1900 occurred during the warm phase of PDO (Mote et al., 1999b, p. 32). Each oscillation's cool phase has the opposite effect, bringing average stream flows about 10% higher than the mean and the highest incidence of flooding. Four or five of the five highest-flow years recorded occurred when PDO was in its cool phase, three of them when ENSO was also in its cool phase.[4] When the two cycles are out of phase, streamflow tends to be near its long-term mean. A study of historical flooding in five smaller river basins found parallel results, a higher probability of flooding in both cool ENSO and cool PDO years, although the pattern was weaker than for the Columbia and not uniform. Snowmelt-dominated rivers, whose floods reflect season-long snow accumulation, showed it more than rivers with strong rainfall components, whose floods more typically reflect less predictable, individual extreme precipitation events. Moreover, the pattern was present for the likelihood but not the intensity of flooding. Still, the pattern suggests a limited ability to predict seasonal flood risk, which most management agencies in the region are not presently exploiting (Mote et al., 1999b, p. 31 and Table 4; Jones and Grant, 1996).

---

[4]The fifth of these years was 1997, the second-highest flow year. This year's status is ambiguous, because its extreme flow is one piece of evidence for the not yet resolved claim that PDO shifted back to cool phase in the mid 1990s.

Understanding the effects of projected warmer temperatures, wetter winters, and drier summers on the region's hydrology requires finer-scale analysis than GCMs alone can provide. To obtain this, two analyses were conducted linking a GCM projection to finer-scale models. In an analysis of snowpack for the entire Northwest region using the PNNL-RCM model described above, projected snowpack declined about 30% over the Northern Rockies and 50% over the Cascades by roughly 2050, the time that CO2 concentration had doubled (Leung and Ghan, 1999a, 1999b). Projected rise of the snowline and earlier spring melt are corroborated by broadly similar results from earlier GCM-driven studies of fine-scale hydrology models elsewhere in the American West (Giorgi and Bates, 1989; Giorgi et al., 1994; Matthews and Hovland, 1997). In a separate study of the Columbia basin alone, several climate models were used to drive detailed models of hydrology, and dam and reservoir operating rules, allowing an integrated examination of interactions between natural and managed responses to climate change. A striking result of these simulations was widespread loss of moderate-elevation snowpack, as Figure 10 shows. Typical snowcover on March 1 in the 2090s is projected to be about equal to present

snowcover on in mid-May, although deep snowcover in the upper basin still persists through June.

On snowmelt-dominated rivers like the Columbia, the very likely effect of these linked changes in temperature, precipitation, and snowpack will be to increase winter flow and decrease summer flow. Both precipitation and temperature matter. Winter flow increases both because there is more winter precipitation and because more of it falls as rain; summer flow decreases both because there is less snowpack and because it melts earlier in the spring. When changes in temperature and precipitation are considered separately, temperature — which climate models project with greater confidence than precipitation — has the larger effect on streamflows (Mote et al., 1999b, Figure 28; Nijssen et al., 1997).

While all the climate models studied agree on these seasonal effects on Columbia streamflow, they differ in the relative size of winter and summer changes, and consequently in whether total annual flow increases or decreases. Projections for annual flow in the 2020s in four models range from a 22% increase to a 6% decrease, with a mean 5% increase; for the 2050s, projected changes range from a 10% increase to a 19% decrease, with a mean 3% decrease. Figure 11 illustrates the range of projected seasonal flow shifts for all seven models in the 2050s.

To assess the socioeconomic effects of these changes in stream flow, a reservoir operations model was used to project how the reliability of different water-management objectives (i.e., the probability of meeting the objective in any year) changes under climate change. The model was first used to examine how present climate variability affects reliability of five uses, under two different sets of system operation rules: present rules, under which the practice is to grant highest priority to ensuring availability of hydroelectric energy sold on "firm" contracts,[5] and a set of alternative, "fish-first" rules that would give highest priority to maintaining minimum flows to protect fish. The effect of alternative operation rules for the system is to distribute risks of shortage among uses. The results showed that reliability is high for all objectives in favorable, high-flow years (i. e., cool PDO/La Niña years) and that one top-priority objective can be maintained at or near 100% reliability even in unfavorable, low-flow years (i. e., warm PDO/El Niño years), but that other uses suffer large reliability losses in unfavorable years. For

Projected Reduction in Columbia Basin Snowpack

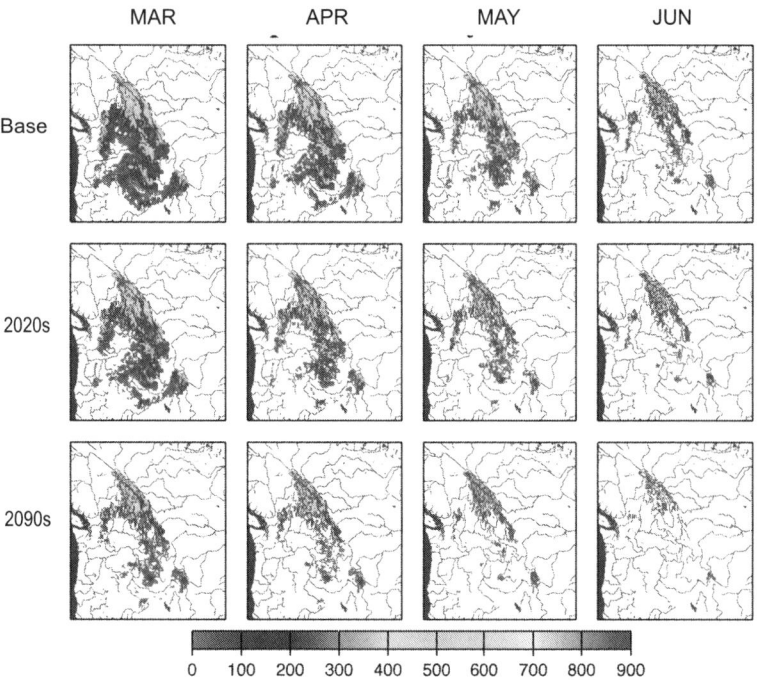

Figure 10: By the 2090s, projected Columbia Basin snowpack on March 1 will be only slightly greater than present snowpack on June 1. Simulations use the VIC hydrology model under the Hadley scenario. Units in millimeters. Source: Hamlet and Lettenmaier, 1999. – See Color Figure Appendix

---

[5]Despite recent policy changes to protect salmon, firm energy contracts still receive highest priority in practice and insufficient reservoir capacity is available to increase late-summer flows for fish.

example, the reliability of both the fish-flow objective under present rules, and the firm energy objective under "fish-first" rules, fall to about 75% in warm PDO/El Niño years. (Mote et al., 1999b, figure 32) Of the five objectives considered, the most sensitive were fish flows and recreational demand for full summer reservoirs, which both drop below 85% reliability when annual flow is only 0.25 standard deviations below its mean, a condition that could occur as often as four years out of ten. (Mote et al., 1999b, section 2.4.2, pp. 39-41.)

Though projected changes in total annual flows are relatively small, the projected seasonal shifts are likely to bring large effects. Smaller, rainfed, and mixed rain/snow basins on the west slope are already susceptible to winter flooding, and have significantly increased risk of flooding in the wetter winters of La Niña years (Mote et al., 1999b, Table 4). The historical severity of flooding, which is more influenced by single extreme events and less by season-long precipitation, shows no such climate signal. Projected warmer, wetter winters are likely to bring further increases in winter flooding risk in these basins, while continuing growth in coastal population will very likely increase the property vulnerable to such flooding. Impacts on human health are also possible, particularly where urban storm-sewer systems are inadequate to handle the increased runoff. No systematic assessment of changes in westside winter flooding risk under climate change, and potential consequences for property damage and human health, has yet been conducted.

In contrast, in large, snowmelt-dominated systems like the Columbia, even large increases in winter flows pose little increased risk of flooding, since existing management systems are adequate to respond to floods and even the highest projected

Projected Seasonal Shift in Columbia River Flow

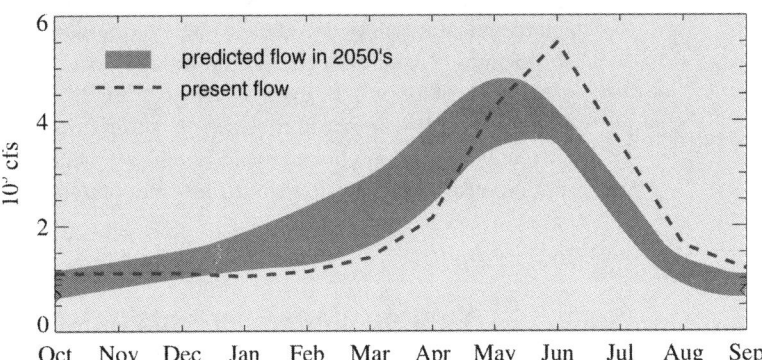

Figure 11: While only small changes are projected in annual Columbia flow, seasonal flow shifts markedly toward larger winter and spring flows, and smaller summer and fall flows. The blue band shows the range of projected monthly flows in the 2050s under the Hadley and Canadian scenarios and the two other 1998-vintage climate models used in the Northwest assessment (MPI and GFDL). Source: Mote et al. (1999), Summary, Figure 7. – See Color Figure Appendix

winter flows remain well below present spring peaks. But the system is much less robust to low flows than high. Reduced summer flows are likely to bring substantial reductions in both hydropower and freshwater availability by mid-century, exacerbating already-sharp allocation conflicts driven by population growth, expansion of irrigated farmland, and increasing priority for maintaining salmon habitat (Cohen et al., 1999).

Projections of changes in reliability for six objectives under present operational rules, using two climate models for the 2020s, and one for the 2090s, are shown in Table 2 (Hamlet and Lettenmaier, 1999). Under present rules, reliability of firm energy is projected to remain near 100%, while other uses suffer reliability losses up

Table 2: Changes in Reliability of Various Columbia Management Objectives, Assuming Present Operating Rules.
Source: Mote et al. (1999b), Table 6.

Objective	Base Case	2020s		2090s
		Hadley	Max-Planck	Hadley
Flood Control	98%	92%	96%	93%
Firm Energy	100%	100%	98%	99%
Non-firm Energy	94%	98%	87%	90%
Snake River Irrigation	81%	88%	76%	75%
Lake Roosevelt Recreation	90%	88%	79%	78%
McNary Fish Flow	84%	85%	79%	75%

to 10%, similar to the effect of PDO. The effects of rule changes, which will interact with both climate change and variability, are likely to be even larger. For example, "fish-first" rules would reduce firm power reliability by 10% even under present climate, and by 17% in warm-PDO years. Adding the projected long-term climate trend would very likely reduce reliability even more, but these interactions have not yet been quantified.

Increasing stresses on the system are highly likely to coincide with increased water demand, principally from regional growth but also induced by climate change itself. For example, an analysis of Portland's municipal water demand for the 2050s projected that climate change would impose an additional 5-8% increase in total summer demand (5% - 10% in peak day demand) on top of a 50% increase in sum-

Yakima Valley Irrigated Agriculture: Rigidity and Vulnerability

The Yakima Valley in south central Washington is one of the driest places in the United States, and contains some of its most fertile farmland. With annual precipitation of only about seven inches, the valley produces annual agricultural revenue of $2.5 billion, with the largest share from tree fruit. Fully 80% of the farmed area of 578,000 acres is irrigated. The basin's hydrology is strongly snowmelt-dominated with its main reservoirs in the mountains. With reservoir capacity equal to about half of annual demand, the region can tolerate a moderate one-year drought. Most farmers also have wells and pump groundwater in times of shortage, subject to state permits.

Drought in the valley, agricultural expansion, and water policy have all followed the PDO cycle. Water shortages have occurred eight times since 1945, all but once in warm-PDO years. Moreover, during the last cool phase (1945-1976) expectations of continued abundance of water led to sharp expansion of agriculture. In the subsequent warm phase farmers experienced substantial hardships, but nearly no contraction occurred. The first major decision allocating water among users was made in 1945, at the end of a warm phase period. Thereafter, no further controls were enacted until after 1979, when the warm phase had returned (Glantz, 1982). Shortage years since then have seen many hundreds of new wells dug, under emergency state permits that typically allow pumping only for the duration of the drought.

Several economic and institutional factors systematically increase the valley's vulnerability to drought. Division of water-right holders into "senior" users (who receive their full allocation every year) and "junior" users (who bear all risk of shortages) imposes large losses on junior users in shortage years – e.g., a 63% proration of water allocation and $140 million of losses in 1994 — and gives strong incentives for unsustainable pumping, depleting the region's groundwater and perpetuating the myth that the region has enough water. A progressive shift from annual crops to more lucrative perennials, such as tree fruit, grapes, and hops, has fur-

Figure 12: Irrigated agriculture in the Yakima Valley, where annual precipitation is about seven inches. Source: P. Mote, University of Washington

ther exacerbated vulnerability: the perennials consume more water, reduce farmers' flexibility to alter their planting for projected dry years, and put many years' investment at risk from a single drought, further increasing incentives to pump groundwater. Some promising initiatives to improve management of the region's water are underway, including state-subsidized investments to increase efficiency of junior users' irrigation systems, and a novel partnership between one junior and one senior irrigation district, but the valley still lacks a coherent basin-wide drought strategy. Moreover, if PDO is now re-entering its cool phase, making continuance of the ample supplies of the past two years likely, the region is likely to face renewed pressure to expand irrigated cropland. As happened during previous cool-PDO periods, such expansion would further increase the region's vulnerability to future recurrence of dry conditions (Gray, 1999).

mer demand from population growth (Mote et al., 1999b, section 2.4.3, p. 43). Such climate-related demand increase, which is also highly likely for other demands such as electric generation, irrigation, and salmon habitat preservation, would compound the climate-related supply decreases discussed above. Further assessment is required to quantify risks of shortfall under the interaction of regional growth, climate-driven demand increase, and climate-driven shifts in seasonal supply.

Adaptation Options

While the Columbia's infrastructure and institutional authority are able to manage high flows, at least up to some fairly high thresholds (Hamlet and Lettenmaier, 1999; Miles et al., 2000), the means for dealing with low flows are rigid and inadequate. The Pacific Northwest Coordinating Agreement, charged with allocating water under scarcity on the basis of defined priorities, is a weak and fragmented body with no one clearly in charge – Idaho is not

Seattle Public Utilities: Learning From Water Shortages

Seattle Public Utilities (SPU) experienced summer droughts and potential shortages in 1987, 1992, and 1998. Its responses to the three events illustrate institutional flexibility and learning. Summer 1987 began with full reservoirs, but a hot dry summer and a late return of autumn rains created a serious shortage in which water quality declined, inadequate flows were maintained for fish, and the main reservoir fell so low that an emergency pumping station had to be installed. In response, the City developed a plan to manage anticipated shortages with four progressive levels of response: advising the public of potential shortages and monitoring use; requesting voluntary use reductions; prohibiting inessential, high-consumption uses such as watering lawns and washing cars; and rationing.

Another drought came in 1992, following a winter with low snowpack but during which SPU had spilled water from its reservoirs to comply with flood-control rules. With a small snowmelt, reservoirs were low by the spring, and SPU invoked mandatory restrictions during the hot dry summer that followed. The resultant low demand caused water quality in the distribution system to decline. Low reservoir levels also caused a decline in the quality of source water, prompting a decision to build a costly ozone-purification plant.

Figure 13: Since 1992, Seattle Public Utilities has used information on year-to-year climate variability to help guide its seasonal forecasting and reservoir operations." Source: Seattle Public Utilities.

Their regrettable spilling of early 1992 alerted SPU to the risks inherent in following rigid reservoir rule curves. Since then, they have used a model that includes both ENSO and PDO to generate probabilistic projections of supply and demand six to twelve months in advance. Using this model during the strong El Niño of 1997-1998, they undertook conservation measures early in the year, including both weekly public announcements of supply conditions and allowing higher than normal winter reservoir fill. When 1998 brought a small snowmelt and a hot dry summer, these measures allowed the drought to pass with the public experiencing no shortage.

In integrating seasonal forecasts into its operations, SPU is an uncommonly adaptable resource-management agency. But it still has a long way to go in adapting to longer-term climate variability and change. SPU presently projects that new conservation measures will keep demand at or below present levels until at least 2010, while conservation measures and planned system expansion (including a connection with a neighboring system) will maintain adequate supply until at least 2030 (A. Chinn, Seattle Public Utilities, personal communication, 2000). Over this period, climate change is likely to have significant effects on both supply and demand, but is not yet included in planning. The warmer drier summers projected under climate change are likely to stress both supply and demand, requiring earlier capacity expansion and triggering the more restrictive conservation provisions more often (Gray, 1999). Moreover, it is possible that the recently suggested shift to cool PDO could mask this effect for a couple of decades, risking sudden appearance of shortages when PDO next shifts back to its warm phase.

even a member. Moreover, the persistence of the prior appropriation doctrine under western water law rigidly maintains large allocations to low-value uses, hindering attempts to rationalize use under present and increasing scarcity.

Managing under projected future scarcity is highly likely to require some combination of reducing demand, increasing supply, and reforming institutions to increase flexibility and regional problem-solving capacity. Demand for water can be reduced through various technical means (e.g., more efficient irrigation methods, changes in agricultural land management, or high-efficiency plumbing fixtures in new construction) and through various policy approaches (e.g., tax incentives for conservation investments, revision of rate structures). The most promising approaches to encourage conservation, however, would be development of institutions to allow reallocation of water to higher-value uses, particularly in times of shortage. Although the barriers to such a major shift of ownership rights are formidable, the Northwest could gain insights from prior western experience with water banks and contingent marketing (Huffaker, Whittlesey, and Wandschneider, 1993; Miller, 1996; Miller, Rhodes, and MacDonnell, 1997). Supply can likewise be increased through various technical and policy means, such as developing groundwater sources or using groundwater recharge for water storage; improving system management by using seasonal climate forecasts (Callahan et al., 1999); promoting optimal use of existing lower-quality supplies, e.g., by delivering reclaimed non-potable water for some uses; or developing non-hydro electric generating capacity; or if permitted by both governments, negotiating water purchase from Canada.

Increasing institutional flexibility is an essential component of response, but is rendered difficult by the fragmentation of the current system. In a survey of water managers on their use of climate forecasts in planning, most stated that they had limited flexibility to use even ENSO forecasts to take advantage of predictably higher-flow years, and most were completely unaware of PDO, whose effect on flows appears to be as large as that of ENSO (Mote et al., 1999b, Section 2.5.3, pp. 47-49.). Indeed, long-term assessment of institutional responses suggests that in at least some cases, responses have served to increase, rather than decrease, vulnerability to inter-decadal variability (see Yakima Valley box). Moreover, long-term climate change is not yet used in planning decisions, even for investment in infra-structure with expected lifetimes of many decades.

2. Salmon

Salmon are anadromous fish, meaning that they swim upstream to spawn after spending most of their adult lives at sea. After hatching, young salmon remain in the stream for a few weeks to several years, depending on the species, then swim downstream to the ocean. Most species make this trip in spring or early summer, taking advantage of the high streamflows that accompany peak spring melting (in snowfed rivers) and arriving at roughly the onset of coastal and estuarine upwelling that fuels marine food-chain productivity. In the ocean the salmon grow to adulthood and live for several months to six years before returning to their spawning grounds. Most die in their natal streams after spawning, thereby delivering marine-derived nutrients that are now recognized as important inputs to stream and riparian ecosystems.

Northwest salmon stocks have been highly stressed for decades by intense fishing pressure and threats to their stream habitats including urbanization, sedimentation and pollution of streams, wetland draining, and dam building. Construction of the Grand Coulee and Hell's Canyon dams eradicated all salmon stocks above these points on the Columbia and Snake Rivers respectively. Fish ladders on other dams are only partly effective at allowing fish to pass, and dams also degrade salmon habitat by changing free-running rivers into chains of lakes, warming in-stream temperatures, reducing dissolved oxygen, and altering sediment loads.

These factors have brought regional salmon stocks to widespread decline (Myers et al., 1998), despite massive efforts to restore habitat and to supplement wild stocks with hatcheries, which reflect salmon's status not just as a commercially important fish but as a regional cultural icon. Over the past century, Pacific salmon have disappeared from about 40% of their historical breeding range in Washington, Idaho, Oregon, and California, and many remaining populations are severely depressed. The decline is not universal: populations from coastal rather than interior streams, from more northerly ranges, and with relatively short freshwater rearing periods have fared better than others. In many cases, the populations that have not declined are now composed largely or entirely of hatchery fish (National Research Council, 1996; Beechie et al., 1994; Bottom, 1995).

In March 1999, eight new salmon stocks were listed under the Endangered Species Act (ESA) as threatened and one as endangered, bringing the number of salmonid stocks listed to 26. The new listings included Puget Sound Chinook, the first-ever ESA listing of a species inhabiting a highly urbanized area. Just as prior listings of Columbia and coastal Oregon stocks severely constrained forest and water management, it is possible that the impacts of this new listing on the Seattle economy will be large, but it will be some years before these impacts are known (Klahn, 1999).

Salmon are sensitive to various climate-related conditions, both inshore and offshore, at various times of their life cycle. Eggs are vulnerable to stream scouring from floods, and migrating juveniles must make the physiological transition from fresh to salt water and require food immediately on reaching the ocean. According to a long-standing but unconfirmed hypothesis, their fate is keenly sensitive to the timing of their arrival relative to the onset of summer northerly winds and the resultant upwelling of nutrient-rich deepwater and spring phytoplankton bloom: either too early or too late, and their survival is imperiled from insufficient food (Pearcy, 1992). More recently, it has been suggested that their survival is predominantly controlled by the balance between predator and baitfish populations at the time of their arrival, which determines predation pressure on salmon (Emmet and Brodeur, 2000).

Although the relative contributions of and interactions between climate and non-climate factors, and inshore and offshore conditions are highly uncertain, salmon stocks throughout the North Pacific show a strong association with PDO (Figure 14) (Hare et al., 1999; Mantua et al., 1997; Francis, 1997; Francis et al., 1998; Reeves et al., 1989). Salmon in the Northwest are more abundant in the cool PDO phase, less abundant in the warm phase, while Alaska salmon show the opposite pattern. When the PDO shifted from cold to warm in 1977, catches in the Northwest dropped sharply while Alaskan catches soared.[6] The mechanisms for this observed climate effect on stocks are poorly known, and probably include some effects of both freshwater and marine changes. It is speculated that coastal waters off Washington and Oregon during the warm phase are warmer and more thermally stratified, and consequently poorer in nutrients. Although ENSO and PDO have similar effects on ocean and terrestrial environments in the Northwest, the signal of PDO in salmon stocks is much stronger than that of ENSO, suggesting that the mechanism involves persistent warm-water conditions over several years. The PDO signal is much weaker for Puget Sound salmon than for stocks that exit directly from rivers into the open

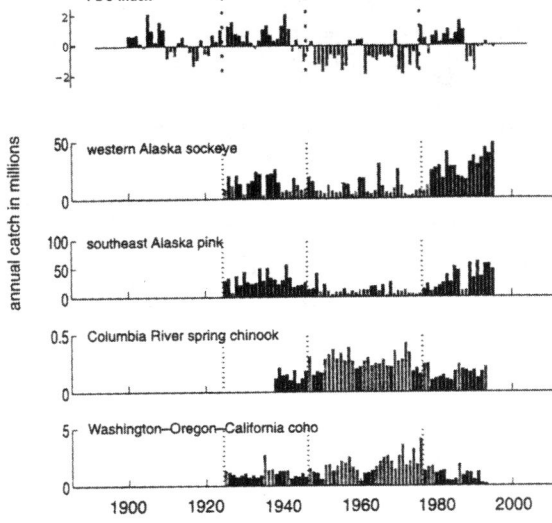

Salmon Catches and Inter-decadal Climate Variability

Figure 14: 20th century catches of Northwest and Alaska salmon stocks show clear influence, in opposite directions, of the Pacific Decadal Oscillation. Source: Mote et al. (1999), Figure 36, pp. 56. – See Color Figure Appendix

ocean, suggesting that the gradual increase of salinity experienced by juveniles passing through an estuarine environment may increase their resilience (Pinnix, 1998).

Climate models presently lack the detail to project changes in many specific factors in the marine environment that are most important for salmon, such as the timing of seasonal coastal upwelling, variations in coastal currents, and vertical stability of the water column. But where climate models are more informative, in salmon's inshore and estuarine habitats, their projections are uniformly unfavorable. Increased winter flooding, reduced summer and fall flows, and warmer stream and estuary temperatures are all harmful for salmon (Baker et al., 1995; Bottom, 1995. Earlier snowmelt and peak spring streamflow will likely deliver juveniles to the ocean before there is adequate food for them, unless the onset of summer northerly winds is also advanced under climate change, but climate models cannot yet address this question. One study has suggested that oceanic warming from even a CO_2 doubling may push the range of some salmon north out of the Pacific entirely (Welch et al., 1998), although recent studies suggest the notion of direct ocean thermal limits to salmon survival is too simplistic (Emmet and Schiewe, 1997; Bisbal and McConnaha, 1998; Hargreaves, 1997). Data from temperature-recording tags show that salmon move hourly and

---

[6]Catches are a good indicator of stocks, because catch variation since the 1930s is almost entirely due to stock fluctuations, not variation of fishing effort (Beamish and Bouillon, 1993)

Figure 15: Old-growth Ponderosa Pine forest with open, grassy understory near Bend, Oregon. Source: S. Garrett and the McKay Collection, Bend, OR.

daily through wide temperature ranges, presumably by moving between surface and deep waters. This result, and high-seas sampling of salmon, suggests that the effect of ocean temperature on salmon is not primarily direct, but operates through changes in food supply (Walker et al., 1999; Pearcy et al., 1999).

Salmon are already beset with a long list of man-made problems, to which climate change is a potentially important addition. At sea, fishing pressure has recently been sharply curtailed, but wild salmon face intense competition from hatchery fish, some of which are being released at ten times the rate of natural smolt migrations. Onshore, the effects of present climate variability on salmon in streams are swamped by clear-cutting, road building, and dams. The effects of future climate trends, and their potential to interact with other stresses, are not known. Neither is the extent to which current and proposed measures to protect salmon by restoring stream habitat and changing dam operations will restore depleted stocks and increase their resilience to climatic stresses. If endangered-species listings and public concern prompt a strong restoration response, it would likely take the form of far-reaching restrictions on land use near rivers and streams. Such measures would very likely have far-reaching social and economic impacts, which could greatly exceed the direct effects of decreases in salmon abundance.

Adaptation Options
At present, the only climatic effects on salmon that are sufficiently understood to allow an informed response involve warming of streams (McCullough, 1999). Maintaining forest buffers for shade along

banks and operational changes on managed rivers can reduce present warming (about 4.5° F, largely due to dams) (Quinn and Adams, 1996), and could potentially slow future warming under climate change, although such measures would ultimately be overwhelmed by continued climate warming. Though other mechanisms of climatic effect are less known, observed fluctuations of stocks with PDO suggest applying a conservative bias to allowable-catch decisions during warm-PDO years. Measures to reduce general stress on stocks, such as changing dam operations to provide adequate late-summer streamflows, might increase salmon's resilience to other stresses, including climate, although maintaining such flows is highly likely to become increasingly difficult under regional warming that shifts peak streamflows to earlier in the year. More forceful measures would include removing existing dams, as now proposed for four dams on the lower Snake River, and accepting the resultant reduced ability to manage summer water shortages. Salmon have evolved a great diversity of life histories and behaviors to thrive in a highly variable and uncertain environment. Maintaining that diversity, which is highly likely to require even greater efforts to preserve healthy, complex freshwater and estuarine habitat, is likely among the most effective options to enhance salmon's resilience to climate change.

3. Forests

Evergreen coniferous forests dominate the landscape of much of the Northwest. In the west, coniferous forests cover about 80% of the land, including some of the world's largest trees and most productive forests, and about half the world's temperate rainforest. Belying their lush appearance, on most sites these forests are constrained by moisture deficit during the warm dry summers, which limits seedling establishment and summer photosynthesis, and creates favorable conditions for insect outbreaks and fires (Agee, 1993; Franklin, 1988). Forests of the dry interior operate under an even more severe summer soil-moisture deficit (Law et al., 1999), which is consequently the dominant factor controlling the species distribution and productivity of forests throughout the region (Zobel et al., 1976; Grier and Running, 1977; Waring and Franklin, 1979; Gholz, 1982).

Forests throughout the Northwest have been profoundly altered by human intervention over the past 150 years. West of the Cascades, forests have been cleared for conversion to other land uses or clear-cut for timber and replanted, replacing massive old-growth forests with young, even-aged managed forests. By various estimates, 75 to 95% of original old-growth forest has been logged, and much of what remains is in small fragmented stands (Mac et al., 1998, p. 646). The transition is estimated to have released to the atmosphere 2 billion metric tons of carbon over the century (Harmon et al., 1990). East of the Cascades, decades of intensive grazing, fire suppression, and selective harvesting of mature trees have transformed the former open, park-like forest of ponderosa pine, Douglas fir, and western larch into a dense mixed forest overstocked with shade-tolerant pines and firs. The new forest mix is highly susceptible to insect outbreaks, disease and catastrophic fire (Mason and Wickman, 1988; Lehmkuhl et al., 1994). Of the ponderosa pine forest that formerly comprised three quarters of east-side forests, more than 90% has been logged or lost (Figure 15). While the former open forest structure was maintained by frequent, low-intensity fires (in contrast with western forests, whose natural regime was of catastrophic fires at intervals of several centuries), fire suppression in the east has allowed large accumulation of fuel. These high fuel loads increase the risk of extreme fires that replace stands over large areas (Quigley et al., 1996). The effects of these human activities overwhelmed climatic effects on Northwest forests during the 20th century, and continued forest management is highly likely to interact strongly with climate effects during the 21st century.

Tree growth can show a clear effect of climate variability (which is why tree rings can provide a useful record of past climate), most pronounced in stands near their climatic limits, e.g. at the upper (cold) or lower (hot and dry) timberline. Figure 16 shows the effect of 20th century climate variability on three Northwest conifer populations, one at high and two at low elevation (Peterson and Peterson, 2000). Near the upper timberline, where trees are not moisture-constrained, growth shows strong positive correlation with PDO. This is because positive PDO periods tend to have lighter snowpack and consequently an earlier start to the high-elevation growing season, promoting tree growth and upward expansion of forests to colonize sub-alpine meadows (Franklin et al., 1971; Rochefort et al., 1994). Near the lower timberline, the opposite relationship is present. Growth is negatively correlated with PDO, because the warm dry winters and light snowpack of positive-PDO periods increase summer drought stress, which is the limiting factor at these elevations (Little et al., 1994; Peterson, 1998). In other regions, such as intermediate elevation stands in the interior, and the western hemlock and pacific silver fir zones west of the Cascade crest, present

Tree Growth and Inter-Decadal Climate Variability

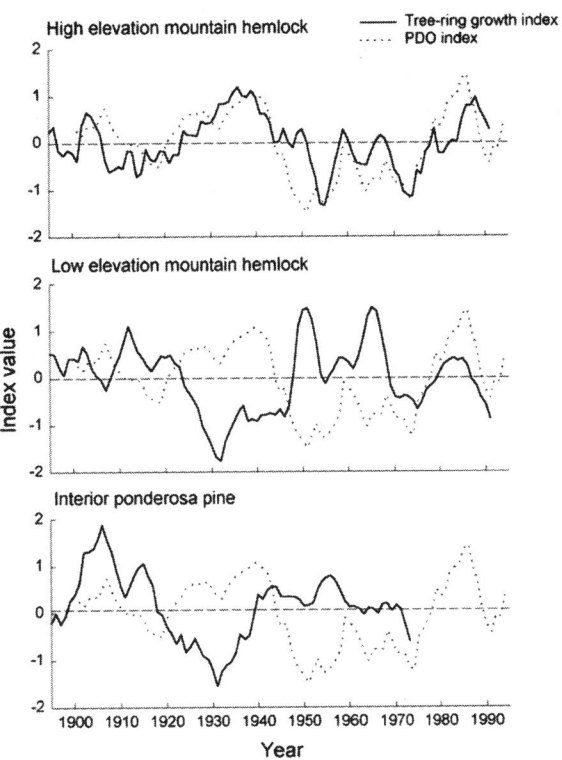

Figure 16: Trees near their climatic limits show strong signals of inter-decadal climate variability. Those near the upper treeline grow best in warm-PDO years because snowpack is lighter, while those near the dry lower treeline grow worst in warm-PDO years, because of summer moisture deficit. Source: Peterson and Peterson, 2000.

climate variations have little discernible influence on the structure and composition of most mature stands, which – once established – have substantial ability to buffer themselves against climate variation. In such stands, competition and other factors obscure present climate signals in individual trees (Brubaker, 1986; Dale and Franklin, 1989).

The principal effect of climate on these forests has come not through direct effects but indirectly, through changes in disturbance by fire, insect infestation, and disease (Overpeck et al., 1990; Fosberg et al., 1992; Ryan, 1991). Major disturbances can reset forests to their establishment stage, when trees are the most sensitive to adverse environmental conditions (Gardner et al., 1996). For insect and disease mortality, the available 20th-century data are inadequate to quantify the region-wide effects of climate variability, but smaller-scale studies have shown strong correlations of both bark beetle and defoliator outbreaks with severe drought conditions (Swetnam and Lynch, 1993). For fire, total

Annual Northwest Area Burned in Forest Fires over the 20th Century

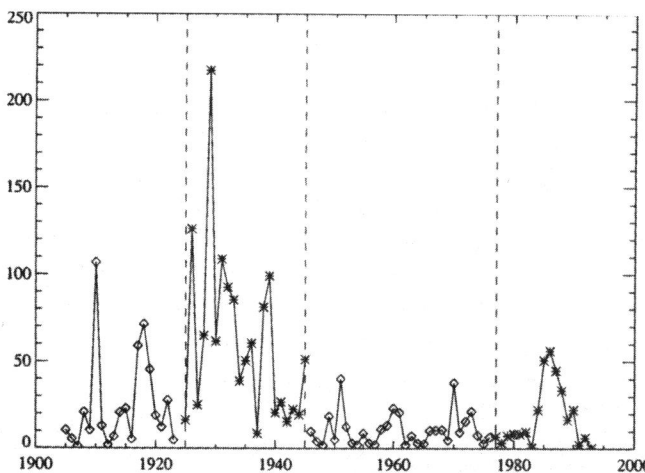

Figure 17: Prior to modern fire suppression, annual Northwest area burned in forest fires showed a clear association with inter-decadal climate variability. Dashed lines show PDO regime shifts. Source: Mote et al. (1999), p. 65.

Projected Northwest Vegetation Changes under two Ecosystem Models, 2100

MAPSS: Percent Change in Leaf Area

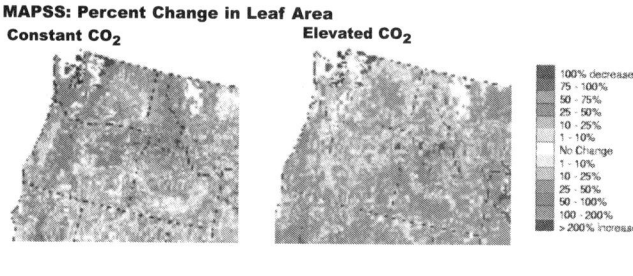

MC1: Percent Change in Vegetation Carbon

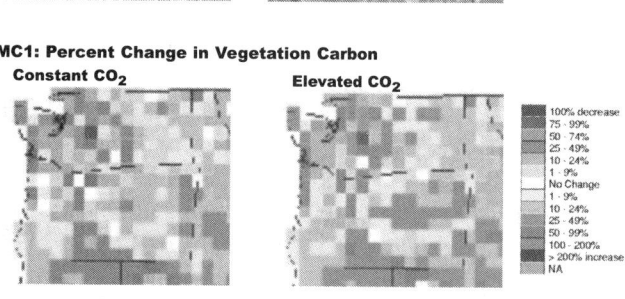

Figure 18: Under the Hadley scenario, the MAPSS (top row) and MC1 (bottom row) models project expansion of forests east of the Cascades and contractions to the west, assuming increased water-use efficiency under elevated atmospheric CO_2 (right column). When no such increase is assumed (left column), projections are nearly unchanged in the MC1 model, but change to a large contraction region-wide in the MAPSS model. Source: Bachelet et al (2000). – See Color Figure Appendix

Northwest forest area burned shows a significant region-wide association with PDO and with the Palmer Drought Severity Index, especially before the introduction of widespread fire suppression. For example, as Figure 17 shows, the warm-PDO period of 1925-1945 had much more area burned than the

cool-PDO periods immediately preceding and following it (Mote et al., 1999a). The effect of ENSO on fire is less clear. In contrast with the strong ENSO signal observed in wildfire in the Southwest US (Swetnam and Betancourt, 1990), no region-wide effect is evident in the Northwest. At smaller scales, however, one study of historical fire data and ongoing tree-ring studies suggest a significant ENSO effect (Heyerdahl 1997; T. Swetnam, personal communication, 1999).

Under projected future climate change, the effect on Northwest forests is highly likely to reflect complex interactions between several factors, some of which vary with particular sites. The direct effect of projected warmer summers without substantial increase in rainfall would be to increase summer soil moisture deficit, resulting in reduced net photosynthesis and tree growth, increased stress and tree mortality, and decreased seedling survival. Reduced snowpack has different effects at different sites: it extends the growing season and facilitates seedling establishment where snowpack is presently heavy, but reduces growing-season moisture availability and consequently increases drought stress in dry areas (Peterson, 1998). One early empirical study used observed associations between existing forest communities and local climate to project impacts of future climate change on Northwest forests. This study projected that forested area in the Northwest would contract, principally through forest dieback and sagebrush-steppe expansion at the dry lower treelines east of the Cascades (Franklin et al., 1991).

It is likely, however, that the effects of increased summer drought will be offset to some degree by wetter winters, or by the direct effects of elevated atmospheric CO_2 concentration. Though drought stress occurs in late summer, its severity depends principally on winter and spring precipitation, because forests rely on moisture stored seasonally in deep soil layers to offset summer water deficits. Under climate model projections, the forest growing season begins several weeks earlier in the spring than at present, making some increased winter precipitation immediately available to forests. Beyond this water put to use immediately, wetter winters could mitigate summer drought stress still further, if the soil has enough storage capacity to hold the additional water until it is needed in the late summer. Although there are some indications that Northwest forest soils below the snow line are already fully recharged under present winter rains, so precipitation increases would be lost as runoff (Harr, 1977; Jones and Grant, 1999; Perkins, 1997), the advance of growing season makes the significance of these results for seasonal water storage under climate change ambiguous (Bachelet et al., 1998). Elevated CO_2 concentration may also possibly mitigate productivity losses from summer drought stress, by increasing photosynthetic efficiency or – likely more important in water-limited Northwest forests – by increasing trees' water-use efficiency through reduced stomatal conductance. Laboratory and field studies have shown increases in both net carbon uptake and water-use efficiency in many plants (Bazzaz et al., 1996), but most such studies have examined agricultural crops, grasses, or tree seedlings. The evidence available on mature trees is quite limited. One experimental study in a young pine plantation in North Carolina found that elevated CO_2 brought increased carbon assimilation with no change in water use - an increase in water-use efficiency, but not through the expected mechanism of increased stomatal resistance (Ellsworth, 1999). The applicability of this result to projecting responses of mature connifer forests in the more moisture-stressed Northwest remains unknown. Cool coniferous forest systems appear to be among the least responsive in aggregate to elevated CO_2 (Ellsworth, 1999; Mooney et al., 1999).

Process-based models are needed to represent and quantify these effects, because empirical studies can only observe forests' responses to the present range of climatic conditions with present CO_2 concentration (VEMAP Members, 1995). With assumptions of increased water-use efficiency, ecological models driven by the Hadley, Canadian, and other climate-change scenarios project opposite effects east and west of the Cascade Crest. Cool coniferous forests to the west are projected to contract, with reductions in vegetation carbon or leaf area exceeding 50% in some areas and replacement by mixed temperate forests over substantial areas. Dry forests to the east are projected to expand, with increases in vegetation carbon or leaf area also exceeding 50% in some locations (Neilson and Drapek, 1998; Daly et al., 2000; Neilson et al., 1998). The increases in the east are relative to a smaller present biomass, and so are smaller in absolute terms. While these model results reflect substantial recent progress, they have significant remaining weaknesses and uncertainties. For example, none adequately represents fog, which can comprise as much as half the water input to forests on some coastal and hillside sites. Moreover, results of different models in the Northwest differ strongly in their sensitivity to the water-use efficiency assumption. When no increase in efficiency is assumed, model results range from a very similar pattern of expansion in the east and contraction in the west, to a region-wide contraction that resembles the projection of the earlier empirical analysis (Bachelet et al., 2000).

The magnitude and consequences of future changes in water-use efficiency represent the most important uncertainties in projecting the climate response of Northwest forests over the next century. While the balance of preliminary evidence suggests at most a small water-use efficiency increase in Northwest coniferous forests due to enhanced CO_2, model results differ substantially in whether the effect of this unknown on the extent, density, and species distribution of Northwest forests is large or small. Over the longer term, there are preliminary indications that, increased evapotranspiration under warmer temperatures could possibly overwhelm even the largest assumptions of increased water-use efficiency, leading to substantial forest dieback.

Northwest Forest Fire Projections

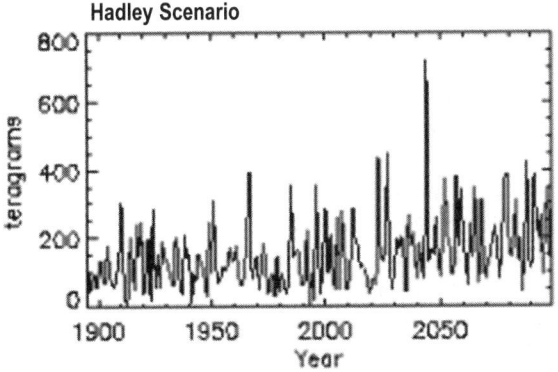

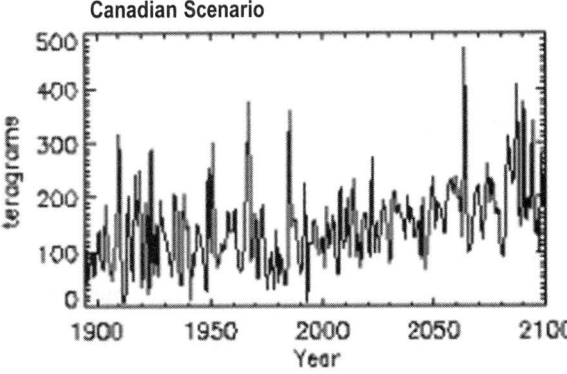

Figure 19: The MC1 ecosystem model projects substantial increases in the amount of forest biomass burnt annually in the Northwest under both the Hadley and Canadian scenarios. Source: Bachelet et al. (2000)

Climate Change Projected for 2050 vs Observed 20th Century Variability

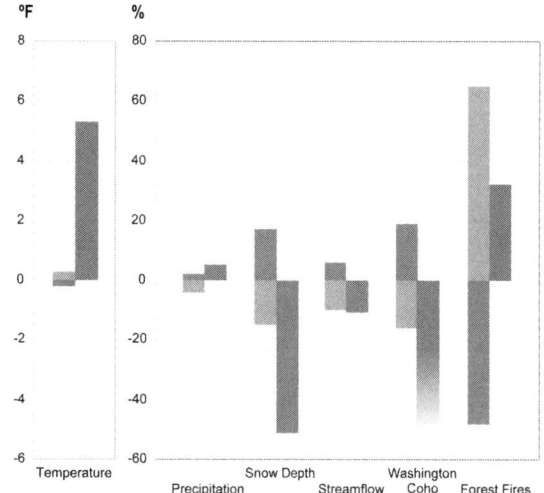

Temperature	Change in annual average regional temperature (°F)
Precipitation	Change in annual average regional precipitation (%)
Snow depth	Change in average winter snow depth at Snoqualmie Pass, WA (%)
Streamflow	Change in annual streamflow at The Dalles on the Columbia River (corrected for changing effects of dams) (%)
Salmon	Change in annual catch of Washington Coho salmon (%)
Forest fires	Change in annual area burned by forest fires in WA and OR (%)

Figure 20: This chart compares possible Northwest impacts from climate change by the 2050s with the effects of natural climate variations during the 20th century. The orange bars show the effects of the warm phase of the Pacific Decadal Oscillation (PDO), relative to average 20th century values. During warm-PDO years, the Northwest is warmer, there is less rain and snow, stream flow and salmon catch are reduced, and forest fires increase. The blue bars show the corresponding effects of cool-phase years of the PDO, during which opposite tendencies occurred.

The pink bars show projected impacts expected by the 2050s, based on the Hadley and Canadian scenarios. Projected regional warming by this time is much larger than variations experienced in the 20th century. This warming is projected to be associated with a small increase in precipitation, a sharp reduction in snowpack, a reduction in streamflow, and an increase in area burned by forest fires. Although quite uncertain, large reductions in salmon abundance ranging from 25 to 50%, are judged to be possible based on projected changes in temperature and streamflow. Source: based on Mote et al., 1999, pp. 27. — See Color Figure Appendix

The largest effects of future climate variability or change on Northwest forests are likely to arise from changes in disturbances (McKenzie et al, 1996; Torn and Fried, 1992). Two dynamic vegetation models now incorporate fire models, which project significant increases in forest area and biomass burned in the Northwest interior by 2100. Changes in other disturbances, such as wind, insects, and disease are also highly likely under climate change and biomass burned in the Northwest interior by 2100 (e.g., results from the MC1 model in Figure 19). The

potential character of these disturbances under climate change is not yet adequately understood. Some likely effects have been suggested for individual disturbance processes, based on present disturbance patterns and tree-ring studies. For example, general warming is likely to encourage northward expansion of southern insects, while longer growing seasons are highly likely to allow more insect generations in a season. Forests that are moisture stressed are more susceptible to attack by sap-sucking insects such as bark beetles. It is also possible that drought stress encourages attack by foliage eaters such as spruce budworm, although the evidence for this is divided. Some studies show attack by foliage eaters increasing when foliage is lush, others when it is stressed (Swetnam and Betancourt, 1998; Thomson et al., 1984; Kemp et al., 1985; Swetnam and Lynch, 1993; Larsson, 1989; Price, 1991). Very little is understood, however, about crucial questions of the interactions between multiple disturbances, e.g., between insect attack and fire, under projected climate change and the changes in forest character that follow (Neilson et al., 1994). Finally, it is crucial to note that ecosystem models only project potential vegetation, the vegetation that would be present on a site in the absence of human intervention. In the Northwest, forest management and land-use change are presently, and are likely to remain, predominant factors shaping the structure, species mix, and extent of forest ecosystems (Franklin and Forman, 1987). Interactions between these human-driven factors and the multiple direct and indirect pathways of climate influence on Northwest forests are essentially unexamined, and are key areas for research (Sohngen et al., 1998).

Adaptation Options

In contrast to water and salmon, where management already reflects at least limited awareness of climate variability, a survey of Northwest forest managers suggests they regard climate as unimportant, because mature stands are resilient to wide climatic ranges and because 40-70 year timber harvest rotations average out the effects of shorter-term climate variation (Mote et al., 1999b, Section 4.5, pp. 73). Long-term climate trends are highly likely to make these assumptions invalid. Trees are likely to mature in a climate substantially different than when they were planted, possibly requiring changes to many dimensions of forest management even beyond those recently adopted due to endangered species concerns. Required adaptations might include: planting species best adapted to projected rather than present climate (e.g., planting Douglas fir on suitable sites in the silver fir zone), or with known broad climatic resilience;

further measures to restore and maintain complexity of forest structure and composition within intensively managed areas; managing forest density for reduced susceptibility to drought stress; and using precommercial thinning, prescribed burning, and other means to reduce the risk of large, high-intensity disturbances and to facilitate adaptation to changed climate regimes.

Managing forests effectively under these conditions is very likely to require increased capacity for long-term monitoring and planning. In addition, the importance of seasonal and interannual climate variability is highly likely to increase when forests are also stressed by long-term trends. Increased understanding of climate variability and predictive skill can allow projected periods of drought stress and fire risk to be factored into short-term forest-management decisions such as timing and species of planting, and use and timing of prescribed burning.

Maintaining forest ecosystem services and biological diversity is also likely to grow increasingly challenging under climate change. Options for doing so would include establishment of further protected areas, incorporating the maximum possible diversity of topography and landscape; active measures to promote species migration and maintain diversity, even in non-commercial forests; and further reduction in the intensity of commercial harvest.

4. Coasts

The Pacific Northwest has three distinct coasts. The inland marine waters of Puget Sound and the Strait of Juan de Fuca are bounded by narrow beaches backed by steep bluffs, and contain large areas of intensive shoreline development. The open Pacific coast has rocky bluffs and headlands punctuated by small pocket beaches, with wide beaches and sandspits in southern Washington and coastal sand dunes in central Oregon. This coast has generally low-intensity development, with no major cities and many stretches in parks, Indian reservations, and large undeveloped parcels, but contains a few pockets of increasing development. Finally, the coastal estuaries of Oregon and southern Washington are principally bordered by farmlands and small towns at river-mouths, and support extensive shellfish aquaculture and harvesting in their shallow waters and on their broad mudflats.

Present stresses in coastal regions include: bluff landsliding from heavy winter rains, principally on the steep hillsides around Puget Sound (Tubbs, 1974); erosion of beaches, barrier islands, sandspits and dunes on the open Pacific coast, principally due to winter storm waves; coastal flooding near river-mouths, particularly in areas that have neither upstream protection nor sufficient height above high tide; loss of wetlands to development or erosion; and invasion by exotic species, particularly in the coastal estuaries. Extensive development on coastal bluffs and near beaches, mainly in Puget Sound but increasingly also along the Pacific coast, has placed considerable valuable property at risk from erosion and landslides. Coastline near major river-mouths is sensitive both to ocean erosion and to variation in the rivers' sediment loads, which have increased on some rivers from adjacent clear-cutting, and decreased on others (including the Columbia) from damming (Canning, 1991; Canning and Shipman, 1994; Field and Hershman, 1997; Park et al., 1993).

Little long-term data are available on coastal effects of climate variability, but a few effects are evident. Severe storm surges and erosion events on the open coast occur on average every five years. These appear to be associated with El Niño events, which increase erosion both by raising sea level for several months and by changing the direction of winds and waves from westerly to southwesterly (Komar and Enfield, 1987). The 1997-1998 El Niño, for example, brought rapid erosion to a 1,000-foot built up segment of the sandspit on which Ocean Shores, Washington is situated, reversing several centuries of slow growth and requiring emergency construction of an armored beach fill. One Oregon study found that construction of shore protection measures follows an ENSO cycle, increasing sharply in the years immediately following a strong El Niño (Good, 1994). As well as causing extreme erosion events, the elevated sea level during El Niño events also increases inundation risk in the low-lying areas of southern Puget Sound such as the city of Olympia, where it reaches 2-4 inches (5-10 cm). In contrast, La Niña events bring reduced erosion on the open coastline, but their heavier than normal winter rainfalls increase soil saturation and landsliding risk on coastal bluffs (Gerstel et al., 1997). For example, the four years with highest landslide incidence in the Seattle area were all La Niña winters (1933-4, 1985-6, 1996-7, 1998-9).

Suggestive indications of a PDO signal in coastal phenomena have also been noted, although these remain speculative. For example, the warm decade of the 1980s, following the shift to warm-phase PDO in the late 1970s, was marked by two striking shifts in the ecosystem of Willapa Bay in southern Washington. The exotic cordgrass (Spartina), which was introduced to the bay nearly 100 years earlier, began a rapid expansion that threatened local species for the first time (Feist and Simenstad, 2000). After several productive decades, the condition of commercially important oysters in Willapa Bay also began a substantial decline in the late 1970s, though other factors such as pollution in the bay could also be responsible (Mote et al., 1999b, section 5.2, pp. 77-78).

Several coastal effects of future climate change have been identified, although detailed assessment of these remains to be done. Future climate warming is likely to raise mean sea level 10 to 35 inches in the 21st century, as opposed to the 4 to 8 inch rise of the 20th century. The apparent rise will differ from place to place, partly due to regional differences in ocean circulation and heat content — for example, the Hadley model projects a larger sea-level rise on the Pacific than the Atlantic coast of North America — and partly due to local variation in the rate of uplift or subsidence of the land surface. In the Pacific Northwest, regions of uplift are centered at the mouth of the Strait of Juan de Fuca, rising at 0.1 inch (2.5 mm) per year, and the mouth of the Columbia River, rising by 0.06 inch (1.7 mm) per year, while southern Puget Sound is subsiding at up to 0.08 inch (2 mm) per year (Shipman, 1989). Consequently, risks of sea-level rise are greatest in southern Puget Sound. Here, low-lying settlements are already at risk of inundation and existing shoreline protection is inadequate even for the high end of projected mean sea level rise, when the far greater risk is from mean rise combined with storm surge (Craig, 1993).

Higher mean sea level is also likely to increase sediment erosion and redistribution on the open coast, which possibly may be further amplified by projected shifts in prevailing wind direction to resemble sustained El Niño conditions. In addition, projected heavier winter rainfall is likely to increase soil saturation, landsliding, and winter flooding. All these changes would increase the risk to property and infrastructure on bluffs and beachfronts, and beside rivers. Climate change could also bring continued changes in coastal and estuarine ecosystems through changes in runoff and warmer water temperatures, with possible increased risk of exotic species introduction or health risks from shellfish contamination.

Adaptation Options

The most effective adaptation strategies for coastal climate impacts involve conserving remaining natural coastal areas and placing less property at risk in low-lying or flood-prone areas, on beaches, or on and below unstable slopes. Although these general adaptation strategies are well known, a series of interviews with coastal-zone managers about how they use climate information suggested that the coastal management system is not particularly adaptable even to current climate variability and its associated risks (Mote et al., 1999b, section 5.5, pp. 80-81). Most managers reported that they are seriously constrained in their ability to incorporate climate in planning, indeed that any climate considerations are overwhelmed in their planning by the present and potential endangered listings of various salmonid species. Many managers did not even view the long-range threat of flooding or inundation as a significant risk to the resources they managed. As for restricting development in vulnerable locations, there appears to be little inclination to move in that direction. A weaker and perhaps more feasible alternative would be to assign more of the risk of living in a coastal zone to property-owners, through incorporating geological assessment into property-insurance rates.

ADDITIONAL ISSUE

Agriculture

Due to limited time and resources in this first Assessment, the Northwest regional study was restricted to the four critical issues discussed above. The choice of these four reflected the concerns of stakeholders in the regional workshops, but was emphatically not intended to imply that these are the only important areas of impact in the Northwest. Other areas of potentially significant impact not covered in this Assessment might include, for example, human health, urban quality of life, recreation, and agriculture. Agriculture, because of its importance in many parts of the region, is a particularly conspicuous omission, and an obvious priority for subsequent assessment. A very preliminary discussion is provided here, based on work conducted by the agriculture sector team of the National Assessment.

The Agriculture Sector Assessment included modeling of climatic effects on several crops at five locations in the Northwest: Boise, Idaho; Medford and Pendleton, Oregon; and Yakima and Spokane, Washington. The studies examined dryland and irrigated yields, and water use for irrigation, under several climate scenarios and using several crop models. Under all scenarios, dryland yields for most crops were projected to increase through the 21st century. The exception was potatoes, whose dryland yields by 2090 declined by as much as 30-35%, with the largest declines in Idaho. Changes in irrigated yields and irrigation water requirements were more mixed. Wheat and potatoes showed large reductions in irrigated yields (7-25% and 35-40%, respectively by 2090), while hay increased 50-70% and tomatoes showed a mixed trend, increasing 15-20% by 2030, then declining to roughly present yields by 2090. Irrigation water needs declined by 20-40% for wheat, and increased by 10-40% for hay, potatoes and tomatoes. To the extent that aggregate irrigation water needs increase under the combined effects of climate change, CO_2 enrichment, and agricultural response, this would exacerbate the summer water shortages and allocation conflicts discussed above. Interactions between climate and related impacts on agriculture and forest lands, and linkages between them as mediated by human land-use conversion and management, are little understood and remain important knowledge needs (Alig et al., 1998).

CRUCIAL UNKNOWNS AND RESEARCH NEEDS

Although this Assessment has produced and synthesized significant advances in our understanding of climate impacts in the four key areas examined in detail, clear needs for additional knowledge are evident for each of them, in order to understand impacts and vulnerabilities more thoroughly and assess potential responses.

For freshwater, the analysis here has concentrated primarily on the response of the Columbia River system to climate variability and change, and on the consequences of projected summer shortages on the reliability with which various present management objectives could be met under potential future flow regimes. Further analysis is required to: project future shifts in demand for multiple uses, and to examine the joint effects of climate variability and change on seasonal supply and demand; to identify the degree of vulnerability to scarcity, assess the effects of alternative operational rules and allocation schemes, and identify and evaluate specific response options. In addition, this Assessment has made only preliminary investigation of other basins than the Columbia. In particular, further examination is needed to characterize the effects of future climate variability and change on winter flooding risk in the low-elevation rainfed and mixed basins west of the Cascades. In view of the rapid development of west-side metropolitan areas, assessment of these climate risks, of how alternative future development patterns may compound or mitigate it, and of potential responses, are of high priority.

For salmon, research priorities include further characterization of historical climate influences on salmon and identification of the mechanisms by which they operate. In particular, very little is known of the effect of climate variability and change on the open-ocean phase of salmon's life cycle. For forests, the top priority is further work on the effects of climate and atmospheric CO_2 on Northwest forests' water use efficiency, which is necessary to understand even the direction of future climate effects on forests. Other priorities include interactions between climate and forest disturbances, including fire, insects, and disease, as well as interactions between these disturbances. Finally, little work has been done quantifying the effects of climate variability on the region's coasts.

Assessment of other sectors and issues than the four examined in detail here will be required. Particularly important areas for investigation in subsequent Northwest climate assessments will be agriculture, energy, and urban issues (e.g., infrastructure, hazards, and quality of life).

More broadly, there are several areas of climate research that are important for better understanding Northwest impacts. This assessment has been based on a single run each of two primary climate models, each using the same emissions scenario, with some additional results drawn from single runs of other models. A more reliable regional assessment would require controlled regional-level comparison of several state-of-the-art models, each with a statistical ensemble of multiple similar runs under each of several emissions scenarios. Ensembles of multiple runs for each model are necessary to allow examination of patterns of climate variability projected by each model. For example, several results in this assessment were strongly influenced by the fact that in the particular Hadley model run employed, the 2020s were an unusually wet decade. Studies of such variability over ensembles of multiple model runs are necessary to interpret such excursions. Comparisons of multiple models with several emission scenarios would allow useful model comparison and explanation of significant regional-scale disparities, and examination of how major impacts vary with higher or lower future emissions. Understanding the dynamics of longer-term variability, in particular whether and how presently observed patterns would shift under a global greenhouse trend, requires further development of climate models. Models are now beginning to represent ENSO, but cannot yet reproduce interdecadal variability of the size and character observed, either in the PDO or elsewhere in the world.

More accurate fine-scale modeling of climate-topography interactions in the Northwest is also required, and their effects on the region's rivers, estuaries and coasts, to understand several processes that strongly affect multiple areas of regional impacts. These processes include, for example, coastal upwelling, the interaction of fresh and salt water in estuaries, windstorms, and rain-on-snow events.

Because interannual and interdecadal climate variability exert such strong influences on the Northwest, better understanding of their dynamics and their relationship to long-term climate trends is also required. Their effects on summer climate are potentially important and have been little studied. An immediate uncertainty is whether the PDO has recently re-entered its cool phase, and indeed, whether the PDO will continue to exhibit more or less regular phase changes. If the PDO continues to behave for the next few decades as it has over the 20th century, and it has re-entered a cool phase, the resultant cooler wetter regional climate would be likely to partly offset greenhouse warming over the next two or three decades, until the PDO reverses again. While this could delay the onset of significant impacts from climate change, it could also obscure the need to undertake long-lead adaptation measures when they can be done with the least disruption.

Finally, a large but crucial area for further research concerns interactions between climate changes, the ecological and hydrological impacts discussed here, and human responses. Land-use change has been the dominant source of environmental stresses in the Northwest over the 20th century, and further population and economic growth are highly likely to bring more pressure for conversion, with complex interactions between metropolitan development, forestry, and other land uses. Coherent scenarios of socioeconomic futures in the Northwest that elaborate more climate-relevant aspects, especially land use and development, are needed. So are methods to couple socioeconomic projections to climatic, ecological, and hydrological models to allow more precise examination of impacts and potential responses. These methods should permit the examination of alternative sets of technological and institutional assumptions, and should support comprehensive examination of uncertainties across socioeconomic, climatic, and ecological domains.

Furthermore, more insight is needed into the feasibility and likely consequences of various adaptation strategies, addressing sectors both singly and jointly. This Assessment has made only a preliminary identification of potential adaptation measures for each sector, and has not attempted to assess their costs, benefits, efficacy, or ancillary effects. Systematic examination of technological, managerial, and institutional adaptation options should be conducted, based on partnerships between researchers, stakeholders, and experienced resource managers. Such partnership will be necessary to identify, develop, and assess adaptation options that are effective, low in cost, and are practical to implement in the context of present management practices, which strongly shape climate impacts. For example, future forest-management and agricultural practices are very likely to strongly mediate climate effects on these sectors, and may also contribute to mitigation of climate change through adjustments to management that increase sequestration of carbon in forests and soils. Where the present assessment has identified that climate variability and change are inadequately considered even in long-lived investment, resource management and infrastructure design decisions, research is needed to identify the cause and potential approaches to increase the time-horizon of planning. Two aspects of adaptation will be particularly important: investigation of interactions with impacts and responses in Canada, since many of the affected resources are shared; and linkages, whether conflicting or complementary, between adaptation measures in multiple sectors. For example, increasing water storage to manage increased summer drought suggests maintaining present dams or building more, while restoring salmon habitat to reduce their vulnerability to climate change would suggest the opposite. The most useful approach to assessment would examine impacts and adaptation measures together with emission scenarios and mitigation measures, to identify the most cost-effective strategy for dealing with climate-change in aggregate.

LITERATURE CITED

Agee, J. K., *Fire Ecology of Pacific Northwest Forests,* Island Press, Washington, DC, 1993.

Alig, R. J., D. M. Adams, B. A. McCarl, Impacts of incorporating land exchanges between forestry and agriculture in sector models, *Journal of Agricultural and Applied Economics, 30,* 2 (December), 389-401, 1998.

Bachelet, D., M. Brugnack, and R. P. Neilson, Sensitivity of a biogeography model to soil properties, *Ecological Modeling, 109,* 77-98, 1998.

Bachelet, D., R. P. Neilson, J. M. Lenihan, and R. J. Drapek, Climate change effects on vegetation distribution and carbon budget in the US, *Ecosystems,* in review, 2000.

Baker, P. F., T. P. Speed, and F. K. Ligon, Estimating the influence of temperature on the survival of chinook salmon smolts migrating through the Sacramento San Joaquin River delta of California, *Canadian Journal of Fisheries and Aquatic Sciences, 52,* 855-863, 1995.

Baker, W. L., Long-term response of disturbance landscapes to human intervention and global change, *Landscape Ecology, 10,* 143-159, 1989.

Bazzaz, F. A., S. L. Bassow, G. M. Berntson, and S. C. Thomas, Elevated CO_2 and terrestrial vegetation: Implications for and beyond the global carbon budget, in *Global Change and Terrestrial Ecosystems,* edited by B. Walker and W. Steffen, pp. 43-76, Cambridge University Press, New York, NY, 1996.

Beamish, R. J., and D. R. Bouillon, Pacific salmon production trends in relation to climate, *Canadian Journal of Fisheries and Aquatic Sciences, 50,* 1002-1016, 1993.

Beechie, T., E. Beamer, and L. Wasserman, Estimating Coho salmon rearing habitat and smolt production losses in a large river basin, and implications for habitat restoration, *North American Journal of Fisheries Management, 14,* 797-811, 1994.

Bisbal, G. A., and W. E. McConnaha, Consideration of ocean conditions in the management of salmon, *Canadian Journal of Fisheries and Aquatic Sciences, 55,* 2178-2186, 1998.

Boer, G. J., N. A. McFarlane, R. Laprise, J. D. Henderson, and J.-P. Blanchet, The Canadian Climate Centre spectral atmospheric General Circulation Model, *Atmosphere-Ocean, 22*(4), 397-429, 1984.

Boer, G. J., G. M. Flato, M. C. Reader, and D. Ramsden, A transient climate change simulation with historical and projected greenhouse gas and aerosol forcing: Experimental design and comparison with the instrumental record for the 20th century, *Climate Dynamics, 16,* 405-426, 1999a.

Boer, G. J., G. M. Flato, and D. Ramsden, A transient climate change simulation with historical and projected greenhouse gas and aerosol forcing: projected climate for the 21st century, *Climate Dynamics, 16,* 427-450, 1999b.

Bottom, D. L., Restoring salmon ecosystems: Myth and reality, *Restoration and Management Notes, 13,* 162-170, 1995.

Brubaker, L. B., Responses of tree populations to climatic change, *Vegetation, 67,* 119, 1986.

Bury, R. B., Vertebrates in the Pacific Northwest: Species richness, endemism, and dependency on old-growth forests, in *Biological Diversity: Problems and Challenges,* edited by S. K. Majumdar, F. J. Brenner, J. E. Lovich, J. F. Schalles, and E. W. Miller, pp. 392-404, Pennsylvania Academy of Science, Easton, Pennsylvania, 1994.

Callahan, B., E. Miles, and D. Fluharty. Policy implications of climate forecasts for water resources management in the Pacific Northwest. *Policy Sciences* 32:269-293, 1999.

Canning, D. J., Sea level rise in Washington State: State-of-the-knowledge, impacts, and potential policy issues, Washington Department of Ecology, Olympia, Washington, 1991.

Canning, D. J., and H. Shipman, Coastal erosion management studies in Puget Sound, Washington: Volume 1, executive summary, report 94-94, Washington Department of Ecology, Olympia, Washington, 1994.

Carter, M. F., and K. Barker, An interactive database for setting conservation priorities for western neotropical migrants, in *Status and Management of Neotropical Migratory Birds,* edited by D. M. Finch and P. W. Stangel, *US Forest Service General Technical Report RM-GTR-229,* pp. 120-144, Rocky Mountain Forest and Range Experiment Station, Ft. Collins, Colorado, 1993.

Cohen, S. J., K. A. Miller, A. F. Hamlet, and W. Avis, Climate change and resource management in the Columbia River Basin, *Water International,* in press, 2000.

Craig, D., Preliminary assessment of sea level rise in Olympia, Washington: Technical and policy implications, City of Olympia Public Works Department Policy and Program Development Division, Olympia, Washington, 1993.

Dale, V. H., and J. F. Franklin, Potential effects of climate change on stand development in the Pacific Northwest, *Canadian Journal of Forest Resources, 19*, 1581, 1989.

Daly C, D. Bachelet J. M. Lenihan, R. P. Neilson, W. Parton and D. Ojima, Dynamic simulation of tree-grass interactions for global change studies, *Ecological Applications, 10*(2), 449-469, 2000

dell'Arciprete, P., N. Mantua, and R. C. Francis, The instrumental record of climate variability in the Pacific Northwest, Progress report for Year 1, JISAO Climate Impacts Group, University of Washington, Seattle, Washington, 1996.

dell'Arciprete, P., D. L. Peterson, R. C. Francis, and N. Mantua, Climate reconstruction through tree growth indices, unpublished manuscript, available from the University of Washington Climate Impacts Group, Seattle, Washington, 1998.

Detling, L. E., Relic islands of xeric flora west of the Cascade Mountains in Oregon, *Madrono, 12*, 39-47, 1953.

Detling, L. E., Historical background of the flora of the Pacific Northwest, *Bulletin No. 13*, Museum of Natural History, University of Oregon, Eugene, Oregon, 1968..

Ellsworth, D. S., CO_2 enrichment in a maturing pine forest: Are CO_2 exchange and water status in the canopy affected?, *Plant, Cell, and Environment, 22,* 461-472, 1999

Emmett, R. L., and M. H. Schiewe (Eds.), Estuarine and ocean survival of Northeast Pacific salmon: Proceedings of the workshop, US Department of Commerce, NOAA Technical Memo., NMFS-NWFSC-29, 1997.

Emmett, R. L., and R. D. Brodeur, Recent changes in the pelagic nekton community off Oregon and Washington in relation to some physical oceanographic conditions, North Pacific Anadromous Fish Commission, Bull. No. 2, in press, 2000.

Feist, B. E., and C. A. Simenstad, Expansion rates and recruitment frequency of exotic smooth cordgrass, (Spartina alterniflora) (Loisel) colonizing unvegetated littoral flats in Willapa Bay, Washington, *Estuaries, 23*(2), 268-275, 2000.

FEMAT (Forest Ecosystem Management Assessment Team), Forest ecosystem management: An ecological, economic, and social assessment, US Department of the Interior, US Department of Agriculture, US Department of Commerce, and US Environmental Protection Agency, Washington, DC, 729 pp., 1993.

Field, J., and M. Hershman, Assessing coastal zone sensitivity and vulnerability to regional climate variability and change in the Pacific Northwest, Climate Impacts Group, University of Washington, Seattle, Washington, 1997.

Flato, G. M., G. J. Boer, W. G. Lee, N. A. McFarlane, D. Ramsden, M. C. Reader, and A. J. Weaver, The Canadian Centre for Climate Modelling and Analysis global coupled model and its climate, *Climate Dynamics*, in press, 2000.

Fosberg, M. A., L. O. Mearns, and C. Price, Climate change: Fire interactions at the global scale: Predictions and limitations of methods, in *Fire in the Environment*, edited by P. J. Crutzen and J. G. Goldammer, Wiley and Sons, New York, 1992.

Francis, R. C., Sustainable use of salmon: Its effects on biodiversity and ecosystem function, in *Harvesting Wild Species —- Implications for Biodiversity Conservation*, edited by C. H. Freeze, pp. 626-670, Johns Hopkins University Press, Baltimore, Maryland, 1997.

Francis, R. C., S. R. Hare, A. B. Hollowed, and W. S. Wooster, Effects of interdecadal climate variability on the oceanic ecosystems of the Northeast Pacific, *Fisheries Oceanography, 7*, 1-21, 1998.

Franklin, J. F., Pacific Northwest forests, in *North American Terrestrial Vegetation*, edited by M. G. Barbour and W. D. Billings, pp. 103-130, Cambridge University Press, Cambridge, United Kingdom, 1988.

Franklin, J. F., and C. T. Dyrness, *Natural Vegetation of Oregon and Washington*, Oregon State University Press, Corvallis, Oregon, 1973.

Franklin, J. F., and R. T. T. Forman, Creating landscape patterns by forest cutting: Ecological consequences and principles, *Landscape Ecology*, 1, 5-18, 1987.

Franklin, J. F., W. H. Moir, G. W. Douglas, and C. Wiberg, Invasion of subalpine meadows by trees in the Cascade Range, *Arctic and Alpine Research, 3*, 215-224, 1971.

Franklin, J. F., et al., Effects of global climatic change on forests in northwestern North America, *Northwest Environmental Journal*, 7, 233-254, 1991.

Gardner, R. H., W. W. Hargrove, M. G. Turner, and W. H. Romme, Climate change, disturbances, and landscape dynamics, in *Global Change and Terrestrial Ecosystems*, edited by B. Walker and W. Steffen, pp., 149-172, Cambridge University Press, Cambridge, United Kingdom, 1996.

Gerstel, W. J., M. J. Brunengo, W. S. Lingley, Jr., R. L. Logan, H. Shipman, and T. J. Walsh, Puget Sound bluffs: The where, why, and when of landslides following the holiday 1996-97 storms, *Washington Geology, 25*(1), 17-31, 1997.

Gholz, H. L., Environmental limits on aboveground net primary production, leaf area, and biomass in vegetation zones of the Pacific Northwest, *Ecology, 63,* 469, 1982.

Giorgi, F., and G. Bates, On climatological skill of a regional model over complex terrain, *Monthly Weather Review, 117,* 2325-2347, 1989.

Giorgi, F., C. S. Brodeur, and G. T. Bates, Regional climate change scenarios over the United States produced with a nested regional climate model, *Journal of Climate, 7,* 375-399, 1994.

Glantz, M. H., Consequences and responsibilities in drought forecasting: The case of Yakima, 1977, *Water Resources Research, 18,(*1), pp. 3-13, 1982.

Good, J. W., Shore protection policy and practices in Oregon: An evaluation of implementation successes, *Coastal Management, 22,* 325-352, 1994.

Graumlich, L. J., Precipitation variation in the Pacific Northwest (1675-1975) as reconstructed from tree rings, *Annals of the Association of American Geographers, 77,* 19-29, 1987.

Graumlich, L. J., Subalpine tree growth, climate, and increasing CO_2: An assessment of recent growth trends, *Ecology, 72,* 1-11, 1991.

Gray, K. N., The impacts of drought on Yakima Valley irrigated agriculture and Seattle municipal and industrial water supply, master's thesis, School of Marine Affairs, University of Washington, Seattle, Washington, 1999.

Grier, C. C., and S. Running, Leaf area of mature northwestern coniferous forests: Relation to site water balance, *Ecology, 58,* 893, 1977.

Hamlet, A. F., and D. P. Lettenmaier, Effects of climate change on hydrology and water resources objectives in the Columbia River Basin, *Journal of American Water Resources Association, 35*(6), 1597-1623, 1999.

Hare, S. J., N. J. Mantua, and R. C. Francis, Inverse production regimes: Alaskan and West Coast Pacific salmon, *Fisheries, 24,* 6-14, 1999.

Harmon, M. E., W. K. Ferrell, and J. F. Franklin, Effects on carbon storage of converting old-growth forests to young forests, *Science, 247,* 699-702, 1990.

Hargreaves, B. N., Early ocean survival of salmon off British Columbia and impacts of the 1983 and 1991-1995 El Niño events, in *Estuarine and Ocean Survival of Northeast Pacific Salmon: Proceedings of the Workshop*, edited by R.L. Emmett and M. H. Schiewe, pp., 197-211, NOAA Tech. Memo., NMFS-NWFSC-29, Seattle, Washington, 1997.

Harr, R. D., Water flux in soil and subsoil in a steep forested slope, *Journal of Hydrology, 33,* 37-58, 1977.

Haynes, R. W., D. M. Adams, and J. R. Mills, The 1993 RPA Timber Assessment update, USDA Forest Service, Rocky Mountain Forest and Range Experimental Station, *General Technical Report RM-GTR.* March 1995.

Henjum, M. G., J. B. Karr, D. L. Bottom, D. Perry, J. C. Bednarz, S. G. Wright, S. A. Beckwitt, and E. Beckwitt, Interim protection for late-successional forests, fisheries, and watersheds: National forests east of the Cascade crest, Oregon, and Washington, *The Wildlife Society Technical Review 94-2*, Bethesda, Maryland, 245 pp., 1994.

Heyerdahl, E. K., Spatial and temporal variation in historical fire regimes of the Blue Mountains, Oregon, and Washington: The influence of climate, Ph.D. dissertation, College of Forest Resources, University of Washington, Seattle, Washington, 1997.

Huffaker, R., N. K. Whittlesey, and P. R. Wandschneider, Institutional feasibility of contingent water marketing to increase migratory flows for salmon on the Upper Snake River, *Natural Resources Journal, 33,* 671-696, Summer 1993.

Hulme, M., E. M. Barrow, N. W. Arnell, P. A. Harrison, T. C. Johns, and T. E. Downing, Relative impacts of human-induced climate change and natural variability, *Nature, 397,* 688-691, 1999.

Jackson, P. L., and A. J. Kimmerling, (Eds.), *Atlas of the Pacific Northwest, 8th edition*, Oregon State University Press, Corvallis, Oregon, 1993.

Johns T. C., R. E. Carnell, J. F. Crossley, J. M. Gregory, J. F. B. Mitchell, C. A. Senior, S. F. B. Tett, and R. A. Wood, The Second Hadley Centre coupled ocean-atmosphere GCM: Model description, spinup, and validation, *Climate Dynamics, 13*, 103-134, 1997.

Jones, J. A., and G. E. Grant, Peak flow responses to clear-cutting and roads in small and large basins, western Cascades, Oregon, *Water Resources Research, 32,* 959-974, 1996.

Jones, J. A., and G. E. Grant, Hydrologic processes and peak discharge response to forest harvest, regrowth, and roads in ten small experimental basins, western Cascades, Oregon, *Water Resources Research*, in review, 2000.

Keane, R. E., and S. F. Arno, Rapid decline of whitebark pine in western Montana: Evidence from 20-year remeasurements, *Western Journal of Applied Forestry, 8*(2), 44-47, 1993.

Kellogg, E., (Ed.), Coastal temperate rain forests: Ecological characteristics, status, and distribution worldwide, *Ecotrust Occasional Paper Series 1*, Ecotrust and Conservation International, Portland, Oregon, and Washington, DC, 64 pp., 1992.

Kemp, W. P., D. O. Everson, and W. G. Wellington, Regional climatic patterns and western spruce budworm outbreaks, USDA Forest Service Cooperative State Research Service, *Technical Bulletin Number 1693*, USDA, Corvallis, Oregon, 1985.

Kendall, K. C., Whitebark Pine: Ecosystem in peril, in *Our living resources: A report to the Nation on the distribution, abundance, and health of US plants, animals, and ecosystems*, edited by E. T. LaRoe, G. S. Farris, C. E. Puckett, P. D. Doran, M .J. Mac, pp. 228-230, US National Biological Service, Washington, DC, 1995.

Klahn, J., This time, city folks will bear cost of restoring species, (AP), *Seattle Times*, pg. 1, September 5, 1999.

Komar, P. D., and D. B. Enfield, Short-term sea-level changes and coastal erosion, in *Sea-level Fluctuations and Coastal Evolution, Special Publication 41*, edited by D. Nummedal, O. Pilkey, and J. D. Howard, pp. 17-27, Society of Economic Paleontologists and Mineralologists, 1987.

Laprise, R., D. Caya, M. Giguere, G. Bergeron, H. Cote, J.-P. Blanchet, G. J. Boer, and N. A. McFarlane, Climate and climate change in western Canada as simulated by the Canadian Regional Climate Model, *Atmosphere-Ocean, 36*, 119-167, 1998.

Larsson, S., Stressful times for the plant stress-insect performance hypothesis, *Oikos, 56*, 277-283, 1989.

Law, B. E., R. H. Waring, P. M. Anthoni, and J. D. Abers, Measurements of gross and net ecosystem productivity and water vapor exchange of a Pinus ponderosa ecosystem, and an evaluation of two generalized models, *Global Change Biology, 5,* 1-15, 1999.

Lehmkuhl, J. F., P. F. Hessburg, R. L. Everett, M. H. Huff, and R. D. Ottmar, Historical and current forest landscapes of eastern Oregon and Washington, Part 1: Vegetation pattern and insect and disease hazards, *General Technical Report PNW-GTR-328*, USDA Forest Service, Corvallis, Oregon, 1994.

Leung, L. R., and S. J. Ghan, Pacific Northwest climate sensitivity simulated by a regional climate model driven by a GCM. Part I: Control simulations, *Journal of Climate, 12*(7), 2010-2030, 1999a.

Leung, L. R., and S. J. Ghan, Pacific Northwest climate sensitivity simulated by a regional climate model driven by a GCM. Part II: 2 times CO_2 simulations, *Journal of Climate, 12*(7), 2031-2053, 1999b.

Little, R. L., D. L. Peterson, and L. L. Conquest, Regeneration of subalpine fir (*Abies lasiocarpa*) following fire: Effects of climate and other factors, *Canadian Journal of Forest Research, 24*, 934-944, 1994.

Mac, M. J., P. A. Opler, C. E. Puckett Haecker, and P. D. Doran (Eds.), Status and trends of the Nation's biological resources, *US Geological Survey*, US Department of the Interior, Biological Resources Division, Reston, Virginia, 1998.

Mantua, N. J., S. R. Hare, Y. Zhang, J. M. Wallace, and R. C. Francis, A Pacific interdecadal climate oscillation with impacts on salmon production, *Bulletin of the American Meteorological Society, 78*, 1069—1079, 1997.

Mantua, N. J., P. dell'Arciprete, and R. C. Francis, Patterns of climate variability in the Pacific Northwest: A regional 20th century perspective, in Impacts of climate variability and climate change in the Pacific Northwest: An integrated assessment, edited by E. L.

Miles, Climate Impacts Group, University of Washington, Seattle, Washington, 1999.

Marcot, B. G., R. S. Holthausen, J. Teply, and W. D. Carrier, Old-growth inventories: Status, definitions, and visions for the future, in *Wildlife and Vegetation in Unmanaged Douglas-Fir Forests*, edited by L. F. Ruggiero, K. Aubry, and M. H. Huff, *US Forest Service General Technical Report PNW-GTR-285*, pp. 47-60, Pacific Northwest Research Station, Portland, Oregon, 1991.

Mason, R. R., and B. E. Wickman, The Douglas-fir tussock moth in the interior Pacific Northwest, chapter 10, in *Dynamics of Forest Insect Populations*, edited by A. A. Berryman, Plenum Press, New York, 1988

Matthews, D. A., and T. Hovland, Nested model simulations of regional orographic precipitation, Global Climate Change Response Program Report, Bureau of Reclamation, Denver, Colorado, 61 pp., 1997.

McCullough, D. A., A review and synthesis of effects of alterations to the water temperature regime on freshwater life stages of salmonids, with special reference to Chinook Salmon, EPA Region 10 Report 910-R-99-010, July 1999.

McFarlane, N. A., G. J. Boer, J. P. Blanchet, and M. Lazare, The Canadian Climate Centre second-generation general circulation model and its equilibrium climate, *Journal of Climate, 5*, 1013-1044, 1992.

McKenzie, D., D. L. Peterson, and E. Alvarado, Predicting the effect of fire on large-scale vegetation patterns in North America, USDA Forest Service Research Paper PNW-489, Fort Collins, Colorado, 1996.

Miles, E. L., A. Hamlet, A. K. Snover, B. Callahan, and D. Fluharty. 2000. Pacific Northwest regional assessment: The impacts of climate variability and climate change on the water resources of the Columbia River Basin. *Journal of American Water Resources Association, 36*(2): 399-420, 2000.

Miller, A. J., D. R. Cayan, T. P. Barnett, N. E. Graham, and J. M. Oberhuber, The 1976-1977 climate shift in the Pacific Ocean, *Oceanography*, 7, 21-26, 1994.

Miller, A. J., D. R. Cayan, and W. B. White, A westward-intensified decadal change in the North Pacific thermocline and gyre-scale circulation, *Journal of Climate, 11*, 3112-3127, 1998.

Miller, K. A., Water banking to manage supply variability, in *Advances in the Economics of Environmental Resources*, vol. 1., edited by D. C. Hall, pp. 185-210, JAI Press, Greenwich, Connecticut, 1996.

Miller, K. A., S. L. Rhodes, L. J. MacDonnell, Water allocation in a changing climate: Institutions and adaptation, *Climatic Change, 35*, 157-177, 1997.

Miller, R. F., and J. A. Rose, Western juniper expansion in eastern Oregon, *Great Basin Naturalist, 55*, 37-45, 1995.

Miller, R. F., and P. E. Wigand, Holocene changes in semi-arid pinyon-juniper woodlands, *BioScience, 44(7)*, 465-474, 1994.

Mitchell J. F. B., T. C. Johns, J. M. Gregory, and S. Tett, Climate response to increasing levels of greenhouse gases and sulphate aerosols, *Nature, 376*, 501-504, 1995.

Mitchell J. F. B., and T. C. Johns, On modification of global warming by sulfate aerosols, *Journal of Climate, 10*(2), 245-267, 1997.

Mooney, H. A., J. Canadell, F. S. Chapin, III, J. R. Ehleringer, C. H.. Korner, R. E. McMurtrie, W. J. Parton, L. F. Pitelka, and E.-D. Schulze, Ecosystem physiology responses to global change, in, *The Terrestrial Biosphere and Global Change: Implications for Managed and Unmanaged Ecosystems*, edited by B. Walker, W. Steffen, J. Canadell, and J. Ingram, pp. 141-189, Cambridge University Press, Cambridge, UK, 1999.

Morse, L. E., J. T. Kartesz, and L. S. Kutner, Native vascular plants, in Our living resources: A report to the Nation on the distribution, abundance, and health of US plants, animals, and ecosystems, edited by E. T. LaRoe, G. S. Farris, C. E. Puckett, P. D. Doran, M. J. Mac, US National Biological Service, Washington, DC, 1995.

Mote, P. W., W. S. Keeton, and J. F. Franklin, Decadal variations in forest fire activity in the Pacific Northwest, 11th Conference on Applied Climatology, American Meteorological Society, Boston, Massachusetts, 1999a.

Mote, P. W., et al., Impacts of climate variability and change: Pacific Northwest, A report of the Pacific Northwest Regional Assessment Group for the US Global Change Research Program, Climate Impacts Group, University of Washington, Seattle, Washington, November 1999b.

Myers, J. M., et al., Status review of chinook salmon from Washington, Idaho, Oregon, and California, NOAA Tech. Memo. NMFS-NWFSC-35, 1998.

National Marine Fisheries Service (NMFS), Northwest Region, West Coast Salmon and the Endangered Species Act, (http://www.nwr.noaa.gov/1salmon/salmesa/index.htm), 2000.

Neilson, R. P., G. A. King, and J. Lenihan, Modeling forest response to climatic change: The potential for large emissions of carbon from dying forests, in *Carbon Balance of the World's Forested Ecosystems: Towards a Global Assessment*, edited by J. Kanninen, pp. 150-162, Academy of Finland, Joensuu, Finland, 1994.

Neilson, R. P., and R. J. Drapek, Potentially complex biosphere responses to transient global warming, *Global Change Biology, 4*, 101-117, 1998.

Neilson R. P., I. C. Prentice, B. Smith, T. G. F. Kittel, and D. Viner, Simulated changes in vegetation distribution under global warming, in *The Regional Impacts of Climate Change: An Assessment of Vulnerability*, edited by R. T. Watson, M. C. Zinyowera, R. H. Moss, and D. J. Dokken, pp. 439-456, Cambridge University Press, Cambridge, United Kingdom, 1998.

National Research Council, *Upstream*, National Academy Press, Washington, DC, 1996.

Nijssen, B., D. P. Lettenmaier, X. Liang, S. W. Wetzel, and E. F. Wood, Streamflow simulation for continental-scale river basins, *Journal of Water Resources Research*, 33, 4, pp 711ff, 1997.

Noss, R. F., E. T. LaRoe, III, and J. M. Scott, Endangered ecosystems of the United States: A preliminary assessment of loss and degradation, *National Biological Service Biological Report 28*, Washington, DC, 58 pp., 1995.

Overpeck, J. T., D. Rind, and R. Goldberg, Climate-induced changes in forest disturbance and vegetation, *Nature, 343*, 51-53, 1990.

Park, R. A., J. K. Lee, and D. J. Canning, Potential effects of sea-level rise on Puget Sound wetlands, *Geocarto International*, 99-110, 1993.

Parsons, G. L., G. Cassis, A. R. Moldenke, J. D. Lattin, N. H. Anderson, J. C. Miller, P. Hammond, and T. D. Schowalter, Invertebrates of the H. J. Andrews experimental forest, western Cascade Range, Oregon, Part V: An annotated list of insects and other arthropods, *US Forest Service General Technical Report PNW-GTR-290*, Pacific Northwest Research Station, Portland, Oregon, 168 pp., 1991.

Pearcy, W. G., *Ocean Ecology of North Pacific Salmonids*, University of Washington Press, Seattle, Washington, 1992.

Pearcy, W. G., K. Y. Aydin, and R. D. Brodeur, What is the carrying capacity of the North Pacific Ocean for salmonids?, *Pisces Press*, 7, 2, 17-22, 1999.

Perkins, R., Climatic and physiographic controls on peak-flow generation in the western Cascades, Oregon, Ph.D. dissertation, Oregon State University, Corvallis, Oregon, 1997.

Peterson, D. L., Climate, limiting factors and environmental change in high-altitude forests of western North America, in *Climatic Variability and Extremes: The Impact on Forests*, edited by M. Benistion and J. L. Innes, Springer-Verlag, Heidelberg, Germany, 1998.

Peterson, D. W., and D. L. Peterson, Growth responses of mountain hemlock (Tsuga mertensiana) to interannual and interdecadal climate variability, *Ecology*, in review, 2000.

Pinnix, W., Climate and Puget Sound, in JISAO/SMA Year 3 report, Climate Impacts Group, University of Washington, Seattle, Washington, 1998.

Price, P. W., The plant vigor hypothesis and herbivore attack, *Oikos, 62*, 244-251, 1991.

Quigley, R. W., R. W. Haynes, and R. T. Graham, Integrated scientific assessment for ecosystem management in the interior Columbia Basin, *USDA Forest Service General Technical Report PNW-382*, USDA Forest Service, Corvallis, Oregon, 1996.

Quinn, T. P., and D. J. Adams, Environmental changes affecting the migratory timing of American shad and sockeye salmon, *Ecology*, 77, 1151-1162, 1996.

Reeves, G. H., F. H. Everest, and T. E. Nickelson, Identification of physical habitats limiting the production of Coho salmon in western Oregon and Washington, *USDA Forest Service General Technical Report PNW-245*, US Forest Service, Corvallis, Oregon, 1989.

Rochefort, R. M., R. L. Little, A. Woodward, D. L. Peterson, Changes in subalpine tree distribution in western North America: A review of climate and other factors, *The Holocene, 4*, 89-100, 1994.

Ryan, K. C., Vegetation and wildland fire: Implications of global climate change, *Environment International, 17*, 169-178, 1991.

Shipman, H., Vertical land movements in coastal Washington: Implications for relative sea level changes, Washington Department of Ecology, Olympia, Washington, 1989.

Sohngen, B., R. Mendelsohn, and R. Neilson, Predicting CO_2 emissions from forests during climatic change: A comparison of natural and human response models, *Ambio, 27,* 509-513, 1988.

Swetnam, T. W., and J. L. Betancourt, Fire-Southern Oscillation relations in the southwestern United States, *Science, 249,* 1017-1020, 1990.

Swetnam, T. W., and J. L. Betancourt, Meso-scale disturbance and ecological response to decadal climate variability in the American southwest, *Journal of Climate, 11,* 3218-3247, 1998.

Swetnam, T. W., and A. M. Lynch, Multicentury regional-scale patterns of western spruce budworm outbreaks, *Ecological Monographs, 63,* 399-422, 1993.

Terleckyj, N. E., Analytic documentation of three alternate socioeconomic projections, 1997—2050, NPA Data Services, Washington, DC, 1999a.

Terleckyj, N. E., Development of three alternate national projections scenarios, 1997—2050, NPA Data Services, Washington, DC, 1999b.

Tett, S. B., P. A. Stott, M. R. Allen, W. J. Ingram, and J. F. B. Mitchell, Causes of twentieth century temperature change, *Nature, 399,* 569-572, 1999.

Thomson, A. R., R. F. Sheperd, J. W. E. Harris, and R. J. Silversides, Relating weather to outbreaks of western spruce budworm in British Columbia, *Canadian Entomologist, 116,* 375-381, 1984.

Torn, M. S., and Fried, J. S., Predicting the impacts of global warming on wildland fire, *Climate Change, 21,* 257-274, 1992.

Trenberth, K. E., The definition of El Niño, *Bulletin of the American Meteorological Society, 78,* 2771-2777, 1997.

Tubbs, D. W., Landslides in Seattle, Washington Division of Geology and Earth Resources Information Circular 52, Washington Department of Natural Resources, Olympia, Washington, 1974.

US Census Bureau, Historical state population series, 2000, at: http://www.census.gov/population/www/estimates/st_stts.html

US Department of Agriculture, Economic Research Service, US State Fact Sheets, 2000 (http://www.ers.usda.gov/epubs/other/usfact/), 2000.

VEMAP Members, Vegetation/ecosystem modeling and analysis project: Comparing biogeography and biogeochemistry models in a continental-scale study of terrestrial ecosystem responses to climate change and CO_2 doubling, *Global Biogeochemical Cycles, 4,* 407-437 1995.

Walker, R. V, K. W. Myers, N. D. Davis, K. Y. Aydin, K. D. Friddland, H. R. Carlson, G. W. Boehlert, S. Urawa, Y. Ueno, and G. Anma, Diurnal variation in the thermal environment experienced by salmonids in the North Pacific as indicated by data storage tags. *Fisheries Oceanography,* in review, 2000.

Waring, R. H., and J. F. Franklin, Evergreen coniferous forests of the Pacific Northwest, *Science, 204,* 1380, 1979.

Weaver, J. L., P. C. Paquet, and L. F. Ruggiero, Resilience and conservation of large carnivores in the Rocky Mountains, *Conservation Biology, 10,* 964-976, 1996.

Welch, D. W., Y. Ishida, and K. Nagasawa, Thermal limits and ocean migrations of sockeye salmon (*Oncorhynchus nerka*): Long-term consequences of global warming, *Canadian Journal of Fisheries and Aquatic Sciences, 55,* 937-948, 1998.

West, N. E., and M. A. Hassan, Recovery of sage-brush-grass vegetation following wildfire, *Journal of Range Management, 38,* 131-134, 1985.

Zobel, D. B., A. McKee, G. M. Hawk, and C. T. Dyrness, Relationships of environment to composition, structure, and diversity of forest communities of the central western cascades of Oregon, *Ecological Monographs, 46,* 135, 1976.

ACKNOWLEDGMENTS

Many of the materials of this chapter are based on contributions from participants on and those working with the

Pacific Northwest Workshop and Assessment Team
Philip Mote*, University of Washington
Douglas Canning, Department of Ecology, State of
    Washington
David Fluharty, University of Washington
Robert Francis, University of Washington
Jerry Franklin, University of Washington
Alan Hamlet, University of Washington
Blair Henry, The Northwest Council on Climate
    Change
Marc Hershman, University of Washington
Kristyn Gray Ideker, Ross and Associates
William Keeton, University of Washington
Dennis Lettenmaier, University of Washington

Ruby Leung, Pacific Northwest National Laboratory
Nathan Mantua, University of Washington
Edward Miles, University of Washington
Ben Noble, Battelle Memorial Institute
Hossein Parandvash, Portland Bureau of Water Works
David W. Peterson, US Geological Survey
Amy Snover, University of Washington
Sean Willard, University of Washington

Comments by the following reviewers are gratefully acknowledged: Ralph Alig, Robert L. Alverts, William Clark, Robert Emmett, Josh Foster, Steven Ghan, Michael Haske, John Innes, Linda Joyce, Kai Lee, L. Ruby Leung, Susan M. Marcus, Steve McNulty, Ron Nielson, Claudia Nierenberg, Michael Scott, Francis Zweirs; Coordinated comments by U.S. Department of Agriculture; Coordinated comments by US Department of Interior; Coordinated comments by US Department of Energy. Remaining errors are the sole responsibility of the coordinating author.

* Assessment Team chair/co-chair

CHAPTER 10

POTENTIAL CONSEQUENCES OF CLIMATE VARIABILITY AND CHANGE FOR ALASKA

Edward A. Parson[1,2] with contributions from:
Lynne Carter[3], Patricia Anderson[4], Bronwen Wang[5], Gunter Weller[4]

Contents of this Chapter

[1]John F. Kennedy School of Government, Harvard University; [2]Coordinating author for the National Assessment Synthesis Team; [3]Office of the U. S. Global Change Research Program, University Corporation for Atmospheric Research; [4]University of Alaska, Fairbanks, co-chair, Alaska Regional Assessment Team; [5]U. S. Geological Survey, Anchorage, co-chair, Alaska Regional Assessment Team.

CHAPTER SUMMARY

Regional Context

Spanning an area nearly a fifth the size of the entire lower 48 states, Alaska includes extreme physical, climatic, and ecological diversity in its rainforests, mountain glaciers, boreal spruce forest, tundra, peatlands, and meadows. Lightly populated and growing about 1.5% per year, Alaska has the nation's highest median household income, with an economy dominated by government and natural resources. In contrast to other regions, the most severe environmental stresses in Alaska at present are already climate-related. Recent warming has been accompanied by several decades of thawing in discontinuous permafrost, which is present in most of central and southern Alaska, causing increased ground subsidence, erosion, landslides, and disruption and damage to forests, buildings, and infrastructure. Sea ice off the Alaskan coast is retreating (by 14% since 1978) and thinning (by 40% since the 1960s), with widespread effects on marine ecosystems, coastal climate, human settlements, and subsistence activities.

Climate of the Past Century

- Alaska's climate has warmed about 4°F since the 1950s, 7°F in the interior in winter, with much of this warming occurring in a sudden regime shift around 1977.
- Alaska's warming is part of a larger Arctic trend corroborated by many independent measurements of sea ice, glaciers, permafrost, vegetation, and snow cover.
- Most of the state has grown wetter, with a 30% average precipitation increase between 1968 and 1990.
- The growing season has lengthened by about 14 days.
- Dramatic reductions in sea ice and permafrost have accompanied the recent warming.

Climate of the Coming Century

- Models project continued strong warming in Alaska, reaching 1.5-5°F by 2030, and 5-18°F by 2100, strongest in the interior and north and in winter.
- Continued precipitation increases are projected, reaching 20-25% in the north and northwest, with areas of decrease along the south coast.
- Increased evaporation from warming is projected to more than offset increased precipitation, making soils drier in most of the state.

Key Findings

- As much as the top 30 feet of discontinuous permafrost is projected to thaw over the 21ˢᵗ century, causing increased ground subsidence, erosion, landslides, and disruption and damage to forests, buildings, and infrastructure.
- The melting of sea ice is projected to continue, with the Canadian climate model projecting a complete loss of summer Arctic sea ice by 2100. Loss of sea ice allows larger storm surges to develop, increasing the erosion and coastal inundation, and also threatens populations of marine mammals and polar bears that depend on ice, and the subsistence livelihoods that depend on them.
- Recent warming has been accompanied by unprecedented increases in forest disturbances, including insect attacks. A sustained infestation of spruce bark beetles, which in the past have been limited by cold, has caused widespread tree deaths over 2.3 million acres on the Kenai Peninsula since 1992, the largest loss to insects ever recorded in North America.
- Increases in blow-downs in forests due to intense windstorms, and in canopy breakage from the heavy snows typical of warm winters, have increased vulnerability of forests to insect attack. Projected further warming is likely to increase the risk of insect attack.
- Significant increases in fire frequency and intensity, both related to summer warming, have

occurred. Simultaneously, the potential damage from forest fires has increased due to an increase in dispersed human settlement in forests. The projected further warming is likely to increase near-term risk of fire.
- In the longer term, large-scale transformation of landscapes is likely, including expansion of boreal forest into the tundra zone, shifts of forest types due to fire and moisture stress, northward expansion of some commercially valuable species, and the appearance of significant fire risk in the coastal forest for the first time since observations began.
- The Gulf of Alaska and Bering Sea support marine ecosystems of great diversity and productivity, and the nation's largest commercial fishery. The effect of projected climate change on these ecosystems could be large.
- Present climate change already poses drastic threats to subsistence livelihoods, practiced mainly by Native communities, as many populations of marine mammals, fish, and seabirds have been reduced or displaced due to retreat and thinning of sea ice and other changes. Projected climate changes are likely to intensify these impacts. In the longer term, projected ecosystem shifts are likely to displace or change the resources available for subsistence, requiring communities to change their practices or move.

POTENTIAL CONSEQUENCES OF CLIMATE VARIABILITY AND CHANGE FOR ALASKA

PHYSICAL SETTING AND UNIQUE ATTRIBUTES

Alaska spans 20 degrees of latitude and 42 of longitude, with a land area of 570,000 square miles - nearly a fifth of the lower 48 states - and a coastline of more than 34,000 miles, longer than those of the other 49 states combined. This enormous expanse embraces extreme physical, climatic, and ecological diversity. In the south, a series of mountain ranges parallel the coast, where intense precipitation produces lush cool rainforests and large mountain glaciers. These southern ranges culminate in the long arc of the Alaska Range, which includes McKinley (Denali), the highest peak in North America. Beyond, in the interior, lie the wide valleys of the Yukon River and its tributary the Tanana. Further north, the Brooks Range divides the interior from the cold and arid Arctic slopes. Alaska contains about 75% of US national parklands and 90% of national wildlife refuge lands (USGS BRD, 1999). It contains roughly 40% of the nation's surface water resources (Lamke, 1986), 63% of its wetlands (Hall et al., 1994), essentially all of its permafrost, and more glaciers and active volcanoes than all other states combined.

SOCIOECONOMIC CONTEXT

Alaska is lightly populated, with 614,000 people in 1998 distributed between the small cities of Anchorage (260,000), Fairbanks (84,000) and Juneau (30,000), a few smaller towns, and many villages and rural settlements. Average annual population growth was more than 3% per year in the 1980s declining to about 1.5% per year in the 1990s, where it is projected to remain, giving a projected total state population that reaches 885,000 by 2025 (US Census Bureau, 2000). Native peoples comprise about 16% of Alaska's population. The state's median household income, nearly $52,800 in 1996, is the highest in the nation.[1] The economy is dominated by government and natural resources, with Federal civilian and military payrolls, and the State's Permanent Fund, contributing 44% of total incomes. The North Slope oil fields, which provide 19% of US crude oil production (8% of US consumption) provide a further 35% of the state's incomes, while fisheries provide an additional 7%. Other significant income shares include tourism (5%), timber (2%), and mining (2%), with the remainder miscellaneous (agriculture contributes 0.1%) (Goldsmith, 1997). In addition, diverse forms of subsistence livelihood are practiced throughout the state, primarily but not exclusively by native communities. These activities depend on fish, marine mammals, and wildlife - including partly commercial reindeer herding - and play a social and cultural role vastly greater than their contribution to monetary incomes.

ECOLOGICAL CONTEXT

Alaska's ecosystem types encompass an extraordinary diversity, reflecting the state's vastness and extreme variety of climates. Ecosystem types include cool Sitka spruce and Western hemlock forest in the southeast and south-central coastal regions; boreal forest of white and black spruce with hardwoods on well-drained uplands through the south-central region and interior; Alpine tundra and meadows at higher elevations on interior mountain ranges; maritime tundra along the west coast from the Alaskan Peninsula and Aleutians to the Seward Peninsula, including vast coastal wetlands in the Yukon-Kuskokwim delta; and Arctic tundra and barrens on the northwest coast and north of the Brooks Range. Only about 30,000 acres is in agricultural production, principally in the Tanana and Matanuska valleys and on the Kenai Peninsula; larger areas are used for pasture (185,000 acres) and reindeer grazing (12 million acres, mostly on the Seward Peninsula).

The Alaskan terrestrial landscape has been altered less by direct human intervention than anywhere else in the United States. The most significant current environmental pressures in Alaska include heavy stress on fish stocks and marine ecosystems from large commercial fisheries, both Alaskan and international; local impacts from the mining and petroleum industries, including the aftermath of the Exxon Valdez spill and its cleanup, as well as smaller ongoing impacts from routine operations; and strain

[1] Median household money income, in 1996 dollars. US Census Bureau (2000).

on fragile ecosystems and communities from rapidly growing summer tourism throughout the state. In addition, like most of the Arctic, Alaska is already experiencing much more rapid climate warming than the lower 48 states, with major ecological and socioeconomic impacts and the early signs of climate-related landscape change increasingly evident. With the exception of direct fishing pressure on marine ecosystems, the greatest present environmental stresses in Alaska are climate-related. Experimental manipulations and model studies both suggest that present and future climate change is highly likely to profoundly alter the range, species mix and functioning of Alaskan ecosystems.

CLIMATE VARIABILITY AND CHANGE

Alaska encompasses extreme climatic differences. The southern coastal margin, including the panhandle and Aleutians, has a maritime climate with cool summers, relatively mild winters, and heavy precipitation, up to 200 inches (500 cm) annually in parts of the southeast, forming large glaciers on the southern mountains. North of the Alaska Range the climate is continental, with moderate summers (July average 59°F or 15°C), very cold winters (January average -13°F or -25°C), rapid seasonal transitions, and annual precipitation of 8-16 inches (20-40 cm). The North Slope beyond the Brooks Range has an Arctic semi-arid climate, with annual precipitation less than 8 inches (20 cm), average July temperature around 39°F (3°C), and snow on the ground nine months of the year. Permafrost is present in all of the state except a narrow belt along the southern coast.

Alaska's climate shows significant interannual and interdecadal variability, associated with large-scale shifts in ocean temperature and salinity regimes, ice conditions, and marine ecosystems in the surrounding seas (Proshutinsky and Johnson, 1997; Groisman and Easterling 1994; Serreze et al., 1995b; Thompson and Wallace, 1998; Parker et al., 1994; Royer, 1993; Francis et al., 1998; Trenberth and Hurrell, 1994; NRC, 1996).

Major Ecological Regions of Alaska

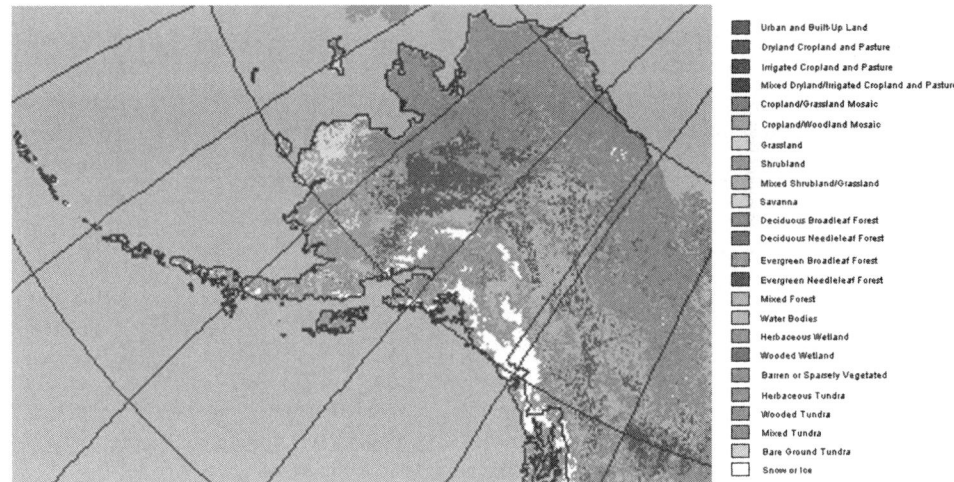

Figure 1: Major Ecological Regions of Alaska. Source: National Atlas of the United States. See Color Figure Appendix

Observed Climate Trends

Alaska's recent climate has shown a strong warming trend. General Arctic warming began in the mid-19th century, but has accelerated in the past few decades (Overpeck et al., 1997). Alaska has warmed 4°F (2°C) since the 1950s on average, with the largest about 7°F (4°C) in the interior in winter (Chapman and Walsh, 1993; Weller et al., 1998). Local weather records show that the growing season in Alaska lengthened by 13 days since 1950 (Keyser et al., 2000). Much of the recent warming, occurred suddenly around 1977, coincident with the most recent of the large-scale Arctic atmosphere and ocean regime shifts (Weller and Anderson, 1998).

Alaska: 20th Century Annual-average Temperature

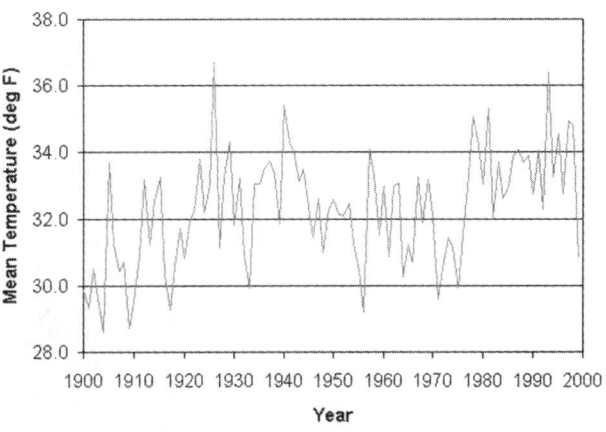

Figure 2: Average temperatures in Alaska have increased over the 20th century, with about 4°F warming since the 1950s. Source: Historical Climate Network, National Climate Data Center. See Color Figure Appendix.

Alaska: 20ᵗʰ Century Annual Total Precipitation

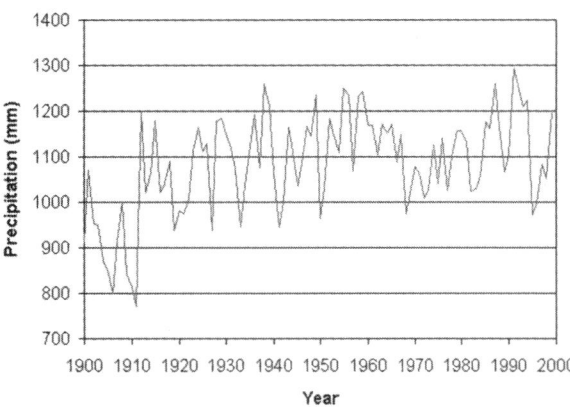

Figure 3: Over the 20th century, precipitation in Alaska has increased. Source: Historical Climate Network, National Climate Data Center. See Color Figure Appendix

Alaska has also grown substantially wetter over the 20th century. The sparse historical record since 1900 shows mixed precipitation trends, with increases of up to 30% in the south, southeast and interior, and smaller decreases in the northwest and over the Bering Sea. The trend to higher precipitation has been stronger recently, a 30% average increase over the region west of 141 degrees longitude (i.e., all of Alaska except the panhandle) between 1968 and 1990 (Groisman and Easterling, 1994).

Alaska's recent warming is part of a strong trend observed throughout the circumpolar Arctic, except for one large region of cooling over eastern Canada and Greenland. This broad Arctic warming has been accompanied and corroborated by extensive melting of glaciers, warming and thawing of permafrost, and retreat and thinning of sea ice (Echelmeyer et al., 1996; Sapiano et al., 1998; Lachenbruch and Marshall, 1986; Osterkamp, 1994; Osterkamp and Romanovsky, 1996; Wadhams 1990; Cavalieri et al., 1997; Serreze et al., 2000; Krabil et al., 1999; Dowdeswell et al., 2000). Paleoclimatic evidence suggests the Arctic is now warmer than at any time in the past 400 years (Overpeck et al., 1997). The start of Arctic warming in the mid-19th century indicates a contribution from natural factors (Overpeck et al., 1997). Of the stronger high-latitude warming of the past three decades, roughly half can be explained by changes in storm track patterns associated with natural patterns of climate variability, although it is possible that anthropogenic changes in radiative forcing may be shifting these patterns so they tend to favor high-latitude warming (Hurrell, 1995, 1996). The remaining share of recent high-latitude warming, roughly half, is broadly consistent with model predictions of the consequences of anthropogenic greenhouse forcing (Serreze et al., 2000). Observations of vegetation and snowcover from satellites, and of the annual fluctuation of atmospheric CO_2 concentration, further corroborate the broad warming trend over northern mid to high latitudes. Mean annual snowcover of the Northern Hemisphere decreased 10% from 1972 to 1992 (Groisman and Easterling, 1994), while the growing season over northern mid- to high latitudes increased by 7 to 14 days (Myneni et al., 1997).

Scenarios of Future Climate

All climate models project the largest warming to occur in the Arctic region, principally because of

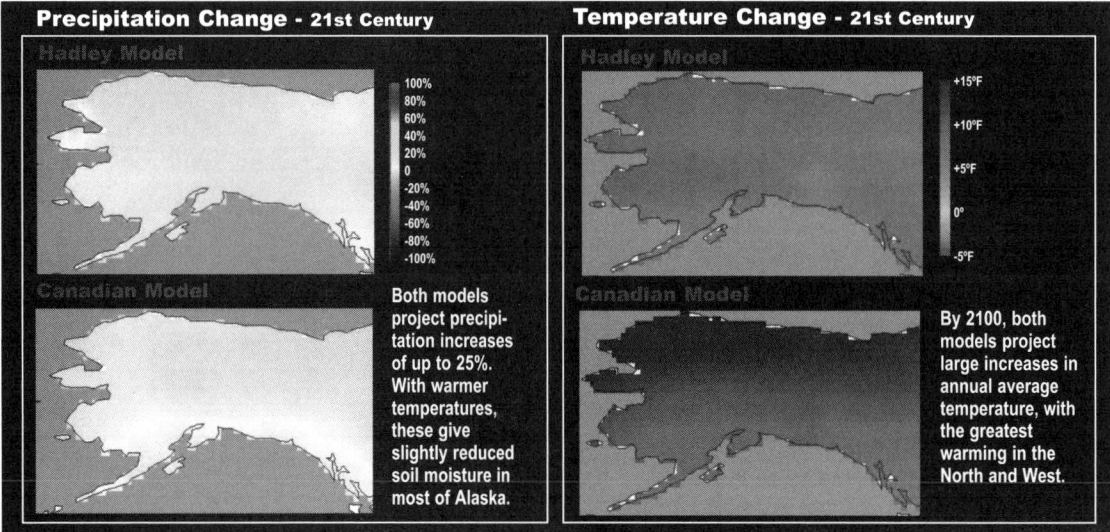

Figure 4: Precipitation and temperature change projected in the 21st century by two climate models. See Color Figure Appendix.

ice-albedo feedback (Kattenberg et al., 1996). Ice and snow are more reflective than the land or water they cover, so after melting, the exposed surface absorbs more solar radiation, accelerating further warming. In both the Canadian and Hadley scenarios (Boer et al., 1984; 1999a,b, McFarlane et al., 1992; Flato et al., 1999; Mitchell et al., 1995; Mitchell and Johns, 1997; Johns et al., 1997), warming in Alaska increases from the southeast to the northwest, and is strongest in winter. In the Canadian model, Alaskan warming ranges from 2 to 5°F (1.1 to 2.8°C) by 2030, and from 7 to 18°F (4 to 10°C) by 2100, accompanied by complete loss of summer Arctic sea ice. In the Hadley model, warming is 1.5 to 3.6°F (0.8 to 2°C) by 2030, 5 to 12°F (3 to 6.5°C) by 2100, with smaller but still extensive loss of sea ice. Comparing these projected future changes to the 4°F (2°C) temperature change already experienced in the last few decades, they range from half as much again to a doubling by 2030, and from a doubling to a quadrupling by 2100.

The Hadley and Canadian scenarios also both project that annual precipitation will increase in most of Alaska, with the largest increases reaching 20-25% in the north and northwest, with some areas of up to 10% decrease along the south coast. The models differ more strongly in projecting seasonal patterns of precipitation changes, particularly in winter. Winters are wetter in the Hadley scenario except in the extreme west and the panhandle, while the Canadian scenario has drier winters everywhere except the Seward Peninsula and northwest coast. Summers have more precipitation in both scenarios except for regions along the south coast. In the Hadley scenario, this region of reduced summer precipitation is confined to the extreme southeast and part of the Alaska Peninsula, while in the Canadian scenario it covers a broad swath along the entire southern coast. Increased evaporation due to warmer summer temperatures exceeds projected precipitation increases, however, so both scenarios project soil moisture decreasing throughout the state, except for an interior region centered on Fairbanks in the Hadley scenario.

KEY ISSUES

The climate changes underway in Alaska have already had major impacts. This synthesis focuses on four key issues - thawing and melting of the cryosphere, particularly permafrost and sea ice; forest and tundra ecosystems; marine ecosystems and fisheries; and subsistence livelihoods. The approach

Winter Maximum Temperature Change

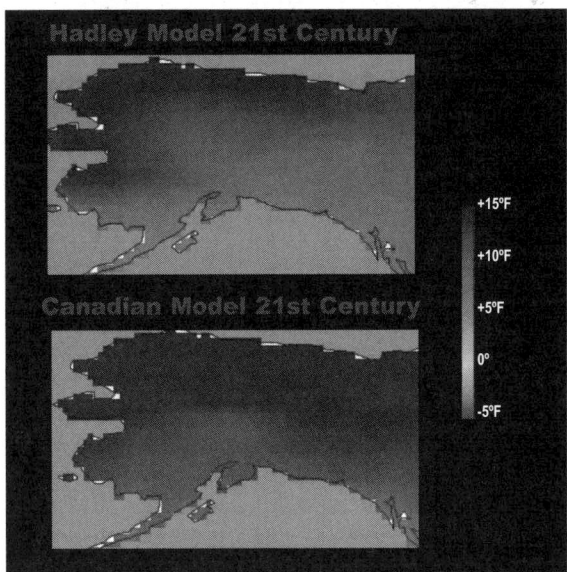

Figure 5: The largest projected warming is in winter, when both models show average daily-high temperatures increasing more than 15°F over the northern half of the state. Source: B.Felzer, UCAR. See Color Figure Appendix.

Summer Soil Moisture Change

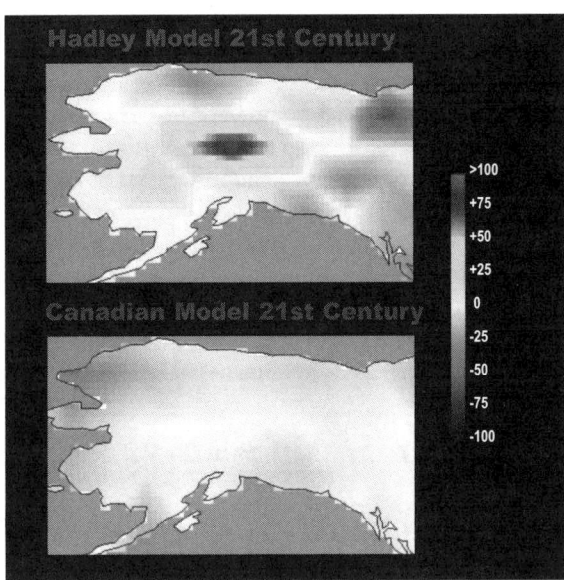

Figure 6: The Hadley model projects increased summer soil moisture in central Alaska and decreases in the north and south, while the Canadian model projects moderate decreases throughout the state. Source: B.Felzer, UCAR. See Color Figure Appendix

combines reviewing projections of future impacts with describing impacts that are already occurring, which are likely to provide insights into the character of the larger projected future impacts.

Thawing and Melting:

Thawing of permafrost, retreat and thinning of sea ice, and reduction of the river and lake ice season are underway and are projected to continue. These changes are likely to bring widespread changes in ecosystems, increased erosion, harm to subsistence livelihoods, and damage to buildings, roads, and other infrastructure (including sanitary systems). In the longer term, longer ice-free seasons are likely to bring substantial benefits to marine transport and offshore operations in the petroleum industry, and will likely have major implications for trade and national defense.

Effects on Forests and Tundra Ecosystems:

Recent warming appears to have brought increased productivity in the boreal forest zone, offset to an uncertain degree by increases in summer moisture stress, fire and insect outbreaks. Future warming is likely to continue increasing both productivity and stresses, and eventually bring large-scale landscape transformation as boreal forest advances into present tundra and mixed forest into present boreal forest. Changes in these ecosystems will possibly have large effects on the global carbon cycle.

Marine Ecosystems and Fisheries:

Alaskan and Bering Sea marine ecosystems show strong signals of climate-driven variation, although their mechanisms are not known. Further climate change is likely to bring large changes in marine ecosystems including stocks important for both the commercial and subsistence catch, but knowledge of their specific character is very limited.

Subsistence Livelihoods:

Subsistence hunting and fishing have been significantly harmed by present climate changes, through

stresses on fish, marine mammals, and wildlife driven by present thawing, sea ice retreat, and ecosystem shifts. While some specific subsistence resources are likely to grow more abundant (e.g., salmon near the northern limit of their range), these stresses are likely to grow more intense, even in the near term.

1. Thawing of Permafrost and Melting of Sea Ice

Throughout Alaska, the landscape and human activities are fundamentally affected by the presence of ice, snow, and permafrost. Because annual average temperatures in much of the southern portion of the state are near 32°F (0°C) – e.g., 28°F (-2°C) in Fairbanks – a small warming can transform the landscape through thawing of permafrost, melting of ice, and reduction of snow cover. The ecological, hydrological, economic, and social effects of these changes to the cryosphere[2] will be large, and will profoundly affect every other domain of impact considered. All components of the cryosphere in the Arctic are experiencing change, including snow cover, mountain and continental glaciers, permafrost, sea ice, and lake and river ice. For example, glaciers in Alaska, as throughout the Arctic, have retreated through most of the 20th century. Estimated losses in Alaskan glaciers are of the order of 30 feet (10 meters) in thickness over the past 40 years, while some have gained thickness in their upper regions, consistent with recent increases in both temperature and precipitation (BESIS, 1997). Melting of glaciers is contributing to rising sea levels worldwide (Meier, 1993), while melting of Alaskan glaciers may have pronounced regional effects through the contribution of their runoff to ocean currents and marine ecosystems in the Gulf of Alaska and Bering Sea, as discussed below. The discussion here concentrates on permafrost and sea ice, whose impacts on people and ecosystems were judged to be most direct and important. General warming would also reduce the ice season on lakes and rivers, impairing transport on ice roads (Cole et al., 1999).

Permafrost

Permafrost underlies about 85% of Alaska, the entire state except for a narrow belt along the southern coast. Its character varies widely, in depth, continuity, and ice content. In the interior and south of the state most permafrost is discontinuous and relatively shallow, reaching depths of 10 to 300 feet (3 to 100 meters). From the Brooks Range north and along the northern and northwestern coasts, it becomes

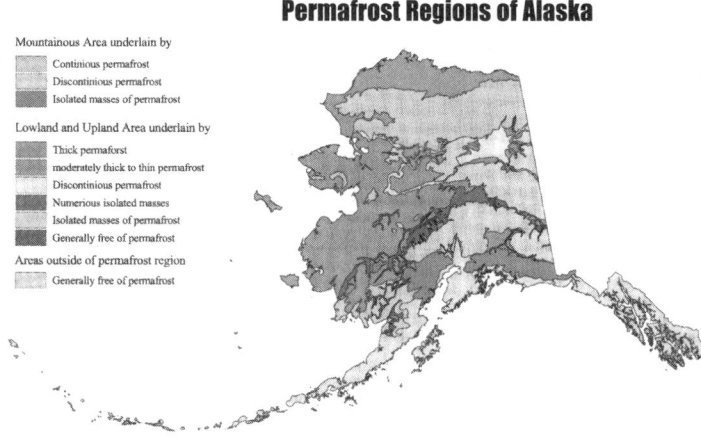

Permafrost Regions of Alaska

Mountainous Area underlain by
  Continuous permafrost
  Discontinuous permafrost
  Isolated masses of permafrost

Lowland and Upland Area underlain by
  Thick permafrost
  moderately thick to thin permafrost
  Discontinuous permafrost
  Numerous isolated masses
  Isolated masses of permafrost
  Generally free of permafrost

Areas outside of permafrost region
  Generally free of permafrost

Figure 7: Permafrost regions of Alaska. Source: O.J. Ferrains, 1965. See Color Figure Appendix

[2] The cryosphere consists of the frozen components of the Earth's surface: ice, snow, and permafrost.

thicker and continuous, reaching depths of 2,200 feet (670 meters) in some locations on the North Slope (Ferrians, 1965; Osterkamp et al., 1985; Brown et al., 1997). Permafrost has profound effects on hydrology, erosion, vegetation, and human activities. It limits movement of ground water and the rooting depth of plants. On slopes, it allows characteristic fluid-like movement of surface soil and deposits. Seasonal thawing over continuous permafrost creates a saturated surface layer in which pools of meltwater accumulate, conducive to marsh and tundra ecosystems and peat formation. Building on permafrost requires that structures be stabilized in permanently frozen ground below the active layer, and that they limit their heat transfer to the ground, usually by elevating them on piles. For example, to prevent thawing of permafrost from transport of heated oil in the Trans-Alaska pipeline, 400 miles of pipeline were constructed elevated on thermosyphons, at an additional cost of $800 million (Cole et al., 1999).

Permafrost in Alaska has been warming for more than a century. Continuous permafrost on the North slope of Alaska has warmed 4-7°F (2-4°C) over the last century (Lachenbruch and Marshall, 1986). Since temperatures at the upper surface of continuous permafrost are still low, typically below 23°F (-5°C), no significant loss of continuous permafrost is projected over the 21st century, although thickening of the active layer may cause active layer detachment, local subsidence, damage to structures, and hydrological changes (Osterkamp and Romanovsky, 1996). The discontinuous permafrost to the south is warmer, usually above 28°F (-2°C). Here, Osterkamp (1994) reported recently increased warming at multiple sites, 1 to 3°F (0.5 to 1.5°C) since the late 1980s, and inferred from this and other evidence that much of the discontinuous permafrost south of the Yukon River and on the south side of the Seward Peninsula must already be thawing. Many reports of localized thawing and associated surface disruptions support this inference (Osterkamp, et al., 1998; Jorgenson et al., 2000; Osterkamp et al., 2000). In the central Canadian Arctic, a general northward retreat of the southernmost margin of discontinuous permafrost by about 60 miles (100 km) over the 20th century has been reported (Kwong and Gan, 1994; French and Egorov, 1998). It is highly likely that recent climate changes have contributed to permafrost warming and thawing. The beginning of permafrost warming pre-dates the recent sharp increase in surface temperature, however, suggesting that other factors than warming – such as natural long-term variability or changes in snow depth and vegetation cover

Figure 8: For much of its length, the Trans-Alaska Pipeline is elevated on refrigerated pilings to prevent local thawing and ground instability. Source: ©David Marusek.

(which alter heat transfer from the surface to the ground) – have also contributed. Moreover, because much of the observed thawing is associated with human disturbance of the surface, and because systematic large-scale observations of changes in permafrost are lacking, the degree of contribution of surface warming to the observed thawing is not yet known.

Continued climate warming is highly likely to bring accelerated thawing of warm discontinuous permafrost. Over the 21st century the top 30 feet (10 meters) is likely to thaw throughout much of the ice-rich discontinuous permafrost zone, although complete thawing is likely to take centuries even in discontinuous permafrost (Osterkamp and Romanovsky, 1999). Canadian studies have projected that even present surface temperatures will cause an eventual further retreat of the southernmost permafrost fringe in the Canadian subarctic by 60 to 100 miles (100 to 160 km) (Dyke et al., 1997), with further retreat of 200 to 300 miles (300 to 500 km) under doubled-CO_2 equilibrium (Woo et al., 1992). The actual pattern of loss of permafrost will, of course, be more complex than a simple uniform retreat. Model studies using three transient climate-model scenarios have projected a 20 - 30% increase in depth of the active layer (Anisimov et al., 1997), and a 12 to 22% reduction in total Arctic permafrost area by 2100 (Anisimov and Nelson, 1997).

Thawing is likely to benefit some activities (e. g., construction, transport, and agriculture) after it is completed, but the intervening transitional period of decades or longer is likely to bring many disrup-

Figure 9: Houses near Fairbanks, Alaska use jacks for support as permafrost thawing causes uneven settling. Source: ©1999, Gary Braasch.

tions and few benefits. Thawing of any permafrost increases groundwater mobility, reduces soil bearing strength, increases susceptibility to erosion and landslides, and can affect soil storage of CO_2, thereby increasing release if the thawed soil drains and dries, or increasing storage if the soil remains flooded. Warming greatly reduces permafrost's bearing strength even if it remains below the freezing point, e.g., by 70% for a pile in permafrost that warms from 25 to 30°F (-4 to -1°C) (Nixon, 1990; Cole et al., 1999). Where permafrost has a high ice content, typically in about half the area of discontinuous permafrost, thawing can induce severe, uneven subsidence of the surface, called thermokarst, observed in some cases to exceed 16 feet (5 m). Human-induced thawing of ice-rich discontinuous permafrost has already damaged houses, roads, airports, pipelines, and military installations; required costly road replacements and increased maintenance expenditures for pipelines and other infrastructure; and increased landscape erosion, slope instabilities and landslides.

Present costs of thaw-related damage to structures and infrastructure in Alaska have been estimated at about $35 million per year,[3] of which repair of permafrost-damaged roads is the largest component (Cole et al., 1999). Longer seasonal thaw of the active layer could disrupt petroleum exploration and extraction and increase associated environmental damage in the tundra, by shortening the season for minimal-impact operations on ice roads and pads. The near-term risk of disruption to operations of the Trans-Alaska pipeline is judged to be small, although costly increases in maintenance due to increased ground instability are likely. The pipeline's support structures are designed for specific ranges

of ground temperatures, and are subject to heaving or collapse if the permafrost thaws. Replacing them, if required, would cost about $2 million per mile. Subsidence from thawing can also destroy the substrate of present ecosystems, destroying them or transforming them to other types of ecosystems, for example changing forests to grasslands or bogs (Jorgenson et al., 2000; Osterkamp et al., 2000). Where large-scale thawing of ground ice has occurred, the landscape has been transformed through mudslides, formation of flat-bottomed valleys, and formation of melt ponds, which can enlarge for decades to centuries (Everett and Fitzharris, 1998, p. 94).

Sea Ice

As permafrost is a prominent feature of the Alaskan landscape, sea ice is a prominent feature of its coasts and the adjoining marine ecosystems. Present for six months along the Bering Sea coast and ten months along most of the Chukchi and Beaufort Seas, sea ice strongly influences coastal climate, ecosystems, and human activities. The area of Arctic sea ice varies up to 50% seasonally, and also shows strong interannual variation. Large and statistically significant reductions in summer sea ice, which have been proposed as an early signal of global climate change (Walsh, 1991), are evident in recent decades despite this variability (Cavalieri et al., 1997; Maslanik et al., 1996, 1999; Wadhams, 1997). Recent reports show area declines of about 3% per decade since the late 1970s, with the largest declines 3.6% per decade in August and September (Serreze et al., 2000). The area of multi-year ice has declined by 14% since 1978 (Johannesen et al., 1999). Model calculations indicate that recent sea-ice trends are consistent with the estimated effects of present greenhouse warming, and are highly unlikely to be accounted for by natural climate variability (Vinnikov et al., 1999). Comparison of two satellite records suggests that the rate of area loss increased from 2.8% per decade in the 1980s to 4.5% per decade in the 1990s (Johannesen et al., 1995). Record low values of summer ice extent have been set repeatedly since 1980 (Chapman and Walsh, 1993; Serreze et al., 1995a), while September 1998 ice area in the Beaufort and Chukchi seas (the western part of the Arctic Basin) was 25% below the prior minimum value over the 45-year record (Maslanik et al., 1999).

Arctic sea ice has also grown thinner over the past few decades. Local observations of sea ice thinning by 3.3 to 6.5 feet (1 to 2 meters) have been reported for several years (Wadhams 1990, 1997; McPhee et al., 1998), but the limited spatial and temporal

[3] This sum represents about 1.4% of the State budget.

coverage of these measurements prevented drawing conclusions about Arctic-wide trends until recently. A recent analysis of submarine ice data, however, has provided the first persuasive evidence of large-scale thinning over the entire Arctic basin. Mean September ice draft[4] observed in six trans-Arctic submarine cruises from 1958 to 1976 was 10 feet (3.1 meters), while mean draft in three similar cruises between 1993 and 1997 was 5.9 feet (1.8 meters). In addition to the 4.1 feet (1.3 meters) of average thinning between the two sets of cruises, the recent cruises also found continued thinning at a rate of 4 inches per year (10 cm/year) from 1993 to 1997 (Rothrock et al., 1999). Evidence of widespread sea-ice melting is corroborated by substantial recent increase in freshwater content of the Arctic Ocean, from depth equivalent of 0.8 meters in 1975 and 2.4 meters in 1997 (McPhee et al., 1998).

Sea-ice retreat allows larger storm surges to develop in the increased open-water areas, increasing erosion, sedimentation, and the risk of inundation in coastal areas. Moreover, coastline where permafrost has thawed is made more vulnerable, which in combination with increased wave action can cause severe erosion (Brown and Solomon, 2000; Forbes and Taylor, 1994; Shaw et al., 1998; Weller, 1998; Wolfe et al., 1998.). Local coastal losses to erosion of the order of 100 feet (40 meters) per year have been observed in some locations in both Siberia and Canada (Semiletov et al., 1999; Solomon and Covill, 1995). Aerial photo comparison has revealed total erosive losses up to 1,500 feet (600 meters) over the past few decades along some stretches of the Alaskan coast (Weller and Anderson, 1998). Several villages on Alaska's west coast are sufficiently threatened by increased erosion and inundation that they must be protected or relocated. Present plans include constructing a $4-6 million sea wall in Shishmaref (a 10-15 year interim solution), and relocating Kivalina on higher ground at an estimated cost of $54 million (US Army Corps of Engineers, 1998).

Under further climate change, further large reductions in sea ice are projected, although there is substantial variation in estimates of the magnitude and timing. Analysis with one transient climate-model scenario projected 60% loss of Arctic summer sea ice area by the time of CO_2 doubling, accompanied by an increase in the duration of the open-water season from 60 to 150 days. The same climate scenario also suggests an increase in the offshore distance of the ice pack from 90 to 125 miles (150 - 200 km) at present, to 300 - 500 miles (500 - 800 km) (Gordon and

Erosion on the Arctic Coast of Siberia

Figure 10: Coastal thawing and sea ice retreat have allowed extreme coastal erosion in both North America and Eurasia, with some local losses of up to 100 feet per year. Source: Igor Semiletov, Pacific Oceanological Institute, Vladivostok.

Projected Summer Sea Ice Change
Canadian Model: An Ice-free Arctic Summer

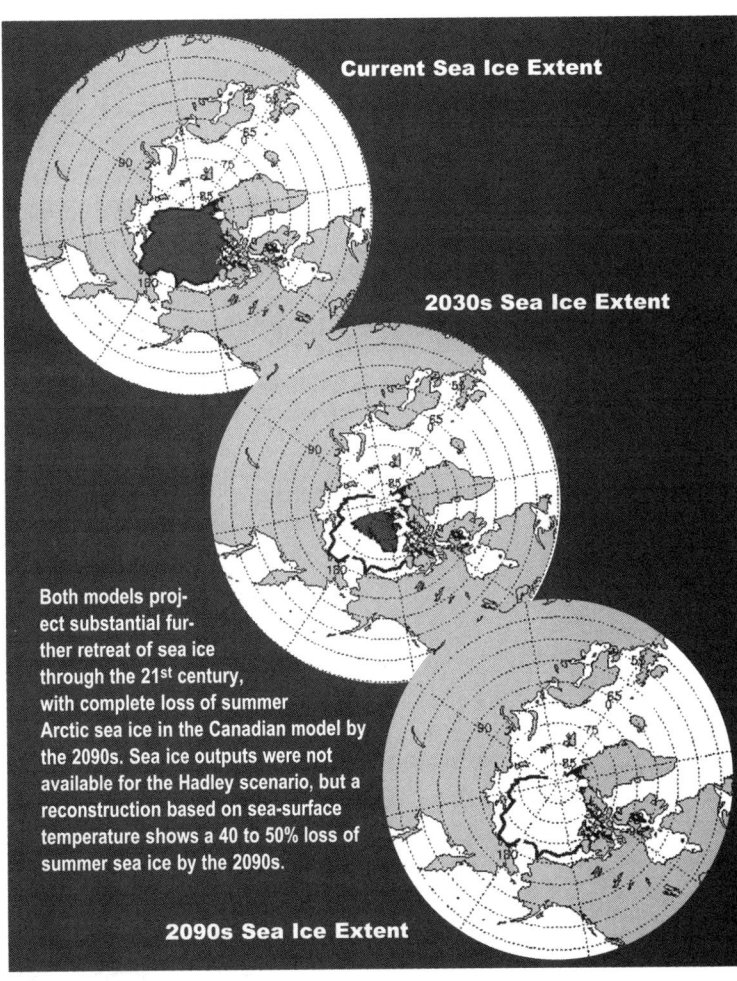

Current Sea Ice Extent

2030s Sea Ice Extent

Both models project substantial further retreat of sea ice through the 21st century, with complete loss of summer Arctic sea ice in the Canadian model by the 2090s. Sea ice outputs were not available for the Hadley scenario, but a reconstruction based on sea-surface temperature shows a 40 to 50% loss of summer sea ice by the 2090s.

2090s Sea Ice Extent

Figure 11: Canadian model projections of future Arctic sea-ice retreat.
Source: B. Felzer, UCAR, 2000. See Color Figure Appendix

---

[4] Ice draft is the depth of sea ice below the water line, equal to roughly 90% of total ice thickness.

O'Farrell, 1997). Both the Hadley and Canadian models project large reduction in summer sea ice by 2100. The Canadian model projects the more rapid loss, as shown in Figure 11, with complete disappearance of summer sea ice by 2100.

Loss of sea ice threatens large-scale change in marine ecosystems, threats to populations of marine mammals and polar bears that depend on the ice, and to the subsistence livelihoods that depend on them. Further retreat may also bring some benefits, principally by facilitating water transport and oil exploration and extraction. Expanded transport possibilities from greatly reduced Arctic sea ice extent and increased open-water season, including the possibility of routine summer navigation through both the Northeast and Northwest Passages (North of the Eurasian and North American continents), are likely to have major implications for both trade and national security.

Adaptation Options

Where sufficient information is available, vulnerability of structures to permafrost thawing can be reduced by careful site selection to avoid permafrost with high ice content and favor permafrost with high gravel content. Unfortunately, local information on permafrost characteristics is often unavailable or inaccurate, and many siting and development decisions fail to consider the information that is available, or the likely future development of the site and its surroundings (Smith and Johnson, 2000). When site or route modifications are not undertaken or not feasible, the effects of permafrost thawing on building and infrastructure can still be reduced, although at substantial cost and difficulty, through several approaches. Local contributions to thawing can be reduced by minimizing physical disturbance of the surface, and through insulation and heat transfer measures to reduce local thermal dis-

Betting on Spring Breakup: The Nenana Ice Classic

The town of Nenana is located about 65 miles southeast of Fairbanks on the Tanana river, a major tributary of the Yukon. In 1917, when Nenana was a construction base for the Alaska Railroad, railroad workers ran a betting pool on when the river ice would break up in the spring. Sufficiently popular to be repeated in subsequent years, the pool became a local tradition that now has been repeated every year for 84 years. Entry tickets cost two dollars, and represent a bet on a single one-minute interval. The jackpot, $800 in 1917, was more than $330,000 in 2000. The high stakes and long continuous history of this contest make it a unique local record of Alaska's 20th century climate history.

The same procedure is used to define the moment of breakup each year. In early March, a large log structure is frozen into the ice about 300 feet from shore, and later joined by a cable to a watchtower on the shore. A strong enough pull on the cable, which occurs when the ice has shifted enough to move the structure about 100 feet downriver, stops the clock.

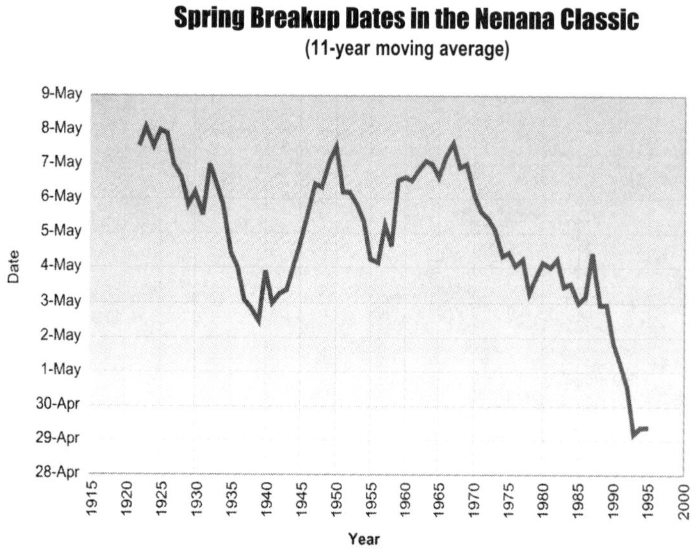

Spring Breakup Dates in the Nenana Classic
(11-year moving average)

Figure 12: The average date of spring breakup of ice on the Tanana River at Nenana has advanced by eight days between the 1920s and the 1990s. Source: Historical data from Nenana Ice Classic, http://www.ptialaska.net/~tripod/breakup.times.html. See Color Figure Appendix.

Over the contest's history the earliest breakup has been on April 20 (in 1940 and 1998), the latest on May 20 (in 1964). Although breakup dates vary greatly from year to year, the past few decades have seen a strong trend toward earlier breakup. Removing some of the year-to-year variation by calculating an 11-year average (for each year, the average of the eleven breakup dates from five years before to five years after) reveals an advance of eight days between the 1920s and the 1990s, from May 7 to April 29. Nenana is a major shipping center for summer barge traffic, so the earlier breakup brings significant local economic benefits. It is also a concrete local indicator of the strong warming trend that has occurred across Alaska over the past few decades.

turbance. Piles used to support structures can be sunk deeper in the permafrost or refrigerated, to maintain their bearing strength longer as the permafrost warms and active layer thickens. With enough advance planning, local thawing can be actively induced before construction, by stripping vegetation and surface soil from the site five years or more in advance (Osterkamp et al., 1998). Which of these types of measures is most promising will depend on site characteristics, the type of project, and its intended lifetime. For roads and runways, the consequences of thawing can be reduced by building with gravel rather than paved surfaces, as they can be more readily repaired after subsidence. Coastal settlements threatened by increased storm surge or erosion can be protected with sea walls or other fortification, or relocated further inland. No adaptation options are likely to be available for terrestrial ecosystems threatened by permafrost thawing, or marine ecosystems threatened by sea-ice retreat.

2. Effects on Forest and Tundra Ecosystems

Forests cover 129 million acres of Alaska, about one third of the state (Powell et al., 1993). Various forms of tundra cover another third, in mountainous and coastal regions and north and west of the Brooks Range. Of the forested land, about 10% is temperate coastal rainforest, the remainder interior boreal forest. About 21 million acres, or 16% of total forest, is classified as productive, capable of average growth of 20 cubic feet per acre per year. About 4 million acres, nearly all of it in the coastal forest, is outside protected areas and has the productivity of 50 cubic feet/acre-year necessary to support commercial harvest with road construction (Berman et al., 1999). The state's timber harvest increased from 600 to 1,100 million board feet from 1986 to 1990, and has since declined to about 500 million board feet. Employment and income in the industry followed the same pattern, peaking at 4,000 jobs and $200 million in 1990, declining to 2,500 jobs and $130 million by 1997. The decline of the 1990s principally reflects two economic causes: the closure of two pulp mills in southeast Alaska, and the depletion of Native Corporation timber inventories, which were exported in large quantities as round logs to convert assets to cash during a period of high world prices (Berman et al., 1999). In addition to their commercial value, Alaskan forests provide various ecosystem services

and support subsistence livelihoods and recreation activities.

Recent warming in Alaska has increased average growing degree-days by about 20% over the state, bringing apparent increases in forest productivity on sites that are not moisture limited – principally in the southern coastal forest, but also including some regions of the boreal zone (Ciais et al., 1995; Myneni et al., 1997). At the northern margin of the boreal forest, the present climate already favors forest expansion into the tundra zone, particularly on the Seward Peninsula, with the potential for such expansion estimated as 35 miles per °F of climate warming (100 km per °C) (Weller and Lange, 1999). On sites that are moisture-limited, which occur through much of the interior, recent warming has apparently increased moisture stress and reduced productivity (Barber et al., 2000). Near Fairbanks the average number of days exceeding 80°F (27°C) annually has tripled since 1950, imposing moisture stress on white spruce stands that can be observed in clear negative correlation of productivity with warm, dry summers (Juday et al., 1998, Fig. 3.13). It has been suggested that the past 20 years have seen the greatest moisture stress and lowest productivity of the 20th century through much of the interior boreal forest (Juday and Barry, 1996) .

The 1990s Outbreak of Spruce Bark Beetles on the Kenai Peninsula

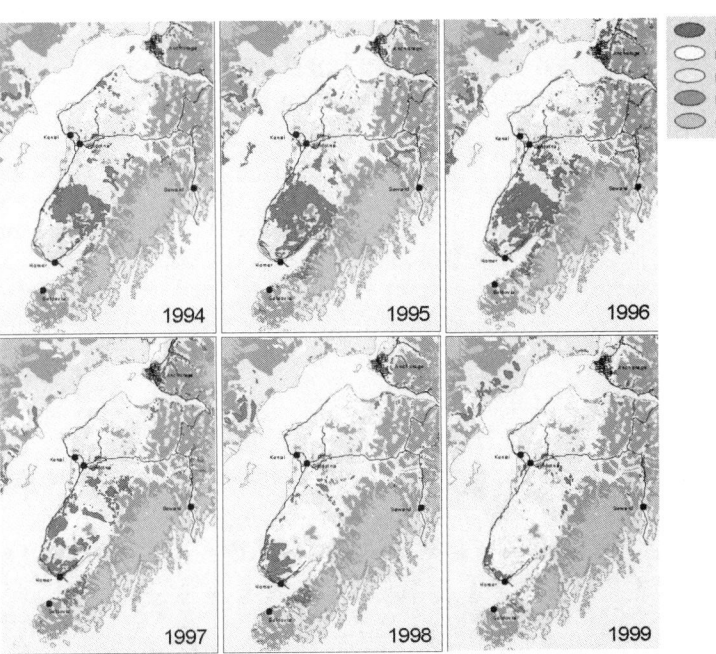

Figure 13: Since 1992, the largest outbreak of forest insects ever recorded in North America has caused widespread tree mortality over 2.3 million acres. Source: USDA Forest Service. See Color Figure Appendix

Substantial changes in patterns of forest disturbance, including insect outbreaks, blowdown, and fire, have also been observed in both the boreal and southeast coastal forest. Although systematic large-scale observations have not been made, localized observations appear to support the hypothesis that these changes are climate-driven. A sustained outbreak of spruce bark beetles since 1992 has caused over 2.3 million acres of tree mortality on the Kenai Peninsula, the largest loss from a single outbreak documented in the history of North America (Werner, 1996). The association of warmer temperatures with both accelerated beetle development times and increased tree vulnerability through moisture stress makes it likely that recent warming contributed to the outbreak (Juday et al., 1998). Outbreaks of defoliating insects in the boreal forest, including spruce budworm, coneworm, and larch sawfly, have also increased sharply in the 1990s, affecting a cumulative total of 800,000 acres (Holsten and Burnside, 1997). Susceptibility of interior forests to insect attack may also have increased due to canopy breakage from the heavy snow loads typical of warmer winters.

In Southeast forests, warmer winters since the 1970s with more precipitation falling as rain have reduced the frequency of low and moderate-elevation avalanches, allowing mountain hemlock to colonize alpine tundra (Veblen and Alaback, 1996). Reduced low-elevation snowpack has also likely

contributed to the extensive decline of Yellow Cedar in the coastal forests, due to freezing of their shallow root systems during winter cold spells with no insulating snow cover (Hennon and Shaw, 1997). Over the same period, the southern coastal forests have also seen a marked increase in the frequency of gale-force winds, which are the primary disturbance agent in these forests (Veblen and Alaback, 1996), and outbreaks of the defoliating western black-headed budworm that appear to be triggered by warm dry summers (Holsten et al., 1985; Furniss and Carolyn, 1977).

Forest fire frequency and intensity have also increased markedly since 1970. As Figure 14 shows, the 10-year average of boreal forest burned in North America, after several decades of around 2.5 million acres (1 million hectares), has increased steadily since 1970 to more than 7 million acres (3 M ha). Boreal forest fire reached extreme values in both Eurasia and North America in 1998, with over 27 million acres burned (11 M ha) in total, 10 million (4.7 M ha) in North America (Kasischke et al., 1999). Analysis of historical Canadian fire data shows a strong association between area burned and anomalous patterns of mid-tropospheric circulation that tend to bring extended warm dry periods (Skinner et al., 1999).

A major change in Alaskan settlement geography since 1970, promoted by policies including large-scale private transfer of public lands and extensive road-building, has greatly increased dispersed settlement in forest land. At the same time, other policies to transform native villages into permanent communities created more than 60 communities with significant costly infrastructure surrounded by boreal forest (Leask, 1985). These trends have greatly increased the vulnerability of people and settlements to forest fires. A single major fire in June 1996, for example, burned 37,000 acres of forest and peat, causing $80 million in direct losses and destroying 450 structures including 200 homes. As many as 200,000 Alaskan residents may now be at risk from such fires, with the number increasing further as outlying suburban development continues to expand (Nash and Duffy, 1997).

Continued increases in CO_2 concentrations and projected further climate warming are likely to bring continuing increases in forest productivity (Keyser et al., 2000), although these are likely to be limited by accompanying increases in summer moisture deficit, fire, and insect outbreaks (Fleming and Volney, 1995; Fleming and Candau, 1997; Hogg and Schwarz, 1997; Hogg, 1999; Oechel et al., 1997b;

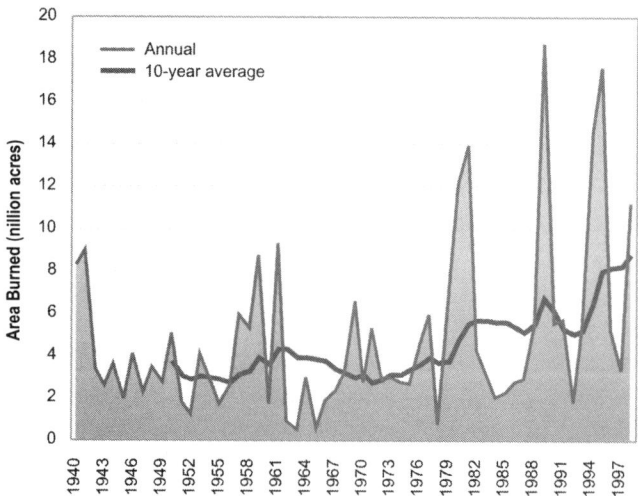

Annual Area of Northern Boreal Forest Burned in North America

Figure 14: The Alaskan boreal forest is a small part of an enormous forest that extends continuously across the northern part of North America. The average area of this forest burned annually has more than doubled since 1970. Source: Kasischke and Stocks, 2000. See Color Figure Appendix

Sieben et al., 1997; Volney, 1996; Kasischke and Stocks, 2000). Increased fire risk is likely, even in the near term. One study projecting fire risk under doubled-CO_2 equilibrium scenarios found a large increase in the area facing extreme fire risk in Canada and Siberia, very similar in both size and spatial pattern under four different climate models (Stocks et al., 1998). Substantial climate-related changes have also been conjectured for the coastal forest over several decades, including the appearance of new fungi and the appearance of significant fire risk for the first time in the observed record (Juday et al., 1998).

Over the longer term, climate change is likely to bring large landscape-level vegetation changes to both forest and tundra regions. Experimental studies in both boreal forest and tundra have shown that warming increases nitrogen availability (Van Cleve et al., 1990; Lukewille and Wright, 1997). At one tundra site, a decade of experimental 6°F (3°C) warming brought major reorganization of the species mix, principally due to increased nutrient availability through changes in nitrogen mineralization (Chapin et al., 1995). Shrubs increased in dominance, while mosses, forbs and lichens were reduced or eliminated. Because shrubs transpire but the declining species do not, such a reorganization would increase evapotranspiration, with large impacts on surface water budgets at many sites likely, including reduced pond formation and runoff and drying of wetlands (Rouse et al., 1997). Moreover, the declining species include some that are critical for lactation and winter nutrition of caribou. Although there exist different views of how sensitive caribou are to such climate-driven changes, it is possible that they could greatly reduce herds, with serious consequences for native communities that depend on them (Gunn, 1995; Callaghan et al., 1998).

Equilibrium studies using biogeochemistry and biogeography models have projected large increases in vegetation carbon under climate change in both boreal and tundra ecosystems (McGuire et al., 1995; McGuire and Hobbie, 1997). These equilibrium studies exclude dynamics of ecosystem response and provide very limited treatment of disturbance, however, and may consequently either over or underestimate the effects of climate change on ecosystems.

Boreal forests and tundra ecosystems also contain large stores of carbon in their soils. Worldwide estimates of carbon content are about 50 gigatonnes (GtC) in tundra soils, and 200 - 500 GtC in boreal

soils (McGuire et al., 1995; Melillo et al., 1995; McGuire and Hobbie, 1997; Oechel et al., 1993; Post et al., 1982; Robinson and Moore, 1999). These soils can act as either sources or sinks of greenhouse gases, depending on temperature and moisture conditions. As temperatures rise and soils thaw and dry, they become more susceptible to oxidation and release of CO_2. Where drainage is poor and the soil remains wet after thawing, emissions of methane

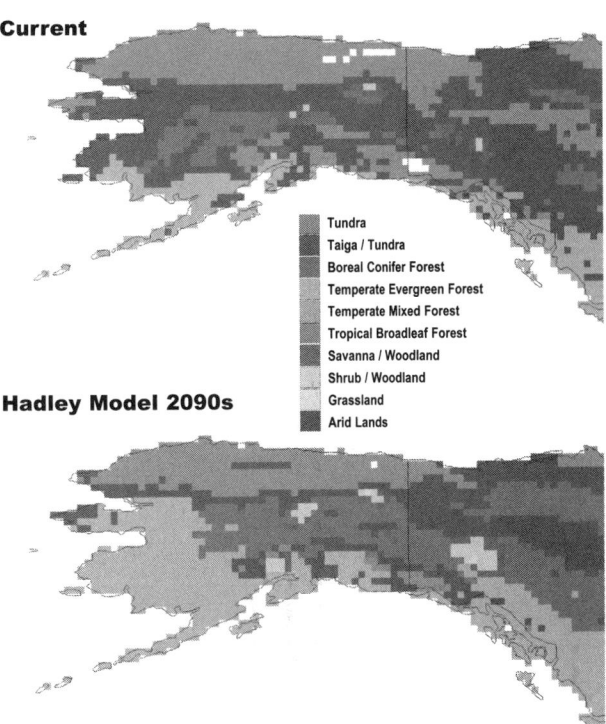

Simulated Vegetation Distribution

Current

Tundra
Taiga / Tundra
Boreal Conifer Forest
Temperate Evergreen Forest
Temperate Mixed Forest
Tropical Broadleaf Forest
Savanna / Woodland
Shrub / Woodland
Grassland
Arid Lands

Hadley Model 2090s

Figure 15: Under the Hadley scenario, the MAPSS biogeography model projects large-scale loss of tundra and taiga ecosystems as forests expand north and west. Likely consequences include disruption of wildlife migration and associated subsistence livelihoods, as well as the potential for large releases of soil carbon. Source: R. Neilson et al., 1998. See Color Figure Appendix

may increase (Gorham, 1995; Rivkin, 1998). Studies of both boreal forest and tundra have observed significant increases in soil carbon release following seasonal warming and thawing (Goulden et al., 1998). Growing-season observations of specific tundra sites have found them to operate as net CO_2 sinks in the cool, wet 1970s, net sources in the warmer, drier 1980s, and net sinks in the warm, wetter 1990s (Oechel et al., 1993, 1995; Vourlitis and Oechel, 1999). The seasonality and mechanism of carbon storage and release in tundra ecosystems have been called into question, however, by recent evidence that carbon release in winter may predominate (Oechel et al., 1997a).

In addition to increased carbon storage and changes in nutrient cycling, biogeography models consistently project large-scale transformation of Arctic landscapes, in which the northern edge of the boreal forest advances into the tundra (Melillo et al., 1996; Everett and Fitzharris, 1998). In Alaska, northward forest advance is likely to be constrained by the Brooks Range, but substantial westward expansion on the Seward Peninsula is possible, as occurred during a warm climatic period 6,000 years ago (Chapin and Starfield, 1997; Foley et al., 1994). Tundra, constrained by the coastline, is likely to both change in composition and shrink in area, by as much as two thirds, worldwide (Neilson et al., 1998; Everett and Fitzharris, 1998).

In southern Alaska, temperate coniferous and mixed forests are likely to advance into the boreal zone. One model study found that an 8°F (4.5°C) warming imposed on the Seward Peninsula induced two landscape transformations over a century, from tundra to boreal spruce forest and subsequently - principally because of fire - to a mixed, deciduous-dominated forest (Rupp et al., 2000). Over one to two centuries, other possible landscape changes include expansion of the coastal forest westward on the Alaska Peninsula (Chapin and Starfield, 1997); an expansion of forests to higher elevations, including colonization of some formerly glaciated lands; and a shift of interior regions with the greatest precipitation deficit to Aspen parkland.

Adaptation Options

Projected increases in forest productivity, including the possibility of northward expansion of commercially valuable species, would likely bring commercial benefits if not offset by increased moisture stress, fire, and insect outbreaks. Various adaptation measures could help to offset climate-induced increases in fire risk in commercially valuable forests or near settlements. These might include expanded road networks to increase fire-suppression capability and facilitate salvage and sanitation logging - assuming that subsequent increase in settlement in forested areas can be discouraged; periodic controlled burns around settled areas to create buffers; and increased investment and staffing in fire suppression. Any strategy based on expanded fire suppression, however, will carry its own ecological costs and also risks being ineffective in the long term, because by removing risk from property owners it would sustain incentives to build in fire-prone areas.

An alternative approach would create incentives to reduce private risk, e.g., by creating rural fire-protec-tion districts in high-risk areas supported by special property taxes; requiring risk-adjusted assessment of fire insurance rates; or encouraging rural residents at risk to form volunteer fire and emergency-response cooperatives at their own expense. This approach would represent a radical departure from historical policies. A related strategy might reverse present policies that encourage dispersed development, by providing infrastructure only in present or designated densely settled areas.

For the projected larger-scale ecological and landscape transformations, no adaptation strategies are likely to be available.

3. Marine Ecosystems and Fisheries

The Gulf of Alaska and Bering Sea support marine ecosystems of great diversity and productivity. The Bering Sea supports at least 450 species of fish, crustaceans, and mollusks, and 25 species of marine mammals. The population of seabirds in Alaska is the largest and most diverse of any similar-sized region in the Northern Hemisphere, with 66 species present at some time of the year, and 38 species - over 50 million individuals - that breed there (Piatt and Anderson, 1996; Meehan et al., 1999).

Roughly 25 species of fish, crustaceans, and mollusks are commercially exploited in the Alaskan fishery. In 1995, Alaska's fisheries landed 2.1 million tons with an ex-vessel value (the amount paid to fisherman) of $1.45 billion, representing 54% of the landings and 37% of the value of all US fisheries. Of this total, pollock were the largest share of volume (1.3 million tons, $297 million) while salmon were the largest share of value (497,000 tons, $490 million, of which sockeye contributed 175,000 tons for $321 million, and pink 218,000 tons for $80 million). A notable contributor to the value of the fishery is the Tanner crab or *Opilio*, for which the volume harvested was only 37,000 tons but the ex-vessel value was $175 million, more than any species except sockeye salmon and pollock. The Alaskan fishery employs about 20,000 people in harvesting and processing (Knapp et al., 1999, Table 1).

The Bering Sea and Gulf of Alaska have shown marked fluctuations in their physical and ecological characteristics over time. Observed ecological fluctuations have included large-scale shifts in the abundance and distribution of many important fish, invertebrates, and marine mammals. Many ecosystem components show clear association with interannual and interdecadal climate variability, with the influence of interdecadal variability apparently stronger than that of interannual variability (Royer, 1982; Parker et al., 1994; NRC, 1996; Francis et al., 1998; Brodeur et al., 1996). While there is climate-driven variability at many different time scales, a few inter-decadal climate variations during the 20th century have apparently caused rapid and extreme shifts in the organization of these marine ecosystems, most recently in 1977. Previous shifts occurred in 1924 and 1946 (NRC, 1996, p. 197), and some data suggest another may have occurred in the mid-1990s (Mantua et al., 1997; NOAA, 1999).

The regime shift of 1977 brought warmer sea-surface temperatures and a sharp reduction in sea ice in the Bering Sea (NRC, 1996; BESIS, 1997). Salmon runs soared, and have largely remained high since then. The catches of 1997 and 1998 were of average volume but dropped sharply in value, principally because of large declines in the lucrative Bristol Bay sockeye run, roughly offset in volume but not in value by huge runs of Pink salmon (Kruse, 1998). Groundfish species including pollock, Pacific cod, Arrowtooth flounder and Yellowfin sole dropped to low levels in 1977 and 1978, then began a sustained climb to record levels in the mid 1980s, after which they have stabilized or declined (Witherell, 1999). Greenland turbot, the groundfish in the region most adapted to a cold climate, declined (NRC 1996, Fig 4.18). As pollock and other predators increased, several species of forage fish with high nutritional value, such as capelin and herring, declined sharply. Various marine mammals and seabirds that fed on these species changed their diets to other less fatty species, and have in turn declined sharply (NRC 1996, Fig. 4.27).

Populations of many species of seabirds, including kittiwakes, murres, cormorants, larus gulls, guillemots, puffins, and murrelets, have declined by 50 to 90% since the 1970s (NRC 1996, pp. 118-120). Marine mammals show similar signs of food stress. In the Gulf of Alaska, both Stellar Sea Lions and Harbor Seals have declined by more than 80%. The extreme decline of Stellar Sea Lions has prompted significant restrictions on the pollock fishery since 1998, to increase the sea lions' food supply. Northern Fur Seals declined by about 35% from 1970 to 1986, then rebounded somewhat through 1990. Sea otters have declined as much as 80% since 1990 over much of the west coast, but this decline has been attributed to predation rather than food shortage. Estes et al. (1998) suggest that a few *Orca* whales, perhaps only a single pod, could account for the observed decline if they began to prey on sea otters following a decline in their usual

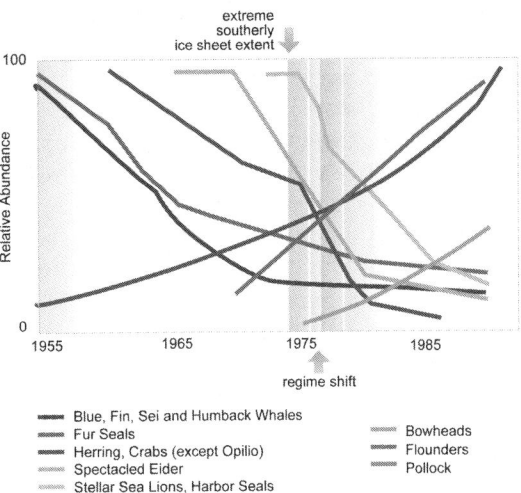

State of Bering Sea Ecosystem

Legend:
— Blue, Fin, Sei and Humback Whales
— Fur Seals
— Herring, Crabs (except Opilio)
— Spectacled Eider
— Stellar Sea Lions, Harbor Seals
— Bowheads
— Flounders
— Pollock

Figure 16: The climatic regime shift of the late 1970s caused large-scale reorganization of the Bering Sea ecosystem. **Source: simplified from NRC (1996). See Color Figure Appendix**

prey.

While climatic effects on larger-bodied, longer-lived species such as marine mammals through changes in food supply are most pronounced at longer time-scales, shorter-term changes can affect them in other ways, such as changes in the extent of sea ice that provides habitat for some species and excludes others (Fay, 1974). For example, the light ice year of 1979 brought unusually large overlap in the distributions of seals and walruses, and saw a high rate of walrus predation on seals (Lowry and Fay, 1984). Large changes in numbers and location of other commercially important fish species have been reported in other Arctic waters, in response to climate regional warmings or coolings of the order of 2°F (1°C) (Buch et al., 1994; Vilhjalmsson, 1997).

These ecosystem changes reflect the joint effects of decadal-scale climate fluctuations and human harvesting of fish and marine mammals, but the complex time and space scales of both climate variation and human pressure prevents separating the contributions of each (NRC, 1996). Moreover, the pathways of climatic influence on these systems are not known. They likely reflect combined effects of changes in streamflow, and the nutrient content, temperature, and vertical stability of coastal waters.

For the Bering Sea and Gulf of Alaska, one likely influence involves changes in the Alaskan coastal current. The intense storm systems generated by the Aleutian Low drop as much as 30 feet (10 meters) of snow annually on the coastal mountains that ring the Gulf of Alaska, forming a large glacier system. The runoff from these glaciers is roughly 800,000 cubic feet per second, comparable to the mean annual discharge from the Mississippi, and contributes more than 40% of the freshwater input to the northeast Pacific. This runoff forms the swift narrow Alaska coastal current, whose low-salinity waters flow westward along the coast and through the Aleutian passes into the Bering Sea (Royer, 1981; Royer, 1982). Significant future changes in these coastal storm systems, which may be associated with either climate variability or anthropogenic climate change, could consequently cause large changes in the temperature, salinity and nutrient content of these waters, and hence in the organization of their ecosystems.

While it is possible that the responses of these ecosystems to future climate change will be large, their specific character is highly uncertain. Effects of stream warming on salmon can be projected with more confidence than any oceanic effects on any species: salmon are likely to benefit in the northern end of their range and be harmed in the south (Berman et al., p. 15). One preliminary study conjectured that 21st century climate change could increase or decrease particular Alaskan fisheries by as much as a factor of two (Knapp et al., 1999).

Adaptation Options
Any substantial change in the abundance, age-class distribution, or location of a commercially exploited species can bring large socioeconomic effects. Fisheries have tended to develop on stocks that are abundant, as the Alaskan pollock fishery has grown from minimal levels over the past 30 years. They

Figure 17: Diverse subsistence livelihoods based on fish, marine mammals and other wildlife, are practiced throughout Alaska. Source: D. Schmitz, National Park Service archive photo

have also tended to over-capitalize, leading to intense competition and rent-seeking when catches decline (Marasco and Arom, 1991; NMFS, 1996, p. 34). The effect of fluctuating catch on fishery revenue depends on the elasticity of demand. The Pacific halibut and salmon fisheries have been estimated to operate in the inelastic region of their demand curves: increases in catch reduce prices so much that revenue falls, as when much of the record 1991 Pink Salmon catch was dumped at sea (Knapp et al., 1999, p. 12). Demand for pollock is presently estimated to be elastic, but would become inelastic with modest increases in catch (Criddle et al., 1998).

In the face of such extreme uncertainty about the direction and magnitude of future climate effects on fisheries, the most useful adaptation options will be measures that increase the robustness of human activities and communities to shifts in the location and abundance of different species. The present system is quite vulnerable to climate change, because specialization of capital and the regulatory structure limit its robustness. Many communities specialize strongly in one or a few species (e. g., Bristol Bay is highly dependent on sockeye salmon, while Dutch Harbor is highly dependent on pollock and crab). In extreme cases like the Bristol Bay salmon fleet, equipment is so specialized that it is only useful for one fishery in one location. Regulatory measures that favor Alaskan shore-based processors over off-shore processing provide jobs and secondary economic benefits in Alaska, but reduce the ability of a fishery to respond efficiently to climate-driven shifts in the distribution of stocks (Huppert, 1991). One important aspect of the present regulatory system that does promote robustness is the use of a limited-entry program. The ability of this program to respond to stock fluctuations could be further improved by allowing buyback of quotas, or by denominating allocations in terms of shares of a variable total harvest, rather than in terms of specific quantities of catch.

4. Subsistence Livelihoods

Subsistence makes an important contribution to livelihoods in many isolated rural communities in Alaska, especially but not exclusively for native peoples. While subsistence is practiced to gather food, subsistence resources and the activities associated with their harvest also make important contributions to health, culture, and identity (Callaway et al., 1999; Berkes, 2000; Wenzel, 1995).

Alaska's 117,000 rural residents are entitled to practice subsistence hunting and fishing on state, federal, and private lands and waters, while urban residents also quality for subsistence activities on state and private lands. The subsistence harvest by rural residents is about 43 million pounds of food annually (20 million kg), or about 375 pounds (170 kg) per rural resident. The subsistence harvest is largest in the most remote communities, about 500 - 800 pounds (225 - 350 kg) per person annually. Fish comprise 60% of total subsistence food, but there is substantial inter-regional variation: west coast communities rely predominantly on fish, interior ones on fish and land mammals, and northern communities principally on marine mammals (Wolfe and Bosworth, 1994).

The links between subsistence harvest and commercial activity are complex. If subsistence food were not available, communities would have to substitute purchased food. With an assumed cost of $3 - 5 per pound of purchased food, a study of four rural communities with large wild food harvests (590 to 760 pounds, or 270 to 350 kg, per person) found that the cost of replacing the wild food harvest would be $1,800 to $3,800 per person, or 13% to 77% of community per capita income (Callaway et al., 1999, p. 70). Moreover, practicing subsistence requires cash income to buy the required equipment, such as guns, boats, and snowmobiles. In one surveyed community (Unalakleet in 1982), the cost of practicing subsistence was about $10,000, nearly half of mean household income (Callaway et al., 1999, p. 65). Consequently, particularly for fishing in coastal communities, the subsistence and commercial harvests may be closely linked: profits from the commercial catch may help pay for required subsistence equipment, and subsistence fish may also be taken during commercial fishing.

Many aspects of the climate change already occurring, and its consequences for forests, marine ecosystems, permafrost, and sea ice discussed above, are already causing multiple serious harms to subsistence livelihoods. Many populations of marine mammals, wildlife, and seabirds have been reduced or displaced. Reduced snow cover, a shorter river ice season, and thawing of permafrost all obstruct travel to harvest wild food. Declines in some fish stocks have harmed subsistence as well as commercial harvesters (Weller and Lange, 1999; Mulvaney, 1998).

The most extreme effects of recent changes on subsistence livelihoods have been from changes in sea ice, which have obstructed hunting of marine mammals. The ice is further from shore, thinner, and present for less of the year. These factors, and the rougher seas encountered in the larger open-water areas between the shore and the ice, have made hunting more difficult, more dangerous, and less productive. Retreat of the ice is also likely to directly harm some species on which subsistence hunters rely, including bearded seals and walrus. Walrus are particularly at risk because they need ice strong enough to hold their weight over water shallow enough that they can reach the bottom to feed (Callaway et al., 1999). Polar bears need sea ice to hunt seals, and recent ice reductions have been associated with declining health and birth rate. Projected further large reductions in ice duration and extent are likely to threaten them with extinction (Stirling et al., 1999).

Some subsistence harvests, such as salmon stocks near the northern end of their range, are likely to benefit from projected climate change. Still, most projected near-term climate changes are likely to intensify existing unfavorable impacts through further loss of sea ice, river ice, and permafrost. In addition, shifts in the composition of tundra vegetation may decrease nutrition available for caribou and reindeer, and invasion of tundra by boreal or mixed forest is likely to curtail the range of caribou and musk-ox (Gunn, 1995). As changes in the cryosphere and both terrestrial and marine ecosystems continue, continuing large changes or displacements of the resources available for subsistence are likely, requiring subsistence communities to make major changes in their practices, or move.

Adaptation Options

Subsistence cultures have historically exhibited substantial adaptability to year-to-year fluctuations in abundance of different species by shifting practices and target species, which likely implies some ability to adapt to effects of near-term climate change (Sabo, 1991). Subsistence practices are now both extensively regulated and hotly contested, however, posing challenges to traditional means of adaptation. Moreover, for many subsistence-dependent communities, particularly northern coastal communities that rely on hunting marine mammals, few adaptation options are likely to be available. Consequently, it is possible that projected climate change will overwhelm the available responses. Some communities may be forced to reduce their dependence on the wild harvest, or relocate. General measures to increase the income and wealth of subsistence-dependent communities, and consequently their ability to adapt to large-scale changes in the subsistence resources on which they depend, would likely mitigate the impacts of lost subsistence resources on nutrition, health, and incomes, but would likely have little effect in mitigating the associated social and cultural impacts (Nuttall, 1998).

ADDITIONAL ISSUES

Due to limited time and resources in this first Assessment, the Alaskan regional study has focused principally on the four critical issues discussed above. The choice of these four reflected the judgments and concerns of participants in Alaskan workshops from 1997 to 1999, but does not imply that these are the *only* important areas of climate impacts in Alaska. Other areas of potentially significant impact include, e.g., freshwater, agriculture, tourism, recreation, and human health. Very preliminary discussions are provided here of freshwater, agriculture, and tourism, based on contributions to workshops from 1997 to 1999, and the scientific literature.

Freshwater

Present and projected climate warming is likely to alter both seasonal and annual river flows in the Yukon and other Alaskan rivers, but the aggregate effects are quite uncertain. Over the entire Arctic basin, a recent analysis of runoff based on streamflow gauges found a significant increase in winter runoff with the largest increases in Alaska and Siberia, consistent with recent winter warming. The larger spring and summer flows show a complex spatial mix of increases and decreases, with significant decreases in Western Canada (both Hudson Bay and the McKenzie Basin) and small increases in Alaska (Lammers et al., 2000).

Model projections of future climate effects on Arctic river flows are also spatially mixed. Total Arctic basin runoff is projected to increase (Van Blarcum et al., 1995; Shiklomanov, 1997; Hagemann and Dumenil, 1998; Walsh et al., 1998), by 10 - 20% annually and 50 - 100% in the small winter flows by the time of CO_2 doubling (Clair et al., 1998; Shiklomanov et al., 2000). One analysis of the Mackenzie basin projected a decrease in annual flow under climate change, however, consistent with the observed spatial distribution of recent flow changes (Kerr, 1997). Uncertainties in these projections for spring, summer, and annual flows are large, however.

Changes in total and seasonal river runoff are likely to interact with other changes in the oceans and cryosphere to yield complex patterns of ecological and socioeconomic impacts. Combined with increases in sea level and storm surge, they would alter the hydrology of coastal wetlands and deltas, possibly impairing seabird and shorebird breeding (Meehan et al., 1999). Spring flooding due to river ice jams is likely to reduce, bringing reduced flood risk to riverside communities but possibly drying out lakes and riparian and delta ecosystems that depend on periodic flooding (Beltaos and Prowse, 2000; Prowse and Conly, 1998). In contrast, smaller rivers and streams that presently freeze solid will likely retain some flowing water beneath the ice, enhancing fish habitat but also making these rivers liable to ice-jam flooding for the first time.

Agriculture

A lengthened growing season could possibly bring a substantial increase in Alaskan agricultural land and production, including the potential for introduction of new crops and animals. Permafrost thawing will likely impair agricultural potential in moisture-limited regions, however, by allowing drying of surface soils. Thawing can also exacerbate soil erosion and loss of organic materials, or obstruct agriculture in regions of ice-rich permafrost through thermokarst formation. Some of these effects could be mitigated through irrigation and soil conservation measures. In the long term, projected changes in tundra ecosystems are likely to seriously harm reindeer herding, through increased snow or ice cover of forage during warmer wetter winters, reduction of forage quality in dry summers, tundra fires, and expansion of forest into tundra. A climatic contribution to the large recent decline in Russia's domestic reindeer herds is possible (Weller and Lange, 1999).

Tourism

Alaskan tourism has increased in recent decades, in parallel with climate warming, although it is not clear how much of the increase can be attributed to the warming climate. Continued warming could possibly bring significant further expansion of tourism, with associated economic benefits and increased risks to sensitive ecosystems and communities (Nuttall, 1998; Weller and Lange, 1999).

CRUCIAL UNKNOWNS AND RESEARCH NEEDS

Despite the strong evidence of present impacts of climate change in Alaska, there remain substantial uncertainties regarding all major domains of future impacts. Even for many impacts that are presently developing, there is insufficient systematic observation and continuing uncertainty about important causal processes. Many near-term research needs follow from this lack.

Cryosphere

Systematic observations of changes presently underway in permafrost are needed over a large scale, to specify the rate and character of present warming and thawing more accurately and to resolve present uncertainties regarding the contribution of climate warming to the observed thaw. Further understanding of the dynamics of permafrost are also needed, in order to identify likely rate and character of future loss under continued climate warming, and thereby to help identify the most appropriate responses.

Better understanding is needed of the interaction between thawing, thermokarst formation, surface hydrology, and ecosystem and site characteristics, in order to understand the effects of permafrost thawing on surface plant communities and carbon storage in vegetation and soils. For instance, recent observations in the Brooks Range foothills, where the observed depth of thawing on adjacent sites facing identical climates was strongly dependent on the acidity of the tundra, illustrate the importance of these investigations (Walker et al., 1998).

Continued monitoring of changes in Arctic sea ice is needed, both in area and thickness, and of coastal erosion and its relationship to both sea ice and storm conditions. In addition, better modeling of Arctic ice dynamics is needed to improve climate models, particularly as regards their projections in Arctic regions.

Forest and Tundra Ecosystems

More systematic large-scale observation of changes in the productivity, range, species mix, and disease and insect activity underway in the boreal and coastal forest are needed, as are studies (observational, experimental, and model-based) of potential interactions between climate change, forest productivity increase, species shift, and disturbances by fire, insects, and disease.

Better understanding is needed of how carbon storage in boreal and tundra ecosystems is controlled. It is possible that warming and thawing of boreal and tundra systems will release large quantities of CO_2 and CH_4 (Anisimov et al., 1997; Bockheim et al., 2000; Goulden et al., 1998; Lindroth et al., 1998; Chase et al., 2000), but the magnitude and even the sign of these fluxes will depend on accompanying hydrological changes and the rate of decomposition of exposed peat under warm temperatures (Oechel et al., 1993; McKane et al., 1997 a, b; Moore et al., 1998).

Models of potential future changes in these ecosystems under climate change should also consider the effect of increased surface ultraviolet (UV) radiation, which has increased 15 - 30% in Arctic regions over the past 20 years (Taalas et al., 1997). In initial experimental studies, enhanced UV has harmed several Arctic plant species with an apparent cumulative effect (Bjorn et al., 1997; Callaghan et al., 1998), and strongly stimulated the growth of one moss species (Gehrke et al., 1996).

Marine Ecosystems

While the history of major regime shifts in the Bering Sea and Gulf of Alaska ecosystems is increasingly well documented, the suite of factors causing them – and in particular, the extent of climatic influence in the shifts and the mechanisms by which they operate – are not yet understood. Further study is needed to understand both historical regime shifts, and the likely effects of future climate change.

One very likely consequence of global climate change will be significant changes in seasonal and annual runoff from the glaciers of Southeast Alaska. The potential effects of these runoff changes on the Gulf of Alaska and Bering Sea ecosystems is large but not well understood, and requires investigation.

Alaska's climate is strongly influenced by existing patterns of climate variability, but little is known about how these are likely to behave in a greenhouse-warmed world. Climate models are now beginning to reproduce ENSO, but do not reproduce observed patterns of interdecadal variability. If these patterns continue to behave as they did during the 20th century, then the changes projected from climate models must be modified by these observed patterns of variability. But whether these cycles will behave as they have in a greenhouse-warmed world, or will show coupled changes, is a critical unknown (Fyfe et al., 1999).

Human Dimensions

The possible interactions between climate-driven changes to natural systems and human responses are strong, diverse, and not well understood. Research is needed in the following areas:

- Develop and refine techniques to assess vulnerability of communities, what determines vulnerability, and strategies to reduce it;
- Investigate interactions between policy, fishing pressure, and the state of marine ecosystems, and how these are likely to adjust under climate change;
- Investigate how changes in economic conditions and forest management practices are likely to interact with climate-driven changes in forest ecosystems;
- Investigate response processes of subsistence-reliant communities to changes in the character of subsistence resources available; and
- Study the potential longer-term influence of climate change on aggregate prospects for Alaskan population and economic growth, including, e.g., large changes in the level and distribution of population, large-scale conversion of forested land to agriculture or settlement, or large shifts in the distribution of economic activity.

Arctic Feedbacks

The Arctic regions, including Alaska, are the site of several key uncertainties in modeling the global climate. These include the potential role of changes in the temperature, salinity, and flow regime of the Arctic Ocean, and of changes in Arctic sea ice, in changes in the global thermohaline circulation, possibly including large or rapid changes (Stocker and Schmittner, 1997; Manabe and Stouffer, 1993; Broecker et al., 1990). They also include a number of potentially important climate-change feedbacks, processes whereby climate change can cause more climate change, either by increasing absorption of solar radiation or by increasing emissions of the greenhouse gases that drive climate change. Gaining further understanding of these processes are key research priorities. While the influence of these processes on regional climate in the Arctic may be especially large, they are also of much wider importance for their contribution to driving climate at the global scale.

One major feedback, the ice-albedo feedback, is included in climate model projections. Present limitations in modeling sea-ice dynamics introduce substantial uncertainty to the representation of this feedback in climate models, however. A second feedback can change albedo through ecological processes, and is not presently represented in climate models. Since forests are darker than tundra, the expansion of boreal forest into the tundra zone as climate warms can reduce the reflectivity of the Earth's surface. It has been suggested that this process could amplify regional climate change by up to 50% (Foley et al., 1994). Both these processes need further study.

Other potential feedbacks would operate through climate change altering patterns of natural greenhouse gas emissions. Both boreal and tundra ecosystems are large carbon stores, particularly in their soils. As discussed above, the controls on carbon storage or release from these systems are not yet well understood, and it is possible that climate change could produce large increases in carbon sequestration, or large increases in release of either CO_2 or methane. Advancing understanding of these controls, and projecting future release under climate change, are key priorities.

A larger, though likely more remote uncertainty, concerns the possibility of methane release from hydrates. Hydrates are crystal structures, in which methane molecules are held at high density by being encased in an ice lattice at high pressure or low temperature. Methane hydrates occur worldwide in enormous quantity. Estimated world reserves are 400 million trillion cubic feet (TCF), versus 5,000 TCF of conventional gas reserves - in ocean sediments at high pressure or low temperature, and at substantial depths in continuous permafrost. Hydrates are of interest as a fuel source, although technical challenges to their exploitation are serious. They are also of interest in the long term for the risk of atmospheric release. While the prospect of significant releases over the next century is presently judged to be highly speculative (Kvenvolden, 1999), long-term Arctic warming could eventually release methane by warming coastal waters to shift the depth at which hydrates become stable, or by thawing hydrate-rich permafrost.

LITERATURE CITED

Anisimov, O. A., N. I. Shiklomanov, and F. E. Nelson, Global warming and active-layer thickness: Results from transient general circulation models, *Global and Planetary Change, 15*(3-4), pp. 61-77, 1997.

Anisimov, O. A., and F. E. Nelson, Permafrost zonation and climate change in the Northern Hemisphere: Results from transient general circulation models, *Climatic Change, 35*(2), 241-258, 1997 .

Barber, V. A., G. P. Juday, and B. P. Finney, Reduced growth of Alaskan white spruce in the 20th century from temperature-induced drought stress, *Nature, 405,* 6787, 668-673, 2000.

Beltaos, S., and T. D. Prowse, Climate impacts on extreme ice jam events in Canadian rivers, *Hydrological Sciences Journal,* in press, 2000.

Berkes, F., Indigenous knowledge and resource management systems in the Canadian sub-Arctic, in *Linking Social and Ecological Systems,* edited by F. Berkes, C. Folke, and J. Colding, Cambridge University Press, Cambridge, United Kingdom, pp. 98-128, 2000.

Berman, M., G. P. Juday, R. Burnside, Climate change and Alaska's forests: People, problems, and policies, in Proceedings of a workshop, Assessing the consequences of climate change for Alaska and the Bering Sea region, Fairbanks, 29-30 October 1998, edited by G. Weller and P. A. Anderson, Center for Global Change and Arctic System Research, University of Alaska Fairbanks, 1999.

BESIS, The impacts of global climate change in the Bering Sea region, an assessment conducted by the International Arctic Science Committee under its Bering Sea Impacts Group (BESIS), International Arctic Science Committee, Oslo, Norway, 1997.

Bjorn, L. O., T.V. Callaghan, C. Gehrke, D. Gwynne Jones, B. Holmgren, U. Johanson, and M. Sonesson, Effects of enhanced UV-B radiation on sub-Arctic vegetation, In *Ecology of Arctic Environments,* edited by S. J. Woodin, and M. Marquiss, Blackwell Science, Oxford, United Kingdom, pp. 241-253, 1997.

Bockheim, J. G., L. R. Everett, K. M. Hinkel, F. E. Nelson, and J. Brown, Soil organic carbon storage and distribution in Arctic tundra, Barrow, Alaska, *Soil Science Society of America Journal, 63*(4), 934-940, 1999.

Boer, G. J., N. A. McFarlane, R. Laprise, J. D. Henderson, and J.-P. Blanchet, The Canadian Climate Centre spectral atmospheric general circulation model, *Atmosphere-Ocean, 22*(4), 397-429, 1984.

Boer, G. J., G. M. Flato, M. C. Reader, and D. Ramsden, A transient climate change simulation with historical and projected greenhouse gas and aerosol forcing: Experimental design and comparison with the instrumental record for the 20th century, *Climate Dynamics,* 16, 405-426, 1999a.

Boer, G. J., G. M. Flato, and D. Ramsden, A transient climate change simulation with historical and projected greenhouse gas and aerosol forcing: projected climate for the 21st century, *Climate Dynamics,* 16, 427-450, 1999b.

Brodeur, R. D., B. W. Frost, S. R. Hare, R. C. Francis, and W. J. Ingraham, Interannual variations in zooplankton biomass in the Gulf of Alaska, and covariation with California current zooplankton biomass, *CalCOFI Report 37,* 80-89, 1996.

Broecker, W. S., G. Bond, and M. A. Klas, A salt oscillator in the glacial Atlantic? 1. The concept. *Paleoceanography, 5,* 469-477, 1990.

Brown, J., O. J. Ferrians, Jr., J. A. Heginbottom, and E. S. Melnikov, Circum-Arctic map of permafrost and ground ice conditions, International Permafrost Association, published as *Map CP45,* US Geological Survey, Washington, DC, 1997.

Brown, J., and S. Solomon, Arctic coastal dynamics: Report of an international workshop, Woods Hole, Massachusetts, A geological survey of Canada open file report, in press, 2000. Available at Web site, http://www.awi-potsdam.de/www-pot/geo/acd.html

Buch, E., S. A. Horsted, and H. Hovgard, Fluctuations in the occurrence of Cod in Greenland waters and their possible causes, *ICES Marine Science Symposium, 198,* 158-174, 1994.

Callaghan, T. V., C. Korner, S. E. Lee, and J. H. C. Cornelison, Part 1: Scenarios for ecosystem responses to global change, in *Global Change in Europe's Cold Regions,* edited by O. W. Heal, T. V. Callaghan, J. H. C. Cornelissen, C. Korner and S. E. Lee, pp. 11-63, European Commission Ecosystems Research Report 27, Luxembourg, 1998.

Callaway, D., J. Eamer, E. Edwardwen, C. Jack, S. Marcy, A. Olrun, M. Patkotak, D. Rexford, and A. Whiting, Effects of climate change on subsistence communities in Alaska, in Proceedings of a workshop, Assessing the consequences of climate change for Alaska and the Bering Sea region, Fairbanks, 29-30 October 1998, edited by G. Weller and P. A. Anderson, Center for Global Change and Arctic System Research, University of Alaska Fairbanks, 1999.

Cavalieri, D. J., P. Gloersen, C. L. Parkinson, J. C. Comiso, and H. J. Zwally, Observed hemispheric asymmetry in global sea ice changes, *Science, 278,* 1104-1106, 1997.

Chapin, F. S., III, G. R. Shaver, A. E. Giblin, K. J. Nadelhoffer, and J. A. Laundre, Responses of Arctic tundra to experimental and observed changes in climate, *Ecology 76,* 3, 694-711, 1995.

Chapin, F. S., III, and A. M. Starfield, Time lags and novel ecosystems in response to transient climatic change in arctic Alaska, *Climatic Change, 35,* 449-461, 1997.

Chapman, W. L., and J. E. Walsh, Recent variations of sea ice and air temperature in high latitudes, *Bulletin of the American Meteorological Society, 74,* 1, 33-47, 1993.

Chase, T. N., R. A. Pielke, Sr., T. G. F. Kittel, R. R. Nemani, and S. W. Running, Simulated impacts of historical land cover changes on global climate in northern winter, *Climate Dynamics, 16*(2-3), 93-105, 2000.

Ciais, P., P. P. Tans, M. Trolier, J. W. C. White, and R. R. J. Francey, A large Northern Hemisphere terrestrial CO_2 sink indicated by the C_{13}/C_{12} ratio for atmospheric CO_2, *Science*, 269, 1098-1102, 1995.

Clair T. A., J. Ehrman, and K. Higuchi, Changes to the runoff of Canadian ecozones under a doubled CO_2 atmosphere, *Canadian Journal of Fisheries and Aquatic Sciences,* 55(11) 2464-2477, 1998.

Cole, H., V. Colonell, and D. Esch, The economic impact and consequences of global climate change on Alaska's infrastructure, in Proceedings of a workshop, Assessing the consequences of climate change for Alaska and the Bering Sea region, Fairbanks, 29-30 October 1998, edited by G. Weller and P. A. Anderson, Center for Global Change and Arctic System Research, University of Alaska Fairbanks, 1999.

Criddle, K. R., M. Hermann, J. A. Greenberg, and E. M. Feller, Climate fluctuations and revenue maximization in the eastern Bering Sea fishery for walleye pollock, North American Journal of Fisheries Management *18*(1), 1-10, 1998.

Dowdeswell, J., J. O. Hagen, H. Bjornsson, A. Glazovsky, P. Holmlund, J. Jania, E. Josberger, R. Koerner, S. Ommanney, and B. Thomas, The mass balance of circum-Arctic glaciers and recent climate change, *Quaternary Research, 48,* 1-14, 1997.

Dyke, L. D., J. M. Aylsworth, M. M. Burges, F. M. Nixon, and F. Wright, Permafrost in the Mackenzie Basin, its influence on land-altering processes, and its relationship to climate change, in Mackenzie Basin Impact Study Final Report, edited by S. J. Cohen, Atmospheric Environment Service, Environment Canada, Downsview, Ontario, 1997.

Echelmeyer K. A, W. D. Harrison, C. F. Larsen, J. Sapiano, J. E. Mitchell, J. DeMallie, B. Rabus, G. Adalgeirsdottir, and L. Sombardier, Airborne surface profiling of glaciers: A case-study in Alaska, *Journal of Glaciology, 42*(142), 538-547, 1996

Estes, J. A., M. T. Tinker, T. M. Williams, and D. F. Doak, Killer whale predation on sea otters linking oceanic and nearshore ecosystems, *Science, 282,* 473-476, 1998.

Everett, J. T., and B. B. Fitzharris, The Arctic and the Antarctic, chapter 3, in *The Regional Impacts of Climate Change: An Assessment of Vulnerability*, edited by R. T. Watson, M. C. Zinyowera, and R. H. Moss, Special report of IPCC Working Group II, Cambridge University Press, Cambridge, United Kingdom, 1998.

Fay, F. H., The role of ice in the ecology of marine mammals of the Bering Sea, in Oceanography of the Bering Sea, edited by D. W. Hoos and E. J. Kelley, University of Alaska Institute of Marine Sciences, Occasional Publication No. 2, 1974.

Ferrians, O. J., Permafrost map of Alaska, *US Geological Survey Miscellaneous Geologic Investigations Map I-445*, 1:2,500,000, 1965.

Flato, G. M., G. J. Boer, W. G. Lee, N. A. McFarlane, D. Ramsden, M. C. Reader, and A. J. Weaver, The Canadian Centre for Climate Modelling and Analysis Global Coupled Model and its climate, *Climate Dynamics,* 16(6), 451-467, 2000.

Fleming, R. A., and J-N Candau, Influences of climatic change on some ecological processes of an insect outbreak system in Canada's boreal forests and the implications for biodiversity, *Environmental Monitoring and Assessment, 49*, 235-249, 1997.

Fleming, R. A., and J. A. Volney, Effects of climate change on insect defoliator population processes in Canada's boreal forest: Some plausible scenarios, *Water, Air, and Soil Pollution, 82*, 445-454, 1995.

Foley, J. A., J. E. Kutzbach, M. T. Coe, S. Levis, Feedbacks between climate and boreal forests during the Holocene epoch, *Nature, 371*(6492), 52-54 (September 1), 1994.

Forbes, D. L., and R. B. Taylor, Ice in the shore zone and the geomorphology of cold coasts, *Progress in Physical Geography, 18*, 59-89, 1994.

Francis, R. C., S. R. Hare, A. B. Hollowed, and W. S. Wooster, Effects of interdecadal climate variability on the oceanic ecosystems of the Northeast Pacific, *Fisheries Oceanography 7*, 1-21, 1998.

French, H. M., and I. E. Egorov, 20th century variations in the southern limit of permafrost near Thompson, northern Manitoba, Canada, in Proceedings of the Seventh International Conference on Permafrost, edited by A. G. Lewkowica and M. Allard, Centre d'Études Nordiques, Université Laval, Quebec, pp. 297-304, 1998.

Furniss, R. L., and V. M. Carolin, Western forest insects, *USDA Forest Service Miscellaneous Publication 1339*, Washington, DC, 1977.

Fyfe, J. C., G. J. Boer, and G. M. Flato, The Arctic and Antarctic oscillations and their projected changes under global warming, *Geophysical Research Letters, 26*(11), 1601-1604 (June 1), 1999.

Goldsmith, O. S., Structural analysis of the Alaska economy: A perspective from 1997, prepared for Alaska Science and Technology Foundation by Institute of Social and Economic Research, University of Alaska, Anchorage, 1997.

Gordon, H. B., and S. P. O'Farrell, Transient climate change in the CSIRO coupled model with dynamic sea ice, *Monthly Weather Review, 25*(5), 875-907, 1997

Gorham, E., The biogeochemistry of northern peatlands and its possible responses to global warming, in Biotic feedback in the global climatic system: Will the warming feed the warming?, edited by G. M. Woodwell and F. T. MacKenzie, pp. 169-187, Oxford University Press, Oxford, United Kingdom, 1995.

Goulden, M. L., et al., Sensitivity of boreal forest carbon balance to soil thaw, *Science, 279,* 214-217, 1998.

Groisman, P.Y., and D.A. Easterling, Variability and trends of precipitation and snowfall over the United States and Canada, *Journal of Climate, 7,* 184-205, 1994.

Gunn, A., Responses of Arctic ungulates to climatic change, in *Human Ecology and Climate Change*, edited by D. L. Peterson and D. R. Johnson, pp. 90-104, Taylor and Francis, Washington DC, 1995.

Hagemann, S., and L. Dümenil, A parameterization of the lateral waterflow for the global scale, *Climate Dynamics, 14,* 17-31, 1998.

Hall, J.V., W. E. Freyer, and B. O. Wilen, Status of Alaska wetlands Alaska Region, US Fish and Wildlife Service, Anchorage, Alaska, 1994.

Hennon P. E., and C. G. Shaw, The enigma of yellow-cedar decline: What is killing these long-lived, defensive trees?, *Journal of Forestry, 95*(12), 4-10, December 1997.

Hogg, E. H., Simulation of inter-annual responses of trembling aspen stands to climatic variation and insect defoliation in western Canada, *Ecological Modeling, 114* 2-3, 175-193, 1999.

Hogg, E. H., and A. G. Schwarz, Regeneration of planted conifers across climatic moisture gradients on the Canadian prairies: Implications for distribution and climate change, *Journal of Biogeography, 24* 4, 527-534, July 1997.

Holsten, E. H., and R. Burnside, Forest health in Alaska: An update, *Western Forester, 24* (4), 8-9, 1997.

Holsten, E. H., P. E. Hennone, and R. A. Werner, Insects and diseases of Alaska forests, USDA Forest Service, *Alaska Region Report Number 181*, revised October 1985, Anchorage, Alaska, 1985.

Huppert, D., Managing the groundfish fisheries of Alaska: history and prospects. *Reviews in Aquatic Sciences.* 4(4), 339-373. 1991.

Hurrell, J.W., Influence of variations in extratropical wintertime teleconnections on Northern Hemisphere temperature, *Geophysical Research Letters, 23,* 665-668, 1996.

Hurrell, J.W., Decadal trends in the North American oscillations: Regional temperatures and precipitation, *Science, 269,* 676-679, 1995.

Johannessen, O. M., M. Miles, and E. Bjorgo, The Arctic's shrinking sea ice, *Nature, 376,* 126-127, 1995.

Johannessen, O. M., E.V. Shalina, and M.W. Miles, Satellite evidence for an Arctic sea ice cover in transformation, *Science, 286* 5446, 1937-1939, 1999.

Johns T. C., R. E. Carnell, J. F. Crossley, J. M. Gregory, J. F. B. Mitchell, C. A. Senior, S. F. B. Tett, and R. A. Wood, The Second Hadley Centre coupled ocean-atmosphere GCM: Model description, spinup and validation, *Climate Dynamics, 13,* 103-134, 1997

Jorgenson, M. T., C. H. Racine, J. C. Walters, and T. E. Osterkamp, Widespread and rapid permafrost degradation and associated ecological changes caused by a warming climate on the Tanana Flats, Central Alaska, *Climatic Change*, in press, 2000.

Juday, G. P. and R. Barry, Growth response of upland white spruce in Bonanza Creek LTER in Central Alaska, *Proceedings of Ecological Society of America, Annual Combined Meeting*, 10-14 August 1996, Providence Rhode Island, 222, 1996.

Juday, G. P, R. A. Ott, D. W. Valentine, and V. A. Barber, Forests, climate stress, insects, and fire, in *Implications of Global Change in Alaska and the Bering Sea Region*, Proceedings of a workshop, University of Alaska, Fairbanks, June 1997, edited by G. Weller and P. A. Anderson, Center for Global Change and Arctic System Research, 23-49, April 1998.

Kasischke, E. S., K. Bergen, R. Fennimore, F. Sotelo, G. Stephens, A. Janetos, and H. H. Shugart, Satellite imagery gives clear picture of Russia's boreal forest fires, *Eos, 80* 13, 141-147, 1999.

Kasischke, E. S., and B. J. Stocks (Eds.), *Fire, Climate Change, and Carbon Cycling in the Boreal Forest, Ecological Studies Series*, Springer-Verlag, New York, 2000.

Kattenberg, A., F. Giorgi, H. Grassl, G. A. Meehl, J. F. B. Mitchell, R. J. Stouffer, T. Tokioka, A. J. Weaver, and T. M. L. Wigley, Climate models: Projections of future change, in Climate Change 1995: *The Science of Climate Change, Contribution of Working Group I to the Second Assessment Report of the Inter-governmental Panel on Climate Change*, edited by J.T. Houghton, L. G. Meira Filho, B.A. Callander, N. Harris, A. Kattenberg, and K. Maskell, pp. 285-357, Cambridge University Press, Cambridge, United Kingdom, 1996.

Keeling, C. D., J. F. S. Chin, and T. P. Whorf, Increased activity of northern vegetation inferred from atmospheric CO_2 measurements, *Nature, 382*(6587), 146-149, 1996.

Kerr, J.A., Future water levels and flows for Great Slave and Great Bear Lakes, Mackenzie River, and Mackenzie Delta, in Mackenzie Basin Impacts Study (MBIS), Final Report, edited by S. J. Cohen, Environment Canada, Ottawa, 1997.

Keyser, A. R., J. S. Kimball, R. R. Nemani, and S. W. Running, Simulating the effects of climate change on the carbon balance of North American high latitude forests, *Global Change Biology, 6,* 1-11, 2000.

Knapp, G., P. Livingston, and A. Tyler, Human effects of climate-related changes in Alaska commercial fisheries, in Proceedings of a workshop, Assessing the consequences of climate change for Alaska and the Bering Sea region, Fairbanks, 29-30 October 1998, edited by G. Weller and P.A. Anderson, Center for Global Change and Arctic System Research, University of Alaska Fairbanks, 1999.

Krabil, W., E. Frederick, S. Manizade, C. Martin, J. Sonntag, R. Swift, R. Thomas, W. Wright, and J. Yungel, Rapid thinning of parts of the Southern Greenland ice sheet, *Science, 283*, 1522-524, 1999.

Kruse, G. H., Salmon run failures in 1997-1998: A link to anomalous ocean conditions?, *Alaska Fishery Research Bulletin, 5*(1), 55-63, 1998.

Kvenvolden, K. A., Potential effects of gas hydrate on human welfare, *Proceedings of the National Academy of Sciences, 96*(7), 3420-3426, March 30, 1999.

Kwong, Y. T. J., and T. Y. Gan, Northward migration of permafrost along the Mackenzie Highway and climatic warming, *Climatic Change 26*, 399-419, 1994.

Lachenbruch A. H., and V. Marshall, Changing climate: Geothermal evidence form permafrost in the Alaskan Arctic, *Science, 234*, 689-696, 1986

Lamke, R. D., Alaska surface-water resource, in National Water Summary 1985, Hydrologic events and surface-water resources: *US Geological Water-Supply Paper 2300*, D. W. Moody, E. B. Chase, and D. A. Aronson, (Compilers), pp. 137-144, 1986.

Lammers, R. B., A. I. Shiklomanov, C. J. Vorosmarty, B. M. Fekete, and B. J. Peterson, Assessment of contemporary Arctic river runoff based on observational discharge records, *Journal of Geophysical Research-Atmospheres,* in press, 2000.

Leask, L., Changing ownership and management of Alaska lands, *Alaska Review of Social and Economic Conditions, 22*:4, October 1985.

Lindroth, A., A. Grelle, and A-S. Moren, Long-term measurements of boreal forest carbon balance reveal large temperature sensitivity, *Global Change Biology, 4*(4), 443-450, 1998.

Lowry, L. F. and F. H. Fay, Seal eating by Walruses in the Bering and Chukchi Seas, *Polar Biology, 3*, 11-18, 1984.

Lükewille, A., and R. F. Wright, Experimentally increased soil temperature causes release of nitrogen at a boreal forest catchment in southern Norway, *Global Change Biology, 3*, 13-21, 1997.

McFarlane, N. A., G. J. Boer, J. P. Blanchet, and M. Lazare, The Canadian Climate Centre second-generation general circulation model and its equilibrium climate, *Journal of Climate, 5*, 1013-1044, 1992.

Manabe, S. and R. J. Stouffer, Simulation of abrupt climate change induced by freshwater input to the North Atlantic Ocean, *Nature, 378* (6553), 165-167, 1995.

Mantua, N. J., S. R. Hare, Y. Zhang, J. M. Wallace, and R. C. Francis, A Pacific interdecadal climate oscillation with impacts on salmon production, *Bulletin of the American Meteorological Society, 78*, 1069—1079, 1997.

Marasco, R. J. and W. Arom, Explosive evolution: the changing Alaska groundfish fishery, *Reviews in Aquatic Sciences, 4*(4), 299-315, 1991.

Maslanik, J. A., M. C. Serreze, and R. G. Barry, Recent decreases in Arctic summer ice cover and linkages to atmospheric circulation anomalies, *Geophysical Research Letters, 23*(13), 1677-1680, 1996.

Maslanik, J. A., M. C. Serreze, and T. Agnew, On the record reduction in 1998 Western Arctic Sea-ice cover, *Geophysical Research Letters, 26*(13), 1905-1908, 1999.

McGuire, A. D., and J. E. Hobbie, Global climate change and the equilibrium responses of carbon storage in Arctic and subArctic regions, in Modeling the Arctic system: A workshop report on the state of modeling in the Arctic System Science Program, pp. 53-54, The Arctic Research Consortium of the United States, Fairbanks Alaska, 1997.

McGuire, A. D., J. M. Melillo, D. W. Kicklighter, and L. A. Joyce, Equilibrium responses of soil carbon to climate change: Empirical and process-based estimates, *Journal of Biogeography, 22*, 785-796, 1995.

McKane, R. B., E. B. Rasetter, G. R. Shaver, K. J. Nadelhoffer, A. E. Giblin, J. A. Laundre, and F. S. Chapin, III, Climate effects of tundra carbon inferred from experimental data and a model, *Ecology, 78*, 1170-1187, 1997a.

McKane, R. B., E. B. Rasetter, G. R. Shaver, K. J. Nadelhoffer, A. E. Giblin, J. A. Laundre, and F. S. Chapin, III, Reconstruction and analysis of historical changes in carbon storage in Arctic tundra, *Ecology 78*, 1188-1198, 1997b.

McPhee, M. G., T. P. Stanton, J. H. Morison, and D. G. Martinson, Freshening of the upper ocean in the Arctic: Is perennial sea ice disappearing?, *Geophysical Research Letters, 25*(10), 1729-1732, 1998.

Meehan, R., V. Byrd, G. Divoky, and J. Piatt, Implications of climate change for Alaska's seabirds, in Proceedings of a workshop, Assessing the consequences of climate change for Alaska and the Bering Sea region, Fairbanks, 29-30 October 1998, edited by G. Weller and P. A. Anderson, Center for Global Change and Arctic System Research, University of Alaska Fairbanks, 1999.

Meier, M. F., Ice, climate, and sea level: Do we know what is happening?, in *Ice in the Climate System, NATO ASI Series, Vol. 112*, edited by W. R. Peltier, Springer-Verlag, Berlin, pp. 141-160, 1993.

Melillo, J. M., D. W. Kicklighter, A. D. McGuire, W. T. Peterjohn, K. M. Newkirk, Global change and its effect on soil organic carbon stocks, in Role of nonliving organic matter in the Earth's carbon cycle, edited by R. G. Zepp and C. H. Sontaff, pp. 175-189, John Wiley and Sons: New York, 1995.

Melillo, J. M., I. C. Prentice, G. D. Farquhar, E.-D Schulze, and O. E. Sala, Terrestrial biotic responses to environmental change and feedbacks to climate, chapter 9, in *Climate Change 1995: The Science of Climate Change, Contribution of Working Group I to the Second Assessment Report of the IPCC*, edited by J. T. Houghton, L. G. Meira Filho, B. A. Callander, N. Harris, A.

Kattenberg, and K. Maskell, Cambridge University Press, Cambridge, United Kingdom, 1996.

Mitchell J. F. B., T. C. Johns, J. M. Gregory, and S. Tett, Climate response to increasing levels of greenhouse gases and sulphate aerosols, *Nature, 376*, 501-504, 1995.

Mitchell J. F. B., and T. C. Johns, On modification of global warming by sulfate aerosols, *Journal of Climate, 10*(2), 245-267, 1997.

Moore T. R., N. T. Roulet, J. M. Waddington, Uncertainty in predicting the effect of climatic change on the carbon cycling of Canadian peatlands, *Climatic Change, 40*(2), 229-245, 1998.

Mulvaney, K., Arctic voices: Global warming is changing the traditional Eskimo environment, *New Scientist*, pp. 55, November 14, 1998.

Myneni, R. B., C. D. Keeling, C. J. Tucker, G. Asrar, and R. R. Nemani, Increased plant growth in the northern high latitudes from 1981 to 1991, *Nature, 386*, 698-702, 1997.

Nash, C. E., and Associates, D. Duffy, Miller's Reach Fire Strategic Economic Recovery Plan, Matanuska-Susitna Borough Department of Planning, October 1997.

National Marine Fisheries Service (NMFS), Economic status of US fisheries, 1996, US Government Printing Office, Washington, DC, 1996.

NRC (National Research Council), *The Bering Sea Ecosystem*, Committee on the Bering Sea Ecosystem, Polar Research Board, National Academy Press, Washington, DC, 1996.

Neilson, R. P., I. C. Prentice, B. Smith, T. G. F. Kittel, and D. Viner, Simulated changes in vegetation distribution under global warming, Annex C, in *The Regional Impacts of Climate Change: An Assessment of Vulnerability*, Special Report of IPCC Working Group II, edited by R. T. Watson, M. C. Zinyowera, R. H. Moss, and D. J. Dokken, pp. 439-456, Cambridge University Press, Cambridge, United Kingdom, 1998.

Nixon, J. F., Effect of climatic warming on pile creep in permafrost, *Journal of Cold Regions Engineering, 4*, 1, pp. 67-73, 1990.

National Oceanic and Atmospheric Administration (NOAA), FOCI International Workshop on Recent Conditions in the Bering Sea, NOAA-Pacific Marine Environmental Laboratory, Seattle, Washington, 1999.

Nuttall, M., Protecting the Arctic: *Indigenous Peoples and Cultural Survival*, Harwood Academic Publishers, Amsterdam, 1998.

Oechel, W. C., S. J. Hastings, G. Vourlitis, M. Jenkins, G. Richers, and N. Gruike, Recent change of Arctic tundra ecosystems from a net carbon dioxide sink to a source, *Nature, 361*, 520-523, 1993.

Oechel, W. C., G. L. Vourlitis, S. J. Hastings, and S. A. Bochkarev, Change in Arctic CO_2 flux over two

decades: Effects of climate change at Barrow, Alaska, *Ecological Application, 5*(3), 846-855, 1995.

Oechel W. C., G. L. Vourlitis, and S. J. Hastings, Cold season CO_2 emission from Arctic soil, *Global Biogeochemical Cycles, 11*, 163-172, 1997a.

Oechel, W. C., A. C. Cook, S. J. Hastings, and G. L. Vourlitis, Effects of CO_2 and climate change on Arctic ecosystems, in *Ecology of Arctic Environments*, edited by S. J. Woodin and M. Marquiss, Blackwell Science, Oxford, United Kingdom, pp. 255-273, 1997b.

Osterkamp, T. E., Evidence for warming and thawing of discontinuous permafrost in Alaska, *Elsevier Oceanography Series, 75*(44), 85, 1994.

Osterkamp, T. E., D. C. Esch, and V. E. Romanovsky, Permafrost, chapter 10, in Implications of global change in Alaska and the Bering Sea region, Proceedings of a workshop, June 3-6 1997, edited by G. A. Weller and P. A. Anderson, Center for Global Change and Arctic System Research, University of Alaska Fairbanks, 1998.

Osterkamp, T. E., J. K. Peterson, and T. S. Collett, Permafrost thicknesses in the Oliktok Point, Prudhoe Bay, and Mikkelsen Bay areas of Alaska, *Cold Regions Science and Engineering, 11*, 99-105, 1985.

Osterkamp, T. E. and V. E. Romanovsky, Evidence for warming and thawing of discontinuous permafrost in Alaska, *Permafrost and Periglacial Processes, 10*, 17-37 (January-March), 1999.

Osterkamp, T. E. and V. E. Romanovsky, Impacts of thawing permafrost as a result of climatic warming, *EOS*, 77(46), 29, 1996

Osterkamp, T. E., L. Viereck, Y. Shur, M. T. Jorgenson, C. H. Racine, A. P. Doyle, and R. D. Boone, Observations of thermokarst and its impact on boreal forests in Alaska, *Arctic, Antarctic, and Alpine Research*, in press, 2000.

Overpeck, J., et al., Arctic environmental change of the last four centuries, *Science, 278*, 1251-1256, 1997

Parker, D. E., P. D. Jones, C. K. Folland, and A. Bevan, Interdecadal changes of surface temperature since the late 19th century, *Journal of Geophysical Research – Atmosphere*, 99(D7), 14373-14399, 1994

Piatt, J. F., and P. Anderson, Response of common murres to the Exxon Valdez oil spill and long-term changes in the Gulf of Alaska marine ecosystem, *American Fisheries Society Symposium, 18*, 720-737, 1996.

Post, W. M., W. R. Emanuel, P J. Zinke, and A. G. Stangenberger, Soil carbon pools and world life zones, *Nature, 298*, 156-159, 1982.

Powell, D. S., J. L. Falukner, D. R. Darr, Z. Zhu, and D. W. MacCleery, Forest Resources of the United States, 1992, USDA Forest Service, Rocky Mountain Forest and Range Experiment Station, Fort Collins Colorado, General Technical Report RM-234, 1993.

Proshutinsky, A. Y., and M. A. Johnson, Two circulation regimes of the wind-driven Arctic Ocean, *Journal of Geophysical Research, 102*, No. C6, 12493-12514, 1997.

Prowse, T. D., and M. Conly, Impacts of climatic variability and flow regulation on ice jam flooding of a northern delta, *Hydrological Processes, 12*, 1589-1610, 1998.

Rivkin, F. M., Release of methane from permafrost as a result of global warming and other disturbances, *Polar Geography, 22*, 105-118, 1998.

Robinson, S. D., and T. R. Moore, Carbon and peat accumulation over the past 1200 years in a landscape with discontinuous permafrost, northwestern Canada, *Global Biogeochemical Cycles, 13,* 591-601, 1999.

Rothrock, D. A., Y. Yu and G. A. Maykut, Thinning of the Arctic Sea ice cover, *Geophysical Research Letters, 26*(23), 3469-3472 (December 1), 1999.

Rouse, W. R., et al., Effects of climate change on the freshwaters of Arctic and sub-Arctic North America, *Hydrological Processes, 11*(8), 873-902 (June 30), 1997.

Royer, T. C., Baroclinic transport in the Gulf of Alaska: A freshwater driven coastal current, *Journal of Marine Research, 39*, 251-266, 1981.

Royer, T. C., Coastal freshwater discharge in the Northeast Pacific, *Journal of Geophysical Research, 87*, 2017-2021, 1982.

Royer, T. C., High-latitude oceanic variability associated with the 18.6-year nodal tide, *Journal of Geophysical Research – Oceans, 98* (C3), 4639-4644 1993

Rupp, T. S., F. S. Chapin, III, and A. M. Starfield, Response of sub-Arctic vegetation to transient climatic change on the Seward Peninsula in northwest Alaska, *Global Change Biology, 6*(5), 541-555, 2000.

Sabo, G., III., *Long-term Adaptations Among Arctic Hunter-Gatherers*, Garland Publishing, London, United Kingdom, 1991.

Sapiano, J.J., W. D. Harrison, and K. A. Echelmeyer, Elevation, volume, and terminus changes of nine glaciers in North America, *Journal of Glaciology, 44*(146), 119-135, 1998,

Semiletov, I. et al., The dispersion of Siberian River flows into coastal waters: Hydrological and hydrochemical aspects, in The Freshwater Budget of the Arctic Ocean, NATO Arctic workshop, 25 April - 1 May 1998, Tallinn, Estonia, edited by E. L. Lewis, 1999.

Serreze, M.C., J. A. Maslanik, J. R. Key, R. F. Kokaly, and D. A. Robinson, Diagnosis of the record minimum in Arctic Sea-ice area during 1990 and associated snow cover extremes, *Geophysical Research Letters 22*(16), 2183-2186 (August 15), 1995a.

Serreze, M.C., R. G. Barry, M. C. Rehder, and J. E. Walsh, Variability in atmospheric circulation and moisture flux over the Arctic, *Philosophical Transactions of the Royal Society of London, Series A, 352*(1699), 215-225 (August 15), 1995b.

Serreze, M. C., J. E . Walsh, F. S. Chapin, III, T. Osterkamp, M. Dyurgerov, V. Romanovsky, W. C. Oechel, J. Morison, T. Zhang, and R. G. Barry, Observational evidence of recent changes in the northern high-latitude environment, *Climatic Change, 46*, 159-207, 2000.

Shaw, J., R. B. Taylor, S. Solomon, H. A. Christian, and D. L. Forbes, Potential impacts of global sea-level rise on Canadian coasts, *The Canadian Geographer, 42*, 365-379, 1998.

Shiklomanov, I. A., On the effect of anthropogenic change in the global climate on river runoff in the Yenisei basin, in Runoff computations for water projects, edited by A. V. Rozhdestvensky, International Hydrological Program, IHP-V, *Technical Documents in Hydrology No. 9*, UNESCO, Paris, pp. 113-199, 1997.

Shiklomanov, I. A., A. I. Shiklomanov, R. B.Lammers, B. J.Peterson, and C. J.Vorosmarty, The dynamics of river water inflow to the Arctic Ocean, in The Freshwater Budget of the Arctic Ocean, edited by E. L. Lewis, E. P.Jones, P. Lemke, T. D. Prowse, P. Wadhamns, Kluwer Academic Publishers, Dordrecht, in press, 2000.

Sieben, B. G., D. L. Spittlehouse, J. A. McLean, and R. A. Benton, White Pine Weevil hazard under GISS climate change scenarios in the Mackenzie Basin using radiosonde derived lapse rates, in Mackenzie Basin Impact Study Final Report, Proceedings of a conference in Yellowknife, Northwest Territories, May 1996, edited by S. J. Cohen, Environment Canada, Atmospheric Environment Service, 1997.

Skinner, W. R., B. J. Stocks, D. L. Martell, B. Bonsal, and A. Shabbar, The association between circulation anomalies in the mid-troposphere and area burned by wildland fire in Canada, *Theoretical and Applied Climatology, 63*, (1-2), 89-105, 1999.

Smith, O. P., and L. Johnston, The warming world: Strategy for Alaskan response, Summary of a workshop, January 5-6, 2000, University of Alaska, Anchorage, 2000.

Solomon, S. M., and R. Covill, Impacts of the September 1993 storm on the coastline at four sites along the Canadian Beaufort Sea coast, in Proceedings of the 1995 Canadian Coastal Conference, Dartmouth Nova Scotia, 2, 779-795, 1995.

Stirling, I., N. J. Lunn, and J. Iacozza, Long-term trends in the population ecology of polar bears in western Hudson Bay in relation to climate change, *Arctic*, 52, 294-306, 1999.

Stocker, T. F., and A. Schmittner, Influence of CO_2 emission rates on the stability of the thermohaline circulation, *Nature, 388*, 862-865, 1997.

Stocks, B. J., M. A. Fosberg, T. J. Lynham, L. Mearns, B. M. Wotton, Q. Yang, J-Z. Jin, K. Lawrence, G. R. Hartley, J. A. Mason, and D. W. McKenney, Climate change and forest fire potential in Russian and Canadian boreal forests, *Climatic Change, 38*(1),1-13, 1998.

Taalas, P., J. Damski, E. Kyro, M. Ginzburg, and G. Talamoni, The effect of stratospheric ozone variations on UV radiation and on tropospheric ozone at high latitudes, *Journal of Geophysical Research, 102*(D1), 1533-1543, 1997.

Thompson, D. W. J., and J. M. Wallace, The Arctic oscillation signature in the wintertime geopotential height and temperature fields, *Geophysical Research Letters, 25*, 1297-1300, 1998.

Trenberth, K. E., and J. W. Hurrell, Decadal atmosphere-ocean variation in the Pacific, *Climate Dynamics, 9*(6), 303-319, 1994.

US Army Corps of Engineers, Community improvement feasibility report, Alaska District, Anchorage, Alaska, (April), 1998.

US Census Bureau, State Population Projections, at http://www.census.gov/population/www/projections/stproj.html, 2000.

USGS BRD (US Geological Service, Biological Resources Division), *Status and trends of the Nation's biological resources*, US Government Printing Office, Washington, DC, 1999.

Van Blarcum, S. C., J. R. Miller, and G. L. Russell, High-latitude river runoff in a doubled CO_2 climate, *Climatic Change, 30*(1), 7-26, 1995.

Van Cleve K., W. C. Oechel, and J. L. Hom, Response of black spruce (Picea mariana) ecosystems to soil temperature modifications in interior Alaska, *Canadian Journal of Forest Research, 20*, 1530-1535, 1990.

Veblen, T. T., and P. B. Alaback, A comparative review of forest dynamics and disturbance in the temperate rainforests of North and South America, in *High-Latitude Rainforests and Associated Ecosystems of the West Coast of The Americas: Climate, Hydrology, Ecology, and Conservation*, edited by R. G. Lawford, P. B. Alaback, and E. Fuentes, pp. 173-213, Springer-Verlag, New York, 1996.

Viljhamsson, H., Climatic variations and some examples of their effects on the marine ecology of Icelandic and Greenland waters, in particular during the present century, *Journal of the Marine Research Institute, Reykjavik, 15*(1), 9-29, 1997.

Vinnikov, K. Y., A. Robock, R. J. Stouffer, J. E. Walsh, C. L. Parkinson, D. J. Cavalieri, J. F. B. Mitchell, D. Garrett, and V. F. Zakharov, Global warming and Northern Hemisphere sea ice extent, Science *286*(5446), 1934-1937, 1999. .

Volney, W. J. A., Climate change and management of insects defoliators in boreal forest ecosystems, in *Forest Ecosystems, Forest Management, and the Global Carbon Cycle,* edited by M. J. Apps and D. T. Price, *NATO ASI Series I (*Global Environmental Change), *Vol. 40*, Springer-Verlag Academic Publishers, Heidelberg, Germany, pp. 79-88, 1996.

Vourlitis, G. L., and W. C. Oechel, Eddy covariance measurements of CO_2 and energy fluxes of an Alaskan tussock tundra ecosystem, *Ecology, 80*(2), 686-701, 1999.

Walsh, J. E., Climate change: The Arctic as a bellwether, *Nature, 352*(6330), 19-20 (July 4), 1991.

Walsh, J. E., V. Kattsov, D. Portis, and V. Meleshko, Arctic precipitation and evaporation: Model results and observational estimates, *Journal of Climate, 11*, 72-87, 1998.

Wadhams P., Ice thickness in the Arctic Ocean: The statistical reliability of experimental data, *Journal of Geophysical Research – Oceans, 102*(C13), 27951-27959 (December 15), 1997.

Wadhams, P., Evidence for thinning of the Arctic ice cover north of Greenland, *Nature, 345*(6278), 795-797 (June 28), 1990.

Walker, D. A., et al., A major Arctic soil pH boundary: Implications for energy and trace gas fluxes, *Nature, 394*, 469-472, 1998.

Weller, G. A., Regional impacts of climate change in the Arctic and Antarctic, *Annals of Glaciology, 27*, 543-552, 1998.

Weller, G. A., and P. A. Anderson (Eds.), Implications of global change in Alaska and the Bering Sea region, Proceedings of a Workshop, June 3-6 1997, Center for Global Change and Arctic System Research, University of Alaska Fairbanks, 1998.

Weller, G. A., and M. Lange (Eds.), Impacts of global change in the Arctic Regions, report from a workshop, Tromso Norway, 25-26 April, International Arctic Science Committee, 1999.

Weller, G., A. Lynch, T. Osterkamp, and G. Wendler, Climate trends and scenarios, in Implications of Global Change in Alaska and the Bering Sea Region: Proceedings of a Workshop, June 3-6 1997, edited by G. A. Weller and P. A. Anderson, Center for Global Change and Arctic System Research, University of Alaska, Fairbanks, Alaska, 1998.

Wenzel, G. W., Warming the Arctic: Environmentalism and the Canadian Inuit, in *Human Ecology and Climate Change*, edited by D. L. Peterson and D. R. Johnson, Taylor & Francis, Washington, DC, 1995.

Werner, R. A., Forest health in boreal ecosystems of Alaska, *The Forestry Chronicle, 72*(1), 43-46, 1996.

Witherell, D., Status and trends of principal groundfish and shellfish stocks in the Alaska EEZ, North Pacific Fishery Management Council, Anchorage Alaska, February 26, 1999.

Wolfe, R. J., and R. G. Bosworth, Subsistence in Alaska: 1994 update, Alaska Department of Fish and Game, Division of Subsistence, Fairbanks, Alaska, 1994.

Wolfe, S.A., S. R. Dallimore, and S. M. Solomon, Coastal permafrost investigations along a rapidly eroding shoreline, Tuktoyaktuk Northwest Territories, Canada, in Permafrost: Proceedings of Seventh International Conference, edited by A. G. Lewkowicz and M. Allard, Centre d'Études Nordiques, Université Laval, Quebec, Canada, 1998.

Woo, M.-K., A. G. Lewkowicz, and W. R Rouse, Response of the Canadian permafrost environment to climatic change, *Physical Geography*, *13*(4), 287-317, 1992.

ACKNOWLEDGMENTS

Many of the materials for this chapter are based on contributions from workshop participants and members of the

Alaska Workshop and Assessment Teams

Gunter Weller*, University of Alaska Fairbanks
Patricia Anderson*, University of Alaska Fairbanks
Bronwen Wang*, US Geological Survey, Anchorage, AK
Matthew Berman, University of Alaska Anchorage
Don Callaway, National Park Service
Henry Cole, Hydro Solutions & Purification LLC
Keith Criddle, Utah State University
Merritt Helfferich, Innovating Consulting Inc.
Glenn Juday, University of Alaska Fairbanks
Gunnar Knapp, University of Alaska Anchorage
Rosa Meehan, U. S. Fish and Wildlife Service
Thomas Osterkamp, University of Alaska Fairbanks

Comments by the following reviewers are gratefully acknowledged: Jerry Brown, F. Stuart Chapin III, Robert W. Corell, Henry P. Huntington, Fae Korsmo, Manfred Lange, Daniel Lashof, Susan M. Marcus, A. David McGuire, Steven P. McNulty, M. Granger Morgan, Susanne Moser, Daniel R. Muhs, Thomas D. Newbury, Thomas Osterkamp, Thomas Pagano, Eve S. Sprunt, Soroosh Sorooshian, Roger B. Street, Francis Zweirs. Remaining errors are the sole responsibility of the coordinating author.

* Regional Assessment co-chair

CHAPTER 11

POTENTIAL CONSEQUENCES OF CLIMATE VARIABILITY AND CHANGE FOR THE US-AFFILIATED ISLANDS OF THE PACIFIC AND CARIBBEAN

Lynne M. Carter[1, 2], Eileen Shea[3], Mike Hamnett[4], Cheryl Anderson[4], Glenn Dolcemascolo[3], Charles 'Chip' Guard[5], Melissa Taylor[2,6], Tony Barnston[7], Yuxiang He[7], Matthew Larsen[8], Lloyd Loope[9], LaShaunda Malone[2], Gerald Meehl[10]

Contents of this Chapter

[1]Coordinating author for National Assessment Synthesis Team (from April, 2000); [2]USGCRP National Assessment Coordination Office; [3]East-West Center; [4]University of Hawaii; [5]University of Guam; [6]Coordinating author for National Assessment Synthesis Team (through March, 2000); [7]NOAA, National Centers for Environmental Prediction; [8]USGS, Guaynabo, Puerto Rico; [9]USGS, Pacific Island Ecosystems Research Center, Haleakala Field Station; [10]National Center for Atmospheric Research

CHAPTER SUMMARY

Regional Context

This section deals with the US-affiliated islands of the Caribbean and Pacific. Included are Puerto Rico and the US Virgin Islands in the Caribbean and the Hawaiian Islands, American Samoa, the Common-wealth of the Northern Mariana Islands, Guam, the Federated States of Micronesia, the Republic of the Marshall Islands, and the Republic of Palau in the Pacific. The latter three are independent states in free association with the United States. Hawaii became the 50th state of the US in 1959. The Virgin Islands, Guam, and American Samoa are US territories. The Northern Mariana Islands and Puerto Rico are commonwealths.

Islands contain diverse and productive ecosystems, and include many specialized and unique species. Many islands are facing the stresses of rapid human popula-tion growth, increasing vulnerability to natural disas-ters, and degradation of natural resources. Droughts and floods are among the climate extremes of most concern as they affect the quality of water supplies in island communities and thus can have significant health consequences. Due to their small size and isola-tion, many islands face chronic water shortages and problems with waste disposal. Some are facing a species extinction crisis; for example, the Hawaiian Islands have the highest extinction rate of any state in the nation. Biological invasion and habitat destruction seem to be the primary causes of the extinction crisis. For most island communities, transportation and other social infrastructure and economic activities are locat-ed near the coast, making them highly vulnerable to storm events and sea-level fluctuations. Over-harvest-ing of reef organisms is a serious issue in parts of the Pacific.

Climate of the Past Century

Over the 20th century, average annual air temperatures in the Caribbean islands have increased by more than 1°F (0.6°C). Average annual air temperatures in the Pacific Islands have increased by about 0.4°F (0.2°C).

Globally, sea level has risen by 4 to 8 inches (10-20 cm) in the past 100 years, with significant local variation. The current rate of sea-level increase in the Caribbean and Gulf of Mexico is about 3.9 inches (10 cm) per 100 years. In the Pacific, the absolute sea level is also rising. However, long-term changes in sea level relative to land vary considerably within the main Hawaiian Islands and some of the South Pacific islands due to geologic uplift and subsidence. Trends in relative sea level thus vary greatly from island to island, such that there is no discernible long-term average trend for sea-level rise for the Pacific islands as a group. There is also extreme variability in relative sea level associated with transient events such as ENSO, storm surges, and extreme lunar tides.

Climate of the Coming Century

The Hadley, Canadian, and other global climate models suggest it is possible that the Pacific and Caribbean islands will be affected by: changes in patterns of natu-ral climate variability (such as El Niño); changes in the frequency, intensity, and tracks of tropical cyclones (hurricanes and typhoons); and changes in patterns of ocean circulation. These islands are also very likely to experience increased ocean temperatures and changes in sea level (including storm surges and sustained rise).

For the Caribbean and the Pacific, both of the primary models used in this Assessment project increasing air and water temperatures. For the Caribbean, both mod-els project slightly wetter conditions in winter, while the Hadley model suggests slightly drier conditions in summer and the Canadian model suggests that in sum-mer parts of the Caribbean will be wetter and parts drier. For the Pacific, both models show greater warm-ing in the central and eastern equatorial Pacific and an attendant eastward shift of precipitation.

Any changes in ENSO patterns would affect precipita-tion, sea level, and the formation and behavior of cyclones (typhoons and hurricanes), with conse-quences for both the Caribbean and Pacific islands.

Key Findings

- Islands are uniquely vulnerable to many of the potential impacts of climate change: changes in hydrologic cycles; temperature increases; and sea-level fluctuations.
- On many islands, water resources are already limited. Water supplies could be adversely affected by changes in hydrologic cycles that result in more frequent droughts and/or floods and changes in sea level that result in saltwater intrusion into freshwater lenses.
- In Puerto Rico, the US Virgin Islands, and mountain islands of the Pacific, climate changes that result in increased extreme precipitation events are of particular concern since flooding and landslides often result from such extreme events today. Model scenarios frequently project increases in rainfall intensity as global temperatures rise.
- Both coastal and inland island populations and infrastructure are already at risk from climate extremes. Model projections suggest that the islands could possibly be affected by changes in patterns of natural variability in climate (such as El Niño) and changes in the frequency, intensity, and tracks of tropical cyclones (hurricanes and typhoons). Both the Pacific and Caribbean regions are familiar with severe cyclones, which have caused billions of dollars in damage from the destruction of housing, agriculture, roads and bridges, and lost tourism revenue. Any increases in frequency or intensity would have negative impacts on island ecosystems and economics.
- Islands are home to rare ecosystems such as tropical forests, mangrove swamps, coral reefs, and seagrass beds, with many unique and specialized endemic species (those that occur nowhere else). These resources are already threatened by invasive non-native plant and animal species, as well as from the impacts of tourism, urban expansion, and various industrial activities, giving the islands the highest rates of extinction of all regions of the US. Many of these ecosystems are also highly sensitive to changes in temperature and are likely to change or be destroyed by changes in climate. For example, coral reefs are sensitive to changes in water temperature, and increases in the frequency or severity of El Niño

events could result in increased bleaching and possible die-off of corals.
- Other ecological concerns include increased extinction rates of mountain species that have limited opportunities for migration, and impacts on forests due to floods, droughts, or increased incidence of pests, pathogens, or fire. Any increase in the frequency or intensity of hurricanes could influence forest succession, and would generally favor invasive species. In addition, the unique cloud forests located on some of these islands occupy a narrow geographical and climatological niche that would be threatened by increases in temperature and changes in the hydrologic cycle.
- Episodic variation in sea level, and the associated erosion and inundation problems are already extremely important issues for many of the US-affiliated islands. Saltwater intrusion into freshwater lenses and coastal vegetation, and coastal erosion would only become greater problems should climate changes result in changes to normal weather patterns, changes in ocean circulation patterns, and increased sea-level rise.
- Tourism remains a significant contributor to the economies of many of these island jurisdictions. Should climate change in these regions negatively impact ecosystems or freshwater supplies, tourism would suffer. As has happened in the past, tourists are likely to select alternate destinations to avoid the impacts of climate changes and extreme weather events.
- For island communities to become more resilient to a changing climate they may consider implementing any number of adaptive strategies. Some of the strategies for the islands involve: increased use of climate forecasting capabilities for planning protection and mitigation measures; consideration of changing land use policies and building codes; use of improved technologies; increased monitoring, scientific research, and data collection; and improved management policies. In addition, broad public awareness campaigns and communication with user groups are important for improving public health and safety and protecting natural resources.

POTENTIAL CONSEQUENCES OF CLIMATE VARIABILITY AND CHANGE FOR THE US-AFFILIATED ISLANDS OF THE PACIFIC AND CARIBBEAN

PHYSICAL SETTING AND UNIQUE ATTRIBUTES

This chapter deals with the US-affiliated islands of the Caribbean and Pacific. Included are: the Commonwealth of Puerto Rico (PR) and the US Virgin Islands (USVI) in the Caribbean: and the Hawaiian Islands, American Samoa, the Commonwealth of the Northern Mariana Islands (CNMI), Guam, the Federated States of Micronesia (FSM), the Republic of the Marshall Islands (RMI), and the Republic of Palau (RP) in the Pacific. The latter three are independent nations in free association with the United States; Hawaii became the 50th state of the US in 1959; the Virgin Islands, Guam, and American Samoa are US territories; and Puerto Rico and the Northern Mariana Islands are commonwealths.

Islands contain diverse and productive ecosystems, and include many specialized and unique species. Many islands are facing the stresses of rapid human population growth, increasing vulnerability to natural disasters, and degradation of natural resources. Droughts and floods are among the climate extremes of most concern as they affect the quality of water supplies in island communities and thus can have significant health consequences. Due to their small size and isolation, many islands face chronic water shortages and problems with liquid and solid waste disposal. Some are facing a species extinction crisis; for example, the Hawaiian Islands have the highest extinction rate of any state in the nation. Biological invasions and habitat loss seem to be the primary causes of the extinction crisis in Hawaii. For most island communities, infrastructure including, transportation aspects, and economic activities are concentrated near the coast, making them highly vulnerable to storm events and sea-level fluctuations.

The islands differ in geologic type, as well as by size, elevation, soil composition, drainage characteristics, and natural resources. There are barrier islands, continental islands, coral islands, and volcanic islands. Islands can also be of mixed type, such as continental islands with raised reefs or volcanic islands sur-

rounded by coral reefs. Some low-lying coral islands or atolls rise only a few feet above sea level whereas some volcanic islands have mountains thousands of feet high. Island entities also may be single islands or groups of islands; for example, American Samoa consists of five volcanic islands and two coral atolls, whereas Guam is a single mountain island. The islands discussed here also vary considerably in terms of the types of human communities, their economic structure, and their lifestyles and infrastructure, ranging from large densely populated cities (e.g., Honolulu, San Juan) to small villages, dispersed populations, and unpopulated islands (Rapaport, 1999; Lobban and Schefter, 1997; Low et al., 1998; Wiley and Vilella, 1998; Loope, 1998).

SOCIOECONOMIC CONTEXT

The islands of the Pacific and Caribbean are host to diverse communities, many of which are especially vulnerable to climate variability and change. While there are many commonalities among island communities, an appreciation of social, political, economic, and cultural diversity can help decision-makers involved in the planning of appropriate strategic responses to climatic perturbations.

American-affiliated islands in the Pacific and Caribbean share a common history of transition from indigenous to colonial and, most recently, to post-colonial society; but these histories have resulted in an array of political structures including US state, US territory, commonwealth, and freely associated state. Many of these islands have held special significance for US national security interests. On many islands, macro-scale political systems operate in conjunction with traditional sociopolitical structures. In some of the Pacific islands, for instance, the chief-based systems still influence political action, particularly at the local level. These diverse political organizations are instrumental in the planning and implementation of climate-related policies and programs.

The US-affiliated islands of the Pacific and Caribbean reflect considerable economic diversity which includes subsistence agriculture and fishing;

tourism; tuna processing and transshipment; and export-led agricultural production of crops such as sugar cane, bananas, pineapple, coffee, spices, and citrus fruits. In addition, some islands are developing the infrastructure to support high-tech telecommunications and computer industries. Some islands, such as CNMI and Puerto Rico, have established substantial manufacturing sectors. Despite this, many of the islands continue to depend on federal subsidies and military spending. While tourism, agriculture, fishing, and transshipment are common in most all the islands, the relative contribution of each sector to the overall economy varies from island to island, and most islands continue to rely on a mix of techno-industrial and subsistence strategies. Further, many island economies are intricately linked with wider economic structures and, as such, are especially vulnerable to fluctuations in the international market, which affect the prices of both imported and exported goods. Climate change and variability, whether through extreme events or gradual change, can potentially affect all island industries.

Culturally, the islands are remarkably diverse as well. Witness the strong Boriqua culture of Puerto Rico in the Caribbean, and the Samoan and Chamorro cultures of the Pacific. The Federated States of Micronesia is home to several different language and culture groups. Further, ethnic diversity is compounded by historical and contemporary migration patterns, which include not only internal migration of indigenous island peoples, but also some migration to and from the continental US and other countries. Growing manufacturing industries in the Commonwealth of the Northern Mariana Islands for instance, have attracted migrants from the Philippines, China, and Korea.

Cultural values and belief systems can significantly influence people's responses to climate and climate changes as well as their responses to preventive or ameliorative policies and strategies. In many Pacific islands, land has inherent cultural value to the populations and therefore any potential threats are regarded as especially serious. Patterns of land ownership/tenure and resource use in many of the island jurisdictions addressed in this Assessment reflect a mix of traditional practices with strong and varying cultural roots combined with more recent policies that derived from periods of colonial occupations. In the Federated States of Micronesia, for example, land ownership remains the most valued right, reflecting the short supply and traditional importance of the land and the natural resources it supports (FSM National Communication, 1999). In the traditional economy of the FSM, land is not a

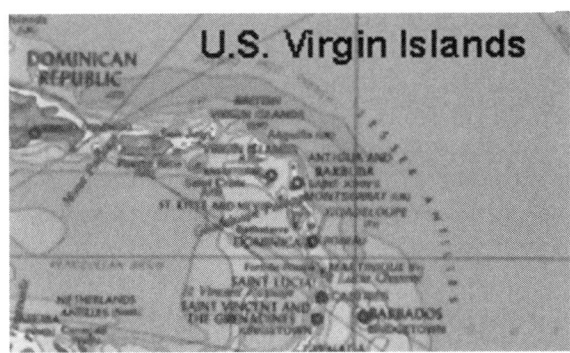

Figure 1: The scope of this section includes the US-affiliated islands of the Caribbean and Pacific. In the Caribbean, this includes Puerto Rico and the US Virgin Islands. In the Pacific, it includes the Hawaiian Islands, American Samoa, the Commonwealth of the Northern Mariana Islands, Guam, the Federated States of Micronesia, the Republic of the Marshall Islands, and the Republic of Palau.

commodity to be sold or traded. While the states have legal authority over land for "eminent domain and condemnation," use of that power is strongly avoided (FSM National Communication, 1999).

These complex patterns of land ownership create challenges for many climate change response (adaptation) options. The FSM National Communication to the United Nations Framework Convention on Climate Change (UNFCCC) notes, for example, that the relocation of a large proportion of the coastal population to elevated lands on high islands can only occur with the concurrence of the owners of those elevated lands and requires long-term land use planning accompanied by changes in traditional patterns of land tenure (FSM National Communication, 1999). The FSM National Communication to the

319

UNFCCC also highlights the fact that such challenges are even greater when considering off-island re-settlement in response to sea-level rise impacts on low-lying atolls. The FSM National Communication notes that pursuit of such an option would "have to be achieved without disruption to host communities and with sensitivity to and careful regard for the traditional values and practices of both the displaced and host communities" (FSM National Communication, 1999).

Demographically, Pacific and Caribbean islands include densely populated urban centers such as Honolulu, Hawaii and San Juan, Puerto Rico to sparsely populated outer islands of the Federated States of Micronesia. Some of the highest population densities in the Pacific are found on Majuro and Ebeye in the Marshall Islands. While future population growth rates in the Caribbean appear to be attenuated, some Pacific island populations are growing rapidly. For example, from 1982 to 1998, the population of Majuro virtually doubled from about 15,000 to over 30,000. Pacific Islanders represent one of the fastest growing groups of migrants to the mainland United States. As the impact of climate change is felt more strongly, this trend is likely to continue.

ECOLOGICAL CONTEXT

The ecosystems of the islands discussed in this chapter can be characterized as tropical, rich in biological uniqueness, and susceptible to disturbance and biological invasions.[1] The isolation of islands has resulted in the evolution of unique floras and faunas with large numbers of endemic species — those found nowhere else on Earth (Low et al., 1998; Wiley and Vilella, 1998; Loope, 1998). Islands in both the Pacific and Caribbean are facing an extinction crisis as a result of steady habitat destruction, and competition and predation by introduced species. Small population sizes and lack of room for migration exacerbates the vulnerability of island species. The situation on many islands is becoming critical as the area of undisturbed natural habitat diminishes (Low et al., 1998; Wiley and Vilella, 1998; Loope, 1998). Fortunately, unique terrestrial and marine resources are still found in the tropical US on islands which have few or no human inhabitants such as Baker and Howland Islands, Jarvis Island, the Johnston Atoll, and the Kingman Reef among others (OIA, 1999).

The Caribbean has a limited resource base, yet contains some of the most biologically productive and complex ecosystems found in the world. These include coral reefs, beaches, seagrass beds, mangrove forests, and coastal estuarine ecosystems. Native tropical rainforest (including mamey, sierra palm, and tree fern) and moist forest once dominated much of Puerto Rico, but much of this land has given way to cultivation, pasture, and other development. Drier areas support dry grasslands and cacti as well as forests of royal palm, acacia, and other tropical trees. Mangrove swamps dominate much of the coastline. The US Virgin islands are hilly due to their volcanic origin with land cover consisting primarily of cactus, grasses, and sparse tropical woodland and shrubland. Some native vegetation has been lost to stock grazing and the cultivation of fruit and vegetable crops and sugarcane.

The native vegetation cover of the Hawaiian islands includes tropical coastal and lowland shrublands, grasslands, wet forests, tropical montane rain forests, and shrublands (Loope, 1998). Much of the coastal and lowland vegetation, and increasingly, the interior montane vegetation, have been lost to urban, recreational, and agricultural development (grazing, sugarcane, and pineapple plantations), and to introduced species.

American Samoa features rain forests in the interior mountains. Much of the coastal zone is under cultivation for coconut, taro, and other crops. Guam's principal land cover consists of tropical shrubland and limestone forest in the north and sword grass and dense jungles along river valleys in the volcanic south. The Marianas are not considered to be tropical rain forest. The flora consist of vines, shrubs, ferns, and grasses, including savanna and trees. The more common trees include: coconut, flame tree, Formosan koa, ironwood (Casuarina), banyan, papaya, tangan, mangrove and a few other varieties. In general, the variety of botanical species is limited on the CNMI and there are no vegetation zones (CNMI website, 2000).

Over 97% of US coral reef resources are in the territorial waters of Hawaii and the US territories and commonwealths. Hawaii has the widest range of reef habitats among the islands discussed in this report, with a significant portion of US reefs in the northwest Hawaiian Islands. The US territories, commonwealths, and freely associated states, because of their location in tropical waters, have extensive reef habitats that are important culturally, economically, and biologically. For example, the abundant and wide variety of sea life within CNMI waters includes

---

[1] This section outlines information on current island ecosystems and biodiversity. The VEMAP modeling system providing information about potential future changes to vegetation and ecosystems was not run for the islands as it was for the conterminous US.

US Affiliated Islands of the Caribbean and Pacific

Name/ US Affiliation	Geologic Type	Location/Description	Current Population (Approx.)	Major Economic Sectors
Commonwealth of Puerto Rico	One volcanic island composed of uplifted sedimentary rocks.	• Located SE of Miami on the northern edge of the archipelago of islands on the Caribbean plate. • 3,425 sq mi; 8,871 sq km • Abundant rain in the North; dry climate in the South.	4 million.	Manufacturing Services Government Agriculture Tourism
US Virgin Islands Territory	Three volcanic islands (St. Croix, St. John, and St. Thomas) composed of uplifted sedimentary rocks.	• Located 1,000 miles SE of Miami on the northern edge of the archipelago of islands on the Caribbean plate. • 133 sq. mi.; 344 sq. km • No lakes, rivers or streams.	120,000	Tourism Government
Hawaiian Islands State	Eight volcanic islands (Kauai, Oahu, Molokai, Lanai, Maui, Kahoolawe, Niihau and Hawaii). Also consists of NW Hawaiian Islands and Johnston Atoll.	• Located 2,400 miles SW of California and on the Tropic of Cancer • 6,450 sq. mi.; 16,706 sq. km. • Sea area: 833,171 sq. mi. • Majority of US coral reefs.	1.2 million	Tourism Agriculture Military spending Manufacturing Government
American Samoa Territory	Five volcanic islands and two coral atolls (Ofu, Ta'u, Swains Island, Tutuila, Olosega, Rose Island, and Aunu'u)	• Located 2,700 miles NE Australia • 76 sq. mi.; 197 sq. km. • Rainforests in interior; mountainous regions	60,000	U.S. federal expenditures Canning industry Small tourism industry Government
Commonwealth of the Northern Mariana Islands	Fourteen volcanic islands	• Located at the edge of the Philippine plate. • 185 sq. mi.; 479 sq. km. • Formed by underwater volcanoes along Mariana Trench.	66,000	Tourism Garment manufacturing Government Fish transshipment
Republic of the Marshall Islands Freely Associated State	29 coral atolls (each made up of many islets) and 5 low-lying volcanic islands	• Located about 2,136 miles SW of Hawaii • 70 sq. mi.; 181 sq. km. • Each atoll is a cluster of many small islands encircling a lagoon • None of these islands is more than a few meters above sea level.	60,000	US aid Tourism Fisheries Subsistence Agriculture Craft items (wood carvings, shell and woven baskets)
Republic of Palau Freely Associated State	Several hundred volcanic islands and a few coral atolls (only 8 of the islands are inhabited).	• Located in the North Pacific Ocean, SE of the Philippines. • 192 sq. mi.; 497 sq. km. • The 8 islands that are inhabited include Kayangel Island, Babelthaup, Urukthapel, Koror, Peleliu, Eli Malk, Angaur, Sonsoral and Tobi.	17,000	Tourism Craft items (wood, shell pearl) Commercial fishing Agriculture (subsistence) Government
Federated States of Micronesia Freely Associated State	A group of 607 small islands consisting of volcanic islands and coral atolls.	• Located in the Western Pacific about 2,500 miles SW of Hawaii. • 271 sq. mi.; 702 sq. km • 4 states: Chuuk, Pohnpei, Yap and Kosrae.	127,000	Tuna fishing Tourism Government Subsistence Craft items (wood carvings, shell and woven baskets)
Guam Territory	One volcanic island composed of uplifted sedimentary rocks.	• Guam is the largest Micronesian Island. • 209 sq. mi.; 541 sq. km. • Formed by the union of two volcanoes, northern Guam is a flat limestone plateau while the southern part is mountainous.	150,000	Tourism Military/government Tuna and other cargo transshipment Airline hub

coral gardens that harbor myriad mollusks, plant life, and tropical fish. The near-shore reef and lagoon areas are important for subsistence fishing, particularly net throwing. Many sea cucumbers occur in the shallow lagoon waters (CNMI website, 2000).

The Freely Associated States contain a wide range of terrestrial and marine habitats, such as coral reefs and mangrove forests. The Exclusive Economic Zone of the Federated States of Micronesia (FSM) covers over one million square miles (2.6 million sq. km) of ocean and contains the world's most productive tuna fishing grounds. The estimated market value of tuna caught within FSM waters is valued at over $200 million per year. Pohnpei State, the location of the FSM's capital, receives an average of over 180 inches of rain per year and is covered by lush rainforests, and surrounded on the southern side with some of the most extensive mangrove wetlands in the Pacific Islands region. The rock islands in the Republic of Palau are uplifted fossil coral reefs that have been rounded by erosion. Coral reef ecosystems in Palau, which suffered from extensive bleaching during the 1997-1998 El Niño, are regarded as among the best diving areas in the world.

Unique terrestrial and marine resources of the tropical US are found in insular areas in the Pacific which have few, if any human inhabitants. These include Baker and Howland islands which are national wildlife refuges (Howland: birds and marine life; Baker: marine life). Jarvis Island, another national wildlife refuge, serves as habitat for seabirds and shorebirds. Johnston Atoll is home to 194 species of inshore fishes and sea turtles, and is a site for roosting and nesting of 500,000 seabirds. Kingman Reef has a rich marine fauna and the Northwest Hawaiian Islands National Wildlife Refuge includes over 100 islands with an estimated 616,000 pairs of Laysan albatrosses and 21 other species of seabirds (Low et al., 1998; Harrison, 1990). The CNMI islands of Asuncion, Guguan, Maug, Managaha, Sarigan, and Uracas (Farallon De Pajaros) are maintained as uninhabited places. No permanent structures may be built and no persons may live on the islands except as necessary for the purposes for which the islands are preserved. The islands are preserved as habitats for birds, fish, wildlife, and plants. The three islands that are collectively known as Maug have been given permanent status as wildlife preserves (CNMI website, 2000).

CLIMATE VARIABILITY AND CHANGE

The islands of the Pacific and Caribbean experience relatively high air temperatures and low seasonal variations in air temperature throughout the year compared to most of the US mainland. However, other climate variables exhibit distinct seasonal patterns, particularly rainfall distribution, which results in wet and dry seasons. For example, precipitation patterns differ greatly over the Pacific, with a pronounced winter rainfall maximum in Hawaii and a late-summer/early fall maximum in Guam. In the Caribbean, the wet season is during the summer. In the Pacific, tropical storms and typhoons are common between May and December, although intense tropical cyclones can affect Guam and the CNMI during any month of the year. In the Caribbean, the hurricane season extends from June to November.

In addition to the seasonal variations, there are strong year-to-year fluctuations. For example, the El Niño/Southern Oscillation (ENSO) causes fluctuations in sea level, rainfall, and cyclone activity (hurricanes or typhoons, depending on the region) in both the Pacific and Caribbean islands. In the Caribbean, Atlantic hurricanes are suppressed during El Niño events, while their numbers increase during La Niña events. In the Pacific, during El Niño events, Hawaii, Micronesia, and the islands of the southwest tropical Pacific often receive below normal rainfall. Despite this, El Niño can also increase the risk of intense tropical cyclones for Hawaii, American Samoa, and the eastern half of Micronesia. Additionally, areas of above normal precipitation, along with greater tropical cyclone activity, typically shift eastward towards French Polynesia. During La Niña events, parts of Micronesia can receive very heavy rainfall during what is normally their dry season.

Observed Climate Trends

Over the past century, average annual air temperatures in the Caribbean islands have increased by more than 1°F (0.5 °C), while average annual air temperatures in the Pacific Islands have increased by about half this amount. Globally, sea level has risen by 4 to 8 inches (10-20 cm) in the past 100 years but with significant local variation. Relative sea level, which takes into account natural or human-caused changes in the land elevation such as tectonic uplifting and land subsidence (sinking), is showing an upward trend at sites monitored in the Caribbean and Gulf of Mexico. The current rate of relative sea-level increase in the Caribbean and Gulf of Mexico is about 3.9 inches (10 cm) per 100 years. While absolute sea level is also rising in the Pacific, geologic uplift is exceeding or keeping pace on many islands. As a result, across the set of islands, there is no consistent sea-level trend for the Pacific. Although showing no average net rise, relative sea level fluctuates widely, due to effects such as the ENSO cycle.

The highest frequency of cyclones in the Caribbean occurs in the western Bahamas and along the US eastern seaboard and adjacent Atlantic Ocean. There has been considerable decadal variability of hurricanes during this century (see Reading, 1990; Diaz and Pulwarty, 1997; Landsea et al., 1995). Among other factors, decadal variability is affected by ENSO as well as changes in the thermohaline circulation (THC), the large-scale oceanic circulation in the Atlantic that includes the Gulf Stream. For example, Gray et al. (1997), show that a weakening of the THC in the latter half of this century resulted in cooling of the Atlantic Ocean along the coast of northwest Africa, an increase in the low-level pressure gradient between this region and the central Sahel, and a resulting reduction in the strength of the easterly waves that originate off the coast of

Africa and propagate into the tropical Atlantic. These easterly waves are responsible for hurricane activity in the Caribbean. Therefore, changes in THC help regulate hurricane occurrence in the Atlantic.

Hurricane activity in the western Atlantic is reduced during the season following the onset of an El Niño event in the Pacific. The reason for El Niño occurrence leading to decreased incidence of hurricanes is because of the presence of anomalously strong westerly winds in the upper troposphere over the Caribbean and equatorial Atlantic (Gray, 1984). The strengthened winds result from extra-deep cumulus convection in the eastern Pacific Ocean that occurs during El Niño events. Hurricanes in the Atlantic tend to return to normal in the second summer following such an event. In spite of the relationship between hurricanes and El Niño events, there is little net effect on the precipitation in the region.

Scenarios for Future Climate

Model projections suggest it is possible that the Pacific and Caribbean islands will be affected by: changes in patterns of natural climate variability (such as El Niño); changes in the frequency, intensity, and tracks of tropical cyclones (hurricanes and typhoons); and changes in patterns of ocean circulation. These islands are also very likely to experience

The El Niño/Southern Oscillation (ENSO) Phenomenon

The term "El Niño" refers to the warm phase of the ENSO cycle when ocean temperatures in the central and eastern tropical Pacific tend to increase. The term "La Niña" refers to a phase of the ENSO cycle when waters in the central and eastern Pacific tend to be cooler. The ENSO cycle, normally about one year in duration, is a continuously evolving pattern of ocean-atmosphere behavior and represents one of the dominant patterns of natural variability in the climate system worldwide. Its frequency of occurrence is approximately every 2 - 7 years.

El Niño is thus one phase of the ocean half of a coupled ocean-atmosphere phenomena in the tropical Pacific (see Pielke and Landsea, 1999). Its atmospheric partner is known as the Southern Oscillation, a periodic oscillation of atmospheric pressure between the western and eastern Pacific Ocean. Under normal, non-El Niño conditions, atmospheric pressure is lower over the western Pacific and higher over the eastern Pacific. This pressure gradient helps drive the (surface) trade winds from east to west or from high pressure to low pressure and results in a tendency to maintain higher sea levels in the western Pacific Ocean than in the east. During an El Niño event, the atmospheric pressure gradient lessens, resulting in surface winds that diminish or, sometimes, reverse direction. The see-saw in atmospheric pressure, coupled with the eastward movement of warm water, results in a drop in sea level in the western Pacific Ocean and a migration of higher than normal sea-level conditions to the east (along with the warm water due to reduced upwelling). In fact, this eastward migration of sea level is one of the observational signals that an El Niño event is underway. Images of El Niño events from satellites like TOPEX-POSEIDON show these changes in sea level as changes in the height of the ocean's surface. When this occurs, islands in the far western Pacific experience lower sea level while those in the central and eastern Pacific experience a transient increase in sea level.

El Niños and La Niñas as a Function of Observed and Projected Sea Surface Temperatures

Index of Monthly Simulated Observed El Niños (Positive) and La Niñas (Negative)

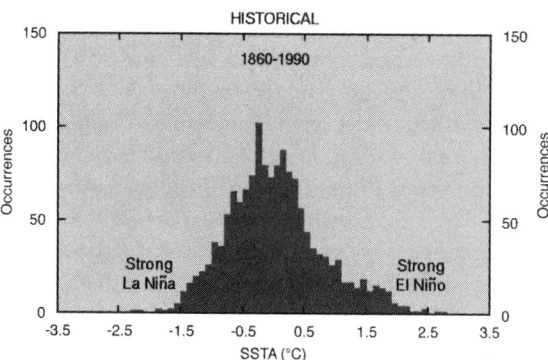

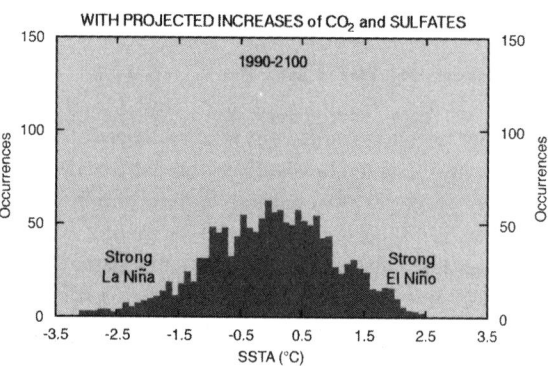

Figure 2. These model projections suggest stronger and more frequent El Niños and La Niñas as a result of climate change. Sea Surface temperature anomalies (SSTA) in the equatorial Pacific are used to measure the strength of El Niños and La Niñas. These model projections by the **Max Planck Institute** suggest a wider range of SST deviations from normal and thus more extreme El Niños and La Niñas in the future. The high bars in the center are occurrences of normal SSTs. In the projections in the bottom graph, these normal temperatures occur less frequently, while lower (La Niña) and higher (El Niño) SSTs occur more frequently. The **Max Planck** model is used here because it has been able to reproduce the strength of these events better than other models due to its physics and ability to resolve fine scale structure in the ocean. Source: Timmermann et al., 1999. See Color Figure Appendix.

increased ocean temperatures and changes in sea level (including storm surges and sustained rise).

For the Caribbean, most model simulations project increasing air and water temperatures, slightly wetter conditions in winter, and slightly drier conditions in summer. Models that suggest more persistent El Niño-like conditions across the Pacific imply a reduction in Atlantic hurricane frequency in the future. However, not all models concur on this point.

For the Pacific, the model simulations project a gradual increase of air and water temperatures. Some recent climate model studies also project that ENSO extremes are likely to increase with increasing greenhouse gas concentrations (Timmermann et al., 1999; Collins, 2000). Additionally, recent model results (Knutson and Manabe, 1998; Timmermann et al., 1999) have agreed with earlier studies (Meehl and Washington, 1996) in showing that as global temperatures rise due to increased greenhouse gases, the mean Pacific climate tends to more resemble an El Niño-like state, with greater relative surface warming in the equatorial eastern Pacific than the west. This implies a reduction of fresh water resources in areas of the western Pacific, Micronesia, and southwest tropical Pacific (Meehl, 1996). The existence of higher sea surface temperatures in the central and eastern Pacific could also result in an expansion in the area of the Pacific where tropical cyclones will form and migrate, as is the case during an El Niño. Other models show uniform warming across the Pacific or somewhat greater warming in the western equatorial Pacific.

Apart from the linkage with ENSO, there is significant uncertainty about how increasing global temperatures will affect hurricane or typhoon frequency and preferred tracks in both the Caribbean and the Pacific regions (see Landsea, et al., 1996; Henderson-Sellers, et al., 1998; Emanuel, 1997, 1999; Meehl, et al., 2000). Studying the effect of global warming at a time of doubled CO_2, Bengtsson et al. (1997) conclude that there are significantly fewer hurricanes (especially in the Southern Hemisphere), but no significant change in the spatial distribution of storms. The model used by Bengtsson, et al. (1997), however, was barely adequate to resolve tropical cyclones. More ocean surface warming could create more opportunity for storm development, given the right atmospheric conditions. Other results, using a nested high-resolution model (resolution of up to 1/6 degree or 18 km) indicate a 5-11% increase in surface winds and a 28% increase in near-storm precipitation (Knutson et al., 1998; Knutson and Tuleya, 2000). Another recent overall assessment regarding changes in hurricane strength suggests a 5-10% increase in tropical storm wind speed by the end of the 21st century (Henderson-Sellers, et al., 1998).

It is also a fact, however, that until capabilities are more developed for accurately simulating ENSO and changes in the thermohaline circulation in global climate models, the possibility of significant changes in the expected frequency, intensity, and duration of ENSO as well as tropical storms cannot be ruled out. It is even conceivable that changes in the strength of monsoon circulations in the South Pacific could have unanticipated effects on ENSO and related tropical storm frequency, paths, and strength.

KEY ISSUES

Two regional workshops were held to identify stakeholders' concerns regarding climate change and climate variability in the context of other current stresses for island and coastal environments and communities. The first workshop was held in Honolulu, Hawaii in March 1998, and participants identified issues of importance to the US-affiliated islands in the Pacific. The second workshop took place in July 1998 and addressed concerns of the south Atlantic coast of the US and the US-affiliated Caribbean islands. This chapter is based on presentations and discussions at these workshops, extant literature, and ongoing research into the consequences of climate variability and change to islands.

While the workshop participants identified numerous issues for consideration, four of the critical climate-related issues for the islands now and in the future include:
• ensuring adequate freshwater supplies
• protecting public safety and infrastructure from climate extremes
• protecting rare and unique ecosystems
• responding to sea-level fluctuations.

It should be noted that participants at the regional workshops did not identify long-term sea-level rise as the highest priority issue, focusing instead on water resource issues and storm impacts. There is a sense that the nature of the sea level issue for islands has been somewhat misunderstood in discussions of climate change. It is important to recognize that episodic variations in sea-level (like those associated with the ENSO cycle in the Pacific) and storm surges are already extremely important issues. In addition to considering only the consequences of a gradual, long-term rise in sea level, island communities will continue to face the consequences of episodic events (e.g., extreme lunar tides, ENSO related changes, and storm-related wave conditions) some of which themselves will be affected by climate change. Importantly, changes in sea level associated with natural variability exceed long-term projections of sea-level rise in some jurisdictions.

Many of the projected consequences of long-term sea-level rise, such as salt water intrusion into fresh-

water lenses and coastal erosion are already problems in some if not most island jurisdictions. According to USGS (1999a), for example, about 25% of the sand beaches on Oahu, Hawaii have been lost or severely degraded during the last 60 years. Climate-related changes of these conditions are seen, therefore, as magnifying existing problems rather than as problems in isolation from other stresses. Island communities must deal with these problems today and, in so doing, can develop important insights into how they might most effectively respond to climate-related changes in sea level over both the short- and long-term. Because the communities will act to mitigate the increased climate change problems periodically, they will not be adjusting to 100 years of climate change effects at one time, but rather to smaller effects more frequently during the century. This allows for staggered adaptation.

1. Freshwater Resources

The availability of freshwater resources is already a problem for island communities, as a result of their unique geography and the growth of population, tourism, and urban centers. Many islands suffer from frequent droughts and chronic water scarcity due to lack of adequate precipitation. In other cases, rainfall is abundant but access to freshwater is limited by lack of adequate storage facilities and delivery systems, or a geographic mismatch between the source and the site of the need. For the Pacific Islands, both a present and future key issue is drought or water scarcity conditions. For the Caribbean, drought is also a problem, but floods and landslides associated with hurricanes and heavy rainfall are an important addition to this region's list of key water-related issues. According to Larsen and Torres-Sanchez (1998) rainfall-triggered landslides are the most common type of landslide in the central mountains and foothills of Puerto Rico.

Future climate changes in the islands regions could include: changes in natural variability in weather patterns, ocean temperature, and currents (e.g., patterns of El Niño); changes in the frequency, intensity, and tracks of tropical cyclones (hurricanes and typhoons) and their resulting precipitation; and/or changes in sea level. Any of these changes would impact the amount, timing, or availability of freshwater, such that freshwater issues will be increasingly serious concerns for the US affiliated Islands.

On the islands, demands for water are from multiple sources including agriculture, industry (including fish processing), households, and natural ecosys-

Freshwater Lens Effect in Island Hydrology

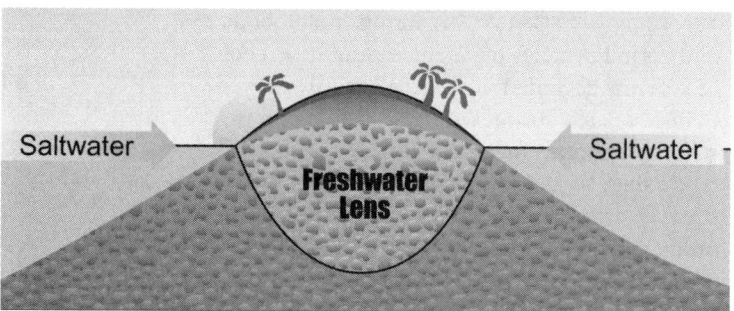

Figure 3. On many islands, the underground pool of freshwater that takes the shape of a lens is a critical water source. The freshwater lens floats atop salt water. If sea level increases, and/or if the lens becomes depleted because of excess withdrawals, salt water from the sea can intrude, making the water unsuitable for many uses. The size of the lens is directly related to the size of the island: larger islands have lenses that are less vulnerable to tidal mixing and have enough storage for withdrawals. Smaller island freshwater lenses shrink during prolonged periods of low rainfall, and water quality is easily impaired by mixing with salt water. Short and light rainfall contributes little to recharge of these sources. Long periods of rainfall are needed to provide adequate recharge. Source: Illustration by Melody Warford. See Color Figure Appendix.

tems. Tourism is extremely water-intensive and is a major industry for the Commonwealth of the Northern Mariana Islands, Guam, Hawaii, Palau, Puerto Rico, and the Virgin Islands. Tourism development continues in these areas and the Federated States of Micronesia, the Republic of the Marshall Islands, and American Samoa all see tourism as one of their best options for economic development.

Many island communities rely upon groundwater resources called freshwater lenses. The size of the groundwater lens is directly related to the size of the island as well as incident rainfall, island recharge characteristics (controlled by land cover, evaporation, and geology), and the lateral leaking of lens water into the ocean. Larger islands have larger lenses and are more buffered from drought conditions while smaller islands have no lenses or shallow lenses that easily become depleted or contaminated with salt water. During drought conditions, there is no recharge to the lens, and the fresh water is depleted rapidly, especially if consumption is high. Low sea levels associated with El Niño lower the water table even more, making it more difficult to access the water and easier to damage the fragile interface between the fresh water lens and the underlying salt water. Low-lying atolls rely heavily on rainwater collection and are therefore least buffered against drought conditions.

The geology of the island controls the amount, quality, and seasonal availability of the water supply. Islands that are continental or of old volcanic origin exhibit complex movement of groundwater through layers of rock or clay. New volcanic islands generally have more dynamic movement of water through tunnels, holes, and layers of ash. Similarly, islands made up of coral or sand are quite permeable, so the rainwater infiltrates very quickly. Water quality is also an issue: many volcanic islands have highly permeable rock, which increases the potential for groundwater contamination and in some Pacific islands contamination problems reduce the "capacity" of the system (USGS Hawaii State Fact Sheet, 1996). In addition, seawater can contaminate the freshwater lens with salt if sea level rises, the lens is depleted, or the delicate freshwater-saltwater interface is damaged by excessive extraction.

Patterns of precipitation are important in determining whether islands have an adequate freshwater supply. Long periods of rainfall are needed to recharge the freshwater lenses since short and light rainfall tends not to contribute to aquifer recharge. Land use is an important determinant of infiltration; how much water infiltrates into the ground or flows into rivers and streams depends upon the land cover that catches the rain. If the land is covered by forest, the forest holds the rainwater for drier periods, but if the forest has been displaced by urban development, for example, the rain runs off faster leaving less for use during dry conditions. On some islands, destruction of forest cover has caused many formerly perennial streams to stop flowing in the dry seasons and has contributed to landslides during periods of heavy rain.

Pacific Islands. Rainfall, stream flow, and groundwater are fairly abundant on the big island of Hawaii, but many areas are withdrawing water at rates close to the estimated yield of aquifers near the populated areas (USGS Hawaii Fact Sheet, 1999a). Water use in many of the other Pacific Islands is now near the limit of the known freshwater resources, and demand exceeds the existing supply on many islands. In some cases, the problem is not the lack of adequate rainfall, but distribution, storage, and maintenance. Guam and the island of Hawaii are examples of the distribution problem due to their "geographic mismatch" between supply and demand. For example, the "Hilo side" of the big island of Hawaii consistently receives abundant rainfall and is one of the wettest places in the Hawaiian Islands, while the "Kona side" is where the majority of tourism development is occurring and has experienced significant periods of drought. Other islands, that more easily access their adequate rainfall, face storage, distribution, and maintenance problems. Pohnpei normally has an abundance of water, including waterfalls, springs, and rivers in the central areas. However, the island lacks storage and distribution systems, causing severe impacts under drought conditions.

Water system development in American Samoa, Guam, and Hawaii has thus far kept pace with growth and almost all residents have access to reliable supplies of quality water. On Saipan, the Commonwealth Utilities Corporation has struggled for decades to make improvements in an inadequate water distribution system and to keep pace with population growth and tourism development. On Pohnpei, despite its abundant rainfall, the public water system in Kolonia, the largest town, cannot provide an adequate supply of quality water to residents under "normal" circumstances. In rural areas, people take water directly from rivers and streams, many of which dry up during severe droughts, and while groundwater is available, there is often no system to access it. In the outer islands of the Marshall Islands and Federated States of Micronesia, residents rely on small rooftop catchment systems with only a few thousand gallons of storage capacity. On Majuro, the public water system is fed primarily from the international airport runway and that system has been supplemented by wells on Laura Islet, some 17 miles from the storage system at the airport. Colonia, Yap (FSM) depends on a small reservoir for water, but the reservoir, even if full at the beginning of a severe drought, dries up.

Currently the Pacific Islands experience drought conditions every two to six years, usually associated with the El Niño/Southern Oscillation (ENSO) cycle. The most extreme droughts appear to be associated with very strong El Niño events like those that occurred in 1982-1983 and 1997-1998. Rainfall during the 1997-1998 event in some areas was well below normal for the period October 1997 through September 1998. In some months, rainfall was as low as 3% of normal. These low monthly totals meant trouble for islands relying on catchment systems.

The water resources for many of the Pacific Islands could be most adversely affected by any increase in El Niño or El Niño-like conditions. These increases could take many forms such as more frequent and/or severe El Niño events, medium-sized El Niños that persist for multi-year periods, or other conditions that diminish precipitation over these islands, such as the "El Niño-like conditions" hypothesized to result

from increased ocean temperatures. Due to overlapping time-scales, these changes could result in almost chronic drought conditions for some islands in the Pacific. Such a sequence would not allow islands critical recovery time from one drought before another sets in, or would stress their emergency preparedness capabilities if they were faced with an additional natural disaster on top of any protracted drought condition. A secondary consequence is migration: if such enhanced drought conditions occur over wide regions of the tropical Pacific, atoll populations may migrate in significant numbers to high-island population centers with more stable freshwater resources (Meehl, 1996).

The Caribbean. In Puerto Rico and the US Virgin Islands, high population densities and the conversion of tropical forest to other uses have affected hydrology and water resources. These changes have contributed to overuse of existing water supply (even with reduced per capita use due to conservation programs, USGS, 1999b), in-filling of public-supply reservoirs with sediment, and contamination of groundwater and surface water (Zack and Larsen, 1993). In the US Virgin Islands, and to a lesser extent in Puerto Rico, leaky septic tanks and inadequate sewage treatment facilities have degraded the quality of near-surface groundwater supplies (Zack and Larsen, 1993). In addition, both Puerto Rico and the US Virgin Islands experience flash flooding and landslides that result from the combination of steep slopes and heavy rainfall typical of this region. Deforestation and development have contributed to increased flash flooding and landslide potential particularly under steep, denuded slopes. Poor drainage on floodplains has increased the vulnerability of these areas to various flooding events, while land-use patterns and inadequate construction increase the exposure of the population (Zack and Larsen, 1993).

In the Caribbean, tropical storms/hurricanes are significant natural hazards, through the mechanisms of storm surge and high winds. In coastal locations of Puerto Rico and the US Virgin Islands, saltwater intrusion threatens the continued use of fresh groundwater and limits groundwater withdrawals.

Puerto Rico has abundant ground-and surface-water resources due to relatively heavy rainfall over the mountainous interior of the island and receptive, sedimentary rocks around the island's periphery (Zack and Larsen, 1993). These form an extensive artesian aquifer system on the north coast. Water-table aquifers overlie the north coast artesian aquifer and occur along most of Puerto Rico's coastline. Artificial reservoirs on principal water courses col-

Responding to Drought in the Pacific Islands

The 1997-1998 El Niño event offers a vivid example of how forecasts and information about potential climate phenomena and their consequences can be used to support decision making and planning that benefit society. In September 1997, the Pacific ENSO Applications Center (PEAC) predicted a near-record drought for Micronesia, beginning in the November-December timeframe and ending in the May-June timeframe.

When severe drought was projected for the Northern Marianas (where there are not regular sources of good water at any time) beginning in February 1998, the government implemented a tightened water-rationing schedule. Each venue was assigned "water hours" of 2 - 4 hours each day. By February, El Niño had already reduced the island's rainfall level to 75% below normal, and the shortage was to continue for many months. The Commonwealth Utilities Corporation, the government entity responsible for the island's water system, also outlined a "severe drought scenario" in which commercial water users, such as hotels and garment factories, would be cut off from the island's water system for a few hours each day. A task force was established which pushed for legislation requiring every home and building to install a rainwater catchment system (this would, however, require loans or subsidies for low-income families).

For Majuro, PEAC provided a drought and typhoon forecast in September 1997 that indicated an increased typhoon risk and a sharp decline in rainfall beginning in December 1997. In November, the Government of the Republic of the Marshall Islands convened a task force to develop a drought response plan. Typhoon Paka struck the Marshall Islands in December 1997 and brought the last significant rainfall until June 1998. The government requested assistance from the US Commander-in-Chief, Pacific Command to repair pumps at the wells on Laura Islet. The water utility on Majuro developed a conservation plan, a public education program, and imposed water hours. At the height of the drought, residents of Majuro were getting 7 hours of water every 15 days. The Japanese government provided two 20,000 gallon per day (gpd) and one 16,000 gpd reverse osmosis (RO) units in February 1998. Following the US Presidential disaster declaration in March, six larger RO units were provided with funding from the US Federal Emergency Management Agency in April. In May 1998, the new pumps at Laura came on line and in June rainfall increased and the RO units were shut down.

The drought in Samoa was delayed until April because of its location in the Southern Hemisphere. For the Samoas, PEAC anticipated an increase in the risk of tropical cyclones and drought as a result of the El Niño. In April, May, and June of 1997, rainfall in Pago Pago was only 64% of normal. In July 1997, it was only 59% of normal. Their summer was closer to normal, but PEAC forecasted that rainfall for the period April through August 1998 would be well below normal and this was borne out. The American Samoa government organized a drought response that included such activities as establishing a drought task force and a public information campaign and has continued to rely on the PEAC and National Weather Service forecasts.

In May of 1998, American Samoa's single largest private employer, StarKist Samoa committed to implementing several in-house conservation

El Niño billboard used as a public information tool in the Pacific Islands - US National Weather Service, Pacific Region Office.

measures which would eventually reduce the canneries' daily water usage by 25%. The company had also utilized scenario-based planning, and acknowledged that the worst case would involve importing water from off-island by tanker in the amount of 50 million gallons.

Agriculture throughout the US-affiliated Pacific Islands during 1997-1998 suffered everywhere from the droughts, except on Guam. There, farmers used public water for irrigation, and the delay of heavy rains toward the end of the drought resulted in one of the most productive harvests in recent history. In the Commonwealth of the Northern Mariana Islands (CNMI), citrus and garden crops were most affected by the drought, and the hospital had to buy imported fruits and vegetables rather than rely on local suppliers. A limited damage assessment was done on Pohnpei and serious losses of both food and cash crops were sustained. Over half the banana trees evaluated had died or were considered seriously stressed. The loss of sakau (kava) was probably the most serious economic loss because it had become a major cash crop for Pohnpei. On Yap, taro losses were estimated at 50-65%, and betel nut prices increased more than 500%, although only 15-20% of the trees were lost. In Palau, imported food shipments increased from twice a month to once a week. Other climate-related consequences felt throughout the islands included changes in the migratory patterns of economically-significant fish stocks like tuna; stresses on coral reefs associated with increased water temperature; and increased sedimentation from erosion; and reduced local air quality in areas affected by wildfires. Guam, for example, experienced some 1400 fires and grass fire burned 20% of Pohnpei.

Relief food was supplied beginning in May 1998 to islands in the Marshalls and was also supplied to the outer islands of Pohnpei, Chuuk, and Yap. The Pohnpei Agriculture and Trade School, the US Natural Resources Conservation Service, and Pohnpei State Agriculture Department provided assistance to the farmers as the drought began to subside. Subsistence crops required 6-10 months for rejuvenation after the rains returned. The disaster management offices used PEAC and National Weather Service forecasts to plan food relief and replanting programs for the droughts in the Republic of the Marshall Islands (RMI) and the Federated States of Micronesia (FSM).

Overall, islands in the Pacific implemented many of the following measures when faced with drought conditions in 1997 and 1998:

- expanded public information and education efforts;
- continued emphasis on and funding for household rainwater collection systems;
- accelerated drought planning efforts;
- expanded drilling efforts and other activation of new water sources, such as the emergency use of monitoring wells to augment production;
- tightened pre-existing water restrictions and delivery scheduling schemes;
- more use of water-haulers to fill existing storage facilities;
- shifted farm location and range management (where livestock were pastured) to take advantage of more reliable, groundwater-based irrigation supplies;
- increased attention to water conservation, for example in concentrated leak detection and repair efforts;
- more extensive and effective use of ENSO forecast information;
- accelerated maintenance on water pumps and lines - acquired spare parts; and
- installed RO units.

Many of these improvements have long lasting positive effects – such as repair of pipes and pumps and increased storage capacity, alleviating the need for conservation until the next drought.

lect runoff and are used for water supply, flood control, and limited hydroelectric power generation. Ground water accounts for ~30% of the total amount of water used in Puerto Rico and surface water accounts for ~70% (Zack and Larsen, 1993).

The US Virgin Islands are much smaller and have a lower maximum elevation than Puerto Rico. These islands receive less rainfall and retain less fresh water. The US Virgin Islands have no perennial streams and only limited ground water resources. Retention dams have been used on ephemeral streams to promote recharge of coastal aquifers (Zack and Larsen, 1993). In the US Virgin Islands, 65% of drinking water supplies are provided by desalinated seawater, making it the most expensive publicly supplied water in the United States (Zack and Larsen, 1993). Another approximately 22% of the drinking water supply originates from groundwater and 13% from rooftop catchments.

Flooding is a critical issue for Puerto Rico and the US Virgin Islands due to the topography and rainfall patterns of the islands. In the US Virgin Islands, human infrastructure (including buildings and roads) has covered many of the ephemeral channels that allowed for infiltration. This not only diminishes water supply, but greatly increases flash flood hazard. In Puerto Rico, flash floods typically result from rainfall that is intense in the upper basins but sparse or nonexistent on the coast. These events can trigger hundreds of landslides, which are common in the mountainous areas of Puerto Rico where mean annual rainfall and the frequency of intense storms are high and hillslopes are steep. One effect of these landslides is adding sediment to river channels, which is carried into downstream reservoirs. Combined with other factors such as heavy rainstorms, urban development, and agricultural practices, this sediment reduces the storage capacity of major reservoirs, and diminishes their efficiency at reducing flood peaks (Zack and Larsen, 1993). Large quantities of sediment are also washed into the ocean, where it is deposited on the reefs and deteriorates reef health.

Droughts are frequent and severe in the US Virgin Islands. Any minor depletions in rainfall dramatically affect agriculture and require water rationing (Zack and Larsen, 1993). Droughts are infrequent in Puerto Rico. However, mandatory water rationing was implemented six times during the 1990s resulting in significant agricultural and other economic losses (Larsen, 2000). A drought in 1994-95 affected more than one million people who endured mandatory rationing for more than a year (Larsen, 2000).

Although Puerto Rico had seen lower mean annual rainfall in previous decades, the population growth changed the nature of the needed response. Public works projects were initiated including a $60 million project to dredge the sediment from the Loiza reservoir and $350 million for a new aqueduct system to interconnect north coast water supply systems (Larsen, 2000). Diminished streamflow has also severely affected biotic communities and ecosystem processes in stream channels in Puerto Rico (Covich et al., 1998).

Adaptation Options
Strategies for providing adequate water resources under changing climate conditions for island communities vary. Options could include: improved rainfall catchment systems; improved storage and distribution systems; development of under-utilized or alternative sources; better management of supply and infrastructure; increased water conservation programs; construction of groundwater recharge basins for runoff; more effective use of ENSO forecast information; and application of new/improved technology, such as desalinization. It would also be useful for island communities to prepare water needs assessments for the future and prepare, test, and improve emergency/contingency plans for periods of water shortage.

For key sectors, strategies include integration of climate considerations and projections of sea-level rise into community planning, water management decision making, and tourism development. Accurate assessments of current water budgets are critical for effective management of water resources, especially on small, densely populated islands with limited storage capacity (Larsen and Concepcion, 1998). Strategies for potential flood conditions are considered in the next section.

2. Public Health and Safety

Both coastal and inland island populations and infrastructure are at risk from climate-related extreme events. As reported earlier, model projections suggest it is possible that the Pacific and Caribbean islands will be affected by: changes in patterns of natural climate variability (such as El Niño); changes in the frequency, intensity, and tracks of tropical cyclones (hurricanes and typhoons); and changes in ocean currents; and that these islands are very likely to experience increased ocean temperatures and changes in sea level (including storm surges and sustained rise). For the Caribbean, models indicate increasing air and water temperatures, slightly wetter conditions in winter,

Potential Consequences of Climate Extremes on Infrastructure and Human Health

Disruption of lifeline systems, including water supply, energy supply, waste management, and sanitation due to fluctuations in temperature and precipitation, as well as storm events. Medical services can also be disrupted during storm events;

Effects on people and public health, due to decreased water quality and sanitation, impact on agriculture and consequently nutrition, and ecological changes that may increase the risk of disease transmission, and direct impacts from severe storms; and

Damage or destruction of infrastructure, including housing, lifelines, transportation, economic, and industrial infrastructure due to storms, related storm surge, and sea-level rise.

and slightly drier conditions in summer. Some models project persistent El Niño-like conditions across the Pacific caused by increased ocean temperatures. This also suggests that Atlantic hurricanes may decrease in the future because of the relationship of reduced hurricane activity in the Atlantic following the season of El Niño event onset in the Pacific. Nevertheless, there seems to be little net effect on total precipitation in the region because rain clouds still form and rain still falls, even in the absence of hurricanes.

For the Pacific, models indicate a gradual warming of air and water temperatures with an accompanying eastward shift of precipitation. This would be somewhat similar to what happens during El Niño events, and while not exactly the same, it would likely cause changes in local rainfall patterns similar to those experienced during the El Niño phase of the ENSO cycle. While this is still uncertain, one result suggests precipitation amounts associated with tropical storms will likely increase by more than 25% on average (Knutson et. al, 1998); other results suggest a 5-11% increase in surface winds and a 28% increase in near-storm precipitation (Knutson and Tuleya, et al., 2000).

Globally, sea level is projected to rise two to five times faster over the 21st century than over the 20th century as increased warming causes glacial melting and thermal expansion of ocean water. The US-affiliated islands would be affected by variations in sea level, such as storm surge associated with tropical storms, and fluctuations associated with El Niño events.

With these types of projected climate changes, many current stresses that islands face will be exacerbated. Storms can directly damage structures, interfere with the provision of services to communities,

cause deaths, and increase disease transmission. In American Samoa, for example, the majority of existing roads are located along the shoreline and are extremely vulnerable to damage by wind driven waves. Each section of new or rehabilitated roadway construction must include the additional costs of sea walls, resulting in extremely expensive highway construction. However, without properly designed shore protection, American Samoa could lose a major portion of the shoreline, as well as the entire length of adjacent roadway (OIA, 1999) to storm surge. Both the Pacific and Caribbean regions are familiar with severe hurricanes and tropical cyclones, which have caused billions of dollars in damage from the destruction of housing, agriculture, roads and bridges, and lost tourism revenue (Walker et al., 1991; Scatena and Larsen, 1991).

The unique topography of Puerto Rico and the Virgin Islands makes them susceptible to floods and landslides usually resulting from the extreme precipitation associated with these storms. On most islands, much of the population, transportation and social infrastructure, and economic activity is located near the coast, leading to dense areas of vulnerability. Diaz and Pulwarty (1997) suggest that the societal impacts of hurricane activity are potentially more worrisome in the future because of several socioeconomic changes that have occurred during this century. Population increases put more people in the path of potential hurricane devastation. Since transportation infrastructure has not kept pace with this population increase, it takes longer to evacuate people (e.g., Hurricane Floyd and East Coast evacuations), even though hurricane-forecasting ability has improved. Finally, the value of insured property at risk is increasing, so the ability of insurance companies to keep pace with potential damage is diminishing.

Estimated Damages for Hurricane Marilyn in the US Virgin Islands

Category of Damage	Estimated Costs
Sewage Treatment Facilities	$1,000,000
Roads and Bridges	1,000,000
Damage to Manufacturing	1,000,000
Agriculture	1,000,000
Water	3,000,000
Protective Measures	10,000,000
Debris Removal	18,000,000
Telephones	30,000,000
Electrical	70,000,000
Lost Employment	80,000,000
Public Buildings	210,000,000
Damage to Hotels	253,000,000
Lost Tourist Revenue	293,000,000
Private Housing	1,300,000,000
Total	**2,271,000,000**

Source: The Virgin Islands Natural Hazard Mitigation Plan, B. Potter et al., 1995.

Leatherman (2000) also suggests that impacts from extreme events on the human community are a function of complex interactions between meteorological phenomena and the environment, including infrastructure (e.g., transportation and other parts of the built environment), land-use patterns, and social, political, and economic systems. He also asserts that any future increases in destructive meteorological events, as are suggested by some of the climate change scenarios, will negatively impact the islands economically as well as socially.

Severe Storms. Hurricanes and typhoons are among the most socially devastating natural disasters, affecting more people on a yearly basis than earthquakes. Cumulative losses are also greater for hurricanes than earthquakes (Leatherman, 1998). Hurricanes and typhoons cause billions of dollars in losses due to destruction of transportation and other infrastructure, life, and property. Increases in population density, changes in age structure and population health, urban sprawl, insufficient transportation infrastructure, and human occupation of coastal and flood-prone areas have all increased the vulnerability of populations and the infrastructure they depend on to hurricanes and typhoons. Deforestation has increased the amount of runoff and erosion that result from precipitation events contributing to the negative impacts of severe storms. One of those negative impacts is the increased risk of some diseases (e.g., *leptospirosis* and other water-borne diseases) following floods caused by storms (even storms that are far less intense than hurricanes). For example, data collect-

ed in certain areas of Florida show that increased flow resulting from storms carries 1,000 times the normal concentration of fecal sanitary bacteria (Alvarez, 1998).

The Caribbean's worst hurricane season since 1933 came in 1995. Hurricanes Luis and Marilyn hit the Virgin Islands in September 1995 (see table for specific damages from Hurricane Marilyn). Damage was greatest in St. Thomas. Every telephone pole and 80% of the homes and businesses were destroyed. It took eight weeks to get desalinization plants running. Every cistern was contaminated by salt water and there was no electricity and no hand pumps to get water out of cisterns. Critical facilities that were affected included hospitals, power and desalinization facilities, sewer systems, fire houses, police stations, public shelters, ports, communication systems, and docks. Notably, hospital roofs were destroyed; sewage distribution systems were disrupted, leading to surface water and marine bay contamination; and power and telephone distribution systems required months of repairs and curtailed service. The US Virgin Islands Bureau of Economic Research estimated the economic loss at $3 billion. (Potter et al., 1995; National Hurricane Center, 2000).

The Pacific Islands have also suffered from severe storms. According to Pielke (2000), between the years 1957 and 1995, Hawaii suffered over $2.4 billion in hurricane damages. For the Pacific islands, 1992 and 1997 were particularly active years. In 1992 FEMA responded to "major disasters" in the

Hurricane Georges of 1998: Effects on Puerto Rico

On September 15, 1998, a tropical weather system off the coast of West Africa was upgraded to a tropical depression. Within 24 hours, the tropical depression intensified to be the tropical storm that would become the fourth hurricane of the 1998 season, Hurricane Georges. At its strongest, Hurricane Georges had sustained winds of 150 mph, becoming a category 4 hurricane, the second most intense level of hurricane on the five-level scale used to measure intensity.

On September 21, Hurricane Georges (reduced to a category 3) began to sweep across Puerto Rico. It was to be the first hurricane since the 1932 San Ciprian hurricane to cross the entire island. The eye of the hurricane, measuring 25-30 miles wide, passed within an estimated 15 miles of San Juan, Puerto Rico's capital, leaving a trail of devastation in its wake.

While rainfall across the island varied greatly, some areas received up to 26 inches of rain within a twenty-four hour period. The rainfall resulted in flooding, landslides, and catastrophic losses in infrastructure. Twelve people on the island of Puerto Rico were killed; three as a direct result of the storm, and nine others as an indirect result. Infrastructure impacts were enormous. The island lost 75% of water and sewage service, 96% of electrical power, and suffered damage to 50% of utility poles and cables. An estimated 33,100 homes were destroyed. Road damage was estimated at $22 million and damage to public schools was estimated at $20-25 million. Agricultural damages were also significant with 75% of the coffee crop, 95% of the plantain and banana crops, and 65% of all poultry destroyed (USGS, 1999; NOAA, 1999).

Overall costs to the US mainland and island territories reached $5.9 billion, with an estimated $2 billion or more in Puerto Rico alone. On October 15, 1998, the Red Cross announced that Hurricane Georges had in fact been "the most expensive disaster relief effort in the organization's 117-year history." They estimate that solely within the US mainland and island territories, about 187,300 families were affected (American Red Cross, 1998, 1999).

Marshall Islands and Micronesia after Typhoon Axel struck in March; then in Guam in the aftermath of Typhoon Omar ($500 million dollars in damages). September 1992 saw Hurricane Iniki hit the Hawaiian island of Kauai. It resulted in 7 deaths, approximately $1.8 billion in damages, and $259.7 million in FEMA disaster relief costs. In December 1996, Typhoon Fern struck Yap and in April of 1997, Typhoon Isa struck Micronesia. Later that year, in December, Supertyphoon Paka ($650 million in damages) pummeled Guam and the Northern Marianas, as well as the Marshall Islands causing widespread destruction (Hamnett et al., 2000; FEMA website, 2000).

Landslides. Each year in Puerto Rico, landslides cause extensive damage to property and occasionally result in loss of life (Larsen and Torres-Sanchez, 1998). Although landslides can be triggered by seismic activity and construction on hillslopes, the leading cause of landslides in Puerto Rico is intense and/or prolonged rainfall (Larsen and Simon, 1993). Population density in Puerto Rico is high, about 1,036 people per square mile (400 people per square kilometer), and is increasing. This increase is accompanied by the use of less desirable construction sites. As a result, human populations are becoming more vulnerable to landslide hazards (Larsen and Torres Sanchez, 1998).

Path of Hurricane Georges in Relation to Puerto Rico with Precipitation Totals

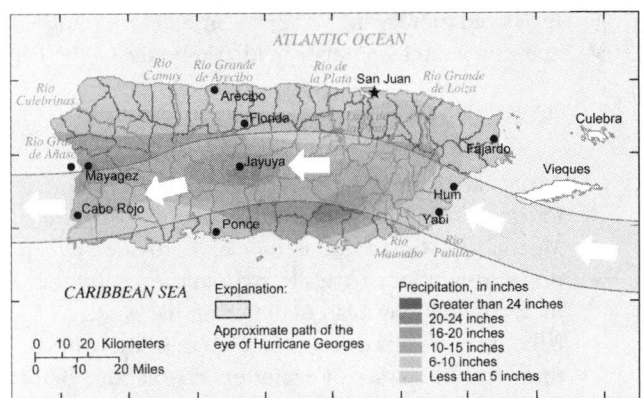

Figure 4. On September 21, 1998, Hurricane Georges swept across Puerto Rico. The eye of the hurricane was 25-30 miles wide and passed within 15 miles of the capital, San Juan, leaving a trail of devastation in its wake. The path of the hurricane and rainfall totals are shown here. Some areas received up to 26 inches of rain within 24 hours. Flooding, landslides, and catastrophic losses in infrastructure resulted. Hurricane Georges Map —USGS: http://water.usgs.gov/pubs/FS/FS-040-99/images/PR_fig01.gif. See Color Figure Appendix

People tend to live in the coastal zone (vulnerable to storm surge), on flood plains, or on hillsides prone to landslides. In the meantime, human modifications are increasing the frequency of landslides. A study conducted in Puerto Rico in 1998 found that, although mean annual rainfall is high, intense storms are frequent, and hillslopes are steep, forested hillslopes are relatively stable as long as they are not modified by humans. However, the greater the modification of a hillslope from its original forested state, the greater the frequency of landslides (Larsen and Torres-Sanchez, 1998).

Adaptation Options.
Strategies for improving public health and safety under changing climate conditions could include numerous options. For example: 1) upgrading and protecting infrastructure (including transportation infrastructure); 2) initiating comprehensive disaster management (and avoidance) programs (including mapping and risk analysis); 3) changing and enforcing land use policies to avoid hazards/hazardous areas; 4) adopting and enforcing increasingly stringent building codes; 5) improving sanitation and health care infrastructure; 6) improving emergency plans; and 7) increasing public and official information and outreach programs related to the range of historical disaster experiences and to future risks associated with climate change.

For potential flood conditions, adaptation measures include changes in land use policies that discourage construction in flood plains or areas at risk for landslides, and that allow for natural patterns of runoff. Improved monitoring, observation, alert, warning, and evacuation systems would also be desirable.

3. Island Ecosystems

Islands are extremely valuable as living laboratories for understanding species adaptation and evolution. Much as the Galapagos archipelago provided exceptional insight for Charles Darwin in the 1830s, the more ancient Hawaiian island chain has since become recognized as a premier site in the world for scientific studies of evolution in isolation. Before human influence, the Hawaiian Islands had no ants, reptiles, amphibians, or mammals (except for one bat species). The roughly 1,500 species of animals and plants that successfully colonized this climatically and topographically complex archipelago over thousands of miles of open ocean — on the winds, by floating, or attached to storm-driven birds (Carlquist, 1980) — gave rise to roughly 15,000 species, over 90% of them endemic (found nowhere else).

Since the time of Darwin, it has been recognized that organisms of oceanic islands are notoriously vulnerable to extinction. Having been isolated for millions of years from the continual challenge of some of the selective forces that shape continental organisms, oceanic island fauna and flora are uniquely vulnerable to the human introduction of previously absent predators, diseases, and competitors (Loope and Mueller-Dombois, 1989). Natural ecosystems also appear to be most vulnerable to many effects of climate changes (e.g., temperature and carbon dioxide increases, precipitation changes, sea-level rise, and changes in the frequency and intensity of extreme events) since climatic conditions strongly determine where particular plants and animals can live, grow, and reproduce. Some species and ecosystems are so strongly influenced by the climate to which they are adapted, that they are very vulnerable to even modest climate changes. Some island ecosystems are already constrained by climate as well as geography and are likely to face extreme stress from projected climate changes while some species may disappear entirely. For example, increased ocean temperatures and possible changes in ocean circulation patterns will affect coastal ecological systems, such as mangrove forests and coral reefs and important marine resources such as fisheries. Often there is very little that can be done to assist ecosystems in adapting to the projected speed and amount of change.

Pacific Islands. The Hawaiian Islands are larger, have more topographical and climatic diversity, and have higher biological uniqueness than most other Pacific Islands. Hawaii also has the highest proportion of extinct and endangered species of anywhere in the US (USGS, 1999a). Hawaii has lost over 70% of its original (pre-Polynesian) 140 land bird and 1200+ land snail species. Of 1,302 known taxa (89% endemic) in the Hawaiian vascular flora, 106 (8%) are extinct and an additional 373 (28%) are considered at risk of becoming extinct in the near future. The "at risk" taxa include over 200 that are federally listed as endangered, roughly one-third of the total listed plants nationwide (Loope, 1998). Loss of about two-thirds of the area of Hawaii's natural habitats has been a primary reason for the losses to date; however, at present, biological invasions pose the greatest threats to further dismemberment of the biota.

Although invasive non-indigenous species are problematic throughout the Pacific, Hawaii, being in the mainstream of commerce, has suffered significantly (Holt, 1998). Hawaii is inherently vulnerable to invasion due to great isolation from source areas of naturally colonizing species and high endemism (Mueller-Dombois and Fosberg, 1998). This is not to say that other Pacific Islands have not suffered devastation by humans and their invasive species introductions. For example, Steadman (1995) estimates that more than 2,000 bird species may have been eliminated by early human settlers in Polynesia. The example of the brown-tree snake (*Boiga irregularis*) on Guam (Fritts and Rodda, 1995) should serve as a cautionary note for all oceanic islands that non-indigenous species may explode in island habitats with unexpected consequences. According to Fritts (1999), the brown tree snake probably arrived on Guam via ship cargo from the New Guinea area and the first inland sighting was in the early 1950s. Some forested areas now have populations up to 13,000 snakes per square mile. The brown tree snake is responsible for essentially wiping out the native forest birds of Guam. Twelve species of endemic birds have disappeared from the island, and several others survive in such low numbers that they are close to extinction. The snake is responsible for economic impacts as well. For example, snakes crawling on electrical lines frequently cause power outages. More than 1,200 power outages have been attributed to snakes since 1978. The power interruptions have reportedly resulted in numerous problems ranging from food spoilage to computer failures and all include significant expenses for the population.

A comparable botanical example of the unintended consequences of non-indigenous species invasions and impact is seen in the invasion of the South American tree *Miconia calvescens* in Tahiti (Loope, 2000). *Miconia* now dominates 65% of Tahiti's forest canopy and its presence threatens some 40-50 endemic plant species to the point they are now on the verge of extinction (Meyer and Florence, 1997). *Miconia* is a serious issue in Hawaii as well. Probably because of its attractive purple and green foliage, it was introduced to Hawaii as an ornamental — a single tree in the Oahu Wahiawa Botanical Garden in early 1960. By 1964, it had reached Hawaii and by the late 1960s or early 1970s it had reached Maui. Based on a 1995 discovery of a fruiting tree over 10 meters tall, it is estimated that it reached Kauai by the early 1980s. Its presence in Hawaii poses a threat to endemic plant species in habitats receiving 1800-2000 mm (75-80 inches) or more of annual precipitation (Loope, 2000). The vulnerability of oceanic islands to invasions cannot be underestimated.

The USGS (2000) Hawaiian Ecosystems at Risk (HEAR) project asserts that "the silent invasion of Hawaii by insects, disease organisms, snakes, weeds, and other pests is the single greatest threat to Hawaii's economy and natural environment and to the health and lifestyle of Hawaii's people." According to HEAR, millions of dollars in crop losses, the extinction of native species, the destruction of native forests, and the spread of disease can already be attributed to alien pest (plant and animal) invasions, with many more harmful pests now threatening to invade Hawaii and cause further damage. For example, an invasion of ant species such as the "red imported fire ant" *(Solenopis invicta)* and the little fire ant *(Wasmannia auropunctata)* may pose a threat to Pacific islands' biodiversity, tourism, agriculture, and quality of life that rivals the threat of global climate change (R. Thaman, University of the South Pacific, personal communication, 1999).

In spite of profound human modification of the Hawaiian landscape and biota (Cuddihy and Stone, 1990), large tracts of near-pristine ecosystems remain at high-elevation. Even with the high incidence of extinction and endangerment, Hawaii has more non-endangered endemic species of vascular plants, birds, and insects than any other state except California (Loope, 1998). In Samoa, as well as on other high islands of the Pacific (e.g., the Society and Marquesas, Tonga, the Cook Islands, and the Hawaiian islands) relatively intact biodiversity is centered in high-elevation cloud forests, which may be exceptionally vulnerable to climate change (see box on Montane Cloud Forests).

Caribbean Tropical Forests. The islands in the Puerto Rican Bank have been largely denuded of native vegetation, have extremely dense human populations, and face a formidable array of environmental problems, including extinction of native plants and animals (Wiley and Vilella, 1998). The introduction of non-indigenous species, particularly vertebrate animals, is yet another difficulty facing the native plant and animal communities. Still, with increasingly aggressive conservation efforts, Puerto Rico and the US Virgin Islands may retain the best-preserved examples of certain natural ecosystems of the West Indies. Being showcases of natural ecosystems is part of the image of these tropical islands, and is deemed vital to their tourist economies (Wiley and Vilella, 1998).

Past policy considerations for tropical forest conservation have generally dismissed the threat of climate change as quite insignificant in comparison to land-use change and other human impacts. However, a better understanding of tropical ecology is now leading many scientists to the conclusion that tropical forests may be very sensitive to climate change. It is being increasingly recognized that factors other than warming, including changes in hydrology, rainfall patterns, and the frequency and intensity of storms and fires, may have far-reaching consequences (Markham, 1998). In comparison to Hawaiian forests, Caribbean forests are well-adapted to disturbance, but increased frequency of hurricanes, floods, droughts, and fires could lead to unprecedented stresses and drastic changes in forest structure and composition. For example, a computer model run for the Luquillo reserve of Puerto Rico showed that increasing intensity of hurricanes could reduce the density of trees in the forest and their total biomass, and favor the development of fast-growing, short-lived and weedy species, including invasive species (O'Brien, et al., 1992).

Mangrove Forests. Coastal wetlands and mangroves of the Caribbean and the Pacific are vital areas for sustaining populations of birds, juvenile fishes, and invertebrates. They also provide important storm and flood buffers between the sea and coastal communities, are coastal stabilizers, and nutrient sinks. Mangrove forests have the capacity to adapt to sea-level rise, although this is being limited by human activities that interfere with forest regeneration. One study suggests that mangrove communities will do better in macrotidal, sediment-rich environments where strong tidal currents redistribute sediment rather than in microtidal, sediment-starved environments, like most small islands (Parkinson et al., 1994). In the pine rocklands (e.g., Sugarloaf Key) in the lower Florida Keys is an area

Mangrove trees grow at the land-ocean interface, helping to protect coastal landsfrom erosion. - Source: ©P. Grabhorn

Coral reefs are among the most sensitive ecosystems to long-term climate changes. Source: ©P. Grabhorn

where pine trees grow directly on rock on very low-lying islands. These pines depend on lenses of freshwater that are being gradually eliminated by salt water as sea level rises. The result has been that mangroves have replaced the pines. This transformation will likely continue as sea level rises (Alvarez, 1998). On the other hand, Snedekar (1993) has argued that mangroves in the Caribbean are more likely to be affected by changes in precipitation than by higher temperatures and rising sea levels because they require large amounts of fresh water to reach full growth potential. A decrease in rainfall in the Caribbean could thus reduce mangrove productive potential and increase their exposure to full-strength seawater. (For more information see the Coastal chapter.)

Coral Reefs. It is projected that coral reefs are among the most sensitive ecosystems to long-term climate change (IPCC, 1995). Widespread coral bleaching has already been observed in the Pacific and Caribbean in association with ENSO events and would be expected to continue and accelerate with increased ocean temperatures. Bleaching occurs when the coral animal expels all or part of its symbiotic algae, when the pigments in the algae decline drastically, or when there is some combination of the two. Bleaching is a stress reaction that can be induced by many individual or combinations of conditions: high or low water temperature, high fluxes of ultraviolet radiation, prolonged aerial exposure, freshwater dilution, high sedimentation, or various pollutants (Glynn and de Weerdt, 1991). While low temperatures have generally been considered a limiting factor for the development of coral reefs in cooler climates, reefs are also susceptible to increased temperatures because they are already near their maximum threshold temperatures in summer. While various species and populations may respond differently, in general, coral are likely to bleach but survive if high temperature anomalies persist for less than a month. Such a short-term exposure usually allows corals to recover. However, sustained high temperatures can result in chronic stress that can cause physiological damage that may be irreversible (Wilkinson et al., 1999; Glynn, 1996). Even a sub-lethal stress may make corals highly susceptible to infection by a variety of opportunistic pathogens.

In the late 1980s and 1990s, after localized bleaching in 1982-1983, bleaching became a regular and pervasive problem in the Caribbean and began to appear in the Pacific and Indian Oceans as well. Bleaching occurred in American Samoa during the warm event of spring 1993. During the El Niño of

1997-1998, bleaching began in the eastern Pacific but then expanded to an unprecedented depth and region across to and including the Indian Ocean, and in the Pacific Ocean from Australia to Polynesia. Bleaching was widespread in the Republic of Palau during the 1997-1998 El Niño.

Corals in the Caribbean are currently impacted by enormous numbers of international tourists, high volume of ship traffic, and fishing. Pacific Ocean coral reefs are some of the world's healthiest overall; about 70% are rated in good-to-excellent condition. But 30% are rated fair-to-poor and many are dying, while human impacts are growing. The region's extremely diverse corals, mangroves, and sea-grasses are pressured by deforestation, agricul-

ture, construction, pollution, and fishing. Climate-related changes could affect coral through more than simply increased water temperatures: in some Pacific jurisdictions, changes are expected in aerial exposure due to sea-level alterations associated with ENSO. In the Caribbean, erosion/sedimentation may be caused either directly through floods or indirectly through increased erosion following fires associated with drought conditions. In addition, hurricanes and typhoons can damage coral reefs. For example, Hurricane Hugo in 1989 caused extensive damage to reefs outside of Puerto Rico by fragmenting, overturning, or destroying virtually all elkhorn coral colonies that were over one meter in size (USGS, 1999). Elkhorn coral are a principal reef builder. Unhealthy coral reefs are the most suscepti-

Montane Cloud Forests: Unique and Vulnerable Ecosystems

The tropical montane cloud forests of islands are rare, diverse, and vulnerable natural habitats, occurring in such Pacific islands as the Society and Marquesas islands, Cook Islands, Tonga, Samoa, and the Hawaiian Islands as well as elsewhere in the world (e.g., Madagascar, Sri Lanka, Greater and Lesser Antilles). The upper and lower limits of cloud forest vegetation are determined by the altitude of a persistent cloud zone. The upper limit of this zone is generally set by the elevation of the trade wind inversion and the lower limit is determined by temperature and humidity. Cloud forests are likely to be extremely sensitive to climate change. These unique forest environments would be affected by any potential changes in the altitudinal level of the trade wind inversion, as well as changes in temperature, precipitation zones, increased drought, or hurricanes. On high islands such as Maui and Hawaii, the two largest of the Hawaiian islands, the cloud forest veg-

The Hawaiian anianiau (*Hemignathus parvus*) an endangered species, is now restricted to high elevation montane rain forest on Kauai. Like other native Honeycreepers, it is highly susceptible to avian pox and malaria, and as a result of mosquito transmitted bird diseases the Anianiau has disappeared from its former low elevation habitats. Source: (http://water.usgs.gov/pubs/FS/FS-012-99/)

etation now changes abruptly to grassland or shrubland above the mean elevation of the inversion layer.

Loope and Giambelluca (1998) have identified Pacific island cloud forests as particularly vulnerable to human-induced climate change, since relatively small climate shifts are likely to trigger major local changes in rainfall, cloud cover, and humidity. Such climatic disruptions will undoubtedly favor the penetration of invasive non-indigenous species into previously intact ecosystems. Pounds et al. (1999) have already demonstrated a constellation of population crashes of native birds, reptiles, and amphibians in biological communities in the montane cloud forest of Monteverde, Costa Rica, apparently resulting from oceanic and atmospheric warming and drying in the extreme eastern tropical Pacific.

The cloud forests of East Maui, Hawaii (Haleakala National Park and neighboring reserves), for example, are home to numerous endemic species, with approximately 90% of the native flowering plants and invertebrates endemic to Hawaii. Nine endemic bird species in the Hawaiian honeycreeper family occupy these cloud forests. One reason these birds survive in the higher forest is because the mosquitoes that carry avian malaria are limited at higher elevations by cooler temperatures. Therefore, any warming that allows the mosquitoes to move farther up the mountain may threaten the birds, five species of which are already listed as endangered by the US Fish and Wildlife Service.(From Loope, personal communication.)

ble to destruction by hurricanes and typhoons. Effects of increasing atmospheric carbon dioxide concentrations could also possibly have adverse effects on reef building corals (see Coastal chapter for details). Without the reefs, low-lying islands would be susceptible to rapid erosion and destined to eventual disappearance.

Adaptation Options

There are a limited number of specific options that can be employed to assist island ecosystems to cope with a changing climate. Some of those options include the following: attempting to slow biological invasions; strengthening and enforcing policies that protect critical habitats; improving understanding of the local effects of climate variability and change; and increasing awareness of tourists and the public concerning the value of

species and biodiversity.

4. Sea-Level Variability

Sea-level rise and the associated erosion and inundation problems are currently extremely important issues for many of the US affiliated islands. There are two factors that affect the impacts of sea-level rise on islands: the rate and extent of global sea-level rise and the occurrence of episodic events, such as extreme lunar tides, ENSO related changes, and storm-related wave conditions. Rising sea levels over the past century have already resulted in salt water intrusion into freshwater lenses, inundation of coastal vegetation (such as taro, pulaka, and yams), and coastal erosion, since a typical beach erosion

Pacific Island Observed Sea-Level Trends
(based on Merrifield, personal communication)

Location/Name	Observed Rise in Relative Sea Level (averaged for the 20th century)	Years of Data Collection
Hawaii		
Hilo	13.6 +/- 2 in (34 +/- 5 cm)	1927- 1999
Honolulu	6 +/- 0.8 in (15 +/- 2 cm)	1905- 1999
Nawiliwili	6 +/- 1.6 in (15 +/- 4 cm)	1954- 2000
Kahului	8.4 +/- 2 in (21 +/- 5 cm)	1950- 1999
Mokuoloe	4 +/- 2 in (10 +/- 5 cm)	1957- 1999
Guam	0 +/- 2.4 in (0 +/- 6 cm)	1948- 1999
American Samoa		
Pago Pago	6.4 +/- 2.4 in (16 +/- 6 cm)	1948- 1999
Commonwealth of Northern Marianas		
Saipan	-0.4 +/- 8.8 in (-1 +/- 22 cm)	1978- 1999
Republic of the Marshall Islands		
Wake	7.2 +/- 2 in (18 +/- 5 cm)	1950- 1999
Kwajalein	3.6 +/- 1.6 in (9 +/- 4 cm)	1946- 1999
Republic of Palau		
Malakal	- 1.6 +/- 7.2 in (-4 +/- 18 cm)	1969- 1999
Federated States of Micronesia		
Kapingamarangi	-6.4 +/- 9.2 in (-16 +/- 23 cm)	1978- 1999
Pohnpei	6.4 +/- 7.2 in (16 +/- 18 cm)	1974- 1999
Yap	-5.6 +/- 7.2 in (-14 +/- 18 cm)	1969- 1999

A typical beach erosion rate can be 150 times the amount of sea-level rise. Source: ©P. Grabhorn

rate can be 150 times the amount of sea-level rise (Alvarez, 1998). The Gulf and Caribbean region experienced on average a relative sea-level rise of 3.9 inches (10 cm) during the 20th century. In the Pacific region, as around the globe, absolute sea level is also rising. There is, however, a great deal of inter-island variability in relative sea level (see table). This occurs because a number of Pacific Islands are rising due to geologic uplift. As a result of this uplift, relative sea-level changes for those islands can appear to be negative. The Pacific regional averages can, therefore, tend toward zero depending on which islands are included in the average; hence, the global sea-level trend is not evident in the long-term average trend in relative sea-level in the Pacific (Merrifield, personal communication).

Long-term global rates of sea-level rise are projected to be 2 to 5 times faster in the 21st century than during the 20th century. Future sea-level rise, both global and episodic (since increasing global sea level will also raise the level from which episodic events occur), will increasingly contribute to negative consequences for populations and ecosystems (Titus et al., 1991).

Risk of flooding, inundation, and coastal erosion on particular islands depends on both physical properties of the island (elevation, rock and soil-type, location) and on biological properties (presence or absence of biotic protection such as coral reefs and mangroves). The most at-risk island types are low-lying coral atolls because of both their elevation and composition. The vulnerability of specific sites depends on the value of what is at risk, including such aspects as natural resources, transportation and social infrastructure, businesses, and human populations. Among other factors, sensitivity increases because of dependence on coastal activities for their economies, dependence on coastal aquifers for freshwater supply, the presence of culturally valued structures along the coast, or because of previous

land-use changes. Ultimately, vulnerability to sea-level rise is a function of both the natural resilience of the shoreline and biological resources, as well as the value of the resource at risk, and the ability of human populations to respond and implement adaptation measures and strategies. Vulnerability also depends on the timeframe over which recovery is possible and the resources that are available.

Islands will be particularly vulnerable to sea-level changes and consequent impacts if they are low-lying and have limited resources for protecting their coastline. In the Pacific, potentially vulnerable island groups include the atolls of the Republic of the Marshall Islands and the Federated States of Micronesia. In the Caribbean, much of the metropolitan area of San Juan in Puerto Rico is already close to sea level. Islands that are high volcanic or limestone, mountainous, naturally rising, or lack resources and populations in coastal areas will be less vulnerable.

Adaptation Options

Coping with sea-level change and the resulting consequences (e.g., flooding, inundation of freshwater and agricultural systems, erosion, destruction of transportation and other built infrastructure) will require a variety of strategies. Some of the options that are likely to be needed could include: protecting coastal infrastructure (both social and transportation), water systems, agriculture, and communities; integrated coastal zone management; and crop shifts and the use of salt-resistant crops. Retreat from risk-prone, low-lying areas is also likely to be necessary in some cases, but will be complicated due to land ownership, and could have significant consequences for social and cultural identity.

For more details on projected sea-level rise and its impacts, see the Coastal chapter in this volume and the additional volume: The Potential Impacts of Climate Change on Coastal and Marine Resources: Report of the Coastal and Marine Resource Sector Team of the US National Assessment.

ADDITIONAL ISSUES

Climate variability and change present island governments, businesses, and communities with challenges and opportunities in a number of other key areas including:

Tourism

Tourism remains a significant contributor to the

Northern Mariana Islands, and is considered to offer economic growth potential for many of the jurisdictions addressed in this chapter. Unique ecosystems (both terrestrial and marine) and attractive coastal areas are among the natural assets that draw tourists to the islands of the Pacific and Caribbean. These natural assets are already under stress as a result of pollution and the growing demands of an increasingly coastal population. Many of the projected climate changes for the islands would exacerbate numerous of the stresses that the islands are already under. For example increasing rates of sea-level rise would contribute to: sea water inundating low-lying areas and threatening key infrastructure (including airports and roads); increasing the risks of damage from tropical storms (particularly damages associated with storm surge); threatening coastal water supplies through salt water intrusion; and reducing the extent and quality of sandy beaches through both inundation and increased erosion associated with storm surge. The South Atlantic Coast-Caribbean workshop report suggests that at the present rate of sea-level rise, there is already a loss of 9 feet (2.7 meters) of coastline due to erosion every decade, in some areas. The projected increased rate of sea-level rise would produce over 33 feet (10 meters) of erosion per decade since a typical beach erosion rate can be 150 times sea-level rise.

Other climate-related changes of concern to tourism include: changes in tropical storm patterns; changes in temperature and rainfall patterns with attendant consequences for terrestrial ecosystems and animals; and changes in ocean temperature, circulation, and productivity that could affect important marine resources like coral reefs and the fish they support. In addition, the climate itself is a magnet for tourists. Locations could become undesirable to tourists because of detrimental temperature and comfort-level changes. It is also evident that just the threat of some impacts has caused tourists to plan vacations or conventions elsewhere.

Fisheries

Climatic conditions influence where specific species can live, grow, and reproduce. The appropriate temperature ranges of some fisheries are extremely narrow and as a result they are sensitive to even small temperature variability within their ranges. Recent scientific studies are demonstrating an important link between patterns of natural climate variability, such as the ENSO cycle, and the migratory patterns of important pelagic species like tuna. Hamnett and Anderson (1999), for example, suggest that the eastward expansion of warm water

in the Pacific during an El Niño is associated with an eastward displacement of Skipjack tuna stocks. As global patterns of water temperature change, species shifts can be expected to follow these changes.

Climate variability and change present a number of challenges for coastal and marine fisheries that remain significant components of the culture and economies of many of the island jurisdictions addressed in this chapter. The movement of the commercially-important Skipjack tuna noted above resulted in the stocks being closer to the Marshall Islands and away from the waters of Micronesia. Industry representatives in FSM noted reduced catches in late 1997 and early 1998 (coinciding with the 1997-1998 El Niño), while government officials in the Marshall Islands noted an increase in fishing activity within their waters in late 1997, resulting in greater access fees from vessels within their Exclusive Economic Zone.

Changes in the ENSO cycle or other climate-related changes in ocean circulation and productivity could bring significant changes to the location of tuna stocks, thus providing opportunities for those jurisdictions that find themselves close to large stocks, and problems for other jurisdictions that find themselves far away from commercially-important species. For many of the Pacific island jurisdictions addressed in this National Assessment, the development of a viable tuna industry is considered an important component of their economic growth. StarKist Samoa and Samoa Packing, for example, are two of the largest tuna canneries operating within the US and they are the largest private sector employers in American Samoa. Fishing access/license fees from Distant Water Fishing Nations catching tuna in the FSM's Exclusive Economic Zone are a major source of revenue for the country (constituting a 17% share of the national GDP in 1994). Further development of that industry is currently the primary focus for economic development in FSM.

In addition to their importance to the tourism industry, many inshore species, including a number of reef fish, remain important contributors to the subsistence diet in small island communities. Any climate-related changes in the habitats that support these fisheries would have consequences for those communities.

Agriculture

Agriculture, for both commercial and subsistence purposes, remains an important part of the economies of many of the island jurisdictions addressed in this National Assessment. Much of the most productive agricultural land on these islands is located in low-lying coastal areas that are at-risk from climate-related changes in sea level. In addition to problems associated with inundation, salt-water intrusion associated with sea-level rise would also present challenges to agricultural production unless appropriate salt-tolerant species could be utilized. Climate-related changes in rainfall and tropical storm patterns also present problems for agriculture in island communities as evidenced by the impacts of the 1997-1998 El Niño event in the Pacific. For example, agriculture in all Pacific Island jurisdictions affiliated with the US except Guam suffered as a result of the droughts associated with the 1997-1998 El Niño event. In Yap, taro losses were estimated at 50-65% and over half of the banana trees evaluated on Pohnpei had died or were seriously stressed. Agriculture in Guam, on the other hand, suffered substantially from the El Niño-induced Typhon Paka that occurred in December 1997.

Data Collection and Availability

One other issue of concern to the islands relates to the complexities of data collection, availability, and reliability, both for climate data and impact studies. These data are necessary to enhance the ability of scientists and decision makers throughout the island regions to understand and respond to the challenges and opportunities presented by changes in climate and other critical environmental conditions.

Climate prediction data availability is at times limited for islands because most climate models in use today are unable to effectively capture island-scale or sub-island scale processes. Also, geographic isolation and dispersion makes data collection and research in island settings, particularly the Pacific, costly and difficult. The harsh environments affecting most islands pose related difficulties in maintenance of monitoring stations. At the same time resources for small-scale, locally-based research focused on individual islands or island ecosystems are difficult to secure, especially since many of these ecosystems are unique and, therefore, research results tend to have more local than global significance. Another difficulty is related specifically to impacts research where many impacts investigations are undertaken by non-local scientists who too often fail to capture/integrate valuable information

associated with traditional knowledge of specific island cultures. In the case of many of the islands, the communication and information management infrastructure is limited and already stressed. Establishing and maintaining an interactive dialogue between scientists and beneficiaries of their work outside the scientific community will require both technological upgrading and institution of a new paradigm of collaboration and integration of information related to climate changes and their impacts. These challenges point to a significant need for local capacity building (e.g., technical training in data gathering, monitoring technology, information distribution, impacts research, and communication technologies).

ADAPTATION STRATEGIES

Overall, strategies helpful for islands to cope with climate variability and change involve a wide range of options. While identifying many of the possible options is an important part of this Assessment, it is only a first step. It was not possible at this time to evaluate any of these potential options for practicality, effectiveness, or cost. Nevertheless, some potential adaptation options for the islands include: increased use of climate forecasting capabilities; consideration of changes in land use policies and building codes; use of new and emerging technologies to both mitigate the causes and reduce the impacts of climate change; increased monitoring, scientific research, and data collection abilities; and improved management policies. In addition, broad public awareness campaigns and communication with user groups are important mechanisms for improving public health and safety, and protecting natural resources.

Land-use policies and building codes can mitigate flood conditions (from sea-level rise as well as from extreme precipitation events) and can encourage construction in areas where life and property will not come into danger from floods, landslides, and severe storms including storm surge. Implementing such land-use policies could be an important coping option. Planting of trees and reforestation encourages natural infiltration, which increases underground water resources and prevents runoff that can lead to flood conditions. Building and transportation infrastructure can be renovated or replaced in the natural renewal cycle with higher standards. It is important to protect the species and ecosystems that currently exist and find strategies to limit sprawl and population growth that are impacting presently healthy ecosystems.

Possible Climate Change Adaptation Options

as identified by the Federated States of Micronesia

Water Resources	• Conduct a comprehensive inventory of existing water resources • Assess the status of storage and distribution systems and secure resources for necessary improvements. • Encourage improvements to residential and commercial catchment systems and identify/support new technology. • Identify opportunities to adjust water conservation and management policies to incorporate information about climate variability and change. • Document the experience gained during the 1997-1998 El Niño and build on the concept of drought management task force(s) to assist governments, communities, and businesses in responding to climate-related events. • Identify opportunities to improve watershed management.
Coastal Resources	• Identify buildings, infrastructure, and ecosystems at risk and explore opportunities to protect critical facilities. • Develop and implement integrated coastal management objectives that enhance resilience of coastal systems to climate change and sea-level rise. • Consider the need for beach nourishment and shoreline protection programs in high-risk areas. • Integrate considerations of climate change and sea-level rise in planning for future construction and infrastructure.
Agriculture	• Document the experience gained during the 1997-1998 El Niño and build on the concept of drought management task forces to assist governments, communities and businesses in responding to climate-related events. • Develop policies that protect both subsistence and commercial crops during extreme events. • Explore opportunities to diversify crops and select drought and/or salt-tolerant species where appropriate. • Document low-lying agricultural areas at-risk from the effects of sea-level rise and consider protection measures where appropriate and necessary.
Fisheries	• Enhance data collection and analyses required to improve understanding of the impacts of El Niño and La Niña events on tuna and other critical fisheries. • Identify and protect critical habitats for key inshore and near-shore species, particularly those important for subsistence fisheries. • Support monitoring and monitoring programs designed to improve understanding of the regional and local consequences of climate variability and change for tuna and other important fisheries.

From the National Communication to the United Nations Framework Convention on Climate Change, October 1999.

Improved technologies and water conservation are critical for protection of water resources. These include more and better rainfall catchment systems, improved storage and distribution systems, artificial recharge of aquifers, and desalinization. In addition, it is important for communities to take advantage of new capabilities in climate forecasting which, among other advantages, can allow for preparation for water scarcity conditions.

Improved management is critical for both ecosystem and human system protection. This includes more efficient management of existing water supplies, active management of natural areas, attention to policies that attempt to slow biological invasions, increased disaster response planning, and appropriate land uses.

At the regional workshops, participants advocated for effective, unified planning and response systems, and for incorporating climate change in growth management regulations. In particular, they stressed the need to design solutions that will be accepted by the local communities, and the need for coordinated institutional structures to manage the coastal resources. An example of a regional response plan for the Federated States of Micronesia is shown in the table opposite.

CRUCIAL UNKNOWNS AND RESEARCH NEEDS

To more effectively evaluate the potential consequences of climate variability and change impacts on Islands, a number of research needs have been identified by the Pacific and Caribbean workshop participants. The following is a brief listing of some of the identified research needs:

• Regional-scale information on changes in water availability, frequency, and intensity of extreme events such as hurricanes or typhoons, and interannual variability of climatic factors such as ENSO; specifically, downscaling by nesting regional climate models within global models.

• How to build local capacity to use such regionally-focused nested models and other data collection techniques and to interpret the information into useful products for island decision makers and island populations;

• How identified key parameters affect communities, the built infrastructure including transportation aspects, businesses, and economic sectors, as well as critical habitats and natural resources;

• The value of forecasting and understanding climate variability and change and how that understanding supports stakeholders in decision making processes;

• Environmental and socioeconomic changes on islands and how the two are related to one another; the interactions of multiple stresses; and identification of the specific parts of society that are most vulnerable to climate change and variability;

• An assessment of the status quo related to water resource management, health care systems, and emergency systems;

• Implications of potential climate-change-response policies for islands and island economies;

• How to improve risk analysis, monitoring systems, and evacuation systems on islands; and

• The impacts of changes in sea level, both short-term variations and long-term trends, on freshwater supply and other potential threats to coastal communities and ecosystems;

• How to incorporate appropriate traditional knowledge and response strategies into response options.

343

LITERATURE CITED

Alvarez, R. (Ed.), Proceedings of the Climate Change and Extreme Events Workshop for the US South Atlantic Coastal and Caribbean Region, a regional workshop, hosted by the International Hurricane Center, Florida International University, Miami, November 1998.

American Red Cross website: (http://205.214.45.11/news/inthenews/98/10-16-98.html) and (http://205.214.45.11/news/inthenews/98/10-16c-98.html) Accessed June 1, 1999.

Bengtsson, L., M. Botzet, and M. Esch, Numerical simulation of intense tropical storms, in *Hurricanes: Climate and Socioeconomic Impacts,* edited by H. F. Diaz and R. S. Pulwarty, Springer-Verlag, New York, 292 pp., 1997

Carlquist, S., Hawaii: *A natural history,* published by Pacific Tropical Botanical Garden, Lawai, Kauai, Hawaii, 1980.

CNMI (Commonwealth of the Northern Mariana Islands) official website: (http://www.saipan.com/cnmi-info/info_pg4.htm#sec100), Accessed May 19, 2000

Collins, M., The El Niño Southern oscillation in the second Hadley Centre coupled model and its response to greenhouse warming, *Journal of Climate,* in press, 2000.

Covich, A. P., T. A. Crowl, S. L. Johnson, and F. N. Scatena, Drought effects on pool morphology and neotropical stream benthos in *Proceedings, Third International Symposium on Tropical Hydrology and Water Resources,* San Juan, Puerto Rico, July 13-17, 1998, edited by R. I. Segarra-Garcia, American Water Resources Association, Middleburg, VA. 1998.

Cuddihy, L. W., and C. P. Stone, Alteration of native Hawaiian vegetation: Effects of humans, their activities, and introductions, University of Hawaii National Park Studies Unit, Honolulu, 138 pp., 1990.

Diaz, H. F., and R. S. Pulwarty, Decadal climate variability, Atlantic hurricanes, and societal impacts: An overview, in *Hurricanes: Climate and Socioeconomic Impacts,* edited by H. F. Diaz and R. S. Pulwarty, Springer-Verlag, New York, 292 pp., 1997.

Emanuel, K. A., Climate variations and hurricane activity: Some theoretical issues, in *Hurricanes: Climate and Socioeconomic Impacts,* edited by H. F. Diaz and R. S. Pulwarty, Springer-Verlag, New York, 292 pp., 1997.

Emanuel, K. A., Thermodynamic control of hurricane intensity, *Nature, 401,* 665-669, 1999.

FSM (Federated States of Micronesia), National communication to the United Nations Framework Convention on Climate Change, Palikir, Pohnpei, FM. Country Coordinator for national climate change program — John Mooteb, PICCAP Coordinator - Department of Economic Affairs, PO Box PS-Palikir, Pohnpei, Federated States of Micronesia, (email: jemooteb@hotmail.com), 1999.

FEMA (Federal Emergency Management Administration) Web site: http://www.fema.gov/library/cat4o5. Cited Aug 1, 2000.

Fritts, T. H., The brown tree snake of Guam: Fact Sheet, USGS, 1999, (http://159.189.24.10/btreesnk.htm)

Fritts, T. H., and G. H. Rodda, Invasions of the brown tree snake, in *Our living resources: A Report on the distribution, abundance, and health of US plants, animals, and ecosystems,* edited by E. T. LaRoe, G. S. Farris, C. E. Puckett, P. D. Doran, and M. J. Mac, pp. 454-456, US Department of the Interior, National Biological Service, Washington, DC, 1995

Glynn, P., Coral reef bleaching: Facts, hypotheses, and implications, *Global Change Biology 2,* 495-509, 1996.

Glynn, P. W., and W. H. de Weerdt, Elimination of two reef-building hydrocorals following the 1982-1983 El Niño warming event, *Science 253*(5015), 69-71, 1991.

Gray, W. M., Atlantic seasonal hurricane frequency, Part 1: El Niño and 30 mb Quasi-biennial oscillation influences, *Monthly Weather Review, 112,* 1649-1668, 1984.

Gray, W. M., J. D. Sheaffer, and C. W. Landsea, Climate trends associated with multidecadal variability of Atlantic hurricane activity, in *Hurricanes: Climate and Socioeconomic Impacts,* edited by H. F. Diaz and R. S. Pulwarty, Springer-Verlag, New York, 292 pp., 1997.

Hamnett, M. P., and C. L. Anderson, Impact of ENSO events on tuna fisheries in the US-affiliated Pacific Islands, Report of research supported by the Office of Global Programs, National Oceanic and Atmospheric Administration, Washington, D.C., October 1999.

Hamnett, M. P., C. L. Anderson, C. Guard, and T. A. Schroeder, Cumulative ENSO impacts: Tropical cyclones followed by drought, in The Pacific ENSO Applications Center: Lessons learned for regional climate forecasting (revised report), Pacific ENSO Applications Center, Honolulu, HI, 2000.

Harrison, C. S., *Seabirds of Hawaii: Natural History and Conservation,* Cornell University Press, Ithaca, NY, 249 pp., 1990.

Henderson-Sellers, A., et al., Tropical cyclones and global climate change: A post- IPCC assessment, *Bulletin of the American Meteorological Society, 79,* 19-38, 1998.

Holt, R. A., An alliance of biodiversity, health, agriculture, and business interests for improved alien species management in Hawaii, in *Invasive Species and Biodiversity Management,* edited by O. T. Sandlund, P. J. Schei, and A. Viken, pp. 65-78, Kluwer Academic Publishers, The Netherlands, 1998.

(IITF) International Institute of Tropical Forestry website (http://www.fs.fed.us/global/iitf/cnf/Georges.html) Accessed June 1, 1999.

IPCC, Intergovernmental Panel on Climate Change *Climate Change 1995: Impacts, Adaptations and Mitigation of Climate Change: Scientific-Technical Analyses,* edited by R. T. Watson, M.C. Zinyowera, and R. H. Moss, pp 289-324, Cambridge University Press, NY, 1995,

Knutson, T. R., and S. Manabe, Model assessment of decadal variability and trends in the tropical Pacific Ocean, *Journal of Climate, 11,* 2273-2296, 1998.

Knutson, T. R., and R. E. Tuleya, Increased hurricane intensities with CO_2-induced warming as simulated using the GFDL hurricane prediction system, *Climate Dynamics,* in press, 2000.

Knutson, T. R., R. E. Tuleya, and Y. Kurihara, Simulated increase of hurricane intensities in a CO_2-warmed climate, *Science, 279,* 1018-1020, 1998.

Landsea, C. W., G. D. Bell, W. M. Gray, and S. B. Goldberg, The extremely active 1995 Atlantic Hurricane season: Environmental conditions and verification of seasonal forecasts, *Monthly Weather Review, 126,* 1174-1193, 1995.

Landsea, C. W., N. Nicholls, W. M. Gray, and L. A. Avila, Downward trends in the frequency of intense Atlantic hurricanes during the past five decades, *Geophysical Research Letters, 23,* 1697-1700, 1996.

Landsea, C. W., R. A. Pielke, Jr., A. M. Mestas-Nunez, and J. A. Knaff, Atlantic basin hurricanes: Indices of climatic changes, *Climatic Change, 42,* 89-129, 1999.

Larsen, M. C., Drought, rainfall, streamflow, and water resources during the 1990s in Puerto Rico, *Physical Geography,* in press, 2000.

Larsen, M. C., and I. M. Concepcion, Water budgets of forested and agriculturally developed watersheds in Puerto Rico, in Proceedings, *Third International Symposium on Tropical Hydrology and Water Resources,* San Juan, Puerto Rico, July 13-17, 1998, edited by R.I. Segarra-Garcia, American Water Resources Association, Middleburg, VA, 199-204, 1998.

Larsen, M. C., and A. Simon, Rainfall-threshold conditions for landslides in a humid-tropical system, Puerto Rico, *Geografiska Annaler 75A*(1-2), 13-23, 1993.

Larsen, M.C., and A. J. Torres Sánchez, The frequency and distribution of recent landslides in three montane tropical regions of Puerto Rico, *Geomorphology, 24,* 309-331, 1998.

Leatherman, S., Opening plenary session speaker #3 pg. 16 in Proceedings of the Climate Change and Extreme Events Workshop for the US South Atlantic Coastal and Caribbean Region, A regional workshop, hosted by the International Hurricane Center, Florida International University, Miami, Florida, November 1998.

Leatherman, S. International Hurricane Center, Florida International University, home page – (http://www.ihc.fiu.edu/ihc/research/researchagenda.html), 2000.

Lobban, C. S., and M. Schefter, *Tropical Pacific Island Environments,* University of Guam Press, Mangilao, Guam, US, 399 pp., 1997.

Loope, L.L., (http://www.hear.org/MiconiaInHawaii/MiconiaSummaryByLLL.htm), 2000.

Loope, L. L., Hawaii and Pacific islands in *Status and Trends of the Nation's Biological Resources,* edited by M. J. Mac, P.A. Opler, C. E. Puckett Haecker, and P. D. Doran, pp. 747-774, US Geological Survey, Reston, Virginia, 1998.

Loope, L. L., and T. W. Giambelluca, Vulnerability of island tropical montane cloud forests to climate change, with special reference to East Maui, Hawaii, *Climatic Change, 39,* 503-517, 1998.

Loope, L. L., and D. Mueller-Dombois, Characteristics of invaded islands, in *Ecology of Biological Invasions: A Global Synthesis,* edited by H. A. Mooney, et al., pp. 257-280, John Wiley & Sons, Chichester, United Kingdom, 1989.

Low, L.-L., A. M. Shimada, S. L. Swartz, and M. P. Sissenwine, Marine resources, in *Status and Trends of the Nation's Biological Resources,* edited by M. J. Mac, P. A. Opler, C. E. Puckett Haecker, and P. D. Doran, pp. 775-866, US Geological Survey, Reston, Virginia, 1998.

Markham, A. (Ed.), Potential impacts of climate change on tropical forest ecosystems, *Climatic Change, 39*, 1-603, 1998.

Meehl, G. A., Vulnerability of fresh water resources to climate change in the tropical Pacific region, *Journal of Water, Air, and Soil Pollution, 92,* 203-213, 1996.

Meehl, G. A., and W. M. Washington, El Niño-like climate change in a model with increased atmospheric CO2 concentrations, *Nature, 382,* 56-60, 1996.

Meehl, G. A., F. Zwiers, J. Evans, T. Knutson, L. Mearns, and P. Whetton, Trends in extreme weather and climate events: Issues related to modeling extremes in projections of future climate change, *Bulletin of the American Meteorological Society, 81(3),* 427-436, 2000.

Meyer, J.-Y., and J. Florence, Tahiti's native *flora endangered by the invasion of Miconia calvescens DC. (Melastomataceae), Journal of Biogeography, 23,* 775-781, 1997.

Mueller-Dombois, D., and F. R. Fosberg, *Vegetation of the Tropical Pacific Islands,* Springer-Verlag, New York, 733 pp., 1998.

National Hurricane Center, (http://www.nhc.noaa.gov/1995marilyn.html), accessed September, 2000.

(NOAA) National Oceanic and Atmospheric Administration (http://www.ncdc.noaa.gov/ol/reports/georges/georges.html) Accessed May 26, 1999.

O'Brien, S. T., B. P. Hayden, and H. H. Shugart, Global climate change, hurricanes, and a tropical forest, *Climatic Change, 22,* 175-190, 1992.

OIA (Office of Insular Affairs), US Department of Interior, A Report on the State of the Islands, URL (http://www.doi.gov/oia/report99.html), 1999.

Parkinson, R. W., R. D. de Laune, and J. R. White, Holocene sea-level rise and the fate of mangrove forests within the Wider Caribbean region, *Journal of Coastal Research, 10,* 1077-1086, 1994.

Pielke, R. A., Jr., (http://www.esig.ucar.edu/HP_roger/sourcebook/hurricane.html), 2000.

Pielke, R. A., Jr., and C. W. Landsea, La Nina, El Niño, and Atlantic hurricane damages in the United States, *Bulletin of the American Meteorological Society, 80*(10), 2027-2033, 1999.

Potter, B., E. L. Towle, and D. Brower, The Virgin Islands Natural Hazard Mitigation Plan, by Island Resources Foundation for the Virgin Islands Territorial Emergency Management Agency (VITEMA). St. Thomas, December, 1995.

Pounds, J. A., M. P. L. Fogden, and J. H. Campbell, Biological response to climate change on a tropical mountain, *Nature, 398,* 611-615, 1999.

Rapaport, M. (Ed.), *The Pacific Islands: Environment and Society,* The Bess Press, Honolulu, Hawaii, 442 pp., 1999.

Reading, A. J., Caribbean tropical storm activity over the past four centuries, *International Journal of Climatology, 10,* 365-376, 1990.

Scatena, F. N., and M. C. Larsen, Physical aspects of Hurricane Hugo in Puerto Rico, *Biotropica, 23,* 317-323, 1991.

Snedaker, S. C., Impact on mangroves, in *Climatic Change in the Inter-Americas Sea,* edited by G. A. Maul, United Nations Environment Programme and Intergovernmental Oceanographic Commission Edward Arnold, London, United Kingdom, pp. 282-305, 1993.

Steadman, D. A., Prehistoric extinctions of Pacific islands' birds: Biodiversity meets zooarcheology, *Science, 267,* 1123-1131, 1995.

Timmermann, A., J. Oberhuber, A. Bacher, M. Esch, M. Latif, and E. Roeckner, Increased El Niño frequency in a climate model forced by future greenhouse warming, *Nature, 398,* 694-696, 1999.

Titus, J. G., R. A. Park, S. P. Leatherman, J. R. Weggel, M. S. Greene, P. W. Mausel, S. Brown, C. Gaunt, M. Trehan, and G. Yohe, Greenhouse effect and sea-level rise: The cost of holding back the sea, *Coastal Management,* 19, 171-204, 1991. On web at (http://www.epa.gov/global-warming/publications/impacts/sealevel/cost_of_holding.html)

US Department of State, Coral Bleaching, Coral Mortality, and Global Climate Change: cited on 5/16/00 (http://www.state.gov/www/global/global_issues/coral_reefs/990305_coralreef_rpt.html)

US Geological Survey (USGS), 2000, Hawaiian Ecosystems at Risk (HEAR) project: (http://www.hear.org/)

US Geological Survey (USGS), 1999a, USGS Hawaii Fact Sheet: (http://water.usgs.gov/pubs/FS/FS-012-99/)

US Geological Survey (USGS), 1996, USGS Hawaii State Fact Sheet: (http://water.usgs.gov/pubs/FS/FS-011-96/)

US Geological Survey (USGS), 1999b, USGS Fact Sheet 040-99: Puerto Rico, (http://water.usgs.gov/pubs/FS/FS-040-99/).

Walker, L. R., D. J. Lodge, N. V. L. Brokaw, and R. B. Waide, An introduction to hurricanes in the Caribbean, *Biotropica, 23*, 313-316, 1991.

Wiley, J. W., and F. J. Vilella, Caribbean islands, in *Status and Trends of the Nation's Biological Resources,* edited by M. J. Mac, P. A. Opler, C. E. Puckett Haecker, and P. D. Doran, pp. 315-349, US Geological Survey, Reston, Virginia, 1998.

Wilkinson, C., O. Linden, H. Cesar, G. Hodgson, J. Rubens, and A. E. Stong, Ecological and socioeconomic impacts of 1998 coral mortality in the Indian Ocean: An ENSO impact and a warning of future change?, *Ambio, 28*, 188-196, 1999.

Zack, Allen, and M. C. Larsen, Puerto Rico and the US Virgin Islands: Research & exploration, *National Geographic Society, Water Issue, 9*, 126-134, 1993.

BACKGROUND REFERENCES

Ackerman, J. D., L. R. Walker, F. N. Scatena, and J. Wunderle, Ecological effects of hurricanes, *Bulletin of the Ecological Society of America, 72*, 178-180, 1991.

Basnet, K., Controls of environmental factors on patterns of montane rain forest in Puerto Rico, *Tropical Ecology, 34*, 51-63, 1993.

Basnet, K., G. E. Likens, F. N. Scatena, and A. E. Lugo, Hurricane Hugo: Damage to a tropical rain forest in Puerto Rico, *Journal of Tropical Ecology, 8*, 47-55, 1992.

Brokaw, N. V. L., and J. S. Grear, Forest structure before and after Hurricane Hugo at three elevations in the Luquillo Mountains, Puerto Rico, *Biotropica, 23*, 386-392, 1991.

Brokaw, N. V. L., and L. R. Walker, Summary of the effects of Caribbean hurricanes on vegetation, *Biotropica, 23*, 442-447, 1991.

Felzer, B., and P. Heard, Precipitation differences amongst GCMs used for the US National Assessment, *Journal of the American Water Resources Association, 35(6)*, 1327-1340, 1999.

Frangi, J. L., and A. E. Lugo, Hurricane damage to a floodplain forest in the Luquillo Mountains of Puerto Rico, *Biotropica, 23*, 324-335, 1991.

Guariguata, M. R., Landslide disturbance and forest regeneration in the upper Luquillo Mountains of Puerto Rico, *Journal of Ecology, 78*, 814-832, 1990.

Herbert, P. J., J. D. Jarrell, and M. Mayfield, The deadliest, costliest, and most intense hurricanes of this century (and other frequently requested hurricane facts). NOAA Technical Memorandum NWS TPC-1, National Hurricane Center, Miami, Florida, 1996. (http://www.nhc.noaa.gov/pastcost.html)

Karl, T. R., R. W. Knight, and N. Plummer, Trends in high-frequency climate variability in the twentieth century, *Nature, 377*, 217-220, 1995.

Larsen, M. C., and A. Simon, Landslides triggered by Hurricane Hugo in eastern Puerto Rico, September 1989, *Caribbean Journal of Science, 28*(3-4), 113-125, 1992.

Lobban, C. S., and M. Schefter, *Tropical Pacific Island Environments,* University of Guam Press, Mangilao, Guam, US, 399 pp., 1997.

Lodge, D. J., and W. H. McDowell, Summary of ecosystem-level effects of Caribbean hurricanes, *Biotropica, 23*, 373-378, 1991.

Loope, L. L., F.G. Howarth, F. Kraus, and T.K. Pratt, Newly emergent and future threats of alien species to Pacific landbirds and ecosystems. in *Studies in Avian Biology, No. 22. Ecology, conservation, and management of Hawaiian birds: a vanishing avifauna,* edited by J. M. Scott et al., published by Cooper Ornithological Society, Allen Press, Lawrence, Kansas, in press, 2000.

Lugo, A. E., Reconstructing hurricane passages over forests: A tool for understanding multiple-scale responses to disturbance, *Trends in Ecology and Evolution, 10,* 98-99, 1995a.

Lugo, A. E., Tropical forests: Their future and our future, in *Tropical forests: Management and ecology. Ecological Studies,* edited by A. E. Lugo and C. Lowe, Springer-Verlag, New York, 1995b.

Lugo, A. E., Mangrove forests: A tough system to invade but an easy one to *rehabilitate, Marine Pollution Bulletin, 37*(8-12), 427-430, 1998.

Lugo, A. E., Comparison of island and continental ecosystems with a focus on disturbance, *Acta Científica, 9*(2-3), 129-130, 1999a.

Lugo, A. E., Ecological aspects of catastrophes in Caribbean islands, *Acta Científica,* 2(1): 24-31, 1999b.

Pringle, C. M., and F. N. Scatena, Aquatic ecosystem deterioration in Latin America and the Caribbean, in *Managed Ecosystems: The Mesoamerican Experience,* edited by L. U. Hatch and M. E. Swisher, Oxford University Press, New York, 104-113, 1999a.

Pringle, C. M., and F. N. Scatena, Freshwater resource development: Case studies from Puerto Rico and Costa Rica, *Managed Ecosystems: The Mesoamerican Experience,* edited by L. U. Hatch and M. E. Swisher, Oxford University Press, New York, Oxford, NY, chapter 13, 1999b.

Rapaport, M. (Ed.), *The Pacific Islands: Environment and Society,* The Bess Press, Honolulu, Hawaii, 442 pp., 1999.

Scatena, F. N., The management of Luquillo cloud forest ecosystems: Irreversible decisions in a non-substitutable ecosystem, in *Tropical Montane Cloud Forests. Proceedings of an International Symposium,* edited by L. S. Hamilton, F. N. Scatena, and J. Juvik, pp. 191-198, East-West Center Publications, Honolulu, Hawaii, 1993.

Silver, W. L., The potential effects of elevated CO_2 and climate change on tropical forest biogeochemical cycling, *Climatic Change, 39,* 337-361, 1998.

Steudler, P. A., J. M. Melillo, R. D. Bowden, M. S. Castro, and A. E. Lugo, The effects of natural and human disturbances on soil nitrogen dynamics and trace gas fluxes in a Puerto Rican wet forest, 356-363, 1991.

Vitousek, P. M., Biological invasions and ecosystem properties: Can species make a difference?, in *The ecology of biological invasions of North America and Hawaii,* edited by H. A. Mooney and J. Drake, pp. 163-176, Springer-Verlag, New York, 1986.

Walker, L. R., J. Voltzow, J. D. Ackerman, D. S. Fernández, and N. Fetcher, Immediate impact of Hurricane Hugo on a Puerto Rican rain forest, *Ecology,* 73, 691-694, 1992.

Wiley, J. W., and F. J. Vilella, Caribbean islands, in *Status and Trends of the Nation's Biological Resources,* edited by M. J. Mac, P. A. Opler, C. E. Puckett Haecker, and P. D. Doran, pp. 315-349, US Geological Survey, Reston, Virginia, 1998.

Willig, M., and L. R. Walker, Disturbance in terrestrial ecosystems: Salient themes, synthesis, and future directions, in edited by L. R. Walker, pp. 747-767, Elsevier Science, Amsterdam, 1999.

Zack, A., and M. C. Larsen, Island hydrology: Puerto Rico and the US Virgin Islands, *National Geographic Research & Exploration. Water Issue,* 126-134, 1994

ACKNOWLEDGMENTS

Much of the material for this chapter is based on contributions from participants on and those work ing with the:

**Pacific Islands Workshop and Assessment
    Teams**
Eileen Shea*, East-West Center
Michael Hamnett*, University of Hawaii
Cheryl Anderson, University of Hawaii
Anthony Barnston, NOAA, National Centers for
    Environmental Prediction,
Climate Prediction Center
Joseph Blanco, Office of the Governor (State of
    Hawaii)
Kelvin Char, Office of the Governor (State of
    Hawaii) and NOAA National Marine Fisheries
    Service, Pacific Islands Area Office
Delores Clark, NOAA National Weather Service,
    Pacific Region Office
Scott Clawson, Hawaii Hurricane Relief Fund
Tony Costa, Pacific Ocean Producers
Margaret Cummisky, Office of the Honorable Daniel
    K. Inouye, United States Senate
Tom Giambelluca, University of Hawaii
Charles "Chip" Guard, University of Guam
Richard Hagemeyer, NOAA National Weather
    Service, Pacific Region Office
Alan Hilton, NOAA Pacific ENSO Applications Center
Honorable Daniel K. Inouye, United States Senate
David Kennard, FEMA Region IX, Pacific Area Office
Roger Lukas, University of Hawaii
Fred Mackenzie, University of Hawaii
Clyde Mark, Outrigger Hotels and Resorts-Hawaii
Gerald Meehl, National Center for Atmospheric
    Research
Jerry Norris, Pacific Basin Development Council
David Penn, University of Hawaii
Jeff Polovina, NOAA National Marine Fisheries
    Service
Roy Price, Hawaii State Civil Defense (retired)
Barry Raleigh, University of Hawaii
Kitty Simonds, Western Pacific Regional Fishery
    Management Council
Peter Vitousek, Stanford University
Diane Zachary, Maui Pacific Center

**South Atlantic Coast and Caribbean Workshop
    Team**
Ricardo Alvarez, International Hurricane Center
Krishnan Dandapani, Florida International University
Shahid Hamid, Florida International University
Stephen Leatherman, International Hurricane Center
Richard Olson, International Hurricane Center
Walter Peacock, International Hurricane Center and
    Laboratory for Social and Behavioral Research
Paul Trimble, South Florida Water Management
    District

Additional Contributors
Robert Cherry, USGCRP National Assessment
    Coordination Office
Benjamin Felzer, University Corporation for
    Atmospheric Research
Mark Lander, University of Guam
Katherin Slimak, USGCRP National Assessment
    Coordination Office

* Assessment Team chair/co-chair

CHAPTER 12

POTENTIAL CONSEQUENCES OF CLIMATE VARIABILITY AND CHANGE FOR NATIVE PEOPLES AND HOMELANDS

Schuyler Houser[1], Verna Teller[2], Michael MacCracken[3,4], Robert Gough[5], and Patrick Spears[6]

Contents of this Chapter

[1] Lac Courte Oreilles Ojibwa Community College, Hayward, Wisconsin, [2] Isleta Pueblo, [3] Lawrence Livermore National Laboratory, on assignment to the National Assessment Coordination Office of the U. S. Global Change Research Program, [4] Coordinating author for the National Assessment Synthesis Team , [5] Intertribal Council on Utility Policy, [6] Lakota

CHAPTER SUMMARY

Context

Native peoples, including American Indians and the indigenous peoples of Alaska, Hawaii, and the Pacific and Caribbean Islands, currently comprise almost 1% of the US population. Formal tribal enrollments total approximately two million individuals, more than half of whom live on or adjacent to hundreds of reservations throughout the country, while the rest live in cities, suburbs, and small rural communities outside the boundaries of reservations. The federal government recognizes the unique status of more than 565 tribal and Alaska Native governments as "domestic dependent nations." The relationships between tribes and the federal government are determined by treaties, executive orders, tribal legislation, acts of Congress, and decisions of the federal courts. These actions cover a range of issues that will be important in adapting to climate change, from responsibilities and governance to use and maintenance of land and water resources.

Tribal land holdings in the 48 contiguous states currently total about 56 million acres, or about 3% of the land. This area is approximately the size of the state of Minnesota. Additionally, Alaska Native corporations hold approximately 44 million acres of land. Despite the relatively extensive total land holdings, most individual reservations are small, supporting communities with populations of less than 2,000. Larger reservation populations, while unusual, do reach as high as 200,000 on the Navajo Reservation.

The federal government has recognized that tribes and tribal governments also have legal rights in territories that lie beyond the boundaries of their respective reservations. Treaties in the Pacific Northwest and the north-central states of Minnesota, Wisconsin, and Michigan recognize rights of tribes to fish, hunt, and gather off-reservation. Further, federal legislation has recognized tribal interests in historical and cultural interest areas beyond reservation boundaries. These interest areas cover a significant fraction of the 48 contiguous states, generally matching the "Native Homelands" that Native peoples inhabited prior to or since European settlement.

With the beginning of clearly observable climate change, and because of the relationships of plants, water, and migrating wildlife with ecosystems outside reservation boundaries, the potential consequences of climate change create significant interest among Native peoples. These interests arise because the consequences will affect both their reservation lands and the much larger land areas encompassed in the concept of Native homelands. While each tribe will face its own challenges, this chapter focuses on a few general issues facing large numbers of Native peoples, particularly American Indians. More region-specific issues are covered in the various regional sections of the report, notably in those dealing with the Pacific Northwest and Alaska.

Climate of Past Century

- During the 20th century, much of the western US, where most reservation lands are located, has warmed several degrees Fahrenheit, contributing to apparent changes in the length of the seasons.
- After relatively dry periods in the central and western US during the first half of the 20th century, the second half has been wetter, with more runoff in key river basins.

Climate of the Coming Century

- During the 21st century, the Hadley and Canadian model scenarios both project that temperatures in the western US are likely to rise significantly (typically 5 to 10°F) with even larger increases possible in the Great Plains and Alaska.
- Greater precipitation is projected for the southwestern US, although projections are uncertain as to how far inland the increase is likely to extend.
- Although rainfall is likely to increase in many regions, the higher temperatures will increase evaporation, likely causing a soil moisture deficit in some regions that will tend to dry out forests and grasslands.

Key Findings

Sustaining economic vitality will require thoughtful planning because many reservation economies and tribal government program budgets depend heavily on agriculture, forest products, and tradition- and recreation-based tourism, which are all likely to be affected as the climate shifts and warm extremes become more frequent. For example, hotter and drier summer conditions are likely to affect recreational use of forest campgrounds and lakes. Economic diversification has the potential to reduce the existing vulnerability.

Preparing for the health and welfare implications of unusual climate episodes is likely to become more important as climatic patterns shift and landscapes change. This is likely to be particularly important because Native housing is typically more sensitive to the prevailing climatic conditions than national average for housing. As a result, increasing use of air-conditioning is not as ready a means of addressing an increasing frequency of very hot and dry conditions. In addition, increased dust and wildfire smoke may well exacerbate respiratory conditions.

Ensuring stable water supplies for tribal lands and surrounding users will require consideration of climate variability and change in management of water resources and in negotiations concerning Indian water rights. Financial resources are also likely to be needed to address infrastructure issues. It is possible that increased precipitation in the Southwest could help to diminish water concerns in that area while other areas are likely to face decreased water availability.

Although only a few tribal economies in Alaska and other regions are primarily based on subsistence, many tribal communities depend on their environment for many types of resources. A changing environment puts such resources at risk, which will affect both sustenance and cultural dependence on environmental resources.

Sacred and historically significant sites (and the experiences associated with them) and cultural traditions are likely to be significantly affected. Because some sites are located in vulnerable locations, changes in climate and ecosystems are likely to alter the site environment. Changes in the timing of animal migrations and changes in the seasonal appearance and abundance of both plants and animals are also likely.

Coping Options

Increasing education about basic science issues and health and wellness as well as increasing scientific and technical expertise could help to build economic resilience. Increased monitoring of ongoing changes and improved projections of future changes are needed to improve the quality of planning and preparing for climate change. Enhancing access to information and technology has the potential to alleviate some of the stresses from climate variability and change.

Promoting and enabling local land use and natural resource planning are likely to help create the conditions for prudent preparations for and responses to climate change. Increasing participation of Native peoples in regional and national decision-making is needed in recognition of the connectedness of tribal reservations, surrounding lands, and larger regional and even national landscapes.

POTENTIAL CONSEQUENCES OF CLIMATE VARIABILITY AND CHANGE FOR NATIVE PEOPLES AND HOMELANDS

The Earth and Nature are inseparable from Indians themselves. Land sustains the lifestyle of countless tribes, even when little acreage is in production or has little productivity. Land sustains far more than subsistence, and indeed many Indians recognized decades ago the folly in attempting to sustain their daily needs on acreage that is marginal, both in resources and in per capita size. But land has emotional meaning, a psychological significance for the Indian that is far more intense than our nostalgic longing for the family farm and a rural way of life.

James Rattling Leaf,
Sicangu Lakota

INTRODUCTION

With projections of significant changes in the climate and in the character of the land and its resources over the next century, American Indians and the indigenous peoples of Alaska, Hawaii, and the Pacific and Caribbean islands that are part of the US will face special challenges. Although counting methods vary, Native peoples currently comprise almost 1% of the US population (see end note 1). Tribal enrollments total approximately two million individuals, of which about 1.2 million live on or adjacent to hundreds of reservations throughout the country. The other 40% of Native peoples live in cities, suburbs, and small rural communities outside the boundaries of reservations (BIA, 2000a). The federal government recognizes the unique status of more than 565 tribal and Alaska Native governments as "domestic dependent nations." The relationships between tribes and the federal government are determined by treaties, executive orders, tribal legislation, acts of Congress, and decisions of the federal courts. These agreements cover a range of issues that will be important in adapting to climate change, from responsibilities and governance, to use and maintenance of land and water resources.
It important to note that many of the phenomena of a changing climate addressed in this chapter are precisely those which affect rural areas throughout the United States. At least half of the Native peoples of the US live in reservations in rural areas. Why, then, if climatological phenomena are so similar, should there be a separate chapter addressing the issues of Native peoples? The answer is at least twofold.

First, the institutional, legal, economic, and political structures in Indian country differ significantly from those of other rural communities. These structures determine and constrain, to a great degree, the resources that Native peoples can deploy in addressing issues that arise from changes in climate.

Secondly, different cultures provide their members with different kinds of tools for addressing life's situations: language, preferences of technology, responses to innovation, tastes in food, attitudes towards strangers, relationships to the physical environment. Native peoples, with traditional cultures as different from each other as they might be from those of Swedish-Americans, bring a wide variety of cultural tools and experience to issues of climate change and adaptation. This chapter examines a number of those tools and their implications for adaptation to significant environmental changes.

The chapter is based on the Native Peoples/Native Homelands portion of the Assessment (NPNH, 2000), which is examining the potential for climate-related impacts that are likely to affect Native peoples and the interactions of these impacts with contemporary Native cultures, communities, and future generations. Although the diversity of land areas and tribal perspectives and situations makes generalization difficult, Native peoples recognize that becoming more resilient to variations in the climate and preparing for and adapting to climate change will require special attention. The preceding regional sections of this report, for example the chapters on Alaska, the West, and the Pacific Northwest, highlight a number of region-specific issues. This section reports on aspects that reach beyond what citizens of particular regions will experience.

HISTORICAL CONTEXT

Over the last 500 years, essential environmental balances that sustained Native peoples in North America for many millennia began to shift rapidly. Forests were cut for homesteads and farming. Alien plants replaced native grasslands. Dry lands flooded, rivers changed their courses, and ponds and swamps drained away as watercourses were dammed and channeled. Important providers of nourishment and protection – buffalo, salmon, and shad – were harvested to near extinction. New and strange creatures – horse, cow, pig, sheep, and pheasant – shoved aside nature's important spiritual guardians and came to dominate local economies. Exotic new diseases eradicated whole villages. Tribal governments and relationships fell. Spiritual leaders lost their followers. Communities – even entire tribal nations – were forced to relocate.

Five hundred years ago, the population of Native peoples in North America is thought to have ranged between about 10 and 18 million. Between 1500 and 1890, the population of Native peoples on the continent was declining at an average rate of between 500,000 and 850,000 individuals each 20-year generation, and by 1890 had dropped to only 228,000 (Snipp, 1991). Some opinion leaders predicted that Native people would soon disappear. However, those who predicted the "vanishing of the Red Man" substantially underestimated the endurance and adaptability of Native peoples, and the strength of Native perspectives and values. Over the last 100 years, the population of Native peoples has grown almost ten-fold as Native communities have been rebuilt; artists, craftworkers, and writers have created a renaissance of beauty and meaning; and economic development has accelerated (Cornell, 1998).

In that the global temperatures have varied relatively little over the past millennium, the environmental changes that drastically altered the lives and circumstances of Native peoples as a whole since 1492 seem to have been more influenced by changes in climate at the continental and regional levels. In addition, as Native peoples were displaced and national development occurred (Brown, 1991), Native peoples experienced continental-scale changes in their surroundings that are not unlike the types of changes that all Americans may face in coming decades. The changes were substantial in magnitude, surprising in their occurrence, unmanageable by available technologies and existing forms of government, and generally irreversible. In those respects, the changes may provide insights of the kinds of transformations – cultural, economic, and social – that global changes in climate may bring, both for Native peoples and for America as a whole.

GEOGRAPHICAL AND SOCIOECONOMIC CONTEXT

The lands held by Native peoples are extensive. In addition to the 40 million acres of land held by Alaska Natives, tribal lands in the rest of the US currently total about 56 million acres (Department of the Interior, 1996). The lands outside Alaska amount to about 3% of the land area of the 48 contiguous states, or approximately the size of the state of Minnesota (see Figure 1). The largest portion of Indian lands are held on reservations, so named because they consist of lands that were reserved for

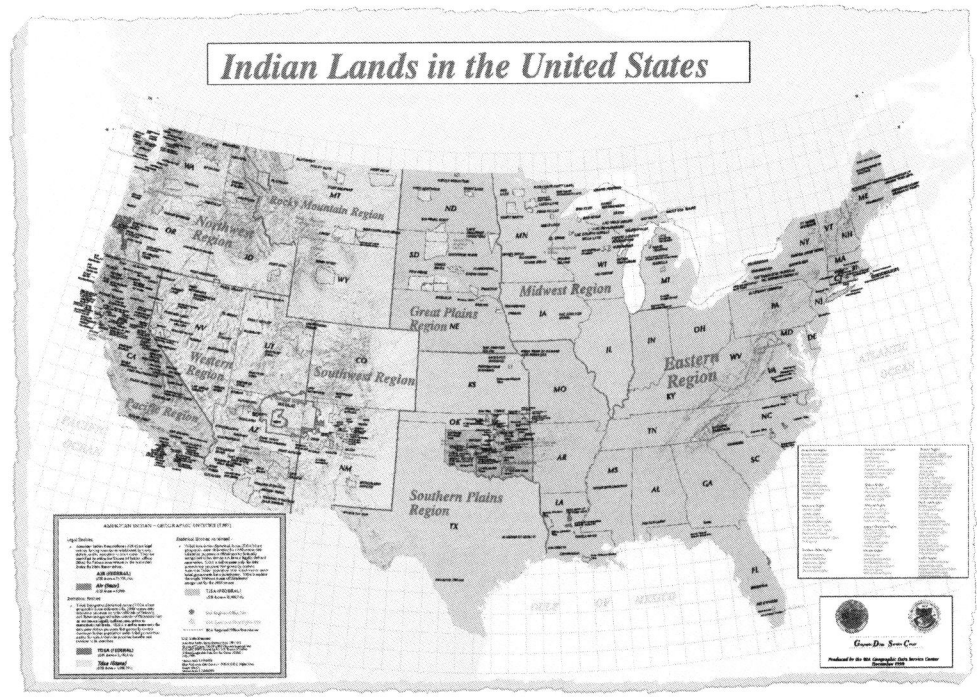

Indian Lands in the United States

Figure 1: Map of Indian lands in the conterminous United States (BIA, 2000b). The largest areas are located in the central to western US. See Color Plate Appendix

the sole use and occupancy of Indian peoples from the vast expanses of land which were ceded to the United States government (Brown, 1991). Property ownership by Native peoples of the Pacific and Caribbean islands varies greatly because of a variety of circumstances, including traditional rights and leadership and historical legal rights. As indicated in the Islands chapter, however, some island lands are overseen by clans with responsibility for steward-ship on behalf of their members, whereas on other islands there are no longer reserved land rights.

By far the majority of reservations are small, both geographically and demographically, with popula-tions less than 2,000 (Tiller, 1996). These lands, although they are owned by tribes or individual Indian people, are held in trust for the owners by the federal government, in much the same way that banks or other trustees hold property for heirs until they come of age and can assume personal manage-ment of the property. One result of this system of trusteeship is that tribes and individual Indian peo-ple have had very limited control over the use, envi-ronmental management, or profits of their own lands. For much of the 20th century, in fact, many of the decisions over these matters rested with the fed-eral government, not with the tribes themselves. Only in the last several decades have tribal govern-ments taken over more control of and responsibility for their lands.

From the most basic perspectives of the American legal system, reservations may be viewed as jurisdic-tional islands, largely exempt from the laws of the states that surround them. Tribal governments hold the authority to levy taxes, regulate commerce, pass and enforce civil and criminal codes and, in princi-ple, regulate the use of tribal lands and water. While federal laws prevail, state authorities generally have no rights of enforcement within these jurisdictional islands.

However, from the perspective of tribal environmen-tal and land management policies and practices, the paradigm of reservations as islands is inadequate. First, the paradigm is inadequate environmentally because these "islands" are surrounded not by oceans, but by land, and so these lands are intimate-ly tied to the forests, grasslands, watersheds, and other ecosystems surrounding them; thus, the changes on Native and surrounding lands are closely coupled. Second, because many reservations have considerable populations of non-Indians residing within the exterior borders of reservations (see end note 2), the paradigm is inadequate administratively. Third, throughout the country, non-Indians also

work on Indian lands because of the leasing of tribal lands to non-Indian farmers and ranchers – or, for example, in the case of Agua Calienta, near Palm Springs, California, for commercial development. The leasing of reservation lands is a long- standing practice and a vital source of income to the landowners (Lawson, 1982). Complicating matters further, a major portion of the lands that were allot-ted to Indian heads of household are now managed either by the Bureau of Indian Affairs (BIA) or by the appropriate tribal government. This land is also frequently leased to non-Indian farmers or ranchers with the proceeds from the leases then being divid-ed among the descendents of the original allottee. Maps of land ownership and tribal jurisdiction on many individual reservations thus resemble checker-boards, greatly complicating planning efforts. At the same time, judicial decisions have sharply limited the jurisdiction of tribal governments and tribal courts over non-Indians.

As a result, many tribes face severe legal difficulties in creating or enforcing comprehensive plans for land use or natural resource management, a situa-tion that will complicate planning for climate change. For example, if a tribal government creates an environmental code, enforcement over an entire watershed or forest may be impossible without the voluntary consent of non-Indian owners of property within and outside of reservation boundaries. If a tribe leases cropland, grazing rights, or timber to non-Indians, environmental regulations can conceiv-ably be written into the terms of the leases, although long-term traditions are likely to be diffi-cult to change and the practical job of enforcing new regulations is likely to stretch the resources of small and understaffed tribal governments (Getches et al., 1998; Pevar, 1992).

Tribal governments also have some legal rights in lands beyond the boundaries of reservations – rights that may establish precedents for collaboration on issues involving climate and environmental changes. For example, the federal government has recognized historical and cultural interests of tribes and tribal governments concerning broader regions, often called "Native Homelands," which include lands occupied by Native peoples at present or in the past. Within the pre-determined boundaries of his-torical and cultural interest areas (generally home-land areas inhabited by a particular Native people prior to contact with Europeans), tribes are entitled, for example, to establish claims to human remains, if evidence of kinship or ancestry can be established.[1] These historical and cultural interest areas cover a

---

[1] For further information, see http://www.doi.gov/oait/docs/eo13007.htm .

significant fraction of the 48 contiguous states, widening greatly the areas of interests of Native peoples.

CLIMATE AND ECOLOGICAL CONTEXT

Reservations are present in all of the major ecosystems across the US, including the unique environments represented by Alaska and the islands of the Pacific and Caribbean regions. Native peoples have been experiencing the vagaries of climate on this continent for many thousands of years. The resource-rich environment created by the woodlands of the northeastern, southeastern, and Great Lakes regions, especially the presence of deer, rabbit, beaver, fish, berries, and many other resources, allowed tribes to occupy particular regions for long periods. The Great Plains provided a source of grains and buffalo, along with fish and other resources, but the wide range of climate extremes caused these tribes to be relatively mobile in order to survive. The western US provided a wide array of environments, from coastlines to mountains and river valleys to deserts, and is now home to the greatest number of Indian reservations. The Native peoples of Alaska have developed a lifestyle that depends, in large part, on very cold winters. Those living on islands depend on the reliability of the rains, and are adversely affected by either too much or too little precipitation.

Their adaptability, and the histories of the experiences and the lessons that have been learned about coping with climate fluctuations, have sustained Native cultures through many generations. Native oral histories are now being linked with past climate data derived from tree rings and other sources in ways that are enriching understanding of past climatic conditions. Oral histories often correlate with events identified in the geological record, such as periods with high or low rainfall, periods of warm or cold winters, and periods of flooding or drought (e.g., Deloria, 1997). What makes these histories especially valuable is that they often record not only the consequences of these climate fluctuations for people and for the environment around them, but also the responses that helped the communities to adjust and survive. Thus, the retelling of these events by tribal elders has created a populace that is relatively well informed about how to adapt to external stresses.

There are, however, two key changes that will limit the application of some of the lessons to the issue of climate change, and are likely to create greater vulnerability than in the past. First, the changes in climate are likely to be larger and more long lasting than the fluctuations experienced in the past. Second, earlier coping strategies of Native peoples, on which many of their histories and traditions are based, relied on shifting and moving, sometimes from one food source to another, sometimes from one place to another, or sometimes to find alternative sources of food and water, or to intersect with the annual migrations of wildlife. In the Southwest, archeological evidence and Native oral histories indicate that the great regional drought of the 13th century caused the ancestral Pueblo people to abandon their permanent homes in the mesas and valleys of marginal areas (see box, "Lessons from the Relocation of the Ancestral Pueblo People in the 13th Century"). When the ability to cope in one place was exceeded, Native peoples moved, later returning if and when climate permitted.

Over recent decades, Native peoples have been observing that changes in the environment have been occurring, some due to regional or global-scale changes in the climate and some due to changing practices of land management and use. These changes are noteworthy, both because Native peoples are changing their practices and because of the nature of the observations themselves. In northwestern Alaska, for example, elders lament that winter temperatures have become so warm (now typically only -20°F instead of -70°F) that the traditional ecosystem on which they have depended for generations is deteriorating and is no longer able to provide the needed resources. In the Southwest, recollections by elders (corroborated by Army records from the early 1800s) are of valleys full of tall sacaton grasslands, whereas the region now is scarred by deep arroyos and supports only sparse vegetation, likely as a result of overgrazing and subsequent drought. All across North America, tribal histories indicate that change is occurring.

Native peoples today feel vulnerable to significant environmental changes because they are no longer able to cope easily with changes by relocating. Few contemporary tribes can afford the purchase of large tracts of new land, and federal laws hinder the transfer or expansion of tribal jurisdiction. Tribes therefore see their traditional cultures directly endangered by the magnitude of the projected climate change. Had the ancient Anasazi been compelled to remain in place, the culture and way of life of an indigenous people that can be traced back

357

Lessons from the Relocation of the Ancestral Pueblo People
in the 13th Century

About 2000 years ago, ancestors of today's Pueblo people inhabited expansive areas of the Southwest, including much of what is now known as New Mexico, Arizona, Utah, Nevada, and Colorado. They developed advanced architecture, moved water via extensive and complex irrigation systems to grow crops, made baskets and pottery, and created sophisticated social and religious structures. Today, those ancestral homelands are dotted with the ruins of large villages that show deserted centers of trade and commerce, abandoned road systems and multi-story houses, forsaken art, and elaborate ceremonial chambers.

What happened to these communities is believed to have been a result, quite possibly in large part, of changes in the climate of this region. In the late 1200s, evidence from the growth rings of trees documents significant changes in the weather and climate. In particular, proxy evidence indicates that droughts became more frequent in southwestern North America. These conditions in turn would have led to disruption of the agricultural lifestyle of the communities dependent on the runoff from the southern Rocky Mountains. Although these regional climate changes were not particularly evident in the global climate record (see the Climate chapter's description of the 1000-year climate record), the regional changes were apparently so large that they contributed to the abandonment of these ancient cities. Cross-sections of soil combined with pollen records tell a similar story of changing conditions, with increasing erosion, ruined farmlands, and possibly a prehistoric "dust bowl." As communities became unable to adapt their abilities to grow food to the deteriorating environment, these changes apparently triggered a mass migration of what had been a typically stationary, agriculture-based people.

Pueblo people living today pass on migration stories that recall the early journeys that brought them to the villages they still occupy, mostly along the Rio Grande in New Mexico. Some clans migrated as far to the west as today's Hopi tribe in Arizona. Others settled in the Pueblos of Zuni, Acoma, and Laguna. Since these moves, much has changed that limits the ability of Native peoples to move again to escape the consequences of climate fluctuations and change. For the Pueblo people, Spaniards claimed their ancestral homelands in the early 1500s. Subsequently, tribal land holdings were generally confined to the very limited boundaries that persist to this day. Thus, if another disruptive climate fluctuation or change were to occur, the Pueblo people could no longer easily move as a group to places with better conditions in order to preserve their societies and cultures.

thousands of years would likely have been lost forever. This history provides a context for thinking about the potential consequences of future changes in the climate.

SCENARIOS OF FUTURE CLIMATE

Most of the large Indian reservations in the US are located in the central and western United States. Coincidentally, most climate models project that changes in temperature and precipitation will be greater over the western than over the eastern US. The Canadian and Hadley model scenarios project warming of as much as 5 to 10˚F (3-6˚C) over the 21st century over much of the western US, with more warming during winters than during summers in many areas (see Chapter 1 for more details).

The warming is already being experienced by some Native peoples. Regions in Alaska are already experiencing significant warming, with some melting of permafrost and sea ice. Across northern North America, warming is already lengthening the period of plant growth. Such changes are expected to intensify. These trends will affect agriculture, wildlife migrations, flowering seasons, and other important ecological events or occurrences that are closely tied to seasonal change.

Some models project that, particularly in the Southwest, warmer winters will also bring increasing wintertime precipitation, a rising snowline, and earlier springtime runoff, thereby affecting the timing and volume of river flows. Increased precipitation would be expected to increase river runoff, which, if it occurs in seasons when the water can be used or stored, may help to augment water supplies. Increased precipitation, however, can cause erosion on sparsely vegetated surfaces, distribute contaminants more widely, and create greater potential for flash flooding. Moreover, warmer conditions will also lead to increased evaporation, especially in summer, that will dry summer soils and vegetation in ways that may more than offset the increase in runoff in some regions. For example, these changes will likely lead to lower river and lake levels in the northern Great Plains and Great Lakes.

KEY ISSUES

The national workshop on Native Peoples/Native Homelands identified a wide variety of issues facing Native peoples. Assessment studies of issues of specific interest to Native peoples in particular regions are just beginning; information on the first such study, reporting on issues in the southwestern US, can be found at http://native-peoples.unm.edu/. In this report, issues that are primarily regional are addressed in more detail in the chapters dealing with particular regions (e.g., Alaska) whereas issues that have broader national implications for Native peoples are addressed in this chapter. These issues include:

Tourism and Community Development. In a number of regions, Native economies are strongly dependent on tourism, agriculture, and other environmentally sensitive activities, so that shifts in temperature and precipitation and the ecosystems that are based on the prevailing climate are likely to require adjustments away from traditional activities. For example, hotter and drier summer conditions are likely to limit recreational use of forest campgrounds and lakes at lower elevations while potentially allowing more use at higher elevations.

Human Health and Extreme Events. As a result of insufficient economic development, Native housing is typically more sensitive to the prevailing climatic conditions than the national average. As a result, increasing use of air-conditioning is not as ready a means of addressing an increasing frequency of very hot and dry conditions. In addition, increased dust and wildfire smoke may exacerbate respiratory conditions.

Rights to Water and Other Natural Resources. Native water rights are established in a variety of treaties, agreements, and court decisions. Provided financing is available, significant amounts of potentially irrigable land exist on reservations, and the potential exercising of these rights along with precipitation changes is likely to complicate water resource allocations, negotiations, planning, and management.

Subsistence Economies and Cultural Resources. Although only a few tribal economies in Alaska and other regions are primarily based on subsistence, many tribal communities depend on their environment for many types of resources. A changing environment puts such resources at risk, which will affect both sustenance and cultural dependence on environmental resources.

Cultural Sites, Wildlife, and Natural Resources. The environment, both climate and the landscape, provides an important sense of place for Native peoples, both for historical and cultural reasons. As the climate changes and vegetation patterns and the presence of wildlife and migrating species shift, the cultural context of Native peoples, who view themselves as tightly coupled with and integral to the natural environment, will be disrupted, with little recourse as the shifts occur.

1. Tourism and Community Development

The most urgent priorities for tribal governments and communities over the past thirty years have been economic development and job creation. Many tribes have based their development initiatives around land-based enterprises, including dryland and irrigation-based agriculture in the central and western US; forestry and forest products in the central, western, and sub-Arctic regions; and recreation- and tradition-based tourism in areas ranging from Hawaii and Alaska to the central, western, and southwestern US. All of these activities are dependent on favorable weather and climatic conditions. Although few of these enterprises have generated enough income to develop a strong economic base for entire tribes, they are all vital to economic development for tribal communities. Adverse conditions, from severe winter storms to unusually wet or dry conditions, can have very severe economic effects, especially because tribal communities are already economically stressed. The 1990 census indicated that 31.6% of all Indian people were below the poverty line, compared to 13.1% of the total population (US Bureau of the Census, 1990). From 1969 though 1989, of the 23 reservations which have data for the period, per capita income declined on 18 reservations during the second decade (Trosper, 1996). The sustained growth of the American economy over the past decade has, for the most part, bypassed Native American households and reservations.

Although none of the primary barriers to development on reservations is currently a result of long-term climate change, recent variations and changes in weather patterns are requiring tribes to adapt and adjust their actions and plans. Tribes have already identified many local needs in response to an increasing frequency of disruptions from severe storms, including improving or re-routing roads (many of which are unpaved), flood control and bank stabilization, providing new or more reliable water and drainage services for industrial sites, strengthening communications links and power supplies, and altering schedules and calendars at schools and medical clinics to adapt to changing weather conditions. Tribal communities now experiencing sharp changes in precipitation patterns are already modifying reservation infrastructures to deal with such situations[2]. Projections of increased occurrence of extreme rains are likely to intensify the need to improve infrastructure on reservations. Significant warming and the rise in the heat index are likely to necessitate alterations in community buildings and water supply systems. For the future, particularly for tribes in the Southwest, longer-term changes in water resources on which many tribes depend are likely to have significant consequences for resource-based sectors such as agriculture and industry that depend on stable water supplies.

Many tribes also are basing an increasing share of their economic development on recreation and tourism (Tiller, 1996). Tourism and recreation-based activities take advantage of the attractions of rivers, lakes, mountains, forests, and the other elements of the natural aesthetic beauty of reservations without, in most cases, causing long-term change. Cultural and historical sites and ceremonies of Native peoples can also be used to attract tourists. These activities provide income while also encouraging the re-establishment of customs and traditions that had been suppressed for many decades by federal policies.

The economic viability of many aspects of recreation and tourism, however, is based on natural attractions that depend on the prevailing climate – rivers and lakes provide water-based recreation opportunities, forests provide campsites and trails, and the wildlife, including migrating fish and birds, and flowering of plants, attract many visitors. As the climate changes, these environments will change: reduced winter runoff from reduced snow cover is

likely to reduce the flow in many streams; drier summer conditions are likely to increase the fire risk and require closure of campgrounds; and the combined effects of climate and ecosystem change are likely to disrupt wildlife and plant communities.

Cultural traditions that draw visitors (and their money) are often tied to the cycles of the seasonal rhythms in plant and animal life, with some traditions honoring annual weather-related events that are likely to be significantly affected by climatic change. The willingness to visit reservations for such events is dependent on the existence (and even the perception) of a safe and healthy environment. Such conditions can be disrupted by unusual climatic occurrences. In 1993, for example, a hantavirus outbreak associated with unusually heavy rains induced by El Niño conditions created a perception of an unhealthy environment (Schmaljohn and Hjelle, 1997). The rains were conducive to high production of piñon nuts and other food sources, leading to an explosion in the hanta-bearing mouse population. This high mouse population then encroached on human populations, resulting in a number of virus-caused deaths (Engelthaler et al., 1999). This one event led to a significant reduction in tourist visits to the Southwest, especially to Pueblo country, indicating how vulnerable sensitive tourist-based economies can be in the event of the outbreak of rare, but frightening diseases, even when these outbreaks are not occurring primarily on Indian reservations. It is possible that a change toward more intense El Niño/La Niña variations could increase the likelihood of such conditions in that such variations would likely upset the predator-prey relationships that develop during less variable conditions.

While some economic diversification of reservation economies is underway, and casino gambling is becoming a basis for attracting tourists in a number of regions, tribal economies tend to be more closely tied to their environments than is typical for the economies in regions where the reservations are located. Because of this, tribal economies tend to be more vulnerable to adverse changes and, on the other hand, more likely to benefit from climatic changes that provide opportunities by enhancing water availability.

Adaptation Options
With many tribal lands already under climatic stress, and with the economies of Indian Nations strongly dependent on climatically sensitive activities such as agriculture and tourism, the lands and economies of many Native peoples are vulnerable to climate change. As a result of the less-diversified Native

---

[2] Repeatedly during the assessment process, participants raised issues about the development of rural transportation systems. Reliable transportation is a major concern in many tribal communities, where many individuals cannot afford newer, more reliable, and fuel-efficient vehicles. Providing effective public transportation in rural communities – most particularly communities with limited incomes – improves access to employment, education, and medical services, and has the additional benefit of improving the efficiency of the use of fossil fuels, thus reducing the production of greenhouse gases.

economies, a larger share of financial resources will likely need to be devoted to ongoing adaptation than is the case for society as a whole, thereby possibly making tribal economies less competitive. To address this issue, enhancement and diversification of reservation resources and the strategic integration of tribal economies with local non-Indian economies could help to make the tribal economies more resilient and sustainable.

At the same time, climate change and resulting policy actions are likely to create some economic opportunities. For example, an increased demand for renewable energy from wind and solar energy could create new opportunities because undeveloped tribal lands could be an important resource for such energy in areas that are already developed. For tribes in the Great Plains, this region could utilize its tremendous wind resource and development could help to reduce greenhouse gas emissions as well as alleviate management problems created by demands for Missouri River hydropower, thereby helping to maintain water levels for power generation, navigation, and recreation. In addition, there may be opportunities for carbon sequestration.

2. Human Health and Extreme Events

The rural living conditions of many Native peoples increase the likelihood of impacts due to variations in the weather. Due to the poor economic conditions, housing on many reservations is old and offers limited protection from the environment. Although many traditional structures are designed to take advantage of the natural warmth or coolness of the landscape (e.g., by being located below ground, having thick walls, and being exposed to or sheltered from the sun), acclimation, both physiological and through use of appropriate clothing, is critical because homes in many areas lack effective heating and cooling systems. A recent study of energy consumption on Indian lands (EIA, 2000) found that reservation households are ten times more likely to be without electricity (14.2%) than the national average (1.4%). While warming in colder regions will alleviate some home heating needs, some acclimation has already occurred so this will not significantly reduce stresses. In presently hot regions, however, there is likely to be a significant increase in natural heating that will require new acclimation and responses as new extremes are reached. While an increase in the presence of air-conditioned facilities would help, it would also require changes in behavior toward a more indoor lifestyle along with improved housing and access to electricity.

Climate change is also likely to exacerbate the delivery of and need for health services. The delivery of health care to rural communities throughout the United States has already been affected significantly by widespread changes in demographic and economic patterns. Rates of depopulation in the least densely populated portions of the country are accelerating. As communities lose population and economic conditions stagnate, the numbers of doctors, pharmacists, and other health care professionals attracted to rural communities have been declining. Full-service health care is increasingly concentrated in regional centers, and consultation with a specialist increasingly involves, for rural residents, an extensive trip.

Reservation populations, on the other hand, are continuing to increase, as birth rates remain high, longevity increases, and tribal members move back to their home communities from urban areas. The institutional structure of health care delivery to Native peoples, however, differs sharply from the market-driven system that provides medical care for rural non-Indians. The federal government, as part of its trust responsibility for Indian people, provides the health care systems for reservation residents. The Indian Health Service (IHS), part of the United States Department of Health and Human Services, is the primary provider of medical services to Native peoples. IHS operates clinics, pharmacies, and hospitals in many tribal communities; in others, tribes have contracted with the federal government to operate health care facilities themselves.

Access to these health care facilities, however, is not always easy for reservation residents, particularly under extreme weather conditions. A single hospital, for example, serves the entire Rosebud Sioux reservation in South Dakota, where many roads are unpaved. The external boundaries of the reservation are approximately 120 miles east-to-west, and 60 miles north-to-south. The hospital is located in the southwestern part of the reservation and access can be disrupted or cut-off by extreme precipitation conditions (whether rain or snow). Because of the distances that many Native people must travel to health care facilities and the conditions of the roads, their access to health care is more subject to sharp variations in the weather than those living and working in cities, and extreme weather events are likely to cause significant interruptions in access.

Changes in climate would also create new challenges for community health. Drier summer conditions and the projected increase in forest fire incidence would likely lead to increased lofting of dust and dust-borne organisms and an increase in forest fire incidence. The poorer air quality resulting from increases in smoke and dust would likely increase respiratory illnesses such as asthma. Hypertension and adult-onset diabetes are pandemic in tribal communities. As life spans increase, larger portions of the populations are becoming increasingly vulnerable to extreme temperatures, and increasingly dependent on uninterrupted access to such therapeutic interventions as kidney dialysis. Further, direct and side effects of many standard medications are affected by climatic conditions.

Adaptation Options

Health care delivery systems that serve Native communities will need to educate health professionals, paraprofessionals, and patients on the potentials for dehydration, overexposure to sunlight, and the need to restrict activity during hot weather. This need for education and appropriate care is intensified because many Indian people live in substandard housing that exposes them to heat and cold. Much of the most effective health education in Native communities is carried out by Community Health Representatives (CHRs), paraprofessionals employed and trained by the Indian Health Service or individual tribal health programs. Specific training, focusing on climate-related health and wellness issues, and targeted towards CHRs could enhance use of a system that has already demonstrated its abilities to reduce infant mortality rates and improve care for the elderly among many Native peoples.

In addition, consideration should be given to the design and construction of appropriate housing for Native peoples. Most housing construction on many reservations is financed through the federal Department of Housing and Urban Development. Tribal housing authorities, however, have taken over much of the responsibility for design, construction, and management of both rental and mutual self-help (purchasable by the occupant) housing units. Given the likelihood of significant changes in average temperatures, precipitation levels, and severe weather events, training and regular updates for decision-makers and technical specialists within tribal housing authorities are needed to assist in improving the design, construction, or remodeling of homes to increase resilience to extreme weather conditions.

3. Rights to Water and Other Natural Resources

For Native peoples, water is recognized as a cultural as well as a physical necessity. Water is vital for life and livelihood, especially for those relying on the resources provided by natural ecosystems. Water is necessary for community use and the production of food as well as for fish, riparian plants, and wildlife. Water is particularly valued where it is most scarce, such as in the southwestern US. Prior to European settlement, water was not owned, but was viewed as a gift to be shared by all. Settlement – both by tribes on reservations and by other Americans – brought increasing demands for water and concepts of ownership that were not traditional to Native peoples. As a result, rights to water, including access to sufficient quantity and quality, are now established and guaranteed by treaties, statutes, and decisional law. Changes in the amount, timing, and variability of flows will affect the exercise of these rights.

Despite the many agreements and treaties, quantitative determination of existing Native rights to water remains a contentious legal issue over much of the western US. Access to water was a key issue in many of the treaties negotiated between tribes and the US government, especially when relocation of tribes to reservations restricted or eliminated their access to traditional homelands. These provisions became the subject of litigation as expectations were not met. The concept of federal reserved water rights for Indians originated in the landmark case *Winters v. United States* [207 US 564 (1908)] involving withdrawal of water from the Milk River along the Fort Belknap Reservation. The Winters Doctrine provided that in the treaties the federal government entered into with various tribes, the government had implicitly reserved a quantity of water necessary to supply the needs of the reservation[3]. Under this provision, the federal government's commitment of water rights would be paramount because it had created the Indian reservations and therefore had a fiduciary duty to protect water implicitly reserved by and for the tribes. The doctrine was based on a set of principles underpinning reservation establishment that retained for Indian tribes all rights not explicitly waived.

Indian reserved water rights have become a subject of considerable importance to tribes, states, the federal government, and private water users due to: (i) the scarcity of water, particularly in the Southwest and Great Plains; (ii) the reality of drought conditions; and (iii) uncertainties arising because of fully (or even over) appropriated watersheds and international water commitments (some of which do not account for tribal water rights at all). The courts

---

[3]The Winters Doctrine later also formed the basis for federal reserved water rights asserted by the government to obtain water for national forest, wetland, wildlife refuge, military, and other reservations.

have often had to resolve conflicting interests in the allocation of water rights, usually using a standard known as "practicable irrigable acreage" for determining the allotment to Indian reservations. This standard quantifies *Winters* rights by providing that allotments include sufficient water to provide for agriculture, livestock, domestic, recreational, cultural, and other uses. In addition, for some tribes, specific legal language also reserves water to maintain in-stream flows necessary to sustain fish or riparian areas. Figure 2 compares the acreage of Indian lands that are currently being irrigated in eleven western states with the areas that could potentially be irrigated under the *Winters* doctrine. Quite clearly, substantially increasing the areas of irrigated lands would significantly increase the amount of water being withdrawn from current resources, increasing the competition for water among the various users.

In some cases, tribes have not had the financial wherewithal to develop their water rights. In other cases, despite having high priority (i.e., senior) rights to sufficient water, tribes have often had to compete for access to water with non-tribal water users, including federal, state, and local governments. This has required negotiations, which have often proven to be time-consuming, costly, and complex. The potential for changes in the amounts and timing of water flows caused by climate change are likely to add to the complexity of the allocations, negotiations, agreements, and management of water resources.

As an indication of the complexity of the issues involved, consider the Southwest, where climate models project that there is likely to be additional precipitation (most probably as rain) in winter, but with earlier snowmelt and generally warmer and drier conditions during the summer. If this occurs, overall winter runoff is likely to increase while summer runoff is likely to decrease significantly. While an increase in annual precipitation could ease overall management of water resources (at least to the extent that increases in the amount of vegetation do not counterbalance the likely increase in runoff), if the extra runoff in winter is not retained in reservoirs, then the increased needs for water in summer would likely increase water demands. Further, water storage that benefits some could be detrimental to others, an example of which is described in the box on "Shifting Ecosystem Boundaries" later in this chapter. With Native peoples exercising their historical water rights more fully, and with federal policies requiring additional amounts of water to protect the environment, the amount of water available for irrigation, communities, and other uses is likely to

decline, particularly during years in which climate fluctuations lead to reduced overall water flows. Water quality is a major issue that is coupled to the issues of water rights and water quantity; as water quantity changes, water quality is likely to change as well. Water quality affects everything from the environment for fish to the purity of drinking water to the quality of aquifers. In addition, because of the cultural connection of Native peoples, water pollution and poor water quality can have unusual ramifications. In the Northeast, for example, streams have been polluted by persistent organic pollutants and heavy metals, so fishing and fish consumption are not permitted. These prohibitions affect not only the diets of Native peoples who observe such restrictions (many do not for cultural reasons), but the restrictions have also reduced the opportunity for intergenerational connection while fishing. In the Southwest, water purity has become an issue based on water's use in religious ceremonies (see box "Ensuring Water Quality for Religious Ceremonies"). The prospect of diminished summertime water flows, along with more intermittent flows of some streams as a result of an increase in frequency and intensity of extreme rains, introduces the potential for water quality to become an issue in some regions.

Adaptation Options

Where water is scarce, it merits careful stewardship. Improving the efficiency of water capture and use

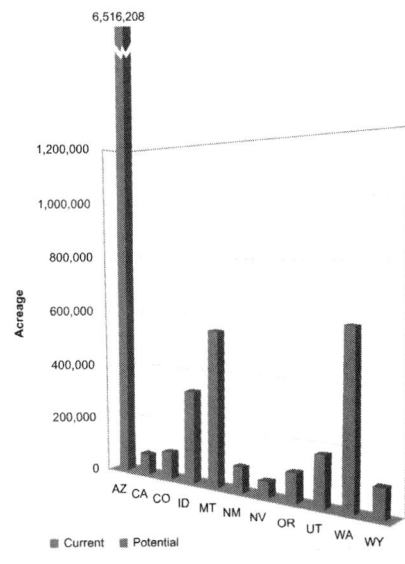

Currently Irrigated vs. Potentially Irrigable Indian Land

Figure 2: Comparison of existing acreage being irrigated on Indian lands in eleven western states with the maximum acreage that could potentially be integrated based on the Winters doctrine that is applied for determining Indian water rights (Riebsame et al., 1997) See Color Plate Appendix

through more efficient irrigation practices and choosing and growing crops that need less water are steps that can be taken now, provided resources are available. As an additional possibility, increased use of cisterns and other water harvesting techniques of the type used several centuries ago by tribes living in the Southwest could also provide a very localized means of conserving water. Increasing in-ground storage of water through artificial recharge is another possibility. Other historical practices of Pueblo cultures may also be applicable; mapping of ancient land use practices show extensive use of water management techniques, including gridding of gardens to slow runoff, pebble mulch fields to reduce evaporation, stone-lined drainage channels to reduce leakage, and terraces on hillsides to retain water and prevent erosion. Quite clearly, however, the climate changes, ensuring the reliability and availability of water resources for tribal lands and surrounding users will require special consideration.

4. Subsistence Economies and Cultural Resources

Native lands have provided a wide variety of resources for Native peoples for thousands of years. Forests, grasslands, streams, coastal zones, and more have provided, and for many groups still provide, substantial amounts of food, fiber, fish, medicines, and culturally important materials. Native traditions are also very closely tied to natural events and resources, including migrating birds and fowl, land animals, fish, and medicinal plants, creating an important cultural link to the land. Indeed, subsistence economies were the predominant form of community organization in North America prior to the colonization of the continent by Europeans. It is clear from oral traditions of Native peoples themselves throughout the continent, and from the accounts of the first Europeans who contacted them, that subsistence economies were able to sustain communities in lives of comfort, relative stability, and abundance, with sophisticated artistic and intellectual traditions.

With the spread of the market economy across the US, especially during the 20th century, subsistence economies began to disappear as the resources necessary to support them were absorbed by commercial markets. By the end of the 20th century, subsistence economies among Native peoples were signif-

Ensuring Water Quality and Quantity for Legally Protected Religious Ceremonies and Cultural Practices

In 1986, the US Environmental Protection Agency (EPA) was authorized to treat tribes as states for purposes of grant and contract assistance, regulatory program development, and permitting and enforcement. This has allowed some tribes to propose water quality standards. As for other states, these standards must then be approved by EPA. One example where this has occurred is the Pueblo of Isleta, which is located immediately south of Albuquerque, New Mexico. This was the first Native government in the US to obtain approval and Treatment-as-State (TAS) status from the EPA, pursuant to Section 518 of the Clean Water Act.

When the Pueblo of Isleta acted to protect its water quality, the City of Albuquerque sued the EPA for approving Isleta's standards (which are more stringent than off-reservation standards) in Federal District Court. In October 1997, the US Supreme Court upheld the tribe's standards as approved by the lower court. The case was decided in the favor of the Isleta Pueblo, with reasoning that has implications for Native governments throughout the country. Isleta Pueblo's standards were based on three significant use designations: use for irrigation, use for recreation, and use for religious ceremonies. The last reason is important to emphasize because no Native government had ever before asserted its right to religious freedom for the protection of its waters. The Pueblo's contention was based on the fact that tribal religious ceremonies (including bathing in the river) were adversely impacted by the contamination of the river water by toxic discharges from Albuquerque's run-off and from the municipal waste treatment facility located four miles north of the Pueblo. Although the city claimed that the tribe's standards were arbitrary and capricious and would create an undue burden on the city to comply, the court upheld the Pueblo's water quality standards.

With projections indicating that climate change is likely to lead to changes in the amounts and timing of water flows, it is possible that changes in water quality and quantity, especially during the summer, could become an issue in other regions. Lower lake levels in the upper Great Lakes region and reduced flows in western rivers may be examples of where this issue could arise.

Premises Supporting Sustainable Subsistence Economies

The most important differences between market economies and the subsistence economies of Native peoples involve concepts of surplus, accumulation, ownership, private property, individuality, and community. Participation in a sustainable subsistence economy demands knowledge, attitudes, behaviors, and expectations that differ significantly from those which work in market economies. The basic premises that have emerged and still generally prevail include recognition that:

Sustainable subsistence activities require traditional ecological knowledge acquired over generations (LaDuke, 1994).

Subsistence communities are fairly, though not completely, closed economic systems. For day-to-day necessities, people rely primarily on goods and services produced within their own community. Other goods might be acquired from external sources through trading (or, in earlier times, raiding), but sustenance depends on work done within the community.

Most subsistence communities use a variety of food sources and methods of food production, including hunting, fishing, gathering, and agriculture.

Success in food production varies from individual to individual within each community, and varies for each individual over time.

Food preservation technologies are often limited primarily to drying and smoking, technologies that do not support the storage of large amounts of foodstuffs over several years.

Food is sought in the amounts needed to sustain the community, with perhaps a small surplus for trading. Hoarding of food makes very little sense because it generally cannot be preserved, and so is actively discouraged.

Subsistence communities are small enough so that everyone within an individual community knows and acknowledges their relationships with everyone else. Kinship ties, in particular, are widely recognized (Akerlof, 1984; Schelling, 1978, 1984).

Family relationships carry with them obligations of mutual support and sharing of resources.

Use of resources by the current generation must take into account the needs of future generations. An individual might gather, harvest, or hunt enough for household and family use, but must leave enough to ensure regeneration of renewable resources.

---

icant as a basis for family life only in the far north of Canada and Alaska, where some communities still support themselves by a combination of subsistence, welfare and market economies; mixed market-subsistence-welfare economies also exist in some tribal communities in the conterminous US.

The values that have supported subsistence economies have persisted, however, and this is a major feature that distinguishes Native peoples from other contemporary rural residents. The ethics embodied in these subsistence systems continue to form the core set of values in many modern Native communities, even as they are also integrated, in varying degrees, into mainstream market economies. These values are incorporated into, and reinforced by spiritual teachings, moral principles, and community and family relationships. To the extent that these values continue to shape contemporary attitudes and relationships of Native peoples, they form a crucial part of the treasury of values on which the Nation may draw in addressing global climate change (see box, "Premises Supporting Sustainable Subsistence Economies").

Relying on a closed system involving primarily local resources rather than on a global network that provides food and goods introduces much greater elements of risk. Membership in a community provides one means of limiting risk, because each household has more or less equal opportunities to share in the community's resources.[4] Under such circumstances, special provisions are often made for individuals such as the elderly and the sick who might be unable to contribute regularly to the community's present food supply (Axelrod, 1984, 1997; Brams, 1996). Prosperity, in this economic system, is measured not exclusively in terms of accumulated possessions – food, fields, and horses – but in terms of personal relationships. Having a large and healthy family, respectful and aware of their obligations, provides assurance of mutual support by

[4]The Lakota (western Sioux) begin each traditional religious ceremony with the phrase "Mitakuye Owasin" which can be translated as "All my relatives" or more freely "We are all related." In the world-view of many Native Peoples, this concept includes not only all of humanity, but all life, and physical objects including rocks, the earth, stars and water. Many traditional Native spiritual leaders support the concept that all life shares a foundation in DNA, and all of life and physical matter shares common structures at the atomic and molecular level.

spreading widely the responsibilities for care and sustenance (Olson, 1965; Hardin, 1982; Houser and White Hat, 1993).

Generosity, in this economy, is a highly prized virtue. When generosity is linked with the obligation for reciprocity, the development of positive human relationships through sharing becomes the most effective way to limit risk for any particular family or individual. Among many Native peoples, there were historically – and continue to be – regular, institutionalized opportunities to show generosity, by giving food and material gifts to members of the community. During dances and special celebrations, Pueblo families open their homes to invited guests and community members for home-cooked feasts, and give away baskets of food and cloth to honor family members whom are participating in ceremonies. Northern Plains' tribes hold giveaways in honor of particular individuals – the honoree is celebrated, not by receiving presents, but by having gifts given to others by the family in his or her name. Families may save for a year or more in order to provide appropriate honor to a family member.

Traditional subsistence economies use a second method of managing risk – the avoidance of over-specialization or over-dependence on single sources of food. In the past, woodlands tribes, for example, raised corn, beans, and squash, while also gathering wild rice, berries, acorns and nuts, wild turnips and onions, and other edible and medicinal roots and plants (Vennum, 1988; Ritzenthaler and Ritzenthaler, 1991). These tribes also fished, and hunted large and small animals and waterfowl. Pueblo communities developed extensive systems of agriculture, but were likewise active hunters. Inuit communities still rely on caribou, whaling, fishing, and gathering of plants and berries for balance in their diets. For each of these economic systems, the likelihood of all sources of food failing simultaneously in a relatively quiescent climatic period is considerably less than the likelihood of one source declining for a particular period (Trosper, 1999).

During the Native Peoples/Native Homelands workshop, Native peoples from the Arctic and sub-Arctic presented substantial evidence that their communities are immediately jeopardized by changes in global climate. In Alaska (see box "A Circle of Life: A Lesson from Alaska"), rapid warming, and the environmental consequences it brings, started about 30 years ago, and the lives of Native peoples there are already being seriously affected (Gibson and

A Circle of Life: A Lesson from Alaska

Caleb Pungoyiwi is a Yupik Eskimo who moves back and forth between Alaska and Siberia in pursuit of walrus and other sea mammals. Gathering food directly from the land and the sea makes the Yupiks very careful observers of their surroundings. In recent years, they have noticed that the walrus are thinner, their blubber less nutritious, and the oil from walrus fat does not burn as brightly in their lamps as in times of old. At the same time, they have noticed that there are fewer and weaker seals. The Yupik hunters have had to go farther and farther from shore to reach the ice pack to find the newborn seals that are being fed fish from nearby waters by their parents. Concurrently, scientists have observed that the sea ice over much of the Arctic is thinner and melting back, with the changes encompassing a broader area than that observed by the Yupiks earlier (Rothrock et al., 1999).

Both the Yupiks and the scientists have come to understand the intertwined chain of events that is occurring (Tynan and DeMaster, 1997). The retreat of the sea ice due to large-scale warming has reduced the platform that seals and walrus have used to rest between searches for fish and mussels; weakened and less productive, they provide less sustenance for both the Yupiks and the whales. The Yupiks have also observed some killer whales eating sea otters, an unusual shift in the whales' diet apparently brought on by the reduced number of seals. The loss of sea otters is important because sea otters control the number of sea urchins. With fewer sea otters, there are more urchins and therefore less kelp (which the urchins eat). And with less kelp, there is reduced habitat for fish; it is fish that would normally be the major food source for the whales as well as the Yupiks, the Inupiats, and other Northern peoples. As another part of this ecological continuum, the sea ice quiets ocean waters during winter storms, helping protect young fish; it also accumulates nutrients that, when the ice melts, create a springtime algal bloom on which the fish feed at a critical stage in their development. What has occurred may seem like only a little warming in a very cold place, but as the Native traditions make clear, everything is woven together – disruption in one place affects everything else (NPNH, 2000; Moreno, 1999; Huntington, 2000).

Schullinger, 1998). Changes in climate, coupled with other human influences, are now becoming more and more rapid in other regions, with projections of much more rapid change in the future. These changes are likely to bring much larger changes in land cover and wildlife than have occurred in the past. As the world warms, as precipitation patterns shift, as sea level changes, as mountain glaciers and sea ice melt (IPCC, 1996a), just as in Alaska, ecosystems elsewhere are likely to change in complex and often unexpected ways, affecting the resources that can be drawn from them (IPCC, 1996b).

In the Plains, warmer winter conditions are already favoring certain types of grasses, thereby changing the mix of vegetation types. The distributions, timing of migrations, and abundances of waterfowl and other birds are also changing (Sorenson et al., 1998; Price, 2001; T. Root, Univ. of Michigan, unpublished data). These changes have been and will be affected not only by temperature and precipitation changes, but also by changes in the timing of the ripening of plants and crops, and other ecosystem factors on which Native peoples depend. How much longer these types of changes can go on without a serious disruption of the services and diversity provided by the various ecosystems that support subsistence economies is not known, although there are signs that some systems are already being seriously stressed.

Adaptation Options

While local environments have changed significantly over past centuries, changes have often been slow enough to allow adjustment or for local environments to be managed. Historically, Native peoples actively worked to manage their environments in ways that led to desired and productive results. Native peoples were not passive inhabitants of their homelands, simply fitting into niches conveniently provided by a supportive environment. Tribes used a variety of consciously developed technologies and culturally based choices to improve opportunities for obtaining resources. The Paiutes, Hopi, Apaches, and Tohono O'odham, all lived in desert environments, but employed significantly different methods of land use. The Paiutes based their subsistence on a wide variety of plants, fish, and animals, and took advantage of whichever food supplies were most abundant, even if this meant making short migrations to take advantage of each season's particular opportunities. Boundaries between the various Paiute bands were relatively flexible, and permitted hunting and gathering over relatively extensive areas of the Great Basin. The Hopi, in contrast, settled in villages on Black Mesa and created permanent fields

that sustained repeated harvests of corn, squash, and beans. The Apaches, on the other hand, used fire as a technique for hunting, driving animals in a single direction for harvest. Regular burning of hunting areas (a technique also used by the Paiutes) had the additional effect of promoting plant growth and improving forage for game animals (White and Cronon, 1988). The ancestors of the modern Tohono O'odham developed a sophisticated system of irrigation to support extensive agriculture. Many of these techniques are no longer available or are not likely to be adequate, however, to sustain subsistence economies in the event of the large changes in climate that are projected.

When changes were rapid and Native peoples were faced with inadequate resources, several approaches were used to adapt. Historically, many tribal communities could rely on migration to adapt to changing resource bases in any particular local area. However, the establishment of reservations has limited the option of entire tribes moving to more hospitable locations to seek water, cropland, forests, or cooler temperatures. While individuals can pursue that option, it is not an option that is available to most tribes because they are tied to where they are by land ownership and governance issues.

When Native peoples were challenged by radical changes in their physical environments, a second approach was to incorporate new technologies. As the Lakota (western Sioux) moved from forests around the Great Lakes to the Great Plains, they found the most suitable agricultural lands – the bottom lands along the Missouri River – already occupied by residents of large and stable villages. Within two generations, however, they adopted the complex technologies of horsemanship and the gun. This rapid adaptation provided the Lakota with both military and economic advantages, and provided the material foundation for a prosperous and vital culture (Houser, 1995). This willingness to absorb new technologies, new materials, and new ways of doing things forms a common theme in the histories of many Native peoples. Cloth (and the complexities of sewing woven textiles) replaced hides. Glass beads from Bohemia replaced porcupine quills. Aniline dyes replaced vegetable colorings. Steel knives replaced stone or bone implements. Cotton thread replaced sinew. For the future, adopting new technologies is likely to be the only means for dealing with the disruptions to the traditional subsistence economies.

5. Cultural Sites, Wildlife, and Natural Resources

Ceremonial and historic sites, graves and archeological locations, special mountain and riverine environments, and seasonal cycles and migrations are central parts of the cultural traditions and traditional indigenous knowledge to Native peoples. Taken together, atmospheric conditions and the character of local landscapes – both the vegetation cover and the wildlife – help to shape people's sense of place and how they relate to what surrounds them. While Native peoples have no monopoly among Americans on love of land, water, wildlife, and the sea, their interests start from different premises and have developed over thousands of years of living on this continent. As a result, the connections of Native peoples to their homelands differ, at fundamental levels, from the kinds of relationships developed in densely populated suburban and urban environments. These differences are frequently explained in spiritual terms, although the differences also include traditional ecological and intellectual knowledge and historical familiarities. These understandings and relationships have been, and continue to be, transmitted orally and through ceremonial forms that carry the interconnections of nature and histories forward to future generations (Brody, 1982; Goodman, 1990; Hiss, 1991; Gallagher, 1993; Basso, 1996; Bordewich, 1996).

Changes in climate and in ecosystems in the decades ahead are likely to have consequences and influences that are both practical and that affect deeper life experiences. A variety of indigenous plants and animals, including migrating fish and waterfowl, provide many tribes with sustenance, as indicated in the previous section, and essential components of many cultural traditions. As climate shifts, the optimal habitats for various plants (including medicinal plants) and fish runs are likely to shift. These changes are, in turn, likely to lead to a declining presence of some plants while other plants become more abundant, altering the resource base and cultural experience for many tribal communities. At deeper levels, humans' whole experience of their environment is likely to diverge from what has been sustained through many generations via historical and religious traditions. For Native peoples, externally driven climate change is likely to disrupt the long history of intimate association with the environment.

Central to the worldviews of Native peoples is an acknowledgement of kinship with all of creation. Through honoring and paying close attention to their relatives, no matter how those relationships are defined, Native peoples have acquired and continue to draw strength from unique insights about the interactions of climate and the environmental health of their homelands. These insights can have very practical significance. For example, in the 1970s and 80s, elders on the Rosebud Sioux Reservation in South Dakota raised many questions about the potential implications of proposed efforts to exterminate prairie dogs in the sparsely settled western portions of their reservation. Although the elders expressed their objections in the language of

SONG OF THE SKY LOOM

Oh our Mother the Earth, oh our Father the Sky
Your children are we, and with tired backs
We bring you the gifts that you love.
Then weave for us a garment of brightness;
May the warp be the white light of morning,
May the weft be the red light of evening,
May the fringes be the falling rain,
May the border be the standing rainbow.
Thus weave for us a garment of brightness
That we may walk fittingly where birds sing,
That we may walk fittingly where grass is green,
Oh our Mother the Earth, oh our Father the Sky!

Found in "Songs of the Tewa"
as translated by Herbert Joseph Spinden, copyright 1933.

Lakota spirituality, which values balance in nature and emphasizes the importance of each part of creation, wildlife biologists have since come to recognize that prairie dogs are a "keystone species" that plays a pivotal role in the maintenance of ecosystems on the Great Plains.[5] Additional examples of the value of this close local knowledge are abundant: the exceptionally successful forestry management of the Menominee Nation; the care of Cree bands in harvesting animals to ensure that all generations of game and fish survive in sufficient quantities to ensure the continuity of all species.

Climate change will bring changes to the landscapes and wildlife that are important to Native peoples, changing the surroundings in ways that will change

Shifting Ecosystem Boundaries and the Sense of Place

Anyone who has ever walked up a mountain has experienced the fact that the climate at a particular altitude determines the vegetation. Because of this, mountaintops have their own particular ecosystems that are isolated like islands from surrounding areas. Although not unique to this region, the very large reservations of southwestern New Mexico and southeastern Arizona are home to many "sky islands" that rise high above the desert floor, some of them reaching to heights exceeding 10,000 feet. These unique regions, which are natural parks in a sea of desert scrub, add greatly to the quality of life, and their biodiversity and genetic richness is among the greatest to be found anywhere on the continent (Arias Rojo et al., 1999). These environments are particularly meaningful, even sacred, to many of the tribes, being environments with particular types of plants and wildlife that are important culturally and medicinally. At the same time, they are particularly vulnerable to climate change because of their unique settings in a largely arid region. As climate warms, ecological communities from the bottom to top will be disrupted as plant species gradually move up-slope, each at their own rate. The existing mountain ecosystems will be forced into ever- diminishing areas of land and those species already at the mountaintops will become extinct because they cannot migrate easily to distant, higher mountains.

Ecosystem shifts will occur not only in the vertical, but also in the horizontal dimension. For centuries, the Anishinaabeg (Ojibway or Chippewa) who live around Lake Superior and along the upper Mississippi River have depended upon the natural resources of the forests, lakes, and rivers of the region (Vennum, 1988). Many of the reservation locations were selected to ensure access to culturally significant resources, such as maple sugar bushes and wild rice beds, whose locations were thought to be fixed. As drier summer conditions cause the western prairies to shift eastward toward the western Great Lakes, the extents of maple, birch, and wild rice habitats in the US are likely to be significantly reduced. Because Ojibway communities cannot, as a whole, move as the ecosystems that are their homes shift, climate change is likely to reduce the resources needed to sustain their traditional culture and impact their economic productivity and the value of established treaty rights unless adjustments are made (e.g., see Keller, 1989).

The wild rice that grows abundantly in shallow lake and marshy habitats of northern Wisconsin and Minnesota is likely to be adversely affected. Wild rice plays a critical role in the economic and ceremonial life of many tribes. The hand-harvested and processed seed is highly prized as a gourmet food and adds significant commercial value to the rural reservation economy. Federal treaties guarantee the right of the Anishinaabeg to gather wild rice in their aboriginal territories, which cover much of the states of Wisconsin and Minnesota. As the climate changes, deep or flooding waters in early spring could delay germination of the seed on lake or river bottoms, leading to crop failure. Lower water levels later in the summer could cause the wild rice stalks to break under the weight of the fruithead or make the rice beds inaccessible to harvesters. Extended drought conditions could encourage greater natural competition from more shallow water species. Climate change will disrupt the current balances, and, as was illustrated during the dry summer of 1988, conflicts over water can pit federal river management policies against tribal treaty rights and state demands for water.

Evidence of significant patterns of change over the past 10,000 years confirms that substantial ecosystem changes can occur as a result of changes in climate. Presuming future changes occur to the same extent as past changes, tribes that trace their ancestry to the wooded regions will slowly become overtaken by grasslands, such that the entire nature of place for many Native peoples is likely to change.

---

[5] Personal communications from Lionel Bordeaux and Ronald Trosper.

human experiences. Mountain environments, edges of ecosystems, and bird populations will be especially vulnerable. For example, the "Sky Islands" in the mountainous west and the prairie-forest interface between the Great Plains and Great Lakes will be places where significant changes seem likely (see box, "Shifting Ecosystem Boundaries and the Sense of Place"). Wildlife, which is central to the cultural life of many tribes, is likely to be significantly affected over coming decades. For example, under the equilibrium climate conditions of the Canadian climate scenario, models of bird distributions (Price, 2000) project a gross loss of up to 27% of the neotropical migratory birds, 32% of the short-distance migratory birds, and 40% of the resident bird species in Arizona (J. Price, draft report prepared for the Environmental Protection Agency). Because some extirpations (local extinctions) are likely to be offset by immigrations, this study suggests that the net changes (6% loss of neotropical migrants, 15% loss of short-distance migrants and 30% gain in resident species) are not likely to be as severe as the gross. Whether colonizing species can "replace" extirpated species in an ecological sense is unknown at this time, as are the overall rates of change. From the point-of-view of Native peoples, what may be as important is the degree to which any of these changes will impact their cultures or religions (see box, "Wildlife and Ceremonies").

Adaptation Options
In some cases, improved or altered land management practices (e.g., fire management) may be able to sustain the presence of particular types of useful plants or animals for at least a while longer. Where ecosystems shift from Native land holdings to nearby non-Native lands, new areas may need to be developed or acquired to allow access to traditional food sources. Increased involvement of Native experts in resource management, particularly of public lands, may improve the quality of the new environments as well as help to sustain traditional plants. Where climatic and ecosystem shifts are significant, new approaches will be needed. In all of these situations, adapting to changing wildlife and land cover on tribal lands will be challenging because options for continued access by Native peoples to traditional ecosystem resources on neighboring lands may be limited.

In planning and working to meet the changing conditions, experience indicates that both traditional knowledge and contemporary scientific knowledge can help to understand and improve the environment. The relationships between Native peoples and their environments provide significant insights and context for the scientific findings about climate change and its implications for human life. Building the bridge will be essential, for many tribal communities use such traditional understandings, developed over many generations, to guide their uses of lands within their immediate political jurisdictions. Because land use decisions on reservations will have influences on, and be influenced by, the health and services provided by wider regional ecosystems, it will be essential for Indian and non-Indian people to

Wildlife and Ceremonies

Many Native peoples use wildlife as integral parts of their cultural and religious ceremonies. Among the various Pueblo peoples (e.g., Hopi, Zuni, Keres, Tewas, Tiwas, and Towas), religious ceremonies are the center of their cultural lives (Tyler 1991). Birds are seen as spiritual messengers and are completely integrated into the traditions of these Native American communities. More than 200 species of birds have unique Native names, and more than 100 are essential to parts of the Pueblo culture. Birds mark the passing of the seasons and are considered to have valuable spiritual properties needed by members of these Pueblos. Among the Zuni, prayer sticks are used as offerings to the spirit realm. Each prayer stick, depending on its purpose, requires a particular combination of feathers drawn from among 72 different species of birds (Tyler 1991). Prayer sticks serve many of the same spiritual purposes in the Zuni religion that rosary beads serve to the Catholic religion.

Zuni also have separate names for the Western and Mountain Bluebirds. These species are only found on the Pueblo in winter and are used as symbols of transitions for fall and spring. Among both the Hopi and the Zuni, bluebirds are associated with puberty rituals surrounding the passage from girlhood into womanhood (Tyler 1991).

Because of such linkages, shifts in climate and consequent shifts in the timing or the distributions of wildlife species are likely to have profound impacts on the cultural and religious lives of these peoples.

work together towards understanding each others' perspectives and choices about climate change, its implications, and how best to adapt.

COPING AND ADAPTATION STRATEGIES

During the assessment process, speakers from tribal communities consistently attributed the endurance of Native peoples, through extreme conditions and brutal transitions, to spiritual and cultural values. Although public and ceremonial expressions of these values differ considerably from tribe to tribe, Native people identify many commonalties across wide geographical, linguistic, and environmental distances. These values form a connector between the past and future, bringing important lessons learned through tens of thousands of years on the continent into the frameworks for choices that will need to be made about the futures of Native peoples and homelands. From the analyses to date, it is clear that responding to substantial changes in climate, while populations remain in fixed and permanent locations, is very likely to require new technologies, skilled personnel, and financial resources. These three necessities, however, are desperately scarce in many tribal communities. Most tribal communities are limited in ways of creating wealth and they continue to rely heavily on transfer payments from the federal government. Adjusting plans for economic and social development to account for climate change may require fresh thinking in federal and tribal policies and budgets. For example, because lands are usually held in trust for Native peoples, it can be very difficult to obtain a mortgage or other loan since the assets are not held personally. Several options, however, are emerging from consideration of adaptation and coping options.

Enhance Education and Access to Information and Technology

Becoming educated on issues concerning climate change will be critical for Native peoples throughout the country. Both those who live in tribal communities and those who make their homes elsewhere need to develop the understanding and skills to deal with a changing climate[6]. This education needs to be both comprehensive and widespread (see Johnson, 1999). It is especially impor-

tant to improve the quality of education in the sciences and technology in the K-12 schools and tribal colleges that serve Native youth. It will be essential to enlist elders and mentors within each Native community (including within each culturally distinct region) to assist in the integration of contemporary information and traditional values, and of Indian students who choose to pursue university degrees and careers in science with their tribal communities.

Promote Local Land-use and Natural Resource Planning

In 1976, with the passage of the Indian Self-Determination and Education Act, the federal government began to encourage tribal governments to take responsibility for developing and implementing plans for use of tribal lands and natural resources in collaboration with local agencies. Since that action, several tribes have received international recognition for success in managing their local resources. The Menominee Nation of Wisconsin regularly trains managers from all parts of the world because of the effectiveness of the tribe's sustainable timber and forestry practices. The Spokane Tribe has built an exemplary water resources program. Cost-effective ways – using existing networks and organizations – need to be developed to inform decision-makers in tribal communities, and to provide shared access to adequate technical resources. Technologies now exist that can assist tribes to make thoughtful and informed choices. Ways need to be found to provide information that will support the abilities of Native peoples and their leaders to make prudent choices based on appropriate knowledge and appropriate values, using appropriate processes aimed at promoting and enhancing diversified and sustainable economies in tribal communities.

Participate in Regional and National Discussions and Decision-making

One result of the trusteeship system has been the tight concentration of the attentions of tribal governments on their relationships with the Bureau of Indian Affairs. Gradually, since the late 1960s, federal agencies have begun to recognize that the trust responsibility to Native peoples extends through all parts of the federal government. Tribes, which once viewed the creation of relationships with federal agencies other than the BIA risky at best and irrelevant at worst, are working increasingly successfully across agency and departmental lines. Federal agencies are also learning how to provide appropriate kinds and levels of service to Native peoples.

---

[6] A variety of professional organizations provide assistance to tribes working on natural resource issues: the Native American Fish and Wildlife Society; Inter-tribal Timber Council; Indian Agriculture Council; National Tribal Environmental Council; and the American Indian Higher Education Consortium, among others.

Although the trust relationships between the federal government and Native peoples are complex, they are not impenetrable. New relationships are essential to address issues of climate change. While creating relationships with agencies of state governments has frequently been viewed by tribes as threatening their sovereignty, serious discussions about climate change – at the regional, state, and national levels – need to include informed stakeholders from every relevant jurisdiction. A pertinent model of interaction and collaboration that provides technical support, advice, and assistance to tribal environmental officers has been developed between tribes in the northern Great Plains and the University of North Dakota. Similarly, the Southwest Strategy, an initiative of all stakeholders in Arizona and New Mexico, serves as a useful framework for strengthening communication and collaboration with tribes and federal agencies. Their success in broadening participation and making knowledge available in useful ways could provide helpful lessons for other states, tribes, and regions.

CRUCIAL UNKNOWNS AND RESEARCH NEEDS

Because there are many hundreds of tribes, there are many hundreds of situations facing Native peoples as

The Johnstone Strait Diversion at Sockeye Salmon

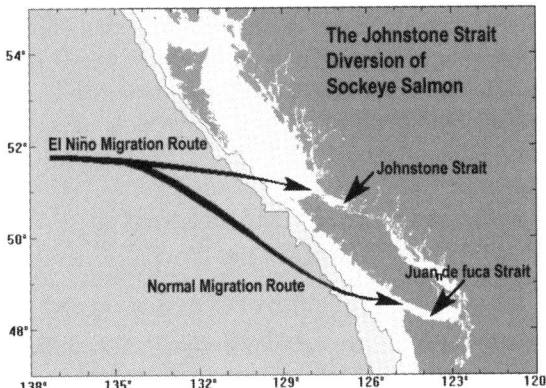

Figure 3: Changes in Pacific Ocean temperatures associated with the Pacific Decadal Oscillation appear to be the cause of changes in the migration path of salmon returning to spawn (Mysak, 1986). Under normal conditions, the salmon return to spawn by passing through the Juan de Fuca Strait and traditional tribal harvesting areas. Under El Niño conditions, however, the salmon take a path to the north of Vancouver Island, passing through Johnstone Strait and not passing through traditional harvesting areas (Groot and Quinn, 1987; Hsieh and Lee, 1989, McKinnel et al., 1999). As the climate warms, the likelihood of the salmon taking one path versus the other is likely to change, and with further warming, the salmon may no longer return to their traditional streams.

they seek to understand and prepare for climate change. Having better estimates of the patterns, magnitudes, and rates of climatic changes to be expected is essential. Accurate weather data from tribal communities – particularly those in remote rural areas – have rarely been compiled. Little systematic effort has been made by federal, tribal, or state agencies to gather information about the microclimatic conditions associated with the various reservations. One significant research priority, therefore, requires the training of members of tribal communities to collect and interpret weather information, including providing for the acquisition, installation, and maintenance of appropriate instrumentation to support the collection and recording of these data.

A second urgent research need requires inventories of the uses and conditions of land and natural resources on each reservation. Such an inventory can integrate information from remote sensing and geographic information systems as well as tribal information on water quantity and quality and first-hand personal observations and culturally based knowledge. While Native oral histories clearly have many insights to offer, data acquisition and sharing of information have become especially complicated issues for Native peoples. Generations of scholarly objectification of Native peoples, and appropriation of knowledge, objects, and human remains have created severe problems of mistrust of scholarly inquiry, and strong resistance to the sharing of sacred or privileged information. The legitimacy of Native concerns over these issues cannot be dismissed, nor can the urgent needs for research to be conducted from within Native communities. As these essential and primary needs are satisfied, Native peoples have consistently demonstrated willingness to take part in broader conversations, as teachers and students, colleagues and leaders, with individual and culturally based perceptions that can enrich discussions and strengthen decisions for all parties involved. Equipping tribes with the capability to scientifically sample and analyze plant and animal populations will enhance the opportunities for broadening tribal participation in such conversations.

Such inventories are needed to provide the basis for establishing baselines of environmental conditions and economic and cultural activities on each reservation. This needs to be followed by assessments of the opportunities and vulnerabilities that changes in climate might bring. These sector analyses will require regional projections of future changes for a variety of climate variables, so that the wide range of potential impacts (and opportunities) can be evaluated by the various tribes. Especially important are predictions of changes in water availability, because water is the fundamental resource for agriculture, tourism, and other vital activities.

Native communities also need to understand how ecosystems are likely to respond to climate changes, both large-scale and small. What will happen to migrations of birds, waterfowl, and fish? What will happen to forests and grasslands? What will happen to the rich mix of flora and fauna on which traditional cultures and subsistence economies depend? How will ecosystems function differently in the future? Will they provide greater or fewer resources, and in what mixtures? That large changes are possible is evident from the changes occurring in response to climate variations. For example, variations in Pacific Ocean temperatures associated with the Pacific Decadal Oscillation have been observed to cause changes in the migration path of salmon returning to spawn, thereby affecting whether fish pass through traditional harvesting areas (see Figure 3). Gaining a better understanding of such variations, and then of the changes that will be brought on by climate change, will be essential.

The rates at which Native communities will be able to respond to major changes vary widely from group to group. Many Native peoples continue to view human actions and community decisions in terms of generations. This perspective sometimes means that reaching conclusions may require more time and discussion than is customary in market-driven societies. Other groups clearly thrive on change and move aggressively to take advantage of new opportunities for education, economic development, and technological innovation. Some of the challenges presented by changes in climate may require relatively rapid social adaptation and swift action. Other challenges may require the capacity to endure discomfort and examine new situations from a variety of perspectives until appropriate responses can be formulated.

Education and the exchanges of information across cultural boundaries, activities that enhance the abilities of Native and non-Native peoples to cope with systemic changes, have been occurring at the individual level for generations. The Native Peoples/Native Homelands assessment process has started to advance these interactions.

Whatever changes befall Earth's climate, there is now fresh ground for hope that Native peoples and non-Natives may be able to address the challenges collaboratively, as relatives. *Mitakuye Owasin* (we are all related)!

END NOTES

1. The number of Native Americans depends on the definition that is used. As a result, the number of those counted as Native Americans can differ. For example, the US Bureau of the Census counts as American Indian anyone who identifies himself or herself as such. As in asking about other ancestral connections, census enumerators require no proof of Indian identity. Thus, census data include individuals who may identify themselves culturally and socially as American Indian, but who are not formally enrolled as members of a particular tribe. As a result, the census produces a comparatively high count of the number of American Indian people in the United States (US Bureau of the Census, 1990). The Bureau of Indian Affairs, on the other hand, counts only individuals who are officially enrolled as members of federally recognized tribes. Each tribe has the right to establish its own criteria for enrollment. Most tribes require that a certain percentage of the individual's ancestors must have been members of that tribe. Some tribes recognize only affiliation through the father's family. Still other tribes have residency requirements indicating that the individual must live on the tribe's reservation for a specified number of years. The BIA's total (see http://www.doi.gov/bia/aitoday/q_and_a.html) thus yields a lower number of American Indians. As a further complication, some tribes are recognized by state governments, but not by the Bureau of Indian Affairs (e.g., the Lumbee of North Carolina). Members of these tribes are, therefore, recognized as Indian by some levels and agencies of government, but not by others. Periodically, a tribe may succeed in completing the BIA's rigorous process for obtaining federal recognition, thus increasing the number of Indian people recognized as such by the Department of the Interior. Further, descendants of the original inhabitants of the Hawaiian Islands have, using the Department of the Interior's own criteria, made credible claim for federal recognition as Native Americans (Bordewich, 1996).

2. The presence of non-Native Americans on reservation lands was largely prevented until the Dawes Severalty Act, commonly called the Allotment Act, was passed by Congress in 1887. Prior to this law, reservation lands were held corporately by an entire tribe and no particular individual held title to any particular tract of land. Furthermore, no outsiders, except government officials and soldiers, were permitted to live within reservation borders. The Allotment Act, however, mandated that each member of a tribe receive an individual allotment of land. The allotments varied in

size from 80 to 1,040 acres, depending on the particular reservation. After each head of household and family member had received an allotment, the remaining unassigned lands within the boundaries of each reservation could be opened to non-Indian homesteaders.[7] These settlers were granted clear title to the lands on which they settled if they fulfilled the normal conditions of homesteading. Land that was conveyed in this way to homesteaders was simply subtracted from the total lands that had been originally reserved for the tribe. As a result, until the passage of the Indian Reorganization Act in 1934, significant amounts of reservation lands passed out of Indian ownership even though they were within the original boundaries of reservation.

LITERATURE CITED

Akerlof, G. A., *An Economic Theorist's Book of Tales,* University of Cambridge Press, Cambridge United Kingdom, 1984.

Arias Rojo, H., J. Bredehoeft, R. Lacewell, J. Price, J. Stromberg, and G. Thomas, *Sustaining and Enhancing Riparian Migratory Bird Habitat on the Upper San Pedro River,* Expert Report prepared for the Commission for Environmental Cooperation, Montreal, Canada, 1999.

Axelrod, R., *The Evolution of Cooperation,* Basic Books, Inc., New York, 1984.

Axelrod, R., *The Complexity of Cooperation,* Princeton University Press, Princeton, New Jersey, 1997.

Basso, K., *Wisdom Sits in Places: Landscape and Language Among the Western Apache,* University of New Mexico Press, Albuquerque, New Mexico, 1996.

BIA (Bureau of Indian Affairs) Website, http://www.doi.gov/bia/aitoday/q_and_a.html), 2000a.

BIA (Bureau of Indian Affairs) Website, http://www.gdsc.bia.gov/images/bia_usa_chip.jpg , 2000b.

Bordewich, F. M., *Killing the White Man's Indian: Reinventing Native Americans at the End of the Twentieth Century,* Anchor Books, New York, 1996.

Brams, S. J., and A. D. Taylor, *Fair Division,* Cambridge University Press, Cambridge, United Kingdom, 1996.

Brody, H., *Maps and Dreams,* Pantheon Books, New York, 1982.

Brown, D., *Bury My Heart at Wounded Knee: An Indian History of the American West,* Henry Holt, New York, 487 pp., 1991.

Cornell, S., *The Return of the Native: American Indian Political Resurgence,* Oxford University Press, Oxford United Kingdom, 1998.

Deloria, V., Jr., *Red Earth, White Lies: Native Americans and the Myth of Scientific Fact,* Fulcrum Publishing, Golden Colorado, 271 pp., 1997.

DOI (Department of the Interior), Lands under the jurisdiction of the Bureau of Indian Affairs, (http://www.doi.gov/bia/realty/area.html.), 1996.

EIA (Energy Information Administration), Energy consumption and renewable energy development potential on Indian lands, Department of Energy, Washington DC, March 2000.

Engelthaler, D. M., et al., Climatic and environmental patterns associated with hantavirus pulmonary syndrome, Four Corners region, United States, *Emerging Infectious Diseases, 5,* 87-94, 1999.

Gallagher, W., *The Power of Place,* Harper Perennial, New York, 1993.

Getches, D. H, C. F. Wilkinson, and R. H. Williams, Jr., *Cases and Materials on Federal Indian Law,* Fourth Edition, West Group, St. Paul, Minnesota, 1998.

Gibson, M.A., and S. B. Schullinger, Answers from the Ice Edge: The consequences of climate change on life in the Bering and Chukchi seas, Arctic Network and Greenpeace USA pamphlet, (see http://www.globalchange.org/monitall/98apr6.htm), 32 pp., June, 1998.

Goodman, R., *Lakota Star Knowledge,* Sinte Gleska University Press, Rosebud, South Dakota, 1990.

Groot, C., and T. P. Quinn, Homing migration of sockeye salmon to the Fraser River, *Fisheries Bulletin, 85,* 455-469, 1987.

Hardin, R., *Collective Action,* Johns Hopkins University Press, Baltimore, Maryland, 1982.

Hiss, T., *The Experience of Place,* Vintage Books, New York, 1991.

---

[7] Many tribes were able to avoid the allotment of tribal lands. The reservations of the Menominee Nation, several Pueblo tribes and a large portion of the Navajo Nation remain undivided and intact within their original borders.

Houser, S., Mending the circle: Peer group lending for microenterprise development in tribal communities, *Rural Development Strategies*, edited by David W. Sears and J. Norman Reid, pp. 204-232, Nelson Hall Publishers, Chicago, Illinois, 1995.

Houser, S., and A. White Hat, Bringing accountability home: Changes in the management of public responsibility within tribal governments, commissioned by the Canadian Federal Department of Indian and Northern Development, Ottawa, 1993.

Hsieh, W. W., and W. G. Lee, A numerical model for the variability of the northeast Pacific Ocean, Canadian Special Publication of *Fisheries and Aquatic Sciences, 108,* 247-254, 1989.

Huntington, H. P., Impacts of changes in sea ice and other environmental parameters in the Arctic, in the *Proceedings of the Workshop on Impacts of Changes in Sea Ice and Other Environmental Parameters in the Arctic,* Girdwood, Alaska, February 15-17, 2000, Marine Mammal Commission, Washingtonn, DC, in press, 2000.

IPCC (Intergovernmental Panel on Climate Change), *Climate Change 1995: The Science of Climate Change, Contribution of Working Group I to the Second Assessment Report of the Intergovernmental Panel on Climate Change,* edited by J. T., Houghton, L. G. Meira Filho, B. A. Callander, N. Harris, A. Kattenberg, and K. Maskell, Cambridge University Press, Cambridge, United Kingdom, 572 pp., 1996a.

IPCC (Intergovernmental Panel on Climate Change), *Climate Change 1995: Impacts, Adaptations, and Mitigation of Climate Change: Scientific-Technical Analyses,. Contribution of Working Group II to the Second Assessment Report of the Intergovernmental Panel on Climate Change,* edited by R. T. Watson, M. C. Zinyowera, and R. H. Moss, Cambridge University Press, Cambridge, United Kingdom, 880 pp., 1996b.

Johnson, T., World out of balance: In a prescient time Native prophecy meets scientific prediction, *Native AMERICAS: Hemispheric Journal of Indigenous Issues,* Volume XVI, Fall/Winter, 8-25, 1999.

Keller, R. H., America's Native Sweet: Chippewa treaties and the right to harvest maple sugar, *American Indian Quarterly,* Spring, 117-135, 1989.

LaDuke, W., Traditional ecological knowledge and environmental futures, *Colorado Journal of International Environmental Law and Policy, 5,* 127, 1994.

Lawson, M. L., *Dammed Indians,* University of Oklahoma Press, Norman, Oklahoma, 1982.

McKinnell, S., H. J. Freeland, and S. D. Grouix, Assessing the northern diversion of sockeye salmon returning to the Fraser River, *Fisheries Oceanography, 8,* 104-114, 1999.

Moreno, F., In the Arctic, ice is life, *Native AMERICAS: Hemispheric Journal of Indigenous Issues, XVI (Fall/Winter),* 42-45, 1999.

Mysak, L. A., El Niño, interannual variability, and fisheries in the northeast Pacific Ocean, *Canadian Journal of Fisheries and Aquatic Sciences, 43,* 464-497, 1986.

NPNH (Native Peoples/Native Homelands Assessment Team), *Report of the Native Peoples/Native Homelands Climate Change Workshop, 28 October – 1 November 1998,* Albuquerque, New Mexico. National Aeronautics and Space Administration, Washington DC, in press, 2000.

Olson, M., *The Logic of Collective Action,* Harvard University Press, Cambridge, Massachusetts, 1965.

Pevar, S. L., *The Rights of Indians and Tribes,* second edition, Southern Illinois University Press, Carbondale, Illinois, 1992.

Price, J., Modeling the potential impacts of climate change on the summer distribution of Michigan's nongame birds, *Michigan Birds and Natural History 7,* 3-13, 2000.

Price, J., Climate change, birds, and ecosystems – Why should we care?, in *Managing for Healthy Ecosystems, Volume II: Issues and Methods, Section 3, Climate Change and Ecosystem Health,* edited by D. Rapport, C. Qualset, W. Lasley, and D. Ralston, in press, 2001.

Riebsame, W., J. Robb, and D. Thoebald (editors), *Atlas of the New West: Portrait of a Changing Region,* W. W. Norton & Co., New York, 192 pp., 1997.

Ritzenthaler, R. E., and P. Ritzenthaler, *The Woodland Indians of the Western Great Lakes,* Waveland Press, Prospect Heights, Illinois, 1991.

Rothrock, D. A., Y. Yu, and G. A. Maykut, Thinning of the Arctic sea-ice cover, *Geophysical Research Letters, 26*(23), 3469, 1999.

Schelling, T. C., *Micromotives and Macrobehavior,* W. W. Norton & Co., New York, 1978.

Schelling, T. C., *Choice and Consequence,* Harvard University Press, Cambridge, Massachusetts, 1984.

Schmaljohn, C., and B. Hjelle, Hantaviruses: A global disease problem, *Emerging Infectious Diseases, 3,* 95-104, 1997.

Snipp, C. M., *American Indians: The First of This Land,* The Russell Sage Foundation, New York, 1991.

Sorenson, L., R. Goldberg, T. L. Root, and M. G. Anderson, Potential effects of global warming on waterfowl populations breeding in the northern Great Plains, *Climatic Change, 40,* 343-369, 1998.

Spinden, H. J., translator, *Songs of the Tewa,* originally published in 1933 by The Exposition of Indian Tribal Arts, Inc., republished by Sunstone Press, Santa Fe, New Mexico, 1993.

Tiller, V. E. V., *American Indian Reservations and Trust Areas,* Economic Development Administration, United States Department of Commerce, Washington, DC, 1996.

Trosper, R. L., American Indian poverty on reservations, in *Changing Numbers, Changing Needs: American Indian Demography and Public Health,* edited by G. Sandefur, R. R. Rindfuss, and B. Cohen, pp. 172-195, National Research Council, Washington, DC, 1996.

Trosper, R. L., Traditional American Indian policy, in *Contemporary Native American Political Issues,* edited by T. R. Johnson, pp. 139-63, Altamira Press, Walnut Creek, California, 1999.

Tyler, H. A., *Pueblo Birds and Myths,* Northland Publishing, Flagstaff, Arizona, 1991.

Tynan, C. T., and D. P. Demaster, Observations and predictions of Arctic climate change: Potential effects on marine mammals, *Arctic,* 50, 308-322, 1997.

US Bureau of the Census, 1990 Census of population, characteristics of American Indians by tribe and language, 1990 CP-3-7, Washington, DC, 1990.

Vennum, T., *Wild Rice and the Ojibway People,* Minnesota Historical Society Press, St. Paul, Minnesota, 1988.

White, R., and W. Cronon, Ecological change and Indian-White relations, in *Handbook of North American Indians, Volume IV,* edited by W. Sturtevant, pp. 417-429, Smithsonian Institution, Washington, DC, 1988.

ADDITIONAL BACKGROUND REFERENCES

Ambler, M., *Breaking the Iron Bonds: Indian Control of Energy Development,* University Press of Kansas, Lawrence, Kansas, 1991.

Berkhofer, R. F., Jr., *The White Man's Indian,* A.A. Knopf, New York, 1978.

Birkes, F., *Sacred Ecology, Traditional Ecological Knowledge and Resource Management,* Taylor & Francis, London, England, 1999.

Braroe, N. W., *Indian & White,* Stanford University Press, Stanford, 1978.

Burger, T. R. *Village Journey: The Report of the Alaska Native Review Commission,* Hill and Wang, New York, 1985.

Cronon, W., *Changes in the Land: Indians, Colonists, and the Ecology of New England,* Hill & Wang, New York, 1983.

Davis, T., *Sustaining the Forest, the People, and the Spirit,* State University of New York Press, Albany, New York, 1999.

Deloria, V., Jr., and C. Lytle, *The Nations Within: The Past and Future of American Indian Sovereignty,* Pantheon Books, New York, 1984.

Diamond, J., *Guns, Germs, and Steel: The Fates of Human Societies,* W. W. Norton & Company, New York, 1999.

Dippie, B. W., *The Vanishing American: White Attitudes and US Indian Policy,* Wesleyan University Press, Middleton Connecticut, 1982.

Gobert, G. G., A. C. Abello, and J. Johnson, The individual economic well-being of Native American men and women during the 1980s: A decade of moving backward, in *Changing Numbers, Changing Needs: American Indian Demography and Public Health,* edited by G. Sandefur, R. R. Rindfuss, and B. Cohen, National Research Council, Washington, DC, pp. 133-171, 1996.

Josephy, A. M. Jr., *Now That the Buffalo's Gone,* Alfred A. Knopf, New York, 1982.

Krech, S., III, (Ed.), *Indians, Animals, and the Fur Trade: A Critique of Keepers of the Game,* University of Georgia Press, Athens, Georgia, 1981.

Martin, C., *Keepers of the Game: Indian Animal Relationships in the Fur Trade,* University of California Press, Berkeley, California, 1978.

Ortiz, A., *The Tewa World: Space, Time, and Becoming in a Pueblo Society,* University of Chicago Press, Chicago, Illinois, 1969.

Oswalt, W. H., *This Land Was Theirs,* John Wiley & Sons, New York, 1978.

Prucha, F. P., *Americanizing the American Indians,* Harvard University Press, Cambridge, Massachusetts, 1973.

Prucha, F. P., *The Indians in American Society: From the Revolutionary War to the Present,* University of California Press, Berkeley, California, 1985.

Scott, C., Property, practice and aboriginal rights among Quebec Cree hunters, *Hunters and Gatherers 2: Property, Power, and Ideology,* edited by T. Ingold, D. Riches, and J. Woodburn, pp. 35-51, Berg Publishing, New York, 1988.

Vogel, V., *This Country Was Ours: A Documentary History of the American Indian,* Harper & Row, New York, 1972.

Weekes, P., *The American Indian Experience,* Forum Press, Arlington Heights, Illinois, 1988.

ACKNOWLEDGMENTS

Many of the materials for this chapter are based on contributions from participants on and those working with the

Native Peoples/Native Homelands—National Workshop Team
Verna Teller, Isleta Pueblo
Robert Gough, Intertribal Council on Utility Policy
Schuyler Houser, Lac Courte Oreilles Ojibwa Community College
Nancy Maynard, NASA
Fidel Moreno, Yaqui/Huichol
Lynn Mortensen, US Global Change Research Program
Patrick Spears, Lakota
Valerie Taliman, Navajo
Janice Whitney, HETF Fiduciary

Native Peoples/Native Homelands—Southwest Assessment Team
Stan Morain*, University of New Mexico
Rick Watson*, San Juan College and University of New Mexico
Diane Austin, University of Arizona
Mark Bauer, Diné College
Karl Benedict, University of New Mexico
Jennifer Bondick, University of New Mexico
Amy Budge, University of New Mexico
Linda Colon, University of New Mexico
Laura Gleasner, University of New Mexico
Jhon Goes In Center, Oglala Lakota Nation
Todd Hinckley, US Geological Survey
Doug Isely, Diné College
Bryan Marozas, DOI Bureau of Indian Affairs
Lynn Mortensen, US Global Change Research Program
Verna Teller, Isleta Pueblo
Carmelita Topaha, Navajo
Ray Williamson, George Washington University

Additional Contributors
Patricia Anderson, University of Alaska
Lynne Carter, National Assessment Coordination Office
Susan Marcus, US Geological Survey
Jeff Price, American Bird Conservancy
James Rattling Leaf, Sinte Gleska University
George Seielstad, University of North Dakota
Eileen Shea, East-West Center, Hawaii
Tony Socci, US Global Change Research Program
Leigh Welling, University of North Dakota

* Assessment Team chair/co-chair

CHAPTER 13

CLIMATE CHANGE AND AGRICULTURE IN THE UNITED STATES

John Reilly[1], Francesco Tubiello[2], Bruce McCarl[3] and Jerry Melillo[4,5]

Contents of this Chapter

[1]Massachussetts Institute of Technology, [2]Goddard Institute for Space Studies and Columbia University, [3]Texas A&M University, [4]The Ecosystems Center, Marine Biological Laboratory, [5]Coordinating author for the National Assessment Synthesis Team

CHAPTER SUMMARY

It is likely that climate changes and atmospheric CO_2 levels, as defined by the scenarios examined in this Assessment, will not imperil crop production in the US during the 21st century. The Assessment found that, at the national level, productivity of many major crops increased. Crops showing generally positive results include cotton, corn for grain and silage, soybeans, sorghum, barley, sugar beets, and citrus fruits. Pastures also showed increased productivity. For other crops including wheat, rice, oats, hay, sugar cane, potatoes, and tomatoes, yields are projected to increase under some conditions and decline under others.

Not all agricultural regions of the United States were affected to the same degree or in the same direction by the climates simulated in the scenarios. In general the findings were that climate change favored northern areas. The Midwest (especially the northern half), West, and Pacific Northwest exhibited large gains in yields for most crops in the 2030 and 2090 timeframes for both of the two major climate scenarios used in this Assessment, Hadley and Canadian. Crop production changes in other regions varied, some positive and some negative, depending on the climate scenario and time period. Yields reductions were quite large for some sites, particularly in the South and Plains States, for climate scenarios with declines in precipitation and substantial warming in these regions.

Crop models such as those used in this Assessment have been used at local, regional, and global scales to systematically assess impacts on yields and adaptation strategies in agricultural systems, as climate and/or other factors change. The simulation results depend on the general assumptions that soil nutrients are not limiting, and that pests, insects, diseases, and weeds, pose no threat to crop growth and yield. One important consequence of these assumptions is that positive crop responses to elevated CO_2, which account for one-third to one-half of the yield increases simulated in the Assessment studies, should be regarded as upper limits to actual responses in the field. One additional limitation that applies to this study is the models' inability to predict the negative effects of excess water conditions on crop yields. Given the "wet" nature of the scenarios employed, the positive responses projected in this study for rainfed crops, under both the Hadley

and Canadian scenarios, may be overestimated.

Under climate change simulated in the two climate scenarios, consumers benefited from lower prices while producers' profits declined. For the Canadian scenario, these opposite effects were nearly balanced, resulting in a small net effect on the national economy. The estimated \$4-5 billion (in year 2000 dollars unless indicated) reduction in producers' profits represents a 13-17% loss of income, while the savings of \$3-6 billion to consumers represent less than a 1% reduction in the consumers food and fiber expenditures. Under the Hadley scenario, producers' profits are reduced by up to \$3 billion (10%) while consumers save \$9-12 billion (in the range of 1%). The major difference between the model outputs is that under the Hadley scenario, productivity increases were substantially greater than under the Canadian, resulting in lower food prices to the consumers' benefit and the producers' detriment.

At the national level, the models used in this Assessment found that irrigated agriculture's need for water declined approximately 5-10% for 2030 and 30-40% for 2090 in the context of the two primary climate scenarios, without adaptation due to increased precipitation and shortened crop-growing periods.

A case study of agriculture in the drainage basin of the Chesapeake Bay was undertaken to analyze the effects of climate change on surface-water quality. In simulations for this Assessment, under the two climate scenarios for 2030, loading of excess nitrogen into the Bay due to corn production increased by 17-31% compared with the current situation.

Pests are currently a major problem in US agriculture. The Assessment investigated the relationship between pesticide use and climate for crops that require relatively large amounts of pesticides. Pesticide use is projected to increase for most crops studied and in most states under the climate scenarios considered. Increased need for pesticide application varied by crop – increases for corn were generally in the range of 10-20%; for potatoes, 5-15%; and for soybeans and cotton, 2-5%. The results for wheat varied widely by state and climate scenario showing changes ranging from approximately

–15 to +15%. The increase in pesticide use results in slightly poorer overall economic performance, but this effect is quite small because pesticide expenditures are in many cases a relatively small share of production costs.

The Assessment did not consider increased crop losses due to pests, implicitly assuming that all additional losses were eliminated through increased pest control measures. This could possibly result in underestimates of losses due to pests associated with climate change. In addition, this Assessment did not consider the environmental consequences of increased pesticide use.

Ultimately, the consequences of climate change for US agriculture hinge on changes in climate variability and extreme events. Changes in the frequency and intensity of droughts, flooding, and storm damage are likely to have significant consequences. Such events cause erosion, waterlogging, and leaching of animal wastes, pesticides, fertilizers, and other chemicals into surface and groundwater.

One major source of weather variability is the El Niño/Southern Oscillation (ENSO). ENSO effects vary widely across the country. Better prediction of these events would allow farmers to plan ahead, altering their choices of which crops to plant and when to plant them. The value of improved forecasts of ENSO events has been estimated at approximately $500 million per year. As climate warms, ENSO is likely to be affected. Some models project that El Niño events and their impacts on US weather are likely to be more intense. There is also a chance that La Niña events and their impacts will be stronger. The potential impacts of a change in frequency and strength of ENSO conditions on agriculture were modeled. An increase in these ENSO conditions was found to cost US farmers on average about $320 million per year if forecasts of these events were available and farmers used them to plan for the growing season. The increase in cost was estimated to be greater if accurate forecasts were not available or not used.

CLIMATE CHANGE AND AGRICULTURE IN THE UNITED STATES

INTRODUCTION

Both weather and climate affect virtually every aspect of agriculture, from the production of crops and livestock, to the transportation of agricultural products to market. Agricultural crop production is likely to be significantly affected by the projected changes in climate and atmospheric CO_2 (Rosenzweig and Hillel, 1998). While elevated CO_2 increases plant photosynthesis and thus crop yields (Kimball, 1983), the projected changes in temperature and precipitation have the potential to affect crop yields either positively or negatively. The negative effects are associated with some climate changes that result in more rapid plant development, and modification of water and nutrient budgets in the field (Long, 1991).

The net effects of increased CO_2 and climate change on crop yields will ultimately depend on local conditions. For example, higher spring and summer air temperatures might be beneficial to crop production at northern temperate latitude sites, where the length of the growing season would increase. However, higher temperatures might have negative effects during crop maturity in those regions where summer temperature and water stress already limit crop production (Rosenzweig and Tubiello, 1997).

The response of agricultural systems to future climate change will additionally depend on management practices, such as the levels of water and nutrient applied. Water limitation tends to enhance the positive crop response to elevated CO_2, compared to well-watered conditions (Chaudhuri et al., 1990; Kimball et al., 1995). The opposite is true for nitrogen limitation: well-fertilized crops respond more positively to CO_2 than less fertilized ones (Sionit et al., 1981; Mitchell et al., 1993).

This Assessment is intended to present our latest understanding of the potential impacts of climate change on the agricultural sector. The Assessment relies on two sources of information: the relevant scientific literature, and new quantitative and qualitative analyses done specifically as part of the Assessment.

Complete documentation of the work of the Agriculture Assessment Team is given in their Sector Report (Reilly et al., 2000; http://www.nacc.usgcrp.gov). This Foundation document, while a stand-alone statement, summarizes the report of the Agriculture Assessment Team.

In this document, we review the major activities undertaken in this Assessment. First, we present a summary of the key findings of the national and international assessments of climate change and agriculture that have been undertaken during the past two decades. Second, we briefly report results of new simulation modeling, done for this Assessment, that considers the consequences of two different climate-change scenarios on crop yield and the economics of agriculture in the US. Third, we set out the essential findings of new analyses on how the impacts of climate change on agriculture may, in turn, affect resources such as water and other aspects of the environment. And finally, we discuss the highlights of a new analysis of climate variability on agriculture.

The agriculture sector Assessment considered crop agriculture, grazing, livestock and environmental effects of agriculture. The focus here is primarily on crop agriculture, which was studied most intensely in the Assessment. Grain production is a major concern, with attention given to vegetables and fruit crops.

The approach used to assess the effects of climate change on crop agriculture involved an "end-to-end" analysis that linked climate-change scenarios for the future derived from general circulation models, with crop models designed to consider the effects of climate change and elevated atmospheric CO_2 on crop yields. The outputs of the crop models were inputs to an economic model that was then used to analyze the economic consequences of changed crop yields on farmers and consumers.

SOCIOECONOMIC CONTEXT

The US is a major supplier of food and fiber for the world, accounting for more than 25% of the total global trade in wheat, corn, soybeans, and cotton.

Cropland currently occupies about 400 million acres, or 17% of the total US land area. In addition, grasslands and permanent grazing and pasturelands, occupy almost 600 million acres, another 26% of US land area. The value of agricultural commodities (food and fiber) exceeds $165 billion at the farm level and over $500 billion, approaching 10% of GDP, after processing and marketing.

Economic viability and competitiveness are major concerns for producers trying to maintain profitability as real commodity prices have fallen by about two-thirds over the last 50 years. Agricultural productivity has improved at over 1% per year since 1950, resulting in a decline in both production costs and prices. This trend maintains intense pressure on individual producers to continue to increase the productivity of their farms and to reduce costs of production. In this competitive economic environment, producers see anything that might increase costs or limit their markets as a threat to their viability. Issues of concern include regulatory actions that might increase costs, such as efforts to control the off-site consequences of soil erosion, agricultural chemicals, and livestock wastes; growing resistance to and restrictions on the use of genetically-modified crops; new pests; and the development of pest resistance to existing pest-control strategies. Future changes in climate will interact with all of these factors.

CLIMATE CONTEXT

This Assessment of climate change is based on climate scenarios derived from climate models developed at the Canadian Centre for Climate Modelling and Analysis and the Hadley Centre in the United Kingdom. While the physical principles driving these models are similar, they differ in how they represent the effects of some important processes. Therefore, these two primary models paint different views of the future. On average over the 21st century the Canadian model projects a greater temperature increase than does the Hadley model, while the Hadley model projects a much wetter climate than does the Canadian model. By using these two models, a plausible range of future temperature conditions is captured, with one model being near the lower end and the other near the upper end of projected temperature changes over the US. Both models project much wetter conditions, compared to present, over many agricultural areas in the US.

Temperature. Average warming in the US is projected to be somewhat greater than for the globe as a whole over the 21st century. In the Canadian

model scenario, increases in annual average temperature of 9°F (5°C) by the year 2100 are common across the central US, with changes about half this large along the East and West coasts. Seasonal patterns indicate that projected changes will be particularly large in winter, especially at night. Large increases in temperature are projected over much of the South in summer. In the Hadley model scenario, the eastern US has temperature increases of 3-5°F (2-3°C) by 2100, while the rest of the nation warms more, up to 7°F (4°C), depending on the region.

In both models, Alaska is projected to experience more intense warming than the lower 48 states, and in fact, this warming is already well underway. In contrast, Hawaii, the other Pacific islands, and the Caribbean islands are likely to experience less warming than the continental US, because they are at lower latitudes and are surrounded by ocean, which warms more slowly than land.

Precipitation. At this time, climate scientists have less confidence in climate model projections of regional precipitation than of regional temperature. For the 21st century, the Canadian model projects the percentage increases in precipitation will be largest in the Southwest and California, while east of the Rocky Mountains, the southern half of the nation is projected to experience a decrease in precipitation. The percentage decreases are projected to be particularly large in eastern Colorado and western Kansas, and across an arc running form Louisiana to Virginia. Projected decreases in precipitation are most evident in the Great Plains during the summer and in the East during both winter and summer. The increases in precipitation projected to occur in the West, and the smaller increases in the Northwest, are likely to occur mainly in winter.

In the Hadley model, the largest percentage increases in precipitation are projected to be in the Southwest and Southern California, but the increases are smaller than those projected by the Canadian model. In the Hadley model, the entire US is projected to have increases in precipitation, with the exception of small areas along the Gulf Coast and in the Pacific Northwest. Precipitation is projected to increase in the eastern half of the nation and in southern California and parts of Nevada and Arizona in summer, and in every region during the winter, except the Gulf States and northern Washington and Idaho.

In both the Hadley and Canadian models, most regions are projected to experience an increase in the frequency of heavy precipitation events. This is especially notable in the Hadley model, but the Canadian model shows the same characteristic.

PREVIOUS ASSESSMENTS – A BRIEF OVERVIEW

Several conclusions are shared among assessments conducted over the past quarter century. Here these are briefly reviewed and a more complete synopsis of some of the important previous assessments is given in the Appendix.

Over the next 100 years and probably beyond, human-induced climate change as currently modeled will not seriously imperil overall food and fiber production in the US, nor will it greatly increase the aggregate cost of agricultural production. Most assessments have looked at multiple climate scenarios. About one-half of the scenarios in any given assessment have shown small losses for the US (increased cost of production) and about one-half have shown gains for the US (decreased cost of production). However, no assessment has adequately included the potential impacts of extreme events, such as flooding, drought, and prolonged heat waves, and the potential effects of increased ranges of pests, diseases, and insects. The result of including these factors could require a reevaluation of this finding.

There are likely to be strong regional production effects within the US, with some areas suffering significant loss of comparative advantage (if not absolute advantage) relative to other regions of the country. With very competitive economic markets, it matters little if a particular region gains or loses absolutely in terms of yield, but rather how it fares relative to other regions. The southern region of the US is persistently found to lose both relative to other regions and absolutely. The likely effects of climate change on other regions within the US are less certain. While warming can lengthen the growing seasons in the northern half of the country, the full effect depends on precipitation, notoriously poorly projected by climate models.

Global market effects and trade dominate in terms of net economic effect on the US economy. Just as climate's effects on regional comparative advantage are important, the relevant concerns are the overall effects on global production and prices and how US producers fare relative to their global competitors or potential competitors. The worst outcome for the US would be severe climate effects on production in most areas of the world, with particularly severe effects on US producers. Consumers would suffer from high food prices, producers would have little to sell, and agricultural exports would dwindle. While an unlikely outcome based on newer climate scenarios, some early scenarios that featured particularly severe drying in the mid-continental US with milder conditions in Russia, Canada, and the northern half of Europe produced a moderate version of this scenario. The US and the world could gain most if climate changes were generally beneficial to production worldwide, but particularly beneficial to US producing areas. Consumers in the US and around the world would benefit from falling prices and US producers would also gain because the improving climate would lower their production costs even more than prices fell, thus increasing their export competitiveness. In fact, most scenarios come close to the middle, with relatively modest effects on world prices. The larger gainers in terms of production are the more northern areas of Canada, Russia, and Northern Europe. Tropical areas are more likely to suffer production losses. The US as a whole straddles a set of climate zones that include gainers (the northern areas) and losers (southern areas).

Effects on producers and consumers often are in opposite directions and this is often responsible for the small net effect on the economy. This result is a near certainty without trade, and reflects the fact that demand is not very responsive to price so that anything that restricts supply (e.g., acreage reduction programs, environmental constraints, climate change) leads to price increases that more than make up for the reduced output. Once trade is factored in, this result depends on what happens to production abroad as discussed above.

US agriculture is a competitive, adaptive, and responsive industry and will likely adapt to climate change; all assessments reviewed have factored adaptation into their analyses. The final effect on producers and the economy after consideration of adaptation may be either negative or positive. The evidence for adaptation is drawn from analogous situations such as the response of production to changes in commodity and input prices, regional shifts in production as economic conditions change, and the adoption of new technologies and farming practices.

KEY ISSUES

Here, we briefly report results of new simulation modeling done for this Assessment that considers the consequences of two different climate change scenarios on crop yield and the economics of agriculture in the US. In addition, we set out the essential findings of new analyses on how the impacts of climate change on agriculture may, in turn, affect resources such as water and other aspects of the environment. And finally, we discuss the highlights of new analyses of climate variability on agriculture.

Four key issues were identified:
- Crop Yield Changes
- Changes in Economic Impacts
- Resource and Environmental Effects
    Changing water demands for irrigation
    Surface water quality
    Increasing pesticide use
- Climate Variability

1. Crop Yield Changes

Approach

The agriculture-sector team investigated the effects of climate change on US crop production, using future climate scenarios generated by two climate models, the Hadley and Canadian models, as input into a family of dynamic crop-growth models. The DSSAT family of models was used extensively in this study to simulate wheat, corn, potato, soybean, sorghum, rice, and tomato (Tsuji et al., 1994). The CENTURY model was used to simulate grassland and hay production (Parton et al., 1994). Finally, the model of Ben Mechlia and Carrol (1989) was used to simulate citrus production. The models were run to simulate yields at 45 sites across the US. These sites were chosen using USDA national and state-level statistics to be in areas of major production.

All models employed have been used extensively to assess crop yields across the US under current conditions as well as under climate change (e.g., Rosenzweig et al., 1995; Parton et al., 1994, Tubiello et al., 1999). Apart from CENTURY, which runs with a monthly time-step, all other models use daily inputs of solar radiation, minimum and maximum temperature, and precipitation to calculate plant phenological development from planting to harvest, photosynthesis and growth, and carbon allocation to grain or fruit. All models use a soil component to calculate water and nitrogen movement, and are thus able to assess the effects of different manage-

ment practices on crop growth. The simulations performed for this study considered: 1) rainfed production; and 2) optimal irrigation, defined as re-filling of the soil water profile whenever water levels fall below 50% of capacity at 30 cm depth. Fertilizer applications were assumed to be optimal at all sites. Atmospheric concentrations of CO_2 assumed in the core analysis were as follows: 350 parts per million by volume (ppmv) for the base, 445 ppmv for the year 2030, and 660 ppmv for 2090. The crop models assumed that crops such as wheat, rice, barley, oats, potatoes, and most vegetable crops, tend to respond favorably to increased CO_2, with a doubling of CO_2 leading to yield increases in the range of 15-20%. Other crops including corn, sorghum, sugar cane, and many tropical grasses, were assumed to be less responsive to CO_2, with a doubling of the gas leading to yield increases of about 5%.

In addition to current practices at each site, simulations were done that included different adaptation techniques. These consisted largely of testing the effects of early planting, a realistic scenario at many northern sites under climate change; and of testing the performance of cultivars better adapted to warmer climates, using currently available genetic stock. In general, early planting was considered for spring crops, to avoid heat and drought stress in the late summer months, while taking advantage of warmer spring conditions. New, better-adapted cultivars were tested for winter crops, such as wheat, to increase the time to maturity (shortened under climate change scenarios) and to increase yield potential.

Two other groups in the US developed additional analyses, independent from the core study described above. Specifically, researchers at the Pacific Northwest National Laboratories (PNNL) developed national-level analyses for corn, winter wheat, alfalfa, and soybean, using climate projections from the Hadley model (Izarraulde et al., 1999). Another group, co-located at Indiana University and Purdue University, focused on corn, soybeans, and wheat, developing a regional analysis for the Midwest, including the states of Indiana, Illinois, Ohio, Wisconsin, and Michigan, using Hadley model projections (Southworth et al., 2000).

In the PNNL study, the baseline climate data were obtained from national records for the period 1961-1990. The scenario runs were constructed for two future periods (2025-2034 and 2090-2099). The Erosion Productivity Impact Calculator (EPIC) was used to simulate the behavior of 204 farms with considerations of soil-climate-management combinations under the baseline climate, the two future periods, and their combinations with two levels of atmospheric CO_2 concentrations (365 and 560 ppmv).

In an independent study by Indiana University and Purdue University, a baseline climate was defined using the period 1961-1990. Several future scenarios were analyzed for the decade of 2050, with atmospheric CO_2 concentration set at 555 ppmv. Crop yields were simulated with the DSSAT model at 10 representative farms in the Corn Belt and Lake States. Adaptations studied included change of planting dates, as well as the use of cultivars with different maturity groups. These results were not included in the economic modeling but provide another source of information.

Although specific differences in time horizons, CO_2 concentrations, and simulation methodologies complicate the comparison of these additional analyses to the work discussed herein, model findings were overall in general agreement with results of the core study.

Dominant Land Uses, 1992

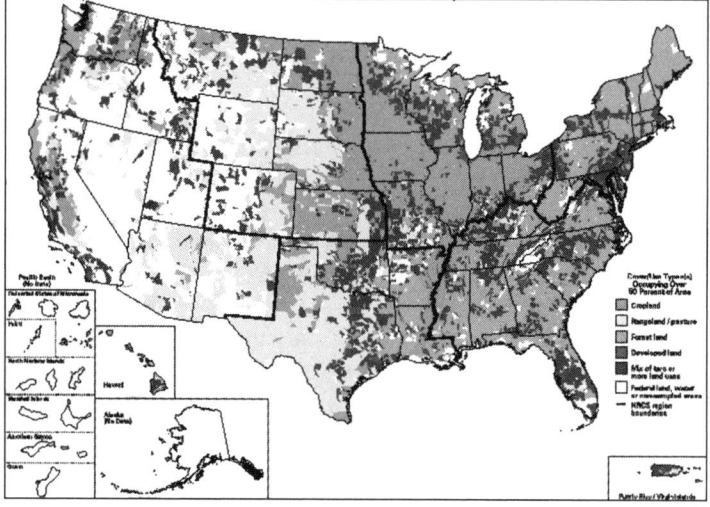

Figure 2. Agriculture Sector Model (ASM) Regions with USDA Regions Overlaid. (ASM regions follow state boundaries except where further disaggregated). The economic analysis in the Assessment is summarized for the 10 USDA regions outlined in the map. Source: USDA, 1997. See Color Plate Appendix

Results

Here we present the results from the models for several major crops. The DSSAT analyses for wheat, corn, alfalfa, and soybean, were integrated with results from two additional independent studies. The national average changes in yields for dryland and irrigated crops with and without adaptation are given in Figures 1a-d. The national averages were calculated by summing regional estimates for the coterminous United States (Figure 2) that are specified in Agriculture Sector Model (ASM). The regional estimates were derived by using crop-model outputs for sites in the region and harvested acreage in each ASM region based on data from the 1992 National Resource Inventory.

Yield Changes for Major Crops

A summary of the changes in simulated crop yields under the Canadian and Hadley scenarios relative to present yields is given below.

Winter wheat. Even with adaptation, rainfed production was reduced by an average of 9% in the 2030 time period under the Canadian scenario. Adaptation techniques helped to counterbalance yield losses in the Northern Plains, but not in the Southern Plains where losses were more severe and due to reductions in precipitation. Yields increased an average of 23% under the Hadley scenario for the 2030 period. Average dryland yields increased under both climate scenarios for the 2090 period, up to 59% in the case of the Hadley scenario. Irrigated wheat production increased under both climate scenarios by up to 16% on average by the end of the 21st century when adaptation strategies were used.

Spring wheat. Dryland production of spring wheat yields increased under both scenarios, either with or without adaptation. Adaptation techniques, including early planting and new cultivars, helped to improve yields under both scenarios, up to 59% for the Hadley scenario in 2090. Irrigated yields were reduced slightly under the Canadian scenario and increased slightly under the Hadley scenario.

Corn. Dryland corn production increased at most sites due to increases in precipitation under both climate scenarios. Average yields were up by between 15 to 40% by the end of the 21st century in much of the Corn Belt region. Larger yield gains were simulated in the Northern Great Plains and in the Northern Lakes Region, where higher tempera-

Figure 1a - Dryland Yields Without Adaptation

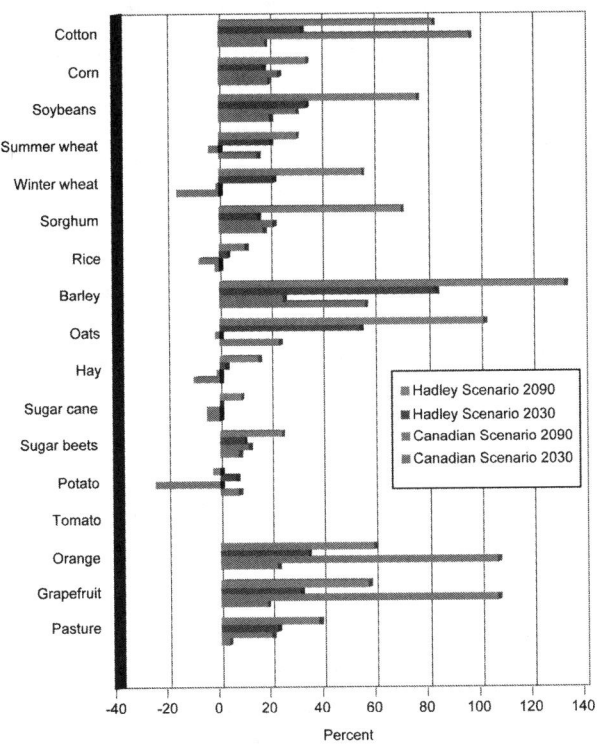

Figure 1b - Dryland Yields With Adaptation

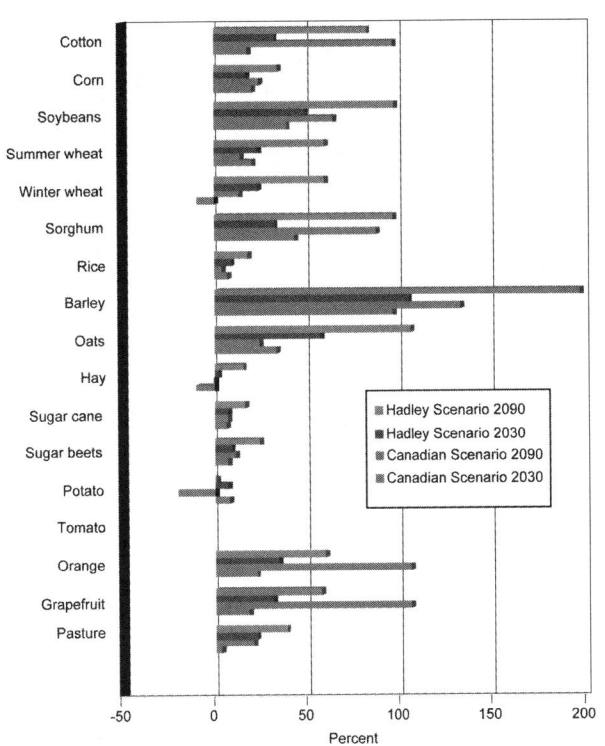

Figure 1c - Irrigated Yields Without Adaptation

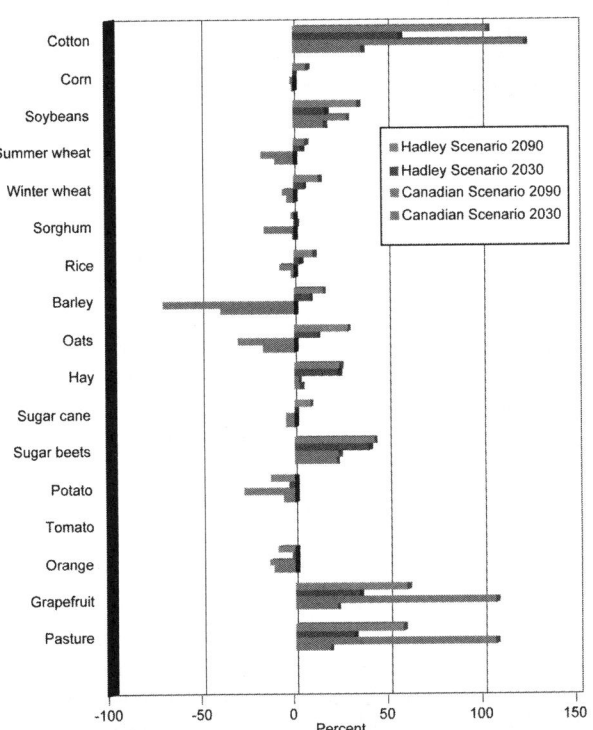

Figure 1d - Irrigated Yields With Adaptation

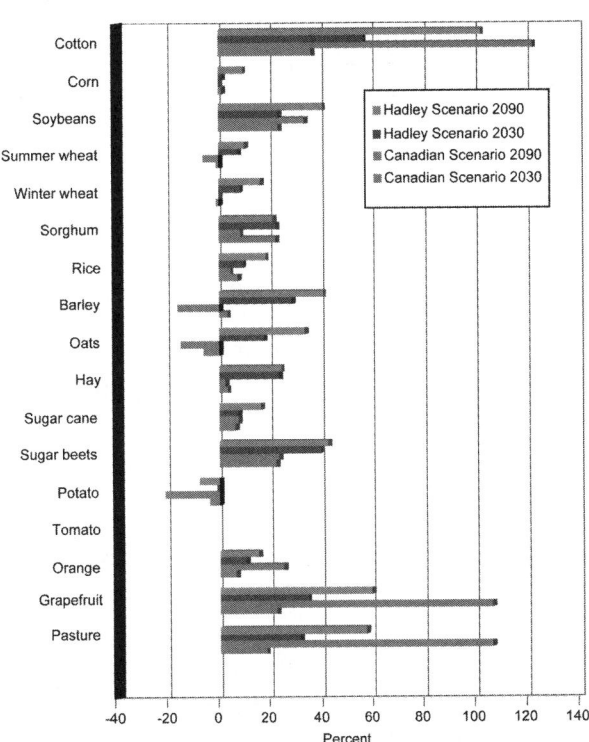

Figure 1a-d. Relative changes (% change relative to present) in crop yield for two time periods, 2030s and 2090s, under the Canadian and Hadley Scenarios. 0 = no change. Under the two climate scenarios, most crops showed substantial yield increase, even without adaptation, under dryland conditions. Irrigated yields increased less or decreased. (Source: Changing Climate and Changing Agriculture: Report of the Agricultural Sector Assessment Team, 2000) See Color Plate Appendix

tures were also beneficial to production. Irrigated corn production was not greatly affected at most sites.

Potato. Potato production decreased across many sites analyzed. While major production areas in the northern US experienced either small increases or small decreases, at some other sites potato yields decreased up to 50% from current levels. At these sites, there was little room for cultivar adaptation, because the projected higher fall and winter temperatures negatively affected tuber formation. Adaptation of planting dates mitigated only some of the projected losses.

Citrus. Production largely benefited from the higher temperatures projected under all scenarios. Simulated fruit yield increased in the range of 60 to 100%, while irrigation water use decreased. Crop losses due to freezing diminished by 65% in 2030, and by 80% in 2090.

Soybean. Soybean production increased at most sites analyzed, in the range 20 to 40% for sites of current major production. Larger gains were simulated at northern sites where cold temperatures currently limit crop growth. The Southeast sites considered in this study experienced large reductions under the Canadian scenario. Losses were reduced by adaptation techniques involving the use of cultivars with different maturity classes. (For regional details see the Southeast chapter).

Sorghum. Sorghum production, especially with adaptation, generally increased under rainfed conditions, due to the increased precipitation projected under the two scenarios considered. Higher temperatures at northern sites further increased rainfed grain yields. By contrast, irrigated production without adaptation was reduced almost everywhere, because of negative effects of higher temperatures on crop development and yield.

Rice. Rice production increased slightly under the Hadley scenario, with the increases in the range 1-10%. Under the Canadian scenario, rice production was 10-20% lower than current levels at sites in California and in the Delta region.

Tomato. Without adaptation, irrigated tomato production decreased at most of the simulated sites due to increased temperatures. Noted exceptions were the northern locations where production is currently limited by low temperatures and by short growing seasons. Reductions were in the 10 to 15% range under the Canadian scenario. Under the

Hadley scenario, reductions in tomato yields were in the 5% range. Adaptation strategies resulted in increased yields of tomatoes under the two climate scenarios.

Even without adaptation, the weighted average yield impact for many crops grown under dryland conditions across the entire US was positive under both the Canadian and Hadley scenarios. In many cases, yields under the 2030 climate conditions improved compared with the control yields under current climate, and improved further under the 2090 climate conditions. These generally positive yield results were observed for cotton, corn for grain and silage, soybean, sorghum, barley, sugar beet, and citrus fruit. The yield results were mixed for other crops (wheat, rice, oats, hay, sugar cane, and potatoes) showing yield increases under some conditions and declines other conditions.

Changes in irrigated yields, particularly for the grain crops, were more often negative or less positive than dryland yields. This reflected the fact that under these climate scenarios precipitation increases were substantial. The precipitation increases provided no yield benefit to irrigated crops because they face no water stress under current conditions since all the water needed is provided through irrigation. Higher temperatures sped development of crops and reduced the grain filling period, thereby reducing yields. For dryland crops, the positive effect of more moisture and CO_2 fertilization counterbalanced the negative effect of higher temperatures.

Water demand by irrigated crops dropped substantially for most crops. The faster development of crops due to higher temperatures reduced the growing period and thereby reduced water demand, more than offsetting increased evapotranspiration due to higher temperatures while the crops were growing. To a large extent, the reduced water use thus reflected the reduced yields on irrigated crops. Increased precipitation also reduced the need for irrigation water.

Adaptations examined in the crop modeling studies contributed small additional gains in yields of dryland crops, particularly for those with large yield increases due to climate change. Adaptation options examined, including shifts in planting times and choice of cultivars adapted to new climatic conditions. For the most part, however, these adaptations had little additional benefit where yields increased from climate change. This suggests that adaptation may be able to partly offset changes in comparative advantage across the US that may result under these scenarios. Other strategies for adaptation, such as whether or not to switch crops or to irrigate, were options available in the economic model. The decisions to undertake these strategies are driven by economic considerations; that is, whether they are profitable under market conditions simulated in the scenario. Adaptations for several crops were not considered because the options available, such as changing planting date, were not applicable to many perennial and tree fruit crops. Adaptation studies were conducted for only a subset of sites considered in the study and these results were extrapolated to other sites.

Adaptation contributed greater yield gains for irrigated crops. Shifts in planting dates were able to reduce some of the heat-related yield losses. With higher yields than in the no-adaptation cases, water demand declines were not as substantial. Again, this reflected the fact that the adaptations considered extended the growing (and grain-filling period) and this extension meant longer periods over which irrigation water was required.

The factors responsible for the positive results of this Assessment varied, but can generally be traced to aspects of the climate scenarios. First, increased precipitation in these transient climate scenarios is an important factor contributing to the more positive effects for dryland crops and explains the difference between dryland and irrigated crop results. The benefits of increased precipitation outweighed the negative effects of higher temperatures for dryland crops, whereas increased precipitation had little yield benefits for irrigated crops because water stress is not a concern for crops already irrigated. In fact, where the climate scenarios projected both higher temperatures and decreases in precipitation, such as for the Central Plains regions of Kansas and Oklahoma, rainfed cereal production, notably winter wheat, was negatively affected.

As noted for the Central Plains, not all agricultural regions of the United States are affected to the same degree or direction by the climates in the scenarios. In general, climate change as projected in the two climate scenarios favored northern areas. The Midwest (especially northern areas), West, and Pacific Northwest exhibited large gains in yields for most crops with both climate scenarios in the 2030 and 2090 time frames. Yield changes in other regions were mixed, depending on the climate scenario and time period. For example in the Southeast, simulated yields for most crops increased under the Hadley scenario in both the 2030 and 2090 time frames. Yield estimates varied widely among crops under the Canadian scenario. Citrus yields increased slightly by 2030, and dramatically by 2090. Dryland soybean yields decreased in the range of 10-30 % in about 2030, and by up to 80 % in about 2090. And rice yields decreased on the order of 5 to 10 % for both time periods.

The potential for within-region differences was highlighted in the Indiana University/Purdue University study of the Midwest. In this study, decreases were found in corn yields across the southern portion of the region's southern states — Indiana, Illinois, and Ohio. In addition, decreases, or only small increases in yields, were found for soybean and wheat across these same southern locations. In the region's northern states, Wisconsin and Michigan, there were simulated increases in yield for all the crops studied, with soybean showing the most dramatic increases. In addition, a variability analysis indicated that a doubling of current climate variability in association with climate change would produce the most detrimental climate conditions for crop growth across this region (Southworth et al., 2000).

Crop models such as those used in this Assessment have been used at local, regional, and global scales to systematically assess impacts on yields and adaptation strategies in agricultural systems, as climate and/or other factors change. The simulation results depend on the general assumption that soil nutrients are not limiting, and that pests, insects, diseases, and weeds pose no threat to crop growth and yield (Patterson et al., 1999; Rosenzweig and Hillel, 1998; Rosenzweig et al., 2000; Strzepek et al., 1999; Tubiello et al., 1999; Walker et al., 1996). One important consequence of these assumptions is that positive crop responses to elevated CO_2, responsible for one-third to one-half of the yield increases simulated in the Assessment studies, should be regarded as upper limits to actual responses in the field. One additional limitation that applies to this study is the models' inability to predict the negative effects of excess water conditions on crop yields. Given the "wet" nature of the scenarios employed, the positive responses projected in this study for rainfed crops, under both the Hadley and Canadian scenarios, may be overestimated.

2. Economic Impacts

Approach

The crop results were combined with impacts on water supply, livestock, pesticide use, and shifts in international production to estimate impacts on the US economy. This allowed the estimation of regional production shifts and resource use in response to changing relative comparative advantage among crops and producing regions. These changes were estimated using a US national agricultural sector model (ASM) (Adams et al., 1990, 1997) that is linked to a global trade model.

The ASM is based on the work of Baumes (1978) which was later modified and expanded by Burton and Martin(1987); Adams et al. (1986); Chang et al. (1992) and Lambert et al. (1995). Conceptually, ASM is a price endogenous, mathematical programming model of the type described in McCarl and Spreen (1980). Constant elasticity curves are used to represent domestic consumption and export demands as well as input and import supplies. Elasticities were assembled from a number of sources including USDA through the USMP modeling team (House, 1987) and prior model versions. ASM is designed to simulate the effects of various changes in agricultural resource usage or resources available on agricultural prices, quantities produced, consumers' and producers' welfare, exports, imports and food processing. In calculating these

effects, the model considers production, processing, domestic consumption, imports, exports and input procurement.

The model distinguishes between primary and secondary commodities, with primary commodities being those directly produced by the farms and secondary commodities being those involving processing. Within ASM, the US is disaggregated into 63 geographical production subregions. Each subregion possesses different endowments of land, labor and water as well as crop yields. Agricultural production is described by a set of regional budgets for crops and livestock. Marketing and other costs are added to the budgets following the procedure described in Fajardo et al.(1981) such that the marginal cost of each budget equals marginal revenue. ASM also contains a set of national processing budgets which uses crop and livestock commodities as inputs (USDA, 1982). There are also import supply functions from the rest of the world for a number of commodities. The demand sector of the model consists of the intermediate use of all the primary and secondary commodities, domestic consumption use and exports.

There are 33 primary crop and livestock commodities in the model. The primary commodities depict the majority of agricultural production, land use and economic value. The model incorporates processing of the primary commodities. There are 37 secondary commodities that are processed in the model. These commodities are chosen based on their linkages to agriculture. Some primary commodities are inputs to the processing activities yielding these secondary commodities and certain secondary products (feeds and by-products) are in turn inputs to production of primary commodities. Three land types (crop land, pasture land, and land for grazing on an animal unit month basis) are specified for each region. Land is available according to a regional price elastic supply schedule with a rental rate as reported in USDA farm real estate statistics. The labor input includes family and hired labor. A region-specific reservation wage and maximum amount of family labor available reflect the supply of family labor. The supply of hired labor consists of a minimum inducement wage rate and a subsequent price elastic supply. Water comes from surface and pumped ground water sources. Surface water is available at a constant price, but pumped water is supplied according to a price elastic supply schedule.

US agricultural sector models typically only deal with aggregate exports and imports facing the total

Economic Impacts of Climage Change on US Agriculture

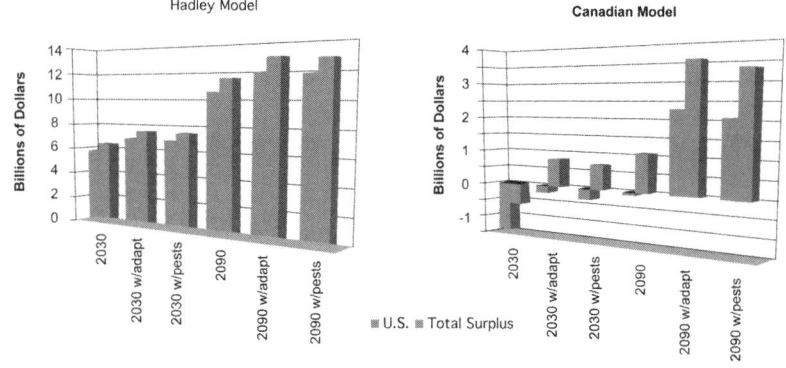

Figure 3a and b. The economic index is change in welfare expressed as the sum of producer and consumer surplus in billions of dollars. There were net economic benefits for the US under most of the scenarios examined in the Assessment. Foreign consumers also gained from lower commodity prices on international markets. Source: Changing Climate and Changing Agriculture: Report of the Agricultural Sector Assessment Team, 2000. See Color Plate Appendix

US without regional trading detail. The ASM includes foreign regions, and shipment among foreign regions modeled as 6 spatial equilibrium models for the major traded commodities (Takayama and Judge, 1971). To portray US regional effects, US markets are grouped into the ten regional definitions used by the USDA. We also added variables for shipment among US regions, and shipment between US regions and foreign regions. The commodities subject to explicit treatment via the spatial equilibrium world trade model components are hard red spring wheat (HRSW), hard red winter wheat (HRWW),soft red winter wheat (SOFT), durum wheat (DURW)), corn, soybeans and sorghum. These commodities are selected based on their importance as US exports. The rest of the world is aggregated into 28 countries/regions. Transportation cost, trade quantity, price and elasticity were obtained from Fellin and Fuller (1998), USDA (1987) statistical sources and the USDA SWOPSIM model (Roningen, 1986).

In the base results, climatic effects on crops and livestock in the rest of the world were assumed to be neutral, that is, no climate change effects on agriculture in the rest of the world were assumed. To test how sensitive the results were to this assumption three scenarios of climate impacts on agriculture in the rest of the world were used. These were developed from previous work reported in Reilly et al. (1993; 1994), and based on a global analysis using a Hadley Centre climate scenario and a global agricultural model developed by Darwin et al. (1995). These climate scenarios were not completely consistent with the new scenarios used for the US, but provide a good test of the sensitivity of US economic results to impacts in the rest of the world.

Results

The net economic effect on the US economy was generally positive, reflecting the generally positive yield effects (Figures 3a,b). The exceptions were simulations under the Canadian climate scenario in 2030, particularly in the absence of adaptation.

Foreign consumers gained in all the scenarios as a result of lower prices for US export commodities. The total effects (net effect on US producers and consumers plus foreign gains) ranged from a $0.5 billion loss to a $12.5 billion gain.

This Assessment found that producers and consumers were affected in opposite ways by climate change (Figures 4a,b). Producers' incomes generally fell due to lower prices. Producer reductions ranged from about $0.1 up to 5 billion. The largest

Producer versus Consumer Impacts of Climate Change

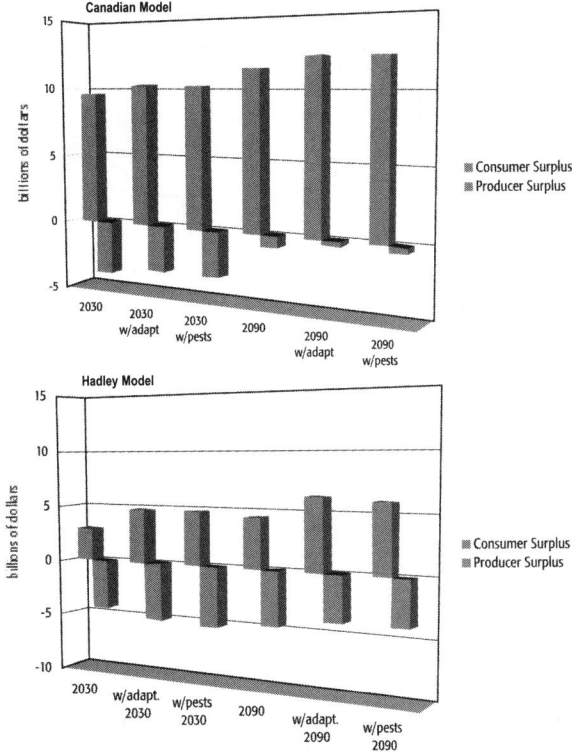

Figure 4a and b. In the model simulations consumers generally benefited from climate change while producers experienced lower income due to lower prices for commodities resulting from increased yields and supply. Source: Changing Climate and Changing Agriculture: Report of the Agricultural Sector Assessment Team, 2000. See Color Plate Appendix

reductions were under the Canadian scenario. Under the Hadley scenario, producers suffered from lower prices, but enjoyed considerable increases in exports such that the net effect was for only very small reductions. Economic gains accrued to consumers through lower prices in all scenarios. Gains to consumers ranged from $2.5 to 13 billion.

Different scenarios of the effect of climate change on agriculture abroad did not change the net impact on the US very much, but redistributed changes between producers and consumers. The direction depended on the direction of effect on world prices. Lower prices increased producer losses and added to consumer benefits. Higher prices reduced producer losses and consumer benefits.

Modeled projections of livestock production and prices were mixed. Increased temperatures directly reduced productivity, but improvements in pasture and grazing and reductions in feed prices due to lower crop prices counter these losses. (For additional comments on livestock, see the Great Plains chapter).

Regional Production Changes Relative to Current Production

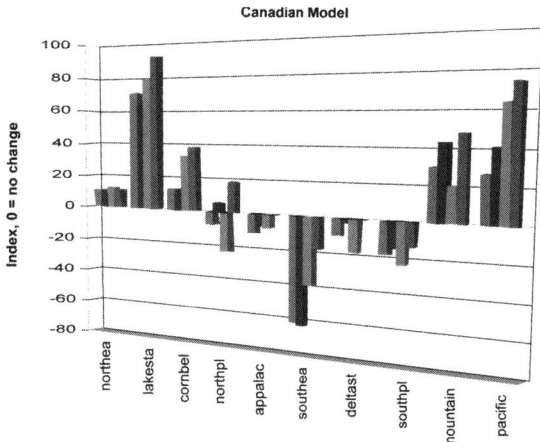

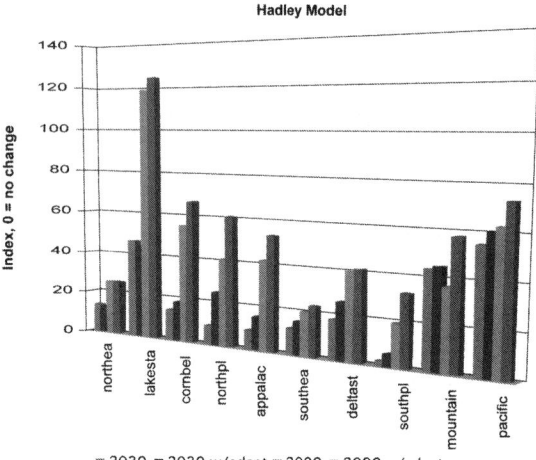

■ 2030 ■ 2030 w/adapt ■ 2090 ■ 2090 w/adapt

Figure 5a and b. In the model simulations production increased in northern regions as a result of longer growing seasons, and in western regions due to increased precipitation. Higher temperatures and increased drought conditions contributed to production declines or smaller increases in southern and plains regions. Source: Changing Climate and Changing Agriculture: Report of the Agricultural Sector Assessment Team, 2000. See Color Plate Appendix

Aggregate regional production changes (Figures 5a, b) were positive for all regions in both the 2030 and 2090 time frames under the Hadley scenario. Adaptation measures had a small additional positive effect. In contrast, aggregate production changes differed among regions under the Canadian scenario in both the 2030s and 2090s. It was positive for most northern regions, mixed for the Northern Plains, and negative for Appalachia, the Southeast, the Delta states, and the Southern Plains. Adaptation measures helped somewhat for the southern regions, but the aggregate production was lower in these regions under both the 2030 and 2090 climates considered. Aggregate production is represented in Figures 5a, b as a price-

weighted index across livestock and crop production. Many of the impacts, because they occur at the regional level, are dealt with in more detail in the regional chapters.

3. Resource and Environmental Effects

In terms of improving the coverage of potential impacts of climate change on agriculture, this study has made advances over previous assessments. Some of the advances were in the area of resource and environmental effects (Figure 6). Details of the studies underlying this summary are given in the Agriculture Sector Report (Reilly, et al. 2000; http://www/nacc.usgcrp.gov).

Demand for land. Agriculture's pressure on land resources generally decreased under both climate scenarios across the 21st century. Area in cropland decreased 5 to 10 % while area in pasture decreased 10 to 15 %.

Grazing pressure. Animal unit months (AUMs) of grazing on western lands decreased on the order of 10% under the Canadian climate scenario and increased 5 to 10% under the Hadley climate scenario.

Demand for water, a national perspective. At the national level, the models used in this Assessment found that irrigated agriculture's need for water declined approximately 5-10% for 2030 and 30-40% for 2090 climate conditions as represented in the two scenarios. At least two factors were responsible for this reduction in water demand for irrigation. One was increased precipitation in some agricultural areas. The other was that faster development of crops due to higher temperatures resulted in a reduced growing period and thereby reduced water demand. In the crop modeling analyses done for the Assessment, shortening of the growing period reduced plant water-use enough to more than compensate the increased water losses from plants and soils due to higher temperatures.

Demand for water, a regional perspective. The competition for water between agriculture and other uses was explored through a case study of the Edwards Aquifer that serves the San Antonio region of Texas. Agriculture uses of water compete with urban and industrial uses and tight economic management is necessary to avoid unsustainable use of the resource. Aquifer discharge is through pumping and artesian spring discharge.

The study found that the Canadian and Hadley scenarios of climatic change caused a slightly negative welfare result in the San Antonio region as a whole, but had a strong impact on the agricultural sector.

The regional welfare loss, most of which was incurred by agricultural producers, was estimated to be between $2.2 and 6.8 million per year if current pumping limits are maintained.

A major reason for the current pumping limits is to preserve the artesian spring flows that are critical to the habitat of local endangered species. To maintain spring flows at the currently specified level to protect endangered species, pumping would need to be reduced in the future with climate change. The study calculated that under the two climate scenarios, pumping would need to be reduced by 10 to 20% below the limit currently set and this would cost an additional $0.5 to 2 million per year. Welfare in the non-agricultural sector was only marginally reduced by the climatic change simulated by the two climate scenarios. The value of water permits rose dramatically.

The agricultural use of water is discussed in several of the regional chapters including the Great Plains.

Surface water quality. As part of the Assessment, a study was undertaken of the linkages between climate change and nitrogen loading of Chesapeake Bay. The Chesapeake Bay is one of nation's most valuable natural resources, but has been severely degraded in recent decades. Soil erosion and nutrient runoff from crop and livestock production have played a major role in the decline of the Bay. Based on simulations done for this Assessment, under the Canadian and Hadley scenarios for the 2030 period, nitrogen loading from corn production increased by 17 to 31% compared with conditions under current climate. Potential effects of climate change on water quality in the Chesapeake Bay must be considered very uncertain because current climate models may not fully represent the effects of extreme weather events such as floods or heavy downpours, which can wash large amounts of fertilizers, pesticides, and animal manure into surface waters.

Surface water quality is also discussed in the Southeast and West chapters.

Pesticide expenditures. The Assessment investigated the relationship between pesticide use and climate for crops that require relatively large

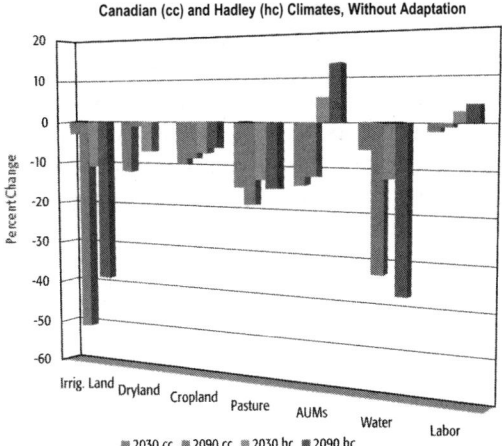

Changes in Resource Use

Canadian (cc) and Hadley (hc) Climates, Without Adaptation

■ 2030 cc ■ 2090 cc ■ 2030 hc ■ 2090 hc

Figure 6. In the simulations resource use generally declined as less crop and grazing land was needed. Use of water and irrigated crop land declined the most because the two climate scenarios used favored dryland over irrigated crops (cc-Canadian, hc=Hadley). Source: Changing Climate and Changing Agriculture: Report of the Agricultural Sector Assessment Team, 2000. See Color Plate Appendix

amounts of pesticide. Pesticide expenditures increased under the climate scenarios considered for most crops studied and in most regions. Increases for corn were generally in the range of 10 to 20%, for potatoes of 5 to 15% and for soybeans and cotton of 2 to 5%. The results for wheat varied widely by state and climate scenario showing changes ranging from approximately –15 to +15%. These projections were based on cross-section statistical evidence on the relationship between pesticide expenditures and temperature and precipitation.

The increase in pesticide expenditures could increase environmental problems associated with pesticide use, but much depends on how pest control evolves over the next several decades. Pests develop resistance to control methods, requiring a continual evolution in the chemicals and control methods used.

The increase in pesticide expenditures resulted in slightly poorer overall economic performance but this effect was quite small because pesticide expenditures are a relatively small share of production costs. The approach used in the Assessment did not consider increased crop losses due to pests, implicitly assuming that all additional losses were eliminated through increased pest control measures. This may underestimate pest losses.

4. Potential Effects of Climate Variability on Agriculture

Ultimately, the consequences of climate change for US agriculture hinge on changes in climate variability and extreme events. Agricultural systems are vulnerable to climate extremes, with effects varying from place to place because of differences in soils, production systems, and other factors. Changes in precipitation type (rain, snow, and hail), timing, frequency, and intensity, along with changes in wind (windstorms, hurricanes, and tornadoes), could have significant consequences. Heavy precipitation events cause erosion, waterlogging, and leaching of animal wastes, pesticides, fertilizers, and other chemicals into surface water and groundwater. While all of the risks associated with these impacts are not known, the system is known to be sensitive to changes in extremes. The costs of adjusting to such changes will likely increase if the rate of climate change is high, although early signals from a rapidly changing climate would reduce uncertainty and encourage early adaptation.

One major source of weather variability is the El Niño/Southern Oscillation (ENSO) phenomenon. ENSO phases are triggered by the movement of warm surface water eastward across the Pacific Ocean toward the coast of South America and its retreat back across the Pacific, in an oscillating fashion with a varying periodicity. Better prediction of these events would allow farmers to plan ahead, planting different crops and at different times. The value of improved forecasts of ENSO events has been estimated at approximately $500 million per year.

ENSO's effects can vary from one event to the next. Predictions of the details of ENSO-driven weather are not perfect. There are also widely varying effects of ENSO across the country. The temperature and precipitation effects are not the same in all regions; in some regions the ENSO signal is relatively strong while in others it is weak, and the changes in weather have different implications for agriculture in different regions because climate-related productivity constraints differ among regions under neutral climate conditions.

As climate warms, ENSO is likely to be affected. Some models project that El Niño and La Niña events and their impacts on US weather will become more intense with climate change. The potential impacts of projected changes in frequency and strength of ENSO conditions on agriculture were modeled in this Assessment. An increase in these ENSO conditions was found to cost US farmers about $320 million on average per year if accurate forecasts of these events were available and farmers used them as they planned for the growing season. The increase in cost was estimated to be greater if accurate forecasts were not available or not used.

ADAPTATION STRATEGIES

Adaptations such as changing planting dates and choosing longer season varieties are likely to offset losses or further increase yields. Adaptive measures are likely to be particularly critical for the Southeast because of the large reductions in yields projected for some crops under the more severe climate scenarios examined. Breeding for response to CO_2 will likely be necessary to achieve the strong fertilization effect assumed in the crop studies. This is an unexploited opportunity and the prospects for selecting for CO_2 response are good. However, attempts to breed for a single characteristic are often not successful, unless other traits and interactions are considered. Breeding for tolerance to climatic stress has already been heavily exploited and varieties that do best under ideal conditions usually also outperform other varieties under stress conditions. Breeding specific varieties for specific conditions of climate stress is therefore less likely to encounter success.

Some adaptations to climate change and its impacts can have negative secondary effects. For example, an examination of use of water from the Edward's aquifer region around San Antonio, Texas found increased pressure on groundwater resources that would threaten endangered species dependent on spring flows supported by the aquifer. Another example relates to agricultural chemical use. An increase in the use of pesticides and herbicides is one adaptation to increased insects, weeds, and diseases that could be associated with warming. Runoff of these chemicals into prairie wetlands, groundwater, and rivers and lakes could threaten drinking water supplies, coastal waters, recreation areas, and waterfowl habitat.

The wide uncertainties in climate scenarios, regional variation in climate effects, and interactions of environment, economics, and farm policy suggest that there are no simple and widely applicable adaptation prescriptions. Farmers will need to adapt broadly to changing conditions in agriculture, of which changing climate is only one factor. Some of the possible adaptations more directly related to climate include:

Sowing dates and other seasonal changes
Plant two crops instead of one or a spring and fall crop with a short fallow period to avoid excessive heat and drought in mid-summer. For already warm growing areas, winter cropping could possibly become more productive than summer cropping.

New crop varieties
The genetic base is very broad for many crops, and biotechnology offers new potential for introducing salt tolerance, pest resistance, and general improvements in crop yield and quality.

Water supply, irrigation, and drainage systems
Technologies and management methods exist to increase irrigation efficiency and reduce problems of soil degradation, but in many areas, the economic incentives to reduce wasteful practices do not exist. Increased precipitation and more intense precipitation will likely mean that some areas will need to increase their use of drainage systems to avoid flooding and waterlogging of soils.

CRUCIAL UNKNOWNS AND RESEARCH NEEDS

Further research is needed in several areas. Broadly, these include: 1) integrated modeling of the agricultural system; 2) research to improve resiliency of the agricultural system to change; and 3) several areas of climate-agriculture interactions that have not been extensively investigated.

Integrated modeling of the agricultural system

- The main methodology for conducting agricultural impacts models has been to use detailed crop models run at a selected set of sites and to use the output of these as input to an economic model. This approach has provided great insights but future assessments will need to integrate these models to consider interactions and feedbacks, multiple environmental stresses (tropospheric ozone, acid deposition, and nitrogen deposition), transient climate scenarios, global analysis, and to allow study of uncertainty where many climate scenarios are used. The present approach of teams of crop modelers running models at specific sites severely limits the number of sites and scenarios that can be feasibly be considered.
- The boundaries of the agricultural system in an integrated model need to be expanded so that more of the complex interactions can be represented. Changes in soils, multiple demands for

water, more detailed analysis and modeling of pests, and the environmental consequences of agriculture and changes in climate are areas that need to be incorporated into one integrated modeling framework. Agricultural systems are highly interactive with economic management choices that are affected by climate change. Separate models and separate analyses cannot capture these interactions.

Resiliency and adaptation

- Specific research on adaptation of agriculture to climate change at the time scale of decades to centuries should not be the centerpiece of an agricultural research strategy. Decision-making in agriculture mostly involves time horizons of one to five years, and long-term climate predictions are not very helpful for this purpose. Instead, effort should be directed toward understanding successful farming strategies and where adaptations to many changes are needed to manage risk.
- There is also great need for research to improve short-term and intermediate term (i.e., seasonal) weather predictions and on how to make better use of these predictions.

New areas of research

- Experimentation and modeling of the interactions of multiple environmental changes on crops (changing temperature, CO_2 levels, ozone, soil conditions, moisture, etc.) are needed. Experimental evidence is needed under realistic field conditions such as Free Air Carbon Dioxide Enrichment (FACE) experiments for CO_2 enrichment.
- Much more work on agricultural pests and their response to climate change is needed.
- Economic analyses need to better study the dynamics of adjustment to changing conditions.
- Climate-agriculture-environment interactions are perhaps one of the more important vulnerabilities, but the existing research is extremely limited. Soil, water quality, and air quality should be included in a comprehensive study of interactions.
- Agricultural modeling must be more closely integrated with climate modeling and modelers to develop better techniques for assessing the impacts of climate variability. This requires significant advances in climate predictions to better represent changes in variability as well as assessment of and improvements in the performance of crop models under extreme conditions.

APPENDIX – REVIEW OF PREVIOUS ASSESSMENTS

Here we provide a short summary of the major assessments of the potential consequences of climate change for agriculture. The summary does not include a detailed scientific literature review that forms the foundation for past assessments as well as this one. A new set of reviews on climate change impacts on crops, livestock, pests, and soils, as well as discussion of global and regional impacts, has been published in a special edition of the journal, *Climatic Change*, Climate Change: Impacts On Agriculture (Reilly, 1999). The 5 articles included in the edition contain over 500 citations, providing a detailed guide to the scientific literature relating climate change and agriculture.

1976-1983: National Defense University

A National Defense University (Johnson, 1983) project produced a series of reports with the 1983 report providing the final report on agriculture, integrating yield and economic effects. It focused on the world grain economy in the year 2000, considering both warming and cooling of up to approximately 1°C (1.8°F) for large warming or cooling and 0.5°C (0.9°F) for moderate changes for the US, with associated precipitation changes on the order of +/- 0-2%. These estimates varied somewhat by region. The base year for comparison purposes was 1975. It relied on an expert opinion survey for yield effects, using the results to create a model of crop-yield response to temperature and precipitation for major world grain regions. There was no explicit account of potential interactions of pests, changes in soils, or of livestock or crops such as fruits and vegetables. No direct effects of CO_2 on plant growth were considered as the study remained agnostic about the source of the climate change (e.g., whether due to natural variability or human-induced). Economic effects were assessed using a model of world grain markets.

Crop yields in the US were estimated to fall by 1.6 to 2.3% due to moderate and large warming and to increase by very small amounts (less than 3%) with large cooling and even smaller amounts with moderate cooling. Warming was estimated to increase crop yields in the (then) USSR, China, Canada, and Eastern Europe, with cooling decreasing crop production in these areas. Most other regions were estimated to gain from cooling and suffer yield losses from warming. The net effect was a very small change in world production and on world prices.

The study assigned subjective probabilities to the scenarios, attempted to project ranges of crop yield improvement in the absence of climate change, and compared climate-induced changes to normal variability in crop yields and uncertainty in future projections of yield. A summary point highlighted the difficulty in ultimately detecting any crop yield changes due to climate given the year-to-year variability and the difficulty in disentangling climate effects from the effects of new varieties and other changing technology that would inevitably be introduced over the 25-year period.

1988-1989: US EPA

US EPA (Smith and Tirpak, 1989) evaluated the impacts of climate change on US agriculture as part of an overall assessment of climate impacts on the US. The agricultural results were published in Adams, et al. 1990. The study evaluated warming and changes in precipitation based on doubled CO_2 equilibrium climate scenarios from 3 widely known General Circulation Models (GCMs), with increased average global surface warming of 4.0 to 5.2°C (7.2 to 9.4°F). In many ways the most comprehensive assessment to date, it included studies of possible changes in pests and interactions with irrigation water supply in a study of California. The main study on crop yields used site studies and a set of crop models to estimate crop yield impacts. These were simulated through an economic model. Economic results were based on imposition of climate change on the agricultural economy in 1985. Grain crops were studied in most detail, with a simpler approach for simulating impacts on other crops. Impacts on other parts of the world were not considered. The basic conclusions summarized in the Smith and Tirpak report were:

* Yields could be reduced, although the combined effects of climate and CO_2 would depend on the severity of climate change.
* Productivity may shift northward.
* The national supply of agricultural commodities may be sufficient to meet domestic needs, but exports may be reduced.
* Farmers would likely change many of their practices.
* Ranges of agricultural pests may extend northward.
* Shifts in agriculture may harm the environment in some areas.

1988-1990: Intergovernmental Panel on Climate Change (IPCC), first assessment report

In the first assessment report of the Intergovernmental Panel on Climate Change (IPCC), (Parry 1990a and in greater detail, Parry, 1990b)

North American agriculture was briefly addressed. The assessment was based mainly on a literature review and, for regional effects, on expert judgement. North American/US results mainly summarized the earlier EPA study. Some of the main contributions of the report were to identify the multiple pathways of effects on agriculture including effects of elevated CO_2, shifts of climatic extremes, reduced soil water availability, changes in precipitation patterns such as the monsoons, and sea-level rise. It also identified various consequences for farming including changes in trade, area farmed, irrigation, fertilizer use, control of pests and diseases, soil drainage and control of erosion, farming infrastructure, and interaction with farm policies. The overall conclusion of the report was that: "on balance, the evidence suggests that in the face of estimated changes of climate, food production at the global level could be maintained at essentially the same level as would have occurred without climate change; however, the cost of achieving this was unclear."

As an offshoot of this effort, the Economic Research Service of USDA (Kane et al., 1991 and subsequently, as Kane et al., 1992, and Tobey et al., 1992) published an assessment of impacts on world production and trade, including specifically the US. The study was based on sensitivity to broad generalizations about the global pattern of climate change as portrayed in doubled CO_2 equilibrium climate scenarios, illustrating the importance of trade effects. A "moderate impacts scenario" brought together a variety of crop model results based on doubled CO_2 equilibrium climate scenarios and the expert judgements for other regions that were the basis for the IPCC. In this scenario, the world impacts were very small (a gain of $1.5 billion 1986 US$). The US was a very small net gainer ($0.2 billion) with China, Russia, Australia, and Argentina also benefiting while other regions lost. On average, commodity prices were estimated to fall by 4%, although corn and soybean prices rose by 9-10%.

1990-1992: US DOE, Missouri, Iowa, Nebraska, Kansas (MINK) Study
In the Missouri, Iowa, Nebraska, Kansas (MINK) Study (Rosenberg, 1993; Easterling et al., 1993) the dust bowl of the 1930s was used as an analogue climate for global change for the four-state region. Unique aspects of the study included consideration of water, agriculture, forestry, and energy impacts, and projection of regional economy and crop variety development to the year 2030. Crop response was modeled using crop models, river flow using historical records, and economic impacts using an

input-output model for the region. Despite the fact that the region was "highly dependent" on agriculture compared with many areas of the country, the simulated impacts had relatively small effects on the regional economy. Climate change losses in terms of yields were on the order of 10 to 15%. With CO_2 fertilization effects, most of the losses were eliminated. Climate impacts were simulated for current crops as well as "enhanced" varieties with improved harvest index, photosynthetic efficiency, pest management, leaf area, and harvest efficiency. These enhanced varieties were intended to represent possible productivity changes from 1990 to 2030 and increased yield on the order of 70%. The percentage losses due to climate change did not differ substantially between the "enhanced" and current varieties.

1992: Council on Agricultural Science and Technology (CAST) Report
The Council on Agricultural Science and Technology (CAST, 1992) report, commissioned by the US Department of Agriculture did not attempt any specific quantitative assessments of climate change impacts, focusing instead on approaches for preparing US agriculture for climate change. It focused on a portfolio approach, recognizing that prediction with certainty was not possible.

1992-1993: Office of Technology Assessment study
The Office of Technology Assessment (OTA, 1993) study, similar to the CAST study for agriculture, focused on steps that could prepare the US for climate change rather than estimates of the impacts. The study's overall conclusions for agriculture were that the long-term productivity and competitiveness of US agriculture were at risk and that market-driven responses may alter the regional distribution and intensity of farming. It found institutional impediments to adaptation, recognized that uncertainty made it hard for farmers to respond, and saw potential environmental restrictions and water shortages as limits to adaptation. It also noted that declining Federal interest in agricultural research and education could impede adaptation. The study recommended removal of institutional impediments to adaptation (in commodity programs, disaster assistance, and water-marketing restrictions), improvement of knowledge and responsiveness of farmers to speed adaptation, and support for both general agricultural research and research targeted toward specific constraints and risks.

1992-1994: US EPA Global Assessment
A global assessment (Rosenzweig and Parry, 1994; Rosenzweig et al., 1995) of climate impacts on world food prospects expanded the method used in the US EPA study for the United States to the entire

world. It was based on the same suite of crop and climate models and applied these to many sites around the world. It used a global model of world agriculture and the world economy that simulates the evolving economy through to 2060, assumed to be the period when the doubled CO_2-equilibrium climates applied. The global temperature changes were +4.0 to +5.2°C (7.2 to 9.4°F). Scenarios with the CO_2 fertilization effect and modest adaptation showed global cereal production losses of 0 to 5.2%. In these scenarios, developed countries showed cereal production increases of 3.8 to 14.2%, while the developing countries showed losses of 9.2 to 12.5%. The study concluded that in the developing world there was a significant increase in the number of people at risk because of climate change. The study also considered different assumptions about yield increases due to technology improvement, trade policy, and economic growth. These different assumptions and scenarios had equal or more important consequences for the number of people at risk of hunger.

Other researchers simulated yield effects estimated in this study through economic models, focusing on implications for the US (Adams et al., 1995) and world trade (Reilly et al. 1993; 1994). Adams et al. (1995) estimated economic welfare gains for the US of approximately $4 and 11 billion (1990 US$) for two climate scenarios and a loss of $16 billion for the other scenario, under conditions reflecting increased export demands and a CO_2 fertilization effect (550 ppmv CO_2). The study found that increased exports from the US, in response to high commodity prices resulting from decreased global agricultural production, led to benefits to US producers of approximately the same magnitude as the welfare losses to US consumers from high prices. Reilly et al. (1993; 1994) found welfare gains to the US of $0.3 billion (1990 US$) under one GCM scenario and up to $0.6 to $0.8 billion losses in the other scenarios when simulating production changes for all regions of the world through a trade model. They also found widely varying effects on producers and consumers, with producers' effects ranging from a $5 billion loss to a $16 billion gain, echoing the general findings of Adams et al. (1995). In particular, Reilly et al. 1994 showed that in many cases, more severe yield effects produced economic gain to producers because world prices rose.

1994-1995: IPCC, Second Assessment Report
The Second Assessment Report of the IPCC included an assessment of the impacts of climate change on agriculture (Reilly et al., 1995). As an assessment based on existing literature, it summarized most of

the studies listed above. The overall conclusions included a summary of the direct and indirect effects of climate and increased ambient CO_2, regional and global production effects, and vulnerability and adaptation. With regard to direct and indirect effects:

The results of a large number of experiments to resolve the effect of elevated CO_2 concentrations on crops have confirmed a beneficial effect. The mean value yield response of C_3 crops (most crops except maize, sugar cane, millet, and sorghum) to doubled CO_2 is +30% although measured responses range from -10 to +80%.

- Changes in soils, e.g., loss of soil organic matter, leaching of soil nutrients, salinization, and erosion, are a likely consequence of climate change for some soils in some climatic zones. Cropping practices including crop rotation, conservation tillage, and improved nutrient management are, technically, quite effective in combating or reversing deleterious effects.
- Livestock production will be affected by changes in grain prices, changes in the prevalence and distribution of livestock pests, and changes in grazing and pasture productivity, as well as the direct effects of weather. Heat stress in particular may lead to significant detrimental effects on production and reproduction of some livestock species.
- The risk of losses due to weeds, insects, and diseases is likely to increase.

With regard to regional and global production effects:

- Crop yields and productivity changes will vary considerably across regions. Thus, the pattern of agricultural production is likely to change in a number of regions, with some areas experiencing significantly lower crop yields and other areas experiencing higher yields.
- Global agricultural production can be maintained relative to base production under climate change as expressed by GCMs under doubled CO_2 equilibrium climate scenarios.
- Based on global agricultural studies using doubled CO_2 equilibrium GCM scenarios, lower latitude and lower income countries are likely to be more negatively affected.

With regard to vulnerability and adaptation:

- Vulnerability to climate change depends on physical and biological response, but also on socioeconomic characteristics. Low-income populations depending on isolated agricultural systems, particularly dryland systems in semi-arid and arid regions, are especially vulnerable to hunger and severe hardship. Many of these at-risk populations are found in Sub-Saharan Africa, South and Southeast Asia, some Pacific island countries, and tropical Latin America.

- Historically, farming systems have responded to a growing population and have adapted to changing economic conditions, technology, and resource availability. It is uncertain whether the rate of change of climate and required adaptation would add significantly to the disruption likely due to future changes in economic conditions, population, technology, and resource availability.
- While adaptation to climate change is likely; the extent depends on the affordability of adaptive measures, access to technology, and biophysical constraints such as water resource availability, soil characteristics, genetic diversity for crop breeding, and topography. Many current agricultural and resource policies are likely to discourage effective adaptation and are a source of current land degradation and resource misuse.
- National studies have shown incremental additional costs of agricultural production under climate change that could create a serious burden for some developing countries.
- Material in the 1995 IPCC Working Group II report was reorganized by region with some updated material in a subsequent special report. Included among the chapters was a report on North America (Shriner and Street, 1998).

1995-1996. The Economic Research Service of the USDA

The Economic Research Service of the USDA (Schimmelpfennig et al., 1996) provided a review and comparison of studies that it had conducted and/or funded, contrasting them with previous studies. The assessment used the same doubled CO_2 equilibrium scenarios of many previous studies (global average surface temperature increases of 2.5 to 5.2°C or 4.5 to 9.4°F). Two of the main new analyses reviewed in the study used cross-section evidence to evaluate climate impacts on production. One approach was a direct statistical estimate of the impacts on land values for the US (Mendelsohn et al., 1994), while the other (Darwin et al., 1995) used evidence on crop production and growing season length in a model of world agriculture and the world economy. Both imposed climate change on

the agricultural sector as it existed in the base year of the studies (e.g., 1990). A major result of the approaches based on cross-section evidence was that impacts of climate were far less negative for the US and world than had previously been estimated with crop modeling studies. While these studies showed similar economic effects as previous studies, they included no direct effects of CO_2 on crops, which in previous studies had been a major factor behind relatively small effects. Hence, if the direct effects of CO_2 on crop yields would have been included, the result would have been significant benefits. The more positive results were attributed to adaptations implicit in cross-section evidence that had been incompletely factored in to previous analyses. The report also contained a crop modeling study (Kaiser et al., 1993) with a complete farm-level economic model that more completely simulated adaptation responses. It also showed more adaptation than previous studies. A summary of this review was subsequently published as Lewandrowski and Schimmelpfennig (1999).

1998-1999: Pew Center Assessment

As part of a series on various aspects of climate change aimed at increasing public understanding, the Pew Center on Global Climate Change completed a report on agriculture (Adams et al., 1999). The report series is based on reviews and synthesis of the existing literature. The major conclusions were:

- Crops and livestock are sensitive to climate changes in both positive and negative ways.
- The emerging consensus from modeling studies is that the net effects on US agriculture associated with doubling of CO_2 may be small; however, regional changes may be significant (i.e., there will be some regions that gain and some that lose). Beyond a doubling of CO_2, the negative effects would be more pronounced, both in the US and globally.
- Consideration of adaptation and human response is critical to an accurate and credible assessment.
- Better climate change forecasts are a key to improved assessments.
- Agriculture is a sector that can adapt, but changes in the incidence and severity of pests, diseases, soil erosion, tropospheric ozone, variability, and extreme events have not been factored in to most of the existing assessments.

LITERATURE CITED

Adams, R. M., C. Rosenzweig, R. M. Peart, J. T. Richie, B. A. McCarl, J. D. Glyer, R. B. Curry, J. W. Jones, K. J. Boote, and L. H. Allen, Global climate change and US agriculture, *Nature, 345,* 219-224, 1990.

Adams, R.M., S.A. Hamilton, and B.A. McCarl, The Benefits of Air Pollution Control: The Case of Ozone and US Agriculture, *American Journal of Agricultural Economics,* 68:886-894, 1986.

Adams, R.M., R.A. Fleming, C.C. Chang, B.A. McCarl, and C. Rosenzweig, A reassessment of the economic effects of global climate change on US agriculture, *Climatic Change, 30,* 147-167, 1995.

Adams, D., R. J. Alig, B.A. McCarl, J. M. Callaway, and S. Winnett, The forest and agricultural sector optimization model: Model structure and applications, USDA Forest Service Research Paper PNW-RP-495, USDA Forest Service, Pacific Northwest Experiment Station, Portland, Oregon, 1997.

Adams, R. M., B. H. Hurd, and J. Reilly, A review of impacts to US agricultural resources, Pew Center on Global Climate Change, Arlington, VA, 36 pp., 1999a.

Baumes, H., A Partial Equilibrium Sector Model of US Agriculture Open to Trade: A Domestic Agricultural and Agricultural Trade Policy Analysis, Ph.D. thesis, Purdue University, 1978.

Ben Mechlia, N., and J. J. Carrol, Agroclimatic modeling for the simulation of phenology, yield, and quality of crop production: 1. Citrus response formulation, *International Journal of Biometeorology, 33,* 36-51, 1989.

Burton, R.O., and M.A. Martin, Restrictions on Herbicide Use: An Analysis of Economic Impacts on US Agriculture." *North Central Journal of Agricultural Economics,* 9:181-194, 1987.

CAST (Council for Agricultural Science and Technology), Preparing US agriculture for global climate change, *Task Force Report No. 119,* CAST, Ames, Iowa, 96 pp., 1992.

Chang, C.C., B.A. McCarl, J.W. Mjelde, and J.W. Richardson, Sectoral Implications of Farm Program Modifications, *American Journal of Agricultural Economics,* 74:38-49, 1992.

Chaudhuri, U. N., M .B. Kirkam, and E.T. Kanemasu, Root growth of winter wheat under elevated carbon dioxide and drought, *Crop Science, 30,* 853-857, 1990.

Darwin, R., M. Tsigas, J. Lewandrowski, and A. Raneses, World agriculture and climate change: Economic adaptations, *Agricultural Economic Report No. 703,* US Department of Agriculture, Natural Resources and Environmental Division, Economic Research Service, Washington, DC, 1995.

Easterling, W.E., III, P. R. Crosson, N. J. Rosenberg, M. McKenney, L. A. Katz, and K. Lemon, Agricultural impacts of and responses to climate change in the Missouri-Iowa-Nebraska-Kansas (MINK) region, *Climate Change, 24,* 23-61, 1993.

Fajardo, D., B.A. McCarl, and R. L. Thompson, A Multi commodity Analysis of Trade Policy Effect: The Case of Nicaraguan Agriculture. *American Journal of Agricultural Economics* 63:23-31, 1981.

Fellin, L., and S. Fuller, Effects of Privatization Mexico's Railroad System on US-Mexico Overland Grain/Oilseed Trade, *Transportation Research Forum* Vol 37, N1:46-64, 1998.

House, R.M., USMP Regional Agricultural Model, United States Department of Agriculture, Economic Research Service, National Economics Division, Draft, 1987.

Izarraulde, C. R., R. A. Brown, and N. J. Rosenberg, US regional agricultural production in 2030 and 2095: Response to CO_2 fertilization and Hadley climate model (HadCM2) projections of greenhouse-forced climatic change, PNNL-12252, Pacific Northwest National Laboratories, Richland, Washington, 1999.

Johnson, D. G., *The World Grain Economy and Climate Change to the Year 2000: Implications for Policy,* National Defense University Press, Washington, DC, 50 pp., 1983.

Kaiser, H. M., S. J. Riha, D. S. Wilks, D .G. Rossier, and R. Sampath, A farm-level analysis of economic and agronomic impacts of gradual warming, *American Journal of Agricultural Economics, 75,* 387-398, 1993.

Kane, S., J. Reilly, and J. Tobey, Climate change: Economic implications for world agriculture, AER No. 647, US Department of Agriculture, Natural Resources and Environmental Division, Economic Research Service, Washington, DC, 1991.

Kane, S., J. Reilly, and J. Tobey, An empirical study of the economic effects of climate change on world agriculture, *Climatic Change, 21*, 17-35, 1992.

Kimball, B.A., Carbon dioxide and agricultural yield: An assemblage and analysis of 430 prior observations, *Agronomy Journal, 75*, 779-786, 1983.

Kimball, B.A., P.J. Pinter, Jr., R. L. Garcia, R. L. LaMorte, G. W. Wall, D .J. Hunsaker, G. Wechsung, F. Wechsung, and T. Kartschall, Productivity and water use of wheat under free-air CO_2 enrichment, *Global Change Biology, 1,* 429-442, 1995.

Lambert, D.K., B.A. McCarl, Q. He, M.S. Kaylen, W. Rosenthal, C.C. Chang, and W.I. Nayda, Uncertain Yields in Sectoral Welfare Analysis: An Application to Global Warming, *Journal of Agricultural and Applied Economics,* 423-435, 1995.

Lewandrowski, J. and D. Schimmelpfennig, Agricultural adaptation to climate change: issues of long-run sustainability, *Land Economics, 75*(1), 39-57, 1999.

Long, S. P., Modification of the response of photosynthetic productivity to rising temperature by atmospheric CO_2 concentrations: Has its importance been underestimated?, *Plant, Cell and Environment, 14*(8), 729-739, 1991.

McCarl, B.A. and T.H. Spreen, Price Endogenous Mathematical Programming As a Tool For Sector Analysis, *American Journal of Agricultural Economics, 62*:87-102, 1980.

Mendelsohn, R., W. D. Nordhaus, and D. Shaw, The impact of global warming on agriculture: A Ricardian analysis, *The American Economic Review, 84*(4), 753-771, 1994.

Mitchell, R.A.C., V.J. Mitchell, S.P. Driscoll, J. Franklin, and D.W. Lawlor, Effects of increased CO_2 concentration and temperature on growth and yield of winter wheat at two levels of nitrogen application, *Plant, Cell, and Environment, 16,* 521-529, 1993.

Office of Technology Assessment, *Preparing for an Uncertain Climate,* Office of Technology Assessment, US Congress, Washington, DC, 1993.

Parry, M. L., Agriculture and forestry, in *Climate Change: The IPCC Impacts Assessment,* edited by W. J. McG. Tegarrt, G. W. Sheldon, and D. C. Griffiths, UN Intergovernmental Panel on Climate Change, Australian Government Printing Office, Canberra, 1990a.

Parry, M. L., *Climate Change and World Agriculture,* Earthscan Publications Ltd., London, England, 1990b.

Parton, W. J., D. S. Schimel, and D. S. Ojima, Environmental change in grasslands: Assessment using models, *Climate Change, 28*(1-2), 111-141, 1994.

Patterson D.T., J. K. Westbrook, R. J. V. Joyce, and P. D. Lindgren, Weeds, insects, and diseases, *Climatic Change, 43*(4), 711-727, 1999.

Reilly, J. (Guest Editor), Climate Change, Impacts on Agriculture, *Climatic Change, 43*(4), 645-793, 1999.

Reilly, J., et al., Changing climate and changing agriculture: Report of the Agricultural Sector Assessment Team, available at: http://www.nacc.usgcrp.gov/sectors/agriculture/working-papers.html, in review, 2000.

Reilly, J., N. Hohmann, and S. Kane, Climate change and agriculture: Global and regional effects using an economic model of international trade, MIT-CEEPR 93-012WP, Center for Energy and Environmental Policy Research, Massachusetts Institute of Technology, Cambridge, Massachusetts, August 1993.

Reilly, J., N. Hohmann, and S. Kane, Climate change and agricultural trade: who benefits, who loses?, *Global Environmental Change, 4*(1), 24-36, 1994.

Reilly, J. et al., Agriculture in a changing climate: Impacts and adaptations, in *Climate Change 1995: Impacts, Adaptations, and Mitigation of Climate Change,* edited by R. T. Watson, M. C. Zinyowera, and R. H. Moss, Cambridge University Press, Cambridge, United Kingdom, pp. 427-469, 1995.

Roningen, V., A Static World Policy Simulation (SWOP-SIM) Modeling Framework, Staff Rep. No. AGE860265, US Department of Agriculture, Economic Research Service, Washington DC, 1986.

Rosenberg, N.J., 1993. Towards an integrated assessment of climate change: the MINK study. *Climatic Change., 24*: 1-175.

Rosenzweig, C., and D. Hillel, *Climate Change and the Global Harvest: Potential Impacts on the Greenhouse Effect on Agriculture,* Oxford University Press, New York, 1998.

Rosenzweig, C., A. Iglesias, X. B. Yang, P. R. Epstein, and E. Chivian, Climate change and US agriculture: The impacts of warming and extreme weather events on productivity, plant diseases, and pests, Center for Health and the Global Environment, Harvard University, Cambridge, Massachusetts, pp. 46, 2000.

Rosenzweig, C., and M. L. Parry, Potential impact of climate change on world food supply, *Nature, 367,*133-138, 1994.

Rosenzweig, C., J. T. Ritchie, J. W. Jones, G. Y. Tsuji, and P. Hildebrand, Climate change and agriculture: Analysis of potential international impacts, *American Society of Agronomy Special Publication No. 59,* American Society of Agronomy, Madison, Wisconsin, 382 pp., 1995.

Rosenzweig, C., and F. N. Tubiello, Impacts of future climate change on Mediterranean agriculture: Current methodologies and future directions, *Climate Change, 1,* 219-232, 1997.

Schimmelpfennig, D., J. Lewandrowski, J. Reilly, M. Tsigas, and I. Parry,. Agricultural adaptation to climate change: issues of long-run sustainability, *Agricultural Economics Report No. 740,* Economic Research Service, USDA, Washington DC, June 1996.

Shriner, D. S., and R. B. Street, North America, in *The Regional Impacts of Climate Change: An Assessment of Vulnerability,* edited by R. T. Watson, M. C. Zinyowera, and R. H. Moss, Cambridge University Press, Cambridge, United Kingdom, pp. 253-330, 1998.

Sionit, N., D. A. Mortensen, B. R. Strain, and H. Hellmers, Growth response of wheat to CO_2 enrichment and different levels of mineral nutrition, *Agronomy Journal, 73,* 1023-1027, 1981.

Smith, J. B., and D. Tirpak (Eds.), The potential effects of global climate change on the United States, EPA-230-05-89-050, US Environmental Protection Agency, Washington, DC, 1989.

Southworth, J., J. C. Randolph, M. Habeck, O. C. Doering, R. A. Pfeifer, G. Rao, and J. Johnston, Consequences of future climate change and changing climate variability on corn yields in the Midwest US, *Agriculture, Ecosystems, and Environment,* in press, 2000.

Strzepek, K. M., D. C. Major, C. Rosenzweig, A. Iglesias, D. N. Yates, A. Holt, and D. Hillel, New methods of modeling water availability for agriculture under climate change: The US Cornbelt, *Journal of the American Water Resources Association, 35,* 1639-1655, 1999.

Takayama, T. and G. G. Judge, *Spatial and Temporal Price and Allocation Models,* North-Holland Publishing Company, 1971.

Tobey, J., J. Reilly, and S. Kane, Economic implications of global climate change for world agriculture, *Journal of Agricultural and Resource Economics, 17*(1),195-204, 1992.

Tsuji, G. Y., G. Uehara, and S. Balas (Eds.), DSSAT v3, University of Hawaii, Honolulu, Hawaii, 1994.

Tubiello, F. N., C. Rosenzweig, B. A. Kimball, P. J. Pinter, Jr., G. W. Wall, D. J. Hunsaker, R. L. Lamorte, and R. L. Garcia, Testing CERES-wheat with FACE data: CO_2 and water interactions, *Agronomy Journal, 91,* 247-255, 1999.

US Department of Agriculture, *Agricultural Statistics,* US Government Printing Office, Washington, D.C., 1987.

US Department of Agriculture, FEDS Budgets, Economic Research Service, 1982.

US Department of Agriculture, Natural Resources Conservation Service, 1997.

Walker, B., and W. Steffen (Eds.), *Global Change and Terrestrial Ecosystems,* IGBP Book Series, Cambridge University Press, Cambridge, United Kingdom, 620 pp., 1996.

ACKNOWLEDGMENTS

Many of the materials for this chapter
are based on contributions from participants on and
those working with the

Agriculture Sector Assessment Team
John Reilly*, Massachusetts Institute of Technology
Jeff Graham*, US Department of Agriculture
(through Sept. 1999)
James Hrubovcak*, US Department of Agriculture
(from October 1999)
David G. Abler, Pennsylvania State University
Robert A. Brown, Battelle-Pacific Northwest National
Laboratory
Roy Darwin, US Department of Agriculture
Steven Hollinger, University of Illinois
Cesar Izaurralde, Battelle-Pacific Northwest National
Laboratory
Shrikant Jagtap, University of Florida-Gainesville
James Jones, University of Florida-Gainesville
John Kimble, US Department of Agriculture
Bruce McCarl, Texas A&M University
Linda Mearns, National Center for Atmospheric
Research
Dennis Ojima, Colorado State University
Eldor A. Paul, Michigan State University
Keith Paustian, Colorado State University
Susan Riha, Cornell University
Norm Rosenberg, Battelle-Pacific Northwest
National Laboratory
Cynthia Rosenzweig, NASA-Goddard Institute for
Space Studies
Francesco Tubiello, NASA-Goddard Institute for
Space Studies

* Assessment Team chair/co-chair

CHAPTER 14

POTENTIAL CONSEQUENCES OF CLIMATE VARIABILITY AND CHANGE FOR THE WATER RESOURCES OF THE UNITED STATES

Katharine Jacobs[1,2], D. Briane Adams[3], and Peter Gleick[4]

Contents of this Chapter

[1]Arizona Department of Water Resources; [2]Coordinating author for the National Assessment Synthesis Team; [3]US Geological Survey, [4]Pacific Institute for Studies in Development, Environment and Security.

CHAPTER SUMMARY

Context

Water supply conditions in all regions and sectors in the US are likely to be affected by climate change, either through increased demands associated with higher temperatures, or changes in precipitation and runoff patterns. Water sector concerns include effects on ecosystems, particularly aquatic systems such as lakes, streams, wetlands, and estuaries. Although competition for water supplies is extremely intense, particularly in the western US, substantial ability to adjust to changing demands for water exists in the current water management system. It is not known whether the effects of climate change will require dramatic changes in infrastructure to control flooding and provide reliable water supplies during drought. However, it is known that precipitation and temperature changes are already increasing runoff volumes and changing seasonal availability of water supply, and that these changes are likely to be more dramatic in the future.

Climate of the Past Century

- Increases in global temperatures have been accompanied by more precipitation in the mid and high latitudes.
- Precipitation has increased an average of 10% across the US, with much of the increase attributed to heavy precipitation events.
- Nationally, streamflow has increased about three times more than the increase in precipitation.
- Regionally, the higher streamflows have increased in many areas, but not in the West where snowmelt dominates peak flows.
- Reductions in areal extent of snowpack in the western mountains have been observed, along with substantial retreat of glaciers.
- In snowpack-dominated streams, a shift has been observed in the timing of the peak runoff to earlier in the season.
- No significant increases in the frequency of droughts or winter-type storms have been observed on a national basis.

Climate of the Coming Century

- Historic trends towards increased precipitation are very likely to continue.
- It is possible that there will be an increase in interannual variability, resulting in more severe droughts in some years.
- The Canadian and Hadley climate models used in the Assessment generally do not agree on precipitation impacts, with the exception of showing an increase in precipitation in the Southwest.
- Increases in temperature, even in the context of increases in precipitation, are likely to result in significant loss of soil moisture in the Northern Great Plains.
- Snowpack is very likely to be reduced even in the context of higher precipitation.
- If the number of high intensity storm events increases, flushing of contaminants into watersheds is likely to increase, causing episodic water quality problems.
- Quality and quantity impacts are very likely to be regionally specific.
- Surprises are likely, since many water-related impacts cannot be predicted.

Key Findings

- More pressure on surface water supplies is likely to come from population shifts and changes in water right allocations to accommodate endangered species and the water rights of Native Americans. Although wetter conditions in the Southwest may alleviate some of these stresses, stress is likely to increase in the Northern Great Plains and in snowpack-dependent watersheds.

- Groundwater supplies are already over-drafted in many parts of the country, and pressure on groundwater supplies is likely to increase to offset changes in surface water supply availability. However, long-term increases in precipitation will possibly increase recharge rates in some areas.

- It is likely that aquatic and riparian ecosystems may be damaged even in the context of higher precipitation, due to higher air temperatures and reduced summer flows. It is also probable that changes in water temperature in lakes and streams will affect species composition.

- Water managers have multiple opportunities to reduce future risks by incorporating "no-regrets" changes into their operating strategies that are appropriate regardless of climate change.

- Institutions governing water rights are generally very inflexible, and are likely to prove to be obstacles to adaptation.

- Improvements are needed in monitoring efforts to identify key impacts related to water quantity and quality, biological conditions of key habitats, snowpack conditions, and groundwater supplies.

POTENTIAL CONSEQUENCES OF CLIMATE VARIABILITY AND CHANGE FOR THE WATER RESOURCES OF THE UNITED STATES

BACKGROUND

Water-related concerns are central to this National Assessment because the hydrologic (water) cycle is a fundamental component of climate and because water plays a role in every sector and region in the US. Despite many remaining uncertainties, a significant amount of research has been done on the connections between climate change and water resources in the US; a searchable bibliography of almost 900 scientific articles is available at http://www.pacinst.org.

The US has a wide variety of tools, institutions and methods for coping with water resource problems, and many of these will be useful for addressing the impacts of climate changes. Water managers already deal with climate variability; reservoirs are designed with some flexibility for extreme high and low flows; techniques and technologies are available for managing water demands. But global climate change raises some unresolved concerns for the water sector. What will be the economic costs of coping with climate changes imposed on top of existing variability? Are existing institutions sufficiently flexible to handle the additional stresses? What might be the nature of unexpected climate "surprises" for the water sector? And will water managers be willing and able to prepare in advance for conditions different from those they are normally faced with?

Figure 1: The Central Arizona Project brings Colorado River water 330 miles uphill to Tucson and Phoenix, Arizona. Source: K. Jacobs

Increases in greenhouse gas concentrations are very likely to affect global temperature and lead to changes in the amount, timing, and geographic distribution of rain, snowfall, and runoff. Changes are also likely in the timing, intensity, and duration of extreme events such as floods and droughts. Such changes possibly will have greater impacts on the regions and sectors than changes in average temperature or precipitation. Higher demand for water is probable in areas where increased temperature results in higher evapotranspiration. Although in most regions it is possible that increased streamflow would relieve current stress, some regions are likely to experience greater difficulty in meeting their water supply needs if precipitation increases do not offset increases in evapotranspiration. Key variables in determining likely impacts and responses include changes in soil moisture and cloud cover, seasonality of precipitation, and the response of vegetation to changes in moisture, temperature, and increased carbon dioxide availability.

While many of the most significant impacts in the agricultural, forestry, ecosystem, energy, and human health sectors relate to the basic issue of water availability, it is likely that there will be some serious impacts on water quality as well. There is a direct relationship between quantity of flows and dilution of pollutants in surface water; higher runoff is likely to improve water quality, but increased intensity of rainfall will probably result in increased erosion and flushing of contaminants into watersheds. Higher water temperature will affect the ecology of wetlands, lakes, and streams. Much less research has been done on impact-related issues.

The primary water resource issue for the US is the distribution of supply and demand, not the total quantity of water available. The nature of water concerns varies by region across the country. For much of the western US, water resources are often separated both by time and distance from water demands. As a result, substantial infrastructure has been developed to store and transport water supplies (for example, from the Colorado River and Northern California to the Southwest and Southern California). There are more than 80,000 dams and reservoirs in the US, and millions of miles of canals, pipes, and tunnels (Schilling et al., 1987, see Figure 1). Although this infrastructure is sophisticated and has allowed the development of urban and agricultural areas, it is also a source of vulnerability to climate change, partially because it has been designed based on the assumption that future conditions will be similar to the historically observed climate. Some argue that there is substantial robustness built into the system that provides some margin of safety, but failure to re-evaluate these assumptions and identify key vulnerabilities may prove to be costly in the future.

Water supply issues in the eastern US relate to aging infrastructure and inadequate storage capacity during times of drought. Flood control issues and environmental impacts of structural solutions are also of concern. In general, local surface water and groundwater supplies are available for domestic and industrial use without major water transfers between basins, but excess reservoir capacity to respond to drought is quite limited. In some areas, such as New York City, reservoir function is threatened by upstream development. New York and other cities also have serious problems on a regular basis with water main breaks causing flooding and other damage.

The initial charge of this Assessment included identifying areas of existing stress and vulnerability and evaluating new problems that climate change may bring. This has necessarily resulted in an identification of negative effects, though certain aspects of climate change are likely to improve conditions in some areas of the US. Even in the absence of climate change, adapting to existing stresses (such as aging infrastructure, inadequate water supplies for areas of rapid growth, etc.) and increased pressures from population dynamics would be expensive. Frederick and Schwarz (1999a) estimate that the annualized water-related costs associated with the demands of an increasing population are likely to approach $13.8 billion by 2030. The impacts of climate change on these costs depend on the nature of the changes. The estimated costs include investments in new water supplies and conservation measures as well as the impacts on streamflows and irrigated lands. The costs would be much higher if climate change were to significantly decrease water availability (as under the Canadian climate model scenario) or increase the magnitude and timing of extreme events. This is because the current infrastructure and management practices are designed based on the historical climate conditions.

Inadequate water and wastewater infrastructure, common along the Mexican border and in some rural areas, leads to high risk for health problems. Because the public has very high expectations regarding water quality, and perceptions of health risks are not always accurate, health issues may have a high profile and require particular attention.

Health-related issues that have been linked to changes in the hydrologic cycle include potential for increases in water-borne pathogens such as *Cryptosporidium* and vector-borne diseases such as encephalitis, as well as outbreaks in marine pathogens associated with red tide (Bernard et al., 1999). Hantavirus, a disease spread by deer mice, has also been linked to extreme climate variability associated with the El Niño Southern Oscillation (ENSO). With higher rainfall, rodent populations tend to increase, which increases the chance of human contact and disease.

Virtually all indices of vulnerability relative to water have identified the over-appropriation of western streams and rivers and over-drafting of groundwater supplies as key issues. Gleick (1990) identified indicators of water resource vulnerability for the US and found that the most vulnerable regions were the high irrigation areas along the eastern drainage of the Rocky Mountains, the Central Valley of California, and Southern California. The overall index prepared by Hurd et al. (1999a) indicates that the most vulnerable watersheds are in the West, Southwest, and Great Plains (see Figure 2).

Current Climate Vulnerability Map, Water Supply, Distribution and Consumptive Use

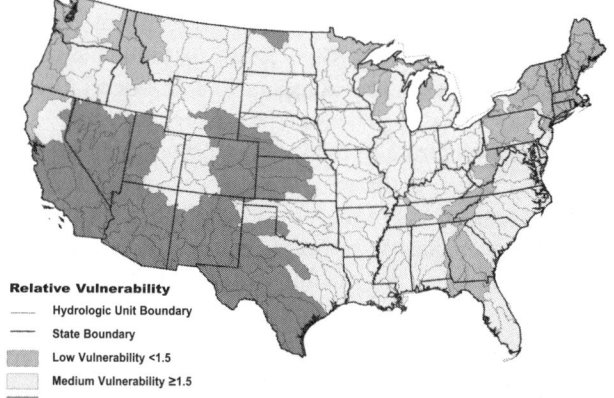

Figure 2: Assessed vulnerability based on current climate and water resource conditions, based on data describing the following: share of streamflow withdrawn for use, streamflow variability, evapotranspiration rate, groundwater overdraft, industrial use savings potential, and water trading potential. Source: Hurd, B.J., N. Leary, R. Jones and J. Smith. (1999a). See Color Plate Appendix

A rise in average temperature, even in the context of higher precipitation, is most likely to impact aquatic systems, including riparian habitat, and freshwater and estuarine wetlands. In some cases, this is because expected changes in precipitation do not offset increased evapotranspiration, though seasonal and regional impacts are likely to vary. Certain coastal systems, prairie potholes (small ponds and lakes formed by glacial deposits), and Arctic and alpine ecosystems are thought to be especially vulnerable. Stresses within the contiguous US are likely to come from changes in the distribution of precipitation as well as increases in its intensity.

SOCIOECONOMIC AND INSTITUTIONAL CONTEXT

Population pressures, including shifts towards western and coastal urban areas, land use practices, and climate change are all likely to increase stress on water supply systems. The need to reserve water for instream uses, endangered species protection, recreation, and American Indian water rights settlements also places new demands on a water rights system that in many parts of the country is already seriously stressed. As society changes, its value system also evolves. Placing more value on protection of fish and wildlife habitat and recreational values is likely to force institutional change at the same time that new stresses are appearing due to climate variability and change. Although there is substantial uncertainty in the projections of changes in runoff that are derived from the climate models, socioeconomic conditions are even less predictable.

There is a need for more flexible institutional arrangements and more effective ways of making water policy decisions in order to adapt to changing conditions (not just changes in climate, but multiple existing stresses). The legal framework for water rights varies from state to state, with nearly infinite permutations at the local level. The one characteristic that is typical of most institutions related to water is inability to respond efficiently to changing socioeconomic and environmental conditions. This is primarily because institutions tend to reflect existing water right holders' interests, and substantial investments are made based on expectations regarding availability of supplies. Devising new legal and related institutions that can introduce the necessary flexibility into water management without destabilizing investors' expectations, while at the same time incorporating public values (ecological, recreational, aesthetic, etc.) is a significant challenge.

The state to state variations among water rights systems results in substantial complexity. In general, water rights in eastern states are not likely to be easily quantifiable, which limits management options. The prior appropriation doctrine of the western US is relatively inflexible in dealing with changing environmental and societal needs (see box, "Major Doctrines for Surface Water and Groundwater").

Some innovative institutions are developing in response to particular problems. For example, the "temporary" water banks in California to respond to drought and in Arizona to respond to long-term supply reliability issues offer some protection to existing water rights while providing much-needed flexibility. Water banks generally provide opportunities for short-term transfers of agricultural water supplies to municipal end users on a willing buyer/willing seller basis. In the case of the Arizona Water Banking Authority, excess Colorado River water is being stored underground through recharge projects to offset future shortfalls in supply. This opportunity is expected to be available on an interstate basis among the Lower Colorado Basin states in the near future. Similar types of contingency planning between jurisdictions and water rights holders could prove beneficial in responding to short-term emergencies. Longer-term changes in climatic conditions that would require permanent changes to legal systems could be more problematic.

Many have argued that an open market in water rights would help resolve conflict and increase efficiency because water would flow to the highest and best use based on willingness to pay (National Research Council, 1992; Western Water Policy Review Advisory Commission, 1998). It is widely acknowledged that market-related solutions may relieve some water supply problems, especially in the West. However, water marketing is an imperfect solution. Of particular concern are third party impacts in water transfers, and overall equity issues. Water markets are developing in many states, but they are generally regulated markets in order to protect the public interest. Mechanisms exist to identify economic values for non-market goods and services, but water rights for non-market values such as ecosystems, aesthetics, and recreation have difficulty competing with major economic forces. There is also a risk that disproportionate burdens will be placed on the social groups that can least afford them (such as rural farming communities, Native Americans, and communities along the Mexican border with inadequate infrastructure) (Dellapenna, 1999b; Gomez and Steding, 1998).

Major Doctrines for Surface Water and Groundwater

Surface Water

Riparian doctrine – Authorization to use water in a stream or other water body is based on ownership of the adjacent land. Each landowner may make reasonable use of water in the stream but must not interfere with its reasonable use by other riparian landowners. The riparian doctrine prevails in the 31 humid states east of the 100th meridian.

Prior appropriation doctrine – Users who demonstrate earlier use of water from a particular source acquire rights over all later users of water from the same source. When shortages occur, those first in time to divert and apply the water to beneficial use have priority. New diversions, or changes in the point of diversion or place or purpose of use, must not cause harm to existing appropriators. The prior appropriation doctrine prevails in the 19 western states.

Groundwater

Absolute ownership – Groundwater belongs to the overlying landowner, with no restrictions on use and no liability for causing harm to other existing users. Texas is the sole absolute ownership state.

Reasonable use doctrine – Groundwater rights are incident to land ownership. Owners of overlying land are entitled to use groundwater only to the extent that the uses are reasonable and do not unreasonably interfere with other users. Most eastern states and California subscribe to this doctrine. Some states, such as Arizona, have modified the reasonable use doctrine by requiring state permits to use groundwater in certain high use areas.

Prior appropriation permit system – Groundwater rights are determined by the rule of priority, which provides that prior uses of groundwater have the best legal rights. States administer permit systems to determine the extent to which new groundwater uses will be allowed to interfere with existing uses. Most western states employ some form of permit system.

Sources: US Army Corps of Engineers, Volume III, Summary of Water Rights – State Laws and Administrative Procedure report prepared for US Army, Institute for Water Resources, by Apogee Research, Inc., June 1992; and US Geological Survey, National Water Summary 1988-89-Hydrologic Events and Floods and Droughts, Water Supply Paper 2375 (Washington, DC, US Government Printing Office, 1991).

Dellapenna (1999a) says true (unregulated) markets have seldom existed for water rights and there are good reasons for believing that they seldom will. This is because water, like air, is viewed as a "public good," which means that people cannot realistically be excluded from using it, at least on a subsistence basis. There is reluctance to pay the full cost of water, including the replacement cost; people are generally charged only the cost for capturing and distributing water. Key factors in developing a workable market are whether the market will enable consumers to meet their needs and whether government regulation and assistance at the margins can correct for market failures.

Numerous institutional issues related to responding to potential climate change stem from the fact that various agencies and levels of government handle water quality and water quantity issues. Water quality regulation derives primarily from federal authorities such as the Clean Water Act and the Safe Drinking Water Act. Water quantity regulation (through rights, allocations, and permits) is primarily handled by the states. An illustration of this problem is found in a study by Eheart et al. (1999), which evaluated the impact of reduced precipitation in the Midwest on ability to meet federal discharge water quality standards. They found that a 25% reduction in precipitation could reduce the critical dilution flow that determines discharge impact on water quality by 63%. They concluded that this has implications for the process of setting Total Maximum Daily Loads (TMDL) for Non-Point Discharge Elimination Permits under the Clean Water Act. Section 303(d) requires states to identify waters that do not meet water quality standards and to establish plans to achieve the TMDL standards. Eheart et al. (1999) noted that the present regulatory scheme in many midwestern states is not sophisticated enough to take into account the interplay between water quality and quantity.

The Tennessee Valley Authority –
Integrated Water Resources Management

A well-known example of integrated water resources management is the Tennessee Valley Authority (TVA), which operates in a seven-state area in the southeastern US. Founded in 1933, TVA pioneered the concept of "unified river basin development" within the Tennessee River Basin, integrating water resources development, social and economic development, power production, and natural resources conservation.

TVA's water management programs focus on the operation of a large, multipurpose reservoir system that includes more than 50 dams and reservoirs. The system is operated as an integrated unit to provide for navigation, flood control, hydropower generation, summer recreation levels, and minimum flows for the maintenance of water quality and aquatic habitat. In one of the most flood-prone areas in the US, TVA has historically taken a dual approach to flood management that combines reservoir system control with a floodplain management program to encourage appropriate shoreline development. Environmental concerns are integrated into reservoir operations, while TVA's Watershed Teams work at grassroots levels to motivate local action to control non-point source pollution. TVA also maintains web-sites and special telephone systems to facilitate public access to streamflow data, dam release information, and other system information.

TVA has a sophisticated streamflow and rainfall data collection and monitoring system, coupled with a state-of-the-art simulation and optimization modeling system. Monitoring and forecasting occur on a continuous, 24-hour basis. Additionally, TVA utilizes 10-day and seasonal weather forecasts to guide reservoir planning. TVA is engaged in joint work with NOAA to better utilize seasonal forecast information. These capabilities will assist TVA in adapting to climate change and variability.

Figure 3: TVA has more than 50 dams in seven states. Source: Tennessee Valley Authority

The institutions that have been successful in managing water resources tend to use an integrated approach to management, and to incorporate natural watershed boundaries rather than political boundaries for management areas. Relatively innovative water management districts have been formed in many states to address specific resource conditions (see box, "Tennessee Valley Authority – Integrated Water Resources Management").

CLIMATE VARIABILITY AND CHANGE

Historic trends show that the surface temperature of the Earth has increased by about 1˚F (just over 0.6˚C) over the 20th century, with 1998 the warmest year on record. Higher temperatures have resulted in reductions in snow cover and sea ice extent.

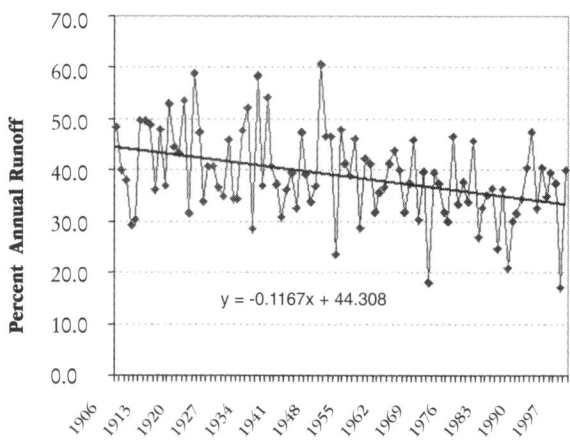

Figure 4: In some western watersheds, runoff timing appears to be shifting from spring to winter, suggesting a change in snowfall and snowmelt dynamics. Source: Gleick, P.H. and E.L. Chalecki. (1999) JAWRA. Dec. pp. 1429-1442.

Trend analysis shows that since WWII, there has been significant retreat of snow cover in spring (Groisman, 1999). At the same time, there appear to be shifts in the seasonality of runoff in some western rivers consistent with what would be expected from changes in snowfall and snowmelt dynamics due to warming (Gleick and Chalecki, 1999; Union of Concerned Scientists, 1999) (see Figure 4).

Increases in global temperatures have been accompanied by more precipitation in the mid and high latitudes (based on Northern Hemisphere land-surface records) and increases in atmospheric water vapor in many regions of North America and Asia where data are adequate for analysis (IPCC, 1996). Karl et al. (1996) show that the meteorological drought indices suggest that there have been more wet spells, but no significant changes in drought on a national basis. The precipitation increase in the US has been attributed primarily to an increase in the heaviest precipitation events (Karl and Knight, 1998; Groisman et al., 1999; Karl et al., 1996). These changes are statistically significant and most apparent during the spring, summer, and autumn months in the contiguous US. Based on recent work by Lins and Slack (1999), the warm season precipitation increases may be responsible for increases in streamflow in the low to moderate range (i.e. the flow values that are most commonly observed during the summer and early autumn months). Using discharge data from a national network of stream gages for the period 1944-1993, Lins and Slack (1999) found statistically significant increases in the annual median streamflow at 29% of the stream gages nationwide and decreases at only 1% of the stream gages. Most trends were even more positive for the lower streamflow quantiles. Fewer significant trends were observed in high streamflows. Only 9% of the gages, for example, had significant trends in the annual maximum streamflow and, of these, more showed decreases than increases. Groisman et al. (2000) show that high streamflow in the mountainous West has not changed despite increases in heavy precipitation events. They attribute this to a trend toward reduced snow cover extent leading to a lower and earlier peak in the annual cycle of runoff. In the East and South however, increasing trends in high and very high streamflow are shown to relate to increases in heavy and very heavy precipitation events. In fact, there is an amplification of the trends in precipitation across the highest precipitation and streamflow rates by a factor of about three. It is well-recognized (Karl and Reibsame, 1989) that small changes in precipitation can be amplified into large changes in streamflow (see Figure 5).

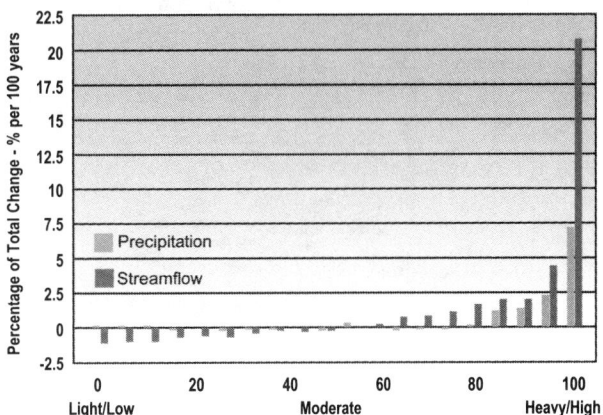

Observed Changes In Streamflow and Precipitation (1939-99)

Figure 5: The graph shows changes in the intensity of precipitation and streamflow, displayed in 5% increments, during the period 1939-99 based on over 150 unregulated streams across the US with nearby precipitation measurements. As the graph demonstrates, the largest changes have been the significant increases in the heaviest precipitation events and the highest streamflows. Note that changes in streamflow follow changes in precipitation, but are amplified by about a factor of 3. Source: Groisman, et.al. (2001).

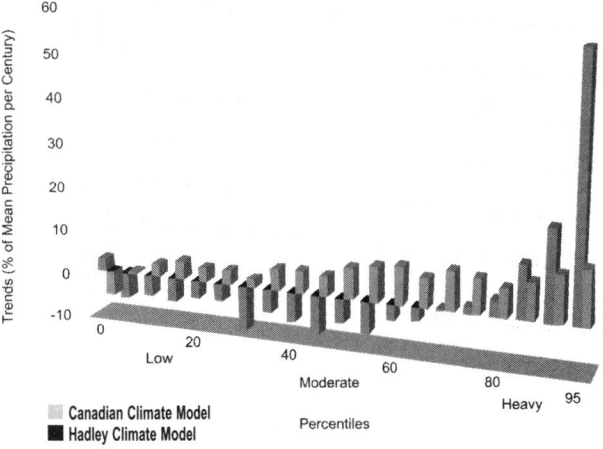

Projected 21st Century Change in US Daily Precipitation

Figure 6: These projections from the Hadley and Canadian models show the changes in precipitation over the 21st century. Each models' projected change in the lightest 5% of precipitation events is represented by the far left bar and the change in the heaviest 5% by the far right bar. As the graph illustrates, both models project significant increases in heavy rain events with smaller increases or decreases in light rain events. Source: National Climatic Data Center. See Color Plate Appendix

Understanding historic changes, or projecting future changes in streamflow conditions will require more evaluation of the complex role of changing precipitation and temperature patterns as well as the role of land-use change on streamflow. These unresolved issues further reinforce the importance of maintain-

Projected Changes in Average Annual Runoff Based on Two GCMs

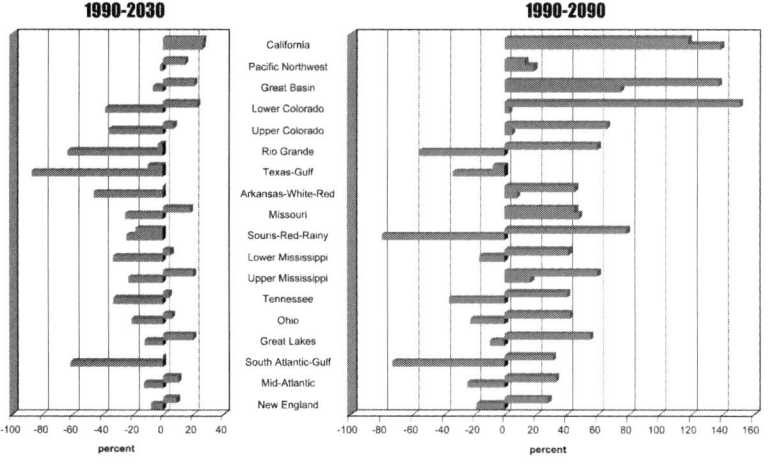

Canadian Climate Model   Hadley Climate Model

Figure 7: The estimated percent changes in average annual runoff based on the Canadian and Hadley models are not well correlated. The Canadian model predicts declines in runoff in all regions except California, while the Hadley model projects increases in most regions, particularly in the Southwest. The models differ in precipitation predictions in part due to underlying model construction. Source: Wolock, D.M. and G.J. McCabe, 1999a. See Color Plate Appendix

ing adequate nationwide networks of precipitation and streamflow gages to help describe and predict changes in average streamflow and, more importantly, streamflow variability.

In addition to a trend analysis of climatic conditions over the past 100 years, this Assessment has evaluated scenarios from two General Circulation Models (GCMs), one from the Canadian Centre for Climate Modelling and Analysis (henceforth referred to as the Canadian model), and the second, the "HadCM2" model from the Hadley Centre for Climate Prediction and Research of the Meteorological Office of the United Kingdom (henceforth referred to as the Hadley model). The Canadian and Hadley models used in this Assessment both project significant warming (5-9°F or 3-5°C) in most parts of the US by 2090. However, with the exception of the southwestern US, where both models show a large increase in precipitation in the future, especially in winter (Felzer and Heard, 1999), the changes in precipitation predicted by the two models are strikingly different. In general, the Hadley model suggests much wetter conditions than the Canadian model. When comparing output of multiple GCMs, the Hadley model increases in precipitation are the most extreme. The precipitation increase in Southern California and Arizona is related to increases in sea surface temperatures in the eastern Pacific and southward shifts in the jet stream that loosely resemble the El Niño pattern. Differences between

the two models are explained in part by differences in the land surface models relating to soil moisture and moisture availability in summer (Felzer and Heard, 1999) (see Figure 6 and figures on precipitation change in the Climate and West Chapters).

Precipitation is a key climatic variable, but it is difficult to predict changes at the local level because they are affected by land surface features that are at smaller scales than the GCM outputs (Felzer and Heard, 1999). Precipitation itself is a sub-grid-scale process, meaning that clouds and convection occur on scales smaller than GCM grids. Both models show increases in heavy precipitation events and increased storminess over the eastern Pacific, off the West Coast of the US (Lambert, 1995; Carnell and Senior, 1998; Felzer and Heard, 1999). However, precipitation patterns will vary regionally. An important issue for improving the utility of GCM output is the ability to downscale the models to a regional or watershed level where the information can be most useful to water managers.

Differences in temperature and moisture levels over land and sea are crucial in determining precipitation levels along the coasts. Over oceans, warming leads to increased evaporation and more precipitation because of the limitless supply of water. In contrast, because of the limited moisture holding capacity of the land, warming may cause drying and less precipitation. Globally, the models show decreased storm frequency, with increases in intensity. Over the US, the models do not produce a consistent projection regarding storm frequency (Lambert, 1995; Carnell and Senior, 1998; Felzer and Heard, 1999). No trends have been identified in North America-wide storminess or in storm frequency variability in the period 1885-1996 (Hayden, 1999a).

Wolock and McCabe (1999a) have used a water-balance model and output from the two GCMs to estimate the effects of climate change on mean annual runoff for the major water resource regions of the US. The model includes the concepts of climatic water supply and demand, seasonality in climatic water supply and demand, and soil-moisture storage. Inputs to the model are monthly precipitation and potential evapotranspiration, which is calculated from monthly temperature using the Hamon equation (Hamon, 1961). To evaluate the model's reliability to estimate mean annual runoff for the 18 water-resources regions in the coterminous US, VEMAP-gridded monthly climate data for 1951-80 were used in conjunction with the water-balance model to estimate mean annual runoff. These estimated runoff data were compared with measured data for the same period. The water-balance model reasonably simulated measured mean annual runoff for most of the water-resources regions. In general, the results from these two GCMs are not well correlated, and project different changes in mean annual runoff (see Figure 7). The difficulty of projecting combined effects of changes in precipitation, temperature, and seasonality of events make projections of impacts based on GCM output uncertain.

On the other hand, both large-scale climate models and regional hydrologic models agree that if changes in temperature of the magnitude identified in the climate models occur, substantial changes in the amount of precipitation that falls as snow versus rain and earlier melting of snowpack are very likely to result in changes in the runoff regime (Frederick and Gleick, 1999; Hamlet and Lettenmaier, 1999; McCabe and Wolock, 1999; Leung and Wigmosta, 1999). Snowpack is very likely to be reduced even in the context of higher precipitation because of the warming trend (see Figure 8). The effects of changes in the timing and volume of runoff will probably be felt in most sectors and regions that are snowpack-dependent (Gleick, 1998), although changes in runoff regimes will probably be highly regionally specific. For example, Leung and Wigmosta (1999) assessed the effects of climate change from the NCAR Community Climate Model (downscaled through a regional climate model) on the American River and the Middle Fork Flathead watersheds in the Pacific Northwest. There was a 61% reduction in snowpack on the American River, accompanied by a major shift in streamflow. On the Middle Fork Flathead there was an 18% reduction and no major shift in runoff. In both watersheds, there was a higher frequency of extreme low and high flow events.

Hamlet and Lettenmaier (1999) used four GCMs to evaluate runoff implications of climate change for various watersheds along the Columbia River. Altered streamflow information was simulated and used to drive a reservoir model to evaluate impacts on water management. Relatively large increases in winter runoff volumes and reductions in winter snowpack resulted in all cases. The March snow water equivalent averaged 75-85% of the base case for 2025, and 55 to 65% of the base by 2045. The earlier snowmelt, coupled with higher temperatures, reduced runoff volumes in spring, summer, and early fall. The researchers found that while higher temperatures increase the potential evapotranspiration, reduced soil moisture in the summer is likely to ultimately limit the actual evapotranspiration.

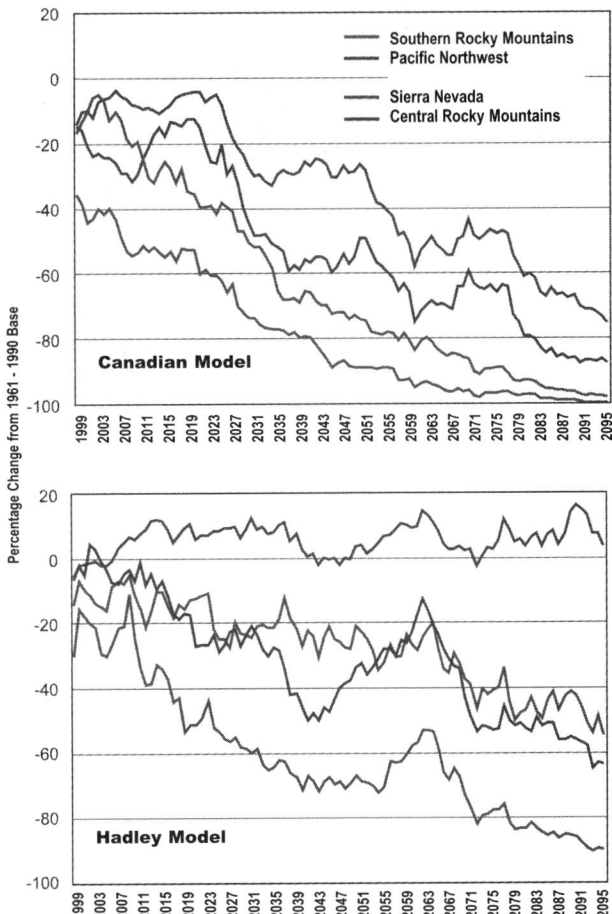

Percentage Change in Snowpack

Figure 8: Percentage change from the 1961-90 baseline in the April 1 snowpack in four areas of the western US as simulated for the 21st century by the Canadian and Hadley models. April 1 snowpack is important because it stores water that is released into streams and reservoirs later in the spring and summer. The sharp reductions are due to rising temperatures and an increasing fraction of winter precipitation falling as rain rather than snow. The largest changes occur in the most southern mountain ranges and those closest to the warming ocean waters. Source: McCabe, G.J. and D.M. Wolock. 1999. See Color Plate Appendix

Hamlet and Lettenmaier also evaluated the impact on various water management objectives of the projected changes in streamflow (see Figure 9). From a water supply perspective, Hamlet and Lettenmaier found that on average, comparing the base case with output from four transient GCMs negatively affected four water resources objectives: non-firm hydropower production, irrigation, instream flow for fish, and recreation at Lake Roosevelt. The Hadley model also showed negative impacts on flood control and navigation, due to the significantly wetter conditions in that model. Hamlet and Lettenmaier noted that an adaptive strategy would be to shift the

Changes in Reliability of Columbia River Water Resources Objectives

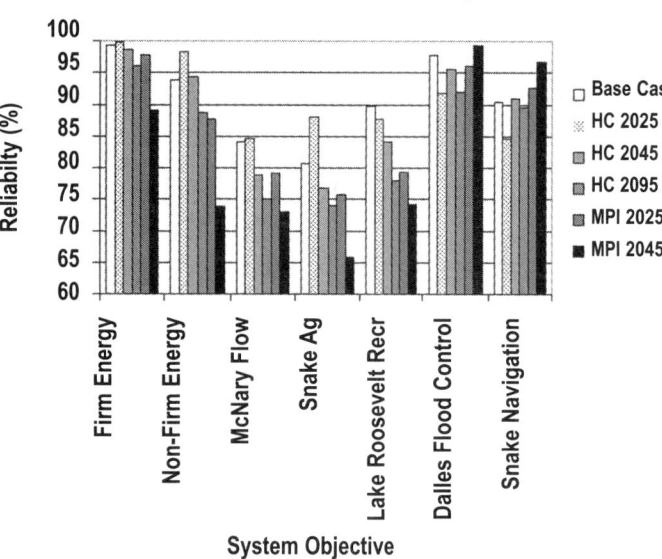

Legend:
- Base Case
- HC 2025
- HC 2045
- HC 2095
- MPI 2025
- MPI 2045

Reliability (%) axis: 60, 65, 70, 75, 80, 85, 90, 95, 100

System Objective categories: Firm Energy, Non-Firm Energy, McNary Flow, Snake Ag, Lake Roosevelt Recr, Dalles Flood Control, Snake Navigation

Figure 9: Four major objectives are impacted by low summer streamflow and reservoir storage: non-firm energy production; irrigation; instream flow; and recreation at Lake Roosevelt. Source: Hamlet, A.F. and D.P. Lettenmaier, 1999. See Color Plate Appendix

Snow Level

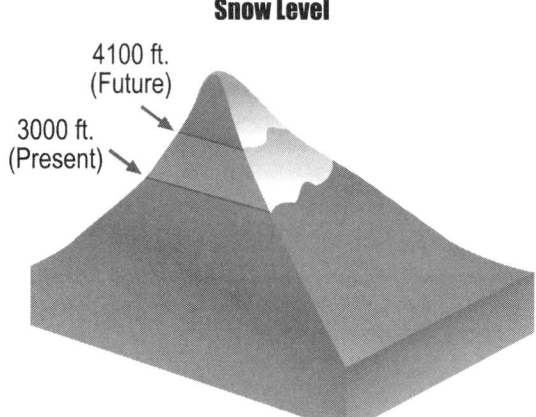

4100 ft. (Future)
3000 ft. (Present)

Figure 10: Rough estimate of how much snowlines in the Pacific Northwest are likely to shift by 2050, assuming about 4°F warming. Source: R. Leung, Pacific Northwest National Laboratory.

hydropower production period in the Columbia to the summer, using stored winter flows. However, an important consideration is that while re-operation of the reservoirs can improve conditions within one management objective, impacts to one or more objectives cannot be avoided unless the total system demands are reduced. This is difficult to accomplish as the regional population increases.

As has been noted by many researchers attempting to model regional impacts, inadequate spatial resolution of climate models to capture the topographic features is a key problem in downscaling from the global models to local hydrologic features. The resulting disparity between GCMs in precipitation predictions must be addressed before water managers will be confident of likely outcomes. However, major advances have been made in the use of regional climate models to drive hydrologic models in the Pacific Northwest (Leung and Ghan, 1999; Leung and Wigmosta, 1999; Georgakakos et al., 1999). In the Southwest, the Regional Climate System Model has been used since 1995 for 48-hour precipitation and streamflow predictions with good success, including during the 1997-1998 El Niño season. This is one of the first global-to-mesoscale-to-watershed-basin scale predictions of this type (Miller et al., 1999).

A key variable in predicting water supply conditions is the impact of CO_2. Higher CO_2 tends to stimulate plant growth, resulting in feedback effects. The water requirements to support more biomass could possibly reduce the runoff associated with a given level of precipitation. However, higher CO_2 levels increase stomatal resistance to water vapor transport, which could decrease water use of plants (Frederick and Gleick, 1999). Under arid conditions, an increase in biomass from elevated CO_2 is likely to reduce runoff to streams, thereby leaving more water on the landscape. Under conditions of ample water for plant growth, elevated CO_2 causes partial stomatal closure with a consequent decrease in transpiration per unit of leaf area. However, leaf area likely will be increased, so it is difficult to predict overall impacts on water use. The CO_2 effect has been observed in several ecosystems, and varies by species (Kimball, 1983).

Natural variability in climate has been traced to a number of phenomena related to ocean temperatures and changes in global circulation patterns. Some of the resulting weather patterns can now be predicted with some accuracy, such as those associated with the El Niño Southern Oscillation (ENSO). The ability to develop forecasts useful to water managers based on these patterns is increasing, allowing for adaptive responses. It is not yet clear how these patterns will be affected by global changes in climate. Both models show more intense storms, but they do not agree on changes in storm frequency (Felzer and Heard, 1999).

Substantial water quality and temperature changes could result from changes in flow regimes. It should be noted that climate change could either increase or decrease the availability of water. While the hydrologic implications of the Canadian model project modest reductions in water supplies (<25%) in some regions, the Hadley model projects relatively large increases in water availability (25-50%) in most regions of the US. However, there are significant regions of precipitation decrease throughout the US in both seasons in the Canadian model and in summer in the Hadley model. The increases in precipitation are greater than the decreases principally because of the large projected increase in the Southwest (Felzer and Heard, 1999).

KEY ISSUES

Five key issues have been identified:
- Competition for water supplies
- Surface water quality
- Groundwater quality and quantity
- Heavy precipitation, floods, and droughts
- Ecosystem vulnerabilities

1. Competition for Water Supplies

Water supply

Changes in water supply availability for economic activities and environmental uses are likely to be affected by changes in average temperature and precipitation as well as by altered frequency of extreme events such as floods and droughts. There is general consensus among climate modelers that a warmer world is very likely to lead to more precipitation at mid and high latitudes as well as an increase in heavy precipitation events in these areas. More precipitation will typically lead to more runoff, but in some regions, higher temperatures and increases in evapotranspiration rates may possibly counteract this effect. Several modeling studies for the western US show that precipitation rates would need to increase by as much as 10-15% just to maintain runoff at historical levels because of increased evapotranspiration (Gleick and Chalecki, 1999).

Changes in the timing of water supply availability are also very likely to occur. Surface water supplies that are dependent on snowmelt are likely to be affected by changes in the amount of precipitation that falls as rain versus snow, changes in snowpack volume, and earlier melting due to warmer temperatures. There is a strong consensus among researchers that there is very likely to be a shift in the peak volume and timing of runoff for water-

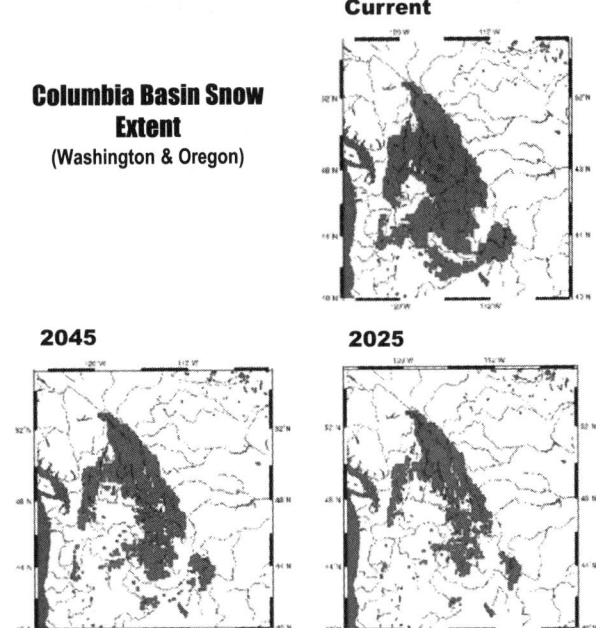

Columbia Basin Snow Extent
(Washington & Oregon)

Figure 11: Complete loss of snow cover is projected at lower elevations. These maps are generated by downscaling output from global to regional climate models. Output shown from these models relates to the Columbia Basin; no projections are included for the blank areas outside the basin. Source: Mote, et.al. (1999) Impacts of climate variability and change in the Pacific Northwest, University of Washington. See Color Plate Appendix

Projected Streamflow Effects from Climate Change in the Pacific Northwest

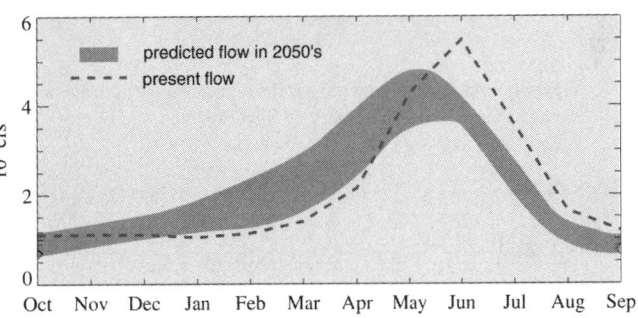

Figure 12: Relative to present flows (dashed), the wetter winters and drier summers simulated by climate models are very likely to shift peak streamflow earlier in the year, increasing the risk of late-summer shortages. Though the Columbia system is only moderately sensitive to climate change, allocation conflicts and a cumbersome network of interlocking authorities restrict its ability to adapt, producing substantial vulnerability to these shortages. Source: Hamlet, A.F. and D.P. Lettenmaier, 1999. See Color Plate Appendix.

sheds that are affected by winter snowpack, resulting in earlier spring runoff, higher winter flows and lower summer flows (Frederick and Gleick, 1999).

Water Withdrawals and Population Trends

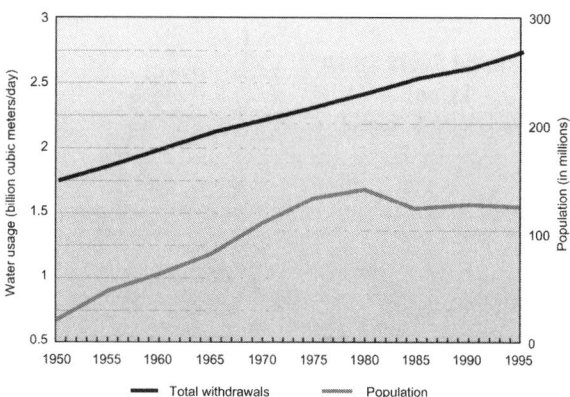

Figure 13: Although US population has continued to increase, withdrawals have declined on a per capita basis. Reductions are due to increased efficiency and recycling in some sectors, and a reduction in acreage of irrigated agriculture. Source: Solley, W.B., R.R. Pierce, and H. A.Perlman, 1998. See Color Plate Appendix.

Consumptive Water Use by Sector

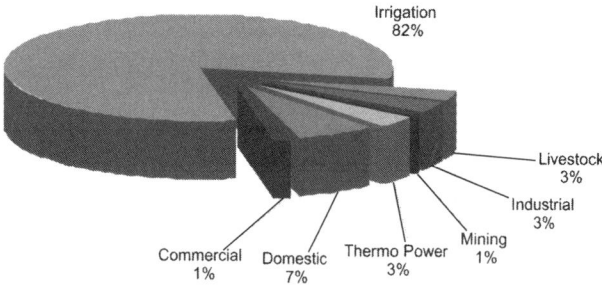

Figure 14: Agricultural water use is the highest consumptive use sector. Source: Data from Solley, W.B., R.R. Pierce, and H. A.Perlman, 1998. See Color Plate Appendix

An important area of vulnerability as a result of climate change is associated with summer water supply from Pacific Northwest (PNW) snow melt and transient snow river basins that are at moderate elevations. Under current climate conditions, summer streamflows in these moderate elevation basins are strongly affected by snow accumulation, which functions as winter storage. As the temperature rises, snow lines move up in elevation (see Figure 10) and overall snow extent is reduced (see Figure 11). As a result, maximum streamflows tend to be shifted towards the winter, with corresponding reductions in summer streamflow volumes (see Figure 12). High-elevation basins, which are below freezing for much of the winter season, are less affected by the changes in temperature, and the timing changes are less pronounced. For regional scale

watersheds like the Columbia River Basin, which integrate both of these responses, the effects are intermediate. The lower summer flows that result from the shift in the hydrograph are likely to exacerbate existing conflicts between summer water supply for human use (e.g., irrigation east of the Cascades and municipal use west of the Cascades), and maintenance of summer instream flow for ecological purposes (such as protecting salmon habitat). In the Pacific Northwest Regional Workshop it was concluded based on the 1995 Max Planck Institute climate model scenario that the most significant vulnerability to climate change is the potential for declining summer water supply in the context of rising demand (Pacific Northwest Regional Report, 1999).

Since spring runoff events are likely to be earlier, reservoir management will need to become more sophisticated in managed watersheds. For example,optimized dynamic reservoir operation rules will likely become more appropriate than traditional rule curves. Relying more on medium and long-term predictions of weather will likely maximize supply and minimize risk of flooding (Georgakakos et al., 1999).

Depending on the degree to which river systems are managed, water supply effects can be dampened by storage and release regimes. However, a study of potential impacts on the Colorado River under current "Law of the River" operating procedures indicates that even small decreases in average runoff could lead to a dramatic decrease in power generation and reservoir levels (Gleick and Nash, 1993) as the system tries to maintain committed deliveries of water. Many storage systems, like the Colorado, can readily handle year-to-year variability but may have more difficulty with long-term change.

Water demand

Water withdrawals increased faster than population growth for most of this century and reached 341 billion gallons per day in 1995. However, since 1975 water use has been decreasing on a per capita basis, and total withdrawals have declined 9% since their peak in 1980 (Solley et al., 1998, see Figure 13). Per capita consumptive use is expected to continue to decline in some areas, due primarily to reductions in irrigated acreage, improvements in water use efficiency, recycling and reuse, and use of new technologies. Brown (1999) developed water-use forecasts to the year 2040 under several scenarios. Total withdrawals would increase only 7% by 2040 with a 41% increase in population under the middle population projection. However, even with reduced per capita use, urban demand is increasing in major metropolitan areas along the coasts and in the Southwest, due to population increases. Agricultural irrigators will likely continue to have competition from municipal users for available supplies. Under drought conditions, competition for water between the agricultural and urban users is likely to intensify (see Figure 14).

Increased water use efficiency is believed to be a key solution to the increasing stress on water supplies. It is widely thought that potential exists to reduce total demand for water without affecting services or quality of life. However, as more and more waste is taken out of the system, future demand is less easily reduced in response to drought or short-term delivery problems. This "hardening of demand" is widely recognized by water managers. Conservation investments generally need to be renewed over time, since the effectiveness of many programs declines over time (including the impacts of conservation pricing and the effectiveness of low water use plumbing devices).

Changes in average temperature, precipitation and soil moisture caused by climate changes are likely to affect demand in most sectors, especially in the agriculture, forestry, and municipal sectors. Increased temperatures and decreased soil moisture are very likely to increase irrigation water needs for some crops. There is a clear linkage between weather patterns and water demand in these sectors (see Figure 15).

In 1995, irrigation accounted for 81% of total consumptive freshwater use and 39% of total water withdrawals in the US (Solley et al., 1998). Total use of water in agriculture has been declining since 1980, with the exception of the Southeast, where a 39% increase in irrigated acreage of row crops was identified between 1970 and 1990 (Irrigation Journal, 1996). McCabe and Wolock (1992) used an irrigation model to demonstrate that increases in mean annual water use in agriculture are more likely to result from increases in temperature than from decreases in precipitation. This finding may be important because runoff is also affected by increased temperatures.

Hydropower and navigation are not consumptive uses, but they are affected by both the volume and the timing of streamflows. Demand for electricity is very likely to increase with higher temperatures due to corresponding demands for summer air conditioning, but the water available for hydropower and

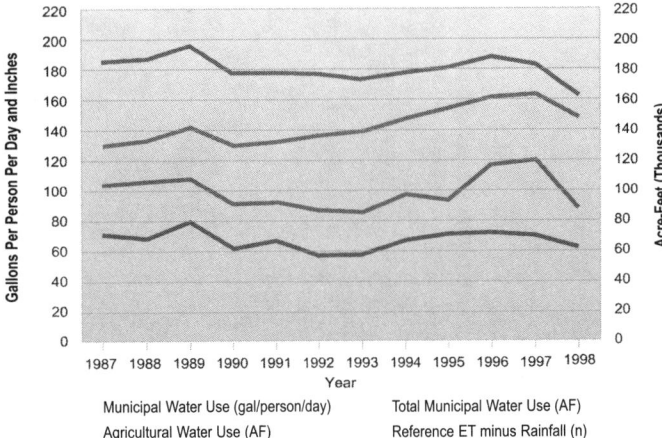

Evapotranspiration and Water Use in Tucson

Municipal Water Use (gal/person/day)   Total Municipal Water Use (AF)
Agricultural Water Use (AF)   Reference ET minus Rainfall (n)

Figure 15: Water demand in the agricultural and municipal water use sectors correlates strongly with evapotranspiration rates. Source: Arizona Deptartment of Water Resources. See Color Plate Appendix

cooling at electric generating plants may decrease because of increased pressure to divert more water for other uses. Climate change could possibly affect navigation by changing water levels in rivers, reservoirs, and lakes (e.g., Great Lakes, Mississippi River, and Missouri River; see Midwest chapter), as well as by changing the frequency of floods and droughts.

In the Pacific Northwest, hydropower and endangered species preservation are in increasing conflict because minimum streamflows must be maintained for habitat or water must be diverted away from turbines to protect migrating fish. Changes in seasonal runoff, even with no change in precipitation, could represent a very serious additional complication. Outflows from some power plants contain waste heat, which affects water temperature in the area of discharge. Although these discharges are regulated, changes in demand for electric power are likely to affect aquatic habitat. Precipitation changes in specific regions, such as the Pacific Northwest and portions of the eastern US, will affect hydropower capacity in the future.

Water issues for Native American communities are particularly critical, in part because of geographic and legal limitations and competition for resources. Significant concerns exist related to fisheries and aquatic habitat, especially with regard to Native subsistence economies. These concerns are particularly important in the Pacific Northwest. Because Indian reservations are found throughout the US, water issues vary substantially by geographic region. In most cases, the tribal culture is tied to a specific place and traditional survival strategies.

419

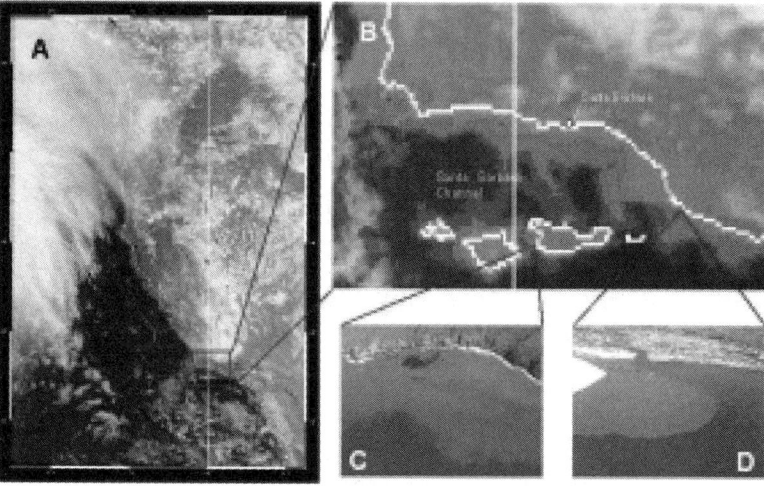

Figure 16: Sediment flow off Santa Barbara caused by El Niño storm runoff. Source: Mertes, L., The Plumes and Blooms Project, ICESS/UCSB. *See Color Plate Appendix*

Figure 17: Coastal mudslide on Highway 50, California following vegetation removal and heavy rainfall. Source: Eplett, R.A., California Governor's Office of Emergency Services.

The viability of the Hopi reservation, for example, is linked to the availability of water on their reservation. This results in added vulnerability to climate change impacts.

Many American Indian communities possess senior, but unexercised water rights. As these rights are put to use, new stresses are very likely to be introduced in the affected watersheds. Quantifying and litigating the water rights claims of Indian communities is a major ongoing issue in many western states.

2. Surface Water Quality

Major improvements have been made in the quality of surface water in the US, largely attributable to the success of the Clean Water Act in reducing industrial pollution and discharge of sewage. In 1994, 83% of the rivers and 87% of the lakes were considered suitable sources for drinking water supply (all sur-

face water must be treated before use), while 95% of the rivers and 82% of the lakes were suitable for fish habitat (US Dept of the Interior, 1997). Remaining water quality problems were attributed primarily to non-point sources of pollution, such as nutrients, bacteria, and siltation deriving from agriculture, and urban runoff (US Dept. of the Interior, 1997).

Water quality issues associated with potential climate change impacts are more subtle than supply issues and include potential impacts on human health and ecosystem function, changes in salinity associated with changes in stream flow, and changes in sediment regimes. Non-point source pollution and agricultural byproducts are likely to become more problematic depending on the change in precipitation patterns; an increase in extreme precipitation events is considered likely (IPCC, 1996), which may increase risk of contamination. A balancing effect is that with some exceptions, higher precipitation will result in lower concentrations of organic and inorganic constituents in surface water, due to dilution. Water quality is greatly influenced by flow variability, and some significant water quality problems are episodic, e.g., episodic acidification from snowmelt, and algal blooms due to nutrient increases (Mulholland and Sale, 1998; Meyer et al., 1999). Increasing salinity related to irrigation return flows (water returning to streams and aquifers after use by agriculture) and greater diversions of surface water are ongoing issues, especially in the West. Changes in streamflow associated with increased precipitation are likely to reduce salinity levels, especially in winter, while lower flows and higher temperatures could exacerbate this problem in summer. Flooding associated with more intense precipitation can also affect water quality by overloading storm and wastewater systems, and damage sewage treatment facilities, mine tailing impoundments, or landfills, thereby increasing risk of contamination.

Many of Santa Barbara's beaches were closed in 1998 due to high bacterial counts from the intense El Niño storm runoff. More winter runoff is likely to bring larger sediment flows to coastal waters, while lower summer streamflow is likely to increase salinity and impact estuarine species (see Figure 16).

A combination of increased precipitation and warmer, drier summers could increase fire hazard in some ecosystems. Sedimentation, landslides, and mudslides frequently follow removal of vegetation by fire (see Figure 17).

Drinking water supplies are very likely to be directly affected by sea-level rise in coastal areas, both through saltwater intrusion into groundwater aquifers and movement of the freshwater/saltwater interface further upstream in river basins. In the case of New York City, if the salt front moves further up the Hudson River, it will threaten emergency water supply intakes (Northeast Regional Report, in preparation). Periodic storm surges can also affect water quality and these are likely to be exacerbated by rising sea levels in a warming climate.

It is likely that climate change will affect lake, reservoir, and stream temperature through direct energy transfer from the atmosphere and changes in dam operations. Increased temperatures in surface water are likely to eliminate some species (such as salmon and trout) that are already near their habitat temperature threshold (see Figure 18). Higher temperatures result in reduced dissolved oxygen in water, which is a measure of ecosystem condition. Hurd et al. (1999) used dissolved oxygen stressed watersheds as an indicator of ecosystem vulnerability, and found that the most vulnerable regions are in Wisconsin, northern Illinois, southern Appalachia, South Carolina, and large portions of east Texas, Arkansas, Louisiana, and Florida (see Figure 19). Changes in temperature regimes are also likely to affect ice cover, and mixing and stratification of water in lakes and reservoirs, conditions that are key to nutrient balance and habitat value (Meyer et al., 1999).

3. Groundwater Quantity and Quality

Groundwater is the source of about 37% of irrigation water withdrawals (Solley et al., 1998), and supplies drinking water to about 130 million Americans (USGS, 1998). Though groundwater supplies are less susceptible to variations in climate than surface water, they may be more affected by long-term trends. More frequent or prolonged droughts are likely to increase pressure on groundwater supplies, which commonly serve as a buffer during shortages of surface water supplies. Depletion of groundwater is significant on the High Plains, the Southwest, parts of the Southeast, and in the Chicago area (USGS, 1998). Groundwater overdraft can cause substantial long-term effects, because in some areas the available groundwater supply is essentially nonrenewable or because land compaction prevents groundwater recharge (see Figure 20). Where the rate of recharge of groundwater aquifers is slower than use, long-term groundwater pumping becomes unsustainable. However, increases in precipitation are likely over a significant portion of the US, and many groundwater aquifers are likely to benefit. Despite the importance

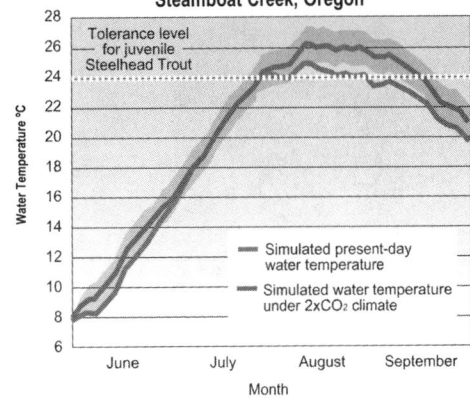

Summer Stream Temperatures
Steamboat Creek, Oregon

Figure 18: Simulated summer stream temperatures under present day climate (blue) and simulated temperatures under about a twice current CO_2 climate (red). The dashed line at 24 °C (75 °F) on the "water temperature" axis indicates the summer temperature tolerance of juvenile steelhead trout. Under doubled CO_2, the model suggests that the length of time within the year when the temperature tolerance limit is exceeded is more than twice as long as under simulated present-day climate conditions. Shaded area surrounding the doubled CO_2 temperature curve indicates an estimate of uncertainty. Source: US Geological Survey Circular 1153, Robert S. Thompson, et al. See Color Plate Appendix

Current Climate Vulnerability Map, Instream Use, Water Quality and Ecosystem Support

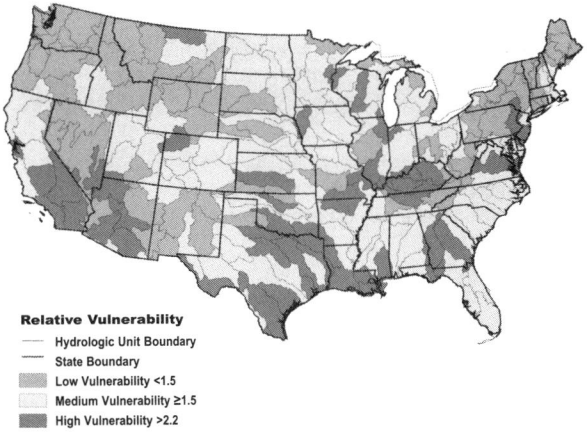

Figure 19: Instream Use, Water Quality, and Ecosystem Support Assessed Vulnerability based on current climate and water resource conditions, based on data describing the following: flood risk population, navigation impacts, ecosystem tolerance to cold and heat, dissolved oxygen stress, low streamflow conditions, and number of aquatic species at risk. Source: Hurd, B.J., N. Leary, R. Jones and J. Smith, 1999a. See Color Plate Appendix

of groundwater to water supply in many regions, effective institutions to manage groundwater are the exception, rather than the rule (Dellapenna, 1999; National Research Council, 1997).

Groundwater is managed through a different mechanism in virtually every state, though there are

Figure 20: Land subsidence fissure, caused by over-pumping of groundwater, can result in earth fissures such as this near Eloy, Arizona. Source: K. Jacobs.

Figure 21: Artificial recharge in Santa Ana riverbed. Artificial groundwater recharge in the Santa Ana Riverbed, Orange County Water District, California. Source: Orange County Water District.

three basic systems of groundwater rights (see box, "Major Doctrines for Surface Water and Groundwater").

Groundwater maintains the base flow for many streams and rivers, and lowering groundwater levels may reduce the seasonal flows and alter the temperature regimes that are required to support critical habitat, especially wetlands. Although conjunctive management (using groundwater and surface water in combination to meet demand) is frequently cited as a solution to water supply problems, this approach is only sustainable if the groundwater supplies are periodically recharged using surplus surface water or other alternative supplies (see Figure 21). In order to ensure a sustainable supply, aquifers may need to be artificially recharged, which involves consideration of multiple issues including changes in water quality in the aquifer. In some

cases, such as portions of the Ogallala Aquifer in the Great Plains, groundwater supplies have already been over utilized and no source of renewable supplies is available.

Groundwater in storage is affected by seasonality, volume, and persistence of surface water inflows, and discharges from groundwater to surface water. Groundwater/surface water interactions are poorly understood in most areas. Changes in precipitation and temperature may have long-term effects on aquifers that are relatively subtle and difficult to identify. Groundwater is frequently found in horizontal layers within an aquifer, separated by relatively impermeable layers of silt and clay or rock. The water quality is affected by the substrate that the water flows in, and the amount of time it has been in storage. In some areas, only the surface zone is contaminated by human activities. An understanding of the aquifer's geology, including such characteristics as location of impermeable layers and the direction of water flow, is necessary in designing appropriate management options.

Industrial pollution is the largest groundwater quality issue in urban areas. Common problems include solvents and petrochemicals. Recently released data from the USGS National Water Quality Assessment indicate that volatile organic compounds (VOCs) were detected in 47% of urban wells tested between 1985 and 1995. The most common VOCs found were the fuel additive MTBE (methyl tertiary butyl ether) and various solvents such as tetrachloroethene, trichloroethene and trichloromethane (Squillace et al., 1999). Contamination of drinking water supplies presents serious challenges to water managers, especially in large urban areas. Cleanup of contaminated aquifers is extremely expensive and in some cases is not practical.

Agricultural chemicals and wastewater treatment byproducts such as nitrates also affect groundwater quality in some areas. Continued degradation is anticipated in some metropolitan areas. It is unclear whether climate change will have a significant effect on groundwater contamination.

Increased pressure on groundwater supplies has resulted from the Safe Drinking Water Act (SDWA) regulations. The Surface Water Treatment Rule now requires filtration of all surface sources. As a result, many small surface water systems are now uneconomical, and have either combined to form larger systems or switched to groundwater, for which filtration is not required. O'Connor et al. (1999) surveyed 506 community water system managers in the Pennsylvania portion of the Susquehanna River Basin, and found that half of all the community water systems in Centre County switched to groundwater or regionalized their surface water systems in response to SDWA regulations.

Another key groundwater quality concern is saltwater intrusion in coastal aquifers. As pumping of groundwater increases to serve municipal demand along the coast, freshwater recharge in coastal areas is reduced, and sea level rises, groundwater aquifers are increasingly affected by infiltration of seawater. Seawater intrusion is already a major issue in Florida, the Gulf Coast, southern California, Long Island, Cape Cod, and island communities. The global sea level is estimated to have risen 4 to 8 inches (10-20 cm) over the past 100 years (Gornitz, 1995); in some areas, the relative sea-level rise has been greater because land surface elevations are sinking in some regions of the coast. Further increases in sea level are very likely to accelerate intrusion of salinity into aquifers and affect coastal ecosystems. The adaptation strategies for dealing with this problem — importation of alternative sources of supply, desalination, and artificial recharge — can be extremely expensive.

Because surface water and groundwater supplies are interconnected and transportation of water across hydrologic and political boundaries is common, issues related to surface water/groundwater interactions will possibly be exacerbated by climate change. Even in the context of higher average precipitation, increased temperatures and changes in seasonality of runoff are likely to reduce streamflows during the warm season, a key period for ecosystem maintenance. Increased urbanization, which generally results in increased channelization of streambeds and higher runoff rates is likely to reduce opportunities for recharge of groundwater near areas of high groundwater demand. Water transfers may increase pressures on areas of origin from the perspective of water supply and ecosystem health. Particularly in arid and semi-arid regions, effects on surface water or groundwater resources resulting from climate change are likely to impact riparian systems that support a high percentage of biological diversity. In many cases, higher precipitation is likely to have a positive effect on groundwater levels and riparian habitat.

Cumulative Number of Large Dams Built in the US

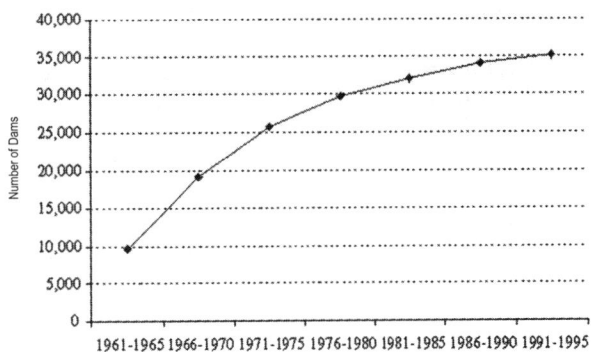

Figure 22: The number of large dams built in the US has declined in recent decades, data yearly. Source: US Army Corps of Engineers. 1996. National Inventory of Dams.

Average Volume of US Reservoirs Built

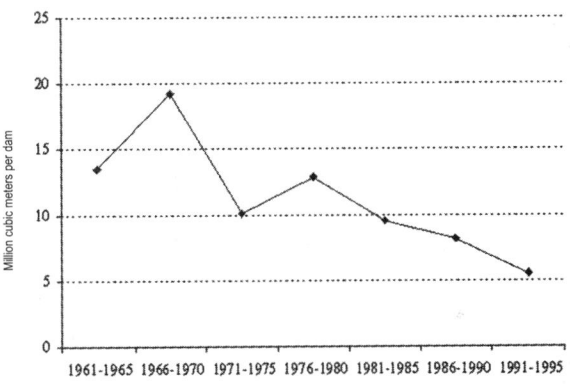

Figure 23: The average volume of reservoirs built in each five-year period since 1960 has declined, data five year interval. Source: US Army Corps of Engineers. 1996. National Inventory of Dams.

4. Heavy Precipitation, Floods and Droughts

Floods, especially those related to flash floods from intense short-duration heavy rains, are likely to increase in magnitude or frequency in many regions. Changes in seasonality of flood flows are very likely to occur in those areas affected by a higher proportion of rain to snow (yielding earlier peak flows of shorter duration). Intensity of droughts is also likely to increase in some areas due to higher air temperatures causing greater evaporation and water use by plants.

There are two types of socioeconomic costs related to floods and droughts: the costs of building and managing the infrastructure to avoid damages, and the costs associated with damages that are not

Impacts of Potential Climate Change on
Aquatic Ecosystem Functioning and Health
(adapted from Meyer et al., 1999)

Region	Potential Climate Effect	Ecosystem Considerations
Great Lakes/ Precambrian Shield	Warmer, more precipitation, but drier soils possible, depending on the magnitude of precipitation increase.	Altered mixing regimes in lakes (e.g., longer summer stratification. Changes in DOC concentration, changes in thermocline depth and productivity. Decreased habitat for cold and cool water fishes, increased habitat for warm water species. Alteration of water supply to wetlands will affect composition of plant communities and carbon storage as peat.
Arctic and sub-Arctic North America	Much warmer, increases in precipitation	Loss or reduction of deltaic lakes. Reduction in area covered by permafrost, leading to drainage of lakes and wetlands, land slumping, sedimentation of rivers. Increased primary productivity, but perhaps not enough to compensate for increased metabolic demands in predatory fish. Shift in carbon balance of peatlands.
Rocky Mountains	Warmer	Changes in timberline would affect stream food webs. Increased fragmentation of cold-water fish habitat. Fishless alpine lakes sensitive to changes in nutrient loading and sedimentation. Current anthropogenic changes are threatening aquatic ecosystems.
Pacific Coast Mountains and western Great Basin	Warmer, less snow but more winter rain, less summer soil moisture	Increases in productivity in alpine lakes. Increased meromixis and decreased productivity in saline lakes. Altered runoff regimes and increased sediment loads leading to decreases in channel stability and negative impact on economically important fish species.
Basin and Range, Arid Southwest	More precipitation, warmer, overall wetter conditions	Aquatic ecosystems highly sensitive to changes in quantity and timing of stream flow. Intense competition for water with rapidly expanding human populations.
Great Plains	Warmer with less soil moisture	Historical pattern of extensive droughts. Reduced water level and extent of open water in prairie pothole lakes with negative effects on waterfowl. Increasing warming and salinity in northern and western surface waters threatening endemic species. Reduction in channel area in ephemeral streams.
Mid-Atlantic England	Warmer and and New somewhat drier	Potentially less episodic acidification during snowmelt. Possible increase in bioaccumulation of contaminants. Bog ecosystems appear particularly vulnerable. Current context: stresses from dense human populations and a long history of land use alterations.
Southeast	Warmer with possible precipitation increases and greater clustering of storms	Increases in rates of primary productivity and nutrient cycling in lakes and streams. More extensive summer deoxygenation in rivers and reservoirs. Loss of habitat for cold-water species like brook trout, which are at their southern limit. Drying of wetland soils. Northward expansion of nuisance tropical exotic species. Increased construction of water supply reservoirs.

avoided, including ecosystem impacts. About $100 billion has been spent by the federal government since 1936 in the US for the construction, operation and maintenance of flood control features, yet damages associated with floods continue to rise (Frederick and Schwarz, 1999b). Flood damage estimates by state are provided by the National Center for Atmospheric Research and the US Army Corps of Engineers. The 1993 flood in the Mississippi and Missouri Rivers caused record damages of over $23 billion. These are only the official damage estimates, and do not take into account total social costs. The 1999 North Carolina flood, resulting from Hurricane Floyd, offers a recent example of the massive dislocations and multi-billion dollar costs that often accompany such events. Dams and levees have also saved billions of dollars of investment, but these facilities, together with insurance programs, encourage development in floodplains, thereby indirectly contributing to damages (Frederick and Schwarz, 1999b). In addition, structural flood control features have high environmental costs. Climate change may affect flood frequency and amplitude, with numerous implications for maintenance and construction of infrastructure and for emergency management. Erosion and deposition rates in rivers and streams are likely to change under different precipitation regimes. The reduction in reservoir construction along with the buildup of sediment in reservoirs will affect the resilience of water supply systems and their ability to handle flood flows (Frederick and Schwarz, 1999b) (see Figures 22 and 23).

Flood risks are ultimately a function of many factors, including populations exposed to floods, the nature and extent of structures within river floodplains and in coastal areas subject to storm surges, the frequency and intensity of hydrologic events, and kinds of protection and warning available. To the extent that each of these factors can be addressed economically and in a timely and integrated manner, future damages can be limited. Wetlands restoration in managed watersheds can reduce the impact of storm water runoff to waterways by slowing down or absorbing excess water. Providing wetland protection including buffer areas beyond the wetland boundary is a viable method of avoiding flood damage or the cost of flood protection.

Palmer Drought Severity Index Change

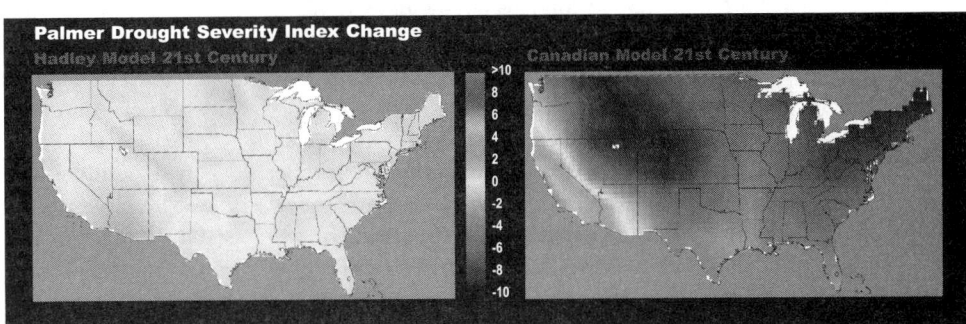

Figure 24: The Palmer Drought Severity Index (PDSI) is a commonly used measure of drought severity taking into account differences in temperature, precipitation, and capacity of soils to hold water. These maps show projected changes in the PDSI over the 21st century, based on the Canadian and Hadley climate scenarios. A PDSI of -4 indicates extreme drought conditions. The most intense droughts are in the -6 to -10 range, similar to the major drought of the 1930s. By the end of the century, the Canadian scenario projects that extreme drought will be a common occurrence over much of the nation, while the Hadley model projects much more moderate conditions. Source: Felzer, B. UCAR. — See Color Plate Appendix

Figure 25: Prairie potholes are considered to be especially vulnerable to drying conditions in the northern Great Plains.

There are many different kinds of droughts, from short-term localized reductions in water availability to long-term and widespread shortages. Severe droughts have had widespread and devastating effects, particularly on agriculture. The drought of the 1930s affected 70% of the US and caused substantial economic dislocations (Woodhouse and Overpeck, 1999). Prolonged droughts affect all sectors of the economy, and may be especially devastating for ecosystems. An evaluation of the paleoclimatic record indicates that droughts of Dust Bowl severity are not unprecedented, at least at a regional level. Agricultural interests in the Great Plains region are particularly concerned about the increased likelihood of drought with global warming (see Figure 24). In the Canadian model, severe and extreme drought becomes the norm for the Great Plains, but in the Hadley model such drying is not evident. Better information is needed on changes in drought risks from climate changes.

Droughts also affect the ability of waterways to support transportation, particularly on the Great Lakes and major river systems like the Mississippi and Missouri. Climate change is likely to affect the volume and timing of streamflows, the amount of sediment carried and deposited in shipping channels, and the extent of ice blockage in the northern waterways (Hurd, et al., 1999).

5. Ecosystem Vulnerabilities

Climate changes are very likely to have a wide variety of effects on ecosystems. Other human-induced changes (such as impacts of changing land use on water quantity and quality, sediment load, and competition from exotic species) are expected to be of greater magnitude in most parts of the country than climate change. However, climate change may add another layer of stress to natural systems that have lost much of their resiliency. From an ecological perspective, the Arctic, Great Lakes, and Great Plains (especially Prairie Potholes, see Figure 25) regions appear most vulnerable (see summary table adapted from Meyer et al., 1999). Aquatic and riparian ecosystems in the arid Southwest are also vulnerable to changing precipitation and runoff regimes, but the nature of predicted climate change in that region may alleviate existing stresses (Meyer et al., 1999), see table on page 424.

Evidence of the current warming trend can be found in Alaska, where the area of sea ice is shrinking, glaciers are melting, and land that has been supported by permafrost for centuries is transitioning to a new ecological regime. Although the impact on river flows in Alaska has not been sufficiently studied, it is clear that major changes are occurring which have already affected species composition and subsistence hunting and fishing. Changes in climatic conditions in Alaska are very likely to be even more dramatic in the future, as albedo (reflectivity) is reduced by reduced ice and snow cover and evaporation rates increase in the summer (Felzer and Heard, 1999).

Impacts on lakes and wetlands from climate change are likely to include changes in water temperature, sedimentation and flushing rates, length of ice cover, amount of mixing of stratified layers, and the inflow of nutrients and other chemicals. Montane and alpine wetlands with temperature-sensitive

species will probably be vulnerable because they have little potential for migration (Kusler and Burkett, 1999). Minor changes in maximum and minimum temperatures and seasonality of precipitation can have significant impacts on wetland habitat (Kusler and Burkett, 1999). Wetlands that are directly dependent on precipitation are likely to be more vulnerable to climate change than those that are dependent on groundwater outflows due to the significant buffering capacity of regional aquifer systems (Winter, 1999). Riparian habitats are of great concern in part because 55% of threatened and endangered species are dependent on them (Herrmann et al., 1999).

Species live in the larger context of ecosystems and have differing environmental needs. A change that is devastating to one species may encourage the expansion of another to fill that niche in the system. It is not possible to determine a single optimum environmental condition for all species in the ecosystem (Meyer et al., 1999). Extreme conditions such as floods, droughts, and fire are critical to sustaining certain ecosystems. Hydrologic conditions affect nutrient cycling and availability in streams and lakes, which affects productivity. Ecological responses to changes in flow regime will depend on the regime to which it is adapted; a system that is historically variable can be severely disrupted by stabilizing the hydrologic regime (which happens when dams are used to regulate flow, as on the Colorado River).

ADAPTATION STRATEGIES

Water management has become more complex as values related to water supply and demand have shifted and regulations have proliferated, both prescribing and proscribing many solutions. Most of the options available for responding to the impacts of climate change and variability on water resources are alternatives that are already used to respond to existing challenges. However, it should be noted that optimizing water resource management under increasing constraints (including regulatory constraints) narrows the options available. In addition, responding to climate changes may require a broader set of information than is usually available to decision-makers.

Current water management practices and infrastructure throughout the country are designed to address problems caused by existing climatic variability. In general, engineering approaches to system design rely on historic data, assuming that future climatic

Key Climate Messages for Water Managers

- Climate is not static and assumptions made about the future based on the climate of the recent past may be inaccurate. Water managers should factor in the potential for climate change when designing major new infrastructure. Assumptions about the probability, frequency, and magnitude of extreme events should be carefully re-evaluated.
- There is substantial stress on the water sector even in the absence of climate change. There are numerous watersheds that are already over-appropriated, and new stresses are coming from population dynamics, land use changes, and changes in international economies. In some areas, the new demands associated with instream flow needs for habitat protection and Indian water rights settlements may cause major shifts in water supply and water rights. Climate change may pose additional stresses and could result in thresholds being reached earlier than currently anticipated.
- Waiting for relative certainty about the nature of climate change before taking steps to reduce risks in water supply management may prove far more costly than taking proactive steps now. (The suggested risk-reducing or "no regrets" steps are those that would have other beneficial effects and so are appropriate regardless of climate change.)
- The types of changes encountered in the future may not be gradual in nature. Non-linearities and surprises should be expected, even if they cannot be predicted.
- The problems that are likely to result from climate change are intergenerational. Decisions made today will commit future generations to certain outcomes. It is important to evaluate benefits of projects over long time frames, and develop an educated citizenry.

Other Key Considerations for Water Managers

- The water delivery, wastewater, and flood control infrastructure, particularly in the eastern US is aging and sometimes inadequately maintained and therefore vulnerable. The likely additional stresses that may result from climate change should encourage upgrading of key infrastructure to limit vulnerability to extreme events.
- As has been observed by many, the days of building large dams and expensive supply-side solutions are nearly over. More innovative solutions will be required in the future. Managers will need to prepare contingency plans to face water quality and supply challenges regardless of changes in climate. Promising options include conservation and efficiency improvements, water banking, water transfers, conjunctive use of surface and groundwater systems, and cooperative arrangements with other jurisdictions and communities.
- There are currently multiple disincentives to efficient utilization of water supplies. Subsidies and failure to reflect the full value of water supplies affect water pricing in virtually every sector. Americans view water as a "public good," believing supplies should be cheap, plentiful, and contain virtually no health risk factors. Agricultural water use is generally the most highly subsidized, but there are few municipal water suppliers that assign a value based on the replacement cost. As stresses increase on the water sector, water costs will definitely increase. Equity issues should be fully evaluated.
- Policies related to floodplain management and insurance currently encourage risky behavior such as rebuilding in floodplains and low-lying coastal areas after floods and storm surges. At both a national and local level, encouraging people to move away from high-risk areas would be beneficial. Incorporating wetland protection in buffer areas beyond current wetland boundaries would offer additional resilience to cope with potential flooding from more intense storms.
- Catastrophic events such as floods and fires are required to sustain some ecosystems over the long term. Management of these ecosystems should allow for continued benefits from these events.
- Hydrologists have developed valuable new models of watershed and regional-level hydrology that are ready for use. Effective use of mid- and long-range forecasts can improve the management of water resources and would be a significant step in developing the flexibility and resilience needed to cope with climate change.
- A key component of either conservation programs or improved water rights administration is metering or measuring of individual uses. This is a basic step to understanding water use and in educating consumers about proper water management. Many large cities in the US (including Fresno and Sacramento, California) currently do not measure water deliveries; most agricultural water use, especially groundwater use, is also not properly monitored.
- Improved management opportunities are available when watersheds are managed as a hydrologic unit.
- A key question that should be considered by policy makers is how much risk is acceptable to the public. It is not reasonable to manage for or expect no risk or zero damages from natural hazards.

conditions won't deviate significantly from those experienced in the recent past. Adaptation strategies for dealing with climate change range from relatively inexpensive options such as revising operating criteria for existing systems, to re-evaluating basic engineering assumptions in facility construction, to building new infrastructure with substantial capital costs. Strategies also include water conservation technologies and policies, use of reclaimed wastewater and other alternative supplies, and improved mechanisms for water transfers. Insufficient work has been done to evaluate the costs and benefits of alternative adaptation strategies. However, improved management of existing systems would certainly be valuable in managing changes in the ranges projected in this Assessment.

Improved efficiency of water use is likely to result from both regulatory requirements and higher water costs, which are expected outcomes of growing demands for clean and adequate water supplies. A move towards marginal-cost pricing (where costs reflect the price of the next available supply) and more extensive water markets may develop.

Water managers are currently not adequately engaged in the process of evaluating the risks of climate change. There is much debate about whether this is because they understand the nature of the risks, but have concluded that the current tools they have are sufficient, or whether they simply do not have the information they need in order to respond more appropriately (Stakhiv and Schilling, 1998; Frederick and Gleick, 1999). Results of a recent survey of western water managers (Baldwin et al., 1999) indicate that water managers who routinely deal with variability that is an order of magnitude greater than predicted climate changes see little reason to respond. Perhaps this is because they do not understand that potential climatic changes may be imposed on top of existing variability. However, availability of information and a greater understanding of the issues are likely to affect management practices. For example, if managers knew that there is a high probability that the magnitude of flood events is likely to increase in their region even if frequency remains the same, this information could be incorporated into planning activities. Even without this kind of information, however, some water organizations are beginning to push for common-sense actions by water managers. The American Water Works Association published recommendations for water managers calling for a re-examination of design assumptions, operating rules, and contingency planning for a wider range of climatic conditions than traditionally used (American Water

Works Association, 1997). The goal was to identify opportunities for reducing the risks associated with future climatic changes.

Various potential water management adaptations have been suggested to respond to either existing stresses or new stresses associated with climate change. They include the following:

- Increase ability to shift water within and between sectors (including agriculture to urban); this could increase flexibility but may require changes to institutional structures.
- Use pricing and market mechanisms proactively to increase efficiency of water use.
- Incorporate potential changes in demand and supply in long term planning and infrastructure design.
- Create incentives or requirements to move people and structures out of floodplains.
- Identify ways to manage all available supplies, including groundwater, surface water and effluent, in a sustainable manner.
- Restore and maintain watersheds as an integrated strategy for managing both water quality and water quantity. For example, restoring watersheds that have been damaged by urbanization, forestry, or grazing can reduce sediment loads, limit flooding, reduce water temperature, and reduce nutrient loads in runoff.
- Reuse municipal wastewater, improve management of urban stormwater runoff, and promote collection of rainwater for local use to enhance urban water supplies.
- Increase the use of forecasting tools for water management. Some weather patterns, such as El Niños, can now be predicted with some accuracy and can help reduce damages associated with extreme weather events.
- Enhance monitoring efforts to improve data for weather, climate, and hydrologic modeling to aid understanding of water-related impacts and management options.

Communication Strategies

Information on the impacts of climate change is only helpful if it is usable by water managers, landowners, emergency response teams, and other decision-makers. They need to understand the range and probability of potential outcomes. This will require timely, detailed information at the scale needed to address local conditions. In addition, since most adaptation strategies incur costs, whether they are in response to existing or new stresses, it will be important to communicate the risks, costs, and opportunities. The need for better information argues for improved monitoring and modeling to link climate information with hydrologic impacts.

There is significant concern that results from GCMs will be misunderstood. While they do not predict the future, they are useful tools for exploring future scenarios. GCM outputs would be more useful to water managers and researchers if the models' underlying concepts were better known. It has been suggested that converting the outputs into information about weather systems, storm tracks, and likely weather events would be helpful.

Balancing water supply and water quality issues while maintaining natural ecosystems and quality of life even in the absence of any climate change is daunting. Adding the overlay of potential climate change increases the difficulty of achieving these goals. One general conclusion of work done to date is that humans have many options for adapting water supply and demand systems to climate change, while unmanaged ecosystems may be more vulnerable to imposed changes. Human adaptation, however, is very likely to come at substantial economic and social cost. In addition, it should be understood that some impacts are likely to be unpredictable or unavoidable because of the very nature of atmospheric and climatic dynamics.

In many parts of the US, the water supply and demand picture is a complex web of imported and local supplies, interconnected physical infrastructure, and overlapping institutions and jurisdictions. Difficulties in downscaling climate models to a useful level for decision-makers continue to limit the utility of the information produced to date. Depending on the geography of particular regions, local predictions of changes in climate may be nearly impossible in the near term. For example, in much of the Southwest, quantitative knowledge of the current hydrologic cycle is quite limited due to the large temporal and spatial variability in precipitation, runoff, recharge, evaporation, and plant water use within basins. Much of the uncertainty is caused by the high degree of diversity in the basin and range topography. Likewise, when predictions of local water supply conditions are aggregated into regions, the resulting picture may be meaningless. Although generalizations are necessary in order to communicate major concepts, they may have little value when applied to particular circumstances without appropriate caveats.

There are, however, some major messages that need to reach water managers. Despite the inability to provide clear, detailed information about many regional impacts, large-scale climatic changes are likely to occur.

These changes will very likely affect water supply and demand in ways that may not be anticipated by current water managers, and these changes may be imposed on top of existing climatic variability and hydrologic risks. While many different alternatives for coping with impacts on the nation's water resources are available, we do not yet understand how effective these will be, how expensive they will be, or what surprises and unavoidable impacts may occur.

CRUCIAL UNKNOWNS AND RESEARCH NEEDS

Strategic Monitoring Needs

- Sophisticated analysis of climate change or its impacts requires continuous data sets provided through environmental monitoring. Monitoring should be enhanced from a strategic perspective in order to integrate key unknowns, particularly groundwater conditions, surface water quality, and biological factors in key habitats. Existing programs are not adequately integrated, and there are critical gaps in both space and time. Recent decreases in funding for stream gages and water quality sampling activities are especially problematic. Monitoring provides important services for society, such as improved predictive capability for weather events and reservoir management.
- Additional data on snowpack, depth, extent, snow water equivalent, etc., would be helpful to scientists and water managers whose supplies are dependent on snowmelt.
- Tools are needed to interpret water quality data and make data readily accessible to decision-makers.
- Engineering/Management Research Needs
- Improved understanding of the demand side of the water resource equation is needed.
- More quantitative evaluation of costs and effectiveness of adaptation strategies is needed.
- Better analyses are needed of the ability of existing infrastructure's capability to adapt. How much flexibility is there in existing systems to deal with variability? Further investigations into the impact of increased precipitation, flooding, and changes in water levels on the nation's infrastructure are needed. For instance, these climate-related changes may adversely affect air and water transportation.
- Design criteria (e.g., for 100-year floods) should be reevaluated to reduce risk to infrastructure in the context of climate change.

429

- More flexible institutional and legal arrangements should be instituted that facilitate the ability to respond to changing conditions.
- Research is needed to compare and evaluate innovative floodplain management strategies at the local, state, and federal levels to improve resiliency to climate change.

Climate Research/Modeling

- Projecting future changes in streamflow conditions will require more evaluation of the complex role of changing precipitation and temperature patterns as well as the role of land-use change on streamflow.
- There is a need to continue to refine existing GCMs, and improve model validation and comparison. Runoff modeling could be improved if differences between the models were better understood. Output should be tailored to users needs. Key areas for model development include better physically based parameterizations for groundwater/surface water interactions, atmospheric feedbacks, and variability of precipitation and land surface characteristics at a watershed scale.
- Additional research is needed to explore the current causes of climate variability, such as the El Niño/Southern Oscillation and Pacific Decadal Oscillation. This will enable evaluation of impacts if such conditions become more persistent in the future.
- Existing models can be used to explore the vulnerabilities of various regions to changes in climate. These evaluations should lead to improved understanding of critical changes in evapotranspiration and runoff regimes.
- Increased data and analysis of paleoclimatic records will provide substantial insight into the nature and range of climate and hydrologic variability (e.g., the incidence of droughts and floods).

Integrated Assessment Research

- Improved tools are needed for translating climate changes into water resource impacts and issues of public interest. For example, to be useful in river basin management, GCMs must more accurately simulate inter-annual climate variability (persistence of extreme events) and probability of extreme events. This could subsequently be linked to economic costs and potential management decisions, such as land use restrictions.

- There is a need to focus on groundwater implications of climate change. Groundwater recharge rates are controlled by many factors that are poorly understood. The response of deep and shallow aquifers to historic drought should be evaluated, as well as stream/aquifer interactions, the extent of interactions between aquifers, and impacts on riparian habitat.
- Water quality changes that result from existing climatic variability, and the impacts of extreme events on ecosystems, need further evaluation.
- Research is needed on highlighting thresholds of change in natural ecosystems, and key areas of vulnerability including impacts of flooding and drought. There are likely to be time lags as ecosystems respond to change, but these have not been identified or modeled so the indicators are not well understood.
- Biotic responses are not being accounted for adequately in modeling efforts, particularly feedbacks associated with changes in land cover, stomatal resistance due to increased CO_2, etc.
- Better integration is needed of human and ecological risk assessment relative to assessments of climate change. Risk factors and willingness to pay for damages caused by climate change should be evaluated as inputs to decision making.
- There is a need to improve communication between scientists and water managers. For purposes of technology transfer, the value and adequacy of integrated climate, hydrologic, and management systems should be demonstrated in prototype applications. Demonstration projects could engage managers of surface water supply systems in applications of reservoir management for their own systems. This could prove helpful in building relationships between modelers and real-world managers.

LITERATURE CITED

American Water Works Association (AWWA), Climate change and water resources, Committee report of the AWWA Public Advisory Forum, *Journal of the American Water Works Association, 89*(11), 107-110, 1997.

Anonymous, 1995 Irrigation survey reflects steady growth, *Irrigation Journal,* Arlington Heights, IL, 1996.

Baldwin, C. K., U. Lall, and F. H. Wagner, Time scales of climate variability important to Western water managers and their views on climate change, in *Proceedings of the AWRA Specialty Conference on Potential Consequences of Climate Variability and Change to Water Resources of the United States,* American Water Resources Association, Herndon, VA, pp. 23-26, 1999.

Bernard, S. M., J. Rose, B. Sherman, M. McGeehin, J. Scheraga, and J. Patz, Water, health and climate: Assessing the potential consequences of climate change and variability on waterborne disease risk, in *Proceedings of the AWRA Specialty Conference on Potential Consequences of Climate Variability and Change to Water Resources of the United States,* American Water Resources Association, Herndon, VA, pp. 409-413, 1999.

Brown, T. C., Past and future fresh water use in the United States, *General Technical Report RMRS-GTR-39,* US Department of Agriculture, Forest Service, Rocky Mountain Research Station, Fort Collins, Colorado, 1999.

Carnell, R. E., and C. A. Senior, Changes in mid-latitude variability due to increasing greenhouse gases and sulphate aerosols, *Climate Dynamics, 4,* pp. 369-383, 1998.

Chalecki, E. L., and P. H. Gleick, A framework of ordered climate effects on water resources: A comprehensive bibliography, *Journal of the American Water Resources Association, 35*(6), pp. 1657-1665, 1999. URL: www.pacinst.org/CCBib.html

Dellapenna, J. W., Adapting the law of water management to global climate change and other hydropolitical stresses, *Journal of the American Water Resources Association, 35*(6), pp.1301-1326, 1999a.

Dellapenna, J. W., Legal responses to global climate change, *Proceedings of the AWRA Specialty Conference on Potential Consequences of Climate Variability and Change to Water Resources of the United States,* American Water Resources Association, Herndon, VA, pp. 13-16, 1999b.

Eheart, J. W., A. J. Wildermuth, and E. E. Herricks, The effects of climate change and irrigation on criterion low streamflows used for determining total maximum daily loads, *Journal of the American Water Resources Association, 35*(6) pp. 1365-1372, 1999.

Felzer, B., and P. Heard, Precipitation differences amongst GCMs used for the US National *Assessment, Journal of the American Water Resources Association, 35*(6), pp. 1327-1340.

Frederick, K. D., and P. H. Gleick, Water and global climate change: Potential impacts on US water resources, Pew Center on Global Climate Change, Washington, DC, 1999.

Frederick, K. D., and G. E. Schwarz, Socioeconomic impacts of climate change on US water supplies, *Journal of the American Water Resources Association, 35*(6), pp. 1563-1584, 1999a.

Frederick, K. D., and G. E. Schwarz, Socioeconomic impacts of climate variability and change on US water supplies, unpublished manuscript, 1999b.

Georgakakos, A. P., H. Yao, and K. P. Georgakakos, Vulnerability of river basin management to climate variability and change, *Proceedings of the AWRA Specialty Conference on Potential Consequences of Climate Variability and Change to Water Resources of the United States,* American Water Resources Association, Herndon, VA, pp. 49-56, 1999.

Gleick, P. H. et al., *Water: The Potential Consequences of Climate Variability and Change.* Report of the National Water Assessment Group for the U.S. Global Change Research Program. [Pacific Institute, USGS, Department of the Interior], Oakland, California. 151 pp., 2000.

Gleick, P. H., Vulnerability of water systems, in *Climate Change and the US Water Resources,* edited by P. E. Waggoner, pp. 223-240, John Wiley and Sons, New York, 1990.

Gleick, P. H., and L. Nash, The societal and environmental costs of the continuing California drought, Pacific Institute for Studies in Development, Environment, and Security, Berkeley, California, 1993.

Gleick, P. H., and B. Chalecki, The impacts of climatic changes for water resources of the Colorado and Sacramento-San Joaquin River basins, *Journal of the American Water Resources Association,* 35(6), 1429-1441, 1999.

Gomez, S., and A. Steding, California water transfers: An evaluation of the economic framework and a spatial analysis of the potential impacts, Pacific Institute for Studies in Development, Environment, and Security, Oakland, California, 57 pp., 1998.

Gornitz, V., Sea-level rise: A review of recent past and near-future trends, *Earth Surface Processes and Landforms, 20,* pp. 7-20, 1995.

Groisman, P. Ya, et al., Changes in the probability of heavy precipitation: Important indicators of climate change, *Climatic Change, 42,* 243-283, 1999.

Groisman, P. Ya., Trends in the precipitation and snow cover records across the United States, *Proceedings of the AWRA Specialty Conference on Potential Consequences of Climate Variability and Change to Water Resources of the United States,* American Water Resources Association, Herndon, VA, pp. 89-92, 1999.

Groisman, P. Ya., R. W. Knight, and T. R. Karl, Mean precipitation and high streamflow in the contiguous United States: Trends in the twentieth century, *Bulletin of the American Meteorology Society,* in press, 2001.

Hamlet, A. F., and D. P. Lettenmaier, Effects of climate change on hydrology and water resources objectives in the Columbia River basin, *Journal of the American Water Resources Association,* 35(6), pp.1597-1625, 1999.

Hatch, U., S. Jagtap, J. Jones, and M. Lamb, Potential effects of climate change on agricultural water use in the Southeast US, *Journal of the American Water Resources Association,* 35(6), pp. 1561-1563, 1999.

Hayden, B. P., Extratropical storms: Past, present, and future, *Proceedings of the AWRA Specialty Conference on Potential Consequences of Climate Variability and Change to Water Resources of the United States,* American Water Resources Association, Herndon, VA, pp. 93-96, 1999a.

Hayden, B. P., Climate change and extratropical storminess in the United States: An Assessment, *Journal of the American Water Resources Association,* 35(6), pp.1387-1399, 1999b.

Herrmann, R., L. Scherbarth, and R. Stottlemyer, Save the wolves, lose the watersheds: Environmental aspects of water use in the United States, *Journal of the American Water Resources Association,* 35(6), 1999.

Hurd, B. J., N. Leary, R. Jones, and J. Smith, Relative regional vulnerability of water resources to climate change, *Journal of the American Water Resources Association, 35(6),* pp.1399-1410, 1999a.

Hurd, B, R. Jones, N. Leary and J. Smith, A regional assessment and database of water resource vulnerability to climate change, *Proceedings of the AWRA Specialty Conference on Potential Consequences of Climate Variability and Change to Water Resources of the United States,* American Water Resources Association, Herndon, VA, pp. 45-48, 1999b.

Jehl, D., Tampa Bay looks to the sea to quench its thirst, *The New York Times,* March 12, 2000.

Intergovernmental Panel on Climate Change (IPCC), *Climate Change 1995: Impacts, Adaptations and Mitigation of Climate Change: Scientific-Technical Analyses,* Cambridge University Press, New York, 1996.

Jacobs, K. L., Water sector summary comments for inclusion in the synthesis report from the AWRA Specialty Conference on Potential Consequences of Climate Variability and Change to Water Resources of the United States, Atlanta, GA, May 10-12, 1999.

Karl, T. R., R. W. Knight, D. R. Easterling, and R. G. Quayle, Indices of climate change for the United States, *Bulletin of the American Meteorological Society, 77,* pp. 279-292, 1996.

Karl, T. R., and R. W. Knight, Secular trends of precipitation amount, frequency, and intensity in the United States, *Bulletin of the American Meteorological Society, 79(2),* pp. 231-241, 1998.

Karl, T. R., and R. W. Riebsame, The impact of decadal fluctuations in mean precipitation and temperature on runoff: A sensitivity study over the United States, *Climatic Change, 15,* 423-447, 1989.

Kimball, B. A., Carbon dioxide and agricultural yield: An assemblage and analysis of 430 prior observations, *Agronomy Journal, 75,* 779-788, 1983.

Kusler, J., and V. Burkett, Climate Change in Wetlands Areas Part I: Potential Impacts and Interactions, Acclimations, 6, 1999. URL: www.nacc.usgcrp.gov/newsletter/1999.06/wet.html

Lambert, S. J., The effect of enhanced greenhouse warming on winter cyclone frequencies and strengths, *Journal of Climate, 8*(5), pp. 1447-1452, 1995.

Lettenmaier, D. P., E. F. Wood, and J. R. Wallis, Hydro-climatological trends in the continental United States 1948-1988, *Journal of Climate, 7,* pp. 586-607, 1994.

Leung, R. L., and S. J. Ghan, Pacific Northwest climate sensitivity simulated by a regional climate model driven by a GCM, Part I: Control simulations, *Journal of Climate, 12,* pp. 2010-2030, 1999a.

Leung, R. L., and S. J. Ghan, Pacific Northwest climate sensitivity simulated by a regional climate model driven by a GCM, Part II: 2 X CO_2 simulations, *Journal of Climate, 12,* pp. 2031-2053, 1999b.

Leung, R. L., and M. S. Wigmosta, Potential climate change impacts on mountain water resources in the Pacific Northwest, *Journal of the American Water Resources Association, 35*(6), pp. 1463-1473, 1999.

Lins, H. F., and J. R. Michaels, Increasing US streamflow linked to greenhouse forcing, *Eos Transactions, 75,* pp. 281-283, 1994.

Lins, H. F., and J. R. Slack, Streamflow trends in the United States, *Geophysical Research Letters, 26*(2), pp. 227-230, 1999.

Matalas, N. C., Note on the assumption of hydrologic stationarity, *Water Resources Update No. 112,* The Universities Council on Water Resources, Carbondale, IL, pp. 64-72, 1988.

McCabe, G. J., and D. M. Wolock, Sensitivity of irrigation demand in a humid-temperate region to hypothetical climate change, *Climatic Change, 37,* pp. 89-101, 1992.

McCabe, G. J., and D. M. Wolock, General-Circulation-Model simulations of future snowpack in the Western United States, *Journal of the American Water Resources Association, 35*(6), pp. 1473-1484, 1999.

Mekis, E., and W. Hogg, Rehabilitation and analysis of Canadian daily precipitation time series, *Conference on Applied Climatology,* Reno, Nevada, pp. 20-24, October 1997.

Meyer, J. L., M. J. Sale, P. J. Mulholland, and N. L. Poff, Impacts of climate change on aquatic ecosystem functioning and health, *Journal of the American Water Resources Association, 35*(6), pp. 1373-1386, 1999.

Miller, N. L., J. Kim, and R. K. Hartman, Downscaled climate and streamflow study of the southwestern United States, *Journal of the American Water Resources Association, 35*(6), pp. 1525-1538, 1999.

Mulholland, P. J., and M. J. Sale, Impacts of climate change on water resources: Findings of the IPCC regional assessment of vulnerability for North America, *Water Resources Update No. 112,* The Universities Council on Water Resources, Carbondale, IL, pp. 10-16, 1998.

National Research Council, *Water Transfers in the West: Efficiency, Equity, and the Environment,* National Academy Press, Washington, DC, 1992.

National Research Council, *Valuing Groundwater: Economic Concepts and Approaches,* National Academy Press, Washington, DC, 1997.

O'Connor, R. E., B. Yarnal, R. Neff, R. Bord, N. Wiefek, C. Reenock, R. Shudak, C. L. Jacoy, P. Pascale, and C. G. Knight, Weather and climate extremes, climate change, and planning: Views of community water system managers in Pennsylvania's Susquehanna River basin, *Journal of the American Water Resources Association, 35*(6), 1999.

Pacific Northwest Regional Report, *Impacts of Climate Variability and Change, Pacific Northwest,* The JISAO/SMA Climate Impacts Group, University of Washington, Seattle, WA, 2000.

Schilling, K. E., Water resources: The state of the infrastructure, report to the National Council on Public Works Improvement, Washington, DC, May 1987.

Snover, A., E. Miles, and B. Henry, Regional workshop on the impacts of global climate change on the Pacific Northwest final report, University of Washington, Seattle, July 1997.

Solley, W. B., R. R. Pierce, and H. A. Perlman, Estimated use of water in the United States in 1995, *US Geological Survey Circular 1200,* Denver, Colorado, 1998.

Solley, W. B., and H. A. Perlman, Factors affecting water use patterns in the United States, 1950-1995, with projections to 2040, *Proceedings of the AWRA Specialty Conference on Potential Consequences of Climate Variability and Change to Water Resources of the United States,* American Water Resources Association, Herndon, VA, pp. 65-67, 1999.

Squillace, P. J., M. J. Moran, W. W. Lapham, C. V. Price, R. M. Clawges, and J. S. Zagorski, Volatile Organic Compounds in Untreated Ambient Groundwater of the United States, 1985-1995, *Environmental Science and Technology,* pp. 4176-4187, December 1999.

Stakhiv, E., and K. Schilling, What can water managers do about global warming?, *Update Water Resources,* No. 112, The Universities Council on Water Resources. pp. 33-40, 1998.

Union of Concerned Scientists and the Ecological Society of America, *Confronting Climate Change in California: Ecological Impacts on the Golden State,* 62 pp., UCS Publications, Cambridge, November 1999.

US Army Corps of Engineers, National study of water management during drought: The report to Congress, *IWR Report 94-NDS-12,* September 1996.

US Army Corps of Engineers, National inventory of dams, 1996.

US Environmental Protection Agency, National water quality inventory: 1996 report to Congress, *EPA841-R-97-008,* Washington, DC, 1998.

US Water Resources Council, The Nation's water resources 1975-2000, second National Water Assessment, US Government Printing Office, Washington, DC, 1978.

US Department of the Interior, Geological Survey, National water summary 1983: Hydrologic events and issues, *Water Supply Paper 2250,* 1984.

US Department of the Interior, Climate change: Impacts on water resources, 15 pp., June 30, 1997.

US Department of the Interior, Geological Survey, Strategic directions for the US Geological Survey Groundwater Resources Program: A report to Congress, November 30, 1998.

US Department of the Interior, Geological Survey, National Assessment of Water Quality, www.water.usgs.gov/lookup/get?nawqa.html
US National Assessment, draft report of the Water Sector, 2000. http://www.nacc.usgcrp.gov/sectors/water/draft-report/full-report.html

Vandemoer, C., Impacts of climate change on American Indian water rights and resources, *Proceedings of the AWRA Specialty Conference on Potential Consequences of Climate Variability and Change to Water Resources of the United States,* American Water Resources Association, Herndon, VA, pp. 17-21, 1999.

Wang, B., D. C. Trabant, G. Weller, and P. Anderson, Effect of climate variability and change on the water resources of Alaska, *Proceedings of the AWRA Specialty Conference on Potential Consequences of Climate Variability and Change to Water Resources of the United States,* American Water Resources Association, Herndon, VA, pp. 273-276, 1999.

Western Water Policy Review Advisory Commission, *Water in the West: Challenge for the Next Century,* National Technical Information Service, Springfield, Virginia, 1998.

Winter, T. C., The vulnerability of wetlands to climate change: A hydrologic landscape perspective, *Journal of the American Water Resources Association, 35*(6), 1999.

Wolock, D. M., and G. J. McCabe, Explaining spatial variability in mean annual runoff in the conterminous United States, *Climate Research, 11,* 149-159, 1999a.

Wolock, D. M., and G. J. McCabe, Simulated effects of climate change on mean and annual runoff in the conterminous United States, *Proceedings of the AWRA Specialty Conference on Potential Consequences of Climate Variability and Change to Water Resources of the United States,* American Water Resources Association, Herndon, VA, pp. 161-164, 1999b.

Woodhouse, C. A., and J. T. Overpeck, A Paleoclimatic record of drought for the past 2,000 years, *Proceedings of the AWRA Specialty Conference on Potential Consequences of Climate Variability and Change to Water Resources of the United States,* American Water Resources Association, Herndon, VA, pp. 59-64, 1999.

ACKNOWLEDGMENTS

Many of the materials for this chapter are based on contributions from participants on and those working with the

Water Resources Sector Assessment Team
D. Briane Adams*, US Geological Survey
Peter Gleick*, Pacific Institute for Studies in
    Development, Environment, and Security
Beth Chalecki, Pacific Institute for Studies in
    Development, Environment, and Security
Joseph Dellapenna, Villanova University
Ted Engman, NASA Goddard Space Flight Center
Kenneth D. Frederick, Resources for the Future
Aris P. Georgakakos, Georgia Institute of Technology
Gerald Hansler, Delaware River Basin Commission
    (retired)
Lauren Hay, US Geological Survey
Bruce P. Hayden, National Science Foundation
Blair Henry, The Northwest Council on Climate
    Change

Steven Hostetler, US Geological Survey
Katharine Jacobs, Arizona Department of Water
    Resources
Sheldon Kamieniecki, University of Southern
    California
Robert D. Kuzelka, University of Nebraska-Lincoln
Dennis Lettenmaier, University of Washington
Gregory McCabe, US Geological Survey
Judy Meyer, University of Georgia
Timothy Miller, US Geological Survey
Paul C. "Chris" Milly, NOAA Geophysical Fluid
    Dynamics Laboratory
Norman Rosenberg, Battelle-Pacific Northwest
    National Laboratory
Michael J. Sale, Oak Ridge National Laboratory
John Schaake, National Oceanic and Atmospheric
    Administration
Gregory Schwarz, US Geological Survey
Susan S. Seacrest, The Groundwater Foundation
Eugene Z. Stakhiv, US Army Corps of Engineers
David Wolock, US Geological Survey

* Assessment Team chair/co-chair

CHAPTER 15

POTENTIAL CONSEQUENCES OF CLIMATE VARIABILITY AND CHANGE FOR HUMAN HEALTH IN THE UNITED STATES

Jonathan A. Patz[1], Michael A. McGeehin[2], Susan M. Bernard[1], Kristie L. Ebi[3], Paul R. Epstein[4], Anne Grambsch[5], Duane J. Gubler[6], Paul Reiter[7], Isabelle Romieu[2], Joan B. Rose[8], Jonathan M. Samet[9], Juli Trtanj[10], with Thomas F. Cecich[11,12]

Contents of this Chapter

[1]Department of Environmental Health Sciences, Johns Hopkins University School of Hygiene and Public Health; [2]National Center for Environmental Health, U.S. Centers for Disease Control and Prevention; [3]EPRI; [4]Center for Health and the Global Environment, Harvard Medical School; [5]Office of Research and Development, U.S. Environmental Protection Agency; [6]Division of Vector-Borne Diseases, U.S. Centers for Disease Control and Prevention; [7]Division of Vector-Borne Diseases, U.S. Centers for Disease Control and Prevention, Dengue Branch; [8]Department of Marine Sciences, University of South Florida; [9]Department of Epidemiology, Johns Hopkins University School of Hygiene and Public Health; [10]Office of Global Programs, National Oceanic and Atmospheric Administration, [11]Glaxo Wellcome, Inc., [12]Liaison for the National Assessment Synthesis Team.

PREFACE

Projections of the extent and direction of the potential health impacts of climate variability and change are extremely difficult to make because of the many confounding and poorly understood factors associated with potential health outcomes, population vulnerability, and adaptation. In fact, the relationship between weather and specific health outcomes is understood for a relatively small number of diseases, with few quantitative models available for analysis. The costs, benefits and availability of resources to address adaptation measures also require evaluation. Research aimed at filling the priority knowledge gaps identified in this assessment would allow for more quantitative assessments in the future.

CHAPTER SUMMARY

Because human health is intricately bound to weather and the many complex natural systems it affects, it is possible that climate change as projected will have a measurable impact, both beneficial and adverse, on health outcomes associated with weather and/or climate. We identified and assessed five such categories of health outcomes: 1) temperature-related morbidity and mortality; 2) health effects of extreme weather events (i.e., storms, tornadoes, hurricanes, and precipitation extremes); 3) air pollution-related health effects; 4) water- and food-borne diseases; and 5) insect-, tick-, and rodent-borne diseases.

Temperature-related Morbidity and Mortality
The more frequent heat waves projected to accompany climate change would pose a risk, particularly against the backdrop of an aging US population, as the elderly are most susceptible to dying from extreme heat. Beyond individual behavioral changes, adaptation measures include development of community-wide heat emergency plans, improved heat warning systems, and better heat-related illness management plans. Death rates are higher in winter than in summer and it is possible that milder winters could reduce deaths in winter months. However, the relationship between winter weather and mortality has been difficult to interpret. The net effect on winter mortality from climatic changes is uncertain and the overall balance between changes in summer and winter weather-related deaths is unknown.

Extreme weather events-related health effects
Health impacts from weather disasters range from acute trauma and drowning, to more medium- and long-term effects, such as conditions of unsafe water, and post traumatic stress disorder (PTSD). The health impacts of extreme weather events such as floods and storms hinge on the vulnerabilities of the natural environment and the local population, as well as on their capacity to recover. The location of development in high-risk areas increases a community's vulnerability to extreme weather events. Adverse health outcomes in the US are low compared with global figures partly because of the many federal, state, and local government agencies and non governmental organizations (NGOs) engaged in disaster planning, early warning, and response.

Air pollution-related health effects
Quantitative studies of the potential effects of climate change on air quality have primarily focused on the impact of increased temperature and ultraviolet radiation on ozone formation. In general, these few studies find that ozone concentrations increase as temperatures rise. The specific type of change (i.e., local, regional, or global), the direction of change in a particular location (i.e., positive or negative), and the magnitude of change in overall air quality (i.e., for all of the criteria air pollutants) that may be attributable to climate change, however, are not known. Additionally, climate change may alter the distribution and types of airborne allergens.

Emissions scenarios are central to assessing future air quality in addition to assessing the effect of altered weather on specific air pollutant formation and/or transport. Integrated air quality modeling studies will be necessary to assess more quantitatively the potential health impacts of air quality changes associated with global climate change.

Water- and food-borne diseases
Weather influences the transport of microbial agents, via rainfall runoff over contaminated sources. Temperature also influences the occurrence of bacterial agents, toxic algal blooms (red tides), and survival of viral pathogens that cause shellfish poisoning. Management of sewage and other wastes, and watershed protection are important to reducing health risks. Federal and state regulations protect much of the US population. Nonetheless, if climate variability increases, current

deficiencies in watershed protection and storm drainage systems will probably increase the risk of contamination events.

Insect-, tick-, and rodent-borne diseases

The ecology and transmission dynamics of insect- and rodent-borne infections are complex and unique for each disease. Many of these diseases exhibit a distinct seasonal pattern, suggestive of weather sensitivity. But demographic, sociological and ecological factors also play a critical role in determining disease transmission. The moderating effect of these other factors makes it unlikely that increasing temperatures alone will have a major impact on tropical diseases spreading into the US. There is greater uncertainty regarding more indigenous diseases that cycle through animals and can also infect humans. Further studies of transmission dynamics, and of pathogen, vector, and animal reservoir host ecology are required to determine whether these diseases will increase or decrease with climate change.

Key Findings

· Multiple levels of uncertainty preclude any definitive statement on the direction of potential future change for each of the health outcomes assessed.

· Although our report mainly addresses adverse health outcomes, some positive health outcomes were identified, notably reduced cold-weather mortality, which has not been extensively examined.

· At present, much of the US population is protected against adverse health outcomes associated with weather and/or climate, although certain demographic and geographic populations are at increased risk.

· Vigilance in the maintenance and improvement of public health systems and their responsiveness to changing climate conditions and to identified vulnerable subpopulations should help to protect the US population from adverse health outcomes of projected climate change.

POTENTIAL CONSEQUENCES OF CLIMATE VARIABILITY AND CHANGE FOR HUMAN HEALTH IN THE UNITED STATES

INTRODUCTION

This chapter is based on a literature review and consultation with experts, interested researchers, and members of the public health community. Using, as an underlying set of assumptions, climate change projections developed for the national assessment, we analyzed the potential relationships between climate variability, climate change, and human health within a given framework of questions:

- What is the current status of the nation's health and what are current stresses on our health?
- How might climate variability and change affect the country's health and existing or predicted stresses on health?
- What is the country's capacity to adapt to climate change, for example, through modifications to the health infrastructure or by adopting specific adaptive measures?
- What essential knowledge gaps must be filled to fully understand the possible impacts of climate variability and change on human health?

These questions were developed to tailor the mandate of the National Assessment to the health sector. That mandate was to identify, for each particular sector or region 1) the current status; 2) the expected impacts of climate variability and change; 3) the adaptive capacity; and 4) the research gaps. Responding to these questions enabled assessment participants to evaluate a baseline and then to identify adaptation measures and research needs. Consistent with the National Assessment as a whole, the health sector did not address the question of the specific role of anthropogenic (human-caused) contributions to changes to climate or identify measures to reduce emissions of greenhouse gases or the presence of greenhouse gases in the atmosphere (e.g., through carbon dioxide sequestration). These issues are critically important, and are the focus of other past and ongoing research and programs in the United States and elsewhere. However, the extent and success of current and future mitigation measures is uncertain. In addition, climate scientists

project that some degree of projected climate change over the next several decades cannot be prevented, as a result of already elevated concentrations of greenhouse gases in the atmosphere, even if mitigation steps are taken. Thus, it is important to understand what adaptation measures might be desirable, or are feasible, regardless of mitigation, given the current climate projections. Future climate change assessments might choose to link adaptation and mitigation research and impacts.

CONTEXT

Because human health is intricately bound to weather and the many complex natural systems it affects, it is possible that climate change as projected will have a measurable impact, both beneficial and adverse, on health outcomes associated with climate and/or weather (see Figure 1). These outcomes include temperature-related illnesses and deaths, injuries or fatalities from extreme weather events (i.e., storms, tornadoes, hurricanes, and precipitation extremes), air pollution- related health effects, and diseases carried by water (water- and food-borne diseases) and by organisms such as mosquitoes, ticks, mites, and rodents (vector- and rodent-borne diseases).

To establish a baseline for projections of the potential impacts of climate on health, we reviewed the current status and context of health in the US, as reflected in indicators such as life expectancy and the leading causes of death. We also identified possible strains on public health and health care systems, such as cost and population growth. Urbanization, funding for public health infrastructure (e.g., sanitation systems and medical research) and scientific developments contributed to advances in health status in the past and are expected to do so in the future. Environmental conditions, such as air and water quality, are important determinants of health.

Chronic diseases—heart disease, cancer, stroke, and chronic obstructive pulmonary disease are the leading four— accounted for almost 75% of all US deaths in 1996 for the 25- to 64- year old age group (NCHS, 1998). Injuries and infectious diseases remain significant causes of morbidity and mortality in the US; infectious diseases caused one third of the deaths in the US in 1992, primarily because of respiratory tract infections, human immunodeficiency virus (HIV), and septicemia (Pinner et al., 1996). Patterns of illness and death vary substantially by socioeconomic status, geographic region, race, age, and gender (NCHS, 1998).

Certain populations within the US — the poor, the elderly, children, and immunocompromised individuals — may be more vulnerable to many of the health risks that might be initially exacerbated by climate change. Poverty, for example, is a risk factor for heat-related illnesses and deaths, because the poor are more likely to live in urban areas and are less likely to be able to afford air-conditioning systems. Thus, making air-conditioned environments readily available to the poor is an adaptive response strategy to reduce illnesses and deaths in heat waves. Understanding what groups may be the most affected by climate change is critical to effective targeting of prevention or adaptation strategies. For example, air pollution and heat advisory warnings should specifically target children and the elderly, respectively.

It is also important to recognize that there are racial differences in health outcomes, including those associated with weather and/or climate, such as heat waves; these differences may be associated with poverty status, which is disproportionately high among African-Americans. For example, data on the 1995 heat wave in Chicago indicate that mortality among African-Americans was 50% higher than among whites (Whitman et al., 1997). The disparity likely reflects residence in inner- city neighborhoods, poverty, housing conditions, and medical conditions (Applegate et al., 1981; Jones et al., 1982; Kilbourne et al., 1982).

It is important to recognize that the proportion of elderly (65 years of age and older) and very elderly (85 years of age and older) residents is expected to rise in the coming decades. The proportion of the senior population in the very elderly category is growing fast: their numbers rose 274% between 1960 and 1994, while the entire US population grew only 45% (Hobbs and Damon, 1998). Aging can be expected to be accompanied by multiple, chronic illnesses that may result in increased vulnerability to infectious disease or external/environmental stresses such as extreme heat (Hobbs and Damon, 1998). Poverty, which increases with age in the elderly, may add to this vulnerability (Day, 1996).

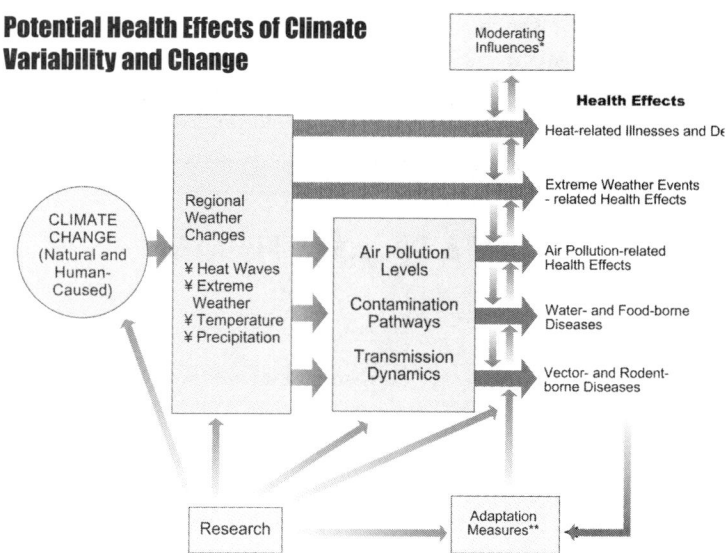

Potential Health Effects of Climate Variability and Change

Figure 1: Schematic diagram of the potential health effects of climate variability and change. (Source, Patz et al., 2000)
* Moderating influences include non-climate factors that affect climate-related health outcomes, such as: population growth and demographic change; standards of living; access to health care; improvements in health care; and public health infrastructure.
** Adaptation measures include actions to reduce risks of adverse health outcomes, such as: vaccination programs; disease surveillance; monitoring; use of protective technologies (e.g., air conditioning, pesticides, water filtration/treatment); use of climate forecasts; and development of weather warning systems; emergency management and disaster preparedness programs; and public education. See Color Figure Appendix

Similarly, although the proportion of children younger than 5 years of age is not expected to grow as significantly as the proportion of the elderly, their number will increase even if immigration levels are kept constant. The variables that may affect children's special vulnerability to the possible impacts of climate change include: poverty [currently approximately 20% of US children are poor (NCHS, 1998)]; access to medical care; and children's susceptibility to environmental hazards because of their size, behavior, and the fact that they are growing and developing (Landrigan et al., 1999).

Finally, it is anticipated that the proportion of immunocompromised people in the US may increase with the aging of the population and the success of medical treatments (e.g., cancer therapy and HIV medications), but data are difficult to obtain. For example, survival has improved for acquired immunodeficiency syndrome (AIDS) patients, resulting in a 12% increase from 1996-1997 in the number of people living with AIDS (CDC, 1998). AIDS patients and other immunocompromised individuals may be more susceptible to water-borne and vector-borne pathogens, to the adverse impacts of exposure to elevated levels of certain air pollutants, and to debilitation due to physical stresses, such as those experienced during heat waves or in adverse emergency weather conditions, unless they can be adequately protected from those stresses with access to air conditioning, sanitation, safe water, and sufficient food.

CLIMATE AND HUMAN HEALTH

The National Assessment climate models project

that over the relevant time period the US climate will be characterized by increased temperatures, an altered hydrologic cycle, and increased variability. These projections are based in part on historical data; however, a detailed systematic record of weather parameters is only available for some places for approximately the last hundred years, although indirect measurements from ice cores, tree rings, other paleo-data and written history extend further (Houghton, 1997). In the past 100 years, the global surface temperature has warmed between 0.7 and 1.4° F (Easterling et al., 1997; Jones et al., 1999;

Table 1. Summary of the Health Sector Assessment

Potential health impacts	Weather factors of interest *	Direction of possible change in health impact	Examples of some specific adaptation strategies	Priority research areas
Heat-related illnesses and deaths	Extreme heat and stagnant air masses	↑	Air conditioning Early warning	Improved prediction, warning and response Urban design and energy systems Exposure assessment Weather relationship to influenza and other causes of winter mortality
Winter deaths	Extreme cold Snow Ice	↓		
Extreme weather events-related health effects	Precipitation variability (heavy rainfall events †) Storms	↑	Early warning Engineering Zoning and building codes	Improved prediction, warning and response Improved surveillance Investigation of past impacts and effectiveness of warnings
Air pollution-related health effects	Temperature Stagnant air masses	↑	Early warning Mass transit Urban planning Pollution control	Relationships between weather and air pollution concentrations Combined effects of temperature/humidity on air pollution Effect of weather on vegetative emissions and allergens (e.g., pollen)
Water- and food-borne diseases	Precipitation Estuary water temperatures	↑	Surveillance Improved water systems engineering	Improved monitoring effects of weather/environment on marine-related disease Land use impacts on water quality (watershed protection) Enhanced monitoring/mapping of fate and transport of contaminants
Vector- and rodent-borne diseases	Temperature Precipitation variability Relative humidity	↑↓	Surveillance Vector control programs	Rapid diagnostic tests Improved surveillance Climate-related disease transmission dynamic studies

* Based on projections provided by the National Assessment Synthesis Team. Other scenarios might yield different changes.
† Projected change in frequency of hurricanes and tornadoes is unknown.

NRC, 2000). In the contiguous US, temperatures have increased by approximately 1°F (Karl et al., 1996), and precipitation has been increasing in the US, with much of this change due to increases in heavy-precipitation events (> 2 inches [5 cm] per day) and decreases in light-precipitation events (Karl et al., 1995; Karl et al., 1996; Karl and Knight, 1998). These historical data are consistent with climate change theory that suggests an altered hydrological cycle accompanying warming of the Earth's surface (Fowler and Hennessey, 1995; Mearns et al., 1995; Trenberth, 1999).

We examined the impact of this projected climate change on five health outcomes: 1) temperature-related morbidity and mortality; 2) health effects of extreme weather events (i.e., storms, tornadoes, hurricanes, and precipitation extremes); 3) air pollution-related health effects; 4) water- and food-borne diseases; and 5) vector- and rodent-borne diseases. Some of these outcomes are relatively direct (e.g., effects of exposure to extreme heat or extreme events); others involve intermediate and multiple pathways, making assessments more challenging (see Figure 1).

Projections of the extent and direction of some potential health impacts of climate variability and change can be made, but there are many layers of uncertainty (see Table 1). First, methods to project changes in climate over time continue to improve, but climate models are unable to accurately project regional-scale impacts. Second, basic scientific information on the sensitivity of human health to aspects of weather and climate is limited. In addition, the vulnerability of a population to any health risk varies considerably depending on moderating factors such as population density, level of economic and technological development, local environmental conditions, preexisting health status, the quality and availability of health care, and the public health infrastructure.

It is also difficult to anticipate what adaptive measures might be taken in the future to mitigate risks of adverse health outcomes, such as vaccines, disease surveillance, protective technologies (e.g., air conditioning or water filtration/treatment), use of weather forecasts and warning systems, emergency management and disaster preparedness programs, and public education (see Figure 1). As they do currently, the need for and the success of adaptation measures can be expected to vary in different parts of the country—for example, Chicago must plan for heat waves, and communities along the southeast coast must be prepared for hurricanes. For the most part,

government organizations fund public health systems within the US. Continued investments in advancing the public health infrastructure are crucial for adapting to the potential impacts of climate variability and change.

KEY ISSUES

- Temperature-related illnesses and deaths
- Health effects related to extreme weather events
- Air pollution-related health effects
- Water- and food-borne diseases
- Insect-, tick-, and rodent-borne diseases

1. Temperature-related Illnesses and Deaths

Heat and heat waves are projected to increase in severity and frequency with increasing global mean

July Heat Index Change - 21st Century

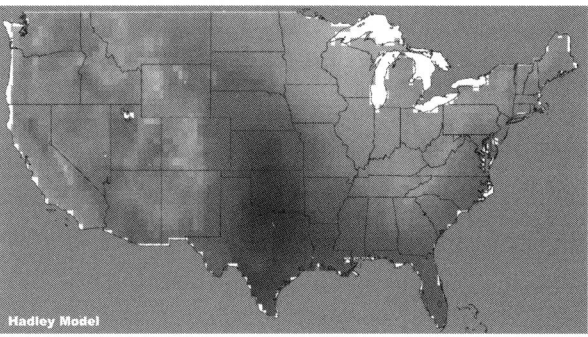

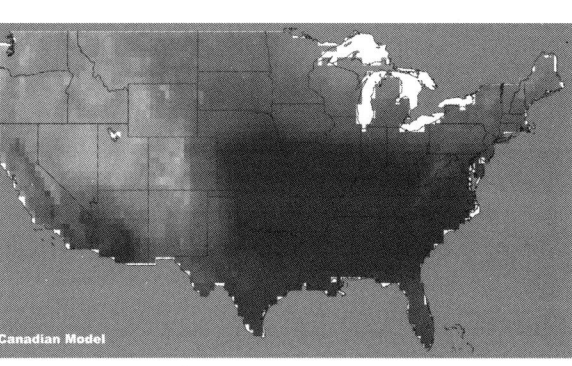

Figure 2: Both models project substantial increases in the July heat index (which combines heat and humidity) over the 21st century. These maps show the projected increase in average daily July heat index relative to the present. The largest increases are in the southeastern states, where the Canadian model projects increases of more than 25°F. For example, a July day in Atlanta that now reaches a heat index of 105°F would reach a heat index of 115°F in the Hadley model, and 130°F in the Canadian model.(Map by Benjamin Felzer, UCAR, based on data from Canadian and Hadley modeling centers.) See Color Figure Appendix

Heat Related Deaths in Chicago in July 1995

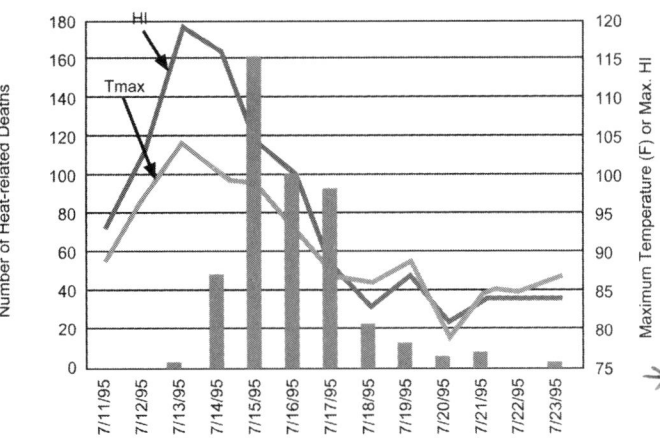

Figure 3: This graph tracks the maximum temperature (Tmax), heat index (HI), and heat-related deaths in Chicago each day from July 11 to 23, 1995. The gray line shows maximum daily temperature, the blue line shows the heat index, and the bars indicate the number of deaths each day. (Source: NOAA/NCDC) See Color Figure Appendix

temperatures (Meehl et al., 2000); see Figure 2. Studies of heat waves in urban areas have shown an association between increases in mortality and increases in heat, measured by maximum or minimum temperature, and heat index (a measure of temperature and humidity). Some of these studies adjust for other weather conditions (Semenza et al., 1996; Kalkstein and Greene, 1997). For example, after a five-day heat wave in 1995 in which maximum temperatures in Chicago, Illinois ranged from 93°F to 104°F, the number of deaths increased 85% over the number recorded during the same period of the preceding year (CDC, 1995); see Figure 3. At least 700 excess deaths (i.e., deaths beyond those expected for that period in that population) were recorded, most of which were directly attributed to heat (CDC, 1995; Semenza et al., 1996; Semenza et al., 1999).

Exposure to extreme and prolonged heat is associated with heat cramps, heat syncope (fainting), heat exhaustion, and heat stroke. These health effects appear to be related to environmental temperatures above those to which the population is accustomed. Models of weather-mortality relationships indicate that populations in northeastern and midwestern US cities are likely to experience the greatest number of heat-related illnesses and deaths in response to changes in summer temperature, and that the most sensitive regions are those where extremely high temperatures occur infrequently or irregularly (Kalkstein and Smoyer, 1993); see Figure 4. For example, Philadelphia, Chicago, and Cincinnati have each experienced a heat wave that resulted in a large number of heat-related deaths. Physiologic and

behavioral adaptations among vulnerable populations may reduce morbidity and mortality due to heat. Although long-term physiologic adaptation to heat events has not been documented, adaptation appears to occur as the summer season progresses; heat waves early in the summer often result in more deaths than subsequent heat waves or than those occurring later in the summer (Kalkstein and Smoyer, 1993). Heat waves are episodic, and although populations may adapt to gradual temperature increases, physiologic adaptation for extreme heat events is unlikely.

Within heat-sensitive regions, populations in urban areas are the most vulnerable to adverse heat-related health outcomes. Heat indices and heat-related mortality rates are higher in the urban core than in surrounding areas (Landsberg, 1981). Urban areas retain heat throughout the nighttime more efficiently than do outlying suburban and rural areas (Buechley et al., 1972; Clarke, 1972). The absence of nighttime relief from heat for urban inhabitants is a factor in excessive heat-related deaths.

The size of US cities and the proportion of US residents living in them are projected to increase over the next century, so it is possible that the population at risk for heat-related illnesses and deaths will increase. High-risk sub-populations include people who live in the top floors of apartment buildings in cities and who lack access to air-conditioned environments (either at home or elsewhere). The elderly (Ramlow and Kuller, 1990; CDC, 1993; Whitman et al., 1997; Semenza, 1999), young children (CDC, 1993), the poor (Schuman, 1972; Applegate et al., 1981), and people who are bedridden or on medications that affect the body's thermoregulatory ability (Kilbourne et al., 1982; Di Maio and Di Maio, 1993; Marzuk et al., 1998) are particularly vulnerable.

There is evidence that heat-related illnesses and deaths are largely preventable through behavioral adaptations, including the use of air conditioning and increased fluid intake (Kilbourne et al., 1982), although the magnitude of mortality reduction cannot be predicted. The proportion of housing units with central and/or room unit air conditioning ranges from below 30% in the Northeast to almost 90% in the South (Bureau of the Census, 1997a). The use of air-conditioning systems in homes, workplaces, and vehicles has increased steadily over the past 30 years and is projected to become nearly universally available in the US by the year 2050 (Bureau of the Census, 1997a; Bureau of the Census, 1997b).

Overall death rates are higher in the winter than in the summer, and it is possible that milder winters could reduce deaths in winter months (Kalkstein and Greene, 1997). However, the relationship between winter weather and mortality is difficult to interpret. For example, many winter deaths are due to infectious diseases such as influenza and pneumonia, and it is unclear how influenza transmission would be affected by higher winter temperatures. In addition, studies indicate an association between snowfall and fatal heart attacks (from winter precipitation rather than cold temperatures) (Spitalnic et al., 1996; Gorjanc et al., 1999). The net effect on winter mortality from climate change is therefore extremely uncertain, and the overall balance between changes in summer and winter weather-related deaths is unknown.

Beyond individual behavioral changes, adaptation measures include the development of community-wide heat emergency plans, improved heat warning systems, and better heat-related illness management plans. Research can refine each of these measures, including which weather parameters are most important in the weather-health relationship, the associations between heat and nonfatal illnesses, the evaluation of implemented heat response plans, and the effectiveness of urban design in reducing heat retention.

2. Health Effects Related to Extreme Weather Events

Climate change may alter the frequency, timing, intensity, and duration of extreme weather events (Fowler and Hennessey, 1995; Karl et al., 1995; Mearns et al., 1995), i.e., meteorological events that have a significant impact on local communities. Increases in heavy precipitation have occurred over the past century (Karl et al., 1995; Karl and Knight, 1998). Future climate scenarios show likely increases in the frequency of extreme precipitation events, including precipitation during hurricanes (Knutson and Tuleya, 1999). This poses an increased risk of floods (Meehl et al., 2000). Frequencies of tornadoes and hurricanes cannot reliably be projected. Whether these changes in climate risk result in increased health impacts cannot currently be assessed.

Injury and death are the direct health impacts most often associated with natural disasters. Secondary health effects have also been observed. These impacts are mediated by changes in ecological systems (such as bacterial and fungal proliferation) and public health infrastructures (such as the availability of safe drinking water). The health impacts of

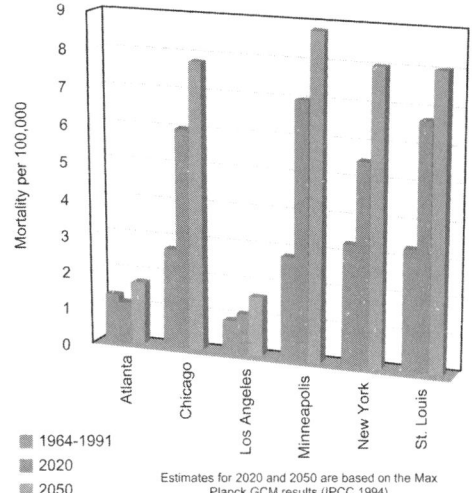

Average Summer Mortality Rates

Attributed to Hot Weather Episodes

Mortality per 100,000

Cities: Atlanta, Chicago, Los Angeles, Minneapolis, New York, St. Louis

■ 1964-1991
■ 2020
■ 2050

Estimates for 2020 and 2050 are based on the Max Planck GCM results (IPCC 1994)

Figure 4: Deaths due to summer heat are projected to increase in US cities, according to a study using time-dependent results (for greenhouse gas increase only) from several climate models (Kalkstein and Greene, 1997). Mortality rates (number of deaths per 100,000 population) are shown from the Max Planck Institute model, the results from which lie roughly in the middle of the models examined (the other climate scenarios used were from Geophysical Fluid Dynamics Laboratory (GFDL) and the Hadley Centre). Because heat-related illness and death appear to be related to temperatures much hotter than those to which the population is accustomed, cities that experience extreme heat only infrequently appear to be at greatest risk. For example, Philadelphia, New York, Chicago, and St. Louis have experienced heat waves that resulted in a large number of heat-related deaths, while heat related deaths in Atlanta and Los Angeles are much lower. In this study, statistical relationships between heat waves and increased death rates are constructed for each city based on historical experience. Deaths under a city s future climate are then projected by applying that city s projected incidence of extreme heat waves to the statistical relationship that was estimated for the city whose present climate is most similar to the projected climate for the city in question. This approach attempts to represent how people will acclimate to the new average climate that they experience. See Color Figure Appendix

extreme weather events such as floods and storms hinge on the vulnerabilities and recovery capacities of the natural environment and the local population. A community's level of preparedness greatly affects the severity of the health impacts of an extreme event.

From 1945 to 1989, 145 natural disasters caused 14,536 deaths in the United States, an average of 323 deaths per year (Glickman and Silverman, 1992). According to the National Weather Service, severe storms caused 600 deaths and 3,799 reported injuries in 1997 (NWS, 1999). Floods are the most frequent natural disaster and the leading cause of death from natural disasters in the US; the average annual loss of life is estimated to be as high as 146 deaths per year (NWS, 1992). Hurricanes also pose an ongoing threat; an average of two each year make landfall on the US coastline (NWS, 1993). The impacts of hurricanes include injuries and deaths resulting from strong winds and heavy rains. ⌐

Depending on the severity and nature of the weather event, people may experience disabling fear or aversion (Drabek, 1996). There is controversy about the incidence and continuation of significant mental problems, such as post traumatic stress disorder (PTSD), after disasters (Quarantelli, 1985). However, an increase in the number of mental disorders has been observed after several natural disasters in the US. Increased psychological problems were reported during a 5-year period after Hurricane Agnes caused widespread flooding in Pennsylvania in 1972 (Logue et al., 1979). More recently, a longitudinal study of local residents who lived through Hurricane Andrew showed that 20-30% of the adults in the area met the criteria for PTSD at 6 months and 2 years after the event (Norris et al., 1999).

A population's ability to minimize the potential health effects associated with extreme weather events is based on a number of diverse and interrelated factors including: building code regulations, warning systems, and disaster policies; evacuation plans; adequate relief efforts; and recovery (Noji, 1997). There are many federal, state, and local government agencies and nongovernmental organizations involved in planning for and responding to natural disasters in the US. For example, the Federal Emergency Management Agency (FEMA) recently launched its National Mitigation Strategy (FEMA, 1996), which is designed to increase public awareness of natural hazard risk and to reduce the risk of death, injury, community disruption, and economic loss. This strategy represents a comprehensive effort to address severe events with a series of initiatives and public-private partnerships.

Future research on extreme weather events and associated health effects should focus on improving climate models to project trends, if any, in regional extreme events. This type of improved prediction capability will assist in public health mitigation and preparedness. In addition, epidemiologic studies of health effects beyond the direct impacts of disaster will provide a more accurate measure of the full health impacts and will assist in planning and resource allocation.

3. Air Pollution-related Health Effects

Air pollutants have many sources—natural (e.g., vegetation and volcanoes), agricultural (e.g., methane and pesticides), commercial (e.g., dry cleaning operations and auto body shops), industrial (e.g. electric power plants and manufacturing facilities), transportation (e.g. truck and automobile emissions), and residential (e.g. home gas and oil burners and wood stoves). Ambient levels of regulated air pollutants (which include particulate matter, ozone, carbon monoxide, and sulfur and nitrogen oxides) have generally dropped since the mid-1970s, but air quality in many parts of the country falls short of health-based air quality standards. In 1997, about 107 million people in the US lived in counties that did not meet the air quality standards for at least one regulated pollutant (USEPA, 1998a).

Air pollution is related to weather both directly and indirectly. Climate change may affect exposures to air pollutants by: 1) affecting weather and thereby local and regional pollution concentrations (Penner et al., 1989; Robinson, 1989); 2) affecting human-caused emissions, including adaptive responses involving increased fuel combustion for power generation; 3) affecting natural sources of air pollutant emissions (USEPA, 1997a; USEPA, 1998a); and 4) changing the distribution and types of airborne allergens (Ahlholm et al., 1998). Local weather patterns, including temperature, precipitation, clouds, atmospheric water vapor, wind speed, and wind direction influence atmospheric chemical reactions. They can also affect atmospheric transport processes and the rate of pollutant export from urban and regional environments to the global scale environments (Penner et al., 1989; Robinson, 1989). In addition, the chemical composition of the atmosphere may in turn have a feedback effect on the local climate.

If the climate becomes warmer and more variable, air quality is likely to be affected. For example, if warmer temperatures lead to more air-conditioning use, power plant emissions could increase without additional air pollution controls. Analyses show that higher surface temperatures are conducive to the formation of ground-level ozone, particularly in urban areas (Morris et al., 1989; NRC, 1991; Sillman and Samson, 1995; USEPA, 1996; USEPA, 1998a); see Figure 5.

Changing weather patterns contribute to yearly differences in ozone concentrations (USEPA, 1998a); for example, the hot, dry, stagnant meteorological conditions in 1995 in the central and eastern US were highly conducive to ozone formation. However, the specific type of change (i.e., local, regional, or global), the direction of change in a particular location (i.e., positive or negative), and the magnitude of change in air quality that may be attributable to climate change are not known.

Because the effect of climate change on all of the air pollutants of concern, especially particulate matter, is unknown, it is difficult to determine the overall effect of climate variability and change on respiratory health. Health effects associated with climate impacts on air pollution will depend on future air pollution levels. Since 1970, emissions and ambient air pollutants have declined overall (USEPA, 1996). However, the majority of regulated air pollutants are from fossil fuel combustion (USEPA, 1997a; USEPA, 1998a) and, as a result, increased energy and fuel-use would increase emissions of air pollutants without additional air pollutant controls. Integrated air quality modeling studies will be necessary to assess more quantitatively the potential health impacts of air quality changes associated with global climate change. These models would need to incorporate variables such as: 1) future human-caused emissions (driven by economic growth, air pollution controls, vehicle usage, and possible changes in use of fuel for heating and cooling); 2) future natural emissions (factoring in possible responses to changing climate); and 3) changes in local meteorology due to global climate change.

Current exposures to air pollutants have serious public health consequences. Ground-level ozone can exacerbate respiratory diseases by damaging lung tissue, reducing lung function, and sensitizing the lungs to other irritants (Romieu, 1999). Short-term drops in lung function caused by ozone are often accompanied by chest pain, coughing, and pulmonary congestion (American Thoracic Society, 1996). Epidemiologic studies have found that exposure to particulate matter can aggravate existing respiratory and cardiovascular diseases, alter the body's defense systems against foreign materials, damage lung tissue, lead to premature death, and possibly contribute to cancer (American Thoracic Society, 1996; Lambert et al., 1998). Health effects of exposures to carbon monoxide, sulfur dioxide, and nitrogen dioxide can include reduced work capacity, aggravation of existing cardiovascular diseases, effects on respiratory function, respiratory illnesses, lung irritation, and alterations in the lung's defense systems (American Thoracic Society, 1996; Lambert et al., 1998).

Maximum Daily Ozone Concentrations versus Maximum Daily Temperature in Atlanta and New York

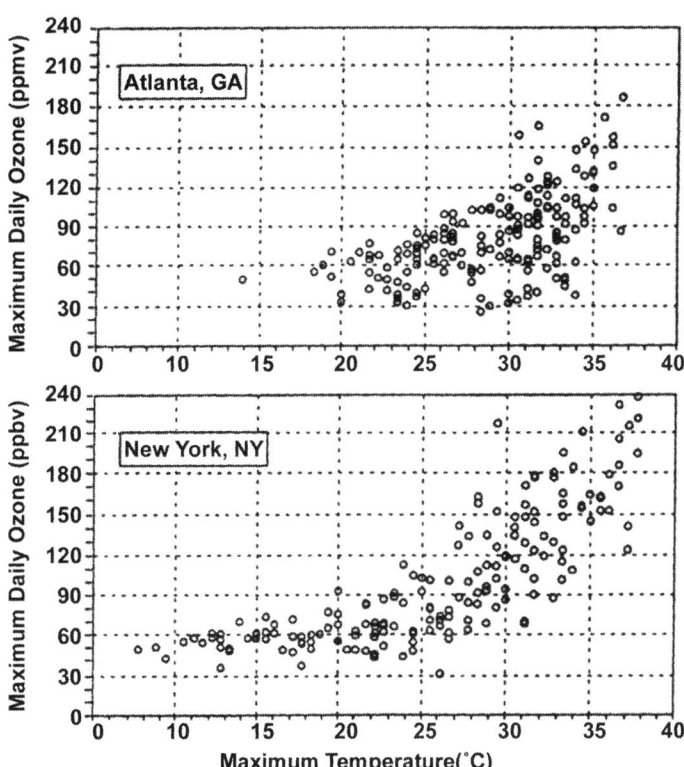

Figure 5: These graphs illustrate the observed association between ground-level ozone concentrations and temperature in Atlanta and New York City (May to October 1988-1990). The projected higher temperature across the US in the 21st century will likely increase the occurrence of high ozone concentrations, especially because extremely hot days frequently have stagnant air circulation patterns, although this will also depend on emissions of ozone precursors and meteorological factors. Ground-level ozone can exacerbate respiratory diseases and cause short-term reductions in lung function. (Maximum Daily Ozone Chart provided by USEPA.) —See Color Figure Appendix

In addition to affecting exposure to air pollutants (whether man-made or naturally emitted), there is some chance that climate change will play a role in human exposure to airborne allergens. Plant species are sensitive to weather, and climate change will possibly alter pollen production in some plants or the geographic distribution of plant species (Ahlholm et al., 1998). Consequently, there is some chance that climate change will affect the timing or duration of seasonal allergies, such as hay fever. The impact of pollen and of pollen changes on the occurrence and severity of asthma, the most common chronic disease of childhood, is currently very uncertain.

There is some chance that climate change will affect the amount of time individuals spend indoors (e.g., individuals may spend more time in air conditioned environments to avoid extreme heat, or may spend more time outdoors if winter temperatures are milder), resulting in changed exposure to indoor air pollutants and allergens. In some cases, these indoor environments may be more dangerous than the ambient conditions.

Adaptation measures include ensuring the responsiveness of federal and state air quality protection programs to changing pollution levels. These standards are designed to protect the public health by limiting emissions of key air pollutants and thus reducing ambient concentrations. The Pollutants Standards Index (Davies and Mazurek, 1998), an EPA-coordinated health advisory system that provides warnings for both the general population and susceptible individuals, could be further strengthened for specific pollutants.

Future research in the area of health effects associated with air pollution should include: basic atmospheric science elucidating the association between weather, ozone, particulate matter, and other air pollutants and aeroallergens; improving existing models (e.g., expanding the spatial domain and lengthening the duration of modeled events) and their linkage with climate change scenarios; and closing the gaps in our understanding of common pollutants, such as particulate matter and ozone, and of individual exposures to these pollutants.

4. Water- and Food-borne Diseases

More than 200 million people in the US have direct access to treated public water supply systems, yet as many as 9 million annual cases of water-borne disease are estimated (Bennett et al., 1987); high uncertainty accompanies this estimate, and reporting is variable by state (Frost et al., 1996). Although most of these cases of water-borne disease involve mild gastrointestinal illnesses, other severe outcomes such as myocarditis (infection of the heart) are now recognized. These infections and illnesses can be chronic and even fatal in infants, the elderly, pregnant women, and people with weakened immune systems (Gerba et al., 1996; ASM, 1998).

In the US, food-borne diseases are estimated to cause 76 million cases of illness, with 325,000 hospitalizations and 5,000 deaths per year (Mead et al., 1999). Microbiologic agents in water (e.g., viruses, bacteria, and protozoa) can contaminate food (e.g. shellfish and fish). In addition, there have been instances of contamination of fresh fruits and vegetables by water-borne pathogens (Tauxe, 1997).

The routes of exposure to water- and food-borne diseases include ingestion, inhalation, and dermal absorption of microbial organisms or algal toxins. For example, people can ingest water-borne microbiologic agents by drinking contaminated water, by eating seafood from contaminated waters, or by eating fresh produce irrigated or processed with contaminated water (Tauxe, 1997). They also can be exposed by contact with contaminated water through commerce (e.g., fishing) or recreation (e.g., swimming) (Coye and Goldoft, 1989). The water-borne pathogens of current concern include viruses, bacteria, and protozoa. Examples include *Vibrio vulnificus*, a naturally occurring estuarine bacterium responsible for a high percentage of the deaths associated with shellfish consumption (Johnston et al., 1985; Shapiro et al., 1998); *Cryptosporidium parvum* and *Giardia lamblia*, protozoa associated with gastrointestinal illnesses (Craun, 1998); and biologic toxins associated with harmful algal blooms (Baden et al., 1996). Many of these were discovered only recently and are the subject of ongoing research.

Between 1980 and 1996, 401 disease outbreaks associated with drinking water were reported, with more than 750,000 associated cases of disease (Craun, 1998). More than 400,000 of those cases (including 54 deaths, primarily of individuals whose immune systems were compromised by HIV infection or other illness) occurred in a 1993 outbreak of *Cryptosporidiosis* that resulted from the contamination of the Milwaukee, Wisconsin, water supply (Hoxie et al., 1997). A contributing factor in the contamination, in addition to treatment system malfunctions, was heavy rainfall and runoff that resulted in a decline in the quality of raw surface water arriving at the Milwaukee drinking water plants (MacKenzie et al., 1994). Studies from other locations in the US found positive correlations between rainfall and *Cryptosporidium* oocyst and *Giardia* cyst concentrations in river water (Atherholt et al., 1998) and human disease outbreaks (Weniger et al., 1983; Curriero et al., 2001). Many water treatment facilities still have difficulty removing these pathogens.

Changes in precipitation, temperature, humidity, salinity, and wind have a measurable effect on the quality of water used for drinking, recreational and commercial use, and as a source of fish and shellfish (see Figure 6). Direct weather associations have been documented for water-borne disease agents such as *Vibrio* bacteria (Motes et al., 1998), viruses

(Lipp et al., 1999), and harmful algal blooms (Harvell et al., 1999). In Florida during the strong El Niño of 1997-1998, high precipitation and runoff greatly elevated the counts of fecal bacteria and infectious viruses in local coastal waters (Harvell et al., 1999). In Gulf Coast waters, *Vibrio vulnificus* bacteria are especially sensitive to water temperature, which dictates their seasonality and geographic distribution (Lipp and Rose, 1997; Motes et al., 1998). In addition, toxic red tides proliferate as seawater temperatures increase (Valiela, 1984). Over the past twenty-five years along the East Coast, reports of marine-related illnesses increased in correlation with El Niño events (Harvell et al., 1999).

For many water-borne diseases, the management and disposal of sewage, biosolids and other animal wastes, and the protection of watersheds and fresh water flows are critical variables that impact water quality and the risk of water-borne disease (ASM, 1998). In September, 1999, the largest reported water-borne associated outbreak of *Escherichia coli* 0157:H7 occurred at a fairground in the state of New York and was linked to contaminated well water (CDC, 1999a). Heavy rains following a period of drought coincided with this major outbreak event (New York Department of Health, 2000). The likelihood of this type of problem occurring could increase under conditions of high soil saturation that enhances the rapid transport of microbiologic organisms (Yates and Yates, 1988). Finally, many communities in the US continue to use combined sewer and storm water drainage systems (Figure 7). These systems may pose a health risk should the frequency or intensity of storms increase, because raw sewage bypasses treatment and is discharged into receiving surface waters during storms (Rose and Simonds, 1998).

Climate changes projected to occur in the next several decades, in particular the likely increase in extreme precipitation events, will probably raise the risk of contamination events. However, whether these increases materialize depends on policy responses and the level of maintenance or improvement of infrastructure. Current adaptations for assessing and preventing water-borne diseases include legal and administrative measures such as water safety criteria, monitoring requirements, and health outcome surveillance, as mandated under the Safe Drinking Water Act, with amendments in 1996 (USEPA, 1997b). Recent legislative and regulatory attention has focused on improved treatment of surface water to address

Seasonality of Shellfish Poisoning in Florida 1981-1994

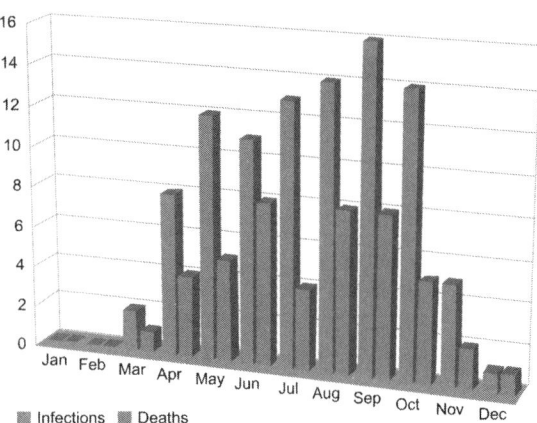

■ Infections ■ Deaths

Figure 6: Monthly distribution of oyster-associated Vibrio vulnificus illness (or shellfish poisoning) and deaths occurring in Florida from 1981-1994. Over the 14-year period, higher numbers of cases occur during summer. Monitoring in Florida shows a statistically significant association between concentrations of this pathogen in estuaries and temperature and salinity, the latter being affected by rainfall and runoff. (Adapted from: Lipp and Rose, 1997.) — See Color Figure Appendix

Locations of Combined Wastewater Systems

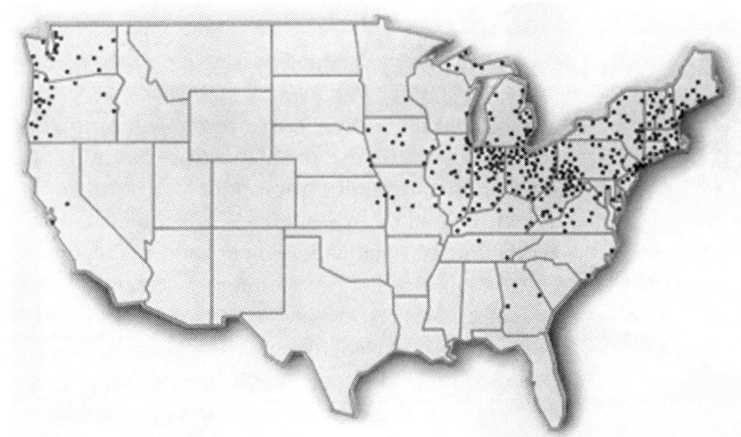

Figure 7: Wastewater systems that combine storm water drainage and sewage and industrial discharges are still in use in about 950 communities in the US, mostly in the Northeast and Great Lakes regions. These combined sewer systems deliver both storm drainage and wastewater to sewage treatment facilities. However, during rain or snowmelt, the volume of incoming water can exceed the capacity of the treatment system. Under those conditions, combined sewer systems are designed to overflow and discharge untreated wastewater into surface water bodies, and are termed as a combined sewer overflow (CSO) event. EPA, in 1994, developed a CSO Control Policy that sets forth a national framework for prevention of combined sewer overflows through the federal Clean Water Act s water discharge permit program. It has been suggested that if they continue to discharge untreated wastewater during storm events, combined sewer systems may pose a greater health risk should the frequency or intensity of storms increase. (Source: USEPA, http://www.epa.gov/owmitnet/cso.htm) See Color Figure Appendix

microbial contaminants and on ground water and watershed protection (ASM, 1998; USEPA, 1998b).

With respect to marine-related human disease outbreaks, protection is provided by measures such as adequate sewage/sanitation systems and safe food storage infrastructures, and beach and recreational water monitoring (USEPA, 1999). However, these measures are inadequate for microbial contaminants. With increasing trends in food importation, improved surveillance and preventive measures are required (Tauxe, 1997), as well as a better understanding of how climate and weather might affect food and water safety outside the US.

Important knowledge gaps must be addressed to improve the assessment of the association of climate with water-borne disease issues. Determinants of transport and fate of microbial pollutants associated with rainfall and snowmelt are not well quantified. Further studies should address the influence of varying land use on the water quality in watersheds. For urban watersheds, much of the current annual load of contaminants is transported into fresh and marine water bodies during storm events. For these reasons, regional and even localized projections of changes in the intensity and frequency of storms and changes in land use are required for improving climate variability/health assessments.

Advances in monitoring are necessary to improve our knowledge base and enhance early warning and prevention capabilities. Application of existing technologies could be expanded, such as molecular fingerprinting to track contaminant sources (CDC, 1999b), improvement of monitoring systems (CDC, 1999c), and the use of satellite remote sensing used to detect coastal algal blooms (Gower, 1995). Coordination and integration of monitoring across the varying agencies responsible for water-borne, food-borne, and coastal surveillance systems could greatly enhance our knowledge and adaptive potential.

5. Insect-, Tick-, and Rodent-borne Diseases

Diseases transmitted between humans by blood-feeding arthropods (insects, ticks, and mites), such as plague, typhus, malaria, yellow fever, and dengue fever were once common in the US and Europe (Philip and Rozeboom, 1973; Beneson, 1995; Reiter, 1996). The ecology and transmission dynamics of these "vector-borne" infections are complex, and the factors that influence transmission are unique to each disease. It is not possible, therefore, to make broad generalizations on the effect of climate on vector-borne diseases (Reiter, 1996; Reiter, 2000). Many of these diseases are no longer present in the US, mainly because of changes in land use, agricultural methods, residential patterns, human behavior, and vector control. However, diseases that may be transmitted to humans from wild animals (zoonoses) continue to circulate in nature in many parts of the country. Humans can become infected with the pathogens that cause these diseases through transmission by insects or ticks. For example, Lyme disease, which is tick-borne, circulates among white-footed mice in woodland areas of the Mid-Atlantic, Northeast, upper Midwest and West Coast of the US, and humans acquire the pathogen when they are bitten by infected ticks (Gubler, 1998). Flea-borne plague incidence increased in conjunction with increasing rodent populations after unseasonal winter-spring precipitation in New Mexico (Parmenter et al., 1999).

Humans may also become infected with pathogens that cause zoonotic diseases by direct contact with the host animals or their body fluids, as occurs with Hantavirus Pulmonary Syndrome (HPS). Hantaviruses are carried by numerous rodent species and are transmitted to humans through contact with rodent urine, droppings, and saliva, or by inhaling aerosols of these products. In 1993, a previously undocumented hantavirus, *Sin Nombre*, emerged in the Four Corners region of the rural southwestern US, causing HPS (Schmaljohn and Hjelle 1997). As of June 2000, 274 cases had been confirmed in the US, 30 in Canada, and 475 in Central and South America. In the US, the mortality rate is currently 39% in otherwise healthy individuals (personal communication, James Mills, CDC).

The impact of weather on rodent populations may affect disease transmission. The Four Corners outbreak was attributed to an explosion in the mouse population caused by an increase in their food supply resulting from unusually prolonged rainfall associated with the 1991-1992 El Niño event (Engelthaler et al., 1999; Glass et al., 2000).

Flooding has also been associated with rodent-borne leptospirosis, as occurred in the 1995 epidemic in Nicaragua. A case-control study showed a 15-fold risk of disease associated with walking through flood waters (Trevejo et al., 1998). In Salvador, Brazil, a large epidemic of leptospirosis peaked two weeks after severe flooding in 1996 (Ko et al., 1999). Although leptospirosis cases are rare in the US, the disease is under-diagnosed (Demers et al.,

1983), and the bacteria has been found in samples from both rats and children from surveys conducted in urban areas (Demers et al., 1983; Childs et al., 1992).

Changes in ecosystems and sociologic factors play a critical role in the occurrence of these diseases. For instance, the increasing numbers of cases and spread of Lyme disease in the US and Europe stemmed from the reversion of large tracts of agricultural land to woodland and the subsequent increase in mouse, deer, and tick populations, combined with the spread of residential areas into undeveloped areas and farmland (IOM, 1992).

Most vector-borne diseases exhibit a distinct seasonal pattern, which clearly suggests that they are weather sensitive. Rainfall, temperature, and other weather variables affect in many ways both vectors and the pathogens they transmit. Rainfall may increase the abundance of some mosquitoes by increasing the number of their breeding sites (Reisen et al., 1995), but excessive rainfall can flush these habitats and thus destroy the mosquitoes in their aquatic larval stages. Increased humidity can extend vector survival times (Reisen et al., 1995). Dry conditions may eliminate the smaller breeding sites, such as ponds and puddles, but create productive new habitats as river flow is diminished. Thus, epidemics of malaria are associated with rainy periods in some parts of the world but with drought in others. High temperatures can increase the rate at which mosquitoes develop into adults, the rate of development of the pathogens in the mosquitoes (Watts et al., 1987), and feeding and egg-laying frequency. The key factor in transmission is the survival rate of the vector (Gilles, 1993). Higher temperatures may increase or reduce survival rate, depending on the vector, its behavior, ecology, and many other factors. Thus, the probability of transmission may or may not be increased by higher temperatures.

In some cases, specific weather patterns over several seasons appear to be associated with increased transmission rates. For example, in the midwestern US, outbreaks of St. Louis encephalitis (SLE, a viral infection of birds that can also infect and cause disease in humans) appear to be associated with the sequence of warm, wet winters, cold springs, and hot dry summers (Monath, 1980). The factors underlying this association are complex and require more investigation (Reeves and Hammon, 1962; Reiter, 1988).

In the western US, one study (Reeves et al., 1994) predicted that a 5.5 to 9°F (3-5°C) increase in average temperature may cause a northern shift in the distribution of both Western Equine Encephalitis (WEE) and SLE outbreaks, and a decreased range of WEE in southern California based on temperature sensitivity of both virus and mosquito carrier.

Many other factors are important in transmission dynamics. For example, dengue fever — a viral disease mainly transmitted by *Aedes aegypti*, a mosquito that is closely associated with human habitation — is greatly influenced by house structure, human behavior, and general socioeconomic conditions. There is a marked difference in the incidence of the disease above and below the US- Mexico border: in the period 1980-1996, 43 cases were recorded in Texas, as compared to 50,333 reported cases in the three contiguous border states in Mexico; Figure 8 shows data updated through 1999 (Reiter, 2001).

The tremendous growth in international travel increases the risk of importation of vector-borne diseases, some of which can be transmitted locally under suitable circumstances at the right time of the year (Gubler, 1998). Key preventive measures must be directed both at protecting the increasing number of US travelers going to disease-endemic areas, as well as preventing importation of disease by US and non-US citizens. The recent importation of West Nile virus encephalitis into New York illustrates the continued need for vigilant surveillance for zoonotic

Reported Cases of Dengue 1980-1999

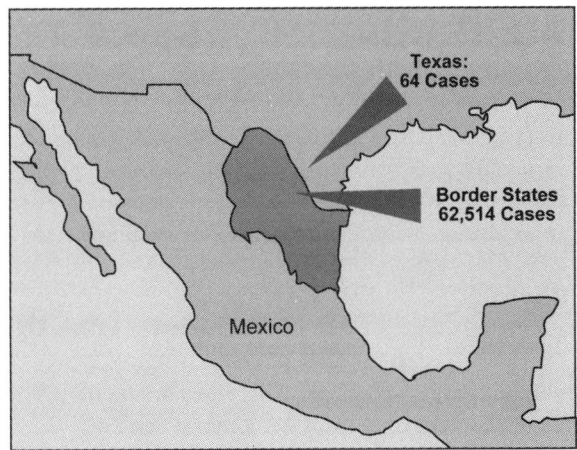

Figure 8: Dengue along the US-Mexico border. Dengue, a mosquito-borne viral disease, was once common in Texas (where there were an estimated 500,000 cases in 1922), and the mosquito that transmits it remains abundant. The striking contrast in the incidence of dengue in Texas versus three Mexican states that border Texas (64 cases vs. 62,514) in the period from 1980-1999 provides a graphic illustration of the importance of factors other than temperature, such as use of air conditioning and window screens, in the transmission of vector-borne diseases. (National Institute of Health, Mexico; Texas Department of Health; US Public Health Service. Unpublished data.) — See Color Figure Appendix

diseases potentially brought in by imported animals or international travelers (Lanciotti et al., 1999). An active survey in Florida recently documented under-reporting for some diseases, such as dengue fever (Gill et al., 2000), further demonstrating the need for improved surveillance to better estimate risk.

Preventive measures from these types of risks include vaccinations and drug prophylaxis for travelers, information for travelers, and the use of repellants and other protective measures. In the US, medical personnel should be made aware of this increased risk to travelers and of the need to improve surveillance of imported vector-borne diseases.

A high standard of living and well-developed public health infrastructure are central to the current capacity to adapt to changing risks of vector- and rodent-borne diseases in the US. Maintaining and improving this infrastructure—including surveillance, early warning, prevention, and control—remain a priority. Integration of climate, environmental, health, and socioeconomic data may facilitate implementing public health prevention measures. For example, climate forecasts can assist in disease prevention by predicting seasonal or interannual events such as El Niño, and early warning from improved vector and disease surveillance can help prevent local transmission of imported vector-borne diseases (Colwell and Patz, 1998).

ADDITIONAL ISSUES

Other health outcomes identified in the literature and by researchers as potentially affected by climate variability and change may warrant future study but are beyond the scope of this current assessment. For example, we did not address the potential impacts

on health of economic losses or gains due to climate variability or attempt to assign a monetary value to the health outcomes of climate change. We did not address the potential impact that changes in the hydrologic cycle might have on crop production and food storage in the US (McMichael et al., 1996). Finally, we did not address stratospheric ozone depletion (McKenzie et al., 1999), although climate change may contribute to the delayed recovery of the stratospheric ozone hole (Shindell et al., 1998; Kirk-Davidoff et al., 1999), and possibly lead to adverse health impacts from increased UV exposure.

ADAPTATION STRATEGIES

If climate change occurs as anticipated, it may have significant impacts on virtually all systems on which human life depends—biologic, hydrologic, and ecologic. The extent of the impacts that climate change may have on human health is very uncertain because it is dependent on multiple interrelated variables as well as on the condition of our public health infrastructure. Climate variability and change will likely have both positive and negative consequences for the health of the US population (see Table 2).

The future vulnerability of the US population to the health impacts of climate change depends on our capacity to adapt to potential adverse changes through legislative, administrative, institutional, technological, educational, and research-related measures. Examples include building codes and zoning to prevent storm or flood damage, severe weather warning systems, improved disease surveillance and prevention programs, improved sanitation systems, education of health professionals and the public, and research addressing key knowledge gaps in climate/health relationships.

Table 2: Human Health: Key Summary Messages

Multiple levels of uncertainty preclude any definitive statement on the direction of potential future change for each of the health outcomes assessed.

Although our report mainly addresses adverse health outcomes, some positive health outcomes were identified, notably reduced cold-weather mortality, which has not been extensively examined.

At present, much of the US population is protected against adverse health outcomes associated with weather and/or climate, although certain demographic and geographic populations are at increased risk.

Vigilance in the maintenance and improvement of public health systems and their responsiveness to changing climate conditions and to identified vulnerable subpopulations should help to protect the US population from adverse health outcomes of projected climate change.

Table 3. Summary of Research Needs and Knowledge Gaps

Temperature-related morbidity and mortality	Improvement of the early prediction of these events by determining the key weather parameters associated with health
	Improvement of urban design to facilitate trees, shade, wind, and other heat-reducing conditions to limit the "urban heat island effect"
	Better personal exposure assessment
	Heat morbidity modeling
	Understanding of weather relationship to causes of winter mortality
Extreme weather events–related health effects	Improvement of warning systems to provide early, easily understood messages to the populations most likely to be affected
	Research on the effectiveness of educational materials and early warning systems
	Long-term health effects from severe events, such as nutritional deficiency and mental health effects
	Standardization of information collection after disasters to better measure morbidity and mortality
	Effects of altered land use on vulnerability to extreme weather
Air pollution-related health effects	Association between weather and pollutants
	Health impacts of chronic exposure to high levels of ozone
	Health effects of exposure to ozone in people with asthma and other lung diseases
	Interaction of ozone with other air pollutants
	Mechanisms responsible for the adverse effects of air pollutants in the general population and within susceptible subgroups
	Measures that can modulate the impact of air pollution on health, such as nutrition and other life-style characteristics
	Urban weather modeling for inversions, etc.
Water- and food-borne diseases	Links between land use and water quality, through better assessment at the watershed level of the transport and fate of microbial pollutants associated with rain and snowmelt
	Methods to improve surveillance and prevention of water-borne disease outbreaks
	Epidemiologic studies
	Molecular tracing of water-borne pathogens
	Links between drinking water, recreational exposure, and food-borne disease monitoring
	Links between marine ecology and toxic algae
	Vulnerability assessment to improve water and waste water treatment systems
Vector- and rodent-borne diseases	Improvement of rapid diagnostic tests for pathogens
	Vaccines
	Improvement of active laboratory-based disease surveillance and prevention systems at the state and local level
	Transmission dynamics (including reservoir host and vector ecology) studies
	Improvement of surveillance systems for the arthropod vector and vertebrate hosts involved in the pathogen maintenance/transmission cycles to allow for more accurate predictive capability for epidemic/epizootic transmission
	More effective and rapid electronic exchange of surveillance data

Many of these adaptive responses are desirable from a public health perspective irrespective of climate change. For example, reducing air pollution obviously has both short- and long-term health benefits. Improving warning systems for extreme weather events and eliminating existing combined sewer and storm water drainage systems are other measures that can ameliorate some of the potential adverse impacts of current climate extremes and of the possible impacts of climate change. Improved disease surveillance and prevention systems at the state and local levels are already needed. Adaptation is a complex undertaking, as demonstrated by the varying degrees of effectiveness of current efforts to cope with climate variability. Considerable work still needs to be done to assess the feasibility (e.g., ability of a community to incur the costs) and the effectiveness of alternative adaptive responses, and to develop improved mechanisms for coping with climate variability and change.

CRUCIAL UNKNOWNS/ RESEARCH NEEDS

We are still learning about the linkages between weather and human health in the present, even as we try to anticipate the health effects of climate variability and change in the future. (Specific knowledge gaps are discussed in the Health Sector report, in press, and are listed in Table 3.)

LITERATURE CITED

Ahlholm J.U., M.L. Helander, J. Savolainen. Genetic and environmental factors affecting the allergenicity of birch (Betula pubescens ssp. czerepanovii [Orl.] Hametahti) pollen. *Clinical and Experimental Allergy* 28:1384-8, 1998.

American Society for Microbiology (ASM), Microbial pollutants in our Nation's water: Environmental and public health issues, American Society for Microbiology, Office of Public Affairs, Washington, DC, 1998.

American Thoracic Society, Health effects of outdoor air pollution, part 2, *American Journal of Respiratory and Critical Care Medicine, 153,* 477-98, 1996.

Applegate, W. B., J. W. Rynyan, L. Brasfield, M. L. Williams, C. Konigsberg, and C. Fouche, Analysis of the 1980 heat wave in Memphis, *Journal of the American Geriatric Society, 29,* 337-42, 1981.

Atherholt, T.B., M. W. LeChevallier, W. D. Norton, and J. S. Rosen, Effect of rainfall on giardia and crypto, *Journal of the American Water Works Association, 90*(9), 66-80, 1998.

Baden, D. G., L. E. Glemming, and J. A. Bean, Marine toxins, in *Handbook of Clinical Neurology, Intoxications of the Nervous System: Part II,* edited by F. de Wolf, Elsevier Science, Amsterdam, 141-75, 1996.

Beneson, A. S., *Control of Communicable Diseases Manual,* 16th edition, American Public Health Association, Washington DC, 577, 1995.

Bennett, J. V., S. D. Homberg, M. F. Rogers, and S. L. Soloman, Infectious and parasitic diseases, *American Journal of Preventive Medicine, 55,* 102-14, 1987.

Buechley, R. W., J. V. Bruggen, and L. E. Truppi, Heat island = death island?, *Environmental Research, 5,* 85-92, 1972.

Bureau of the Census, American housing survey for the United States in 1995, US Bureau of the Census, Washington, DC, 1997a.

Bureau of the Census, *Statistical Abstracts of the United States: 1997* (11th edition), US Bureau of the Census, Washington, DC, 1997b.

Centers for Disease Control and Prevention (CDC), Heat-related deaths — United States, 1993, *Morbidity and Mortality Weekly Report 42,* 558-560, 1993.

Centers for Disease Control and Prevention (CDC), Heat-related mortality—Chicago, July 1995, *Morbidity and Mortality Weekly Report 44,* 577-579, 1995.

Centers for Disease Control and Prevention (CDC), HIV/AIDS surveillance report, December 1998, CDC, Atlanta, Georgia, 10, December 1998.

Centers for Disease Control and Prevention (CDC), FoodNet, 1999c. http://www.cdc.gov/ncidod/dbmd/foodnet/98surv.htm

Centers for Disease Control and Prevention (CDC), Outbreak of Escherichia coli O157:H7 and Campylobacter among attendees of the Washington County Fair – New York, 1999, *Morbidity and Mortality Weekly Report, 48*(36), 803, 1999a.

Centers for Disease Control and Prevention (CDC), PulseNet, 1999b. http://www.cdc.gov/ncidod/dbmd/pulsenet/pulsenet.htm

Childs, J. E., B. S. Schwartz, T. G. Ksiazek, R. R. Graham, J. W. LeDuc, and G. E. Glass, Risk factors associated with antibodies to leptospires in inner-city residents of Baltimore: A protective role for cats, *American Journal of Public Health, 82*(4), 597-9, 1992.

Clarke, J. F., Some effects of the urban structure on heat mortality, *Environmental Research, 5,* 93-104, 1972.

Colwell, R. R., and J. A. Patz, *Climate, Infectious Disease and Health: An Interdisciplinary Perspective American Academy of Microbiology,* Washington, DC, pp. 24, 1998.

Coye, M. J., and M. Goldoft, Microbiological contamination of the ocean, and human health, *New Jersey Medicine, 86*(7), 533-8, 1989.

Craun, G. F., *Waterborne Disease in the United States,* CRC Press, Boca Raton, Florida, pp. 295,,1998.

Curriero, F., J. Patz, J. Rose, and S. Lele, Analysis of the association between extreme precipitation and waterborne disease outbreaks in the United States: 1948-1994, *American Journal of Public Health,* in press, 2001.

Davies, C. J., and J. Mazurek, *Pollution Control in the United States:* Evaluating the system, Resources for the Future, Washington, DC, pp. 366, 1998.

Day, J. C., Population projections of the United States by age, sex, race and Hispanic origin: 1995-2050, Current population reports provided by the US Bureau of the Census, US Government Printing Office, Washington, DC, 1996.

Demers, R. Y., A. Thiermann, P. Demers, and R. Frank, Exposure to Leptospira icterohaemorrhagiae in inner-city and suburban children: A serologic comparison, *Journal of Family Practice, 17*(6), 1007-11, 1983.

Di Maio, D. J., and V. J. M. Di Maio, *The Effects of Heat and Cold: Hyperthermia and hypothermia, Forensic Pathology,* CRC Press, Boca Raton, Florida, pp. 503, 1993.

Drabek, T. E., *The Social Dimensions of Disaster,* Federal Emergency Management Agency (Emergency Management Institute), Emmitsburg, Maryland, 619 pp., 1996.

Easterling, D. R., et al., Maximum and minimum temperature trends for the globe, *Science, 277,* 364-367, 1997.

Engelthaler, D. M., et al., Climatic and environmental patterns associated with Hantavirus pulmonary syndrome, Four Corners region, United States, *Emerging Infectious Diseases, 5*(1), 87-94, 1999.

Federal Emergency Management Agency (FEMA), National mitigation strategy: Partnerships for building safer communities, 1996, http://www.fema.gov/mit/ntmstrat.htm

Fowler, A. M., and K. J. Hennessey, Potential impacts of global warming on the frequency and magnitude of heavy precipitation, *Natural Hazards, 11,* 283-303, 1995.

Frost, F. J., G. F. Craun, and R. L. Calderon, Waterborne disease surveillance, *Journal of the American Water Works Association, 88*(9), 66-75, 1996.

Gerba, C. P., J. B. Rose, and C. N Haas, Sensitive populations: Who is at the greatest risk? *International Journal of Food Microbiology, 30*(1-2), 113-23, 1996.

Gill, J., L. M. Stark, and G. C. Clark, Dengue surveillance in Florida, 1997-1998, *Emerging Infectious Diseases, 6*(1), in press, 2000.

Gilles, H. M., Epidemiology of Malaria, in *Bruce-Chwatt's Essential Malariology* edited by H. M. Gilles and D. A. Warrell, Edward Arnold Division of Hodder & Stoughton, London, England, 124-63, 1993.

Glass G.E., Cheek J.E., Patz J.A., Shields T.M., Doyle T.J., Thoroughman D.A., Hunt D.K., Enscore R.E., Gage K.L., Irland C., Peters C.J., Bryan R. Using remotely sensed data to identify areas at risk for hantavirus pulmonary syndrome, *Emerging Infectious Diseases, 6*(3):238-47.

Glickman, T. S., and E. D. Silverman, *Acts of God and Acts of Man.* Resources for the Future,, CRM 92-02, Washington, DC, 1992.

Gorjanc, M. L., W. D. Flanders, J. VanDerslice, J. Hersh, and J. Malilay, Effects of temperature and snowfall on mortality in Pennsylvania, *American Journal of Epidemiology, 149,* 1152-60, 1999.

Gower, J. F. R.. Detection and mapping of bright plankton blooms and river plumes using AVHRR imagery, in *Third Thematic Conference, Remote Sensing for Marine and Coastal Environments,* Seattle, Washington, pp. 151-62, 1995.

Gubler, D. J., Resurgent vector-borne diseases as a global health problem, *Emerging Infectious Diseases, 4*(3), 442-50, 1998.

Harvell, C. D., et al., Emerging marine diseases—climate links and anthropogenic factors, *Science, 285*(5433), 1505-1510, 1999.

Hobbs, F. B., and B. L. Damon, 65+ in the United States, Bureau of the Census, US Government Printing Office, Washington, DC, 1998.

Houghton, J., *Global Warming: The Complete Briefing,* Cambridge University Press, Cambridge, United Kingdom, 1997.

Hoxie, N. J., J. P. Davis, J. M. Vergeront, R. D. Nashold, and K. A. Blair, Cryptosporidiosis - associated mortality following a massive waterborne outbreak in Milwaukee, Wisconsin, *American Journal of Public Health, 87*(12), 2032-35, 1997.

Institute of Medicine (IOM), Emerging infections: Microbial threats to health in the United States, Institute of Medicine, Washington, DC, 1992. Johnston, J. M., S. F. Becker, and L. M. McFarland, Vibrio vulnificus. Man and the sea, *JAMA, 253*(19), 2850-2853, 1985.

Jones, P. D., D. M. New, D. E. Parker, S. Martin, and I. G. Rigor, Surface air temperature and its changes over the past 150 years, *Review of Geophysics, 37,* 173-200, 1999.

Jones, T. S., A. P. Liang, E. M. Kilbourne, M. R. Griffin, P. A. Patriarca, S. G. Fite-Wassilak, R. J. Mullan, R. F. Herrick, H. D. Donnell, C. Keewhan, et al. et al., Morbidity and mortality associated with the July 1980 heat wave in St. Louis and Kansas City, Missouri, *JAMA, 247,* 3327-3331, 1982.

Kalkstein, L. S., and J. S. Greene, An evaluation of climate/mortality relationships in large US cities and the possible impacts of climate change, *Environmental Health Perspectives, 105*(1), 84-93, 1997.

Kalkstein, L. S., and K. E. Smoyer, The impact of climate change on human health: Some international implications, *Experientia, 49*(11), 969-79, 1993.

Karl, T. R., and R. W. Knight, Secular trends of precipitation amount, frequency, and intensity in the USA, *Bulletin of the American Meteorological Society, 79,* 231-241, 1998.

Karl, T. R., R. W. Knight, D. R. Easterling, and R. G. Quayle, Indices of climate change for the United States, *Bulletin of the American Meteorological Society, 77,* 279-303, 1996.

Karl, T. R., R. W. Knight, and N. Plummer, Trends in high-frequency climate variability in the twentieth century, *Nature, 377*(6546), 217-20, 1995.

Kilbourne, E. M., K. Choi, T. S. Jones, S. B. Thacker, and F. I. Team, Risk factors for heat stroke: A case-control study, *JAMA, 247,* 3332-3336, 1982.

Kirk-Davidoff, D.B., E. J. Hintsa, J. G. Anderson, and D. W. Keith, The effect of climate change on ozone depletion through changes in stratospheric water vapour, *Nature, 402,* 399-401, 1999.

Knutson, T. R., and R. E. Tuleya, Increased hurricane intensities with CO 2 -induced global warming as simulated using the GFDL hurricane prediction system, *Climate Dynamics, 15,* 503-19, 1999.

Ko, A. I., M. Galvao Reis, C. M. Ribeiro Dourado, WD, Jr., Johnson, and L .W. Riley, Urban epidemic of severe leptospirosis in Brazil, Salvador Leptospirosis Study Group, *Lancet, 354*(9181), 820-825, 1999.

Lambert, W. E., J. M. Samet, and D. W. Dockery, Community air pollution, in *Environmental and Occupational Medicine,* edited by W. N. Rome, pp. 1501-1522, Lippincott-Raven, Philadelpia, Pennsylvania, 1998.

Lanciotti, R. S., et al., Origin of the West Nile virus responsible for an outbreak of encephalitis in the northeastern United States, *Science, 286*(5448), 2333-2337, 1999.

Landrigan, P. J., W. Suk, and R. W. Amler, Chemical wastes, children's health, and the Superfund basic research program, *Environmental Health Perspectives, 107*(6), 423-7, 1999.

Landsberg, H. E., *The Urban Climate,* Academic Press, New York, 1981.

Lipp, E. K., and J. B. Rose, The role of seafood in food-borne diseases in the United States of America, *Revue Scientifique et Technique, 16:* 620-40, 1997.

Lipp, E. K., J. B. Rose, R. Vincent, R. Kurz, and C. Rodriguez-Palacios, Assessment of the microbiological water quality in Charlotte Harbor, Florida, Technical report to the Southwest Florida water management district, surface water improvement, and management plan, Tampa: Southwest Florida Water Management District, 1999.

Logue J. N., H. Hansen, and E. Struening, Emotional and physical distress following Hurricane Agnes in Wyoming Valley of Pennsylvania, *Public Health Report 94,* 495-502, 1979.

McKenzie, R., B. Connor, and G. Bodeker, Increased summertime UV radiation in New Zealand in response to ozone loss, *Science, 285,* 1709-1711, 1999 .

MacKenzie, W. R., et al., A massive outbreak in Milwaukee of cryptosporidium infection transmitted through the public water supply, *New England Journal of Medicine, 331*(3), 161-167, 1994.

Marzuk, P. M., K. Tardiff, A. C. Leon, C. S. Hirsch, L. Portera, M. I. Iqbal, M. K. Nock, and N. Hartwell, Ambient temperature and mortality from unintentional cocaine overdose, *JAMA, 279*(22), 1795-800, 1998.

McMichael, A. J., A. Haines, R. Sloof, and S. Kovats, *Climate Change and Human Health,* World Health Organization, Geneva, 297, 1996.

Mead, P. S., L. Slutsker, V. Dietz, L. F. McCaig, J. S. Bresee, C. Shapiro, P. M. Griffen , and R. V. Tauxe, Food-related illness and death in the United States, *Emerging Infectious Diseases 5*(5), 607-25, 1999.

Mearns, L. O., F. Giorgi, L. McDaniel, and C. Shields, Analysis of daily variability of precipitation in a nested regional climate model: Comparison with observations and doubled CO2 results, *Global and Planetary Change, 10,* 55-78, 1995.

Meehl, G. A., F. Zwiers, J. Evans, T. Knutson, L. O. Mearns, and P. Whetton, Trends in extreme weather and climate events: Issues related to modeling extremes in projections of future climate change, *Bulletin of the American Meteorological Society, 81,* 427-436, 2000.

Monath, T. P., Epidemiology, in *St. Louis encephalitis,* edited by T. P. Monath, American Public Health Association, Washington, DC, 1980.

Morris, R. E., M. S. Gery, M. K. Liu, G. E. Moore, C. Daly, and S. M. Greenfield, Sensitivity of a regional oxidant model to variations in climate parameters, in *The Potential Effects of Global Climate Change on the United States,* edited by J. B. Smith and D. A. Tirpak, US EPA, Office of Policy, Planning and Evaluation, Washington, DC, 1989.

Motes, M.L., A. DePaola, D. W. Cook, J. E. Veazey, J. C. Hunsucker, W. E. Garthright, R. J. Blodgett, and S. J. Chirtel, Influence of water temperature and salinity on Vibrio vulnificus in Northern Gulf and Atlantic Coast oysters *(Crassostrea virginica), Applied Environmental Microbiology, 64*(4), 1459-65, 1998.

National Center for Health Statistics (NCHS), Health, United States, 1998 with socioeconomic status and health chartbook, US Department of Health and Human Services, Hyattsville, Maryland, pp. 465, 1998.
National Research Council (NRC), *Rethinking the Ozone Problem in Urban and Regional Air Pollution,* National Academy Press, Washington, DC, 1991.

National Research Council (NRC), Reconciling *Observations of Global Temperature Change,* National Academy Press, Washington, DC, pp. 86, 2000.

National Weather Service (NWS), Summary of natural hazard deaths for 1991 in the US, National Weather Service, Rockville, Maryland, 1992. 21

National Weather Service (NWS), Hurricane Andrew: South Florida and Louisiana, August 23-26, 1992, Natural disaster survey repor, US Dept. of Commerce, National Oceanic and Atmospheric Administration, National Weather Service, Silver Spring, Maryland, 1993.

National Weather Service (NWS), Summary of natural hazard statistics, 1999. http://www.nws.noaa.gov/om/hazstats.htm

New York Department of Health, Health Commissioner releases *E. coli* outbreak report, March 31 2000. http://www.health.state.ny.us/nysdoh/commish/2000/ecoli.htm

Noji, E. K., The nature of disaster: General characteristics and public health effects, in *The Public Health Consequences of Disasters,* edited by E. K. Noji, pp. 3-20, Oxford University Press, New York, 1997. 31

Norris, F. H., J. L. Perilla, J. K. Riad, K. Kaniasty, and E. Lavizzo, Stability and change in stress, resources, and psychological distress following natural disaster: Findings from a longtitudinal study of Hurricane Andrew, *Anxiety, Stress and Coping,* 12, 363-96, 1999.

Parmenter R.R., E.P. Yadav, C.A. Parmenter, P. Ettestad, K.L. Gage. Incidence of plague associated with increased winter-spring precipitation in New Mexico. *American Journal of Tropical Medicine and Hygiene* 61:814-21, 1999.

Patz, J.A., M.A. McGeehin, S.M. Bernard, K.L. Ebi, P.R. Epstein, A. Grambsch, D.J. Gubler, P. Reiter, I Romieu, J.B. Rose, J.M. Samet, J. Trtanj, The Potential Health Impacts of Climate Variability and Change in the United States: Executive Summary of the Report of the Health Sector of the U.S. National Assessment, *Environmental Health Perspectives, 108*(4), 367-376, 2000.

Penner, J. E., P. S. Connell, D. J. Wuebbles, and C. C. Covey, Climate change and its interactions with air chemistry: Perspective and research needs, in *The potential effects of global climate change on the United States,* edited by J. B. Smith and D. A. Tirpak, US EPA, Office of Policy, Planning and Evaluation, Washington, DC, 1989.

Philip, C. B., and L. E. Rozeboom, Medico-veterinary entomology: A generation of progress, in History of Entomology, edited by R. F. Smith, T. E. Mittler, and C. N. Smith, Annual Reviews, Inc., Palo Alto, California, 1973.

Pinner, R.W., S. M. Teutsch, L. Simonsen, L.A. Klug, J. M. Graber, M. J. Clarke, and R. L. Berkelman, Trends in infectious diseases mortality in the United States, *JAMA, 275*(3), 189-93, 1996.

Quarantelli, E. L., An assessment of conflicting views on mental health: The consequences of traumatic events, in *Trauma and Its Wake: The Study and Treatment of Post-Traumatic Stress Disorder,* vol 4 (Figley CR, ed). Brunner/Mazel, New York, 173-215, 1985.

Ramlow, J. M., and L. H. Kuller, Effects of the summer heat wave of 1988 on daily mortality in Allegheny County, Pennsylvania, *Public Health Report 105,* 283-9, 1990.

Reeves, W. C., and W. M. Hammon, Epidemiology of the arthropod-borne viral encephalitides in Kern County, California, 1943-1952, University of California at Berkeley Public Health Report 4, 1-257, 1962.

Reeves, W. C., J. L. Hardy, W. K. Reisen, and M. M. Wilby, Potential effect of global warming on mosquito-borne arboviruses, *Journal of Medical Entomology, 31*(3), 323-32, 1994.

Reisen, W. K., H. D. Lothrop, and J. L. Hardy, Bionomics of *Culex tarsalis* (Diptera: Culicidae) in relation to arbovirus transmission in southeastern California, *Journal of Medical Entomology 32*(3), 316-27, 1995.

Reiter, P., Weather, vector biology, and arboviral recrudescence, in *The Arboviruses: Epidemiology and Ecology,* edited by T. P. Monath, pp. 245-255, CRC Press, Boca Raton, Florida, 1988.

Reiter, P., Global warming and mosquito-borne disease in USA, *Lancet, 348*(9027), 622, 1996.

Reiter, P., Global climate change and mosquito-borne disease, *Environmental Health Perspectives, 109*(supplement 1), in press, Feb. 2001.

Reiter, P., From Shakespeare to Defoe: Malaria in England in the Little Ice Age, *Emerging Infectious Diseases, 6,* 1-11, 2000.

Robinson, P., The effects of climate change, in *Global Climate Change Linkages: Acid Rain, Air Quality, and Stratospheric Ozone,* edited by J. C. White, Elsevier, New York, 1989.

Romieu, I., Epidemiological studies of the health effects arising from motor vehicle air pollution, in *Urban Traffic Pollution,* edited by D. Schwela and O. Zali, pp. 10-69, World Health Organization, New York, 1999.

Rose, J. B., and J. Simonds, King County water quality assessment: Assessment of public health impacts associated with pathogens and combined sewer overflows, Seattle, Washington, Report for Water and Land Resources Division. Department of Natural Resources, 1998.

Schmaljohn, C., and B. Hjelle, Hantaviruses: A global disease problem, *Emerging Infectious Diseases 3*(2), 95-104, 1997.

Schuman, S. H., Patterns of urban heat-wave deaths and implications for prevention: Data from New York and St. Louis during 1966, *Environmental Research, 5,* 59-75, 1972.

Semenza, J. C., Are electronic emergency department data predictive of heat-related mortality? *Journal of Medical Systems, 23*(5), 419-424, 1999.

Semenza, J. C., J. McCullough, D. W. Flanders, M. A. McGeehin, and J. R. Lumpkin, Excess hospital admissions during the 1995 heat wave in Chicago, *American Journal of Preventive Medicine, 16,* 269-77, 1999.

Semenza, J. C., C. H. Rubin, and K. H. Falter, Heat-related deaths during the July 1995 heat wave in Chicago, *New England Journal of Medicine, 335*(2), 84-90, 1996.

Shapiro, R. L., S. Altekruse, and P. M. Griffin, The role of Gulf Coast oysters harvested in warmer months in Vibrio vulnificus infections in the United States, 1988-1996, *Journal of Infectious Diseases, 178*(3), 752, 1998.

Shindell, D. T., D. Rind, and P. Lonergan, Increased polar stratospheric ozone losses and delayed eventual recovery owing to increasing greenhouse-gas concentrations, *Nature, 392,* 589-592, 1998.

Sillman, S., and P. J. Samson, Impact of temperature on oxidant photochemistry in urban, polluted, rural, and remote environments, *Journal of Geophysical Research, 100,* 11497-508, 1995.

Spitalnic, S. J., L. Jagminas, and J. Cox, An association between snowfall and ED presentation of cardiac arrest, *American Journal of Emergency Medicine, 14*(6), 572-3, 1996.

Tauxe, R. V., Emerging foodborne diseases: An evolving public health challenge, *Emerging Infectious Diseases, 3*(4), 425-34, 1997.

Trenberth, K. E., Conceptual framework for changes of extremes of the hydrologic cycle with climate change, *Climatic Change, 42,* 327-39, 1999.

Trevejo, R. T., et al., Epidemic leptospirosis associated with pulmonary hemorrhage - Nicaragua, 1995, *Journal of Infectious Diseases, 178*(5), 1457-63, 1998.

United States Environmental Protection Agency (USEPA), Air quality criteria for ozone and related photochemical oxidant: Volume I of III, USEPA, Office of Research and Development, Washington, DC, 1996.

United States Environmental Protection Agency (USEPA), National air pollutant emission trends update: 1970-1996, USEPA, Washington, DC, 1997a.

United States Environmental Protection Agency (USEPA), Water on tap: A consumer's guide to the nation's drinking water, USEPA, Office of Water, Washington, DC, 1997b.

United States Environmental Protection Agency (USEPA), National air quality and emissions trends report, 1997, USEPA, Office of Air Quality Planning and Standards, Washington, DC, 1998a.

United States Environmental Protection Agency (USEPA), *National Primary Drinking Water Regulations: Interim Enhanced Surface Water Treatment; Final Rule,* USEPA, Washington, DC, 1998b.

United States Environmental Protection Agency (USEPA), Action plan for beaches and recreational waters, USEPA, Office of Research and Development, Office of Water, Washington, DC, 1999.

Valiela, I., *Marine Ecological Processes,* Springer-Verlag, New York, 1984.

Watts, D. M., D. S. Burke, B. A. Harrison, R. E. Whitmire, and A. Nisalak, Effect of temperature on the vector efficiency of *Aedes aegypti* for Dengue 2 virus, *American Journal of Tropical Medicine and Hygiene, 36*(1), 143-52, 1987.

Weniger, B. G., M J. Blaser, J. Gedrose, E. C. Lippy, and D. D. Juranek, An outbreak of waterborne giardiasis associated with heavy water runoff due to warm weather and volcanic ashfall, *American Journal of Public Health, 73*(8), 868-72, 1983.

Whitman, S., G. Good, E. R. Donoghue, N. Benbow, W. Shou, and S. Mou, Mortality in Chicago attributed to the July 1995 heat wave, *American Journal of Public Health, 87,* 9, 1515-8, 1997.

Yates, M. L., and S. R. Yates, Modeling microbial fate in the subsurface environment, *Critical Reviews in Environmental Control, 17*(4), 307-44, 1988.

CHAPTER 16

POTENTIAL CONSEQUENCES OF CLIMATE VARIABILITY AND CHANGE ON COASTAL AREAS AND MARINE RESOURCES

John C. Field[1], Donald F. Boesch[2], Donald Scavia[1], Robert Buddemeier[3], Virginia R. Burkett[4,5], Daniel Cayan[6], Michael Fogarty[2], Mark Harwell[7], Robert Howarth[8], Curt Mason[1], Leonard J. Pietrafesa[10], Denise Reed[11], Thomas Royer[12], Asbury Sallenger[4], Michael Spranger[13], James G. Titus[14]

Contents of this Chapter

[1]National Oceanic and Atmospheric Administration, [2]University of Maryland, [3]University of Kansas, [4]U.S. Geological Survey, [5]Coordinating author for the National Assessment Synthesis Team, [6]Scripps Institution of Oceanography, [7]University of Miami, [8]Eco-Modeling, [10]North Carolina State University, [11]University of New Orleans, [12]Old Dominion University, [13]University of Washington, [14]U.S. Environmental Protection Agency

CHAPTER SUMMARY

Context

The US has over 95,000 miles of coastline and approximately 3.4 million square miles of ocean within its territorial sea, all of which provide a wide range of essential goods and services to human systems. Coastal and marine ecosystems support diverse and important fisheries throughout the nation's waters, hold vast storehouses of biological diversity, and provide unparalleled recreational opportunities. Some 53% of the total US population lives on the 17% of land in the coastal zone, and these areas become more crowded every year. Demands on coastal and marine resources are rapidly increasing, and as coastal areas become more developed, the vulnerability of human settlements to hurricanes, storm surges, and flooding events also increases. Coastal and marine environments are intrinsically linked to climate in many ways. The ocean is an important distributor of the planet's heat, and this distribution could be strongly influenced by changes in global climate. Sea-level rise is projected to accelerate in the 21st century, with dramatic impacts in those regions where subsidence and erosion problems already exist.

Climate of the Past Century

- Sea level has risen by 4 to 8 inches (10-20 cm) in the past century.
- Ocean temperatures have risen over the last 55 years, and there has been a recent increase in the frequency of extreme ocean warming events in many tropical seas.
- Sea ice over large areas of the Arctic basin has thinned by 3 to 6 feet (1 to 2 meters), losing 40% of its total thickness since the 1960s; it continues to thin by about 4 inches (10 cm) per year.
- Marine populations and ecosystems have been highly responsive to climate variability.

Climate of the Coming Century

- Sea level is projected to rise an additional 19 inches (48 cm) by 2100 (with a possible range of 5 to 37 inches [13 cm to 95 cm]) along most of the US coastline.
- Ocean temperatures will continue to rise, but the rate of increase is likely to lag behind temperature changes observed on land.
- The extent and thickness of Arctic sea ice is expected to continue to decline. All climate models project large continued losses of sea ice, with year-round ice disappearing completely in the Canadian model by 2100.
- Increased temperature or decreased salinity could trigger abrupt changes in thermohaline ocean circulation.

Key Findings

- Climate change will increase the stresses already occurring to coastal and marine resources as a result of increasing coastal populations, development pressure and habitat loss, overfishing, excess nutrient enrichment, pollution, and invasive species.

- Marine biodiversity will be further threatened by the myriad of impacts to all marine ecosystems, from tropical coral reefs to polar ecosystems.

- Coral reefs are already under severe stress from human activities, and have experienced unprecedented increases in the extent of coral bleaching, emergent coral diseases, and widespread die-offs.

- The direct impact of increasing atmospheric carbon dioxide on ocean chemistry will possibly severely inhibit the ability of coral reefs to grow and persist in the future.

- Globally averaged sea level will continue to rise, and the developed nature of many coastlines makes both human settlements and ecosystems more vulnerable to flooding and inundation.

- Barrier islands are especially vulnerable to the combined effects of sea-level rise and uncontrolled development that hinders or prevents natural migration.

- Ultimately, choices will have to be made between the protection of human settlements and the protection of coastal ecosystems such as beaches, barrier islands, and coastal wetlands.

- Human development and habitat alteration will limit the ability of coastal wetlands to migrate inland as sea levels rise, however, if sediment supplies are adequate, many wetlands may survive through vertical adjustments.

- Increases in precipitation and runoff are likely to intensify stresses on estuaries in some regions by intensifying the transport of nutrients and contaminants to coastal ecosystems.

- As rivers and streams also deliver sediments, which provide material for soil in wetlands and sand in beaches and shorelines, dramatic declines in streamflows could have negative effects on these systems.

- Changes in ocean temperatures, currents, and productivity will affect the distribution, abundance, and productivity of marine populations, with unpredictable consequences to marine ecosystems and fisheries.

- Increasing carbon dioxide levels could trigger abrupt changes in thermohaline ocean circulation, with massive and severe consequences for the oceans and for global climate.

- Extreme and ongoing declines in the thickness and extent of Arctic sea ice will have enormous consequences for Arctic populations, ecosystems, and coastal evolution.

POTENTIAL CONSEQUENCES OF CLIMATE VARIABILITY AND CHANGE ON COASTAL AREAS AND MARINE RESOURCES

INTRODUCTION

Human caused alterations and impacts to coastal and marine environments must be considered in the context of evaluating the health and viability of these resources. In many instances, it is difficult to assess the effects of climate variability and changes because human activities are often responsible for the greatest impacts on coastal and marine environments. Such disturbances often reduce the capacity of systems to adapt to climate variations and climate stress, and often mask the physical and biological responses of many systems to climate forces. For example, naturally functioning estuarine and coastal wetland environments would typically be expected to migrate inland in response to relative sea-level rise, as they have responded to sea-level variations throughout time. When this natural migration is blocked by coastal development, such habitat is gradually lost by "coastal squeeze" as rising sea levels push the remaining habitat against developed or otherwise altered landscapes. Similarly, estuaries already degraded by excess nutrients could recover more slowly from droughts or floods.

Population Distribution across the US

Projected change in county population (percent), 1970 to 2030

- >+250% (highest +3,877%) +50% to
- +250%
- +5% to +50%
- -5% to +5%
- -20% to -5%
- -40% to -20%
- <-40% (lowest -60%)

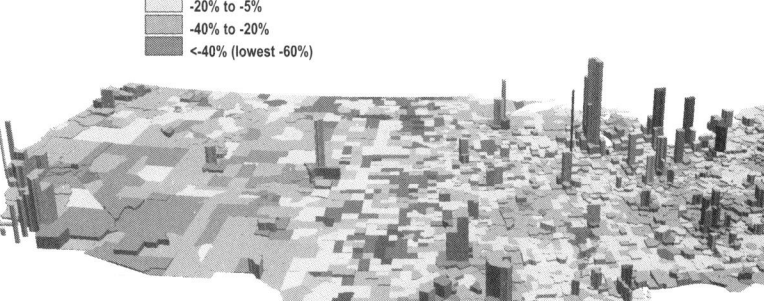

Figure 1: Over the next 25 years, population gains of some 18 million people are projected to occur in the coastal states of Florida, California, Texas, and Washington (NPA, 1999). See Color Figure Appendix.

CONTEXT

The US has over 95,000 miles of coastline and approximately 3.4 million square miles of ocean within its territorial sea, all of which provide a wide range of essential goods and services to society. These include overlapping and often competing uses, including but not limited to tourism, coastal development, commercial and recreational fisheries, aquaculture, biodiversity, marine biotechnology, navigation, and mineral resources.

The coastal population of the US is currently growing faster than the nation's population as a whole, a trend that is projected to continue. Currently some 53% of the total population of the US live in 17% of the land area considered coastal (Culliton, 1998). Over the next 25 years, population gains of approximately 18 million people are projected to occur in the coastal states of Florida, California, Texas, and Washington alone (Figure 1) (NPA, 1999). With this growth, as well as increased wealth and affluence, there are rapidly increasing demands on coastal and marine resources for both aesthetic enjoyment and economic benefits.

This large and growing population pressure in coastal areas is responsible for many of the current stresses to coastal resources. For example, the EPA (1996) estimated that nearly 40% of the nation's surveyed estuaries were impaired by some form of pollution or habitat degradation. Some 30 to 40% of shellfish-growing waters in the nation's estuaries are harvest prohibited or restricted each year, primarily due to bacterial contamination from urban and agricultural runoff and septic systems (Alexander, 1998). Additionally, over 3,500 beach advisories and beach closings occurred in the United States in 1995, primarily due to storm-water runoff and sewage overflows (NOAA, 1998).

Population pressures from further inland can also have detrimental impacts on coastal resources. Effluent discharges as well as agricultural runoff have caused significant nutrient over-enrichment in many coastal areas. Sewage and siltation are significant contributors to coral reef degradation in Hawaii, Florida, and US-affiliated islands of the Pacific and Caribbean. Dams, irrigation projects, and other water management activities have further impacted coastal ecosystems and shorelines by diverting or otherwise altering the timing and flow of water, sediments, and nutrients.

As population and development in coastal areas increase, many of these stresses can be expected to increase as well, further decreasing the resilience of coastal systems. This, in turn, increases the vulnerability of coastal communities and economies which depend upon healthy, functioning ecosystems. The interaction of these ongoing stresses with the expected impacts imposed by future climate change is likely to greatly accentuate the detrimental impacts to coastal ecosystems and communities (Reid and Trexler, 1992, Mathews-Amos and Bernston, 1999).

Despite these ongoing stresses to coastal environments, the oceans and coastal margins provide unparalleled economic opportunities and revenues. One estimate suggests that as many as one out of every six jobs in the US is marine-related, and nearly one-third of the gross domestic product (GDP) is produced in coastal areas (NOAA, 1998; NRC, 1997). In 1996, approximately $590 billion worth of goods passed through US ports, over 40% of the total value of US trade and a much larger percentage by volume. The US is also the world's fifth largest fishing nation and the third largest seafood exporter; total landings of marine stocks have averaged about 4.5 million metric tons over the last decade. Ex-vessel value (the amount that fishermen are paid for their catch) of commercial fisheries alone was estimated at approximately $3.5 billion in 1997, and the total (direct and indirect) economic contribution of recreational and commercial fishing has been estimated at over $40 billion per year (NRC, 1999). The growing field of marine biotechnology has also generated substantial opportunities; for example recent research has yielded five drugs originating from marine organisms with a cumulative total potential market value of over $2 billion annually (NOAA, 1998).

Coastal tourism also generates enormous revenues. Travel and tourism are multi-billion dollar industries in the US, representing the second largest employer in the nation (after health care) and employing some 6 million people (NOAA, 1998). It has been estimated that the US receives over 45% of the developed world's travel and tourism revenues, and oceans, bays, and beaches are among the most popular tourist destinations in the nation (Houston, 1996). As many as 180 million people visit the coast each year for recreational purposes in all regions of the country, and many regions depend upon tourism as a key economic activity. For example one study estimated that in the San Francisco Bay area alone tourism has been estimated to generate over $4 billion a year (EPA, 1997). Clean water, healthy ecosystems, and access to coastal areas are critical to maintaining tourism industries; ironically, however, these industries themselves often pose additional impacts to coastal environments and local communities (Miller and Auyong, 1991).

As coastal populations increase, the vulnerability of developed coastal areas to natural hazards is simultaneously expanding. Disaster losses are currently estimated at about $50 billion annually in the US, compared to just under $4.5 billion in 1970. As much as 80% of these disasters were meteorologically related storms, hurricanes, and tornadoes (as opposed to geologically related disasters such as earthquakes and volcanoes) and many of these had their greatest impacts on coastal communities. The potential increase in such events related to climate change will pose an even greater threat to coastal population centers and development in the future. However, the ongoing trends of population growth and development in coastal areas alone will ensure that losses due to hurricanes, storms, and other disasters in coastal areas will continue to increase.

In addition to direct economic benefits, coastal and marine ecosystems, like all ecosystems, have characteristic properties or processes which directly or indirectly benefit human populations. Costanza et al. (1997) have attempted to estimate the economic value of sixteen biomes, or ecosystem types, and seventeen of their key goods and services, including nutrient cycling, disturbance regulation, waste treatment, food production, raw materials, refugia for commercially and recreationally important species, genetic resources, and opportunities for recreational and cultural activities. For example, the societal value per hectare of estuaries, tidal marshes, coral reefs, and coastal oceans were estimated at $22,832, $9,990, $6,075, and $4,052 per hectare, respectively. On a global basis, the authors suggested that these environments were of a disproportionately higher

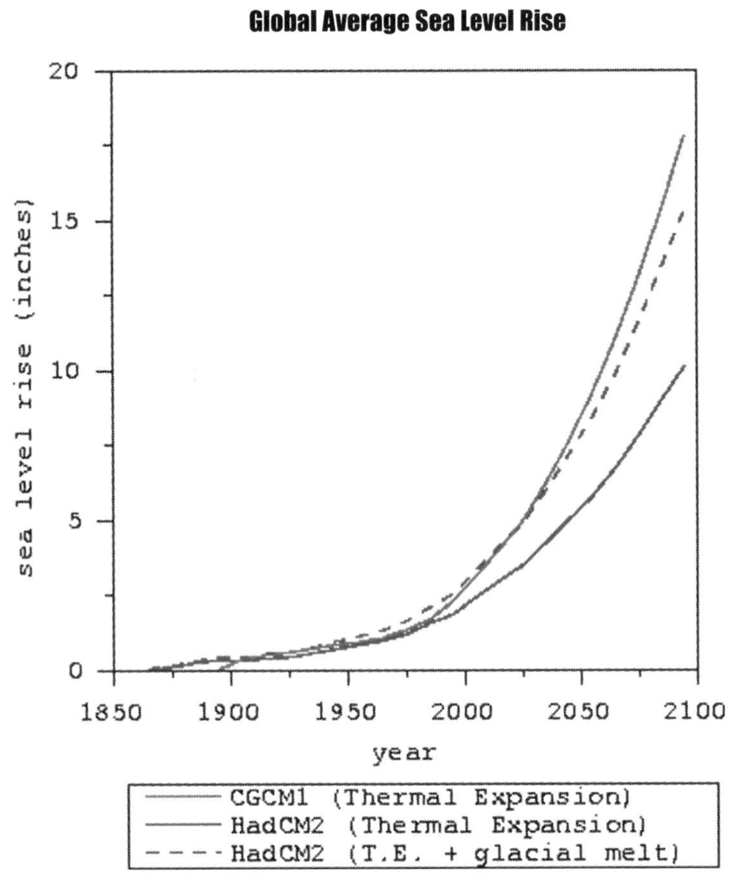

Global Average Sea Level Rise

(Figure: line graph with y-axis "sea level rise (inches)" from 0 to 20, x-axis "year" from 1850 to 2100)

Legend:
— CGCM1 (Thermal Expansion)
— HadCM2 (Thermal Expansion)
- - - - HadCM2 (T.E. + glacial melt)

Figure 2 : Projected rise in global average sea level based on the Hadley and Canadian General Circulation Model (GCM) scenarios. See color figure appendix.

value, covering only some 6.3% of the world's surface area but responsible for some 43% of the estimated value of the world's ecosystem services. These results suggest that the oceans and coastal areas contribute the equivalent of some $21 trillion per year to human activities globally (Costanza, 1999). The approach of Costanza et al. (1997) to valuation is not universally accepted, and the authors themselves agree that ecosystem valuation is difficult and fraught with uncertainties. However, the magnitude of their estimates, and the degree to which coastal and marine ecosystems rank as amongst the most valuable to society, serve to place the importance of the services and functions of these ecosystems in an economic context.

CLIMATE AND COASTAL ENVIRONMENTS

Sea-level Change

Global sea levels have been rising since the conclusion of the last ice age approximately 15,000 years ago. During the last 100 years, globally averaged sea level has risen approximately 4 to 8 inches (10-20 cm, or about 1 to 2 millimeters per year). This represents "eustatic" sea-level change, the change in elevation of the Earth's oceans that has been determined from tidal stations around the globe. Most of

Spatial Distribution Around North America in Sea Level Rise

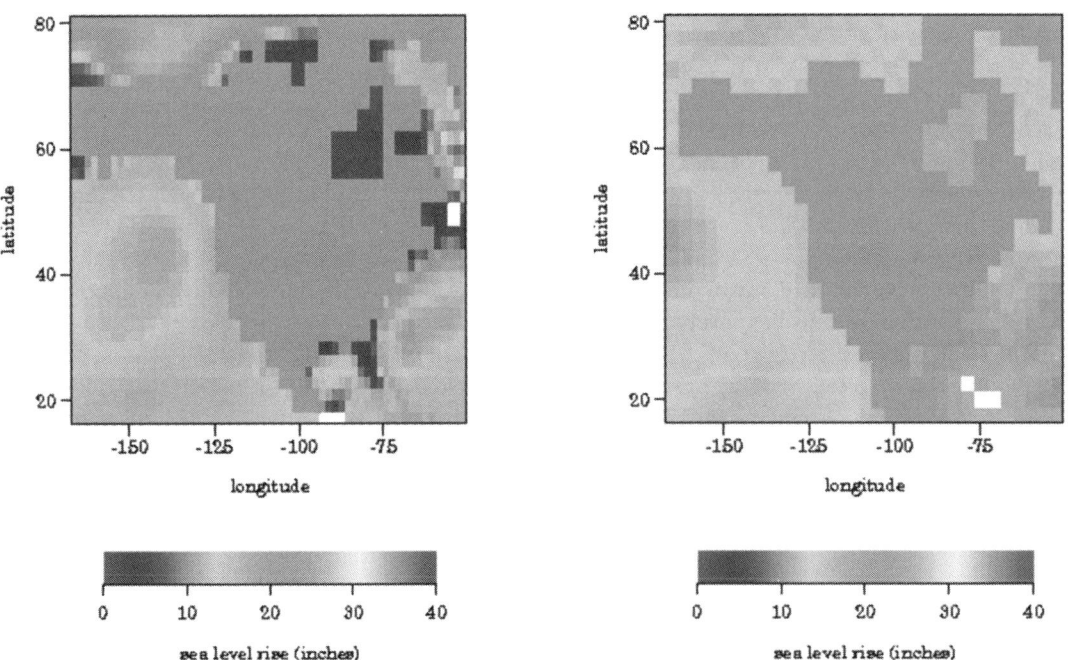

Figure 3 : Projections of the regional pattern of global sea level rise by the year 2100 based on the Canadian (left) and Hadley (right) scenarios. These estimates do not include contributions to sea-level change due to vertical movement of coastal lands. See color figure appendix.

the observed global sea-level change is accounted for by two major variables: the thermal expansion of seawater in the oceans with rising ocean temperatures and changes in the amount of the Earth's water that is locked up in glaciers and ice sheets. The vast majority of landlocked water, enough to raise global sea levels by some 80 meters, is found in the Greenland and Antarctic Ice sheets. Of these, there has been considerable concern regarding the stability of the West Antarctic Ice Sheet (WAIS). Oppenheimer (1998) suggested that the probability of mass wasting (melting) of this ice sheet over the next 100 years is relatively low, but the probability of wasting after 2100 will be considerably greater. Over the next 100 years, most studies predict that the majority of observed sea-level change is expected to come as a result of thermal expansion of the oceans (Gornitz, 1995).

In addition to global changes in mean sea level, there are large regional variations in sea level as measured at the shoreline due to changes in coastal land masses such as subsidence (sinking), isostatic (glacial) rebound and tectonic uplift. The combined effects of eustatic and land mass factors contribute to relative sea-level change in each locality. For example, within the US, portions of the Gulf Coast are experiencing a relative sea-level rise of nearly half an inch (approximately 10 mm) per year due to subsidence. Concurrently, some portions of the southeastern Alaska coastline are experiencing uplift associated with tectonic activity, causing a local relative sea-level fall of slightly less than one half inch (approximately 8 mm) per year.

Finally, there are also regional changes in mean sea level that result from dynamic changes to the ocean geoid, or the "topography" of the sea surface. These result from changes in ocean circulation, wind and pressure patterns, and ocean-water density (IPCC, 1996). The result is that sea-level rise will affect various coastal regions differently, independent of the land motion that contributes to sea-level change. Figures 2 and 3 show the results of Hadley and Canadian General Circulation Model (GCM) projections for regional changes in sea level, independent of land movement. In general, the Hadley model predicts a greater sea-level rise for the Pacific coast than for the Atlantic and Gulf coasts as a result of current and wind patterns. By contrast, the Canadian model predicts a more complex pattern of sea-level rise but with increases relatively similar along all US coasts.

The Hadley climate model projects that sea level will rise between 8 and 12 inches (20 to 30 cen-

timeters) by 2100 for the Atlantic and Gulf Coasts, and 13 to 16 inches (33 cm to 41 cm) for the Pacific Coast (Figure 3). The Canadian model projects a significantly greater estimate of 20 to 24 inches (51 cm to 61 cm) along parts of the US coast (Figure 3). These results are very similar to those found by the Intergovernmental Panel on Climate Change (IPCC, 1996), which estimated that sea levels would most likely increase by approximately 19 inches (48 cm) by 2100, with a range of 5 to 37 inches (13 to 95 cm). Most additional studies conducted since then yield similar estimates, generally projecting increases of about one foot above current trends over the 21st century, for a total sea-level rise of approximately 18 to 20 inches (45 to 52 cm) above their current level by the year 2100 (Titus and Narayanan, 1996; Wigley, 1999). In addition, sea-level rise will continue to occur, even accelerate, beyond 2100 as a result of the long time frame necessary for oceans and ice sheets to approach equilibrium under the long-term perturbations anticipated with climate change.

Hurricanes and Non-Tropical Storms

Storm flooding, wave forces, and coastal erosion are natural processes that pose hazards only when they affect people, homes, and infrastructure that are concentrated in coastal areas. During the 20th century, loss of life and threats to human health during hurricanes has decreased substantially because of improved tracking and early warning systems; however, property losses have increased greatly (Herbert et al., 1996). Yet even if storm intensity and frequency remain the same in the future, continued acceleration in property losses is likely as the increased concentration of people and infrastructure along the coasts continues. This acceleration in property losses will be amplified should global climate change increase storm activity, although the current relationship between climate change and hurricane frequency and intensity is not clear.

Historical records of hurricanes suggest strong annual to decadal variability. For example, the number of hurricanes occurring per year can vary by a factor of three or more for consecutive years, and during El Niño years, hurricanes are less prevalent in the Atlantic Basin (Pielke and Landsea, 1999). Furthermore, during the 25-year period from 1941 to 1965, there were seventeen Category 3 hurricanes landfalling on the US East Coast or peninsular Florida, yet between 1966 and 1990 there were only two (Figure 4). The mid-1990s have seen a recurrence of large numbers of hurricanes in the North Atlantic, perhaps suggesting a return to an active regime, although the long term implications are not yet clear (Landsea et al., 1996). Globally, the historical record shows "no discernible trends in tropical cyclone number, intensity, or location" (Henderson-Sellers et al., 1998). However, regional variability, such as observed in the North Atlantic, can be large.

While interdecadal and interannual variability of hurricane frequency and strength is likely to dominate changes in hurricanes for at least the first half of the next century, increases in hurricane wind strength could result from future elevated sea surface temperatures over the next 50 to 100 years. A recent model investigation shows that increases in hurricane wind strength of 5 to 10% are possible with a sea surface warming of $4°F$ ($2.2°C$) (Knutson et al., 1998). Other research supports the possibility that tropical cyclones could become more intense (Kerr, 1999). For a

moderate hurricane, such an increase in wind strength would translate into approximately a 25% increase in the destructive power of the winds; thus the resulting increase in wave action and storm surge would be greater than the percentage increase in wind speed. Similarly, wave height and storm surge would have a greater percentage increase than wind speed, potentially yielding increased impacts to coasts.

However, it should be noted that recent global climate model investigations have shown that El Niño/Southern Oscillation (ENSO) extremes could become more frequent with increasing greenhouse gas concentrations. For example, work by Timmermann et al. (1999), using a GCM with sufficient resolution, suggests that the tropical Pacific is likely to change to a state similar to present-day El Niño conditions. Since fewer hurricanes occur in the Atlantic during El Niño years, their results suggest that Atlantic hurricanes will likely decrease in frequency in the future. Interestingly, during severe El Niño events such as those in 1982-83 and 1997-98, the jet stream over the North Pacific brought winter storms farther south causing extensive coastal erosion and flooding in California. Hence, a prolonged El Niño state should decrease the occurrence of hurricanes in the Atlantic but lead to an increase in coastal impacts by winter storms on the West Coast. On the other hand, the results of Timmermann et al. (1999) also suggest stronger interannual variability, with relatively strong cold (La Niña) events becoming more frequent. Although these La Niña events would be superimposed upon a higher mean temperature, this could suggest more interannual variability in Atlantic hurricanes with more intense activity during the stronger cold events.

Even if storm magnitudes and frequencies of occurrence remain the same, an important impact of future storms, whether tropical or extratropical, will be their superposition on a rising sea level. This has recently been demonstrated by examining historical storm surge magnitudes calculated from sea-level records (Zhang et al., 1997). These records included surges induced by both winter storms and hurricanes but were dominated by the more frequent winter storms. No significant long term trends were found in storm frequency or severity. When considering that the storms were

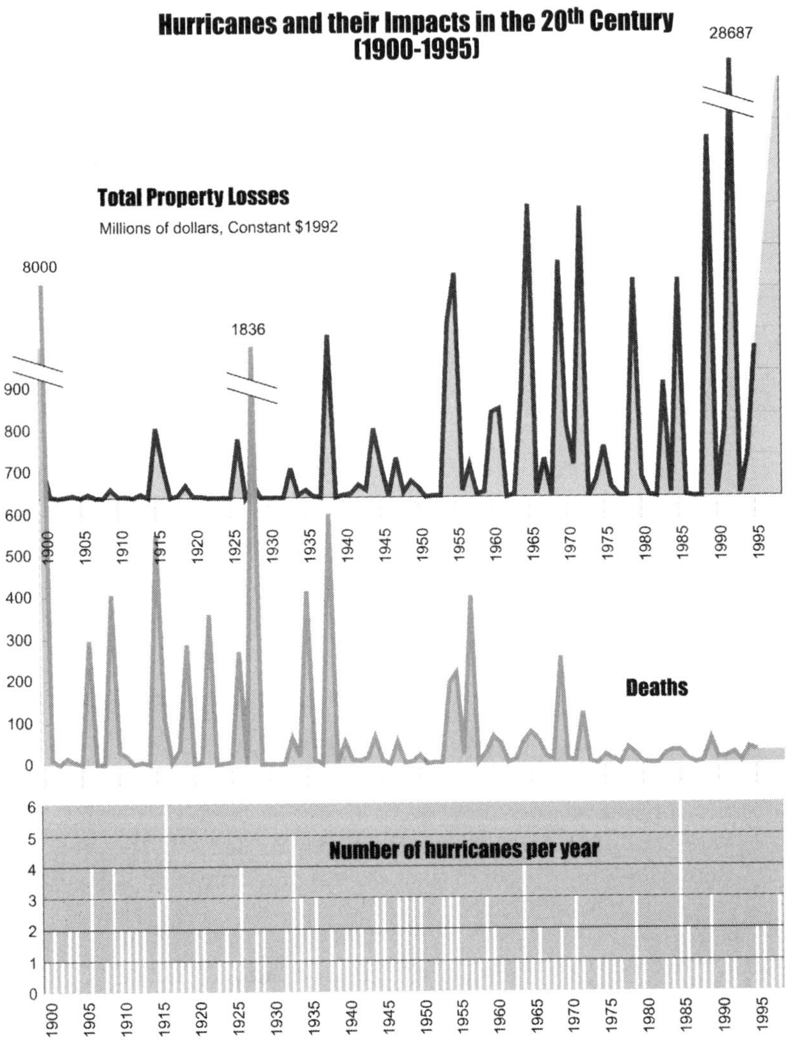

Hurricanes and their Impacts in the 20ᵗʰ Century (1900-1995)

Total Property Losses
Millions of dollars, Constant $1992

28687

Deaths

Number of hurricanes per year

Figure 4: Loss of life and property from hurricanes making landfall in the continental U.S. over the past 20ᵗʰ century Source: National Hurricane Center: NOAA. See color figure appendix.

superimposed on a rising sea level (0.15 inch/year or 3.9 mm/year at Atlantic City), there is an inferred increase in storm impact over an 82-year record. For example, the number of hours of extreme water levels per year increased from less than 200 in the early 1900s to abnormally high values of up to 1200 hours (averaging typically 600 hours) in the 1990s. Thus sea-level rise will increase impacts to the coast by a storm of a given magnitude.

Freshwater Runoff

Hydrological cycles are fundamental components of climate, and climate change is likely to affect both water quality and water availability. In general, the consensus is that the hydrologic cycle will become more intense, with average precipitation increasing, especially at high latitudes. Extreme rainfall conditions, already demonstrated to have increased over the 20th century, are likely to become more common, as could droughts and floods (Karl et al., 1995a; Karl et al., 1995b). Changes in freshwater runoff will result both from climate-related factors and changes in population and land use patterns relating to supply and demand. While the reader is referred to the Water Sector chapter for further discussion of both climatic and human-induced changes in hydrological cycles, the close relationship between precipitation, streamflow, and coastal ecosystems is a topic of special interest to coastal researchers, managers, and planners.

The Canadian and Hadley climate models have been used in concert with hydrology models to estimate the changes in freshwater runoff for three portions of coastline: the Atlantic, the Gulf of Mexico, and the Pacific. In contrast to GCM scenarios of temperature changes under the influence of increased CO_2, the estimates of runoff vary widely. Wolock and McCabe (1999) determined that for some regions the Hadley and Canadian climate models produce opposite results. For the Atlantic coast, the Hadley model projects an increase of more than 60% by the 2090s, whereas the Canadian model projects a runoff decrease of about 80%. The large differences are attributed to the projected increases in precipitation in the Hadley model, versus small changes to significant decreases in precipitation in the Canadian model during the 21st century.

Freshwater runoff affects coastal ecosystems and communities in many ways. The delivery of sediment, nutrients, and contaminants is closely linked to both the strength and timing of freshwater runoff. Salinity gradients are driven by freshwater inputs into estuaries and coastal systems, and have

strong effects on biotic distributions, life histories, and geochemistry. Coastal runoff also affects circulation in estuaries and continental shelf areas, and increases in runoff have the potential to increase the vertical stratification and decrease the rate of thermohaline circulation by adding more freshwater to the system.

In the event that increased river flows result from climate change, more suspended sediments could be transported into the coastal regions, increasing the upper layer turbidity and potentially reducing available light to both plankton and submerged aquatic vegetation. Changes in sediment transport could also alter the amount of sediment available for soil aggradation (accumulation) in wetlands and sands for littoral systems. Increased sediment transport will provide needed material for accretion to coastal wetlands threatened by sea-level rise, whereas a decrease in sediment transport will concurrently diminish the ability of some wetlands to respond to sea-level rise.

Increased river flows could also increase the flux of nutrients and contaminants into coastal systems, which influence eutrophication and the accumulation of toxins in marine sediments and living resources. Both increased temperatures and decreased densities in the upper layers might also reduce the vertical convection enough to prevent oxygenation of the bottom waters, further contributing to anoxic conditions in the near-bottom waters (Justic´ et al., 1996). Decreased freshwater inflows into coastal ecosystems would be likely have the reverse effect, reducing flushing in estuaries, increasing the salinity of brackish waters, and possibly increasing the susceptibility of shellfish to diseases and predators.

Ocean Temperatures

The oceans represent enormous reservoirs of heat; their heat capacity is such that the upper 16 to 33 feet (5 to 10 meters) of the water column generally contain as much heat as the entire column of air above it. The oceans work to distribute heat globally, and atmospheric and oceanic processes work together to control both pelagic and coastal ocean temperatures.

Strong evidence for ocean warming was recently published by Levitus et al. (2000), who evaluated some five million profiles of ocean temperature taken over the last 55 years. Their results indicate that the mean temperature of the oceans between 0 and 300 meters (0 and 984 feet) has increased by

0.31°C (0.56°F) over that same period, which corresponds to an increase in heat content of approximately 1×10^{23} Joules of energy. Furthermore, the warming signal was observable to depths of some 3000 meters (9843 feet), and the total heat content between 300 and 3000 meters increased by an additional 1×10^{23} Joules of energy (see Figure 5). Although Levitus et al. (2000) could not conclude that the signal was primarily one of climate change, as opposed to climate variability, they note that their results are in strong agreement with those projected by many general circulation models. Earlier work done by Cane et al. (1997) also suggest that observed patterns of change in ocean temperature are consistent with warming scenarios.

Understanding how future ocean temperatures will be affected by climate variability and change will depend on improved understanding of the coupling between atmospheric and oceanic processes, and predicting future variations in the forcing functions. The coupling between these processes is regionally or locally controlled, whereas the climate change models are usually more global. As these physical structures and processes, along with their natural modes of variability, have important implications for overall levels of biological productivity in the ocean, an accurate assessment of potential changes in ocean dynamics will necessarily be speculative until GCMs with finer resolution of ocean processes are developed.

The ecosystem responses to increased ocean temperatures can change productivity both directly and indirectly. Ocean temperature changes will affect not only the metabolic rate of organisms but also sea level, currents, movement of larvae, erosion rates, substrate structures, turbidity, water column stratification, nutrient cycling, and subsequently production (McGowan, et al., 1998; Rice, 1995). With population changes, community dynamics will change, altering such things as predator-prey relationships. Ocean temperature increases will affect the distribution of marine species, likely with a poleward migration of tropical and lower latitude organisms. The result could be major changes in the composition, function, and productivity of marine ecosystems, including potential impacts to marine biodiversity. For example, observable changes in the distribution and abundance of intertidal species along the California coast have been documented by Sagarin et al. (1999) which are consistent with warming trends. Similarly, large interannual and interdecadal temperature changes in the North Pacific have already been associated with changes in the mixed layer depth over the North Pacific and diminished biological productivity (McGowan et al., 1998).

Increases in temperature will also result in further melting of sea ice in polar and subpolar regions. Recent observations in the Arctic have already shown significant declines in ice extent, which has been shrinking by as much as 7% per decade over the last 20 years (Johannessen et al., 1999) as well as ice thickness, with Arctic sea ice thinning (and subsequently decreasing in volume) by as much as 15% per decade (Rothrock et al., 1999). Additionally, record low levels of ice extent occurred in the Bering and Chukchi seas during 1998, the warmest year on record (Maslanik et al., 1999). While a possible mechanism could be related to the Arctic Oscillation, a decadal scale mode of atmospheric variability in the Arctic, some comparisons with GCM outputs strongly suggest that these observed declines in sea ice are related to anthropogenically induced global warming (Vinnikov et al., 1999). The reduction and potential loss of sea ice has enormous

Ocean Heat Content in the 0-3000 m Layer

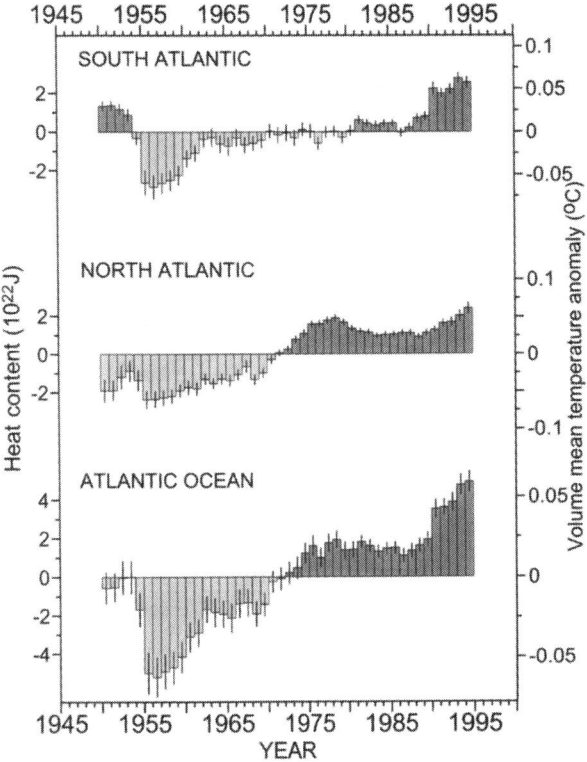

Figure 5: A comprehensive analysis of over 5 million temperature profiles by Levitus, et al. (2000) reveals a pattern of warming in both the surface and the deep ocean over the last 40 years. The largest warming has occurred in the upper 300 meters (984 feet), which have warmed by an average of 0.31°C (0.56°F), with additional warming as deep as 3000 meters (9843 feet). See Color Figure Appendix

feedback implications for the climate system as well; ice and snow are highly reflective surfaces that reflect the vast majority of the Sun's incoming radiative heat back to outer space. By contrast, open oceans reflect only 10 to 20% of the Sun's energy. Thus, the conversion of the Arctic ice cap to open ocean could greatly increase solar energy absorption, and act as a positive feedback to global warming.

Ocean Currents

Major ocean current systems play a significant role in the dispersal of marine organisms and in the production characteristics of marine systems. These current systems are likely to be affected in critical ways by changes in global and local temperatures, precipitation and runoff, and wind fields. Similarly, oceanic features such as fronts and upwelling and downwelling zones will be strongly influenced by variations in temperature, salinity, and winds. These changes will be manifest on scales ranging from the relatively small spatial and temporal scales characteristic of turbulent mixing processes to very large scales characteristic of the deep water "conveyor belt" circulation, with potentially dramatic feedback influences on climate patterns. Changes occurring on this spectrum of spatial and temporal scales have important implications for overall levels of biological productivity in the ocean, and are critical to understanding the implications of global climate change on living marine resources.

Research suggests that the processes of formation and circulation of deep-water through the so-called conveyor-belt circulation (see Figure 6) could be strongly influenced by changes in temperature and salinity, with significant implications for the North Atlantic region and global ocean circulation (Broecker et al., 1999; Taylor 1999). Generally warm, saline surface waters are transported northwards by the Gulf Stream, ultimately feeding the Norwegian, East Greenland, and West Greenland currents. Here winter air-sea interactions cool the already highly saline water masses, resulting in a rapid increase in density. This rapid increase in density drives convection, in which the heavier, denser water sinks and flows away from the polar regions to fill the deep ocean basins, driving deep sea thermohaline circulation.

Increased temperature or decreased salinity (resulting from changes in precipitation patterns or melt-

The Global Ocean Conveyor Belt

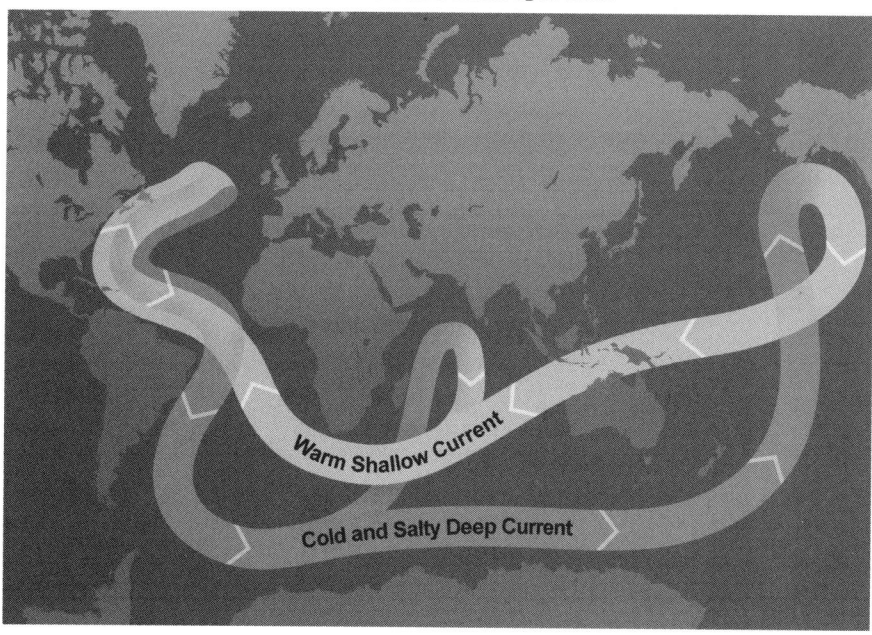

Figure 6: The ocean plays a major role in the distribution of the planet's heat through deep sea circulation. This simplified illustration shows this "conveyor belt" circulation which is driven by differences in heat and salinity. Records of past climate suggest that there is some chance that this circulation could be altered by the changes projected in many climate models, with impacts to climate throughout lands bordering the North Atlantic (Modified from Broecker, 1991). See color figure appendix.

ing ice sheets) at high latitudes could result in a reduction in the deep-water formation by decreasing the density of surface waters, with important consequences for this convective system (Schmittner and Stocker, 1999; Broecker, 1997). Hadley Centre models have suggested a decline in the strength of deep-water circulation of approximately 25% under some scenarios (Wood et al., 1999). Such a decline could lead to a general cooling throughout the North Atlantic region, resulting from a reduction in transport of warmer waters from lower latitudes (Driscoll and Haug, 1998). Additionally, a complete cessation of the conveyor belt circulation is possible, which could cause winter temperatures throughout the North Atlantic to fall abruptly by as much as 9°F (5°C). Sedimentary records in the northern Atlantic suggest that past circulation shutdowns have occurred in extremely short time intervals, associated with abrupt climate shifts and dramatic cooling in Europe. Ironically, the result would likely be further acceleration of global warming trends as a result of a decrease in the oceanic uptake of carbon dioxide that would follow weakening or cessation of thermohaline circulation (Sarmiento and Le Quere, 1996).

KEY ISSUES

1. Shoreline Systems, Erosion, and Developed Coastal Areas
2. Threats to Estuarine Health
3. Coastal Wetland Survival
4. Coral Reef Die-offs
5. Stresses on Ocean Margins and Marine Fisheries

Coastal and marine resources are uniquely influenced by the long-term climate dynamics described above. For the purposes of this Assessment, the potential impacts on five principal ecosystem types were evaluated: shorelines, estuaries, coastal wetlands, coral reefs, and ocean margins/marine fisheries. While there is significant overlap between some of these ecosystem types, and also further divisions that could be made within these ecosystems, this breakdown provides a methodology for understanding what may be the most significant generic impacts on ecological systems. These are consequences which are either ongoing or could be reasonably expected on the basis of past climate variability, forcing scenarios, and change thresholds. In addition, individual case studies were conducted in order to provide specific examples of the complex set of interactions between the effects of climate and human activities; results from several of these case studies have been incorporated here.

1. Shoreline Systems, Erosion, and Developed Coastal Areas

Storms, hurricanes, typhoons, and similarly extreme atmospheric phenomena along coasts produce high winds that in turn generate large waves and currents. Storms and hurricanes produce storm surges that temporarily raise water levels by as much as 23 feet (7 meters) above normal. Although these events are sporadic, they are a primary cause of beach erosion and shoreline impacts throughout the US. Hurricanes account for far more insured losses of property than do other hazards such as earthquakes and wildfires. Sea-level rise increases the vulnerability of shorelines and floodplains to storm damage by increasing the baseline water level for extreme storms and coastal flooding events. In addition to storm and flooding damages, specific impacts associated with sea-level rise could include increased salinity of estuaries and freshwater aquifers, altered tidal ranges in rivers and bays, changes in sediment and nutrient transport which drive beach processes, and the inundation of waste disposal sites and landfills which could reintroduce toxic materials into coastal ecosystems.

Individual hurricanes have resulted in enormous economic impacts, particularly to the Southeast. Hurricane Hugo in 1989 caused an estimated $9 billion in damages; Hurricane Andrew in 1992 caused an estimated $27 billion in damages; and Hurricane Georges in 1998 caused an estimated $5.9 billion in damages (Source: National Climatic Data Center). However, hurricanes are not the only storm threat to coastal inhabitants and property owners. Extratropical winter storms have significant impacts as well, such as the Halloween "nor'easter" of 1991 which caused damages amounting to over $1.5 billion along the Atlantic Coast. Along the Pacific Coast, extratropical storms battered California causing an estimated $500 million in damage during the 1997-98 El Niño (Griggs and Brown, 1999). As coastal population and property construction increases, the economic vulnerability of human developments in coastal areas to hurricane and storm activity will continue to rise, and will be aggravated by any climate-induced changes in the frequencies or intensities of these events.

Coastal erosion is already a widespread problem throughout much of the country (Figure 7), and has significant impacts to both undeveloped shorelines and coastal development and infrastructure. In addition to storms and extreme events, interannual modes of variability such as ENSO have been shown to impact many shorelines. For example, along the Pacific Coast, cycles of beach and cliff erosion have been linked to El Niño events, which elevate average sea levels over the short term and alter the frequency, intensity, and direction of storms impacting the coastline. During the 1982-83 and the 1997-98 El Niños, impacts were especially severe along shorelines throughout California, Oregon, and Washington (Komar, 1999; Kaminsky et al., 1999). Good (1994) has shown that along the central Oregon coast, a rapid buildup of seawalls and revetments routinely followed major El Niño events of the last two decades, as coastal property owners attempted to protect their shorelines from increasing erosion.

Atlantic and Gulf coastlines are especially vulnerable to sea-level rise, as well as to changes in the frequency and severity of storms and hurricanes. Most of the East and Gulf coasts are rimmed by a series of barrier island and bay systems which separate the gently sloping mainland coastal plains from the continental shelf. These islands bear the brunt of forces from winter storms and hurricanes, protecting the mainland from wave action. However, these islands and coastlines are not stable but are instead highly dynamic systems that are highly sensitive and respon-

Classification of Annual Shoreline Change Around the United States

Severely eroding   Moderately eroding   Relatively stable

sive to the rate of relative sea-level rise as well as the frequency and severity of storms and hurricanes.

In response to rising sea levels, these islands typically "roll over" towards the mainland, through a process of beach erosion on their seaward flank and overwash of sediment across the island. Human activities block this natural landward migration through the construction of buildings, roads, and seawalls; as a result, shorelines erode, increasing the threat to coastal development and infrastructure. Subsequent impacts include the destruction of property, loss of transportation infrastructure, increased coastal flooding, negative effects on tourism, saltwater intrusion into freshwater aquifers, and impacts to fisheries and biodiversity. Such impacts will unquestionably further modify the functioning of barrier islands and inland waters that have already been impacted by pollution, physical modification, sediment starvation by dams, and material inputs related to human activities.

Barrier islands are certainly not the only coastlines at risk. Private property on mainland and estuarine shorelines, harbor installations, other water-dependent activities and infrastructure, transportation infrastructure, recreational areas, and agricultural areas are all particularly vulnerable to inundation and increased erosion as a result of climate change, especially in low-lying areas. Assessing the total

Figure 7: A general classification scheme of shoreline erosion rates throughout the US. (modified from Dolan et al., 1985). See Color Figure Appendix

economic impacts from sea-level rise on coastal areas and on a national scale is still somewhat speculative. Nevertheless, a study by Yohe et al. (1996) quantified the economic costs (protection plus abandonment) to coastal structures with a one meter sea-level rise. This analysis estimated costs of as much as $6 billion (in 1990 dollars) between 1996 and 2100. However, this number represents market-valued estimates only, which are derived from property-value appreciation, market adaptation, and protection costs. As such, this is a minimum cost estimate, as it does not include the lost ecosystem services value of non-market resources, such estuaries and tidal wetlands, or the costs to communities resulting from reductions in coastal economic activities such as fishing, tourism, and recreation. A comprehensive review of the potential cost of sea-level rise to developed coastlines was recently completed by Neumann et al. (2000); and should be referred to for a wider range of estimates of the potential economic consequences of sea-level rise to developed areas.

Many coastal structures were designed with the 100- year flood as their basis. This flooding level determines the elevations to which the federal projects are built, such as the Army Corps of Engineers

South Florida Case Study

The natural South Florida ecosystem was largely defined by a high degree of variability in rainfall both within and across years. The region originally supported a mosaic of multiple interactions among many different ecosystem types, including freshwater, wetland, mangrove, estuarine, and coral habitats all intimately coupled by the regional hydrology. All of these ecosystems now severely impacted by a large and very rapidly growing human population perched along a very narrow strip of coastal land. Tourism in South Florida is a multibillion-dollar-a-year industry, and the existence of the Everglades, Biscayne National Park and the Florida Keys National Marine Sanctuary illustrate the high value society places on this unique environment.

The South Florida region is vulnerable to multiple climate change stresses, as sea-level rise, changes in the frequency of freezing events, hurricanes, droughts and associated fires, sea surface temperatures and many others all affect the full diversity of ecosystems. The responses of all of these ecosystems will be significantly modified or constrained by human development. In particular, South Florida has developed one of the world's largest water management systems with the primary objectives of dampening hydrological variability, providing flood protection, and supplying water to urban and agricultural systems (Harwell, 1998; US COE 1994). This alteration of hydrological functions has caused serious degradation to the ecosystems, which were formerly adapted to natural modes of variability but have been made more vulnerable as a result of human disturbance and development.

Any significant change to the precipitation regime could have major consequences to coastal ecosystems; already the Everglades and associated systems are not considered sustainable due to regional hydrologic alterations. While a massive and costly restructuring of the water management system is currently underway, a reduced precipitation regime could further increase multi-year droughts, increase fire frequency and intensity, reduce water supply to urban users, and increase estuarine hypersalinity. Thus, global climate change can be expected to exacerbate the effects of natural and anthropogenic stresses on South Florida, making the current challenge of restoring and sustaining its ecosystems even greater.

levees that protect New Orleans. It is also the level to which coastal structures must be built to qualify for flood insurance through the Federal Emergency Management Agency's (FEMA) Flood Insurance Program. If sea level rises, the statistics used to design these structures change. For example, what was once considered a 50-year flood could become as severe as a 100-year flood following an increase in sea level (Pugh and Maul, 1999). Coastal insurance rates would have to be adjusted to reflect the change in risk. Furthermore, FEMA has estimated that the number of households in the coastal floodplain could increase from 2.7 million currently to some 6.6 million by the year 2100, as a result of the combination of sea-level rise and rapidly growing coastal populations and development (FEMA, 1991). Again, it is clear that the vulnerability of coastal developed areas to natural disasters can be expected to rise throughout the 21st century, even independent of climate-induced changes in risk.

2. Threats to Estuarine Health

Estuaries are among the most productive coastal ecosystems, critical to the health of a great many commercial and recreational fisheries, and are affected in numerous ways by climate variability

and change. Warmer spring and fall temperatures will alter the timing of seasonal temperature transitions, which affect a number of ecologically important processes and will thus impact fisheries and marine populations. As referenced above in the section on freshwater forcing, the amount and timing of freshwater flow into estuaries greatly influence salinity, stratification, circulation, sediment, and nutrient inputs. Thus, increases in winter-spring discharges in particular could deliver more excess nutrients as well as contaminants to estuaries, while simultaneously increasing the density stratification. Both of these would increase the potential for algal blooms (including potentially harmful species) and the development of hypoxic (low oxygen) conditions, which could in turn increase stresses on sea grasses and affect commercial fishing and shellfish harvesting.

Currently, nutrient over-enrichment is one of the greatest threats to estuaries in the US, with over half the nation's estuaries having at least some of the symptoms of moderate to high states of eutrophication (Bricker et al., 1999). Eutrophication has multiple impacts to estuaries, as well as other coastal ecosystems, including more frequent and longer lasting harmful algal blooms, degradation of seagrass beds and coral reefs, alteration of ecological structure, decreased biological diversity, and the loss of fishery resources resulting from low oxygen events. In this instance, climate change is only one aspect of human-accelerated global environmental change, and particularly for estuaries, the broader effects of global change have to be considered. Specifically, human activities have altered the biological availability of nitrogen through the production of inorganic fertilizer, the combustion of fossil fuel, and the management of nitrogen-fixing agricultural crops. As a result, since the 1960s the world has changed from one in which natural nitrogen fixation was the dominant process to a situation in which human-controlled processes are at least as important in making nitrogen available on land.

This increased availability has led to an increase in the delivery of nitrogen to coastal marine systems, particularly estuaries (Vitousek et al., 1997). The delivery of nutrients has been closely linked to variability in streamflow; for example, in the Gulf of Mexico, increases in freshwater delivery from the Mississippi River are closely linked with increases in nutrient delivery and the subsequent development of hypoxia. The spring floods of 1993 resulted in the greatest nitrogen delivery ever recorded in the Gulf of Mexico (Rabalais et al., 1999), and the areal extent of the resulting hypoxic zone was twice as large as the average for the preceding eight years. However, the response of individual estuaries to changes in freshwater and nutrient delivery will differ significantly as related to variation in water residence times, stratification, and the timing and magnitude of inputs. Yet because the solubility of oxygen in warmer waters is lower than in cooler waters, any additional warming of estuaries during summer months would be likely to aggravate the problems of hypoxia and anoxia that already plague many estuaries.

With few exceptions, the potential consequences of climate change to estuarine ecosystems are not yet being considered in long-term estuarine and coastal zone management. In the Mid-Atlantic, estuaries are already in a degraded environmental condition, but are currently the subject of substantial societal commitments for their restoration through pollution reduction, habitat rehabilitation, and more sustainable use of living resources. An increase in winter-spring precipitation that delivers additional nutrients to estuaries such as the Chesapeake Bay would make achieving the management goals of reducing nutrient inputs and improving dissolved oxygen levels more difficult. More efficient nutrient management practices and more extensive restoration of riparian zones and wetlands would be required to meet current nutrient goals in seasonally wetter watersheds. However, it should be recognized that climate change is expected to bring about alterations to many estuaries for which mitigating responses may not exist.

3. Coastal Wetland Survival

Coastal wetlands are some of the most valuable ecosystems in the nation as well as some of the most threatened. By providing habitat, refuge, and forage opportunities for fishes and invertebrates, marshes and mangroves around the coastal US are the basis of many communities' economic livelihoods (Costanza et al., 1997; Mitsch and Gosslink, 1993). The role of wetlands in nutrient uptake, improving water quality, and reducing nutrient loads to the coastal ocean is widely recognized, as is their value in providing recreational opportunities and protecting local communities from flooding — either by dampening storm surges from the ocean or providing storage for river floodwaters. Climate variability and change compound existing stresses from human activities (Markham, 1996), such as dredging and/or filling for development, navigation or mineral extraction, altered salinity and water quality, and the direct pressures of increasing numbers of people living and recreating in close proximity.

These ecosystems are sensitive to a number of climate related variables. Changes in atmospheric temperatures and carbon dioxide concentrations generally result in increased plant production, although the response varies considerably among species. Increases in freshwater discharge would generally benefit many coastal wetlands, and decreases might result in salinity stress for some communities, particularly in the western Gulf of Mexico where already limited freshwater inputs are expected to decrease dramatically. For coastal wetlands facing current or future relative sea-level rise, increased sediment delivery will be necessary for vertical accumulation of the substrate.

Coastal wetlands can cope with changes in sea level when they are capable of remaining at the same elevation relative to the tidal range, which can occur if sediment buildup equals the rate of relative sea-level rise or if the wetland is able to migrate. Migration occurs when the wetland moves upslope along with the tidal range, with the seaward edge of the wetland drowning or eroding and the landward edge invading adjacent upslope communities (Figure 8). However, if wetlands are unable to keep pace with relative sea-level change, or if their migration is blocked by bluffs, coastal development, or shoreline protection structures, then the wetland will become immersed and eventually lost as rising seas submerge the remaining habitat.

Currently, many US coastlines are areas of coastal subsidence where both natural and human-induced processes strongly influence wetlands. Coastal Louisiana has experienced the greatest wetland loss in the nation, where a combination of anthropogenic (human-caused) and natural factors have driven losses between 24 and 40 square miles per year during the last 40 years. As approximately 25% of the nation's brackish and freshwater coastal wetlands are found in Louisiana, this constitutes as much as 80% of the total loss of these wetlands in the US (Boesch et al., 1994). Changes have occurred so rapidly in the many bald cypress forests near New Orleans that they have been converted directly to open water rather than being gradually overtaken by salt marsh. Once lost to open water, these wetlands become extremely difficult to restore. The remaining 3.5 million acres of wetlands in South

Louisiana provide critical nursery areas for finfishes and crustaceans, particularly shrimp and crabs, which make up the bulk of the state's multimillion dollar seafood industry. Additionally, these coastal wetlands serve as important buffers against storm surges, protecting inland residential and commercial infrastructure from severe flooding. Thus, if climate change exacerbates the current rate of sea-level rise, those regions of coastal Louisiana where marshes are on the margin of survival will suffer, and even more extensive land loss will result.

4. Coral Reef Die-Offs

Although coral reefs might seem exotic to many residents of the US, these ecosystems play a major role in the environment and economies of two states (Florida and Hawaii) as well as most US territories in both the Caribbean and the Pacific. Coral reefs are valuable, if often over-utilized, economic resources for many tropical coastal regions, providing numerous fisheries opportunities, recreation, tourism, and coastal protection (Wilkinson and Buddemeier, 1994). It is widely recognized that reefs and related communities are also some of the largest storehouses of marine biodiversity, with untapped resources of genetic and biochemical materials, and of scientific knowledge (Veron, 1995). Further, the living reef communities are not isolated dots on a map, but part of an interacting mosaic of oceanic and terrestrial habitats and communities, and many of the organisms and processes found on reefs are important in a much wider sphere of related environments (Buddemeier and Smith, 1999). In many important ways, the condition and responses of coral reef communities can be seen as diagnostic of the condition of the world's low-latitude coastal oceans.

Degradation of reef communities has been increasing worldwide over the past few decades, and the last few years have seen a dramatic increase in the extent of coral bleaching, new emergent coral diseases and widespread reef die-offs. Coral bleaching, which occurs when symbiotic algae that live in the corals themselves are expelled from their hosts due to high sea surface temperatures, has been occurring with increasing frequency in the last several decades, notably during El Niño events. The 1998 El Niño in particular was associated with unprecedented high sea surface temperatures and

Processes Affecting Wetland Migration

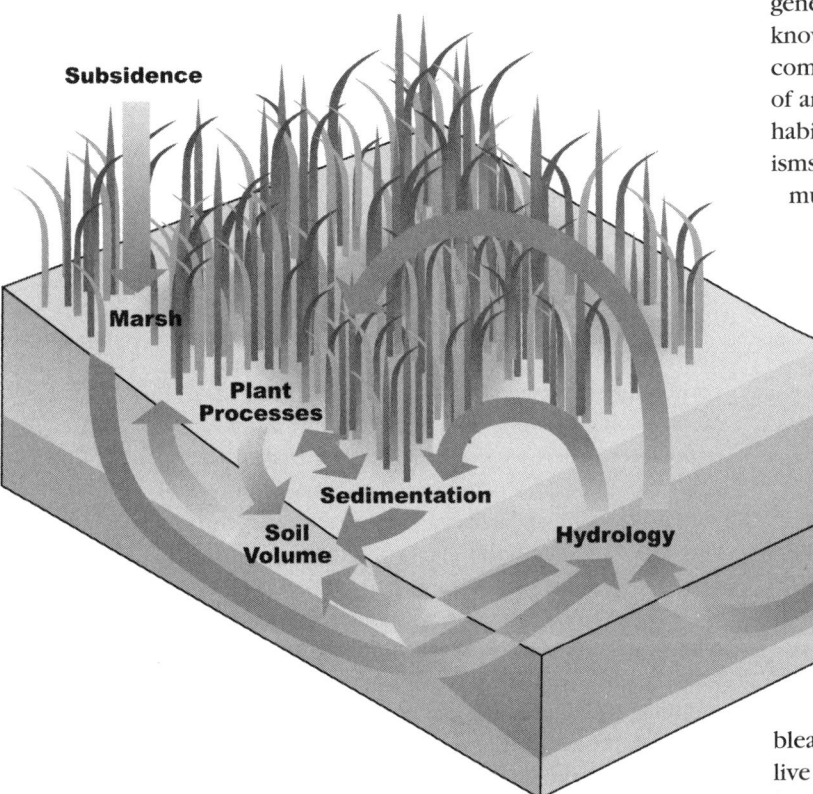

Figure 8: The rate of sea-level rise is projected to accelerate 2 to 5 fold over the next 100 years. The delivery of sediments to coastal wetlands is extremely important in determining the potential of these systems to maintain themselves in the face of current and future sea-level changes (based on Reed, 1995). See Color Figure Appendix

thus the most widespread coral bleaching ever observed (Strong et al., 2000; Wilkinson et. al., 1999; Hoegh-Guldberg, 1999). In many past bleaching events, both the algae and the corals are capable of recovery; however, warming events in 1998 resulted in unusually high levels of mortality from which many reefs have yet to recover. In addition to bleaching effects, there has been an upsurge in the variety, incidence, and virulence of coral diseases in recent years, with major die-offs reported, particularly in Florida and the Caribbean region. The causes of these epizootics are not known, but they suggest that coral ecosystems and organisms are weakened and vulnerable.

One of the most significant unanticipated direct consequences of increased atmospheric carbon dioxide concentration is the reduction of carbonate ion concentrations in seawater. Carbon dioxide acts as an acid when dissolved in seawater, causing seawater to be less alkaline. A drop in alkalinity subsequently decreases the amount of calcium carbonate (aragonite) which can be dissolved in seawater, which in turn decreases the calcification rates of reef-building corals and coraline algae (Figure 9) (Kleypas et al., 1999a; Gattuso, 1999). The impacts on coral reefs are weaker skeletons, reduced growth rates, and an increased vulnerability to erosion, with a wide range of impacts on coral reef health and community function. Additionally, the effects of increasing atmospheric carbon dioxide on the concentration of carbonate ions is greatest at the margins of coral distributions, due to the fact that carbon dioxide is more soluble in cooler waters. Thus, these effects will be most severe at high latitudes, so coral reefs at the margins of their distribution are not expected to expand their ranges as might otherwise be predicted by some ocean warming scenarios (Figure 10).

Human-induced disturbances have already taken a significant toll on coral reef systems, from activities such as over-fishing, the use of destructive fishing techniques, recreational activities, ship groundings, anchor damage, sedimentation, pollution, and eutrophication. Thus, the future for many coral reefs appears bleak. The synergy of stresses and the response time scales involved to restore reefs make integrated management essential but difficult. Given trends already in motion, some of the more marginal reef systems will almost certainly continue to deteriorate, making allocation of resources an important policy issue. The demise or continued deterioration of reef communities will have profound social and economic implications for the US, as well as serious political and economic implications for the world.

5. Stresses on Ocean Margins and Marine Fisheries

Projected changes in the marine environment have important implications for marine populations and fisheries resources. In addition to their economic value, marine fisheries hold a special social and cultural significance in many coastal communities. However, over-fishing and habitat loss have already taken serious tolls on the nation's fisheries and the communities that depend upon them. Currently some 33% of the stocks for which trends are known are either over-fished or depleted (NMFS, 1998), and for the vast majority of stocks, an accurate assessment of their status is unknown. The possible effects of climate change range from shifts in distribution to changes in survival and growth rates, with direct implications for fishery yields.

Effects of CO₂ on Coral

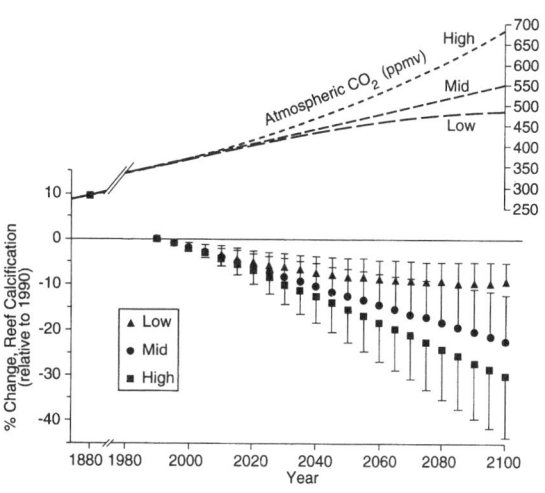

Bleached coral.

Figures 9: Increasing levels of atmospheric CO_2 are projected to decrease carbonate ion concentrations in seawater, which will decrease the calcification rates of many coral building species and impact coral reef health (Gattuso et al., 1999).

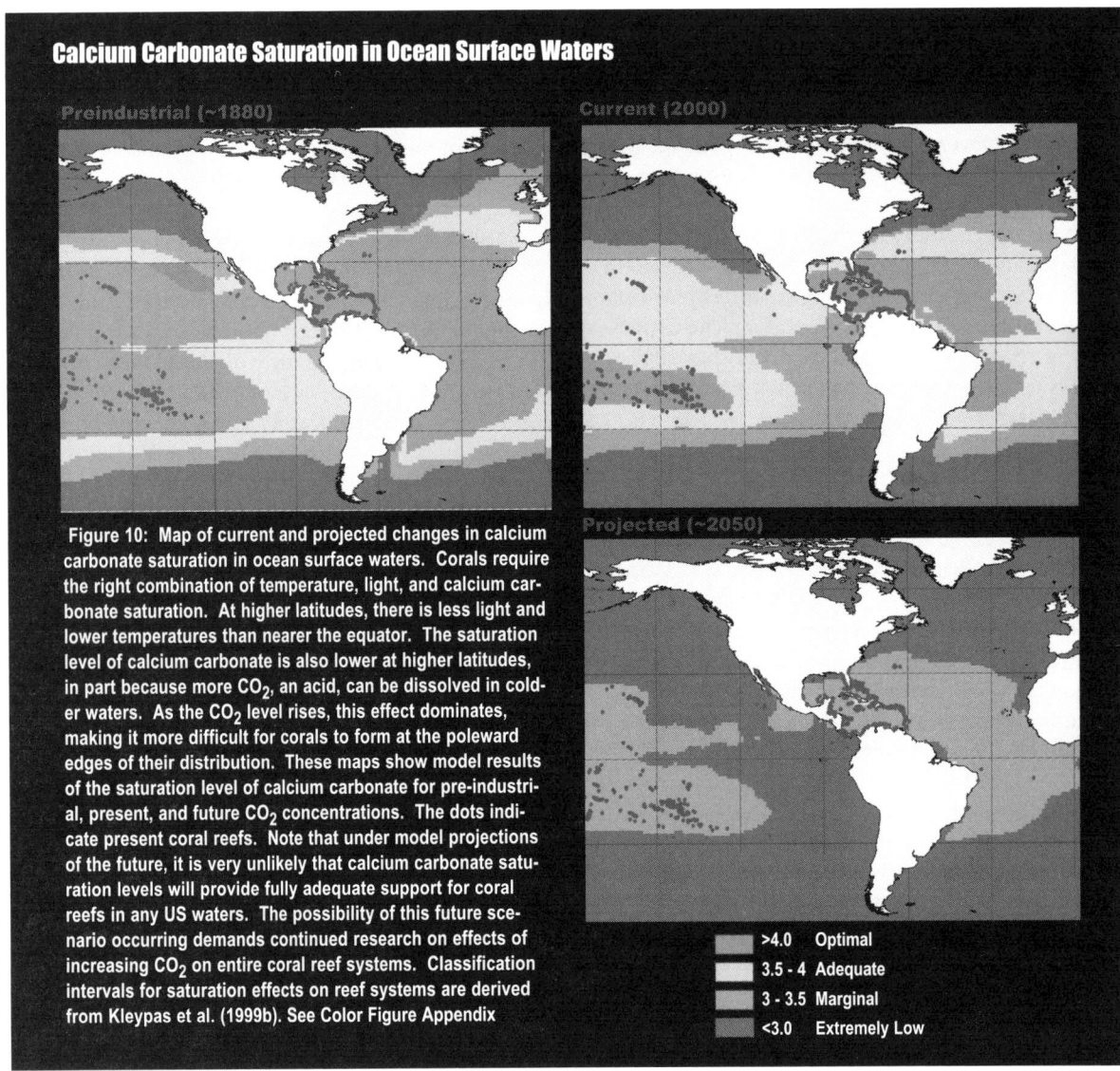

Calcium Carbonate Saturation in Ocean Surface Waters

Preindustrial (~1880)

Current (2000)

Projected (~2050)

Figure 10: Map of current and projected changes in calcium carbonate saturation in ocean surface waters. Corals require the right combination of temperature, light, and calcium carbonate saturation. At higher latitudes, there is less light and lower temperatures than nearer the equator. The saturation level of calcium carbonate is also lower at higher latitudes, in part because more CO_2, an acid, can be dissolved in colder waters. As the CO_2 level rises, this effect dominates, making it more difficult for corals to form at the poleward edges of their distribution. These maps show model results of the saturation level of calcium carbonate for pre-industrial, present, and future CO_2 concentrations. The dots indicate present coral reefs. Note that under model projections of the future, it is very unlikely that calcium carbonate saturation levels will provide fully adequate support for coral reefs in any US waters. The possibility of this future scenario occurring demands continued research on effects of increasing CO_2 on entire coral reef systems. Classification intervals for saturation effects on reef systems are derived from Kleypas et al. (1999b). See Color Figure Appendix

>4.0	Optimal
3.5 - 4	Adequate
3 - 3.5	Marginal
<3.0	Extremely Low

Changes in climate forcing will have important effects in ocean margin ecosystems through expected changes in the distribution and abundance of marine organisms and in fundamental changes in the production characteristics of these systems. Changes in temperature, precipitation, wind fields, and sea level can all be expected to affect oceanographic conditions in the ocean margins with direct ramifications for marine life in these areas. Physiological effects of temperature and salinity changes can also be expected with potentially important consequences for growth and mortality of marine species (Jobling, 1996). Impacts to marine ecosystems and fisheries associated with El Niño events illustrate the extent to which climate and fisheries can interact. For example, the high sea surface temperatures and anomalous conditions associated with the 1997-98 El Niño had a tremendous impact on marine resources off of California and the Pacific Northwest. Landings of market squid, California's largest fishery by volume and second largest in value, fell from over 110,000 metric tons in the 1996-97 season to less than 1000 metric tons during the 1997-98 El Niño season (Kronman, 1999). Amongst the many other events associated with this El Niño were high sea lion pup mortalities in California (Wong, 1997), poor reproductive success in seabirds off of Oregon and Washington, and unexpected catches of warm-water marlin off of the Washington coast (Holt, 1997). Even further north were rare coccolithophore blooms, massive seabird die-offs along the Aleutian Islands, and poor salmon returns in Alaska's Bristol Bay sockeye salmon fishery (Macklin, 1999).

Climate change is likely to cause even greater impacts to marine populations. Poleward shifts in distribution of marine populations can be expected with increasing water temperatures (Murowski, 1993). Species at the southern extent of their range along the East Coast can be expected to shift their distribution to the north with important overall consequences for ecosystem structure. Cod, American plaice, haddock, Atlantic halibut, redfish, and yellowtail flounder all would be expected to experience some poleward displacement from their southerly limits in the Gulf of Maine and off New England under increasing water temperatures (Frank et al., 1990). Thus, the loss of populations or sub-populations due to shifts in temperature under these constraints is likely, with regional impacts to communi-

ties that depend on local populations. An expansion of species commonly occurring in the middle Atlantic region, such as butterfish and menhaden, into the Gulf of Maine could also be expected.

In the Pacific, Welch et al. (1998) suggest that projected changes in water temperatures in the North Pacific could possibly result in a reduction of suitable thermal habitat for sockeye salmon. The most surprising prediction is that under scenarios of future climate, none of the Pacific Ocean is projected to lie within the thermal limits that have defined the distribution of sockeye salmon over the last 40

Understanding Global Climate and Marine Productivity

The US Global Ocean Ecosystems Dynamics (US GLOBEC) research program is operated in collaboration between the NOAA/National Ocean Service (NOS) Coastal Ocean Program and the NSF Biological Oceanography Program. GLOBEC is designed to address the questions of how global climate change may affect the abundance and production of marine animals. Two large marine regions have been the subjects of intensive research, the Northwest Atlantic and the Northeast Pacific. In the Atlantic program, the overall goal is to improve the predictability and management of the living marine resources of the region. This is being done through improved understanding of ecosystem interactions and the coupling between climate change, the ocean's physical environment and the ecosystem components (Figure 11). Particularly crucial physical drivers are the North Atlantic Oscillation and the salinity variations derived from flows from the Labrador Sea. A major objective of this program is to apply the understanding of the physical processes that affect the distribution, abundance and production of target species. This information will be used in the identification of critical variables that will support ecosystem-based forecasts and indicators as a prelude to the implementation of a long-term ecosystem monitoring strategy.

Remarkable changes have been observed in recent decades in the Northeast Pacific. Concurrent changes in atmospheric pressure and ocean temperatures indicate that in 1976 and 1977 the North Pacific shifted from one climate state or regime to another that persisted through the 1980s. Analysis of records of North Pacific sea surface temperature and atmospheric conditions show a pattern of such regime shifts lasting several years to decades. Although the important linkages are poorly understood, there is growing evidence that biological productivity in the North Pacific responds quite strongly to these decadal-scale shifts in atmospheric and oceanic conditions by alternating between periods of high and low productivity. Some of the key issues being addressed in this region concern the evaluation of life history patterns, distributions, growth rates, and population dynamics of higher trophic level species and their direct and indirect responses to climate variability. Additionally, the program seeks to better understand complex ecosystem interactions, such as how the North Pacific ecosystem is structured and whether higher trophic levels respond to climate variability solely as a consequence of bottom-up forcing.

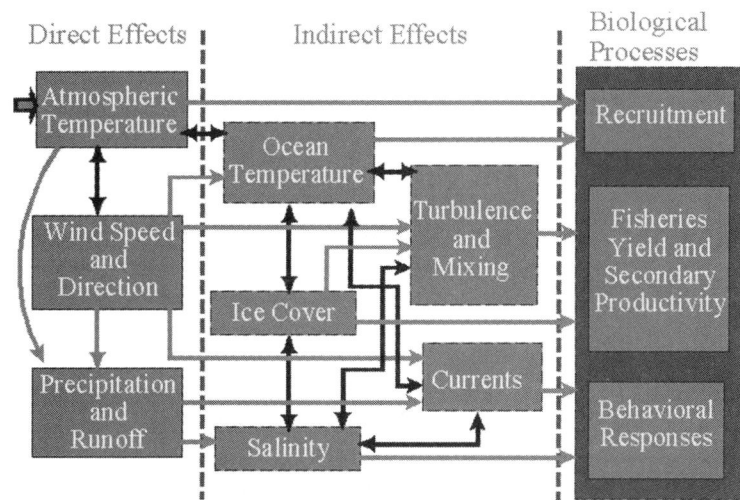

Climatic Pathways Affecting the Abiotic Environment and Biological Processes

Figure 11: Biological processes in the ocean are related to climate in many ways, both directly and indirectly. Improvements in our understanding of the direct and indirect effects of climate on biological processes in the oceans are essential to predicting how marine populations might respond to future change Source: US GLOBEC.

years. Under such scenarios, the distribution of all species of salmonids could be restricted to marginal seas in the North Pacific region (see also the Pacific Northwest chapter and Mantua et al. 1997 for more information on interactions between salmon and climate variability). Conversely, higher sea surface temperatures will be beneficial for other species. For example, changes in the abundance of sardine and anchovy populations appear to be strongly linked to global climate shifts (Lluch-Belda et al., 1992). The California sardine population was

recently declared "recovered," decades after its near disappearance from Pacific waters. The recovery has in part been related to an increase in sea surface temperatures off the Pacific Coast observed over the last 20 years (California Department of Fish and Game, 1999). Sardines are once again found in large numbers as far north as British Columbia (Hargreaves et al., 1994), and their recovery has led to a concurrent resurgence of the fishery long ago immortalized in John Steinbeck's *Cannery Row.*

Predicting Coastal Evolution at Societally-Relevant Time and Space Scales

One of the most important applied problems in coastal geology today is determining the response of the coastline to sea-level rise. Prediction of shoreline retreat and land loss rates is critical to the planning of future coastal zone management strategies, and assessing biological impacts due to habitat changes or destruction. Presently, long-term (~50 years) coastal planning and decision-making has been done piecemeal, if at all, for the nation's shoreline. Consequently, facilities are being located and entire communities are being developed without adequate consideration of the potential costs of protecting them from or relocating them due to sea-level rise-related erosion, flooding and storm damage.

The prediction of future coastal evolution is not straightforward. There is no standard methodology, and even the kinds of data required to make such predictions are the subject of much scientific debate. A number of predictive approaches could be used, including: 1) extrapolation of historical data, (e. g., coastal erosion rates), 2) inundation modeling, 3) application of a simple geometric model (e. g., the Bruun Rule), 4) application of a sediment dynamics/budget model, or 5) Monte Carlo (probabilistic) simulation. Each of these approaches, however, has its shortcomings or can be shown to be invalid for certain applications. Similarly, the types of input data required vary widely, and for a given approach (e. g., sediment budget), existing data may be indeterminate or simply not exist.

The relative susceptibility of different coastal environments to sea-level rise, however, may be quantified at a regional to national scale (e. g., Gornitz et al. 1994) using basic data on coastal geomorphology, rate of sea-level rise, and past shoreline evolution (Figure 12). A pilot project is underway at the US Geological Survey, Coastal and Marine Geology Program to assess the susceptibility of the nation's coasts to sea-level rise from a geologic perspective. The long-term goal of this project is to predict future coastal changes with a degree of certainty useful to coastal management, following an approach similar to that used to map national seismic and volcanic hazards. This information has immediate application to many of the decisions our society will be making regarding coastal development in both the short- and long-term.

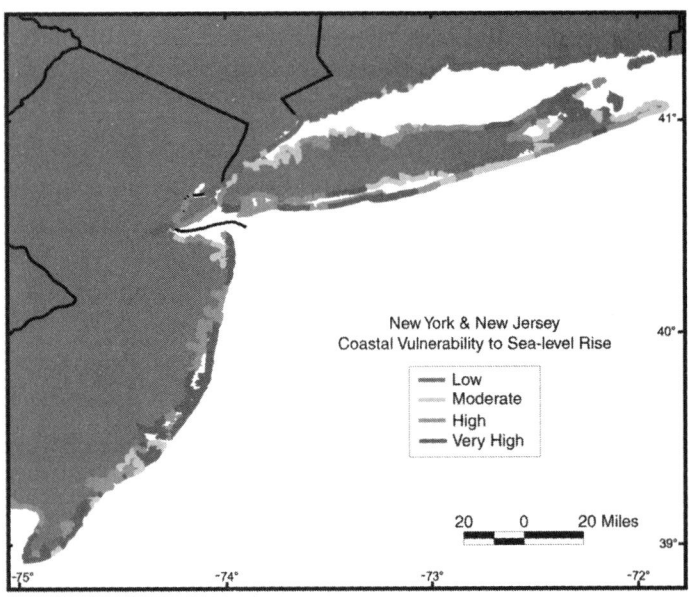

New York & New Jersey
Coastal Vulnerability to Sea-level Rise

— Low
— Moderate
— High
— Very High

20   0   20 Miles

Figure 12: These preliminary results illustrate the relative vulnerability to sea-level rise along the New York and New Jersey coastline as assessed by ongoing USGS research. Note that the vulnerability mapped here is likely to change as methodologies in this pilot program are critically evaluated and improved (Source: USGS). See Color Figure Appendix

ADAPTATION STRATEGIES

Assessing the effects of climate variability and change on coastal and marine resources is especially difficult given that human activities are generally responsible for the greatest impacts on coastal and marine environments. The nature of climate effects — both detrimental and beneficial to resources in question — are likely to vary greatly in the diverse coastal regions of the US. Anthropogenic disturbance often results in a reduction in the adaptive capacity of systems to cope with change and stress, making the real or potential impacts of climate difficult to observe. It is in this context that climate change acts as an increased stress on coastal and marine environments, adding to the cumulative impact of both natural and anthropogenic stress on ecological systems and resources.

This is most abundantly clear in coral reef systems. Coral reefs, both in US waters and worldwide, are clearly stressed, and many are degraded to the point of destruction. The prospects for the future are that in many, if not most, instances, the levels of both climate-related stress and local or regional stress resulting from anthropogenic impacts will increase. The implication for coral reefs is that local and regional reef protection and management efforts must be even more effective in controlling local stresses, to provide some compensation for large scale impacts. If local and regional anthropogenic stresses continue or increase, many of the reefs that are heavily used or affected by humans will be poor candidates for survival. This would result in substantial impacts to the communities and regional economies that depend upon healthy reefs for fisheries, subsistence, recreation, and tourism.

For responding or adapting to sea-level rise, three groups of strategies have been discussed, these being: (planned) retreat, accommodation, and protection (IPCC, 1996). A retreat strategy would prevent or discourage major developments in vulnerable coastal areas and could include rolling easements, which allow development but explicitly prevent property owners from preventing the upland migration of wetlands and beaches. An accommodation strategy might elevate land surfaces or human structures, modify drainage systems, or otherwise change land use practices, thus allowing many coastal ecosystems to be maintained. A protection strategy could utilize beach nourishment and dune stabilization, as well as dikes, seawalls, bulkheads, and revetments, to form a barrier between water and land. This might generally lead to a loss of natu-ral functions for beaches, wetlands, and other inter-tidal zones, but would be capable of roughly maintaining the coastline in place.

In areas where beaches or wetlands must migrate inland in response to sea-level change, it has been shown that planning and implementing protection or retreat strategies for coastal developments can substantially reduce the long-term economic impacts of inundation and shoreline migration. While some regulatory programs continue to permit structures that block the migration of wetlands and beaches, others have tried to decrease economic motivations for developing in vulnerable areas, or have experimented with retreat strategies. For example, coastal management programs in Maine, Rhode Island, South Carolina, and Massachusetts have implemented various forms of rolling easement policies in an attempt to ensure that wetlands and beaches can migrate inland as sea-level rises. Several other states require coastal counties to consider sea-level rise in their coastal management plans. However, the difficulties and obstacles facing coastal managers in effectively implementing setback and rolling easement policies are substantial. Many of the programs initiated to date are not mandatory, and the effectiveness of state coastal zone management programs often varies significantly.

While the potential impacts of sea-level rise have begun to initiate concern and some discussion of potential response strategies, the broader ramifications of changes in temperature, freshwater discharges, and the frequency and intensity of storm events, have scarcely been assessed. For example, in estuaries such as Chesapeake Bay, a particular focus of restoration efforts is the reduction of nutrient over-enrichment, or eutrophication, from both point discharges and diffuse sources throughout the watersheds draining into the estuaries. However, efforts to reduce eutrophication will have to contend with multiple climate change-related problems such as sea-level rise, increased winter-spring discharges but reduced summer runoff, warmer sea surface temperatures, and greater shoreline erosion.

For marine fisheries, adaptations to changes in the production characteristics of exploited populations will include adjustments in the recommended harvest levels and/or exploitation rates and in the size or age at which fish and invertebrate populations are first harvested. The limiting level of exploitation (which is the rate at which the risk of population collapse is high) for a population is directly related to the rate of recruitment at low abundance levels. Thus, environmental changes that result in a reduc-

481

tion in recruitment rates must be countered by reductions in exploitation rates. Conversely, some higher levels of exploitation should be sustainable for some stocks under favorable environmental conditions. Adaptation to changes in the composition of fish stocks will also be necessary; regional markets will undoubtedly have to adjust to shifts in species composition due to changes in the availability of different species. Additionally, changes in distribution are likely to lead to more complex conflicts regarding the management of transboundary stocks and species.

Although much remains to be learned in order to confidently project impacts of climate change on coastal areas and marine resources, the trends and relationships already apparent suggest that the managers, decision makers, and the public must take climate change impacts into account. To be successful, this will have to be done in the context of coastal and resource management challenges already being addressed, for example:

- Strategic adaptation of coastal communities (e.g., barrier islands and other low-lying areas) to sea-level rise and increased storm surge.
- Adaptive management of coastal wetlands to improve their prospects of soil building to keep up with sea-level rise and allow their migration over adjacent lowlands.
- Comprehensive and forward-looking water use and management policies that factor in requirements for coastal ecosystems, such as reduced nutrient and pollutant delivery.
- Control procedures to reduce the risk of invasions by non-indigenous species.
- Fishery management regimes that incorporate knowledge of fluctuations in productivity and populations resulting from varying modes of climatic variability.
- Controls on impacts such as runoff from land and unsustainable fishing pressures that reduce the resilience of coral reef ecosystems.

In general, many of the strategies which might be appropriate for coping with future climate change have been or are currently being discussed or implemented in response to current stressors on coastal and marine environments. The future impacts of climate will be deeply integrated with the ongoing impacts resulting from human activities; thus, attempts to manage or mitigate the effects of climate must be tightly coupled with management of human behavior at all spatial and temporal scales. Most importantly, those who are or will likely be affected by climate impacts must be made aware of

the risks and potential consequences that future change will pose to their communities and their livelihoods.

CRUCIAL UNKNOWNS AND RESEARCH NEEDS

Direct human impacts dominate the changes occurring in most coastal areas, making it difficult to separate climatic effects from these direct stresses. In most areas, study of climate-related coastal impacts is only beginning. Consequently, for nearly all coastal and marine ecosystems considered in this Assessment, knowledge is inadequate to assess potential impacts fully and to determine effective adaptations. The Assessment has identified six important areas for research related to climate impacts on coastal and marine systems.

Coastal Hazards and the Physical Transformations of Coastlines and Wetlands. The significant erosion of beach fronts, barrier islands, and coastal marshes, coupled with accelerated sea-level rise, increases the vulnerability of coastal life and property to storm surge. Regardless of projected changes in the frequency and severity of coastal storms (hurricanes and nor'easters), storms will be riding on a higher sea level in the future. Research and assessments are needed to fully evaluate the vulnerability of human and natural coastal systems to the combined effects of sea-level rise, land subsidence, and storm surge. This information is required for rational responses in coastal protection, setbacks, and mitigation approaches to sustaining coastal wetlands.

Changes in Freshwater Loads to Coastal Ecosystems. Because of the importance of changes in land-use patterns and freshwater inflow to coastal ecosystems, particularly estuaries and wetlands, and to key species like Pacific salmon, considerable effort is needed to improve assessments of the impact of changes in the extent and timing of freshwater runoff. While contemporary GCM estimates of potential runoff vary widely, it is clear that changes are likely to occur and that the impacts could be substantial. Thus, new research is needed to assess the consequences of changes in runoff and the attendant changes in nutrient, contaminant, and sediment supply, circulation, and biological processes.

Decline of Coral Ecosystems. The decline of coral ecosystems is significant and global.

to this decline include changes in ocean temperatures, levels of atmospheric CO_2, and a series of more direct anthropogenic stress (e.g., over-fishing, eutrophication, and sedimentation). Increased effort is needed to adequately understand and predict the cumulative effects of these multiple stresses on coral ecosystems. It is important in this work to recognize the significance of the full ecosystem (including e.g., sand beds, sea grasses, and the water column) associated with corals, and not only the coral reefs alone.

Alterations and Geographic Shifts in Marine Ecosystems. Changes in ocean temperature and circulation (e.g., ENSO, PDO, and NAO), coupled with changes in nutrient supplies (driven by changes in freshwater fluxes and arctic ice dynamics), are likely to modify patterns of primary productivity, the distribution and recruitment success of marine fish, the reproductive success of protected species, and the economic viability of marine fisheries. While research on environmental variability and marine ecosystems is advancing (e.g., in GLOBEC), the current effort is limited to relatively few important regional ecosystems. More research is needed to understand and predict potential changes and regional shifts for all important coastal and US marine ecosystems, including the socioeconomic impacts to fishing communities.

Loss of Arctic Sea Ice. Loss of sea ice in Arctic regions will have widespread regional impacts on coastal environments and marine ecosystems. Recent dramatic reductions in the extent of sea ice in the Arctic Ocean and Bering Sea have led to more severe storm surges because the larger open water areas are capable of generating much larger waves. This has led to unprecedented erosion problems both for Native villages and for oil and gas extraction infrastructure along the Beaufort Sea coast. Reductions in sea ice also result in a loss of critical habitat for marine mammals such as walrus and polar bears, and significant changes in the distribution of nutrients supporting the base of the food web. Research to better understand how changing ice regimes will affect the productivity of polar ecosystems, and to assess the long-term consequences of these impacts is essential, both to sustain marine ecosystems and to develop coping strategies for the Native communities that depend on hunting for their food and other aspects of their culture.

LITERATURE CITED

Alexander, C. E., Classified shellfish growing waters, in NOAA's *State of the Coast Report*, National Oceanic and Atmospheric Administration (NOAA), Silver Spring, Maryland, 1998. URL: http://state-of-coast.noaa.gov/topics/html/state.html

Boesch, B. F., M. N. Josselyn, A. J. Mehta, J. T. Morris, W. K. Nuttle, C. A. Simenstad, and D. J. P. Swift, Scientific assessment of coastal wetland loss, restoration, and management in Louisiana, *Journal of Coastal Research,* Special Issue 20, 1994.

Bricker, S. B., C. G. Clement, D. E. Pirhall, S. P. Orlando, and D. R. G. Farrlow, National estuarine eutrophication assessment: A summary of conditions, historical trends, and future outlook, National Oceanic and Atmospheric Administration (NOAA), Silver Spring, Maryland, 1999.

Broecker, W. S., Thermohaline circulation, the Achilles Heel of our climate system: Will man-made CO_2 upset the current balance?, *Science, 278,* 1582-1588, 1997.

Broecker, W. S., The great ocean conveyor, *Oceanography, 4*(2), 79-89, 1991.

Broecker, W. S., S. Sutherland, and T. H. Peng, A possible 20th century slowdown of southern ocean deep water formation, *Science, 286,* 1132-1135, 1999.

Buddemeier, R. W., and S. V. Smith, Coral adaptation and acclimatization: A most ingenious paradox, *American Zoologist, (39)*1, 66-79, 1999.

California Department of Fish and Game, Pacific sardine resource now officially recovered, press release, January 15, 1999.

Cane, M. A., A. C. Clement, A. Kaplan, Y. Kushnir, D. Pozdnyakov, R. Seager, S. E. Zebiak, and R. Murtugudde, Twentieth-century sea surface temperature trends, *Science, 275,* 957-960, 1997.

Costanza, R., The ecological, economic, and social importance of the oceans, *Ecological Economics, 31,* 199-213, 1999.

Costanza, R., et al., The value of the world's ecosystem services and natural capital, *Nature, 387,* 253-260, 1997.

Culliton, T. J., Population: Distribution, density and growth, NOAA state of the coast report, National Oceanic and Atmospheric Administration (NOAA), Silver Spring, Maryland, 1998. URL: http://state-of-coast.noaa.gov/topics/html/pressure.html

Dolan, R., F. Anders, and S. Kimball, Coastal erosion and accretion in US, *Geological Survey National Atlas* Department of Interior, US Geological Survey, Reston, Virginia, 1985.

Driscoll, N. W., and G. H. Haug, A short circuit in thermohaline circulation: A cause for Northern Hemisphere glaciation?, *Science, 282,* 436-438, 1998.

Environmental Protection Agency (EPA), Natural resource valuation: A report by the Nation's Estuary Program, 1997.

Environmental Protection Agency (EPA), National water quality inventory: 1996 Report to Congress, 1996. URL: http://www.epa.gov/305b/

Federal Emergency Management Agency, Projected impact of relative sea-level rise on the National Flood Insurance Program, Report to Congress, Federal Emergency Management Agency, Washington, DC, Federal Insurance Administration, 1991.

Frank, K. T., R. I. Perry, and K. F. Drinkwater, Predicted response of Northwest Atlantic invertebrate and fish stocks to CO_2-induced climate change, *Transactions of the American Fisheries Society, 119,* 353-365, 1990.

Gattuso, J. P., D. Allemand, and M. Frankignoulle, Photosynthesis and calcification at cellular, organismal, and community levels in coral reefs: A review on interactions and control by the carbonate chemistry, *American Zoologist, 39*(1), 160-183, 1999.

Good, J. W., Shore protection policy and practices in Oregon: An evaluation of implementation success, *Coastal Management, 22,* 335-352, 1994.

Gornitz, V. M., Sea-level rise: A review of past and near-future trends, *Earth Surface Processes and Landforms, 20,* 7-20, 1995.

Gornitz, V. M., R. C. Daniels, T. W. White, and K. R. Birdwell, The development of a coastal risk assessment database: Vulnerability to sea-level rise in the US Southeast, in *Coastal Hazards: Perception, Susceptibility, and Mitigation,* edited by C. W. Finkl, Jr., *Journal of Coastal Research,* Special Issue 12, pp. 327-338, 1994.

Griggs, G. B., and K. M. Brown, Erosion and shoreline damage along the Central California Coast: A comparison between the 1997-1998 and 1982-1983 ENSO winters, *Shore and Beach, 66*(3), 18-23, 1999.

Hargreaves, N. B., D. M. Ware, and G. A. McFarlane, Return of the Pacific sardine *(Sardinops sagax)* to the British Columbia coast in 1992, *Canadian Journal of Fisheries and Aquatic Sciences, 51,* 460-463, 1994.

Harwell, M. A., Science and environmental decision making in South Florida, *Ecological Applications, 8*(3), 580-590, 1998.

Henderson-Sellers, A., H. Zhang, G. Berz, K. Emanuel, W. Gray, C. Landsea, G. Holland, J. Lighthill, S.-L. Shieh, P. Webster, and K. McGuffie. Tropical cyclones and global climate change: A post-IPCC assessment *Bulletin of the American Meteorological Society,* 79:19-38. 1998.

Herbert, P. J., J. D. Jarrell, and M. Mayfield, The deadliest, costliest, and most intense hurricanes of this century (and other frequently requested hurricane facts), NOAA Technical Memorandum NWS TPC-1, National Hurricane Center 1996.

Hoegh-Guldberg, O., Climate change, coral bleaching, and the future of the world's coral reefs, *Marine and Freshwater Research, 50*(8), 839-866, 1999.

Holt, G., Marlin hooked in state's 'tropical' waters, *Seattle Post Intelligencer,* p. C1, Friday, September 5, 1997.

Houston, J. R., International tourism and US beaches, *Shore and Beach, 64*(3), 27-35, 1996.

Intergovernmental Panel on Climate Change (IPCC), *Climate Change 1995: Impacts, Adaptations, and Mitigation of Climate Change: Scientific-Technical Analysis,* Cambridge University Press, Cambridge, United Kingdom, 1996.

Jobling, M. 1996. Temperature and growth: Modulation of growth rate via temperature change, in *Global Warming: Implications for Marine and Freshwater Fish,* edited by C. M. Wood and D. G. MacDonald, pp. 225-253, Cambridge University Press. 1996..

Johannessen, O. M., E. V. Shalina, and M. W. Miles, Satellite evidence for an Arctic sea ice cover in transformation, *Science, 286,* 1937-1939, 1999.

Justic´, D., N. N. Rabalais, and R. E. Turner, Effects of climate change on hypoxia in coastal waters: A doubled CO_2 scenario for the northern Gulf of Mexico, *Limnology and Oceanography, 41,* 992-1003, 1996.

Kaminsky, G. M., P. Ruggeiero, and G. Gelfenbaum, Monitoring coastal change in Southwest Washington and Northwest Oregon during the 1997-1998 El Niño, *Shore and Beach, 66*(3), 42-51, 1999.

Karl, T. R., R. W. Knight, and N. Plummer, Trends in high-frequency climate variability in the twentieth century, *Nature, 377,* 217-220, 1995a.

Karl, T. R., R. W. Knight, D. R. Easterling, and R. G. Quayle, Indices of climate change for the United States, *Bulletin of the American Meteorological Society, 77,* 279-292, 1995b.

Kerr, E. A., Thermodynamic control of hurricane intensity, *Nature, 401,* 665-669, 1999.

Kleypas, J. A., R. W. Buddemeier, D. Archer, J. P. Gattuso, C. Langdon, and B. N. Opdyke, Geochemical consequences of increased atmospheric carbon dioxide on coral reefs, *Science, 284*(2), 118-120, 1999a.

Kleypas, J. A., J. W. McManus, and L. A. B. Menez, Environmental limits to coral reef development: Where do we draw the line?, *American Zoologist, 39,* 146-159, 1999b.

Knutson, T. R., R. E. Tuleya, and Y. Kurihara, Simulated increase of hurricane intensities in a CO_2-warmed climate, *Science, 279,* 1018-1020, 1998.

Komar, P. D., The 1997-1998 El Niño and erosion on the Oregon coast, *Shore and Beach, 66*(3), 33-41, 1999.

Knutson, R. R., R. E. Tuleya, and Y. Kurihara, Simulated increase of hurricane intensities in a CO_2 warmed climate, *Science, 279,* 1018-1020, 1998.

Kronman, M., Market report: Year in review, *National Fisherman,* p. 40, April 1999.

Landsea, C. W., N. Nicholls, W. M. Gray and L. A. Avila, Downward trends in the frequency of intense Atlantic hurricanes during the past five decades, *Geophysical Research Letters, 23,* 1697-1700, 1996.

Levitus, S., J. I. Antonov, T. P. Boyer, and C. Stephans, Warming of the world ocean, *Science, 287,* 2225-2229, 2000.

Lluch-Belda, D. S., H. Vasquez, D. B. Lluch-Cota, C. A. Salinas-Zavela, and R. A. Schwartzlose, The recovery of the California sardine as related to global change, *California Cooperative Oceanic Fisheries Investigations Reports, 33,* 50-59, 1992.

Macklin, S. A., Report on the Fisheries Oceanography Coordinated Investigations International Workshop on Recent Conditions in the Bering Sea, National Oceanic and Atmospheric Administration ERL. Contribution No. 2044 from NOAA/Pacific Marine Environmental Laboratory, Contribution B358 from Fisheries-Oceanography Coordinated Investigations. 1999.

Mantua, N. J., S. R. Hare, Y. Shang, J. M. Wallace, and R. C. Francis, A Pacific interdecadal climate oscillation with impacts on salmon production, *Bulletin of the American Meteorological Society, 78,* 1069-1079, 1997.

Markham, A., Potential impacts of climate change on ecosystems: A review of implications for policymakers and conservation biologists, *Climate Research, 6*(2), 179-191, 1996.

Maslanik, J. A., M. C. Serreze, and T. Agnew, On the record reduction in 1998 Western Arctic Sea-ice cover, *Geophysical Research Letters, 26,* 1905-1908, 1999.

Mathews-Amos, A., and E. A. Berntson, Turning up the heat: How global warming threatens life in the sea, Marine Conservation Biology Institute, 1999. URL: http://www.mcbi.org/

McGowan, J. A., D. R. Cayan, and L. M. Dorman, Climate, ocean variability, and ecosystem response in the Northeast Pacific, *Science, 281,* 210-217, 1998.

Miller, M. L., and J. Auyong, Coastal zone tourism: A potent force affecting environment and society, *Marine Policy, 15*(2), 75-99, 1991.

Mitsch, W. J., and J. G. Gosslink, *Wetlands: Second edition,* Van Norstrand Reinhold, New York, 1993.

Murowski, S. A., Climate change and marine fish distributions: Forecasting from historical analogy, *Transactions of the American Fisheries Society, 122,* 657-658, 1993

National Marine Fisheries Service. 1998. Status of Fisheries of the United States: A Report to Congress. National Oceanic and Atmospheric Administration, Silver Spring, MD. 88pp.

National Research Council, *Sustaining Marine Fisheries,* National Academy Press, Washington, DC, 1999.

National Research Council, *Striking a Balance: Improving Stewardship of Marine Areas,* National Academy Press, Washington DC, 1997.

Neumann, J. E., G. Yohe, R. Nicholls, and M. Maino, *Sea-level rise and global climate change: A Review of Impacts to US Coasts,* Pew Center on Global Climate, 2000. URL: http://www.pewclimate.org/projects/env_sealevel.html

National Oceanic and Atmospheric Administration (NOAA), Year of the Ocean discussion papers, 1998.

NPA Data Services, Inc., Analytic documentation of three alternate socioeconomic projections, 1997-2050, Washington, DC, May 1999.

Oppenheimer, M., Global warming and the stability of the West Antarctic ice sheet, *Nature, 393,* 325-332, 1998.

Pielke, R. A., Jr., and C. W. Landsea, La Niña, El Niño, and Atlantic hurricane damages in the United States, *Bulletin of the American Meteorological Society, 80*(10), pp. 2027-2033, 1999.

Pugh, D. T., and G. A. Maul, Coastal sea level prediction for climate change, in *Coastal and Estuarine Studies 56: Coastal Ocean Prediction,* edited by C.N.K. Mooers, American Geophysical Union, Washington D.C. pp. 377-404, 1999.

Rabalais, N. N., R. E. Turner, D. Justic´, Q. Dortch, and W.J. Wiseman, Jr., Characterization of hypoxia, *NOAA Coastal Ocean Program Decision Analysis Series No. 15,* National Oceanic and Atmospheric Administration, Silver Spring, Maryland, 1999.

Reed, D.J. The response of coastal marshes to sea-level rise: Survival of submergence. *Earth Surface Processes and Landforms,* 20:39-48. 1995.

Reid, W. V., and M. C. Trexler, Responding to potential impacts of climate change on US coastal biodiversity, *Coastal Management, 20*(2), 117-142, 1992.

Rice, J., Food web theory, marine food webs, and what climate change may do to northern marine fish populations, in Climate Change and Northern Fish Populations, edited by R. J. Beamish, *Canadian Special Publication of Fisheries and Aquatic Sciences, 121,* 1995.

Rothrock, D. A., Y. Yu, and G. A. Maykut, Thinning of the Arctic Sea-ice cover, *Geophysical Research Letters, 26*(23), 3469-3472, 1999.

Sagarin, R. D., J. P. Barry, S. E. Gilman, and C. H. Baxter, Climate-related change in an intertidal community over short and long time scales. *Ecological Monographs, 69*(4), 465-490, 1999.

Sarmiento, J. L., and C. Le Quere, Oceanic carbon dioxide uptake in a model of century-scale global warming, *Science, 274,* 1346-1350, 1996.

Schmittner, A., and T. F. Stocker, The stability of the thermohaline circulation in global warming experiments, *Journal of Climate, 12,* 1117-1134, 1999.

Strong, A. E., E. J. Kearns and K. K. Gjovig, Sea surface temperature signals from satellites: An update, *Geophysical Research Letters, 27*(11), 1667-1670, 2000.

Taylor, K., Rapid climate change, *American Scientist, 87,* 320-327, 1999.

Timmermann, A., J. Oberhuber, A. Bacher, M. Esch, M. Latif, and E. Roeckner, Increased El Niño frequency in a climate model forced by future greenhouse warming, *Nature, 398,* 694-696, 1999.

Titus, J. G., and V. K. Narayanan, The risk of sea-level rise, *Climatic Change, 33,* 151-212, 1996.

US Army Corps of Engineers, Central and Southern Florida Project (C&SF) comprehensive review study, Draft integrated feasibility report and programmatic environmental impact statement, US Army Corps of Engineers, Jacksonville District, Jacksonville, Florida, 1998.

Veron, J. E. N., *Corals in Space and Time.* Cornell University Press, Ithaca 1995.

Vinnikov, K. Y., A. Robock, R. J. Stouffer, J. E. Walsh, C.L. Parkinson, D. J. Cavalieri, J. F. B. Mitchell, D. Garrett, and V. F. Zakharov, Global warming and Northern Hemisphere sea ice extent, *Science, 286,* 1934-1937, 1999.

Vitousek, P. M., J. Aber, S. E. Bayley, R. W. Howarth, G. E. Likens, P. A. Matson, D. W. Shindler, W. H. Schlesinger, and G. D. Tilman, Human Alteration of the global nitrogen cycle: Sources and consequences, *Ecological Applications, 7*(3), 737-750, 1997.

Welch, D. W., Y. Ishida, and K. Nagasawa, Thermal limits and ocean migrations of sockeye salmon (*Onchorhynchus nerka*): Long term consequences of global warming, *Canadian Journal of Fisheries and Aquatic Sciences, 55,* 937-948, 1998.

Wigley, T. M. L., *The Science of Climate Change: Global and US Perspectives,* Pew Center on Global Climate Change, Arlington, Virginia, 1999.

Wilkinson, C., O. Linden, H. Cesar, G. Hodgson, J. Rubens, and A. E. Strong, Ecological and socioeconomic impacts of 1998 coral mortality in the Indian Ocean: An ENSO impact and a warning of future change?, *Ambio, 28*(2), 188-196, 1999.

Wilkinson, C. R., and R. W. Buddemeier, Global climate change and coral reefs: Implications for people and reefs, Report of the UNEP-IOC-ASPEI-IUCN Global Task Team on the Implications of Climate Change on Coral Reefs, International Union for the Conservation of Nature. Gland, Switzerland, 1994.

Wolock, D.M. and G.J. McCabe. Simulated effects of climate change on mean annual runoff in the conterminous United States. *Journal of the American Water Resources Association, 35*: 1341-1350, 1999.

Wong, J., Thousands of seals, sea lions starving, *Seattle Times,* pp. A8, Thursday, December 11, 1997.

Wood, R. A., A. B. Keen, J. F.B. Mitchell, and J. M. Gregory, Changing spatial structure of the thermohaline circulation in response to atmospheric CO_2 forcing in a climate model, *Science, 399,* 572-575, 1999.

Yohe, G., J. Neumann, P. Marshall, and H. Ameden, The economic cost of greenhouse induced sea-level rise for developed property in the United States, *Climatic Change, 32,* 387-410, 1996.

Zhang, K., et al., East Coast storm surges provide unique climate record, *Eos, Transactions of the American Geophysical Union, 78*(37), 389, 396-97, 1997.

ACKNOWLEDGMENTS

Special thanks to:

Jim Allen, Paul Smith College
Donald Cahoon, US Geological Survey
Benjamin Felzer, National Center for Atmospheric Research
Michael MacCracken, US Global Change Research Program
LaShaunda Malone, US Global Change Research Program
Melissa Taylor, US Global Change Research Program
Rob Thieler, US Geological Survey
Elizabeth Turner, National Oceanic and Atmospheric Administration
Justin Wettstein, US Global Change Research Program

CHAPTER 17

POTENTIAL CONSEQUENCES OF CLIMATE VARIABILITY AND CHANGE FOR THE FORESTS OF THE UNITED STATES

Linda Joyce[1,2], John Aber[3], Steve McNulty[1], Virginia Dale[4], Andrew Hansen[5], Lloyd Irland[6], Ron Neilson[1], and Kenneth Skog[1]

Contents of this Chapter

Chapter Summary

Background

Climate and Forests

Key Issues

> Forest Processes

> Forests and Disturbance

> Biodiversity

> Socioeconomic Impacts

Adaptations: Forest Management Strategies under Climate Change

Crucial Unknowns and Research Needs

Literature Cited

Acknowledgments

[1]USDA Forest Service; [2]Coordinating author for the National Assessment Synthesis Team; [3]University of New Hampshire, [4]Oak Ridge National Laboratory; [5]Montana State University, [6]The Irland Group

CHAPTER SUMMARY

Context

Forests cover nearly one-third of the US, providing wildlife habitat, clean air and water, cultural and aesthetic values, carbon storage, recreational opportunities such as hiking, camping, fishing, and autumn leaf tours, and products that can be harvested such as timber, pulpwood, fuelwood, wild game, ferns, mushrooms, and berries. This wealth depends on forest biodiversity—the variety of plants, animals, and microbe species, and forest functioning—water flow, nutrient cycling, and productivity. These aspects of forests are strongly influenced by climate and human land use.

Key Findings

Carbon storage in US forests is currently estimated to increase from 0.1 to 0.3 Pg of carbon per year and analyses suggest that carbon dioxide fertilization and land use have influenced this current storage. Within the next 50 years, forest productivity is likely to increase with the fertilizing effect of atmospheric carbon dioxide. Those productivity increases are very likely to be strongly tempered by local environmental conditions (e.g., moisture stress, nutrient availability) and by human land use impacts such as forest fragmentation, increased atmospheric deposition, and tropospheric ozone.

Economic analyses when driven by several different climate scenarios indicate an overall increase in forest productivity in the US that is very likely to increase timber inventory, subject to other external forces. With more potential forest inventory to harvest, the costs of wood and paper products to consumers are likely to decrease, as are the returns to owners of timberland. The changes in climate and consequent impact on forests are likely to change market incentives to harvest and plant trees, and shift land uses between agriculture and forestry. These changes will likely vary within a region. Market incentives for forestry are likely to moderate some of the climate-induced decline in the area of natural forests. International trade in forest products could either accentuate or dampen price effects in the US, depending on whether forest harvest activity increases or decreases outside the US.

Over the next century, changes in the severity, frequency, and extent of natural disturbances are possible under future climate change. These changes in natural disturbances then impact forest structure, biodiversity, and functioning. Analyses of the results from climate and ecological models suggest that the seasonal severity of fire hazard is likely to increase by 10% over much of the US, with possibly larger increases in the southeastern US and Alaska and actual decreases in the Northern Great Plains. Although the interactions between climate change and hurricanes, landslides, ice storms, wind storms, insects, disease, and introduced species are difficult to predict; climate changes, changes in these disturbances and their effects on forests are possible.

Analyses of the results of ecological models when driven by several different climate scenarios indicate changes in the location and area of potential habitats for many tree species and plant communities. For example, alpine and subalpine habitats and the variety of species dependent upon them are likely to be greatly reduced in the conterminous US. The ranges of some trees are likely to contract dramatically in the US and largely shift into Canada. Expansion of potential habitats is possible for oak/hickory and oak/pine in the eastern US and Ponderosa pine and arid woodland communities in the West. How well plant and animal species adapt to or move with changes in their potential habitat is strongly influenced by their dispersal abilities and the disturbances to these environments. Introduced and invasive species that disperse rapidly are likely to find opportunities in newly forming communities.

The effects of climate change on socioeconomic benefits obtained from forests will likely be influenced greatly by future changes in human demands, as determined by population growth, increases in income, changing human values, and consumer preferences.

Outdoor recreation opportunities are very likely to be altered by climate change. For example, warmer waters would increase fish production and opportunities for some species; but decrease opportunities for cold water species. Outdoor recreation opportunities in mountainous areas are likely to be altered. Summer recreation opportunities are likely to expand in the mountainous areas when warmer lowland temperatures attract more people to higher elevations. Skiing opportunities are likely be reduced with fewer cold days and snow events. In marginal climate areas, the costs to maintain downhill skiing opportunities are likely to rise which would possibly result in the closure of some areas.

POTENTIAL CONSEQUENCES OF CLIMATE VARIABILITY AND CHANGE FOR THE FORESTS OF THE UNITED STATES

BACKGROUND

Covering nearly one-third of the US, forests are an integral part of the vegetative cover of the nation's landscape (Figure 1). The total area in forests in the US is 747 million acres, which is about 7% of the forestland in the entire world. US forests are distributed in the eastern and western parts of the US, with small stands of forests in the central part located mainly along rivers and streams. The white-red-jack pine forests of New England have supported a timber industry and the northeastern deciduous forests of maples, beeches, birches, and oaks provide the colorful autumn landscapes for tourists. The Mid-Atlantic region is rich in tree species from pine forests and coastal wetlands to northern upland hardwoods such as oak-hickory and maple-beech-birch forest types. The southern forests are also a mix of conifer and deciduous forests. Western forests are primarily conifer forests such as Douglas-fir, fir-spruce, Ponderosa pine and piñon-juniper.

Ownership of forestland varies by region in the US. Over 63% of US forestland is in private ownership.

Current Distribution of Forests in the United States

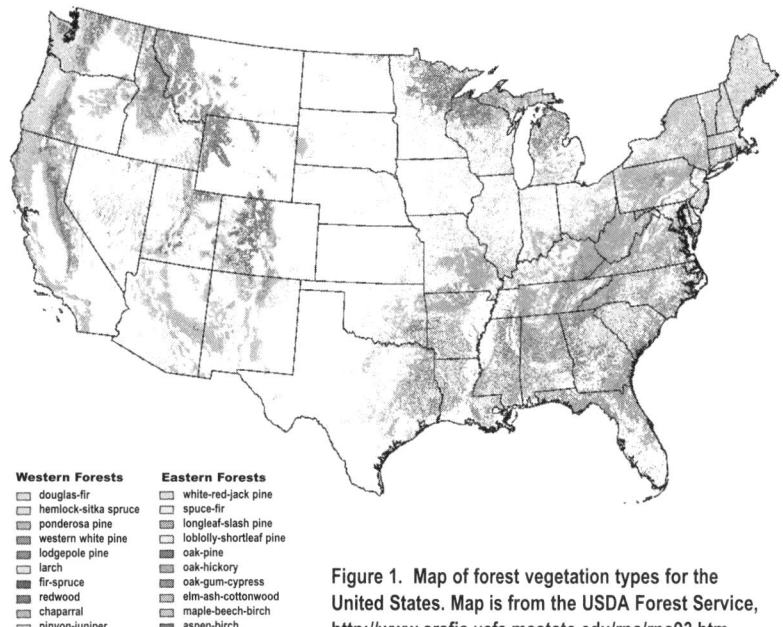

Western Forests
- douglas-fir
- hemlock-sitka spruce
- ponderosa pine
- western white pine
- lodgepole pine
- larch
- fir-spruce
- redwood
- chaparral
- pinyon-juniper
- western hardwoods
- aspen-birch

Eastern Forests
- white-red-jack pine
- spuce-fir
- longleaf-slash pine
- loblolly-shortleaf pine
- oak-pine
- oak-hickory
- oak-gum-cypress
- elm-ash-cottonwood
- maple-beech-birch
- aspen-birch

Figure 1. Map of forest vegetation types for the United States. Map is from the USDA Forest Service, http://www.srsfia.usfs.msstate.edu/rpa/rpa93.htm. See Color Plate Appendix

The remaining forestland is in various federal, state, county, and municipal ownerships. Over 50% of the federal land in forests is managed by the US Forest Service. The largest state owner of forestland is Alaska. County and municipal ownerships comprise less than 4% of the total forestland. However, over 2.5 million acres of forestland are managed by these local governments in Minnesota alone. Over 80% of the forestland in federal ownership is found in the west, while most of the forestland held by states is in the northern part of the US. Of the 472 million acres in private ownership, over 60% is found in the eastern part of the US. Forest industry owners account for 14% of the forestland in private ownership, and these lands are found mainly in the southern part of the US (54%).

Forests are an environment in which people recreate, such as the National Forests and Parks, and an environment in which people live, such as the New England woods, and the conifer forests of Rocky Mountains. Forests provide recreational opportunities such as hiking, camping, fishing, hunting, bird watching, downhill and cross-country skiing, and autumn leaf tours. In addition, many rivers and streams flowing through forests provide fishing, boating, and swimming recreational opportunities. Activities associated with recreation provide income and employment in every forested region of the US and Canada (Watson et al., 1998). Forests provide clean air and water, watershed and riparian buffers, moderate streamflow, and help to maintain aquatic habitats. New York City's water supply is derived from water collected within forested watersheds in a 2,000 square mile area. Forests provide wildlife habitat. Flather et al. (1999) report that at least 90% of the resident or common migrant vertebrate species in the US rely on forest habitats for part of their life requisites. Forests also provide cultural and aesthetic values, carbon storage, and products that can be harvested such as timber, pulpwood, fuelwood, wild game, edible plants, fruits and nuts, mushrooms, and floral products. This wealth from forests depends on forest biodiversity – the variety of plant and animal species – and forest functioning—water flow, nutrient cycling and productivity. These aspects of forests are strongly influenced by climate.

Climate change is one of several pressures on forests encompassed under the broader term, global change. Human activities have altered the vegetation distribution of forests in the US. The arrival of Europeans along the eastern coast initiated the harvest of forests. For example, eastern white pines were highly prized as ship masts for the English Navy. From 1600 to the mid 1800s, the native forests in the eastern US were extensively harvested for wood products as well as to clear land for agriculture and urban uses (Figure 2). While total forestland area has stabilized in the US since the early 1900s, land use shifts still occur where forestland is converted to agricultural or urban use and agricultural or pasture is planted to trees. In some parts of the eastern US, new forests have regrown on abandoned agricultural lands, although forestland in the northeastern part of the US is still less than 70% of its original extent in the 1600s. Forestland in the South occupies less then 60% of the 1600s extent of forestland, and the establishment of pine plantations has placed some 20% of the forestland in this region under intensive management. While western forests have remained relatively constant in area since the 1600s, recent expansion of urban areas and agriculture is fragmenting them. Across the Nation, urban areas have continued to increase (Flather et al., 1999). This expansion of human influences into the rural landscape alters disturbance patterns associated with fire, flooding, and landslides. In addition, human activities can result in the dispersal of pollutants into forests. Increasing atmospheric concentrations of ozone and deposition of nitrogen (N) compounds have profound effects on tree photosynthesis, respiration, water relations, and survival.

Human activities modify the species composition of forests and the disturbance regimes associated with forests. Fire suppression has altered southeastern, midwestern, and western forests. Harvesting methods, such as clearcutting, shelterwood, or individual tree selection, have changed species composition in native forests. The average age of forest stands in the East is less than the average for the West, reflecting the extensive harvesting as the eastern US was settled. Intensive management along with favorable climates in parts of the US has resulted in highly productive forests that are maintained for timber production, such as the southern pine plantations. Native and introduced insects and disease species, such as gypsy moth, chestnut blight, Dutch elm disease, have altered US forests (Ayres and Lombardero, 2000). Trees have been planted far outside their natural ranges for aesthetic and landscaping purposes in urban and rural areas.

Forest Land Coverage over the Past 400 Years

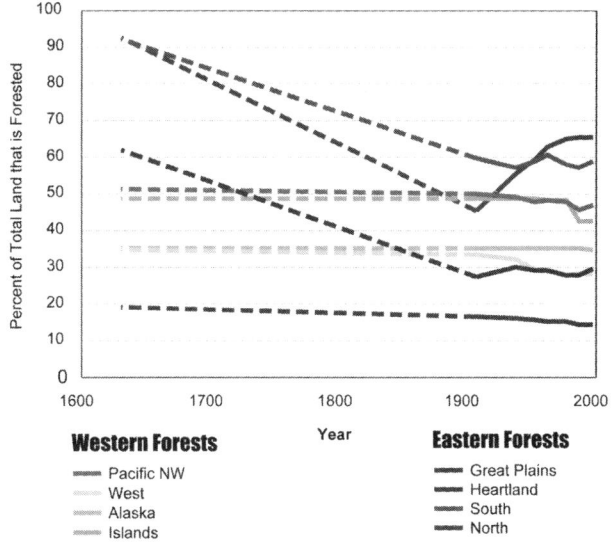

Figure 2. Land area changes in forestland. Data are from Forest Service Resource Bulletin PNW-RB-168, Forest Resource Report No. 23, No. 17, No. 14, the Report of the Joint Committee on Forestry, 77th Congress 1st Session, Senate Document No. 32. Data for 1850 and 1870 were based on information collected during the 1850 and 1870 decennial census; data for 1907 were also based on the decennial census modified by expert opinon, reported by R.S. Kellogg in Forest Service Circular 166. Data for 1630 were included in Circular 166 as an estimate of the original forest area based on the current estimate of forest and historic land clearing information. These data are provided here for general reference purposes only to convey the relative extent of the forest estate in what is now the US at the time of European settlement. See Color Plate Appendix

Population levels, economic growth, and personal preferences influence the socio-economic values associated with forests, and consequently the resources demanded from forests. Per capita consumption of wood in the US has been relatively stable over the last several decades and the future demand for wood products is projected to follow population growth over the next 50 years. Technological development and consumer preferences strongly affect the demand for specific wood and paper products. For example, consumer preference has influenced the increasing amount of recycled material used in fiber over the last several years. Though recreational hunting has been declining, the economic impact is significant; Flather et al. (1999) estimated that, for all wildlife, hunters alone spent nearly 21 billion dollars on equipment and travel in 1996. Other products harvested from forests are more difficult to assign an economic value or to project future demand. Blatner (1997) estimates the harvest of mushrooms from Washington, Oregon, and Idaho forests in 1992 to be valued at about $40 million.

At the global scale, population, economic growth, and personal preferences influence the demands from forests. Wood consumption rose 64% globally since 1961. This increase was strongly influenced by rising wood per capita consumption for fuelwood, paper products, and industrial fiber (Matthews and Hammond, 1999). More than half of the wood fiber produced globally is consumed as fuel in contrast to the US, where fuelwood comprises around 14% of total wood fiber consumed (Brooks, 1993; Matthews and Hammond, 1999). The demands on forests globally are expected to change as the world's populations become increasingly urban. More industrial products and environmental services are likely to influence the management of the world's forests (Brooks, 1993).

CLIMATE AND FORESTS

Climate influences the composition, structure, and function of forest ecosystems, the amount and quality of forest resources, and the social values associated with forests. Native forests are adapted to local climatic features. Where summer drought is typical in the Pacific Northwest, native conifer and hardwood forests have water-conserving leaves (Shriner et al., 1998). Black spruce and white spruce are found in the cold-tolerant boreal forests where winter temperature extremes can reach –62° to –34°C (-79° to –30°F) (Burns and Honkala, 1990). The piñon-juniper forests of the Southwest are drought-adapted.

Changes in the distribution and abundance of plant or animal species reflect the birth, growth, death, and dispersal rates of individuals in a population. Climate and soil are strong controls on the establishment and growth of plants. Climate influences the distribution and abundance of animal species through changes in resource availability, fecundity, and survivorship (Hansen and Rotella, 1999). Changes in disturbance regimes, and competitive and cooperative interactions with other species also affect the distribution of plants and animals. In addition, human activities influence the occurrence and abundance of species on the landscape.

Temperature and Precipitation. The spatial and temporal distribution of water and temperature are the primary determinants of woody plant distributions over the Earth. Air temperature affects physiological processes of individual plants, and over the long-term, the environmental conditions for seed development and germination, population, and com-

munity development. Temperature affects fruit and seed yields and quality by influencing factors such as flowering, bud dormancy, and ripening of fruit and cones (Kozlowski and Pallardy, 1997). Changes in air temperature in autumn and spring can also affect hardening and dehardening of tree needles (Guak et al., 1998). Temperature affects ecosystem-level processes such as soil decomposition and mineralization. Indirect effects associated with warming could be larger than the direct effect on plant growth in subpolar biomes where warming of permafrost is likely (Mooney et al., 1999).

Shortages or excesses of water offset the rates of most important processes controlling the biogeochemistry of the major nutrients. In particular, forested wetlands are sensitive to changes in hydrologic regimes. While forest ecosystems generally occupy those regions with low annual water deficits, water limitation in space and time is still critical for overall carbon balances, and is one of the major drivers embedded in the models used to predict global change effects.

KEY ISSUES

The vulnerability of forests to climate variability and change is examined by looking at four key issues: forest processes, disturbance, biodiversity, and socioeconomic benefits.

- Forest processes regulate the flux and apportionment of carbon, water, nutrients, and other constituents within a forest ecosystem. These processes operate at spatial scales from leaf to landscape and control responses of forest ecosystems, such as forest productivity, to environmental factors such as temperature, precipitation, and atmospheric concentrations of CO_2.
- Forests are subjected to **disturbances** that are themselves strongly influenced by climate. These natural disturbances include fire, hurricanes, landslides, ice storms, wind storms, drought, insects, disease, and introduced species.
- **Biodiversity** refers to the variety of populations, species, and communities. Climate influences the distribution and abundance of plant and animal species through food availability, habitat availability, and survivorship.
- Forest processes and forest biodiversity are uniquely capable of providing goods such as wood products, wild game, and harvested plants, ecological services such as water purification, and amenities such as scenic vistas and wilderness experiences—the **socioeconomic benefits of forests.**

Changes in these goods, services, and amenities are influenced by changes in factors that determine their supply — land area, productivity, management, production technology, quality of amenities — and their demand — human population levels, economic growth, and personal preferences.

1. Forest Processes

Current Environmental Changes Include Deposition of Nitrogen and Ozone.

These environmental changes influence the ability of forests to respond to changes in climate (Aber et al., 2001). Tree species have been shown to be sensitive to air pollutants such as ozone and sulfur (S) (Fox and Mickler, 1996; Taylor et al., 1994; US EPA, 1996) and nitrogen (N) deposition has been linked to soil acidification and cation depletion in forests (Aber et al., 1995; Aber et al., 1998). Total deposition of N and associated acidity in the US have increased as much as 10-fold over global background levels as a result of human activity (Galloway, 1995; Vitousek et al., 1997; Matthews and Hammond, 1999). Combustion of fossil fuels injects N and S into the atmosphere as simple oxides. The N and S oxides are retained in the atmosphere only for days to weeks, whereas carbon dioxide (CO_2) is retained for decades. Some N and S compounds can be re-deposited on the forest surface either in a dry form or, by dissolution into water, in a wet form ("acid rain") (Boubel et al., 1994). The shorter residence time of N and S in the atmosphere results in an intensely regional distribution of deposition, with the eastern US, and especially the Northeast, experiencing the highest levels of both S and N (NADP/NTN, 1997).

With the large reduction in S deposition in the last decade from the controls imposed by the Clean Air Act of 1990, the importance of the acids in precipitation has been reduced. Sources of N vary regionally. Urban areas contribute to increased N deposition in pine forests in southern California (Fenn et al., 1996). Nitrogen deposition as ammonium occurs where fertilizer is applied intensively or where livestock are concentrated in feedlots (Lovett, 1994). Agricultural lands and feedlots along the Front Range of the Rocky Mountains in Colorado contribute to increased N deposition in the alpine and forest ecosystems in Rocky Mountain National Park (Baron et al., 1994; Musselman et al., 1996; Williams et al., 1996).

High levels of N deposition can result in negative effects such as soil acidification, causing depletion of nutrient cations (calcium, magnesium, and potassium). Low availability of N in the soil limits forest production. Increases in forest growth in response to N deposition have been reported both in Scandinavia and the US, although the response varies between deciduous and coniferous species (Magill et al., 1997; Aber et al., 1998; Magill et al., 1999).

Ozone is a highly reactive gas. Closely associated with the combustion of fossil fuels, ozone is formed in the lower atmosphere (ground-level ozone) through chemical reactions between nitrogen oxides and hydrocarbons in the presence of sunlight. High levels of ozone occur, generally in summer, when warm, stagnant air masses over densely populated and highly industrialized regions accumulate large quantities of nitrogen oxides and hydrocarbons. Thus, the distribution of ozone concentrations is very irregular in space and time.

Ozone leaves the atmosphere through reactions with plants and soil surfaces along a number of physical and chemical degradation pathways. Ozone concentrations can dissipate in a matter of days when air masses move away from urban areas throughout the eastern US, or from western cities across more remote forested areas, such as from Los Angeles to the San Bernardino and San Gabriel Mountains, California. Ozone tends to remain in the atmosphere when it cannot react with material (e.g. vegetation or soils). Thus, ozone concentrations in eastern Maine can be as high as over urban Boston because ozone-bearing air masses have reached these remote areas with little loss of ozone as they passed over the ocean. High concentrations can occur in relatively remote mountaintop locations because ozone also accumulates at the top of the atmospheric boundary layer where contact with vegetated surfaces is minimal.

Unlike nitrogen and S deposition, which by their nature are slow and cumulative, the effects of ozone on vegetation are direct and immediate, as the primary mechanism for damage is through direct plant uptake from the atmosphere through stomata (small openings in the plant leaf through which water and gases pass into and out of the plant). Ozone is a strong oxidant that damages cell membranes; the plant must then expend energy to maintain these sensitive tissues. The net effect is a decline in net photosynthetic rate. The degree to which photosynthesis is reduced is a function both of dose and species conductance rates, the rate at which leaves exchange gases with the atmosphere (Musselman and Massman, 1999). Analyses suggest that current ozone levels have decreased production 10% in northeastern forests (Ollinger et al., 1997) and 5% in southern pine plantations (Weinstein et al., 1998). Warming of surface air, a consistent feature of the Hadley and Canadian scenarios used in this assessment, is likely to increase ozone and other air-quality problems (see Watson et al., 1996 for analysis of previous climate scenarios), further increasing the stress on forests in areas where air quality is compromised.

Impacts of Elevated Atmospheric Carbon Dioxide and Climate Change on Forest Processes

Experimental exposure of trees to elevated atmospheric CO_2 has shown significant changes in physiological processes, growth, and biomass accumulation (Aber et al., 2001; Mooney et al., 1999). Over a wide range of CO_2 concentrations, photosynthesis has been increased in plants representative of most northern temperate forests (Eamus and Jarvis, 1989; Bazzaz, 1990; Mohren et al., 1996; Long et al., 1996; Kozlowski and Pallardy, 1997). Reviews of the extensive CO_2-enrichment studies have shown variable but positive responses in plant biomass accumulation (Ceulemans and Mousseau, 1994; Saxe et al., 1998; Mooney et al., 1999). In a review of studies not involving environmental stress, biomass accumulation was greater for conifers (130%) than for deciduous species (only 49% increase) under elevated CO_2 (Saxe et al., 1998). The wide range of plant responses reflects, in part, the interaction of other environmental factors and the CO_2 response (Mooney et al., 1999; Stitt and Krapp, 1999; Morison and Lawlor, 1999; Johnson et al., 1998; Curtis and Wang, 1998). A recent field-scale experiment produced significant (25%) increases in forest growth under continuously elevated concentrations of CO_2 for loblolly pine in North Carolina (Delucia et al., 1999). While positive tree responses to elevated CO_2 are likely to be overestimated if abiotic and biotic environmental factors are not considered, most ecosystems responded positively in terms of net carbon uptake to increases in atmospheric CO_2 above the current ambient level.

A significant question is how long these increased responses can be sustained. The observed acclimation or down-regulation of photosynthetic rates (Long et al., 1996; Lambers et al., 1998; Rey and Jarvis, 1998) has been ascribed to a physiological response (accumulation of photosynthetic reserves, Bazzaz, 1990), or a morphological response (changes in trees, Pritchard et al., 1998; Tjoelker et al., 1998b). Declining photosynthetic rates have also been ascribed to the result of a water or nutrient stress imposed on pot-grown seedlings where root growth

is limited (Will and Teskey, 1997a; Curtis and Wang, 1998). Down-regulation has been shown in low-temperature systems such as the Arctic tundra, where the initial enhancement of net carbon uptake declined after 3 years of elevated CO_2 exposure (Shaver et al., 1992; Mooney et al., 1999). Down regulation is likely in areas where nutrient availability does not increase along with carbon dioxide. A recent review of large-scale field exposures to carbon dioxide suggested that, though variable, tree response to CO_2 was sustained over these short-term studies (Norby et al., 1999). They also examined the CO_2 response of trees growing near surface vents of deep geothermal springs, and concluded that a basal area increase of 26% was sustained through 3 decades of elevated carbon dioxide.

Under enriched CO_2 conditions, water use efficiency (WUE) has been shown to increase, which results in higher levels of soil moisture. These increased levels of soil moisture have been shown to be a significant factor in increased carbon uptake in water-limited ecosystems (Mooney et al., 1999). While there is still uncertainty as to whether stomatal conductance decreases under elevated CO_2 (Long et al., 1996; Will and Teskey, 1997b; Saxe et al., 1998; Curtis and Wang, 1998), WUE increases either with or without changes in stomatal conductance (Aber et al., 2001). With constant conductance, the higher atmospheric CO_2 concentration results in faster carbon (C) uptake with constant water loss. If conductance is reduced, a tradeoff is established between increased C gain (which is partially reduced by decreased conductance) and decreased water loss (also reduced by decreased conductance). Experimental studies have emphasized leaf-level responses. A physiological response observed at one scale (i.e., leaf) does not necessarily imply that a response will be observed at the larger scale (i.e., canopy, watershed). There is some evidence that the reduction in stomatal conduction of tree seedlings is not seen in mature trees (Ellsworth, 1999; Mooney et al., 1999; Norby et al., 1999).

In the near-term, changes in the physiology of plants are likely to be the dominant response to elevated carbon dioxide, strongly tempered by local environmental conditions such as moisture stress, nutrient availability, as well as by individual species responses (Egli and Korner, 1997; Tjoelker et al., 1998a; Berntson and Bazzaz, 1998; Crookshanks et al., 1998; Kerstiens, 1998). Over the long term, plant species changes within forests will have a large influence on the response of forests (Mooney et al., 1999).

Interactions of Multiple Stresses on Forests

It is crucial to understand not only the direct effects of CO_2, ozone, temperature, precipitation, and N and S deposition on forests, but also the interactive effects of these stresses. For example, if canopy conductance in forests is reduced in response to CO_2 enrichment, then ozone uptake is reduced and the effects of this pollutant mitigated. Drought stress has a similar effect by reducing stomatal conductance. However, if N deposition leads to increased N concentrations in foliage and hence higher rates of photosynthesis and increased stomatal conductance, then the positive effect of increased photosynthesis is partially offset by increased ozone uptake. Reductions in production from ozone damage could possibly speed the onset of N saturation in ecosystems and the attendant development of acidified soils and streams. To project the effects of climate change and other stresses on forests, their interactive effects on forest processes must be understood and integrated into ecological models.

Carbon Sequestration

Globally, an estimated 62-78% of the terrestrial carbon is in forests (Shriner et al., 1998). North American (Canada and US) forests hold 14% of this global total, with large amounts contained in the boreal forests. Estimates of the carbon sequestered annually as a result of climate and carbon dioxide fertilization in US forests were analyzed recently (Schimel et al., 2000) at a value of 0.08 Pg (Pg= Petagrams where 1 Pg = 10^{15} grams) of carbon per year. This analysis focused on the 1980-1993 period. Estimates of carbon sequestration based on forest inventory data were 0.28 Pg of carbon per year, and most of this increase in carbon storage was estimated to be on private timberland (Birdsey and Heath, 1995). Houghton et al. (1999) estimated the increase in carbon stored per year ranged from 0.15 to 0.35 Pg of carbon per year. Schimel et al. (2000) suggest that the larger inventory-based estimates imply that the effects of intensive forest management and agricultural abandonment on carbon uptake in the US are probably equal to or larger than the effects of climate and atmospheric carbon dioxide. A comprehensive approach to account for carbon fluxes to and from the atmosphere is the focus of the recent IPCC Special Report on Land Use, Land-Use Change, and Forestry (Watson et al., 2000).

Projecting the Impact of Climate Change on Forest Processes

A number of studies have examined the climate controls on the distribution and productivity of forests using different types of models (Goudriaan et al., 1999; McGuire et al., 1992; VEMAP members, 1995). In this assessment, three types of models were used to project the impact of climate change on forest processes. Biogeochemistry models simulate the gain, loss and internal cycling of carbon, nutrients, and water; with these models, the impact of changes in temperature, precipitation, soil moisture, atmospheric carbon dioxide, and other climate-related factors can be examined for their influence on such processes as ecosystem productivity and carbon storage. Biogeography models examine the influence of climate on the geographic distribution of plant species or plant types such as trees, grasses, and shrubs. Dynamic global vegetation models integrate biogeochemical processes with dynamic changes in vegetation composition and distribution.

An earlier analysis using biogeochemistry models and four climate scenarios (different from this assessment) shows increased net primary productivity (NPP) at the continental scale (VEMAP members, 1995). Carbon storage results vary with the modeled sensitivity to changes in water availability. When the direct effects of carbon dioxide are not included, several analyses indicate a reduction in productivity (biogeochemistry models: VEMAP members, 1995) or in vegetation density (biogeography model: Neilson et al., 1998).

In this assessment, three biogeochemistry models (TEM, Century, and Biome-BGC) and one dynamic global vegetation model (MC1) show increases in total live vegetation carbon storage (3 to 11 Pg C) for forest ecosystems within the conterminous US under the Hadley scenario (Table 1). Under the Canadian scenario, the models project changes in live vegetation carbon for forests from a reduction of 1.6 Pg to an increase of 11 Pg C (Pg= Petagrams where 1 Pg = 10^{15} grams). The MC1 model simulates a decline in vegetation carbon of about 2 Pg (Bachelet et al., 2001; Daly et al., 2000). Results for live vegetation in all ecosystems and for live and dead vegetation in both forest and all ecosystems parallel these results. Modeling the changes in species groups and fire disturbance are the primary reasons for the carbon responses in MC1.

In the MC1 projections, regional changes in carbon storage vary greatly across the conterminous US, and reflect the likelihood of disturbances such as

Table 1. Changes in Carbon Storage for the Conterminous US

These results are based on simulations by the VEMAP models (TEM, BIOME-BGC, CENTURY, MC1) using the two transient climate change scenarios described in the text. Baseline period is 1961-1990. Changes are given at decades centered on 2030 and 2095, and for forest ecosystems, and all ecosystems. Values are expressed in Pg (billions of metric tons).

	Hadley		Canadian	
	2030	2095	2030	2095
Forest Ecosystems				
Live Vegetation	0.4 to 4.0	3.0 to 11.1	-1.5 to 2.9	-1.6 to 10.5
Total (Live + Dead)	0.3 to 3.3	3.2 to 10.5	-2.9 to 2.0	-0.6 to 9.4
All Terrestrial Ecosystems				
Live Vegetation	0.4 to 4.7	3.3 to 13.2	-1.2 to 3.8	-1.8 to 13.9
Total (Live + Dead)	0.2 to 4.8	3.4 to 14.5	-0.2 to 4.4	-1.9 to 15.0

drought or fire (Figure 3). Under the Hadley scenario about 20% of current forest area experience some level of carbon loss, while the remaining 80% experience increased storage. Under the much warmer and generally drier Canadian scenario, close to 80% of current forest area experiences a drought-induced loss of carbon. Reductions in carbon storage are projected in MC1 to be especially severe in the eastern and southeastern US, where losses exacerbated by drought or fire approach 75%.

In summary, synthesis of laboratory and field studies and recent simulation experiments indicate that forest productivity increases with the fertilizing effect of atmospheric carbon dioxide and that these productivity increases are strongly tempered by local conditions such as moisture stress and nutrient availability. It is likely that modest warming could result in carbon storage gains in some forest ecosystems in the conterminous US. Under even warmer conditions, it is likely that drought-induced losses of carbon would occur in some forests, notably in the Southeast and the Northwest. These losses of carbon would possibly be enhanced by increased fire disturbance. These potential gains and losses of carbon are very likely to be subject to changing land-use patterns, such as the conversion of forests to other uses.

2. Forests and Disturbance

Natural Disturbances Impacted by Climate

Natural disturbances, impacted by climate, include insects, disease, introduced species, fires, droughts,

hurricanes, landslides, wind storms, and ice storms. Over geologic time, local, regional, and global changes in temperature and precipitation have influenced the occurrence, frequency, and intensity of these natural disturbances.

Impacts of disturbances are seen over a broad spectrum of spatial scales, from the leaf and tree to the forest and forested landscape. Disturbances can result in: leaf discoloration and reduction of leaf function; deformation of tree structure such as broken branches or crown losses; tree mortality or chronic stress resulting in tree death; altered regeneration patterns including losses of seed banks; disruption of physical environment including soil erosion; alterations in biomass and nutrient turnover; impacts on surface soil organic layers and the underground plant root and reproductive tissues; and increased landscape heterogeneity (patchiness of forest communities). Introduced species (invasives) can affect forest ecosystems through herbivory, predation, habitat destruction, competition, loss of gene pools through hybridization with native species, and disease (either causing or carrying disease). Outbreaks of native insects and disease can result in similar impacts on forests.

At the ecosystem scale, introduced species as well as outbreaks of native insects and diseases can alter natural cycles and disturbance regimes, such as nutrient cycles, and fire frequency and intensity (Mack and D'Antonio, 1998). Some tree species have developed adaptations to survive repeated occurrences of certain disturbances over time. Thick bark on some trees allows their survival in

Patterns of Live Vegetation for Different Times and Climate Scenarios

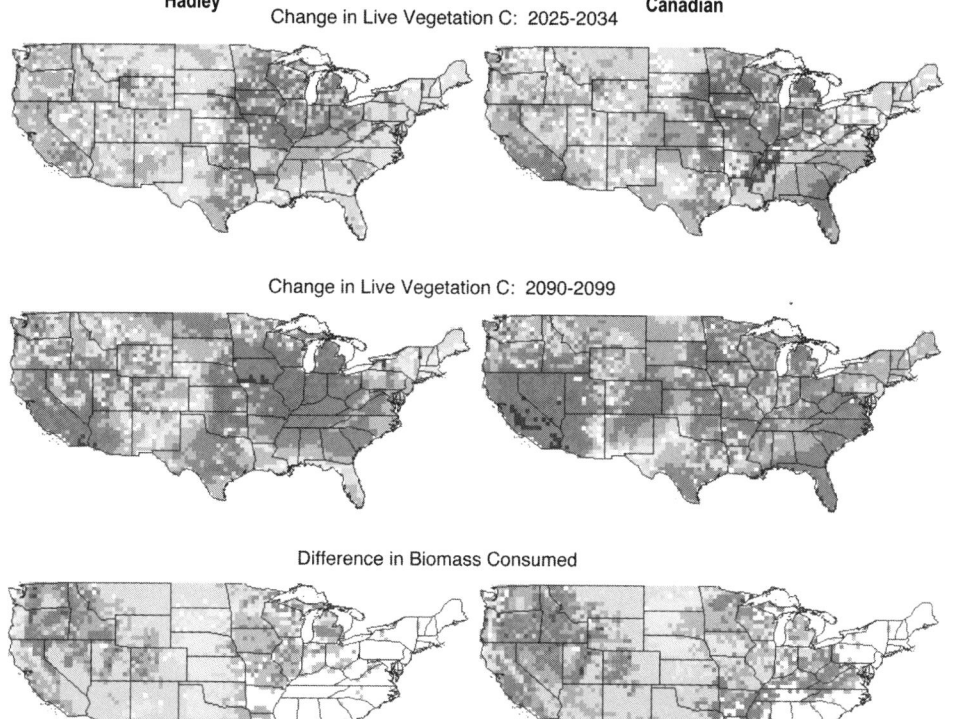

Hadley   Canadian

Change in Live Vegetation C: 2025-2034

Change in Live Vegetation C: 2090-2099

Difference in Biomass Consumed

■ 100% decrease
75 - 100%
50 - 75%
25 - 50%
10 - 25%
1 - 10%
□ No Change
1 - 10% increase
10 - 25%
25 - 50%
50 - 100%
100 - 200%
■ > 200%
NA

Figure 3. Change in live vegetation carbon density from the historical period (1961-1990) to the 2030s (2025-2035) and the 2090s (2090-2099) under two climate scenarios. Change in the biomass consumed by natural fires between the 20th century (1895-1993) and the 21st century 1994-2100) as simulated by the dynamic vegetation model MC1 under two climate scenarios (bottom two panels) (Bachelet et al. 2001) See Color Plate Appendix

ground-level fires. In western forests, repeated ground-level fires reduce intermediate height vegetation that can serve as fuel between the surface and the crown. Thus, these repeated ground-level fires reduce the occurrence of stand-killing crown fires.

Disturbances can be regional or widespread. Landslide processes exhibit very strong geographic patterns. Pacific coastal mountains are particularly prone to sliding because of weak rocks, steep slopes, and high precipitation from frontal storms in these tectonically active areas. Other disturbances, such as insects, pathogens, or introduced species, are widespread across the US. Many disturbances cascade into others. For example, drought often leads to insect outbreaks, disease, or fire. Insects and disease can also create large fuel loads and thereby contribute to increased fire frequency.

Disturbances can be of either natural or anthropogenic origin (e.g., fire). Many large forest areas have been affected by past human activities, and current disturbance regimes are profoundly different from historical regimes. For example, fire suppression in the fire-adapted western forests has led to increased forest density and biomass, changes in forest composition, and increased outbreaks of insects and disease (Flannigan et al., 2000; Ayres and Lombardero, 2000). Some forest types have evolved

to depend on the periodic occurrence of the disturbance. Long-leaf pine forests are a fire-climax ecosystem that would not exist if fire were not a part of the Southeast. Nearly 1.5 million acres of prescribed burning and other fuel reduction treatments in 1998 were used to enhance forest health and diversity by restoring fire-dependent ecosystems that have been affected by long-term fire suppression (USDA Forest Service, 1998). Natural disturbances interact with human activities such as air pollution, harvest, agricultural and urban encroachment, and recreation.

Impact of Climate Change on Forest Disturbances

Climate variability and climate changes alter the frequency, intensity, timing, and/or spatial extent of disturbances. Many potential consequences of future climate change will possibly be buffered by the resilience of forest communities to natural climatic variation. However, the extensive literature on this subject suggests that new disturbance regimes under climate change will likely result in significant perturbations to US forests, with lasting ecological and socioeconomic impacts (Dale et al., 2001).

Hurricanes. Hurricanes seriously impact forests along the eastern and southern coasts of the US as well as the Caribbean Islands (Lugo, 2000). They can inflict sudden and massive tree mortality, complex patterns of tree mortality including delayed mortality, and altered patterns of forest regeneration (Lugo and Scatena, 1996). Because most hurricane damage is from floods, effects can be removed from the actual hurricane in distance (heavy precipitation well inland from the coast) and in time (delays involved in water movement, and in mortality resulting from excessive water). Hurricanes can lead to shifts in successional direction, higher rates of species turnover, opportunities for forest species change, increasing landscape heterogeneity, faster biomass and nutrient turnover, and lower aboveground biomass in mature vegetation (Lugo and Scatena, 1995).

Hurricane location, size, and intensity are influenced by sea surface temperatures and by regional weather features (Emanuel, 1999). Sea surface temperatures (SSTs) are expected to increase, with warmer SSTs expanding to higher latitudes (Royer et al., 1998; Walsh and Pittock, 1998). Climate change is also likely to influence the frequency of regional weather events conducive to hurricane formation, although it is not yet possible to say whether hurricane frequency increases or decreases (Royer et al., 1998; Henderson-Sellers et al., 1998; Knutson et al., 1998; Knutson and Tuleya, 1999).

Fire. Fire frequency, size, intensity, seasonality, type, and severity are highly dependent on weather and climate. An individual fire results from the interaction of ignition agents (such as lightning, fuel conditions, and topography) and weather (including air temperature, relative humidity, wind velocity, and the amount and frequency of precipitation). Over the 1989-1998 period, an average of 3.3 million acres burned annually in the US, varying from 1 million acres a year to over 6 million, mostly in the west and southeast. Most of the burned acres resulted from human-caused fires.

Two modeling approaches were used to look at the impact of climate change on fire: 1) fire severity ratings estimated from future climate (Flannigan et al., 2000), and 2) the interaction of vegetation biomass and climate in establishing conditions for fire (Bachelet et al., 2000).

In the analysis by Flannigan et al. (2000), the future fire severity is projected to increase over much of North America under both climate scenarios and these results are consistent with earlier analyses (Flannigan et al., 1998). The results show great variation for the US and Canada. The warmer and wetter Hadley scenario suggests fire severity increases

near 20% for the Northeast and small decreases for the northern Great Plains, with increases less than 10% over the rest of the continent. The warmer and drier Canadian scenario produces a 30% increase in fire severity for the Midwest, Alaska, and sections of the Southeast, with about 10% increases elsewhere. These results suggest a possible increase of 25-50% in the area burned in the US. Temperature and precipitation are not the only climate-related factors that influence fire regimes; for example, lightning strike frequency was estimated to increase 44% under the Goddard Institute for Space Studies general circulation model scenario (Price and Rind, 1994). Other factors such as length of the fire season, weather conditions after ignition, and vegetation characteristics also influence the fire regimes. Wotton and Flannigan (1993) found that the fire season would be on average 30 days longer in a double carbon dioxide climate as compared to the current climate for Canada. Wildfire severity was at least as sensitive to changes in wind as to changes in temperature and precipitation in a climate change sensitivity analysis for California (Torn and Fried, 1992). Human activities such as fire policies and land use will likely also influence fire regimes in the future (Keane et al., 1998). For example, wildfires on all lands in the western US increased in the 1980s after 30 years of aggressive fire suppression that had led to increases in forest biomass.

The second modeling study examined the influence of climate change on vegetation and the interaction with natural fires (Bachelet et al., 2001). The amount of biomass consumed in fire increased under future climates in analyses with the dynamic global vegetation model, MC1 (Figure 4). In this model, fire occurrence, severity, and size are simulated as a direct function of fuel and weather conditions (Lenihan et al., 1997; Daly et al., 2000, Bachelet et al., 2001). In the western US, increased temperature, steady to increased precipitation, CO_2 fertilization, and increased water use efficiency enhance ecosystem productivity, resulting in more biomass. The highly variable climate of dry years interspersed with wet years and the fuel buildup contributes to more and larger fires in the western landscape. In the eastern US under the Canadian scenario, fires are projected to increase in the Southeast as a result of increased drought stress in forests.

Both approaches suggest the potential for an increased area to be burned with a changing climate. These analyses are based on the physical and biological factors influencing potential fire hazard. Factors such as current land management, land use, and ownership are not considered. Harvest activities and the conversion of forest land to other uses would also alter the amount and kind of fuel.

The rapid response of fire regimes to changes in climate is well established (Flannigan et al., 1998; Stocks et al., 1998), so this response has the potential to overshadow the direct effects of climate change on species distribution, migration, and extinction within fire-prone areas. This possibility of increased fire poses challenges to the management of protected areas such as national parks (Malcolm and Markham, 1998) and to the management of forests for carbon storage.

Drought. Droughts occur in nearly all forest ecosystems (Hanson and Weltzin, 2000). The primary immediate response of trees to drought is to reduce net primary production (NPP) and water use, which are both driven by reduced soil moisture and stomatal conductance. Under extended severe drought conditions, plants die. Seedlings and saplings usually die first and can succumb under moderate drought conditions. Deep rooting, stored carbohydrates, and nutrients in large trees make them susceptible only to longer, more severe droughts. Secondary effects also occur. When reductions in NPP are extreme or sustained over multiple growing seasons, increased susceptibility to insects or disease is possible, especially in dense stands (Negron, 1998). Drought can also reduce decomposition processes leading to a buildup of organic matter on the forest floor.

The consequences of a changing drought regime depend on annual and seasonal changes in climate and whether a plant's current drought adaptations offer resistance and resilience to new conditions. Forests tend to grow to a maximum leaf area that uses nearly all available growing-season soil water (Eagleson, 1978; Hatton et al., 1997; Kergoat, 1998; Neilson and Drapek, 1998). A small increase in growing-season temperature could increase evaporative demand and trigger moisture stress. Using this assumption about leaf area, results from MAPSS, a biogeography model, and MCI, a dynamic global vegetation model, suggest that increased evaporative demands will likely cause future increases in drought stress in forests of the Southeast, southern Rocky Mountains, and parts of the Northwest (Neilson and Drapek, 1998; Bachelet et al., 2000). While earlier forest models, often known as 'Gap' models (Shugart, 1984) also suggested forest declines, all of a tree's current drought adaptations, such as adjusting growth rates or aboveground-belowground allocations of carbon, had not been incorporated into the analysis (Loehle and LeBlanc, 1996). These adaptations buffer the impact of cli-

Biomass Consumed under Two Scenarios of Future Climate

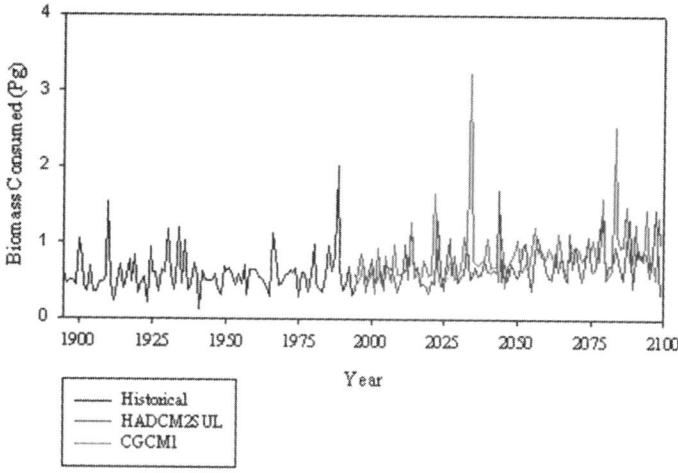

Figure 4. Simulated total biomass consumed by fire over the conterminous US under historic and two future climates; Hadley (HADCM2SUL) and Canadian (CGCM1) scenarios. The fire simulations are for potential vegetation and do not consider historic fire suppression activities. However, grid cells with more than 40% agriculture have been excluded from the calculations (Lenihan et. al., 1997, Daly et. al., 2000, Bachelet et. al., 2001). See Color Plate Appendix

mate, including drought, on individual trees and forest stands. The current generation of ecological models, such as MAPSS, MC1 and others, have improved upon these process algorithms, including the suggestions from Loehle and LeBlanc (1996).

Wind events. Small-scale wind events, such as tornadoes and downbursts, are products of mesoscale weather circumstances (Peterson, 2000). These disturbances can create very large areas of damage. For example, an October 25, 1997 windstorm flattened nearly 13,000 acres of spruce-fir forest in the Routt National Forest of Colorado (USDA Forest Service Routt National Forest, 1998), and a July 4, 1999 windstorm flattened roughly 250,000 acres of forest in the Boundary Waters Canoe Area of Minnesota (Minnesota Dept. of Natural Resources press release 7/12/99). Although small-scale wind events occur throughout the US, the highest concentration of tornadoes occurs across the Central Great Plains states of Oklahoma, Texas, Kansas, and Nebraska.

If climate change increases intensity of all atmospheric convective processes, this change will accelerate the frequency and intensity of tornadoes and hailstorms (Berz, 1993). Karl et al. (1995a) found that the proportion of precipitation occurring in extreme weather events increased in the US from

1910 to 1990. Karl et al. (1995b) further suggest that the US climate has become more extreme (in terms of temperature and precipitation anomalies) in recent decades. Further, Etkin (1995) found a positive correlation between monthly tornado frequency in western Canada and mean monthly temperature, and inferred that this relationship suggests increased tornado frequency under a warmer climate. Despite the above indirect inferences about tornado frequencies, and the direct data on thunderstorm trends, there is still inadequate understanding of tornado genesis to directly forecast how climate change will affect the frequency or severity of windstorms in the next century (Peterson, 2000).

Ice Storms. Ice storms, also known as glaze events, result when rain falling through subfreezing air masses close to the ground is supercooled so that raindrops freeze on impact (Irland, 2000). The National Weather Service (NWS) defines an ice storm as an occurrence of freezing precipitation resulting in either structural damage or at least 0.25 inch of ice accumulation. Ice accumulation can vary dramatically with topography, elevation, aspect, and areal extent of the region where conditions favor glaze formation. While ice storms can occur as far into the southern US as northern Mississippi and Texas, the frequency and severity of ice storm events generally increases toward the northeast (Irland, 2000).

Depending on forest stand composition, amount and extent of ice accumulation, and stand history, damage can range from light and patchy to total breakage of all mature stems (Irland, 1998). Even though the weather conditions producing ice storms are well understood, it is not known how changes in climate will affect the frequency, intensity, location, or areal extent of ice storms.

Introduced species. Climate, as well as human activities, largely determine the potential and realized distributional ranges of introduced species (Simberloff, 2000). Unsuitable climate at points of arrival restricts the survival of a great majority of introduced species (Williamson, 1999). In warmer parts of the US, introduced species comprise a larger fraction of the biota (Simberloff, 1997). Where climate currently restricts invasives, changes in temperature or precipitation may facilitate increased growth, reproduction, or expansion of their ranges. For example, laboratory studies of balsam woolly adelgid (*Adelges piceae*), growing under various temperature conditions, provided the basis for simulations suggesting that temperature-induced changes in the population dynamics of the insect significant-

ly affect Fraser fir (*Abies fraseri*) survival (Dale et al., 1991).

A key feature of most invasive species is that they have the capacity to thrive in disturbed environments through their high reproductive rates, good dispersal abilities, and rapid growth rates (Vitousek et al., 1996). If climate change results in increased disturbances such as fire or drought, these disruptions to ecosystems create just the type of environments in which invasive species are likely to expand rapidly. The interactions among introduced species, native communities, intensively managed forests, human activities that fragment ecosystems, increased atmospheric deposition, and climate change might positively affect the prevalence of invaders, but forecasting specific impacts of invasions remains problematic (Dukes and Mooney, 1999; Williamson, 1999).

Insect and Pathogen Outbreaks. Outbreaks of insects and pathogens can adversely affect recreation, wildlife habitat, wood production and ecological processes. Over the 1986 to 1995 period, these 4 native insect species damaged annually the following acreages: over 4 million acres for western spruce budworm; less than 1 million acres for eastern spruce budworm; less than 2 million acres for mountain pine beetle; and nearly 12 million acres for southern pine beetle (The Heinz Center, 1999). Within this time period, the acreage affected by any one of these insects could vary from less than half of the long-term average to three times the long-term average. Nearly 13 million acres of southern forests are affected by a single disease, fusiform rust, and 29 million acres of western forests are affected by a parasitic plant, dwarf-mistletoe (The Heinz Center, 1999). Disturbances such as drought and fire influence these outbreaks.

An extensive body of scientific literature suggests many pathways through which elevated carbon dioxide and climate change could significantly alter patterns of disturbance from insects and pathogens (Ayres and Lombardero, 2000). Elevated carbon dioxide and climate change could possibly increase or decrease the disturbances of insects and pathogens through direct effects on the survival, reproduction, and dispersal of these organisms. For example, an increase in the interannual variation in minimum winter temperatures could possibly favor more northerly outbreaks of southern pine beetles, while decreasing more southerly outbreaks (Ungerer et al., 1999). It is also possible that climate change would alter insect and pathogen disturbances indirectly through changes in the abundance of their natural enemies and competitors. In addition, climate, and elevated carbon dioxide, influence the susceptibility of trees to insects and pathogens through changes in the chemistry of plant tissues (Ayres and Lombardero, 2000).

The short life cycles, high mobility, reproductive potential, and physiological sensitivity to temperature suggest that even modest climate change will possibly have rapid impacts on the distribution and abundance of many forest insects and pathogens. Beneficial impacts could possibly result where decreased snow cover increases winter mortality. Detrimental impacts could possibly result when warming accelerates insect development and facilitates dispersal of insects and pathogens into areas where tree resistance is less. Detrimental impacts could also possibly result from interactions of disturbances. For example, warming could increase outbreaks in boreal forests that would tend to increase fire frequency (Ayres and Lombardero, 2000). Already, the impact of insects and disease in forests is widespread. In 1995, over 90 million acres of forestland in the US were affected by a few species: southern pine beetle, mountain pine beetle, spruce budworm (eastern and western), spruce beetle, dwarf mistletoe, root disease, and fusiform rust. Thus, there are potentially important ecological and socio-economic consequences to these beneficial and detrimental impacts (Ayres and Lombardero, 2000; Ayres and Reams, 1997).

3. Biodiversity

Land Use Impacts Species, Communities, and Biomes

Global change encompasses a number of events occurring at the continental scale, including climate change, land use change, species invasion, and air pollution. Global change has and will likely continue to affect the abundance and distribution of plants and animals which, in turn, will have considerable ramifications for human economics, health, and social well-being. Organisms provide goods including material products, foods, and medicines. In addition, the number and kinds of species present affect how ecosystems respond to global change. It is the responses of individual organisms that begin the cascade of ecological processes that then manifest themselves as changes across landscapes, biomes, and the globe (Hansen et al., 2001, Walker et al., 1999).

Humans modify the quality, amount, and spatial configuration of habitats. A number of natural

community types now cover less than 2% of their pre-settlement ranges (Noss et al., 1994). Examples include: spruce-fir forests in the southern Appalachians; Atlantic white-cedar in parts of Virginia and North Carolina; red and white pine in Michigan; longleaf pine forests in the southeastern coastal plains; slash pine rockland habitat in southern Florida; loblolly/shortleaf pine forests in the west gulf coastal plains; and oak savannas in Oregon. For the species dependent upon ecosystems that have declined in area, such habitat loss can reduce effective population sizes, genetic diversity, and the ability of species to evolve adaptations to new environments (Gilpin, 1987). The area involved in land use shifts can dwarf the land area involved in natural disturbances. For example, the area of harvested cropland went from 292 million acres in 1964 to 347 million acres in 1982, and then down to 293 million in 1987. For comparison, the total area burned in fires at 1 to 6 million acres annually is much less than these land area shifts in and out of agriculture. Though the forest remains standing, only five species of insects are estimated to defoliate about 21 million acres each year, on average (The Heinz Center, 1999).

Land use change alters the spatial pattern of habitats by creating new habitats that are intensively used by humans or by reducing the area and fragmenting natural habitats. These changes increase the distance between habitat patches and reduce overall habitat connectivity. Native forests have been converted to agricultural and urban uses, notably in the eastern and midwestern parts of the US. In some cases, forests have regrown on abandoned agricultural lands. Recent expansion of urban areas and agriculture are fragmenting western forests. Nationally, urban areas have doubled in area between 1942 and 1992 (Flather et al., 1999). While urban areas increased most rapidly in the South and Pacific Coast regions, the most influential land use change has been the increase in human population density in rural areas, particularly in the Rocky Mountain and Pacific Coast regions. This expansion alters disturbance patterns associated with fire, flooding, and landslides. Roadways and expansion of urban areas have fragmented forests into smaller, less-contiguous patches.

Loss of habitat and degradation of habitat quality can reduce population size and growth rates, and elevate the chance of local extinction events (Pulliam, 1988). The steadily increasing number of species listed as threatened and endangered in the US is currently at 1,232 (USDI, Fish and Wildlife Service, 2000). Factors contributing to species endangerment include habitat conversion, resource extraction, and exotic species (Wilcove et al., 1998). The spatial distribution of such factors results in species at risk being concentrated in particular regions of the US, especially the southern Appalachians, the arid Southwest, and coastal areas (Flather et al., 1998), which can be traced back to human population growth and attendent land-use intensification. Land-use intensification has also been found to affect animal communities broadly. Along a gradient of increasing land-use intensification in forested regions of the eastern US, native species of breeding birds are increasingly under-represented and exotic species over-represented (Flather, 1996).

Species that take advantage of anthropogenic habitats, such as some deer, goose, and furbearer species, have greatly expanded in recent years (Flather et al., 1999). Some species of deer (e.g., white-tailed deer) are now so abundant that the primary concern is controlling their populations. In addition, exotic species have become established and greatly expanded their ranges in the US (Drake et al., 1989).

Impact of Climate Change on Biodiversity

Changes in the distribution and abundance of plant or animal species reflect the birth, growth, death, and dispersal rates of individuals in a population. When aggregated, these changes manifest as local extinction and colonization events, which are the mechanisms that determine a species' range. While climate and soils are strong controls on the establishment and growth of plants, the response of plant and animal species to climate change will be the result of many interrelated processes operating over several scales of time and space. Migration rates, changes in disturbance regimes, and competitive and cooperative interactions with other species will affect the distribution of plants and animals. Because of the individualistic response of species, biotic communities are not expected to respond as intact units to climate change. Community composition responds to a complex set of factors including the direct effects of climate, differential species dispersal, and indirect effects associated with changes in disturbance regimes, land use, and interspecific interactions (Peters, 1992).

Some of the best evidence that species' distributions are correlated with changing climates is found for plants (Webb, 1992). Recent observations of some species suggest a response to historical changes in climate. Breeding ranges of some mobile species, such as waterfowl, have expanded northward in

association with climate warming (Abraham and Jefferies, 1997). The breeding dates of both amphibians and birds in Great Britain have shifted one to three weeks earlier since the 1970s in association with increasing temperatures (Beebee, 1995; Crick et al., 1997).

The primary focus for this analysis is on the continental-scale response of forest vegetation as reflected in climate-induced changes in the distributions of biomes, community types, species richness, and individual tree and shrub species. Vegetation responses to projected future climate change were assessed by reviewing the available literature and by using the climate predictions of global climate models as input to a set of different vegetation simulation models. Paleoecological studies of climate change impacts on forests provide information about past responses. However, their results are limited with respect to the future because the current size, age, and species composition of temperate forests has been strongly impacted by human activities, and second, global temperatures are predicted to increase at an unprecedented rate (Dale, 1997). Although vegetation models incorporating land-use dynamics are under rapid development at local scales, the current state of knowledge does not allow for integrating the effects of land use at the continental scale.

The vegetation models used in this analysis include the following. Two statistical models project the distribution of individual tree species with results aggregated into community types. For the eastern US, the DISTRIB model, which projects potential changes in suitable habitat of 80 tree species, was developed from 33 environmental variables and the current distribution of each tree species (Iverson et al., 1999). The results for 80 species under each climate change scenario were also aggregated to examine potential changes in forest types in the eastern US (Iverson and Prasad, 2001). Shafer et al. (2001) developed local regression models that estimate the probability of occurrence of 51 tree and shrub species across North America based on 3 bioclimatic variables: mean temperature of the coldest month, growing degree days, and a moisture index. Changes in species richness of trees and terrestrial vertebrates based on energy theory (Currie, 1991) were also analyzed under the different climate scenarios. Species interactions and the physiological response of species to carbon dioxide are not included in these statistical models. In contrast, the MAPSS biogeography model (Neilson, 1995) projects biome response to climate as change in vegetation structure and density based on light, energy, and water limitations (VEMAP Members, 1995). The potential natural vegetation is coupled directly to climate and hydrology, and rules are applied to classify vegetation into biome types. The model considers the effects of altered carbon dioxide on plant physiology. The analyses for species, communities, and biomes used an equilibrium climate scenario based on the transient Canadian and transient Hadley scenarios. The baseline scenario was the average climate for the 1961-1990 period. The "climate change" scenario was the average of the projected climate for 2070 to 2100. The results of these analyses are given below. The implications of species dispersal and land use change are discussed in the context of these analyses.

Tree Species. The potential distribution of tree species under climate change is modeled using both statistical models (Iverson and Prasad, 2001; Shafer et al., 2001) and both climate scenarios used in the National Assessment. These statistical approaches assume that there are no barriers to species migration. Results should be viewed as indications of the potential magnitude and direction of range shifts under a changed climate and not as predictions of change.

For many of the eastern tree species, their possible ranges shift north (Iverson et al., 1999; Iverson and Prasad, 2001). Under both climate scenarios, the range of sugar maple (*Acer saccharum*) shifts out of the United States entirely (Figure 5). White oak (*Quercus alba*) remains within its current range but is reduced in importance in the southern parts and increases in the northern parts of its range. A total of 7 of the 80 eastern tree species are projected to decline in regional importance by at least 90%: bigtooth aspen (*Populus grandidentata*), aspen (*Populus tremuloides*), sugar maple, northern white-cedar (*Thuja occidentalis*), balsam fir (*Abies balsamea*), red pine (*Pinus resinosa*), and paper birch (*Betula papyrifera*). The ranges of most species are projected to move to the north, with the ranges of several species moving north by 60 to 330 miles (100 to 530 km). For some species, such as aspen, paper birch, northern white-cedar, balsam fir, and sugar maple, the optimum latitude for their occurrence moves north of the US-Canadian border.

When integrated into community types, southern forest types expand while higher elevation and northern forest types decline in area (Figure 6). The oak-hickory type is projected to expand in area by 34% primarily to the north and east (Iverson and Prasad, 2001). The oak-pine type is

projected to expand in area by 290% throughout the Southeast. Area of spruce-fir and aspen-birch types is projected to decline by 97% and 92% respectively. These types are replaced mainly by oak-hickory and oak-pine forests. The loblolly-shortleaf pine type is also projected to be reduced by 32% and shifts north and west, being replaced in its current zone by the oak-pine type. The longleaf-slash type is projected to be reduced by 31% in area.

In the western US, the potential future ranges for many tree species are simulated to change, with some species' ranges shifting northward into Canada (Shafer et al., 2001). Simulated future ranges for Western hemlock (*Tsuga heterophylla*) and Douglas-fir (*Pseudotsuga menziesii*) (Figure 7) are projected to decrease west of the Cascade

Projected Changes in Distribution of Sugar Maple

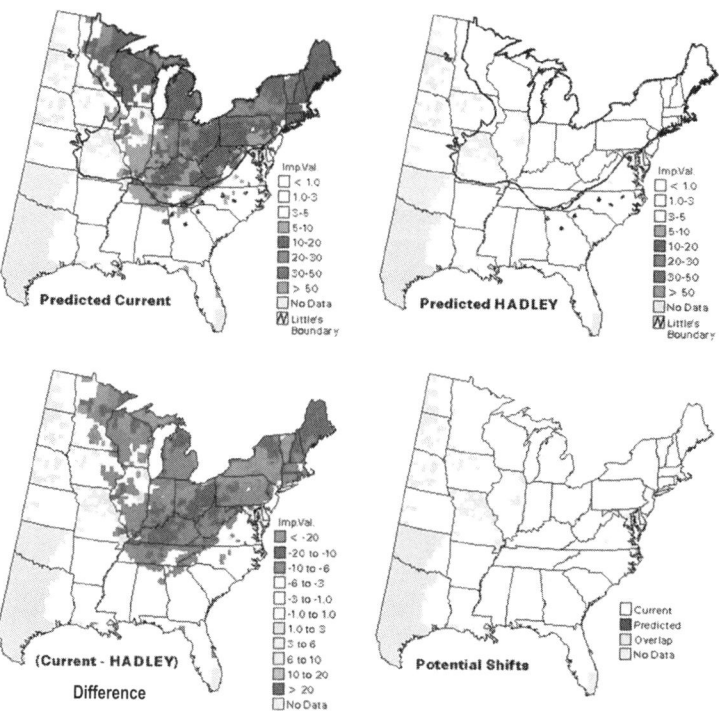

Figure 5. Projected distribution for sugar maple under current climate and the Hadley climate scenario and for the eastern United States, using statistical models developed by Iverson et al. (1999). The Predicted Current map is the current distribution and importance value of sugar maple, as modeled from the regression tree analysis. Importance value is an index based on the number of stems and basal area of both the understory and the overstory. Predicted Hadley is the potential suitable habitat for sugar maple under the Hadley climate scenario. These potential maps imply no barriers to migration. The Difference map represents the difference between Modeled Current and Predicted Hadley maps. The Potential Shifts map displays the modeled current distribution, along with predicted potential future distribution (using the Hadley scenario) and the overlap where the species is now and is projected to be in the future. See Color Plate Appendix

Current and Projected Forest Communities in the Eastern US

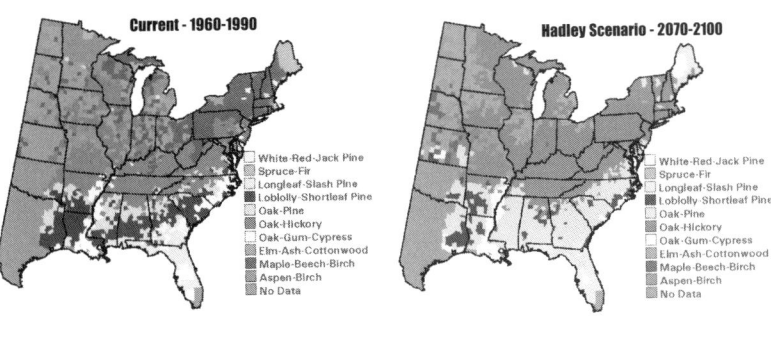

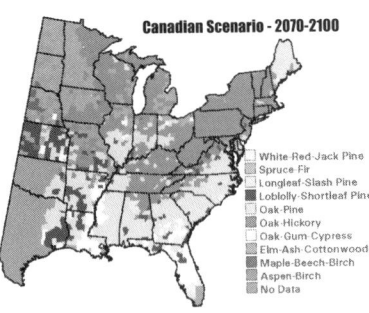

Figure 6. Projected forest communities under (a) current climate, (b) the Hadley climate scenario, and (c) the Canadian climate scenario, based on the results of individual analyses of 80 tree species shifts (see Prasad and Iverson, 1999-ongoing http://www.fs.fed.us/ne/delaware/atlas/index.html) See Color Plate Appendix

Mountains. The potential future range of Western hemlock extends into mountain ranges throughout the interior west while Douglas-fir expands east of the Cascades and Sierras as well as northward along the west coast of Canada into Alaska. The potential future ranges for subalpine conifers such as Engelmann spruce (*Picea engelmannii*), Mountain hemlock (*Tsuga mertensiana*), and several species of fir (*Abies species*) are much reduced in the western parts of the US; however these subalpine species expand to the north along the west coast of Canada and into Alaska. While many of the more mesic and higher elevation species shift northward into Canada, the potential future range of Ponderosa pine (*Pinus ponderosa*) expands within the interior western US.

The complex topography of the western US, combined with its seasonal and regional variations in climate, strongly influences potential future shifts in the ranges of tree species and their likely future abilities to successfully disperse to new habitat (Hansen et al., 2001; Shafer et al.,;2001). Species range shifts in the western US are simulated to occur in all directions whereas in the eastern US, the shifts tend to be northward as temperature increases and westward as precipitation increases. In the western US, several conifer species associated with moderately moist climates shift south and east along the Rocky

Mountains with, for example, species typical today of Glacier National Park expanding to the southeast into Yellowstone National Park (Bartlein et al., 1997). Contrasts between eastern and western range shifts can be seen in the simulated future range of Paper birch (*Betula papyrifera*) whose eastern US range limit contracts northward but whose western US range limit expands southward (Figure 7).

Community Richness. The relationship between spatial patterns of climate and richness of trees and terrestrial vertebrate species were used to predict changes in species richness under climate change across North America (Hansen et al., 2001; Currie, 1991; Currie, 2001). The climate relationships with species richness were stronger for temperature than precipitation. Because the climate-richness relationships, including where maximal richness occurs, differ across taxonomic groups, the impacts of climate change will likely also vary across the groups.

Across all scenarios, tree richness is projected to increase in the cooler regions: northern US, the western mountains, and near the Canadian Border (Currie, 2001). Because species richness of birds and mammals is currently highest in moderately warm areas and decreases in hotter areas, the model projected species richness for these taxa declines as air temperatures increase. Those scenarios where warming is projected to occur results in a decrease of bird and mammal richness—a 25% decrease in low-elevation areas in the Southeast, for example. Warming in colder areas (such as mountainous areas) is projected to result in highly variable increases in richness (11% to >100%). Cold-blooded animals such as reptiles and amphibians would benefit from increased air temperatures. Consequently, under the warming climate scenarios, species richness for these taxa is projected to increase from about 11% to 100% over the entire conterminous US.

Flather et al. (1998) identified regions in the US that have high concentrations of the currently designated threatened and endangered species. Species richness was used to explore the potential impact of climate change on taxonomic groups within these hotspots of species endangerment. The projected richness results were overlaid on maps of current hot spots for threatened and endangered species, and the species richness changes within each taxonomic group were evaluated. Within these hotspots, species richness for reptiles and amphibians increased, whereas bird and mammal richness appeared to be much reduced in many of them, especially in the East.

Biome Shifts. The impact of potential climate change on the biogeography of the US is examined using six scenarios that varied in degree of warming (Bachelet et al., 2001). Across the scenarios, forest area is projected to decrease by an average of 11%, with a range of +23% under the moderately warming scenarios to –45% under the hottest scenarios. Northeast mixed forests (hardwood and conifer) are projected to decrease by 72% in area in the US, on average, as they shift into Canada (Neilson et al., 1998). The range of eastern hardwoods is projected to decrease by an average of 34%. Although these deciduous forests shift north, replacing Northeast mixed forests, they are squeezed from the south by Southeast mixed forests or from the west by savannas and grasslands, depending on the scenario. Southeast mixed forests are projected to increase under all scenarios (average 37%). This biome would remain intact under moderately warm scenarios, but would be converted to savannas and grasslands in the South under the hotter scenarios.

Alpine ecosystems are projected to all but disappear from the western mountains, being overtaken by encroaching forests. Under the climate scenarios studied, wet coniferous forests in the northwest decrease in area by 9% on average, while the extent of interior western pines change little. Responses in both wet conifer and interior pine forests show wide variations among scenarios, with expansions under newer transient scenarios (Bachelet et al., 2001). Arid woodlands also are projected to expand in the interior West and Great Plains, encroaching on some grasslands.

Likely Species Shifts, Dispersal, and Land Use Impacts. The locations and areas of potential habitats for many plant and animal species are likely to shift as climate changes. Potential habitats for trees favored by cool environments are likely to shift northward. The habitats of alpine, subalpine spruce/fir, and aspen communities are likely to contract dramatically in the US and largely shift into Canada. Potential habitats that are likely to increase in the US are oak/hickory, oak/pine, ponderosa pine, and arid woodland communities. Most of these results were evident in the results from three independent models: the mechanistic model MAPSS and two statistical models. These results were also relatively robust across the several climate scenarios

Paper Birch and Douglas Fir Tree Distributions under Future Climate Change Scenarios

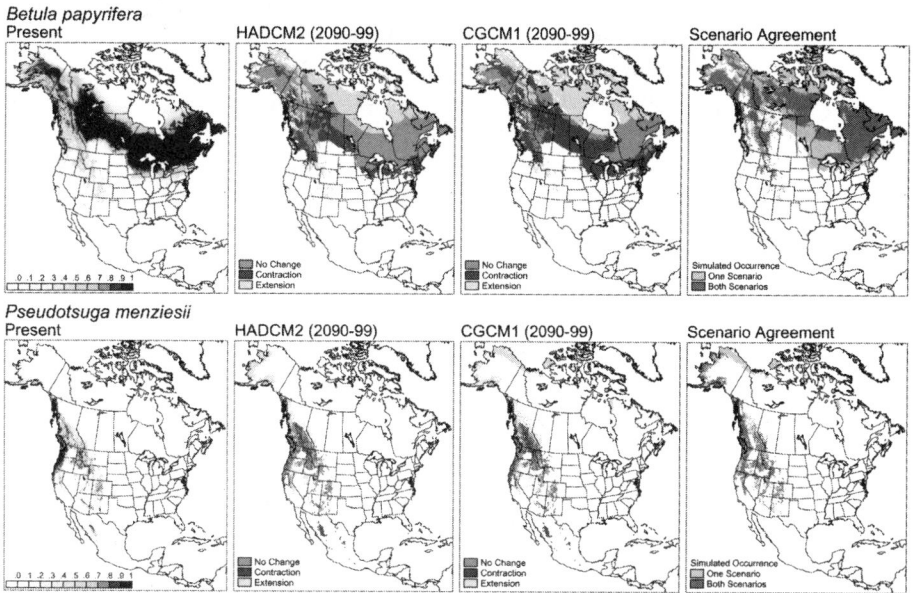

Figure 7. Simulated distributions and scenario agreement for *Betula papyrifera* and *Pseudotsuga menziesii* (after Hansen et al., 2001). Estimated probabilities of occurrence for each taxon simulated with observed modern climate (left panel). Comparison of the observed distributions with the simulated future distributions under future climate conditions as generated by the Hadley (HADCM2) and Canadian (CGCM1) scenarios for 2090-2099 (middle panels). Gray indicates locations where the taxon is observed today and is simulated to occur under future climate conditions; red indicates locations where the taxon is observed today but is simulated to be absent under future climate conditions; and blue indicates locations where the taxon is absent today but is simulated to occur under future climate conditions. Scenario agreement (right panel). Light purple indicates locations where the species is simulated to be present under the future climate of either the HADCM2 or CGCM1 scenario; dark purple indicates locations where the species is simulated to be present under both future climate scenarios. See Color Plate Appendix

analyzed, including the Hadley and the Canadian scenarios. The statistical models are highly defensible for this application because of the strong and extremely well-documented relationship between tree species performance and environmental conditions, especially climate and soil. However, statistical models do not incorporate a direct CO_2 effect, which enhances water-use efficiency. Even so, it is well accepted that climate and soils are strong controls on the establishment, growth, and reproduction of many tree species.

How well these species actually track changes in their potential habitats will very likely be strongly influenced by their dispersal abilities and the disturbances to these environments. Good analytical models for dispersal are few. Some native species will very likely have difficulty dispersing to new habitats because of the rapid rate of climate change and the varying land uses along alternate migration routes. For example, aspen communities are currently being reduced by conifer encroachment, grazing, invasive species, and urban expansion. Weed species that disperse rapidly will likely to be well-represented in these new habitats. Hence, the species composition of newly forming communities will likely differ substantially from those occupying similar habitats today.

4. Socioeconomic Impacts

Current Supply and Demand for Amenities, Services, and Goods from Forests

Forests and the forestry sector provide the resource base and flow of amenities, goods, and services that provide for needs of individuals, communities, regions, the nation, and export customers. Harvested products include timber, pulpwood, fuelwood, wild game, ferns, mushrooms, floral greenery, and berries. Recreational opportunities in forests range from skiing, swimming, hiking and camping to birding, and autumn leaf tours. Forest land is also managed to provide clean water and habitat for wildlife.

Environmental conditions and available technology as well as social, political, and economic factors influence the supply of and the demand for forest amenities, goods, and services. Factors that influence supply of these services include forest area, forest productivity, forest management, production technology, and the quality of amenities. Factors that influence demand include population, economic growth and structure, and personal preferences. Changes in these factors also influence supply of and demand for forest amenities, goods, and services. Land-use pressures alter the amount and type of land available for forest reserves, multiple-use forests, commercial recreation, forest plantations, and carbon storage (Alig, 1986; Leemans et al., 1996; Solomon et al., 1993). Changes in forest species composition, growth, and mortality alter the possible supply of specific types of wood products, wildlife habitat, and recreation. Assumptions about changes in human needs in the US and overseas also affect the socioeconomic impacts from climate change on US forests (Joyce et al., 1995; Perez-Garcia et al., 1997). Clearly, forest changes caused by human use of forests could exceed those impacts from climate change (Dale, 1997). However, climate change could impact many of the amenities, goods, and services from forests (Bruce et al., 1996), includ-

ing productivity of locally harvested plants such as berries or ferns; local economies through land use shifts from forest to other uses; forest real estate values; and tree cover and composition in urban areas, and associated benefits and costs. In addition, climate-change impacts on disturbances such as fire could increase fire suppression costs and economic losses due to wildfires (Torn et al., 1999). In this assessment, it was only possible to explore the impact of climate change on wood products and recreation in more detail.

North America is the world's leading producer and consumer of wood products. The US has substantial exports of hardwood lumber, wood chips, logs, and some types of paper (Haynes et al., 1995). The US also depends on Canada for 35% of its softwood lumber and more than half of its newsprint. Past, current, and future land uses and management as well as environmental factors such as insects, disease, extreme climate events, and other disturbances affect the supply of forest products. These factors can alter the cost of making and using wood products, and associated jobs and income in an area. Changes in economic viability of wood production influence whether owners keep land in forests or convert it to other uses. Demand for wood products, consumption, and trade is strongly influenced by US and overseas population, economic growth, and human values. While demand for timber needed to make US wood products is projected to grow at about the same rate as population over the next 50 years, the demands for products and the kinds of timber harvested will be affected by technology change and consumer preferences. For example, use of recycled paper slows the increase in the amount of timber harvest needed to meet increasing demands for various paper products.

The combinations of resources, travel behavior, and population characteristics vary uniquely across the regions of the US. Participation in outdoor activities is strongly related to age, ability and disability, race, education, and income. These factors influence the types of recreation opportunities available and in which people participate. Approximately 690 million acres of federal lands are used for recreation, of which 95% are in the West. State and local governments manage over 54 million acres, of which 30 million (55%) are in the East (Cordell et al., 1990). Most of the downhill skiing capacity is in the western US while most of the cross-country skiing capacity is in the East. The number of people participating in recreation is expected to continue to increase for many decades. The importance of recreation opportunities near urban areas is rising as US preference shifts from long-distance vacations to frequent close-to-home trips (Cordell et al., 1990). A little over 14% of private lands are open to recreation, but this amount is declining as lands are converted to other uses or access is restricted. Projected increases in per capita income will likely contribute to higher demand for snow-related recreation, although land- and water-based activities will continue to dominate total recreation patterns (USDA Forest Service, 1994).

Potential Impact of Climate Change on Timber and Wood Product Markets, and Recreation

Adaptation in forest land management and timber markets. The possible degree and uncertainty in the flow of value from forests — goods, services, and amenities — is influenced by the combined (and individual) uncertainty in changes in climate, forest productivity, and the economy. Comparisons of earlier with more recent analyses indicate that differences in assumptions, modeling structure, and scope of analysis such as spatial scale significantly affected the socioeconomic results. For example, an early study of Scandinavian boreal forests concluded that increased warming would benefit higher latitude regions (Binkley, 1988); yet when the global trade patterns were analyzed, other regions with lower production costs benefited most (Perez-Garcia et al., 1997). Early analyses (e.g., Smith and Tirpak, 1989) that did not include economic forces concluded there would be significant damage to US forests. When active forest management was included in a previous economic analysis, timber markets adapted to short-term negative effects of climate change by reducing prices, salvaging dead and dying timber, and replanting species appropriate to the new climate (Sohngen and Mendelsohn, 1998). The total of consumer and producer benefit (surplus) remained positive.

For this assessment, the dynamic optimization model FASOM (Adams et al., 1996, 1997; Alig et al., 1997, 1998) was used to evaluate the range of possible projected changes in forest land area, timber markets, and related consumer/producer impacts associated with climate change (Irland et al., 2001). The range of scenarios considered alternate assumptions about 1) climate (the Hadley and the Canadian scenarios), 2) forest productivity (the TEM and CENTURY biogeochemistry models), and 3) timber and agricultural product demand (determined by population growth and economic growth).

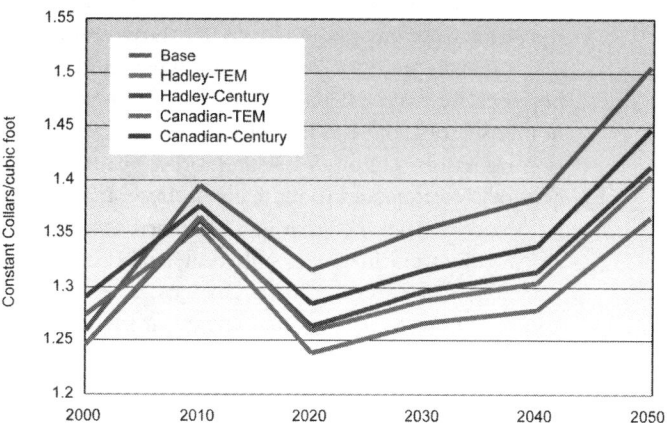

Average Price for Standing Timber in US Forests

Figure 8. Prices for standing timber under all climate change scenarios remain lower than a future without climate change (baseline). Prices under the Canadian scenario remain higher than prices under the Hadley scenario when either the TEM or the CENTURY model are used. (Irland et al., 2001). See Color Plate Appendix

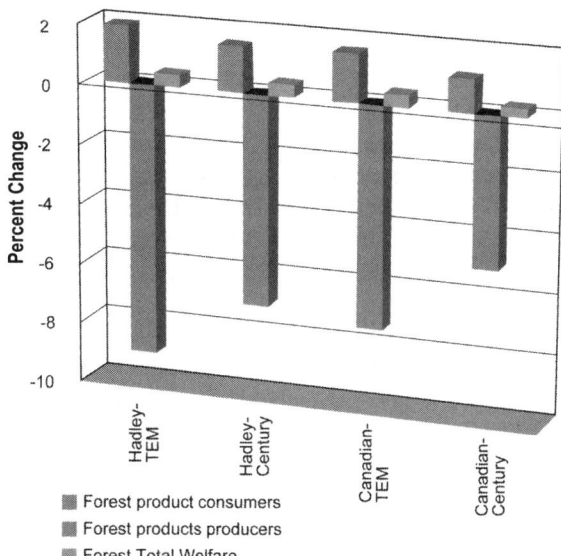

Change in Timber Product Welfare from 2001 to 2100

Figure 9. Increased forest growth overall leads to increased wood supply; reductions in log prices decrease producers welfare (profits), but generally benefit consumers through lower wood-product prices. Welfare is present value of consumer and producer surplus discounted at 4% for 2000-2100. (Irland et al., 2001). See Color Plate Appendix

Analyses of these particular scenarios indicate that forest productivity gains increase timber inventories over the next 100 years. This increased wood supply results in reductions in log prices, which, in turn, decrease producers' profits. At the same time, lower forest-product prices mean that consumers generally benefit (Figure 8). The net effect on the economic welfare of participants in both timber and agricultural markets was projected to increase in all scenarios from between 0.4 to 0.7% above the current values (Figure 9). Land would likely shift between forestry and agricultural uses as these economic sectors adjusted to climate-induced changes in production. Although US total forest production generally is projected to increase in these analyses, hardwood output is higher in all scenarios whereas softwood output increases only under moderate warming. The extent of these changes varies by region. In these analyses, timber output increases more in the South than in the North, and sawtimber volume increases more than pulpwood volume.

While previous studies and this analysis differ in the degree of market and human adaptation, one general conclusion is that timber and wood product markets will likely be able to adjust and adapt to climate change (Irland et al., 2001). Assumptions about changes in population, land use, trade in wood products, consumption of wood products, recreation patterns, and human values are highly uncertain on a century time scale. For example, if human needs from forests increase over the next 100 years and imports are limited, the socioeconomic impacts of climate change on forests would be greater than if needs are low or products can be imported from areas where climate increases forest growth. Thus, assumptions about change in human needs in the US and overseas, and about climate change effects in other parts of the world, are likely to be the major factors that determine socioeconomic impacts on the US.

Recreation. Outdoor recreation will likely be altered as a result of changes in seasonality of climate, and air and water temperatures (Irland et al., 2001). Secondary impacts of environmental changes, such as increased haze with increased temperatures, and degraded aquatic habitats under changing climates, will also likely affect outdoor recreation opportunities. Because recreation is extremely broad and diverse in its environmental requirements (Cordell et al., 1999), it is difficult to generalize about the impact of climate change across recreation as a whole (Wall, 1998). Change in benefits to consumers, as measured by aggregate days of activities and economic value, vary by type of recreation and location (Loomis and Crespi, 1999; Mendelsohn and Mackowiki, 1999). In some cases, recreation in one location will be substituted for recreation in other locations. For example, temperature increases will likely extend summer activities such as swimming and boating in some forest areas, and substitute to some degree for such activities in

more tropical areas. Effects on fishing opportunities will likely vary with warming waters increasing fish production and opportunities for some species while decreasing habitat and opportunities for cold water species.

Recreation is likely to expand in mountainous areas where warmer temperatures attract more people to higher elevations. Skiing is an important use of forested mountain landscape and is sensitive to the climate in the mountains. Competition within the skiing industry is strong, with successful ski areas attracting customers by providing high-speed lifts, overnight accommodations, modern snowmaking and grooming equipment, and other amenities. Small ski areas often cannot compete and many have closed or have been annexed by adjacent larger areas (Irland et al., 2001). Climate change will likely alter the primary factors influencing the ability of a ski area to make snow; namely temperature, water availability, and energy. Higher winter temperatures could possibly increase the amount of snow melting, the number of rain events, and decrease the opportunities for snowmaking while increasing the need for machine-made snow. The efficiency of snowmaking declines as temperature warms. The cost of making snow is 5 times greater at 28°F (-2 °C) as at 10°F (-12°C). The annual electricity usage at Maine's Sunday River ski area is approximately 26 million kilowatt-hours (Hoffman, 1998), at a cost of nearly 2 million dollars per year and most of this energy is used for snowmaking. On the other hand, higher winter temperatures and more rain events together could possibly result in increased water availability for snowmaking and perhaps increased visitation with fewer extremely cold days. Changes in the geographic line of persistent subfreezing winter temperature could possibly alter the location of winter recreation by affecting the feasibility of snow-making, such as in the southern Appalachian Mountains. While the impacts of climate change on the US ski industry remain speculative, ski areas operating in marginal climates are likely to be seriously affected.

ADAPTATIONS: FOREST MANAGEMENT STRATEGIES UNDER CLIMATE CHANGE

A major challenge in developing strategies for coping with the potential effects of climate change on forest processes and subsequent values is that the magnitude and direction of such changes at the local level remain highly uncertain. In addition, potential climate-induced changes in forest processes must also be put into context of other human-induced pressures on forests, which will likely change significantly over future decades and centuries. Finally, such strategies must be considered in terms of their economic viability. Strategies for coping with climate change could include: 1) active forest management to promote forest adaptation to climate change, and 2) assistance to urban and rural communities to adapt to changing forest conditions.

Active Forest Management under Climate Change

Current forest management capabilities provide initial guidance on how to manage forests under a future changing climate. The value of the forest, any changes in natural disturbance regimes, and the available environmentally and economically acceptable management options influence the coping strategies for forests. One way forests may be aided in adapting to climate change is to take steps to decrease other forest stresses, such as atmospheric deposition. Strategies for coping with disturbances in forests will vary regionally.

If climate change results in alteration of such disturbance regimes as fire, drought, or insects and disease, managers could try to cope with these impacts by influencing forest ecosystems prior to the disturbance, mitigating the forest disturbance itself, manipulating the forest after the disturbance, or facilitating the recovery process. Prior to the disturbance, the ecosystem could be managed in ways that alter its vulnerability or ability to enhance recovery from a disturbance. For example, trees susceptible to ice or wind storms could be removed, as is common in cities. Density and spacing of tree planting could be altered to reduce susceptibility to drought. Species composition could be changed to reduce vulnerability of forests to fire, drought, wind, insects, or pathogens. Management could be designed to reduce the opportunity for disturbance to occur. Examples include: limiting the introductions of non-native species; burning restrictions; and prescribed fires to reduce fuel loads. While manipulations of fuel type, load, and arrangement could protect local areas of high value, fuel management may not be complicated for larger landscapes. Some disturbances, such as fire, insects, disease, and drought can be managed through preventive measures, or through manipulations that affect the intensity or frequency of the disturbance. For example, fire, insect, and disease controls are examples of managing to reduce the impact of a disturbance for high-valued forests.

Mitigation measures, such as irrigation, might temporarily support specific gene pools until new and stable environments are identified. Recovery efforts can focus either on managing the state of the system immediately after the disturbance, or managing the ongoing process of recovery. Recovery can be enhanced by adding structural elements that create shade or other safe sites necessary for reestablishing vegetation or that serve as perches for birds (and thus places where seeds would be dispersed). Alternatively, late successional species might be planted to speed up succession.

There will likely be surprises in how changes in climate alter the nature of forest disturbances and the forests themselves. A monitoring program could determine the influence of climate change on forests and the natural disturbance regimes. Programs for monitoring the impact of forest disturbances currently exist for insects, pathogens, and fire. However, few surveys quantify the extent and severity of damage from wind, ice storms, and landslides. Further, reserve areas such as wildernesses, are not currently monitored. Information from monitoring programs could then be used to update risk assessments in management plans and prescriptions in an adaptive management sense (Walters, 1997) or to assess the regional vulnerability of landscapes (O'Neill et al., 2000). A risk ranking system could identify aspects of the forest most susceptible to severe repercussions from disturbance under a changing climate.

Human Adaptation to Forest Changes under an Altered Climate

While recent studies have suggested that human activities associated with markets will likely allow for adaptation to ecological changes by changing land use practices, production and use technologies, and consumption patterns, it may be important to examine the breadth of possible adaptations. As forest conditions change costs of goods, services, and amenities, investors, producers, and consumers will likely shift investments and consumption decisions. However, it is possible that separate effects on producers and consumers of timber products, especially in different regions, could be large and in opposite directions.

Helping human communities to adapt to changing forest conditions could include reducing potential socioeconomic impacts by mitigating carbon buildup in the atmosphere. Carbon buildup in the atmosphere can be slowed by changing forest management and forest industry technology to increase net carbon sequestration. Carbon storage in forests could be increased by reducing the conversion of forests to other land uses, setting aside existing forests from harvest, or reducing forest fires (Birdsey et al., 2000; Sampson and Hair, 1996). Carbon storage could also be increased by converting other land uses to forests and by enhancing forest management. Improvements in forest industry technology could enhance sequestration by allowing substitution of wood products where appropriate for products requiring more energy and carbon emissions in production and use, and increasing the use life of products and recycling of products allowing more carbon accumulation in forests (Sampson and Hair, 1996).

CRUCIAL UNKNOWNS AND RESEARCH NEEDS

Linkages between Forest Processes, Air Quality, and Climate Change. It is crucial to begin to understand not only the direct effects of CO_2, ozone, temperature, precipitation, and nitrogen and sulfur deposition on forests, but also the interactive effects of these stresses. Physiological impacts propagate through forest ecosystems by altering competitive interactions among individuals and species, litter characteristics, and soil processes. These impacts and interactions feed back to affect physiological processes, nutrient cycling, and hydrology. These interactions stress the need for integrated ecological research to improve our understanding of these forest dynamics.

Responses of Terrestrial Animals to Changing Climate, and Trace Gases. The response of trees to climate, soils, and other biophysical controls is relatively well-documented, in comparison to other plants or animals. For native species, and the interaction of introduced species and natives, understanding the combined effects of climate, carbon dioxide concentration, and nutrients such as nitrogen on terrestrial vertebrates and invertebrates is severely lacking. The response of vertebrates and invertebrates to soils and other biophysical controls is also lacking. Research is needed to combine models of impacts of climate change and other factors on forest pests (insects and pathogens, introduced species) with models of range and abundance changes of host species.

Integration of Climate and Land Use Change into Biodiversity, Ecosystem Structure, and Higher Trophic Level Models. Climate and land use interact in ways that influence biodiversity, implying that these factors cannot be considered in isolation (Hansen et al., 2001). Climate and land use jointly influence species distributions through alteration of dispersal routes and changes in habitat through changes in disturbances such as wildfire, flooding, and landslides. Land use may modify the local climate through changes in transpiration, cloud formation and rainfall, and increased levels of drying. Land use models include socio-economic variables such as human population size, affluence, and culture (Alig, 1986); however, the influence of climate on land use and land use impacts on climate are not considered. Global and local climate projections simplify the feedbacks between land use, vegetation, and climate. Integrated models of land use and climate are needed to predict the interactions of these two influences on biodiversity.

Understanding the Interactions of Current Forest Disturbances and Climate. Basic information on the disturbance — frequency, intensity, spatial extent — and the climatological conditions that initiate these current forest disturbances are currently not available for a number of forest disturbances. For example, the genesis of ice storms is well-understood, but the nature of the ice accumulation in relation to storm characteristics is not well-understood. Similarly, the understanding of what specific climatological conditions lead to the formation and occurrence of small-scale wind events is inadequate. Analyses suggest that future climate variability may push fire and hurricanes outside of the well-studied historical conditions that initiate them and control their intensity and frequency. For most disturbances, there is a need for better understanding of the interactions between climate variability, disturbance frequency, and spatial patterns. Once the relationships between weather, climate, and forest disturbances have been quantified, the occurrences of such disturbances could be predicted to help minimize their impact.

Quantifying the Impact of Disturbances on Forest Structure and Function. A key aspect of managing the forest before, during, or after disturbances is an understanding of the disturbance impacts on forests. For disturbances such as fire, the impact has been studied extensively. For others such as ice storms, insects, diseases, and introduced species, a more complete understanding of impacts is necessary before we can manage for the disturbance, or explore the interactions of multiple distur-bances. Research might identify herbivores and pathogens that are likely to be key agents of forest disturbance in the next 50 years. Integrated conti-nental surveys would help to determine the sensitiv-ity of different types of pests and diseases to envi-ronmental change and their potential for increased outbreaks at the margins of their existing ranges.

Trees and forests have developed adaptations by co-evolving with disturbances. Research exploring the utility of these adaptations under a changing climate would be valuable. For example, more work in understanding the inherent genetic variation within particular species is needed to assess the resilience of both unmanaged and managed forests. Research on how genetic traits can be manipulated to increase resistance to specific stresses through biotechnology, expanded tree breeding programs, and seed banks would also be useful for adapting to change in areas where extensive planting is used.

Information to Manage Forests under Multiple Disturbances. Capabilities to manage forests cur-rently, as well as under future potential climate change, rest on the understanding of how forests respond to multiple disturbance events, and how one event affects the response to subsequent distur-bance. Fire and biological disturbances such as insects, pathogens, and introduced species have sig-nificant interactions. A better understanding of these interactions would improve the ability to make long-range predictions about the fate of forest succession and ecosystem dynamics, and would lead to better predictions of conditions under which one event affects the response to a subsequent one.

Human Adaptation in Response to Climate Change Impacts on Forests. Research needs include better links between ecological models and economic models, and improvements in ecological models to capture human adaptation activities. Enhancing these linkages may require additional information about the interface between ecological processes and management objectives. Continued evaluation is needed of climate change impacts on a wide range of forest goods and services such as water supply, carbon storage, nonwood forest prod-ucts, as well as timber and recreation. For example, an improved understanding is needed of how extreme weather events affect timber yields, salvage activity, and timber markets. When stakeholders associated with forest operations in the Mid-Atlantic region were surveyed, more than 20% reported major impacts on their forests from heavy rains,

high winds, or ice storms (Fisher et al., 2000). Increased forest operation costs associated with extreme weather events were noted by all man-agers regardless of their forest management objec-tive. New generations of combined vegetation/economic models should incorporate a range of adaptation mechanisms – for example investments in new/altered management strategies and in new technologies.

Research needs also include more applied research on how adaptation could occur based on how it has occurred in the recent and historical past (McIntosh et al, 2000). Previous marketing and recreational studies typically examined human behavior against many attributes including a set of environmental attributes. These studies have shown that people adapt behavior to environmen-tal events such as fish advisories, smog alerts, beach closing, fish catch rates, and nitrogen depo-sition. This wealth of information could be used to examine market and recreational choices to cli-mate change phenomena. Further, this informa-tion would be the basis to begin testing the sensi-tivity of human adaptation to different climate sce-narios.

US Trade Flows. The US forest sector operates within a global market. Information is needed on how climate change will alter US trade flows in wood and harvest in the US, versus harvest over-seas. As mitigation alternatives are identified, the costs, benefits, and the distribution of costs and benefits must be evaluated. Trade flows can be sensitive to fairly modest shifts in exchange rates and price levels, to a degree that can be important for regions within a nation.

Longer-term Socioeconomic Impacts on Consumer and Producer Benefit/Costs. Improvements in methods are needed that value long-term changes in resource conditions and con-sumer and producer benefits/costs. While it has been shown that human migration patterns are sensitive to climate and climate change, a larger social question remains: will climate change phe-nomena cause larger socioeconomic changes than demographic or macroeconomic phenomena?

LITERATURE CITED

Aber, J., A. Magill, S. G. McNulty, R. D. Boone, K. J. Nadelhoffer, M. Downs, and R. Hallett, Forest biogeo-chemistry and primary production altered by nitrogen saturation, *Water, Air, and Soil Pollution, 85,* 1665-1670, 1995.

Aber, J., W. H. McDowell, K. J. Nadelhoffer, A. Magill, G. Berntson, M. Kamakea, S. G. McNulty, W. Currie, L. Rustad, and I. Fernandez, Nitrogen saturation in temper-ate forest ecosystems: Hypotheses revisited, *BioScience, 48,* 921-934, 1998.

Aber, J., R. P. Neilson, S. McNulty, J. Lenihan, D. Bachelet, and R. Drapek, Forest processes and global environ-mental change: Predicting the effects of individual and multiple stresses, *BioScience,* in press, 2001.

Abraham, K. F., and R. L. Jefferies, High goose popula-tions: Causes, impacts, and implications, in *Arctic Ecosystems in Peril: Report of the Arctic Goose Habitat Working Group,* edited by B. D. J. Batt, Arctic Goose Joint Venture Special Publication, US Fish and Wildlife Service, Washington, DC, and Canadian Wildlife Service, Ottawa, Ontario, 1997.

Adams, D. M., R. J. Alig, B. A. McCarl, J. M. Callaway, and S. Winnett, An analysis of the impacts of public timber harvest policies on private forest management in the United States, *Forest Science, 42,* 343-357, 1996.

Adams, D., R. J. Alig, B. A. McCarl, J. M. Callaway, and S. Winnett, The forest and agricultural sector optimization model: Model structure and applications, in *USDA Forest Service Research Paper PNW-RP-495,* USDA Forest Service, Pacific Northwest Experiment Station, Portland, Oregon, 1997.

Alig, R. J., Econometric analysis of forest acreage trends in the Southeast, *Forest Science, 32,* 119-134, 1986.

Alig, R. J., et al., Assessing effects of mitigation strategies for global climate change with an intertemporal model of the US forest and agriculture sectors, *Environmental and Resource Economics, 9,* 259-274, 1997.

Alig, R. J., D. M. Adams, and B. A. McCarl, Impacts of incorporating land exchanges between forestry and agriculture in sector models, *Journal of Agricultural and Applied Economics, 30,* 389-401, 1998.

Ayres, M. P, and M. J. Lombardero, Assessing the conse-quences of global change for forest disturbances from herbivores and pathogens, *The Science of the Total Environment, 262,* 263-286, 2000.

Ayres, M. P, and G. A. Reams, Global change and distur-bance in southern forest ecosystems, in *Global Change and Disturbance in Southern Ecosystems,* edited by R. A. Mickler and S. Fox, Springer-Verlag, New York, 741-752, 1997.

Bachelet, D., R. P. Neilson, J. M. Lenihan, and R. J. Drapek, Climate change effects on vegetation distribution and carbon budget in the US, *Ecosystems,* in press, 2001.

Baron, J. S, D. S. Ojima, E. A. Holland, and W. J. Parton, Analysis of nitrogen saturation potential in Rocky Mountain tundra and forest: Implications for aquatic systems, *Biogeochemistry, 27,* 61-82, 1994.

Bartlein, P. J., C. Whitlock, and S. I. Shafer, Future climate in the Yellowstone National Park region and its poten-tial impact on vegetation, *Conservation Biology, 11,* 782-792, 1997.

Bazzaz, F. A., The response of natural ecosystems to the rising global CO_2 levels, *Annual Review of Ecology and Systematics, 21,* 167-196, 1990.

Beebee, T. J. C., Amphibian breeding and climate, *Nature, 374,* 219-220, 1995.

Berntson, G. M., and F. A. Bazzaz, Regenerating temperate forest mesocosms in elevated CO_2: Belowground growth and nitrogen cycling, *Oecologia, 113,* 115-125, 1998.

Berz, G. A., Global warming and the insurance industry, *Interdisciplinary Science Reviews, 18,* 120-125, 1993.

Birdsey, R. A., and L. S. Heath, Carbon changes in US forests, in *Productivity of America's Forests and Climate Change, General Technical Report RM-GTR-271,* edited by L. A. Joyce, Rocky Mountain Forest and Range Experiment Station, Fort Collins, Colorado, 1995.

Birdsey, R. A., R. Alig, and D. Adams, Mitigation activities in the forest sector to reduce emissions and enhance sinks of greenhouse gases, in *The Impact of Climate Change on America's Forests, General Technical Report RMRS-GTR-59,* edited by L. A. Joyce and R. A. Birdsey, Rocky Mountain Research Station, Fort Collins, Colorado, 2000.

Binkley, C. S., A case study of the effects of CO_2-induced climatic warming on forest growth and the forest sector: B. Economic effects on the world's forest sector, in *The Impact of Climatic Variations on Agriculture*, edited by M. L. Parry, T. R. Carter, and N. T. Konijn, Kluwer Academic Publishers, Dordrecht, the Netherlands, 1988.

Blatner, K., Special forest products markets in the Pacific Northwest with global implications, in *Special Forest Products – Biodiversity Meets the Marketplace*, edited by N. C. Vance and J. Thomas, US Department of Agriculture, Washington, DC, 1997.

Boubel, R. W, D. L. Fox, D. B. Turner, and A. C. Stern, *Fundamentals of Air Pollution*, Academic Press Limited, San Diego, California, 1994.

Brooks, D. J., US forests in a global context, *General Technical Report RM-228*, Rocky Mountain Forest and Range Experiment Station, Fort Collins, Colorado, 1993.

Bruce, J. P., H. Lee, and E. Haites, *Climate Change 1995: Economic and Social Dimensions of Climate Change. Contribution of Working Group III to the Second Assessment Report of the Intergovernmental Panel on Climate Change*, Cambridge University Press, Cambridge, United Kingdom, 1996.

Burns, R. M., and B. H. Honkala, *Silvics of North America, Volume 1, Conifers, Agriculture Handbook 654*, US Department of Agriculture Forest Service, Washington, DC, 1990.

Ceulemans, R., and M. Mousseau, Effects of elevated atmospheric CO_2 on woody plants, *New Phytologist, 127*, 425-446, 1994.

Cordell, H. K., J. C. Bergstrom, L. A. Hartmann, and D. B. K. English, An analysis of the outdoor recreation and wilderness situation in the United States: 1989-2040, *General Technical Report RM-189*, USDA Forest Service Rocky Mountain Forest and Range Experiment Station, Fort Collins, Colorado, 1990.

Cordell, H. K., B. L. McDonald, R. J. Teasley, J. C. Bergstrom, J. Martin, J. Bason, and V. R. Leeworthy, Outdoor recreation participation trends, in *Outdoor Recreation in American Life: A National Assessment of Demand and Supply Trends*, edited by H. K. Cordell, Sagamore Publishing, Champaign, Illinois, 1999.

Crick, H. Q. P, C. Dudley, D. E. Glue, and D. L. Thomson, UK birds are laying eggs earlier, *Nature, 388*, 526, 1997.

Crookshanks, M., G. Taylor, and M. Broadmeadow, Elevated CO_2 and tree root-growth contrasting responses in *Fraxinus excelsior, Quercus petraea, and Pinus sylvestris, New Phytologist, 138*, 241-250, 1998.

Currie, D. J., Energy and large scale patterns of animal and plant species richness, *American Naturalist, 137*, 27-39, 1991.

Currie, D. J., Tree and vertebrate species richness, *Ecosystems*, in review, 2001.

Curtis, P. S, and X. Z. Wang, A meta-analysis of elevated CO_2 effects on woody plant mass, form, and physiology, *Oecologia, 113*, 229-313, 1998.

Dale, V. H., The relationship between land-use change and climate change, *Ecological Applications, 7*, 753-769, 1997.

Dale V. H, R. H. Gardner, D. L. DeAngelis, C. C. Eagar, and J. W. Webb, Elevation-mediated effects of balsam wooly adelgid on southern Appalachian spruce-fir forests, *Canadian Journal of Forest Research, 21*, 1639-1648, 1991.

Dale, V. H., et al., Forest disturbances and climate change, *BioScience*, in press, 2001.

Daly, C., D. Bachelet, J. M. Lenihan, R. P. Neilson, W. Parton, D. Ojima, Dynamic simulation of tree-grass interactions for global change studies, *Ecological Applications, 10*(2), 449-469, 2000.

Deluccia, E. H., et al., Net primary production of a forest ecosystem with experimental CO_2 enrichment, *Science, 284*, 1177-1179, 1999.

Drake, J. A., H. A. Mooney, F. diCastri, R. H. Groves, F. J. Kruger, M. Rejmanek, M. Williamson (Eds.), *Biological Invasions. A Global Perspective.* SCOPE 37, John Wiley and Sons, Chichester, United Kingdom, 1989.

Dukes, J. S., and H. A. Mooney, Does global change increase the success of biological invaders?, *Trends in Ecology and Evolution, 14*, 135-139, 1999.

Eagleson, P. S., Climate, soil, and vegetation 1. Introduction to water balance dynamics, *Water Resources Research, 14*, 705-712, 1978.

Eamus, D., and P. G. Jarvis, The direct effects of increase in the global atmospheric CO_2 concentration on natural and commercial temperate trees and forests, *Advances in Ecological Research, 19*, 1-55, 1989.

Egli, P., and C. Korner, Growth responses to elevated CO_2 and soil quality in beech spruce model ecosystems, *ACTA Oecologica- International Journal of Ecology,* 18, 343-349, 1997.

Ellsworth, D. S., CO_2 enrichment in a maturing pine forest: Are CO_2 exchange and water status in the canopy affected?, *Plant, Cell, and Environment,* 22, 461-472, 1999.

Emanuel, K. A., Thermodynamic control of hurricane intensity, *Nature, 401,* (6754), 665-669, 1999.

Etkin, D. A., Beyond the Year 2000, more tornadoes in western Canada?, Implications from the historical record, *Natural Hazards, 12,* 19-27, 1995.

Fenn, M. E., M. A. Poth, and D. W. Johnson, Evidence for nitrogen saturation in the San Bernardino Mountains in southern California, *Forest Ecology and Management, 82,* 211-230, 1996.

Fisher, A., et al., Mid-Atlantic Overview, Pennsylvania State University, 2000, http://www.essc.psu.edu/mara/results/foundations_report/index.html#report.

Flannigan, M. D., et al. Future wildfire in circumboreal forests in relation to global warming, *Journal of Vegetation Science, 9,* 469-476, 1998.

Flannigan, M. D., B. J. Stocks, and B. M. Wotton, Climate change and forest fires, *The Science of the Total Environment, 262,* 221-230, 2000.

Flather, C. H., Fitting species-accumulation functions and assessing regional land use impacts on avian diversity, *Journal of Biogeography, 23,* 155-168, 1996.

Flather, C. H., M. S. Knowles, and I. A. Kendall, Threatened and endangered species geography: Characteristics of hot spots in the conterminous United States, *BioScience, 48,* 365-376, 1998.

Flather, C. H., S. J. Brady, and M. S. Knowles, An analysis of wildlife resources in the United States: A technical document supporting the 1999 RPA Assessment, USDA Forest Service, Rocky Mountain Research Station, Fort Collins, Colorado, 1999.

Fox, S., and R. A. Mickler, (Eds.), *Impacts of Air Pollutants on Southern Pine Forests,* Springer-Verlag, New York, 1996.

Galloway, J. N., Acid deposition: Perspectives in time and space, *Water, Air, and Soil Pollution, 85,* 15-24, 1995.

Gilpin, M. E., Spatial structure and population vulnerability, in *Viable Populations for Conservation,* edited by M. Soule, 125-140, Cambridge University Press, Cambridge, United Kingdom, 1987.

Goudriaan, J., et al., Use of models in global change studies, in *Implications of Global Change for Natural and Managed Ecosystems: A Synthesis of GCTE and Related Research,* edited by B. H. Walker, W. L. Steffen, J. Canadell, and J. S. I. Ingram, IGBP Book Series No. 4, Cambridge University Press, Cambridge, United Kingdom, pp. 190-220, 1999.

Guak, S., D. M. Olsyzk, L. H. Fuchigani, and D. T. Tingey, Effects of elevated CO_2 and temperature on cold hardiness and spring bud burst and growth of Douglas-fir *(Pseudotsuga menziesii), Tree Physiology, 18,* 671-679, 1998.

Hansen, A. J., and J. J. Rotella, Environmental gradients and biodiversity, in *Managing Forests for Biodiversity,* edited by M. Hunter, Cambridge University Press, London, 161-209, 1999.

Hansen, A. J., R. P. Neilson, V. H. Dale, C. Flather, L. Iverson, D. J. Currie, S. Shafer, R. L. Cook, and P. Bartlein, Global change in forests: Interactions among biodiversity, climate, and land use, *BioScience,* in press, 2001.

Hanson, P., and J. Weltzin, Drought, forests, and climate change, *The Science of the Total Environment, 262,* 206-220, 2000.

Hatton, T. J., G. D. Salvucci, and H. I. Wu, Eagleson's optimality theory of an ecohydrological equilibrium: Quo vadis?, *Functional Ecology, 11,* 665-674, 1997.

Haynes, R. W., D. M. Adams, and J. R. Mills, The 1993 RPA Timber Assessment Update, USDA Forest Service Rocky Mountain Forest and Range Experiment Station, Fort Collins, Colorado, 1995.

Henderson-Sellers, A., H. Zhang, G. Berz, K. Emanuel, W. Gray, C. Landsea, G. Holland, J. Lighthill, S-L. Shieh, P. Webster, and K. McGuffie, Tropical cyclones and global climate change: A post-IPCC assessment, *Bulletin of the American Meteorological Society, 79,* 19-38, 1998.

Hoffman, C., Let it snow, *Smithsonian, 29,* 50-58, 1998.

Houghton, R. A., J. L. Hackler, and K. T. Lawrence, The US carbon budget: Contributions for land-use change, *Science, 285,* 574-578, 1999.

Irland, L. C., Ice storm 1998 and the forests of the Northeast, *Journal of Forestry, 96,* 32-30, 1998.

Irland, L. C., Ice storms and forest impacts, *The Science of the Total Environment, 262,* 231-242, 2000.

Irland, L. C., D. Adams, R. Alig, C. J. Betz, C. Chen, M. Mutchins, B. A. McCarl, K. Skog, and B. W. Sohngen, Assessing socioeconomic impacts of climate change on US forests, wood product markets, and forest recreation, BioScience, in press, 2001.

Iverson, L. R., and A. M. Prasad, Predicting potential future abundance of 80 tree species following climate change in the eastern United States, *Ecological Monographs, 68,* 465-485, 1998.

Iverson, L. R., A. M. Prasad, B. J. Hale, and E. K. Sutherland, An atlas of current and potential future distributions of common trees of the eastern United States, *General Technical Report,* USDA Forest Service, Northeastern Forest Experiment Station, Radnor, Pennsylvania, 1999.

Iverson, L. R., and A. M. Prasad, Potential changes in tree species richness and forest community types following climate change, *Ecosystems,* in press, 2001.

Johnson, D. W., R. B. Thomas, K. L. Griffin, D. T. Tissue, J. T. Ball, B. R. Strain, and R. F. Walker, Effects of carbon dioxide and nitrogen on growth and nitrogen uptake in ponderosa and loblolly pine, *Journal of Environmental Quality, 27,* 414-425, 1998.

Joyce, L. A., J. R. Mills, L. S. Heath, A. D. McGuire, R. D. Haynes, and R. A. Birdsey, Forest sector impacts form changes in forest productivity under climate change, *Journal of Biogeography, 22,* 703-713, 1995.

Karl T. R., R. W. Knight, and N. Plummer, Trends in high-frequency climate variability in the twentieth century, *Nature, 377,* 217-220, 1995a.

Karl T. R., R. W. Knight, D. R. Easterling, and R. G. Quayle, Indices of climate change for the United States, *Bulletin of the American Meteorological Society, 77,* 279-292, 1995b.

Keane, R. E., K. C. Ryan, and M. A. Finney, Simulating the consequences of fire and climate regimes on a complex landscape in Glacier National Park, Montana, in *Fire in Ecosystem Management: Shifting the Paradigm from Suppression to Prescription,* edited by T. L. Pruden and L. A. Brennan, Tall Timbers Fire Ecology Conference Proceedings, No. 20, Tall Timbers Research Station, Tallahassee, Florida, 1998.

Kellogg, R. S., The timber supply of the United States, *Forest Service Circular 166,* USDA Forest Service, Washington, DC, 1907.

Kergoat, L., A model for hydrological equilibrium of leaf area index on a global scale, *Journal of Hydrology, 213,* 268-286, 1998.

Kerstiens, G., Shade tolerance as a predictor of responses to elevated CO_2 in trees, *Physiologia Plantarum, 102,* 472-480, 1998.

Knutson, T. R., T. R. Tuleya, and Y. Kurihara, Simulated increase of hurricane intensity in a CO_2-warmed climate, *Science, 279,* 1018-1020, 1998.

Knutson, T. R., and R. E. Tuleya, Increased hurricane intensities with CO_2-induced warming as simulated using the GFDL hurricane prediction system, *Climate Dynamics, 15,* 503-519, 1999.

Kozlowski, T. T., and S. G. Pallardy, *Growth Control in Woody Plants,* Academic Press, San Diego, California, 1997.

Lambers, H., F. S. Chapin, III, and T. L. Pons, *Plant Physiologyical Ecology,* Springer-Verlag, New York, 1998.

Leemans, R., A. van Amstel, C. Battjes, E. Kreileman, and S. Toet, The land cover and carbon cycle and consequences of large-scale utilization of biomass as an energy source, *Global Environmental Change, 6,* 335-357, 1996.

Lenihan, J. M., C. Daly, D. Bachelet, and R. P. Neilson, Simulating broad-scale fire severity in a dynamic global vegetation model, *Northwest Science, 72,* 91-103, 1997.

Loehle, C, and D. LeBlanc, Model-based assessments of climate change effects on forests: A critical review, *Ecological Modelling, 90,* 1-31, 1996.

Long, S. P., C. P. Osborne, and S. W. Humphries, Photosynthesis, rising atmospheric carbon dioxide concentration and climate change, in *Global Change: Effects on Coniferous Forests and Grasslands,* edited by A. I. Breymeyer, D. O. Hall, J. M. Melillo, and G. I. Agren, pp. 121-159, John Wiley and Sons, Ltd, New York, 1996.

Loomis, J., and J. Crespi, Estimated effects of climate change on selected outdoor recreation activities in the United States, in *The Impact of Climate Change on the United States Economy,* edited by R. Mendelsohn and J. E. Neumann, Cambridge University Press, Cambridge, England, 1999.

Lovett, G. M., Atmospheric deposition of nutrients and pollutants in North America: an Ecological perspective, *Ecological Applications, 4,* 629-650, 1994.

Lugo, A. E., Effects and outcomes of hurricanes in a climate change scenario, *The Science of the Total Environment, 262,* 243-252, 2000.

Lugo, A. E., and F. N. Scatena, Ecosystem-level properties of the Luquillo Experimental Forest, with emphasis of the tabonuco forest, in *Tropical Forests: Management and Ecology,* edited by A. E. Lugo and C. Lowe, Springer-Verlag, New York, 1995.

Lugo, A. E., and F. N. Scatena, Background and catastrophic tree mortality in tropical moist, wet, and rain forests, *Biotropica, 28,* 585-599, 1996.

Mack, M. C., and C. M. D'Antonio, Impacts of biological invasions on disturbance regimes, *Trends in Ecology and Evolution, 13,* 195-198, 1998.

Magill, A. H., J. D. Aber, J. J. Hendricks, R. D. Bowden, J. M. Melillo, and P. Steudler, Biogeochemical response of forest ecosystems to simulated chronic nitrogen deposition, *Ecological Applications, 7,* 402-415, 1997.

Magill, A., J. Aber, G. Berntson, W. McDowell, K. Nadelhoffer, J. Melillo, and P. Steudler, Long-term nitrogen additions and nitrogen saturation in two temperate forests, *Ecosystems, 3,* 2000.

Malcolm, J. R, and A. Markham, Climate change threats to the national parks and protected areas of the United States and Canada, World Wildlife Fund, Washington, DC, 1998.

Matthews, E., and A. Hammond, Critical consumption trends and implications, World Resources Institute, Washington, DC, 1999.

McGuire, A. D., J. M. Melillo, L. A. Joyce, D. W. Kicklighter, A. L. Grace, B. Moore, and C. J. Vorosmarty, Interactions between carbon and nitrogen dynamics in estimating net primary productivity for potential vegetation in North America, *Global Biogeochemical Cycles, 6,* 101-124, 1992.

McIntosh, R. J., J. A. Tainter, and S. K. McIntosh, (Eds.), *The Way the Wind Blows, Climate, History and Human Action,* Columbia University Press, 2000.

Mendelsohn, R., and M. Mackowiki, The impact of climate change on outdoor recreation, in *The Impact of Climate Change on the United States Economy,* edited by R. Mendelsohn and J. E. Neumann, Cambridge University Press, Cambridge, England, 1999.

Mohren, G. M. J., K. Kramer, and S. Sabate, *Impacts of Global Change on Tree Physiology and Forest Ecosystems,* Kluwer Academic Publishers, Boston, Massachusetts, 1996.

Mooney, H. A., J. Canadell, F. S. Chapin, J. Ehleringer, C. Korner, R. McMurtrie, W. J. Parton, L. F. Pitelka, and E. D. Schulze, Ecosystem physiology responses to global change, in *Implications of Global Change for Natural and Managed Ecosystems: A Synthesis of GCTE and Related Research,* edited by H. B. Walker, W. L. Steffen, J. Canadell, and J. S. I. Ingram, IGBP Book Series No. 4, pp. 141-189, Cambridge University Press, 1999.

Morison, J. I. L, and D. W. Lawlor, Interactions between increasing CO_2 concentration and temperature on plant growth, *Plant, Cell, and Environment, 22,* 659-682, 1999.

Musselman, R. C., and W. J. Massman, Ozone flux to vegetation and its relationship to plant response and ambient air quality standards, *Atmospheric Environment, 33,* 65-73, 1999.

Musselman, R. C., L. Hudnell, M. W. Williams, and R. A. Sommerfeld, Water chemistry of Rocky Mountain Front Range aquatic ecosystems, *Research paper RM-RP-325,* USDA Forest Service Rocky Mountain Forest and Range Experiment Station, Fort Collins, Colorado, 1996.

NADP/NTN, NADP/NTN Wet deposition of the United States. 1996. National Atmospheric Deposition Program Office, Champaign, Illinois, 10 pp., 1997.

Negron, J. F., Probability of infestation and extent of mortality associated with the Douglas-fir beetle in the Colorado Front Range, *Forest Ecology & Management, 107,* 71-85, 1998.

Neilson, R. P., A model for predicting continental-scale vegetation distribution and water balance, *Ecological Applications 5,* 362-385, 1995.

Neilson, R. P., I. C. Prentice, B. Smith, T. Kittel, and D. Viner, Simulated changes in vegetation distribution under global warming, Appendix C, in *The Regional Impacts of Climate Change: An Assessment of Vulnerability. Intergovernmental Panel on Climate Change,* edited by R. T. Watson, M. C. Zinyowera, and R. H. Moss, Cambridge University Press, New York, 1998.

Neilson, R. P., and R. J. Drapek, Potentially complex biosphere responses to transient global warming, *Global Change Biology, 4,* 505-521, 1998.

Norby, R. J., S. D. Wullschleger, C. A. Gunderson, D. W. Johnson, and R. Ceulemans, Tree responses to rising CO_2 in field experiments: Implication for the future forest, *Plant, Cell and Environment, 22,* 683-714, 1999.

Noss, R. F., E. T. LaRoe, and J. M. Scott, Endangered ecosystems of the United States: A preliminary assessment of loss and degradation, US Fish And Wildlife Service Report, Washington, DC, 1994.

Ollinger, S. V., J. D. Aber, and P. B. Reich, Simulating ozone effects on forest productivity: Interactions between leaf,- canopy-, and stand-level processes, *Ecological Applications, 7,* 1237-1251, 1997.

O'Neill, R. V., K. H. Riitters, J. D. Wickham, and K. Bruce Jones, Landscape patterns and regional assessment, *Ecosystem Health, 5,* 225-233, 1999.

Perez-Garcia, J., L. A. Joyce, A. D. McGuire, and C. S. Binkley, Economic impact of climatic change on the global forest sector, in *Economics of Carbon Sequestration in Forestry,* edited by R.A. Sedjo, R. N. Sampson, and J. Wisniewski, Lewis Publishers, Boca Raton, Florida, 1997.

Peters, R. L., Conservation of biological diversity in the face of climate change, in *Global Warming and Biological Diversity*, edited by R. L. Peters and T. E. Lovejoy, Yale University Press, New Haven, Connecticut, 15-30, 1992.

Peterson, C. J., Catastrophic wind damage to North American forests and the potential impact of climate change, The Science of the Total Environment, 262, 287-312, 2000.

Price, C., and D. Rind, The impact of a 2xCO_2 climate on lightning-caused fires, Journal of Climate, 7, 1484-1494, 1994.

Pritchard, S. G., C. Mosjidis, C. M. Peterson, G. B. Runion, and H. H. Rogers, Anatomical and morphological alterations in longleaf pine needles resulting from growth in elevated CO_2: Interactions with soil resource availability, International Journal of Plant Sciences, 159, 1002-1009, 1998.

Pulliam, H. R., Sources, sinks, and population regulation, *American Naturalist, 132,* 652-661, 1988.

Rey, A., and P. G. Jarvis, Long-term photosynthetic acclimation to increased atmospheric CO_2 concentration in young birch (*Betula pendula*) trees, *Tree Physiology, 18,* 441-450, 1998.

Royer, J. F., F. Chauvin, B. Timbal, P. Araspin, and D. Grimal, A GCM study of the impact of greenhouse gas increase on the frequency of occurrence of tropical cyclones, *Climatic Change, 38,* 307-343, 1998.

Sampson, R. N., and D. Hair, Forests and *Global Change, Volume 2: Forest Management Opportunities for Mitigating Carbon Emissions,* American Forests, Washington, DC, 1996.

Saxe, H., D. S. Ellsworth, and J. Heath, Tree and forest functioning in an enriched CO_2 atmosphere, *New Phytologist, 139,* 395-436, 1998.

Schimel, D, J., et al., Contribution of increasing CO_2 and climate to carbon storage by ecosystems in the United States, *Science, 287,* 2004-2006, 2000.

Shafer, S. L., P. J. Bartlein, and R. S. Thompson, Potential changes in the distributions of western North America tree and shrub taxa under future climate scenarios, *Ecosystems,* in review, 2001.

Shaver, G. R., W. D. Billings, F. S. Chapin, III, A. E. Giblin, K. J. Nadelhoffer, W. C. Oechel, and E. B. Rastetter, Global change and the carbon balance of arctic ecosystems, *BioScience, 42,* 433-441, 1992.

Shriner, D.S., et al., North America, in *The Regional Impacts of Climate Change,* edited by Robert T. Watson, M. C. Zinyowera, and R. H. Moss, pp. 253-330, Cambridge University Press, Cambridge, England, 1998.

Shugart, H. H., *A Theory of Forest Dynamics: The Ecological Implications of Forest Succession Models*, Springer-Verlag, New York, 1984.

Simberloff, D., The biology of invasions, in *Strangers in Paradise: Impact and Management of Nonindigenous Species in Florida,* edited by D. Simberloff, D. Schmitz, and T. Brown, pp. 3-17, Island Press, Washington, DC, 1997.

Simberloff, D., Global climate change and introduced species in forests, *The Science of the Total Environment, 262,* 253-262, 2000.

Sohngen, R., and R. Mendelsohn, Valuing the impact of large-scale ecological change in a market: The effect of climate change on US timber, *American Economic Review, 88,* 686-710, 1998.

Solomon, A. M., I. C. Prentice, R. Leemans, and W. P. Cramer, The interactions of climate and land use in future terrestrial carbon storage and release, *Water, Air, and Soil Pollution 70,* 595-614, 1993.

Stitt, M., and A. Krapp, The interaction between elevated carbon dioxide and nitrogen nutrition: The physiological and molecular background, *Plant, Cell, and Environment, 22,* 583-621, 1999.

Stocks, B. J., M. A. Fosberg, et al., Climate change and forest fire potential in Russian and Canadian boreal forests, *Climatic Change, 38,* 1-13, 1998.

Taylor, Jr., G. E., D. W. Johnson, and C. P. Andersen, Air pollution and forest ecosystems: A regional to global perspective, *Ecological Applications, 4,* 662-689, 1994.

Tjoelker, M. J., J. Oleksyn, and P. B. Reich, Seedlings of five boreal tree species differ in acclimation of net photosynthesis to elevated CO_2 and temperature, *Tree Physiology, 18,* 715-726, 1998a.

Tjoelker, M. J., J. Oleksyn, and P. B. Reich, Temperature and ontogeny mediate growth response to elevated CO_2 in seedlings of five boreal tree species, *New Phytologist, 140,* 197-210, 1998b.

The H. John Heinz III Center, Designing a report on the state of the Nation's ecosystems: Selected measurements for cropland, forests and coasts and oceans, The H. John Heinz III Center, Washington, DC, 1999.

Torn, M. S., and J. S. Fried, Predicting the impact of global warming on wildfire, *Climatic Change, 21,* 257-274, 1992.

Torn, M. S., E. Mills, and J. S. Fried, Will climate change spark more wildfire damages, *LBLN Report No. LBNL-42592,* Lawrence Berkeley National Laboratory, California, 1999.

Ungerer, M. J., M. P. Ayres, and M. J. Lombardero, Climate and the northern distribution limited of *Dendroctonus frontalis* Zimmerman (Coleoptera: Scolytidae), *Journal of Biogeography 26,* 113-1145, 1999.

USDA Forest Service, Forest Statistics of the United States, Forest Service Resource Bulletin PNW-RB-168, USDA Forest Service Pacific Northwest Research Station, Portland, Oregon, 1989.

USDA Forest Service, RPA assessment of the forest and rangeland situation in the United States, *Forest Resource Report No. 27,* USDA Forest Service, Washington, DC, 1994.

USDA Forest Service, Report of the Forest Service, Fiscal Year 1998, USDA Forest Service, Washington DC, 171 pp., 1998.

USDA Forest Service, Routt National Forest, Draft Environmental Impact Statement South Fork Salvage Analysis, USDA Forest Service, Routt National Forest, Hahns Peak/Bears Ears Ranger District, Steamboat Springs, Colorado, 159 pp., 1998.

United States Department of Interior, Fish and Wildlife Service, Box score, *Endangered Species Bulletin, 25,* 40, 2000.

US Environmental Protection Agency, Air quality criteria for ozone and related photochemical oxidants, *EPA Report No. EAP/600/P-93/004af, Vol. II,* US EPA, Research Triangle Park, North Carolina, 1996.

VEMAP members, Vegetation/ecosystem modeling and analysis project: Comparing biogeography and biogeochemistry models in a continental-scale study of terrestrial ecosystem responses to climate change and CO_2 doubling, *Global Biogeochemical Cycles, 9,* 407-437, 1995.

Vitousek, P. M., J. D. Aber, R. W. Howarth, G. E. Likens, P. A. Matson, D. W. Schindler, W. H. Schlesinger, and D. G. Tilman, Human alteration of the global nitrogen cycle: Sources and consequences, *Ecological Applications, 7,* 737-750, 1997.

Vitousek, P. M., C. M. D'Antonio, L. L. Loope, and R. Westbrooks, Biological invasions as global environmental change, *American Scientist, 84,* 468-478, 1996.

Wall, G., Implications of global climate change for tourism and recreation in wetland areas, *Climatic Change, 40,* 371-389, 1998.

Walker, B. H., W. L. Steffen, and J. Langridge, Interactive and integrated effects of global change on terrestrial ecosystems, in *The Terrestrial Biosphere and Global Change,* edited by B. Walker, W. Steffen, J. Canadell, and J. Ingram, Cambridge University Press, 1999.

Walsh K., and A. B. Pittock, Potential changes in tropical storms, hurricanes, and extreme rainfall events as a result of climate change, *Climatic Change, 39,* 199-213, 1998.

Walters, C., Challenges in adaptive management of riparian and coastal ecosystems, *Conservation Ecology, 1*(2), 1997.

Watson, R. T., M. C. Zinyowera, and R. H. Moss, (Eds.), *Climate Change 1995. Impacts, Adaptations, and Mitigation of Climate Change: Scientific-Technical Analyses,* Cambridge University Press, Cambridge, United Kingdom, 1996.

Watson, R. T., M. C. Zinyowera, R. H. Moss, (Eds.), *The Regional Impacts of Climate Change: An Assessment of Vulnerability. Intergovernmental Panel on Climate Change,* Cambridge University Press, New York, 1998.

Watson, R. T., I. R. Nobel, B. Bolin, N. H. Ravindranath, D. J. Verardo, and D. J. Dokken (Eds.), *IPCC, 2000: Land Use, Land-Use Change, and Forestry. A Special Report of the Intergovernmental Panel on Climate Change,* Cambridge University Press, Cambridge, United Kingdom, 2000.

Webb, T., III, Past changes in vegetation and climate: Lessons for the future, in *Global Warming and Biological Diversity,* edited by R. L. Peters and T. E. Lovejoy, Yale University Press, New Haven, Connecticut, 59-75, 1992.

Weinstein, D. A., W. P. Cropper, Jr., and S. G. McNulty, Summary of simulated forest response to climate change in the southeastern United States, in *The Productivity and Sustainability of Southern Forest Ecosystems in a Changing Environment,* edited by R. A. Mickler and S. Fox, Springer-Verlag, New York, pp. 479-500, 1998.

Wilcove, D. S., D. Rothstein, J. Dubow, A. Phillips, and E. Losos, Quantifying threats to imperiled species in the United States, *BioScience, 48,* 607-615, 1998.

Will, R. E., and R. O. Teskey, Effect of elevated carbon dioxide concentration and root restriction on net photosynthesis, water relations, and foliar carbohydrate status of loblolly pine seedlings, *Tree Physiology, 17,* 655-661, 1997a.

Will, R. E, and R. O. Teskey, Effect of irradiance and vapor-pressure deficit on stomatal response to CO_2 enrichment of four tree species, *Journal of Experimental Botany, 48,* 2095-2102, 1997b.

Williams, M. W., J. S. Baron, N. Caine, R. Sommerfeld, and J. R. R. Sanford, Nitrogen saturation in the Rocky Mountains, *Environmental Science and Technology, 30,* 640-646, 1996.

Williamson, M., Invasions, *Ecography, 22,* 5-12, 1999.

Wooton, B. M., and M. D. Flannigan, Length of the fire season in a changing climate, *Forestry Chronicle, 69,* 198-192, 1993.

ACKNOWLEDGMENTS

Many of the materials for this chapter are based on contributions from participants on and those working with the:

Forest Sector Assessment Team
John Aber*, University of New Hampshire
Steven McNulty*, US Department of Agriculture, Forest Service
Darius Adams, Oregon State University
Ralph Alig, US Department of Agriculture, Forest Service
Matthew P. Ayres, Dartmouth College
Dominique Bachelet, Oregon State University
Patrick Bartlein, University of Oregon
Carter J. Betz, US Department of Agriculture, Forest Service
Chi-Chung Chen, Texas A&M University
Rosamonde Cook, Colorado State University
David J. Currie, University of Ottawa, Canada
Virginia Dale, Oak Ridge National Laboratory
Raymond Drapek, Oregon State University
Michael D. Flannigan, Canadian Forest Service
Curt Flather, US Department of Agriculture, Forest Service
Andy Hansen, Montana State University

Paul J. Hanson, Oak Ridge National Laboratory
Mark Hutchins, Sno-Engineering, Inc
Louis Iverson, US Department of Agriculture, Forest Service
Lloyd Irland, The Irland Group
Linda Joyce, US Department of Agriculture, Forest Service
María Lombardero, Universidad de Santiago, Lugo, Spain
James Lenihan, Oregon State University
Ariel E. Lugo, US Department of Agriculture, Forest Service
Bruce McCarl, Texas A&M University
Ron Neilson, US Department of Agriculture, Forest Service
Chris J. Peterson, University of Georgia
Sarah Shafer, University of Oregon
Daniel Simberloff, University of Tennessee
Ken Skog, US Department of Agriculture, Forest Service
Brent L. Sohngen, Ohio State University
Brian J. Stocks, Canadian Forest Service
Frederick J. Swanson, US Department of Agriculture, Forest Service
Jake F. Weltzin, University of Tennessee
B. Michael Wotton, Canadian Forest Service

* signifies Assessment team chairs/co-chairs

CHAPTER 18

CONCLUSIONS AND RESEARCH PATHWAYS

National Assessment Synthesis Team

Contents of this Chapter

Conclusions

Research Pathways

CONCLUSIONS

Large Impacts in Some Places

The impacts of climate change will be significant for Americans. The nature and intensity of impacts will depend on the location, activity, time period, and geographic scale considered. For the nation as a whole, direct economic impacts are likely to be modest. However, the range of both beneficial and harmful impacts grows wider as the focus shifts to smaller regions, individual communities, and specific activities or resources. For example, while wheat yields are likely to increase at the national level, yields in western Kansas, a key US breadbasket region, are projected to decrease substantially under the Canadian climate model scenario. For resources and activities that are not generally assigned an economic value (such as natural ecosystems), substantial disruptions are likely.

Multiple-stresses Context

While Americans are concerned about climate change and its impacts, they do not think about these issues in isolation. Rather they consider climate change impacts in the context of many other stresses, including land-use change, consumption of resources, fire, and air and water pollution. This finding has profound implications for the design of research programs and information systems at the national, regional, and local levels. A true partnership must be forged between the natural and social sciences to more adequately conduct assessments and seek solutions that address multiple stresses.

Urban Areas

Urban areas provide a good example of the need to address climate change impacts in the context of other stresses. Although large urban areas were not formally addressed as a sector, they did emerge as an issue in most regions. This is clearly important because a large fraction of the US population lives in urban areas, and an even larger fraction will live in them in the future. The compounding influence of future rises in temperature due to global warming, along with increases in temperature due to local urban heat island effects, makes cities more vulnerable to higher temperatures than would be expected due to global warming alone. Existing stresses in urban areas include crime, traffic congestion, compromised air and water quality, and disruptions of personal and business life due to decaying infrastructure. Climate change is likely to amplify some of these stresses, although all the interactions are not well understood.

Impact, Adaptation, and Vulnerability

As the Assessment teams considered the negative impacts of climate change for regions, sectors, and other issues of concern, they also considered potential adaptation strategies. When considered together, negative impacts along with possible adaptations to these impacts define vulnerability. As a formula, this can be expressed as vulnerability equals negative impact minus adaptation. Thus, in cases where teams identified a negative impact of climate change, but could not identify adaptations that would reduce or neutralize the impact, vulnerability was considered to be high. A general sense emerged that American society would likely be able to adapt to most of the impacts of climate change on human systems but that the particular strategies and costs were not known.

Widespread Water Concerns

A prime example of the need for and importance of adaptive responses is in the area of water resources. Water is an issue in every region, but the nature of the vulnerabilities varies, with different nuances in each. Drought is an important concern in every region. Snowpack changes are especially important in the West, Pacific Northwest, and Alaska. Reasons for the concerns about water include increased threats to personal safety, further reductions in potable water supplies, more frequent disruptions to transportation, greater damage to infrastructure, further degradation of animal habitat, and increased competition for water currently allocated to agriculture. The table below illustrates some of the key concerns related to water in each region.

Health, an Area of Uncertainty

Health outcomes in response to climate change are highly uncertain. Currently available information suggests that a range of health impacts is possible. At present, much of the US population is protected against adverse health outcomes associated with weather and/or climate, although certain demographic and geographic populations are at greater risk. Adaptation, primarily through the maintenance

and improvement of public health systems and their responsiveness to changing climate conditions and to identified vulnerable subpopulations, should help to protect the US population from adverse health outcomes of projected climate change. The costs, benefits, and availability of resources for such adaptation need to be considered, and futher research on the relationships between climate/weather and health is needed.

Vulnerable Ecosystems

Many US ecosystems, including wetlands, forests, grasslands, rivers, and lakes, face possibly disruptive climate changes. Of everything examined in this Assessment, ecosystems appear to be the most vulnerable to the projected rate and magnitude of climate change, in part because the available adaptation options are very limited. This is important because, in addition to their inherent value, they also supply Americans with vital goods and services, including food, wood, air and water

Ecosystem Goods and Services

Ecosystem	Goods	Services
Forests	timber, fuelwood, food (such as honey, mushrooms, and fruits)	purify air and water, generate soil, absorb carbon, moderate weather extremes and impacts, and provide wildlife habitat and recreation
Freshwater Systems	drinking and irrigation water, fish, hydroelectricity	control water flow, dilute and carry away wastes, and provide wildlife habitat, transportation corridors, and recreation
Grasslands	livestock (food, game, hides, fiber), water, genetic resources	purify air and water, maintain biodiversity, and provide wildlife habitat, employment, aesthetic beauty, and recreation
Coastal Systems	fish, shellfish, salt, seaweeds, genetic resources	buffer coastlines from storm impacts, maintain biodiversity, dilute and treat wastes, and provide harbors and transportation routes, wildlife habitat, employment, beauty, and recreation
Agro-ecosystems	food, fiber, crop genetic resources	build soil organic matter, absorb carbon, provide employment, and provide habitat for birds, pollinators, and soil organisms

WATER ISSUES

Region	Floods	Droughts	Snowpack/ Snowcover	Groundwater	Lake, river, and reservoir levels	Quality
Northeast	X	X	X	X		X
Southeast	X	X		X		X
Midwest	X	X	X	X	X	X
Great Plains	X	X	X	X	X	X
West	X	X	X	X	X	X
Northwest	X	X	X		X	
Alaska		X	X			
Islands	X	X		X		X

This table identifies some of the key regional concerns about water. Many of these issues were raised and discussed by stakeholders during regional workshops and other Assessment meetings held between 1997 and 2000.

purification, and protection of coastal lands. Ecosystems around the nation are likely to be affected, from the forests of the Northeast to the coral reefs of the islands in the Caribbean and the Pacific.

Agriculture and Forestry Likely to Benefit in the Near Term

In agriculture and forestry, there are likely to be benefits due to climate change and rising CO_2 levels at the national scale and in the short term under the scenarios analyzed here. At the regional scale and in the longer term, there is much more uncertainty. It must be emphasized that the projected increases in agricultural and forest productivity depend on the particular climate scenarios and assumed CO_2 fertilization effects analyzed in this Assessment. If, for example, climate change resulted in hotter and drier conditions than projected by these scenarios, both agricultural and forest productivity could possibly decline.

Potential for Surprises

Some of the greatest concerns emerge not from the most likely future outcomes but rather from possible "surprises." Due to the complexity of Earth systems, it is possible that climate change will evolve quite differently from what we expect. Abrupt or unexpected changes pose great challenges to our ability to adapt and can thus increase our vulnerability to significant impacts.

A Vision for the Future

Much more information is needed about all of these issues in order to determine appropriate national and local response strategies. The regional and national discussion on climate change that provided a foundation for this first Assessment should continue and be enhanced. This national discourse involved thousands of Americans: farmers, ranchers, engineers, scientists, business people, local government officials,

and a wide variety of others. This unique level of stakeholder involvement has been essential to this process, and will be a vital aspect of its continuation. The value of such involvement includes helping scientists understand what information stakeholders want and need. In addition, the problem-solving abilities of stakeholders have been key to identifying potential adaptation strategies and will be important to analyzing such strategies in future phases of the assessment.

The next phase of the assessment should begin immediately and include additional issues of regional and national importance including urban areas, transportation, and energy. The process should be supported through a public-private partnership. Scenarios that explicitly include an international context should guide future assessments. An integrated approach that assesses climate impacts in the context of other stresses is also important. Finally, the next assessment should undertake a more complete analysis of adaptation. In the current Assessment, the adaptation analysis was done in a very preliminary way, and it did not consider feasibility, effectiveness, costs, and side effects. Future assessments should provide ongoing insights and information that can be of direct use to the American public in preparing for and adapting to climate change.

The following section offers suggestions about new approaches, new knowledge, and new capabilities that would improve future assessments and thus provide more effective guidance for responding to the challenges posed by climate change.

ECOSYSTEMS VULNERABILITIES

The table below gives a partial list of potential impacts for major ecosystem types in various regions of the US. While the impacts are often stated in terms of what is likely to happen to plant communities, it is important to recognize that plant-community changes will also affect animal habitat.

Ecosystem Type

Forests

Impacts	NE	SE	MW	GP	WE	PNW	AK	IS
Changes in tree species composition and alteration of animal habitat	X	X	X		X	X	X	X
Displacement of forests by open woodlands and grasslands under a warmer climate in which soils are drier		X						

Grasslands

Displacement of grasslands by open woodlands and forests under a wetter climate					X			
Increase in success of non-native invasive plant species				X	X	X		X

Tundra

Loss of alpine meadows as their species are displaced by lower-elevation species	X				X	X	X	
Loss of northern tundra as trees migrate poleward							X	
Changes in plant community composition and alteration of animal habitat							X	

Semi-arid and Arid

Increase in woody species and loss of desert species under wetter climate					X			

Freshwater

Loss of prairie potholes with more frequent drought conditions				X				
Habitat changes in rivers and lakes as amount and timing of runoff changes and water temperatures rise	X	X	X	X	X	X		

Coastal & Marine

Loss of coastal wetlands as sea level rises and coastal development prevents landward migration	X	X			X	X		X
Loss of barrier islands as sea-level rise prevents landward migration	X	X						
Changes in quantity and quality of freshwater delivery to estuaries and bays alter plant and animal habitat	X	X			X	X	X	X
Loss of coral reefs as water temperature increases		X						X
Changes in ice location and duration alter marine mammal habitat							X	

RESEARCH PATHWAYS

New Approaches, New Knowledge, and New Capabilities for Our Nation

The National Assessment has defined a new vision for climate-impacts research. This vision has at its core a focus on integrated regional analysis and a close partnership of natural and social scientists with local, regional, and national stakeholders. Integrated analysis refers to considering the full range of stresses that affect a resource or system, including climate change and variability, land use change, air and water pollution, and many other human and natural impacts. For example, in studying water quality in a particular place, the direct and indirect effects of urban development, agricultural runoff, industrial pollution, and climate change-induced increases in heavy precipitation events all would need to be considered, along with many other factors. Integrated analysis also refers to integrating across all the relevant spatial scales for an issue, and these may extend from local to regional to national and even to global, depending on the issue. In the example of the local water quality study, this would mean integrating the effects of large-scale weather patterns on precipitation, as well as pollution inputs on both large and small scales, some of which originate far from the study area. Such integration across both multiple stresses and multiple scales is needed to provide the type of comprehensive analysis that decision-makers seek.

Guided by this vision, the first National Assessment has identified a range of regional and sector vulnerabilities to climate variability and change that call for the attention of the American people and their leaders. In identifying vulnerabilities, the authors of the Assessment took great care to evaluate the likelihood of various climate-related outcomes. The likelihood of a number of outcomes was considered to be high. However, the likelihood of some of the other important potential outcomes was more difficult to judge due to lack of appropriate methods, uncertainties in knowledge, or shortcomings in research infrastructure such as computer power.

As our nation considers its strategies for dealing with climate-related vulnerabilities, scientists must work to reduce uncertainties underlying the vulnerability estimates. To assure that efforts to reduce uncertainties are efficient and to the point, the authors of the Assessment have identified a short list of priority research steps that are outlined below. These steps are organized into three categories: New Approaches, New Knowledge, and New Capabilities. It is vital to our national interest that we meet these research needs so that we can, with increasing certainty, address the critical question: How vulnerable or resilient are the nation's natural and human resources and systems to the changes in climate projected to occur over the decades ahead? With the new vision of regional analysis and scientist-stakeholder partnerships developed in the National Assessment, we have a powerful approach to effectively address this complex question.

Expanding New Approaches

This first National Assessment experimented with new approaches for linking the emerging findings and capabilities of the scientific community with the real-time needs of stakeholders who manage resources, grow food, plan communities, sustain commerce, and ensure public welfare. Teams were established in various regions across the country to look at how climate variability and change would affect particular locations. Other teams were established on particular topics that focused on how climate variability and change would affect issues of national significance. These types of efforts need to be sustained and expanded. In doing so we can build on a number of existing federal programs.

Recommendation 1: Develop a More Integrated Approach to Examining Impacts and Vulnerabilities to Multiple Stresses

The key requirement is to develop a truly integrated and widely accessible approach at regional and national scales appropriate for the examination of regional and national problems associated with biologic, hydrologic, and socioeconomic systems. A number of the regional and sector studies supporting the National Assessment have made important progress in such efforts but substantial additional efforts are required.

Expand the national capability to develop integrated, regional approaches for assessing impacts and vulnerabilities.

The regional teams supporting the National Assessment have produced new and innovative partnerships among a wide variety of scientists and stakeholders. In the process, they have catalyzed new modes of research and have demonstrated the potential of an integrated approach to assessing the consequences of climate variability and change. If the nation is to have improved projections of the impacts of and vulnerabilities to multiple stresses, we must accelerate the process of integrating research capabilities across the spectrum of natural and social sciences with the needs of public and private decision-makers. The importance of multiple stresses on specific environments and the importance of linkages between physical, biological, chemical, and human systems, require enhanced capabilities for regional analysis. The key elements of this strategy must be to: (a) integrate observations at a regional level; (b) develop a comprehensive system designed to make the enormous amounts of data and information more accessible and useful to the public; (c) enable field and experimental studies that focus on solving regional problems; and (d) develop a foundation for building tractable, high-resolution coupled models that can address the outcomes associated with multiple stresses unique to each area of the country. Regional assessments add impetus for developing a comprehensive integrated approach, and this integration will engender substantial new capabilities to address the relationships between climate and air quality, energy demand, water quality and quantity, species distribution, ecosystem character, ultraviolet radiation, and human health indices in specific regions.

Perform integrated national investigations of additional sectors and issues.

The choice of major sectors in this Assessment (water, forests, human health, coastal regions, and agriculture) reflected the fact that they were likely to be informative and important. Some of the themes were represented by a strong foundation of methods and models (e. g., agriculture) while others represented capabilities at very early stages of development (human health). These themes yielded a number of future research

needs (e.g., the need to better understand human health relationships to extreme temperatures, other extreme weather events, and air quality, and to better characterize the relationships between climate and disease vectors and then between vector distribution and disease). In addition, a large number of important sectors and themes were not addressed, including climate impacts on transportation, energy, urban areas, and wildlife. These will require future investment in supporting research and assessment.

Consider international linkages in assessing national impacts.

In some cases, the vulnerability and resiliency of the US to climate impacts are highly dependent on the nature of the changes in other countries. For example, in the case of agriculture, the nature of the impacts strongly depends on international markets, and therefore the production and distribution of major crops around the world. These markets will reflect the extent of temperature and precipitation changes in other nations and the ability of these nations to cope with climate change and variability. Assessment of US impacts, including potential benefits, will increasingly require an examination of changes and response strategies around the world, and the manner in which these are translated to the global marketplace and environment.

Recommendation 2: Develop New Ways to Assess the Significance of Global Change to People

New methods for examining the potential impacts of climate change, adaptation options, and the vulnerability of communities, institutions, and sectors are essential to improving the assessment process. Research on these issues would result in a far greater ability to anticipate possible surprises, incorporate socioeconomic data in our analyses, and provide information that is useful for public and private decision-makers. The key research requirements involve improving methods to:

Understand and assign value to large and non-market impacts (e.g., on communities, resources, and ecosystems).

Changes that occur in natural and managed ecosystems, natural resources, and the other sectors are important because people assign them value, either in market or non-market terms. It is crucial to develop new ways to assign values to possible future changes in resources and ecosystems, especially in the cases of very large impacts and of processes and services that do not produce marketable goods. A focus on large impacts and non-market systems should provide insights that would enable decision-makers to understand the potential consequences of environmental change, as well as the potential consequences of particular adaptation or mitigation decisions.

Represent, analyze, and report uncertainties.

Describing scientific uncertainty is a task that faces every assessment of impacts or vulnerability. Findings in this Assessment, and their associated uncertainties, are based on the considered judgment of the NAST and on the peer-reviewed literature. The assessment process should extend the capabilities of decision-makers to understand potential uncertainties. For this reason, a range of additional methods for representing, analyzing, and reporting uncertainties should become a research focus.

Assess potential thresholds and breakpoints.

Some ecosystems and human institutions do not respond to rapid changes or stresses in continuous ways; if the stress exceeds certain thresholds, the system changes very rapidly, and sometimes in irreversible ways. It is very important to understand these types of responses because they raise particularly difficult challenges for adaptation. Using climate scenarios to help determine the conditions under which such changes might occur is therefore extremely important to pursue, because it can provide information of direct utility to decision-makers.

Develop and apply internally consistent socioeconomic futures for use in assessing impacts.

In order to consider adaptation responses and ultimate vulnerabilities to climate change and other environmental stresses, assessments need to consider alternative possible socioeconomic and climatic futures using scenarios, probability distributions, or other methods. This Assessment began this process, but new methods to develop and apply such futures will improve the quality of the evaluation of the potential for adaptation in sector, regional and national analyses.

Developing New Knowledge

Determining how climate change will affect us necessarily builds on a wide array of scientific knowledge, not just of how the atmosphere works, but of how land ecosystems, the oceans, society, and many other aspects of the Earth system interact. Building this base of knowledge was the reason for establishing the US Global Change Research Program, and for the many programs and projects that it supports. Findings of this research have been essential to the overall undertaking of the Assessment. However, in the course of this Assessment, a number of areas have been identified where specific types of new knowledge are needed to assist society in preparing for the changing conditions of the 21st century. Improving projections of the responses of ecosystems, societal and economic systems, and climate, would improve scientists' ability to answer questions that are important to decision-makers.

Recommendation 3: Improve Projections of How Ecosystems Will Respond

The nature of the response of complex natural and managed ecosystems to multiple stresses is one of the most important challenges to providing more certain projections of the impacts of climate variability and future climate change. Scientific studies are needed to extend our knowledge of many types of interrelationships between climate and ecosystems. These complex, interdependent interactions determine how organisms will respond to climate and other stresses and determine the potential vulnerability and/or resilience of these systems. Areas requiring intensified research to address this challenge include:

Terrestrial and aquatic natural ecosystem responses to multiple stresses, including the consequences for productivity, biodiversity, and other ecosystem processes and services.

Information is lacking on local, watershed, and continental scales to evaluate the responses of terrestrial and aquatic ecosystems to combinations of environmental stresses. Experimentally and observationally based investigations are needed of combinations of important environmental changes (such as changes in: CO_2 concentration; climate variability, temperature, land use, air and water quality, and species composition) as they affect important ecological processes (including productivity and nutrient cycling), attributes (such as species diversity, the responses and interactions of important individual plant and animal species, and the interactions among plant and animal communities), and services (such as regulating runoff). Results from such field experiments are needed as inputs for more sophisticated generations of ecosystem models to provide the information needed to project the responses of ecosystems to combinations of stresses, rather than separately treating individual stresses.

Managed ecosystem responses to multiple stresses, including their consequences for water quality and runoff, soil fertility, agricultural and forest productivity, and pest, weed, crop, and pathogen interactions through the development of integrated observations, process studies, and models.

Effective management of agricultural lands, forested ecosystems, and watersheds is closely coupled to the influences of pests, pathogens, climate, and other environmental variables. These interactions have not been addressed adequately in an experimental fashion. Research is needed on the interactions of these factors as they affect crop and forest productivity, soil fertility, water quality and quantity, the spread of pathogens and weeds, etc. Results from such experiments would provide insight into potential vulnerabilities from the combinations of environmental stresses, and input to better management responses to changes in those stresses.

The importance and interactions of climate, land cover, and land use in nutrient cycling, water supply and quality, runoff, and soil fertility.

Current land use models include socioeconomic variables such as human population size, affluence, and culture. However, the influences of climate change on land use, and of land use changes on regional climate, are not adequately considered. Available analyses of ecological change tend to make one of two simplifications: either that a region is covered by its potential natural vegetation (the vegetation that would exist in the absence of human activities) or that the land-cover and land-use will remain as it is now, independent of climatic or other stresses. Both of these perspectives are limiting because they simplify the feedbacks between land use, ecological systems, and climate. New models of land-use change that integrate actual land-cover and land-use information with ecological and economic processes would provide a crucial context for examining the potential consequences of human land-use decisions on a wide variety of ecosystem goods and services.

Better observations and models of ecosystem disturbance, and species dispersal and recruitment.

The ability to project changes in the ranges of important tree and plant species, and therefore changes in the make-up of forests and other ecosystems, is critically limited by information about the frequency of fires and other disturbances, the ability of seeds to disperse across current landscapes, and the factors that determine the success of plants in establishing themselves in new habitats and locations. Field observations, experimental studies, and historical analyses of past ecosystem changes are needed in order to understand both what is possible and the distances and rates of range changes that important species might actually achieve as climate changes. Such results would help fill crucial gaps in knowledge and enhance our ability to project the future ranges and distribution of important species.

Recommendation 4: Enhance Knowledge of How Societal and Economic Systems Will Respond to a Changing Climate and Environment

A greater understanding of the vulnerability and resilience of societal and economic systems is essential to addressing key uncertainties. Human roles in and responses to climate change and other environmental stresses are among the most important features of impact assessments. To assess real impacts on people and their societies, we should improve our understanding of how people and institutions will adapt to change and of the factors that will determine their vulnerability. Gaining such knowledge will require investing in the following key areas:

Understanding the resilience of communities, institutions, regions, and sectors (e.g., human health, urban areas, transportation, and international linkages).

The ability of communities, institutions, regions, and sectors to adapt has only begun to be addressed in this Assessment. Understanding how the capacity to adapt to a changing climate might be exercised, and therefore what vulnerabilities to climate change and other environmental stresses might remain, is an important next step in the human dimensions research agenda. The results of this research would enable a more integrated evaluation of both natural and social science aspects of human responses.

Improving understanding of how people and institutions have adapted to past climate variability and extreme events.

There is a wealth of information available on how people and institutions have responded to climate variability and other environmental changes in the past. New research that documents these responses, analyzes the underlying reasons for them, and explains how individual and institutional decisions were actually made will provide important insights into the feasibility of coping and adaptation options that might be available and considered in the future.

Greater information and analysis of specific potential adaptation options (e.g., costs, efficacy, time horizons, feasibility, and other impacts).

One of the critical unknowns in this Assessment's consideration of adaptation options stems from a lack of information about their potential costs, efficacy, time horizons required for implementation, other consequences, and feasibility. This type of information should be gathered as decision-makers consider specific adaptation options.

Recommendation 5: Refine our Ability to Project How Climate Will Change

This first National Assessment has revealed a number of key uncertainties in projecting climate change and variability at global, national, and regional scales. These uncertainties limit our ability to assess the responses of natural and managed ecosystems and societal and economic systems. Of greatest significance to decision-makers will be reducing uncertainties in several key areas by pursuing research that will lead to:

Improved understanding and analysis of the potential for future changes in severe weather, extreme events, and seasonal to interannual variability.

Many of the results in this Assessment demonstrate that changes in climate variability across a wide range of spatial and time scales have very important impacts on ecosystems, natural resources, and human systems. Long-term climate variations are strongly affected by how the oceans store and transport heat from warm regions near the equator to cold regions at high latitudes. For example, the El Niño-Southern Oscillation (ENSO) influences extreme weather and climate events by affecting the paths, frequency, and severity of winter and tropical storms. While model projections of ENSO and other sources of variability have advanced greatly over the past few years, much additional work is needed on how human-caused climate change might affect these patterns of variability. Much greater understanding is also needed about how climate change will influence the frequency, intensity, and likely locations of severe weather and climate events such as droughts, hurricanes, tornadoes, severe thunderstorms, and meteorological events that produce severe flooding.

Improved understanding of the spatial and temporal character of hydrologic processes, including precipitation, soil moisture, and runoff.

Some of the most important differences among climate model simulations involve projections of precipitation, soil moisture, and runoff, and these are of the greatest significance for ecosystems, agriculture, and water quality and quantity. Despite their importance, there is incomplete understanding of the physical processes that govern the water cycle and the extent to which these processes will be modified by climate change. The differences are sufficiently large in some regions of the country that we cannot even project whether there will be an increase or a decrease in soil moisture and runoff in these regions. A critical element of climate research must be an improved ability to simulate all aspects of the water cycle. Results of this research would substantially improve the estimates of potential vulnerabilities to climate change and other stresses.

Increased information on the nature of past climate, including its spatial and temporal character.

Model-generated scenarios of climate change and variability are only one way to examine potential futures. Another important method is to reconstruct the regional record of past changes in climate and their consequences in order to improve our understanding of how the natural world has operated in the past. These records illustrate the nature of past variability, provide an opportunity to assess climate model sensitivity, and offer insights into the response of ecological systems to past climate change. Results of this research would raise our confidence in the application of climatic information to evaluate impacts and vulnerability.

Adding New Capabilities

The nation needs a stronger capability for providing climate information that serves the national requirements for assessing vulnerabilities and impacts. A stronger national capability could deliver climate projections, including increased access to reliable model outputs and observational information, improved understanding of limita-

tions, and greater availability of the specialized products required by increasingly sophisticated assessment science. Climate modeling and analysis are the foundation for developing climate scenarios that describe alternative futures for analysis of potential impacts of climate change, adaptation options, and vulnerability. Several steps can be taken in the nearterm to enhance our capabilities to provide and use scenarios.

Recommendation 6: Extend Capabilities for Providing Climate Information

Addressing the broad spectrum of future societal needs will require continued improvements in observations, analysis, and the ability to forecast a wide variety of environmental variables. The elements required to develop comprehensive capability include:

A national modeling and analysis capability designed to provide long-term simulations, analysis of limitations and uncertainties, and specialized products for impact studies.

The nation's climate modeling expertise is widely recognized throughout the world and this expertise is dedicated to developing state-of-the-science model capability. Ensembles of long-term simulations, extending from the start of the robust historical record to at least the next 100 years, would provide important information to the nation. Further, the regional and sector teams in this Assessment have been requesting a host of specialized climate products that more directly tie future climate projections to specific decisions or vulnerabilities. The assessment process requires greater access to and a greater understanding of the limitations inherent in future projections in order to weigh the advantages and risks of alternative courses of action. Substantially higher model resolution is required to link climate with the scales of human decisions. The demand for these climate services exceeds the capabilities of the research functions of the nation's climate modeling centers.

Dedicated computer capability for developing ensemble climate scenarios, high-resolution models, and multiple emission scenarios for impact studies.

There is a need for ensemble climate simulations based on multiple-emission scenarios devoted to

535

studies of climate impacts, vulnerabilities, and responses. The investment that is needed is to enhance the capacity of the climate modeling community to generate and analyze model runs that are dedicated for use by impact analysts. Similarly, future assessments need to investigate a range of plausible emissions and atmospheric concentrations of carbon dioxide and other greenhouse gases. The results of enhancing the capability to generate dedicated scenarios of emissions and climate would be a dramatic improvement in the range of outcomes that future assessments of vulnerability could analyze.

Reliable long-term observations and data archives.

One of the most often encountered limitations to the conduct of this Assessment, and one of the most often expressed needs of participants in the regional and sector assessment process, has been the lack of databases that truly reflect changes and variations in the environment, as opposed to those that unduly reflect uncertainties in observing methods. A commitment by the nation to provide integrated databases and information on multiple environmental conditions and trends, and indicators/measures of climate and related environmental changes, is essential to support and implement the research agenda. The US has tremendous potential to create more efficient and comprehensive measurement, archive, and data access systems that would provide greater scientific benefit to society by building upon existing weather and hydrologic stations and remote sensing capability, and integrating current efforts of local, state, and federal agencies. Improved data and information archives will substantially enhance future assessments.

Addressing the Full Agenda

While the proposed research activities are all important individually, it will not be possible to substantially reduce our uncertainties and gaps in knowledge without consideration of their interconnections and interdependencies.
The National Assessment Synthesis Team is convinced that the nation will benefit from multi-year investments in this focused program of research. Benefits will include major enhancement in our knowledge of the impacts of and vulnerabilities to global change on scales appropriate to the national interest and in our capacity to assess their importance.

We must accept the challenge to learn more and to conduct future assessments of multi-dimensional changes in climate, other environmental processes, and socioeconomic conditions. Meeting this challenge will require the new approaches, new knowledge, and new capabilities outlined above that can help reduce uncertainties while taking advantage of the great amount that we already know. Many of the building blocks of scientific knowledge, analytical capability, and commitment to the required integration are now in place. Through its regional, sector, and integrated approach, this Assessment has taken an important first step toward that future.

Areas with High Potential for Providing Needed Information in the Near-Term

Expand the national capability to develop integrated, regional approaches for assessing the impacts of multiple stresses, perhaps beginning with several case studies.

Develop capability to perform large-scale (over an acre) whole-ecosystem experiments that vary both CO_2 and climate.

Incorporate representations of actual land cover and land use into models of ecosystem responses.

Identify potential adaptation options and develop information about their costs, efficacy, side effects, practicality, and implementation.

Develop better ways to assign values to possible future changes in resources and ecosystems, especially for large changes and for processes and service that do not produce marketable goods.

Improve climate projections by providing dedicated computer capability for conducting ensemble climate simulations for multiple emission scenarios.

Focus additional attention on research and analysis of the potential for future changes in severe weather, extreme events, and seasonal to interannual variability.

Improve long-term data sets of the regional patterns and timing of past changes in climate across the US, and make these data-sets more accessable.

Develop a set of baseline indicators and measures of environmental conditions that can be used to track the effects of changes in climate.

Develop additional methods for representing, analyzing, and reporting scientific uncertainties related to global change.

APPENDIX: USGCRP LEADERSHIP

About the National Science and Technology Council

President Clinton established the National Science and Technology Council (NSTC) by Executive Order on November 23, 1993. This cabinet-level council is the principal means for the President to coordinate science, space, and technology policies across the Federal Government. The NSTC acts as a "virtual" agency for science and technology to coordinate the diverse parts of the Federal research and development enterprise. The NSTC is chaired by the President. Membership consists of the Vice President, the Assistant to the President for Science and Technology, Cabinet Secretaries and Agency Heads with significant science and technology responsibilities, and other senior White House officials.

An important objective of the NSTC is the establishment of clear national goals for Federal science and technology investments in areas ranging from information technology and health research, to improving transportation systems and strengthening fundamental research. The Council prepares research and development strategies that are coordinated across Federal agencies to form an investment package that is aimed at accomplishing multiple national goals.

To obtain additional information regarding the NSTC, contact the NSTC Executive Secretariat at 202-456-6100 (voice).

About the Office of Science and Technology Policy

The Office of Science and Technology Policy (OSTP) was established by the National Science and Technology Policy, Organization, and Priorities Act of 1976. OSTP's responsibilities include advising the President on policy formulation and budget development on all questions in which science and technology are important elements; articulating the President's science and technology policies and programs; and fostering strong partnerships among Federal, State, and local governments, and the scientific communities in industry and academia. To obtain additional information regarding the OSTP, contact the OSTP Administrative Office at 202-456-6004 (voice).

About the Committee on Environment and Natural Resources

The CENR is charged with improving coordination among Federal agencies involved in environmental and natural resources research and development, establishing a strong link between science and policy, and developing a Federal environment and natural resources research and development strategy that responds to national and international issues. To obtain additional information about the CENR, contact the CENR Executive Secretary at 202-482-5917 (voice).

The Committee on Environment and Natural Resources (CENR) is one of five committees under the NSTC.

D. James Baker, Co-Chair
National Oceanic and Atmospheric Administration

Rosina Bierbaum, Co-Chair
White House Office of Science and Technology Policy

Ghassem Asrar
National Aeronautics and Space Administration

James Decker
Department of Energy

Roland Droitsch
Department of Labor

Albert Eisenberg
Department of Transportation

Delores Etter
Department of Defense

Terrance Flannery
Central Intelligence Agency

George Frampton
Council on Environmental Quality

Kelley Brix
Department of Veteran s Affairs

Charles Groat
Department of the Interior

Len Hirsch
Smithsonian Institution

Kathryn Jackson
Tennessee Valley Authority

Eileen Kennedy
Department of Agriculture

Margaret Leinen
National Science Foundation

Paul Leonard
Housing and Urban Development

Norine Noonan
Environmental Protection Agency

Kenneth Olden
Department of Health and Human Services

David Sandalow
Department of State

Wesley Warren
Office of Management and Budget

Craig Wingo
Federal Emergency Management Agency

Samuel Williamson
Office of the Federal Coordinator for Meteorology

SUBCOMMITTEES

Air Quality
Dan Albritton (NOAA), Chair
Bob Perciasepe (EPA), Vice Chair

Ecological Systems
Mary Clutter (NSF), Co-Chair
Don Scavia (NOAA), Co-Chair

Global Change
D. James Baker (NOAA), Chair
Ghassem Asrar (NASA), Vice Chair
Margaret Leinen (NSF), Vice Chair

Natural Disaster Reduction
Mike Armstrong (FEMA), Chair
John Filson (USGS), Vice Chair
Jaime Hawkins (NOAA), Vice Chair

Toxics and Risk
Norine Noonan (EPA), Chair
Bob Foster (DOD), Vice Chair
Kenneth Olden (HHS), Vice Chair

About the Subcommittee on Global Change Research

The Subcommittee on Global Change Research (SGCR) is one of five subcommittees under the Committee on Environment and Natural Resources (CENR). The SGCR is charged with improving coordination among Federal agencies participating in the U. S. Global Change Research Program (USGCRP), which was established by Congress in 1990 "to provide for development and coordination of a comprehensive and integrated United States research program which will assist the Nation and the world to understand, assess, predict, and respond to human-induced and natural processes of global change." The NAST is grateful for the SGCR establishing the NAST and providing oversight for its activities.

To obtain additional information regarding the SGCR, contact the Office of the USGCRP at 202-488-8630 (voice) or see http://www.usgcrp.gov.

D. James Baker, Chair
(from January 2000)
National Oceanographic and Atmospheric
  Administration
Department of Commerce

Robert W. Corell, Chair (through December, 1999)
National Science Foundation

Ghassem Asrar, Vice Chair
National Aeronautics and Space Administration

Margaret Leinen, Vice Chair
National Science Foundation

William Sommers
U. S. Forest Service
Department of Agriculture

Warren Piver
National Institute of Environmental Health
  Sciences,
Department of Health and Human Services

Charles (Chip) Groat
Department of the Interior

J. Michael Hall
National Oceanic and Atmospheric
  Administration,
Department of Commerce

Patrick Neale
Smithsonian Institution

Mark Mazur
Department of Energy

Margot Anderson
Department of Energy

Jeff Miotke
Department of State

Aristides A. Patrinos (SGCR liaison to the NAST)
Department of Energy

Fred Saalfeld
Department of Defense

Michael Slimak
Environmental Protection Agency

Executive Office Liaisons

Rosina Bierbaum
Office of Science and Technology Policy

Peter Backlund
Office of Science and Technology Policy

Steven Isakowitz
Office of Management and Budget

Sarah G. Horrigan
Office of Management and Budget

Ian Bowles
Council on Environmental Quality

About the National Assessment Working Group

The National Assessment Working Group is charged by the SGCR with overseeing and facilitating the coordination and preparation of national-scale assessments to document the current state of knowledge of the consequences of global change and their implications for policy and management decisions for the Nation. As such, they were the organizers and sponsors of the regional and sectoral assessments.

Paul Dresler
(through December 1999), Chair
Department of the Interior

Joel Scheraga, Vice Chair
Environmental Protection Agency

Richard Ball (through November 1999), Vice Chair
Department of Energy

Department of Agriculture
Margot Anderson (through March 2000)
Jeff Graham (through Sept. 1999)
Robert House
James Hrubovcak (from October 1999)
Fred Kaiser

Department of Defense and US Army Corps of Engineers
Thomas Nelson
Eugene Stakhiv

Department of Energy
Mitchell Baer (from November, 1999)
Jerry Elwood

Department of the Interior
Dave Kirtland
Ben Ramey

Environmental Protection Agency
Janet Gamble

National Aeronautics and Space Administration
Anne Carlson (from Jan. 2000)
Nancy Maynard (through Jan. 2000)
William Turner (from April 2000)
Louis Whitsett (through December 1999)

National Institute of Environmental Health Sciences
Mary Gant
Warren Piver

National Oceanic and Atmospheric Administration
Claudia Nierenberg
Roger Pulwarty
Caitlin Simpson

National Science Foundation
Thomas Spence

Office of Science and Technology Policy
Peter Backlund

COLOR FIGURE APPENDIX

Figure 2: Records of CO₂ emissions, CO₂ concentrations, and Northern Hemisphere average surface temperature for the past 1000 years: (a) Reconstruction of past emissions of CO₂ as a result of land clearing and fossil fuel combustion since about 1750 (in billions of metric tons of carbon per year) [data from CDIAC, 2000; Andres et al., 2000; Marland et al., 1999; Houghton, 1995; Houghton and Hackler, 1995]; (b) Record of the CO₂ concentration for the last 1000 years, derived from measurements of CO₂ concentration in air bubbles in the layered ice cores drilled in Antarctica, a location that has been found to be representative of the global average concentration [data from Etheridge et al., 1998; Keeling and Whorf, 1999]; (c) Reconstruction of annual-average Northern Hemisphere surface air temperatures based on paleoclimatic records (Mann et al., 1999). For the Mann et al. data, the zero change baseline is based on the average conditions over the period 1902-80. The error bars for the estimate of the annual-average anomaly increase somewhat going back in time, with one standard deviation being about 0.25˚F (0.15˚C). Although this record comes mostly from the Northern Hemisphere, it is likely to be a good approximation to the global anomaly based on comparisons of recent patterns of temperature fluctuations.

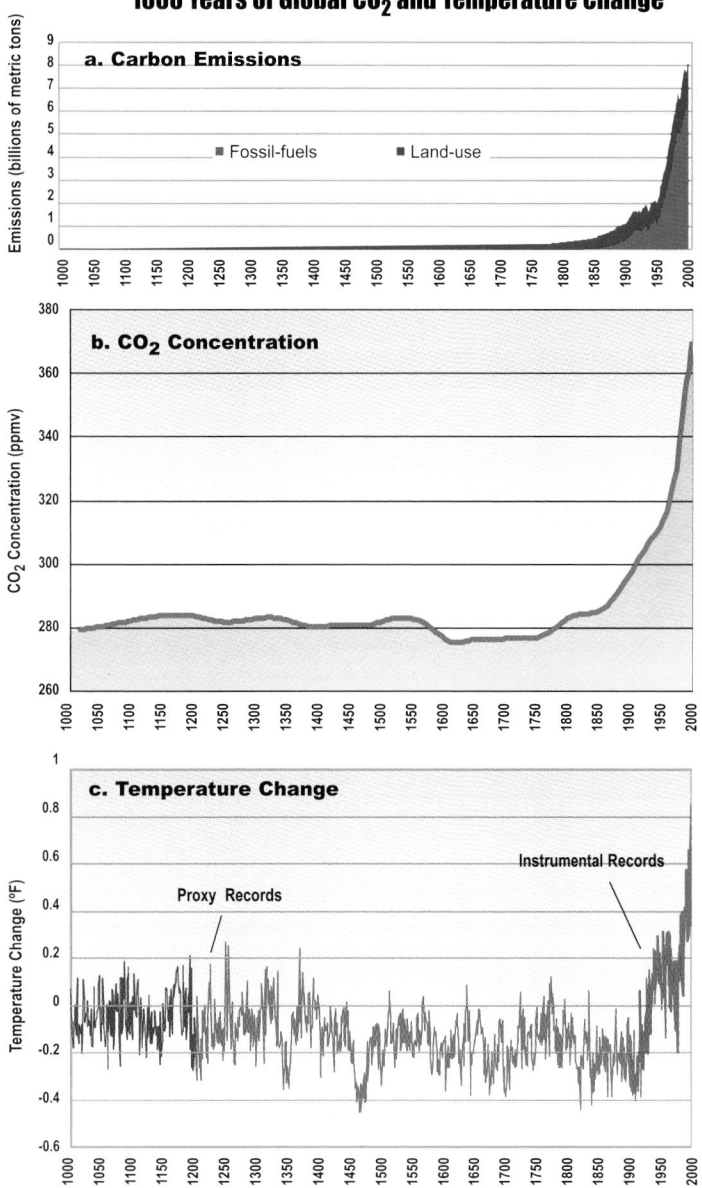

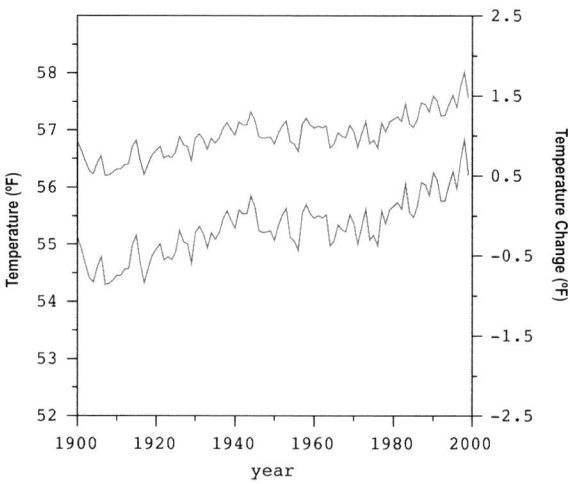

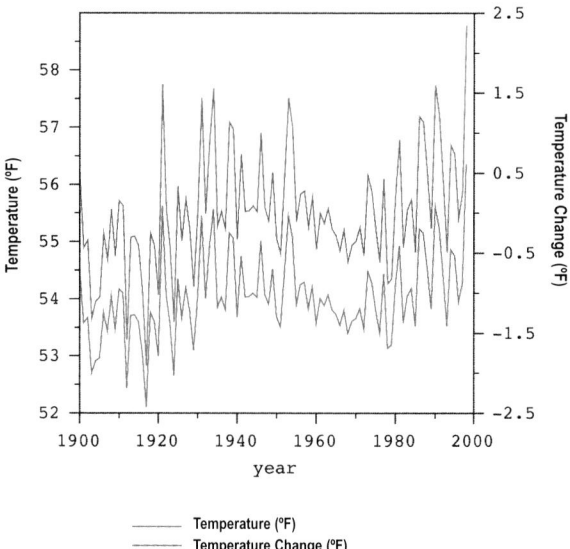

Figure 3: (a) Global annual-average surface temperature and temperature change for combined land and ocean regions for the period 1900-1999 based on the method of Quayle et al. (1999); (b) US annual-average surface temperature and temperature change for the period 1900-1999 using the USHCN data set (Easterling et al., 1996).

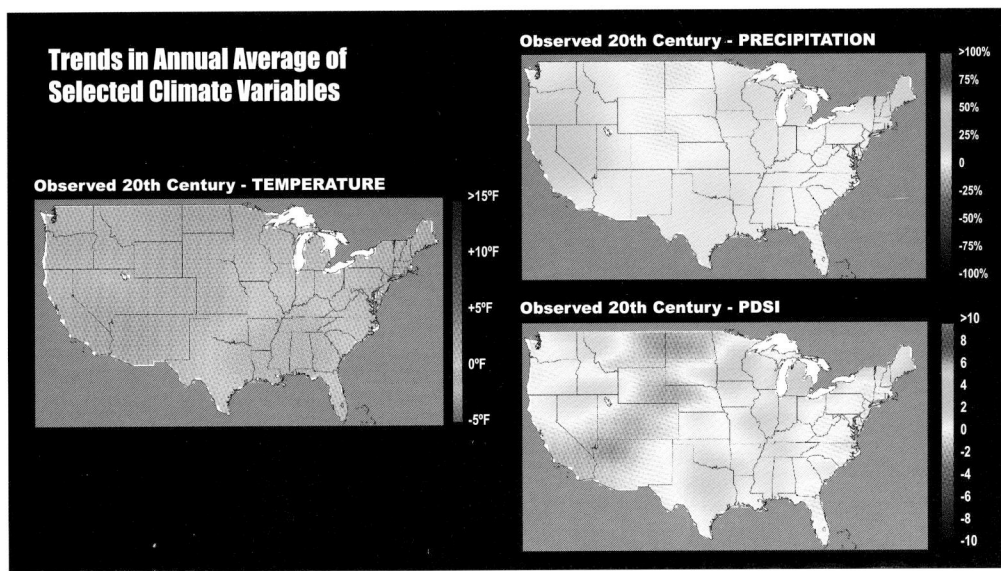

Figure 4: Trends in the annual average of selected climatic variables over the US during the 20th century as derived from observations compiled in the USHCN data set (Easterling et al., 1996). (a) Temperature (°F/century); (b) Precipitation (percent change/century); (c) Palmer Drought Severity Index (percent change/century).

Global Precipitation Anomalies

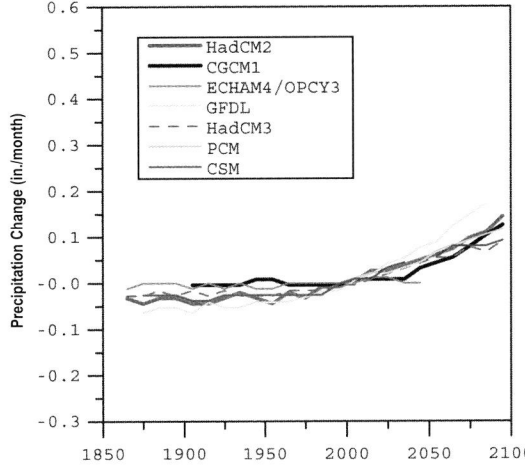

Observed US Trends in Daily Precipitation Intensity

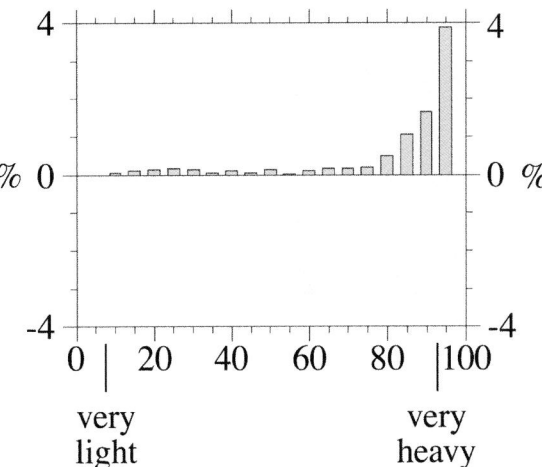

Figure 5: US trends (1910-1996) in mean precipitation (in percent change per century) for various categories of daily precipitation intensity. Values are plotted for each 5%, such that 5 represents from the lowest to 5th percentile and 95 represents the 95th to highest values of precipitation intensity. The lowest to 5th percentile are the lightest daily precipitation amounts and the 95th to highest are the heaviest daily amounts (Karl and Knight, 1998).

US Precipitation Anomalies

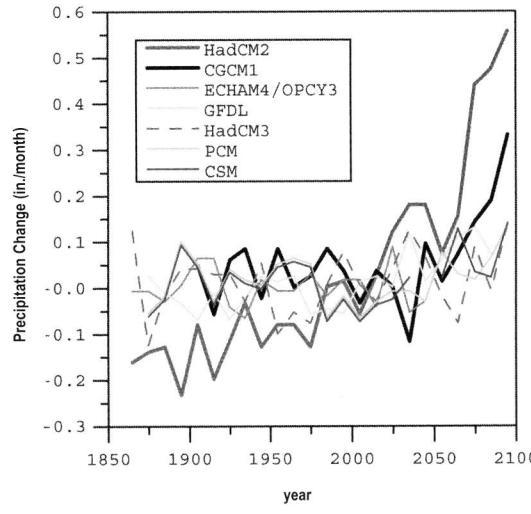

Figure 8: Comparison of the annual average changes in (a) global average precipitation (inches per month), and (b) average precipitation over the US from the Canadian model scenario and Hadley model scenario simulations used in the National Assessment and from the simulations of other groups (same as for Figure 7). The baseline period is assumed to be 1961-1990. Although decadal means have been applied to suppress year-to-year fluctuations, the greater variability of precipitation than temperature still reveals significant variations due to natural factors; the magnitude, although not the timing, of the remaining fluctuations may be considered plausible. The anomalies are with respect to the year 2000, calculating the values from a 2nd order polynomial fit.

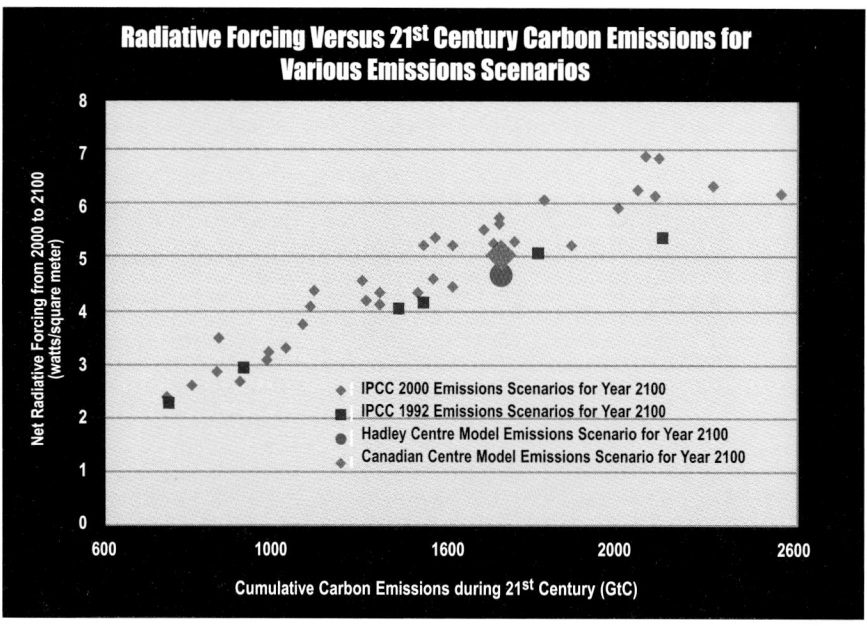

Figure 6: Comparison of the projections of total carbon emissions and overall human-induced radiative forcing for the six emissions scenarios prepared by the IPCC in 1992 (IS92 scenarios; IPCC, 1992) and the 35 emissions scenarios prepared by the IPCC in 2000 for which radiative forcing could be estimated (SRES scenarios; IPCC, 2000). These scenarios are based, although in different ways, on projected changes in emissions resulting from changes in population, economic development, energy use, efficiency of energy use, the mix of energy technologies, etc. The horizontal axis gives the total emissions of fossil fuel-derived carbon dioxide projected for the 21st century (in billions of tonnes of carbon, GtC). For reference, if the current level of global carbon emissions is maintained from 2000 to 2100, cumulative emissions over the 21st century would be roughly 650 GtC. Assuming no climate-related controls on emissions are introduced, this value is near the lowest value projected by any of the scenarios for the 21st century. The vertical axis gives the projected change in net radiative forcing at a pressure level approximating the tropopause (in watts per square meter) for all human-induced changes in greenhouse gases and aerosols (both direct and indirect contributions) over the 21st century using relationships employed in the IPCC Second Assessment Report (IPCC, 1996a; Smith et al., 2000), including the uptake of CO_2 by the oceans and land. Radiative forcing is important because it is the driving force for global warming; for reference, the projected change in radiative forcing up to the year 1992 is about 1.6 watts per square meter (IPCC, 1996a). The figure also shows the net radiative forcing and the approximate emissions of carbon used in the Hadley and Canadian scenarios. For these scenarios, which increase the equivalent CO_2 concentration by 1% per year, the carbon emissions are estimated by calculating the emissions needed to match the net radiative forcing after subtracting the radiative effects of other greenhouse gases and aerosols based on the average of IS92a and IS92f scenarios, and is an amount between the IS92a and IS92f scenarios. Based on these calculations, the Canadian and Hadley scenarios lie near the mid-range of the proposed scenarios in terms of both carbon emissions and net radiative forcing.

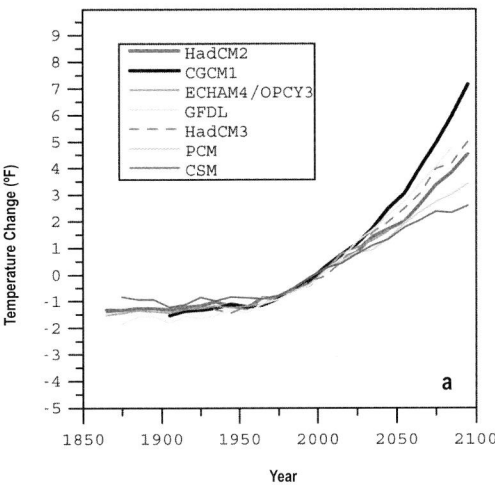

Global Mean Temperature Anomalies (a)

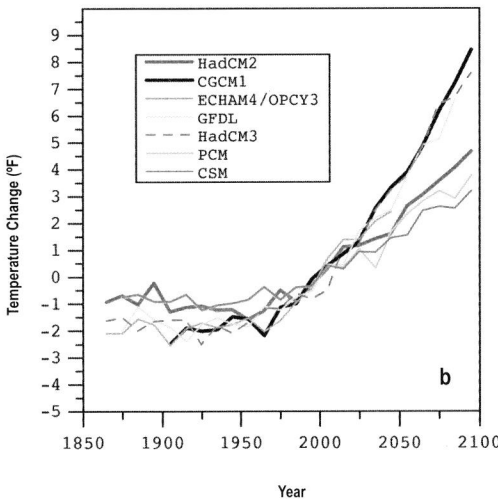

U.S. Mean Temperature Anomalies (b)

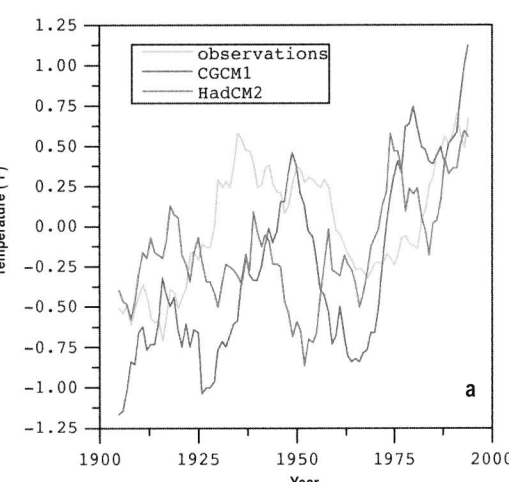

Mean Temperature Change

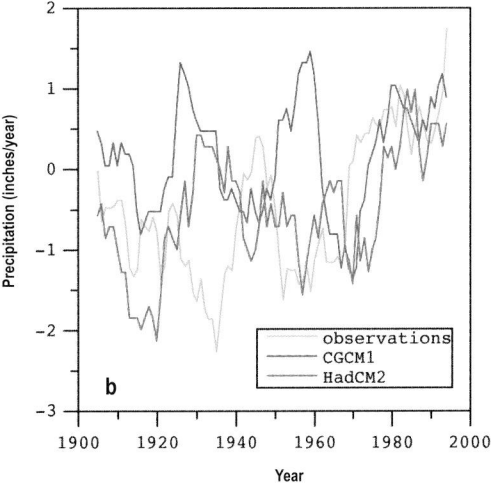

Precipitation Change

Figure 7: Comparison of the annual average changes in (a) global average surface air temperature (°F), and (b) US average surface air temperature (°F) from the Canadian model scenario and Hadley model scenario simulations used in the National Assessment and from the simulations of other modeling groups, including a very recent result from the Hadley Centre model version 3, Germany's Max Planck Institute/German Climate Computing Center (DKRZ), NOAA's Geophysical Fluid Dynamics Laboratory, and from the Parallel Climate and the Climate System models from the National Center for Atmospheric Research (which used a slightly lower greenhouse gas emission scenario and a significantly lower sulfate emissions scenario than the other models). Decadal means have been plotted to suppress the natural year-to-year variability. The baseline period is 1961-1990. The anomalies are with respect to the year 2000, calculating the values from a 2nd order polynomial fit over adjacent decades.

Figure 11: Time histories of the changes in (a) annual average temperature (°F), and (b) annual total precipitation (inches per year) for the 20th century based on observations and on simulations from the Canadian and Hadley models, calculated as 10-year running means from 1900 to 2000. Mean temperature is the actual mean temperature from the models, rather than the mean of the minimum and maximum temperatures. Anomalies are shown with respect to 1961-1990. In these simulations, unlike in intercomparisons of the atmospheric models as in the AMIP project (Gates et al., 1999), the ocean temperatures are freely calculated and the concentrations of greenhouse gases and aerosols are imposed; natural forcings, such as changes in solar radiation and volcanic eruptions that are likely affecting the observed climate are not, however, being treated in the models because observations of their precise radiative influences are not available.

Scenarios for Climate Variability and Change

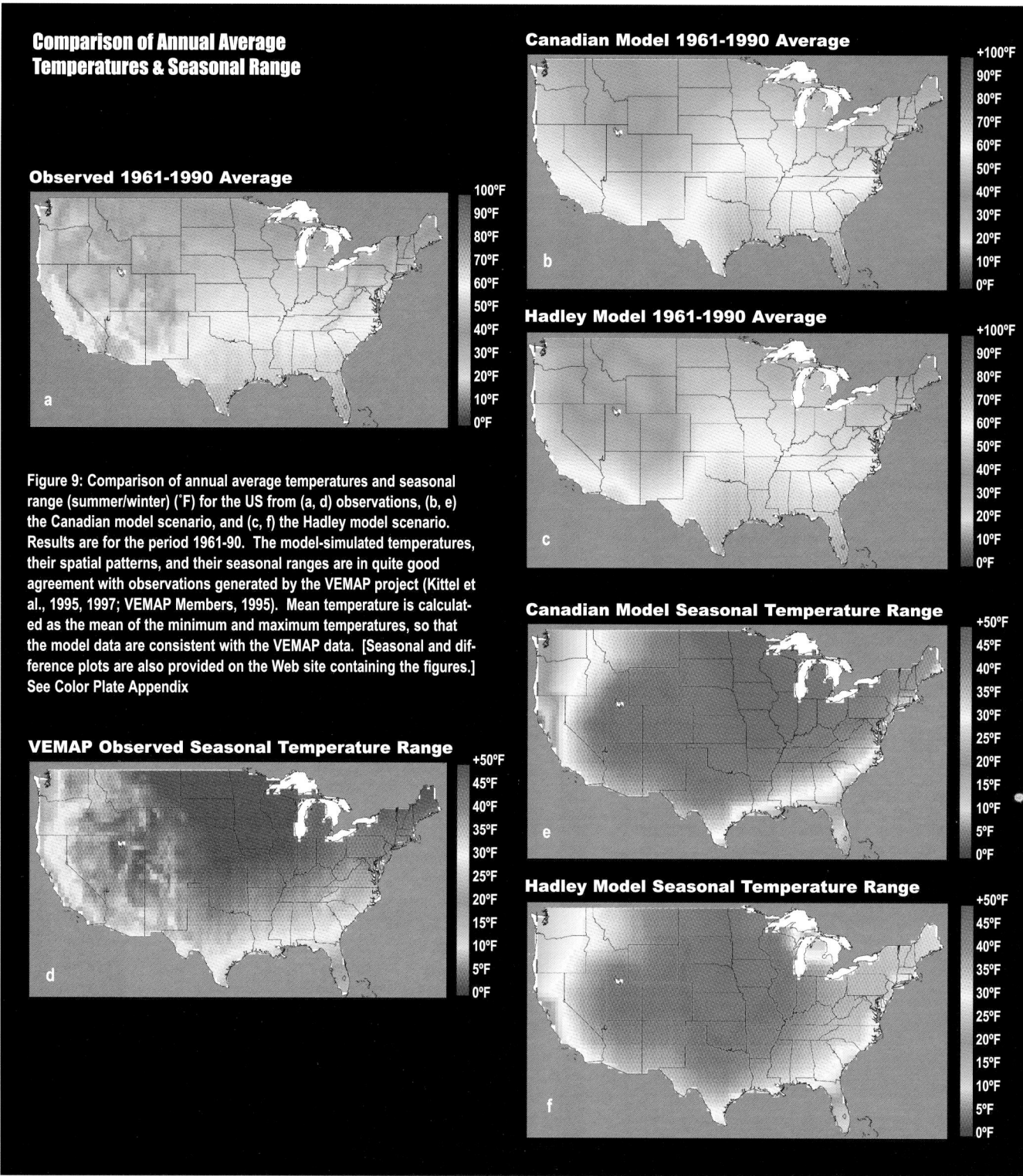

Comparison of Annual Average Temperatures & Seasonal Range

Observed 1961-1990 Average

a

Canadian Model 1961-1990 Average

b

Hadley Model 1961-1990 Average

c

Canadian Model Seasonal Temperature Range

e

Hadley Model Seasonal Temperature Range

f

VEMAP Observed Seasonal Temperature Range

d

Figure 9: Comparison of annual average temperatures and seasonal range (summer/winter) (°F) for the US from (a, d) observations, (b, e) the Canadian model scenario, and (c, f) the Hadley model scenario. Results are for the period 1961-90. The model-simulated temperatures, their spatial patterns, and their seasonal ranges are in quite good agreement with observations generated by the VEMAP project (Kittel et al., 1995, 1997; VEMAP Members, 1995). Mean temperature is calculated as the mean of the minimum and maximum temperatures, so that the model data are consistent with the VEMAP data. [Seasonal and difference plots are also provided on the Web site containing the figures.] See Color Plate Appendix

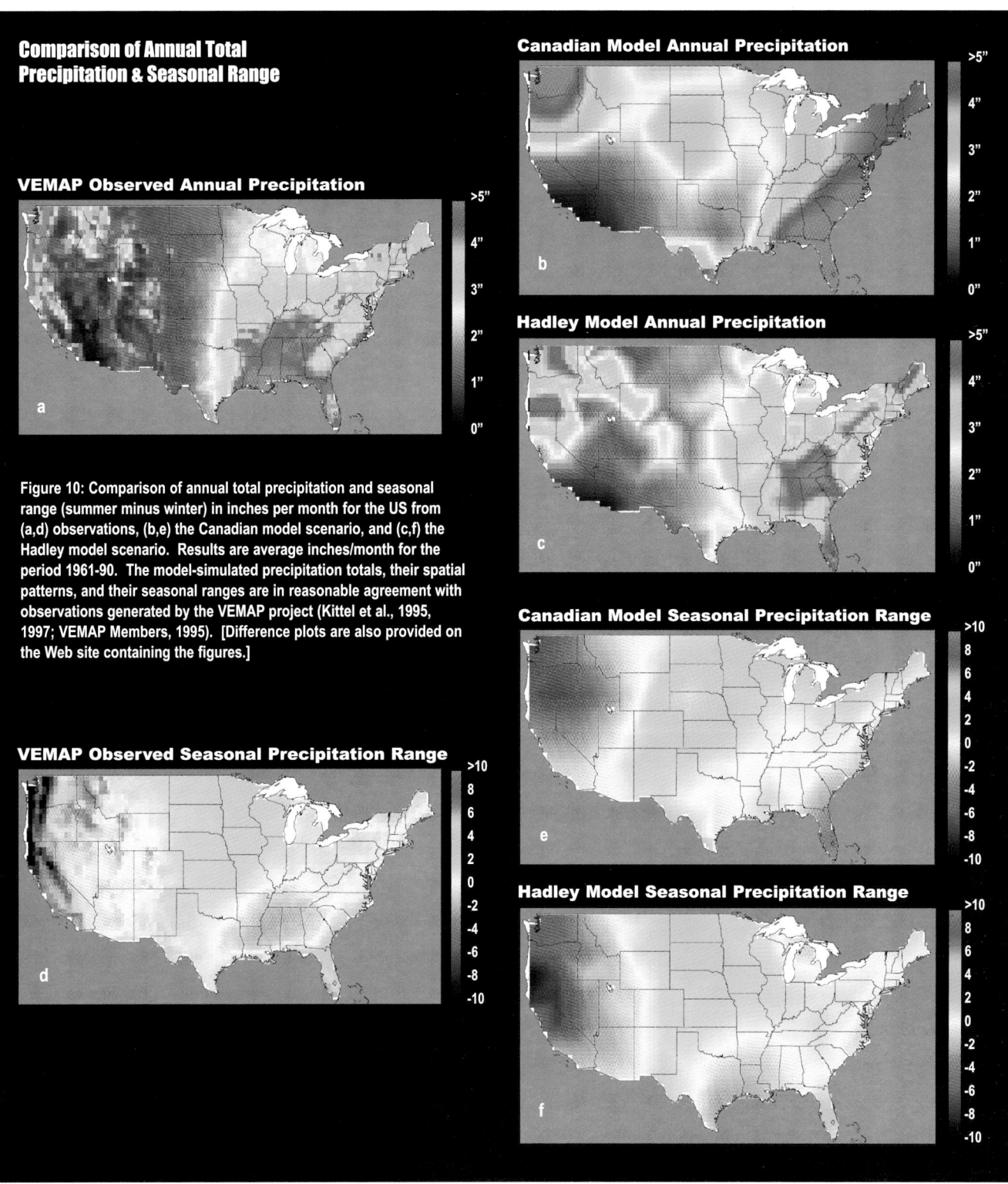

Comparison of Annual Total Precipitation & Seasonal Range

VEMAP Observed Annual Precipitation

Figure 10: Comparison of annual total precipitation and seasonal range (summer minus winter) in inches per month for the US from (a,d) observations, (b,e) the Canadian model scenario, and (c,f) the Hadley model scenario. Results are average inches/month for the period 1961-90. The model-simulated precipitation totals, their spatial patterns, and their seasonal ranges are in reasonable agreement with observations generated by the VEMAP project (Kittel et al., 1995, 1997; VEMAP Members, 1995). [Difference plots are also provided on the Web site containing the figures.]

VEMAP Observed Seasonal Precipitation Range

Canadian Model Annual Precipitation

Hadley Model Annual Precipitation

Canadian Model Seasonal Precipitation Range

Hadley Model Seasonal Precipitation Range

Scenarios for Climate Variability and Change

549

Forcing Scenarios

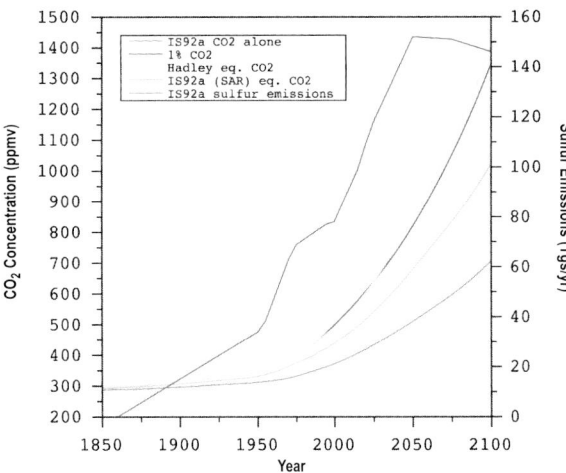

Radiative Forcing

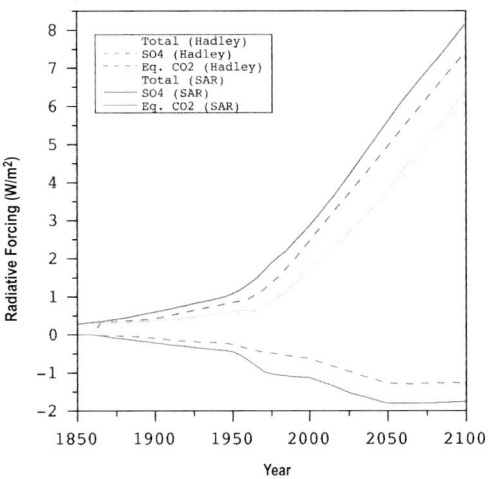

Figure 12: Comparison of different projections for aerosol effects and for (a) the CO_2 and equivalent CO_2 concentrations, and (b) the associated radiative forcings for the period 1850-2100. In the top figure, the lavender line shows the IPCC's IS92a scenario estimate of the CO_2 concentration; values prior to 1990 are based on observations. Based on this projection, the CO_2 concentration would rise to about 705 ppmv in 2100 from a level of about 353 ppmv in 1990. Because many of the climate models treat the effects of the set of human-affected greenhouse gases by use of an equivalent CO_2 concentration, the green line shows the scenario for the equivalent CO_2 concentration, which rises to about 1022 ppmv in 2100 from a value of about 410 ppmv in 1990. For this curve, the equivalent CO_2 concentration is calculated so as to incorporate the radiative effects of changes in the concentrations of all greenhouse gases using the IPCC radiative forcing equivalents (the conversion factor is 6.3 based on Appendix 2 in IPCC, 1997). The light blue line shows the equivalent CO_2 concentration that results from using the Hadley radiative forcing equivalents to approximate the IS92a scenario; the conversion factor used is 5.05 (John Mitchell, personal communication). Using the Hadley conversion factor, the equivalent CO_2 concentration for the IS92a scenario would rise to about 1409 ppmv in 2100. The red line shows that the Hadley IS92a equivalent CO_2 scenario is quite well fitted by use of a 1% per year compounded increase in the Hadley equivalent CO_2 concentration. In this case, the CO_2 equivalent concentration in 2100 reaches about 1346 ppmv. The deep blue line shows the IPCC IS92a scenario for sulfur emissions, which shows a rise until about 2050, when emissions roughly level off. While there are some differences in the projected concentrations of equivalent CO_2 between the IPCC (1996a) and the Hadley model scenario, the bottom figure shows that these differences are mostly overcome when comparing the radiative forcings that are projected by the IPCC and are actually used in the Hadley model scenario. The red and blue lines, respectively show the radiative forcings as projected by the IPCC (solid lines) and as included in the Hadley model (dotted lines). For both forcings, the Hadley model projects slightly less influence than the projections using the IPCC conversion factors. When these forcings are combined, as shown by the green lines, the net radiative forcings projected by the IPCC and used in the Hadley model 1% per year scenario are very close.

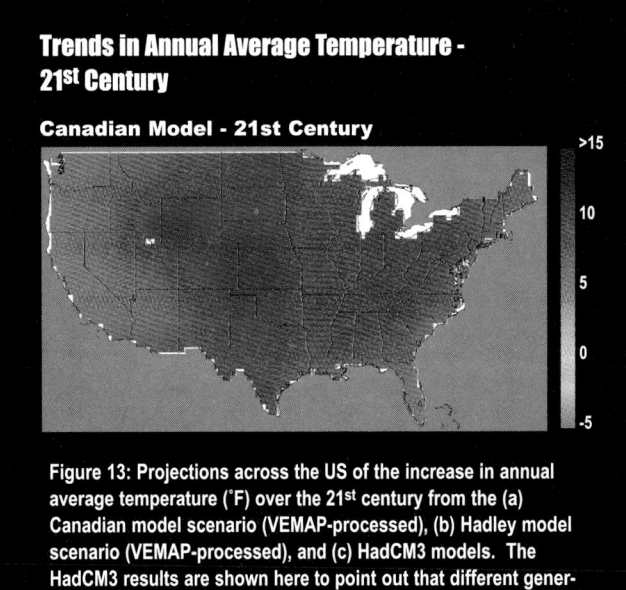

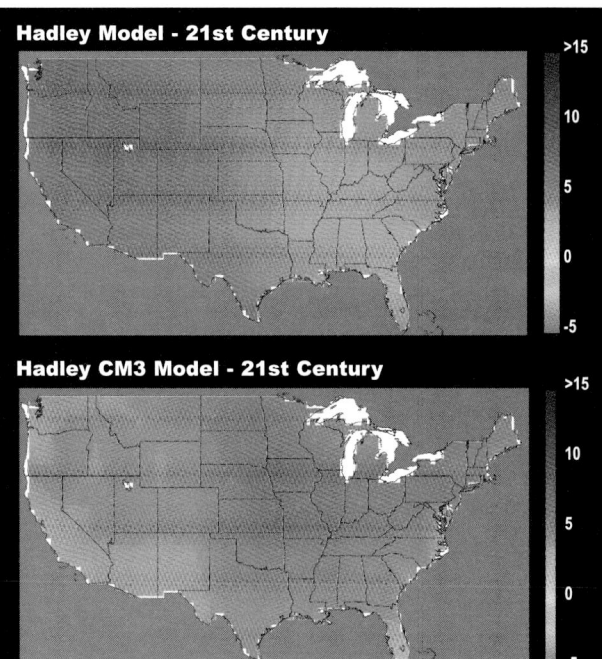

Figure 13: Projections across the US of the increase in annual average temperature (°F) over the 21st century from the (a) Canadian model scenario (VEMAP-processed), (b) Hadley model scenario (VEMAP-processed), and (c) HadCM3 models. The HadCM3 results are shown here to point out that different generations of the same basic model can yield results that are as different as results of different models.

Maximum Temperature in the US (annual average)

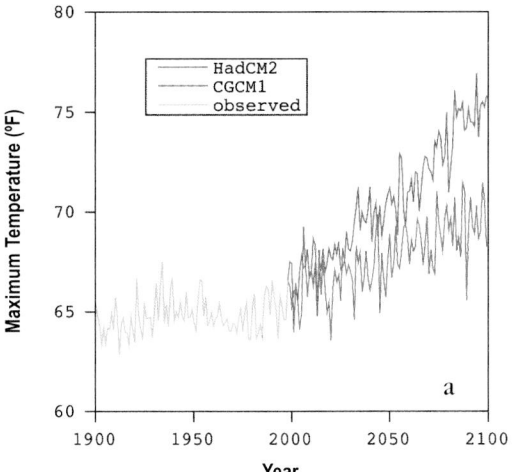

Minimum Temperature in the US (annual average)

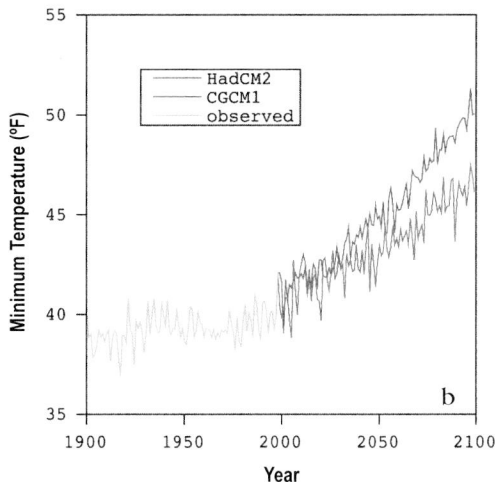

Figure 14: Time histories of (a) maximum and (b) minimum temperature over the US (°F). The values prior to the present are based on observations from 1900-1998 (the HCN data set) and values for the future are based on the VEMAP version of the Canadian and Hadley model scenarios (i.e., in the VEMAP data sets, model projections of climate change are added to the observed 1961-90 baseline climate).

July Heat Index Change - 21st Century

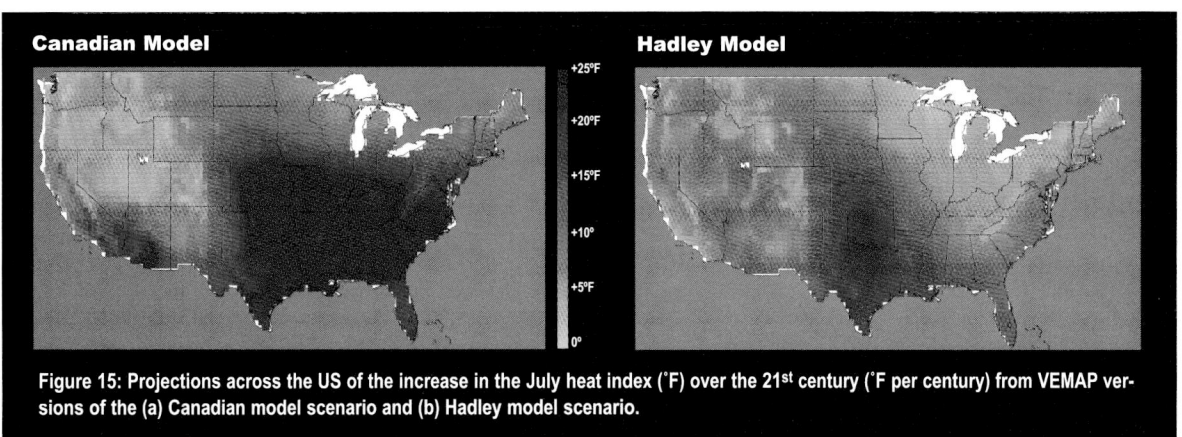

Figure 15: Projections across the US of the increase in the July heat index (°F) over the 21st century (°F per century) from VEMAP versions of the (a) Canadian model scenario and (b) Hadley model scenario.

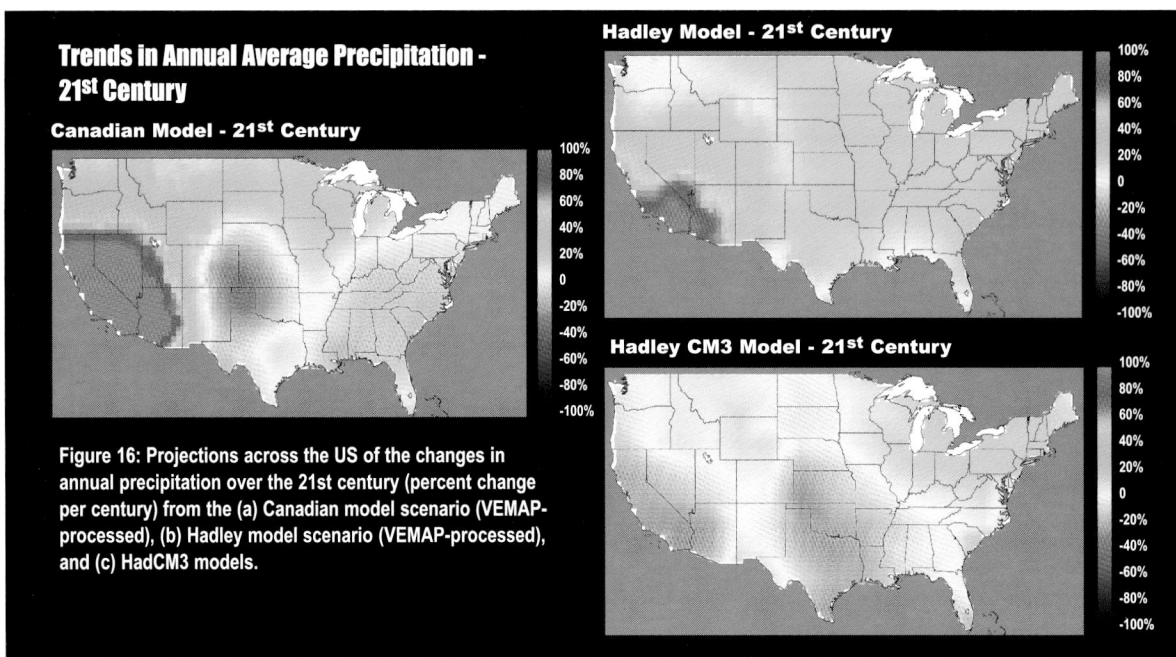

Trends in Annual Average Precipitation - 21st Century

Canadian Model - 21st Century

Hadley Model - 21st Century

Hadley CM3 Model - 21st Century

Figure 16: Projections across the US of the changes in annual precipitation over the 21st century (percent change per century) from the (a) Canadian model scenario (VEMAP-processed), (b) Hadley model scenario (VEMAP-processed), and (c) HadCM3 models.

Summer Soil Moisture - 21st Century

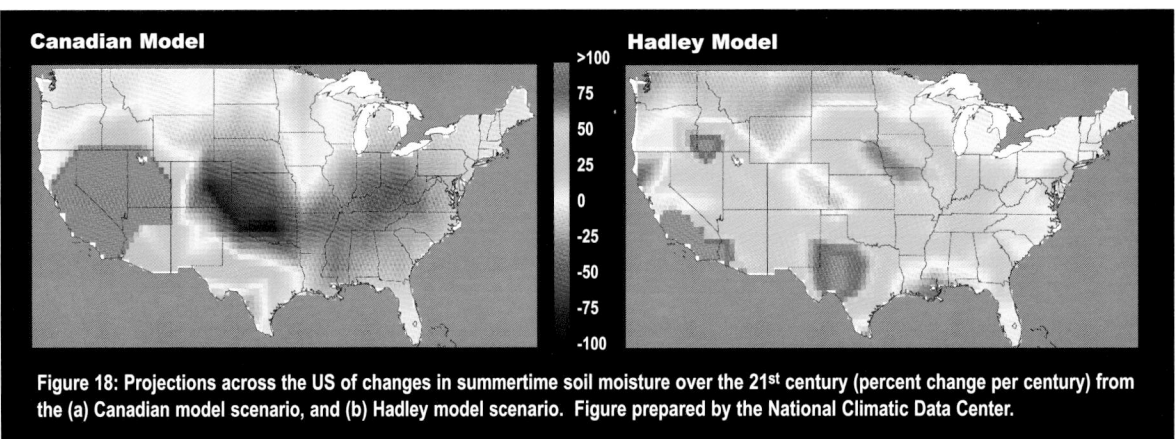

Canadian Model

Hadley Model

Figure 18: Projections across the US of changes in summertime soil moisture over the 21st century (percent change per century) from the (a) Canadian model scenario, and (b) Hadley model scenario. Figure prepared by the National Climatic Data Center.

Annual Precipitation in the US

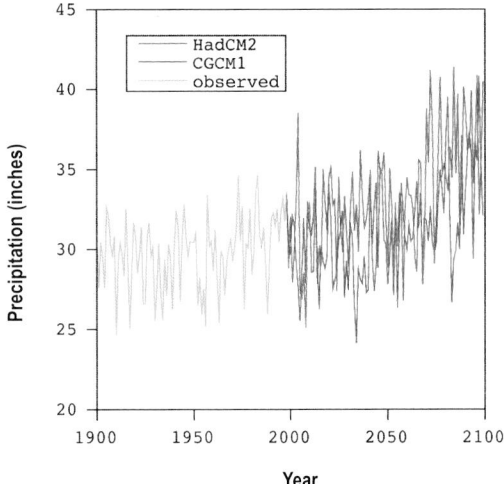

Figure 17: Time history of model projected changes in precipitation over the US (inches per year). The values prior to the present are based on observations from 1900-1998 (the HCN data set) and values for the future are based on the VEMAP version of the Canadian and Hadley model scenarios.

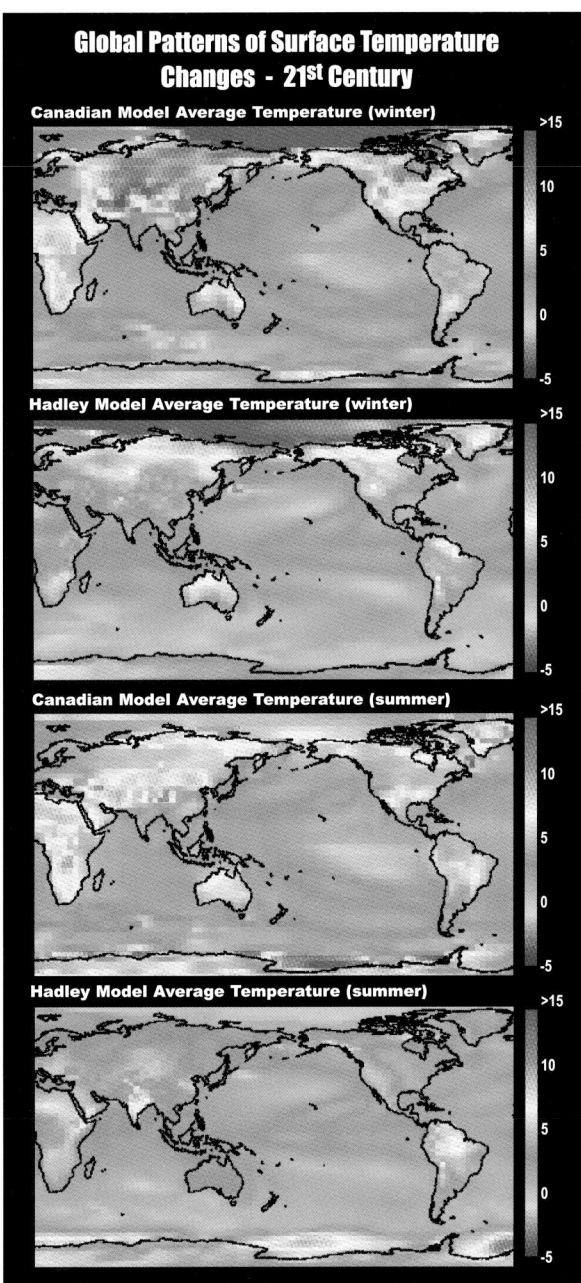

Figure 20: Global patterns of projected changes in surface temperature (°F) over the 21st century [future (2090-2099) and modern (1961-1990)] for (a) December, January, February (DJF) from the Canadian model scenario, (b) DJF from the Hadley model scenario, (c) June, July, August (JJA) from the Canadian model scenario, and (d) JJA from the Hadley model scenario.

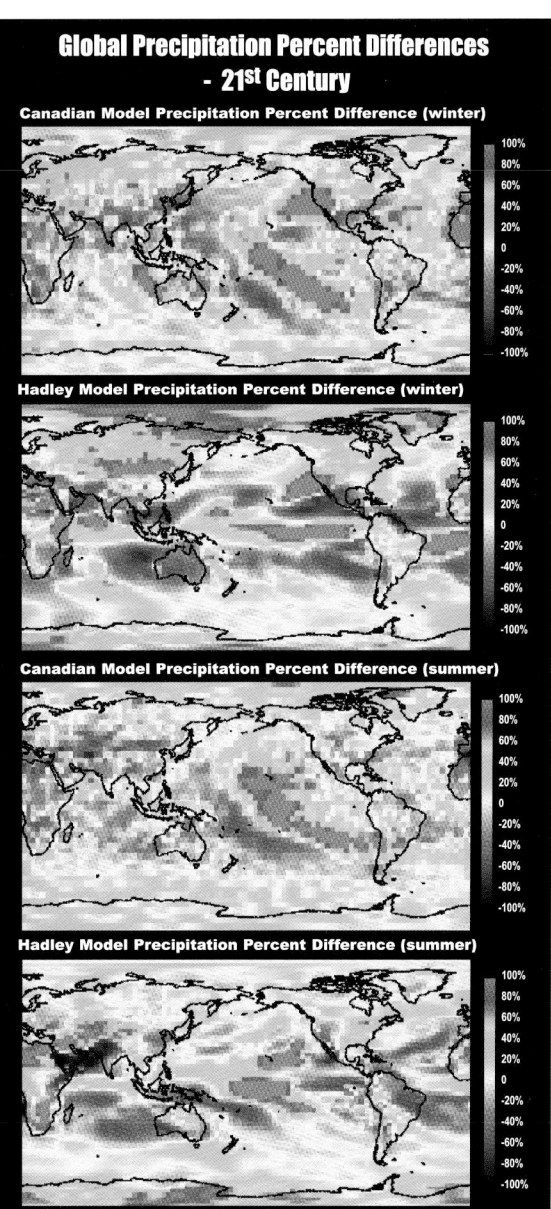

Figure 21: Global precipitation percent differences [(future - modern)/modern) x 100] for (a) December, January, February (DJF) from the Canadian model scenario, (b) DJF from the Hadley model scenario, (c) June, July, August (JJA) from the Canadian model scenario, and (d) JJA from the Hadley model scenario.

Wintertime Changes in Jet Stream and Atmospheric Circulation

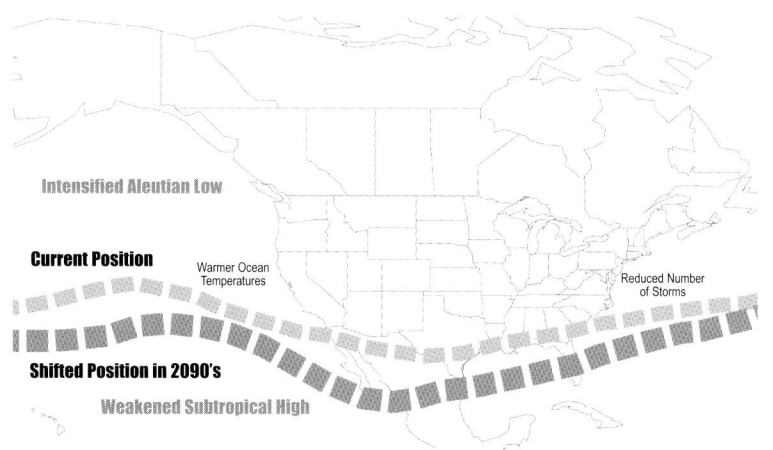

Figure 22: Schematic illustrating wintertime changes in the jet stream, pressure systems, sea surface temperatures, and storm tracks over and adjacent to North America. The Canadian and Hadley model scenarios both show: a southward-shifted jet stream over the eastern Pacific and Southwest; a southward-shifted and intensified Aleutian Low and weakened subtropical High in the West; and warmer ocean surface temperatures off the coast of California. The Canadian model scenario also shows a reduction in the number of storms along the East Coast storm track; however, the Hadley model scenario does not show this reduction nor did it develop this observed storm center in its control simulation. For more details, see Sousounis (1999).

Sea Level Rise

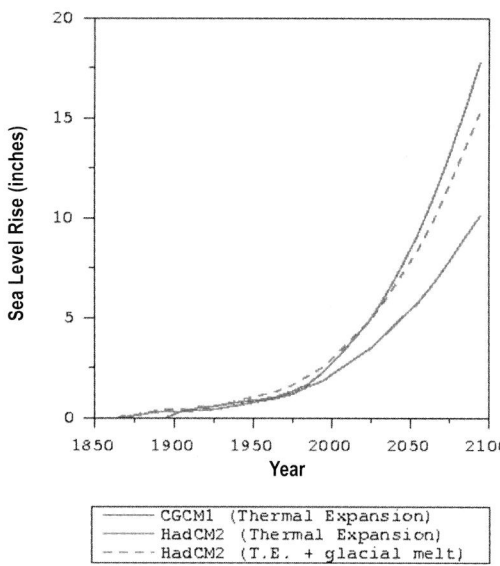

Figure 19: Historic and projected changes in sea level (inches above baseline) based on the Canadian and Hadley model scenarios. The Canadian model projection includes only the effects of thermal expansion of warming ocean waters (F. Zwiers, personal communication). The Hadley model simulation adds on the sea level increment of melting of mountain glaciers (Gregory and Oerlemans, 1998). Neither model includes consideration of possible changes of sea level (upward or downward) due to melting or accumulation of snow on Greenland and Antarctica.

Projected Changes in Intensity of National Daily Precipitation

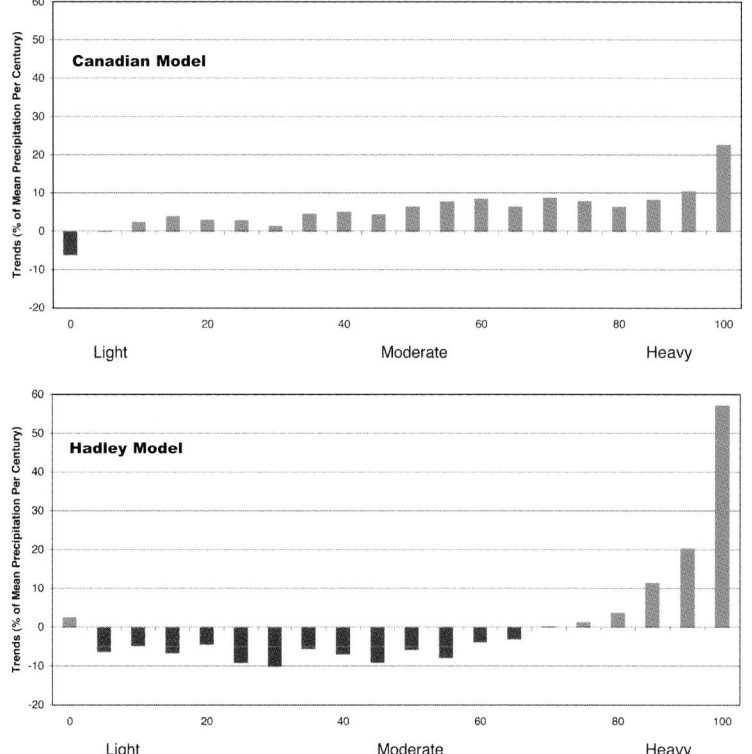

Figure 25: Bar chart showing projected changes in frequency of various types of precipitation. Both the (a) Canadian and (b) Hadley model scenarios project increases in the frequency of heavy precipitation events, intensifying the trend observed for the 20th century. Figure prepared by Byron Gleason of the National Climatic Data Center based on the methods described in Karl and Knight (1998).

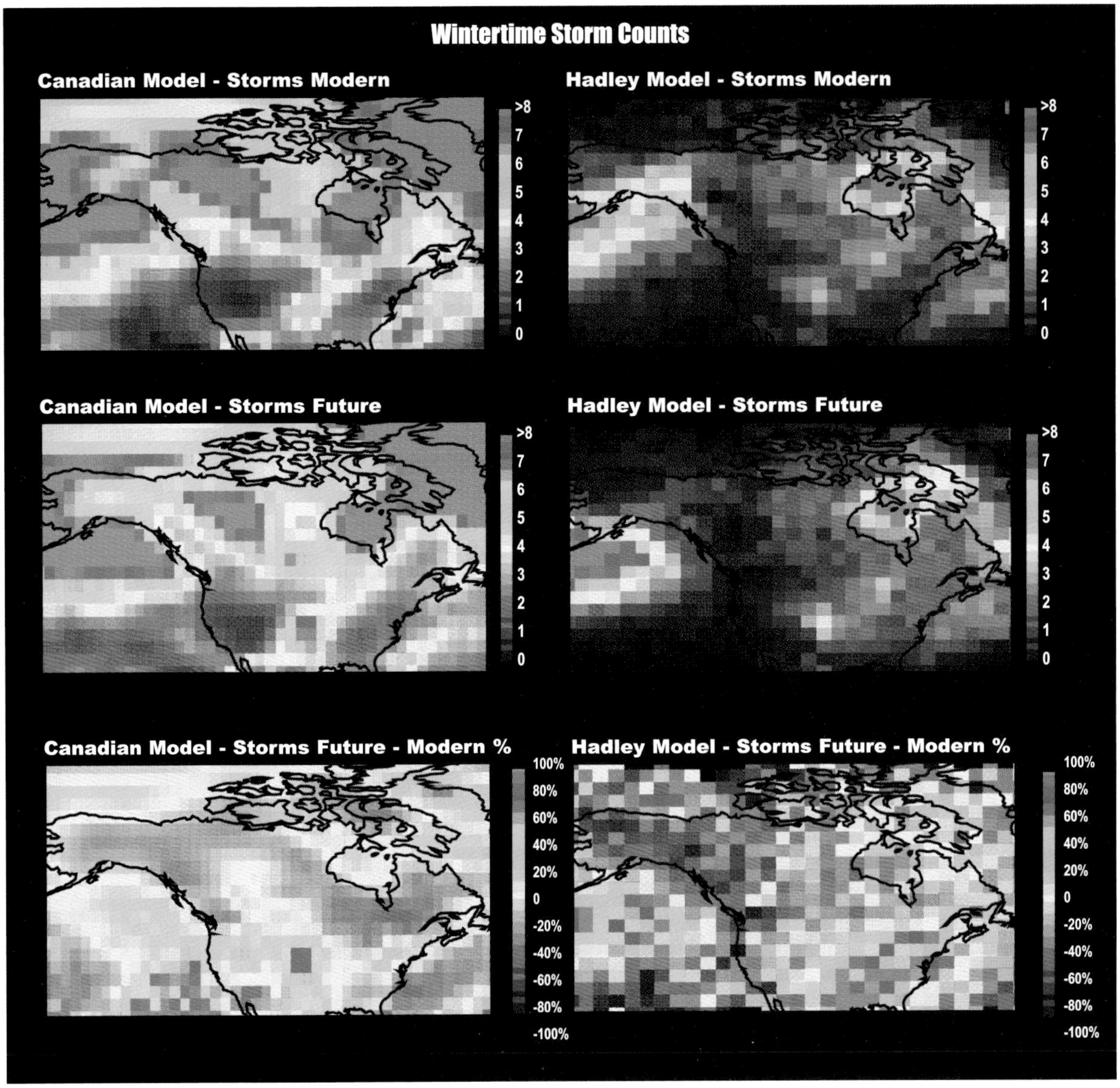

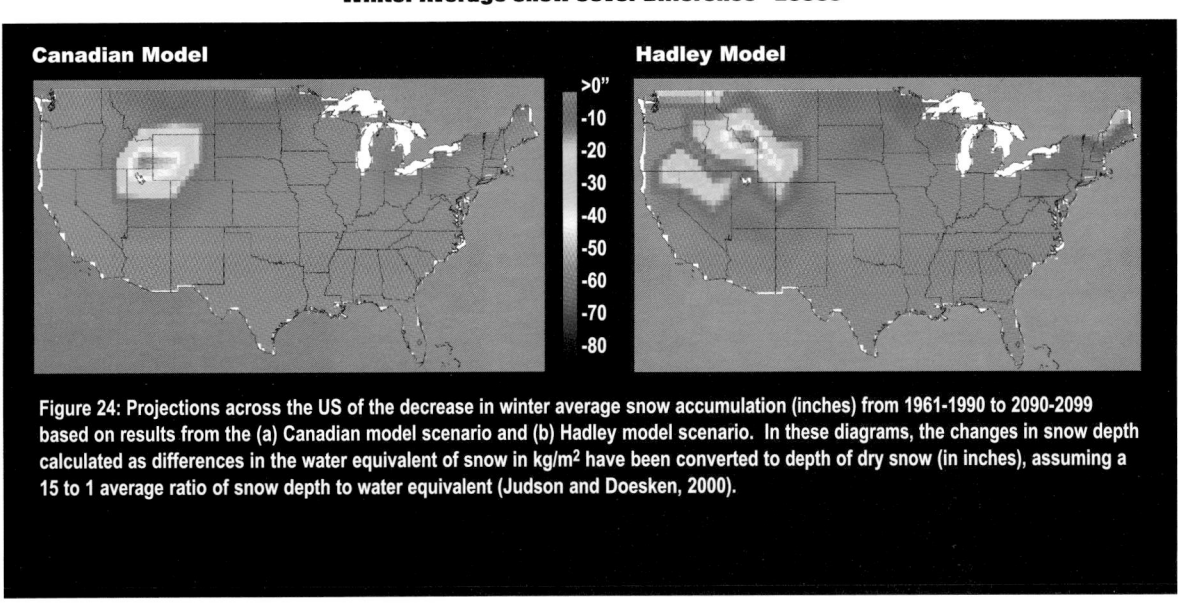

Figure 23: Wintertime (DJF) storm counts (Carnell and Senior, 1998; Lambert, 1995) from the (a) Canadian model scenario (1901-1910 total); (b) Hadley model scenario (1990-2110 mean from unforced control run); (c) Canadian model scenario (2091-2100 total); (d) Hadley model scenario (2070-2100 mean from transient run); (e) Canadian model scenario delta (c-a); and (f) Hadley model scenario delta (d-b). Units are number of winter storms per 145,000 km².

Figure 24: Projections across the US of the decrease in winter average snow accumulation (inches) from 1961-1990 to 2090-2099 based on results from the (a) Canadian model scenario and (b) Hadley model scenario. In these diagrams, the changes in snow depth calculated as differences in the water equivalent of snow in kg/m² have been converted to depth of dry snow (in inches), assuming a 15 to 1 average ratio of snow depth to water equivalent (Judson and Doesken, 2000).

Scenarios for Climate Variability and Change

Global Mean Temperature Anomolies

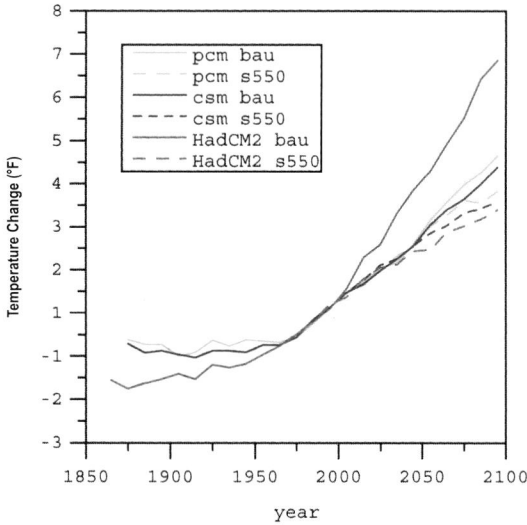

US Mean Temperature Anomolies

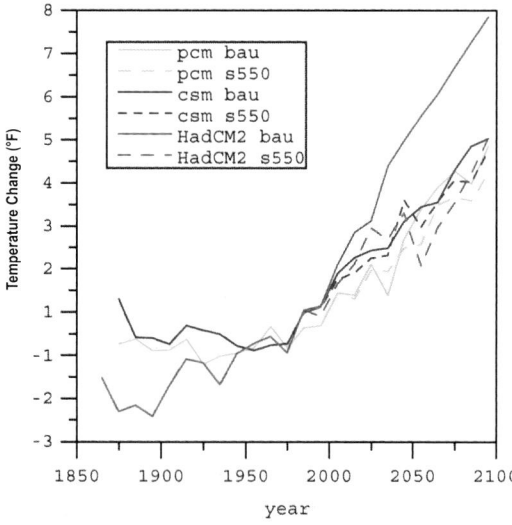

Figure 26: Comparison of the time history of the increase in annual-average surface temperature for (a) the globe and (b) the US as projected by two related models developed at the National Center for Atmospheric Research for an emission scenario where the greenhouse gas concentrations are allowed to rise without restriction (baseline) and for a case (stabilization) where steps are taken to limit the rise in the CO_2 concentration to 550 ppmv (Dai et al., 1999; Washington et al., 2000). Results are also shown for a recent Hadley model simulation (Mitchell et al., 2000). See Color Plate Appendix.

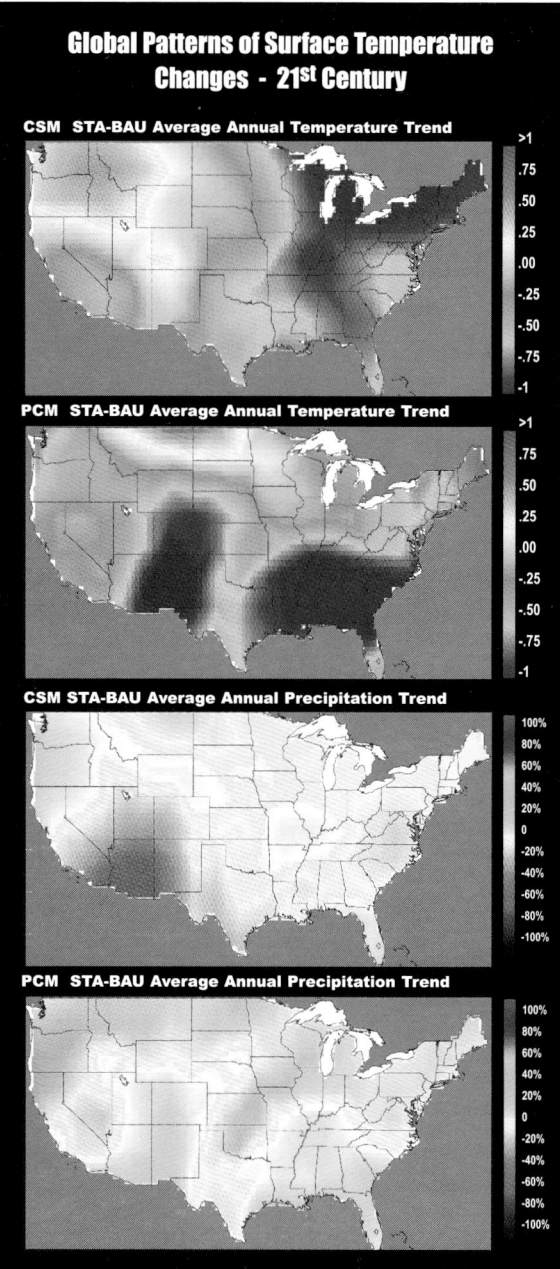

Figure 27: Patterns across the US of projected changes in the trends of annual mean surface temperature and precipitation for the 21st century assuming an emissions profile that moves toward stabilization of the CO_2 concentration at 550 ppmv in the 22nd century (STA) as compared to the baseline case (roughly case IS92a, or BAU, except projections in sulfur emissions are reduced in the CSM scenario). The projected differences in the changes that would generally be projected (case STA minus BAU) are based on results from: (a) NCAR CSM for annual mean temperature; (b) PCM for annual mean temperature: (c) NCAR CSM annual average monthly precipitation; and (d) PCM annual average monthly precipitation. Temperature trend differences are given as °F per 100 years. Precipitation trend differences are given in percent, with both trends calculated using a 1980-1999 baseline. Trends are derived based on a linear regression through each grid point. Results are described in Dai et al. (1999) and Washington et al. (2000). See Color Plate Appendix.

Changes in Vegetation Carbon

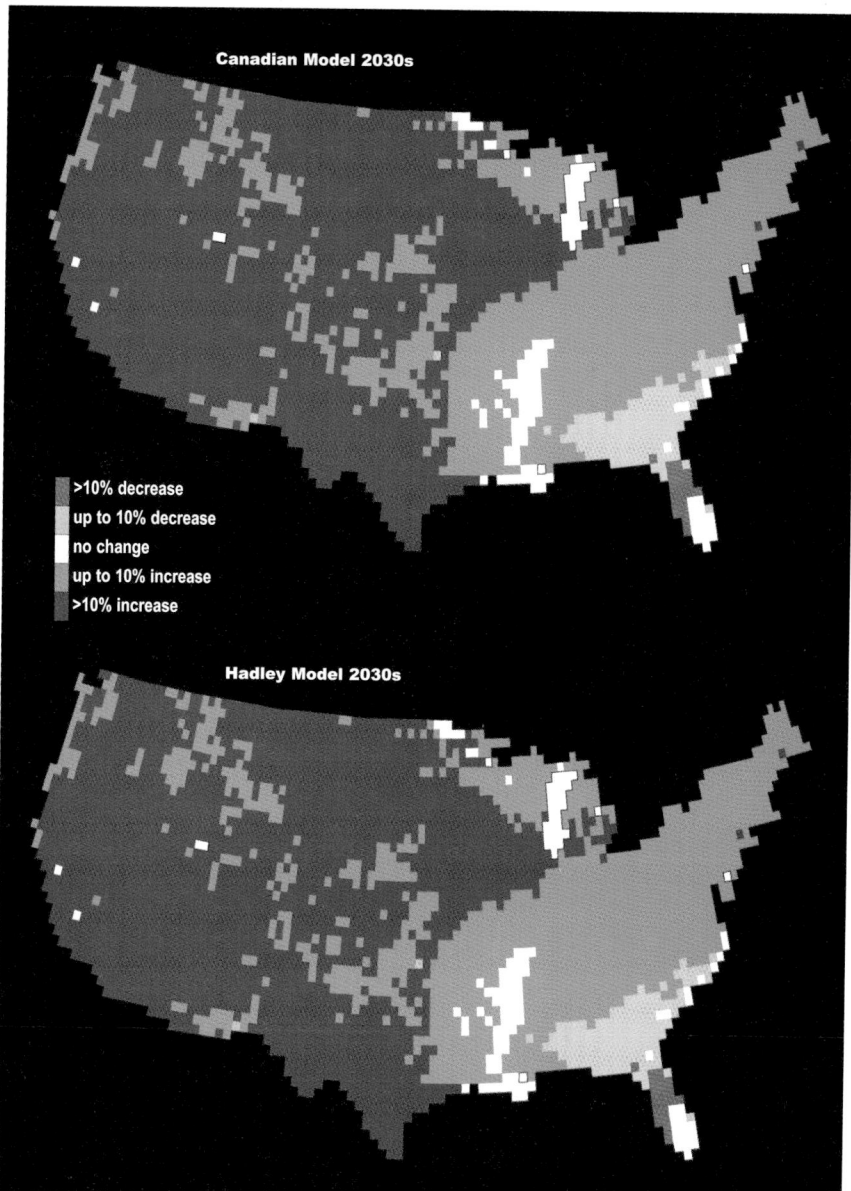

Figure 1. The maps above show projections of relative changes in vegetation carbon between 1990 and the 2030s for two climate scenarios. Under the Canadian model scenario, vegetation carbon losses of up to 20% are projected in some forested areas of the Southeast in response to warming and drying of the region by the 2030s. A carbon loss by forests is treated as an indication that they are in decline. Under the same scenario, vegetation carbon increases of up to 20% are projected in the forested areas in the West that receive substantial increases in precipitation. Output from TEM (Terrestrial Ecosystem Model) as part of the VEMAP II (Vegetation Ecosystem Modeling and Analysis Project) study.

Ecosystem Models

Current Ecosystems

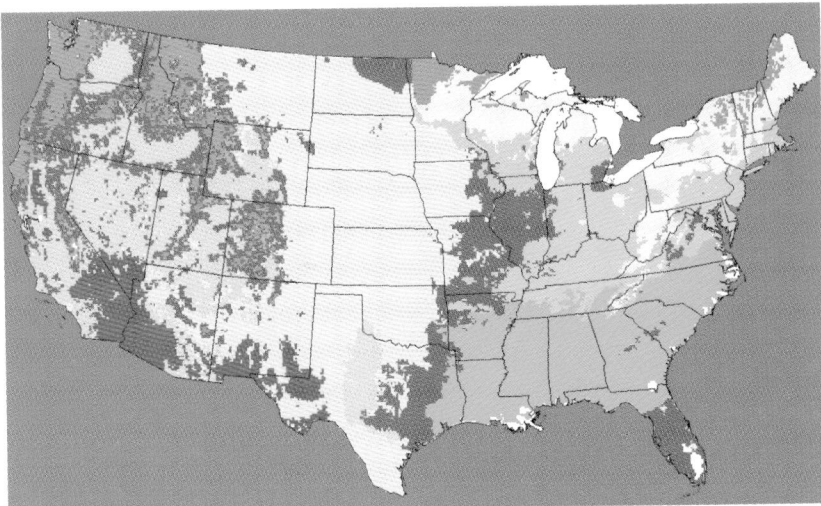

Canadian Model

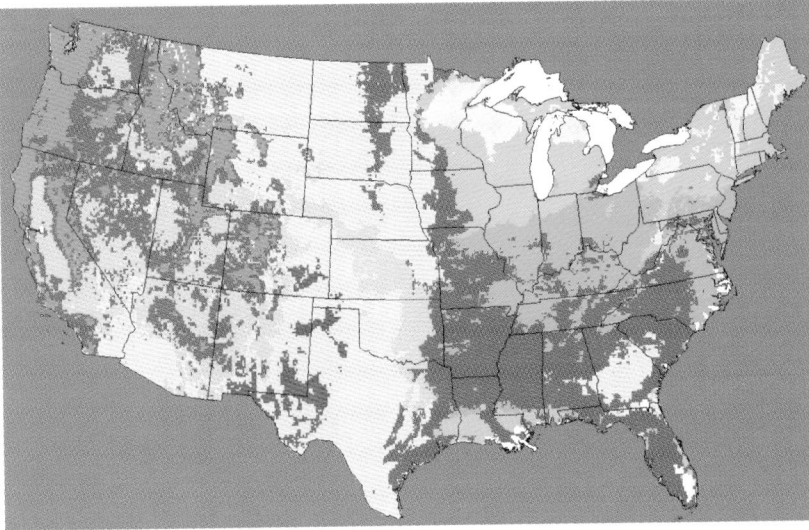

Tundra
Taiga / Tundra
Conifer Forest
Northeast Mixed Forest
Temperate Deciduous Forest
Southeast Mixed Forest
Tropical Broadleaf Forest
Savanna / Woodland
Shrub / Woodland
Grassland
Arid Lands

Hadley Model

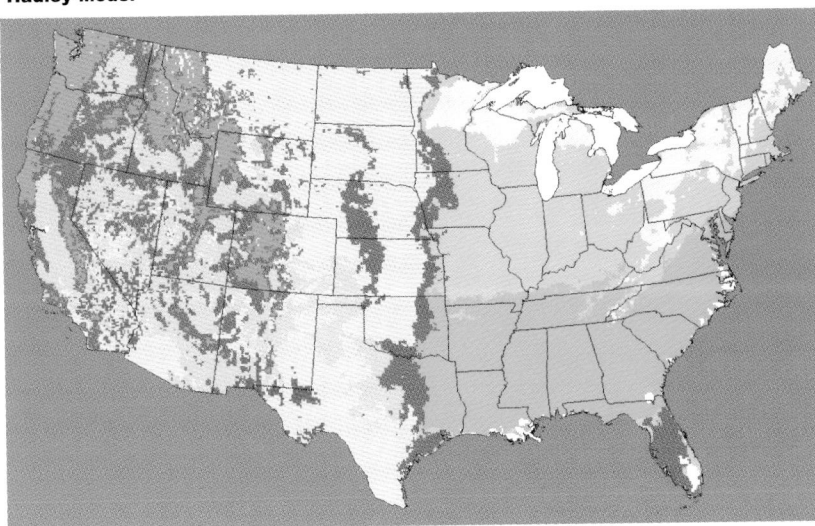

Figure 2. The models used to estimate biogeographic responses to climate change in VEMAP
II include LPJ, MAPSS, and MC1. These three
models predict the local dominance of various
terrestrial vegetation forms based on: (1) ecophysiological constraints, which determine the
broad distribution of major categories of woody
plants; and (2) response limitations, which determine specific aspects of community composition, such as the competitive balance of trees
and grasses. Though similar in some respects,
these models simulate potential evapotranspiration and direct CO_2 effects differently, and as a
result they show varying sensitivities to temperature, CO_2 levels, and other factors. Two of the
model models, LPJ and MC1 have biogeochemistry modules, while the third, MAPSS, does not.
For both the Hadley and Canadian climate scenarios, the biogeography models project shifts
in the distribution of major vegetation types as
plant species move in response to climate
change. The projected changes in vegetation
distribution with climate change vary from
region to region. (Source: VEMAP, 1998).

Vegetation and Biogeochemical Scenarios

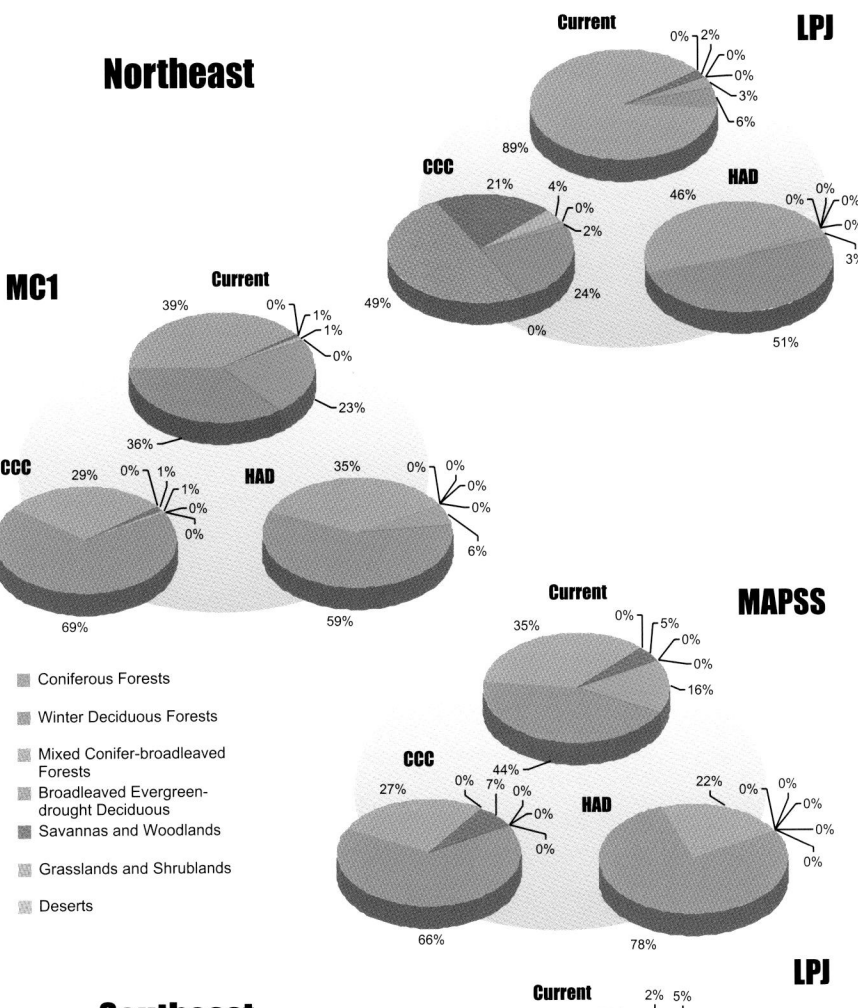

LPJ, MC1 and MAPSS Estimates

Figure 3(a) Under both simulated climates, forests remain the dominant natural vegetation, but the mix of forest types changes. For example, winter-deciduous forests expand at the expense of mixed conifer-broad-leaved forests. Under the climate simulated by the Canadian model, there is a modest increase in savannas and woodlands.

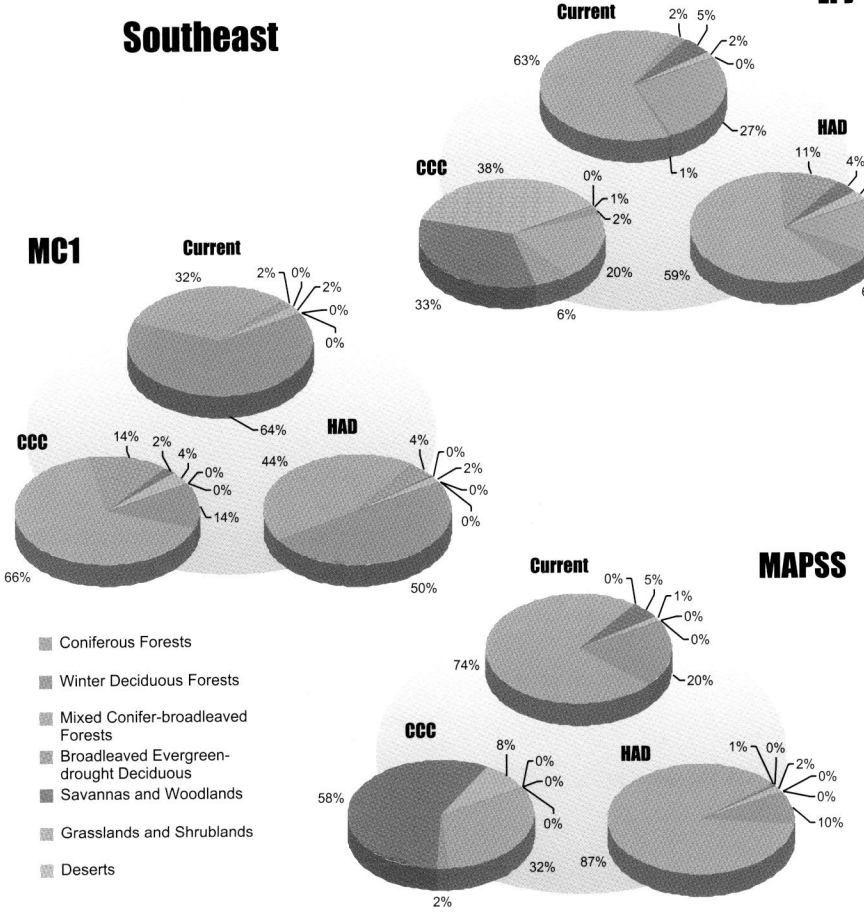

Figure 3(b) Under the climate simulated by the Hadley model, forest remains the dominant natural vegetation, but once again the mix of forest types changes. Under the climate simulated by the Canadian model, all three biogeography models show an expansion of savannas and grasslands at the expense of forests. For two of biogeography models, LPJ and MAPSS, the expansion of these non-forest ecosystems is dramatic by the end of the 21st century. Both drought and fire play an important role in the forest breakup.

559

Mid-West

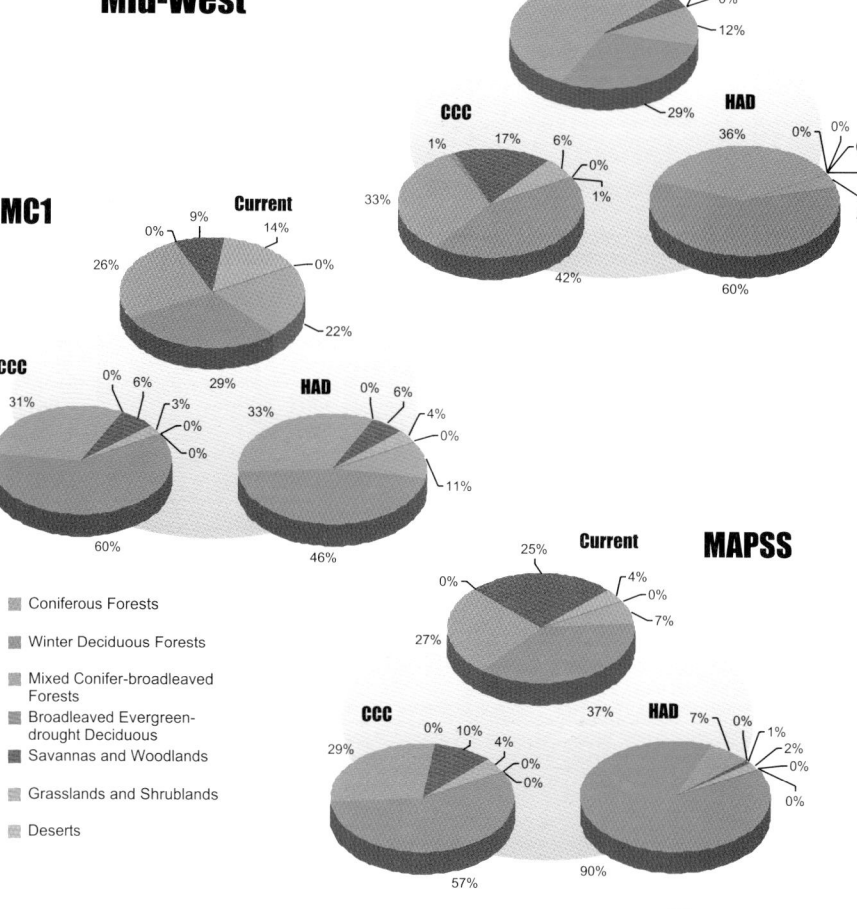

Figure 3(c) Under both simulated climates, forests remain the dominant natural vegetation, but the mix of forest types changes. One biogeography model, LBJ, simulates a modest expansion of savannas and grasslands.

- Coniferous Forests
- Winter Deciduous Forests
- Mixed Conifer-broadleaved Forests
- Broadleaved Evergreen-drought Deciduous
- Savannas and Woodlands
- Grasslands and Shrublands
- Deserts

Great Plains

Figure 3(d) Under the climate simulated by the Hadley model, two biogeography models project an increase in woodiness in this region, while the third projects no change in woodiness. Under the climate simulated by the Canadian Model, the biogeography models project either no change in woodiness or a slight decrease.

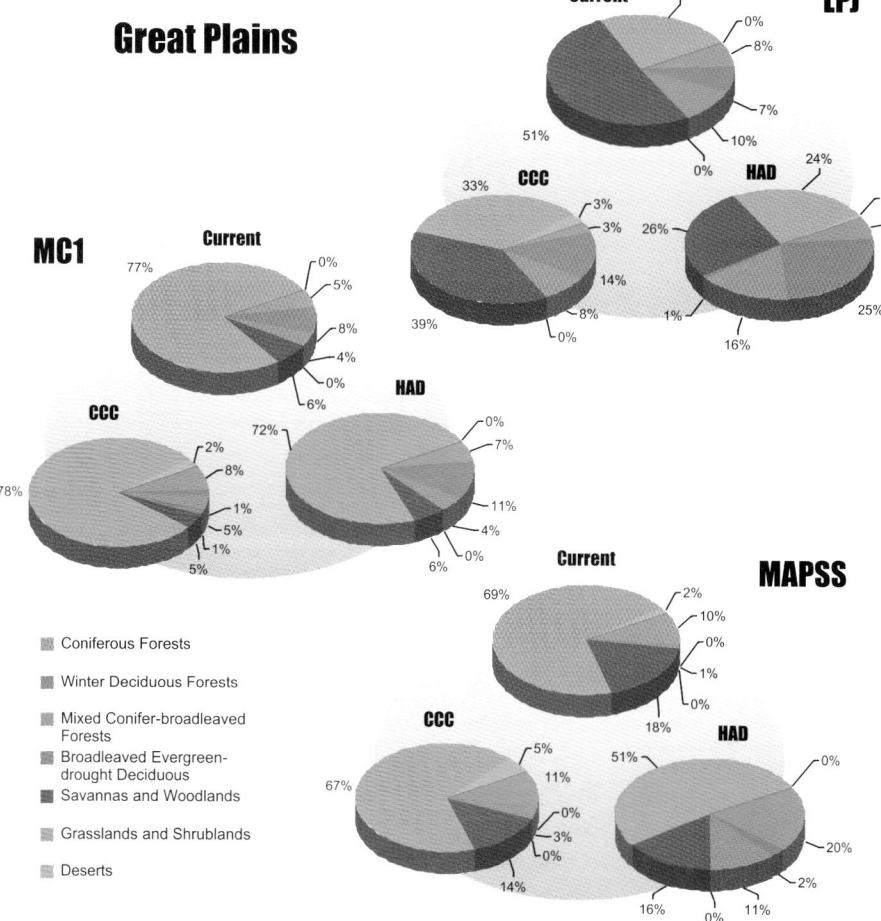

- Coniferous Forests
- Winter Deciduous Forests
- Mixed Conifer-broadleaved Forests
- Broadleaved Evergreen-drought Deciduous
- Savannas and Woodlands
- Grasslands and Shrublands
- Deserts

LPJ, MC1 and MAPSS Estimates

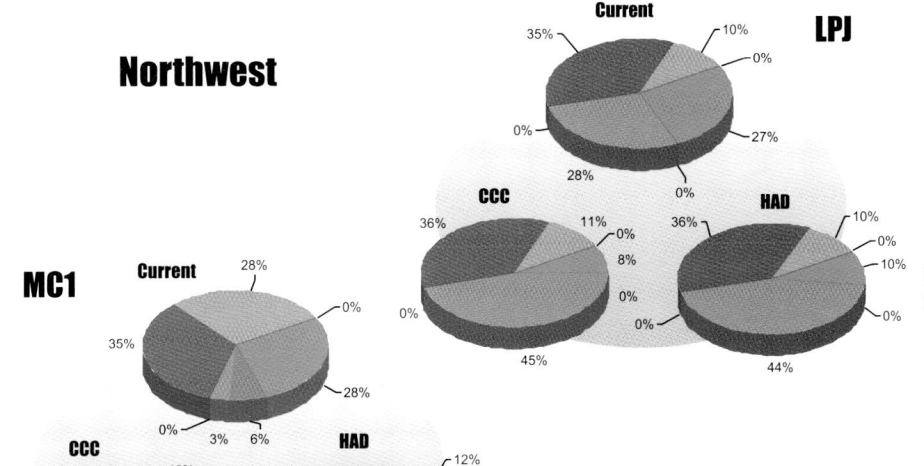

Figure 3(e): Under both simulated climates, the forest area grows slightly.

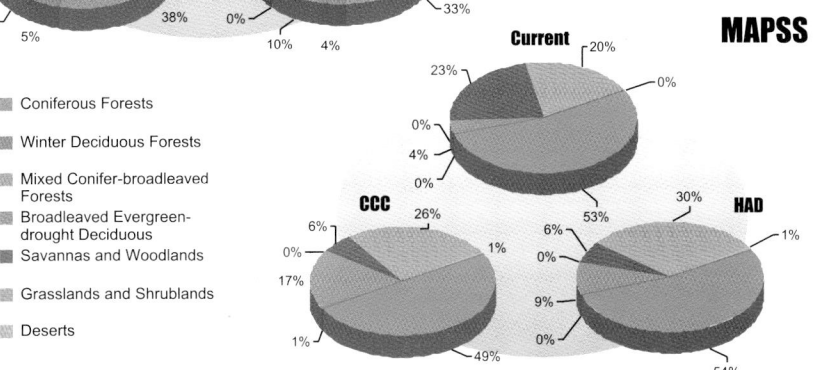

Figure 3(f). Under the climate simulated by both the Hadley and Canadian models, the area of desert ecosystems shrinks and the area of forest ecosystems grows.

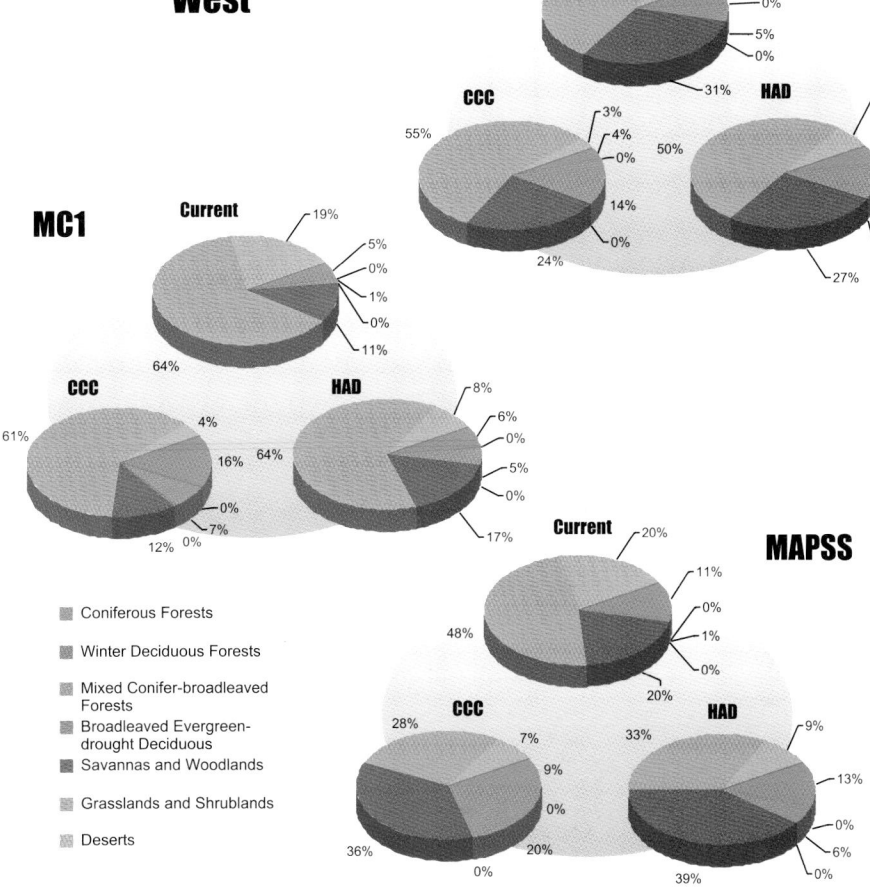

Coniferous Forests

Winter Deciduous Forests

Mixed Conifer-broadleaved Forests

Broadleaved Evergreen-drought Deciduous

Savannas and Woodlands

Grasslands and Shrublands

Deserts

561

Scenarios of 21st Century Growth in America

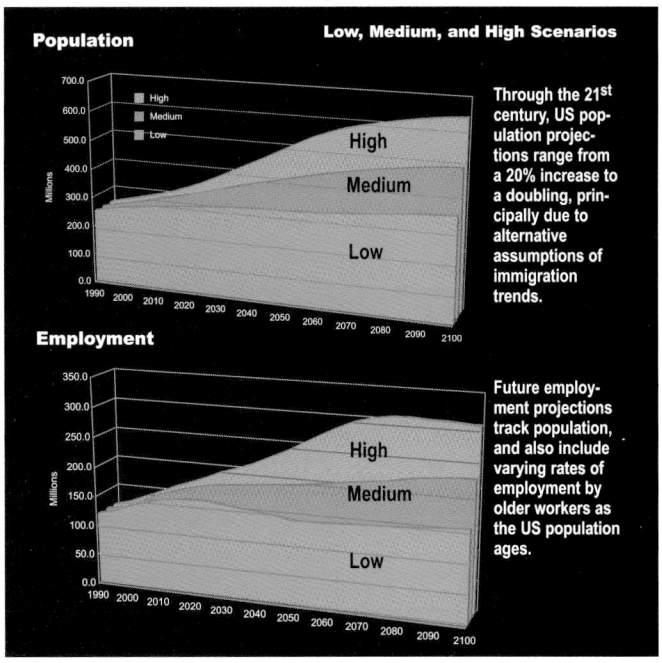

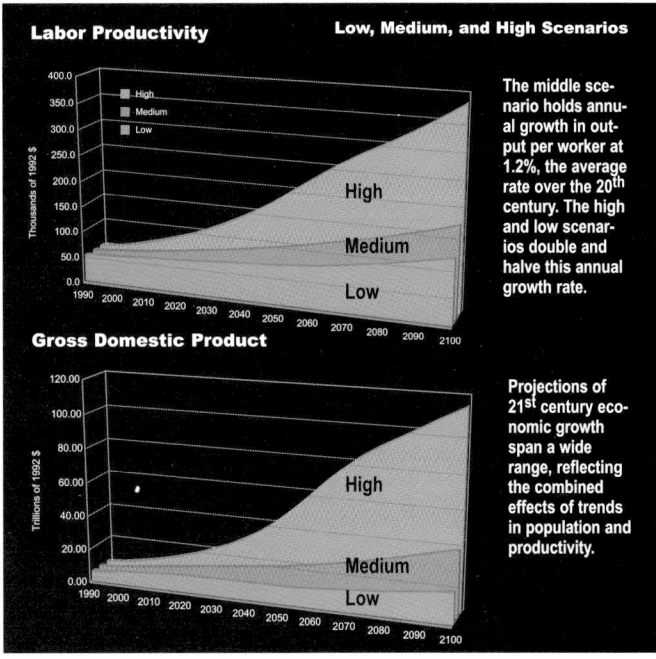

Figure 1. The Assessment considered high, medium, and low scenarios of future US population and economic growth. Future trends in population, economic growth, and technological change will all shape our contribution to climate change, our vulnerability to it, and our ability to adapt.

Changes in Storm Tracks

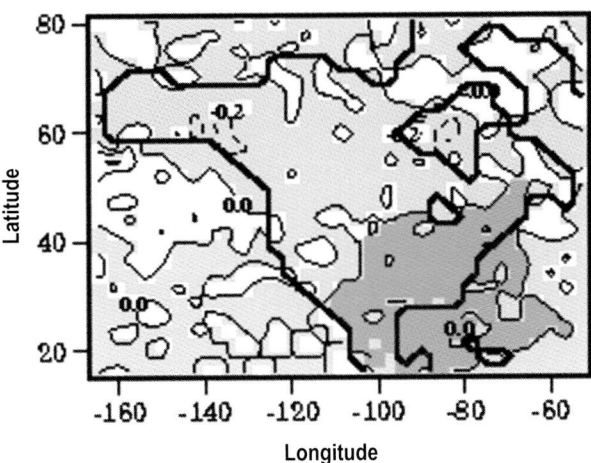

Figure 1. A storm track analysis from the Hadley climate model scenario projects a slightly strengthened wintertime storm track through the Northeast in the 2020s, because the jet stream has a more north-to-south position along the East Coast. This scenario projects a slightly stronger winter storm area (dark shaded region). The Canadian climate scenario has a more east-west jet, and in general indicates slightly weaker storminess.

Palmer Drought Severity Index Change

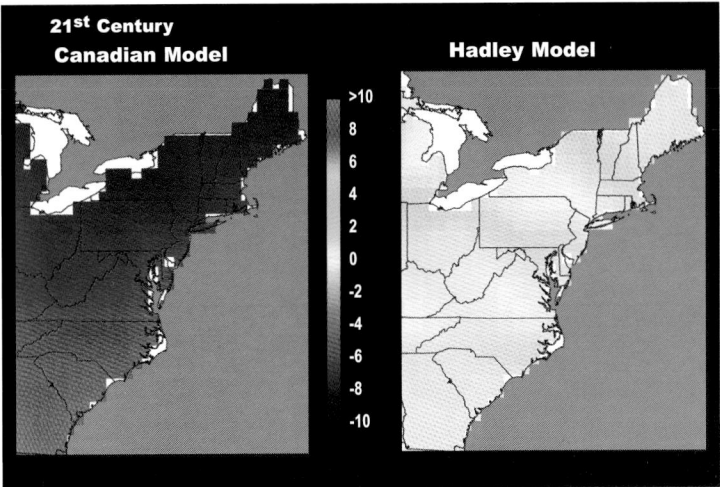

Figure 2. The projected trends in the Palmer Drought Severity Index (PDSI) are dependent on the projections of temperature and precipitation. Large increases in drought tendencies occur in the Northeast in the Canadian model associated with substantial warming and small changes in precipitation. In contrast, the Hadley model yields larger increases in precipitation and a more modest warming, conditions under which the drought tendency tends to decline.

Ecosystem Models

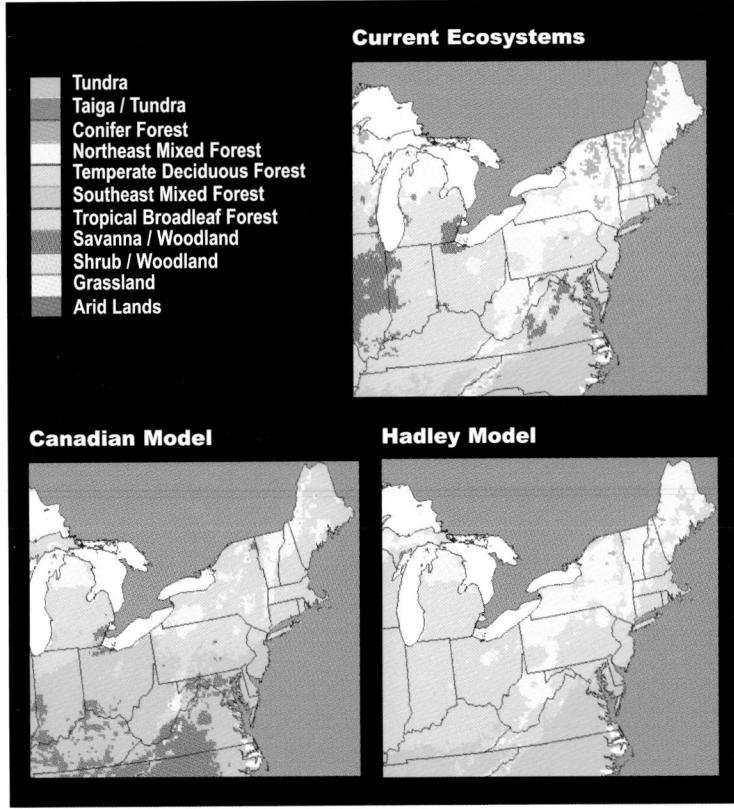

Figure 3. The projected changes in vegetation character using output of the Canadian (a) and the Hadley (b) models indicates a substantial northward shift in the vegetation types. These changes are significantly larger in the Canadian model scenario, which projects a greater warming trend with little change or a decrease in precipitation. Based on the model of Neilson and Drapek (1998).

Northeast

Percent Salinity Change in the Chesapeake Bay

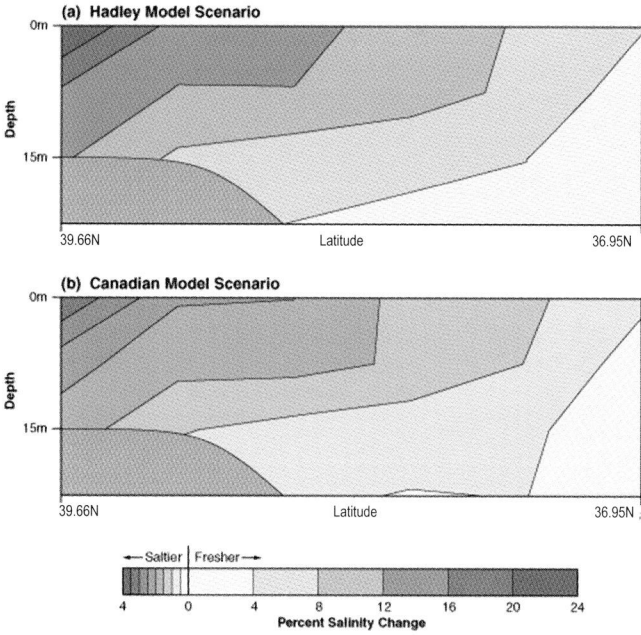

Figure 7. Calculated salinity within the Chesapeake given the run-off calculated from the Hadley (top) and Canadian (bottom) climate scenarios by Gibson and Najjar (2000). The distribution of salinities ranges from the upper reaches of the Bay (39.66N) to the Lower Chesapeake near its Atlantic opening (36.95N).

Potential Changes In Severe Weather

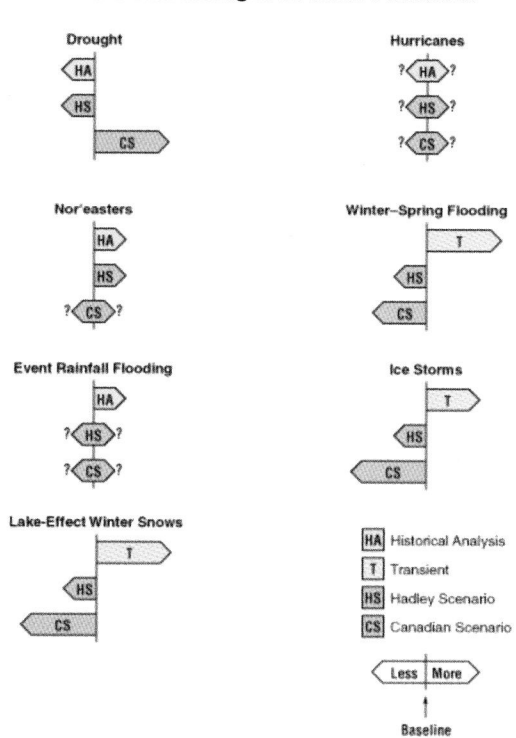

Figure 5. Schematic of the potential changes in severe weather for the Northeast based on historical data (H), the Hadley model scenario (HS), the Canadian model scenario (CS) or an assessment of possible transient effects (T).

Nor'easter of December 1992

The December 11-12, 1992 nor'easter produced some of the worst flooding and strongest winds on record for the area. It resulted in a near shutdown of the New York metropolitan transportation system and evacuation of many seaside communities in New Jersey and Long Island. This storm should have provided a "wake-up" call, heralding the vulnerability of the transportation system to major nor'easters and hurricanes. Had flood levels been only 1 to 2 feet above the actual high water level of 8.5-foot above mean sea level, massive inundation of rail and subway tunnels could have resulted in loss of life. With rising sea levels, even a weaker storm would produce comparable damage. While hurricanes are much less frequent than nor'easters in this area, they can be even more destructive because the geometry of the New Jersey and Long Island coasts amplifies surge levels toward the New York City harbor. For a worst-case scenario category-3 hurricane, surge levels could rise 25 feet above mean sea level at JFK airport and 21 feet at the Lincoln tunnel entrance.

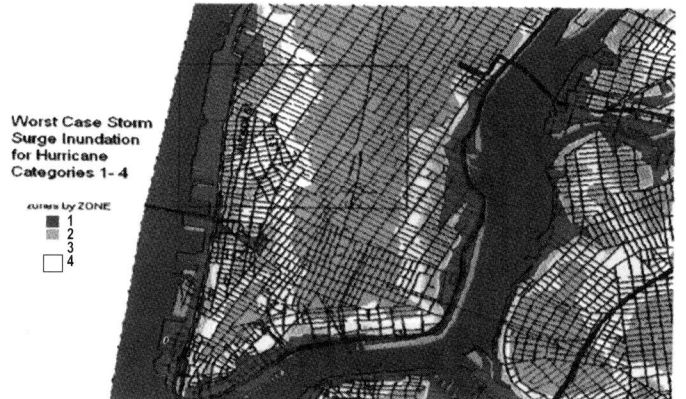

Figure 8. Vulnerable coastal areas for Manhattan, based on a 20-foot high flooding zone for the year 2100, derived by Klaus Jacob of Lamont-Doherty Earth Observatory for the Metroeast workshop.

Dominant Forest Types

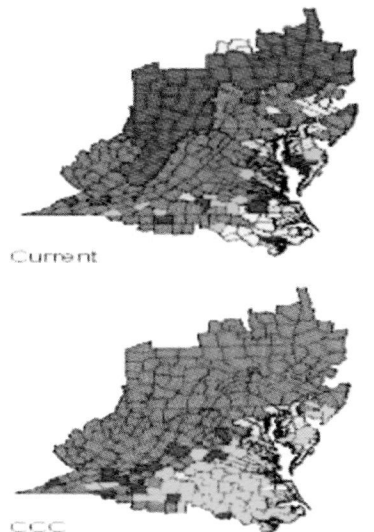

Current

Maple-Beech-Birch
Elm-Ash-Cottonwood
Oak-Gum-Cypress
Oak-Hickory
Oak-Pine
Loblolly-Shortleaf Pine
Longleaf-Slash Pine
No data

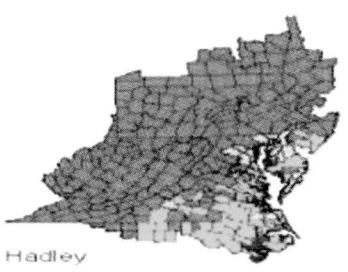

CCC   Hadley

Figure 12. Dominant forest types for the mid-Atlantic region for current climate, and the potential distribution of these forest types for the Canadian and Hadley climate scenarios based on the Mid-Atlantic Assessment. Based on the model of Iverson and Prasad (1998).

Figure 11. On warm humid days when temperatures exceed 90°F, ozone problems are exacerbated across the region. The top figure shows the view on a clear day at the Great Gulf of Mount Washington, New Hampshire. The bottom figure shows the same view when temperatures exceed 90°F and air quality problems occur.

Figure 10. Outdoor recreation is of major economic importance in the Northeast, and it is tightly coupled to climatic conditions.

Northeast

Land Cover Map of the Southeast Region

Figure 1. The Southeast region includes all of nine states (Alabama, Florida, Georgia, Kentucky, Louisiana, North Carolina, Mississippi, South Carolina, and Tennessee), the southern portion of Virginia, and 50 counties in east Texas. Four subregional workshops were conducted in the Southeast region.

Southeast US Annual Mean Temperature

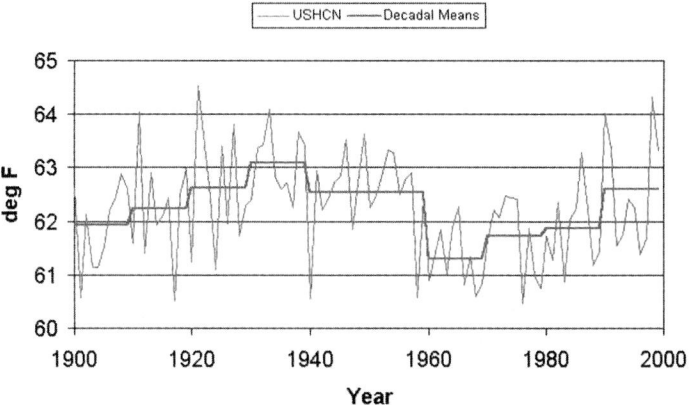

Figure 2. Decadal average temperatures in the Southeast. Source D. Easterling, NOAA National Climatic Data Center.

Gulf Landfalling Hurricanes By Decade

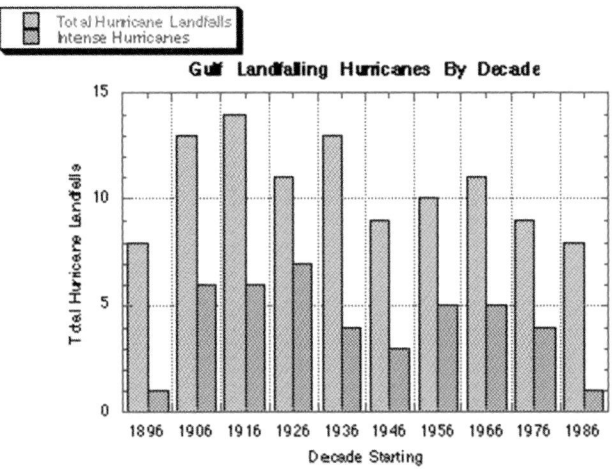

Figure 3(a). US Hurricane Landfall Trends in the Gulf of Mexico. This figure shows the number of US hurricanes making landfall in the Gulf of Mexico by decade for the past 100 years. There were peaks in activity during the 1910s and 20s, as well as a lower peak in the 1960s. The past 30 years have shown a decrease in the number and intensity of Gulf hurricanes making landfall.

US Hurricanes

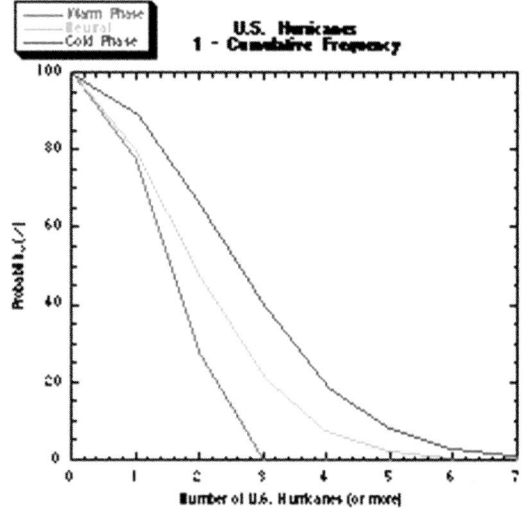

Figure 3(b). Effect of ENSO Phase on Hurricane Landfall
This figure shows the probability of the number of hurricane landfalls on the US in a given hurricane season and ENSO phase (El Niño, Neutral, La Niña). Based on the past 100-year record, the probability of at least 1 hurricane landfall is similar for all three phases, with probabilities ranging from 78% for El Niño to 90% for La Niña. For multiple landfalls, however, the differences caused by ENSO phase become apparent. The probability of at least 2 landfalls during El Niño is 28%, but is 48% in neutral years, and 66% during La Niña. The probability of at least 3 landfalling hurricanes is near 0% for El Niño, 20% for neutral years, and 50% for La Niña. It is clear that El Niño years have few multiple hurricane strikes on the US, while neutral years and La Niña years often see multiple hurricane strikes on the US coast. Source: Florida State University, Center for Ocean-Atmosphere Prediction.

July Heat Index Change - 21st Century

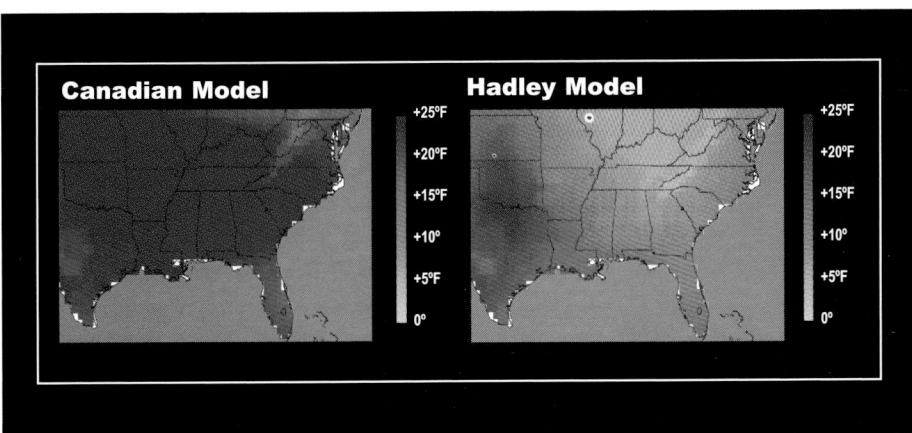

Figure 4. The changes in the simulated heat index for the Southeast are the most dramatic in the nation with the Hadley model suggesting increases of 8 to 15°F for the southern-most states, while the Canadian model projects increases above 20°F for much of the region. Heat indices simulated for the Southeast by 2100. Source, NOAA National Climatic Data Center.

Summer and Winter Climate Changes from Hadley Centre Scenario

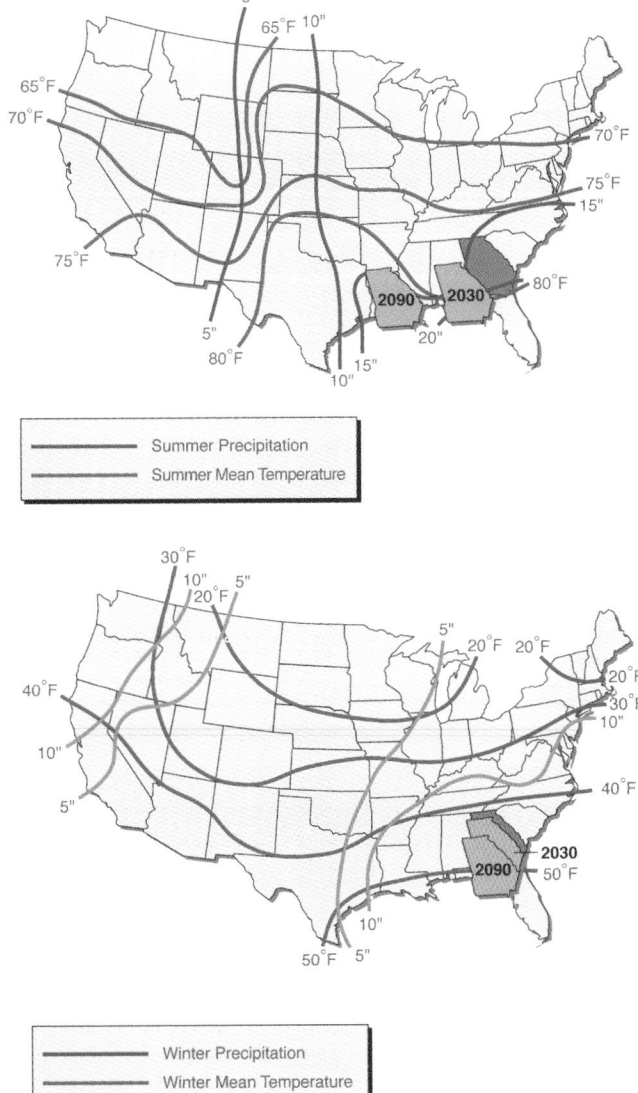

Figure 5. Illustration of how the summer and winter climates in Georgia would shift under the Hadley climate scenario (HADCM2). For example, the summer climate in Georgia in the 2030s would be more like the current climate of the Florida panhandle. Source: NOAA, National Climatic Data Center.

Dryland Crop Yield Changes

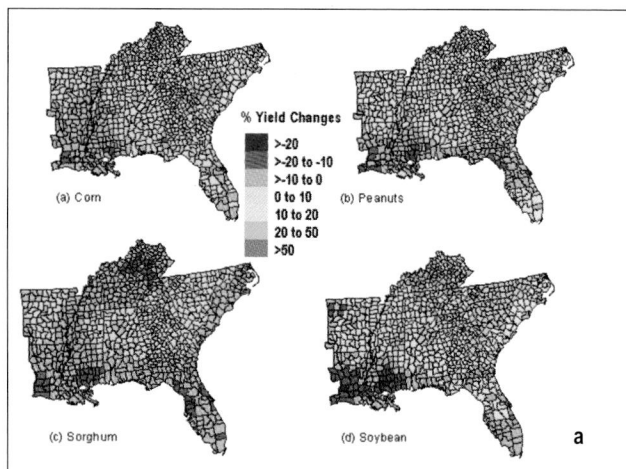

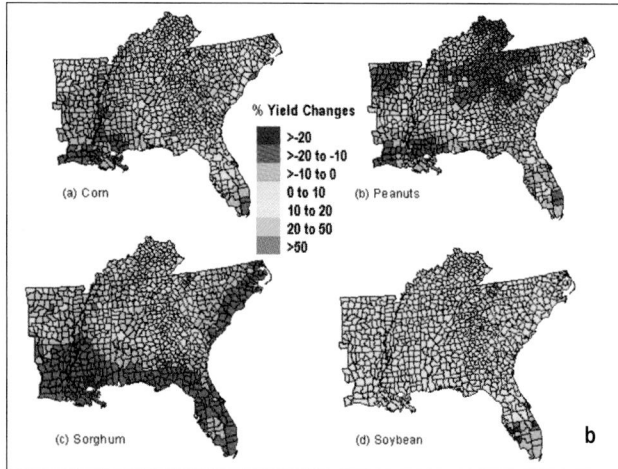

Figure 6. Dryland crop yield changes in 2030 (a) and 2090 (b) without adaptation for various climate sensitivity scenarios Source: Auburn University, Global Hydrology and Climate Center; University of Florida, Agricultural and Biological Engineering Department.

Simulated Changes in Dryland Yields for Southeastern Crops based on the Hadley (HADCM2) Scenario

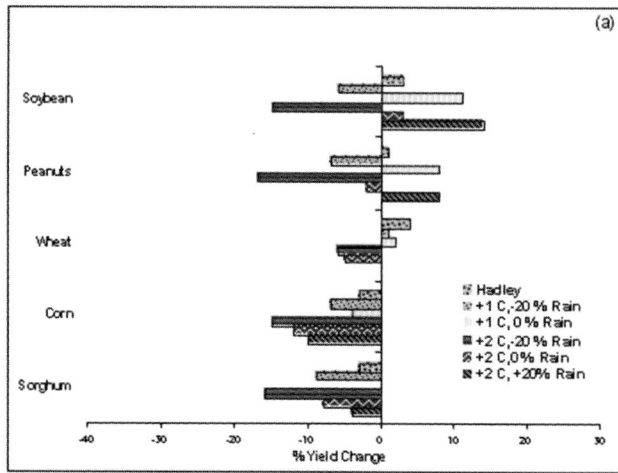

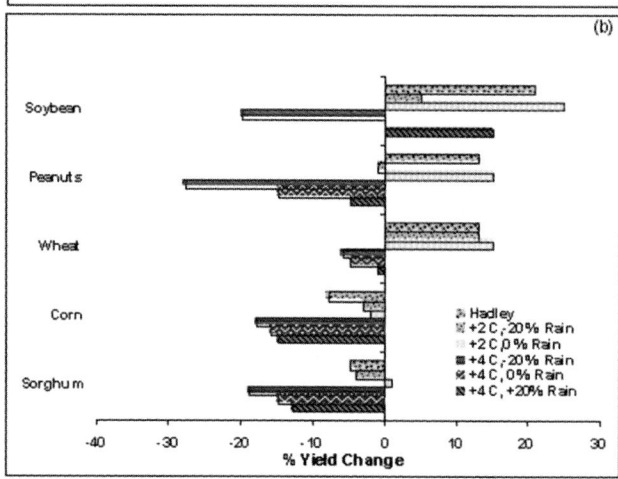

Figure 7. Dryland yields changes in 2030(a) and 2090(b) without adaption for various climate sensitivity scenarios. Source: Auburn University, Global Hydrology and Climate Center; University of Florida, Agricultural and Biological Engineering Department.

Changes in Florida's Big Bend

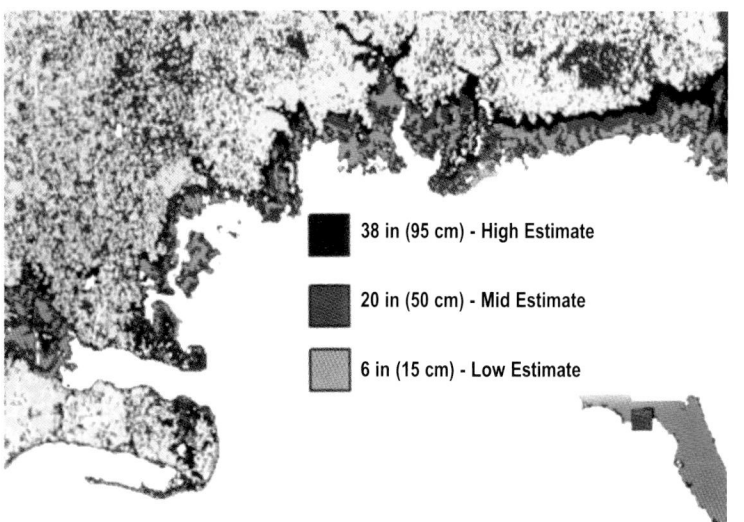

Figure 11. Changes in Florida's Big Bend region forest, marshes, and open water under IPCC (1998) sea-level rise scenarios. Source: Doyle, 1998.

Timberland Acreage Shift

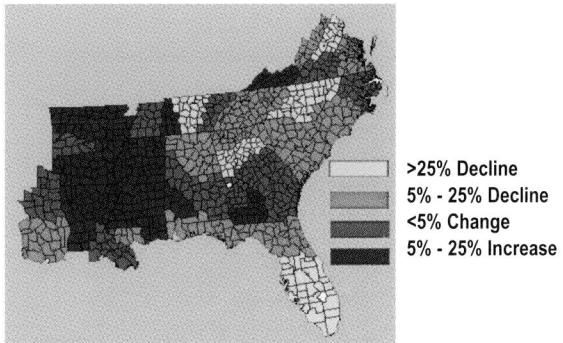

>25% Decline
5% - 25% Decline
<5% Change
5% - 25% Increase

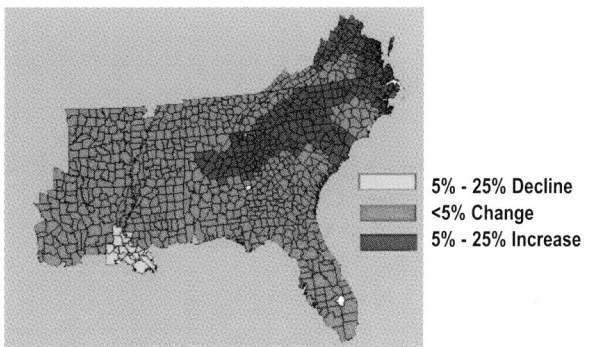

5% - 25% Decline
<5% Change
5% - 25% Increase

Figure 9 (a). Changes in land use based on Timberland Acreage Shift 1993-2040: Baseline Without Climate Change. Forestland losses are projected in the more urbanized areas of the Southeast, from northern Virginia through the Georgia piedmont and southern Florida. The movement of land from agriculture to forest is projected in many parts of the mid-South.

Figure 9(b). Timberland acreage shifts by 2040 due to Hadley climate change. In 2040, forestland is projected to be slightly higher with Hadley base climate change than without climate change in some of the northern reaches of the Southeast, but slightly lower under climate change in parts of the deep South. Year 2040 land allocation effects in most of the region are fairly neutral. Source: North Carolina State University, Department of Forestry; Research Triangle Institute, Center for Economics Research.

Potential Southern Pines and Hardwoods Net Primary Productivity (NPP)

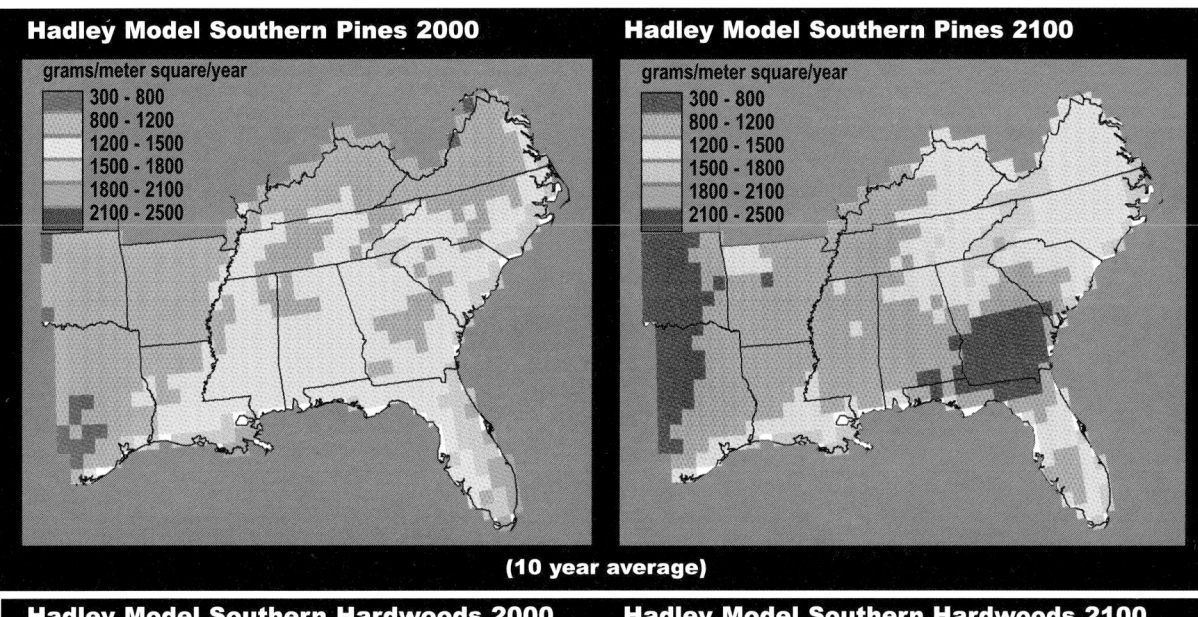

Figure 8. Potential net primary productivity (NPP) of loblolly pine and southern hardwoods simulated by the PnET model with the Hadley climate scenario (HADCM2). Source: USDA Forest Service, Southern Global Change Program.

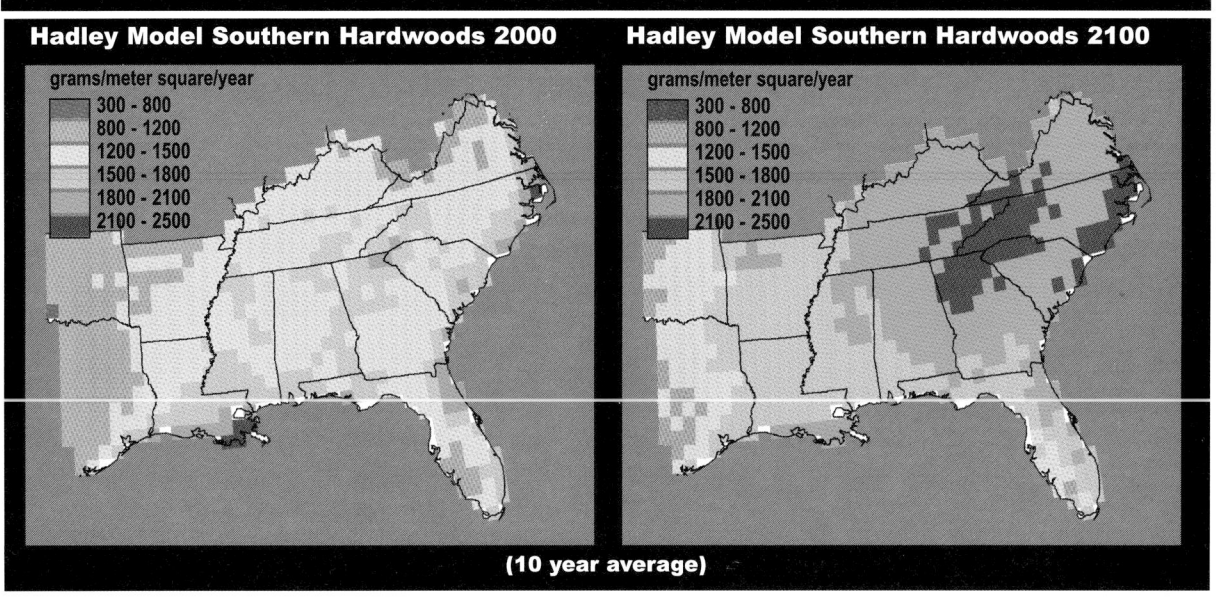

Southeast

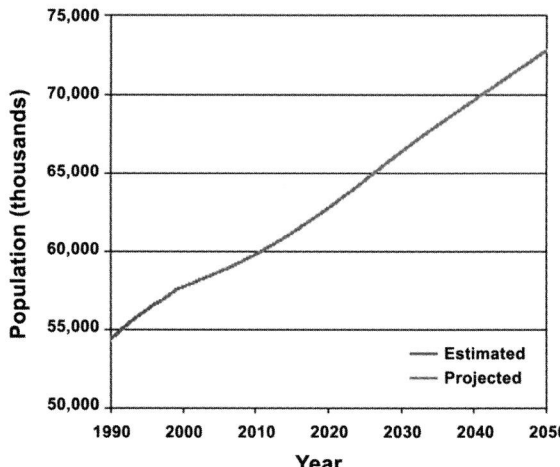

Midwest Population Estimates and Projections

Figure 2. Population trend estimate for the Midwest region using the baseline assumptions from the NPA Data Services estimates. Under this scenario, the population of the Midwest is expected to increase by about 30% by 2050.

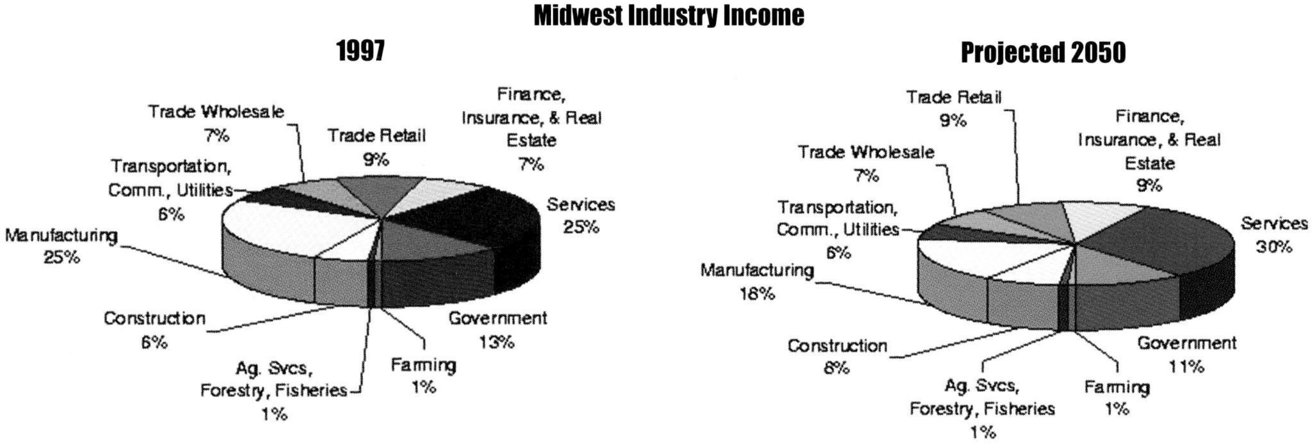

Figure 1. Map of Midwest region.

Midwest Industry Income

1997

Projected 2050

Figure 3. Percentage of economy by sector for (a) 1997, and (b) 2050, estimated from the NPA Data Services using baseline assumptions (NPA 1999b). Under these estimates, by 2050 the manufacturing percentage of the economy decreases by 7%, and the service sector increases by 5%.

Great Lakes Water Level Change

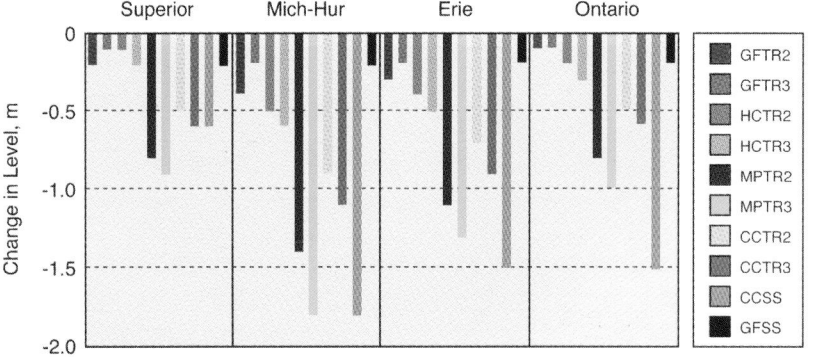

Figure 5. Change in water level for each of the Great Lakes under a number of climate change scenarios, from Chao (1999).

Midwest Daily Precipitation/HadCM2

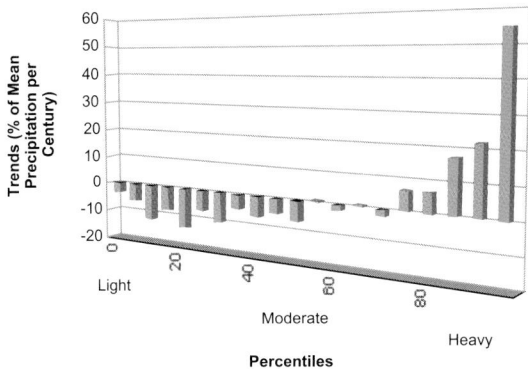

Midwest Soybean Yield and Precipitation

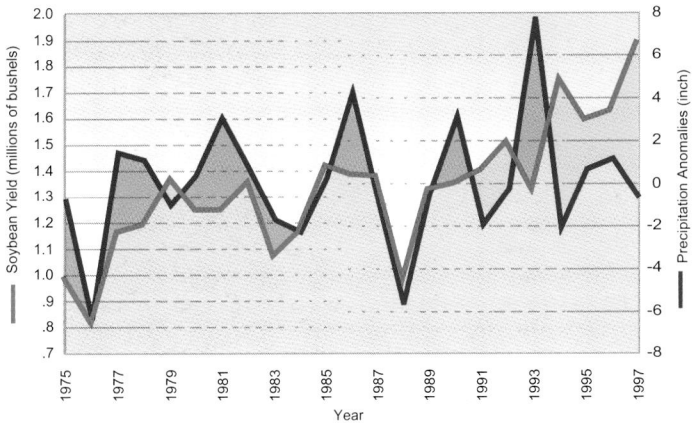

Figure 8. The relationship between Midwest soybean yield and precipitation is shown here. Soybean yields in thousands of bushels are shown as the differences from the average yield in recent decades. Precipitation is the difference from the 1961-90 average precipitation. Note that lower yields result from both extreme wet and extreme dry conditions. Soybean yields from National Agricultural Statistics Service, USDA

Midwest Daily Precipitation/CGCM1

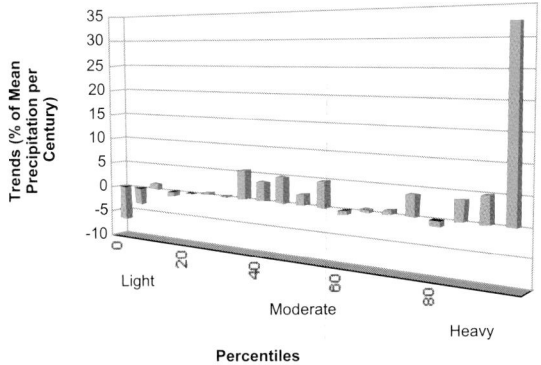

Figure 4. Annual trends in daily precipitation by percentile for the (a) CGCM1 (Canadian model) and (b) HadCM2 (Hadley model) scenarios. Notice the largest trend is in the heaviest daily precipitation amount for both model simulations indicating that most of the increase in annual precipitation is due to an increase in precipitation on days already receiving large amounts (analysis based on method in Karl and Knight, 1998).

Summer Climate Shifts

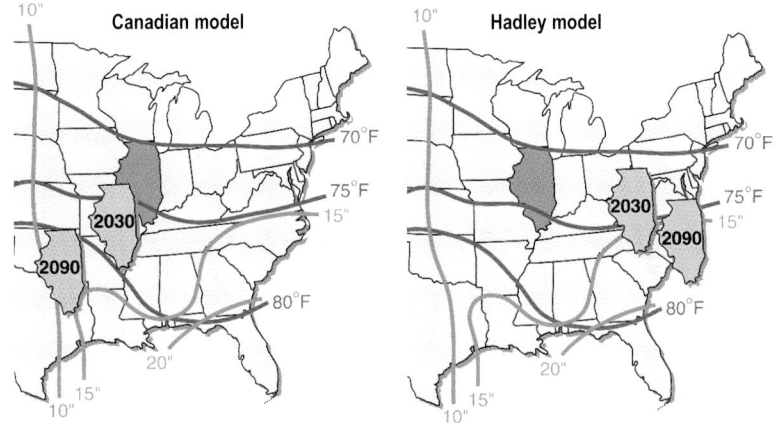

Figure 7. Illustration of how the summer climate of Illinois would shift under the (a) CGCM1 (Canadian model) scenario, and (b) HadCM2 (Hadley model) scenarios. For example, under the CGCM1 Canadian scenario, the summer climate of Illinois would become more like the current climate of southern Missouri in 2030 and more like Oklahoma's current climate in 2090.

Shipping Cost Change

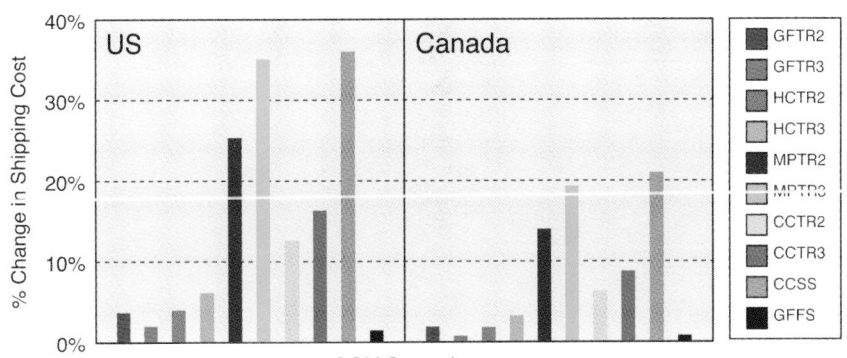

Figure 6. Change in shipping costs under a number of climate change scenarios, from Chao (1999).

Great Plains Vegetation Map

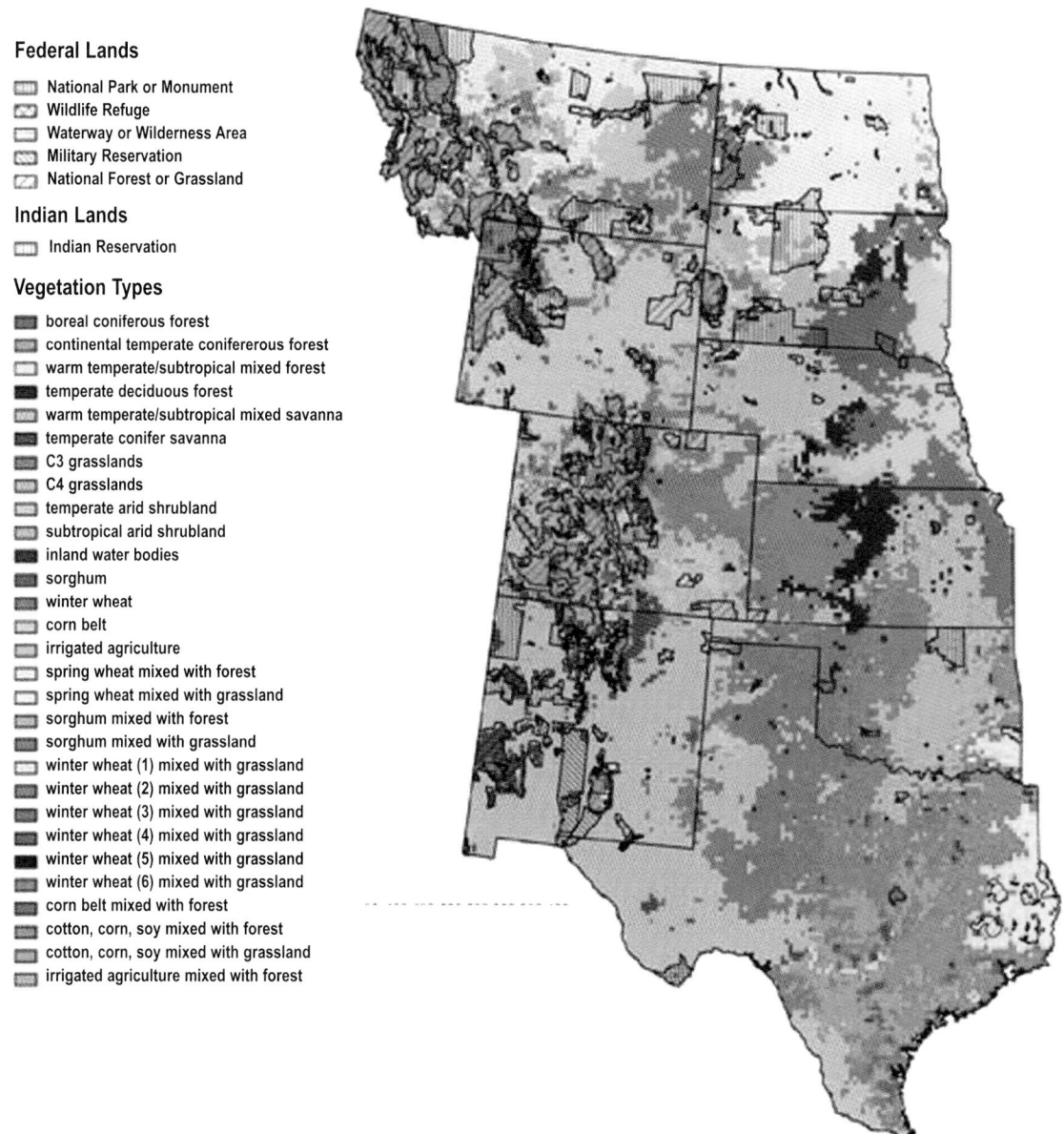

Federal Lands

National Park or Monument
Wildlife Refuge
Waterway or Wilderness Area
Military Reservation
National Forest or Grassland

Indian Lands

Indian Reservation

Vegetation Types

boreal coniferous forest
continental temperate coniferous forest
warm temperate/subtropical mixed forest
temperate deciduous forest
warm temperate/subtropical mixed savanna
temperate conifer savanna
C3 grasslands
C4 grasslands
temperate arid shrubland
subtropical arid shrubland
inland water bodies
sorghum
winter wheat
corn belt
irrigated agriculture
spring wheat mixed with forest
spring wheat mixed with grassland
sorghum mixed with forest
sorghum mixed with grassland
winter wheat (1) mixed with grassland
winter wheat (2) mixed with grassland
winter wheat (3) mixed with grassland
winter wheat (4) mixed with grassland
winter wheat (5) mixed with grassland
winter wheat (6) mixed with grassland
corn belt mixed with forest
cotton, corn, soy mixed with forest
cotton, corn, soy mixed with grassland
irrigated agriculture mixed with forest

Figure 1. Distributions of the naturally occurring vegetation and the current planted agricultural crops are strongly linked to the gradients of temperature (north to south) and precipitation (west to east) within the Great Plains. Outlines show the federal land holdings in the region (vegetation map from Natural Resource Ecology Lab, Colorado State University. Potential natural vegetation according to VEMAP members, 1995).

Great Plains Climate

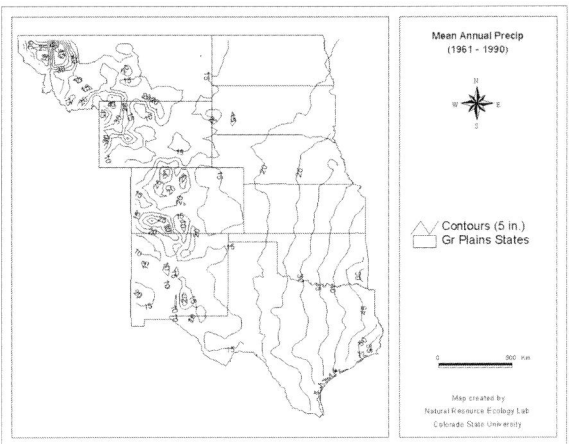 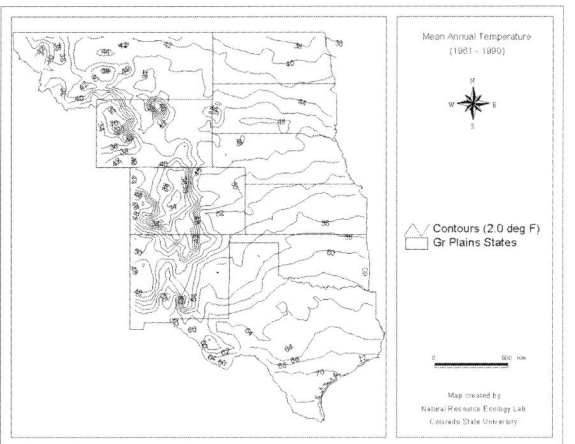

Figure 2. Great Plains climate is characterized by a strong north-south temperature gradient and a strong east-west precipitation gradient (averages based on the 1961-1990 period; data from Dennis Ojima, VEMAP climate).

Great Plains Agicultural Exports

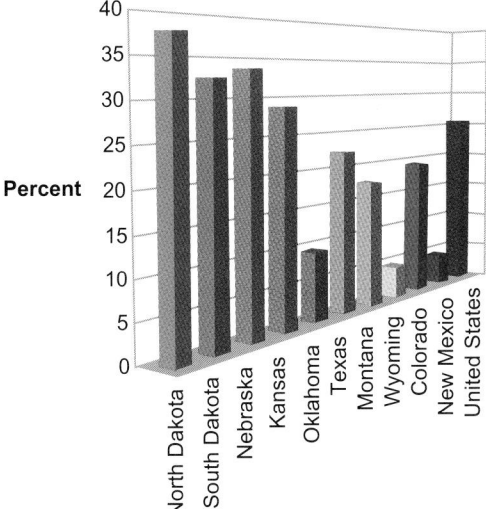

Figure 3. Agricultural exports are an important percentage of the total agricultural production within each state. (USDA, 1997 Census of Agriculture).

Agricultural Land Comparisons

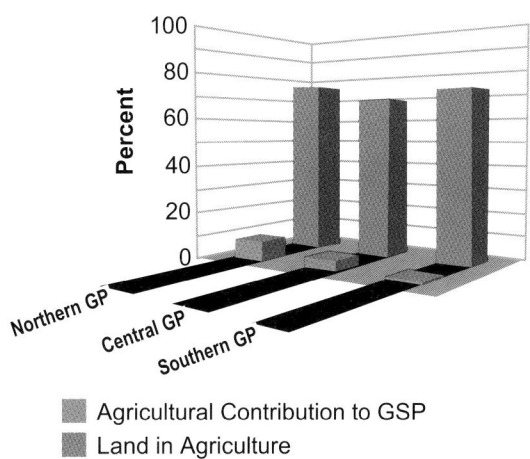

Agricultural Contribution to GSP
Land in Agriculture

Figure 4. The Northern Great Plains are more dependent on agriculture than the Central which is more dependent on agriculture than the Southern Great Plains, yet agriculture dominates land use in all regions of the Great Plains. (Economic data from US Dept of Commerce, Bureau of Economic Analysis, Regional Economic Analysis Division, June 1998, and land use data from USDA, 1997 Census of Agriculture.)

Great Plains

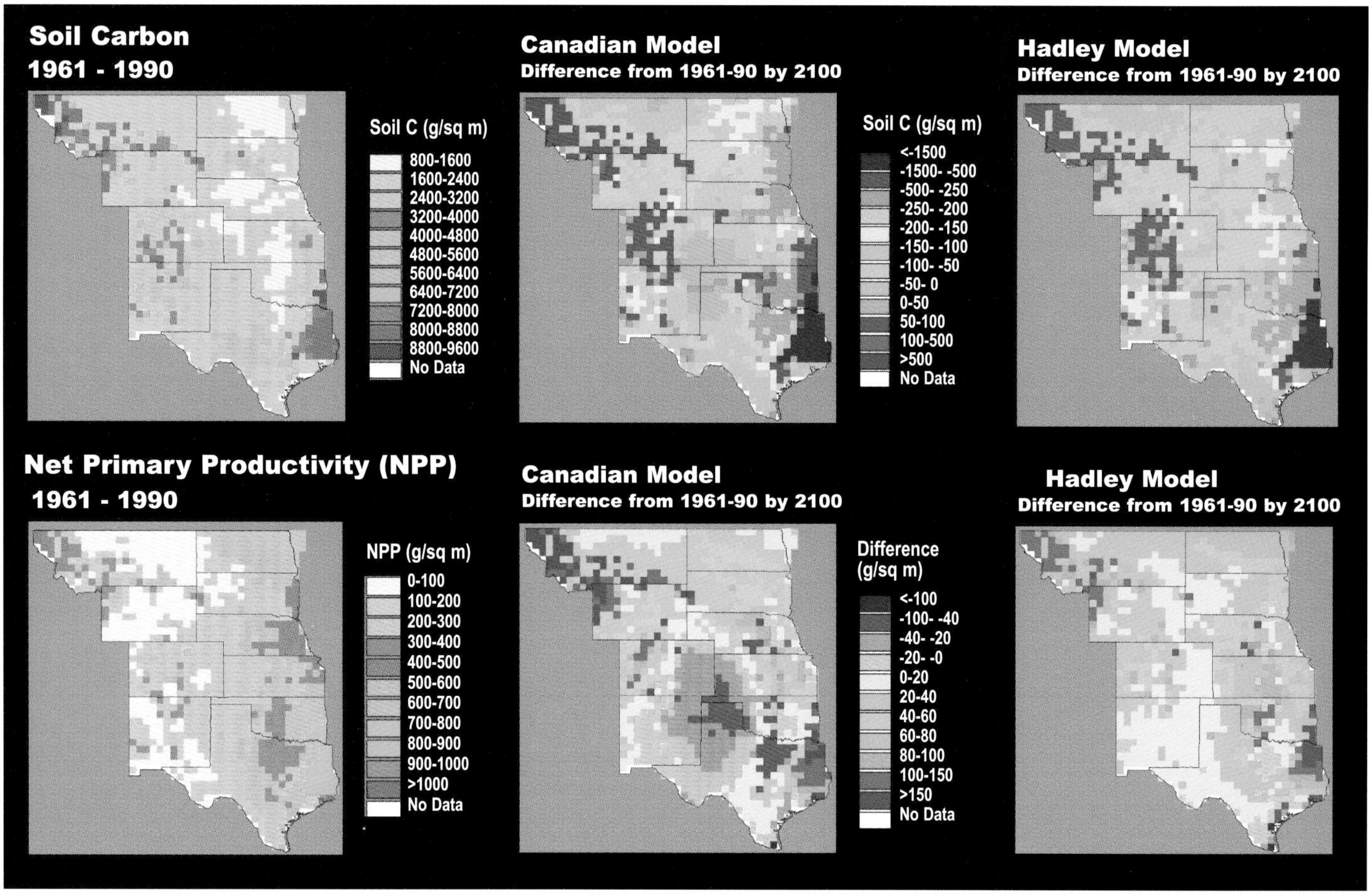

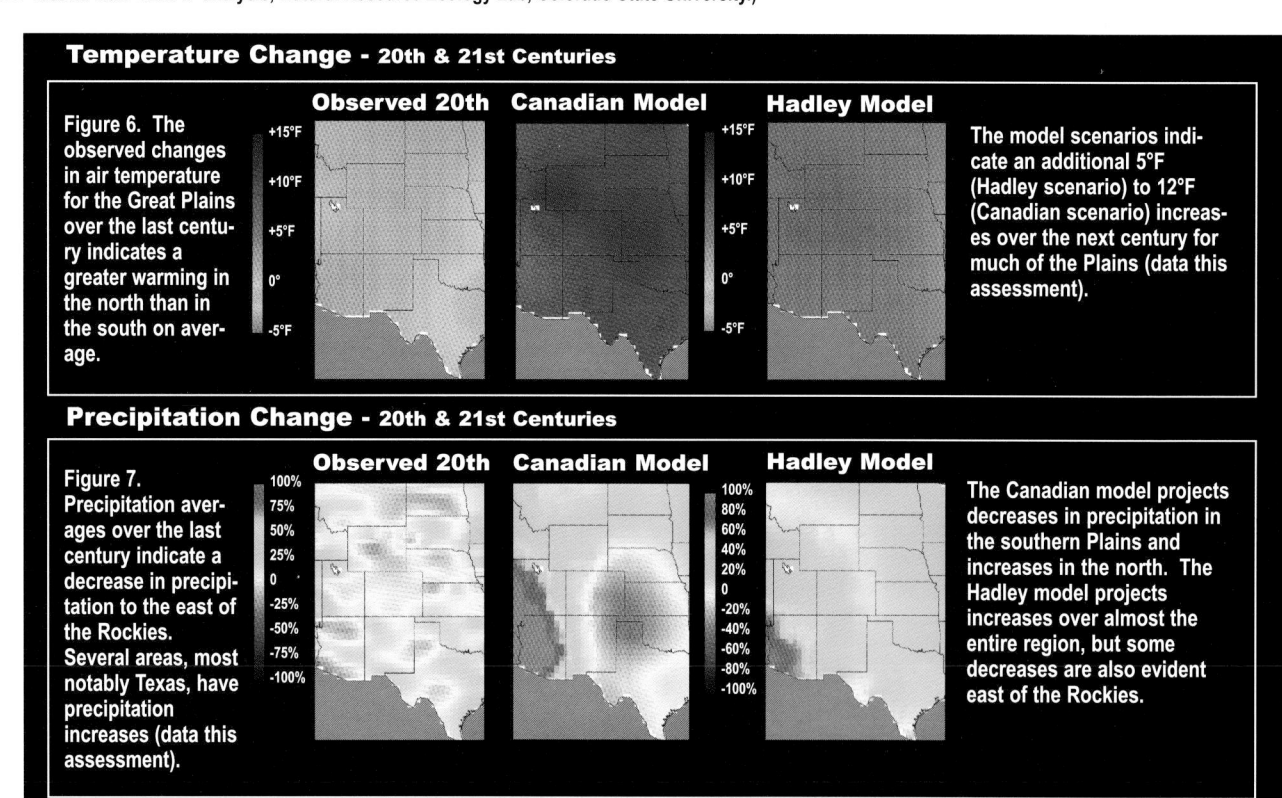

Soil Carbon
1961 - 1990

Soil C (g/sq m)
- 800-1600
- 1600-2400
- 2400-3200
- 3200-4000
- 4000-4800
- 4800-5600
- 5600-6400
- 6400-7200
- 7200-8000
- 8000-8800
- 8800-9600
- No Data

Canadian Model
Difference from 1961-90 by 2100

Soil C (g/sq m)
- <-1500
- -1500- -500
- -500- -250
- -250- -200
- -200- -150
- -150- -100
- -100- -50
- -50- 0
- 0-50
- 50-100
- 100-500
- >500
- No Data

Hadley Model
Difference from 1961-90 by 2100

Net Primary Productivity (NPP)
1961 - 1990

NPP (g/sq m)
- 0-100
- 100-200
- 200-300
- 300-400
- 400-500
- 500-600
- 600-700
- 700-800
- 800-900
- 900-1000
- >1000
- No Data

Canadian Model
Difference from 1961-90 by 2100

Difference (g/sq m)
- <-100
- -100- -40
- -40- -20
- -20- -0
- 0-20
- 20-40
- 40-60
- 60-80
- 80-100
- 100-150
- >150
- No Data

Hadley Model
Difference from 1961-90 by 2100

Figure 5. The productivity of the Great Plains increases from west to east and from north to south, following the precipitation and the temperature gradients. Land uses are strongly influenced by productivity. Both climate scenarios increase the moisture stress on the central parts of the Great Plains and productivity declines in this region. Soil organic matter in the Great Plains is an important reservoir of terrestrial carbon. The amount of carbon stored in the soil is strongly influenced by past and present land management practices and weather patterns. Where moisture levels and productivity decline, soil carbon may actually increase as decomposition processes become limiting. Where soil moisture levels increase from increased water use efficiency, soil carbon levels may decline. (CENTURY results from VEMAP analysis, Natural Resource Ecology Lab, Colorado State University.)

Temperature Change - 20th & 21st Centuries

Observed 20th   **Canadian Model**   **Hadley Model**

Figure 6. The observed changes in air temperature for the Great Plains over the last century indicates a greater warming in the north than in the south on average.

+15°F
+10°F
+5°F
0°
-5°F

The model scenarios indicate an additional 5°F (Hadley scenario) to 12°F (Canadian scenario) increases over the next century for much of the Plains (data this assessment).

Precipitation Change - 20th & 21st Centuries

Observed 20th   **Canadian Model**   **Hadley Model**

Figure 7. Precipitation averages over the last century indicate a decrease in precipitation to the east of the Rockies. Several areas, most notably Texas, have precipitation increases (data this assessment).

100%
75%
50%
25%
0
-25%
-50%
-75%
-100%

100%
80%
60%
40%
20%
0
-20%
-40%
-60%
-80%
-100%

The Canadian model projects decreases in precipitation in the southern Plains and increases in the north. The Hadley model projects increases over almost the entire region, but some decreases are also evident east of the Rockies.

Annual Average Palmer Drought Severity Index (PDSI)

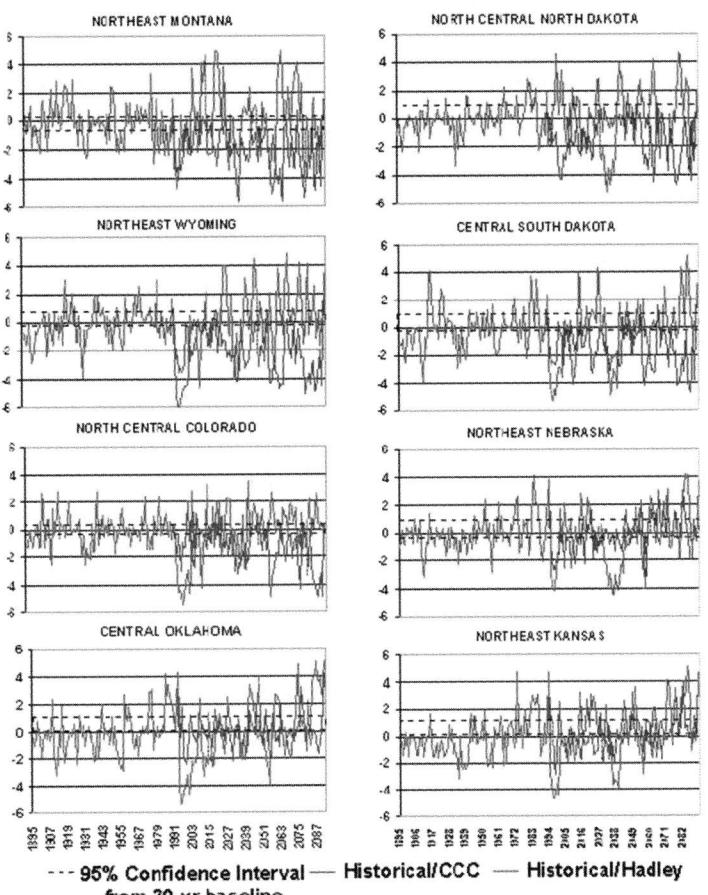

Figure 8. The droughts of the 1930s and 1950s are shown as years or periods where the Palmer Drought Severity Index (PDSI) was less then –2 and both climate scenarios (CCC = Canadian, Hadley) suggest future periods in each of these 8 climate divisions in the Great Plains where drought conditions appear likely. The 95% confidence interval for the historical period of 1960 to 1990 is shown as two dashed lines (VEMAP data, Natural Resource Ecology Lab, Colorado State University.)

3+ Consecutive Days exceeding 90°F (32° C)

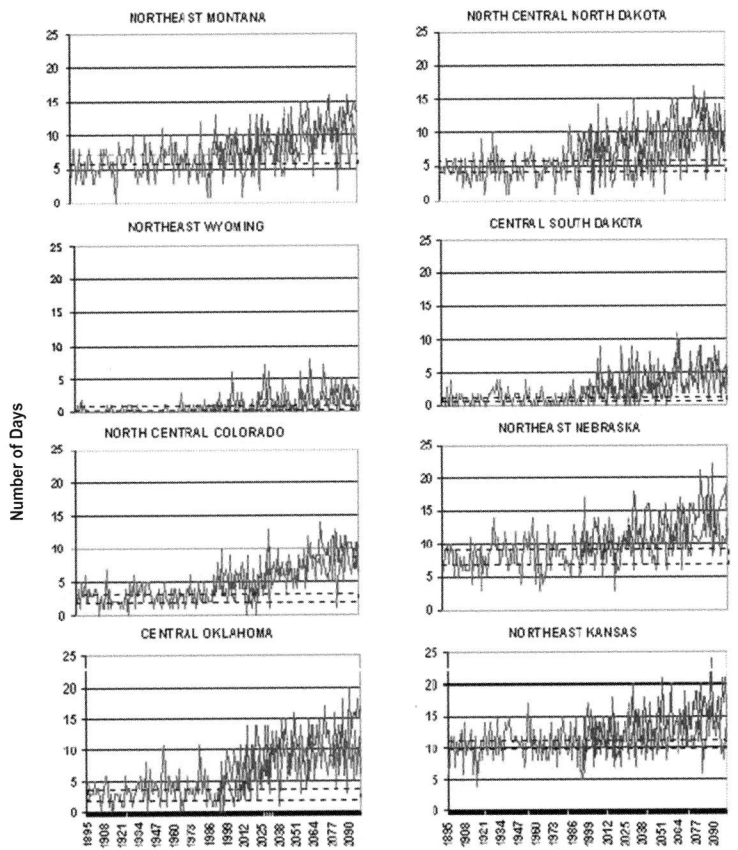

Figure 9. Heat stress events can be triggered for livestock and for humans when the temperature exceeds 90˚F (32˚C) for three or more consecutive days. The number of times that a climate division in each of the 8 Great Plains states experiences three consecutive days where temperatures exceed 90˚F (32˚C) increases in both scenarios. The 95% confidence interval for the historic period of 1960 to 1990 is shown as two dashed lines (VEMAP data, Natural Resource Ecology Lab, Colorado State University).

Great Plains Water Use

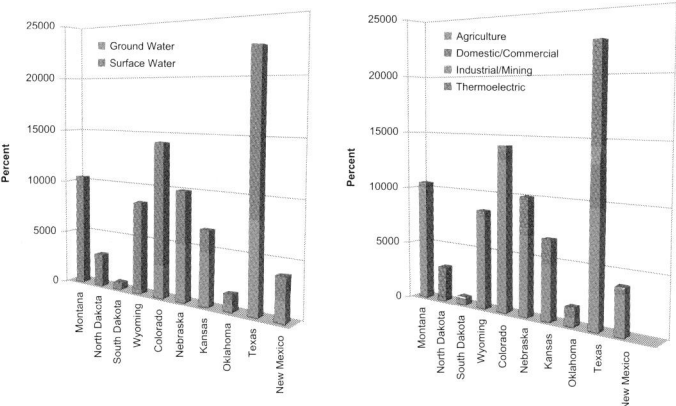

Figure 10. Surface waters are important sources for the western and northern Great Plains. Ground water, such as the Ogalalla aquifer, supplies large shares of the water for Nebraska, Kansas, and Oklahoma. Although the total amount of water withdrawal varies across the Great Plains, agriculture is the dominant consumptive use in all states (Solley, 1997).

Consumptive Water Use

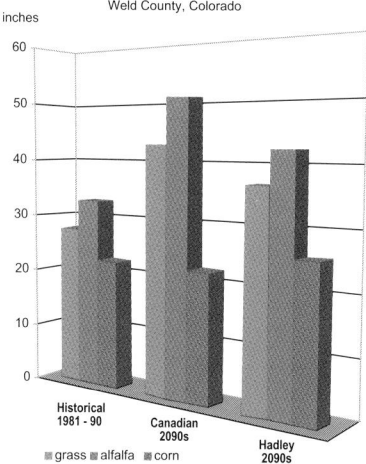

Figure 11. Lack of soil moisture can greatly reduce yield of crops and forage. Under both climate scenarios, the consumptive demand for water on grass pasture increases more than 50% while the water needs for irrigated corn change little. Perennial crops experience an increase in consumptive demand for water; the size of the increase depends on the climate scenario (Ojima et al., 1999).

Historic and Estimated Population for the West

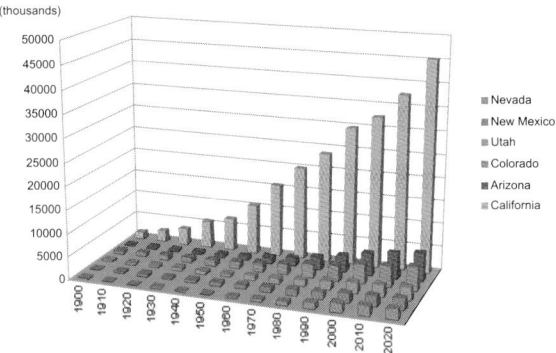

Figure 1: The West's population grew from less than 10 million in 1940 to 46.2 million in 1998 (US Census Bureau, 1998). California's population mushroomed from less than 7 million in 1940 to more than 33 million in 1998 (California Trade and Commerce Agency, 1997; California Department of Finance, 1998). Although more than two-thirds of the West's population lives in California, in recent decades, the intermountain states have become the fastest-growing in the nation. For example, Arizona's population grew from 1.3 million in 1960 to 4.5 million in 1998 (CLIMAS, 1998). Six of the 10 fastest-growing states in the US are projected to be in this region, with Arizona, Nevada, and Utah being the fastest. California's population is projected to rise from its 1998 level of 33 million to about 45 million (NPA Data Services, Inc., 1999).

Urban Population Growth in the West

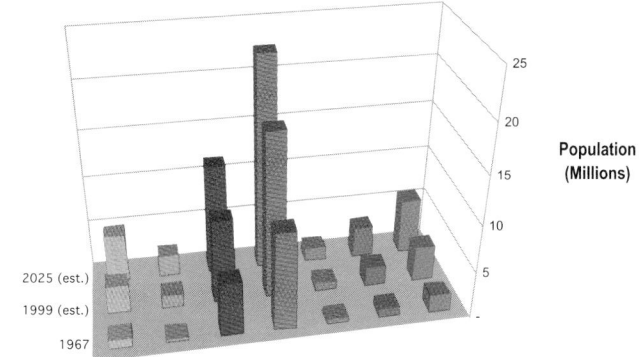

Population (Millions)

▪ Denver-Boulder-Greeley; CO-KS-NE   ▪ Salt Lake City-Ogden; UT-ID
▪ Albuquerque; NM-AZ   ▪ Los Angeles-Riverside-Orange County; CA-AZ
▪ San Francisco-Oakland-San Jose; CA   ▪ Las Vegas; NV-AZ
▪ Phoenix-Mesa; AZ

Figure 2: Over 93% of California's residents live in cities, including San Francisco, Los Angeles, San Diego, and Sacramento, and their surrounding metropolitan areas. In intermountain areas, population growth is also largely concentrating in cities, such as Denver, Salt Lake City, Albuquerque, Phoenix, Las Vegas, Santa Fe and Provo. Much of the future population growth is expected to occur in urban areas. Source: NPA Data Services, 1999.

Relative Water Use in the West

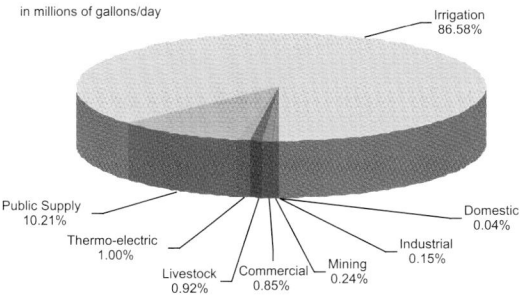

Figure 4: In 1995, 87% of the water consumed in the West was for irrigation (Solley et al., 1998; see Figure 4). However, water use for irrigation has declined slightly since 1980, while municipal uses have grown (Diaz and Anderson, 1995). For example, agriculture accounts for 81% of all water used in Arizona, down from 93% in 1963, while municipal demand currently accounts for 14% of water used, up from 5% in 1963 (CLIMAS, 1998). In addition, irrigated land in the region fell by 8% from 1982 to 1992, although acreage may have increased in recent years (USDA, 1997). Total water use in the region appears to have been declining since 1980 (Templin, 1999).

The Relative Value of Economic Activitiy in the West

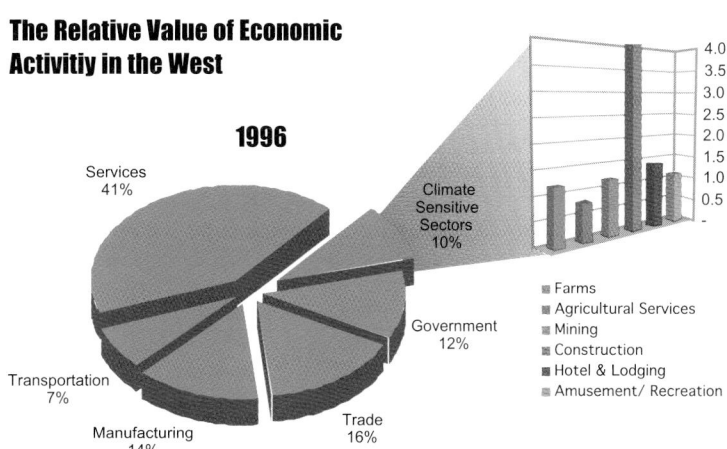

Figure 3: The West produces 18% of US Gross National Product. The region has a slightly greater share of its economy in relatively climate-sensitive sectors such as agriculture, mining, construction, and tourism, than the nation as a whole. While 1.8% of the nation's economic output is from agriculture (which includes forests and fisheries), 2.0% of the West's economic output is from the agriculture sector. The West has 4.1% of its gross product from hotels, amusement/recreation, restaurants, and museums, which are strongly affected by tourism, while the nation as a whole has 1.6% (US BEA, 1999a). With its Gross State Product of $962 billion, California comprises 72% of the total Regional Product of $1.3 trillion in 1996 (US BEAa, 1999). Ranked as a nation, California would be the seventh largest economy in the world (California Trade and Commerce Agency, 1997).

West

El Niño and Events 1997-1998

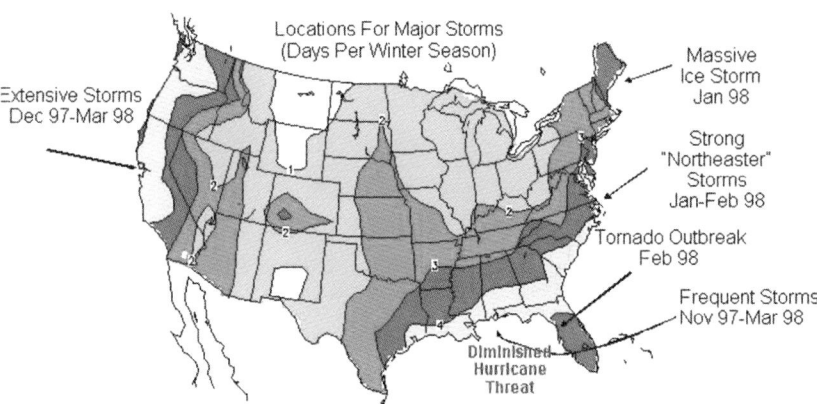

Figure 5: The 1997-1998 El Niño had quite strong effects in the West, with particularly large winter precipitation events. The heavy precipitation lead to such localized consequences as flooding and landslides.

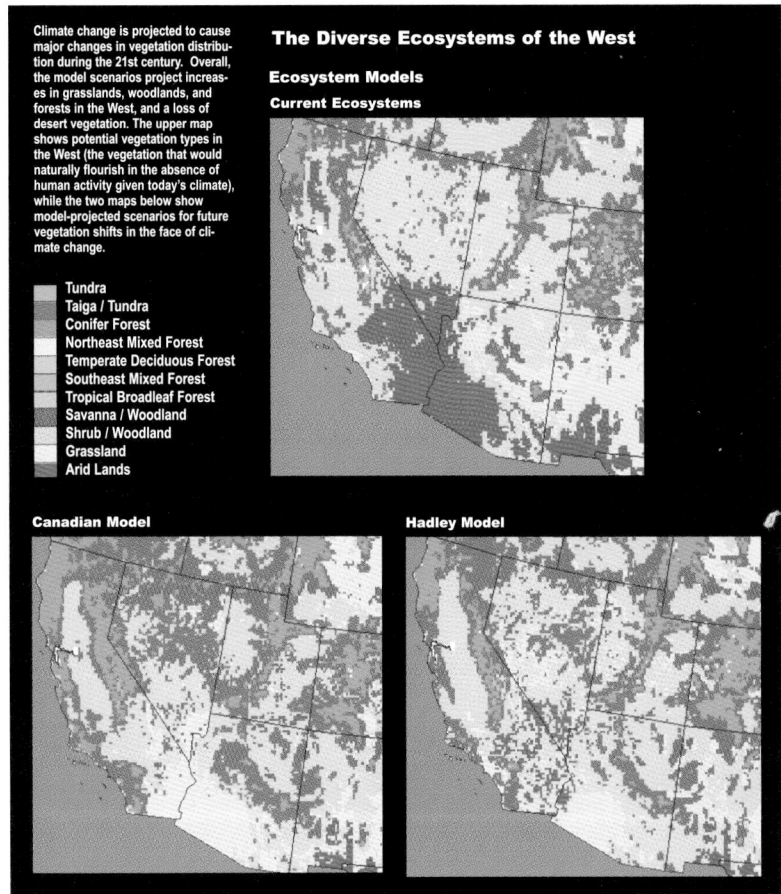

Figure 8: Currently the West has a large diversity of ecosystems. Under the two climate change scenarios, the area in arid and grassland ecosystems would decrease and the area in forest ecosystems would increase.

Observed Shift in Range of Edith's Checkerspot Butterfly: 1900 to 1990s

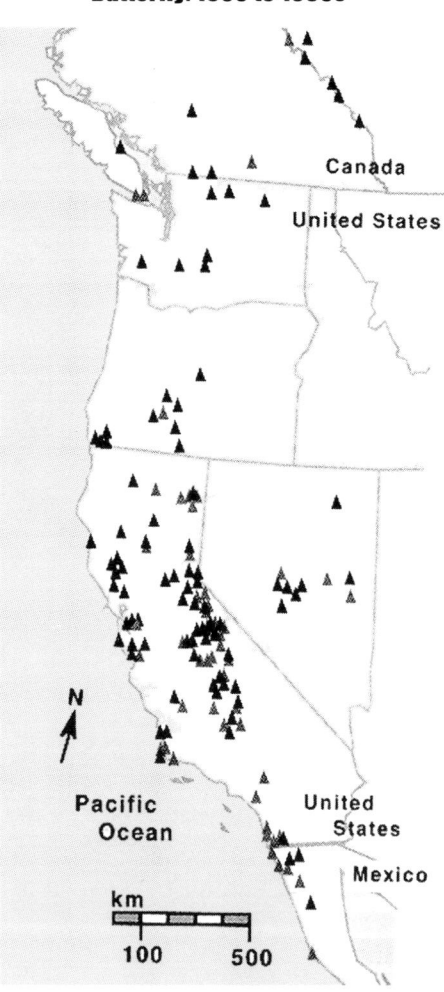

Figure 10: On this map of studied sites, the lighter triangles represent extinct populations of Edith's Checkerspot butterfly, while the darker triangles represent present populations. The mean location of populations of this butterfly has shifted northward by 57 miles (92 kilometers) and upward in altitude by 407 feet (124 meters) since 1900. This is an indication that climate change is already having an affect on the some species ranges. Source: Parmesan, 1996.

Relative Share of Crop and Livestock Output in the West.

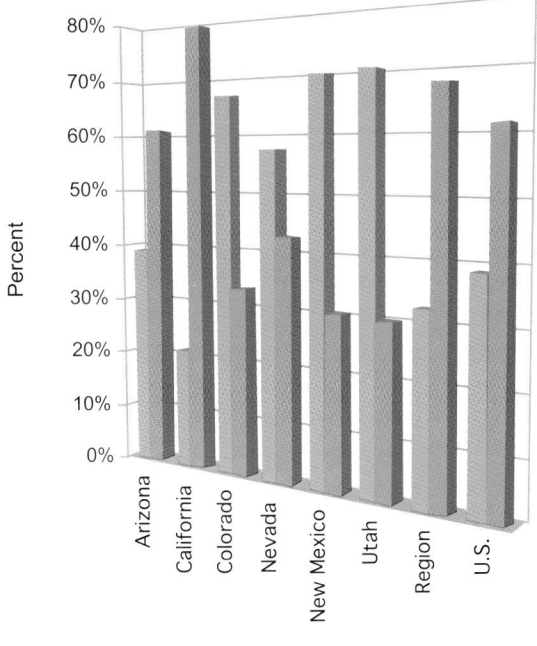

Livestock (dairy & meat) ■ Crops

Figure 11: For most of the states in the West, the majority of value-added agriculture production comes from livestock and dairy production. However, because California's agricultural production is dominated by crops (75% of total agricultural output for the state), and because California dominates regional agricultural output (84% of regional crop production, 51% of regional livestock and dairy production), the majority of the region's total agricultural production comes from crops. This difference between the dominant types of agricultural production on a state level and on a regional level highlights the heterogeneity of agriculture in the West. Source: USDA Economic Research Service State Farm Sector Value-Added Data; (http://www.econ.ag.gov/briefing/fbe/fi/fivadmu.htm). August 30, 1999.

Current and Projected Wetlands in South San Francisco Bay

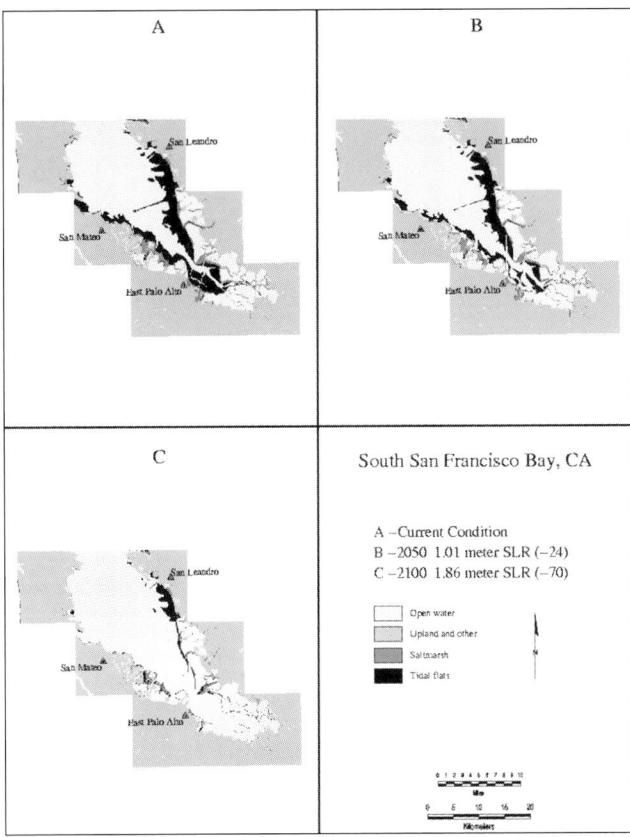

Figure 12: This figure shows the spatial extent and distribution of current and projected wetland habitat types in southern San Francisco Bay (derived from US Fish and Wildlife, National Wetlands Inventory data) following sea-level rise as calculated using the Sea Level Affecting Marshes Model (SLAMM4) (Galbraith et al., In prep.). The sea-level rise scenarios use historic rates that include local subsidence (obtained from tide gages at or close to each of the sites), superimposed on the median estimate of the likely rate of sea-level change due to climate change (Titus and Narayanan, 1996). The historic rate of sea-level rise in the southern part of San Francisco Bay is estimated to be 3.0 feet (0.9 meter) by 2050 and 5.3 feet (1.6 meter) by 2100. This could be due to tectonic movements resulting in land subsidence and/or crustal subsidence due to the depletion of subterranean aquifers. When combined with the projected median estimate of 13.4 inches (34 cm) eustatic (global) sea-level rise by 2100 from climate change, sea-level rise is estimated to be 3.3 feet (1.0 meter) by 2050 and 6.1 feet (1.9 meters) by 2100. The numbers shown in parenthesis on the figure indicate that approximately 57.7% of tidal flat habitat will be lost by 2050 and 62.1% by 2100, compared to the current condition. Using only the historic rate of local sea level rise, approximately 58.9% (2050) and 61.1% (2100) of tidal flat habit.

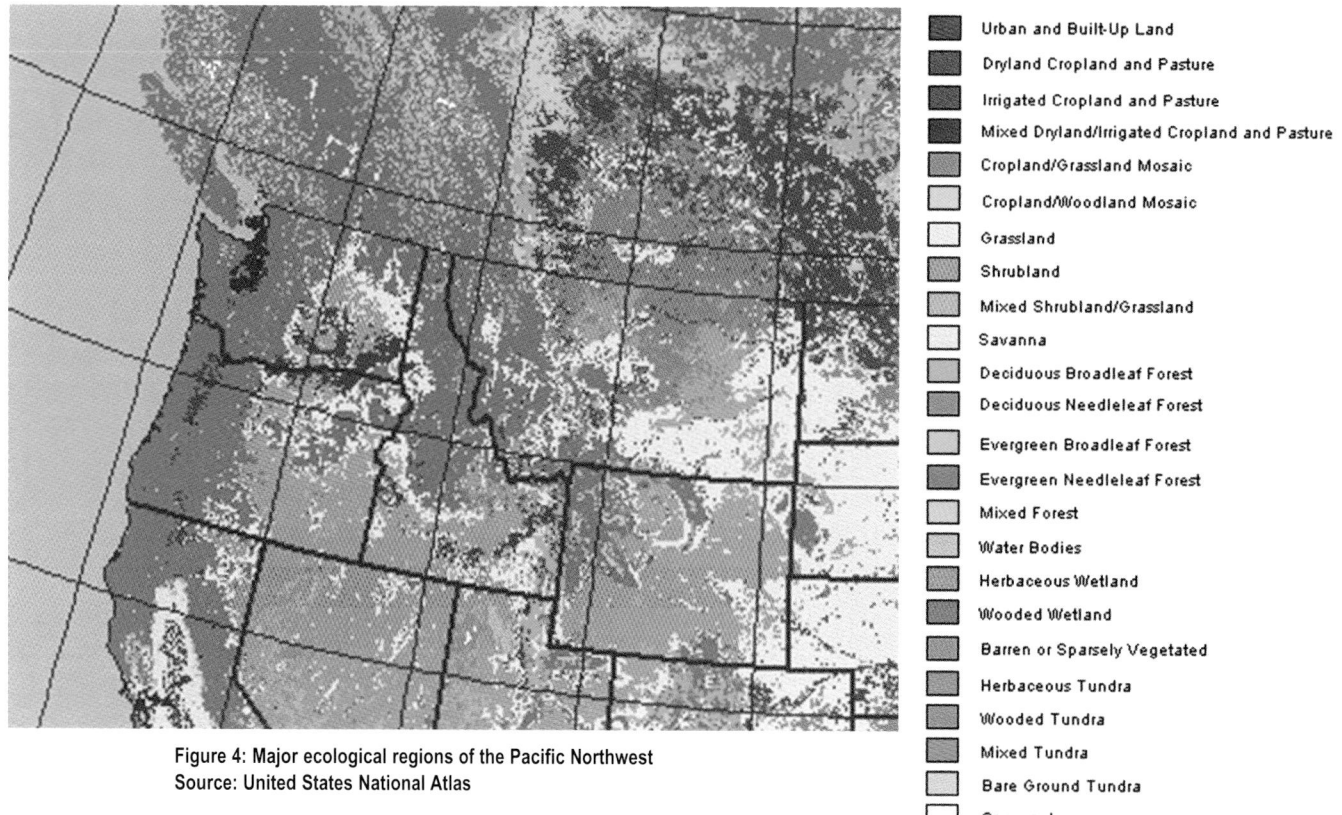

Urban and Built-Up Land
Dryland Cropland and Pasture
Irrigated Cropland and Pasture
Mixed Dryland/Irrigated Cropland and Pasture
Cropland/Grassland Mosaic
Cropland/Woodland Mosaic
Grassland
Shrubland
Mixed Shrubland/Grassland
Savanna
Deciduous Broadleaf Forest
Deciduous Needleleaf Forest
Evergreen Broadleaf Forest
Evergreen Needleleaf Forest
Mixed Forest
Water Bodies
Herbaceous Wetland
Wooded Wetland
Barren or Sparsely Vegetated
Herbaceous Tundra
Wooded Tundra
Mixed Tundra
Bare Ground Tundra
Snow or Ice

Figure 4: Major ecological regions of the Pacific Northwest
Source: United States National Atlas

Average Annual Precipitation, Pacific Northwest, 1961-1990

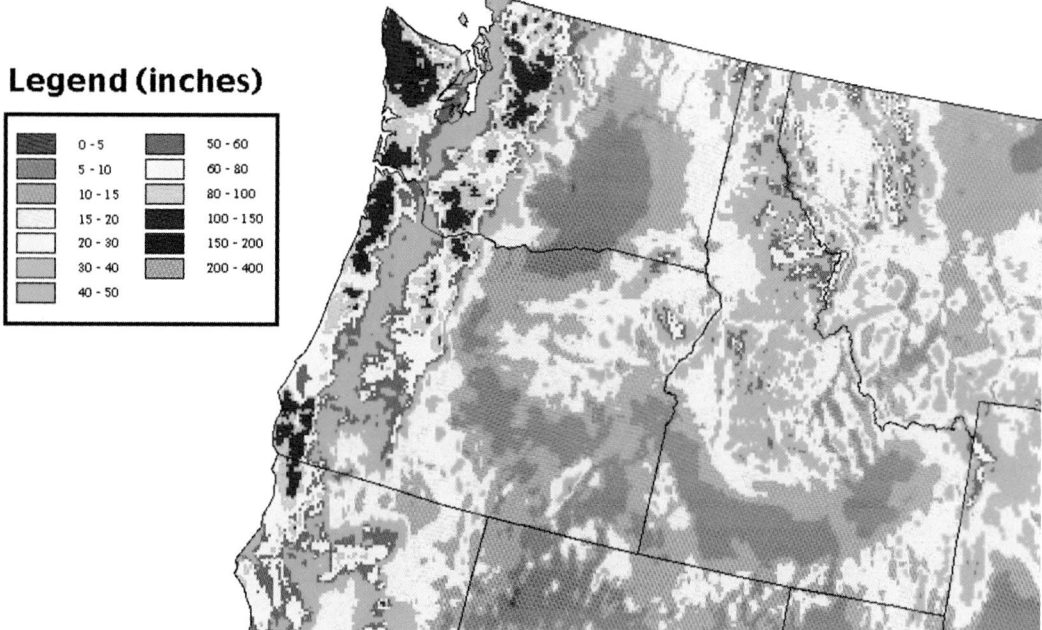

Legend (inches)

0 - 5	50 - 60
5 - 10	60 - 80
10 - 15	80 - 100
15 - 20	100 - 150
20 - 30	150 - 200
30 - 40	200 - 400
40 - 50	

Figure 5: The Cascade mountains divide the wetter west from the drier east. Source: Mapping by C. Daly, graphic by G. Taylor and J. Aiken, copyright © 2000, Oregon State University.

Northwest Average Temperature, Observed and Modeled

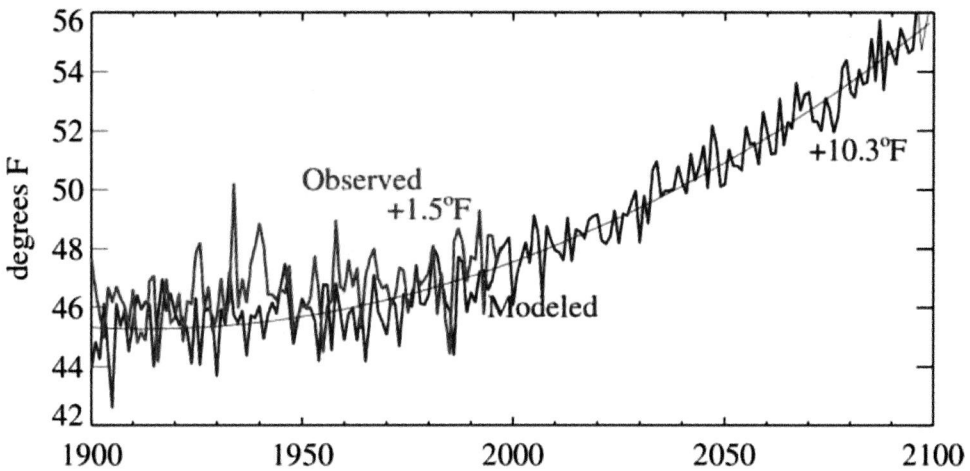

Figure 7: The red line shows annual-average temperature in the Northwest in the 20th century, observed from 113 weather stations with long records. The blue line shows the historical Northwest average temperature calculated by the Canadian model from 1900 to 2000, and projected forward to 2100. Source: Mote et al (1999), Summary (p. 6).

Temperature Change 20th and 21st Centuries

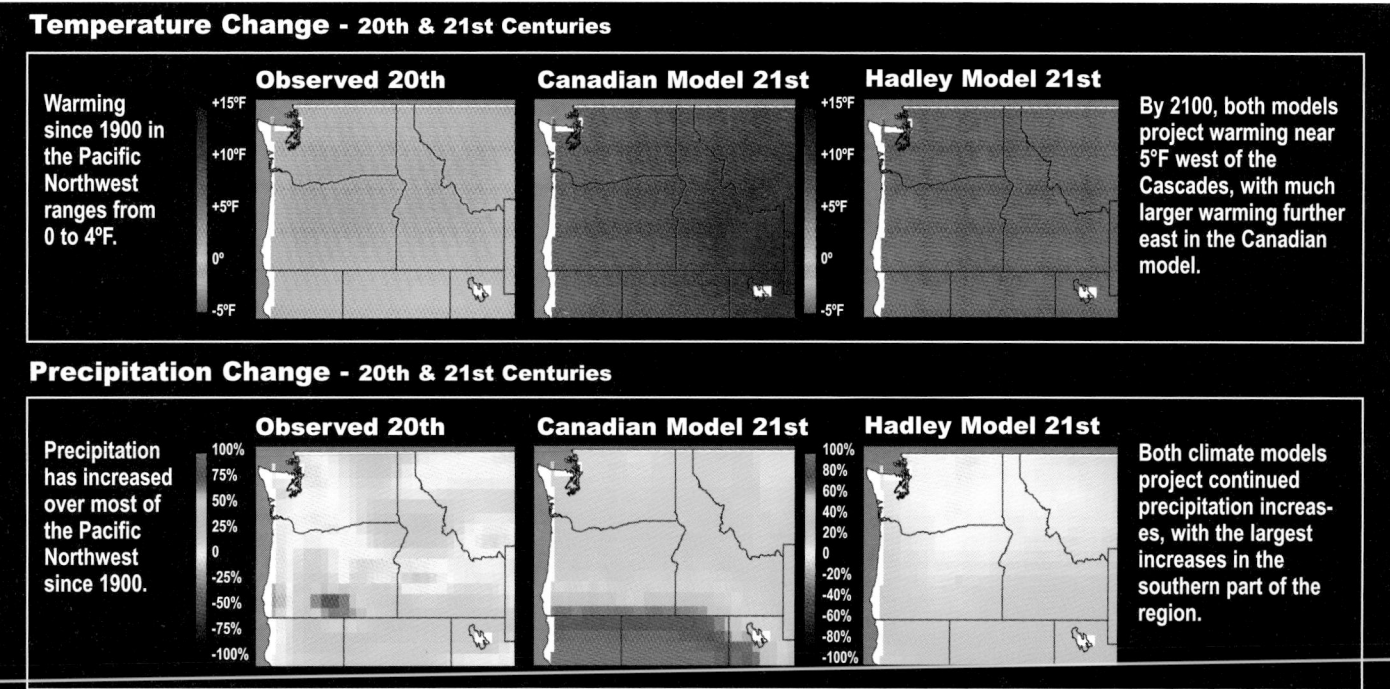

Figure 8: Temperature change observed in the 20th and projected for the 21st centuries.

Projected Northwest Climate Change, Compared to 20th Century Variability

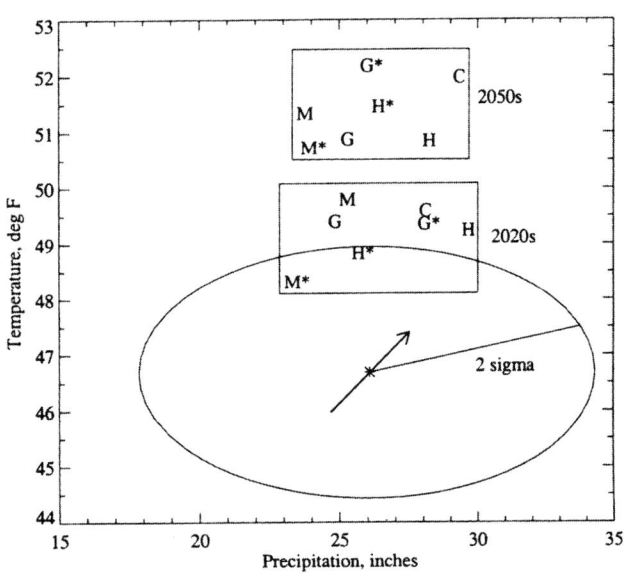

Figure 9: Climate change by the 2020s and 2050s over the Northwest Region from seven climate model scenarios. Any point on the graph shows a particular combination of regional annual-average temperature and total annual precipitation. The asterisk and arrow through it show the average climate over the 20th century and its trend, warming about 1.5°F (0.8°C) with a 2.5" (6 cm) precipitation increase. The oval illustrates how much the region's climate varied over the 20th century, enclosing all combinations of temperature and precipitation that were more than 5% likely to occur. Each letter shows one model's projection of the region's average climate, either in the 2020s or the 2050s. The models project that regional precipitation changes will lie within the range of 20th century variability, but projected temperature changes lie outside it. By the 2050s, all models project a climate so much warmer in the Northwest that it lies well outside the range of 20th century variability (*=1995-vintage model; H=Hadley; M=Max-Planck; G=GFDL; C=Canadian). Source: Regional report, Mote et al. (1999), fig. 12, pg. 19.

Projected Reduction in Columbia Basin Snowpack

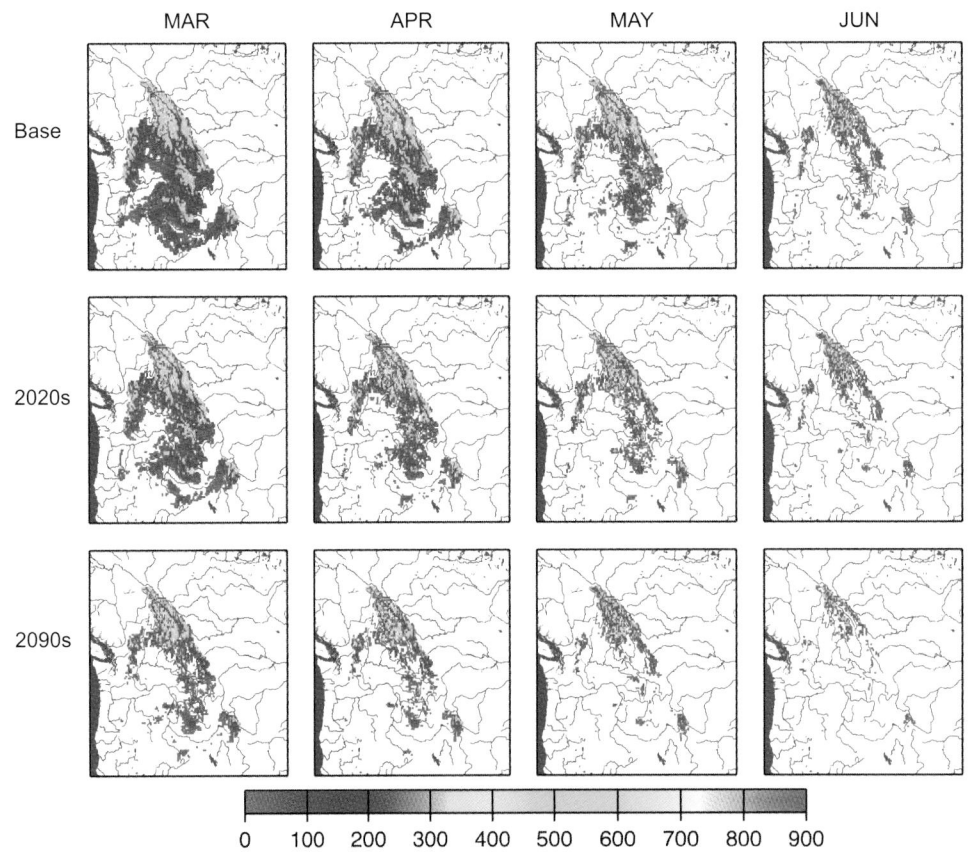

Figure 10: By the 2090s, projected Columbia Basin snowpack on March 1 will be only slightly greater than present snowpack on June 1. Simulations use the VIC hydrology model under the Hadley scenario. Units in millimeters. Source: Hamlet and Lettenmaier, 1999.

Pacific Northwest

Projected Seasonal Shift in Columbia River Flow

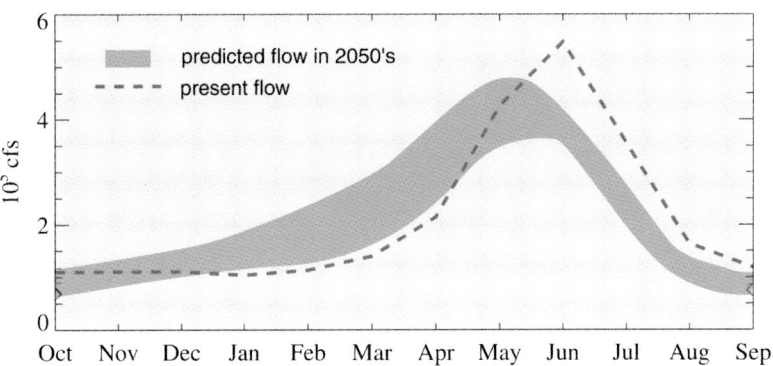

Figure 11: While only small changes are projected in annual Columbia flow, seasonal flow shifts markedly toward larger winter and spring flows, and smaller summer and fall flows. The blue band shows the range of projected monthly flows in the 2050s under the Hadley and Canadian scenarios and the two other 1998-vintage climate models used in the Northwest assessment (MPI and GFDL). Source: Mote et al. (1999), Summary, Figure 7.

Salmon Catches and Inter-decadal Climate Variability

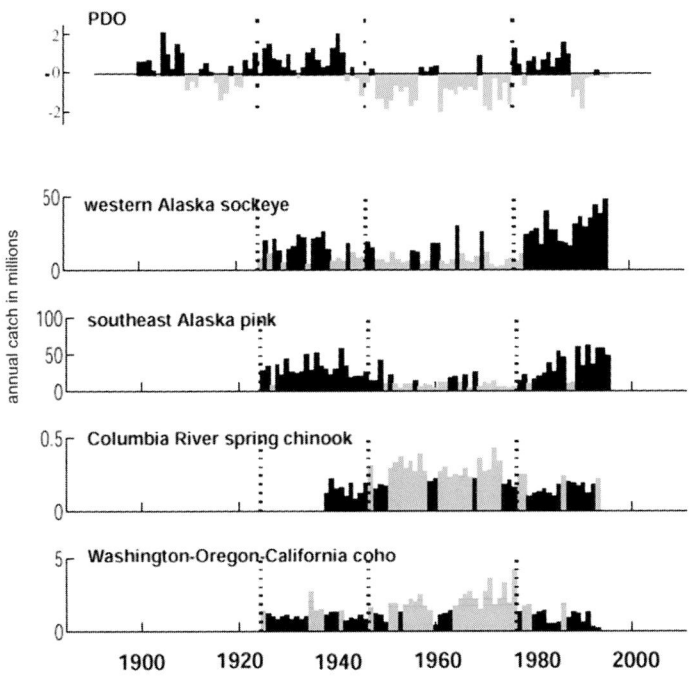

Figure 14: 20th century catches of Northwest and Alaska salmon stocks show clear influence, in opposite directions, of the Pacific Decadal Oscillation. Source: Mote et al (1999), Figure 36, p. 56.

Tree Growth and Inter-Decadal Climate Variability

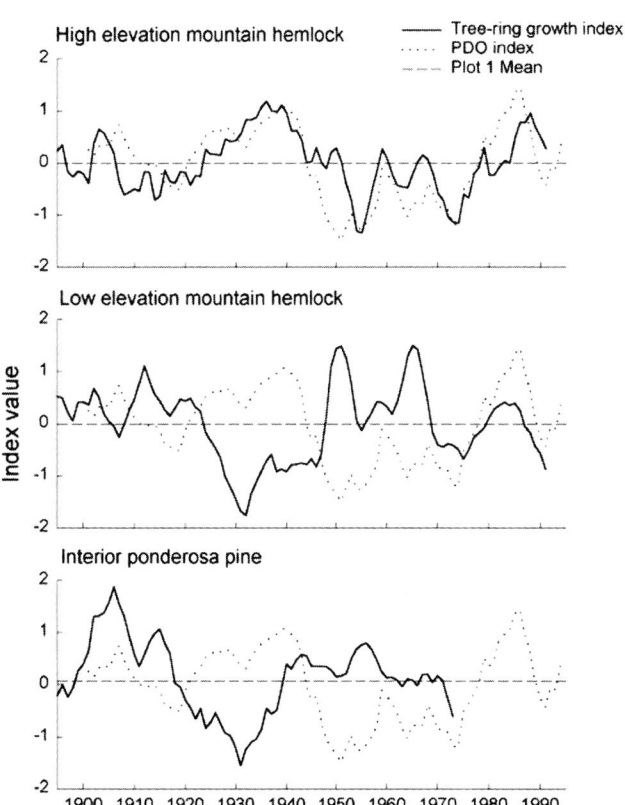

Figure 16: Trees near their climatic limits show strong signals of inter-decadal climate variability. Those near the upper treeline grow best in warm-PDO years because snowpack is lighter, while those near the dry lower treeline grow worst in warm-PDO years, because of summer moisture deficit. Source: Peterson and Peterson, 2000.

Pacific Northwest

Projected Northwest Vegetation Changes under two Ecosystem Models, 2100

MAPSS: Percent Change in Leaf Area

Constant CO$_2$ Elevated CO$_2$

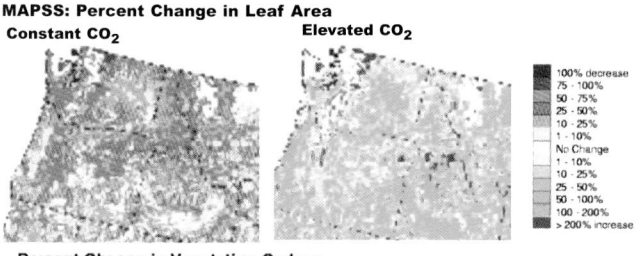

	100% decrease
	75 - 100%
	50 - 75%
	25 - 50%
	10 - 25%
	1 - 10%
	No Change
	1 - 10%
	10 - 25%
	25 - 50%
	50 - 100%
	100 - 200%
	> 200% increase

Percent Change in Vegetation Carbon

MC1: Percent Change in Vegetation Carbon

Constant CO$_2$ Elevated CO$_2$

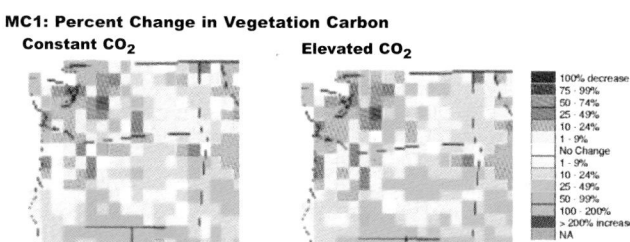

	100% decrease
	75 - 99%
	50 - 74%
	10 - 24%
	1 - 9%
	No Change
	1 - 9%
	10 - 24%
	25 - 49%
	50 - 99%
	100 - 200%
	> 200% increase
	NA

Figure 18: Under the Hadley scenario, the MAPSS (top row) and MC1 (bottom row) models project expansion of forests east of the Cascades and contractions to the west, assuming increased water-use efficiency under elevated atmospheric CO$_2$ (right column). When no such increase is assumed (left column), projections are nearly unchanged in the MC1 model, but change to a large contraction region-wide in the MAPSS model. Source: Bachelet et al (2000).

Climate Change Projected for 2050 vs Observed 20th Century Variability

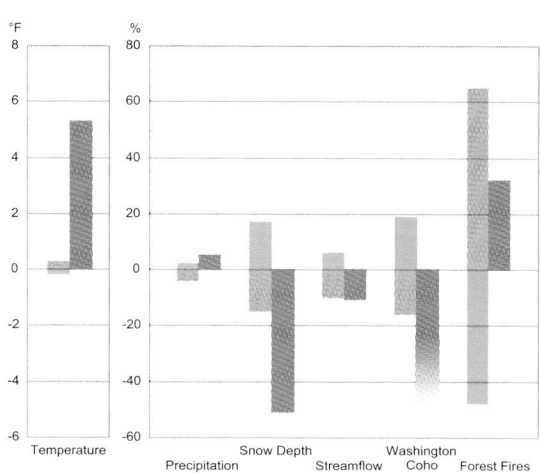

Warm PDO Years   Cool PDO Years   Regional impacts of climate change in 2050s
(1925-1945, 1977-95)   (1900-1924, 1946-76)

Temperature	Change in annual average regional temperature (°F)
Precipitation	Change in annual average regional precipitation (%)
Snow depth	Change in average winter snow depth at Snoqualmie Pass, WA (%)
Streamflow	Change in annual streamflow at The Dalles on the Columbia River (corrected for changing effects of dams) (%)
Salmon	Change in annual catch of Washington Coho salmon (%)
Forest fires	Change in annual area burned by forest fires in WA and OR (%)

Figure 20: This chart compares possible Northwest impacts from climate change by the 2050s with the effects of natural climate variations during the 20th century. The orange bars show the effects of the warm phase of the Pacific Decadal Oscillation (PDO), relative to average 20th century values. During warm-PDO years, the Northwest is warmer, there is less rain and snow, stream flow and salmon catch are reduced, and forest fires increase. The blue bars show the corresponding effects of cool-phase years of the PDO, during which opposite tendencies occurred.

The pink bars show projected impacts expected by the 2050s, based on the Hadley and Canadian scenarios. Projected regional warming by this time is much larger than variations experienced in the 20th century. This warming is projected to be associated with a small increase in precipitation, a sharp reduction in snowpack, a reduction in streamflow, and an increase in area burned by forest fires. Although quite uncertain, large reductions in salmon abundance ranging from 25 to 50%, are judged to be possible based on projected changes in temperature and streamflow. Source: based on Mote et al., 1999, pg. 27.

Major Ecological Regions of Alaska

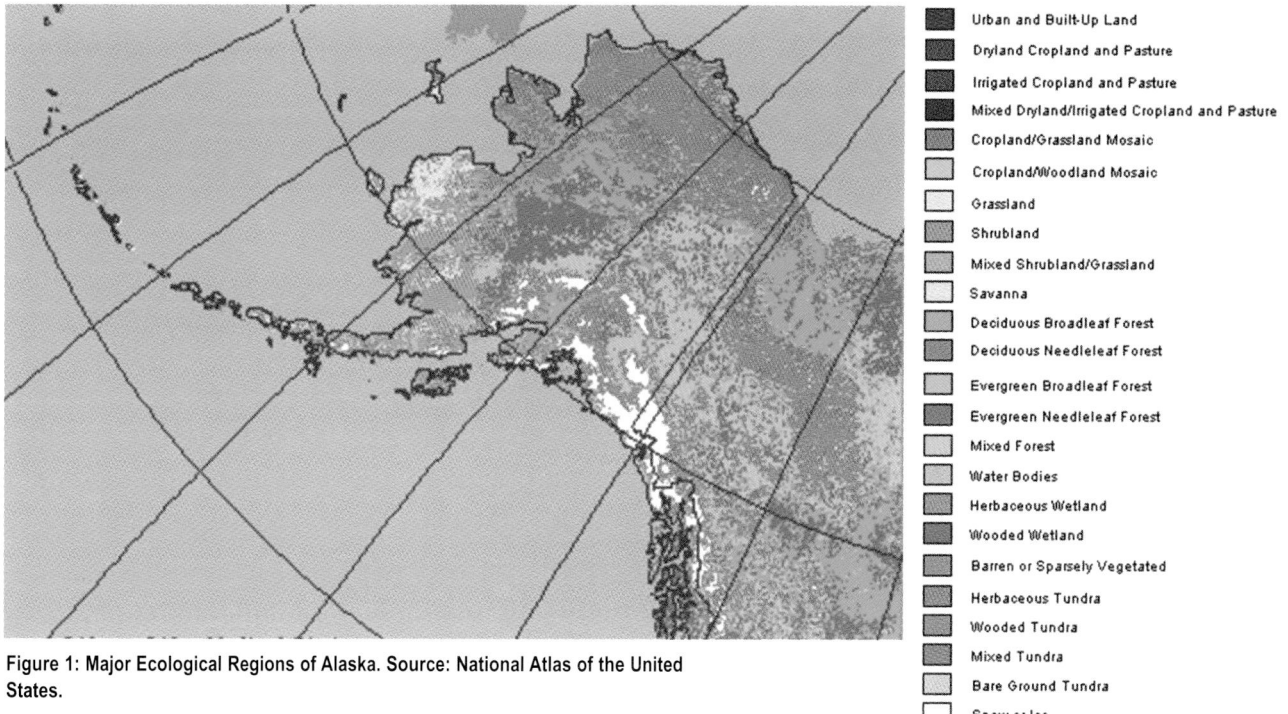

Urban and Built-Up Land

Dryland Cropland and Pasture

Irrigated Cropland and Pasture

Mixed Dryland/Irrigated Cropland and Pasture

Cropland/Grassland Mosaic

Cropland/Woodland Mosaic

Grassland

Shrubland

Mixed Shrubland/Grassland

Savanna

Deciduous Broadleaf Forest

Deciduous Needleleaf Forest

Evergreen Broadleaf Forest

Evergreen Needleleaf Forest

Mixed Forest

Water Bodies

Herbaceous Wetland

Wooded Wetland

Barren or Sparsely Vegetated

Herbaceous Tundra

Wooded Tundra

Mixed Tundra

Bare Ground Tundra

Snow or Ice

Figure 1: Major Ecological Regions of Alaska. Source: National Atlas of the United States.

Alaska: 20th Century Annual-average Temperature

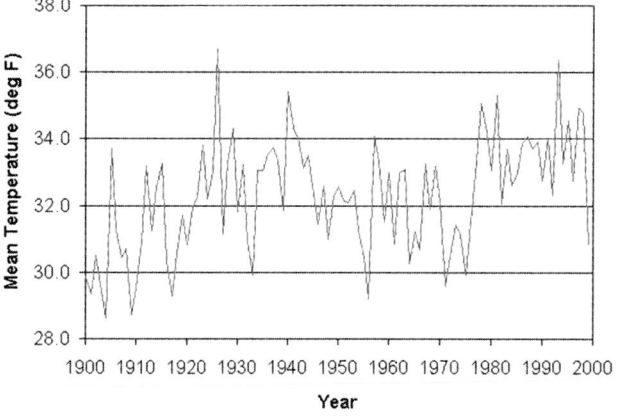

Figure 2: Average temperatures in Alaska have increased over the 20th century, with about 4°F warming since the 1950s. Source: Historical Climate Network, National Climate Data Center.

Alaska: 20th Century Annual Total Precipitation

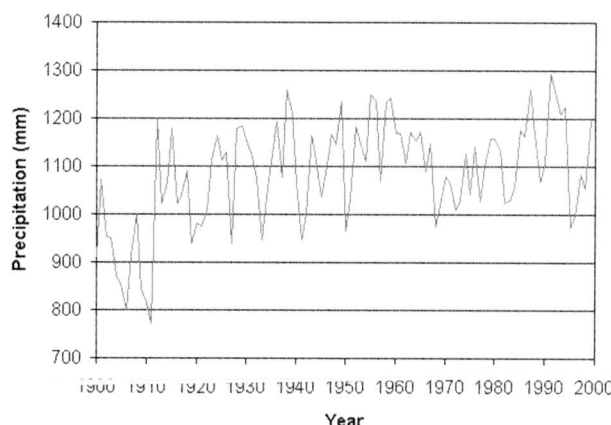

Figure 3: Over the 20th century, precipitation in Alaska has increased. Source: Historical Climate Network, National Climate Data Center.

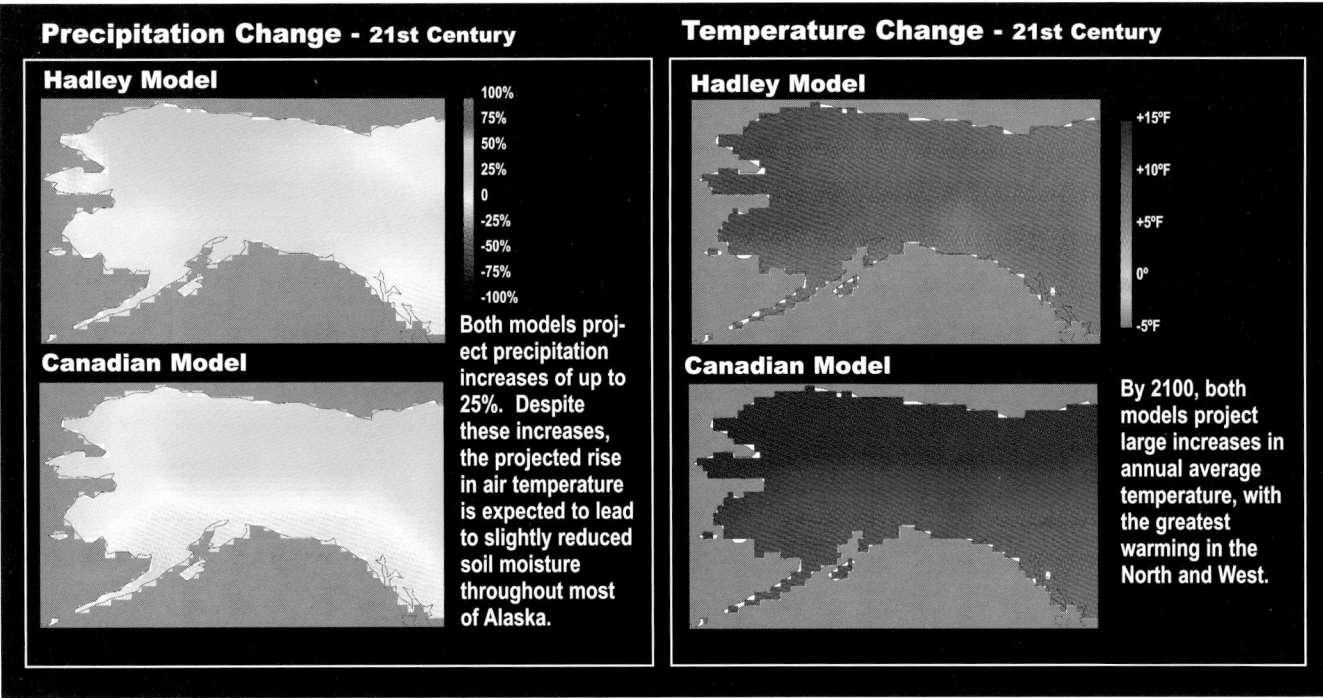

Figure 4: Precipitation and temperature change projected in the 21st century by two climate models.

Winter Maximum Temperature Change

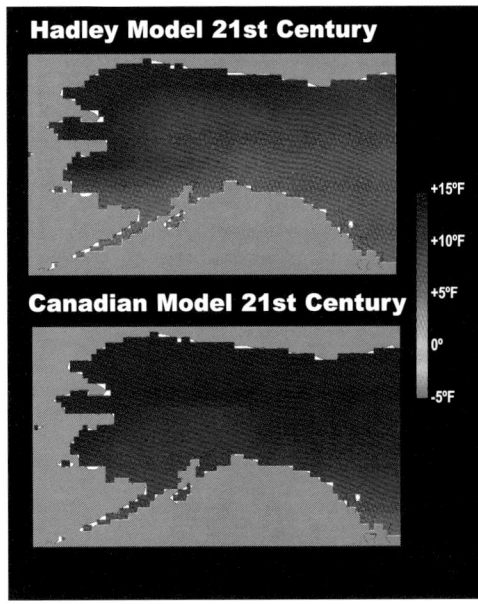

Figure 5: The largest projected warming is in winter, when both models show average daily-high temperatures increasing more than 15°F over the northern half of the state. Source: B.Felzer, UCAR.

Summer Soil Moisture Change

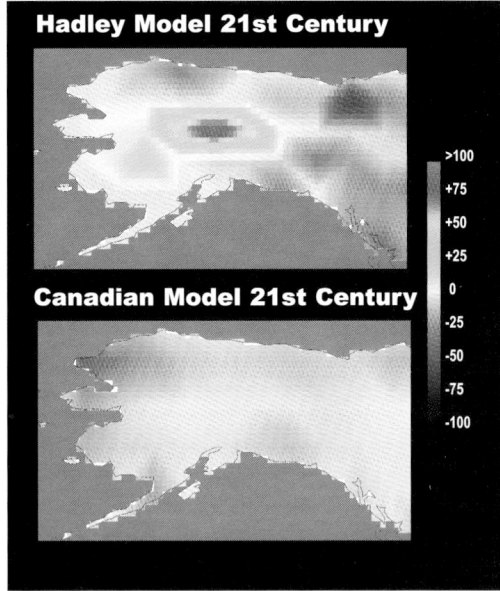

Figure 6: The Hadley model projects increased summer soil moisture in central Alaska and decreases in the north and south, while the Canadian model projects moderate decreases throughout the state. Source: B. Felzer, UCAR.

Permafrost Regions of Alaska

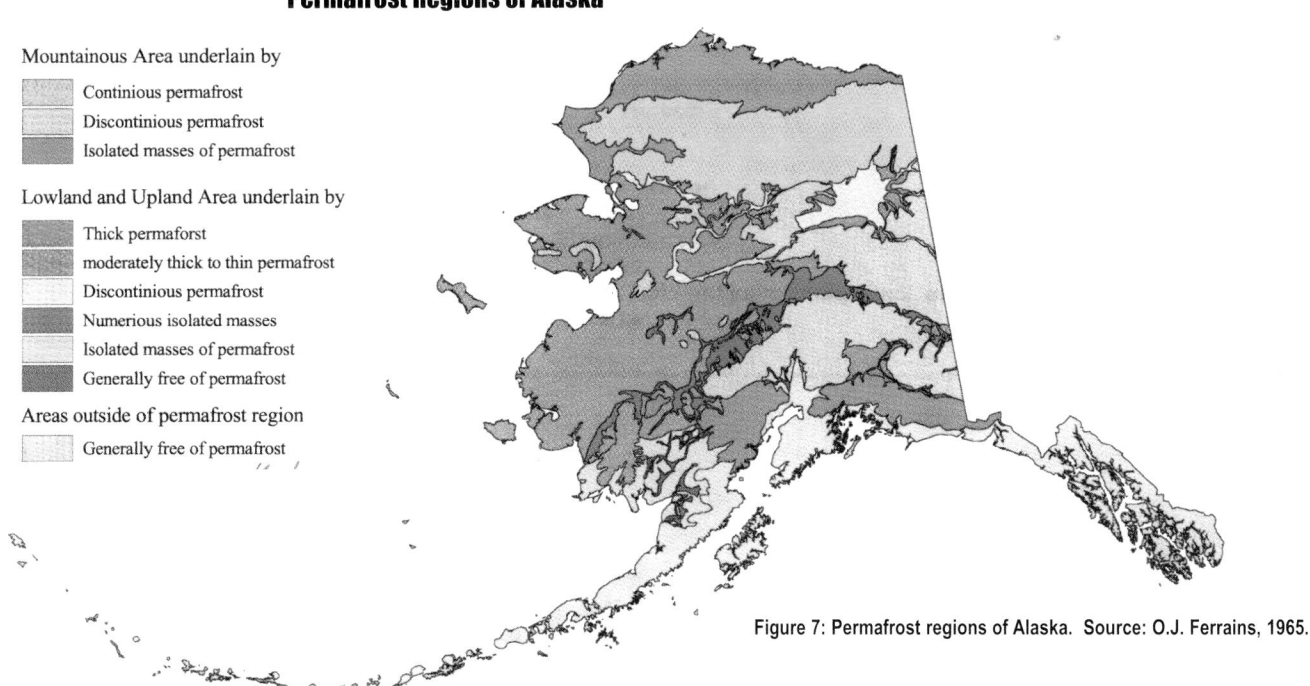

Mountainous Area underlain by

    Continious permafrost

    Discontinious permafrost

    Isolated masses of permafrost

Lowland and Upland Area underlain by

    Thick permaforst

    moderately thick to thin permafrost

    Discontinious permafrost

    Numerious isolated masses

    Isolated masses of permafrost

    Generally free of permafrost

Areas outside of permafrost region

    Generally free of permafrost

Figure 7: Permafrost regions of Alaska. Source: O.J. Ferrains, 1965.

Alaska

Projected Summer Sea Ice Change
Canadian Model: An Ice-free Arctic Summer

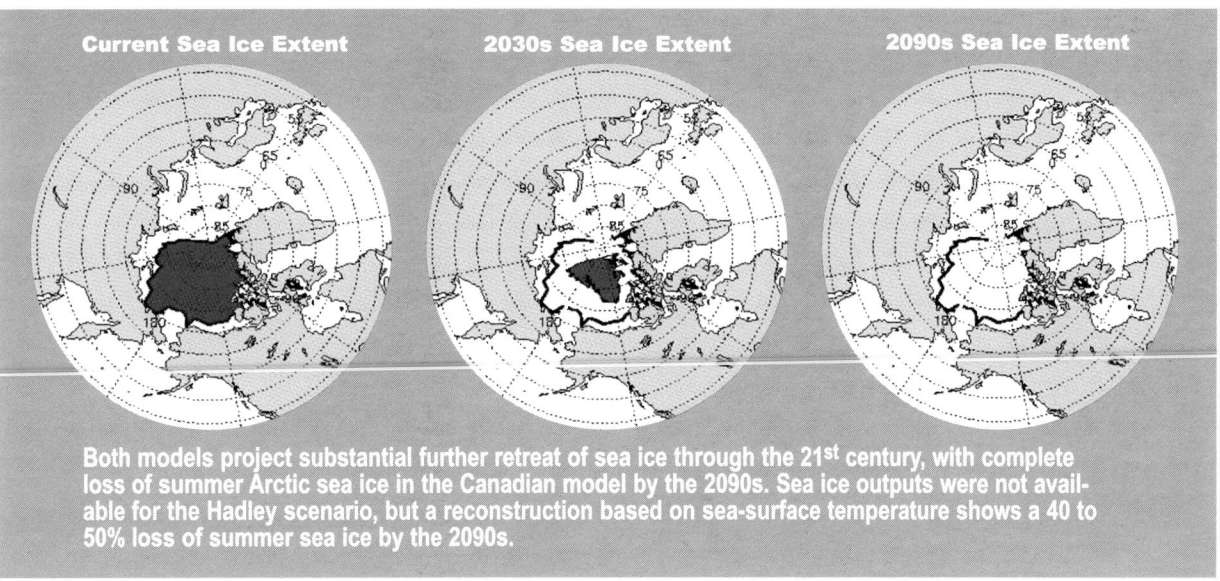

Current Sea Ice Extent   2030s Sea Ice Extent   2090s Sea Ice Extent

Both models project substantial further retreat of sea ice through the 21st century, with complete loss of summer Arctic sea ice in the Canadian model by the 2090s. Sea ice outputs were not available for the Hadley scenario, but a reconstruction based on sea-surface temperature shows a 40 to 50% loss of summer sea ice by the 2090s.

Figure 11: Canadian model projections of future Arctic sea-ice retreat. Source: B. Felzer, UCAR, 2000.

Spring Breakup Dates in the Nenana Classic
(11-year moving average)

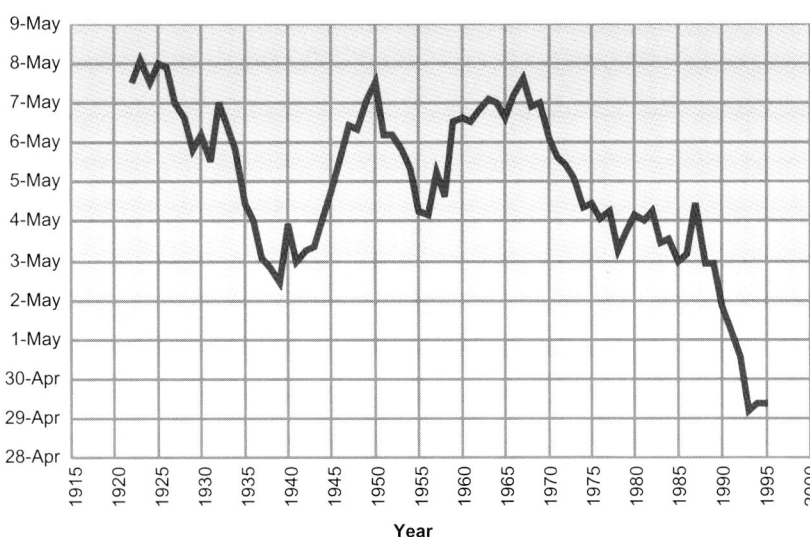

Figure 12: The average date of spring breakup of ice on the Tanana River at Nenana has advanced by eight days between the 1920s and the 1990s. Source: Historical data from Nenana Ice Classic, http://www.ptialaska.net/~tri-pod/breakup.times.html.

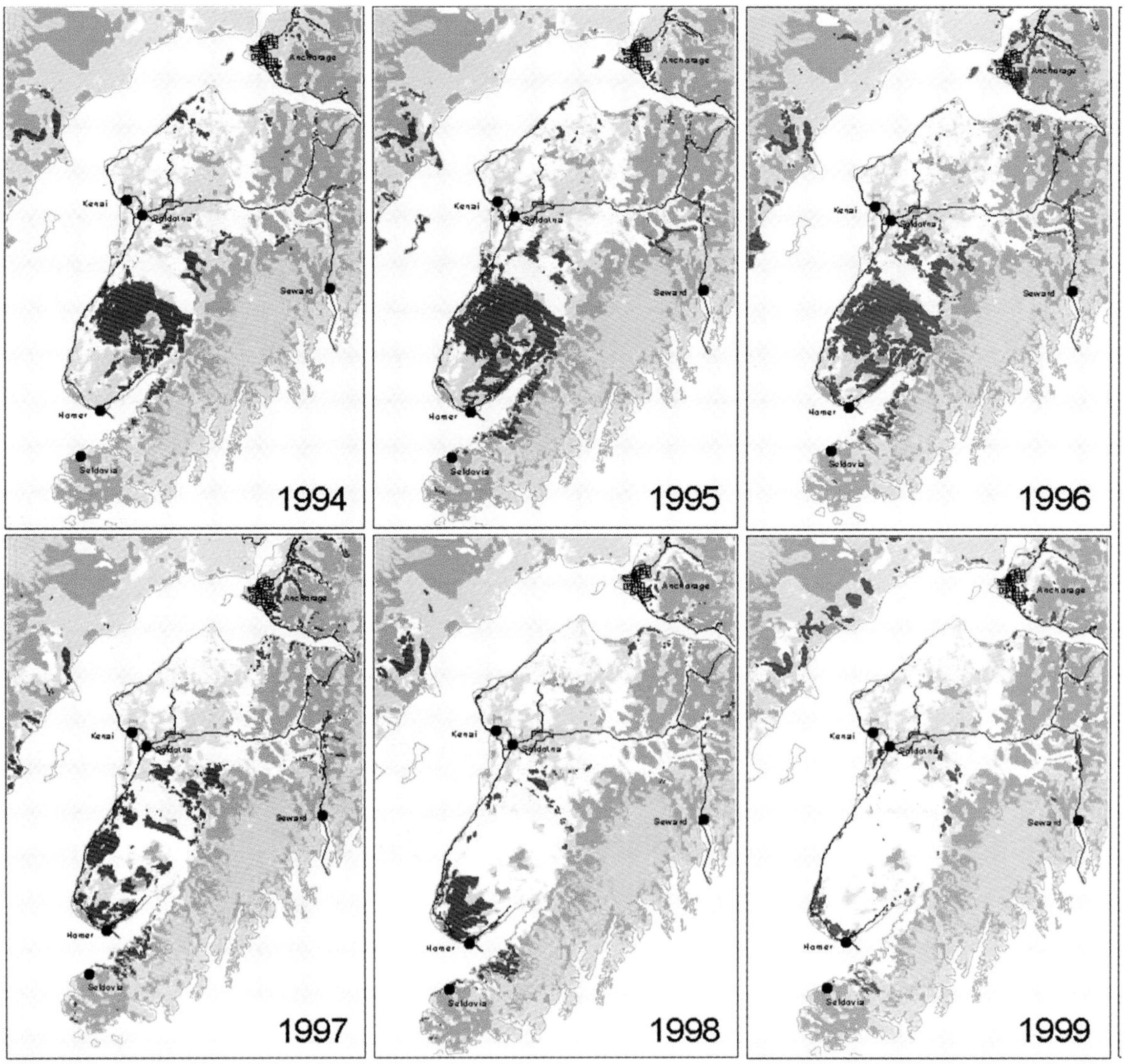

Alaska

The 1990s Outbreak of Spruce Bark Beetles on the Kenai Peninsula

Legend:
- Spruce Beetle Mortality
- Prior SPB Mortality
- Forest
- Non-Forest
- Glacier

Figure 13: Since 1992, the largest outbreak of forest insects ever recorded in North America has caused widespread tree mortality over 2.3 million acres. Source: USDA Forest Service.

1994

1995

1996

1997

1998

1999

Annual Area of Northern Boreal Forest Burned in North America

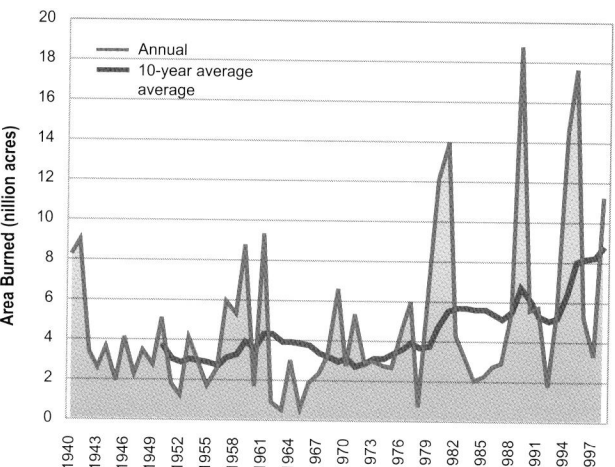

Figure 14: The Alaskan boreal forest is a small part of an enormous forest that extends continuously across the northern part of North America. The average area of this forest burned annually has more than doubled since 1970. Source: Kasischke and Stocks, 2000.

State of Bering Sea Ecosystem

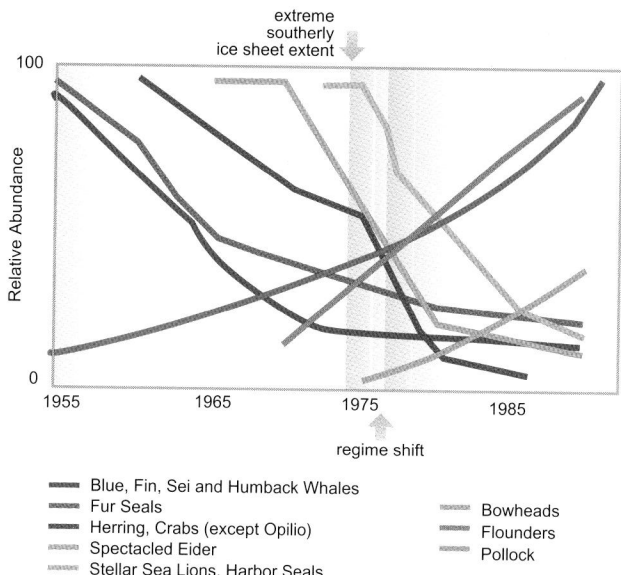

Blue, Fin, Sei and Humback Whales
Fur Seals
Herring, Crabs (except Opilio)
Spectacled Eider
Stellar Sea Lions, Harbor Seals

Bowheads
Flounders
Pollock

Figure 16: The climatic regime shift of the late 1970s caused large-scale reorganization of the Bering Sea ecosystem. Source: simplified from NRC (1996).

Simulated Vegetation Distribution

Current

Tundra
Taiga / Tundra
Boreal Conifer Forest
Temperate Evergreen Forest
Temperate Mixed Forest
Tropical Broadleaf Forest
Savanna / Woodland
Shrub / Woodland
Grassland
Arid Lands

Figure 15: Under the Hadley scenario, the MAPSS biogeography model projects large-scale loss of tundra and taiga ecosystems as forests expand north and west. Likely consequences include disruption of wildlife migration and associated subsistence livelihoods, as well as the potential for large releases of soil carbon.   Source: R. Neilson et al, 1998.

Hadley Model 2090s

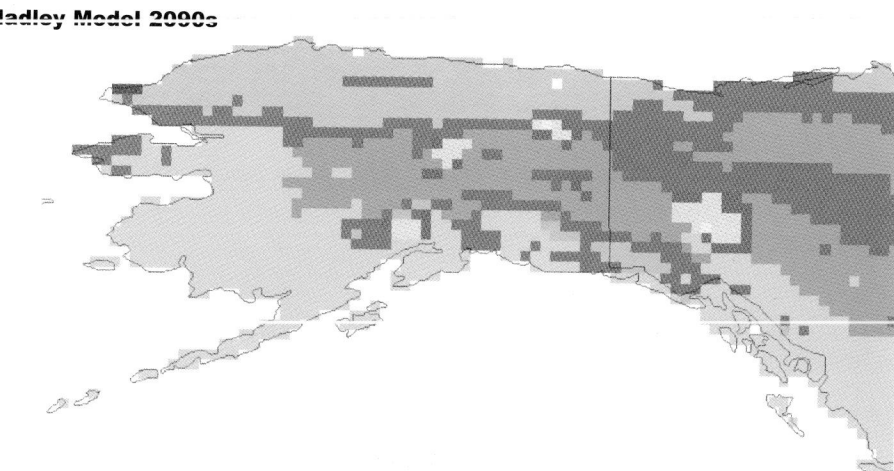

Alaska

El Niños and La Niñas as a Function of Observed and Projected Sea Surface Temperatures

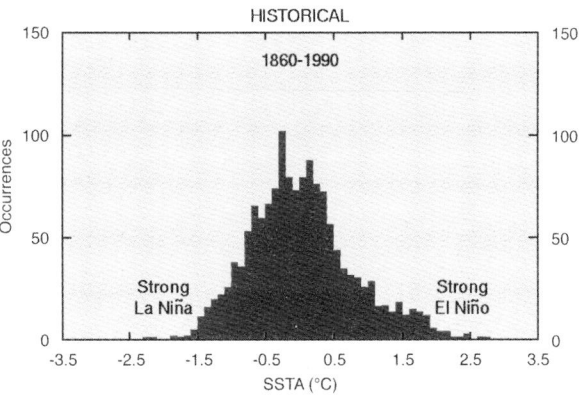

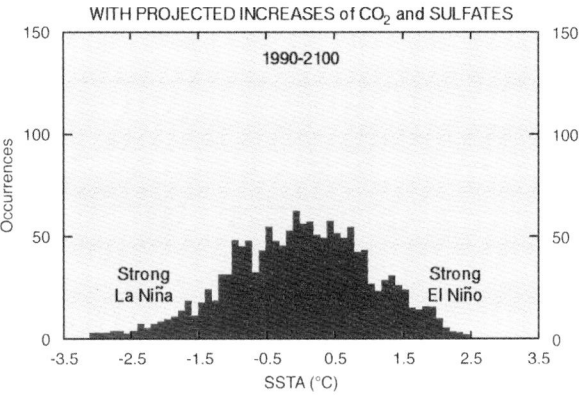

Figure 2. These model projections suggest stronger and more frequent El Niños and La Niñas as a result of climate change. Sea Surface temperature anomalies (SSTA) in the equatorial Pacific are used to measure the strength of El Niños and La Niñas. These model projections by the Max Planck Institute suggest a wider range of SST deviations from normal and thus more extreme El Niños and La Niñas in the future. The high bars in the center are occurrences of normal SSTs. In the projections in the bottom graph, these normal temperatures occur less frequently, while lower (La Niña) and higher (El Niño) SSTs occur more frequently. The Max Planck model is used here because it has been able to reproduce the strength of these events better than other models due to its physics and ability to resolve fine scale structure in the ocean. Source: Timmermann et al., 1999

Freshwater Lens Effect in Island Hydrology

Figure 3. On many islands, the underground pool of freshwater that takes the shape of a lens is a critical water source. The freshwater lens floats atop salt water. If sea level increases, and/or if the lens becomes depleted because of excess withdrawals, salt water from the sea can intrude, making the water unsuitable for many uses. The size of the lens is directly related to the size of the island: larger islands have lenses that are less vulnerable to tidal mixing and have enough storage for withdrawals. Smaller island freshwater lenses shrink during prolonged periods of low rainfall, and water quality is easily impaired by mixing with salt water. Short and light rainfall contributes little to recharge of these sources. Long periods of rainfall are needed to provide adequate recharge. Source: Illustration by Melody Warford.

Path of Hurricane Georges in Relation to Puerto Rico with Precipitation Totals

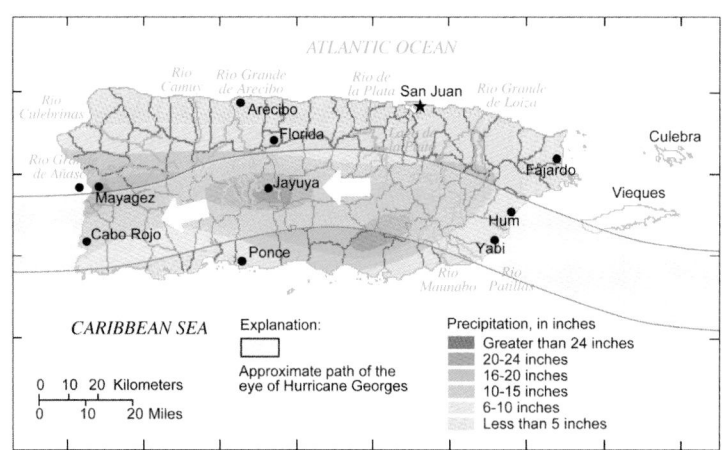

Figure 4. On September 21, 1998, Hurricane Georges swept across Puerto Rico. The eye of the hurricane was 25-30 miles wide and passed within 15 miles of the capital, San Juan, leaving a trail of devastation in its wake. The path of the hurricane and rainfall totals are shown here. Some areas received up to 26 inches of rain within 24 hours. Flooding, landslides, and catastrophic losses in infrastructure resulted. Hurricane Georges Map –USGS: http://water.usgs.gov/pubs/FS/FS-040-99/images/PR_fig01.gif

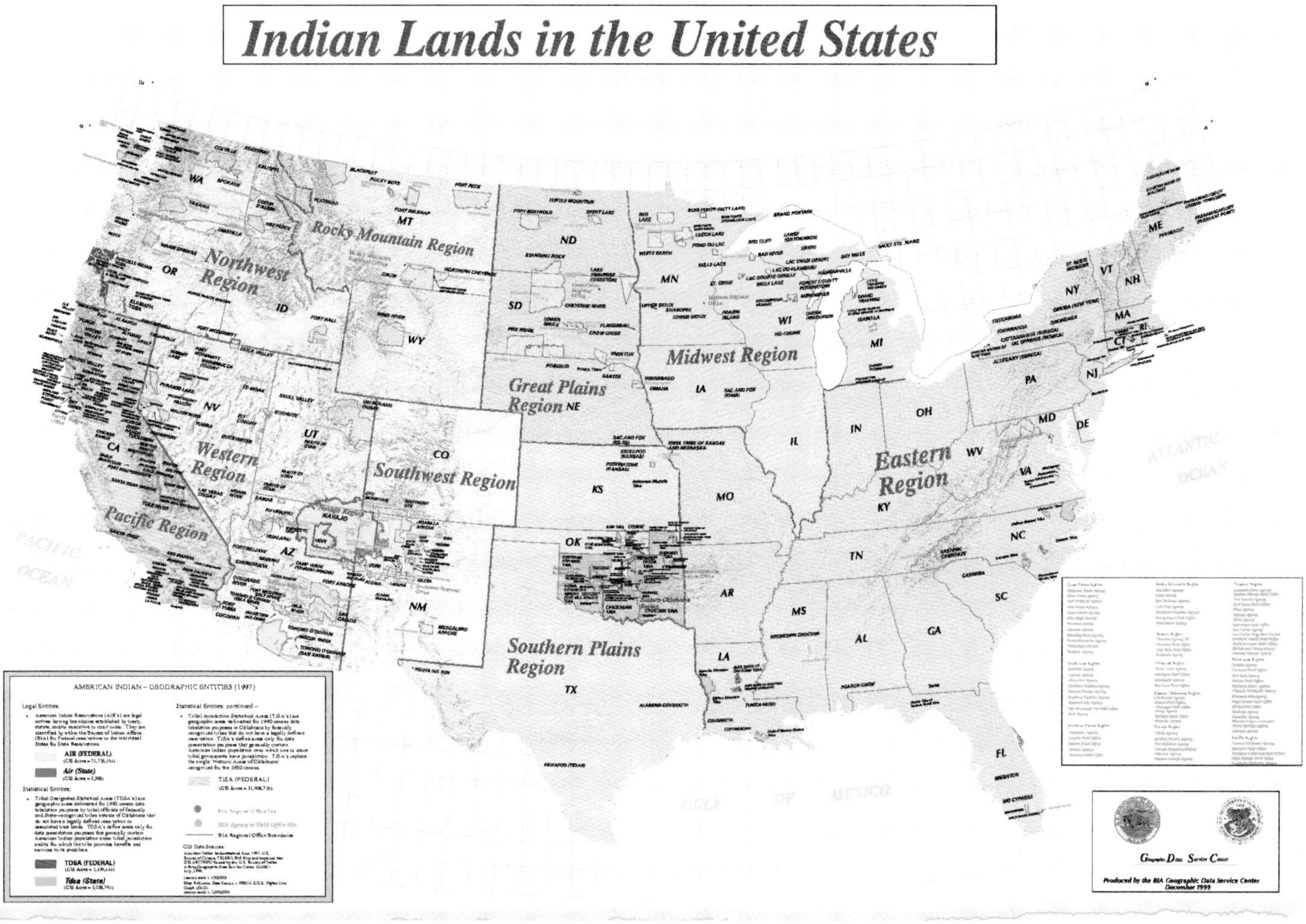

Indian Lands in the United States

Figure 1: Map of Indian lands in the conterminous United States (BIA, 2000b). The largest areas are located in the central to western US.

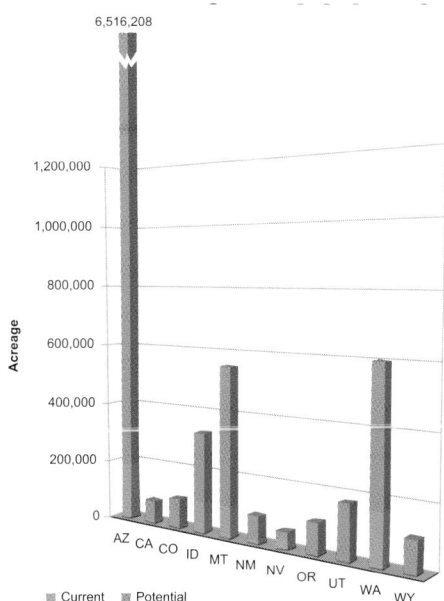

Currently Irrigated vs. Potentially Irrigable Indian Land

6,516,208

Figure 2: Comparison of existing acreage being irrigated on Indian lands in eleven western states with the maximum acreage that could potentially be integrated based on the Winters doctrine that is applied for determining Indian water rights (Riebsame et al., 1997)

▪ Current ▪ Potential

591

Dominant Land Uses, 1992

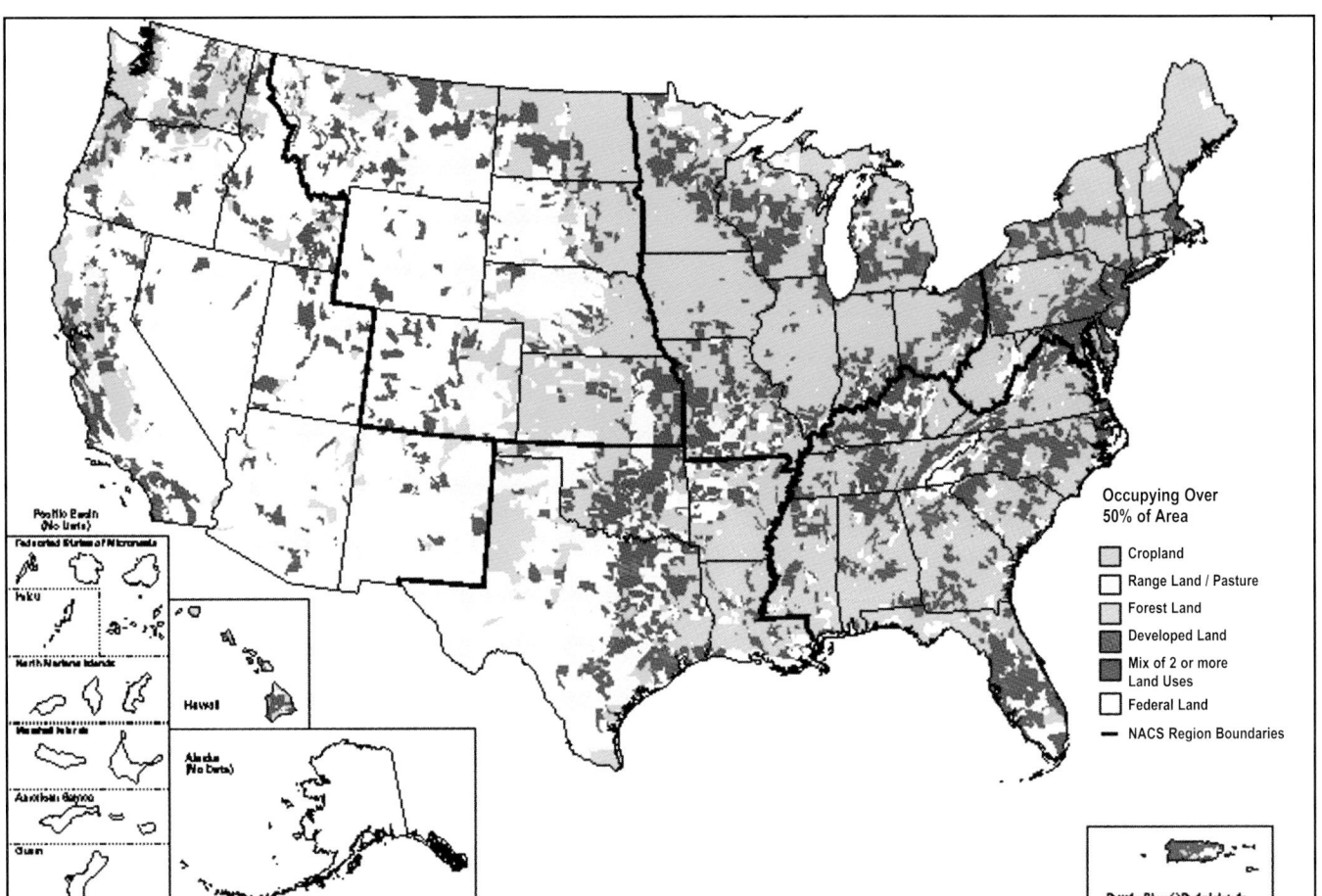

Figure 2. Agriculture Sector Model (ASM) Regions with USDA
Regions Overlaid. (ASM regions follow state boundaries except
where further disaggregated). The economic analysis in the
Assessment is summarized for the 10 USDA regions outlined in the
map. Source: USDA, 1997.

Figure 1a-d. Relative changes (% change relative to present) in
crop yield for two time periods, 2030s and 2090s, under the
Canadian and Hadley Scenarios. 0 = no change. Under the two climate scenarios, most crops showed substantial yield increase,
even without adaptation, under dryland conditions. Irrigated yields
increased less or decreased. Source: Changing Climate and
Changing Agriculture: Report of the Agricultural Sector Assessment
Team, 2000.

Figure 1a - Dryland Yields Without Adaptation

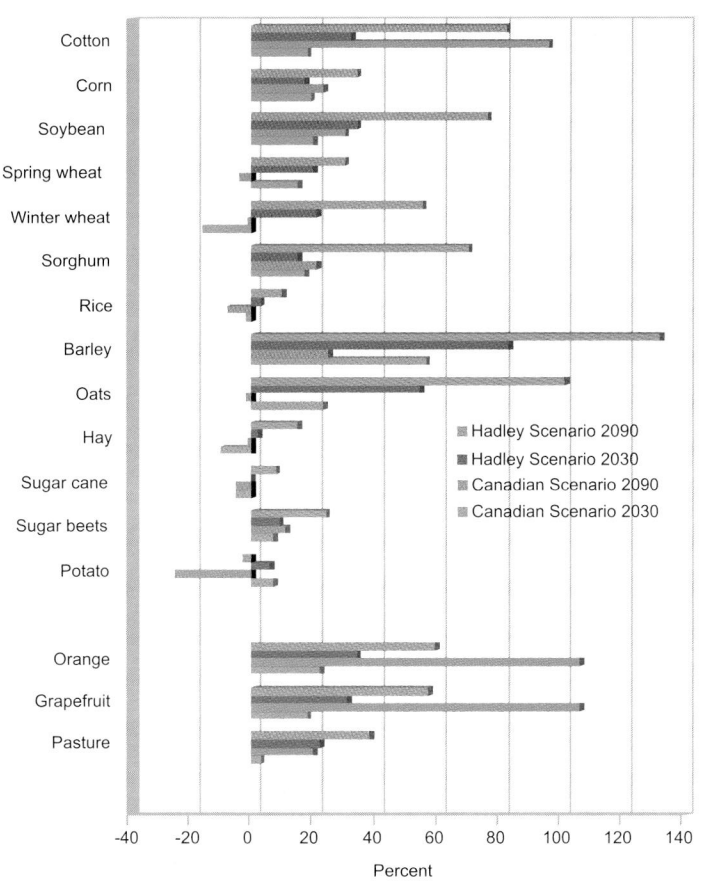

Figure 1b - Dryland Yields With Adaptation

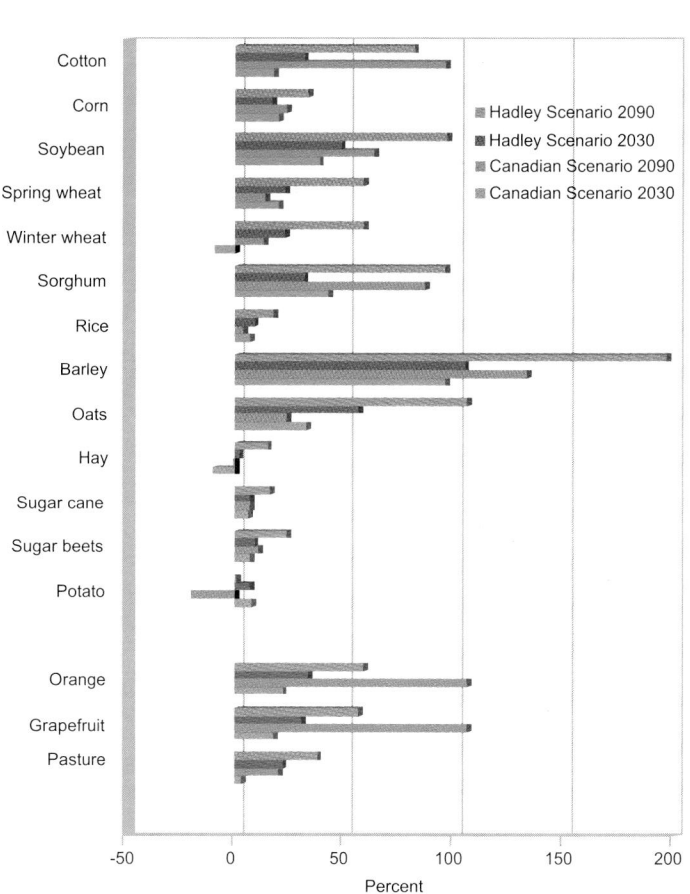

Figure 1c - Irrigated Yields Without Adaptation

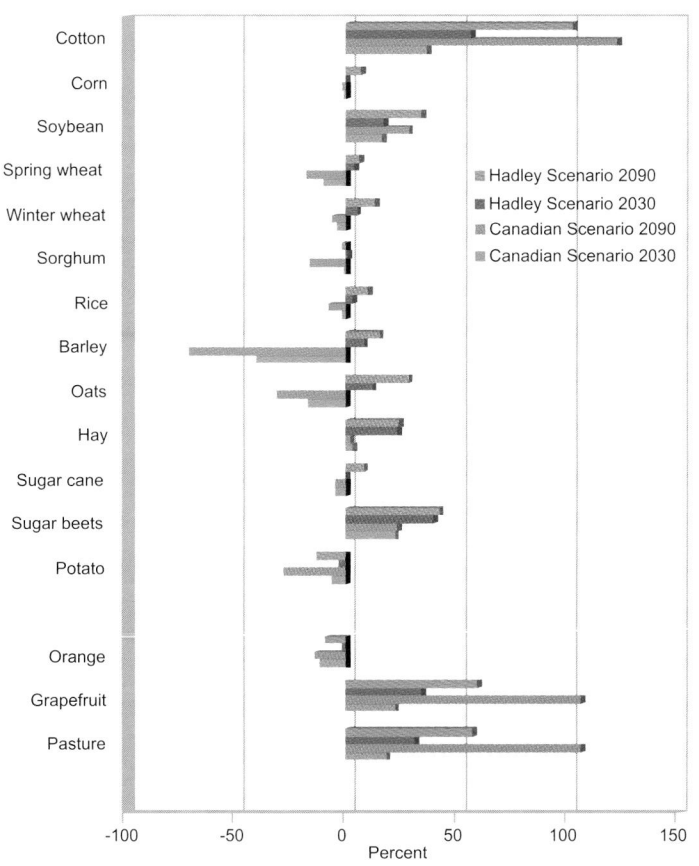

Figure 1d - Irrigated Yields With Adaptation

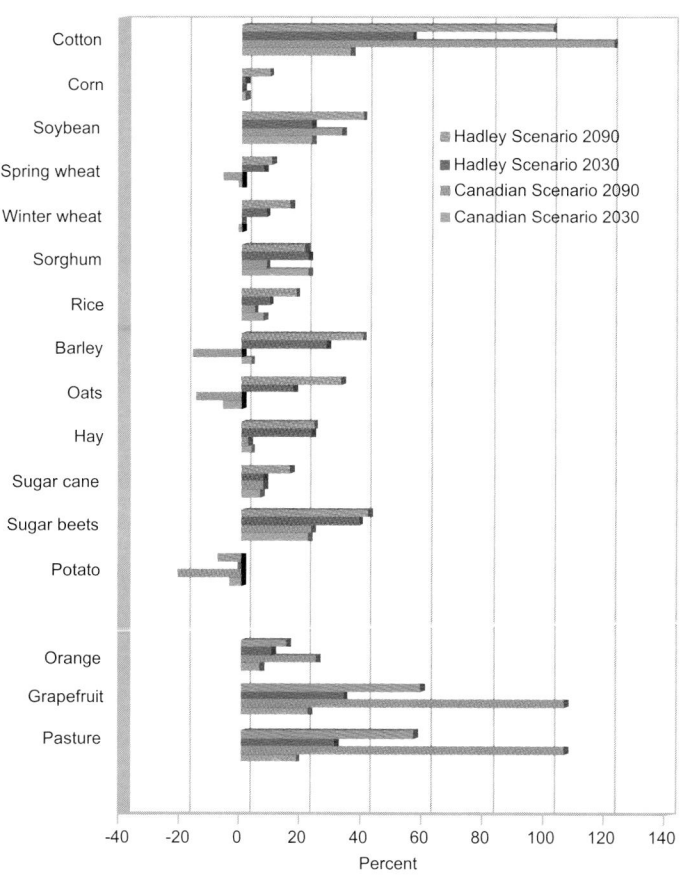

Agriculture

Economic Impacts of Climage Change on US Agriculture

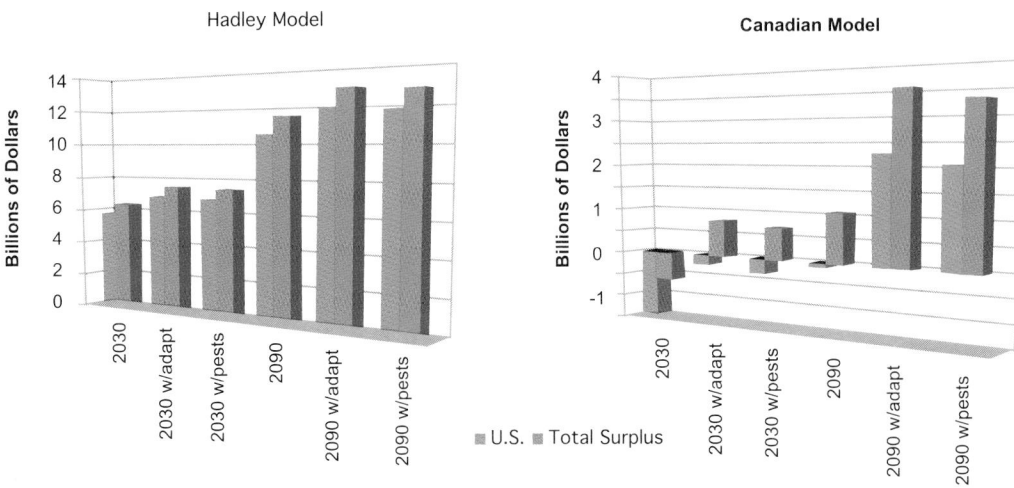

Figure 3a and b. The economic index is change in welfare expressed as the sum of producer and consumer surplus in billions of dollars. There were net economic benefits for the US under most of the scenarios examined in the Assessment. Foreign consumers also gained from lower commodity prices on international markets. Source: Changing Climate and Changing Agriculture: Report of the Agricultural Sector Assessment Team, 2000.

Producer versus Consumer Impacts of Climate Change

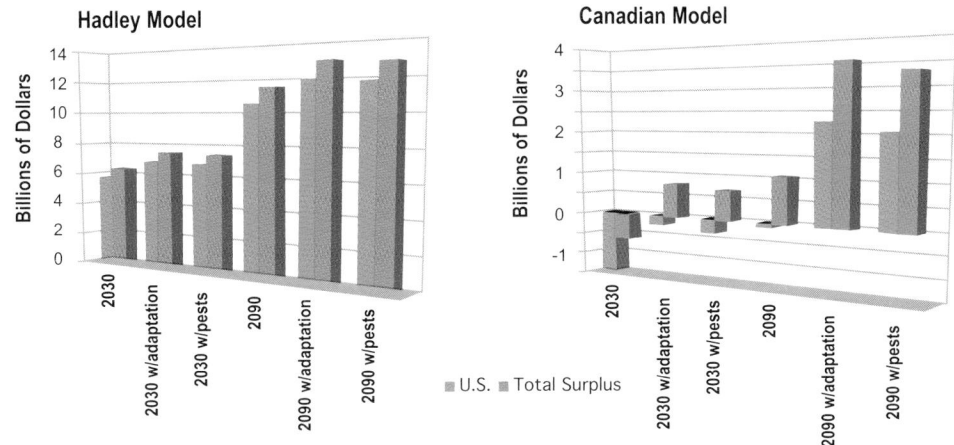

Figure 4a and b. In the model simulations consumers generally benefited from climate change while producers experienced lower income due to lower prices for commodities resulting from increased yields and supply. Source: Changing Climate and Changing Agriculture: Report of the Agricultural Sector Assessment Team, 2000.

Regional Production Changes Relative to Current Production

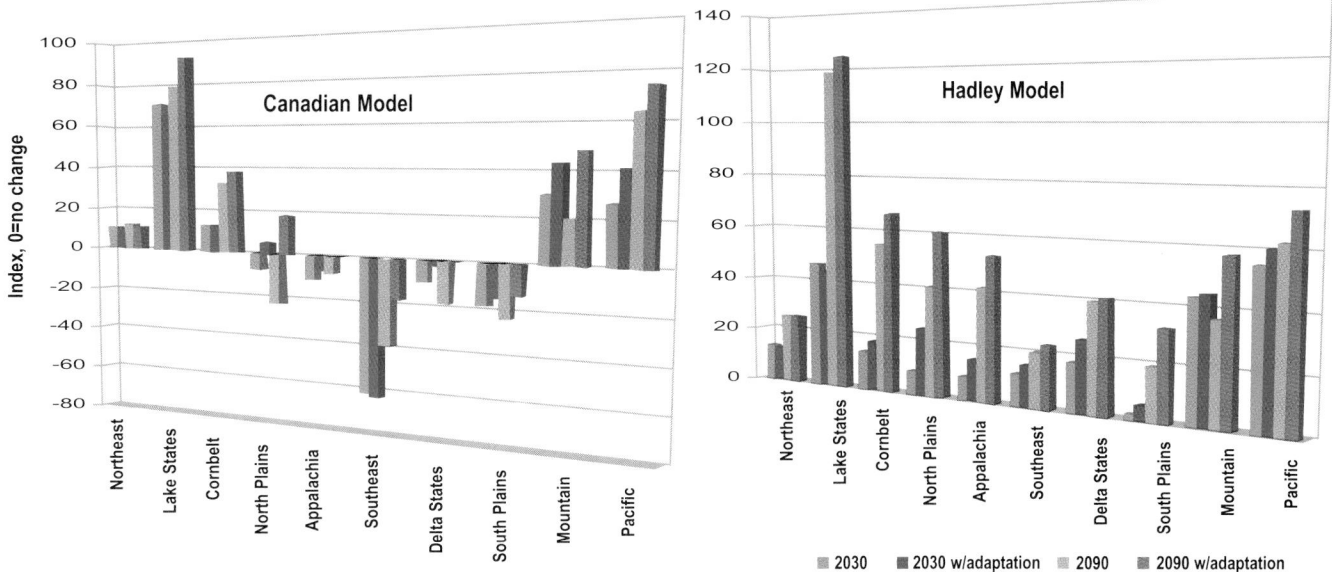

Figure 5a and b. In the model simulations, production increased in northern regions as a result of longer growing seasons, and in western regions due to increased precipitation. Higher temperatures and increased drought conditions contributed to production declines or smaller increases in southern and plains regions. Source: Changing Climate and Changing Agriculture: Report of the Agricultural Sector Assessment Team, 2000.

Changes in Resource Use

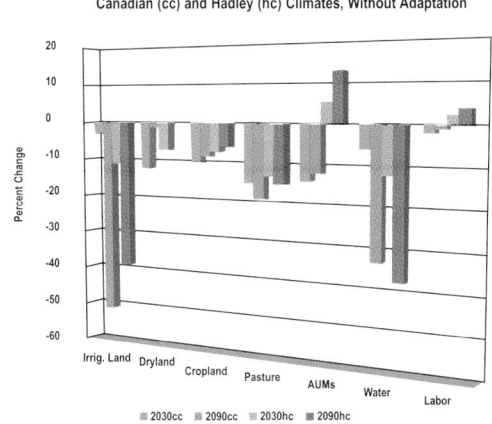

Figure 6. In the simulations, resource use generally declined as less crop and grazing land was needed. Use of water and irrigated crop land declined the most because the two climate scenarios used favored dryland over irrigated crops(cc-Canadian, hc=Hadley). Source: Changing Climate and Changing Agriculture: Report of the Agricultural Sector Assessment Team, 2000.

Agriculture

Current Climate Vulnerability Map, Water Supply, Distribution and Consumptive Use

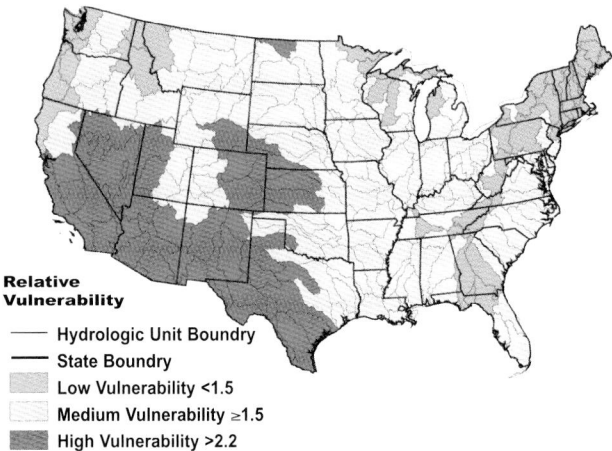

Relative Vulnerability

—— Hydrologic Unit Boundry

—— State Boundry

▢ Low Vulnerability <1.5

▢ Medium Vulnerability ≥1.5

▨ High Vulnerability >2.2

Figure 2: Assessed vulnerability based on current climate and water resource conditions, based on data describing the following: share of streamflow withdrawn for use, streamflow variability, evapotranspiration rate, groundwater overdraft, industrial use savings potential, and water trading potential. Source: Hurd, B.J., N. Leary, R. Jones and J. Smith. (1999a).

Observed Changes In Streamflow and Precipitation (1939-99)

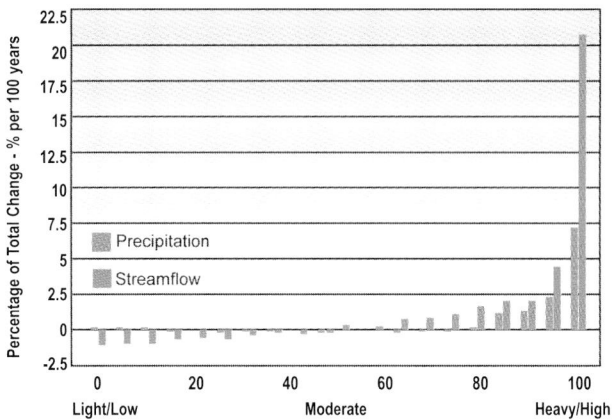

Figure 5: The graph shows changes in the intensity of precipitation and streamflow, displayed in 5% increments, during the period 1939-99 based on over 150 unregulated streams across the US with nearby precipitation measurements. As the graph demonstrates, the largest changes have been the significant increases in the heaviest precipitation events and the highest streamflows. Note that changes in streamflow follow changes in precipitation, but are amplified by about a factor of 3. Source: Groisman, et.al. (2001).

Current Climate Vulnerability Map, Instream Use, Water Quality and Ecosystem Support

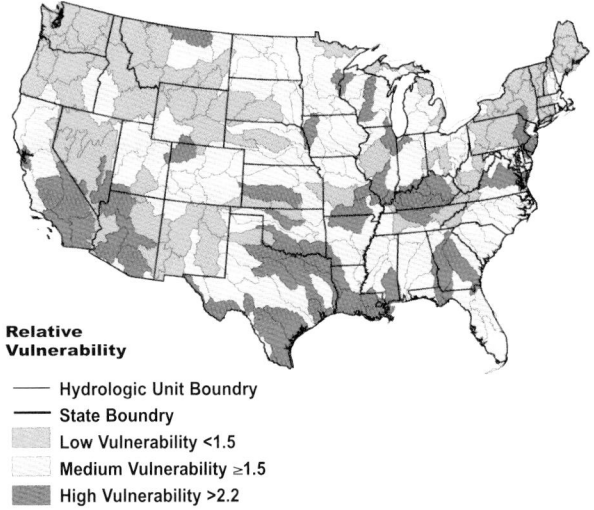

Relative Vulnerability

—— Hydrologic Unit Boundry

—— State Boundry

▢ Low Vulnerability <1.5

▢ Medium Vulnerability ≥1.5

▨ High Vulnerability >2.2

Figure 19: Instream Use, Water Quality, and Ecosystem Support Assessed Vulnerability based on current climate and water resource conditions, based on data describing the following: flood risk population, navigation impacts, ecosystem tolerance to cold and heat, dissolved oxygen stress, low streamflow conditions, and number of aquatic species at risk. Source: Hurd, B.J., N. Leary, R. Jones and J. Smith, 1999a. (Duplicated on Page 600)

Projected 21st Century Change in US Daily Precipitation

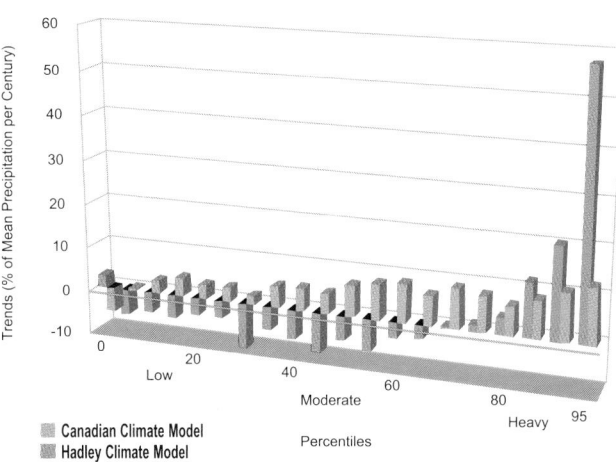

■ Canadian Climate Model
■ Hadley Climate Model

Figure 6: These projections from the Hadley and Canadian models show the changes in precipitation over the 21st century. Each models' projected change in the lightest 5% of precipitation events is represented by the far left bar and the change in the heaviest 5% by the far right bar. As the graph illustrates, both models project significant increases in heavy rain events with smaller increases or decreases in light rain events. Source: National Climatic Data Center.

Water

Projected Changes in Average Annual Runoff
Based on Two GCMs

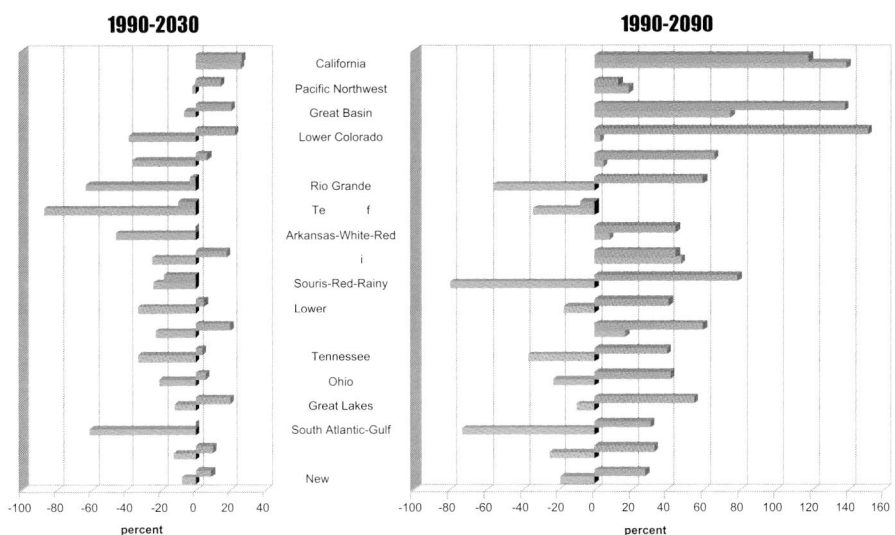

1990-2030

1990-2090

California
Pacific Northwest
Great Basin
Lower Colorado

Rio Grande
Te   f
Arkansas-White-Red
i
Souris-Red-Rainy
Lower

Tennessee
Ohio
Great Lakes
South Atlantic-Gulf

New

Figure 7: The estimated percent changes in average annual runoff based on the Canadian and Hadley models are not well correlated. The Canadian model predicts declines in runoff in all regions except California, while the Hadley model projects increases in most regions, particularly in the Southwest. The models differ in precipitation predictions in part due to underlying model construction. Source: Wolock, D.M. and G.J. McCabe, 1999a.

Percentage Change in Snowpack

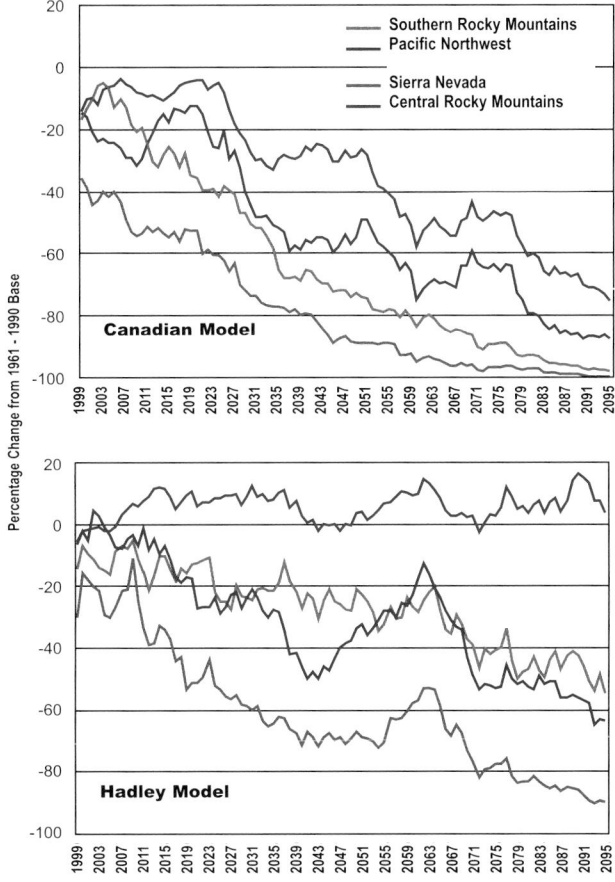

Southern Rocky Mountains
Pacific Northwest
Sierra Nevada
Central Rocky Mountains

Canadian Model

Hadley Model

Figure 8: Percentage change from the 1961-90 baseline in the April 1 snowpack in four areas of the western US as simulated for the 21st century by the Canadian and Hadley models. April 1 snowpack is important because it stores water that is released into streams and reservoirs later in the spring and summer. The sharp reductions are due to rising temperatures and an increasing fraction of winter precipitation falling as rain rather than snow. The largest changes occur in the most southern mountain ranges and those closest to the warming ocean waters. Source: McCabe, G.J. and D.M. Wolock. 1999.

Changes in Reliability of Columbia River Water Resources Objectives

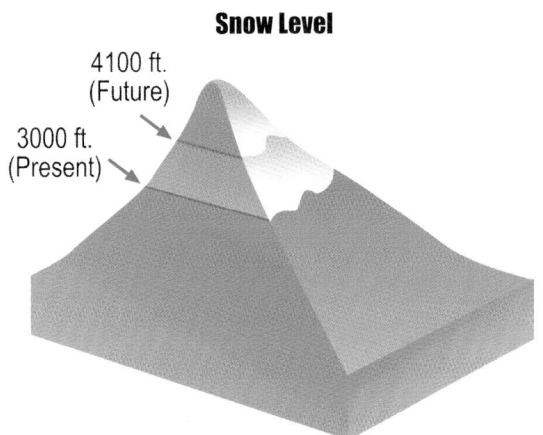

	Base Case
	HC 2025
	HC 2045
	HC 2095
	MPI 2025
	MPI 2045

Figure 9: Four major objectives are impacted by low summer streamflow and reservoir storage: non-firm energy production; irrigation; instream flow; and recreation at Lake Roosevelt. Source: Hamlet, A.F. and D.P. Lettenmaier.

Columbia Basin Snow Extent
(Washington & Oregon)

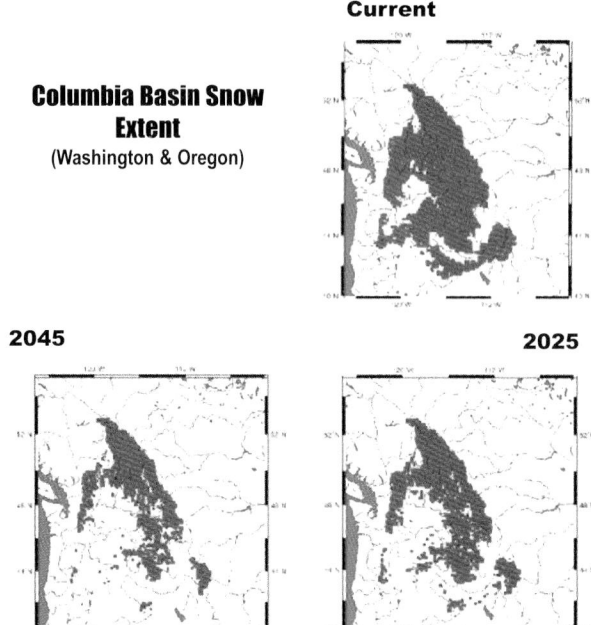

Figure 11: Complete loss of snow cover is projected at lower elevations. These maps are generated by downscaling output from global to regional climate models. Output shown from these models relates to the Columbia Basin; no projections are included for the blank areas outside the basin. Source: Mote, et.al.,(1999) Impacts of climate variability and change in the Pacific Northwest, University of Washington.

Snow Level

4100 ft. (Future)

3000 ft. (Present)

Figure 10: Rough estimate of how much snowlines in the Pacific Northwest are likely to shift by 2050, assuming about 4°F warming. Source: R. Leung, Pacific Northwest National Laboratory.

Projected Streamflow Effects from Climate Change in the Pacific Northwest

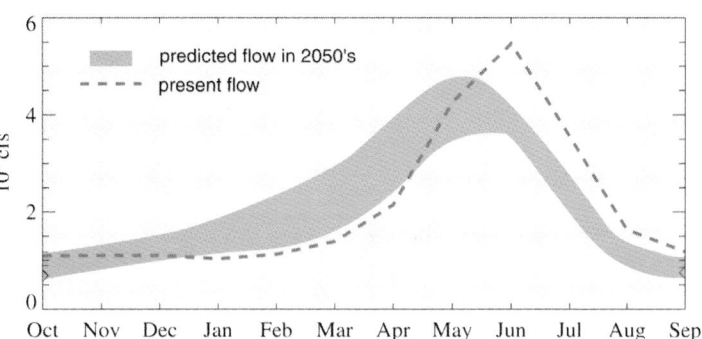

Figure 12: Relative to present flows (dashed), the wetter winters and drier summers simulated by climate models are very likely to shift peak streamflow earlier in the year, increasing the risk of late-summer shortages. Though the Columbia system is only moderately sensitive to climate change, allocation conflicts and a cumbersome network of interlocking authorities restrict its ability to adapt, producing substantial vulnerability to these shortages. Source: Hamlet, A.F. and D.P. Lettenmaier. 1999.

Water Withdrawals and Population Trends

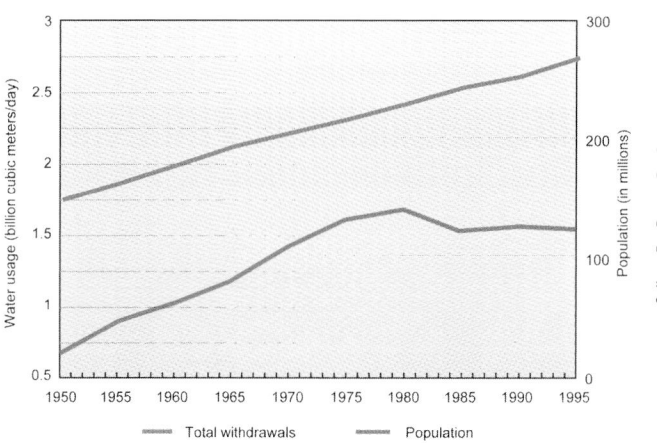

Figure 13: Although US population has continued to increase, withdrawals have declined on a per capita basis. Reductions are due to increased efficiency and recycling in some sectors, and a reduction in acreage of irrigated agriculture. Source: Solley, W.B., R.R. Pierce, and H. A.Perlman, 1998.

Evapotranspiration and Water Use in Tucson

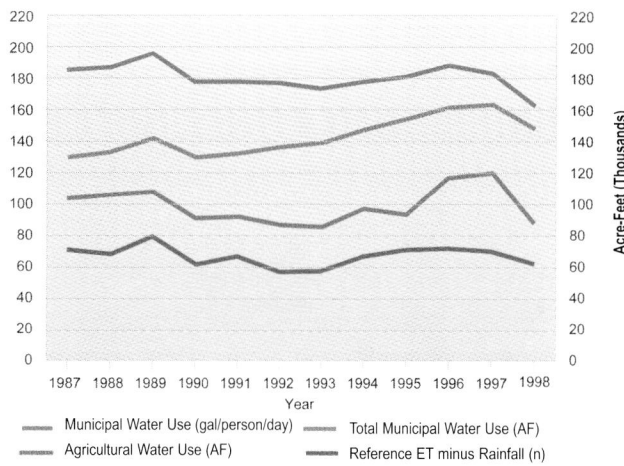

Figure 15: Water demand in the agricultural and municipal water use sectors correlates strongly with evapotranspiration rates. Source: Arizona Deptartment of Water Resources.

Consumptive Water Use by Sector

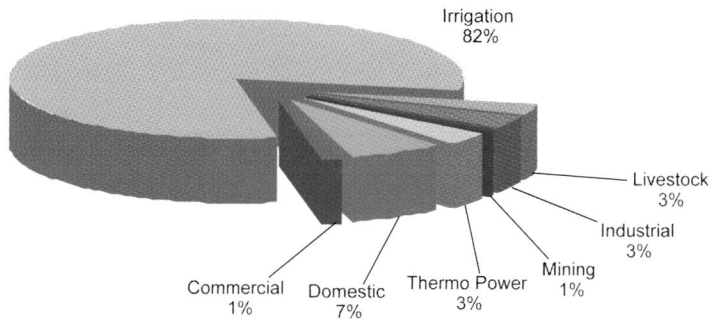

Figure 14: Agricultural water use is the highest consumptive use sector. Source: Data from Solley, W.B., R.R. Pierce, and H. A.Perlman, 1998.

Summer Stream Temperatures
Steamboat Creek, Oregon

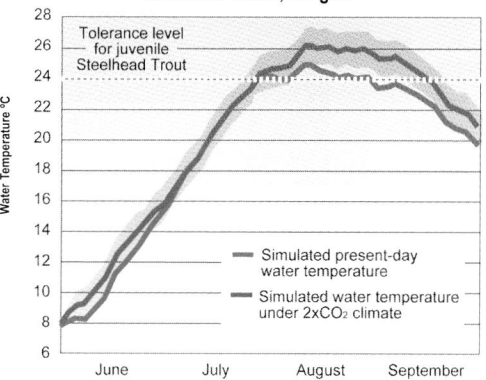

Figure 18: Simulated summer stream temperatures under present day climate (blue) and simulated temperatures under about a twice current CO_2 climate (red). The dashed line at 24 °C (75 °F) on the "water temperature" axis indicates the summer temperature tolerance of juvenile steelhead trout. Under doubled CO_2, the model suggests that the length of time within the year when the temperature tolerance limit is exceeded is more than twice as long as under simulated present-day climate conditions. Shaded area surrounding the doubled CO_2 temperature curve indicates an estimate of uncertainty.

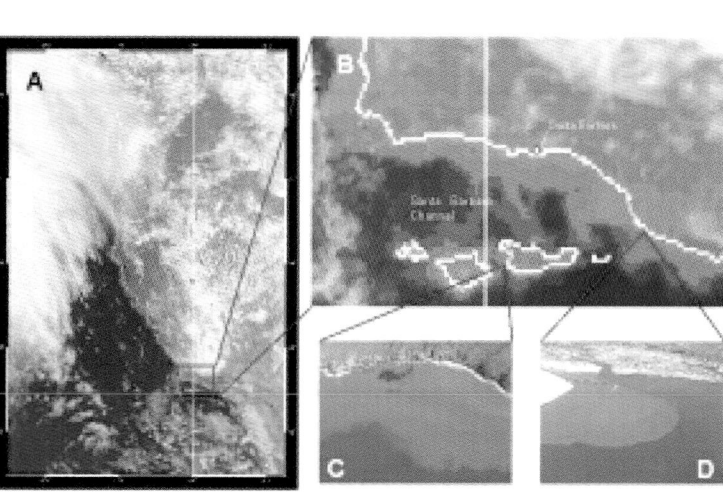

Figure 16: Sediment flow off Santa Barbara caused by El Niño storm runoff. Source: Mertes, L., The Plumes and Blooms Project, ICESS/UCSB.

Water

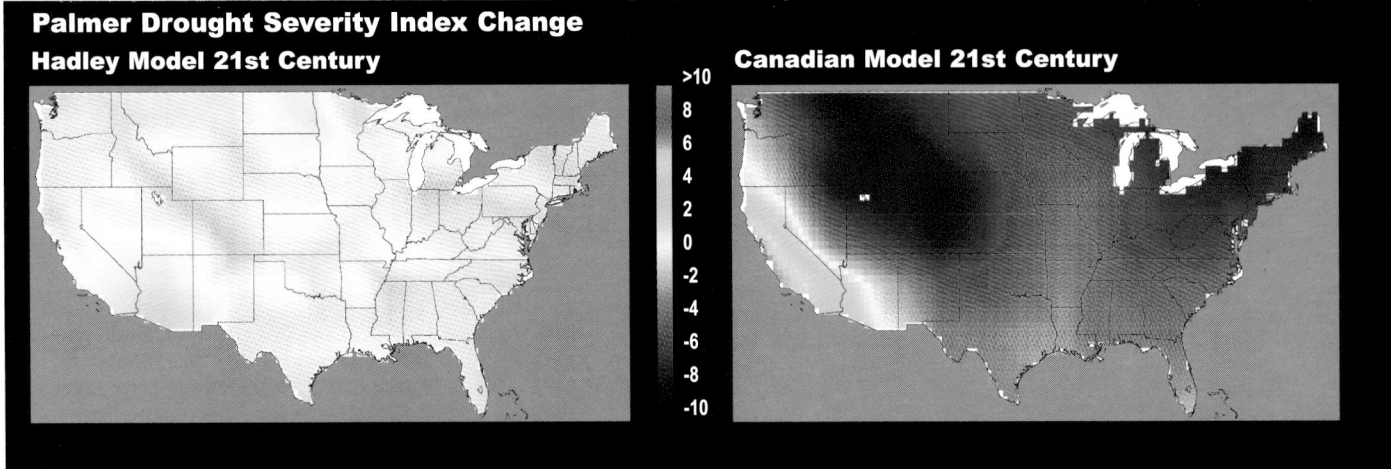

Figure 24: The Palmer Drought Severity Index (PDSI) is a commonly used measure of drought severity taking into account differences in temperature, precipitation, and capacity of soils to hold water. These maps show projected changes in the PDSI over the 21st century, based on the Canadian and Hadley climate scenarios. A PDSI of -4 indicates extreme drought conditions. The most intense droughts are in the -6 to -10 range, similar to the major drought of the 1930s. By the end of the century, the Canadian scenario projects that extreme drought will be a common occurrence over much of the nation, while the Hadley model projects much more moderate conditions. Source: Felzer, B. UCAR.

Current Climate Vulnerability Map, Instream Use, Water Quality and Ecosystem Support

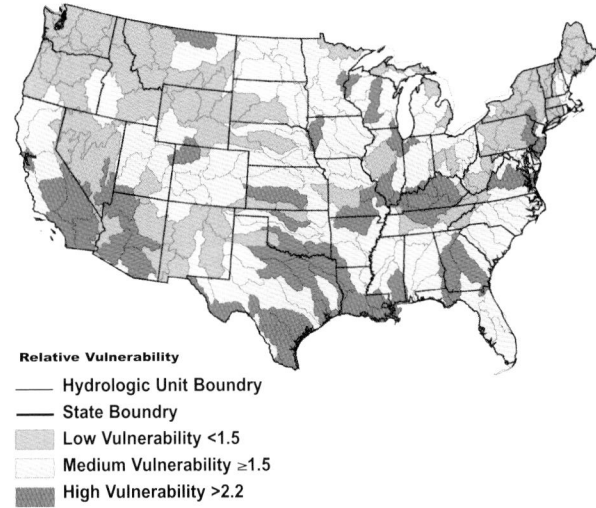

Relative Vulnerability

—— Hydrologic Unit Boundry

—— State Boundry

Low Vulnerability <1.5

Medium Vulnerability ≥1.5

High Vulnerability >2.2

Figure 19: Instream Use, Water Quality, and Ecosystem Support Assessed Vulnerability based on current climate and water resource conditions, based on data describing the following: flood risk population, navigation impacts, ecosystem tolerance to cold and heat, dissolved oxygen stress, low streamflow conditions, and number of aquatic species at risk. Source: Hurd, B.J., N. Leary, R. Jones and J. Smith, 1999a.

Potential Health Effects of Climate Variability and Change

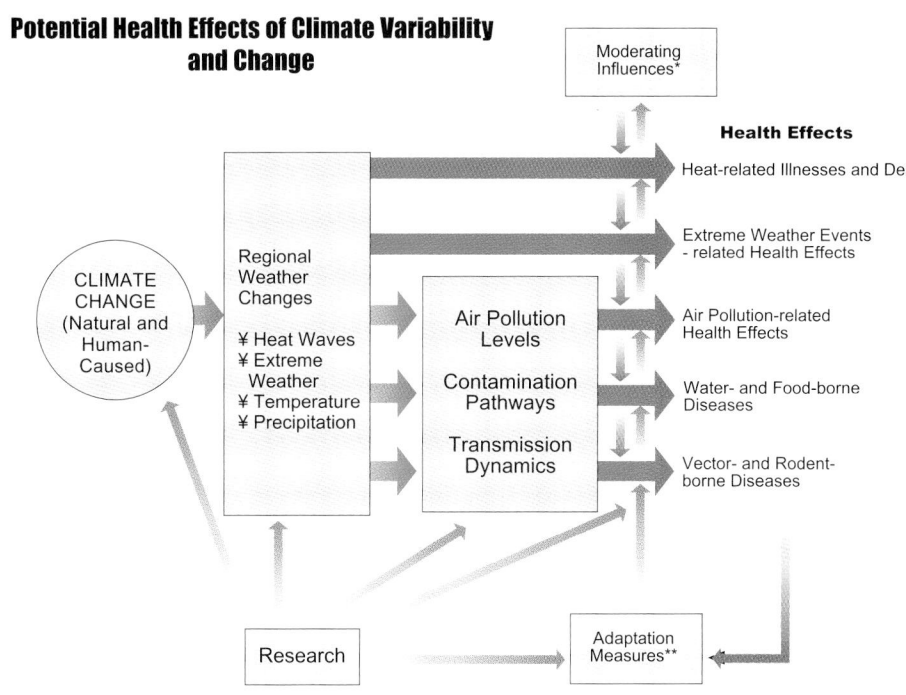

Figure 1: Schematic diagram of the potential health effects of climate variability and change. (Source, Patz et al., 2000)
 * Moderating influences include non-climate factors that affect climate-related health outcomes, such as: population growth and demographic change; standards of living; access to health care; improvements in health care; and public health infrastructure.
 ** Adaptation measures include actions to reduce risks of adverse health outcomes, such as: vaccination programs; disease surveillance; monitoring; use of protective technologies (e.g., air conditioning, pesticides, water filtration/treatment); use of climate forecasts; and development of weather warning systems; emergency management and disaster preparedness programs; and public education.

July Heat Index Change - 21st Century

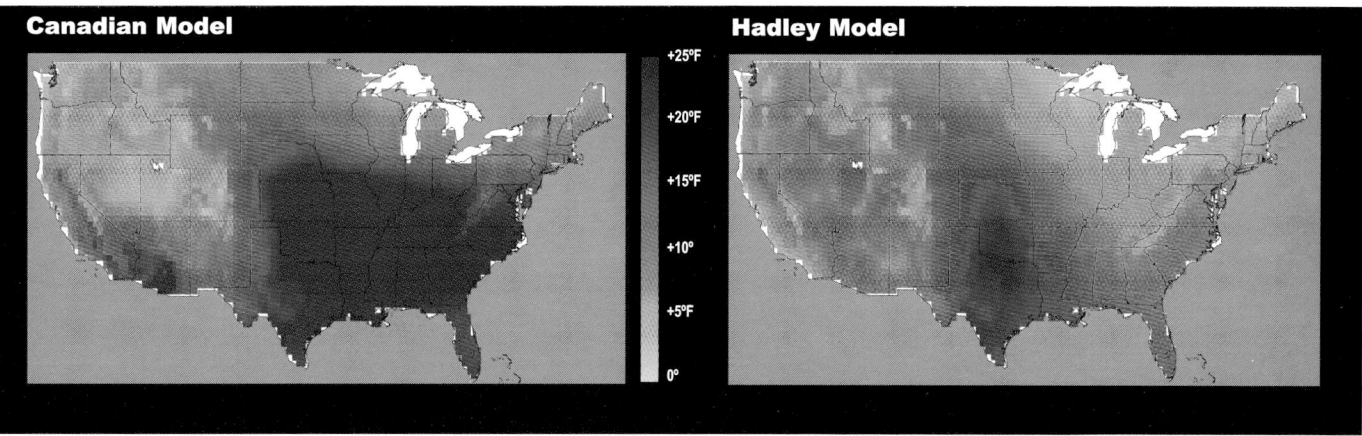

Figure 2: Both models project substantial increases in the July heat index (which combines heat and humidity) over the 21st century. These maps show the projected increase in average daily July heat index relative to the present. The largest increases are in the southeastern states, where the Canadian model projects increases of more than 25°F. For example, a July day in Atlanta that now reaches a heat index of 105°F would reach a heat index of 115°F in the Hadley model, and 130°F in the Canadian model. Map by B. Felzer, UCAR, based on data from Canadian and Hadley modeling centers.

Heat Related Deaths in Chicago in July 1995

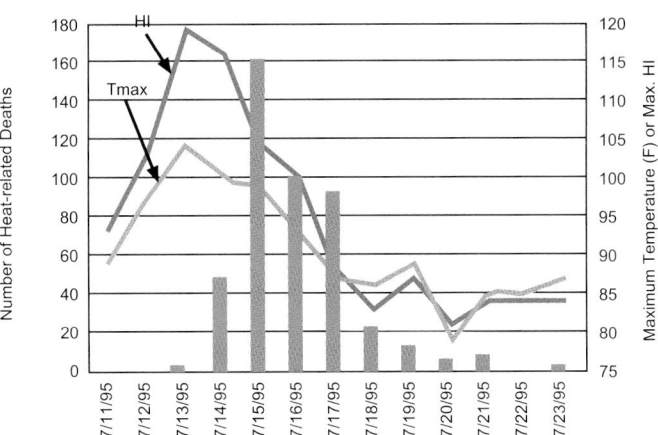

Figure 3: This graph tracks the maximum temperature (Tmax), heat index (HI), and heat-related deaths in Chicago each day from July 11 to 23, 1995. The gray line shows maximum daily temperature, the blue line shows the heat index, and the bars indicate the number of deaths each day. Source: NOAA/NCDC.

Maximum Daily Ozone Concentrations versus Maximum Daily Temperature in Atlanta and New York.

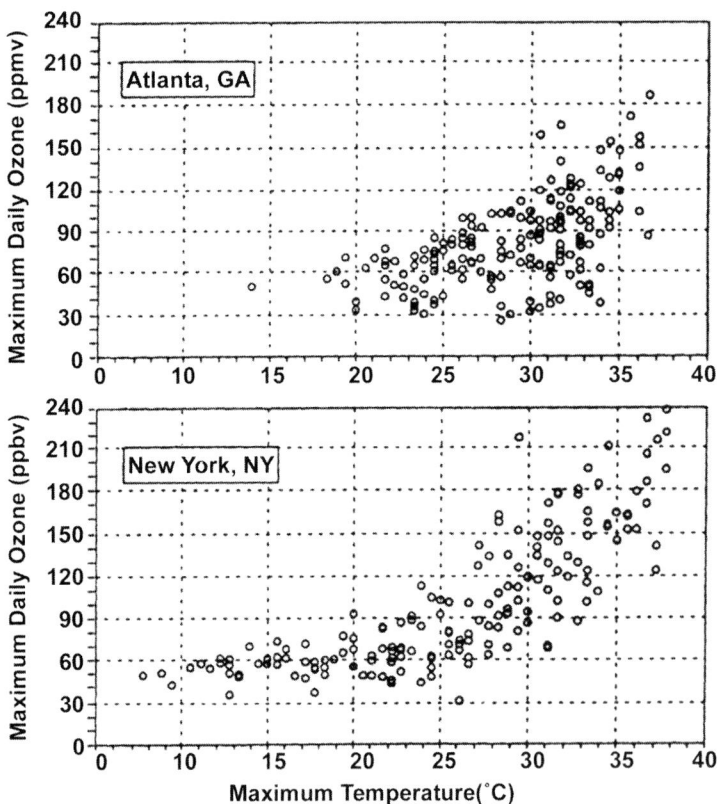

Figure 5: These graphs illustrate the observed association between ground-level ozone concentrations and temperature in Atlanta and New York City (May to October 1988-1990). The projected higher temperature across the US in the 21st century will likely increase the occurrence of high ozone concentrations, especially because extremely hot days frequently have stagnant air circulation patterns, although this will also depend on emissions of ozone precursors and meteorological factors. Ground-level ozone can exacerbate respiratory diseases and cause short-term reductions in lung function. (Maximum Daily Ozone Chart provided by USEPA.)

Average Summer Mortality Rates

Attributed to Hot Weather Episodes

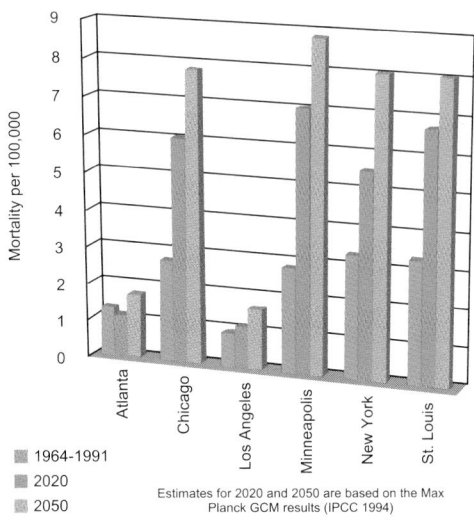

■ 1964-1991
■ 2020
■ 2050

Estimates for 2020 and 2050 are based on the Max Planck GCM results (IPCC 1994)

Figure 4: Deaths due to summer heat are projected to increase in US cities, according to a study using time-dependent results (for greenhouse gas increase only) from several climate models (Kalkstein and Greene, 1997). Mortality rates (number of deaths per 100,000 population) are shown from the Max Planck Institute model, the results from which lie roughly in the middle of the models examined (the other climate scenarios used were from Geophysical Fluid Dynamics Laboratory (GFDL) and the Hadley Centre). Because heat-related illness and death appear to be related to temperatures much hotter than those to which the population is accustomed, cities that experience extreme heat only infrequently appear to be at greatest risk. For example, Philadelphia, New York, Chicago, and St. Louis have experienced heat waves that resulted in a large number of heat-related deaths, while heat related deaths in Atlanta and Los Angeles are much lower. In this study, statistical relationships between heat waves and increased death rates are constructed for each city based on historical experience. Deaths under a city's future climate are then projected by applying that city's projected incidence of extreme heat waves to the statistical relationship that was estimated for the city whose present climate is most similar to the projected climate for the city in question. This approach attempts to represent how people will acclimate to the new average climate that they experience.

Human Health

Seasonality of Shellfish Poisoning in Florida 1981-1994

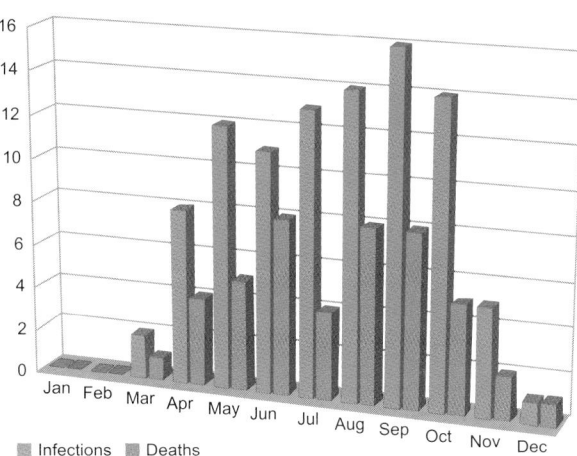

■ Infections ■ Deaths

Figure 6: Monthly distribution of oyster-associated *Vibrio vulnificus* illness (or shellfish poisoning) and deaths occurring in Florida from 1981-1994. Over the 14-year period, higher numbers of cases occur during summer. Monitoring in Florida shows a statistically significant association between concentrations of this pathogen in estuaries and temperature and salinity, the latter being affected by rainfall and runoff. Adapted from: Lipp and Rose, 1997.

Reported Cases of Dengue 1980-1999

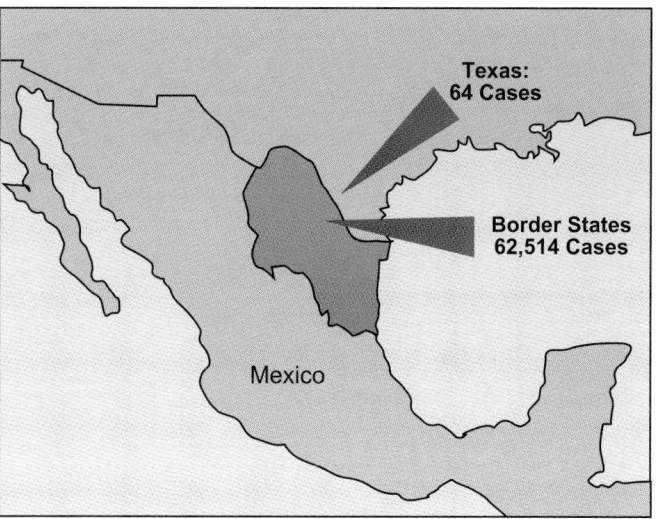

Figure 8: Dengue along the US-Mexico border. Dengue, a mosquito-borne viral disease, was once common in Texas (where there were an estimated 500,000 cases in 1922), and the mosquito that transmits it remains abundant. The striking contrast in the incidence of dengue in Texas versus three Mexican states that border Texas (64 cases vs. 62,514) in the period from 1980-1999 provides a graphic illustration of the importance of factors other than temperature, such as use of air conditioning and window screens, in the transmission of vector-borne diseases. National Institute of Health, Mexico; Texas Department of Health; US Public Health Service. Unpublished data.

Locations of Combined Wastewater Systems

Figure 7: Wastewater systems that combine storm water drainage and sewage and industrial discharges are still in use in about 950 communities in the US, mostly in the Northeast and Great Lakes regions. These combined sewer systems deliver both storm drainage and wastewater to sewage treatment facilities. However, during rain or snowmelt, the volume of incoming water can exceed the capacity of the treatment system. Under those conditions, combined sewer systems are designed to overflow and discharge untreated wastewater into surface water bodies, and are termed as a combined sewer overflow (CSO) event. EPA, in 1994, developed a *CSO Control Policy* that sets forth a national framework for prevention of combined sewer overflows through the federal Clean Water Act's water discharge permit program. It has been suggested that if they continue to discharge untreated wastewater during storm events, combined sewer systems may pose a greater health risk should the frequency or intensity of storms increase. Source: USEPA, http://www.epa.gov/owmitnet/cso.htm.

Human Health

Population Distribution across the US

Projected change in county population
(percent), 1970 to 2030

- >+250% (highest +3,877%)
- +50% to +250%
- +5% to +50%
- -5% to +5%
- -20% to -5%
- -40% to -20%
- <-40% (lowest -60%)

Each block on the map illustrates one county in the US. The height of each block is proportional to that county's population density in the year 2000, so the volume of the block is proportional to the county's total population. The color of each block shows the county's projected change in population between 1970 and 2030, with shades of orange denoting increases and blue denoting decreases. The patterns of recent population change, with growth concentrated along the coasts, in cities, and in the South and West, are projected to continue.

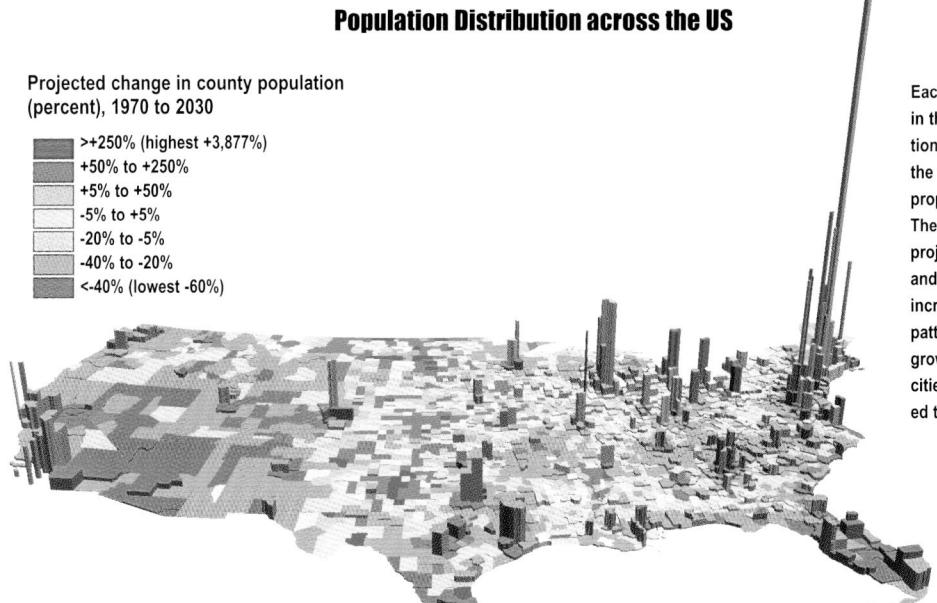

Figure 1: Over the next 25 years, population gains of some 18 million people are projected to occur in the coastal states of Florida, California, Texas, and Washington (NPA, 1999).

Global Average Sea Level Rise

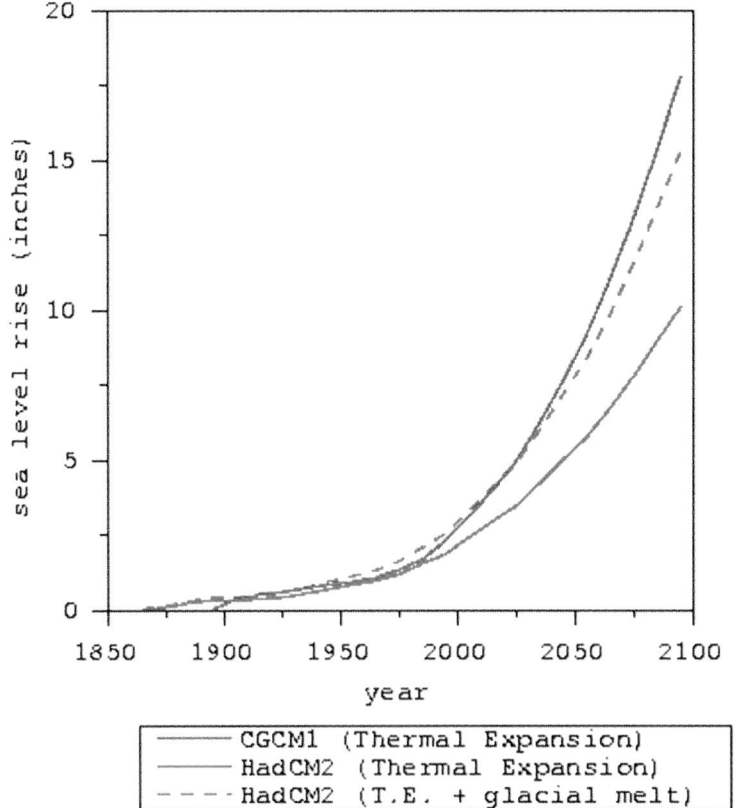

CGCM1 (Thermal Expansion)
HadCM2 (Thermal Expansion)
HadCM2 (T.E. + glacial melt)

Figure 2 : Projected rise in global average sea level based on the Hadley and Canadian General Circulation Model (GCM) scenarios.

Spatial Distribution Around North America in Sea Level Rise

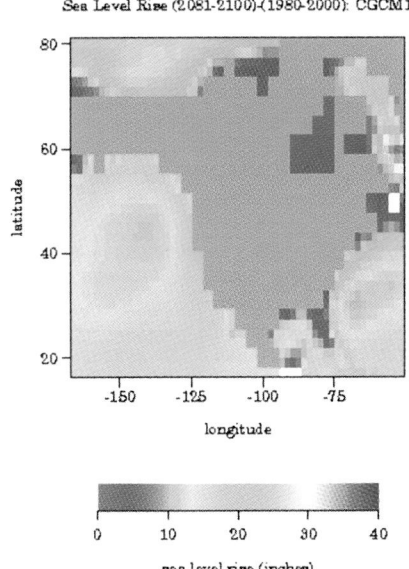

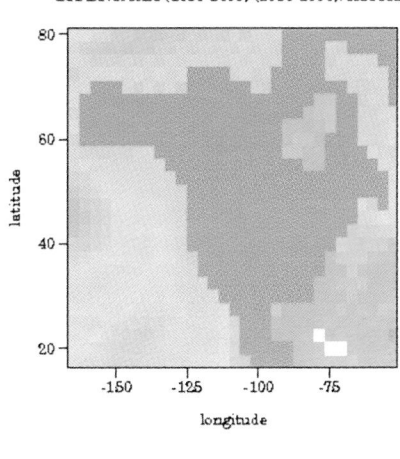

Figure 3 : Projections of the regional pattern of global sea level rise by the year 2100 based on the Canadian (left) and Hadley (right) scenarios. These estimates do not include contributions to sea-level change due to vertical movement of coastal lands.

Hurricanes and their Impacts in the 20ᵗʰ Century (1900-1995)

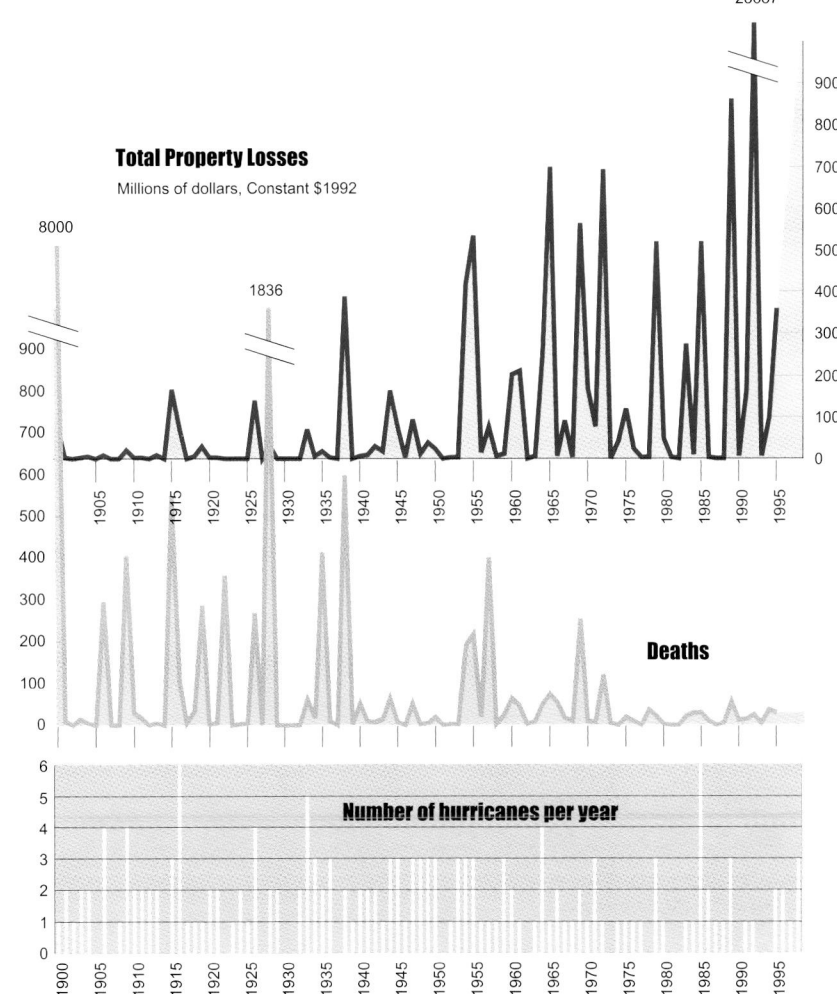

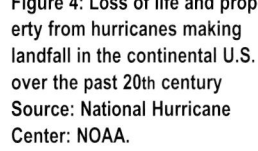

Figure 4: Loss of life and property from hurricanes making landfall in the continental U.S. over the past 20th century Source: National Hurricane Center: NOAA.

605

Ocean Heat Content in the 0-3000 m Layer

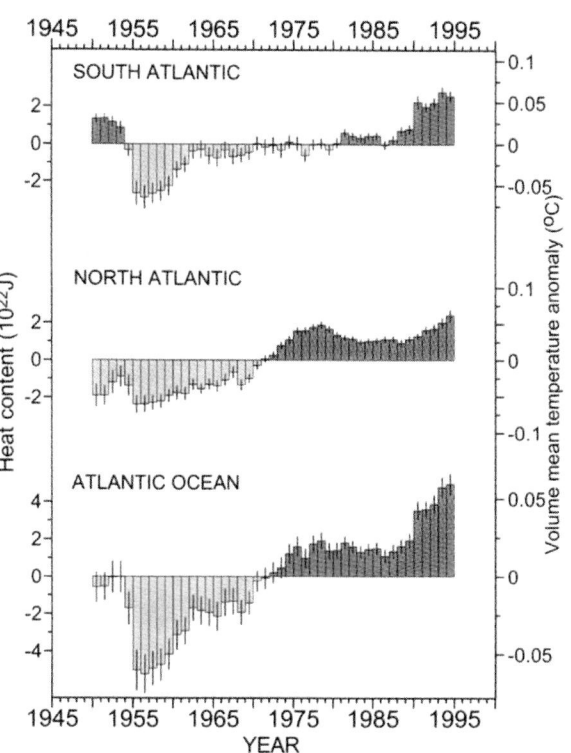

Figure 5: A comprehensive analysis of over 5 million temperature profiles by Levitus, et al. (2000) reveals a pattern of warming in both the surface and the deep ocean over the last 40 years. The largest warming has occurred in the upper 300 meters (984 feet), which have warmed by an average of 0.31°C (0.56°F), with additional warming as deep as 3000 meters (9843 feet).

The Global Ocean Conveyor Belt

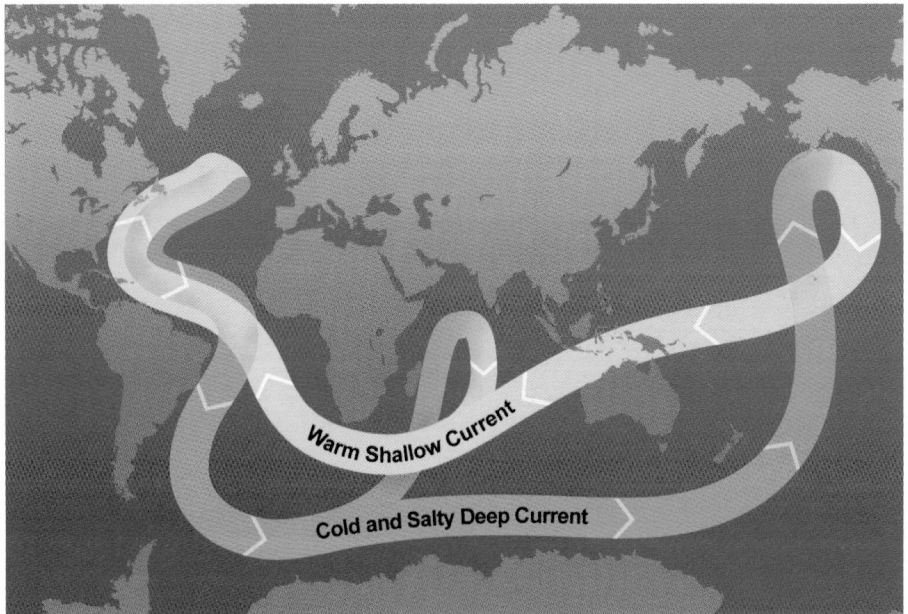

Figure 6: The ocean plays a major role in the distribution of the planet's heat through deep sea circulation. This simplified illustration shows this "conveyor belt" circulation which is driven by differences in heat and salinity. Records of past climate suggest that there is some chance that this circulation could be altered by the changes projected in many climate models, with impacts to climate throughout lands bordering the North Atlantic (Modified from Broecker, 1991).

Classification of Annual Shoreline Change Around the United States

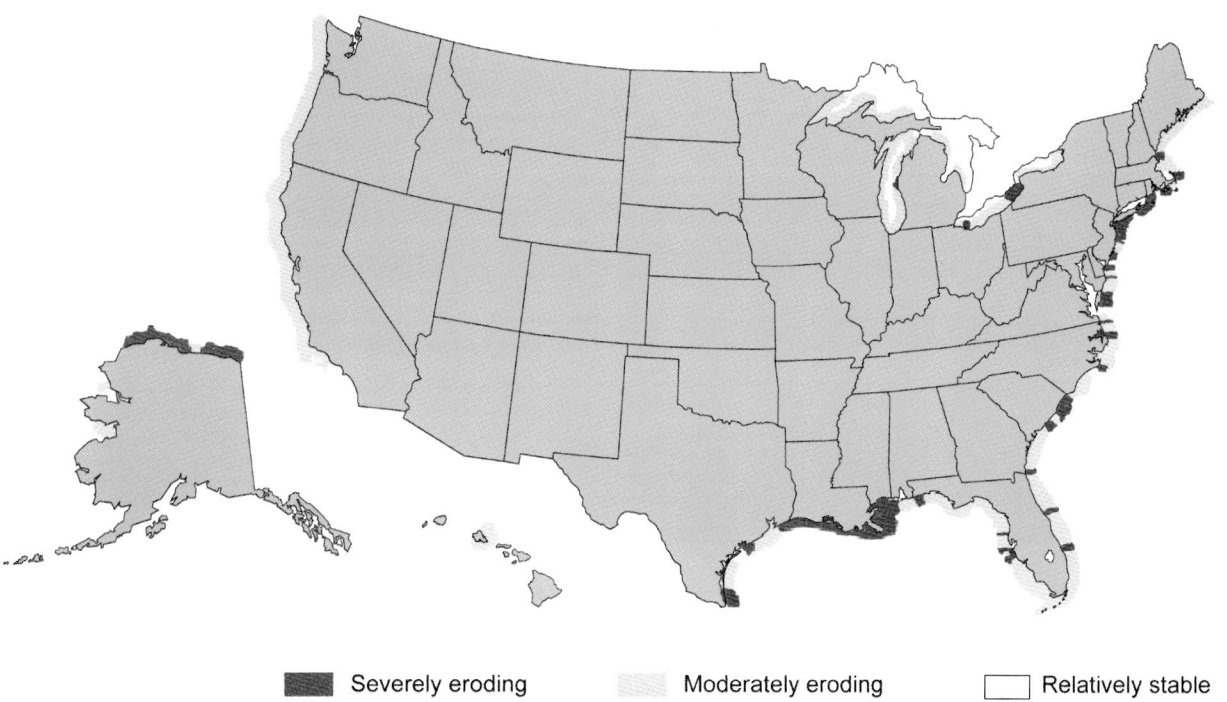

■ Severely eroding   ▨ Moderately eroding   ☐ Relatively stable

Figure 7: A general classification scheme of shoreline erosion rates throughout the US. (modified from Dolan et al., 1985).

Processes Affecting Wetland Migration

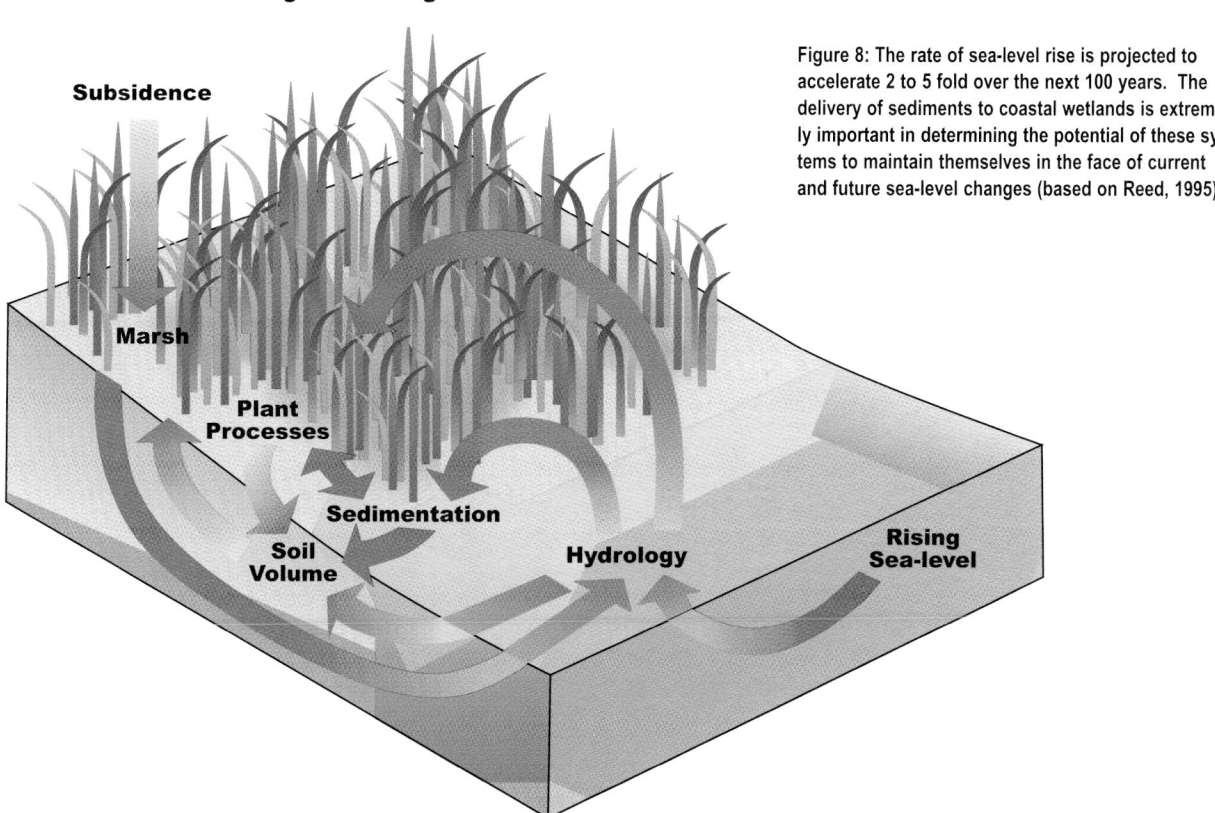

Figure 8: The rate of sea-level rise is projected to accelerate 2 to 5 fold over the next 100 years. The delivery of sediments to coastal wetlands is extremely important in determining the potential of these systems to maintain themselves in the face of current and future sea-level changes (based on Reed, 1995).

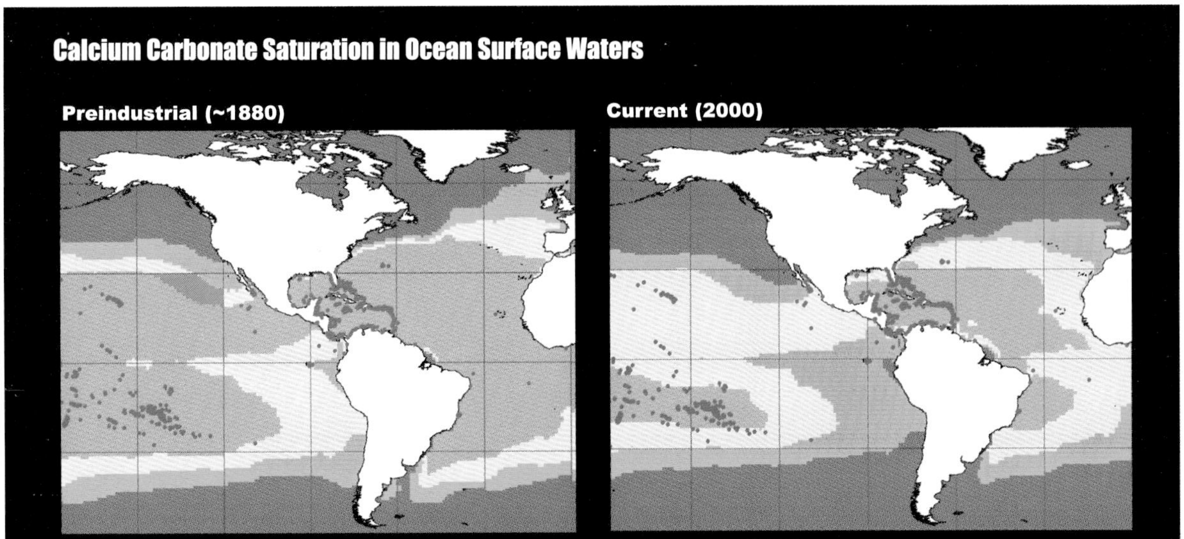

Calcium Carbonate Saturation in Ocean Surface Waters

Preindustrial (~1880)

Current (2000)

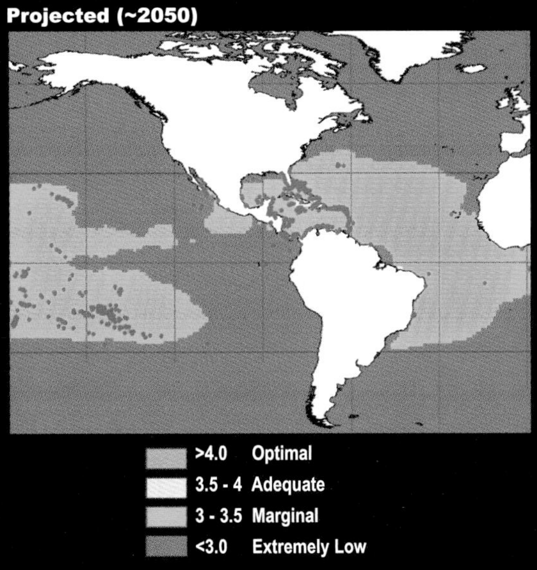

Projected (~2050)

Figure 10: Map of current and projected changes in calcium carbonate saturation in ocean surface waters. Corals require the right combination of temperature, light, and calcium carbonate saturation. At higher latitudes, there is less light and lower temperatures than nearer the equator. The saturation level of calcium carbonate is also lower at higher latitudes, in part because more CO_2, an acid, can be dissolved in colder waters. As the CO_2 level rises, this effect dominates, making it more difficult for corals to form at the poleward edges of their distribution. These maps show model results of the saturation level of calcium carbonate for pre-industrial, present, and future CO_2 concentrations. The dots indicate present coral reefs. Note that under model projections of the future, it is very unlikely that calcium carbonate saturation levels will provide fully adequate support for coral reefs in any US waters. The possibility of this future scenario occurring demands continued research on effects of increasing CO_2 on entire coral reef systems. Classification intervals for saturation effects on reef systems are derived from Kleypas et al. (1999b).

>4.0   Optimal
3.5 - 4   Adequate
3 - 3.5   Marginal
<3.0   Extremely Low

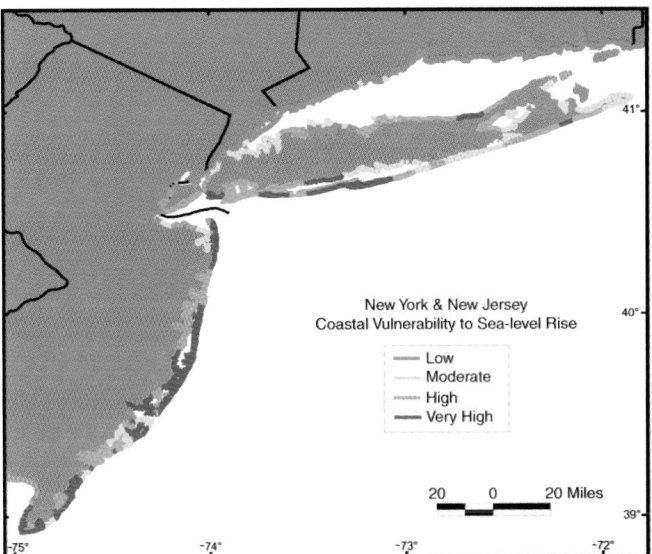

New York & New Jersey
Coastal Vulnerability to Sea-level Rise

— Low
— Moderate
— High
— Very High

20   0   20 Miles

Figure 12: These preliminary results illustrate the relative vulnerability to sea-level rise along the New York and New Jersey coastline as assessed by ongoing USGS research. Note that the vulnerability mapped here is likely to change as methodologies in this pilot program are critically evaluated and improved (Source: USGS).

Current Distribution of Forests in the United States

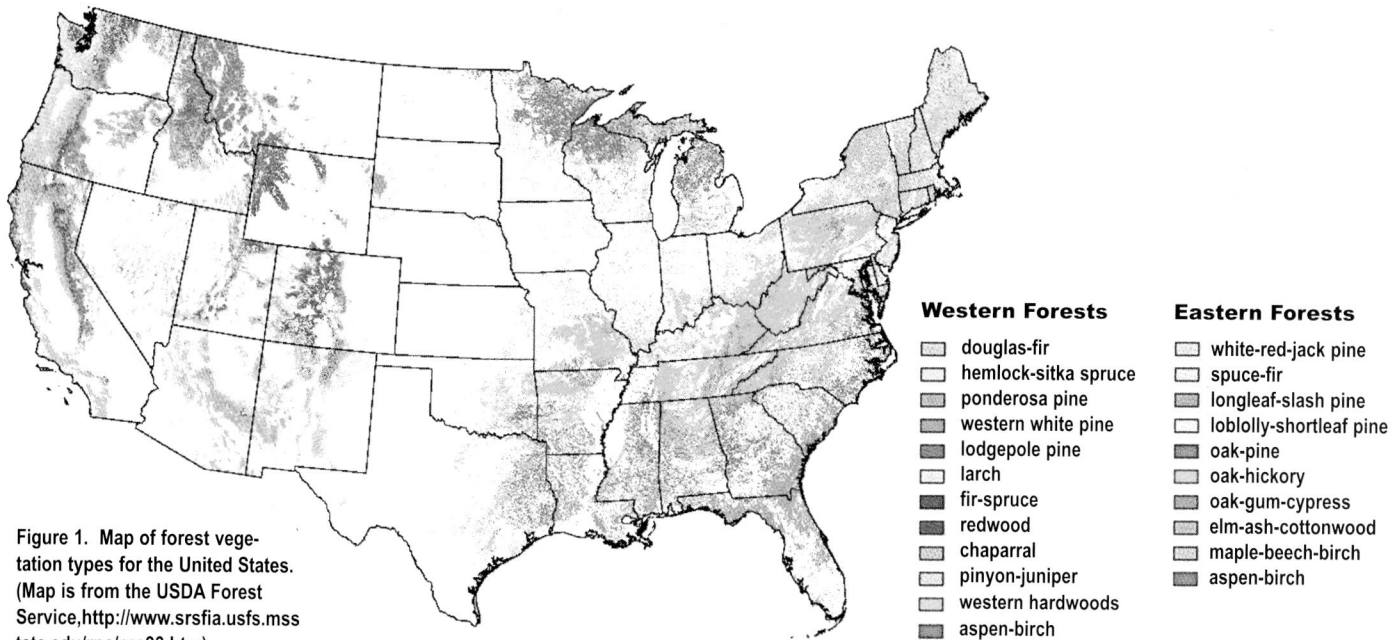

Western Forests
- douglas-fir
- hemlock-sitka spruce
- ponderosa pine
- western white pine
- lodgepole pine
- larch
- fir-spruce
- redwood
- chaparral
- pinyon-juniper
- western hardwoods
- aspen-birch

Eastern Forests
- white-red-jack pine
- spuce-fir
- longleaf-slash pine
- loblolly-shortleaf pine
- oak-pine
- oak-hickory
- oak-gum-cypress
- elm-ash-cottonwood
- maple-beech-birch
- aspen-birch

Figure 1. Map of forest vegetation types for the United States. (Map is from the USDA Forest Service,http://www.srsfia.usfs.mss tate.edu/rpa/rpa93.htm)

Forest Land Coverage over the Past 400 Years

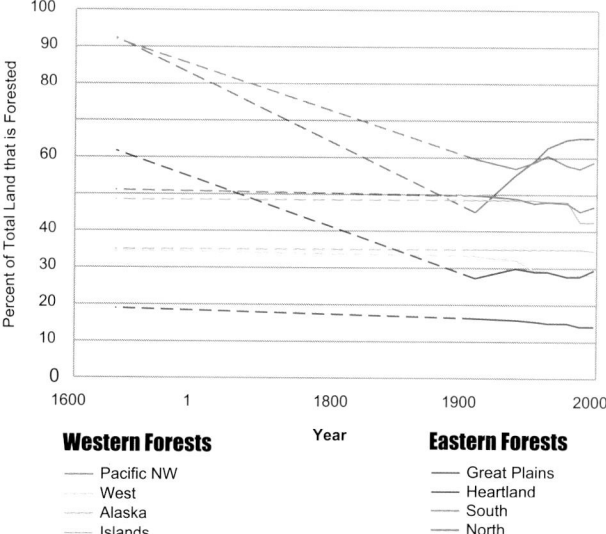

Western Forests
- Pacific NW
- West
- Alaska
- Islands

Eastern Forests
- Great Plains
- Heartland
- South
- North

Figure 2. Land area changes in forestland. Data are from Forest Service Resource Bulletin PNW-RB-168, Forest Resource Report No. 23, No. 17, No. 14, the Report of the Joint Committee on Forestry, 77th Congress 1st Session, Senate Document No. 32. Data for 1850 and 1870 were based on information collected during the 1850 and 1870 decennial census; data for 1907 were also based on the decennial census modified by expert opinon, reported by R.S. Kellogg in Forest Service Circular 166. Data for 1630 were included in Circular 166 as an estimate of the original forest area based on the current estimate of forest and historic land clearing information. These data are provided here for general reference purposes only to convey the relative extent of the forest estate in what is now the US at the time of European settlement.

Biomass Consumed under Two Scenarios of Future Climate

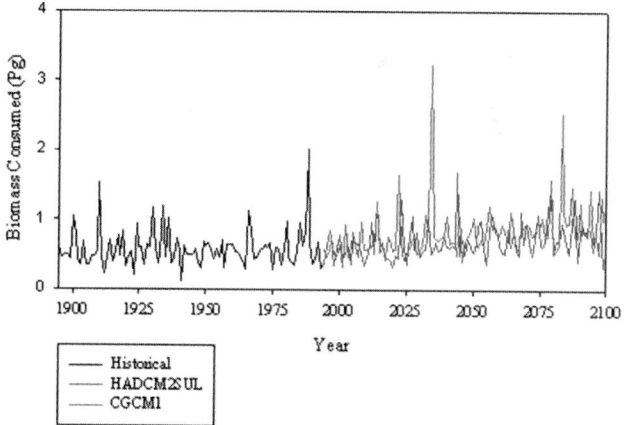

- Historical
- HADCM2SUL
- CGCM1

Figure 4. Simulated total biomass consumed by fire over the conterminous US under historic and two future climates; Hadley (HADCM2SUL) and Canadian (CGCM1) scenarios. The fire simulations are for potential vegetation and do not consider historic fire suppression activities. However, grid cells with more than 40% agriculture have been excluded from the calculations (Lenihan et. al., 1997, Daly et. al., 2000, Bachelet et. al., 2001).

Forests

Patterns of Live Vegetation for Different Times and Climate Scenarios

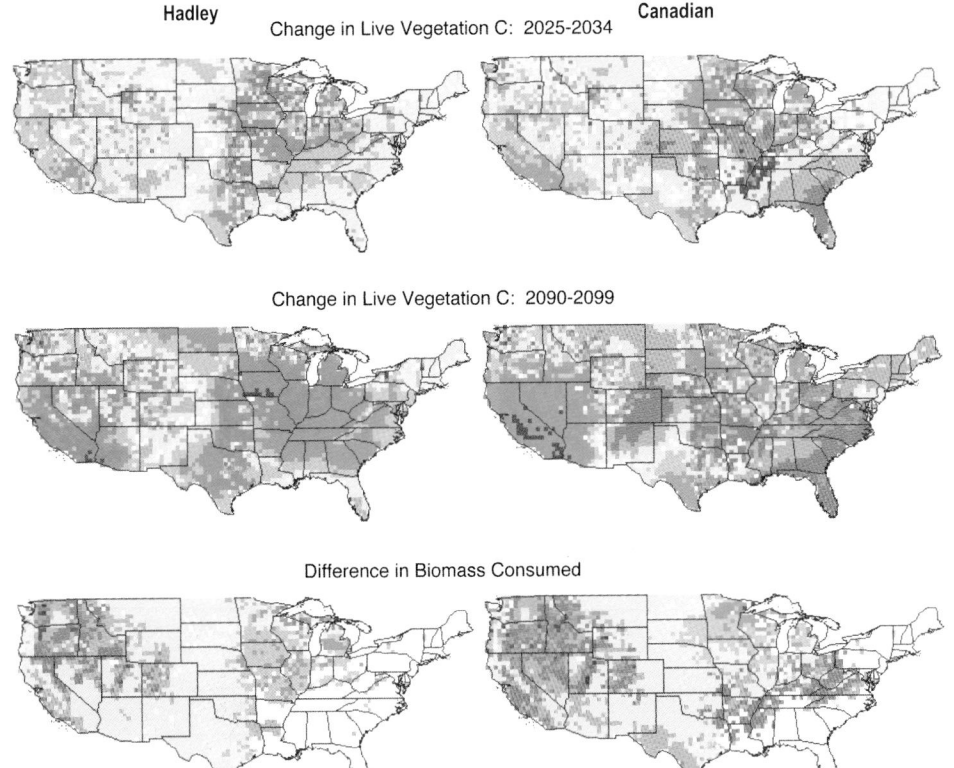

Figure 3. Change in live vegetation carbon density from the historical period (1961-1990) to the 2030s (2025-2035) and the 2090s (2090-2099) under two climate scenarios. Change in the biomass consumed by natural fires between the 20th century (1895-1993) and the 21st century 1994-2100) as simulated by the dynamic vegetation model MC1 under two climate scenarios (bottom two panels) (Bachelet et al. 2001)

Projected Changes in Distribution of Sugar Maple

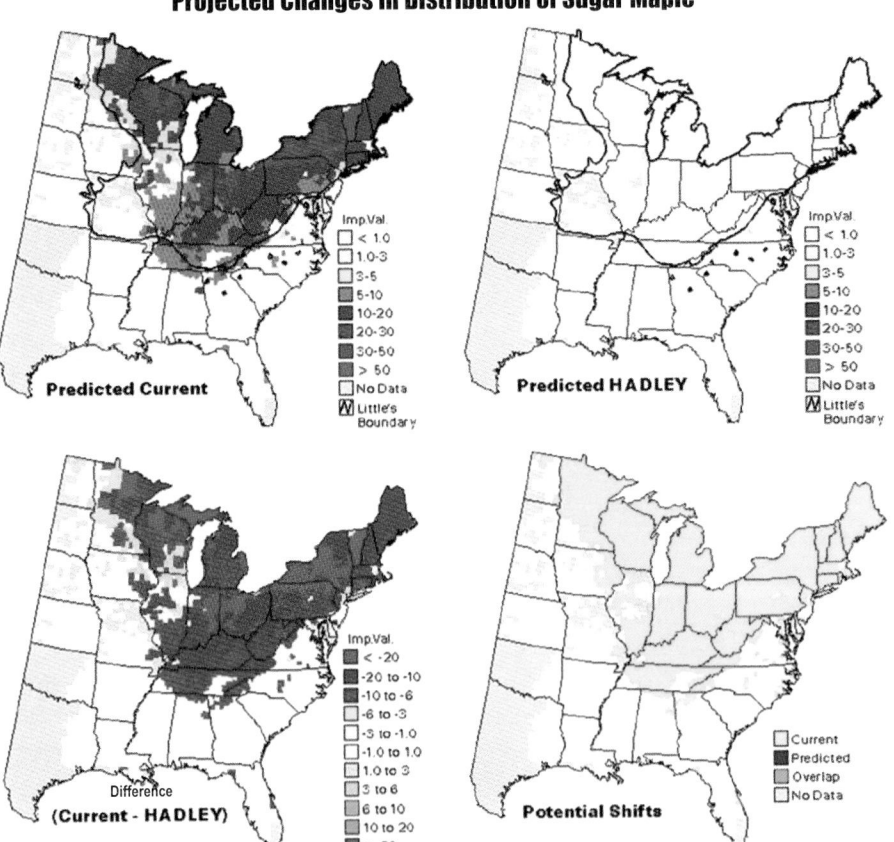

Figure 5. Projected distribution for sugar maple under current climate and the Hadley climate scenario and for the eastern United States, using statistical models developed by Iverson et al. (1999). The Predicted Current map is the current distribution and importance value of sugar maple, as modeled from the regression tree analysis. Importance value is an index based on the number of stems and basal area of both the understory and the overstory. Predicted Hadley is the potential suitable habitat for sugar maple under the Hadley climate scenario. These potential maps imply no barriers to migration. The Difference map represents the difference between Modeled Current and Predicted Hadley maps. The Potential Shifts map displays the modeled current distribution, along with predicted potential future distribution (using the Hadley scenario) and the overlap where the species is now and is projected to be in the future. As these maps indicate, very little sugar maple is likely to remain in the US by the late 21st century.

Current and Projected Forest Communities in the Eastern US

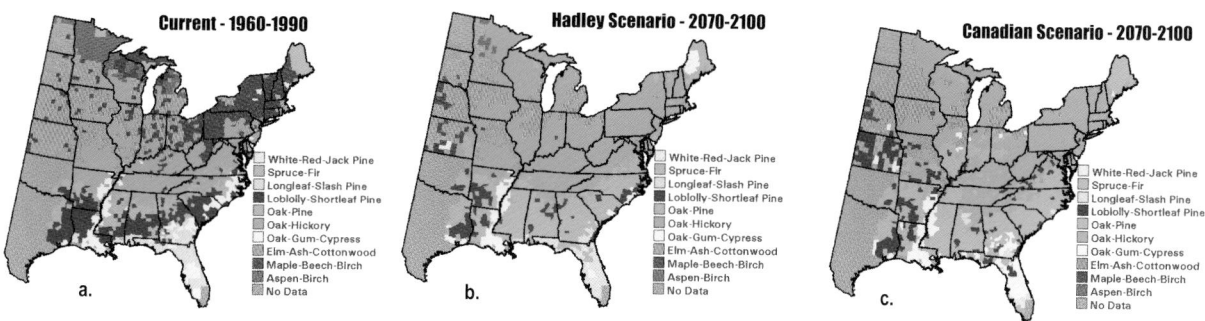

Figure 6. Projected forest communities under (a) current climate, (b) the Hadley climate scenario, and (c) the Canadian climate scenario, based on the results of individual analyses of 80 tree species shifts (see Prasad and Iverson, 1999-ongoing http://www.fs.fed.us/ne/delaware/atlas/index.html)

Paper Birch and Douglas Fir Tree Distributions under Future Climate Change Scenarios

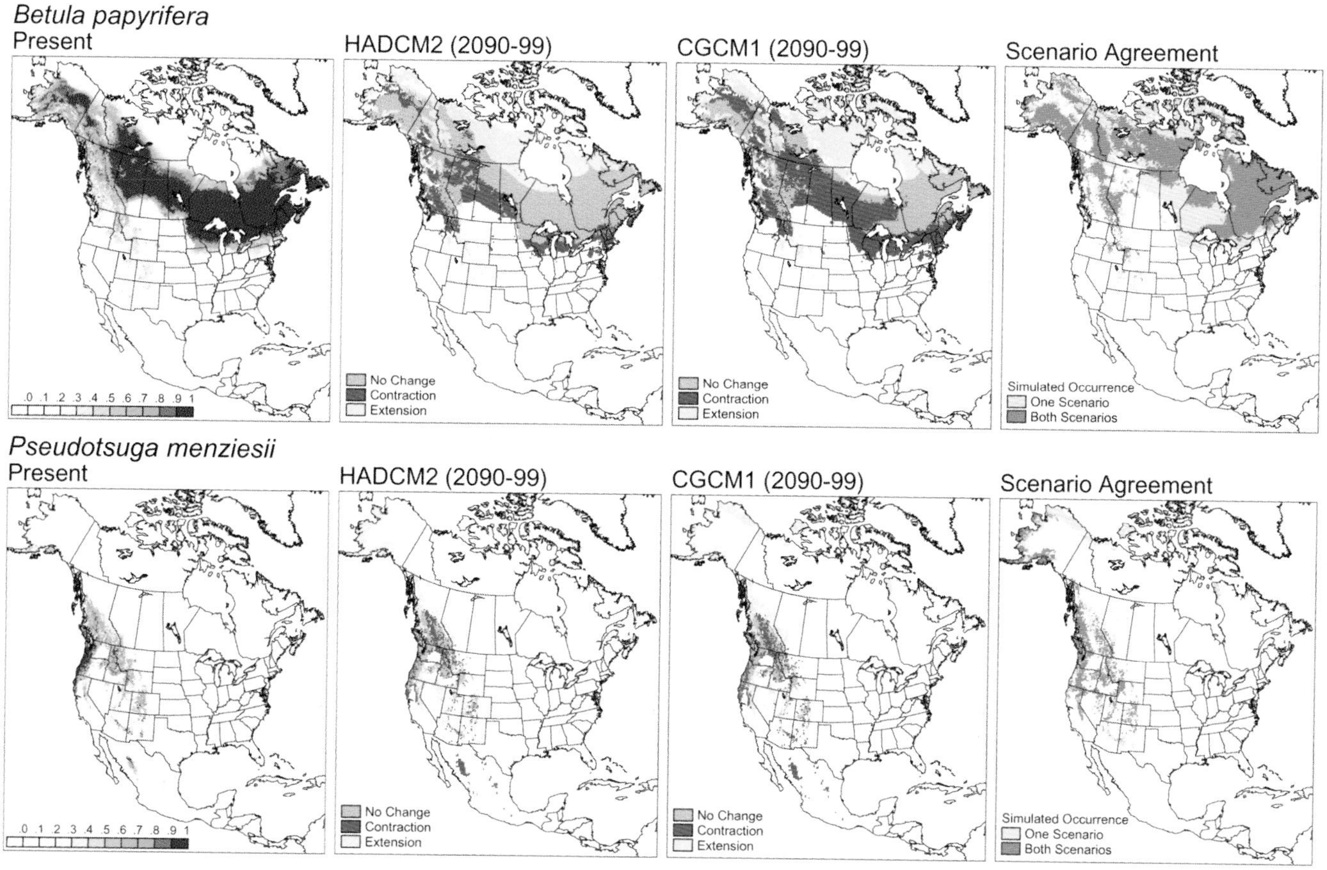

Betula papyrifera

Present	HADCM2 (2090-99)	CGCM1 (2090-99)	Scenario Agreement

Pseudotsuga menziesii

Present	HADCM2 (2090-99)	CGCM1 (2090-99)	Scenario Agreement

Figure 7. Simulated distributions and scenario agreement for *Betula papyrifera* and *Pseudotsuga menziesii* (after Hansen et al., 2001). Estimated probabilities of occurrence for each taxon simulated with observed modern climate (left panel). Comparison of the observed distributions with the simulated future distributions under future climate conditions as generated by the Hadley (HADCM2) and Canadian (CGCM1) scenarios for 2090-2099 (middle panels). Gray indicates locations where the taxon is observed today and is simulated to occur under future climate conditions; red indicates locations where the taxon is observed today but is simulated to be absent under future climate conditions; and blue indicates locations where the taxon is absent today but is simulated to occur under future climate conditions. Scenario agreement (right panel). Light purple indicates locations where the species is simulated to be present under the future climate of either the HADCM2 or CGCM1 scenario; dark purple indicates locations where the species is simulated to be present under both future climate scenarios.

Average Price for Standing Timber in US Forests

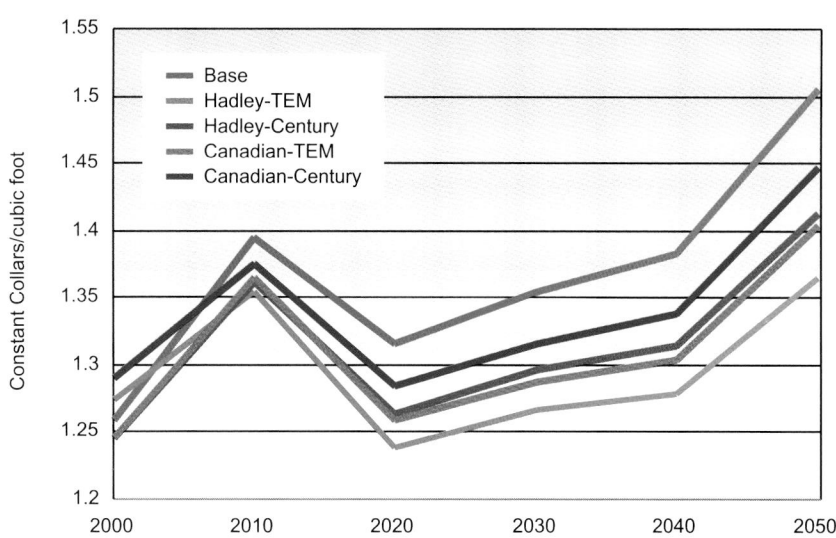

Figure 8. Prices for standing timber under all climate change scenarios remain lower than a future without climate change (baseline). Prices under the Canadian scenario remain higher than prices under the Hadley scenario when either the TEM or the Century model are used. (Irland et al., 2001).

Change in Timber Product Welfare from 2001 to 2100

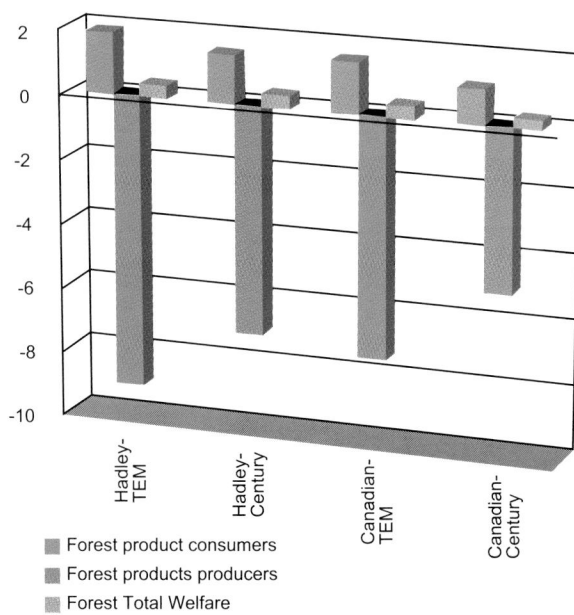

Figure 9. Increased forest growth overall leads to increased wood supply; reductions in log prices decrease producers' welfare (profits), but generally benefit consumers through lower wood-product prices. Welfare is present value of consumer and producer surplus discounted at 4% for 2000-2100. (Irland et al., 2001).